AA002268

# International Exhibition & Conference for Power Electronics, Intelligent Motion, Renewable Energy and Energy Management (PCIM Europe 2024)

Nuremberg, Germany
11 – 13 June 2024

Volume 2 of 5

ISBN: 978-1-7138-9966-2

**Printed from e-media with permission by:**

Curran Associates, Inc.
57 Morehouse Lane
Red Hook, NY 12571

**Some format issues inherent in the e-media version may also appear in this print version.**

Copyright© (2024) by Mesago Messe Frankfurt GmbH
All rights reserved.

Printed with permission by Curran Associates, Inc. (2024)

For permission requests, please contact VDE VERLAG GMBH
at the address below.

VDE VERLAG GMBH
Bismarckstr. 33
P.O.B. 12 01 43
10625 Berlin, Germany

Phone:   +49 30 34 80 01 - 0
Fax:       +49 30 34 80 01 - 9088

kundenservice@vde-verlag.de

**Additional copies of this publication are available from:**

Curran Associates, Inc.
57 Morehouse Lane
Red Hook, NY 12571 USA
Phone:  845-758-0400
Fax:       845-758-2634
Email:   curran@proceedings.com
Web:     www.proceedings.com

# TABLE OF CONTENTS

## VOLUME 1

### KEYNOTE

K01    AI BETWEEN HYPE AND INDUSTRIAL-GRADE - THE IMPACT OF AI ON THE ENTIRE POWER ELECTRONICS LIFECYCLE.................................................................................. 1
*Rolf Hellinger*

K02    INFRASTRUCTURE REQUIREMENTS FOR ELECTRIFIED HEAVY GOODS TRANSPORT IN GERMANY AND THE EU.................................................................................. 7
*Martin Wietschel*

K03    CHALLENGES AND SOLUTIONS TO POWER LATEST PROCESSOR GENERATIONS FOR HYPER SCALE DATACENTERS ............................................................. 15
*Gerald Deboy*

### GAN RUGGEDNESS

OP001   AN IMPROVED ULTRAFAST DESATURATION-BASED PROTECTION SCHEME FOR GAN HEMT ............................................................................................................... 19
*Juncheng Lu*

OP002   THE PERFORMANCE OF A GAN EMODE HEMT IN SURGE CURRENT SCENARIOS SUCH AS THE ACTIVE SHORT CIRCUIT.................................................... 24
*Dominik Nehmer*

OP003   GATE RESISTANCE EFFECT ON SHORT-CIRCUIT ROBUSTNESS OF P-GAN HEMTS ....................................................................................................................................... 34
*Mohamed Lemine Dedew*

### ADVANCED PACKAGING TECHNOLOGIES

OP004   NEURAL NETWORK ASSISTED NUMERICAL SIMULATION BENCHMARKING FOR ELECTRIC VEHICLE THERMAL MANAGEMENT SYSTEM .......................................... 40
*Ekin Alp Bicer*

OP005   RELATIONSHIP BETWEEN POROSITY IN CU SINTERED BONDING AND BONDING RELIABILITY.......................................................................................................... 49
*Hideo Nakako*

OP006   HIGH THERMAL DURABILITY OF THIN COPPER DIE-ATTACH LAYERS AND FINITE ELEMENT MODEL SIMULATION................................................................................ 56
*Takaaki Eyama*

### THERMAL CYCLING RELIABILITY

OP007   THERMAL SHOCK TEST LIFETIME IMPROVEMENT WITH OPTIMIZED ADHESIVE STRENGTH BETWEEN EPOXY RESIN AND COPPER .......................................... 62
*He Kangjia*

OP008    POWER CYCLING RELIABILITY AND FAILURE MODE ANALYSIS OF POL...... 67
*Kenichi Koi*

OP009    ACCELERATED POWER CYCLING OF GAN HEMTS USING SWITCHING LOSS
AND FAST TEMPERATURE MEASUREMENT ...... 74
*Wing Tai Leung*

## HIGH POWER CONVERTERS

OP010    CONTROL OF AN MMC-BASED HYBRID TRANSFORMER WITH STAR-POINT
VOLTAGE INJECTION ...... 84
*Rui Wang*

OP011    PROTECTION AND CONTROL OF A DUAL MMC MEDIUM VOLTAGE SUPPLY ...... 93
*Max Dupont*

OP012    STATION POWER ELECTRONICS CONVERTER WITH HIGH THERMAL
ENDURANCE TO POLE-TO-POLE SHORT CIRCUITS FOR LVDC DISTRIBUTION GRID ...... 103
*Frédéric Reymond-Laruina*

## GATE DRIVERS

OP013    SUPPRESSION OF OSCILLATIONS IN A SIC BRIDGE-LEG USING A CUSTOM
SINGLE-CHIP DIGITAL ACTIVE GATE DRIVER WITH 2×255 STRENGTH LEVELS......113
*Qilei Wang*

OP014    SIC MOSFET SHORT-CIRCUIT PROTECTION: A FASTER SOFT SHUT DOWN
METHOD FOR GATE DRIVERS ...... 121
*Julien Weckbrodt*

OP015    PARAMETER IDENTIFICATION: GATE SENSOR FOR POWER TRANSISTOR
TOLERANCE COMPENSATION IN ADVANCED GATE DRIVER ICS ...... 128
*Christopher Wille*

## ADVANCED CONTROL TECHNIQUES ON ELECTRICAL DRIVES I

OP016    AN INNOVATIVE HIGH-SPEED TRACK RANGE RESTART STRATEGY FOR
PERMANENT MAGNET SYNCHRONOUS MOTOR...... 135
*Anna Corbitt*

OP017    STEADY-STATE ERROR REDUCTION OF REINFORCEMENT LEARNING BASED
INDIRECT CURRENT CONTROL OF PERMANENT MAGNET SYNCHRONOUS
MACHINES ...... 140
*Tobias Schindler*

OP018    PERFORMANCE COMPARISON OF USING SHUNT-BASED AND INTEGRATED
CURRENT SENSING FOR SENSORLESS FIELD-ORIENTED CONTROL ...... 150
*John Emmanuel Tan*

## GAN CONVERTERS

OP019    DESIGN OF HIGH-POWER INVERTER WITH 12 PARALLEL GAN DEVICES ...... 161
*Takashi Sawada*

OP020   OVER 99.7% EFFICIENT GAN-BASED 6-LEVEL CAPACITIVE-LOAD POWER
CONVERTER ............................................................................................................................ 167
Stefan Mönch

OP021   CASCADED PRIMARY-SIDE-ONLY CONTROL OF A COMPACT 2 MHZ 500 W
WIRELESS POWER TRANSFER SYSTEM............................................................................ 174
Tim Krigar

## ADVANCED MATERIALS AND TECHNOLOGIES

OP022   POWER MODULE EVALUATION USING ULTRA HIGH HEAT DISSIPATION AND
HIGH HEAT RESISTANCE RESIN SHEET CONTAINING CARD HOUSE TYPE BORON
NITRIDE FILLER ..................................................................................................................... 180
Ayano Imai

OP023   INVESTIGATING TEMPERATURE DEPENDENT WARPAGE IN METAL CERAMIC
SUBSTRATES FOR POWER ELECTRONICS DEVICES ......................................................... 190
Benjamin Fabian

OP024   DEGRADATION MODE ANALYSIS OF DIFFERENT BONDING TECHNOLOGIES
OF SIC POWER SEMICONDUCTORS STRESSED BY ACTIVE POWER CYCLING............................ 197
Rasched Sankari

## CHARGING STATION TECHNOLOGY

OP025   IMPLEMENTATION AND VERIFICATION OF A 50KW OPPORTUNITY WIRELESS
CHARGER DESIGN .................................................................................................................. 205
Carlos Costas Sos

OP026   PERFORMANCE EVALUATION OF SILICON-BASED 3-LEVEL VIENNA
RECTIFIER IN ISOPLUS SMPD PACKAGE ............................................................................. 214
Karsten Haehre

OP027   PERFORMANCE ANALYSIS OF A 25-KW SIC-BASED DUAL ACTIVE BRIDGE
CONVERTER BASED ON PARALLEL-CONNECTED DEVICES ............................................. 222
Francesco Porpora

## MODELLING AND MONITORING

OP028   SEMICONDUCTOR CHIP MODELS ARE THE KEY FOR ENABLING VIRTUAL
DESIGN AND OPTIMIZATION WORKFLOWS OF POWER ELECTRONIC SYSTEMS......................... 230
Stefan Haensel

OP029   IMPROVED RESONANT FREQUENCY-BASED PARASITIC INDUCTANCE
ESTIMATION METHOD FOR SIC MOSFET HALF-BRIDGE CIRCUIT ................................... 238
Hongpeng Zhang

OP030   FAST SIMULATOR WITH INVERTER TEMPERATURE ESTIMATION FOR
TRACTION EDRIVES IN VEHICLES SUBJECTED TO DRIVING CYCLES............................ 248
Simone Giuffrida

## SOLID STATE TRANSFORMERS

OP031 A NEW FAMILY OF THREE-PHASE-UNFOLDER-BASED MVAC-LVDC SOLID-STATE TRANSFORMERS ........................................................................................................ 254
*Jonas Huber*

OP032 VOLTAGE BALANCING OF A SPLIT-CAPACITOR IGCT 3L-NPC LEG FOR THE RESONANT DC TRANSFORMER ................................................................................................. 264
*Renan Pillon Barcelos*

OP033 COMPARATIVE ANALYSIS OF UNIDIRECTIONAL HIGH STEP-UP CONVERTERS FOR MEDIUM VOLTAGE APPLICATIONS ............................................................................... 274
*Stefan Subotic*

## ADVANCED CONTROL TECHNIQUES ON ELECTRICAL DRIVES II

OP034 STARTUP BEHAVIOR OF HARMONIC SUPPRESSION IN ELECTRICAL MACHINES USING ITERATIVE LEARNING CONTROL AND NEURAL NETWORKS ......................... 284
*Annette Mai*

OP035 ANALYTICAL APPROACH OF THE VECTOR CURRENT CONTROL FLUX-WEAKENING STRATEGY FOR PERMANENT MAGNET SYNCHRONOUS MACHINES .................... 290
*Oriol Subirats Rillo*

## POWER ELECTRONICS FOR E-MOBILITY

OP036 INVESTIGATION ON DIRECT LIQUID COOLING DESIGN OF POWER MODULES WITH FLAT BASEPLATE FOR AUTOMOTIVE APPLICATION ............................................... 298
*Nobuhide Arai*

OP037 A NOVEL APPROACH FOR AFFORDABLE ELECTRIC VEHICLES BASED ON DUAL 48V BATTERY SYSTEM WITH MULTI-FUNCTIONAL 3-LEVEL CONVERTER ......................... 305
*Radovan Vuletic*

OP038 AN INNOVATIVE 3-LEVEL SOLUTION FOR AUTOMOTIVE APPLICATIONS: EMPACK .................................................................................................................................... 315
*Pranav Panchal*

OP039 GATED RECURRENT UNITS-ASSISTED STATE-SPACE MODELING FOR ELECTRIC VEHICLE TEMPERATURE PREDICTION ............................................................... 322
*Xinyuan Liao*

OP040 NOVEL BIDIRECTIONAL SINGLE-STAGE ISOLATED 600-V GAN M-BDSBASED SINGLE/THREE-PHASE-OPERABLE EV ON-BOARD CHARGER ............................................. 330
*Sven Weihe*

## ENCAPSULATION MATERIALS

OP041 APPLICATION-SPECIFIC INVESTIGATION OF INORGANIC POTTING MATERIAL IN DRIVE TRAINS ......................................................................................................................... 338
*Soenke Fleck*

OP042   THE INFLUENCE OF THE GLASS TRANSITION TEMPERATURE OF EPOXY
MOLD COMPOUNDS ON THE RELIABILITY OF A SEMICONDUCTOR DEVICE .............................. 343
   *Stefan Schwab*

OP043   CORROSION RESISTANT PACKAGING FOR POWER SEMICONDUCTOR
MODULES - MODIFIED INSULATION MATERIALS FOR CONTAMINATED
ENVIRONMENTS  ................................................................................................................................ 351
   *Michael Hanf*

OP044   INVESTIGATION OF INORGANIC ENCAPSULATION MATERIALS IN POWER
ELECTRONIC SYSTEMS FOR HIGH POWER DENSITY APPLICATIONS ............................................ 361
   *Stefan Behrendt*

OP045   CHARACTERIZATION OF THERMALLY AGED SILICONE GELS FOR POWER
SEMICONDUCTOR MODULES .................................................................................................................. 369
   *Sonja Madloch*

## POWER QUALITY

OP046   A COORDINATED CONTROL OF HYBRID SINGLE-PHASE AC/DC MICROGRIDS
BASED ON THE NATURAL HARMONIC INJECTION CONCEPT .......................................................... 378
   *Mehdi Baharizadeh*

OP047   A HIGH-POWER DENSITY SIC BASED TP PFC WITH HIGH-FREQUENCY RIPPLE
CANCELLATION LEG ................................................................................................................................ 383
   *Serkan Dusmez*

OP048   HIGH FREQUENCY ACTIVE FILTER FOR AC-DC HIGH POWER CONVERTERS ................ 390
   *Sarah Sifoune*

OP049   LABORATORY SETUP FOR ACCURACY INVESTIGATION OF ELECTRICITY
METERS AND MONITORS UNDER INDUSTRY-TYPICAL OPERATING CONDITIONS ...................... 397
   *Matthias Schmidt*

## GRID CONNECTED CONVERTERS

OP050   REAL-TIME EVALUATION OF WEIGHTING FACTORLESS PREDICTIVE
CONTROL OF LCL FILTER EQUIPPED GRID-SIDE CONVERTERS USING SORTING
NETWORKS .................................................................................................................................................. 403
   *Kristóf Bándy*

OP051   RELAXED ROBUST CONTROL WITH PRAGMATIC SHORTAGE OF PASSIVITY
FOR WIND, STORAGE AND PV POWER CONVERTERS ...................................................................... 411
   *Sergio De Lopez Diz*

OP052   AN EFFECTIVE DC VOLTAGE REGULATION OF ACTIVE FRONT-END
RECTIFIER THROUGH MODEL PREDICTIVE CONTROL ..................................................................... 419
   *Mobina Pouresmaeil*

OP053   BI-DIRECTIONAL 11KW MULTI-LEVEL ACTIVE-NEUTRAL-POINT-CLAMPED
AC-DC CONVERTER USING 600V/750V SI SUPER-JUNCTION AND SIC MOSFETS FOR
HIGH-EFFICIENCY AND HIGH-DENSITY APPLICATIONS .................................................................. 424
   *Mengxing Chen*

OP054    A STUDY OF GRID-FORMING INVERTER CONTROL STRATEGY FOR FAULT-RIDE-THROUGH CAPABILITY ........................ 433
Hirofumi Uemura

## PASSIVE COMPONENTS

OP055    FILM CAPACITORS FOR HIGH TEMPERATURE AC-DC INVERTER APPLICATIONS ........................ 440
Adel Bastawros

OP056    LOSS REDUCTION IN HF-TRANSFORMERS USING LAMINATED FERRITE E-CORES ........................ 447
Lukas Reißenweber

OP057    MULTIGAP TOROIDAL TRANSFORMER AND INDUCTORS FOR OVERCOMING FRINGING LOSSES IN HIGH FREQUENCY CONVERTERS ........................ 456
Pau Colomer

OP058    STUDY ON SAMPLE GEOMETRIES FOR FERRITE CHARACTERISATION IN THE MHZ RANGE ........................ 463
Till Piepenbrock

OP059    FEM-SUPPORTED AND NON-DESTRUCTIVE MAGNETIC CHARACTERIZATION METHOD FOR NON-LAMINATED STEEL ........................ 472
Stefan Tobler

## DRIVES FOR HIGH DEMANDING APPLICATIONS

OP060    HIGHLY-COMPACT BEARINGLESS AXIAL-FLUX MOTOR FOR A PEDIATRIC IMPLANTABLE FONTAN BLOOD PUMP ........................ 480
Andreas Horat

OP061    A NOVEL PERMANENT MAGNET SYNCHRONOUS MOTOR DRIVE FOR REACTION WHEELS IN SATELLITES ........................ 490
Baris Colak

OP062    EXPLORING HIGH FREQUENCY OPERATION OF MOTOR DRIVES: PRACTICAL INSIGHTS ON EFFICIENCY AND LOSS ........................ 497
Asantha Kempitiya

OP063    HIGH POWER DENSITY SYSTEM DESIGN FOR GAN-BASED LV MOTOR DRIVES ........................ 502
Marco Cannone

OP064    DESIGN OF GAN TRANSISTOR BASED VARIABLE SPEED DRIVE INVERTER WITH OUTPUT VOLTAGE FILTERING ........................ 510
Kaspars Kroics

## IGBT

OP065    THE 8TH GENERATION LV100 IGBT MODULE WITH HIGHER CURRENT RATING ........................ 518
Daichi Otori

LOSS REDUCTION BY LAMINATIGN FERRITE E CORES.................................................................... 525
Lukas Reißenweber

OP066   NEW PLANAR 4.5 KV SPLIT-GATE (SG) SI-IGBT DEVICE FOR IMPROVED
SWITCHING CHARACTERISTICS AND HIGH FREQUENCY OPERATION ......................................... 534
Gaurav Gupta

OP067   4.5 KV DOUBLE-GATE REVERSE-CONDUCTING PRESS-PACK IEGT .................................. 543
Satoshi Yoshida

## DEVICE CONCEPTS

OP068   EVALUATION OF A 3 KV POLARIZATION SUPERJUNCTION GAN HEMT.......................... 549
Alireza Sheikhan

OP069   MORE THAN 1200 V BREAKDOWN AND LOW AREA-SPECIFIC ON STATE
RESISTANCES BY PROGRESS IN LATERAL GAN-ON-SI AND GAN-ON-INSULATOR
TECHNOLOGIES.......................................................................................................................... 557
Richard Reiner

OP070   NOVEL 200 V MOSFET TECHNOLOGY PUSHES MOTOR DRIVE INVERTER
EFFICIENCY TO AN UNPRECEDENTED LEVEL .................................................................... 564
Mark Thomas

## DEGRADATION MECHANISMS

OP071   MOISTURE ROBUST CHIP DESIGN - IMPROVED EDGE-TERMINATIONS FOR
HIGH LIFETIME UNDER HIGH HUMID CONDITIONS........................................................... 571
Michael Hanf

OP072   METHOD FOR MEASURING THE INITIAL STATE OF A SOLDER JOINT
DELAMINATION IN A 3D PCB INTEGRATION ASSEMBLY OF SIC MOSFETS.................................. 581
Souhila Bouzerd

OP073   GENERIC LIFETIME MODEL FOR WIRE BONDS DEGRADATION IN IGBT
MODULES BASED ON A FRACTURE MECHANICS PARAMETER.......................................... 589
Merouane Ouhab

## ADVANCED CONVERSION CONCEPTS

OP074   MODULAR COAXIAL POWER CONVERTER FOR HIGH-DENSITY INTEGRATION
INTO MEDIUM-VOLTAGE CABLES ....................................................................................... 599
Mark Cairnie

OP075   CONTROLLED INDUCTOR BASED BCM BUCK CONVERTERS ............................................ 608
Ziv Gellman

OP076   INFLUENCE OF VARYING COMMON MODE CHOKE SIZES ON THE
PERFORMANCE AND STABILITY OF AN ACTIVE EMI FILTER............................................ 615
Patrick Körner

## PHOTOVOLTAIC SYSTEMS

OP077    A HIGH EFFICIENCY BATTERY CHARGER WITH MAXIMUM POWER POINT
TRACKING FOR MAGNETIC ENERGY HARVESTERS ............................................................. 625
*Antonio Miguel Munoz Gomez*

OP078    SYMMETRIC FLYING-CAPACITOR BOOST CONVERTER FOR MEDIUM-
VOLTAGE PHOTOVOLTAIC APPLICATIONS ................................................................. 635
*Luis Alves Rodrigues*

OP079    COMPARISON OF SI IGBT, SIC MOSFET AND ADJUSTABLE HYBRID SWITCH
PV INVERTERS FOR DIFFERENT GEOGRAPHICAL LOCATIONS ....................................... 645
*Tanya Thekemuriyil*

## MODEL BASED SYSTEM ANALYSIS

OP080    OPTIMISING A POWER MODULE FOR ELECTRICAL AND THERMAL
PERFORMANCE AND SYMMETRY USING EDA TOOLS ................................................... 655
*Wilfried Wessel*

OP081    CONDUCTOR-BASED MODELING OF VOLTAGE DISTRIBUTION ALONG A
SINGLE-TOOTH WINDING OF ELECTRICAL MACHINES ............................................... 665
*Hujun Peng*

OP082    REDUCTION OF PWM HARMONICS WITH CARRIER PHASE SHIFTING IN A
DUAL-STATOR PMSM WITH MAGNETIC COUPLED WINDINGS ....................................... 672
*Bünyamin Tekir*

## VOLUME 2

## SIC DEVICES

OP083    THE NEW COOLSIC MOSFET 1200 V G2: ELECTRICAL PERFORMANCE AND
COMPACT MODELLING ............................................................................................. 681
*Andreas Huerner*

OP084    PARALLELING SIC-POWER-MOSFET BODY DIODES UNDER HARSH
SWITCHING CONDITIONS .......................................................................................... 690
*Michael Rauh*

OP085    3.3KV SBD-EMBEDDED SIC-MOSFET MODULE FOR TRACTION USE ............................. 699
*Yoichi Hironaka*

OP086    DEAD TIME OPTIMIZATION FOR HIGH POWER SIC MOSFET MODULE IN
CONSIDERATION OF PARASITIC COMPONENTS ........................................................... 707
*Pham Ha Trieu To*

## WBG RELIABILITY

OP087    PERFORMANCE INSTABILITY OF 650 V P-GAN GATE HEMT DEVICE UNDER
TEMPERATURE-RELATED POSITIVE GATE BIAS STRESSES ............................................. 717
*Renze Yu*

OP088  GATE OXIDE RELIABILITY OF CURRENT GENERATION 1.2 KV SIC MOSFETS
UNDER STEP-WISE INCREASED GATE VOLTAGE............................................................. 723
  *Roman Boldyrjew-Mast*

OP089  AN ACCELERATED DYNAMIC GATE SWITCHING STRESS TEST CONCEPT OF
SIC MOSFETS AT HIGH DRAIN-SOURCE VOLTAGE (HV-GSS) ............................................ 731
  *Clemens Herrmann*

OP090  SILICON CARBIDE POWER DEVICE USE IN SPACECRAFT AND AIRCRAFT ..................... 739
  *Akin Akturk*

## POWER ELECTRONICS FOR E-MOBILITY/ CONTROL

OP091  CURRENT RIPPLE REDUCTION BY COMBINATION OF SI IGBT AND SIC
MOSFETS IN HEAVY DUTY FUEL CELL TRUCKS.............................................................. 745
  *Yavuz Gürlek*

OP092  EVALUATION OF ACTIVE GATE DRIVERS WITH SWITCHABLE GATE
RESISTORS AND INTERMEDIATE VOLTAGE LEVELS FOR SIC MOSFETS IN WLTC ...................... 754
  *Michael Frank*

OP093  PERFORMANCE EVALUATION OF TCM-BASED, ZERO-VOLTAGE SWITCHING
(ZVS) THREE-PHASE INVERTER FOR ELECTRIC VEHICLE DRIVE SYSTEMS ................................ 764
  *Khizra Abbas*

OP094  A PARTIAL LOAD THREE-PHASE TRIANGULAR CURRENT MODE
MODULATION CONCEPT WITH AN OPTIMIZED FILTER INDUCTOR FOR HIGH
EFFICIENCY TRACTION DRIVES  ................................................................................. 774
  *Bhaskar Chatterjee*

## DC-DC CONVERTERS I

OP095  GAN VS SI SYNCHRONOUS RECTIFIER FOR LLC CONVERTER ........................................ 784
  *Gokhan Sen*

OP096  CO-SIMULATION DESIGN OF A GAN-BASED THREE-PHASE LLC CONVERTER
WITH INTEGRATED THREE-PHASE MAGNETICS .............................................................. 791
  *Jhih-Cheng Hu*

OP097  SWITCHING ASSISTING CIRCUIT IMPROVING THE EFFICIENCY OF DC-DC
CONVERTERS BASED ON PIEZOELECTRIC RESONATORS.................................................... 797
  *Ghislain Despesse*

OP098  TRANSFORMER-BASED FIXED-RATIO RESONANT DC-DC CONVERTERS FOR
48V DATA CENTERS ............................................................................................... 803
  *Xufu Ren*

## PFC CONVERTERS

OP099  HIGH-DENSITY 3.3 KW GAN RECTIFIER FOR SERVER APPLICATIONS
COMPRISING A 130 KHZ TOTEM-POLE PFC AND A 500 KHZ LLC.......................................... 812
  *Manuel Escudero Rodriguez*

OP100    ADDRESSING POWER SWITCH TECHNOLOGY SELECTION SI/SIC/GAN IN
HIGH EFFICIENCY ZVS-PFC RESONANT CONVERTERS .................................................... 822
    Marco Torrisi

OP101    BUCK-TYPE CURRENT UNFOLDING CONVERTER WITH DISCONTINUOUS
CONDUCTION MODE IN ULTRA-LOW POWER-FACTOR OPERATION ................................. 831
    Tomoyuki Mannen

OP102    GAN BASED BI-DIRECTIONAL 6.6KW INTERLEAVED TOTEM-POLE PFC WITH
13KW/L POWER DENSITY AND HIGH EFFICIENCY ........................................................... 837
    Juncheng Lu


## SIC MODULES

OP103    THE DESIGN OF A 2KV 1700A SIC MOSFET DUAL MODULE ............................... 843
    Jorge Mari

OP104    TECHNOLOGICAL APPROACHES TO HIGH-POWER DENSITY SIC POWER
MODULE FOR AUTOMOTIVE ....................................................................................... 849
    Takeshi Tokorozuki

OP105    EXTREMELY COMPACT SIC POWER MODULE FOR EV TRACTION INVERTERS
IN THE 250 KW CLASS .................................................................................................. 855
    Raffael Schnell

OP106    BENEFITS OF .XT INTERCONNECTION TECHNOLOGY FOR 3.3 KV XHP 2
MODULE WITH 3.3 KV COOLSIC MOSFET .................................................................. 863
    Matthias Bürger


## ADVANCED COOLING

OP107    LARGE-AREA BONDING WITH LMEE: SUPPRESSION OF THE DEGRADATION
OF THE JUNCTION-TO-WATER THERMAL RESISTANCE IN POWER MODULES ............................. 870
    Yo Mochizuki

OP108    ACTIVE THERMAL CONTROL OF SIC MOSFETS UTILIZING TRANSIENT
THERMAL CHARACTERIZATION .................................................................................. 875
    Varaha Satya Bharath Kurukuru

OP109    THERMAL MANAGEMENT SOLUTIONS BY ADDITIVE MANUFACTURING –
POWDER BED FUSION AND DIFFUSION BONDING .......................................................... 883
    Simon Jahn

OP110    ADVANCED PUMPED TWO-PHASE COLD PLATE FOR COOLING POWER
ELECTRONICS ......................................................................................................... 888
    Elizabeth Seber


## DC-DC CONVERTERS II

OP111    FEASIBILITY STUDY OF HIGH-POWER DENSITY ISOLATED CLLC DC-DC
INTERFACE WITH WIDE RANGE OF VOLTAGE/CURRENT REGULATION ......................................... 893
    Oleksandr Husev

OP112   DC-BIAS REDUCTION IN HIGH-FREQUENCY DUAL ACTIVE BRIDGE DC-DC CONVERTERS THROUGH SLOW DC MEASUREMENTS................................................................... 903
*Patrick Lenzen*

OP113   OPTIMIZED CURRENT SHARING TECHNIQUE FOR INTERLEAVED CLLC CONVERTERS FOR MINIMAL OUTPUT CURRENT DISTORTION ......................................... 909
*Martin Gendrin*

OP114   PRIMARY-SIDE OUTPUT REGULATION PRINCIPLES IN DYNAMIC MULTI-MHZ INDUCTIVE POWER TRANSFER SYSTEMS AND ISOLATED DC/DC CONVERTERS ....................... 916
*Ioannis Nikiforidis*

## SMART GRID

OP115   LOW VOLTAGE DC-GRIDS WITH GALVANIC ISOLATION: SYSTEM DISCUSSION, EFFICIENCY AND PERFORMANCE COMPARISON TO AC-FEEDING....................... 926
*Lukas Fräger*

OP116   IMPLEMENTATION AND EXPERIMENTAL EVALUATION OF AN ADAPTIVE DC GRID CONTROLLER FOR DECENTRALISED GRID CONTROL ............................................. 933
*Steffen Menzel*

OP117   DEMONSTRATING THE EFFECTIVENESS OF A DC SOLID-STATE CIRCUIT BREAKER'S FAST RESPONSE TIME.............................................................................. 942
*Ehab Tarmoom*

OP118   MODELLING AND SIZING SENSITIVITY ANALYSIS OF A FULLY RENEWABLE ENERGY-BASED ELECTRIC VEHICLE CHARGING STATION MICROGRID ....................................... 949
*David A. Stone*

## MEASUREMENT TECHNIQUES AND METHODS

OP119   LED POWERED ROTOR TELEMETRY SYSTEM.................................................................. 958
*Raphael Beyerle*

OP120   'INFINITY GATE SENSOR': A DIFFERENTIAL MAGNETIC FIELD SENSOR FOR MEASURING GATE CURRENT OF SIC POWER TRANSISTORS............................................. 966
*Yushi Wang*

OP121   CHARACTERISING WIDE BANDGAP POWER MODULES: VALIDATING THE M-SHUNT CONCEPT FOR HIGH-POWER APPLICATIONS IN THE KILOAMPERE RANGE ................... 976
*Hauke Lutzen*

OP122   CHARACTERIZATION OF POWER-MODULE PARASITICS: SUB-NANOSECOND LARGE SIGNAL PULSING VS. DOUBLE-PULSE TESTING .................................................. 986
*Gerhard Groos*

STATISTICAL VARIATIONS IN THE PARASITIC CAPACITANCE OF A COIL...................................... 997
*Kevin Talits*

## HIGH VOLTAGE SWITCHES

PP001   A 4.5 KV FAST RECOVERY DIODE PLATFORM FOR HIGH-CURRENT IGBTS .................. 1002
*Jan Vobecky*

PP002    6.5 KV INNOVATIVE SILICON POWER DEVICE (I-SI) MODULE WITH HIGH POWER DENSITY AND LOW LOSS BY STORED CARRIER CONTROL ............................................. 1007
Takashi Hirao

PP003    HIGH CURRENT DENSITY 4.5KV PRESSPACK IGBTS PUSH SOA LIMITS ........................ 1013
Hossein Davoodi

PP004    2.5KV IGBT MODULE WITH HIGH RELIABILITY FOR RENEWABLE APPLICATIONS ............................................................................................................................ 1018
Akiyoshi Masuda

PP005    NEW GENERATION 4.5KV IGCT AND FAST RECOVERY DIODE FOR RAILWAY POWER SUPPLY APPLICATIONS .............................................................................................. 1025
Umamaheswara Reddy Vemulapati

PP006    NEXT GENERATION 4.5 KV IGBT-ONLY STAKPAK MODULE WITH REDUCED LOSSES AND HIGH TEMPERATURE CAPABILITY ............................................................................ 1031
Jeremy Jones

## THERMAL MODELLING AND SIMULATIONS

PP007    FINITE ELEMENT ANALYSIS OF THE UPSCALING OF WARPAGE AND BIFURCATION HYSTERESIS LOOPS: FROM CU/SI DIE TO LARGE WAFERS ................................ 1039
Vincenzo Vinciguerra

PP009    MAXIMUM JUNCTION TEMPERATURE SIMULATION AND VALIDATION FOR THE HOT SPOT IN MULTI-CHIP SIC POWER MODULE ................................................................... 1046
Wonjin Dylan Cho

PP010    INTEGRATION OF CFD-SIMULATION RESULTS IN PLECS USING LOOKUP TABLES ........................................................................................................................................... 1051
Simon Cepin

PP011    PCB ONLY THERMAL MANAGEMENT TECHNIQUES FOR EGAN FETS IN A HALF-BRIDGE CONFIGURATION ................................................................................................... 1057
Adolfo Herrera

## HIGH POWER DENSITY DESIGNS

PP013    FROM 4X TO 3X STPAK – OPTIMIZATION FOR A MORE COMPACT EV TRACTION INVERTER SOLUTION ................................................................................................... 1065
Vittorio Giuffrida

PP014    A MULTI-OBJECTIVE STRUCTURAL OPTIMIZATION METHOD BASED ON MULTI-PHYSICS SIMULATIONS FOR POWER MODULE ............................................................... 1072
Baihan Liu

PP015    HOLISTIC APPROACH TO MAXIMIZE LIFETIME AND POWER DENSITY IN HIGH POWER SEMICONDUCTOR MODULES ................................................................................... 1077
Martin Schulz

PP016    REGULATED HIGH DENSITY SWITCH CAPACITOR TOPOLOGY ....................................... 1082
Pierrick Ausseresse

PP017   SILICON INTERPOSER AS A SUBSTRATE FOR POWER MODULES WITH HIGH
POWER DENSITY AND SUPERIOR THERMAL PERFORMANCE ........................................................ 1087
   *Ahmed Ammar*

## SPECIAL CONVERTER APPLICATIONS

PP018   ANALYTICAL MODELING AND STABILITY CHARACTERIZATION OF A
DAMPED VSCC CM ACTIVE EMI FILTER FOR SINGLE- AND THREE-PHASE AC-DC
APPLICATIONS ........................................................................................................................................ 1092
   *Timothy Hegarty*

PP020   A REPETITIVE HIGH VOLTAGE NANOSECOND PULSE GENERATOR: FIRST
PROTOTYPE DESIGN AND TEST RESULTS ...................................................................................... 1101
   *Serge Gavin*

PP021   FREQUENCY SHIFT KEYED DUAL SIDE CONTROL OF INDUCTIVE POWER
TRANSFER: AN APPLICATION OF TALKATIVE POWER CONVERSION .................................... 1105
   *Hamzeh Beiranvand*

PP022   STUDY OF A MULTI-ACTIVE BRIDGE CONVERTER FOR A DOMESTIC
ELECTRICAL GRID .............................................................................................................................. 1113
   *Abdennour Merrouche*

## INTEGRATION TECHNOLOGIES AND RELIABILITY DESIGN

PP023   FABRICATION DEVELOPMENT FOR GATE DRIVER EMBEDDED DOUBLE-
SIDED COOLING SIC POWER MODULE FOR ELECTRIC VEHICLE APPLICATION ...................... 1123
   *Anna Corbitt*

PP024   PRINTED CIRCUIT EMBEDDING OF PREPACKAGED 150V POWER MOSFETS IN
A PORTABLE WELDING APPLICATION ............................................................................................ 1128
   *Thomas Gebhard*

PP025   PROCESS CHALLENGES AND PROGRESS TOWARDS DIRECT CONNECTION OF
AUTOMOTIVE POWER MODULES (TMM) TO HEATSINK .............................................................. 1133
   *Indrajit Paul*

PP026   OPTIMIZING PCB STACKUPS FOR ENHANCED GAN TRANSISTOR
PERFORMANCE IN HIGH-POWER APPLICATIONS .......................................................................... 1139
   *Philipp Czerwenka*

PP027   NEW GENERATION CERAMIC SUBSTRATES – KEY COMPONENTS FOR POWER
ELECTRONIC APPLICATIONS: PROCESSING AND CHARACTERIZATION .................................... 1147
   *Stefanie Schindler*

PP028   AI-ENHANCED VACUUM REFLOW OVEN: PRECISION CONTROL FOR
RELIABLE LARGE-AREA SOLDERING .............................................................................................. 1152
   *Chih Hui Lee*

PP030   CORROSION-COMPATIBLE DRIVE ELECTRONICS FOR ELECTRIC VEHICLES
AND INDUSTRIAL POWER MODULES .............................................................................................. 1158
   *Tom Petzold*

PP031   EVALUATING THE SAFETY ISOLATION OF THE PACKAGE IN AN INTEGRATED
POWER DEVICE ...........................................................................................................................1168
    *Thomas Anthony Capobianco*


## CONTROL METHODS I

PP032   FLEXIBLE CONTROL SYSTEM FOR MODULAR ONE-PHASE INTERLEAVED
GAN-BASED TOTEM POLE PFC USING REAL-TIME HARDWARE ......................................1174
    *Oleksandr Solomakha*

PP033   A PEAK CURRENT MODE CONTROL METHOD FOR PFC ...................................................1180
    *Sean Yu*

PP034   ADAPTIVE RESONANT CONTROLLER FOR A THREE-PHASE PFC CONVERTER
FOR AN ON-BOARD CHARGE APPLICATION ........................................................................1185
    *Rami Troudi*

PP035   SYNTHESIS OF A FIELD ORIENTED CONTROL ALGORITHM BY USING TWO
DIFFERENT POLE-ZERO COMPENSATION APPROACHES ...................................................1192
    *Marco Denk*

PP037   AVERAGE CURRENT MODE CONTROL AND ITS LOOP DESIGN ......................................1200
    *Niklas Schwarz*

PP038   NOVEL POWER FEED-FORWARD REGULATION FOR DUAL STAGE PFC+DCDC
CONVERTERS ................................................................................................................................1207
    *Alfredo Medina-Garcia*


## HIGH POWER AC-DC AND DC-AC CONVERTER

PP039   22 KW BI-DIRECTIONAL WALL-BOX CHARGER WITH 1200 V SIC MOSFET....................1212
    *Sanbao Shi*

PP040   DYNAMIC SWITCHING FREQUENCY SELECTION FOR EFFICIENCY
OPTIMIZATION IN ON-BOARD CHARGER PFC STAGE BASED ON NOVEL SIC MOSFET
POWER MODULE ...........................................................................................................................1217
    *Giuseppe Aiello*

PP041   DESIGN AND OPTIMIZATION OF SIC-BASED 11KW MOTOR DRIVE WITH HIGH
EFFICIENCY ...................................................................................................................................1222
    *Iris Liu*

PP042   MODEL DESIGN DEVELOPMENT FOR FALSE TURN-ON CHARACTERIZATION
IN SIC-BASED ACTIVE T-TYPE CONVERTER CONSIDERING ALL PARASITICS ..........1227
    *Amir Babaki*

PP043   EFFICIENCY INVESTIGATIONS OF AN AUXILIARY RESONANT COMMUTATED
POLE INVERTER..............................................................................................................................1233
    *Markus Zocher*

PP044   A NOVEL HYBRID TWO-STAGE AC-DC CONVERTER WITH SOFT-SWITCHED
CCM PFC STAGE FOR EVS CHARGING APPLICATIONS..................................................1242
    *Lei Wang*

PP045    A METHOD FOR TUNING LEAKAGE INDUCTANCE IN TRANSFORMERS ...... 1249
Rosemary O'Keeffe

PP046    LOW COST HIGH DENSITY 300W/20V AC-DC CONVERTER ENABLED BY GAN
POWER ICS.................................................................................................................. 1254
Tom Ribarich

PP047    25KVA GRID-TIED BI-DIRECTIONAL T-TYPE INVERTER WITH HIGH-
EFFICIENCY AND HIGH-POWER DENSITY USING SIC MOSFETS..................................... 1259
Tamanna Bhatia

PP048    COST-EFFECTIVE EFFICIENCY ENHANCEMENT IN AC-DC CONVERTERS: A
STUDY ACROSS THE FULL LOAD CYCLE ......................................................... 1264
Sebastian Gick

## E-MOBILITY TRACTION I

PP049    NEXT GENERATION POWER MODULE WITH PARALLEL CONNECTED SIC
MOSFETS FOR BEV TRACTION INVERTERS......................................................... 1272
Kohei Tanikawa

PP051    INVESTIGATION OF COMMON SOURCE FEEDBACK IN SIC POWER MODULES
REGARDING PERFORMANCE AND SHORT CIRCUIT ROBUSTNESS............................... 1277
Dominik Ruoff

PP052    HYBRIDPACK DRIVE POWER MODULES WITH SIC-MOSFET'S AND
MONOLITHIC RC- SNUBBER CHIPS FOR OPTIMIZED POWER DENSITY....................... 1283
Andre Uhlemann

PP053    ROBUST AUXILIARY POWER SUPPLY FOR EVS BASED ON INNOVATIVE
STI2GAN 650V IC...................................................................................................... 1289
Federica Cammarata

PP054    IMPACT OF VARIOUS SILICON DIODES ON THE HYBRID SWITCH INVERTER ............ 1297
Michael Walter

PP055    ADVANCED PULSE SEQUENCE FOR SALIENCY-BASED HIGH-ACCURATE
ROTOR POSITION ESTIMATION OF RAILWAY TRACTION LOCOMOTIVE MOTORS .................. 1307
Markus Vogelsberger

## CONTROL TECHNIQUES

PP056    OPTIMIZED HALF-BRIDGE GATE-DRIVE WITH LOW TIME-SKEW FOR RC-
IGBTS AND SIC-MOSFET DEAD-TIME CONTROL ......................................................... 1315
Jan Fuhrmann

PP057    DESIGN OF A TRACTION INVERTER BASED ON PCB-EMBEDDED GAN
DEVICES ..................................................................................................................... 1322
Maurizio Tranchero

PP058    OPTIMIZING ELECTRIC VEHICLE PERFORMANCE WITH GAN DESIGN....................... 1330
Andrew Patterson

PP059   FAST ANALYTICAL CALCULATION OF THE MAGNETIC FIELD IN PERMANENT MAGNET SYNCHRONOUS MACHINES WITH FLUX BARRIERS INCLUDING SATURATION ........................................................................................................ 1336
*Martin Ackermann*

PP060   MODELING AND CONTROL OF LCL FILTERED 3L-VSCS IN INTERLEAVED TOPOLOGY ........................................................................................................ 1346
*Adeel Jamal*

PP062   ENHANCING SAFETY AND EFFICIENCY FOR ISOLATED PLC I/O DESIGNS WITH SPI DAISY CHAIN ...................................................................................... 1352
*Travis Lenz*

## VOLUME 3

PP063   COST-EFFECTIVE METHOD TO DISCHARGE DC LINK CAPACITORS WITH SIC POWER MODULES ....................................................................................... 1361
*Paul Kanatzar*

## POWER QUALITY

PP064   A STUDY ON CIRCULATION CURRENT IN PARALLEL OPERATION OF TRANSFORMER LESS UPS ..................................................................................... 1368
*Koji Kato*

PP065   DESIGN CHALLENGES AND CONSIDERATIONS FOR GATE DRIVERS OF SIC MOSFETS AND THEIR TESTING .............................................................................. 1374
*Niranjan Hegde*

PP066   A PORTABLE EFFICIENCY CHARACTERIZATION SETUP FOR TECHNOLOGY DEMONSTRATION OF POWER MODULES ........................................................... 1380
*Sebastian Tengvall*

PP067   FAST EME CHARACTERIZATION OF BARE-DIE SIC MOSFETS ........................... 1385
*Robert Kragl*

PP068   THEORETICAL COMPARISON OF COMPONENT-RELATED MEASUREMENT METHODS OF PHOTOVOLTAIC INVERTERS FOR LONG-TERM TESTING ...................... 1393
*Niclas Reitz*

DYNAMIC TRANSIENTS AND RELIABILITY OF HIGH-VOLTAGE SILICON & 4H-SIC BIPOLAR JUNCTION TRANSISTORS UNDER AVALANCE AND SHORT-CIRCUITS ....................... 1402
*Mana Hosseinzadehlish*

PP069   POWER CYCLING TEST OPTIMIZATION TOWARD RELIABILITY ASSESSMENT OF SINTERED POWER MODULES .................................................................... 1410
*Robert Graham*

PP070   REAL-TIME ESTIMATION AND SENSITIVITY ANALYSIS OF PARASITIC CAPACITANCES IN ELECTRIC DRIVE SYSTEMS .................................................... 1418
*Mohammadreza Bagheribavaryani*

## MODELLING AND TESTING

PP071 PARASITIC COMPONENT EFFECTS OF INTERNAL AND EXTERNAL PACKAGE LEVEL ON SWITCHING PERFORMANCE OF SIC POWER MODULE .................................................... 1428
*Nguyen Nghia Do*

PP072 A MULTI-PHYSICS ITERATIVE APPROACH FOR TEMPERATURE ESTIMATION IN SIC POWER MODULE FOR ELECTRIC VEHICLE ......................................................................... 1434
*Stefano Orlando*

PP073 VOLTAGE BALANCING METHOD FOR SERIES CONNECTION OF 50 SIC MOSFETS ......................................................................................................................................... 1441
*Antoine Philippe*

VOLTAGE BALANCING METHOD FOR SERIES-CONNECTION OF 50 SIC MOSFETS .................... 1449
*Antoine Philippe*

PP074 A LABORATORY-SCALE MMC-BASED DC SYSTEM WITH RCP AND PHIL SIMULATION CAPABILITIES ................................................................................................................ 1457
*Marc René Lotz*

PP075 FILM CAPACITOR STANDARD SERIES DIGITALIZATION: ELECTROMAGNETIC & THERMAL MODELLING IMPLEMENTATION IN CLARA WEB TOOL ........................................... 1467
*Fernando Aunon*

PP076 ACCURACY EVALUATION AND PROPOSED DYNAMIC TUNING PROCEDURE OF A COMPACT SIC SPICE MODEL ....................................................................................................... 1475
*Austin Curbow*

PP077 INVESTIGATION OF USE-CASE-DEPENDENT MODELING APPROACH FOR SWITCHED-MODE POWER CONVERTER FOR LVDC GRID EVALUATION ..................................... 1485
*Melanie Lavery*

PP078 AVERAGED MODEL WITH BLOCKING CAPABILITY FOR SOLID-STATE TRANSFORMERS ........................................................................................................................................... 1495
*Ahmed Meligy*

## ADVANCED COMPONENTS

PP080 SURFACANT-MODIFIED NANOCOMPOSITE THIN-FILM CAPACITORS .......................... 1504
*Bartosz Gackowski*

PP081 INCREASING ENERGY STORAGE CAPABILITIES OF POWDER CORES BY ADAPTING THE WINDING AND THE USE OF FRINGING FLUX ......................................................... 1511
*Paul Winkler*

PP082 PEEC-BASED THERMAL MODELING OF PASSIVE COMPONENTS .................................... 1516
*Sascha Langfermann*

PP083 GALVANICALLY ISOLATED POWER SUPPLY FOR GATE DRIVERS IN HIGH VOLTAGE APPLICATIONS ..................................................................................................................... 1523
*Priyanka Ghosh*

PP084    FABRICATION TECHNIQUE FOR NOVEL NANOCRYSTALLINE CORES WITH
HIGH SATURATION POLARIZATION AND LOW LOSSES .................................................. 1532
*Merlin Thamm*

PP085    EXCITATION-DEPENDENT TEMPERATURE BEHAVIOR OF THE QUASI-STATIC
HYSTERESIS LOSS ENERGY DENSITY OF N87 FERRITE MATERIAL............................. 1538
*Jeremias Kaiser*

PP087    PASSIVE METHODS LIMITING LEAKAGE CURRENT IN METAL-OXIDE
VARISTOR AS VOLTAGE CLAMPING DEVICE USED DC LOW VOLTAGE POWER
ELECTRONICS-BASED CIRCUIT BREAKERS ....................................................................... 1545
*Kenan Askan*

## GAN DEVICES AND APPLICATIONS

PP088    ESD SOLUTIONS FOR 650V NORMALLY-OFF ALGAN/GAN HEMTS ..................... 1555
*Thanh Hai Phung*

PP089    A SIMULATIVE STUDY OF MEASUREMENT ERRORS DURING DOUBLE PULSE
TESTING OF GAN DEVICES ...................................................................................................... 1561
*Severin Klever*

PP090    PARALLEL CONNECTION OF GAN FETS: AN EXPERIMENTAL INVESTIGATION
APPROACH...................................................................................................................................... 1568
*Marco Palma*

PP091    REPETITIVE SHORT CIRCUITS ON 650 V GAN ...................................................... 1574
*Adrien Lambert*

PP092    COMPARISON OF SWITCHING LOSSES AND DYNAMIC ON RESISTANCE OF
600 V-CLASS GAN HEMTS......................................................................................................... 1584
*André Thönnessen*

PP093    PERFORMANCE EVALUATION OF DEADTIME AND GATE RESISTANCE FOR
PARALLEL CONNECTED GAN HEMTS .................................................................................... 1590
*Junhyeok Jegal*

PP094    REACHING BEYOND 1200V: LATERAL GAN HEMTS FOR HIGH-RELIABILITY
EV AND INDUSTRIAL APPLICATIONS ..................................................................................... 1598
*Kamal Varadarajan*

## SIC DEVICES AND TECHNOLOGIES

PP095    SMARTSIC 150 & 200MM ENGINEERED SUBSTRATE: INCREASING SIC POWER
DEVICE CURRENT DENSITY UP TO 30%................................................................................. 1604
*Eric Guiot*

PP096    DYNAMIC TRANSIENTS IN HIGH-VOLTAGE SILICON AND 4H-SIC NPN
BIPOLAR JUNCTION TRANSISTORS ....................................................................................... 1610
*Mana Hosseinzadehlish*

PP097    AN ADVANCED MULTI-ASPECT PERFORMANCE ANALYSIS OF PLANAR-GATE
1.2 KV SIC POWER MOSFETS ................................................................................................... 1613
*Anja Katerina Brandl*

PP098    SIC MOSFET DIE SORTING AND PARALLEL FOR OPTIMAL MODULE DESIGN ............. 1621
Zhong Ye

PP099    SIMULATION APPROACH FOR RADIATED ELECTRO-MAGNETIC FIELDS
ESTIMATION ON ACEPACK DRIVE SIC POWER MODULE .................................................................. 1627
Andrea Cusumano

## CONTROL METHODS II

PP100    EXACT ANALYSIS OF CONTROL-TO-OUTPUT TRANSFER FUNCTIONS OF
PWM-CONVERTERS - A COMPARISON OF TWO METHODS ............................................................. 1634
Daniel Breidenstein

PP101    3-LEVEL FLYING CAPACITOR MULTILEVEL TOPOLOGY WITH DELTA-SIGMA
MODULATION ...................................................................................................................................... 1642
Jannik Maier

PP102    MODEL BASED CONTROLLED POWER CONVERTER TEST PLATFORM .......................... 1651
Dawid Koczy

PP103    EDUCATIONAL HARDWARE TRAINER FOR TEACHING THE DUAL ACTIVE
BRIDGE IN A DC GRID ....................................................................................................................... 1658
Peter Van Duijsen

PP104    STUDY OF THE OPERATING PERFORMANCE OF A FCS-MPC-CONTROLLED
MATRIX-CONVERTER FOR PMSM AT DIFFERENT FREQUENCY RATIOS ..................................... 1664
Robert Zipprich

PP105    ENHANCING REACTIVE POWER CAPACITY IN BATTERY-FED POWER
CONDITIONING SYSTEMS ................................................................................................................... 1673
Lucas Araujo

PP106    PULSE SHARING: ACHIEVING HIGH EFFICIENCY AND EXCELLENT REGULA-
TION IN MULTI-OUTPUT FLYBACK POWER SUPPLIES ..................................................................... 1680
Xingda Yan

PP107    RELIABILITY-OPTIMIZED SPACE VECTOR MODULATION (RO-SVM) FOR
SEMICONDUCTORS LIFETIME ENHANCEMENT ................................................................................ 1686
Amin Rezaeizadeh

## INTELLIGENT POWER MODULES

PP108    ANALYSIS AND OPTIMIZATION OF INTERNAL COUPLING INTERFERENCE IN
INTEGRATED SIC POWER MODULE BASED ON DBC .......................................................................... 1693
Chenhang Zeng

PP109    MULTISPECTRAL ELECTROLUMINESCENCE SENSING OF SIC MOSFETS FOR
JUNCTION TEMPERATURE AND CURRENT EXTRACTION ................................................................. 1703
Lukas Ruppert

PP110    SIC-IPM FOR COMPACT AND ENERGY EFFICIENT LOW-POWER MOTOR
DRIVES ............................................................................................................................................... 1712
Jongmu Lee

PP111   CONCEPT FOR A GAN-BASED INTELLIGENT MOTOR CONTROLLER WITH
INTEGRATED FAILURE PREDICTION FOR THE INVERTER AND THE DRIVE ............................... 1717
   *Christoph Blechinger*

PP112   INTRODUCING THE NEW 1200 V CIPOS MAXI IM817 INTELLIGENT POWER
MODULE FOR MOTOR DRIVE APPLICATIONS ...................................................................... 1724
   *Kihyun Lee*

PP113   THERMAL PERFORMANCE OF INFINEON'S NEW 600 V CIPOSTM MICRO IM241
IPM FOR LOW POWER MOTOR DRIVE SYSTEMS WITHOUT HEATSINK ............................... 1732
   *David Jo*

INTRODUCING THE NEEW 1200 V CIPOSTM MAXI IM12BXXXC1 INTELLIGENT POWER
MODULE FOR MOTOR DRIVE APPLICATIONS ...................................................................... 1737
   *Kihyun Lee*

## INTELLIGENT GATE DRIVE UNITS

PP114   AN ADAPTIVE DEAD TIME CONTROL BASED ON SWITCH NODE VOLTAGE
DERIVATIVE ......................................................................................................................... 1745
   *Lukas Knappstein*

PP115   COUPLING COIL DESIGN AND POSITIONING OPTIMIZATION ON NEW HIGH
POWER SEMICONDUCTOR MODULE FOR FAST SHORT CIRCUIT DETECTION ........................... 1751
   *Yannick Dumollard*

PP116   ENABLING ACTIVE THERMAL CONTROL VIA AN ADAPTIVE MULTI-VOLTAGE
GATE DRIVER ......................................................................................................................... 1759
   *Tianlong Albert*

PP117   INNOVATIVE GATE DRIVE METHOD TRIC3 FOR MOTOR ................................................... 1765
   *Hisashi Sugie*

PP118   A NEW CLASS OF SOLID STATE ISOLATORS ENHANCES THE RELIABILITY OF
SOLID STATE RELAYS ............................................................................................................ 1770
   *Wolfgang Frank*

PP119   A SELF-DRIVING 3-LEVEL ACTIVE GATE DRIVER NETWORK TO CONTROL
THE SWITCHING SLEW RATE FOR SIC MOSFETS ............................................................... 1775
   *Vin Loong Choo*

## E-MOBILITY TRACTION II

PP121   ANALYSIS OF LONG-TERM RELIABILITY OF SIC IN TRACTION INVERTER
CONSIDERING VTH INSTABILITY .............................................................................................. 1781
   *Chi Zhang*

PP122   EFFICIENT MAPPING OF ON-DEMAND DRIVE LOAD PROFILES ON INVERTER
STRESS ................................................................................................................................... 1788
   *Zlatko Bosnjic*

PP123   EV TRACTION INVERTER OPTIMAL DESIGN IS DOMINATED BY 3-LEVEL
ANPC ..................................................................................................................................... 1797
   *Timothé Delaforge*

PP124 INTRODUCTION OF POWER SEMICONDUCTOR OPTIONS FOR AN EXCITER OF ELECTRICALLY EXCITED SYNCHRONOUS MOTOR ....................................................... 1804
*Yeriel Bai*

PP125 A NOVEL HIGH POWER DENSITY THREE PHASE TRACTION INVERTER ARCHITECTURE FOR ELECTRIC VEHICLE (EV) APPLICATIONS.................................... 1809
*Yiyang Yan*

PP126 A MODULAR DC-LINK CAPACITOR SOLUTION FOR THE MAIN POWERTRAIN INVERTER OF XEV .................................................................................................. 1814
*David Olalla*

PP127 FAULT IDENTIFICATION TESTING METHODS FOR A COMMERCIAL TRACTION INVERTER ..................................................................................................................... 1821
*Anna Corbitt*

PP128 SHORT CIRCUIT ROBUSTNESS FOR TRACTION INVERTERS FROM AN APPLICATION POINT OF VIEW ....................................................................................... 1828
*Karl Oberdieck*

## INVESTIGATIONS OF PARTICULAR SIC DEVICE PHENOMENON

PP129 THE IMPACT OF THE DEADTIME ON THE STABILITY OF 1.2KV SIC MOSFET BODY DIODE UNDER HARD SWITCHING WITH SYNCHRONOUS RECTIFICATION.................... 1835
*Mohammed Amer Karout*

PP130 RC-DC SNUBBER IMPLEMENTATION FOR SUPPRESSION OF DIODE VOLTAGE PEAK AND RINGING IN A FULL SIC HALF-BRIDGE POWER MODULE ......................... 1844
*Emanuela Alfonzetti*

PP131 SUB-5 SECOND WIDE-BANDGAP POWER DEVICE CALORIMETRIC MEASUREMENTS UTILZIING OPTICAL SENSORS AND PELTIER ELEMENTS ............................. 1851
*Ruben Schnitzler*

PP132 SIC TRENCH MOSFETS IN AVALANCHE MODE WITH RC SNUBBER CIRCUIT.............. 1858
*Sebnem Tuncay*

PP133 HIGH-FREQUENCY OSCILLATIONS IN SIC MOSFET POWER MODULES DURING TURN-ON SWITCHING TRANSIENT – ANALYSIS BASED ON SIMULATIONS AND MITIGATION METHODS.......................................................................................... 1865
*Rajani Kumar Thirukoluri*

PP134 A DYNAMIC CURRENT BALANCING METHOD USING FULL-COUPLED INDUCTORS IN PARALLELED GATE BRANCHES............................................................ 1872
*Jianwei Lv*

PP135 QUANTITATIVE PERFORMANCE COMPARISON OF LARGE-FORMAT SIC MOSFET AND SI IGBT MODULES ...................................................................................... 1878
*Arthur Boutry*

## THERMAL MANAGEMENT AND ADVANCED COOLING

PP136 SOLDER PREFORM TECHNOLOGY FOR IMPROVED THERMOMECHANICAL PERFORMANCE IN MOLDED POWER MODULE PACKAGE-ATTACH .............................. 1886
*Joseph Hertline*

PP138   EFFECT OF FLIP-CHIP DIE-ATTACH ON THE THERMAL BEHAVIOR OF POWER
GAAS DIODES .................................................................................................................... 1891
   Felix Steiner

PP139   INFLUENCES OF SOLDER DELAMINATION ON THE THERMAL
PERFORMANCE IN AUTOMOTIVE TRACTION MODULE .......................................... 1896
   Hansol Seo

PP141   DEVELOPMENT OF A PASSIVE CAPILLARY-PUMPED COOLING SYSTEM FOR
HIGH-PERFORMANCE ELECTRONICS ........................................................................ 1902
   Justin Fey

PP143   ADVANCED COOLING OF POWER ELECTRONICS WITH COPPER COLD
SPRAYED ALUMINIUM HEATSINKS & BUSBARS......................................................... 1907
   Michael Dasch

PP144   COLD PLATE DESIGN FOR COOLING LV100 SILICON CARBIDE POWER
MODULE PACKAGING......................................................................................................... 1910
   Wahid Cherief

PP145   AN IMPROVED DOUBLE-LAYER SPACER IN DOUBLE-SIDED COOLING POWER
MODULE................................................................................................................................ 1917
   Linhao Ren


## RELIABILITY TESTING

PP146   POWER CYCLING OF 1.7KV MULTI-CHIP POWER MODULES – SIC MOSFETS
VS SILICON IGBTS................................................................................................................ 1923
   Nick Baker

PP147   POWER CYCLING CAPABILITY OF DISCRETE SIC MOSFET DEVICES WITH
DIFFERENT DESIGNS.......................................................................................................... 1930
   Luhong Xie

PP148   MODEL-BASED PARAMETER TUNING OF SEMICONDUCTOR DEVICES IN DC
POWER CYCLING TEST....................................................................................................... 1936
   Yi Zhang

PP149   INFLUENCE OF TRANSFER MOLDING ON THE RELIABILITY OF DCM SIC
POW-ER MODULES............................................................................................................... 1942
   Jacek Rudzki

PP150   DAMP HEAT BEHAVIOR OF HIGH HEAT CAPACITORS FOR APPLICATIONS IN
ELECTRIC VEHICLES.......................................................................................................... 1951
   Adel Bastawros

PP151   INFLUENCE OF THE GATE VOLTAGE DURING ON-TIME ON THE POWER
CYCLING CAPABILITY OF SIC MOSFETS ....................................................................... 1955
   Patrick Heimler

PP152   INVESTIGATION OF THE TEMPERATURE MEASUREMENT VIA VSD(T)-
METHOD APPLIED TO PARALLELED SIC MOSFET CHIPS DURING POWER CYCLING .............. 1964
   Kevin Ladentin

PP153   APPROACHES OF TSEP MEASUREMENTS FOR POWER SEMICONDUCTORS ................ 1969
   Philipp Hauenschild

PP154    REALTIME JUNCTION TEMPERATURE ESTIMATION IN SIC POWER MODULES
BASED ON MULTIPLE TSEP ACQUISITION ........................................................................ 1978
    Kevin Muñoz Barón


## HIGH VOLTAGE WBG DEVICES

PP155    ENHANCED CURRENT MEASUREMENT APPROACH FOR NON-ISOLATED 6.5
KV SILICON CARBIDE MOSFETS ............................................................................................. 1987
    Xinyuan Du

PP156    NEW 2KV SIC-MOS TECHNOLOGY FOR APPLICATION FIELDS IN THE
INDUSTRIAL LANDSCAPE.......................................................................................................... 1991
    Igor Kasko

PP157    HIGH TEMPERATURE EXPERIMENTAL CHARACTERIZATIONS OF COSS OF 3.3
KV SIC MOSFET FOR MEDIUM VOLTAGE PV APPLICATIONS............................................. 1999
    Paul Schmidt

PP158    IMPACT OF GATE CONTROL ON THE SWITCHING PERFORMANCE OF 3.3KV
SBD-EMBEDDED SIC-MOSFET................................................................................................... 2006
    Junya Sakai

PP159    COMPARATIVE ASSESSMENT OF OVERLOADABILITY POTENTIAL OF 3.3 KV
SI-IGBTS AND SIC-MOSFET POWER MODULES ..................................................................... 2013
    Muhammad Nawaz

PP160    IMPROVED RELIABILITY OF A 2200 V SIC MOSFET MODULE WITH AN EPOXY-
ENCAPSULATED INSULATED METAL SUBSTRATE................................................................. 2022
    Hiroshi Kono

PP161    PARALLELING 3.3-KV/800-A RATED SIC-MOSFET MODULES – AN
OPTIMIZATION METHOD.......................................................................................................... 2028
    Hiroyuki Irifune

PP162    PERFORMANCE ASSESSMENT OF 10 KV SIC MOSFET AND PIN DIODE IN 3L-
NPC CONVERTER TOPOLOGY ................................................................................................. 2036
    Renato Amaral Minamisawa


## VOLUME 4

PP163    PERFORMANCE EVALUATION OF COOLSIC 2 KV SIC MOSFET DISCRETE IN
1500 V DC LINK SYSTEMS ......................................................................................................... 2041
    Ajith Kumar Sekar

PP164    A NEW 2.3 KV RATED SIC MOSFET MODULE WITH LOW-INDUCTANCE HIGH-
POWER PACKAGE HPNC FOR 1500 VDC APPLICATIONS ................................................... 2049
    Junya Kawabata


## PACKAGING AND INTERCONNECTION MATERIALS

PP166    MECHANISM FOR IMPROVING THE HEAT-RESISTANCE OF ADHESIVE
INTERFACE IN FLEXIBLE PRINTED CIRCUITS...................................................................... 2053
    Keita Suzuki

PP167    A SYSTEMATIC COMPARISON STUDY OF DIFFERENT BONDING
TECHNOLOGIES FOR SUBSTRATE ATTACHMENT OF POWER ELECTRONICS............................ 2060
   Lisheng Wang

PP168    STABILITY OF PRESSURE SINTERED INTERCONNECTS AS A FUNCTION OF
TEMPERATURE AND ENVIRONMENTAL CONDITIONS.................................................................... 2067
   Kentaro Yoshioka

PP169    THE EFFECT OF NANO-CU INTERCONNECTION MATERIALS ON THE
THERMOMECHANICAL PROPERTIES OF SIC DOUBLE-SIDED POWER MODULES ..................... 2074
   Suhang Wei

PP170    ALL-IN-ONE-SINTERING: DIE-ATTACH AND SUBSTRATE-ATTACH ON BARE
COPPER IN A PRESSURE ASSISTED SINTERING ONE-STEP PROCESS............................................ 2082
   Battist Rabay

PP171    SEQUENTIAL MANUFACTURING OF HIGHLY FUNCTIONALIZED THREE-
DIMENSIONAL CERAMIC COMPONENTS FOR POWER ELECTRONICS......................................... 2088
   Lars Rebenklau

PP173    PARAMETRIC STUDY OF DAMAGE EVOLUTION IN SILVER SINTERED
LAYERS OF DOUBLE SIDED POWER ELECTRONICS MODULES OF ELECTRICAL
VEHICLES......................................................................................................................................... 2094
   Saeed Akbari

## DC-DC CONVERTER I

PP174    TRISTATE MODIFIED BOOST CONVERTER..................................................................... 2104
   Johannes Gragger

PP175    COMPARATIVE EVALUATION OF THE CENTER TAPPED BOOST CONVERTER
TOPOLOGY ..................................................................................................................................... 2112
   Bryan Radix

PP176    COMPARISON OF MULTI-LEVEL TOPOLOGIES TO REDUCE THE
COMPONENTS VOLTAGE STRESSES WHEN POWERED FROM INDUSTRIAL DC GRIDS.............. 2119
   Katharina Machtinger

PP177    HARD-SWITCHING HIGH-FREQUENCY GAN-BASED DC-DC CONVERTERS
WITH CONCOMITANT DATA TRANSMISSION FUNCTIONALITY ....................................................... 2128
   Abdelmoumin Allioua

PP178    EFFICIENT DESIGN OF HIGH-CURRENT, LOW-OUTPUT VOLTAGE DC-DC
CONVERTERS USING ARTIFICIAL INTELLIGENCE-BASED TOPOLOGY SELECTION
AND OPTIMIZATION ....................................................................................................................... 2138
   Thomas Harmand

## HIGH POWER DC-DC CONVERTER I

PP180    A SIC BASED 60KW LLC CONVERTER WITH NOVEL TRANSFORMER DESIGN
FOR IMPROVING VOLTAGE BALANCE ........................................................................................... 2146
   Frank Wei

PP181    ANALYSIS OF INVERTER OPERATION MODES OF AN IGBT-BASED ZCS LLC
CONVERTER FOR A 2 KW AUTOMOTIVE ON-BOARD DC-DC ........................................................... 2152
    Daniel Urbaneck

PP182    DUAL OUTPUT HYBRID CONVERTER FOR 48 V DATA CENTERS: M-HSC ....................... 2162
    Simone Mazzer

PP183    3.6KW HIGH EFFICIENCY SIC-BASED HV/LV DC-DC CONVERTER FOR EVS ................. 2167
    Veera Bharath Chandra Reddy Gandluru

PP184    BIDIRECTIONAL DC-DC TOPOLOGIES COMPARISON FOR 800 V AUTOMOTIVE
APPLICATIONS INTEGRATING 650 V GAN-ON-SI DEVICES ......................................................... 2175
    Ilias Chorfi

PP185    ANALYSIS OF PHASE SHIELDING METHOD BASED ON ?-CR-Y THREE-PHASE
INTERLEAVED LLC CONVERTER ................................................................................................. 2182
    Jin Wen

PP186    22KW IMS-BASED BIDIRECTIONAL DC-DC CONVERTER USING SURFACE
MOUNT SIC MOSFETS FOR OBCS ............................................................................................... 2185
    Hamlin Wang

PP187    COMPARATIVE ANALYSIS OF DC-DC CONVERTERS FOR ELECTROLYZERS
USING GEOMETRIC PROGRAMMING ......................................................................................... 2190
    Tim McRae

PP188    DESIGN CONSIDERATION OF BI-DIRECTIONAL CLLLC RESONANT
CONVERTER IN ENERGY STORAGE SYSTEMS ........................................................................... 2200
    Sheng-Yang Yu


## SMART-GRID TECHNOLOGIES

PP189    ADAPTIVE FAST CHARGING SYSTEM WITH SECOND LIFE BATTERIES - AN
OVERVIEW OF A RESEARCH PROJECT ........................................................................................ 2208
    Lukas Böhning

PP190    PARALLEL OPERATION AND SYNCHRONIZATION OF MICROGRIDS BY USING
THE THEVENIN THEOREM ........................................................................................................... 2217
    Marius Block

PP192    21 KA SOLID STATE DC BREAKER FOR SUPERGRID INSTITUTE'S HIGH
POWER TEST FACILITY ................................................................................................................ 2227
    Christophe Conilh

PP193    DESIGN AND ANALYSIS OF A 50KW SIC-BASED ACTIVE FRONT END WITH A
VERY SMALL LINE CHOKE FOR DC-GRIDS .............................................................................. 2234
    Raphael Otte

PP194    INVESTIGATION OF LOAD TRANSITIONS BETWEEN LOADED AND LOAD
FREE CONDUCTOR SEGMENTS IN INDUSTRIAL CONDUCTOR SYSTEMS ..................................... 2240
    Jan-Niklas Koch

PP195    A METHOD TO CONTROL VOLTAGE AND POWER FLOW IN A DC GRID ......................... 2248
    Peter Van Duijsen

## ENERGY STORAGE SYSTEMS

PP196    CONSIDERATIONS ON A HIGH-CELL-COUNT CONVERTER-BASED BATTERY STORAGE SYSTEM WITH REDUCED COMMUNICATION EFFORT ..................................................... 2258
*Paul Aspalter*

PP197    STUDYING CONVERTORS FOR VOLTAGE EQUALIZATION IN ENERGY STORAGE SYSTEM WITH ACTIVE BMS .................................................................................................. 2268
*Dimitar Arnaudov*

PP198    CHALLENGES OF HIGH SIDE GATE DRIVER AND DISCONNECT MOSFET FOR BATTERY PROTECTION UNIT DURING START-UP, TURN-OFF AND OVER CURRENT EVENTS.................................................................................................................................................. 2273
*Niranjan Suravarapu Reddy*

PP199    ELECTRIC INSULATION COORDINATION TO PREVENT ELECTRIC ARCS IN LITHIUMION BATTERIES .................................................................................................................. 2278
*Daniel Chatroux*

PP201    BATTERY CHARGER WITH IMPEDANCE SPECTROSCOPY CAPABILITY FOR LI-ION CELLS................................................................................................................................................. 2286
*Christian Branas*


## EMC

PP202    EFFICIENCY, VOLUME AND CO2 EMISSIONS IMPACT IN A PFC CONVERTER WITH AN ACTIVE FILTER SOLUTION FOR OBC APPLICATION............................................... 2294
*Kelly Ribeiro*

PP203    ANALYTICAL AND EXPERIMENTAL VALIDATION COMMON MODE FEEDBACK LOOP FOR A THREE-PHASE_LEVEL VIENNA RECTIFIER............................................. 2303
*Daniel San Laureano Igartuburu*

PP204    ROBUSTNESS OF FREQUENCY-DOMAIN TERMINAL MODELING OF ELECTROMAGNETIC INTERFERENCES IN STATIC CONVERTERS ............................................. 2309
*Mehyeddine Singer*

PP205    STUDY OF EMI BEHAVIOR OF A 2-LEVEL GAN-INVERTER – SIMULATION AND MEASUREMENT................................................................................................................................... 2316
*Benedikt Kohlhepp*

COMMON MODE CURRENTS IN RESONANT CIRCUITS GENERATED WITH A DELTA-SIGMA MODULATED VOLTAGE SOURCE INVERTER.............................................................. 2326
*Tobias Haas*

PP206    ANALYSIS OF COMMON-MODE NOISE GENERATED DUE TO FAST-SWITCHING GAN DEVICES IN TOTEM-POLE PFCS ........................................................................ 2334
*Serkan Dusmez*

PP207    CONDUCTED EMI FROM GAN-BASED 48V TO 12V DC-DC-CONVERTERS FOR AUTOMOTIVE APPLICATIONS ................................................................................................................ 2342
*Erik Kampert*

## ADVANCED DESIGN

PP208   APPLIED DESIGN AUTOMATION FOR FINDING FEASIBLE DESIGNS FOR HIGH-FREQUENCY PLANAR TRANSFORMERS .................................................................................. 2350
*Rando Raßmann*

PP209   FREQUENCY DEPENDENT AREA PRODUCT METHOD ........................................ 2359
*Alfonso Martínez*

HIGH RESOLUTION MIXED-SIGNAL PULSE WIDTH MODULATOR FOR HIGH-FREQUENCY DC-DC CONVERTERS ................................................................................... 2364
*Tim McRae*

PP210   DESIGNING A CONTROL LIBRARY FOR GRID-FOLLOWING AND GRID-FORMING POWER INVERTERS ................................................................................................ 2370
*Lars Lindner*

PP211   INTELLIGENT OPTIMISATION OF A WIND TURBINE DIGITAL TWIN MODEL ................. 2377
*René Reimann*

PP212   THERMAL TRANSIENT DIGITAL TWIN MODELLING FOR POWER CONVERTERS ............................................................................................................... 2386
*Xianghao Mo*

PP213   A DIGITAL TWIN APPROACH TOWARD LIFETIME ANALYSIS AND PREDICTIVE MAINTENANCE OF POWER SEMICONDUCTORS FOR RAILWAY APPLICATION ........................... 2394
*Emmanuel Batista*

## INDUCTORS

PP214   SATURABLE FERRITE CORE INDUCTORS IN LCL FILTERS OF THREE-PHASE VOLTAGE SOURCE INVERTERS ................................................................................................ 2400
*Marius Kaufmann-Bühler*

PP215   2D COPPER LOSS ANALYTICAL MODEL FOR PLANAR INDUCTOR COMBINING HIGH AND LOW PERMEABILITY MATERIALS ................................................................. 2408
*Idriss Nachete*

PP216   CNC-MANUFACTURED POWER INDUCTORS WITH EXCELLENT BANDWIDTH FOR MULTI-MEGAWATT CONVERTERS ....................................................................... 2416
*Thomas Kreppel*

PP217   ANALYTICAL EVALUATION OF DIFFERENTIAL MODEL DC EMI FILTER INDUCTORS USING MATERIAL SATURATION COEFFICIENT ......................................... 2425
*Lukas Mueller*

PP218   DESIGN AND PERFORMANCE EVALUATION OF AIR CORE INDUCTORS FOR VERY HIGH FREQUENCY POWER CONVERSION ...................................................... 2431
*Florentin Salomez*

PP220   IMPROVING MULTI-PHASE FERRITE MAGNETICS BY COUPLING FOR MV AND UPS CONVERTERS ...................................................................................................... 2438
*Michael Schmidhuber*

## E-MOBILITY CHARGING

PP221    22-KW BIDIRECTIONAL SINGLE-STAGE DIRECT-AC-AC POWER CONVERSION ON-BOARD CHARGER WITH HIGH-POWER-DENSITY IMPLEMENTATION ..................................... 2448
*Oscar Lucia*

PP222    BENCHMARKING DC FAST CHARGERS: A COMPARATIVE ANALYSIS OF POWER CONVERTER STRUCTURES FOR WIDE VOLTAGE RANGE ................................................. 2453
*Sadik Cinik*

PP223    PERFORMANCE OPTIMIZATION OF SINGLE-PHASE ON-BOARD CHARGERS WITH RIPPLE PORT .................................................................................................................... 2461
*Davide Gottardo*

PP224    A REDUCED-SENSOR MODULAR DUAL ACTIVE BRIDGE-BASED BATTERY CHARGING SYSTEM FOR ELECTRIC VEHICLES USING AN IMPROVED LINEAR EXTENDED STATE OBSERVER ................................................................................................. 2469
*Armel Asongu Nkembi*

PP225    BIDIRECTIONAL NON-ISOLATED THREE-PHASE ONBOARD CHARGER WITH A LOW-VOLTAGE LOWER-PHASE OPERATION MODE ............................................................ 2478
*Steffen Frei*

PP226    CONTROL OF A THREE-PHASE INDUCTIVE POWER TRANSFER SYSTEM BASED ON DD²Q COIL TOPOLOGY ............................................................................................ 2488
*Nikola Mirkovic*

PP227    COMPARISON OF TWO BIDIRECTIONAL 11KW 400V CLLC AND CLLLC RESONANT CONVERTERS FOR EV APPLICATIONS ............................................................... 2494
*Hasan Mousavi Somarin*

PP228    DYNAMIC WIRELESS CHARGING SYSTEM DESIGN FOR EXTRA-URBAN AREAS BASED ON RESONANT INDUCTIVE POWER TRANSFER .......................................... 2503
*Irene Maria Torres Alfonso*

PP229    BIDIRECTIONAL ISOLATED 400-12V DC-DC CONVERTER WITH IMPROVED POWER DENSITY AND FULL-RANGE OPERAION FOR EV APPLICATIONS ................................... 2513
*Oscar Lucia*

## HIGH POWER DC-DC CONVERTER II

PP230    GAIN OPTIMIZATION CONTROL METHOD FOR CLLLC RESONANT CONVERTERS UNDER PHASE SHIFT MODE ...................................................................................... 2518
*Sean Yu*

PP231    ANALYSIS OF COMMON AND SPLIT DC-BUS INTERLEAVED H-BRIDGE CONVERTERS FOR HIGH-CURRENT LOW-RIPPLE APPLICATIONS ................................................. 2524
*Bhavana Gudala*

PP232    OPTIMAL FREQUENCY OPERATING POINTS FOR HYBRID SWITCHED CAPACITOR CONVERTERS AND LOSSLESS CURRENT SENSE METHOD ...................................... 2532
*Simone Mazzer*

PP233    DESIGN AND TESTING OF A 250 KW 50 KHZ SIC-BASED HALF-BRIDGE-
SERIES-RESONANT-CONVERTER ................................................................................ 2538
   *Daniel Haake*

PP234    30KW - 97% EFFICIENCY ISOLATED DC-DC CONVERTER WITH LARGE INPUT
VOLTAGE RANGE BASED ON A BOOST DAB ASSOCIATION ........................................ 2547
   *Jean-Jacques Huselstein*

PP235    ANALYSIS OF A FULL-BRIDGE PUSH-PULL FORWARD DUAL ACTIVE BRIDGE
DC-DC CONVERTER ...................................................................................................... 2557
   *Gean Sousa*

## DC-DC CONVERTER II

PP236    SYMMETRICAL OPERATION OF FOUR CHANNEL RESONANT BOOST DC-DC
CONVERTERS IN CONTINUOUS CONDUCTION MODE ................................................ 2566
   *Kristóf Bándy*

PP237    IMPACT OF MAGNETICS TOLERANCE ON THE POWER SHARING OF
PARALLEL DUAL-OUTPUT PHASE-SHIFT FULL-BRIDGE CONVERTERS ...................... 2576
   *Riccardo Mandrioli*

PP238    A BALANCING CONVERTER WITH SERIES CONNECTED MOSFETS FOR +/-
700V BIPOLAR DC GRIDS ............................................................................................ 2583
   *Sachin Yadav*

PP239    OPTIMIZATION AND DESIGN OF LOW-VOLTAGE AND HIGH-CURRENT POINT-
OF-LOAD CONVERTER UNDER 48V BUS ARCHITECTURE ........................................... 2591
   *Jiajia Guan*

PP240    INTERLEAVED BOOST CONVERTER EFFICIENCY AND POWER DENSITY
MODEL FOR ACTIVE AND PASSIVE COMPONENT DESIGN .......................................... 2596
   *Damien Lemaitre*

## NOVEL AND ADVANCED SEMICONDUCTOR DEVICES

PP241    EVALUATION OF A HYBRID POWER SWITCH BASED ON TRENCH CLUSTERED
IGBT AND SIC MOSFET ................................................................................................ 2606
   *Alireza Sheikhan*

PP242    CONTRIBUTIONS FOR BUILDING BLOCKS FOR NORMALLY-OFF 650V GAN-
ON-SI POWER INTEGRATED CIRCUITS ....................................................................... 2612
   *Thanh Hai Phung*

PP243    NEW BIDIRECTIONAL ASYMMETRIC HIGH VOLTAGE TVS (TRANSIENT
VOLTAGE SUPPRESSOR) DIODE ................................................................................. 2620
   *Boris Rosensaft*

PP244    ISO247: HIGH PERFORMANCE CERAMIC BASED ADVANCED ISOLATED
DISCRETE PACKAGE TO FULLY EXPLOIT THE ADVANTAGES OF SIC MOSFET .......... 2627
   *Sachin Shridhar Paradkar*

PP245    IMPACT OF CURRENT RIPPLE REDUCTION USING HIGH SWITCHING
FREQUENCIES ON PMSM EFFICIENCY ....................................................................... 2632
   *Jannik Fuchs-Gade*

PP246   MAXIMIZING COST-EFFICIENCY IN ELECTRIC DRIVETRAINS: A SIC/SI
FUSION SWITCH APPROACH ................................................................................................ 2638
    Matthias Ippisch


## ADVANCED CONTROL

PP247   CONCISE AND RELIABLE SIC MOSFET DRIVER CIRCUITS .............................................. 2646
    Zhong Ye

PP248   ARTIFICIAL INTELLIGENCE ENHANCED RESOLVER SYSTEM FOR
AUTOMOTIVE TRACTION INVERTER APPLICATIONS BASED ON AURIX TC4X ......................... 2651
    David Zipperstein

PP250   MULTIFUNCTIONAL GRID MANAGER TOPOLOGY WITH CONFIGURABLE
OUTPUT ................................................................................................................................ 2657
    Peter Van Duijsen

PP252   CO2 FOOTPRINT OF MEDIUM VOLTAGE DC SOLID STATE
TRANSFORMER   ................................................................................................................... 2663
    Adriana Campos


## SIC MOSFET

PP253   THERMO-ELECTRICAL ANALYSIS AND PERFORMANCE: A COMPARATIVE
STUDY BETWEEN MODULAR AND DISCRETE APPROACHES ......................................... 2673
    Stefano Orlando

PP254   IMPACT OF PARAMETER SPREAD IN PARALLEL-OPERATED SIC MOSFETS
FOR HARD-SWITCHING CONVERSION ................................................................................. 2680
    Andrea Piccioni

PP255   ASSESSMENT OF THE RDS,ON OF SIC MOSFET DIES THROUGH KELVIN WIRE
CONNECTION .......................................................................................................................... 2686
    Philipp Rehlaender

PP256   CHALLENGES IN SCALING SIC SINGLE-CHIP MEASUREMENTS TO
CORRESPONDING POWER MODULES ................................................................................... 2693
    Hao Wang

PP257   SWITCHING PERFORMANCE EVALUATION OF HIGH-POWER 1.7 KV SIC
MOSFET MODULES USING A COMMON BUSBAR DESIGN ............................................... 2700
    Sebastian Neira

PP258   CHARACTERIZING THE SWITCHING BEHAVIOR OF A 1.2 KV MIXED SIC JFET
AND MOSFET HALF BRIDGE ................................................................................................ 2708
    Tim Ringelmann


## VOLUME 5

## WBG HIGH FREQUENCY APPLICATION

PP259   PERFORMANCE EVALUATION OF THE PACKAGING OF SIC DIODES IN A 6.78
MHZ WIRELESS POWER TRANSFER SYSTEM ....................................................................... 2718
    Ioannis Nikiforidis

PP260    VOLTAGE WAVEFORM GENERATION FOR SAWYER-TOWER COSS LOSS MEASUREMENTS USING A HYBRID POWER CONVERTER .............................................................. 2724
*Malachi Hornbuckle*

PP261    EVALUATION OF SIC DEVICES FOR OVER 500KHZ APPLICATION BASED ON BUCK CIRCUIT ................................................................................................................................ 2730
*Minli Jia*

PP262    LINEARIZATION OF DRAIN-SOURCE CAPACITANCES FOR ANTISERIAL CONFIGURATED SIC MOSFETS IN HIGH FREQUENCY SOLID STATE SWITCHES ........................ 2737
*Lars Dresel*

## SIC RUGGEDNESS

PP263    EFFECTS OF NON-KILLER DEFECTS ON SIC MOSFET SHORT-CIRCUIT RUGGEDNESS AND RELIABILITY ..................................................................................................... 2745
*Sara Kuzmanoska*

PP264    DYNAMIC REVERSE BIAS TEST: ELECTRO-THERMAL CHARACTERIZATION OF SIC MOSFETS ................................................................................................................................. 2751
*Giuseppe Mauromicale*

PP266    RADIATION HARDNESS OF SIC BASED INVERTERS BASED ON AN EV MISSION PROFILE ........................................................................................................................... 2758
*Hadiuzzaman Syed*

PP267    RAPID SHORT CIRCUIT PROTECTION USING DIDT DETECTION FOR SIC POWER MODULES ............................................................................................................................... 2764
*Koki Samura*

PP268    COMPARISON OF DYNAMIC GATE STRESS TEST RESULTS OF SIC MOSFETS ............. 2769
*Mathias Gebhardt*

PP279    EXTENDING SIC MOSFET SHORT-CIRCUIT WITHSTANDING TIME BY TWO-LEVEL TURN-OFF GATE DRIVING ........................................................................................................ 2778
*Kwokwai Ma*

PP270    EXPERIMENTAL INVESTIGATIONS ON PARASITIC TURN-ON OF 1.2KV SIC MOSFET DISCRETE DEVICES ....................................................................................................... 2786
*Thanh-Toan Pham*

PP271    BEHAVIOR MODELLING THE SHORT CIRCUIT CHARACTERISTICS OF SIC MOSFETS USING COMPACT MODELS ....................................................................................... 2791
*Qing Sun*

## THERMAL CHARACTERIZATION

PP273    THERMAL ANALYSIS AND MODELLING OF CHARGING STATIONS FOR ELECTRIC VEHICLES ............................................................................................................................ 2796
*Ruben Kopischke*

PP274    JUNCTION TEMPERATURE MEASUREMENT OF A 3.3 KV SILICON CARBIDE MOSFET POWER MODULE ................................................................................................................ 2803
*Michael Gleissner*

PP275 INNOVATIVE 3D POWER MODULE DEFAULTS DETECTION VIA THERMAL IMPEDANCE ANALYSIS AND SIMULATIONS ..................................................................2811
*Louis Alauzet*

PP276 THERMAL CHARACTERIZATION OF AN AIR-COOLED PEBB BASED ON SIC MOSFET POWER MODULES ...................................................................................... 2819
*Alexandre Marie*

PP277 THERMAL BEHAVIOUR OF SIC MOSFET WITH PLANAR PACKAGING TECHNOLOGY ................................................................................................................ 2826
*Yijun Ye*


## RELIABILITY AND AVAILABILITY

PP279 IMPLEMENTING MODULE HEALTH MONITORING IN EV TRACTION INVERTERS ................................................................................................................... 2831
*Karol Rendek*

PP280 RELIABILITY TESTS OF COPPER THICK-FILM SUBSTRATES FOR POWER ELECTRONIC APPLICATIONS ............................................................................... 2838
*Henry Barth*

PP281 POWER MODULE SOLUTIONS WITH IMPROVED RELIABILITY FOR ELEVATOR DRIVE APPLICATIONS ......................................................................... 2843
*Tiago Jappe*

PP282 FAIL-OPERATIONAL LLC TOPOLOGIES WITH FAULT-TOLERANCE INTEGRATED REDUNDANT CAPABILITIES ................................................................ 2850
*Aswathy M. Prince*

PP283 THERMAL AND RELIABILITY OPTIMIZATION OF CLIPS IN SIC MOSFET POWER MODULES ........................................................................................................ 2860
*Zexiang Zheng*

PP284 CONDITION MONITORING OF A GAN FULL-BRIDGE BY MEANS OF FORWARD VOLTAGE IN CONTINUOUS OPERATION ........................................................ 2866
*Michael Vogt*

PP285 A SIMPLE AND LOW COST OVERCURRENT PROTECTION SYSTEM BASED ON COMMERCIAL SHUNT FOR WIDE-BANDGAP DEVICES .................................... 2874
*Emanuele Martano*

PP286 SVM-BASED FAULT-TOLERANT CONTROL FOR A CASCADED H-BRIDGE MULTILEVEL CONVERTER UNDER MULTIPLE OPEN-CIRCUIT SWITCH FAULTS ....................... 2880
*Dong Xie*

PP287 REVOLUTIONIZING MOBILITY: THE SECOND LIFE OF ONBOARD CHARGING SYSTEMS IN COMMERCIAL VEHICLES ...................................................................... 2886
*Ajay Krishna Voppu Muralikrishna*


## LOW VOLTAGE SWITCHES

PP288 A BEHAVIORAL TRANSIENT MODEL FOR IGBT DEVICE WITH ANTI PARALLEL FREEWHEELING DIODE ..................................................................................................... 2893
*Shiwu Zhu*

PP289    PARAMETER EXTRACTION FOR AN ANN-ASSISTED IGBT MODEL IN
TRANSIENT SIMULATIONS ............................................................................................... 2901
    *Huaiyuan Zhang*

PP290    FABRICATION OF 600V RC-IGBT USING 300MM WAFER ...................................... 2909
    *Masaki Ueno*

PP291    NEXT LEVEL OF POWER MODULE SOLUTION FOR PV C&I STRING INVERTER
WITH 1200V H7 TECHNOLOGY IN EASY3B PACKAGE ............................................... 2914
    *Tilo Poller*

PP292    ANALYSIS OF MOSFET SWITCHING LOSSES IN RESONANT CONVERTERS
USING ELECTRICAL AND THERMAL MEASUREMENTS AND LOSS TRENDS WITH
MOSFET SIZE VARIATION ................................................................................................. 2921
    *Alfio Scuto*

PP293    OPTIMOS 6 135V FOR HIGH POWER MOTOR DRIVES ........................................... 2930
    *Kunal Jha*

PP294    AUTO POWER-SOI: SHAPING THE FUTURE OF BATTERY MONITORING
TECHNOLOGY ...................................................................................................................... 2937
    *Alex Lim*

## LIFETIME MODELLING AND CONDITION MONITORING

PP295    UNDERSTANDING THE IMPACT OF IEC60747-17 ON CAPACITIVE AND
MAGNETIC COUPLERS ....................................................................................................... 2942
    *Shu Ee Ong*

PP296    PARIS LAW APPLIED TO WIRE BONDS DEGRADATION USING CRACK
GROWTH MEASUREMENT ................................................................................................. 2948
    *Merouane Ouhab*

PP297    CONDITION MONITORING TECHNIQUE OF POWER ELECTRONIC MODULES
VIA SQUARE-WAVE GATE SIGNAL EXCITATION ........................................................ 2956
    *Isabel Austrup*

PP298    STATISTICS-BASED LIFETIME SIMULATION ENVIRONMENT FOR POWER
MODULES INCORPORATING DEGRADATION MODELS .................................................. 2963
    *Karthik Debbadi*

PP299    POWER CYCLING RESULTS FOR RELIABILITY STUDIES OF SIC-INVERTERS ............. 2972
    *Robert Keilmann*

PP300    GAN CASCODE IN HIGH SPEED DRIVEN AIR COMPRESSORS FOR
AUTOMOTIVE FUEL CELLS................................................................................................ 2981
    *Florian Lippold*

PP301    PROGNOSTIC ANALYSIS OF IGBT HEALTH: REAL-TIME ON-STATE VOLTAGE
PREDICTION THROUGH MACHINE LEARNING .............................................................. 2986
    *Tanya Thekemuriyil*

PP302    ROBUSTNESS ANALYSIS OF TEMPERATURE-SENSITIVE ELECTRICAL
PARAMETERS OF IGBTS ..................................................................................................... 2995
    *Laurids Schmitz*

PP303 OBSERVATION OF THERMAL-RESISTANCE INCREASE OF DEGRADED IGBT MODULES BY VCE (SAT) MEASUREMENT IN A CHOPPER CIRCUIT ............................................. 3002
*Kazunori Hasegawa*

## PULSE WITH MODULATION METHODS

PP304 MODULATION TECHNIQUE FOR REDUCED AC CONTENT OF THE DC LINK CURRENT IN THREE-PHASE TWO-LEVEL INVERTERS ................................................. 3007
*Steffen Frei*

PP305 COMMON MODE CURRENTS IN RESONANT CIRCUITS GENERATED WITH A DELTA-SIGMA MODULATED VOLTAGE SOURCE INVERTER ............................................. 3017
*Tobias Haas*

PP306 EVALUATION OF NEW MODULATION SCHEME FOR 3L-ANPC USING BOTH CURRENT PATHS IN ZERO STATE ................................................................................. 3020
*Felix Eichler*

PP307 AN INNOVATIVE SYNCHRONOUS RECTIFICATION METHOD FOR 11KW CLLC CONVERTER .................................................................................................................. 3029
*Sanbao Shi*

PP308 INTERLEAVED ASYNCHRONOUS DELTA-SIGMA MODULATION CONCEPT FOR DYNAMIC POWER CONVERTERS .......................................................................... 3034
*Philipp Czerwenka*

PP309 HIGH RESOLUTION MIXED-SIGNAL PULSE WIDTH MODULATOR FOR HIGH-FREQUENCY DC-DC CONVERTERS ........................................................................... 3042
*Tim McRae*

PP310 IMPLEMENTATION AND CONTROL OF OPTIMIZED PULSE PATTERNS FOR SALIENT PERMANENT MAGNET SYNCHRONOUS MACHINES IN ELECTRIC VEHICLES ........... 3045
*Maximilian Hepp*

PP311 A 3-LEG INTERLEAVED TP PFC WITH A 90° PHASE-SHIFTED ASYMMETRIC LEG FOR REDUCED MAGNETICS ........................................................................... 3060
*Serkan Dusmez*

PP312 FAULT-TOLERANT OPERATION ANALYSIS OF A FIVE-PHASE THREE-LEVEL TNPC INVERTER FOR ELECTRIC AIRCRAFT PROPULSION SYSTEMS ............................. 3067
*Chanuch Chaisakdanugull*

## AC-DC AND DC-AC CONVERTER

PP313 CCM TOTEM-POLE PFC FOR ULTRA-HIGH POWER DENSITY USB-PD CHARGERS ................................................................................................................... 3077
*Manuel Escudero Rodruigez*

PP314 COMPARISON OF HYBRID SI/SIC AND SIC TWO-LEVEL AND THREE-LEVEL CONVERTERS FOR LOW-VOLTAGE LOW-POWER APPLICATIONS ........................... 3086
*Tim Augustin*

PP315 ANALYSIS OF ANALOGUE CURRENT AND FLUX BALANCING FOR THE DUAL-ACTIVE-BRIDGE CONVERTER ........................................................................... 3096
*Christophe Basso*

PP316 DESIGN AND OPTIMIZATION OF A SINGLE-STAGE PHOTOVOLTAIC
MICROINVERTER WITH INTEGRATED MAGNETICS ............................................................... 3103
*Jin Wen*

PP317 EXPERIMENTAL INVESTIGATION OF CLASS F INVERTER UNDER VARIOUS
LOAD CONDITIONS............................................................................................................................3110
*Baptiste Daire*

PP318 ANALYSIS, MODELING, DESIGN, AND LIMITATIONS OF CURRENT INJECTION
BASED UPF RECTIFIER WITH SMALL DC-LINK CAPACITOR...............................................3118
*Ramkrishan Maheshwari*

PP319 HIGH-EFFICIENT ISOLATED AC-DC CONVERTER WITH CIRCULATING
CURRENT REDUCTION FOR AC ADAPTERS ............................................................................ 3125
*Hiroki Watanabe*

PP320 A PHASE-LOCKED LOOP (PLL) BASED STRATEGY FOR ACCURATE BLANKING
TIMES IN BRIDGELESS TOTEM-POLE PFCS............................................................................ 3130
*Sandu Tigira Tigira*

PP321 CIRCULATING CURRENTS IN COUPLED MULTI-TERMINAL HYBRID AC-DC
GRIDS............................................................................................................................................... 3136
*Fabian Herzog*

## ADVANCED CONVERTER TOPOLOGIES

PP322 COMPARISON OF 4500V STATE-OF-THE-ART XHP3 IGBT AND CONVENTIONAL
IHV IGBT FOR 3300V 3-LEVEL ANPC MEDIUM VOLTAGE DRIVES ................................... 3142
*Martin Knecht*

PP323 GENERALIZED SWITCHING SEQUENCE FOR VOLTAGE BALANCING IN A
FLYING CAPACITOR DC-DC CONVERTER WITH QUASI-2-LEVEL MODULATION........................ 3150
*Jose Andres Aguilar Croston*

PP324 OPTIMIZATION-BASED SIZING OF A MODULAR MULTILEVEL CONVERTER
BASED ON 650 V GAN MODULES FOR NEW LVDC/MVDC GRIDS...................................... 3160
*Gregoire Le Goff*

PP325 A NOVEL THREE-PHASE LOW-SWITCH-COUNT AC-DC GRID CONVERTER
TOPOLOGY WITH GALVANIC ISOLATION................................................................................ 3169
*Liska Steenbock*

PP326 SINGLE-STAGE LED DRIVER BASED ON COUPLED INDUCTOR POWER
FACTOR CORRECTION AND LLC CONVERTER....................................................................... 3175
*Alireza Ramezan Ghanbari*

PP327 A INVERSE COUPLED DC-DC BOOST INDUCTOR WITH 2-KV SIC MOSFET
MODULE FOR 1500V SOLAR INVERTER MPPT....................................................................... 3181
*Yusi Liu*

PP328 ENVIRONMENTAL IMPACT OF MODULAR POWER ELECTRONICS SYSTEMS
CONSIDERING DIAGNOSTIC-DRIVEN UNIT REPLACEMENT ............................................ 3187
*Briac Baudais*

## POWER ELECTRONICS FOR RAILWAY APPLICATIONS

PP329   SWITCHING PERFORMANCE COMPARISON OF 3.3 KV SIC MOSFET AND SI IGBT POWER MODULES FOR RAILWAY TRACTION SYSTEMS ......................................... 3197
*Yue Zhao*

PP330   COMPARISON OF THREE-LEVEL INVERTER TOPOLOGIES FOR MVDC REVERSIBLE RAILWAY SUBSTATIONS ............................................................................ 3206
*Luc Bimmel*

PP331   CONTROL OF BIDIRECTIONAL POWER FLOW IN RAILWAY CATENARY OVERHEAD LINES.......................................................................................................... 3213
*Peter Van Duijsen*

PP332   A RAIL TRACTION CONVERTER PLATFORM BASED ON POWER MODULE IMPLEMENTATIONS WITH 450 A, 600 A AND 800 A 3.3 KV IGBT MODULES .................................. 3221
*Ekrem R. Gunes*

PP333   COMPARISON OF SELECTED MEGAWATT-LEVEL TRACTION CONVERTER POWER MODULE IMPLEMENTATIONS IN TERMS OF COMMUTATION INDUCTANCE AND PRACTICALITY.................................................................................................................... 3229
*Abdulkerim Ugur*

## CURRENT RELATED TESTING

PP334   PITFALLS AND THEIR AVOIDABILITY IN THE DOUBLE-PULSE TEST ............................ 3237
*Nikolas Förster*

PP335   MODELING AND SIMULATION OF FLUXGATE BASED CURRENT SENSOR ................... 3247
*Yunus Çay*

PP336   SIGMA-DELTA BASED CURRENT ACQUISITION WITH REDUCED SETTLING TIME ........................................................................................................................................... 3256
*Joschka Randerath*

PP337   CHARACTERISATION OF WIDE-BANDGAP SEMICONDUCTORS IN DOUBLE PULSE TESTING USING OPTICALLY ISOLATED PROBES............................................................ 3264
*Lennart Hoffmann*

PP338   NON-INVASIVE BATTERY CONDITION TESTING USING ELECTRICAL SIGNALS AND OSCILLOSCOPES...................................................................................................................... 3269
*Srikrishna N. H*

PP339   INSTRUMENTATION REQUIREMENTS FOR FAST 130 V/NS SWITCHING OF 1700 V, 35 M? SIC MOSFETS ............................................................................................................ 3276
*Matthew Appleby*

## POWER ELECTRONICS FOR AEROSPACE APPLICATIONS

PP340   CONCEPTUALIZATION AND EXPERIMENTAL ASSESSMENT OF DESIGN ASPECTS FOR 3-LEVEL ANPC INVERTERS ................................................................................. 3286
*Lukas Radomsky*

PP341    DESIGN OF A HIGH POWER DENSITY INVERTER AND FOC IMPLEMENTATION
FOR UAVS............................................................................................................................ 3296
    Matthias Neuner

PP342    HIGHLY-INTEGRATED, FLEXIBLE POWER SOLUTION FOR AEROSPACE 5KVA –
20 KVA MOTOR DRIVE APPLICATIONS .......................................................................... 3305
    Alain Calmels

PP343    DATABASE-SUPPORTED PRELIMINARY DESIGN, SIMULATION AND
EVALUATION OF POWER CONVERTERS IN ELECTRIC AIRCRAFT PROPULSION
SYSTEMS ............................................................................................................................ 3315
    Jeff Kugener

PP344    DESIGN AND ANALYSIS OF GATE-DRIVER FOR SIC-BASED INVERTER FOR
MEGAWATT SCALE ALL ELECTRIC AIRCRAFT............................................................ 3318
    Jeff Kugener

## MEASUREMENT TECHNIQUES AND METHODS

PP345    ADDRESSING TESTING CHALLENGES FOR POWER MODULES AND THREE-
LEVEL INVERTERS.............................................................................................................. 3328
    Oleg Fotteler

PP346    CHARACTERIZATION OF THE BONDING QUALITY OF SILVER SINTERED
COMPOUNDS BY MEANS OF LASER-INDUCED BREAKDOWN SPECTROSCOPY ........ 3334
    Yannick Bockholt

PP347    INVERTER-INTEGRATED MEASUREMENT OF THE FREQUENCY-DEPENDENT
WINDING IMPEDANCE OF ELECTRIC MACHINES ........................................................ 3340
    Christian Mühlfeld

PP348    COMPENSATION TECHNIQUES FOR BANDWIDTH-DISTORTED
MEASUREMENTS OF FAST TRANSIENTS IN DOUBLE PULSE TESTS............................. 3347
    Christian Lottis

PP349    AN AERODYNAMIC LOAD MEASUREMENT TECHNIQUE FOR AUTONOMOUS
AERIAL VEHICLES ............................................................................................................ 3353
    Mehmet Oguz Girgin

COMPENSATION TECHNIQUES FOR BANDWIDTH-DISTORTED MEASUREMENTS OF
FAST TRANSIENTS IN DOUBLE PULSE TESTS .................................................................. 3358
    Christian Lottis

PP350    A HIGH-BANDWIDTH MULTILEVEL COUNTER CIRCUIT FOR BEARING
CURRENT EVALUATION..................................................................................................... 3364
    Felix Schulte

## TRANSFORMERS

PP351    CORE LOSS MODEL FOR CONSIDERING ANISOTROPY AND TEMPERATURE
EFFECTS ON ELECTRICAL STEEL UNDER POWER ELECTRONIC CONDITIONS ........... 3371
    Michael Owzareck

PP353    CIRCULAR ECONOMY ORIENTED AND RECONFIGURABLE PLANAR
TRANSFORMER DESIGN FOR ISOLATED DC-DC CONVERTERS ........................................................ 3380
    *Fabian Groon*

PP354    CONTROLLABLE MAGNETICS: VARIABLE TRANSFORMERS AND VARIABLE
INDUCTORS, THEORY – PRODUCTION – APPLICATION ..................................................................... 3390
    *Florian Fenske*

PP355    A THREE-PHASE INTERLEAVED LLC INTEGRATED TRANSFORMER USING
PCB WINDINGS FOR FUEL CELL DCDC CONVERTERS ....................................................................... 3395
    *Jiajia Guan*

PP356    TESTING THE PRIMARY-SECONDARY COIL COUPLING OF HIGH-FREQUENCY
TRANSFORMER IMPLEMENTED ON ETD AND TOROIDAL CORES .................................................. 3400
    *Alexis Gioda*

**Author Index**

PCIM Europe 2024, 11– 13 June 2024, Nuremberg          DOI: 10.30420/566262083

# The New CoolSiC™ MOSFET 1200 V G2: Electrical Performance and Compact Modelling

Andreas Huerner[1], Qing Sun[1], Rudolf Elpelt[1]

[1] Infineon Technologies AG, Sieglitzhoferstr. 9, 91054 Erlangen, Germany

Corresponding author:   Andreas Hürner, Andreas.huerner@infineon.com
Speaker:                Andreas Hürner, Andreas.huerner@infineon.com

## Abstract

In this paper, the electrical performance, and the corresponding compact model of the Cool-SiC™ MOSFET 1200 V G2 portfolio are presented with a special focus on IMBG120R026M2H. A comparison between the measured and simulated static, dynamic, and switching characteristics under various conditions is also presented. Additionally, this paper also discusses how customers can utilize the full potential of compact models to implement the CoolSiC™ MOSFET 1200 V G2 portfolio in their systems.

## 1   Introduction

To support circuit designers in implementing the new CoolSiC™ MOSFET 1200 V G2 portfolio in D2PAK package, Infineon released a compact model describing its typical device features.

Generally, compact models are provided by device manufacturers to support customers in designing and optimizing their systems [1] [2] [3] [4] [5]. Thus, compact models should reflect major device features such as the dependence of forward and reverse conduction characteristics on gate voltage and temperature, the device capacitances, the internal gate resistance, and the body diode's reverse recovery behavior.

There are three different approaches for compact modelling:

- Behavioral compact models
- Physical compact models
- Physics-based behavioral models

### Behavioral models

Behavioral compact models normally use polynomial functions without any physical background or look-up tables. With this approach a very high accuracy between measurement and simulation can be achieved with lesser effort. However, the accuracy of this model depends on the quality of the data entered, e.g., measurement results, because no post-processing is possible. These models can be highly accurate only under the conditions covered by the data entered and any extrapolation can be very difficult.

Also, parameter variations such as different gate source threshold voltages cannot be simulated in these models because they are not a specific element of the model.

Therefore, the influence of device-to-device variations on the final product cannot be investigated.

### Physical models

Physical compact models use physical device equations that are parametrized by process and layout information [6] [7]. The main benefit of this modelling approach is that it enables circuit designers to simulate process variations and investigate their influence on the results during production.

Although this approach is beneficial, the development time for such models is quite high, making it unlikely that the benefits justify the efforts. Secondly, physical equations require many sub-equations using limit functions that, typically, result in slow simulation speed and high convergence issues.

### Physics-based behavioral compact models

Considering all advantages and disadvantages of other compact models, the physics-based behavioral compact modelling approach was used to create the compact models for the newly released

CoolSiC™ MOSFET 1200 V G2 portfolio in D2PAK package [1] [2] [3]. This approach provides the advantages of both the behavioral and physical compact modelling approach. For instance, for the output characteristic, an equation very close to the textbook equation is used in this approach. With this equation, it is possible to extract the parameters from measurement results by pure fitting or by considering the key performance values of the device, such as the gate source threshold voltage, the breakdown-voltage, and the transistor factor.

This approach combines the high predictivity of compact models with high simulation speed and full parametrization capability.

# 2 Compact model approach and verification flow

In [1], [2] and [3], the physics-based behavioral compact modelling approach is described in detail. In this approach, as shown in Fig.1, the device is separated in single elements that are described through mathematical equations. These equations are very close to well-known textbook equations, and along with the main key performance values only a limited number of fitting parameters are required. This enables the creation of very accurate and stable models for circuit designers, to support them in resolving development issues.

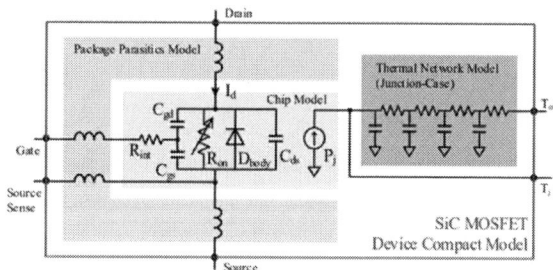

**Fig. 1** Simplified electrical structure of a compact model for SiC MOSFETs in a D2PAK package.

In the following sections, the core elements of this model, such as the MOSFET current source, device capacitances, internal gate resistance, reverse recovery behavior of the body diode, and the thermal impedance network are discussed in detail.

Hereby, the focus is on explaining how the models for these core elements are verified.

## 2.1 Thermal impedance network

As described in the previous section, the focus of electrical simulation is to derive a deeper understanding of the interaction between chip-technology and circuitry to avoid, for instance, ringing and oscillations.

For thermal considerations, normally PLECS models are used, but the virtual junction temperature can be calculated with the released compact model also.

To do this, the dissipated power is calculated and injected into the Cauer thermal impedance network shown in Fig. 1 via a current source. Then, the coefficients of the thermal impedance network are taken from measurement results using the dual thermal interface method [8].

Although, a very high accuracy can be achieved with this method, time constants lower than 10 µs are challenging and should be used carefully. Therefore, e.g., short circuits cannot be simulated accurately. For such conditions, as described by Sun et al., in [9], advanced methods to determine the coefficients of the thermal impedance network must be carried out.

## 2.2 MOSFET current source

From the perspective of global devices, it is highly important that the MOSFET current source correctly describes the typical drain-to-source on-resistance and its dependence on drain current, gate voltage, and temperature. For this, the linear part of the output characteristic is measured for many devices and a typical device behavior is derived.

Fig. 2 shows the dependence of drain-to-source on-resistance on gate voltage and temperature characteristic of IMBG120R026M2H – a product in the 1.2 kV CoolSiC™ MOSFET portfolio in D2PAK. Furthermore, in Fig. 3, the drain-to-source on-resistance's dependence on drain-current is given for a junction temperature of 25°C, whereas Fig. 4 demonstrates the drain-to-source on-resistance's dependence on drain-current at 175°C.

Therefore, with these models, circuit designers can investigate the on-state characteristics for several conditions potentially occur in application.

Although, understanding the device's on-state characteristics are very important for properly designing a circuit, compact models normally describe the switching characteristics of the device correctly under several conditions. For this, as described in [1], the output characteristics in the saturation region are more relevant.

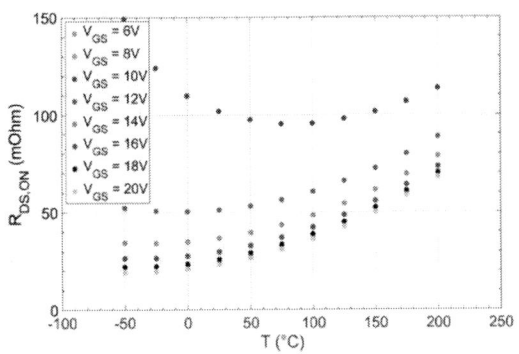

**Fig. 2** Simulated drain-to-source on-resistance in dependence on temperature and gate-voltage.

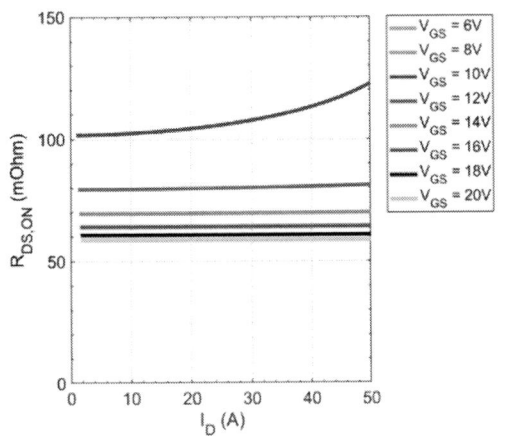

**Fig. 3** Simulated on-resistance in dependence on drain-current and gate-voltage @ Tj = 25°C.

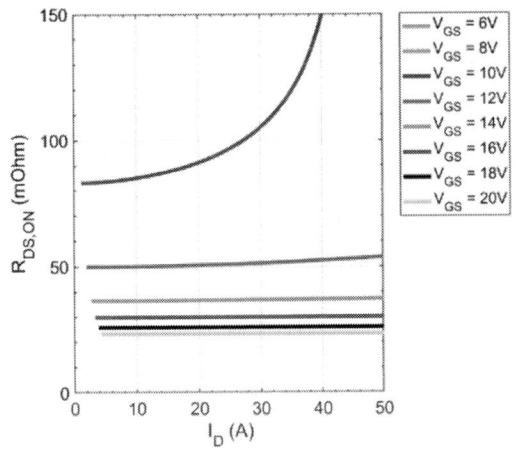

**Fig. 4** Simulated on-resistance in dependence on drain-current and gate-voltage @ T = 175°C.

However, due to the limitations of the output power in state-of-the art curve tracers, measuring the saturation current region is not easy. Although, there are methods discussed in [11] [12] and [13], for example, extracting the dynamic transfer characteristics from double-pulse measurements, calibrating a compact model's accuracy in the saturation current regime is one of the most challenging aspects in compact modelling.

To overcome this challenge and to calibrate this region for the output characteristic of the Cool-SiC™ 1200 V G2 compact models, a special measurement setup called the short circuit IV-characterization tool was developed.

In principle, this setup is comparable to a typical short circuit type 1 measurement setup, but this board was endued with a stray inductance lower than 1 nH.

As shown exemplarily in Fig. 5, it enabled the measurement of short circuits at very high turn-on speed, mandatory for avoiding significant self-heating effects.

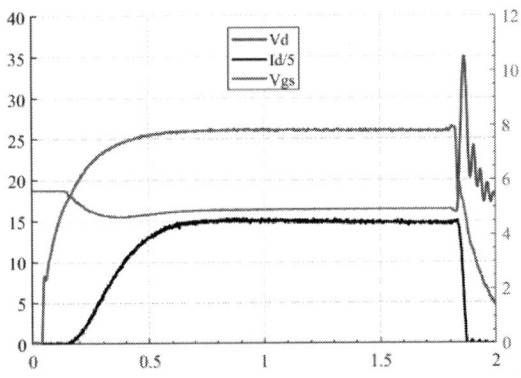

**Fig. 5** Typical high-voltage IV characteristic test pulse @ Tj = 25°C

With this measurement method, the output characteristics of several devices was measured for drain-voltages up to 800 V, gate-voltages in the range from 0 V to 18 V and different temperatures. Based on the experimental results, the parameters of the MOSFET current source equation were calibrated. In Fig. 6, a comparison between the measured and simulated output characteristic is shown.

Simulation = straight line

Measurement = dots

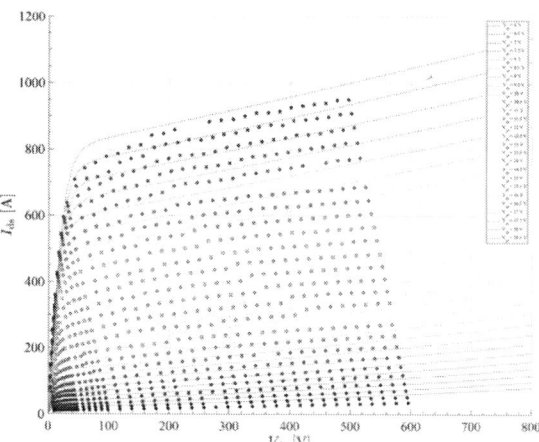

**Fig. 6** Simulated and measured high voltage output characteristic @ Tj = 25°C.

Furthermore, the devices were measured with state-of-the art curve tracers. As shown in Fig. 7, in contrast to the short circuit IV-board, with state-of-the art curve tracers the current-slope in the saturation region is significantly overestimated. Mainly, this must be attributed to the limited measurement pulse length of state-of-the-art curve tracer being in the range of min. 20 µs. Therefore, the power dissipated during the measurement cannot be neglected and results, especially in the saturation region of the output characteristics into significant self-heating.

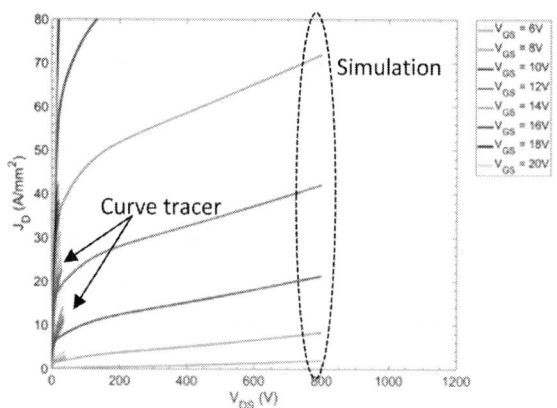

**Fig. 7** Simulated and measured high voltage output characteristic @ T = 25°C.

This slope has a significant influence on switching behavior and implementing it incorrectly in a SiC MOSFET compact model can reduce the accuracy of simulated switching events significantly [1].

### 2.3 Device capacitances

Device capacitances are highly non-linear and endued with a strong dependence on drain source and gate source voltage because of the SiC MOSFET cell structure and technology [10]. Therefore, to avoid severe discrepancies between simulation and measurement, and convergence issues the device capacitance was implemented using a pure behavioral model.

Devices from the entire portfolio were measured and the extracted device capacitances were nominated. Fig. 8 shows a comparison between the simulated and measured device capacitances.

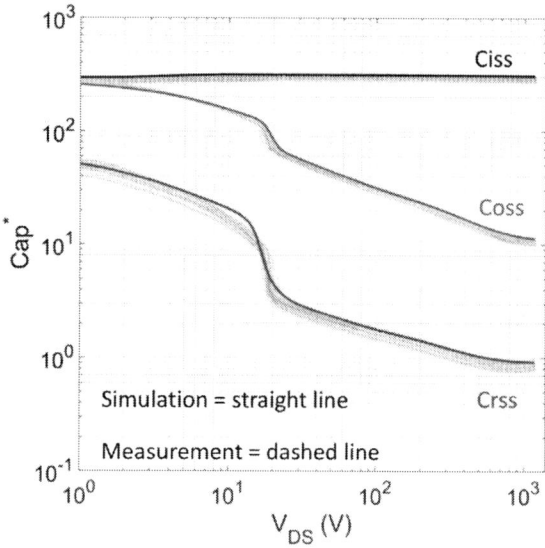

**Fig. 8** Simulated and measured device capacitances.

A detailed investigation was also carried out to determine the dependence of the gate source capacitance on the gate voltage. Based on this investigation, a highly accurate gate charge characteristic, shown in Fig. 9, was achieved.

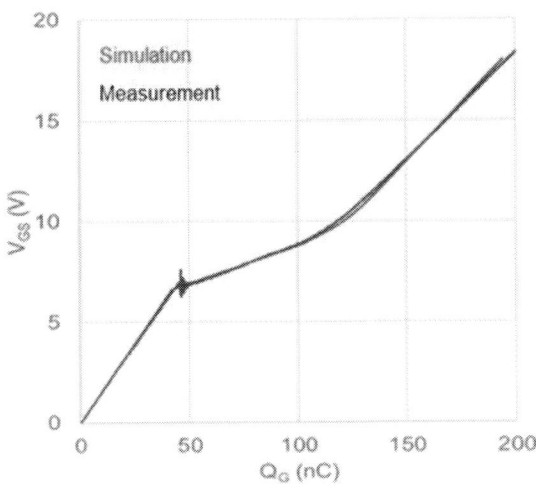

**Fig. 9** Simulated and measured gate charge characteristic

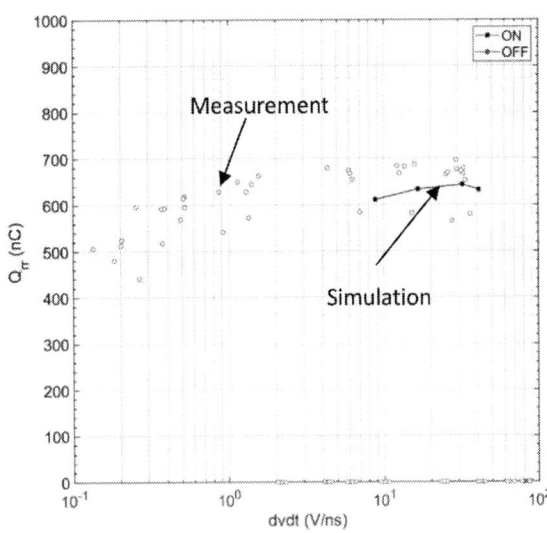

**Fig. 10** Measured and simulated reverse recovery charge.

### 2.4 Reverse recovery behavior

As well known, the switching characteristics of SiC MOSFETs are strongly influenced by device capacitances and output characteristics.

For higher temperatures, however, the reverse recovery behavior of the body diode is one of the most important device features, the interaction between chip behavior and circuitry of which cannot be overlooked.

Therefore, a physics-based behavioral model describing the reverse recovery behavior has been developed and implemented into the CoolSiC™ MOSFET 1200 V G2 models [5]. In these models, the influence of drain current, deadtime, turn-off gate voltage, and temperature on the body diode's reverse recovery behavior have been considered.

For highlighting the accuracy of the model, in Fig. 10 the dependence of measured and simulated reverse recovery charge on drain voltage slope is shown. For this, the reverse recovery charge was extracted by the method proposed by Sochor et al., in [14]. It represents the reverse recovery charge without considering the influence of the circuitry.

In addition, as discussed in the next section, the reverse recovery current and the body diode overvoltage can be simulated with a very high accuracy using this model. With this functionality, circuit designers can virtually investigate the interaction between the specific switching characteristics of the chip and their circuitry.

## 3 Switching characteristics

For demonstrating compact-model's high accuracy in terms of switching characteristics, the switching characteristics of IMBG120R026M2H, a product in the 1.2 kV CoolSiC™ MOSFET portfolio in D2PAK (shown in Fig.11), was measured exemplarily through a double-pulse setup endued with a low stray inductance. The equivalent circuit of the double-pulse setup is shown in Fig. 12. The double-pulse setup consisted of a DC link capacitor that enabled DC link voltages from 200 V up to 1.8 kV, a stray inductance in the range of 6 nH, a load inductance of 600 nH, and SiC MOSFETs on both the low side and high side.

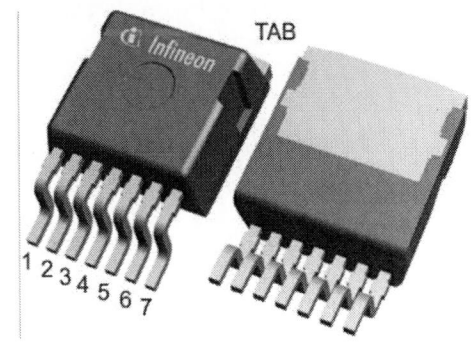

**Fig. 11** Devices under test

PCIM Europe 2024, 11– 13 June 2024, Nuremberg    DOI: 10.30420/566262083

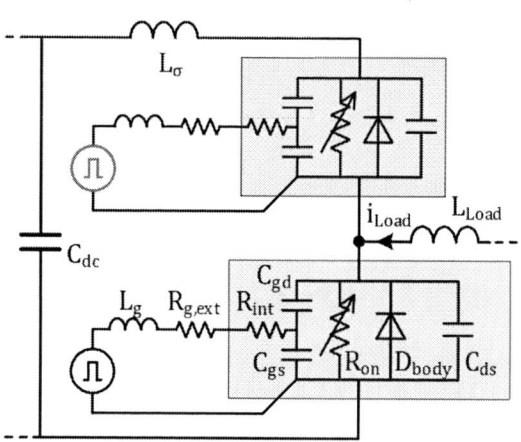

**Fig. 12**    Double-pulse test setup.

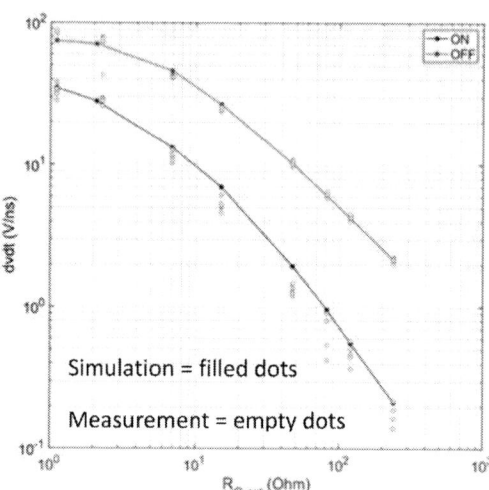

**Fig. 14**    Dependence of simulated and measured drain voltage slopes on external gate resistance

For switching measurements, discussed in this section, the DC link voltage was 800 V and the drain current 27.3 A. The SiC MOSFET at the low side was driven with different external gate resistances, the turn-on gate voltage was 18 V, and the turn-off gate voltage was 0 V. The turn-off gate voltage of the SiC MOSFET at the high side was -5 V. The case temperature was 175°C.

The dependence of measured and simulated switching energy loss on the drain voltage slope is shown in Fig.13. In addition to this, Figure 14 shows the dependence of drain voltage slopes on external gate resistance.

Hence, using these advanced compact models, the simulation shows a particularly good agreement of switching losses as a function of dV/dt and for the controllability of dV/dt by the external gate resistance.

This underlines the ability of the models to accurately simulate switching energy losses giving the circuit designer a strong tool to optimize his own system. But, as previously discussed, for compact models the focus is not the switching energy loss, but the switching transients, corresponding oscillations and the over-voltages at the high- and low-side device.

Normally, for calibrating and optimizing the switching transients, the measured and simulated switching transients are compared, and, in case of differences, some adjustments of output characteristics, device capacitances and package parasitic are derived. To do this, the influence of both the chip model and the double-pulse setup on the measured and simulated switching characteristic need to be understood and to highlight this aspect, different double-pulse setups were used to measure just one device. Figure 15 shows the measured turn-on characteristics and Fig. 16 shows the measured turn-off characteristics.

Based on this comparison, it is clear that for evaluating compact model's accuracy in terms of switching characteristics, a detailed understanding of the used double-pulse setup is required. Therefore, a lot of effort was spent in improving and understanding the described double-pulse setups.

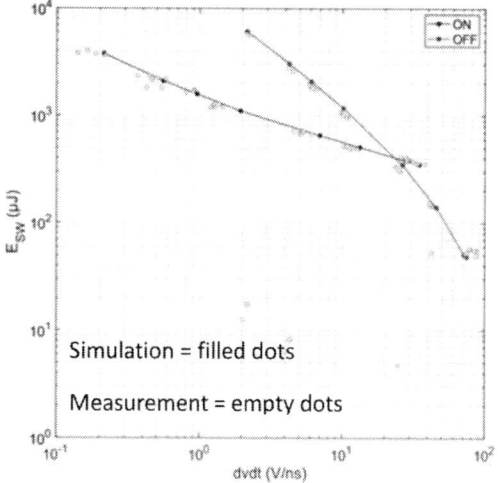

**Fig. 13**    Dependence of simulated and measured switching energy losses on drain voltage.

PCIM Europe 2024, 11– 13 June 2024, Nuremberg    DOI: 10.30420/566262083

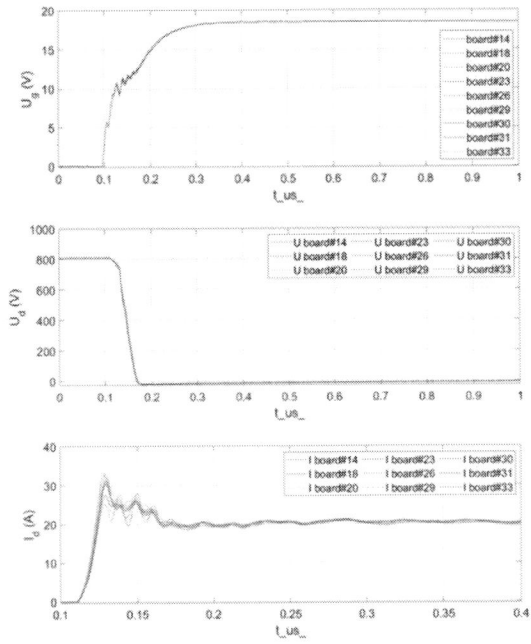

**Fig. 15**    Measured turn-on switching characteristics.

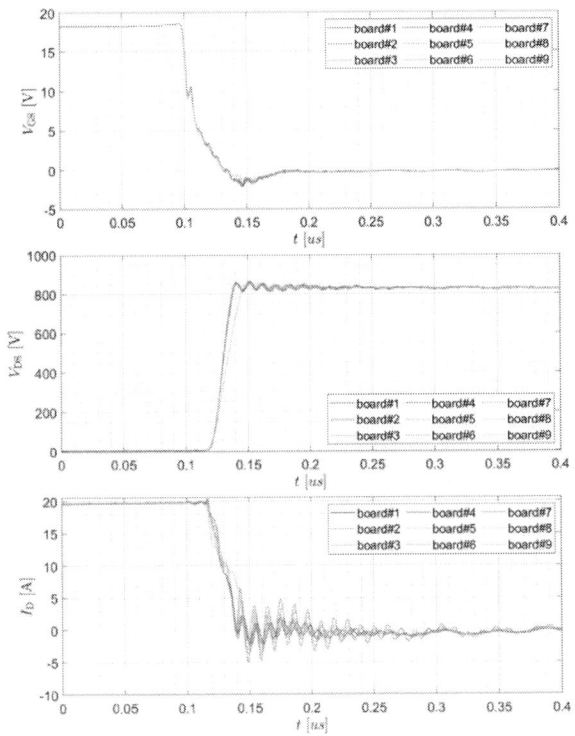

**Fig. 16**    Measured turn-off switching characteristics.

Besides iterative development cycles in which the parasitic capacitance and stray inductance of the double-pulse setup can be reduced significantly, modelling techniques were carried out to translate the characterization setup's properties into an equivalent circuit usable in the simulation environment.

With this detailed understanding of the double-pulse setup, it was possible to distinguish between the influence of the chip-model and the circuitry in the optimization loop of the compact-model. With this approach a very high accuracy between measured and simulated switching characteristics, also in terms of switching transients and oscillations was achieved.

In Fig. 17 a comparison between the measured and simulated turn-on switching characteristics is shown. Fig. 18 shows the corresponding turn-off switching characteristics. For both simulation and measurement, an external gate resistance of 2 Ohm was chosen.

The comparison between the simulated and the measured turn-on switching characteristics shows an extremely high accordance the gate voltage, the drain current, and the drain voltage's dependence on time. Especially, the reverse recovery behaviour of the body diode, and the corresponding turn-on behaviour of the low side device, including the reverse recovery current and body diode overvoltage can be simulated with extremely high accuracy.

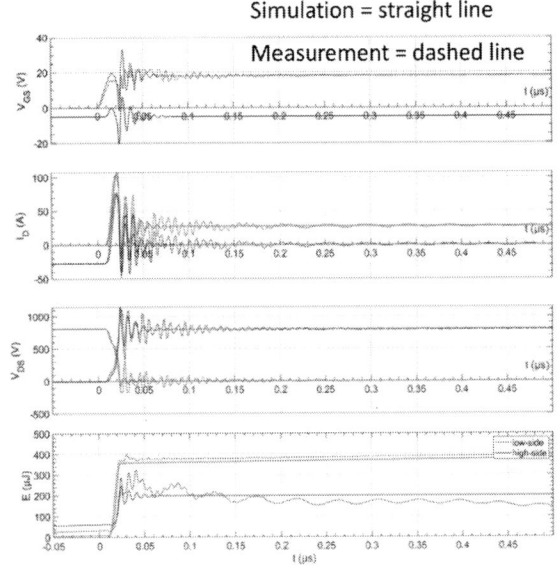

**Fig. 17**    Simulated and measured turn-on switching characteristics.

In Fig. 18, a comparison between the measured and simulated turn-off switching characteristic is shown. In accordance to the turn-on switching characteristics, for the turn-off switching characteristic also an extremely high accordance the gate voltage, the drain current, and the drain voltage's dependence on time was achieved.

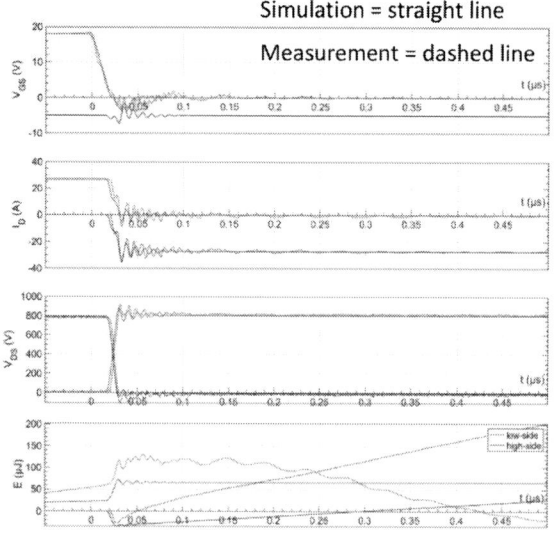

**Fig. 18** Simulated and measured turn-on switching characteristics.

This underlines that the released compact models are suitable for virtually optimizing the turn-on switching characteristics to, for example, avoid significant body diode overvoltage.

## 4 Conclusion

In this study, the recently developed compact model for the CoolSiC™ MOSFET 1200 V G2 portfolio in D2PAK package was discussed with a special focus on the verification flow. With this model, typical device parameters such as the drain-to-source on-resistance, output characteristics, and device capacitances can be considered correctly. Using these models, a detailed understanding of the characterization setup, switching energy losses, and switching transients can also be simulated with very high accuracy. In conclusion, it was proven that compact models have the potential to support circuit designers in optimizing their systems.

## References

[1] P. Sochor, A. Huerner, Q. Sun and R. Elpelt, "Characteristics of SiC MOSFET Compact Models Suitable for Virtual Prototyping of Power Electronic Circuits," *2023 11th International Conference on Power Electronics and ECCE Asia (ICPE 2023 - ECCE Asia)*, Jeju Island, Korea, Republic of, 2023, pp. 112-119, doi: 10.23919/ICPE2023-ECCEAsia54778.2023.10213920.

[2] P. Sochor, A. Huerner, and R. Elpelt, "Commutation loop design for optimized switching behavior of CoolSiC™ MOSFETs using compact models," *PCIM Europe digital days 2020; International Exhibition and Conference for Power Electronics, Intelligent Motion, Renewable Energy and Energy Management*, 2020, pp. 1–8

[3] P. Sochor, A. Huerner and R. Elpelt, "A Fast and Accurate SiC MOSFET Compact Model for Virtual Prototyping of Power Electronic Circuits," *PCIM Europe 2019; International Exhibition and Conference for Power Electronics, Intelligent Motion, Renewable Energy and Energy Management*, Nuremberg, Germany, 2019, pp. 1-8.

[4] P. Sochor, A. Huerner, Q. Sun and R. Elpelt, "Understanding the Switching Behavior of Fast SiC MOSFETs," *PCIM Europe 2022; International Exhibition and Conference for Power Electronics, Intelligent Motion, Renewable Energy and Energy Management*, Nuremberg, Germany, 2022, pp. 1-8, doi: 10.30420/565822153.

[5] A. Huerner, P. Sochor, Q. Sun and R. Elpelt, "Reverse Recovery Behavior in SiC MOSFETs: Characterization and Modelling," *PCIM Europe 2023; International Exhibition and Conference for Power Electronics, Intelligent Motion, Renewable Energy and Energy Management*, Nuremberg, Germany, 2023, pp. 1-7, doi: 10.30420/566091108.

[6] J. Victory *et al.*, "A Physically Based Scalable SPICE Model for High-Voltage Super-Junction MOSFETs," *PCIM Europe 2014; International Exhibition and Conference for Power Electronics, Intelligent Motion, Renewable Energy and Energy Management*, Nuremberg, Germany, 2014, pp. 1-8.

[7] C. He, J. Victory, Y. Xiao, H. D. Vleeschouwer, E. Zheng and Z. Hu, "SiC MOSFET Corner and Statistical SPICE Model Generation," *2020 32nd International Symposium on Power Semiconductor Devices and ICs*

*(ISPSD)*, Vienna, Austria, 2020, pp. 154-157, doi: 10.1109/ISPSD46842.2020.9170091.

[8] D. Schweitzer, H. Pape, L. Chen, R. Kutscherauer and M. Walder, "Transient dual interface measurement — A new JEDEC standard for the measurement of the junction-to-case thermal resistance," *2011 27th Annual IEEE Semiconductor Thermal Measurement and Management Symposium*, San Jose, CA, USA, 2011, pp. 222-229, doi: 10.1109/STHERM.2011.5767204.

[9] Q. Sun, A. Huerner, R. Elpelt, „Behavior Modelling the Short Circuit Characteristics of SiC MOSFETs using Compact Models", to be presented at PCIM 2024

[10] R. Siemieniec *et al.*, "A SiC Trench MOSFET concept offering improved channel mobility and high reliability," *2017 19th European Conference on Power Electronics and Applications (EPE'17 ECCE Europe)*, Warsaw, Poland, 2017, pp. P.1-P.13, doi: 10.23919/EPE17ECCEEurope.2017.8098928.

[11] P. Hofstetter, R. W. Maier and M. -M. Bakran, "Influence of the Threshold Voltage Hysteresis and the Drain Induced Barrier Lowering on the Dynamic Transfer Characteristic of SiC Power MOSFETs," *2019 IEEE Applied Power Electronics Conference and Exposition (APEC)*, Anaheim, CA, USA, 2019, pp. 944-950, doi: 10.1109/APEC.2019.8721772.

[12] A. Huerner, P. Sochor, Q. Sun, M. W. Feil and R. Elpelt, "Influence of Turn-Off Gate-Voltage Undershoots on the Turn-On Behavior of SiC MOSFETs," *PCIM Europe 2022; International Exhibition and Conference for Power Electronics, Intelligent Motion, Renewable Energy and Energy Management*, Nuremberg, Germany, 2022, pp. 1-8, doi: 10.30420/565822038.

[13] A. Huerner, P. Sochor, M. Feil and R. Elpelt, "Influence of the Threshold-Voltage Hysteresis on the Switching Properties of SiC MOSFETs," *PCIM Europe digital days 2021; International Exhibition and Conference for Power Electronics, Intelligent Motion, Renewable Energy and Energy Management*, Online, 2021, pp. 1-8.

[14] P. Sochor, A. Huerner, M. Hell and R. Elpelt, "Understanding the Turn-off Behavior of SiC MOSFET Body Diodes in Fast Switching Applications," *PCIM Europe digital days 2021; International Exhibition and Conference for Power Electronics, Intelligent Motion, Renewable Energy and Energy Management*, Online, 2021, pp. 1-8.

PCIM Europe 2024, 11– 13 June 2024, Nuremberg    DOI: 10.30420/566262084

# Paralleling SiC-Power-MOSFET Body Diodes under Harsh Switching Conditions

Michael Rauh [1], Matthias Bürger[2], Mark-M. Bakran [1]

[1] University of Bayreuth, Department of Mechatronics, Center of Energy Technology, Germany
[2] Infineon Technologies AG, Warstein, Germany

Corresponding author:    Michael Rauh, michael.rauh@uni-bayreuth.de
Speaker:                 Michael Rauh, michael.rauh@uni-bayreuth.de

## Abstract

To reduce switching losses of SiC-Power-MOSFETs, one approach is to maximize the switching speed by using dead-time optimization. However, if the driver fails to meet the required dead-time, the body diode may be operated under avalanche conditions due to high switching stress. Since power modules consist of numerous semiconductors operating in parallel, the energy distribution between the separate devices during a possible avalanche event of the body diodes has to be taken into consideration, which will be addressed in this paper. The main focus is the parallel operation of semiconductors with different breakdown voltages under avalanche conditions due to high switching stress as well as the impact of the inductive coupling among the parallel devices during avalanche operation.

## 1   Introduction

Dead-time optimization, as described in [1] and [2], is a well-known approach to reduce the switching losses in highly efficient power inverters fit with silicon carbide metal-oxide-semiconductor field-effect transistors (SiC-MOSFETs). By minimizing the dead-time between the active and the passive switch of a half-bridge setup, the turn-on speed of the active switch can be increased which comes along with lower turn-on losses. Pushing this method to its limits, the semiconductor's safe operating area (SOA) is only met in the presence of the optimized dead-time. If the driver does not meet the required dead-time, the overvoltage across the body diode during diode turn-off will increase. In the worst case, the body diode may be operated under avalanche conditions due to high switching stress. One scenario in which the desired dead-time control is not effectively applicable is when the inverter stops while the electric motor is still in motion. In this case, the inverter operates as a diode rectifier and the conduction time of the body diodes is much longer compared to the optimized dead-time. This results in a significantly higher overvoltage at the diode turn-off event as the inverter restarts its operation. Nevertheless, such events

are seldom in the application but must be dealt with by the SiC-Power-MOSFETs. [3]

In addition to reducing losses, decreasing the dead-time also results in an overall softer reverse recovery behaviour of the body diode, as mentioned in [4]. Furthermore, if the body diode has a snappy behaviour due to fast-switching, a lower dead-time can also mitigate a harsh current snapback and therefore prevent the body diode from a possible avalanche operation, according to [5]. Looking at the avalanche robustness of the body diode of a single SiC-MOSFET, the overcurrent turn-off robustness has been investigated by [6]. The effect of repetitive avalanche events caused by high switching stress on a single body diode has been investigated in previous work [3]. The findings indicate that a single device can endure substantial repetitive stress for a significant number of events without destruction, only experiencing a reasonable degradation of the electrical parameters. Looking at high power modules consisting of multiple semiconductors in parallel, the distribution of losses between individual dies plays a crucial part in the durability of the module, according to [7].

That said, the reliability during an avalanche event of the body diode caused by high switching stress when paralleling SiC-MOSFETs has also to be taken into consideration. The behaviour of parallel

SiC-MOSFETs with different avalanche breakdown voltages during unclamped inductive switching was already investigated by [8], [9]. This publication addresses the parallel operation of two SiC-MOSFET body diodes, especially an avalanche operation due to high switching stress, with the main focus on a mismatch of the breakdown voltage $V_{BD}$. Another aspect will be the impact of the inductive loop among the two devices operating in parallel since an asymmetric connnection significantly influences the dynamic current sharing during switching transients [7], [10].

## 2 Characterization of devices under test

For the evaluation of the energy distribution between the devices in parallel during the avalanche event due to high switching stress, the devices under test (DUT; Infineon's 3.3 kV CoolSiC trench MOSFET) undergo a characterization of the static breakdown voltage $V_{BD,st}$ at a temperature of $T_J = 25\,°C$. Out of all the devices characterized, the three devices with the highest $V_{BD,st}$ (D1, D2 and D3) as well as the three with the lowest $V_{BD,st}$ (D4, D5 and D6) are extracted. The results, normalized to the DC-link voltage $V_{DC}$, are shown in Fig. 1 (blue). The devices D3 and D4 have the highest disparity in $V_{BD,st}$ with $\Delta V_{BD,st} \approx \frac{6}{100} \cdot V_{DC}$. After the identification of $V_{BD,st}$, the dynamic breakdown voltage $V_{BD,dyn}$ for each of the six DUTs body diode will be determined in the standard double pulse test setup. The DUTs are only operated as freewheeling diodes throughout all the upcoming dynamic measurements and therefore are actively clamped to the rated negative gate voltage $V_{GS,off,nom}$. One SiC-MOSFET of the same kind as the DUTs is used as an active switch and remains the same for all single chip measurements. For the determination of $V_{BD,dyn}$, the turn-on gate resistance $R_{G,on,DUT}$ was adjusted separately for each DUT to meet the maximum rated voltage at the diode turn-off event at a temperature of $T_J = 150\,°C$. For the single chip measurement, a stray inductance of $L_{\sigma,sc} = L_{\sigma,nom}$ and a switching current $I_{SW,sc} = 2 \cdot I_{nom}$ are used, where $I_{nom}$ is the rated current of the DUT and $L_{\sigma,nom}$ is the stray inductance for the typical application scaled for the single chip measurement according to [11]. Figure 2 and Fig. 3 show the switching waveforms for the diode turn-off event and the active turn-on event for the DUTs with highest $V_{BD,st}$ (D3 - blue) and lowest $V_{BD,st}$ (D4 - red), respectively, including the drain-to-source voltage $V_{DS}$ (solid) as well as the drain- and source-currents $I_D$ and $I_S$ (dashed). As illustrated by the switching waveforms, the devices with higher $V_{BD,st}$ have a worse diode switching behaviour than the devices with lower $V_{BD,st}$ and therefore require a higher turn-on gate resistance for the same operating point, which results in $R_{G,on,D3} > R_{G,on,D4}$. This circumstance directly influences the switching energies for the different devices. An evaluation of the switching energies for the DUTs D3 and D4, illustrated in Fig. 4, shows, that the different body diode behaviours result in a difference of 12 % in the total switching energy $E_{TOT}$ and a difference of 22 % in turn-on and reverse recovery switching energies $E_{ON+RR}$.

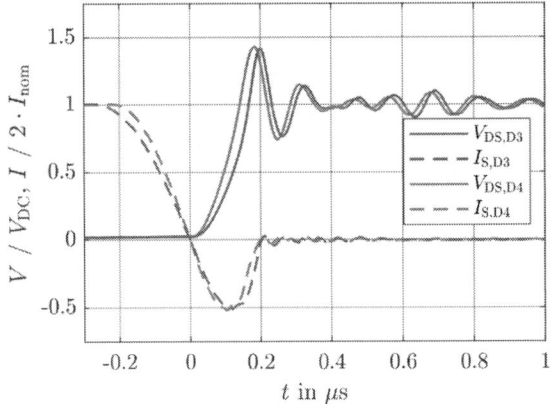

**Fig. 1:** Results for static and dynamic breakdown voltages $V_{BD,st}$ and $V_{BD,dyn}$ for the six devices with highest (D1, D2 and D3) and lowest $V_{BD}$ (D4, D5 and D6)

**Fig. 2:** Diode turn-off event for D3 and D4 at $T_J = 150\,°C$, $V_{DC}$, $2 \cdot I_{nom}$, $L_{\sigma,nom}$ as well as $R_{G,on,D3}$ and $R_{G,on,D4}$

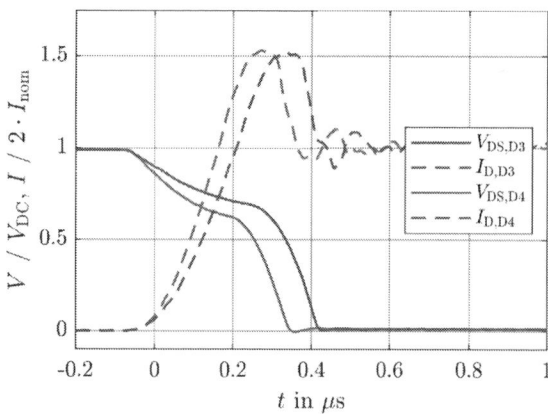

**Fig. 3:** Active turn-on event for D3 and D4 at $T_J = 150\,^\circ$C, $V_{DC}$, $2 \cdot I_{nom}$, $L_{\sigma,nom}$ as well as $R_{G,on,D3}$ and $R_{G,on,D4}$

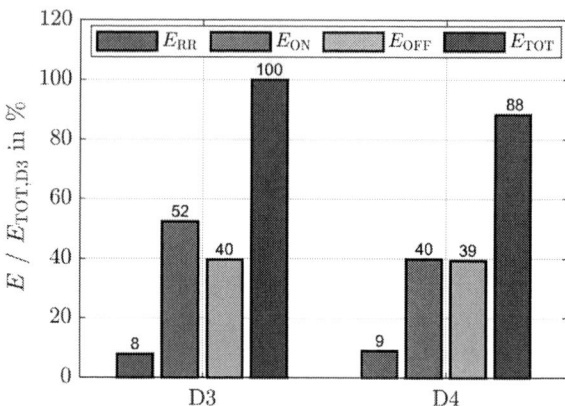

**Fig. 4:** Switching losses for D3 and D4 at $T_J = 150\,^\circ$C, $V_{DC}$, $2 \cdot I_{nom}$, $L_{\sigma,nom}$ as well as $R_{G,on,D3}$ and $R_{G,on,D4}$

To cause the avalanche operation due to high switching stress, the stray inductance is increased to $L_{\sigma,sc,aval} = 4 \cdot L_{\sigma,nom}$. Figure 5 shows the switching waveform during the avalanche operation for the two DUTs with the largest disparity in $V_{BD,dyn}$ D3 (blue) and D4 (orange) with $\Delta V_{BD,dyn} \approx \frac{6}{100} \cdot V_{DC}$ which also matches $\Delta V_{BD,st}$. The results of the determination of $V_{BD,dyn}$ are also depicted in Fig. 1 (red) and show the same trend for each DUT as for $V_{BD,st}$, which means, that the device D3 with highest $V_{BD,st}$ also has the highest $V_{BD,dyn}$ and the device D4 with the lowest $V_{BD,st}$ also has the lowest $V_{BD,dyn}$, as to be expected.

## 3 Evaluation of the parallel operation

After the characterization process, the six DUTs are divided into three pairs for the evaluation of the parallel operation, summarized in table 1. Since all of the upcoming measurements are dynamic switching operations, the breakdown voltage $V_{BD,DUT}$ always refers to the dynamic breakdown voltage $V_{BD,dyn,DUT}$ from Fig. 1, which was determined in the previous section. Pair one (P1) includes two DUTs with a similar breakdown voltage at the upper end (D1 and D2). The second pair (P2) consists of the two DUTs with the largest disparity in breakdown voltage (D3 and D4). The third pair (P3) includes two DUTs with a similar breakdown voltage at the lower end (D5 and D6). The DUTs are only operated as freewheeling diodes throughout all the upcoming dynamic measurements and therefore are actively clamped to the rated negative gate voltage $V_{GS,off,nom}$.

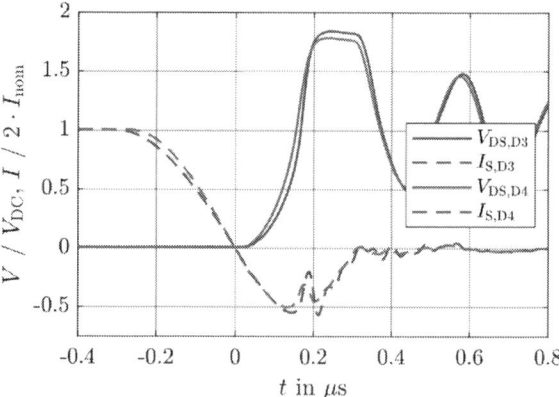

**Fig. 5:** Determination of the avalanche breakdown voltage at $T_J = 150\,^\circ$C, $V_{DC}$, $2 \cdot I_{nom}$, $4 \cdot L_{\sigma,nom}$ as well as $R_{G,on,D3}$ and $R_{G,on,D4}$

### 3.1 Measurement setup

The equivalent electric circuit of the measurement setup for the parallel operation is shown in Fig. 6, the DUTs are connected to the printed circuit board (PCB) as illustrated in Fig. 7. The setup itself is a standard double pulse test setup. Two MOSFETs of the same kind as the DUTs are also used in parallel as an active switch and remain the same for all parallel measurements. The setup itself consists of three Rogowski coils, one coil for each DUT ($I_{LS,1}$ and $I_{LS,2}$) as well as one coil as a reference to measure the overall current flowing back to the DC-Link ($I_{LS}$).

For the evaluation of the influence of the inductive loop for the two devices operating in parallel in the later stage of this publication, two different connection inductances $L_h$ and $L_l$ towards the PCB are

| pair | DUT1 | $V_{\mathrm{BD,DUT1}}$ | DUT2 | $V_{\mathrm{BD,DUT2}}$ |
|------|------|------|------|------|
| P1 | D1 | high | D2 | high |
| P2 | D3 | high | D4 | low |
| P3 | D5 | low | D6 | low |

**Tab. 1:** Summary of the different pairs for the evaluation of the parallel operation

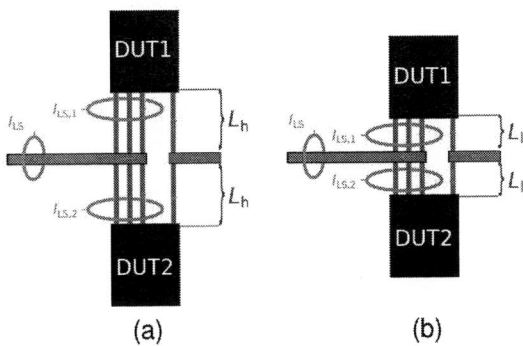

**(a)**            **(b)**

**Fig. 7:** Rogowski-Setup and PCB connection for different connection inductances $L_{\mathrm{h}}$ (a)) and $L_{\mathrm{l}}$ (b))

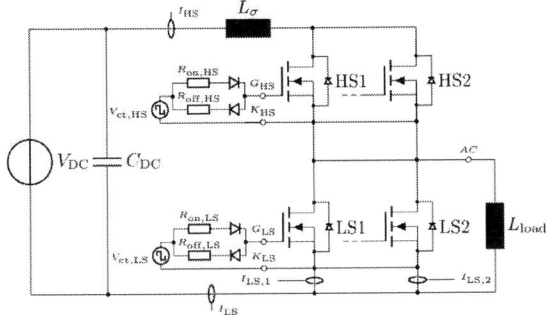

**Fig. 6:** Equivalent circuit of the measurement setup

used, as shown in Fig. 7. The inductance for each DUT operating in parallel is approximated from the drain PCB-connection to the power source PCB-connection. The ratio of those two inductances is $L_{\mathrm{h}} \cdot L_{\mathrm{l}}^{-1} \approx 1.39$. $L_{\mathrm{h}}$ is realized by soldering the TO247-4L package at the end of the package legs, $L_{\mathrm{l}}$ by soldering the package as close as possible to the PCB. Looking at power modules, the low inductive configuration $L_{\mathrm{l}}$ is meant to be as close as possible to the loop inductance of parallel SiC-Power-MOSFETs in high power modules.

### 3.2 Regular parallel operation with symmetric loop inductance

To evaluate the parallel avalanche operation for each of the three pairs mentioned above, the same operating point is scaled for the parallel operation as for the previous single chip characterization. With the correct scaling, $L \cdot I$ in the commutation loop remains the same for both single chip and parallel operation [11]. This results in $L_{\sigma,\mathrm{par}} = 0.5 \cdot L_{\mathrm{nom}}$ and $I_{\mathrm{SW,par}} = 4 \cdot I_{\mathrm{nom}}$. The turn-on gate resistance $R_{\mathrm{G,on,PAIR}}$ for each pair is adjusted for $L_{\sigma,\mathrm{par}}$ at $150\,^{\circ}\mathrm{C}$, $V_{\mathrm{DC}}$ and $I_{\mathrm{SW,par}}$ to reach $V_{\mathrm{rated}}$ at diode turn-off. The devices are connected to the PCB as illustrated in Fig. 7 a) with $L_{\mathrm{h}}$ for both DUTs for a symmetric connection. Figure 8 and Fig. 9 show the switching waveforms for the diode turn-off event and the active turn-on event, respectively, for the three pairs P1 (blue), P2 (red)

and P3 (yellow) at $T_{\mathrm{J}} = 150\,^{\circ}\mathrm{C}$, $L_{\sigma,\mathrm{par}}$, $V_{\mathrm{DC}}$ and $I_{\mathrm{SW,par}}$. Looking at the switching wavesforms, the pairs P1 and P3 show the same behaviour as for the single chip measurement. P1, consisting of two devices with higher $V_{\mathrm{BD}}$, shows a worse reverse recovery behaviour than P3, consisting of two devices with lower $V_{\mathrm{BD}}$. Therefore, the turn-on gate resistance $R_{\mathrm{G,on,P1}}$ has a higher value than $R_{\mathrm{G,on,P3}}$ to set up the same operating point. For the pair P2, consisting of one device with high $V_{\mathrm{BD}}$ and one device with low $V_{\mathrm{BD}}$, the reverse recovery behaviour is somewhere in between P1 and P3 and therefore the value of $R_{\mathrm{G,on,P2}}$ also lies in between $R_{\mathrm{G,on,P1}}$ and $R_{\mathrm{G,on,P3}}$ to set up the same operating point.

The same procedure for setting up the turn-on gate resistances has also been conducted for junction temperatures of $T_{\mathrm{J}} = 125\,^{\circ}\mathrm{C}$ and $T_{\mathrm{J}} = 175\,^{\circ}\mathrm{C}$ to determine a possible influence of $T_{\mathrm{J}}$ on the per-

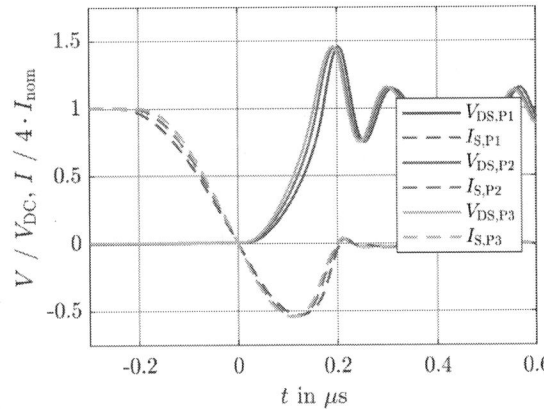

**Fig. 8:** Diode turn-off event for the regular, parallel switching operation of all three pairs at $T_{\mathrm{J}} = 150\,^{\circ}\mathrm{C}$, $V_{\mathrm{DC}}$, $4 \cdot I_{\mathrm{nom}}$, $0.5 \cdot L_{\mathrm{nom}}$ as well as $R_{\mathrm{G,on,P1}}$, $R_{\mathrm{G,on,P2}}$ and $R_{\mathrm{G,on,P3}}$

PCIM Europe 2024, 11– 13 June 2024, Nuremberg DOI: 10.30420/566262084

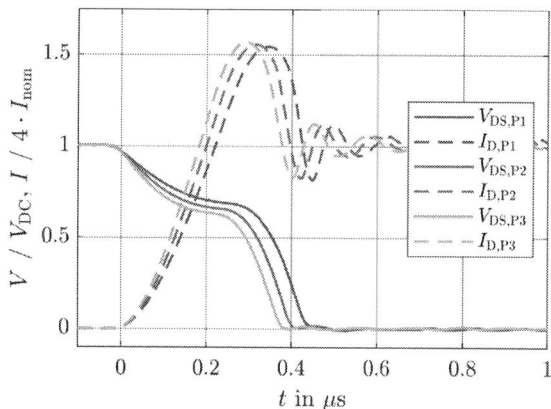

**Fig. 9:** Active turn-on event for the regular, parallel switching operation of all three pairs at $T_J = 150\,°C$, $V_{DC}$, $4·I_{nom}$, $0.5·L_{nom}$ as well as $R_{G,on,P1}$, $R_{G,on,P2}$ and $R_{G,on,P3}$

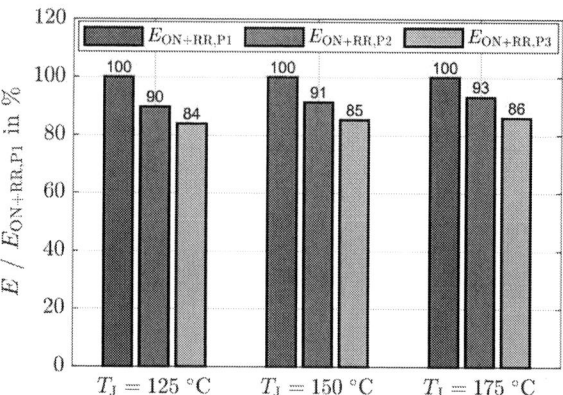

**Fig. 10:** Turn-on and reverse recovery losses for the regular, parallel switching operation for different temperatures $T_J$ at $V_{DC}$, $4·I_{nom}$, $0.5·L_{nom}$ as well as $R_{G,on,P1}$, $R_{G,on,P2}$ and $R_{G,on,P3}$

formance of P2 in comparison to P1 and P3. Figure 10 illustrates the turn-on and reverse recovery losses $E_{ON+RR,PAIR}$ for P1 (blue), P2 (red) and P3 (yellow). An evaluation of $E_{ON+RR,PAIR}$ shows, that for $T_J = 125\,°C$, the performance of the mixed pair P2 is closer to P3. At higher $T_J$, the value of $E_{ON+RR,P2}$ shifts towards P1. At $T_J = 175\,°C$, $E_{ON+RR,P2}$ has the mean value of $E_{ON+RR,P1}$ and $E_{ON+RR,P3}$.

### 3.3 Parallel avalanche operation with symmetric loop inductance

For the evaluation of the parallel avalanche operation due to high switching stress, the stray inductance is increased to four times the value used for the regular parallel switching operation, which results in $L_{\sigma,par,aval} = 2 · L_{nom}$. The devices are connected to the PCB as illustrated in Fig. 7 a) with $L_h$ for both DUTs. The turn-on gate resistances $R_{G,on,PAIR}$ for each pair are the same as for the regular parallel switching operation, which was evaluated in the previous section. The waveforms for the parallel avalanche operation for P1, P2 and P3 are shown in Fig. 11 to Fig. 13, respectively, consisting of the drain to source voltage $V_{DS}$ (blue), the currents for both DUTs $I_{S,DUT1}$ and $I_{S,DUT2}$ (dark red and yellow), the sum of both individual currents $I_{S,Sum} = I_{S,DUT1} + I_{S,DUT2}$ (black, dashed), the current $I_S$ flowing back to the DC-Link (black, solid) as well as the interval, in which the avalanche energy was calculated (light red). The calculation of the avalanche energy, according to [12], follows Eq. (1), where $t_{AV}$ specifies the interval in which $V_{DS}$ is at or above $V_{BD}$.

$$E_{AV} = \int_{t_{AV}} V_{BD} · i\,(t)\,\mathrm{d}t \qquad (1)$$

A comparison of $I_{S,Sum}$ and $I_S$ validates the measurement setup for the individual currents for each DUT. Looking at the avalanche operation of the pairs P1 and P3 with similar $V_{BD}$ (Fig. 11 and Fig. 13), the current distribution between the two devices operating in parallel stays close to equal during the avalanche event. For pair P2, illustrated in Fig. 12, the device with the significant lower $V_{BD}$ (D4) takes over the majority of the reverse recovery current as soon as its breakdown voltage $V_{BD,D4}$ is reached and therefore is exposed to the majority of the avalanche energy throughout the whole avalanche event.

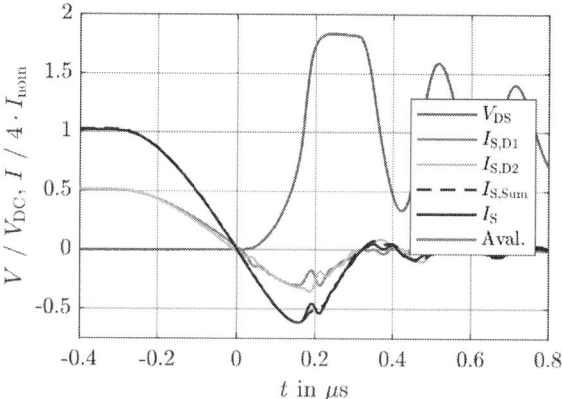

**Fig. 11:** Results for the parallel avalanche operation for P1 at $T_J = 150\,°C$, $V_{DC}$, $4·I_{nom}$, $2·L_{nom}$ and $R_{G,on,P1}$

694

PCIM Europe 2024, 11– 13 June 2024, Nuremberg    DOI: 10.30420/566262084

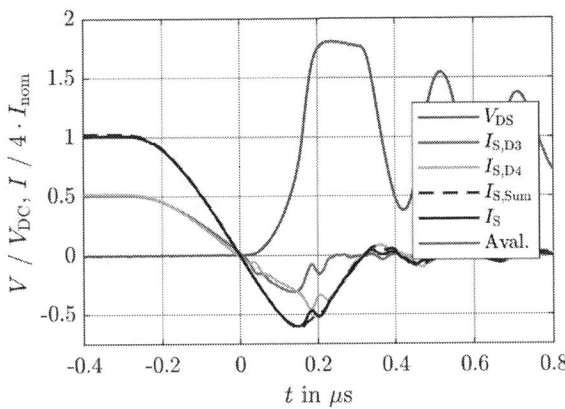

**Fig. 12:** Results for the parallel avalanche operation for P2 at $T_J = 150\,°C$, $V_{DC}$, $4 \cdot I_{nom}$, $2 \cdot L_{nom}$ and $R_{G,on,P2}$

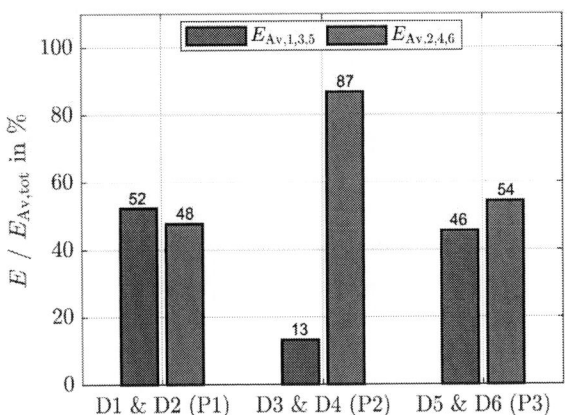

**Fig. 14:** Avalanche energy distribution between the two parallel devices for each pair at $T_J = 150\,°C$, $V_{DC}$, $4 \cdot I_{nom}$, $2 \cdot L_{nom}$ and pair specific $R_{G,on,PAIR}$

Figure 14 summarizes the avalanche energy distribution between the two parallel devices for each pair. The findings indicate that when two DUTs with comparable $V_{BD}$ are used in parallel, as it is the case for P1 and P3, the avalanche operation results in an almost equal energy distribution between the two devices. For P1, D1 is exposed to slightly more avalanche energy than D2 which matches the fact that $V_{BD,D1} < V_{BD,D2}$. Looking at P3, D6 is exposed to slightly more avalanche energy than D5 which matches $V_{BD,D5} > V_{BD,D6}$. Nonetheless, when two devices with a significant difference in $V_{BD}$ are paired, as it is the case for P2, the device with the lower $V_{BD}$ (D4) takes over the majority of the current during the avalanche operation and, in this case, is exposed to more than $85\,\%$ of the total avalanche energy.

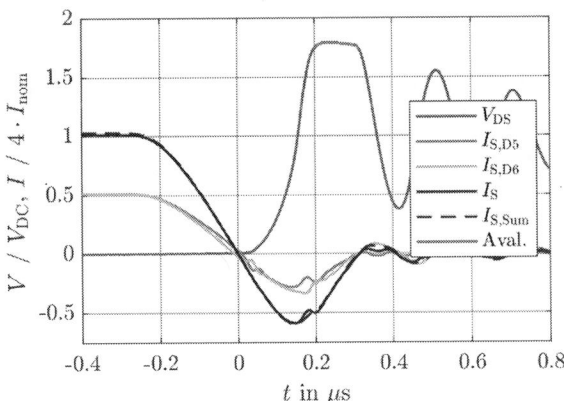

**Fig. 13:** Results for the parallel avalanche operation for P3 at $T_J = 150\,°C$, $V_{DC}$, $4 \cdot I_{nom}$, $2 \cdot L_{nom}$ and $R_{G,on,P3}$

The same procedure has also been applied at temperatures of $T_J = 125\,°C$ and $T_J = 175\,°C$ to evaluate, if a different junction temperature has an influence on the energy distribution during the avalanche event for the parallel operation of two devices with a significant difference in $V_{BD}$. The results, depicted in Fig. 15, show, that the avalanche energy distribution has the same ratio for P2 at $125\,°C$ and $175\,°C$ compared to the measurement at $T_J = 150\,°C$.

## 3.4 Influence of the inductive loop on the parallel avalanche operation

The last aspect that will be addressed in this paper is the influence of the inductive loop between the two devices operating in parallel during the avalanche event due to high switching stress concerning the avalanche energy distribution. Since the results in the previous section showed, that two DUTs with similar $V_{BD}$ have an almost equal energy distribution during the avalanche event, the upcoming investigation of the influence of the loop inductance will be performed for P2 in the first instance. For the evaluation of the loop inductance, a total number of four different loop configurations will be evaluated, as illustrated in Fig. 16. The four different loop setups include two configurations with a symmetric connection for both DUTs operating in parallel, one with high inductance $L_h$ (Fig. 16 a)) and one with low inductance $L_l$ (Fig. 16 b)) for both DUTs towards the common PCB connection. The last two configurations address an asymmetric connection of the two DUTs operating in parallel towards the common PCB connection, one with

695

PCIM Europe 2024, 11– 13 June 2024, Nuremberg    DOI: 10.30420/566262084

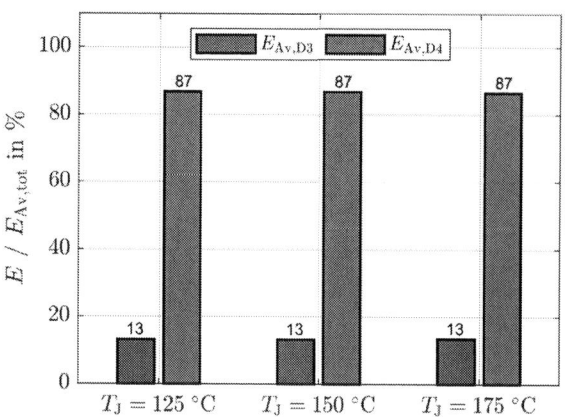

**Fig. 15:** Influence of the junction temperature $T_\mathrm{J}$ for P2 on the avalanche energy distribution at $V_\mathrm{DC}$, $4 \cdot I_\mathrm{nom}$, $2 \cdot L_\mathrm{nom}$ and $R_\mathrm{G,on,P2}$

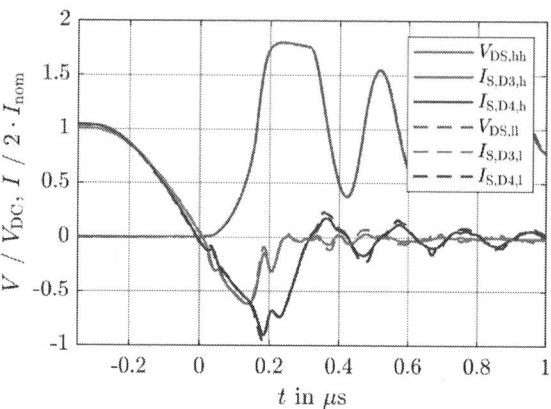

**Fig. 17:** Symmetric high and low loop inductance at $T_\mathrm{J} = 150\,°\mathrm{C}$, $V_\mathrm{DC}$, $4 \cdot I_\mathrm{nom}$, $2 \cdot L_\mathrm{nom}$ and $R_\mathrm{G,on,P2}$

high inductance $L_\mathrm{h}$ for the device with higher $V_\mathrm{BD}$, depicted in Fig. 16 c), as well as one with lower inductance $L_\mathrm{l}$ for the device with higher $V_\mathrm{BD}$ (D3), illustrated in Fig. 16 d).

Figure 17 and Fig. 18 show the switching waveforms for the avalanche event during diode turn-off for symmetric (Fig. 16 a) and b)) and asymmetric (Fig. 16 c) and d)) loop inductance, respectively, with the drain-source voltage $V_\mathrm{DS}$ (blue) and the individual currents of both DUTs $I_\mathrm{S,D3}$ (red) and $I_\mathrm{S,D4}$ (violet). Looking at the symmetric loop inductance (Fig. 17), the configuration with high inductance $L_\mathrm{h}$ (solid - Fig. 16 a)) as well as the configura-

tion with low inductance $L_\mathrm{l}$ (dashed - Fig. 16 b)) show the same switching behaviour throughout the whole diode turn-off event, including the current commutation, the reverse recovery behaviour and the avalanche event.

The waveforms of the avalanche operation of P2 with asymmetric loop inductance are depicted in Fig. 18, including the configuration with $L_\mathrm{h}$ for D3 and $L_\mathrm{l}$ for D4 (solid - Fig. 16 c)) as well as the configuration with $L_\mathrm{l}$ for D3 and $L_\mathrm{h}$ for D4 (dashed - Fig. 16 d)). Looking at the measurement results, the ratio for the two current slopes is $\left(\frac{\mathrm{d}i}{\mathrm{d}t}\right)_\mathrm{l} \cdot \left(\frac{\mathrm{d}i}{\mathrm{d}t}\right)_\mathrm{h}^{-1} \approx 1.41$ which is close to the ratio of the two approximated inductances $L_\mathrm{h} \cdot L_\mathrm{l}^{-1} \approx 1.39$. Although the difference in the inductance towards the PCB causes a significant difference in the $\frac{\mathrm{d}i}{\mathrm{d}t}$ of the two DUTs operating in parallel, the behaviour during the avalanche stress is the same. Until the zero

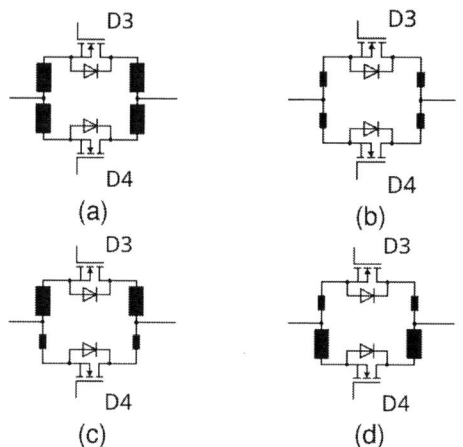

**Fig. 16:** Variation of the loop inductance: Symmetric with high inductance (a), symmetric with low inductance (b), asymmetric with high inductance for the device with higher $V_\mathrm{BD}$ (c) and asymmetric with high inductance for the device with lower $V_\mathrm{BD}$ (d)

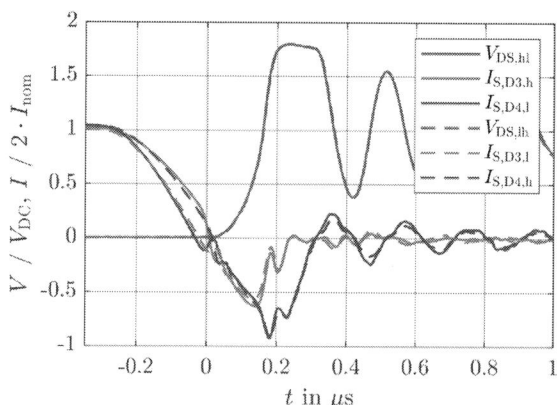

**Fig. 18:** Asymmetric distribution of loop inductance at $T_\mathrm{J} = 150\,°\mathrm{C}$, $V_\mathrm{DC}$, $4 \cdot I_\mathrm{nom}$, $2 \cdot L_\mathrm{nom}$ and $R_\mathrm{G,on,P2}$

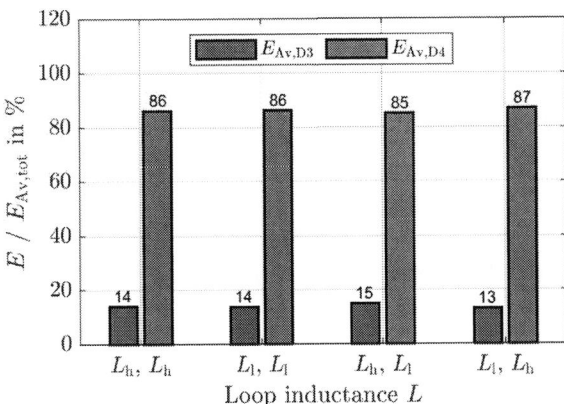

**Fig. 19:** Avalanche energy distribution for different loop inductance setups at $T_J = 150\,°C$, $V_{DC}$, $4 \cdot I_{nom}$, $2 \cdot L_{nom}$ and $R_{G,on,P2}$

crossing of the current at diode turn-off event, the commutation process is current driven and therefore the difference in the inductances causes the difference in $\frac{di}{dt}$. As soon as the diodes start to take over the voltage, the commutation process is voltage driven and the reverse recovery process is not affected by the different commutation inductances since the voltage across the diode is much higher than the voltage across the connection inductance. The energy distribution of all four different loop configurations is illustrated in Fig. 19. The avalanche energy distribution is identical for both symmetric layouts. Looking at the asymmetric layouts, the energy distribution slightly shifts towards the device with the higher inductance towards the PCB, as to be expected, but the device with the lower $V_{BD}$ is still exposed to the majority of the total avalanche energy. The results show, that the loop inductance has close to no influence on the energy distribution during the avalanche event and the reverse recovery process itself for the given setup. Therefore, an evaluation of the loop inductance for P1 and P3 will not be performed.

## 4 Conclusion

This paper investigated the parallel operation of SiC-Power-MOSFET body diodes under harsh switching conditions with the main focus on the pairing of devices with a significant difference in the breakdown voltage $V_{BD}$ for a regular switching operation as well as an avalanche operation due to high switching stress. Another aspect was the influence of the inductive loop between the two devices operating in parallel, including symmetric

and asymmetric loop layouts. The regular switching operation shows, that SiC-MOSFETs with a significant difference in $V_{BD}$ exhibit a different reverse recovery behaviour of the internal body diode which results in different switching losses. Looking at the parallel avalanche operation of SiC-MOSFET body diodes, it can be said that the device with the lower breakdown voltage $V_{BD}$ is exposed to the majority of the total avalanche energy. The variation of the loop inductance, especially an asymmetric connection of the DUTs towards the PCB, has close to no influence on the avalanche energy distribution. Considering multiple parallelization of chips in a power module, strong asymmetries in the avalanche energies between the chips can occur. Operating beyond SOA limits of the datasheet is not guaranteed by power module manufacturers. For upcoming research, a verification of the results from the parallel operation of two SiC-MOSFET body diodes is planned by increasing the number of parallel devices for the regular switching operation as well as the avalanche operation. Another aspect will be the evaluation of the influence of a difference in the junction temperature $T_J$ of the individual DUTs operating in parallel concerning a regular switching operation as well as an avalanche operation due to high switching stress.

## References

[1] R. Horff, A. Maerz, and M.-M. Bakran, "Analysis of reverse-recovery behaviour of sic mosfet body-diode - regarding dead-time," in *Proceedings of PCIM Europe 2015; International Exhibition and Conference for Power Electronics, Intelligent Motion, Renewable Energy and Energy Management*, 2015, pp. 1–8.

[2] A. März, T. Bertelshofer, and M.-M. Bakran, "Improving the performance of sic trench mosfets under hard switching operation," in *2017 IEEE 12th International Conference on Power Electronics and Drive Systems (PEDS)*, 2017, pp. 553–558. DOI: 10.1109/PEDS.2017.8289177.

[3] M. Rauh, A. Maerz, S. Schoenewolf, and M.-M. Bakran, "Impact of operating a sic-mosfets body diode beyond its soa," in *PCIM Europe 2023; International Exhibition and Conference for Power Electronics, Intelligent Motion, Renewable Energy and Energy Management*, 2023, pp. 1–8. DOI: 10.30420/566091076.

[4] A. Maerz, S. Schoenewolf, A. Nagel, M. Rauh, and M.-M. Bakran, "Deadtime optimization eliminating snap-off of 3.3kv sic mosfet bodydiodes," in *2023 25th European Conference on Power Elec-*

*tronics and Applications (EPE'23 ECCE Europe,* 2023, pp. 1–7.

[5] X. Liu, X. Li, C. Herrmann, and T. Basler, "The impact of the dead-time on the reverse recovery behavior of sic-mosfet body diodes," in *2023 35th International Symposium on Power Semiconductor Devices and ICs (ISPSD)*, 2023, pp. 322–325. DOI: 10.1109/ISPSD57135.2023.10147719.

[6] S. Palanisamy, T. Basler, X. Liu, C. Herrmann, R. Elpelt, and P. Sochor, "Overcurrent turn-off robustness and stability of the switching behavior of sic mosfet body diodes," in *2022 IEEE 34th International Symposium on Power Semiconductor Devices and ICs (ISPSD)*, 2022, pp. 257–260. DOI: 10.1109/ISPSD49238.2022.9813611.

[7] H. Li, S. Zhao, X. Wang, L. Ding, and H. A. Mantooth, "Parallel connection of silicon carbide mosfets—challenges, mechanism, and solutions," *IEEE Transactions on Power Electronics*, vol. 38, no. 8, pp. 9731–9749, 2023. DOI: 10.1109/TPEL. 2023.3278270.

[8] C. Herrmann, M. He, M. Alaluss, T. Basler, and J. Lutz, "Avalanche robustness of sic mosfets in parallel connections," in *PCIM Europe 2023; In-*

*ternational Exhibition and Conference for Power Electronics, Intelligent Motion, Renewable Energy and Energy Management*, 2023, pp. 1–10. DOI: 10.30420/566091077.

[9] C. Herrmann, M. He, M. Alaluss, T. Basler, R. Elpelt, and G. Zeng, "Energy balancing in paralleled sic mosfets during an avalanche event," in *CIPS 2024; 13th International Conference on Integrated Power Electronics Systems*, 2024.

[10] R. Werner, J. da Cunha, and H.-G. Eckel, "Mutual influence of quasistatic and dynamic current imbalances of paralleled igbts," in *2019 21st European Conference on Power Electronics and Applications (EPE '19 ECCE Europe)*, 2019, P.1–P.8. DOI: 10.23919/EPE.2019.8915438.

[11] R. W. Maier and M.-M. Bakran, "Switching sic mosfets under conditions of a high power module," in *2018 20th European Conference on Power Electronics and Applications (EPE'18 ECCE Europe)*, 2018, P.1–P.9.

[12] J. Lutz, H. Schlangenotto, U. Scheuermann, and R. D. Doncker, *Semiconductor Power Device*, 2nd ed. Springer, 2018. DOI: 10.1007/978-3-319-70917-8.

PCIM Europe 2024, 11– 13 June 2024, Nuremberg

DOI: 10.30420/566262085

# 3.3kV SBD-Embedded SiC-MOSFET Module for Traction Use

Yoichi Hironaka[1], Shigeru Okimoto[1], Mamoru Matsuo[1], Shota Saito[1], Kenji Hatori[1], Nils Soltau[2]

[1] Mitsubishi Electric Corporation, Japan

[2] Mitsubishi Electric Europe B.V., Germany

Corresponding author:  Yoichi Hironaka, Hironaka.Yoichi@da.MitsubishiElectric.co.jp
Speaker:                       Yoichi Hironaka, Hironaka.Yoichi@da.MitsubishiElectric.co.jp

## Abstract

We have developed a 3.3kV Schottky-barrier-diode-embedded (SBD-embedded) silicon-carbide metal-oxide-semiconductor field-effect transistor (SiC-MOSFET) module for traction use. It achieves 60% lower switching loss compared to the conventional 3.3 kV full-SiC module. It also outperforms the conventional module in output current across all frequencies, with 40% improvement at 3 kHz. The SBD-embedded SiC-MOSFETs ensure high reliability by preventing bipolar degradation. We have introduced a novel structure, the bipolar mode activation (BMA) cell, to enhance surge current capability, achieving a similar level to body-diode-operated SiC-MOSFET modules. The required BMA area   is 0.2% of the chip area and does not affect electrical characteristics within the safe operating range. Continuous current and repetitive surge tests confirm the robustness to bipolar degradation of our SBD-embedded SiC-MOSFETs. Additionally, our module exhibits excellent moisture resistance according to the High Voltage High Humidity High Temperature Reverse Bias (HV-H3TRB) test based on ECPE guidelines.

## 1    Introduction

Approximately 25 years ago, silicon-based (Si-based) Insulated Gate Bipolar Transistors (IGBTs) were initially utilized as power modules in railcar traction systems. Since then, significant advancements have been made in semiconductor chip technology for power modules, including the reduction of losses and improvements in package technologies, resulting in notable enhancements in the performance of railcar traction systems.

However, Si-based chips for power modules are gradually approaching their physical limits, and further dramatic improvements are not expected. Consequently, wide bandgap semiconductors, particularly silicon carbide (SiC), have garnered increased attention due to their ability to overcome the limitations of Si. In railcar traction systems, the desired features include not only loss reduction, but also high reliability achieved through improved semiconductor chips and package technologies.

Considering these requirements, we have developed a new 3.3kV Schottky-barrier-diode-embedded (SBD-embedded) SiC-metal-oxide-semiconductor field-effect transistor (SiC-MOSFET) module. This module enables more efficient and reliable traction systems.

## 2    3.3kV/800A SBD-embedded SiC-MOSFET module

### 2.1   Structure

The newly developed module is rated for 3.3 kV / 800 A, and its specifications and appearance are presented in Table 1 and Fig. 1, respectively. To achieve low commutation inductance and fast switching, the LV100 package is utilized, which has become the established new standard for high-voltage power modules. This package is identical to the conventional 3.3kV full-SiC power module (FMF750DC-66A) [1]. Fig. 2 illustrates the structure of the SBD-embedded SiC-MOSFET module. Our previous study indicated that the active area of the SBD chip in the 3.3 kV full-SiC power module needs to be approximately 1.3 times larger than that of the MOSFET chip to keep the external SBD chips inactive [2]. This is because the voltage applied to the body diode must be lower than the built-in potential of the P-N junction. However, the SBD-embedded SiC-MOSFET allows for a significant reduction in chip area, as it avoids voltage drop across the drift layer [1]. To ensure high-temperature operation (maximum junction temperature $T_{jmax}$ = 175 °C), the chips are bonded using silver (Ag) sintering, and the terminals are connected to the substrate through ultrasonic bonding. Furthermore, to enhance heat dissipation, materials with lower thermal resistance ($R_{th}$) for substrates and system solder are employed compared to the conventional 3.3 kV full-SiC power module. As a result, the $R_{th}$ of the newly developed module is reduced by approximately 30% compared to the FMF750DC-66A.

**Table 1.** 3.3kV/800A SBD-embedded SiC-MOSFET module

| Type name | Rated voltage | Rated current | Isolation voltage | $T_{jmax}$ |
|---|---|---|---|---|
| FMF800DC-66BEW | 3.3 kV | 800 A | 6.0 kV$_{rms}$ | 175 °C |

**Fig. 1** 3.3kV/800A SBD-embedded SiC-MOSFET module.

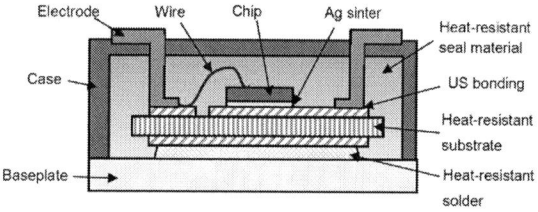

**Fig. 2** Cross-sections of SBD-embedded SiC-MOSFET module.

## 2.2 Features of SBD-embedded SiC-MOSFET

SBD-embedded SiC-MOSFETs exhibit two significant features. Firstly, they offer lower switching losses compared to conventional 3.3 kV full-SiC power modules. The switching waveforms, as depicted in Fig. 4(a), 4(b), demonstrate a decrease in switching delay time and an increase in switching speed. These improvements are achieved through optimized chip design, resulting in a reduction in input capacitance ($C_{iss}$) and an increase in mutual conductance ($g_m$). $C_{iss}$ is measured under the following conditions: drain-source voltage $V_{DS}$ = 10 V, gate-source voltage $V_{GS}$ = 0 V, frequency $f$ = 100 kHz. As shown in Fig. 3, $g_m$ is measured by calculating $dI_D/dV_{GS}$ close to rated current. Table 2 presents the measurements of $C_{iss}$ and $g_m$, showing a 52.6% reduction in $C_{iss}$ and a 41.9% increase in $g_m$ comparing the new FMF800DC-66BEW with the conventional FMF750DC-66A. Consequently, the new module exhibits reduced turn-on and turn-off switching delay times, leading to lower switching losses.

Furthermore, SBD-embedded SiC-MOSFETs suppress the superposition of recovery current in the diode when the MOSFET turns on. This reduces turn-on and recovery losses. The turn-on and recovery switching waveforms, illustrated in Fig. 4 indicate that the newly developed module experiences fewer charge carriers during turn-on compared to the conventional module. In the conventional module, an external SBD was connected, but it did not completely suppress recovery charge by minority carrier injection. This results in a small amount of reverse recovery current (Fig. 4(a)) [3]. However, in the SBD-embedded MOSFET, the recovery current is suppressed. Only a displacement current for charging and discharging the output capacitance is evident. This leads to reduced recovery losses. Additionally, since recovery current influences the increase in turn-on current of the opposing switching arm, the suppression of recovery current also contributes to reducing $E_{on}$, which constitutes a significant portion of switching losses.

As a result, the SBD-embedded SiC-MOSFET and optimized design reduce switching losses by 66% compared to the conventional 3.3kV full-SiC power module (FMF750DC-66A).

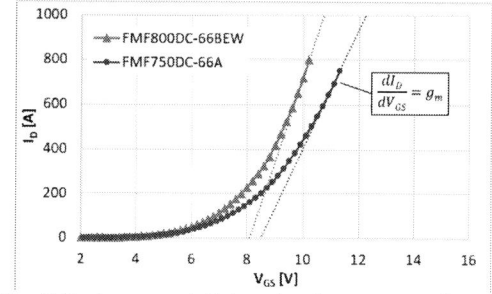

**Fig. 3** Drain current ($I_D$) and gate-source voltage ($V_{GS}$) characteristics.
(Drain-source voltage ($V_{DS}$) = 20V, $T_j$ = 25 °C).

**Table 2** Comparison of capacitance and conductance characteristics.

| | $C_{iss}$ | $g_m$ |
|---|---|---|
| Conventional module (FMF750DC-66A) | 209 nF | 267 $\Omega^{-1}$ |
| New module (FMF800DC-66BEW) | 110 nF | 379 $\Omega^{-1}$ |

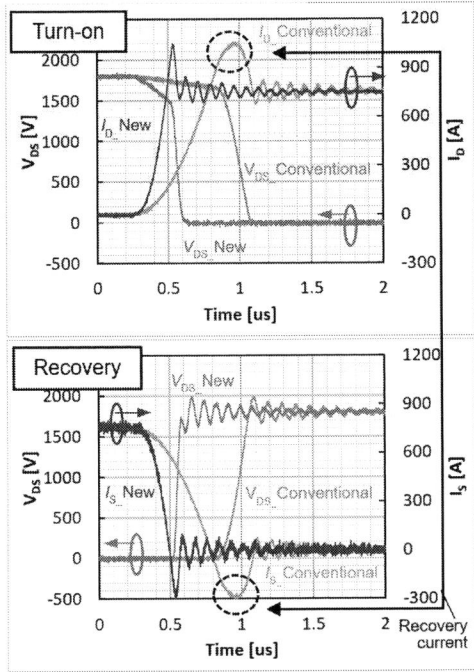

**Fig. 4(a)** Turn-on and recovery switching waveforms ($V_{DD}$=1800V, $I_D$=$I_S$=750A, $L_S$=40nH, $T_j$=175 °C)

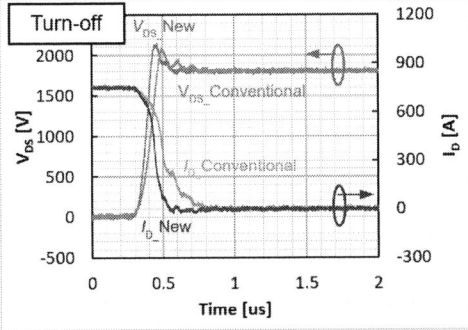

**Fig. 4(b)** Turn-off switching waveform. ($V_{DD}$=1800V, $I_D$=750A, $L_S$=40nH, $T_j$=175 °C)

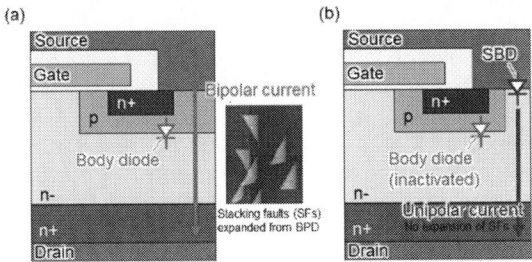

**Fig. 5** Cross-sections of (a) a conventional MOSFET (b) an SBD-embedded MOSFET.

Secondly, SBD-embedded SiC-MOSFETs offer high reliability by preventing device degradation caused by the inherent body diode current. Typically, SiC-MOSFETs can experience a permanent increase in on-state voltage due to bipolar degradation (refer to Fig. 5) [3-5]. However, with SBD-embedded SiC-MOSFETs, the threshold voltage of the SBD is lower than that of the body diode. This effectively suppresses current flow through the body diode.

On the other hand, it has been believed that SBD-embedded MOSFETs have lower surge current capability compared to bipolar devices, because the embedded SBDs hinder conductivity modulation during a surge. Moreover, our previous reports have shown that current crowding in parallel-connected chips during a surge is caused by variations in the snapback voltage among chips. The snapback voltage is defined as the voltage at which a body diode becomes active [6-7]. To address this issue, we have developed a novel structure called the bipolar mode activation cell (BMA cell) to enhance the surge current capability of the newly developed module (refer to Fig. 6) [6-8]. By implementing BMA cells, the snapback voltages of different chips become more uniform, resulting in improved surge current capability by preventing current crowding in parallel-connected chips. Since bipolar operation spreads outward from the BMA cells and covers the entire chip, the required area for BMA cells is only 0.2% of the total chip area. Therefore, bipolar operation occurs in the SBD-embedded MOSFETs only when a large accident current flows during irregular operation, without affecting the electrical characteristics of the chips during normal operation.

Fig. 7 illustrates the I-V characteristics at $T_j$ = 175°C of SBD-embedded SiC-MOSFETs. We have confirmed that new module operates in unipolar within the safe operating area, both with and without BMA cells. Thus, the application of BMA cells does not induce bipolar degradation caused. Fig. 8 presents the results of the surge current capability. With BMA cells implemented, the surge current capability of the SBD-embedded SiC-MOSFET module reaches a similar level to that of the body-diode-operated SiC-MOSFET module.

Additionally, we have included Fig. 9 to showcase the appearance of the chip surface after surge-current capability tests. Without BMA cells, we observe melted aluminum chip metallization of single chips only. This indicates that the current concentrates on specific chips due to variations in the snapback voltage. However, with BMA cells, we observed melted aluminum chip metallization in all chips. This indicates that the application of BMA cells leads to a more uniform current distribution by narrower snapback voltage variation. Consequently, we have confirmed that the application of BMA cells allows for the distribution of current in parallel-connected chips as intended, resulting in an improvement in surge current capability.

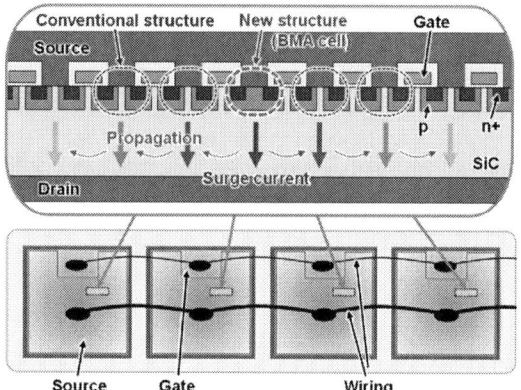

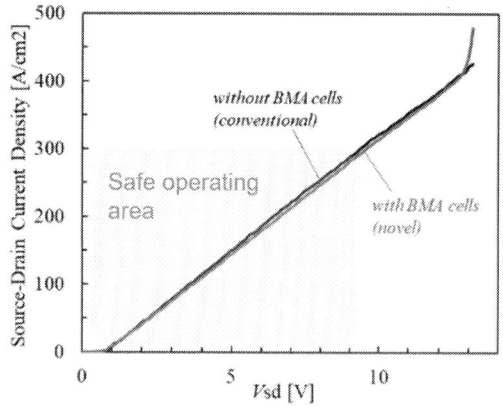

**Fig. 6** Cross-section of BMA cell.

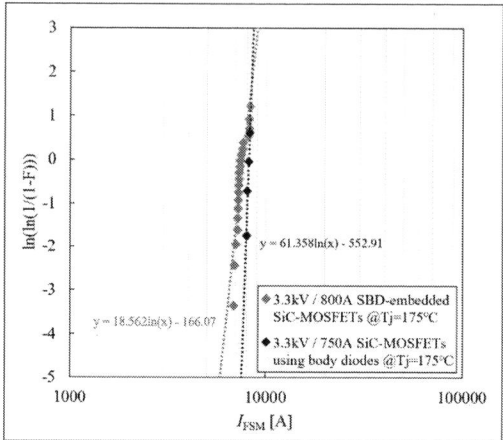

**Fig. 8** Weibull plots of surge forward current ($I_{FSM}$) measurement results.

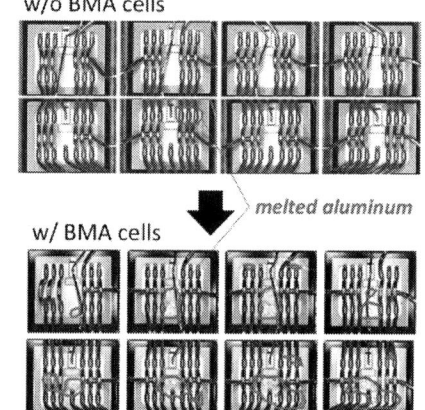

**Fig. 9** Chip appearance after test of surge current capability (w/ and w/o BMA cells).

**Fig. 7** Source-drain current density and source-drain voltage ($V_{SD}$) waveforms at $T_j$=175 °C of SBD-embedded SiC-MOSFETs.

# 3 Application Benefits

## 3.1 Inverter Loss Simulation

To evaluate the advantages of the new module for traction applications, the inverter loss is calculated using the Melcosim power loss simulator [9].

The simulation is conducted under the following conditions: supply voltage $V_{DD}$ = 1800 V, output current $I_O$ = 400 $A_{rms}$, carrier frequency $f_c$ = 1 kHz, and power factor P.F. = ±0.85. For each power module the respective recommended gate-drive condition are considered. The simulation results are presented in Fig 10. As mentioned in section 2.2, turn-on, turn-off, and recovery losses are all reduced. Consequently, it is confirmed that the new module (FMF800DC-66BEW) achieves a 59% reduction in switching losses during inverter operation compared to the conventional full-SiC module (FMF750DC-66A). Furthermore, when compared to the conventional Si module (CM600DA-66X), the reduction is 92%. These results indicate that the new product can significantly contribute to the improvement of efficiency in traction systems.

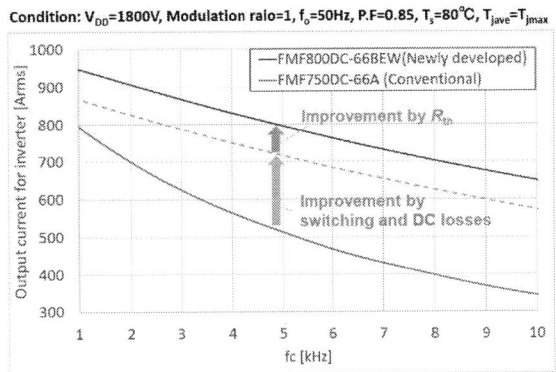

Fig. 11 The switching carrier frequency characteristics of the inverter output current.

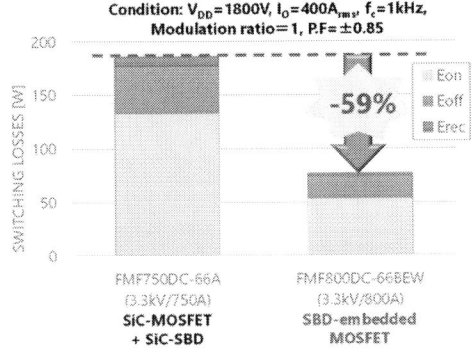

Fig. 10 Comparison of result of inverter switching losses.

## 3.2 Inverter Output Current Characteristics

The switching carrier frequency characteristics of the inverter output current are calculated using Melcosim [9]. These characteristics are commonly used as a benchmark for assessing overall device performance. The simulation results are presented in Fig. 11. For Fig. 11, the following conditions are applied: $V_{DD}$ = 1800 V, P.F. = 0.85, heat sink temperature $T_S$ = 80 °C, modulation ratio = 1. As before, respective recommended gate-drive conditions are considered for each power module. It can be observed that the inverter output current of the new module surpasses that of the conventional full-SiC module across the entire frequency range. Especially, at a frequency of 3 kHz, the inverter output current of the developed module is 40% higher than that of the conventional full-SiC module. These results confirm that FMF800DC-66BEW delivers exceptional inverter output, particularly in the high-frequency range where SiC devices can showcase their performance.

## 4 Robustness

### 4.1 Bipolar Degradation

In section 2.2, we have demonstrated that the device does not exhibit bipolar conduction within the safe operating region. However, for long-term reliability considerations, it is crucial to verify if this trend remains consistent during prolonged continuous conduction. To address this concern, we have conducted continuous conduction tests specifically on the embedded SBD to assess its robustness against an increase in on-state voltage. These tests were performed on individual SBD-embedded SiC-MOSFET chips, and the test conditions are outlined in Table 3.

The waveforms of the drain current density ($J_D$) and drain-source voltage ($V_{DS}$) before and after the tests are depicted in Fig. 11. If bipolar degradation occurs, it is expected that the on-state voltage will increase by approximately 10% [8]. It is evident from the graphs that there has been no observable increase in the on-state voltage, thus confirming the absence of bipolar degradation even during prolonged continuous conduction.

Table 3 Continuous current test conditions with SBD-embedded SiC-MOSFET chip.

| Source current | Gate voltage | Test duration | Number of samples |
|---|---|---|---|
| Rated current | -10 V | 100 h | N=10 |

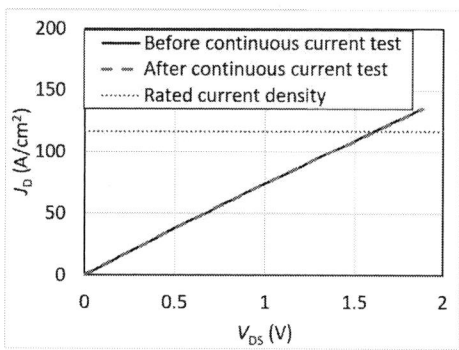

**Fig. 11** Drain current density ($J_D$) and drain-source voltage ($V_{DS}$) waveforms of before and after continuous current test at $T_j$=25 °C.

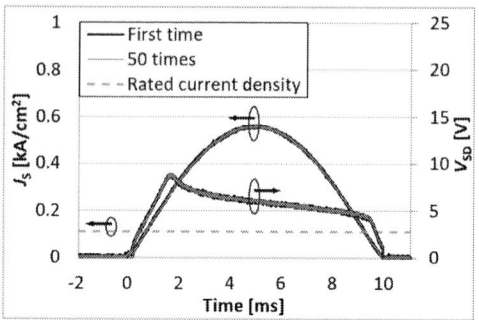

**Fig. 12** Source current density ($J_S$) and source-drain voltage ($V_{SD}$) waveforms of first time and 50 times of surge capability tests.

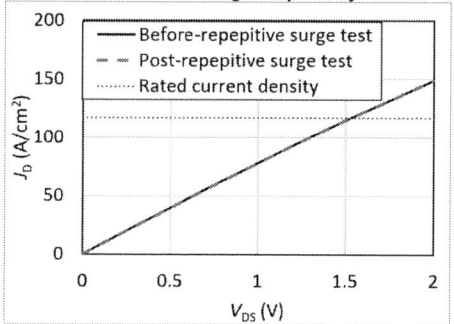

**Fig. 13** $J_D$ - $V_{DS}$ waveforms of before and after repetitive surge test at Tj=25 °C.

Furthermore, as mentioned in section 2.2, if a surge current flows due to irregular operation, the device will undergo bipolar operation during the duration of the surge current flow. To assess robustness in the event of multiple instances of irregular operation, we have conducted a repetitive surge capability test where surge current has been applied continuously. The test conditions are detailed in Table 4, with the source current set to five times the rated current.

Fig. 12 illustrates the waveforms of the first and 50th test. Even after 50 times of surge current events, no degradations is observed in the waveforms. This indicates that SBD-embedded SiC-MOSFET chips maintained their performance and integrity throughout the repetitive surge capability test.

Furthermore, Fig. 13 presents the $J_D$-$V_{DS}$ waveforms before and after the repetitive surge capability test. Despite approximately 500ms of bipolar operation, there is no increase in the on-state voltage. These findings demonstrate that the new module exhibits a high level of robustness against bipolar degradation.

Based on these results, it can be concluded that the SBD-embedded SiC-MOSFET module demonstrates exceptional resilience in the face of multiple instances of irregular operation involving surge currents.

**Table 4** Continuous surge capability test conditions with SBD-embedded SiC-MOSFET chip.

| Source current | Pulse width | Number of repetitions | Number of samples |
|---|---|---|---|
| 5-times of rated current | 10 ms | 50 times | N=10 |

## 4.2 Humidity Robustness

Humidity robustness remains a significant concern for power modules, as these modules are not hermetically sealed. Measurements and simulations have shown that high voltage power modules are exposed to humid conditions within a traction converter cabinet, which can be even more severe than the ambient conditions [10-11].

To confirm the humidity robustness, the HV-H3TRB (High Voltage High Humidity High Temperature Reverse Bias) test is typically conducted. For railway applications, the HV-H3TRB test guideline "*ECPE guideline PSRRA 01*" has been defined [12].

For this paper, we have performed the HV-H3TRB test based on the ECPE guideline on 3.3kV/800A SBD-embedded SiC-MOSFET modules. The test conditions are presented in Table 5. According to the ECPE guidelines, the test voltage for a voltage class of 3300V is typically set to 1950V. However, to ensure an extensive quality margin, we have decided to use a more rigorous test voltage of 2100V. Throughout the 2000-hour test, no increase in leakage current is observed, as shown in Fig. 14. Additionally, Table 6 displays the characteristic data before and after the HV-H3TRB test, which remained equivalent. These results indicate that the HV-H3TRB test has not caused any degradation in performance. Thus, it can be concluded that the developed module exhibits excellent moisture resistance.

**Table 5** HV-H3TRB test conditions

| Temperature | Relative Humidity | Voltage | Test duration | Number of samples |
|---|---|---|---|---|
| 85 °C | 85% RH | 2100 V | 2000 h | N=5 |

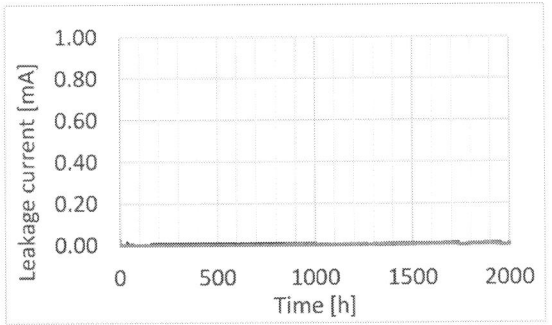

**Fig 14.** Leakage current waveform during HV-H3TRB test

**Table 6** Comparison of characteristics between before and after 1000h the HV-H3TRB test ($T_j$=25 °C)

| | | Before HV-H3TRB test | After 1000h HV-H3TRB test |
|---|---|---|---|
| $I_{DSX}$ | $V_{DS}$=3300V, $V_{GS}$=-7V | 1.40 μA | 0.90 μA |
| $I_{GSS(\pm)}$ | $V_{DS}$=0V, $V_{GS}$=±20V | ≤0.1 μA | ≤0.1 μA |
| $V_{DS(on)}$ | $I_D$=800A, $V_{GS}$=17V | 1.88 V | 1.89 V |
| $V_{GS(th)}$ | $I_D$=80mA, $V_{DS}$=10V | 2.22 V | 2.21 V |
| $V_{SD}$ | $I_S$=800A, $V_{GS}$=17V | 2.32 V | 2.34 V |

## 5 Conclusion

High Voltage power modules in railcar traction system need to reduce power losses, enable system-size reduction, and achieve higher system reliability. To satisfy these needs, we have developed the 3.3kV SBD-embedded SiC-MOSFET module. The inverter switching loss of the developed module can be reduced by about 60% compared to the conventional 3.3 kV full-SiC module. We also calculated the switching carrier frequency characteristics of the inverter output current. We confirmed that the new module surpasses the conventional 3.3 kV full-SiC module in output current in all frequency ranges, with a 40% improvement in output current at a frequency of 3 kHz.

Moreover, the SBD-embedded SiC-MOSFETs provide high reliability by preventing bipolar degradation. Furthermore, we have developed a novel structure to improve surge current capability, called BMA cell. By implementing BMA cells, the surge current capability of the new module reaches similar levels to that of a body-diode-operated SiC-MOSFET module. The BMA cell promotes bipolar conduction during exceptionally high current flow. For this, the required area for the BMA cells is only 0.2% of the total chip area. Within the safe operating range, there is no bipolar conduction, and it does not affect the electrical characteristics. We also conducted continuous current tests and repetitive surge tests on SBD-embedded SiC-MOSFET to confirm its strong robustness to bipolar degradation.

Furthermore, we have conducted the HV-H3TRB (High Voltage High Humidity High Temperature Reverse Bias) test based on the ECPE guidelines. The results confirm that the developed module exhibits excellent moisture resistance.

## References

[1] T. Negishi, R. Tsuda, K. Ota, S. Iura, H. Yamaguchi, "3,3kV All-SiC Power Module for Traction use", *PCIM Europe 2017*, Conference proceedings, pp.51-56

[2] K. Kawahara, S Hino, K. Sadamatsu, Y. Nakao, T. Iwamatsu, S. Nakata, S. Tomohisa, S. Yamakawa, "Impact of Embedding Schottky Barrier Diodes into 3.3 kV and 6.5 kV SiC MOSFETs", *Materials Science Forum 2018*, vol. 924, pp. 663-666

[3] T. Tominaga, S. Hino, Y. Mitsui, J. Nakashima, K. Kawahara, S. Tomohisa, N. Miura, "Investigation on the Effect of Total Loss Reduction of HV Power Module by Using SiC-MOSFET Embedding SBD", *Materials Science Forum 2020*, vol. 1004, pp. 801-807

[4] M. Furukawa, H. Kono, K. Sano, M. Yamaguchi, H. Suzuki, T. Misao, G. Tchouangue, "Improved reliability of 1.2kV SiC MOSFET by preventing the intrinsic body diode operation", *PCIM Europe digital days 2020*, pp. 1- 5

[5] T. Ishigaki, T. Murata, K. Kinoshita, T. Morikawa, T. Oda, R. Fujita, K. Konishi, Y. Mori, A. Shima, "Analysis of Degradation Phenomena in Bipolar Degradation Screening Process for SiC-MOSFETs", Proceedings of *ISPSD'19*, pp. 259-262

[6] A. Iijima, K. Kawahara, K. Sugawara, S. Hino, K. Fujiyoshi, Y. Oritsuki, T. Murakami, T. Takahashi, Y. Kagawa, Y. Hironaka, K. Nishikawa, "Improving Surge Current Capability of SBD-Embedded SiC-MOSFETs in Parallel Connection by Applying Bipolar Mode Activation Cells", Proceedings of *ISPSD'23*, pp. 238-241

[7] S. Okimoto, Y. Hironaka, K. Hatori, A. Iijima, K. Kawahara, K. Sugawara, N. Soltau, "Improvement of Surge Current Capability of 3.3 kV SBD-Embedded SiC-MOSFET Module", Proceedings of *EPE'23 ECCE Europe*, No. 042

[8] K. Kawahara, K. Sugawara, A. Iijima, K. Fujiyoshi, Y. Oritsuki, T. Murakami, T. Takahashi, Y. Kagawa, Y. Hironaka, K. Nishikawa, "Comparison of the Surge Current Capabilities of SBD-Embedded and Conventional SiC MOSFETs", Proceedings of *ICSCRM 2023*

[9] Melcosim download page: https://www.mitsubishielectric.com/semiconductors/simulator/index.html

[10] O. Schuster, A Nagel, B Laska, "Observation and simulation of dynamic humidity in power converters for railway applications due to moisture diffusion in plastics", Proceedings of *EPE'21 ECCE Europe*, No. 149

[11] K. Hatori, K. Ebihara, K. Ishimoto, R. Tsuda, N. Soltau, S. Idaka, E. Wiesner, S. Schönewolf, O.Schuster, M. Hanf, "Investigation of acceleration factors of the HV-H3TRB test on 3.3kV SiC SBDs", Proceedings of *EPE'23 ECCE Europe*, No. 173

[12] ECPE Working Group Railway Reliability: https://www.ecpe.org/research/working-groups/railway-reliability/

PCIM Europe 2024, 11– 13 June 2024, Nuremberg     DOI: 10.30420/566262086

# Dead Time Optimization for High-Power SiC MOSFET Module in Consideration of Parasitic Components

Pham Ha Trieu To [1], Hao Wang[1], Florian Sawallich[1], Felix Kayser[1], Hans-Günter Eckel[1]

[1] University Rostock, Germany

Corresponding author: Pham Ha Trieu To, pham.to2@uni-rostock.de
Speaker: Pham Ha Trieu To, pham.to2@uni-rostock.de

## Abstract

The parasitic components influence the gate voltage of the high power SiC MOSFET module significantly during di/dt and dv/dt phases of the switching transients. In some particular situations, a decrease in dead time can potentially introduce unexpected turn-on and reverse-recovery behaviors. This paper investigates the change of dead time under various operating conditions and their effects on the turn-on, reverse-recovery losses and reverse-recovery oscillation. From this perspective, the author recommends an optimization method for dead time that takes into consideration the influence of parasitic components.

## 1 Introduction

SiC MOSFET's body diode is well known for snappy recovery, especially at high temperature and high load current [2, 3]. By reducing deadtime, the bipolar charges in the drift region have not yet reached their equilibrium between recombination and generation while it is turned off, leading to a lower amount of storage charge in the drift region and a softer reverse-recovery behavior which benefits lower reverse-recovery loss and smaller overvoltage [1]. In high power applications, SiC MOSFET modules are usually operated under high di/dt and dv/dt conditions, which makes their gate's voltage sensible to the parasitic components. Consequently, the optimal dead time selection in this scenario, is influenced and dependent on the operating conditions of the MOSFET.

This paper first presents the feedback mechanisms of dv/dt and di/dt to the gate loops then the gate's behaviors at different dead time under those circumstances are investigated. From that point of view, the dead time optimization process is proposed.

## 2 Parasitic components in a high-power module

Different from discrete chip, high-power SiC MOSFET module has its unique parasitic

components due to the parallel network inside the module.

In a half-bridge module, there are five terminals with varying copper surface areas: -DC, +DC, AC, high-side gate loop and low-side gate loop. These correspond to five parasitic capacitors connected to the base plate: $C_{sn}$, $C_{dn}$, $C_{an}$, $C_{ghn}$, and $C_{gln}$ as can be seen in Fig.1.

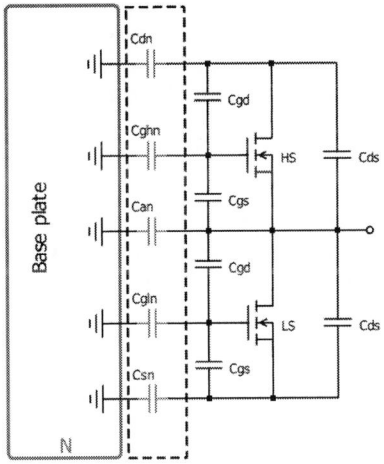

**Fig. 1**     Parasitic capacitors to base plate inside a high-power SiC MOSFET module.

To handle large currents effectively, the copper areas are typically made large, which in turn increases the value of the parasitic capacitors.

During rapid transient voltage events in SiC MOSFETs, the dv/dt is fed back to the high-side and low-side gate loops not only by the internal capacitors of the MOSFET but also by the series capacitors $C_{dn}$, $C_{ghn}$ and $C_{an}$, $C_{gln}$.

Moreover, a high-power module is typically equipped with a Kelvin source terminal connected to the gate network. Due to a large amount of parallel chips, there is a parasitic inductance $L_s$ between the Kelvin terminal and the real source terminal of the MOSFET. Apart from that, the proximity of the power terminals and the gate networks inside the module create a magnetic coupling $M_g$ between the power loops and gate loops during the current transients [10] (Fig. 2.).

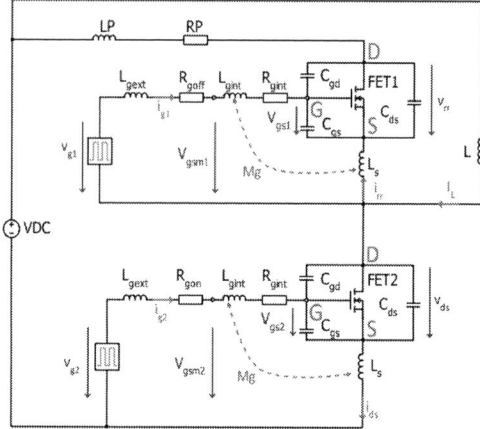

**Fig. 2**    Double pulse setup with details parasitic components.

It should be noticed that the measured gate voltages ($v_{gsm1}$, $v_{gsm2}$) are different from the real chip terminal voltages ($v_{gs1}$, $v_{gs2}$) during the current transient because of those $L_s$ and $M_g$.

$$v_{gs} = v_{gsm} - \left(L_s + M_g\right)\frac{di_{ds}}{dt} \qquad (1)$$

The actual value of the combination $L_s+M_g$ can be estimated by comparing the gate voltage waveform in three different cases which are displayed in Fig. 3.

## 3    The gate voltage under effect of di/dt and dv/dt.

The static gate voltage ($v_{gs}^*(t)$) is the gate voltage under zero voltage and zero current switching. In this switching condition, there are no feedback from dv/dt or di/dt to the gate loop. The gate voltage in this case is purely the RC charging and discharging behavior which is defined by the gate loop's resistance and the MOSFET's input

capacitance $C_{iss}$. The static turn-off gate voltage is depicted in Fig. 3 (black dashed curve).

To investigate how dv/dt is fed back to the gate loop during the transient, a double pulse test with high load inductance at nominal voltage $V_n$ is set up. At a very small load current, di/dt can be considered as zero, the gate voltage during dv/dt has a deviation from the static gate voltage (Fig.3 red curve). It is the result of the dv/dt feedback to the gate loop by $C_{gd}$ capacitor [4]. In case of high-power module, the effect also involves the base plate parasitic capacitors.

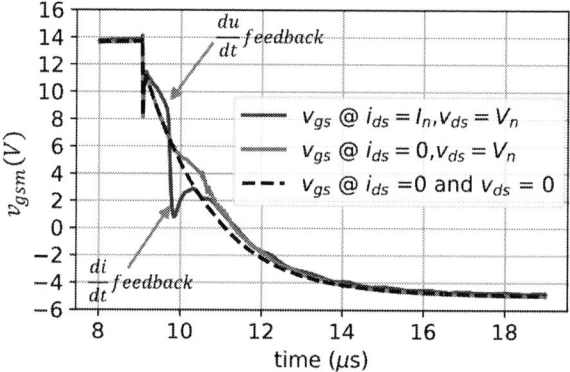

**Fig. 3**    Measured gate voltage under different di/dt and dv/dt.

When the MOSFET is switched off under nominal voltage and current conditions, the impact of both dv/dt and di/dt on its gate voltage can be observed at the blue curve in Fig 3.

## 4    The reverse-recovery of SiC MOSFET's body diode under parasitic turn-on effect.

Different from IGBT, MOSFET has its own intrinsic body diode. In a half-bridge configuration like Fig. 2, when FET2 is turned on, the load current through FET1's body diode's is cut off and the diode enter its reverse-recovery process.

The reverse-recovery process of the body diode is complicated because the reverse-recovery charge is a combination of the bipolar charges ($Q_{bi}$) and the output capacitor charges ($Q_{oss}$) [1]. The portions of those two types of charges are significantly changed with the junction temperature and load current. The portion of bipolar charge is dominated at high temperature and high load current due to longer life time of the charge carriers [5, 6]. The more bipolar charges, the higher reverse current peak and the higher

$di_{rr}/dt$. The consequence is the large overvoltage and severe oscillation.

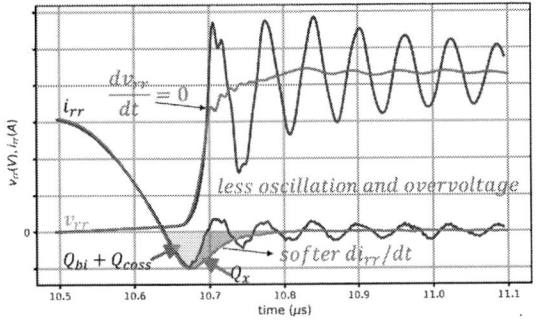

**Fig. 4** MOSFET's body diode voltage and current during reverse-recovery with parasitic turn-on (red lines) and without parasitic turn-on (blue lines).

Moreover, the process may involve parasitic turn-on effect especially during its dv/dt phase [4]. When the channel is on during the body diode's reverse-recovery, the short-circuit current starts flowing across the half-bridge. There are some major consequences:

- The total reverse-recovery current now has three components: the recovery current of the body diode $i_{bd}$, the output capacitor discharging current $i_{coss}$ and the short-circuit current in the channel $i_x$.

$$i_{rr} = i_{bd} + i_{coss} + i_x \qquad (2)$$

  Because of the additional current, the total reverse-recovery current slope $di_{rr}/dt$ is softer, which leads to lower oscillation and overvoltage [7].

- The total measured reverse-recovery charge $Q_{rr}$ is a combination of the body diode's bipolar charge $Q_{bi}$, the output capacitor's charge $Q_{coss}$ and the short-circuit current's charge $Q_x$.

$$Q_{rr} = Q_{bi} + Q_{coss} + Q_x \qquad (3)$$

- The short-circuit current injects more electrons into the space charge region and prevents its expansion. Consequently, the rising voltage $dv_{rr}/dt$ is clamped to zero.

The influence of the parasitic turn-on effect on the reverse-recovery process is depicted in Fig. 4.

## 5 The reverse-recovery of SiC MOSFET's body diode under short dead time.

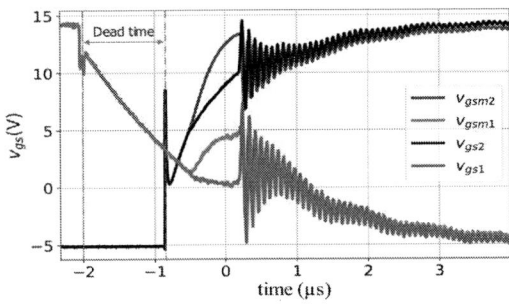

**Fig. 5** Measured gate voltages (red, blue curves) and the estimated chip's gate voltages (green, black curves) and dead time definition.

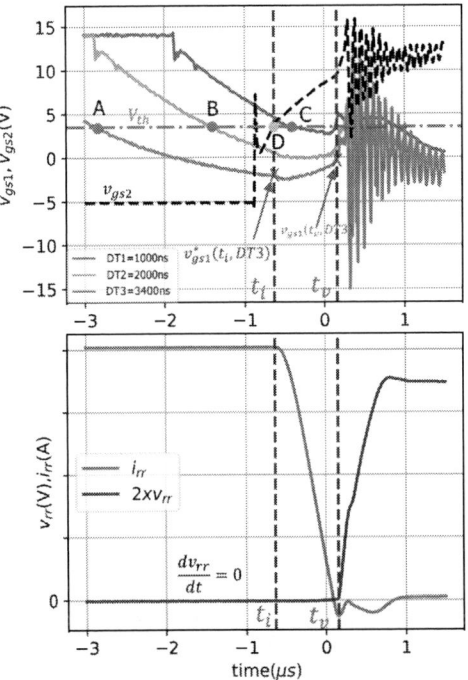

**Fig. 6** Estimated $v_{gs1}$ (solid lines) and $v_{gs2}$ (black dashed line) at different dead times DT. Points A, B, C are $v_{gs1}= V_{th}$, point D is at $v_{gs2}= V_{th}$. Distance AD, BD are body diode's on-times.

In the double pulse test shown in Fig. 2, when FET2 is turned on, the reverse-recovery occurs at FET1's body diode. The dead time in this context refers to the duration between FET1's turn-off and FET2's turn-on. To ensure consistency across various test setups with differing transport delay times between the controller and gate drivers, this paper defines dead time as the interval from the turn-off gate voltage spike to the turn-on gate voltage spike (Fig.5). These voltage spikes arise due to the voltage drop across the gate loop's inductances during turn-on and turn-off.

To make it easy for understanding, the gate terminal voltages are estimated from the measured gate voltages by equation (1). The difference between them are depicted in Fig.5.

When reducing dead time from 3400ns to 1000ns, the estimated gate voltages of the upper (index 1) and lower (index 2) MOSFETs can be observed in Fig.6.

The red dots: A, B, C mark the moment when the FET1's channel is off and its body diode starts conducting or $v_{gs1} = V_{th}$. The green dot D presents the moment when FET2's channel starts conducting and FET1's body diode is off or $v_{gs2} = V_{th}$. The times $t_i, t_v$ mark the beginning of $di_{rr1}/dt$ and $dv_{rr1}/dt$. Distances between the red dots and green dot: AD and BD are the body diode's on-times correspondingly to 3400ns and 2000ns dead time. When dead time is 1000ns, point C is after point D, FET1 and FET2 channels are overlapping, FET1's body diode doesn't conduct is this case.

Before $t_i$, $dv_{rr1}/dt = 0, di_{rr1}/dt = 0$, FET1's gate voltage is the static gate voltage $v_{gs1}^*(t)$. When the dead time (DT) is varied: DT3 > DT2 > DT1, correspondingly, FET1's static gate voltages at $t_i$ are:

$$v_{gs1}^*(t_i, DT3) < v_{gs1}^*(t_i, DT2) < v_{gs1}^*(t_i, DT1) \quad (4)$$

In the period between $t_i$ and $t_v$, $dv_{rr1}/dt = 0$, FET1's gate voltage is mainly influenced by $di_{rr1}/dt$ and it can be modeled:

$$v_{gs1}(t) \approx v_{gs1}^*(t, DT) + \left(M_g + L_{sin}\right)\frac{di_{rr}}{dt} \quad (5)$$

Since $R_{gon}$ is not changed, $\left(M_g + L_{sin}\right)\frac{di_{rr}}{dt}$ is constant at different dead time DT.

At time $t_v$, when $v_{rr1}$ starts rising, from equations (4) and (5), it can be concluded that:

$$v_{gs1}(t_v, DT3) < v_{gs1}(t_v, DT2) < v_{gs1}(t_v, DT1) \quad (6)$$

The higher $v_{gs1}(t_v, DT)$ is the more possibility of parasitic turn-on during $dv_{rr1}/dt$ due to the feedback from $C_{gd}$ to the upper gate loop [4].

After $t_v$, FET1's terminal gate voltage is under effect of both di/dt and dv/dt transient:

$$v_{gs1}(t) \approx v_{gs1}(t) + R_{goff} C_{gd} \frac{dv_{rr}}{dt} \quad (7)$$

$$+ \left(M_g + L_{sin}\right)\frac{di_{rr}}{dt}$$

When reducing dead time, there are three different regions (I, II, III) with distinct reverse-recovery behaviors can be observed in Fig.7:

- Region I: This is long dead time region, where $Q_{rr}$ are constant with varying dead time values. In this region, FET1's body diode has long on-time before FET2 is turned on. The body diode enters its reverse-recovery with full bipolar charge (Fig.6, DT3).

- Region II: In this region, FET1's body diode on-time is reduced before the bipolar charges reach equilibrium of carriers' generation and recombination [1, 11, 12, 13, 14], resulting less bipolar charge $Q_{bi}$ (Fig.6, DT2).

- Region III: In this region, when $v_{gs1}(t_v, DT)$ is high enough to parasitically turn on FET1 during its dv/dt, the short-circuit current in FET1's channel increase the total $Q_{rr}$ charge exponentially.

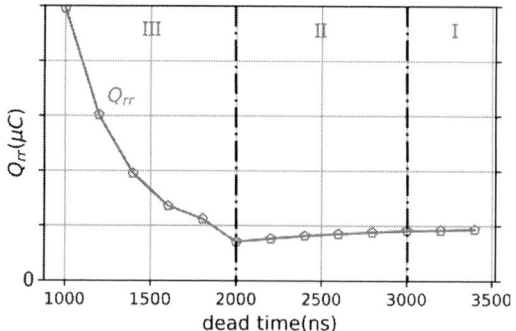

**Fig. 7** Reverse-recovery charge $Q_{rr}$ at different dead times. IL = 2x$I_n$, $T_j$ = 150°C.

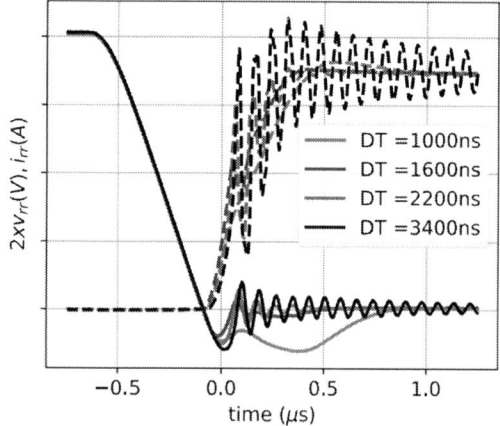

**Fig. 8** FET1's reverse-recovery voltages and currents at different dead times. IL = 2x$I_n$, $T_j$ = 150°C.

As shown in Fig. 8, with a dead time of 2200ns, the reverse-recovery process occurs in region II, resulting in a small reduction in $Q_{rr}$. Because there is less bipolar charge, the reverse current slope is slightly softer, resulting slightly reducing the switching oscillation. When the dead time is set at 1600ns, the reverse-recovery process is in region III. Although the total $Q_{rr}$ is just slightly higher compare to region I, there is a significant reduction in switching oscillation. When the dead time is too short at 1000ns, the short-circuit current is clearly observed at the end of the reverse-recovery process, which leads to a high reverse-recovery loss.

# 6 Effect of $L_s+M_g$ on dead time selection

The regions outlined in Fig.7 based on the measurements at load current double nominal current $I_n$. When the load current increase from $0.2 \times I_n$ to $2 \times I_n$ as depicted in Fig.9, The reverse-recovery tends to get more parasitic turn-on at low load current.

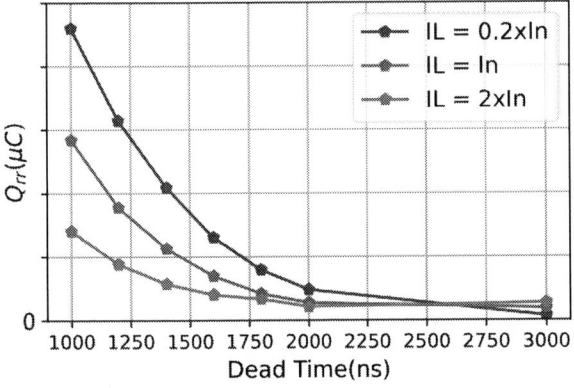

**Fig. 9** $Q_{rr}$ at different load currents. $T_j = 150°C$.

The experiments show that, for a selected dead time, the reverse-recovery behaviors may be different depending on the load current. For instance, if the dead time is selected at 2000ns, at $IL = 2 \times I_n$, the reverse-recovery behavior is in region II, but for $IL = I_n$ and $0.2 \times I_n$, the reverse-recovery is already in region III. The load current dependency can be explained by investigating FET1's gate voltage $v_{gs1}(t)$ during the reverse-recovery process.

Fig. 10 depicts the estimated gate terminal voltage of FET1 when FET2 is on. During the dv/dt phase, the gate-source voltage ($v_{gs1}$) is increased above the threshold voltage ($V_{th}$), when $v_{gs1}$ is high enough, the channel is fully on and its rising voltage is clamped to zero (dv/dt = 0). At the same

time, $i_{rr}$ is at its negative slope, resulting in a drop in the gate voltage (Fig. 10 black dashed line). This dip voltage is the voltage across $L_s+M_g$ during the negative current slope.

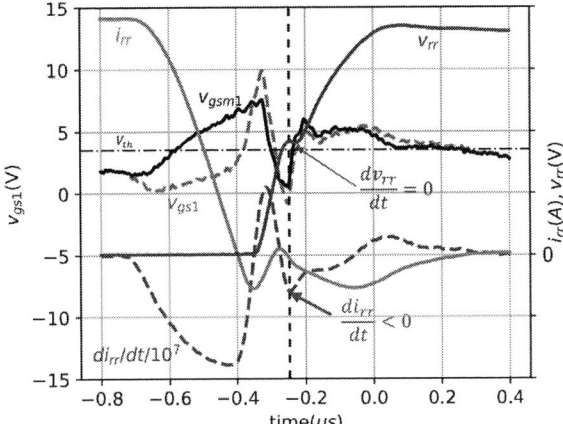

**Fig. 10** FET1's estimated gate voltage($v_{gs1}$) during the reverse-recovery and the voltage dip due to negative di/dt when dv/dt is clamped to zero.

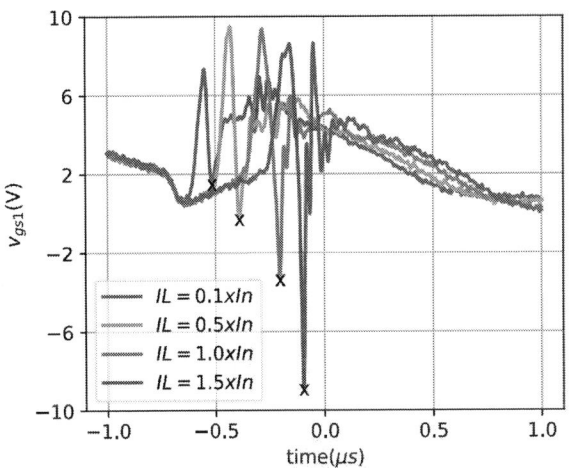

**Fig. 11** FET1's estimated gate voltage($v_{gs1}$) during the reverse-recovery and the voltage dip (marked with X) due to negative di/dt at different load current IL.

The higher IL is, the lower $v_{gs1}$ is dropped, which can be observed in Fig. 11. The gate charges accumulated during previous dv/dt phase can be faster discharged by the larger voltage dip, leading to lower short-circuit current.

After the voltage dip, FET1's channel is off, $v_{rr}$ rises again and lifts $v_{gs1}$ back above $V_{th}$. When $v_{rr}$ reaches DC link voltage, dv/dt becomes zero, $v_{gs1}$ reduces back below $V_{th}$ and turns off FET1's channel. When $i_{rr}$ turns back to zero, $v_{gs1}$ returns to its static RC discharge.

Due to $L_s + M_g$ feeding back the di/dt to the gate loop, a selected dead time can exhibit varied reverse-recovery behaviors at different load currents. Consequently, the dead time selection must consider the change of load current accordingly.

## 7 Effect of junction temperature on dead time selection

When selecting the reverse-recovery in region II or III, it becomes evident that the reverse-recovery behaviour may vary with different threshold voltages.

In region II, with a fixed dead time, when the threshold voltage decreases, the on time of the body diode also decreases, leading to lower $Q_{rr}$.

Conversely, in region III, with a fixed dead time, the occurrence of parasitic turn-on at a lower threshold voltage can lead to higher $Q_{rr}$ charges.

It is widely documented in the literature that the threshold voltage of SiC MOSFETs decreases at high junction temperatures [8, 9]. As a result, the reverse-recovery behaviour at a constant dead time will vary junction temperatures.

In Fig.12, when dead time equal to 1600ns, at 25°C the reverse-recovery is in region II while at 150°C, it is in region III.

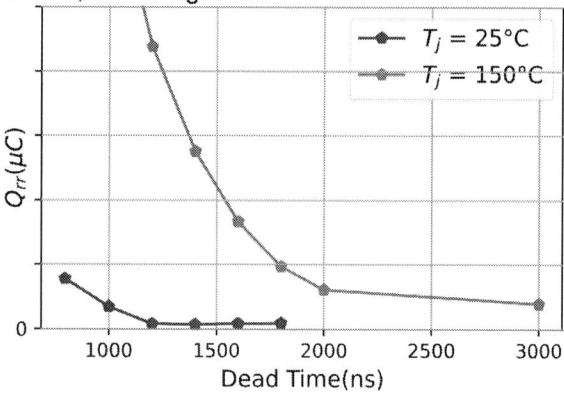

**Fig. 12** $Q_{rr}$ at different junction temperature $T_j$. IL = $I_n$.

## 8 Dead time optimization

### 8.1 Dead time optimization targets

As discussed in sections 7 and 8, it is evident that for a constant dead time, the expected reverse-recovery behaviours can change with different load currents and junction temperatures. Therefore, it is crucial to clearly define the optimization target before initiating the selection process.

There are two possible optimization targets achievable when reducing dead time:

1. Reduce reverse-recovery oscillation and overvoltage.

2. Reduce reverse-recovery oscillation, overvoltage, and total turn-on and reverse-recovery losses.

In the first target, $R_{gon}$ and $R_{goff}$ are adjusted under worst-case operating conditions and a long dead time. The dead time is then selected to minimize oscillations. With $R_{gon}$ fixed under worst-case conditions, the overvoltage remains within safe limits despite variations in reverse-recovery behaviours.

In the second optimization target, after tuning $R_{gon}$ and $R_{goff}$ under worst-case conditions, the dead time is decreased to minimize oscillations. With lower overvoltage, there's potential to increase turn-on speed, further reducing turn-on losses. However, a key challenge arises: overvoltage now relies on reverse-recovery behaviors. Any unforeseen variations in these behaviors can result in device failure or destruction.

### 8.2 Reducing oscillation and overvoltage

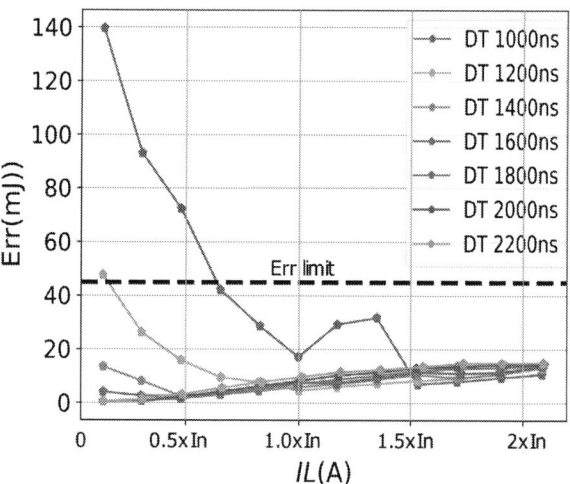

**Fig. 13** $E_{rr}$ at different dead times and load current. $T_j$ = 150°C.

For a fixed dead time, the critical scenario of the reverse-recovery behavior is the high reverse-recovery energy $E_{rr}$ due to parasitic turn-on. The situation happens at low current and highest junction temperature as illustrated in Fig.9 and Fig.12. Limiting this energy loss should be considered as a dead time selection criterion. There is flexibility in selecting $E_{rr}$ limit. For instance, Fig. 13 shows the $E_{rr}$ at different load current and dead time. If the maximum allowed $E_{rr}$ is selected at 45mJ, a dead time of 1200ns can be used. The maximum loss happens at the lowest load current.

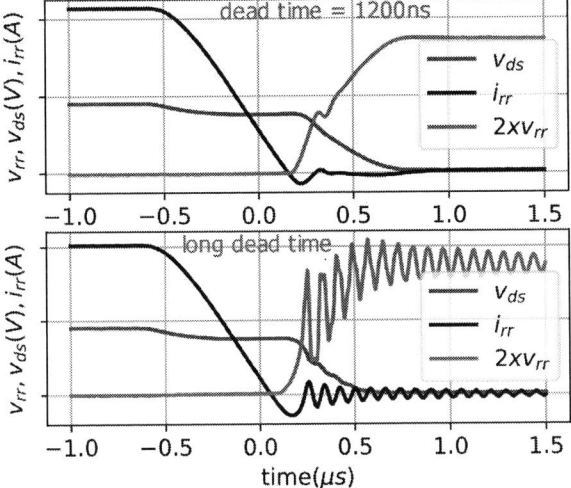

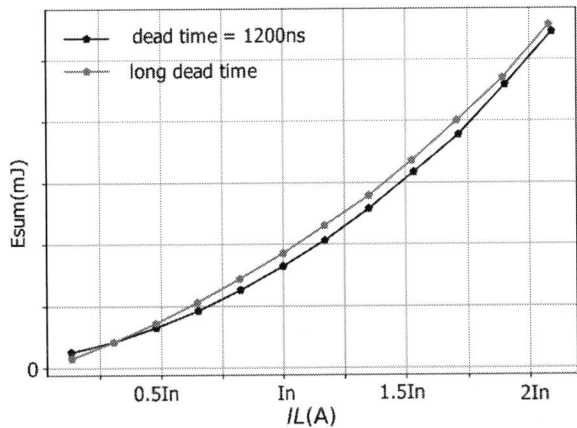

**Fig. 14** Compare reverse-recovery waveforms at long dead time and at 1200ns, $R_{gon}$ = 2.9Ω, IL = 2xI$_n$, T$_j$ = 150°C.

**Fig. 15** Compare total turn-on and reverse-recovery energy $E_{sum}$ = $E_{rr}$ + $E_{on}$ of the optimal dead time at 1200ns and long dead time. $R_{gon}$ = 2.9Ω, T$_j$ = 150°C.

When most of the time the converter is operated at IL = I$_n$, with the same $E_{rr}$ limit, a dead time of 1000ns can be selected, provided that a temporary $E_{rr}$ =140mJ at the lowest current is acceptable.

The main difference between the two approaches is the operating current range, where the reverse-recovery behaviour is located in region III. With a dead time of 1600ns, reverse-recovery is in region III when IL < I$_n$, whereas with a dead time of 1000ns, this occurs when IL < 1.5xI$_n$.

Fig. 14 shows the reverse-recovery voltage and current of FET1 during the turn-on of FET2 with the optimal dead time of 1200ns and with a long

dead time. The optimal dead time results in a softer reverse-recovery and clean switching waveform.

The total turn-on and reverse-recovery energy $E_{sum}$ is just slightly lower than the long dead time case because of less oscillation, which can be seen in Fig.15.

## 8.3 Reducing oscillation, overvoltage and turn-on losses.

When oscillation and overvoltage are reduced, it becomes possible to use smaller $R_{gon}$ values to decrease turn-on losses. The primary objective in this scenario is to find the smallest total turn-on and reverse-recovery energy ($E_{sum}$ = $E_{rr}$ + $E_{on}$). Since $E_{sum}$ depends on multiple variables such as $R_{gon}$, dead time, load current, and junction temperature, optimizing dead time selection becomes a more complex process.

Fig. 16 illustrates $E_{sum}$ at various load currents and dead times with a junction temperature of 150°C. At low load current ranges, a dead time of 2000ns clearly has the smallest $E_{sum}$. However, when IL > 1.5I$_n$, there is overlap between the curves. To identify the best dead time, the author suggests using the optimization function $F_{op}$:

$$F_{op} = \int_{ILmin}^{ILmax} E_{sum} \cdot dIL \qquad (8)$$

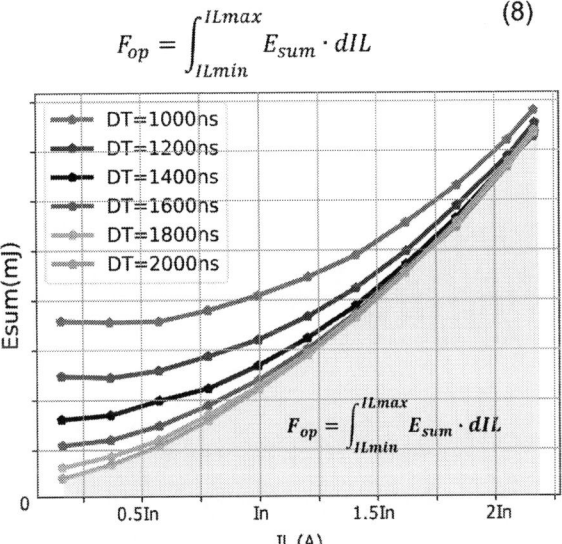

**Fig. 16** Total turn-on and reverse-recovery energy ($E_{sum}$ = $E_{rr}$+$E_{on}$) at different load current and dead time. $R_{gon}$ =2.9Ω, T$_j$ = 150°C.

The function actually represents the area below the dead time curve. The dead time that yields the smallest $F_{op}$ corresponds to the smallest average $E_{sum}$ over the load current range from IL$_{min}$ to IL$_{max}$. One can flexibly adjust these two current limits to align with the application's target requirements. For example, the target can be optimized either for the entire range of operating currents or focused specifically around the nominal current.

Another advantage of this optimization function is that it transforms a 2D variable plane ($E_{sum}$, IL) into a 1D variable $F_{op}$, thereby simplifying the multi-dimensional optimization space. The 3D optimization space in Fig. 16 is converted to 2D space in Fig.17.

It should be noticed that all the measurement points in Fig.16 are at the worst-case scenario of the operating conditions and all the reverse-recovery overvoltage is under the safe limit.

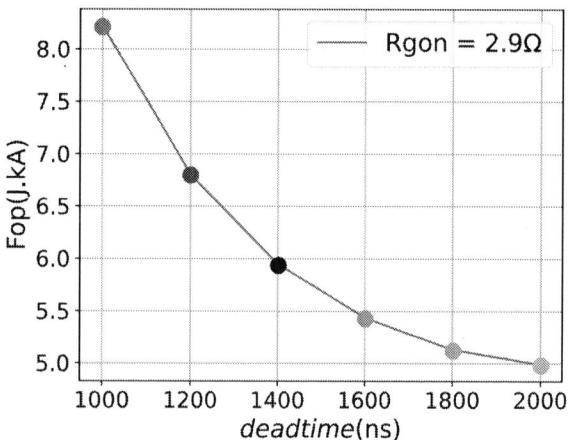

**Fig. 17** $F_{op}$ at different deadtimes, $R_{gon}$ = 2.9Ω, $T_j$ = 150°C.

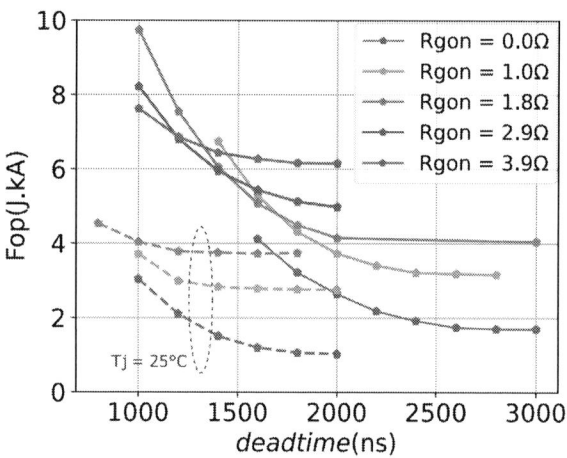

**Fig. 18** $F_{op}$ at different dead times, $R_{gon}$, $T_j$ = 150°C (solid curves), $T_j$ = 25°C (dashed curves).

The optimal function values for different combinations of $R_{gon}$ and dead time at 150°C (solid curves) and 25°C (dashed curves) junction temperatures are shown in Fig.18. The lowest optimal function values are achieved with a dead time of 2000ns and $R_{gon}$ of zero ohms at both 150°C and 25°C junction temperatures.

The voltage and current waveforms across FET1 can be observed in Fig.19 when $R_{gon}$ is set to zero ohms and the dead time is set to 2000ns. Compared to the initial case where $R_{gon}$ is fixed at 2.9Ω and dead time is set at 1200ns, there is a notable reduction in total turn-on and reverse-recovery losses. These results are depicted in Fig.20. At nominal load current, a reduction of 63% of total turn-on and reverse-recovery can be achieved.

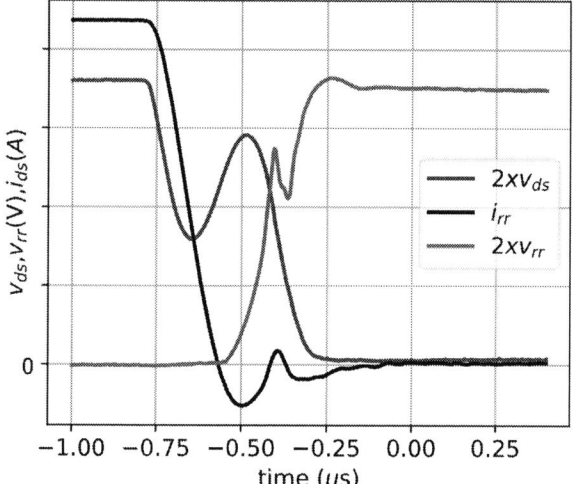

**Fig. 19** Reverse-recovery switching waveform when $R_{gon}$ = 0Ω, dead time = 2000ns, $T_j$ = 150°C.

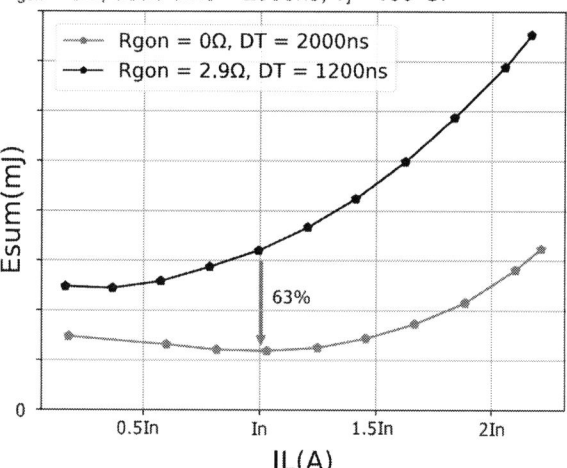

**Fig. 20** Compare total turn-on and reverse-recovery energy $E_{sum}$ between fixed $R_{gon}$ and reduced $R_{gon}$.

# 9 Conclusion

The paper explores the impact of parasitic components in a high-power SiC MOSFET module on the gate voltages during its voltage and current transients. Specifically, during the reverse-recovery of the body diode, the feedback of di/dt and dv/dt to the gate loop alters reverse-recovery behaviors at varying load currents and junction temperatures. For a chosen dead time, these behaviors can be unpredictable. Two primary optimization targets can be set in this scenario.

Firstly, fixed $R_{gon}$ aims for clean reverse-recovery and safe overvoltage across different reverse-recovery behaviors. The key criterion for dead time selection is the acceptable limit of reverse-recovery energy.

Secondly, reducing $R_{gon}$ targets minimizing total turn-on and reverse-recovery energy, a function of multiple variables. The optimization process can be simplified using the suggested optimization functions, reducing the dimensionality of the optimization space.

Since the overvoltage now depends on the reverse-recovery behaviors, any variation outside the optimization zone may lead to overvoltage exceeding the safe limit. However, a huge reduction in total turn-on and reverse-recovery can be achieved if the variation of the reverse-recovery behaviors are predictable.

## Acknowledgements

This work is supported by the Federal Ministry for Economic Affairs and Climate Action.

## Reference

[1] A. März, S. Schönewolf, A. Nagel, M. Rauh and M. -M. Bakran, "Deadtime optimization eliminating snap-off of 3.3kV SiC MOSFET bodydiodes," 2023 25th European Conference on Power Electronics and Applications (EPE'23 ECCE Europe), Aalborg, Denmark, 2023, pp. 1-7, doi: 10.23919/EPE23ECCEEurope58414.2023.10264563.

[2] P. Sochor, A. Huerner, Q. Sun and R. Elpelt, "Understanding the Switching Behavior of Fast SiC MOSFETs," PCIM Europe 2022; International Exhibition and Conference for Power Electronics, Intelligent Motion, Renewable Energy and Energy Management, Nuremberg, Germany, 2022, pp. 1-8, doi: 10.30420/565822153.

[3] S. Jahdi et al., "An Analysis of the Switching Performance and Robustness of Power MOSFETs Body Diodes: A Technology Evaluation," in IEEE Transactions on Power Electronics, vol. 30, no. 5, pp. 2383-2394, May 2015, doi: 10.1109/TPEL.2014.2338792.

[4] A. März, T. Bertelshofer, M. Helsper and M. -M. Bakran, "Comparison of SiC MOSFET gate-drive concepts to suppress parasitic turn-on in low inductance power modules," 2017 19th European Conference on Power Electronics and Applications (EPE'17 ECCE Europe), Warsaw, Poland, 2017, pp. P.1-P.10, doi: 10.23919/EPE17ECCEEurope.2017.8099044

[5] Z. Wang, J. Ouyang, J. Zhang, X. Wu and K. Sheng, "Analysis on reverse recovery characteristic of SiC MOSFET intrinsic diode," 2014 IEEE Energy Conversion Congress and Exposition (ECCE), Pittsburgh, PA, USA, 2014, pp. 2832-2837, doi: 10.1109/ECCE.2014.6953782.

[6] K. Peng, S. Eskandari and E. Santi, "Characterization and modeling of SiC MOSFET body diode," 2016 IEEE Applied Power Electronics Conference and Exposition (APEC), Long Beach, CA, USA, 2016, pp. 2127-2135, doi: 10.1109/APEC.2016.7468161.

[7] P. Hofstetter, R. W. Maier and M. Bakran, "Parasitic Turn-On of SiC MOSFETs – Turning a Bug into a Feature," PCIM Europe digital days 2020; International Exhibition and Conference for Power Electronics, Intelligent Motion, Renewable Energy and Energy Management, Germany, 2020, pp. 1-7.

[8] N. Shiono and C. Hashimoto, "Threshold-voltage instability of n-channel MOSFET's under bias-temperature aging," in IEEE Transactions on Electron Devices, vol. 29, no. 3, pp. 361-368, March 1982, doi: 10.1109/T-ED.1982.20710.

[9] Lei Ren, Qian Shen and Chunying Gong, "Real-time aging monitoring for power MOSFETs using threshold voltage," IECON 2016 - 42nd Annual Conference of the IEEE Industrial Electronics Society, Florence, Italy, 2016, pp. 441-446, doi: 10.1109/IECON.2016.7793625

[10] Jorge Mari, Fabio Carastro, Max-Josef Kell, "Assessing the Presence of Parasitic Turn On in SiC Mosfet Power Modules", EPE'21 ECCE Europe, September 2021.

[11] Lutz, J.: Halbleiter-Leistungsbauelemente, Physik, Eigenschaften, Zuverlässigkeit, Springer Verlag, ISBN 3-540-34206-0, 2006.

[12] Kolessar, R.: Physical study of the power diode turn-on process, Proc. of IEEE Industry Applications Conf. 2000, Vol 5, pp.2934-2940, 2000.

[13] Yamazaki, M., Kobayashi, H., Shinohara, S.: Forward transient behaviour of PiN and super

junction
diodes, Proc. of IEEE ISPSD 2004, Vol 5, pp.197-200, 2004.

[14] Baburske, R., Domes, D., Lutz, J., Hofmann, W.: Passive turn-on process of IGBTs in matrix converter
applications, Proc. of EPE Conf. 2009.

[15] R. Elferich and T. Lopez, "Impact of gate voltage bias on reverse recovery losses of power MOSFETs," Twenty-First Annual IEEE Applied Power Electronics Conference and Exposition, 2006. APEC '06., Dallas, TX, USA, 2006, pp. 6 pp.-, doi: 10.1109/APEC.2006.162072.

PCIM Europe 2024, 11– 13 June 2024, Nuremberg    DOI: 10.30420/566262087

# Performance Instability of 650 V *p*-GaN Gate HEMT Device under Temperature-related Positive Gate Bias Stresses

Renze Yu [1], Saeed Jahdi [1], Phil Mellor [1], Jose Ortiz Gonzalez [2], Olayiwola Alatise [2]

[1] School of Electrical & Mechanical Engineering, University of Bristol, Bristol, UK
[2] School of Engineering, University of Warwick, Coventry, UK

Corresponding author:    Renze Yu, renze.yu@bristol.ac.uk
Speaker:                 Renze Yu, renze.yu@bristol.ac.uk

## Abstract

In this work, the effects of positive gate stresses on threshold voltage ($V_{th}$) and on-state resistance ($R_{on}$) instability of the 650 V Schottky *p*-GaN gate HEMT devices were investigated under different gate bias stresses and a wide temperature range. It is noticed that once the stress was applied, there was an immediate $V_{th}$ jump. The drift magnitude and direction of $V_{th}$ and $R_{on}$ during the stressing tests were largely dependent on the applied $V_{gs}$ bias. The temperature had a significant impact on static parameters. The variations in $V_{th}$ ($\Delta V_{th}$) demonstrated an overall decreasing trend with the progress of experiment under the positive gate stress of 6 V. The initial $\Delta V_{th}$ jump increased with temperature at first and then reduced at certain temperatures. The variation in $R_{on}$ ($\Delta R_{on}$) caused by the positive gate bias of 6 V was minor at temperatures below zero, but it became more pronounced with the increase in temperature.

## 1   Introduction

Wide-bandgap power semiconductor devices, including Gallium Nitride (GaN) and Silicon Carbide (SiC) [1]–[3] are fast becoming a key players in power electronics applications [4], [5]. GaN High Electron Mobility Transistors (HEMTs) have received widespread attention in power electronics systems because the devices could offer lower power losses, higher switching frequency, and increased power density compared with Si-based semiconductor devices [6], [7].   To ensure operational safety and achieve easy gate driving methodology, normally-off GaN HEMT devices are preferred.    Among many approaches to realize normally-off property, implementing a heavily doped *p*-type gate stack between the gate electrode and AlGaN/GaN layer has gained popularity owing to easier manufacturability and adequate electrical performances. Depending on the device technology, the contact between the introduced layer and adjacent layers is either Ohmic type or Schottky type.   The Ohmic *p*-GaN gate HEMT can offer stable device performance, but its current-driven property brings about drawbacks such as large gate leakage current and high power consumption of gate drivers [8]. On the contrary, the Schottky *p*-GaN gate HEMT has advantages including reduced leakage current, enlarged driving margin, and voltage-driven solution [9].

However, the *p*-GaN layer between the gate metal and the barrier layer is electrically floating, which can easily induce the performance instability of the Schottky *p*-GaN gate HEMT device. Besides, two back-to-back connected diodes are formed and connected by the *p*-GaN stack.   The gate metal and the Schottky *p*-GaN stack form a Schottky diode ($J_1$), while the *p*-GaN/AlGaN/GaN is a P-i-N heterojunction ($J_2$).    It is reported that trapping/de-trapping of electrons and holes in the gate structure is responsible for the shift of threshold voltage ($V_{th}$) and on-state resistance ($R_{on}$) [10].    The Schottky *p*-GaN gate HEMT device is sensitive to gate-source voltage ($V_{gs}$), drain-source voltage ($V_{ds}$), bias direction and magnitude, static/dynamic operating conditions, etc [11]–[14].   Here, the impact of positive gate bias stress on the static performance instability of Schottky *p*-GaN gate HEMT device is studied. The tested devices from different manufacturers were stressed and recovered at different $V_{gs}$ levels at various temperatures. The experiment procedures were demonstrated in the next section. The drift patterns were illustrated and the mechanism is explained in Section 3, followed by the conclusion.

## 2 Experiment Methodology

The devices under test (DUTs) were 650 V/15 A and 650 V/18 A Schottky $p$-GaN gate HEMTs with part number GPI65015DFN and GS-065-018-2-L, respectively. The test conformed with the typical measure-stress/recovery-measure sequence [15], as shown in Figure 1. During the positive gate bias stress tests at different $V_{gs}$, the devices suffered from $V_{gs}$ of 1 V to 6 V for 1000 sec. Then, the stress was removed and the devices recovered without any bias applied for another 1000 sec. To evaluate the performance of devices under different temperatures, the devices were placed in the thermal chamber (TAS LTCL600) and tested from -50°C to 150°C, with a 25°C step between tests. The stressing $V_{gs}$ was 6 V and lasted for 1000 sec. The recovery stage was unbiased and lasted for 1000 sec as well. To characterize the instability of the tested devices, the transfer curves were collected by the source measuring unit (B2902A) after stressing/recovering for 1 ms, 10 ms, 100 ms, 1 sec, 10 sec, 50 sec, 100 sec, 500 sec, and 1000 sec. The measuring $V_{gs}$ was swept from 0 V to 6 V, and the measuring $V_{ds}$ was fixed at a low value of 0.1 V to avoid potential injection of electrons/holes during the characterization transient as much as possible. It is noted that the resistance of the wires can be excluded because of the usage of four point probe method in the characterization.

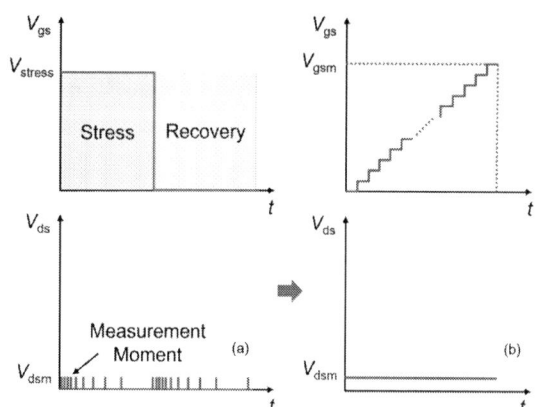

**Fig. 1:** Schematic of the test method (a) Stress and recovery phase, (b) I-V characterization.

## 3 Test results and analysis

### 3.1 Positive Gate Bias Stress Test under Different $V_{gs}$

The variations in $V_{th}$ ($\Delta V_{th}$) of two devices under positive gate bias stress tests at different $V_{gs}$ are shown in Figure 2 and Figure 3, respectively. $\Delta V_{th}$ is defined as the change of $V_{th}$ in the test compared with the initial $V_{th}$ at each group of test. It can be seen that two devices behaved differently at different $V_{gs}$ bias.

For the GPI65015DFN device, $\Delta V_{th}$ kept decreasing throughout the test at 1 V. When $V_{gs}$ was increased to 2-4 V, $\Delta V_{th}$ experienced a fast drop and then started to increase. At $V_{gs}$ of 5 V, $\Delta V_{th}$ was relatively stable and demonstrated an overall increasing trend. Finally, after the initial jump in $\Delta V_{th}$ once the gate stress at 6 V was applied, $\Delta V_{th}$ stayed steady before 1 sec, and then began to decrease.

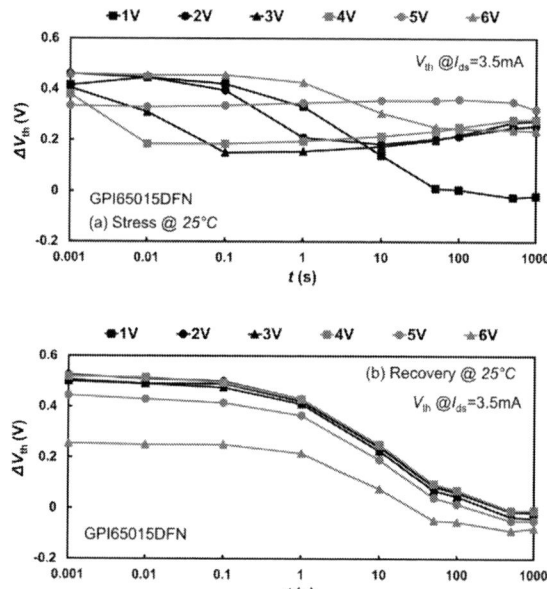

**Fig. 2:** Shift of $\Delta V_{th}$ of GPI65015DFN device in positive gate bias test at different $V_{gs}$: (a) Stress phase, (b) Recovery phase.

Regarding the GS-065-018-2-L device, $\Delta V_{th}$ suffered from a continuous decrease at $V_{gs}$ of 1 V as well. $\Delta V_{th}$ decreased at first and then increased at $V_{gs}$ of 2-5 V. Depending on the gate stress level, the amount of increase and decrease in $\Delta V_{th}$ was different. Lastly, $\Delta V_{th}$ increased marginally and then decreased at 6 V. In the recovery stage, two devices showed similar trend. The devices recovered at a nearly comparable rate under most stressing conditions.

As soon as the positive gate bias was applied on the GaN HEMT device, the metal/$p$-GaN stack Schottky diode was reverse-biased, and the $p$-GaN stack/AlGaN/GaN P-i-N diode was forward-biased. When the stressing $V_{gs}$ was lower than $V_{th}$, such as at $V_{gs}$=1 V, the P-i-N diode cannot be fully turned on. The transport of electrons and holes in $J_2$ was weak. At the same time, the tunneling of electrons and holes in the reverse-biased $J_1$ was also suppressed [16]. Hence, the change in $\Delta V_{th}$ was not caused by the gate bias and resulted from the measuring $V_{gs}$ of 6 V. Considering that the mobility of electrons was higher than the mobility of holes [14], the electrons across the AlGaN layer could be injected into the $p$-GaN during the characterization immediately, causing the deficiency in holes and initial $\Delta V_{th}$ jump. Once the measuring $V_{gs}$ was removed, the device began to recover because the $V_{gs}$ was low enough not to induce significant carrier transport.

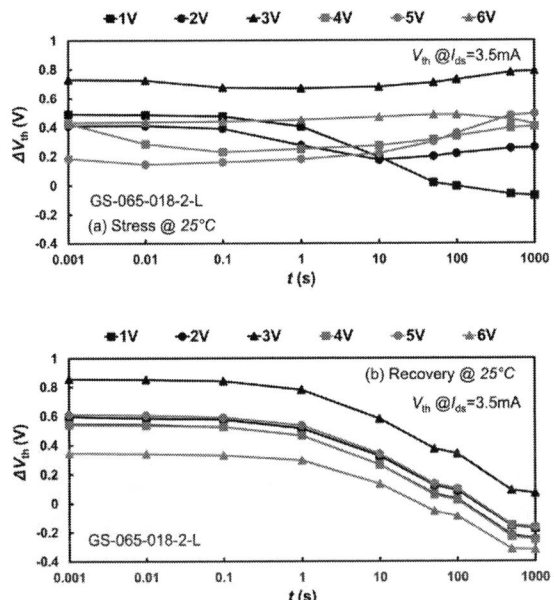

**Fig. 3:** Shift of $\Delta V_{th}$ of GS-065-018-2-L device in positive gate bias test at different $V_{gs}$: (a) Stress phase, (b) Recovery phase.

With the increase in $V_{gs}$, the space charge region in $J_2$ began to shrink and eventually disappeared. The P-i-N diode gradually turned on. The positive $V_{gs}$ attracted electrons, which then recombined with the acceptors in the $p$-GaN stack. Simultaneously, holes were emitted out from the $p$-GaN layer to the AlGaN interface, further inducing the deficiency in holes. Hence, $\Delta V_{th}$ decreased after the initial jump of $\Delta V_{th}$. The reduction in $\Delta V_{th}$ happened at the first few seconds, and

then $\Delta V_{th}$ increased instead. The initial decrease in $\Delta V_{th}$ became faster with the increase in $V_{gs}$ from 2-4 V for the GPI65015DFN device and from 2-5 V for the GS-065-018-2-L device due to hole deficit. When further increasing $V_{gs}$, $J_2$ became saturated. The additional voltage drop was suffered by $J_1$ and the energy band of the $p$-GaN layer became steeper [17], which facilitated the hole tunneling from the metal to the $p$-GaN layer. At the same time, more acceptors could be ionized in the reverse-biased $J_1$. These holes could accumulate at $p$-GaN/AlGaN interface, resulting in the negative shift of $\Delta V_{th}$ again of the GPI65015DFN device at 6 V. When the amount of the tunneling holes became similar to the extent of attracted electrons, the $\Delta V_{th}$ would present a relatively moderate shifting trend, such as the GPI65015DFN device at 5 V and the GS-065-018-2-L device at 6 V.

The changes in $R_{on}$ ($\Delta R_{on}$) under different positive gate bias stress tests are demonstrated in Figure 4 and Figure 5. Similarly, $\Delta R_{on}$ is defined as the change of $R_{on}$ compared with the initial $R_{on}$ measured at the start of each test.

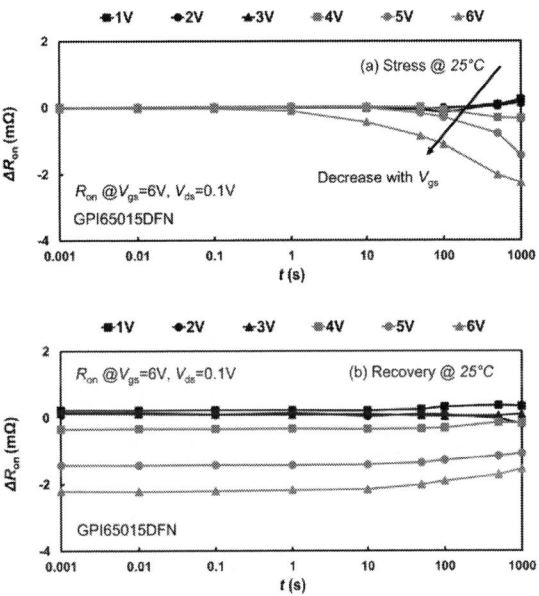

**Fig. 4:** Shift of $\Delta R_{on}$ of GPI65015DFN device in positive gate bias test at different $V_{gs}$: (a) Stress phase, (b) Recovery phase.

For the GPI65015DFN device, $\Delta R_{on}$ did not show obvious shift at 1-3 V in the stressing test, but with further increase in $V_{gs}$, $\Delta R_{on}$ exhibited a minor decrease. As for the GS-065-018-2-L device, the obvious reduction in $\Delta R_{on}$ happened at $V_{gs}$ of 6 V. For both devices, the recovery of $\Delta R_{on}$ was slow.

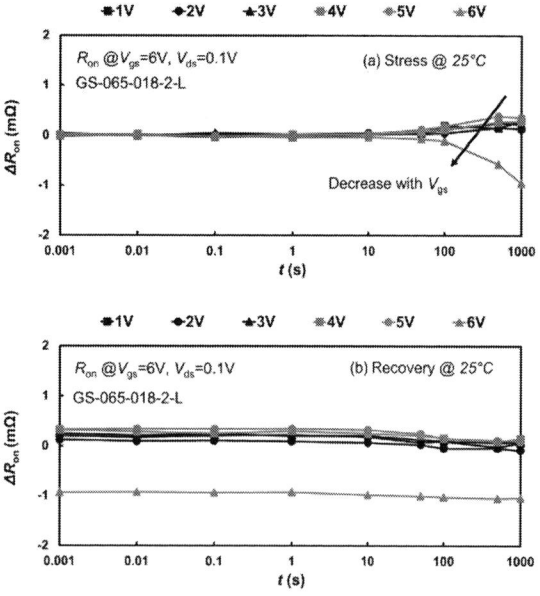

**Fig. 5:** Shift of $\Delta R_{on}$ of GS-065-018-2-L device in positive gate bias test at different $V_{gs}$: (a) Stress phase, (b) Recovery phase.

Under positive bias stress tests, the variation in $\Delta R_{on}$ was different from the $V_{th}$ trend, especially at high $V_{gs}$ stress, suggesting a different mechanism. According to [17], the highest electrical field near the gate structure was distributed at the $p$-GaN gate corner, which was likely to induce the injection of holes into passivation/$p$-GaN interface [18], [19]. The injected holes had certain possibility to access to the passivation layer between gate and drain electrodes, leading to the reduced drain-source resistance and the negative shift of $\Delta R_{on}$. With the increase in $V_{gs}$, the trapping of the holes was more likely to happen. Since the trapping/de-trapping time of holes was long, the recovery of the $\Delta R_{on}$ was slow as well in the recovery phase.

## 3.2 Positive Gate Bias Stress Test under Different Temperatures

The $\Delta V_{th}$ of two GaN HEMT devices under 6 V positive gate bias stress test and 0 V recovery test at different temperatures are shown in Figure 6 and Figure 7. It can be seen for both devices, the initial $\Delta V_{th}$ jump increased with the increase in temperature below 50-75°C, beyond which the $\Delta V_{th}$ jump decreased instead with the increase in temperature for the GPI65015DFN device, while

the $\Delta V_{th}$ jump of the GS-065-018-2-L device hardly changed. Moreover, the $\Delta V_{th}$ of the GPI65015DFN device decreased with the progress of the stressing test, while the $\Delta V_{th}$ of the GS-065-018-2-L device slightly increased at the first 10 sec and then decreased. Compared with the GS-065-018-2-L device, the $\Delta V_{th}$ drift of the GPI65015DFN device was faster. In the recovery phase, the $\Delta V_{th}$ of both devices gradually restored to the initial state. The recovery was slow at low temperatures below 0°C and was accelerated with the increase in temperature. It is worth mentioning that for the GPI65015DFN device, the $V_{th}$ was over-recovered at low temperatures, but after fully recovering at enhanced temperature, $\Delta V_{th}$ slightly increased on the contrary.

For the positive gate bias tests at 6 V, the testing temperature played an important role on the device performance. Although the magnitude of the $\Delta V_{th}$ was different, the trend of each device under different temperatures was close, indicating that the shift mechanism was similar. When the testing temperature was below 0°C, the trapping of electrons and holes was less pronounced. With the increase in temperature, the transport of electrons and holes in $J_2$ was accelerated [20], which led to the initial increase in $\Delta V_{th}$ jump.

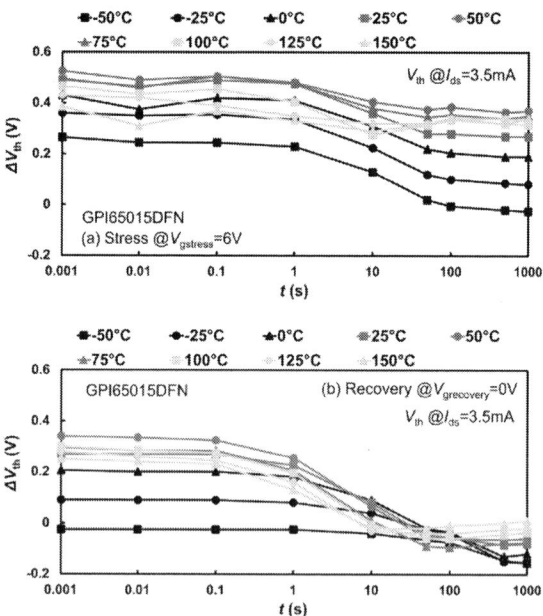

**Fig. 6:** Shift of $\Delta V_{th}$ of GPI65015DFN device under positive gate bias test at different temperatures: (a) Stress phase, (b) Recovery phase.

PCIM Europe 2024, 11– 13 June 2024, Nuremberg    DOI: 10.30420/566262087

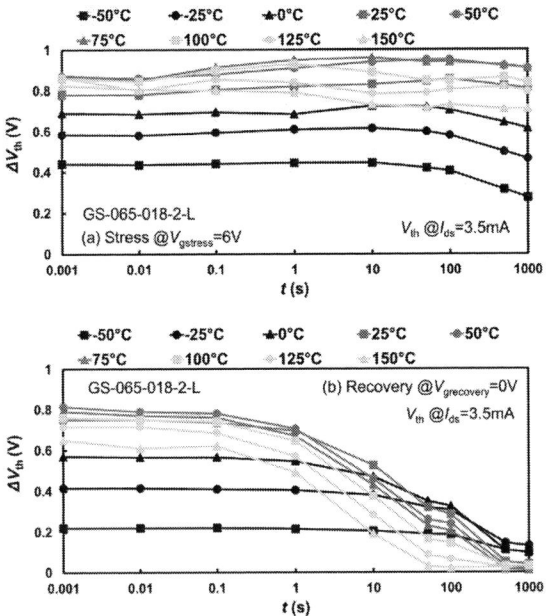

**Fig. 7:** Shift of $\Delta V_{th}$ of GS-065-018-2-L device under positive gate bias test at different temperatures: (a) Stress phase, (b) Recovery phase.

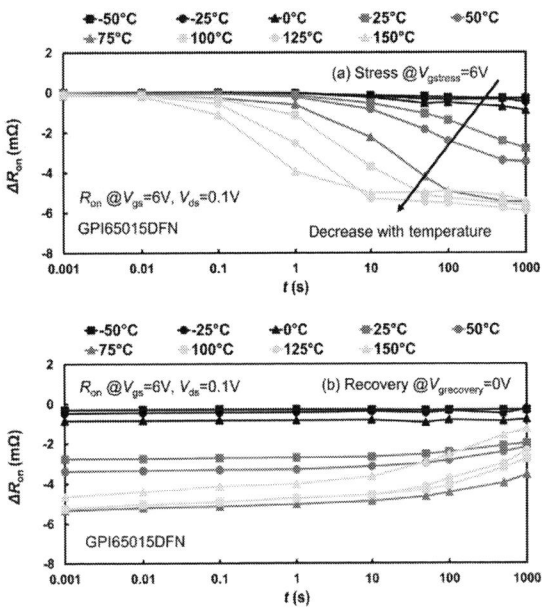

**Fig. 8:** Shift of $\Delta R_{on}$ of GPI65015DFN device under positive gate bias test at different temperatures: (a) Stress phase, (b) Recovery phase.

On the other hand, the thermionic emission across $J_1$ was enhanced as well, which favored the hole injection to the $p$-GaN layer and caused the elevated $\Delta V_{th}$ shift. However, it can be noticed that the decreasing trend of $\Delta V_{th}$ throughout the test became less prominent, especially at high temperatures, which indicated that during the gate bias stress tests, the injection of electrons from AlGaN interface to $p$-GaN stack was reinforced to a greater extent, but the tunneling of holes in $J_1$ was still the dominate mechanism leading to the drift of the threshold voltage as $\Delta V_{th}$.

The $\Delta R_{on}$ of two devices under positive gate bias stress and recovery tests at different temperatures are shown in Figure 8 and Figure 9, respectively. Test results indicate that the positive gate bias stress has caused negative shifts in $\Delta R_{on}$ for both devices. During the stressing phase, the variations were minor when temperature was below 0°C. With the increase in temperature, $\Delta R_{on}$ gradually decreased. This is because the spill over of the holes was intensified at higher temperatures. The holes gained more energy and had higher possibility to be injected into the gate corners under high gate bias stress. When the bias was removed, the devices started to recover. The recovery speed was enhanced at elevated temperatures.

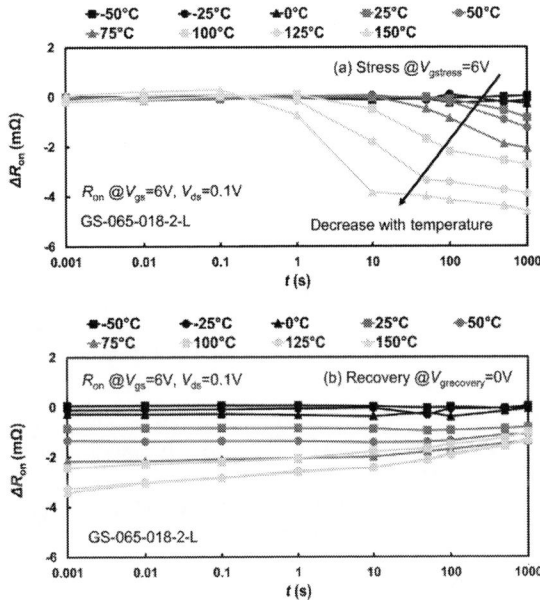

**Fig. 9:** Shift of $\Delta R_{on}$ of GS-065-018-2-L device under positive gate bias test at different temperatures: (a) Stress phase, (b) Recovery phase.

## 4 Conclusion

In this paper, the $V_{th}$ and $R_{on}$ instabilities of 650 V Schottky $p$-GaN gate HEMT devices were characterized under positive gate stresses.

721

According to test results, $\Delta V_{th}$ and $\Delta R_{on}$ behaved significantly different under low and high $V_{gs}$. The drift mechanism of $\Delta V_{th}$ was explained by the carrier transport in the back-to-back diodes, which was dependent on the applied stress value. The shift in $\Delta R_{on}$ was assumed to be caused by the injection of holes in the gate corner/passivation layer. Under different temperatures, the variation in $\Delta V_{th}$ was similar, suggesting similar drift mechanisms. However, the initial $\Delta V_{th}$ jump increased with temperature below 50-75°C, but it decreased or kept unchanged as temperature further increased. With regard to $\Delta R_{on}$, it decreased with time and increasing temperatures, indicating a higher possibility of carrier injection.

# References

[1] S. Jahdi and et al., "Temperature and switching rate dependence of crosstalk in si-igbt and sic power modules," *IEEE Transactions on Industrial Electronics*, vol. 63, no. 2, pp. 849–863, 2016.

[2] R. Wu and et al., "Performance of parallel connected sic mosfets under short circuits conditions," *Energies*, vol. 14, no. 20, 6834, 2021.

[3] R. Wu and et al., "Measurement and simulation of short circuit current sharing under parallel connection: Sic mosfets and sic cascode jfets," *Microelectronics Reliability*, vol. 126, p. 114 271, 2021.

[4] S. Jahdi and et al., "Dg islanding operation detection methods in combination of harmonics protection schemes," in *2nd IEEE PES International Conference and Exhibition on Innovative Smart Grid Technologies*, 2011.

[5] S. Jahdi and et al., "The impact of silicon carbide technology on grid-connected distributed energy resources," in *IEEE PES ISGT Europe*, 2013.

[6] S. Yang and et al., "Dynamic on-resistance in gan power devices: Mechanisms, characterizations, and modeling," *IEEE Journal of Emerging and Selected Topics in Power Electronics*, vol. 7, no. 3, pp. 1425–1439, 2019.

[7] L. Gill and et al., "A review of gan hemt dynamic on-resistance and dynamic stress effects on field distribution," *IEEE Transactions on Power Electronics*, 2023.

[8] Z. Jiang and et al., "Negative gate bias induced dynamic on-resistance degradation in schottky-type *p*-gan gate hemts," *IEEE Transactions on Power Electronics*, vol. 37, no. 5, pp. 6018–6025, 2021.

[9] Y. Cheng and et al., "Gate reliability of schottky-type *p*-gan gate hemts under ac positive gate bias stress with a switching drain bias," *IEEE Electron Device Letters*, vol. 43, no. 9, pp. 1404–1407, 2022.

[10] C. Feng and et al., "Gate-bias-accelerated $V_{TH}$ recovery on schottky-type *p*-gan gate algan/gan hemts," *IEEE Trans. on Electron Devices*, 2023.

[11] F. Yang and et al., "Characterization of threshold voltage instability under off-state drain stress and its impact on *p*-gan hemt performance," *IEEE Journal of Emerging and Selected Topics in Power Electronics*, vol. 9, no. 4, pp. 4026–4035, 2020.

[12] S. Li and et al., "Time-dependent threshold voltage instability mechanisms of *p*-gan gate algan/gan hemts under high reverse bias conditions," *IEEE Transactions on Electron Devices*, vol. 68, no. 1, pp. 443–446, 2020.

[13] J. He and et al., "$V_{TH}$ instability of *p*-gan gate hemts under static and dynamic gate stress," *IEEE Electron Device Letters*, vol. 39, no. 10, pp. 1576–1579, 2018.

[14] R. Wang and et al., "$V_T$ shift and recovery mechanisms of *p*-gan gate hemts under dc/ac gate stress investigated by fast sweeping characterization," *IEEE Electron Device Letters*, vol. 42, no. 10, pp. 1508–1511, 2021.

[15] A. Tallarico and et al., "Pbti in gan-hemt's with p-type gate: Role of the aluminum content on $\Delta V_{TH}$ and underlying degradation mechanisms," 2018.

[16] Y. Shi and et al., "Carrier transport mechanisms underlying the bidirectional $V_{TH}$ shift in p-gan gate hemts under forward gate stress," *IEEE Transactions on Electron Devices*, vol. 66, no. 2, pp. 876–882, 2018.

[17] X. Chao and et al., "Analysis of $V_{TH}$ degradation and recovery behaviors of p-gan gate hemts under forward gate bias," *IEEE Transactions on Electron Devices*, 2023.

[18] A. Tajalli and et al., "Impact of sidewall etching on the dynamic performance of gan-on-si e-mode transistors," *Microelectronics Reliability*, vol. 88, pp. 572–576, 2018.

[19] A. Stockman and et al., "On the origin of the leakage current in p-gate algan/gan hemts," in *2018 IEEE international reliability physics symposium (IRPS)*, IEEE, 2018, 4B–5.

[20] H. Wu and et al., "Time-resolved threshold voltage instability of 650-v schottky type p-gan gate hemt under temperature-dependent forward and reverse gate bias conditions," *IEEE Transactions on Electron Devices*, vol. 69, no. 2, pp. 531–535, 2022.

PCIM Europe 2024, 11– 13 June 2024, Nuremberg    DOI: 10.30420/566262088

# Gate Oxide Reliability of Current Generation 1.2 kV SiC MOSFETs under Step-Wise Increased Gate Voltage

Roman Boldyrjew-Mast, Sven Thiele, Thomas Basler
Chemnitz University of Technology, Reichenhainer Str. 70, D-09126 Chemnitz, Germany

Corresponding author:  Roman Boldyrjew-Mast, roman.boldyrjew-mast@etit.tu-chemnitz.de
Speaker:           Roman Boldyrjew-Mast, roman.boldyrjew-mast@etit.tu-chemnitz.de

## Abstract

The gate oxide reliability of current generation SiC MOSFETs from five different manufacturers divided in seven groups has been analyzed in step-wise increased gate stress tests. All groups show a significantly varying intrinsic lifetime in terms of dielectric breakdown, dominantly resulting from the specific oxide thickness of each group. Low extrinsic failure rates have been found for each group indicating a well-defined screening procedure during the manufacturing process. During read-out interruptions, the threshold voltage has been measured with the hysteresis method in alignment with the JEDEC guideline JEP-184 at room temperature. Only two groups from one manufacturer showed a lifetime comparable to Si IGBTs.

## 1    Introduction

Typically, time-to-breakdown ($t_{BD}$) [1, 2] or charge-to-breakdown ($Q_{BD}$) [3] tests with a high DC gate voltage are being applied on a high number of specimens to analyze the gate oxide integrity of processed MOS-capacitors and/or commercially available MOSFETs. Since a higher extrinsic failure rate is expected for SiC MOSFETs [4] due to the presence of carbon atoms in the lattice and the more complicated manufacturing process of SiC MOSFETs, a fast test to separate intrinsic from potentially present extrinsic oxide failures in these devices is highly desirable. Intrinsic failures are determined by the oxide thickness of the ideal gate dielectric, whereas extrinsic failures result from impurities, particles, enclosures, etc., leading to randomly distributed oxide ruptures during reliability tests [5]. The step-wise increased gate stress is a suitable method to obtain reasonably fast statements about the distribution of extrinsic and intrinsic failures. Hence, it was recently described in a JEDEC guideline (JEP-194) [6].

Eight years after our previous investigations in [7] and the roll-out of new generations of SiC MOSFETs for many manufacturers, it is time to analyze the current gate oxide reliability status. With the implementation of optimized screening approaches [8, 9] a lower extrinsic failure rate can be expected.

## 2    Procedure of step-wise increased gate stress tests

First step-stress tests have been presented by Nelson [10] and Anolick [11]. Both publications describe a similar test procedure, but they focus on the reliability of different test objects, with electrical cable insulation and MOS-capacitors, respectively. Due to an increasing relevance of SiC MOSFETs, the step-stress tests came back into focus as a suitable approach to analyze their gate oxide integrity.

### 2.1    Test procedure

Similar to constant voltage stress tests, the aim of step-wise increased gate stress test is to analyze the gate oxide reliability in power semiconductors with MOS-structures, focusing on the time depend dielectric breakdown (TDDB). However, in contrast to constant voltage stress tests, the applied gate voltage will be increased in steps, raising the electric field in the gate oxide with every stress level, until a dielectric breakdown occurs for all devices.

In the presented step-wise increased gate stress test, all specimens have been tested initially under the recommended gate voltage for a fixed stress time of $t_{step}$ = 18 h at a high case temperature of $T_C$ = 150°C, as shown in Fig. 1. The applied gate voltage is increased in each subsequent stress period by 2 V after the datasheet maximum $V_{GS(max)}$ until all specimens reach dielectric breakdown. With a stress time of 18 h per step it was possible to heat up, perform the stress sequence, cool down and measure the $V_{GS(Th)}$ within 24 h for all specimens, forming a test cycle in this manner.

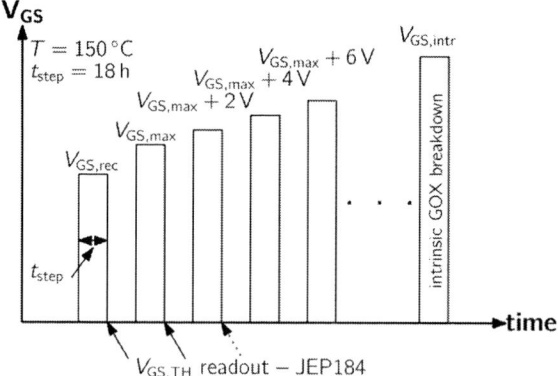

Fig. 1 Procedure of the step-wise increased gate stress test at high case temperature $T_C = 150°C$ with a stress period $t_{step} = 18$ h; increase of stress voltage by 2 V steps until dielectric breakdown occurs for all SiC MOSFETs; before test start and after every stress period, the threshold voltage is recorded at room temperature

## 2.2 Threshold voltage read-out procedure

Typically, the threshold voltage does not belong to the parameters which are recorded during gate stress tests, focusing on time-to-breakdown and gate leakage current. But the periodical interruptions in the step-stress test provide the opportunity to measure this parameter as well.

The threshold voltage has been recorded at room temperature, initially before test start and after every stress period with the *hysteresis method* in alignment to the JEDEC guideline JEP-184 [12] and [13], depicted in Fig. 2. Prior to every $V_{GS(Th)}$-measurement a preconditioning pulse has been applied to every single device under test with positive and negative gate voltage $V_{GS} = \pm 20$ V for a fixed time of $t_{precon} = 100$ ms. The preconditioning pulse is necessary to achieve a stable and comparable condition regarding the electric charge at the SiC/SiO$_2$-interface. For each group, the threshold voltage has been measured in the gated-diode configuration at a drain current $I_D = 10$ mA. Further, a constant acquisition delay time ($t_{aq(del)}$) is mandatory for a precise $V_{GS(Th)}$-analysis, because a variation in $t_{aq(del)}$ can result in significant scattering of measurement results [14]. In the given

measurements, the acquisition delay time was set to $t_{aq(del)} = 1$ µs. For a clear overview, only values for $V_{GS(Th)Down}$ will be discussed, because this parameter reaches more rapidly a stable value during the readout [13].

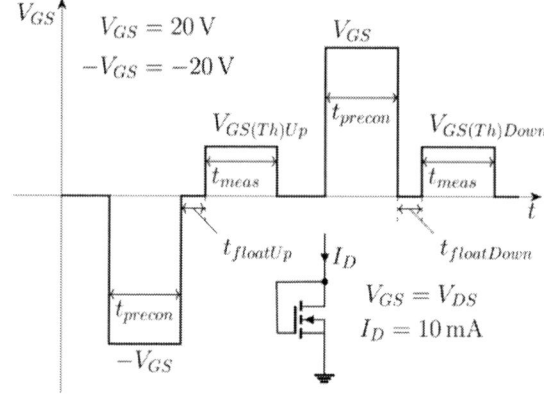

Fig. 2 $V_{GS(Th)}$ read-out with the hysteresis method and gated-diode configuration in alignment to the JEDEC JEP 184 guideline and [13]; preconditioning with $V_{GS} = \pm 20$ V for 100 ms prior to $V_{GS(Th)}$ measurement with $I_D = 10$ mA after an acquisition delay time $t_{aq(del)} = 1$ µs

## 3 Overview of test results

Six groups, containing 100 SiC MOSFETs each, from 5 different manufacturers have been tested in the step-wise increased gate stress test according to the test procedure described in paragraph 2, while the results are depicted in Fig. 3. The diagram shows the failure probability vs. time until the dielectric breakdown in test. The failure probability has been calculated with median ranks according to the approximation of Benard and the corresponding lifetime for each specimen. The lifetime represents the accumulated time in test for every stress level. No extrapolation of test results to nominal operation conditions with the linear E-model or other approaches [15] has been performed at this stage. Solid lines represent the extrapolated Weibull distribution of each group in the corresponding color code, derived from intrinsic failures only. Each group has different scale and shape parameters for the Weibull distribution, characterizing the time until dielectric breakdown under the specified test conditions, as shown in Table 1.

The stress voltage level is shown on top of the diagram, correlating with the corresponding total stress period at the bottom. With the applied gate

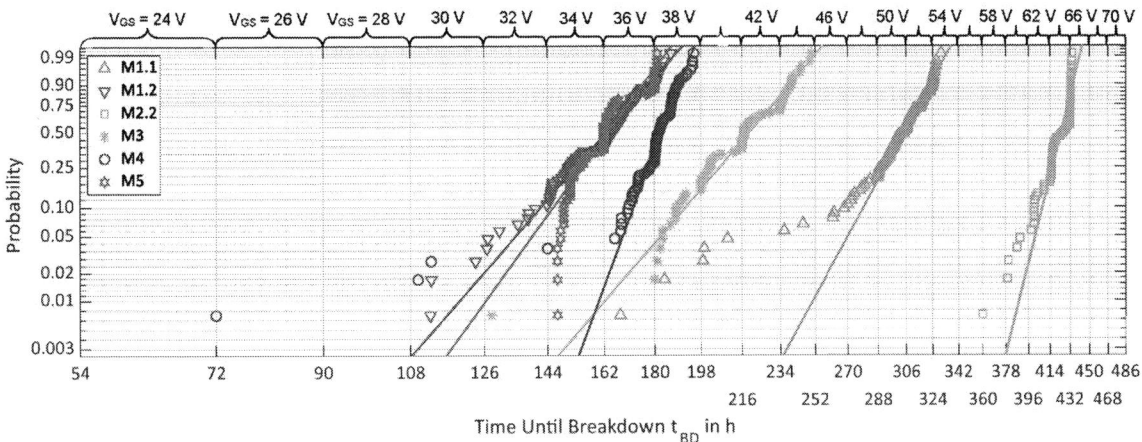

Fig. 3 Overview of test results in step-wise increased gate bias test for 6 groups; each group, represented by different markers with corresponding colors, contains 100 1.2 kV SiC MOSFETs; median ranks (MR) calculated with approximation of Benard are drawn vs. time until dielectric breakdown; solid lines represent extrapolated Weibull distributions and the extracted parameters are reported in Table 1

| Test Group | M 1.1 | M 1.2 | M 2.2 | M 3 | M 4 | M 5 |
|---|---|---|---|---|---|---|
| Scale parameter ($\alpha$) | 310.1 | 167.5 | 427.2 | 225.4 | 184.9 | 167.2 |
| Shape parameter ($\beta$) | 21.7 | 13.7 | 48.6 | 14.0 | 32.1 | 16.5 |

Table 1 Extracted scale ($\alpha$) and shape ($\beta$) parameter for Weibull distribution from Fig. 3; scale parameter $\alpha$ represents the time, where 63.2 % of the test population reached TDDB; shape parameter of $\beta > 1$ represent failure due to wear out (intrinsic branch)

voltage and the assumption of dielectric breakdown of thick $SiO_2$ oxides occurring around 10 MV/cm [5, 16] the gate oxide thickness for each group can be estimated. Different oxide thickness of devices from each manufacturer is the reason for strong deviation between individual groups in the intrinsic segment. Comparing group M 1.1 and M 1.2, which represent specimens from the same manufacturer, but from different chip generations, it becomes evident that some manufacturers have decreased the oxide thickness to reduce the channel resistance $R_{ch}$. For lower and medium voltage class devices, the channel resistance dominates the specific on-state resistance $R_{DS(on)}*A$. Thus, a lower $R_{ch}$ leads directly to a decrease in $R_{DS(on)}*A$.

In comparison to previous publications [7, 17], lower extrinsic failure rates have been found for each manufacturer, indicating a well-defined oxide screening procedure and/or improved manufacturing process. However, for some manufacturers, the screening voltage seems to be very close to the intrinsic branch, whereas for others, extrinsic failures occurred nearby field operation conditions (typically $V_{GS}$ = 15 V).

Finally, a typical behaviour is visible in Fig. 3 for step-wise increased gate stress tests on SiC MOSFETs. In the intrinsic area, many oxide breakdowns typically occur at the beginning of a new stress voltage level, leading to a crowding of failures in the diagram. This can be explained by the step-wise increase of stress voltage. With every new voltage level, the electric field in the oxide raises as well. Assuming a Gaussian distribution for intrinsic oxide thickness across the device population, the resulting electric field will exceed the critical limit for a specific group of devices, leading directly to a dielectric breakdown. For the remaining devices, the aging will be accelerated significantly, and failures will be dominated by TDDB. This behaviour is particularly visible for group M 3 and M 5.

## 4 Detailed discussion of particular test groups

After a general overview of all TDDB test results in the previous paragraph, the following section focuses on four selected groups for a closer discussion of the test results concentrating on the time dependent dielectric breakdown, the threshold voltage drift, and their correlation.

## 4.1 TDDB and $V_{GS(Th)}$-increase for group M 1.2

According to the Weibull distribution, depicted in Fig. 4 a) for group M 1.2, no extrinsic failures occurred during the stress test with step-wise increased gate bias. All specimens reached end of life (EOL) due to a time dependent dielectric breakdown at a gate voltage between $V_{GS}$ = 30 V and 38 V, whereas most devices remained within 95 % confidence intervals of the Weibull distribution (dashed lines). This wide spread led to the lowest shape parameter among all tested groups with $\beta$ = 13.7, reported in Table 1. With 57 out of 100 specimens, the majority reached TDDB at $V_{GS}$ = 36 V.

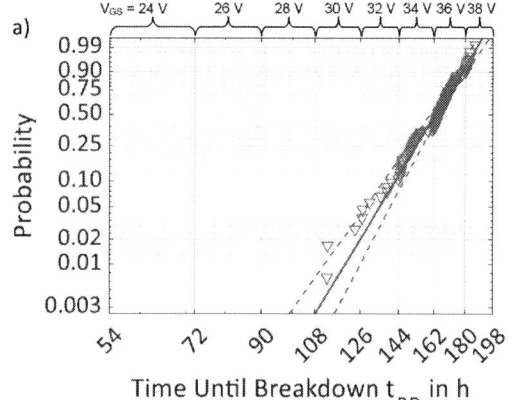

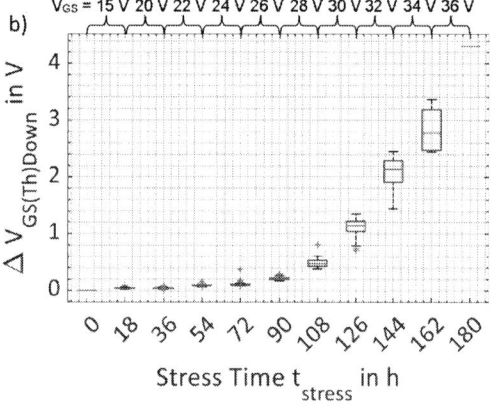

Fig. 4 Depiction of failures in the step-wise increased gate stress test in a) for group M 1.2; solid line represents the fitted Weibull distribution and the dashed lines the 95 % confidence intervals; extrapolated scale and shape parameters are $\alpha$ = 167.5 and $\beta$ = 13.7, respectively; box plot of $V_{GS(Th)Down}$-shift in b) recorded before test start and after each stress period; until $V_{GS}$ = 30 V $V_{GS(Th)Down}$ was measured for all specimens and from $V_{GS}$ = 32 V only 10 specimens per step until TDDB

With increasing stress time and gate voltage in the test, the $V_{GS(Th)Down}$ drift becomes larger, as shown in Fig. 4 b). For moderate stress voltages ($V_{GS}$ < 28 V), the drift is caused mostly by bias temperature instability (BTI) effects. Electrons will be trapped in the oxide layer, accumulating negative charge. Consequently, a net higher positive gate voltage is required, to establish the conducting channel, increasing the threshold voltage. As soon as the gate voltage exceeds $V_{GS}$ = 28 V, the $V_{GS(Th)}$ drift becomes more severe as a result from initiated tunneling mechanisms in the step-wise increased gate stress test. This leads to bond braking mechanisms and enhanced electron capturing at present or newly generated trap states. Up to a gate voltage of $V_{GS}$ = 30 V, the threshold voltage has been recorded for all devices and after stress step with $V_{GS}$ = 32 V, $V_{GS(Th)}$ has been measured for 10 out of 100 specimens until their TDDB.

Assuming the dielectric breakdown at 10 MV/cm [5, 16] for $V_{GS}$ = 36 V, the electric field at $V_{GS}$ = 28 V would be around 7.8 MV/cm, enhancing tunneling currents already. Comparing the $V_{GS(Th)}$ drift with the failure times in Fig. 4 a) the applied screening voltage might be estimated around $V_{GS(scr)} \approx$ 28 – 30 V for group M 1.2. However, the screening voltage must be chosen carefully, without causing further/new irreversible $V_{GS(Th)}$ drift and gate oxide degradation [8, 18].

## 4.2 Extrinsic TDDB failures of group M 4 and assumed screening voltage

In contrast to group M 1.2, four extrinsic failures are clearly visible in Fig. 5 a), for group M 4. Extrinsic failures often occur randomly and are a bigger threat than intrinsic failures in field operation, regarding the gate oxide integrity of SiC MOSFETs. Two out of four failures occurred directly at the beginning of a new stress period after increasing the applied gate voltage by a 2 V step, whereas the other two specimens reached TDDB after a certain time at $V_{GS}$ = 30 V. In particular, the first extrinsic failure, detected at a gate voltage of $V_{GS}$ = 26 V might become a serious threat for the lifetime under field operation. Most devices reached the dielectric breakdown at a gate voltage $V_{GS}$ = 36 V or 38 V, leading to a high shape parameter ($\beta$ = 32.1), as reported in Table 1.

Considering the first extrinsic failure passing the screening procedure, the estimated screening voltage might be around $V_{GS(scr)} \approx$ 26 V. According to Fig. 5 b), the BTI-driven threshold voltage drift is below 0.2 V after stress with $V_{GS}$ = 24 V and remains moderate up to a gate voltage of $V_{GS}$ = 30 V. Hence, the screening voltage could be increased slightly, in order to further reduce the extrinsic population. However, with respect to the sample size of 100 specimens, the rate of extrinsic failures might vary for the complete chip population.

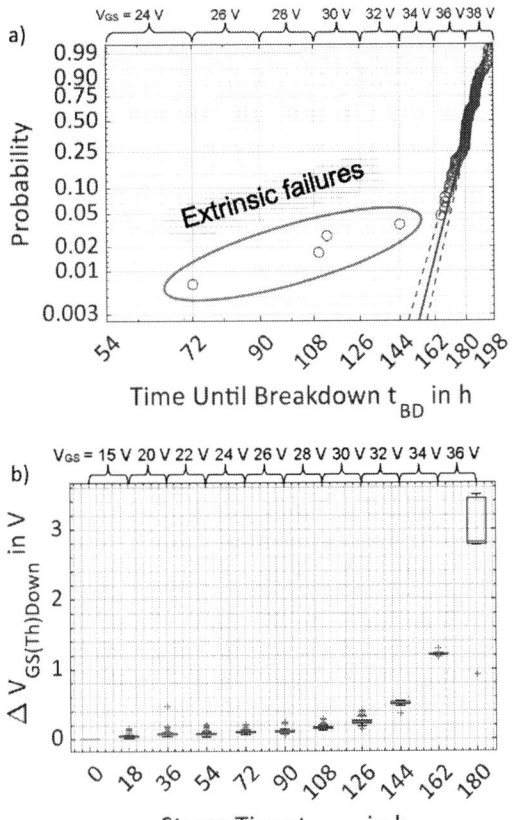

Fig. 5 Depiction of failures in the step-wise increased gate stress test in a) for group M 4; solid line represents the fitted Weibull distribution and the dashed lines the 95 % confidence intervals; extrapolated scale and shape parameters are $\alpha = 184.9$ and $\beta = 32.1$, respectively; box plot of $V_{GS(Th)Down}$-drift in b) recorded before tests start and after each stress period; until $V_{GS} = 30$ V $V_{GS(Th)Down}$ was recorded for all specimens and from $V_{GS} = 32$ V only 10 specimens until TDDB

## 4.3 Lifetime variation for manufacturer M 2

In Fig. 6, the results from the step-wise increased gate stress test are shown for group M 2.1 and M 2.2. Both groups contain specimens from the same manufacturer and the same chip generation but significantly different date code. In group M 2.2 only one specimen could be classified as an extrinsic failure, which occurred after 360 h in test. In

contrast to M 2.2, twelve specimens could be classified as extrinsic for group M 2.1. M 2.2 can be regarded as an improvement, reducing the ppm-rate of extrinsic failures at nominal operation conditions, potentially resulting from an enhanced manufacturing process and/or screening procedure. However, the intrinsic lifetime of group M 2.2 is lower. This is confirmed by the slightly higher scale parameter $\alpha$ for M 2.1 than for M 2.2, reported in Table 2 with $\alpha = 458.6$ and $427.2$, respectively. This deviation can be explained by two influencing factors. Either the gate oxide thickness has been decreased from group M 2.1 to M 2.2, or the applied screening procedure reduced the lifetime, as indicated by the model proposed by Chbili et al. in [18]. As discussed in paragraph 3, the specific on-state resistance $R_{DS(on)}*A$ would clearly benefit from a lower gate oxide thickness. Assuming a lifetime consumption by the screening procedure as further impact, this would illustrate the complex screening and burn-in process. Hence, selecting an adequate screening procedure is crucial to reduce the population of critical extrinsic failures in the application-relevant gate voltage range, without shortening the time until dielectric breakdown below a critical limit. This can be achieved only with a sufficiently high gate oxide thickness with respect to the use gate voltage.

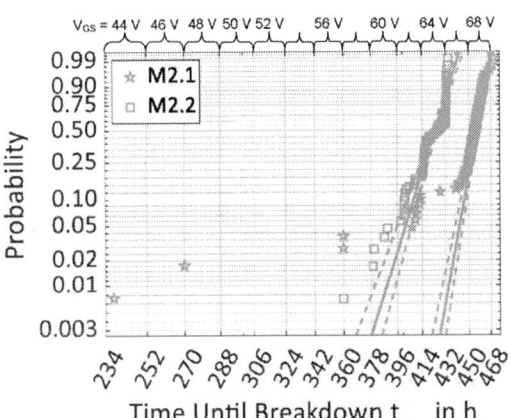

Fig. 6 Depiction of failures in the step-wise increased gate stress test for group M 2.1 and M 2.2; both groups contain specimens from the same chip generation but strongly different manufacturing dates and potentially with deviating screening voltages; solid lines represent the fitted Weibull distribution and the dashed lines the 95 % confidence intervals; extrapolated scale and shape parameters are reported in Table 2

| Test Group | M 2.1 | M 2.2 |
|---|---|---|
| Scale parameter (α) | 458.6 | 427.2 |
| Shape parameter (β) | 83.5 | 48.6 |

Table 2 Extrapolated scale (α) and shape (β) parameter for Weibull distribution from Fig. 6 for group M 2.1 and M 2.2 containing specimens from the same manufacturer and the same chip generation

## 5 Lifetime estimation at application near conditions

Although the presented approach is reasonable to depict results from step-wise increased gate stress tests, and extrinsic failures can be clearly separated from intrinsic breakdown, the lifetime extrapolation to field operation condition is more complicated. Main drawback is the accumulation of test time, as mentioned in paragraph 3, neglecting the influence of stress voltage level on time to dielectric breakdown. Increasing gate voltage with every stress level accelerates degradation of the gate oxide and must be considered for lifetime estimation. Consequently, the extracted Weibull parameter describe only the test results under these specific conditions, but neither scale nor shape parameter can be directly considered for lifetime evaluation or calculation of ppm-rates. This becomes evident, selecting significantly higher or lower stress period $t_{step}$. With $t_{step}$ = 168 h, as applied in [7] and [17], extracted Weibull parameter would vary significantly, although the specimens reached TDDB at similar gate voltage stress.

This circumstance can be overcome, considering the influence of individual stress periods on the lifetime, as proposed by Nelson [10] and Anolick [11]. However, these approaches require not only specific lifetime models (for instance linear or reciprocal E-model [15]), but the corresponding parameters as well, which must be valid for SiC MOSFETs. For a rough estimation, the linear E-model can be applied on the first failed specimen, to calculate the lifetime $t_{use}$ under field application conditions, according to Eq. (1) from [6]:

$$t_{use} = t_{str}\left[\left(\gamma E_{BD}^{1h} - \ln\left(\frac{t_{str}}{3600}\right)\right)\left(1 - \frac{V_{GS(use)}}{V_{GS(str)}}\right)\right] \quad (1)$$

with the voltage acceleration factor γ, the electric field $E_{BD}^{1h}$ where the dielectric breakdown occurs after approximately 1 h, stress time $t_{str}$ and corresponding gate voltages $V_{GS(use)}$ and $V_{GS(str)}$.

Fig. 7 contains bar graphs, with a separation of extrinsic from intrinsic failures for a) group M4 and b) group M 2.2, observed in step-wise increased gate stress tests. Additionally, accumulated failures in test are shown vs. stress voltage level, where they have been detected. According to Eq. (1), the time until breakdown can be extrapolated to field application conditions. Thus, for group M 4, where the first extrinsic failure occurred directly at the beginning of stress with $V_{GS}$ = 26 V, the gate oxide lifetime at $V_{GS}$ = 18 V would be approximately 6 years and can be regarded as critical. The longest time until breakdown in test for group M 2.2 leads consequently to the highest extrapolated lifetime as well, which is comparable to Si IGBTs [17].

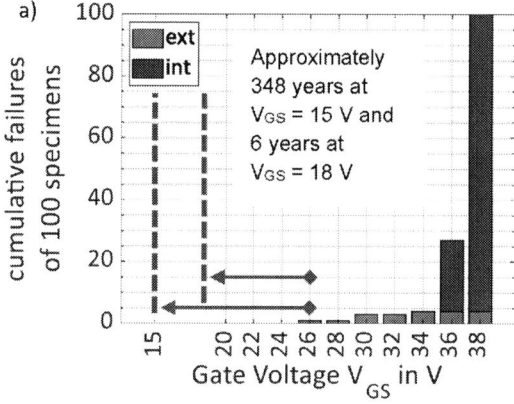

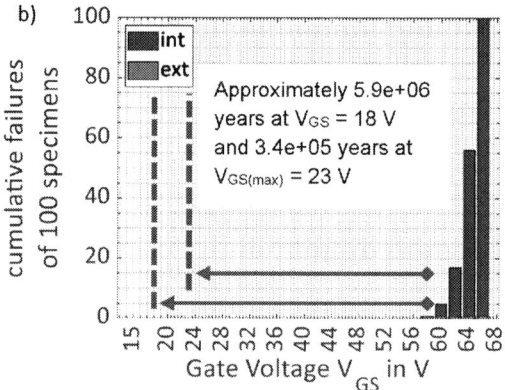

Fig. 7 Bar graph containing classification of extrinsic and intrinsic failures during the step-wise increased gate stress test for a) group M 4 and b) group M 2.2; graphs contain accumulated failures with the corresponding gate voltage levels in test; extrapolation of first recorded failure to field application conditions according to Eq. (1) from the JEDEC guideline JEP 194 [6]

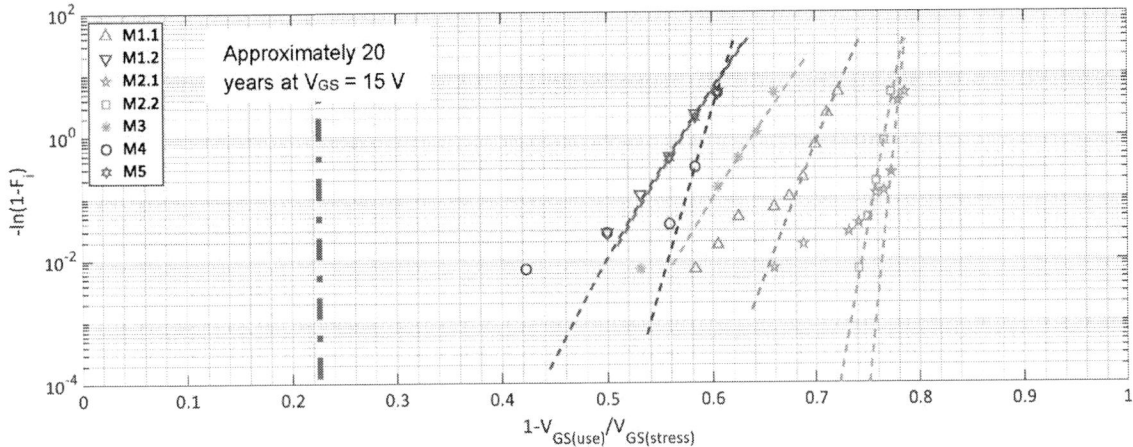

Fig. 8 Overview of test results in step-wise increased gate bias test for 7 groups; each group, represented by different markers with corresponding colors, contains 100 1.2 kV SiC MOSFETs; median ranks (MR) calculated with approximation of Benard are drawn vs. $1 - V_{GS(use)}/V_{GS(str)}$; assumed gate voltage for field application $V_{GS(use)} = 15$ V; dashed lines represent extrapolated Weibull distributions for intrinsic lifetime in test; vertical red dash-dotted line represents targeted lifetime of 20 years

An alternative concept is the analysis of test data focusing on the stress voltage level in step-wise increased gate bias tests, as shown in Fig. 8, based on the JEDEC guideline JEP 194 [6]. In this approach, specimens are accumulated after the dielectric breakdown at the corresponding gate voltage step, in the same manner as in Fig. 7. In the next step, median ranks are calculated with the approximation of Benard, considering only the total number of failures, observed until the corresponding gate voltage step. The resulting failure probability $F_i$ is depicted vs. $1 - V_{GS(use)} / V_{GS(str)}$, assuming $V_{GS(use)} = 15$ V for field applications. The vertical red dash-dotted line represents the targeted chip lifetime of 20 years. It results from the required gate voltage, which is applied during one single stress cycle, representing the application-near voltage applied for 20 years. The main advantage is that this approach considers the stress voltage as major acceleration factor for aging in gate stress tests. Thus, although time until dielectric breakdown is neglected, lifetime can be extrapolated more realistically, than according to Fig. 3.

## 6 Conclusion

Compared to [7, 17], the presented results show a promising gate oxide reliability for the majority of manufacturers. This might have been achieved either with a better screening procedure and/or a significantly improved gate oxide manufacturing process. Most test groups reach their time dependent dielectric breakdown with lower extrinsic failure rates, than observed in previous publications, starting above the maximum allowed gate-voltage level $V_{GS(max)}$, recorded in the corresponding datasheets. Some manufacturers have a larger distance between first extrinsic failures and $V_{GS(max)}$ which will lead to lower ppm rates in the field, whereas others should further reduce the extrinsic failure rate and increase the distance to $V_{GS(max)}$. Thus, aiming on the lowest specific on-state resistance by reducing the gate oxide thickness, reliability aspects must not be neglected. Therefore, a sufficiently high distance between recommended gate voltage levels and the intrinsic dielectric breakdown voltage must be ensured, especially, if an efficient gate oxide screening procedure should be applied.

With an adequate adoption of the existing models for step-stress tests, one open question remains for the lifetime extrapolation to nominal operation conditions in field applications. The model presented by Anolick [11] focuses on a linear E-model [19]. However, recent publications indicate a behavior, which might be characterized either by a two stage E-model [20] or a $Q_{BD}$-approach [3]. Hence, further investigations are required, to establish a model for step-wise increased gate stress tests, valid for SiC MOSFETs, to derive the expected lifetime at the application-near operation conditions directly. Applying the linear E-model on the first failed specimen, with respect to the corresponding acceleration factors, can provide a rough estimation of the lifetime.

# References

[1] T. Liu et al., "Gate Oxide Reliability Studies of Commercial 1.2 kV 4H-SiC Power MOSFETs," in *International Reliability Physics Symposium (IRPS)*, Dallas, TX, USA, 2020, pp. 1–5.

[2] Z. Chbili et al., "Time Dependent Dielectric Breakdown in High Quality SiC MOS Capacitors," *MSF*, vol. 858, pp. 615–618, 2016.

[3] P. Moens, J. Franchi, J. Lettens, L. de Schepper, M. Domeij, and F. Allerstam, "A Charge-to-Breakdown (Q BD ) Approach to SiC Gate Oxide Lifetime Extraction and Modeling," in *Proceedings of the 32nd International Symposium on Power Semiconductor Devices and ICs (ISPSD): virtual conference*, Vienna, Austria, 2020, pp. 78–81.

[4] R. Siemieniec et al., "A SiC Trench MOSFET concept offering improved channel mobility and high reliability," in *2017 19th European Conference on Power Electronics and Applications (EPE)*, Warsaw, 2017, P.1-P.13.

[5] Josef Lutz, Thomas Aichinger, and Roland Rupp, "Reliability evaluation," in *Woodhead Publishing Series in Electronic and Optical Materials Ser, Wide Bandgap Power Semiconductor Packaging: Materials, Components, and Reliability*, K. Suganuma, Ed., San Diego: Elsevier Science & Technology, 2018, pp. 155–197.

[6] *JEDEC Publication No. 194: Guideline for Gate Oxide Reliability and Robustness Evaluation Procedures for Silicon Carbide Power MOSFETs*, JEP 194, JEDEC Solid State Technology Association, Feb. 2023.

[7] M. Beier-Moebius and J. Lutz, "Breakdown of gate oxide of 1.2 kV SiC-MOSFETs under high temperature and high gate voltage," in *PCIM Europe 2016: International Exhibition and Conference for Power Electronics, Intelligent Motion, Renewable Energy and Energy Management*, Nürnberg, Germany, 2016.

[8] J. Berens and T. Aichinger, "A straightforward electrical method to determine screening capability of GOX extrinsics in arbitrary, commercially available SiC MOSFETs," in *International Reliability Physics Symposium (IRPS)*, Monterey, CA, USA, 2021, pp. 1–5.

[9] T. Aichinger and M. Schmidt, "Gate-oxide reliability and failure-rate reduction of industrial SiC MOSFETs," in *International Reliability Physics Symposium (IRPS)*, Dallas, TX, USA, 2020, pp. 1–6.

[10] W. Nelson, "Accelerated Life Testing - Step-Stress Models and Data Analyses," *IEEE Trans. Rel.*, R-29, no. 2, pp. 103–108, 1980.

[11] E. S. Anolick and L.-Y. Chen, "Application of Step Stress to Time Dependent Breakdown," in *International Reliability Physics Symposium (IRPS)*, Las Vegas, NV, USA, 1981, pp. 23–27.

[12] *JEDEC Publication No. 184: Guideline for evaluating Bias Temperature Instability of Silicon Carbide Metal-Oxide-Semiconductor Devices for Power Electronic Conversion*, JEP 184, JEDEC Solid State Technology Association, Mar. 2021.

[13] T. Aichinger, G. Rescher, and G. Pobegen, "Threshold voltage peculiarities and bias temperature instabilities of SiC MOSFETs," *Microelectronics Reliability*, vol. 80, pp. 68–78, 2018.

[14] K. Puschkarsky, H. Reisinger, T. Aichinger, W. Gustin, and T. Grasser, "Understanding BTI in SiC MOSFETs and Its Impact on Circuit Operation," *IEEE Transactions on Device and Materials Reliability*, vol. 18, no. 2, pp. 144–153, 2018.

[15] J. W. McPherson, "Time dependent dielectric breakdown physics – Models revisited," *Microelectronics Reliability*, vol. 52, 9-10, pp. 1753–1760, 2012.

[16] T. Kimoto and J. A. Cooper, *Fundamentals of silicon carbide technology: Growth, characterization, devices and applications*. Singapore, Piscataway, New Jersey: Wiley, 2014.

[17] M. Beier-Moebius and J. Lutz, "Breakdown of Gate Oxide of SiC-MOSFETs and Si-IGBTs under High Temperature and High Gate Voltage," in *PCIM Europe 2017: International Exhibition and Conference for Power Electronics, Intelligent Motion, Renewable Energy and Energy Management*, Nürnberg, Germany, 2017, pp. 1–8.

[18] Z. Chbili et al., "Modeling Early Breakdown Failures of Gate Oxide in SiC Power MOSFETs," *IEEE Trans. Electron Devices*, vol. 63, no. 9, pp. 3605–3613, 2016.

[19] J. W. McPherson and D. A. Baglee, "Acceleration Factors for Thin Gate Oxide Stressing," in *International Reliability Physics Symposium (IRPS)*, Orlando, FL, USA, 1985, pp. 1–5.

[20] S. Zhu, T. Liu, M. H. White, A. K. Agarwal, A. Salemi, and D. Sheridan, "Investigation of Gate Leakage Current Behavior for Commercial 1.2 kV 4H-SiC Power MOSFETs," in *International Reliability Physics Symposium (IRPS)*, Monterey, CA, USA, 2021, pp. 1–7.

PCIM Europe 2024, 11– 13 June 2024, Nuremberg    DOI: 10.30420/566262089

# An Accelerated Dynamic Gate Switching Stress Test Concept for SiC MOSFETs at High Drain-Source Voltage (HV-GSS)

Clemens Herrmann[1] , Sven Thiele[1] , Dezhi Yang[1], Thomas Basler[1]
Matthias Neumeister[2], Markus Pfeifer[2]

[1] Chemnitz University of Technology, Chair of Power Electronics, Germany
[2] Siemens AG, Germany

Corresponding author:   Clemens Herrmann, clemens.herrmann@etit.tu-chemnitz.de
Speaker:           Clemens Herrmann

## Abstract

This paper presents a test concept for a dynamic gate stress test under high drain-source voltage and high acceleration via switching frequency (HV-GSS). Gate-switching stress tests with and without drain voltage (HV-GSS and GSS) were conducted using SiC trench MOSFETs in order to determine the influence of high drain-source voltage on gate reliability in terms of gate-switching instability.

## 1  Introduction

### 1.1  Motivation

The appearance of gate-switching-instability (GSI) of silicon carbide (SiC) MOSFETs is generally accepted and well-known. It was preceded by new test concepts that addressed the threshold voltage instability under traditional DC-stress (e.g. HTGB) and upcoming AC-stress conditions (DGS/GSS) in recent years. AC stress is considered more critical and meaningful regarding an application-compliant assessment of a device´s threshold voltage ($V_{th}$) development [1]. A typical test method is the execution of high-frequency gate switching stress at room temperature (GSS/DGS), as described e.g. in [1] and [2]. However, this test method does not address the influence of a high drain-source voltage on the device as it would be present in the application. Due to the superposition of the electric fields resulting from the negative gate-source voltage and positive drain-source voltage across the gate oxide, the additional influence of the drain voltage is of interest. It is questionable whether a widely used standard gate switching stress test that dispenses high voltage is sufficient to estimate the gate oxide reliability due to GSI over the whole lifetime.

### 1.2  Theoretical Background

The underlying question behind this concern is: does the drain voltage affect gate oxide reliability, mainly the threshold voltage and leakage currents of a SiC MOSFET in addition to the stress resulting from gate-source voltage?

From a theoretical point of view, it is obvious that high drain voltage affects the channel region and the gate oxide, as shown by the following effects.

In [3], Yao et al. investigated the differences between two trench device structures and their ability to protect the gate oxide from high drain voltage via a shielding structure. This is a general concern in all modern SiC MOSFET concepts [4] [5] [6], especially during an avalanche event. It was shown by simulation that the extent to which the device and shielding structure were realized, affects the maximum field strength, the gate oxide is subjected to, which in turn directly affects the measured avalanche ruggedness.

Depending on the effectiveness of the shielding approach, the electric field resulting from the drain-source voltage overlaps with and adds up to the electric field caused by the negative gate-source voltage ($V_{GS-}$). This effect is decisive during e.g. an avalanche event and is one of several reasons why the negative gate voltage in datasheets is limited.

In [7], a simulation of the electric field in the gate oxide during blocking state (high $V_{DS}$, negative $V_{GS}$) of an asymmetric trench MOSFET shows that especially on the edges of the trench bottom, a high electric field strength is present. This aspect may not be responsible for a stronger threshold voltage drift at high voltage, since the channel region is, at least in this example, not directly exposed to these high field strengths. However, the general oxide reliability could be affected, which may be visible apart from $V_{th}$ drifts, e.g. in the increase in gate leakage current.

The general presence of a potential influence of $V_{DS}$ on the channel itself can be seen via the appearance of the drain-induced barrier lowering (DIBL) or short channel effect, which effectively reduces $V_{th}$ with increasing $V_{DS}$. Figure 1 depicts the development of $V_{th}$ at increasing $V_{DS}$. The intensity of the DIBL effect varies for different device manufacturers and their individual device designs - however, its appearance is an indication that the channel region is directly influenced by the presence of drain voltage.

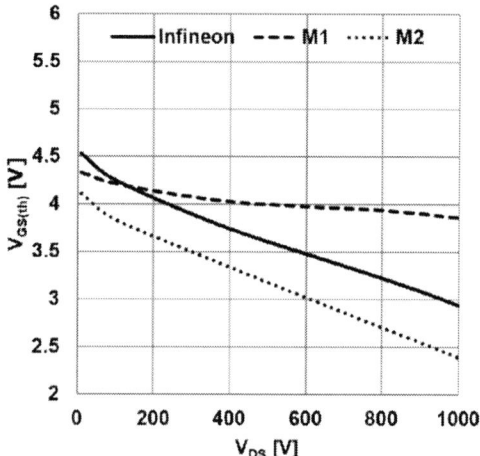

**Fig. 1**    $V_{th}$ (at 10 mA) as a function of $V_{DS}$ for different SiC MOSFET manufacturers. Infineon - trench, M1 - trench, M2 - planar, 1200 V class. Picture taken from [4]

Under consideration of these aspects, it appears reasonable to perform switched gate stress tests which include high drain-source voltage in addition to the gate-source voltage.

Today's standard gate-switching stress tests mainly focus on the device's subjection of gate-source voltage. The drain voltage is zero in these cases since the drain terminal is shorted to the gate in order to not keep it floated at an undefined potential. The development of the $V_{th}$-drift mainly depends on the number of switching events rather than on absolute test duration, which is why a high AC frequency is preferred to gain acceleration [8] [1].

However, gate stress tests with additional high drain-source voltage have also been published as well. In [9] an application test at high drain voltage was performed and compared to a GSS test without drain voltage but otherwise comparable conditions. There was no difference in the development of the device parameters between the two tests. However, this statement only applies to the devices tested (Infineon 1200 V class trench SiC MOSFETs) and must be proven for other manufacturers, voltage classes, and device concepts. Furthermore, a different test approach would be useful in general since an application test is hard to accelerate strongly and demands therefore a lot of test time to attain a high cycle number.

## 2    Test Concept

### 2.1    Technical Requirements

The aim is to find a test concept to perform application-compliant and accelerable high voltage gate switching stress tests. For this purpose, the following requirements must be met:

#### 2.1.1    Application-compliant Drain and Gate Voltage and Junction Temperature

The devices should be subjected to application-compliant gate-source and drain-source voltages concerning their shape, levels, and transients. A test acceleration should not be applied via increased voltage, but via switching frequency, i.e. the number of switching events (see section 2.1.2). To keep the results and statements of the test clearly evaluable, over- and undershoots, especially in gate voltage should be avoided as much as possible since the gate oxide reliability should be investigated concerning drain voltage and aging should not be additionally provoked by transient voltage peaks. The applications of interest are hard-switched applications, such as inverters. Furthermore, the junction temperature should be kept within the datasheet limit.

#### 2.1.2    Acceleration via Frequency

Depending on the application, the switching frequency $f_{switching}$, and the duty cycle, a certain number of switching events $n_{cycles}$ must be expected over the total lifetime. For applications using high switching frequencies, long lifecycles, and high daily duty cycles, such as solar inverters or all-time running DC-DC-converters operating at 100 kHz and more, 12-24 h operation per day over

20 years, switching cycles in the order of magnitude of $10^{13}$ will be obtained. It was also observed that the $V_{th}$-drift shows a saturation before this cycle number is finally attained, i.e. not 100% of the expected $n_{cycles}$ have to be necessarily covered by the test. It can stop earlier, and the rest of the switching cycles can be estimated by linear extrapolation of the $V_{th}$ drift to cover the worst-case estimation.

We aim to reach a cycle number in the range of $10^{12}$. A usual and manageable test time should not exceed 1000 h. Considering these requirements, a switching frequency of at least 200-300 kHz is necessary.

## 2.2 Topology

One of the biggest challenges is to keep power losses per switching event low to ensure a strong acceleration via switching frequency. Application-compliant hard switching with full current and voltage is therefore not the preferred solution as it is too close to an inverter setup with high power-supply requirements and does not offer a significantly higher margin in terms of switching frequency.

Resonant switching could ensure higher frequencies due to lower switching losses, but the waveforms differ strongly from hard-switched applications and are therefore not considered to be representative. Furthermore, the operation of a resonant tank can be more challenging since the switching frequency has to hit a certain operating point and cannot be changed flexibly without adjusting the resonant elements of the topology.

A half-bridge topology without load was found suitable to meet all the requirements. The basic structure and functional principle are already known from dynamic reverse bias (DRB) tests addressing e.g. the stability of edge terminations and other internal structures due to fast charging and discharging processes initiated by a steep $dV_{DS}/dt$ transient. Due to the necessity to focus on the high switching frequency and addressing the gate structure instead of edge termination or p-areas, the test concept has been adapted according to the individual needs of our HV-GSS test.

The topology consists of a half-bridge of two devices under test (DUT) which are complementary switched under consideration of a sufficient dead time. No dedicated load is attached. In contrast to DRB topologies, where several DUTs in parallel could also be passively driven via an active auxiliary switch, the DUTs in the HV-GSS concept must be necessarily switched actively due to the necessity of an AC voltage on both gate and drain terminals.

Due to the missing load, no load current is switched, which ensures that no classical power losses due to hard switching and no forward losses as well are generated. The only losses persisting are due to capacitive displacement currents of the DUTs (output capacitance $C_{oss}$) and the coupling capacitance of the surrounding setup (e.g. via heatsink). Furthermore, LC oscillations can contribute to the losses. In general, these losses can be decreased by a proper design, but not eliminated. Especially those caused by the output capacitance $C_{oss}$ are inevitable. The minimum losses physically possible are determined by $C_{oss}$, if all other shares of losses are neglected:

$$
\begin{aligned}
P_{loss} &= E_{oss} \cdot f_{switching} \\
&= \int_0^{V_R} C_{oss}(V) \cdot V \, dV \cdot f_{switching}
\end{aligned} \tag{1}
$$

However, our experience is that the other loss shares are in the same magnitude of order and can, hence, not be neglected, especially, if small chip sizes (high ohmic classes) are used. Nevertheless, smaller chip sizes should be preferred for such tests since they offer a more beneficial ratio between output capacitance and thermal resistance compared to low ohmic chips.

The following depictions show the circuit of the topology (Fig. 2) and simplified, schematized waveforms of voltage and power (Fig. 3).

The gate voltage duty cycle is freely adjustable by signal control to a maximum of 0.5 supposing the two DUTs are constantly switched at even times. In that case, the drain voltage duty cycle will always remain at 0.5, even if the gate voltage duty cycle is lower, e.g. 0.25 (as shown in Fig. 3). The reason is that the drain voltage will not redistribute in the short period in which the two DUTs are simultaneously in the off state.

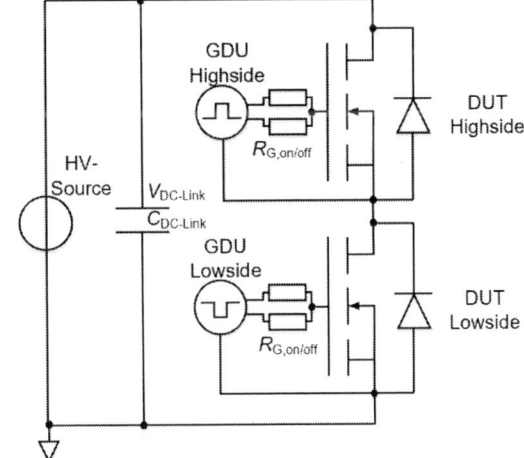

**Fig. 2**  Topology of the HV-GSS test circuit.

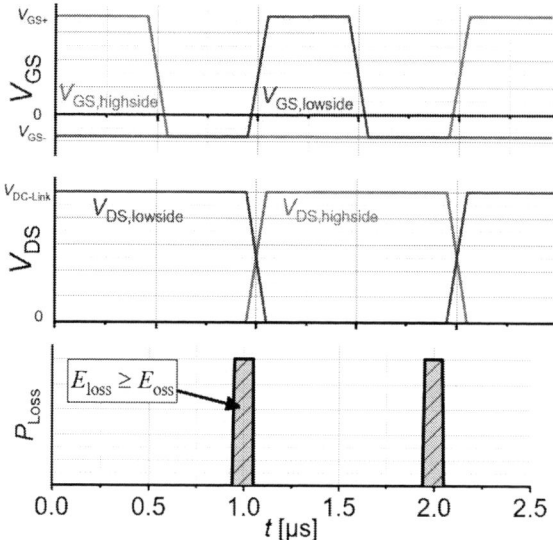

**Fig. 3** Schematic waveforms of the HV-GSS test, $f_{switching}$ = 500 kHz.

One difference compared to the application is that in the HV-GSS test, the negative $V_{GS}$ transient and the positive $V_{DS}$ transient do not coincide in time. Even if the dead time is smaller than shown in Fig. 3, one device will be turned off before it takes over the DC-link voltage since the redistribution of $V_{DS}$ depends on the turn-on of the complementary device. The temporal overlap of these two processes could potentially increase the dynamic stress to the gate oxide and influence therefore the gate oxide reliability. This aspect cannot be fully covered by this HV-GSS approach since a total overlap of the two transients would require the elimination of dead time.

By incurring only minimal possible losses, switching frequencies several times higher than in the application, and therefore corresponding strong test accelerations can be achieved. Due to serious power losses by displacement currents at such high switching frequencies and drain voltages, proper cooling is essential. The test presented in a later section was running at $f_{switching}$ = 500 kHz.

Typically, if the output capacitance $C_{oss}$ is not too large and the cooling is powerful enough, the maximum switching frequency will not be limited thermally - $T_{j,max}$ will not be exceeded - but dynamic limits due to high voltage transients may occur. At shorter period lengths, the d$V_{GS}$/d$t$ and, hence, d$V_{DS}$/d$t$ transients also have to be increased in order to maintain rectangular voltage waveforms. They can be controlled via the gate resistances. However, this causes additional challenges due to stronger oscillation tendencies, gate and drain voltage over- and undershoots as well as Miller

feedback in the gate voltage, which could trigger a parasitic turn-on and a bridge short in the worst case. Therefore, the maximum switching frequency and transients must be limited to a justifiable level.

The test setups of the HV-GSS and GSS test are depicted in Fig. 4 and Fig. 5, respectively.

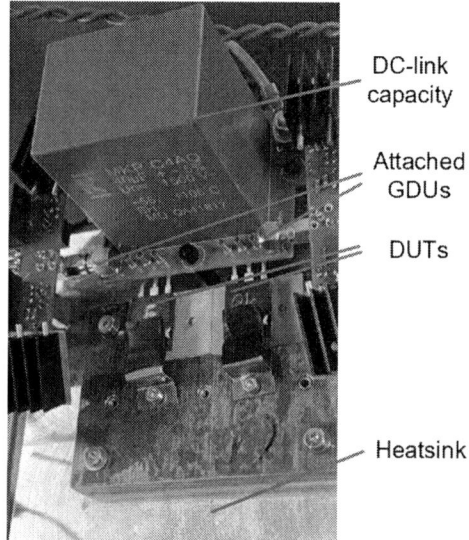

**Fig. 4** One setup (half-bridge) of the HV-GSS test circuit containing two DUTs.

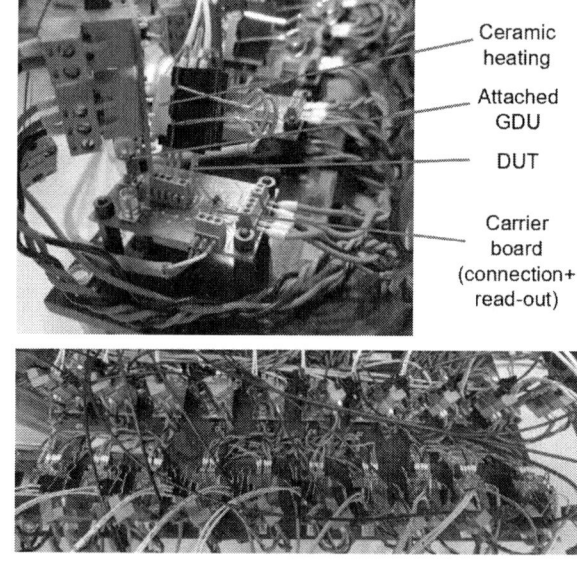

**Fig. 5** One setup of the GSS test (top) with in total 20 channels (bottom). Pictures show exemplary DUTs, both 3- and 4-pin devices are possible.

## 3 Results and Discussion

### 3.1 Test Conditions

In the following, a gate stress test was carried out using the proposed approach to study the influence of the drain voltage on the gate switching instability. As a reference test, a standard GSS test without high drain voltage but otherwise similar test conditions was performed. The DUTs were 1200 V, 60 mΩ class SiC MOSFETs with trench structure. The test conditions are shown in Table 1.

| | HV-GSS | GSS |
|---|---|---|
| $f_{switching}$ | 500 kHz | 500 kHz |
| $T_j$ | 100°C | 100°C |
| $V_{GS+/-}$ (rec.) | 18 V/0 (4 DUTs) | 18 V/0 (4 DUTs) |
| $V_{GS+/-}$ (max.) | 21/-4 V (4 DUTs) | 21/-4 V (4 DUTs) |
| $+dV_{GS}/dt$ | 0.2 V/ns | 0.2 V/ns |
| $R_{G,on/off}$ | 24/15 Ω | 24/15 Ω |
| $V_{DS}$ | 1000 V | - |
| $+dV_{DS}/dt$ | 20 kV/µs | - |
| Test time | 1000 h | 830 h |
| $n_{cycles}$ | 1.8E12 | 1.5E12 |

**Table 1** Test conditions of HV-GSS and GSS test.

Due to self-heating, the DUTs in the HV-GSS test reached a $T_j$ of 100°C. Hence, the maximum test frequency was not thermally limited, but due to the dynamic behavior. As a result, it was kept at 500 kHz, as described before. In the GSS test, no significant self-heating (only slightly via the gate network in interaction with the input capacitance $C_{iss}$, $\Delta V_{GS}$, and $f_{switching}$) occurs due to the missing drain voltage. To reach the same $T_j$ in the GSS test, external heating via ceramic resistors was applied. Two different gate voltage sets were used in both test approaches (GSS and HV-GSS):

- The datasheet recommended conditions $V_{GS,rec}$ (+18 V/0) as well as
- the maximum conditions $V_{GS,max}$ (+21/-4 V).

Furthermore, the HV-GSS test's drain voltage was set to 1000 V, which is considered more than the usual 800 V in the 1200 V device class. This allows the test to cover transient turn-off voltage peaks at least partially. Moreover, potential influences of drain-source voltage on the gate reliability can tendentially be provoked stronger and made more visible.

To limit the transients, the $R_{G,on}$ has been set to a relatively high value of 24 Ω. Since only the turn-on process of one DUT affects the displacement of high drain voltage from one device to another, the $R_{G,off}$ can be set within a higher degree of freedom. The gate and drain voltage waveforms of the HV-GSS and the GSS test are depicted in Fig. 6 and Fig. 7 exemplarily for the $V_{GS,max}$ setting.

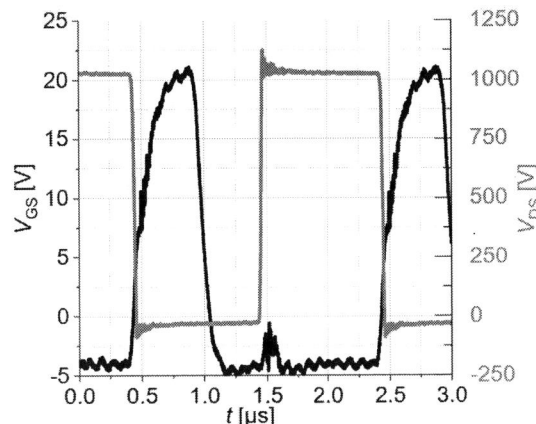

**Fig. 6** Gate-source voltage and drain-source voltage waveforms applied in the HV-GSS test at $V_{GS,max}$ setting.

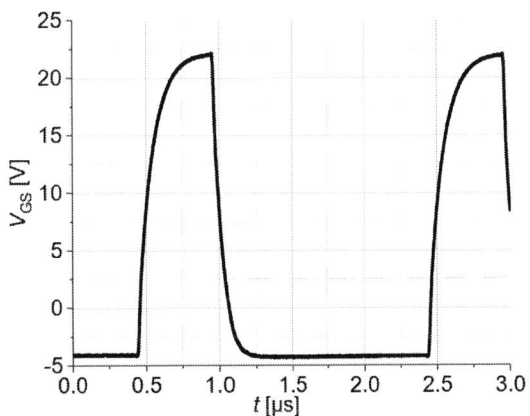

**Fig. 7** Gate-source voltage waveform applied in the GSS test at $V_{GS,max}$ setting.

### 3.2 Readouts of Threshold Voltage and Leakage Currents

The gate threshold voltage $V_{th}$ and gate and drain leakage currents $I_{GSS}$ and $I_{DSS}$ have been recorded before and after as well as throughout the tests. To conduct intermediate readouts, the HV-GSS and GSS tests were interrupted, and the devices cooled down to room temperature.

$V_{th}$ was read out by measuring the threshold voltage hysteresis $V_{th,hyst}$ which is the difference between the $V_{th,down}$ and a $V_{th,up}$ values. The so-called double sense method with positive and negative preconditioning (D-S short) and a $V_{th}$-read-out in gated-diode configuration (G-D short) as described in [8] was used. The preconditioning voltage was set to +/-20 V and the preconditioning time was 100 ms. The threshold drain current was set to 5 mA according to the datasheet.

The leakage current $I_{GSS}$ was measured at room temperature and 20 V using a shielded test fixture, which enabled a measurement with a resolution down to the one-digit pA range. There was a D-S short applied, which means that $I_{GSS}$ contains gate leakage current in source and drain direction.

## 3.3 Test Results

In the following section, the test results of the HV-GSS test and the reference GSS test will be presented, compared, and analyzed. For the analysis of the threshold voltage evolution $\Delta V_{th}(n_{cycles})$, the $V_{th,down}$ value will be evaluated.

The intermediate $V_{th,down}$ readouts of the GSS tests started at a higher switching cycle number than in the HV-GSS tests. Therefore, the progression of $\Delta V_{th,down}(n_{cycles})$ between the premeasurement at $n_{cycle} = 0$ and the first intermediate readout (first data point in Fig. 8 and Fig. 9) will be linearly interpolated and is drawn as a dashed line.

### 3.3.1 HV-GSS vs. GSS at $V_{GS,rec}$

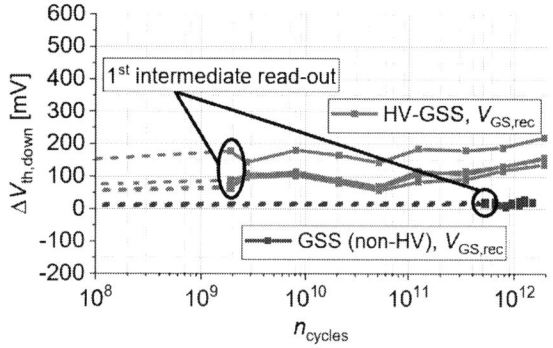

**Fig. 8** Comparison of the development of $V_{th,down}$ for HV-GSS and GSS tests at $V_{GS,rec}$. Dashed lines show regions without intermediate read-out.

Figure 8 depicts the evolution of $V_{th,down}$ over the number of switching cycles at recommended gate-source stress voltage $V_{GS,rec}$. The GSS test shows no significant increase in $V_{th,down}$, while the HV-GSS test leads to a stronger, but in absolute numbers still small drift of $V_{th,down}$. This can be considered as first indication that the GSI is affected by the presence of drain-source voltage.

### 3.3.2 HV-GSS vs. GSS at $V_{GS,max}$

At maximum allowed gate voltage $V_{GS,max}$ the absolute gate threshold voltage drifts are more pronounced compared to $V_{GS,rec}$-conditions (see Fig. 9 vs. Fig. 8) due to higher gate voltage values, presumably especially in $V_{GS-}$ (-4 V instead of 0).

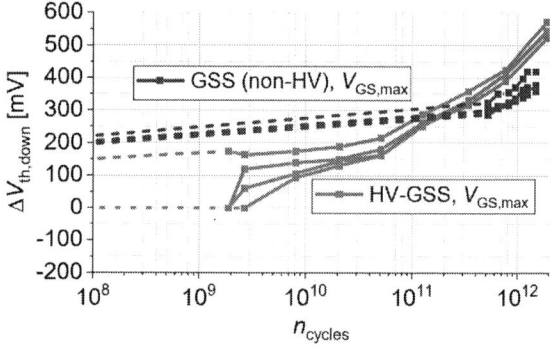

**Fig. 9** Comparison of the development of $V_{th,down}$ for HV-GSS and GSS tests at $V_{GS,max}$. Dashed lines show regions without intermediate read-out.

The differences between HV-GSS and GSS become more pronounced. Fig. 9 shows, analogously to Fig. 8, the development of $V_{th,down}$ in comparison. At the end of the test, the devices subjected to HV-GSS show a significantly higher drift in $V_{th}$ compared to the devices subjected to GSS. While the GSS test results in a drift in $V_{th,down}$ of ~400 mV, the HV-GSS test causes ~500-600 mV of positive threshold voltage drift. This confirms the statement of section 3.3.1 and delivers a clear indication that the threshold voltage development due to gate switching stress is affected by the drain voltage. The discrepancy in $V_{th,drift}$ between HV-GSS and GSS appears to be significant in relative numbers (up to max. ~50% stronger), but it is still tolerable in absolute numbers. All devices stay within the datasheet limits of $V_{th}$ and still have enough margin before exceeding it.

In all tests, no significant changes in the leakage currents have been observed. Table 2 provides an overview of the development of the gate reliability parameters of the tests by comparing the pre- and post-measurement values. The measurements have been performed by TUC and were double-checked and independently confirmed by our project partner Siemens. The change in $V_{th}$ increase of the HV-GSS in contrast to the GSS test (see also Fig. 9) implies that the additional drain voltage modifies the trapping behavior in the channel region. However, since there was no change in leakage currents, the oxide, especially the area apart from the channel region seems not to be further attacked or damaged, at least not in that matter that it would be visible in leakage currents.

| | HV-GSS | GSS |
|---|---|---|
| | $\Delta V_{th,down}$ | $\Delta V_{th,down}$ |
| **test at** $V_{GS,rec}$ (+18 V/0) | ~+100-200 mV | no significant change |
| **test at** $V_{GS,max}$ (+21/-4 V) | ~+500-600 mV | ~+350-400 mV |
| <ul><li>no significant change in $I_{GSS}$ (low pA range)</li><li>no significant change in $V_{th,hyst}$ (=$V_{th,down}$-$V_{th,up}$)</li><li>untested golden (reference) devices that were subjected to the same pre-, intermediate, and post-measurements did not show any significant change in device parameters</li><li>$R_{DS,on}$ was almost unchanged according to low $\Delta V_{th}$</li><li>blocking capability ($I_{DSS}$) unchanged</li></ul> | | |

**Table 2**  Overview of the development of gate reliability parameters during the HV-GSS and GSS tests using the pre- and final post-measurements.

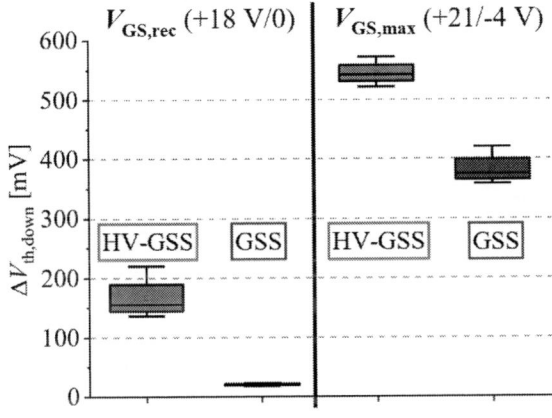

**Fig. 9**  Development of $V_{th,down}$ over the HV-GSS and GSS tests at $V_{GS,rec}$ and $V_{GS,max}$.

## 4  Conclusion and Outlook

In this paper, an HV-GSS test concept was introduced and tested that offers the possibility to perform accelerated gate switching stress tests at high drain voltage. It enables the execution of application-compliant gate-switching stress tests covering most of the important aspects concerning hard-switched applications (application-compliant gate and drain voltage waveforms, levels, and transients as well as junction temperatures). Simultaneously, a strong acceleration via switching frequency is achieved. This ensures that a large part of the device's lifespan can be covered.

A first test demonstrated that the devices subjected to an HV-GSS test show a stronger $V_{th}$ drift at both recommended and maximum datasheet gate voltages. The leakage currents, blocking behavior, on-state resistance, and threshold voltage hysteresis were not affected significantly.

The strong relative additional $V_{th}$ caused by the HV-GSS test in contrast to the GSS test suggests that the gate oxide reliability is influenced by drain voltage and should be considered when performing gate switching stress tests. However, the $V_{th}$ drift was in none of the tests critical in absolute terms. The datasheet limits were never exceeded, and a normal device operation was not threatened.

Nevertheless, all statements made only apply to the tested devices and their device concept, shielding approach, and gate oxide. Tests with different device designs, especially concerning the outcome of [3] should be performed to prove these statements for a wider range of different SiC MOSFETs. A following contribution that we publish at the ISPSD conference will deal with more manufacturers and device concepts.

## References

[1] P. Salmen et al. "A new test procedure to realistically estimate end-of-life electrical parameter stability of SiC MOSFETs in switching operation," *2021 IEEE International Reliability Physics Symposium (IRPS)*, Monterey, CA, USA, 2021, pp. 1-7, doi: 10.1109/IRPS46558.2021.9405207.

[2] Guideline for evaluating gate switching instability of silicon carbide metal-oxide-semiconductor-devices for power electronic conversion, JEP195, 2023.

[3] K. Yao, H. Yano and N. Iwamuro, "Investigations of UIS Failure Mechanism in 1.2 kV Trench SiC MOSFETs Using Electro-Thermal-Mechanical Stress Analysis," *2021 33rd International Symposium on Power Semiconductor Devices and ICs (ISPSD)*, Nagoya, Japan, 2021, pp. 115-118, doi: 10.23919/ISPSD50666.2021.9452281.

[4] T. Basler, D. Heer, D. Peters, T. Aichinger and R. Schoerner, "Practical Aspects and Body Diode Robustness of a 1200 V SiC Trench MOSFET," PCIM Europe 2018; *International Exhibition and Conference for Power Electronics, Intelligent Motion, Renewable Energy and Energy Management*, Nuremberg, Germany, 2018, pp. 1-7.

[5] R. Nakamura, Y. Nakano, M. Aketa, K. Noriaki and K. Ino, "1200V 4H-SiC Trench Devices," *PCIM Europe 2014; International Exhibition and Conference for Power Electronics, Intelligent Motion, Renewable Energy and Energy Management*, Nuremberg, Germany, 2014, pp. 1-7.

[6] D. Peters *et al.*, "Performance and ruggedness of 1200V SiC — Trench — MOSFET," *2017 29th International Symposium on Power Semiconductor Devices and IC's (ISPSD)*, Sapporo, Japan, 2017, pp. 239-242, doi: 10.23919/ISPSD.2017.7988904.

[7] D. Peters et al., "The New CoolSiC™ Trench MOSFET Technology for Low Gate Oxide Stress and High Performance,*" PCIM Europe 2017; International Exhibition and Conference for Power Electronics, Intelligent Motion, Renewable Energy and Energy Management*, Nuremberg, Germany, 2017, pp. 1-7.

[8] Guideline for evaluating Bias Temperature Instability of Silicon Carbide Metal-Oxide-Semiconductor Devices for Power Electronic Conversion, JEP184, 2021.

[9] P. Salmen et al. "Gate-switching-stress test: Electrical parameter stability of SiC MOSFETs in switching operation," Microelectronics Reliability, Volume 135, 2022, doi: 10.1016/j.microrel.2022.114575.

PCIM Europe 2024, 11– 13 June 2024, Nuremberg          DOI: 10.30420/566262090

# Silicon Carbide Power Device Use in Spacecraft and Aircraft

Akin Akturk[1] , Ethan Mountfort[2], Christopher Darmody[3], Mitchell Gross[4], Bryce Galey[5], Usama Khalid[6], Neil Goldsman[7]

[1-7] CoolCAD Electronics, LLC, USA

Corresponding author:    Akin Akturk, akin.akturk@coolcadelectronics.com
Speaker:                 Akin Akturk, akin.akturk@coolcadelectronics.com

## Abstract

To enable high power systems in future spacecraft and aircraft, radiation tolerant high voltage power switches are required. These radiation hardened devices are needed to address the capability of future performance goals of a) developing basic power building blocks for multiple applications, and b) distributing power at increased voltage to lower overall power system mass. Among alternatives, silicon carbide power devices offer a possible solution; however, the safe operating areas of these devices are needed to be quantified. To this end, we present the existing limits of this technology, and our solutions for higher altitude and space use versions of the silicon carbide technology.

## 1    Introduction

Power devices fabricated using the silicon carbide semiconductor are poised to benefit many power electronics applications by offering a low on-resistance high voltage power switch, having the most mature fabrication technology and the high-quality wafer production developed for a wide bandgap semiconductor. One such array of applications is linked to electrical power conversion and regulation systems needed for spacecraft and aircraft; however, unlike similar systems working at the sea level, these higher altitude and outer space applications require components that tolerate various forms of increased radiation relative to the ground-level, present at the point of use.

### 1.1    Spacecraft Power Needs

To achieve massive weight and volume savings in space missions and newly developed electric aircraft, radiation tolerant high voltage and power devices are needed. These devices would pave the way for higher voltages and frequencies in power converters and power distribution systems. For example, the international space station (ISS) has the highest power (~100 kW) for a spacecraft. It also has the largest space power distribution system with eight interleaved micro-grids. The planned space stations such as Gateway in lunar orbit, and the future lunar and planetary surface missions for establishing bases for example on the lunar surface and eventually on Mars require low mass and high efficiency modular / interchangeable power electronic regulators. For these new generation power conversion systems that can result in more efficient scale-up of power systems, high voltage power devices are needed to be developed.

Silicon carbide high voltage power devices would provide a solution for increasing voltage levels in spacecraft, and lunar and planetary base power distribution networks. This is likely to increase overall efficiency and mission lifetimes. It is also likely to decrease design complexity, and lower power distribution losses through use of high voltages. In addition, use of SiC power converters is likely to reduce cooling needs. Through use of high voltages and possibly less cooling, radiation tolerant silicon carbide power devices are likely to lead to overall power system weight and cost reductions, allowing more space and power for the critical instrument payload.

In the case of spacecraft, a recent NASA Technology Roadmap for the Space Power and Energy Storage category [1] states that the current state of the art for a space radiation hardened power distribution component is limited to <200V after derating due to semiconductor limitations. To overcome this limitation and to achieve the technology performance goal of >300V for the derated semiconductor operating voltage, new technologies are required.

An example of an immediate benefit of space hardened high voltage parts with voltage ratings >300V is the enabling of next generation high-power electric propulsion systems beyond their current limit of approximately 5kW. Use of a wide bandgap semiconductor such as silicon carbide is also likely to increase their efficiency beyond that which is currently achieved in these systems, which is approximately >92%. For instance, a theoretical study on a piloted electric vehicle to an asteroid shows that a possible use of 300V solar arrays instead of the 120V option for solar electric propulsion would decrease the payload by approximately 2.5 tons with larger voltage operation resulting in weight cuts, as mentioned in [2].

Roughly speaking, the volume of a conductor that needs to be employed in an electrical harness is inversely proportional to the square of the rated voltage (for the same power). A transition from a 100V to a 300V system would result in a 9x reduction in the attendant electrical harness weight, which can be a substantial part of the overall weight of a high-power system.

Even though power devices such as MOSFETs and diodes with voltage ratings more than 1700V exist for terrestrial high voltage and power applications, their space environment tolerant counterparts lack in availability, as well as voltage and current ratings. The state-of-the-art technology for space ready parts is <200V, which is significantly lower than the voltage rating of a ground based equivalent component. This results in challenges in circuit design and gives rise to less reliable and heavier systems. One of the consequences of the lack of high voltage (and power) devices with approximately >200V is that even the power system of an advanced spacecraft such as that used in International Space Station (ISS) is practically limited to roughly 100-150kW due to circuit complexity, weight, and other concerns. Extending this power level in ISS and other spacecraft necessitates higher voltage ratings. For high voltage applications such as thrusters, electric propulsion systems, and ISS power channels, the current blocking voltage rating needs to be increased to several hundreds of volts or even higher to enable integration and implementation of next generation energy efficient and reliable high voltage and power systems into these applications. The availability of high power and voltage components would obviate the need to build power systems by connecting many modules composed of low voltage rated parts in series or parallel. The current practice of increasing the voltage blocking capability and the current drive by connecting existing modules in series and parallel, respectively, gives rise to increased system failures in time, real estate needs and weight due to a greater circuit complexity and a larger number of parts used.

## 1.2 Aircraft and Sea Level Radiation

In the terrestrial environment, the Earth's atmosphere provides shielding against space radiation. Nevertheless, cosmic showers consisting of cosmic rays and solar particles manage to penetrate the atmosphere, interacting with gases in the atmosphere to generate secondary neutron particles. These secondary terrestrial neutrons pose notable risks to electronics in commercial and military aircraft as well as ground vehicles.

Primary cosmic rays in the Earth's atmosphere are a result of high-energy particles, mostly protons and alphas, coming from outer space or the trapped proton belt, penetrating our atmosphere and colliding with gases in the upper atmosphere. This generates several energy-rich products such as neutrons, and smaller pions and muons. The resulting interactions give rise to a cosmic shower [3]. Among the cosmic shower constituents, neutrons pose the greatest threat to electronics since they are relatively heavy, have relatively large capture cross sections with other atoms, and are highly penetrating.

The high energy neutrons lose their energy through elastic and inelastic collisions, and finally are captured as they travel through materials including a semiconductor device such as a power MOSFET. The high energy neutrons can occasionally collide with lattice atoms and knock them off their sites. These charged knock-on atoms are very efficient in creating ionization trails along their trajectory. These can then in turn result in a power MOSFET failure in the form of a gate rupture or a burn-out.

Wide bandgap (WBG) semiconductor devices such as those made from silicon carbide have the potential to offer a revolution in building high efficiency power and energy systems, due to their lower losses and high temperature operation compared to silicon at high power and voltage levels. The fabrication of silicon carbide (SiC) devices has the most maturity among wide bandgap (WBG) material processes, and therefore it is the most likely candidate in demonstrating the energy efficiency of WBG devices in real world applications. However, unlike silicon, reliability of SiC devices is a big unknown for power and energy system designers since the reliability problems due to their usage in long term applications are relatively undetermined. To enable their integration into energy efficient high-power systems, and therefore reducing waste of energy, money and time in these systems, firstly requires knowing the reliability problems that may arise due to their usage [4, 5].

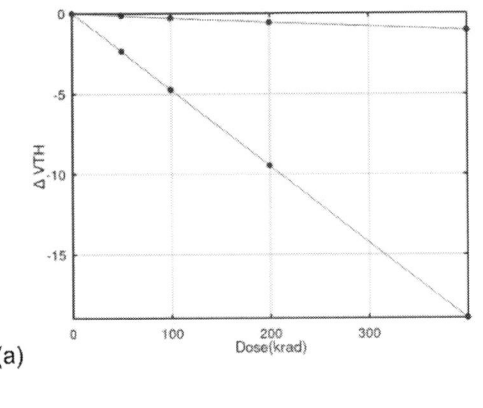

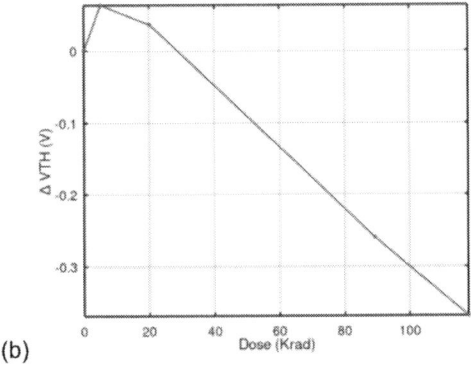

(b)

**Fig. 1** (a) Threshold voltage shift as a function of total ionizing dose: Silicon carbide versus silicon metal-oxide-semiconductor field effect transistor (MOSFET). (b) Total ionizing dose versus threshold voltage shift of our new generation power MOSFETs.

## 2 Spacecraft Use Considerations

Space readiness of a component requires its tolerance to the following three radiation effects:

### 2.1 Total Ionizing Dose (TID)

Total ionizing dose is related to the ionization of the material (creation of electron and hole pairs) due to the absorption of incident high energy electromagnetic waves or charged particles. It is measured by absorbed energy per unit mass. Its unit rad, which is commonly used to describe TID effects in semiconductors, is defined as 100 ergs of energy absorbed by a gram of material.

For power MOSFETs, the main effect on electrical parameters is the decrease in threshold voltage of n-channel devices by absorbed dose. As shown in Fig. 1(a), hole trapping efficiency of the silicon dioxide grown on silicon carbide is less than that of the silicon dioxide grown on silicon, rendering SiC power MOSFET more tolerant of this effect.

Here, silicon carbide power MOSFET threshold voltage shift is modeled after our earlier experiments [6]. As commercial silicon carbide power MOSFETs become available, we report the total ionizing dose response of these devices, and the effects of the radiation on the device parameters at room temperature and 125C. The threshold voltage shifts with radiation are explained in terms of electron/hole generation and trapping as in $Si/SiO_2$. The hole trapping efficiency of $SiC/SiO_2$ is calculated to be approximately 5%.

Later, we perform radiation studies on newer generation SiC power MOSFETs at Cobalt 60 gamma facilities. In these studies, the threshold voltage shifts are observed to decrease compared to those measured using the first-generation devices. We mostly attribute these changes in threshold voltage shifts to changes in gate oxide thickness.

Hole trapping efficiencies of unhardened $Si/SiO_2$ systems are known to be relatively large, indicating large threshold voltage shifts with increasing dose, as shown in Fig. 1(a).

In our new generation MOSFETs, the attendant threshold voltage shift as a function of TID dose is relatively modest even at the highest dose of approximately 120Krad(Si). Here, a Cobalt-60 gamma source is used. The dose rate is measured to be approximately 2rad/s. Furthermore, during measurements, the gate-to-source bias is kept at 20V, and the drain-to-source bias is kept as zero to give rise to maximum threshold voltage shift as a function of dose. At certain dose intervals, we interrupt the test and re-measure the devices to observe the new threshold voltage shift.

Here we see that as TID increases, in general, the threshold voltage of the MOSFET shifts negative, where after approximately 120Krad(Si), the threshold changes by less than 0.4V. Additionally, after finishing dosing the MOSFETs, we allow the devices to anneal at room temperature to determine if the threshold voltages recover. We again measure the devices after several days annealing at room temperature, but the devices show little to no meaningful recovery at room temperature.

We note that these threshold voltage shifts indicate an excellent TID response for a device with a gate insulator, especially when compared to its silicon variants.

## 2.2 Displacement Damage

Displacement damage is due to the displacement of lattice atoms within the target material by incoming particles such as protons. The consequence of this is the creation of defects and trapping sites within the material. The displaced atoms become interstitials and leave behind vacancies. Over time, this phenomenon can significantly affect material properties and the electrical performance of devices.

In the case of silicon carbide, the displacement damage threshold is more than $10^{12}$ protons or neutrons per cm$^2$, which is higher than that of silicon, giving rise to less leakage increase with particle fluence.

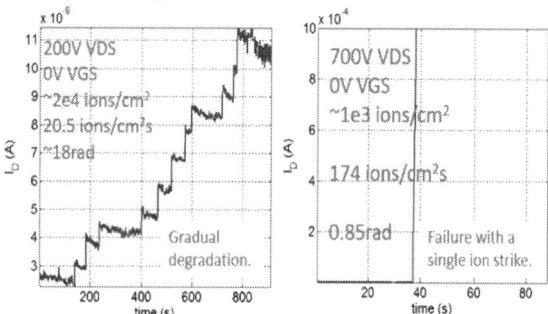

(a)

(b)

**Fig. 2** (a) A heavy ion irradiation test we previously pursued at Texas A&M University (TAMU) cyclotron facility shows that commercially available SiC power devices suffer from terminal current degradations due to ion strikes at relatively low voltages. At high voltages, SiC power devices experience sudden failure events. (b) Heavy ion induced damage is visible at the upper left of the power device.

## 2.3 Heavy Ion Single Event Effects

Silicon carbide power MOSFETs show excellent TID and displacement damage tolerance. However, silicon carbide power MOSFETs and diodes are very susceptible to heavy ions, characterized by deposited energy per unit length (also known as linear energy transfer, or LET). Figure 2 shows some of our previously measured data. We have an extensive experience in performing a variety of such radiation experiments at various facilities.

The metric that is used to predict how the heavy ions affect a device is the LET (linear energy transfer) with units of MeVcm$^2$/mg. LET is a measure of how much differential energy the heavy ion transfers (dE) to the surrounding material for a given incremental distance traveled (dx). Generally, for a given starting energy, the LET of a particle increases the heavier it is, and also increases the denser the material it is traveling through becomes. This increase in LET due to particle mass and surrounding material density also decreases the range of the particle since its energy is lost more quickly compared to lighter particles and/or less dense surrounding material.

Single event burnout (SEB) and single event gate rupture (SEGR) are two different catastrophic failure mechanisms, associated with single event transients. They are both the result of radiation-induced high current states in the device which occur suddenly and can result in a catastrophic failure. The high current state is a consequence of the ionization trail of the heavy ion, triggering positive feedback within the device. This gives rise to a self-sustaining high current state in the presence of high field until device damage.

In silicon carbide power MOSFETs, the attendant high LET heavy ion damage is found to change in character with bias during irradiation. In the high bias range, which starts at approximately larger than one third of the breakdown or avalanche voltage, a single high LET heavy ion strike in the active volume is usually sufficient to irreparably damage the device due to burnout. At medium biases, degradations or increases in drain and gate currents are observed, with or without equal degradation amounts mirroring at both terminals, but nonetheless correlating well with the heavy ion fluence [7, 8].

These degradations remain present even after the removal of the ion beam. For some devices that survive irradiation but exhibit some degradations in current during irradiation, device failures that are usually in the form of increased gate leakage are observed when performing post irradiation gate stress tests. At low biases that are as low as 10-20 % of the rated voltage (~8-16 % of the avalanche voltage), a threshold is found, below which no measurable effects are observed after irradiation. This clearly dramatically limits the safe operating voltage of such silicon carbide power MOSFETs in most space environments.

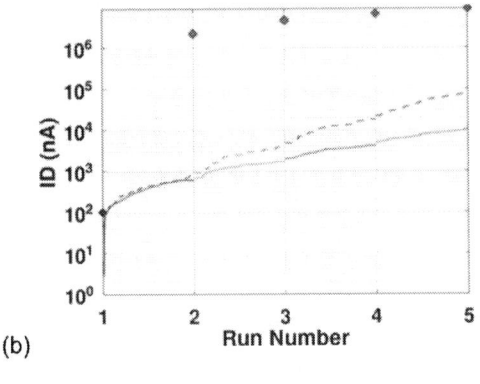

(b)

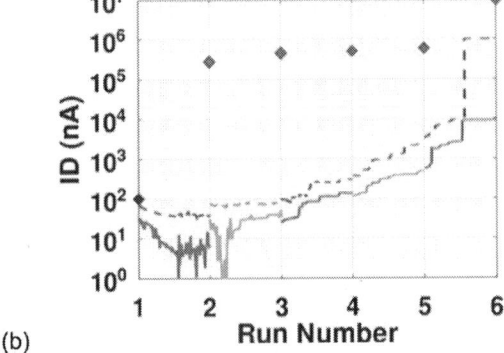

(b)

**Fig. 3** (a) During each run, the bias is increased from 700V to 1kV in 100V steps. LET(SiC) ~ 20MeVcm²/mg (b) During each run, the bias is increased from 800V to 1.1kV in 100V steps. LET(SiC) ~ 10MeVcm²/mg. During each run, the fluence is 100Kions/cm².

Similar degradation mechanisms are also observed in silicon carbide power diodes. As bias is increased beyond a threshold value such as that observed in MOSFETs, degradations in terminal currents become apparent. This is followed by a catastrophic device failure at higher biases.

Comparison between different types of vertical SiC power devices indicate a threshold field above which a catastrophic damage occurs irrespective of the device design. This indicates a damage mechanism that is purely a function of electric field, and possibly driven by electrical-thermal effects. At lower voltages, a drain or anode current degradation can also be related to a thermal initiated process that is large enough to cause permanent damage but small enough to remain localized. In the case of MOSFETs, gate damage is also related to a temporary increase in oxide field,

resulting in physical damages that might be explained by the percolation theory, latent ionizing and trapping induced electrical drifts, and subsequent increased gate leakage currents.

Catastrophic damages due to burnout and gate rupture events cannot be tolerated in a power device. When these failures occur, all terminal voltages short to each other. A typical commercially available 1.2kV silicon carbide power device has a burnout threshold of approximately 500V. It usually shows catastrophic failures even at lower voltages. The experiments also indicate that this threshold is approximately the same in higher voltage rated parts. For example, a previous generation 3.3kV device shows a marginal improvement to 600V [8]. Considering of a conservative derating factor that is as high as 50%, such a device becomes appropriate for use at voltages less than 300V.

To address the need for high power silicon carbide power devices that are radiation tolerant, we work on improving the radiation tolerance, and more specifically the burnout threshold of silicon carbide power devices. The goal is to achieve tolerance to heavy ion LET of at least 40MeVcm²/mg and fluence of at least $10^5$ions/cm² up to the 300V bias after derating. Figure 3 shows our earlier results indicating a burnout threshold more than 1kV for heavy ion LETs as high as 20. The later results indicate a burnout threshold of slightly less than 1kV for higher LET values.

## 3 Aircraft Use Considerations

High altitude readiness of a component requires its tolerance to atmospheric neutrons. The damage neutrons induce on a device is mainly via knocking off lattice atoms, which from that point on behaves similar to a low LET ion strike.

Atmospheric neutrons are known to cause failures in silicon carbide power devices at sea level and higher altitudes. Our earlier investigations show that these failures are a result of fast heating in a small filament, initiated by an inelastic or elastic collision of a neutron-lattice atom. Each of these collisions deposit a charge within the device, and when deposited beyond a critical value for a given bias would give rise to a failure.

As the incident neutron knocks one of the lattice atoms off its site, the knock-on atom deposits charge along its path, giving rise to an ionization trail in a device biased in its off state. The interaction of these deposited electron-hole pairs with high internal electric fields results in a localized fast Joule heating, and consequently lattice damage and device failure.

This kind of failure is the same as that observed during heavy ion burnouts. Burnout events in similarly rated silicon power devices are usually due to a different failure mechanism that involves activation of a parasitic bipolar transistor. Because of this, Si power devices exhibit high failure rates at high biases; however, the failure rates drop rapidly as the parasitic bipolar transistor activation ceases. Figure 4 shows a comparison of failure rates of silicon and silicon carbide power devices. It also shows that a design and process change can suppress failure rates at lower biases.

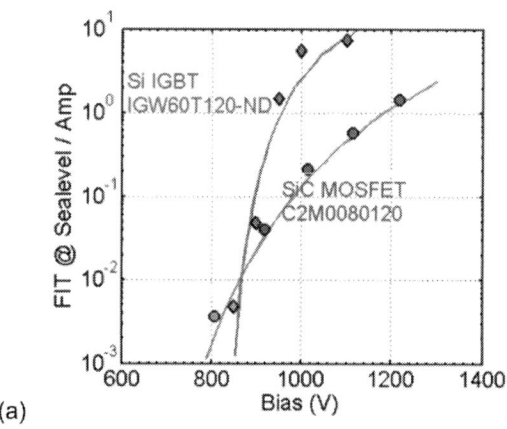

(a)

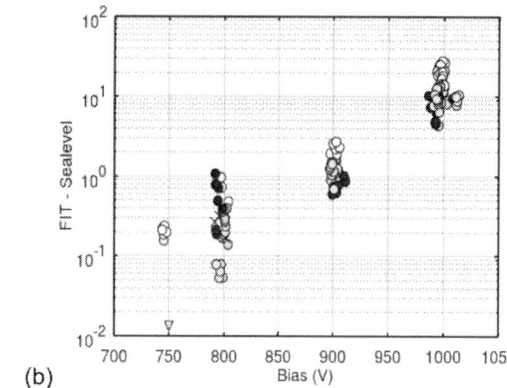

(b)

**Fig. 4** (a) Example failure rate of a silicon carbide MOSFET compared to a similar silicon device. (b) Our designs indicate improved failure rate values at lower voltages. Here circles indicate actual failures. The triangle shows an upper limit on failures. It does not indicate an actual failure.

## 4 Conclusion

To pave the way for the use of silicon carbide power devices in high altitude environments, a thorough understanding of the effects of radiation on silicon carbide power components needs to be achieved. This is for risk assessment and mitigation. Here, we present the limits of silicon carbide power device use in spacecraft and aircraft. We also present our work on improving the tolerance of silicon carbide devices to these effects.

Silicon carbide power devices compared to their silicon parts exhibit excellent TID and displacement damage tolerance. However, the burnout threshold in silicon carbide power devices due to heavy ions is relatively low. Likewise, they exhibit excellent failure rates at high biases due to atmospheric neutrons; however, they have a tail that extends into lower voltages. As shown by our work, there are ways to achieve radiation tolerance in these devices through design and processing.

## References

[1] NASA Space Technology Roadmaps and Priorities: Restoring NASA's Technological Edge and Paving the Way for a New Era in Space. Washington, DC: The National Academies Press. https://doi.org/10.17226/13354.

[2] C. R. Mercer et al, "Benefits of Power and Propulsion Technology for a Piloted Electric Vehicle to an Asteroid", AIAA SPACE 2011 Conference & Exposition, AIAA–2011–7252 (2011). https://doi.org/10.2514/6.2011-7252.

[3] A. Akturk et al, "Terrestrial neutron-induced failures in silicon carbide power MOSFETs and diodes," IEEE Transactions on Nuclear Science 65(6), 1248-1254 (2018).

[4] D. J. Lichtenwalner et al, "Reliability of SiC power devices against cosmic ray neutron single-event burnout," Proceedings of Int. Conference on Silicon Carbide and Related Materials (ICSCRM), 559-562 (2017).

[5] A. Akturk et al, "Predicting cosmic ray-induced failures in silicon carbide power devices," IEEE Transactions on Nuclear Science 66(7), 1828-1832 (2019).

[6] A. Akturk et al, "Radiation effects in commercial 1200 V 24 A silicon carbide power MOSFETs," IEEE Transactions on Nuclear Science 59(6), 3258-3264 (2012).

[7] A. Akturk et al, "Space and terrestrial radiation response of silicon carbide power MOSFETs," Proceedings of IEEE Radiation Effects Data Workshop (REDW), (2017).

[8] J-M. Lauenstein et al, "Space Radiation Effects on SiC Power Device Reliability," 2021 IEEE International Reliability Physics Symposium (IRPS), 1-8, (2021), 10.1109/IRPS46558.2021.9405180.

PCIM Europe 2024, 11– 13 June 2024, Nuremberg    DOI: 10.30420/566262091

# Current Ripple Reduction by combination of Si IGBT and SiC MOSFETs in Heavy-Duty Fuel Cell Trucks

Yavuz Gürlek[1], Firat Yüce[1], Martin Ackerl[1], Roland Dold[1], Martin Neuburger[2], Thomas Basler[3]

[1] Daimler Truck AG, Germany
[2] Esslingen University of Applied Sciences, Germany
[3] Chemnitz University of Technology, Germany

Corresponding author:    Yavuz Gürlek, yavuz.guerlek@daimlertruck.com
Speaker:    Yavuz Gürlek, yavuz.guerlek@daimlertruck.com

The transformation to emission-free heavy-duty transportation on road requires the use of a fuel cell. The heavy-duty fuel cell vehicle is characterized by advantages such as a long range, fast refueling and thus a short charging time as well as environmental friendliness. However, the operation with power electronic components within the high voltage power net of a fuel cell vehicle leads to current ripples in the DC-link that generate high-frequency harmonics. These ripples have a profound impact on power quality, system efficiency, availability and can potentially destabilize critical components. Consequently, this study proposes methodologies for evaluating and mitigating current ripples within the propulsion system of a Heavy-Duty Fuel Cell Truck. In pursuit of this objective, an equivalent model of the fuel cell truck's propulsion system, characterized by real parameters, was created in *Simulink* and is presented in this paper. Utilizing real measurement data such as voltages, currents, and torques collected from various driving cycles of a fuel cell truck, we assessed the distribution of current ripples. Furthermore, this work encompasses techniques for reducing current ripples and optimization methods, along with the corresponding test bench results, validating our theoretical concepts. Furthermore, we investigate the optimization of switching losses in the inverter through the incorporation of various power semiconductor components.

## 1    Introduction

### 1.1    Structure of the HDFCT HV Power Net

sources in the HDFCT are HV battery system and two fuel cell systems, including HV auxiliaries like electric turbo chargers, humidity recirculation blowers and two Fuel Cell Converters - DC/DC

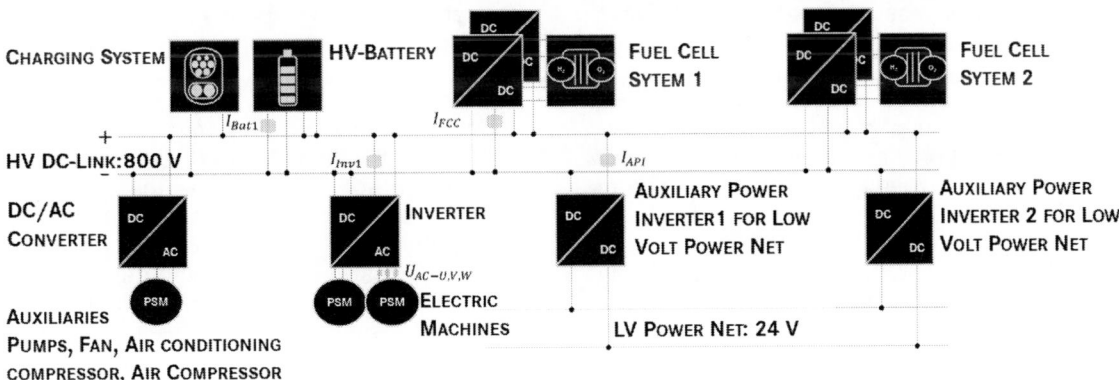

**Fig. 1** HV power net of the HDFCT and measurement setup with transducers (yellow)

The High Voltage (HV) power net of a Heavy-Duty Fuel Cell Truck (HDFCT), which is analyzed in this work, consists of three different energy sources and several consumers as shown in Fig. 1. Energy

boost converters (FCC). The third energy source is the electrical drive system in recuperation mode consisting of two Permanent Magnet Synchronous

Machines (PMSM) and a dual inverter. The remaining HV components support auxiliary functions for the braking, the thermal management system or supply electrical energy - Auxiliary Power Inverter (API – see Fig. 1) - to the low voltage power systems. The operation with power converters within the HV power net leads to current ripples in the DC-link [1], [2]. This was shown in our prior work, where we identified that the inverter generated the most pronounced current ripples. These detrimental effects must be reduced to guarantee a high efficiency and reduced ageing of components for the HDFCT's power net [3]-[6]. In real industrial environment, the hardware design of the entire vehicle takes place in the early phase of the development process. Therefore, a system analysis is essential.

## 1.2 Current Ripples at HDFCT's DC-link

To understand the interaction between the various components of the HV power net, different driving cycles of the HDFCT and measurements of the HV power net must be analyzed. To achieve this, recording the currents within the HV power net is essential to examine and understand the nature of current ripples. Existing literature extensively discusses the injection of current ripples into the DC-link, providing substantial knowledge in this domain. The impact of the switching frequency has not been investigated with regard to drive cycles yet. Battery Electric Vehicles (BEV) are modeled and the results are presented in various works [3]-[6]. In these publications, the focus is more on the impedance characterization and the influence of the power net impedance on current ripples, but the reduction of the current ripples is not considered. As fuel cell vehicles are promising for long haul transportation, deep power net analyses need to be performed also on the inverter level. Our previous analysis on HDFCT [1], [2] is used to establish the HV power net simulation and loss calculation by means of additional losses generated by current ripples in the HV power net. For instance, when considering the mission profile of a route from Stuttgart to Hamburg and back, the drive train inverter generates an additional energy consumption of 0.39 kWh due to current ripples in the high-voltage (HV) power net [2]. The novelty of this work is the reduction of current ripples in a real HDFCT considering the switching frequency by keeping the switching losses as low as possible. Current ripples must be considered in the design process. Besides this, the work also shows the advantages of using higher switching frequencies in the drivetrain inverter. The procedures are shown to illustrate the possibility of current ripple reduction in the DC-link, as they decrease the system efficiency due to increased electrical losses [2] or may cause stronger aging for example in the battery cells [3].

## 1.3 Semiconductors in a HDFCT

The considered vehicle is equipped with Silicon Insulated Gate Bipolar Transistor (Si IGBT) power semiconductors. With other device technologies, lower losses can be realized. The device losses of the IGBT and Silicon Carbide MOSFET (SiC) [5] are modeled and calculated according to measured drive cycles and a procedure is presented for the loss calculation to reduce the current ripples in the HV power net (Fig. 1). This is shown in section 3. SiC MOSFET devices are frequently considered for power electronic applications due to their distinctive material properties, encompassing a wide bandgap and high thermal conductivity. These attributes render SiC devices appealing for high-power, high-frequency applications. For example, several factors contribute to the potential reduction of diode losses in SiC devices due to the reverse channel-on mode possibility and low natural reverse-recovery charge ($Q_{rr}$), compared to traditional silicon devices [8]:

- Faster Switching Speed: SiC devices, along with their integrated diodes, often demonstrate faster switching speeds compared to silicon devices. This leads to shorter transition times during switching.

- Reduced reverse recovery time: Traditionally, silicon diodes have a higher reverse recovery time due to high plasma stored charge.

- Higher temperature stability: SiC devices can operate at higher temperatures than silicon devices. Switching losses will stay almost constant with temperature. This characteristic can result in enhanced overall efficiency in high-temperature applications, but the conduction losses need to be considered carefully, as the $R_{DS,ON}$ is strongly temperature dependent and its temperature coefficient worse compared to Si devices. It is noteworthy that while SiC devices offer advantages in terms of reduced losses, they may pose their own set of challenges and considerations, such as cost, gate drive and package requirements, and specific application needs. The choice between SiC and Si devices depends on the requirements and constraints of a given application.

# 2 Methods

## 2.1 Measurements with HDFCT

Several tests with the HDFCT were carried out on the test road track. A series of driving sessions involving accelerations and braking maneuvers were executed to identify the specific maneuvers leading to the highest current ripples.

| Test | Setup |
|:---:|:---:|
| 1 | **Initial software:** <br> Switching frequency of 6.4 kHz |
| 2 | **Modification of modulation scheme:** <br> Changing from SVPWM to DPWM |
| 3 | **Modification of switching frequency:** <br> Increasing switching frequency to 10 kHz |

**Table 1** Driving cycles with HDFCT

Drive tests, which were carried out for the analysis, are shown in Table 1. The performed measurements of the HDFCT are recorded with 200 kHz resolution transducers at full throttle acceleration and deceleration with maximum recuperation. The first test is made with the initial setup of the HDFCT as a reference. The second and third tests represent the HDFCT with software modifications in the drivetrain inverter: In test 2, the switching pattern is changed from Space Vector Pulse Width Modulation (SVPWM) to Discontinuous Pulse Width Modulation (DPWM) and in test 3, the switching frequency is increased from $f = 6.4\,\text{kHz}$ to $f = 10\,\text{kHz}$.

## 2.2 Simulation of Inverter Losses

For calculating the device losses of the power modules of the drivetrain inverter, an additional simulation model is implemented. An inverter model is characterized, which can analyze the switching, conduction and total losses. Finally, the total losses of the inverter are shown with two different semiconductor types: Si IGBT and SiC MOSFET. Also here, conditions of Table 1 are used. The verification of the simulation model is made by calculating the same operation points with *IPOSIM* simulator, an online power simulation tool offered by *Infineon*.

# 3 Results and Discussion

## 3.1 Excess Losses in HV DC-link

Equation 1 describes the supplementary losses induced by current ripples within the HV power network. These additional losses are defined as excess losses and the equation 1 serves to compute the excess losses within a given mission profile. The excess losses of all test series are calculated according to [2]:

$$P_{\text{V,Excess}} = I_{\text{RMS,AC}}^2 \cdot R \qquad (1)$$

$I_{\text{RMS,AC}}$ represents the current ripple of the DC current and $R$ is the characterized system resistance of the HDFCT, see Fig. 2. Each test of Table 1 corresponds to its excess losses by equation 1 depicted in Fig. 2.

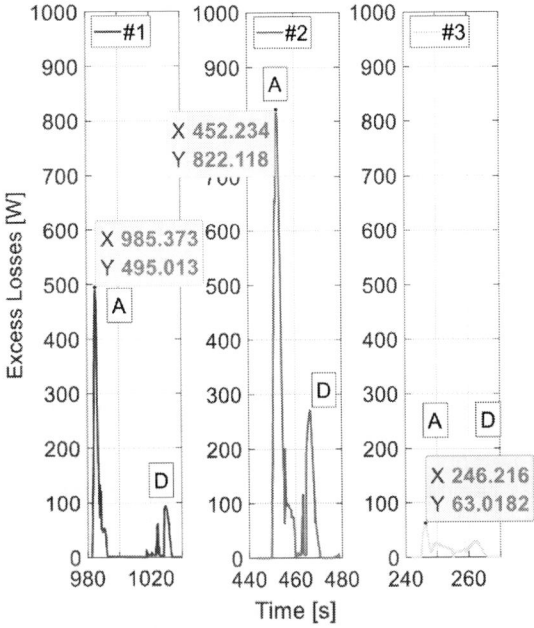

**Fig. 2** Calculated excess losses of test series 1-3 from Table 1, where A presents the acceleration phase and D the deceleration phase

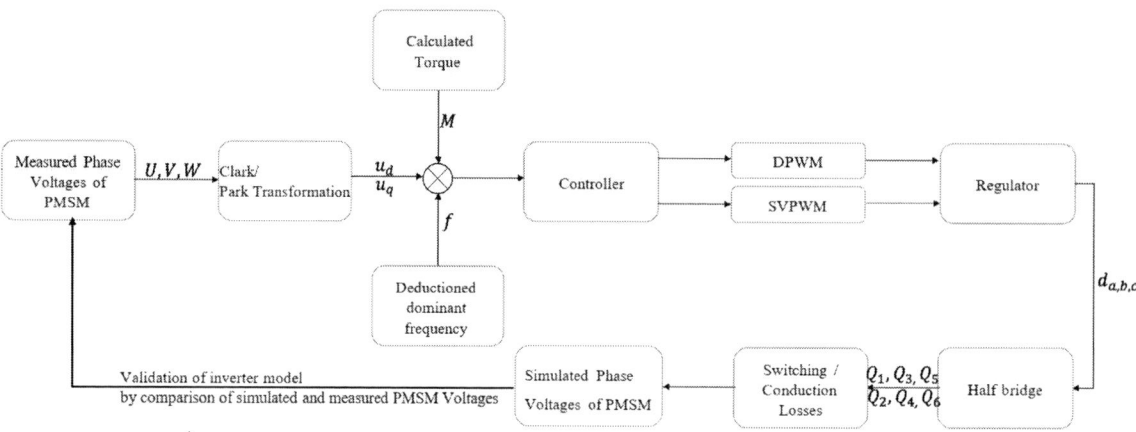

**Fig. 3** Control diagram of proposed method

Based on calculation results, the setup, which was used in test 3, shows the best results in terms of generated excess losses at the HV power net which are the smallest, particularly:

$$P_{V,Excess,Test\ 3} = 63\ W.$$

In comparison, the worst results are generated by test 2 and will be neglected in further investigations:

$$P_{V,Excess,Test\ 2} = 822.1\ W.$$

The physical explanation of these effects are discussed in [1]-[6].

## 3.2 Higher Switching Frequency of Drivetrain Inverter and Device Loss Calculation

This study encompasses the validation of switching and conduction losses in the device, a process carried out through the utilization of *IPOSIM*. *IPOSIM* is designed for the precise computation of losses and thermal characteristics for *Infineon* power modules. The maximum achievable switching frequency of a power device is depending on the switching losses. The total switching loss can be calculated according to equation 2:

$$P_{on} + P_{off} = f \cdot (E_{on} + E_{off}) \qquad (2)$$

where $f$ represents the switching frequency. In practical applications, the energy loss per pulse is typically extracted from oscillograms. Conduction losses, as well as blocking losses, contribute to the overall losses. In the case of power MOSFETs and lower voltage-class IGBTs, the off-state leakage current is typically in the range of a few microamperes allowing us to disregard blocking losses. However, conduction losses cannot be neglected. These losses can be computed by defining the duty cycle, denoted as $d$ which represents the ratio of the time during which the MOSFET conducts relative to the entire switching period. The calculation for total losses is as follows:

$$P_v = P_{cond} + P_{on} + P_{off} = d \cdot R_{on} \cdot I_{DS}^2 + \qquad (3)$$
$$f \cdot (E_{on} + E_{off})$$

whereas $R_{on}$ represents the on-state resistance and $I_{DS}$ the drain source current of a MOSFET device. These losses need to be dissipated as heat flux through the device's package. The maximum permissible losses are determined by the cooling conditions, the acceptable temperature difference, and the thermal resistance. The equation 3 is used in this work to determine the losses in the devices, which are calculated with the presented simulation model and *IPOSIM*. In this section, an analytical loss calculation based on the measurement data is obtained from the propulsion inverter's drive cycle [5]. The data includes parameters such as DC voltage, DC current, and AC voltages. Initially, the loss calculation is focused on the IGBTs, since the given measurements are acquired from an inverter equipped with IGBTs. Subsequently, a simulation model is built up to assess and verify the accuracy of the model's thermal performance, output voltage and pulse pattern in comparison to real-world measurements. Following the IGBT analysis, a further simulation with a comparable module of SiC MOSFET technology is performed. This simulation includes the evaluation of switching losses in accordance with the mission profile. In accordance with the AC voltage deviation method, Fig. 3 illustrates the diagram for predicting DC-link current. The voltage deviation method derives from measured AC voltages of the

PMSM the pulse pattern of the inverter. In this section the Si IGBT (= device in the reference inverter) and the SiC MOSFET loss calculation by the self-built inverter model and *IPOSIM* by *Infineon* is derived. The detailed calculations and validation of switching and conduction losses for two specific devices are presented. Initially, one general challenging operational scenario within the acceleration mode is examined: low speed with high torque, where the peak to peak current ripple in the DC-link is high, according to Fig. 4. Subsequently, two different switching frequencies are analyzed, as per measured data, to comprehensively evaluate and quantify these losses for a mission profile, which is shown in Fig. 4.

| Input requirements | |
|---|---|
| Blocking Voltage | 1200 V |
| Modulation Algorithm | SVPWM [7] |
| DC-link Voltage | 700 V |
| Output Current in Analyzed Time Interval | 203 A |
| Load | 1636 Nm |
| Heat Sink Temperature | 25° C |
| Output Frequency | 40 Hz |
| M | < 1 |
| Switching Frequency | Test 1: 6.4 kHz, Test 3: 10 kHz (see Table 1) |

**Table 2**  Conditions for simulation extracted from Fig. 4 red dot for IGBT and SiC MOSFET device

|  | Switching Losses | | Conduction Losses | | Total Losses | |
|---|---|---|---|---|---|---|
|  | I [W] | II [W] | I [W] | II [W] | I [W] | II [W] |
| IGBT | 138 | 123.9 | 40.4 | 55.6 | 178.4 | 179.5 |
| Diode | 37.9 | 44 | 68.2 | 56.3 | 106.1 | 100.3 |

**Table 3**  Extracted losses from simulation during acceleration phase at given operation point (see red dot Fig. 4) IGBT loss simulation conditions given in Table 2 at 6.4 kHz and $T_J = 40\,°C$
I Simulated with *Matlab Simulink Simscape Electric*
II Simulated with *Infineon IPOSIM*

|  | Switching Losses | | Conduction Losses | | Total Losses | |
|---|---|---|---|---|---|---|
|  | I [W] | II [W] | I [W] | II [W] | I [W] | II [W] |
| IGBT | 183.8 | 197.4 | 67.1 | 55.6 | 250.9 | 253 |
| Diode | 52.1 | 70.5 | 70.1 | 56 | 122.2 | 126.5 |

**Table 4**  Extracted losses from simulation during acceleration phase at given operation point (see red dot Fig. 4) IGBT loss simulation conditions given in Table 2 at 10 kHz and $T_J = 40\,°C$
I Simulated with *Matlab Simulink Simscape Electric*
II Simulated with *Infineon IPOSIM*

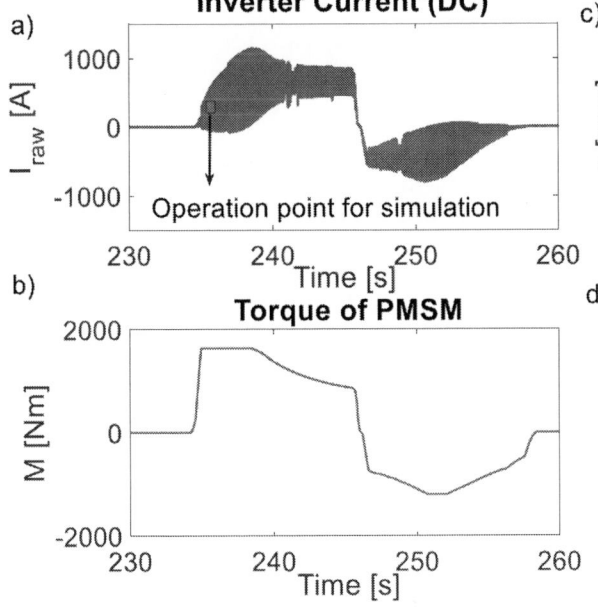

**Fig. 4** Measured data during HDFCT's mission profile

The switching losses and conduction losses are computed as per the specifications outlined in Table 2. The analyzed operation point of Fig. 4 red dot is the beginning of the acceleration phase. Therefore, the junction temperature $T_J$ of the device is at only 40° C. In Table 3 and 4, both types of losses are calculated using the model presented in this study (detailed time frame in Fig. 5) and with *IPOSIM*. As for the comprehensive evaluation of the total losses (diode + IGBT), this is addressed as follows:

$$P_{v,I} = 284.4 \text{ W}$$

and:

$$P_{v,II} = 279.8 \text{ W}.$$

The deviation is: 1.58 %, which is an adequate result. Also, for the calculation of Table 4 with higher switching frequency of 10 kHz, the results are competent:

$$P_{v,I} = 373.1 \text{ W}$$

and:

$$P_{v,II} = 379.5 \text{ W}.$$

The deviation is: 1.6 %, which is an adequate result. Based on the former comparisons, the self-built model is valid, according to *IPOSIM*. Since the examined segment of the mission profile does not encompass the entire mission of the HDFCT, further investigation is necessary. Current ripples in the HV DC-link can be reduced according to Fig. 2 by increasing the switching frequency. Table 4 shows a significant loss increase in the IGBT device for the 10 kHz case, which shows already a limitation for a pure IGBT solution. Therefore, the usage of SiC MOSFETs can be helpful to reduce the losses as already indicated in the right side of Fig. 5, where a pure SiC MOSFET solution with same current rating was used. Particularly under light load conditions, SiC MOSFETs are likely to demonstrate superior performance. Integrating both IGBTs and SiC MOSFETs into a hybrid module can optimize the use of each device according to the varying loads within the mission profile [8].

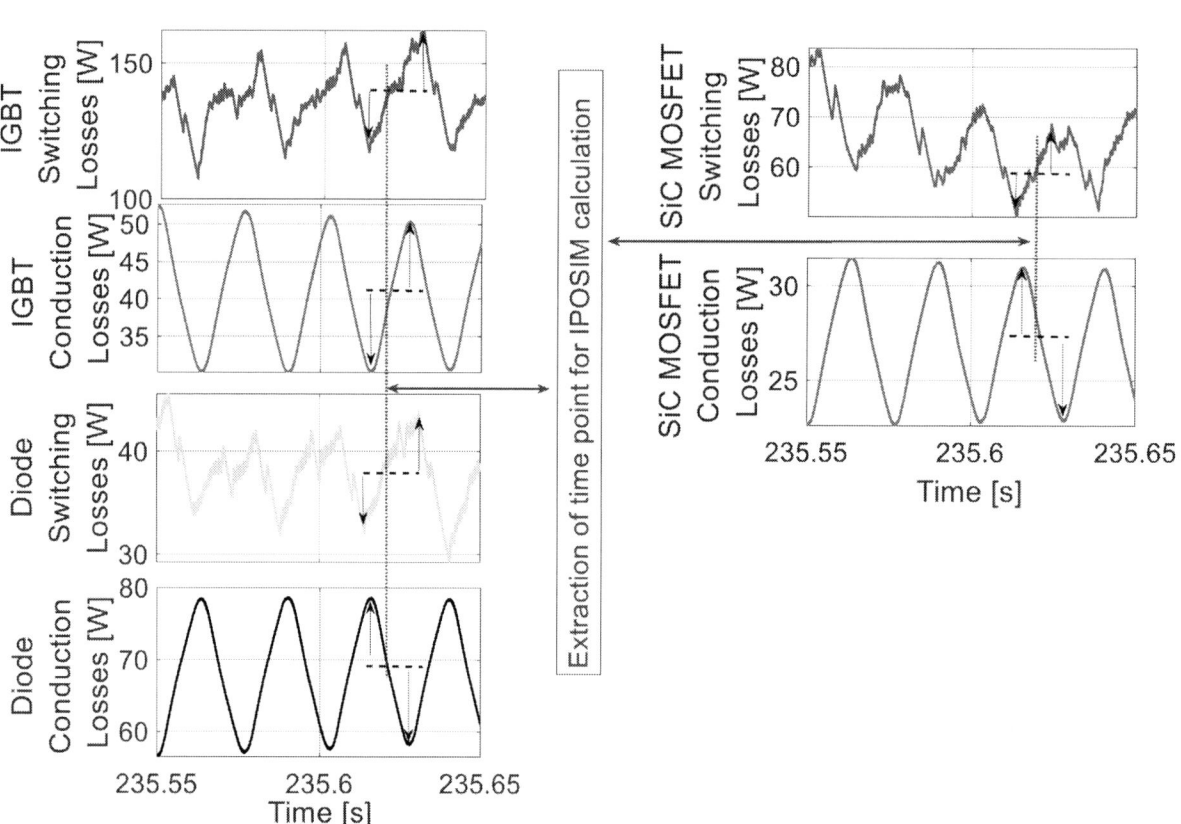

**Fig. 5** Loss calculation with own *Matlab Simulink Simscape Electric* Model for dedicated current during acceleration (red dot in Fig. 4); left: IGBT and diode switching and conduction losses; right: SiC MOSFET switching and conduction losses at $T_J = 40\ °C$

This hybrid strategy will be explored in greater detail in the subsequent section.

## 3.3 Hybrid Approach for HDFCT: Si IGBT and SiC MOSFET

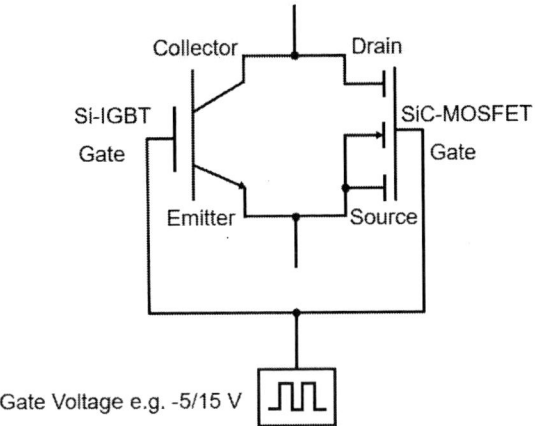

Fig. 6 Hybrid Approach according to [9] (only the Body Diode (BD) of the SiC MOSFET is used for free-wheeling)

In Fig. 6, a schematic of the hybrid module is shown. In the following, a comparison of the total losses for 1200V class devices is performed. The equivalent resistance at room temperature is significant for the used IGBT module with $R_{CE} = 2.25\ \mathrm{m\Omega}$ at the simulated current condition in table 1. For a similar rated full SiC power module the $R_{DSON}$ at room temperature is only $R_{DSON} =$

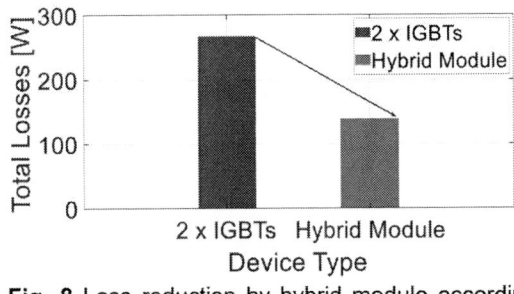

Fig. 8 Loss reduction by hybrid module according to analyzed operation point in Fig. 4

$2.13\ \mathrm{m\Omega}$. Consequently, the distribution of current is determined according to the on-state resistances of the respective devices. For hypothetical PSMSM which needs two power modules with a rating of ca. 600 A for a single phase (which is twice the original rated module from the HDFCT) the calculation between IGBT-only versus hybrid approach was done. That means for the IGBT-only solution 2 x 600 A were calculated and for the hybrid module a 600 A IGBT module in parallel to a similar rated SiC MOSFET module were studied for simplicity. On the left-hand side of Fig. 7, two paralleled Si IGBT modules are simulated. On the right-hand side, the hybrid is simulated and compared. The losses for the 2 IGBT-solution is in total: 266.5 W at the cursor-point of Fig. 7. The losses for the hybrid approach is in total: 138.4 W. The hybrid module has at the regarded operation point 35 % less losses in total [8], which is shown in Fig.

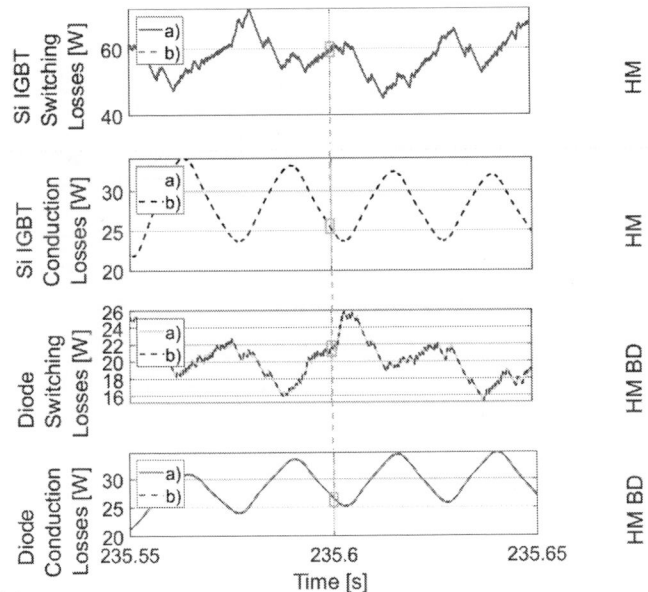

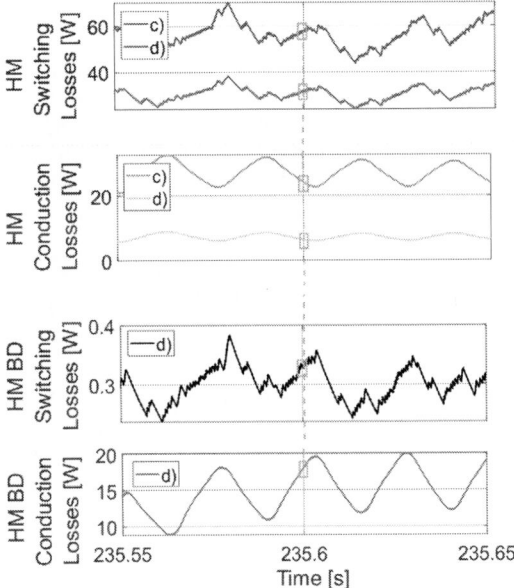

Fig. 7 Loss calculation with own *Matlab Simulink Simscape Electric* Model for dedicated current during acceleration red dot in Fig. 4 at .6.4 kHz switching frequency; left: 2 IGBTs: a) & b); right: Hybrid Module (HM) and its Body Diode (BD): c) IGBT d) SiC MOSFET: d) at $T_j = 40\ °C$ (starting point of mission profile); marked points are used for total loss calculation

7 and Fig. 8. In high-current scenarios, the hybrid module, with its Si IGBT component, can handle higher currents and maintain lower conduction losses compared to a standalone SiC MOSFET, thus boosting overload capacity. Further, we see reduced switching losses. Consequently, the overall switching performance and possible switching frequency of the hybrid module are substantially enhanced thus, potentially enabling lower DC-link current ripples. Another advantage is the lower cost compared to a pure SiC MOSFET solution [10]: In the hybrid module setup, a smaller, current-rated SiC MOSFET can be used.

# 4 Conclusion

Several operation points of the mission profile are calculated in this work with the self-built model and validated with *IPOSIM*. The Si IGBT devices have been analyzed and compared with SiC MOSFETs in the frame of a HDFCT. From this analysis, it is evident that Si IGBTs and SiC MOSFETs exhibit complementary features concerning conduction loss, switching loss, and device cost. Consequently, there's a logical inclination to merge the benefits offered by these two semiconductor devices. This hybrid system presents several advantages over using Si IGBT or SiC MOSFET individually like enhanced conduction characteristics: The hybrid module surpasses the pure Si IGBT or SiC MOSFET in terms of conduction. When compared to a single Si IGBT, the hybrid module does not have a threshold in forward voltage due to its parallel SiC MOSFET. This results in a lower forward voltage at low currents. Thus, through comprehensive analysis and comparison, the motivation to explore the hybrid module in more detail is grounded in its ability to provide a cost-effective solution for achieving high efficiency and power density in power electronic systems for HDFCT to reduce current ripples in the HV DC-link and decrease at the same the device losses even at higher switching frequencies.

# References

[1]   Y. Guerlek, M. Ackerl, M. Neuburger, R. Dold, H. Xu, T. Basler, "Characterization of High Voltage Power Net in Heavy Duty Fuel Cell Truck with Focus on Current Ripples. Berlin", Offenbach: VDE Verlag GmbH, 2023

[2]   Y. Gürlek, O. Joos, C. Lang, M. Ackerl, R. Dold, M. Neuburger, T. Basler, "Analysis on the Impact of Current Ripples in Fuel Cell Electric Heavy-Duty Trucks," in 2023

25th European Conference on Power Electronics and Applications (EPE'23 ECCE Europe), Aalborg, Denmark, 2023, pp. 1–8.

[3]   M. Gentejohann, M. Schlüter, and S. Dieckerhoff, "Analysis and experimental verification of the high voltage DC-link current and voltage ripple in electric vehicles," IEEE.

[4]   M. Gentejohann, M. Schluter, S. Dieckerhoff, and M. Hepp, "Driving Cycle Analysis of the DC-link Current Ripple in Electric Vehicles." Piscataway, NJ: IEEE, 2021. [Online]. Available: https://ieeexplore.ieee.org/servlet/opac?punumber=9570185

[5]   M. Schlüter, M. Gentejohann, and S. Dieckerhoff, "Driving Cycle Power Loss Analysis of SiC-MOSFET and Si-IGBT Drivetrain Inverters for Electric Vehicles," in 2023 25th European Conference on Power Electronics and Applications (EPE'23 ECCE Europe), Aalborg, Denmark, 2023, pp. 1–11.

[6]   C. Sagert, B. Rolle, M. Walter, and O. Sawodny, "Current and Voltage Harmonics in the Powertrain of Electric Vehicles," IEEE Trans. Veh. Technol., vol. 69, no. 10, pp. 10736–10749, 2020, doi: 10.1109/TVT.2020.3008385.

[7]   Q. Li and D. Jiang, "DC-link current analysis of three-phase 2L-VSI considering AC current ripple," IET Power Electronics, vol. 11, no. 1, pp. 202–211, 2018, doi: 10.1049/iet-pel.2017.0133

[8]   H. Liu, J. Zhou, T. Zhao and X. Xu, "Si IGBT and SiC MOSFET Hybrid Switch-Based Solid State Circuit Breaker for DC Applications," 2022 IEEE Energy Conversion Congress and Exposition (ECCE), Detroit, MI, USA, 2022, pp. 1-6, doi: 10.1109/ECCE50734.2022.9948172.

[9]   M. Rahimo et al., "The Cross Switch "XS" Silicon and Silicon Carbide Hybrid Concept," Proceedings of PCIM Europe 2015; International Exhibition and Conference for Power Electronics, Intelligent Motion, Renewable Energy and Energy Management, Nuremberg, Germany, 2015, pp. 1-8.

[10]  F. Kayser, R. Baburske, P. Brandt, U. Queitsch and H. -G. Eckel, "Hybrid Switch with SiC MOSFET and fast IGBT for High Power Applications," PCIM Europe digital days 2021; International Exhibition and

Conference for Power Electronics, Intelligent Motion, Renewable Energy and Energy Management, Online, 2021, pp. 1-6.

PCIM Europe 2024, 11– 13 June 2024, Nuremberg    DOI: 10.30420/566262092

# Evaluation of Active Gate Drivers with Switchable Gate Resistors and Intermediate Voltage Levels for SiC MOSFETs in WLTC

Michael J. Frank ©[1], Mark-M. Bakran ©[1]

[1] University of Bayreuth, Centre for Energy Technology - ZET, Department of Mechatronics

Corresponding author:     Michael J. Frank, Michael.Frank@uni-bayreuth.de
Speaker:                  Michael J. Frank, Michael.Frank@uni-bayreuth.de

## Abstract

This paper explores the impact of employing an intermediate gate voltage level in an active gate driver on the performance of a SiC MOSFET inverter during the Worldwide Harmonized Light Vehicles Test Cycle (WLTC) class 3. It evaluates the effectiveness of this gate driving approach compared to a gate driving method employing switchable gate resistors. The study includes detailed loss calculations that are substantiated through experimental verification using an H-bridge in continuous operation mode.

## 1 Introduction

Wide-bandgap semiconductors, such as SiC MOS-FETs, are particularly valued for their reduced conduction and switching losses [1]. However, the choice of gate resistors ($R_G$) and consequently the switching speed is often dictated by the most demanding operational conditions. This can lead to suboptimal efficiency in applications with fluctuating operating conditions, such as automotive traction inverters. Under typical conditions, like those found in the Worldwide Harmonized Light Duty Test Cycle (WLTC), partial load states are prevalent, further complicating efficiency optimization. Active Gate Drivers (AGD) offer a potential solution by dynamically adjusting the switching speed to align with prevailing load conditions. In AGDs, any parameters influencing switching speed can be adjusted to optimise the switching behaviour. The turn-on and turn-off speeds, represented by the drain current slew-rate $\left(\frac{di_D}{dt}\right)_{on/off}$ and the drain-source voltage slew-rate $\left(\frac{dv_{DS}}{dt}\right)_{on/off}$, assuming a constant Miller voltage ($V_{Mil}$), are described by the following equations [2]:

$$\left(\frac{di_D}{dt}\right)_{on/off} = \frac{v_{GG} - V_{Mil}}{C_{iss} \cdot R_{G,on/off}/g_m + L_{cs}} \quad (1)$$

$$\left(\frac{dv_{DS}}{dt}\right)_{on/off} = \frac{V_{Mil} - v_{GG}}{C_{GD} \cdot R_{G,on/off}} \quad (2)$$

In these relations, $v_{GG}$ represents the applied gate drive voltage, $C_{iss}$ denotes the input capacitance of the device, and $R_{G,on/off}$ signifies the total gate resistance applied during the turn-on and turn-off phases, respectively. The device intrinsic transconductance $g_m$, the device intrinsic common-source inductance $L_{cs}$, and the gate drain capacitance $C_{GD}$ also affect the switching charateristic. Subsequent relationships are given by:

$$C_{iss} = C_{GS} + C_{GD} \quad (3)$$

$$R_{G,on/off} = R_{G,on/off,ext} + R_{G,int} \quad (4)$$

$$V_{Mil} = V_{th} + \frac{i_D}{g_m} \quad (5)$$

Here, $C_{GS}$ is the gate-source capacitance, $R_{G,on/off,ext}$ is the externally applied gate resistance for turn-on and turn-off, respectively, $R_{G,int}$ is the internal gate resistance of the device, $V_{th}$ is the threshold voltage, and $i_D$ represents the drain current. These elements together illustrate how gate resistance, drive voltage, gate-source capacitance, and gate-drain capacitance, along with the gate current, can be manipulated to alter the switching speed [3]. The two most prevalent methods for modulating the switching speed in AGDs are Gate Resistance Manipulation (GRM) and Gate Voltage Manipulation (GVM) [4].

## 2 State-of-the-Art Analysis

This study utilises the intermediate gate voltage approach for GVMAGDs as depicted in Fig. 1, focusing on a genuine intermediate gate voltage level

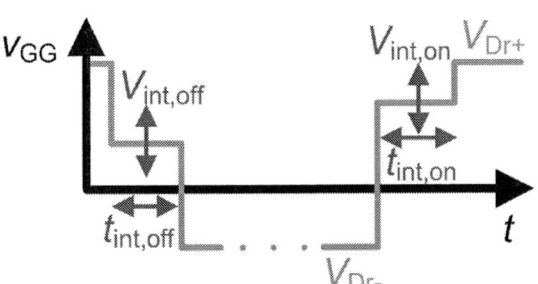

**Fig. 1:** Schematic of the investigated intermediate gate voltage control method [9]

situated between the positive ($V_{Dr+}$) and negative ($V_{Dr-}$) gate drive voltage levels. This strategy simplifies the voltage generation process. Additionally, the gate voltage pattern analysed in this study assumes that the intermediate gate voltage phase immediately follows the onset of the switching process, analogous to the turn-on approach in [5], but without the intermediate phase for turn-off as observed in [5] and [6], which has been demonstrated to reduce the switching delay. Such GVMAGDs are already commercially available [7], capable of independently adjusting the intermediate gate voltage level and duration for both turn-on ($V_{int,on}$, $t_{int,on}$) and turn-off ($V_{int,off}$, $t_{int,off}$), similar to the configurations presented in [8] and [9]. This GVMAGD is utilised throughout this study. Although the current setup does not adjust these parameters ($V_{int,on}$, $t_{int,on}$, $V_{int,off}$, $t_{int,off}$) on a pulse-to-pulse basis, theoretically, such an adaptation is feasible.

Recent literature highlights the influence of various gate control parameters on SiC MOSFET switching behaviour. Nevertheless, direct comparisons between GRMAGDs and GVMAGDs are scarce. For example, [10] broadly examines the impact of gate control parameters on SiC MOSFETs but does not specifically contrast GRMAGDs with GVMAGDs. Another study [6] compares these AGD types but focuses on a distinct gate voltage pattern for GVMAGDs, involving a short phase of regular gate drive voltage post-switching initiation before transitioning to the intermediate voltage level. This research does not explore potential loss reductions during driving cycles such as the WLTC.

Numerous studies have adopted an 'online' approach with AGDs, adapting the gate control of the Device Under Test (DUT) during a single switching transient [2], [11]–[15]. These methods, however, necessitate highly accurate nanosecond-range tim-

ing, thereby increasing hardware complexity and cost. Conversely, AGDs that adjust control parameters between successive switching events offer a balanced approach, with timing requirements in the microsecond range, thus reducing costs and complexity while maintaining effective switching performance. For instance, [16] demonstrated that a GRMAGD achieved significant switching loss reductions during the WLTC, ranging from 27% to 47%, depending on the operating conditions. This shows that even relatively straightforward approaches can yield notable loss savings.

Extensive research has been conducted on the efficiency of automotive electric powertrains during various driving cycles, including the benefits of using SiC [17]–[19]. The focus is increasingly shifting towards intelligent gate control. For example, [20] introduces an adjustable current source driver capable of shaping the gate current for a SiC MOSFET, achieving a 7% reduction in total electrical losses during the WLTC. Moreover, [21] and [22] investigate a GRMAGD with two gate resistors each for turn-on and turn-off, showing significant reductions in switching losses for an IGBT, though these studies do not include full drive cycles or SiC MOSFETs. To the best of the authors' knowledge, no published investigation has yet explored the effect of an active gate driver with an intermediate gate voltage level on the efficiency of a SiC MOSFET inverter during a full drive cycle.

# 3 Operating Principles and Conditions

Figure 2 illustrates the basic control regime and the trajectory of the load current ($i_{Load}$) for a specific vehicle during WLTC class 3. The load current is the most variable parameter in an automotive traction inverter during a drive cycle and primarily dictates the control strategy. Other parameters such as DC link voltage ($V_{DC}$) and junction temperature ($T_J$) also vary, however, to a lesser extent. The overarching objective is to minimise switching losses while ensuring that the maximum drain-source overvoltage $\Delta V_{DS,sw/Diode,max}$ remains within the permissible limit $\Delta V_{DS,allowed}$ for both the active switch during turn-off and the complementary body-diode during turn-on.

As the load current varies, so does the resultant overvoltage. Equations (1) and (2) imply that in the GRM technique, a higher gate resistance is necessary to increase gate impedance and thereby

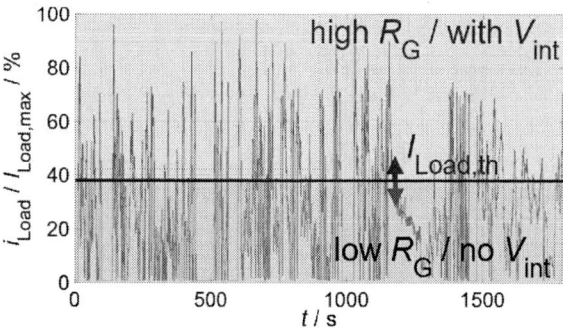

**Fig. 2:** Adaptation of $R_G$ and $V_{int}$ throughout WLTC

reduce switching speed. Conversely, the GVM technique employs an intermediate gate voltage level instead of a constant gate drive voltage to modulate the gate current. Hence, low gate resistances or conventional gate drive levels, instead of intermediate gate voltages, are utilised for the predominantly low load currents during WLTC, facilitating fast switching and low losses. For high load currents, higher gate resistances or intermediate gate voltage levels are employed to keep $\Delta V_{DS,sw/Diode,max}$ within the $\Delta V_{DS,allowed}$ limit, leading to slower switching and increased losses. Consequently, a load current threshold ($I_{Load,th}$) is established, dictating the transition between gate control states, monitored by a current sensor that requires only modest timing and accuracy compared to online methods. Table 1 provides an overview of the operating conditions derived from an automotive inverter specification using SiC power modules. Five current scalings, and thus module utilizations, ranging from $\left( \frac{I_{Load,RMS,max}}{I_{Load,RMS,th.\ limit\ WLTC}} \right) = 20\%$ to $100\%$, have been explored to assess how efficiency improvements vary with module utilization. These scalings are intended to encompass a broad spectrum of electric vehicles; for instance, a low current scaling such as $I_{Load,RMS,max}/I_{Load,RMS,th.\ limit\ WLTC} = 20\%$ typically corresponds to a high-power sports car, which is less utilised during WLTC. Conversely, a high current scaling like $I_{Load,RMS,max}/I_{Load,RMS,th.\ limit\ WLTC} = 100\%$ usually relates to a small city car, which experiences significant load during WLTC class 3. A Cree / Wolfspeed C3M0016120K SiC MOSFET in a TO247-4 package served as the DUT for the measurements. The results were then scaled to the module level for simulation according to the

following equations:

$$L_{\sigma,\text{Module}} \cdot I_{N,\text{Module}} = L_{\sigma,\text{Chip}} \cdot I_{N,\text{Chip}} \quad (6)$$

$$R_{G,\text{Module}} \cdot C_{GS,\text{Module}} = R_{G,\text{Chip}} \cdot C_{GS,\text{Chip}} \quad (7)$$

| Parameter | Value |
|---|---|
| DUT | Wolfspeed C3M0016120K |
| GRMAGD | Infineon 1ED3241MC12H |
| GVMAGD | Microchip 2ASC-12A2HP |
| Maximum DC link voltage | $V_{DC,max} = 830\,V$ |
| Max. allowed drain source overvoltage | $\Delta V_{DS,allowed} = 200\,V$ |
| Switching frequency | $f_{sw} = 10\,kHz$ |
| Interlock time | $t_{lock} = 1\,\mu s$ |
| Maximum junction temperature | $T_{J,max} = 150\,°C$ |
| Coolant temperature | $T_{cool} = 65\,°C$ |
| Maximum load current | $I_{Load,max} = 99\,A$ |
| Stray inductance | $L_{\sigma} = 120\,nH$ [a] |
| (Maximum) Positive gate drive voltage | $V_{Dr+} = +15\,V$ |
| (Minimum) Negative gate drive voltage | $V_{Dr-} = -4\,V$ |

[a] Scaled from $I_{Load,RMS,th.\ limit\ WLTC} = 600\,A$ at $L_{\sigma} = 15\,nH$.

**Tab. 1:** Key parameters and operating conditions

Figure 3a illustrates the schematic setup of the Infineon GRMAGD with switchable gate resistors, while Fig. 3b displays the setup of the Microchip GVMAGD featuring intermediate gate voltage levels. The Infineon GRMAGD includes two output ports: OUT for standard sourcing and sinking, and OUTF for accelerated sourcing and sinking, with the selection of OUTF depending on the state of the /INF control input port [23]. Rather unusual, the Microchip GVMAGD employs a Darlington-like series configuration of output buffers with increasing peak current capability.

# 4 Switching Process Measurements

To investigate the effect of dynamic gate control, it was crucial to measure the switching processes using a double-pulse test setup. The primary processes observed were the turn-on and turn-off of the active switch, as well as the passive diode turn-off. Figure 4a displays exemplary active turn-on operating points, while Fig. 4b illustrates the corresponding passive turn-off switching processes. It is evident that very similar switching behaviours can be achieved using different combinations of $R_{G,on,ext}$ and $V_{int,on}$. These range in Fig. 4 from $V_{int,on} = 15\,V$ and $R_{G,on,ext} = 30\,\Omega$ (typical of regular switching with a conventional gate driver(CGD)) to $V_{int,on} = 9.0\,V$ and $R_{G,on,ext} = 8\,\Omega$.
Figure 5 provides a partial overview of $\Delta V_{DS,Diode,max}$ across different load currents

PCIM Europe 2024, 11– 13 June 2024, Nuremberg   DOI: 10.30420/566262092

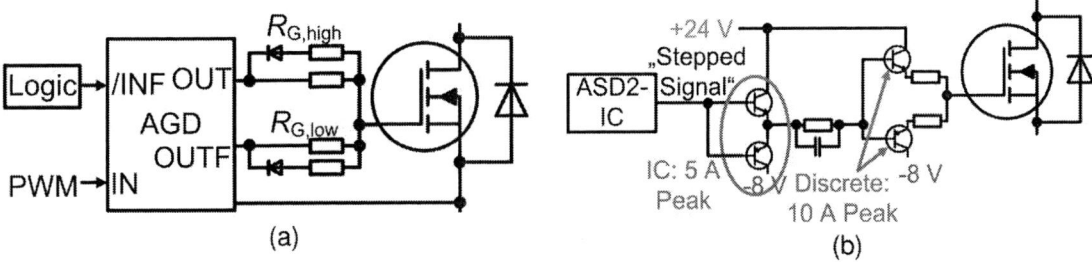

**Fig. 3:** Schematic setup of (a) the Infineon GRMAGD with switchable gate resistors and (b) the Microchip GVMAGD with intermediate gate voltage levels

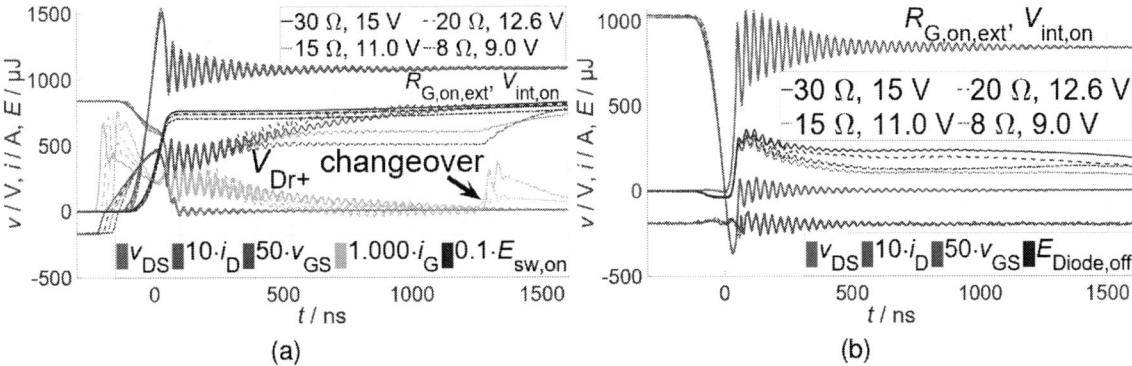

**Fig. 4:** Exemplary (a) active turn-on and (b) passive turn-off measurements for $V_{DC,max} = 830\,\text{V}$, $L_\sigma = 120\,\text{nH}$, $T_{J,max} = 150\,^\circ\text{C}$, and $i_{Load} \approx I_{Load,max}$ under different combinations of $R_{G,on,ext}$ and $V_{int,on}$

and combinations of $R_{G,on,ext}$ and $V_{int,on}$ at a junction temperature of $T_J = T_{J,max} = 150\,^\circ\text{C}$. Double pulse tests were also conducted at further combinations of $R_{G,on,ext}$ and $V_{int,on}$ as well as at junction temperatures of $T_J = 25\,^\circ\text{C}$ and $T_J = 75\,^\circ\text{C}$, yielding over 1,000 turn-on data points in total. Figure 6 displays the corresponding turn-on losses, while Fig. 7 shows the diode turn-off losses. Generally, $\Delta V_{DS,Diode,max}$ increases with rising load current, increasing $V_{int,on}$, and decreasing $R_{G,on,ext}$. Conversely, $E_{sw,on}$ generally rises with increasing load current, decreasing $V_{int,on}$, and increasing $R_{G,on,ext}$. Diode turn-off losses ($E_{Diode,off}$) also increase with higher load currents, higher $V_{int,on}$, and lower $R_{G,on,ext}$.

Figure 8a displays exemplary active turn-off operating points at $i_{Load} \approx I_{Load,max}$. It is evident that various combinations of $R_{G,off,ext}$ and $V_{int,off}$ can achieve similar switching behaviours, ranging from $R_{G,off,ext} = 33\,\Omega$ and $V_{int,off} = -4\,\text{V}$ (representative of conventional gate driving) to $R_{G,off,ext} = 8\,\Omega$ and $V_{int,off} = 4.0\,\text{V}$. Notably, the latter combination results in the MOSFET not being fully turned off during the intermediate gate voltage phase, leading to significantly high switching losses as observed

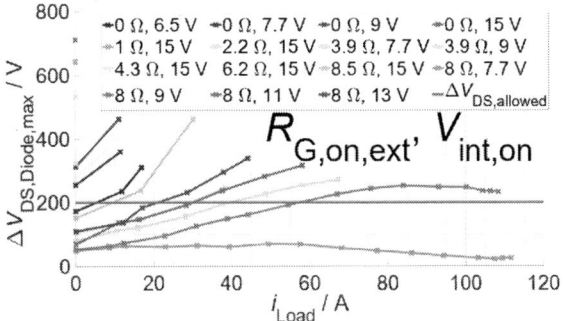

**Fig. 5:** Characterisation of $\Delta V_{DS,Diode,max}$ at $V_{DC,max} = 830\,\text{V}$, $L_\sigma = 120\,\text{nH}$, and $T_J = 150\,^\circ\text{C}$ for various $i_{Load}$ and different combinations of $R_{G,on,ext}$ and $V_{int,on}$

in Fig. 8. These losses are contingent on the duration of the intermediate gate voltage phase, which critically influences the outcomes of subsequent WLTC simulations. For realistic simulation, a phase duration of $t_{int,off} = 1,000\,\text{ns}$ was assumed.

Figure 9a shows an overview of $\Delta V_{DS,sw,max}$ as a function of different load currents as well as different combinations of $R_{G,off,ext}$ and $V_{int,off}$ for $T_J = 25\,^\circ\text{C}$. Double pulse tests were also conducted at further combinations of $R_{G,off,ext}$ and $V_{int,off}$. Therefore,

757

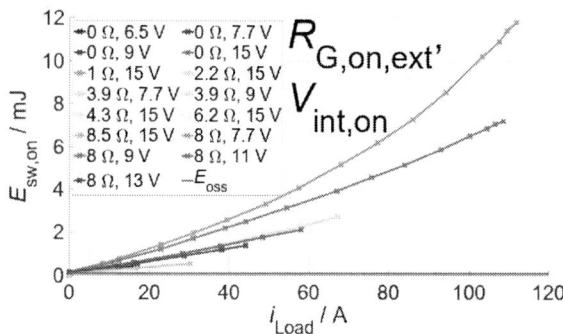

**Fig. 6:** Characterisation of $E_{sw,on}$ at $V_{DC,max} = 830\,V$, $L_\sigma = 120\,nH$, and $T_J = 150\,°C$ for various $i_{Load}$ and different combinations of $R_{G,on,ext}$ and $V_{int,on}$

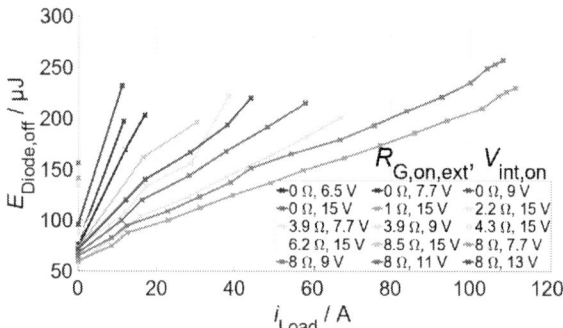

**Fig. 7:** Characterisation of $E_{Diode,off}$ at $V_{DC,max} = 830\,V$, $L_\sigma = 120\,nH$, and $T_J = 150\,°C$ for various $i_{Load}$ and different combinations of $R_{G,on,ext}$ and $V_{int,on}$

over 650 turn-off data points have been considered in total. Figure 9b displays the corresponding turn-off switching losses. In general, $\Delta V_{DS,sw,max}$ rises with increasing load current, decreasing $V_{int,off}$ and decreasing $R_{G,off,ext}$. Conversely, $E_{sw,off}$ tends to rise with increasing load current and $V_{int,off}$ (particularly when $V_{int,off} \geq V_{th}$), and also increases with $R_{G,off,ext}$.

## 5 WLTC Simulation Model

The WLTC class 3 simulation is conducted in MATLAB, utilizing measurement data from conduction and switching processes. These data are applied through multidimensional interpolation to reflect real-world driving conditions accurately.

### 5.1 Structure of Simulation Model

The simulation model is structured to depend dynamically on various parameters for both turn-on and turn-off processes. Specifically, the parameters for active turn-off are modelled based on gate resistance $R_{G,off}$, intermediate gate voltage $V_{int,off}$, and load current $i_{Load}$. Similarly, parameters for

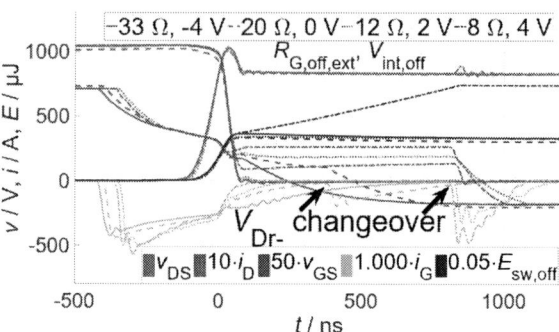

**Fig. 8:** Exemplary turn-off measurements for $V_{DC,max} = 830\,V$, $L_\sigma = 120\,nH$, $T_J = 25\,°C$, $i_{Load} \approx I_{Load,max}$ under different combinations of $R_{G,off,ext}$ and $V_{int,off}$

active turn-on and passive turn-off are modelled depending on gate resistance $R_{G,on}$, intermediate gate voltage $V_{int,on}$, load current $i_{Load}$, and junction temperature $T_J$. The relevant parameters for each sample point are calculated using the following equations:

$$\Delta V_{DS,sw,max} = f(R_{G,off}, V_{int,off}, i_{Load}) \tag{8}$$
$$E_{sw,off} = f(R_{G,off}, V_{int,off}, i_{Load}) \tag{9}$$
$$\Delta V_{DS,Diode,max} = f(R_{G,on}, V_{int,on}, i_{Load}, T_J) \tag{10}$$
$$E_{sw,on} = f(R_{G,on}, V_{int,on}, i_{Load}, T_J) \tag{11}$$
$$E_{Diode,off} = f(R_{G,on}, V_{int,on}, i_{Load}, T_J) \tag{12}$$

To ensure operational safety and efficiency, $R_{G,on,ext}$ and $V_{int,on}$, as well as $R_{G,off,ext}$ and $V_{int,off}$, are determined to satisfy the following conditions:

$$\Delta V_{DS,Diode,max}(R_{G,on,ext}, \\ V_{int,on}, I_{Load,max}, T_{J,max}) = \Delta V_{DS,allowed} \tag{13}$$
$$\Delta V_{DS,sw,max}(R_{G,off,ext}, \\ V_{int,off}, I_{Load,max}) = \Delta V_{DS,allowed} \tag{14}$$

Switching with $V_{Dr+}$ will be selected for turn-on if the maximum drain-source overvoltage during passive turn-off is not exceeded. If it is exceeded, switching with $V_{int,on}$ will be implemented instead. Similarly, for turn-off, $V_{Dr-}$ will be used if the maximum drain-source overvoltage during active turn-off is not exceeded; otherwise, the intermediate gate voltage level $V_{int,off}$ will be employed. It should be noted that, as defined in Eq. (13) and Eq. (14), the usage of $V_{int,on}$ for turn-on and $V_{int,off}$ for turn-off ensures that the maximum permissible drain-source overvoltage is never exceeded. Figure 10 illustrates all possible combinations of $R_{G,on,ext}$ and $V_{int,on}$, as

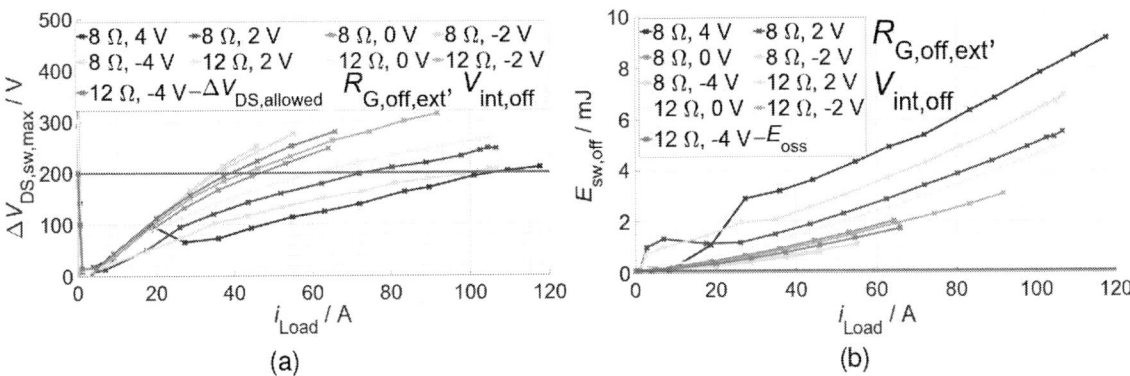

**Fig. 9:** Characterisation of (a) $\Delta V_{DS,sw,max}$ and (b) $E_{sw,off}$ for $V_{DC,max} = 830\,V$, $L_\sigma = 120\,nH$, $T_J = 25\,°C$ for various $i_{Load}$ and different combinations of $R_{G,off,ext}$ and $V_{int,off}$

well as $R_{G,off,ext}$ and $V_{int,off}$, that result in the allowed drain-source overvoltage of $\Delta V_{DS,allowed} = 200$ V for $i_{Load} = I_{Load,max}$. These combinations delineate the boundary of feasible options for slow switching. Fast switching scenarios, then, invariably involve employing the chosen gate resistor with the standard positive gate drive voltage $V_{Dr+}$ or negative gate drive voltage $V_{Dr-}$, as appropriate.

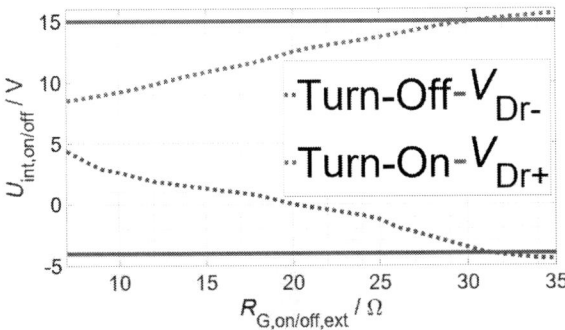

**Fig. 10:** Mapping of potential selections of $R_{G,on/off,ext}$ and $V_{int,on/off}$ that achieve a target overvoltage of $\Delta V_{DS,allowed} = 200$ V under worst-case conditions with $V_{DC,max} = 830$ V, $L_\sigma = 120\,nH$, $T_J = 150\,°C$, $I_{Load} = I_{Load,max}$ for active turn-on and $T_J = 25\,°C$ for active turn-off

Conduction losses were quantified by measuring the on-state resistance $R_{DS,on}$ of four different SiC MOSFET devices across four distinct junction temperatures ($T_J = 25, 75, 125, 175\,°C$), using a curve tracer. To mitigate parameter scatter, the mean of the measurements from the four DUTs at each temperature level was computed. Drain-source resistance in the on-state during the $n$-th simulation step $R_{DS,on,n}$ is then calculated through a

two-dimensional interpolation defined by:

$$R_{DS,on,n} = f(i_{Load,n}, T_{J,n}) \quad (15)$$

Conduction losses, both during regular conduction and interlock periods, are calculated as follows:

$$P_{Cond,n} = 0.5 \cdot R_{DS,on,n}(i_{Load,n}, T_{J,n}) \cdot i_{Load,n}^2 \quad (16)$$

$$P_{Cond,lock,n} = 0.5 \cdot V_{F,n}(i_{Load,n}, T_{J,n}) \cdot i_{Load,n} \quad (17)$$

where $V_{F,n}$ represents the forward voltage across the body diode during the $n$-th simulation step , calculated as the sum of its knee voltage and the product of its differential resistance and the load current. The factor of $0.5$ accounts for the fact that, over a complete sine wave, each switch (high-side (HS) and low-side (LS)) only conducts for half the cycle. It is assumed that the MOSFET is activated for synchronous rectification to minimise reverse conduction losses.

All switching losses $E_{sw,n}$ are scaled by the switching frequency $f_{sw}$ to determine power dissipation $P_{sw,n}$:

$$P_{sw,n} = E_{sw,n} \cdot f_{sw} \quad (18)$$

Total power losses for each simulation step encompass all switching losses and conduction losses during both regular operation and interlock time:

$$P_{Loss,total,n} = P_{sw,n} + P_{Cond,n} + P_{Cond,lock,n} \quad (19)$$

In the simulation, the influence of temperature on switching and conduction losses is iteratively assessed. The junction temperature for each subsequent step is updated using the following relation:

$$T_{J,n} = T_{cool} + P_{Loss,total,n} \cdot R_{th} \quad (20)$$

This updated temperature is then applied in the subsequent calculation of losses. This iterative process continues until the change in temperature between two consecutive cycles is less than a specified threshold $\epsilon$, ensuring stability:

$$|T_{J,n+1} - T_{J,n}| \leq \epsilon \qquad (21)$$

## 5.2 Seven considered Scenarios

In total, seven different gate driver configurationss have been investigated in the WLTC as summarised in Table 2 with the aim of comparing GRMAGD and GVMAGD.

| Scenario | # $R_{G,on}$ | # $R_{G,off}$ | $V_{int,on}$ | $V_{int,off}$ |
|---|---|---|---|---|
| 1 | 1 | 1 | no | no |
| 2 | 2 | 1 | no | no |
| 3 | 1 | 2 | no | no |
| 4 | 2 | 2 | no | no |
| 5 | 1 | 1 | yes | no |
| 6 | 1 | 1 | no | yes |
| 7 | 1 | 1 | yes | yes |

**Tab. 2:** Overview of considered gate driver configurations

Scenario 1 uses a CGD and serves as the reference for loss reduction in all other scenarios. Scenarios 2 to 4 explore variations of GRMAGD, with Scenario 2 implementing it only during turn-on, Scenario 3 only during turn-off, and Scenario 4 during both switching processes. Similarly, Scenarios 5 to 7 test GVMAGD applications, where Scenario 5 applies it only during turn-on, Scenario 6 only during turn-off, and Scenario 7 during both phases. The selection of $V_{int,on}$ and $V_{int,off}$ balances two competing factors, similar to the optimisation problem for $R_{G,low}$ discussed in [16]. On one hand, a more positive $V_{int,on}$ or more negative $V_{int,off}$ can enhance energy savings during each switching process where these voltages are applied, as demonstrated in Fig. 6 and Fig. 9b. On the other hand, more extreme voltage settings restrict the number of switching processes where these can be used due to the limits imposed by $\Delta V_{DS,allowed}$, as shown in Fig. 5 and Fig. 9a. To address this optimization challenge and find the optimum combinations for turn-on and turn-off, respectively,, $V_{int,on/off}$ is finely adjusted in conjunction with $R_{G,on/off,ext}$ as per Fig. 10. The WLTC class 3 is then simulated, and losses are accumulated for comparison.

# 6 Results of WLTC Simulation

Figure 11 illustrates the potential savings in switching losses relative to Scenario 1 (using a CGD) across various module utilisations and gate driver configurations. Meanwhile, Fig. 12 presents the potential total loss savings, which also consider conduction losses. It is apparent that total loss savings are consistently smaller than switching loss savings, since conduction losses are not influenced by AGDs.

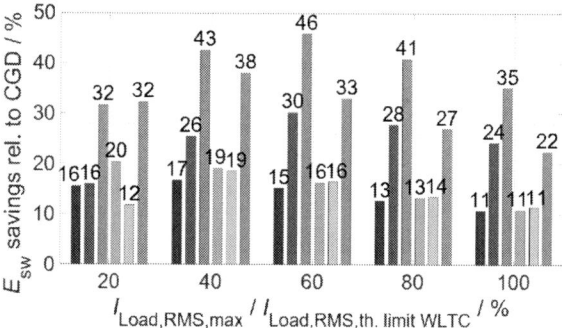

**Fig. 11:** Switching loss savings relative to CGD for different module utilisations and gate driver configurations (GRMAGD vs. GVMAGD) during WLTC

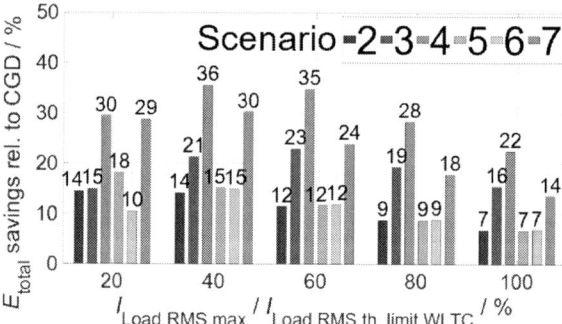

**Fig. 12:** Total loss savings relative to CGD for different module utilisations and gate driver configurations (GRMAGD vs. GVMAGD) during WLTC

Interestingly, both AGD approaches yield approximately the same results for turn-on across the range of module utilisations (comparing scenario 2 for GRM turn-on with scenario 5 for GVM turn-on). However, while the GVM approach for turn-off remains competitive with GRM for low module utilisations, it falls behind as the module utilisations increase. This variance is indicative of the dynamic performance differences between the two AGD types. The comparative analysis of the AGD variants using Fig. 11 and Fig. 12 reveals distinct performance characteristics between GVMAGDs and GRMAGDs. While the two types of AGDs exhibit similar efficiencies during active turn-on, the GRMAGD demonstrates superior performance in terms of $E_{sw,off}$. This systematic difference stems from the relationship between $V_{Mil}$ and $v_{GG}$.

For both turn-on and turn-off, the dimensioning of gate resistors occurs at $I_{Load,max}$. According to Eq. (5), the Miller voltage increases with load current. Therefore, for turn-on with $I_{Load,max}$, $R_{G,on,ext}$ is selected for the minimum voltage difference $V_{Dr+/int,on} - V_{Mil}$. Consequently, $R_{G,on,ext}$ for GV-MAGD can be reduced significantly compared to GRMAGDs. For lower $i_{Load}$, a smaller $V_{Mil}$ leads to a larger voltage difference $V_{Dr+/int,on} - V_{Mil}$. Combined with a relatively low $R_{G,on,ext}$, this results in a high switching speed, low switching losses and competitive performance of the GVMAGD against the GRMAGD.

Conversely, for turn-off design at $I_{Load,max}$, the voltage difference $V_{Mil} - V_{Dr-/int,off}$ is maximised, resulting in the highest switching speed. Hence, the $R_{G,off,ext}$ for GVMAGD needs to be relatively high to limit the switching speed. At lower $i_{Load}$, however, the voltage difference $V_{Mil} - V_{Dr-/int,off}$ decreases as well. This, combined with the relatively high $R_{G,off,ext}$ for the GVMAGD, leads to slower switching speeds and higher switching losses compared to the GRMAGD.

# 7 Experimental Verification using H-bridge in Continuous Operation Mode

To validate the drive cycle simulations derived from double pulse measurements, continuous operation tests were conducted using an H-bridge configuration. The schematic of the H-bridge setup is illustrated in Fig. 13a, while a photograph of the actual setup is shown in Fig. 13b. An Infineon 1ED3241MC12H was utilised as the GRMAGD. The criteria for toggling between fast or slow switching were determined by the magnitude of the measured AC current $i_{AC}(t)$. A logical circuit PCB processed the signal from the LEM current sensor. The AGD was designated to drive the LSS of half-bridge 1, with CGDs employed for the other three SiC MOSFETs. Both the control board, generating the control signals for the AGD, and the AGD itself, are visible on the left side of Fig. 13b.

To evaluate potential loss reductions using the GRMAGD shown in this research and in [16], exper-iments were first conducted with the feature disabled and subsequently enabled. The static operation point was characterised by $V_{DC} = 800\,\text{V}$, $I_{AC,rms} = 16.55\,\text{A}$, $L_\sigma = 120\,\text{nH}$, $L_{Load} = 3\,\text{mH}$, $f_{el} = 400\,\text{Hz}$, and $f_{sw} = 10\,\text{kHz}$.

Figure 14a displays a thermal image of the operation without dynamic slew-rate control, and Fig. 14b with it activated. In both instances, the DUT is located on the left side. Comparing Fig. 14a and Fig. 14b, a relative temperature difference can be calculated based on an ambient temperature of $T_{amb} = 25\,°\text{C}$, resulting in a loss reduction of $-19.3\,\%$. Notably, the temperature difference between the LSS and HSS, where the AGD approach is not used, results in higher losses at the HSS, causing heat flow towards the DUT and reducing temperature reduction at the DUT. Further evaluations accounting for this heat flow showed a total relative temperature difference of $-24.6\,\%$, closely matching the simulated reduction of $-28.2\,\%$ according to the simulation model, thus confirming the potential for the discussed loss reductions.

# 8 Conclusion

This paper explored two prevalent variants of AGDs, the Gate Resistance Manipulation (GRMAGD) and the Gate Voltage Manipulation AGD (GVMAGD), under a pulse-to-pulse control strategy. The results demonstrate that both types of AGDs perform comparably during the turn-on phase under similar operational conditions. However, during turn-off, the GRMAGD shows superior efficiency relative to the GVMAGD, a distinction primarily attributable to the strategic selection of gate resistors. These are optimised for scenarios with high $i_{Load}$ and consequently elevated $V_{Mil}$ conditions. Despite these variances, it is significant to note that both AGD variants markedly outperform conventional gate drivers, highlighting their potential in enhancing the efficiency of power electronics systems. Looking ahead, further research could investigate configurations that vary gate voltage levels and intermediate phase durations more dynamically. An 'online' approach could be particularly promising, where AGDs optimally set gate drive conditions for each individual operating point, potentially leading to even greater efficiency gains in diverse application environments.

# References

[1] J. Millán, P. Godignon, X. Perpiñà, A. Pérez-Tomás, and J. Rebollo, "A Survey of Wide

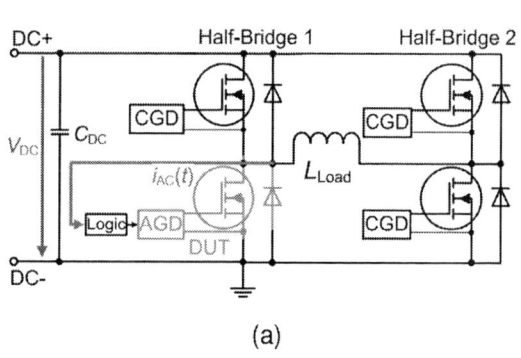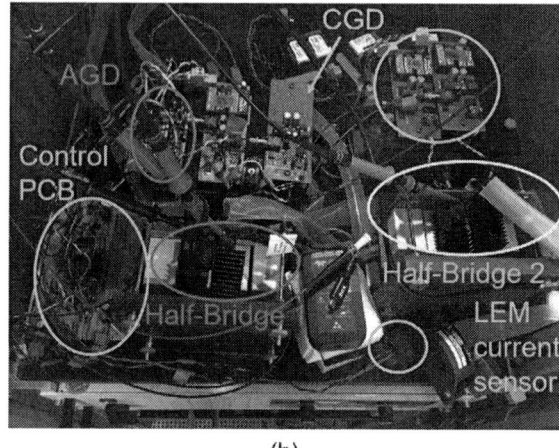

**Fig. 13:** (a) Schematic and (b) real setup of H-bridge with Infineon GRMAGD

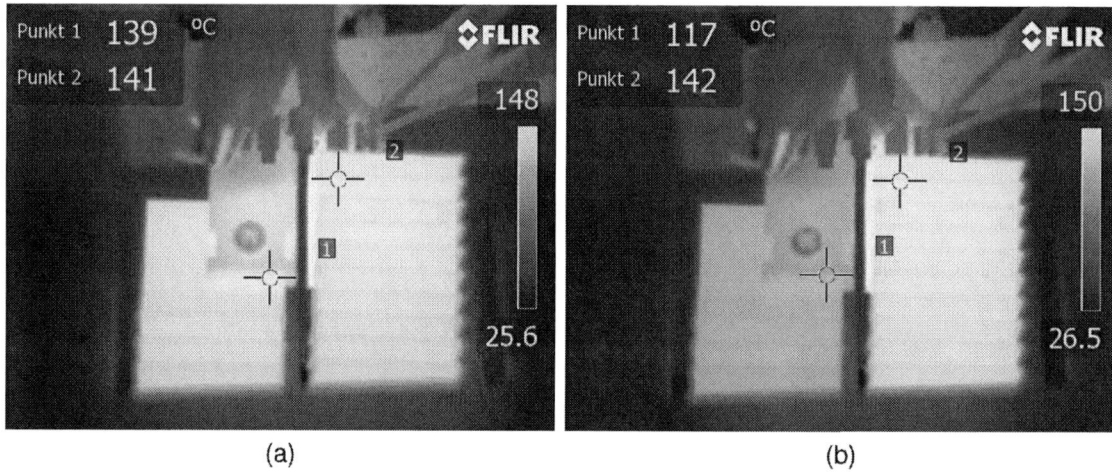

**Fig. 14:** Thermal image of static continuous operation point (a) without and (b) with dynamic slew-rate control on the left-hand side SiC MOSFET

Bandgap Power Semiconductor Devices," *IEEE Transactions on Power Electronics*, vol. 29, no. 5, pp. 2155–2163, 2014. DOI: 10.1109/TPEL.2013. 2268900.

[2] Y. Yang, Y. Wen, and Y. Gao, "A Novel Active Gate Driver for Improving Switching Performance of High-Power SiC MOSFET Modules," *IEEE Transactions on Power Electronics*, vol. 34, no. 8, pp. 7775–7787, 2019. DOI: 10.1109/TPEL.2018. 2878779.

[3] S. Zhao, X. Zhao, Y. Wei, Y. Zhao, and H. A. Mantooth, "A Review of Switching Slew Rate Control for Silicon Carbide Devices Using Active Gate Drivers," *IEEE Journal of Emerging and Selected Topics in Power Electronics*, vol. 9, no. 4, pp. 4096–4114, 2021. DOI: 10.1109/JESTPE. 2020.3008344.

[4] J. Henn, C. Lüdecke, M. Laumen, S. Beushausen, S. Kalker, *et al.*, "Intelligent Gate Drivers for Future Power Converters," *IEEE Transactions on Power Electronics*, vol. 37, no. 3, pp. 3484–3503, 2022. DOI: 10.1109/TPEL.2021.3112337.

[5] N. Idir, R. Bausiere, and J. J. Franchaud, "Active Gate Voltage Control of Turn-On di/dt and Turn-Off dv/dt in Insulated Gate Transistors," *IEEE Transactions on Power Electronics*, vol. 21, no. 4, pp. 849–855, 2006. DOI: 10.1109/TPEL.2007.876895.

[6] A. Dearien, S. Zhao, C. Farnell, and H. A. Mantooth, "Slew Rate Control of High-Voltage SiC MOSFETs Using Gate Resistance vs. Intermediate Voltage Level," in *2019 10th International Conference on Power Electronics and ECCE Asia (ICPE 2019 - ECCE Asia)*, 2019, pp. 2146–2152. DOI: 10.23919/ICPE2019-ECCEAsia42246.2019. 8797027.

[7] Microchip Technology Inc. "2ASC-12A2HP - 1200v Dual-Channel Augmented High Performance SiC Core 2." Accessed: 2024-04-02. (2021), [Online]. Available: https://shorturl.at/iKV26.

[8] X. Du, Y. Wei, A. Stratta, L. Du, V. S. Machireddy, and A. Mantooth, "A Four-level Active Gate Driver with Continuously Adjustable Intermediate Gate Voltages," in *2022 IEEE Applied Power Electronics Conference and Exposition (APEC)*, 2022, pp. 1379–1386. DOI: 10.1109/APEC43599.2022.9773689.

[9] H. B. Ekren, D. A. Philipps, G. Lyng Rødal, and D. Peftitsis, "Four Level Voltage Active Gate Driver for Loss and Slope Control in SiC MOSFETs," in *2022 IEEE 13th International Symposium on Power Electronics for Distributed Generation Systems (PEDG)*, 2022, pp. 601–606. DOI: 10.1109/PEDG54999.2022.9923113.

[10] Z. Zeng and X. Li, "Comparative Study on Multiple Degrees of Freedom of Gate Drivers for Transient Behavior Regulation of SiC MOSFET," *IEEE Transactions on Power Electronics*, vol. 33, no. 10, pp. 8754–8763, 2018. DOI: 10.1109/TPEL.2017.2775665.

[11] P. Nayak and K. Hatua, "Active Gate Driving Technique for a 1200 v SiC MOSFET to Minimize Detrimental Effects of Parasitic Inductance in the Converter Layout," *IEEE Transactions on Industry Applications*, vol. 54, no. 2, pp. 1622–1633, 2018. DOI: 10.1109/TIA.2017.2780175.

[12] G. Engelmann, T. Senoner, and R. W. De Doncker, "Experimental Investigation on the Transient Switching Behavior of SiC MOSFETs Using a Stage-Wise Gate Driver," *CPSS Transactions on Power Electronics and Applications*, vol. 3, no. 1, pp. 77–87, 2018. DOI: 10.24295/CPSSTPEA.2018.00008.

[13] Y. Lobsiger and J. W. Kolar, "Closed-Loop di/dt and dv/dt IGBT Gate Driver," *IEEE Transactions on Power Electronics*, vol. 30, no. 6, pp. 3402–3417, 2015. DOI: 10.1109/TPEL.2014.2332811.

[14] A. P. Camacho, V. Sala, H. Ghorbani, and J. L. R. Martinez, "A Novel Active Gate Driver for Improving SiC MOSFET Switching Trajectory," *IEEE Transactions on Industrial Electronics*, vol. 64, no. 11, pp. 9032–9042, 2017. DOI: 10.1109/TIE.2017.2719603.

[15] Z. Li, R. W. Maier, M.-M. Bakran, F.-J. Niedernostheide, and D. Domes, "A Simulation Model for SiC MOSFET Switching Transients Controlled by an Adaptive Gate Driver with the Capability of Reducing Switching Losses and EMI across the Full Operating Range," in *2022 24th European Conference on Power Electronics and Applications (EPE'22 ECCE Europe)*, 2022, pp. 2744–2753.

[16] M. J. Frank and M.-M. Bakran, "Effect of Dynamic Gate Control Driver on SiC MOSFET Power Module Performance in WLTC," in *2023 25th European Conference on Power Electronics and Applications (EPE'23 ECCE Europe)*, 2023, pp. 1–10. DOI: 10.23919/EPE23ECCEEurope58414.2023.10264362.

[17] S. Hain, M. Meiler, and M. Denk, "Evaluation of 800V Traction Inverter with SiC-MOSFET versus Si-IGBT Power Semiconductor Technology," in *PCIM Europe 2019; International Exhibition and Conference for Power Electronics, Intelligent Motion, Renewable Energy and Energy Management*, 2019, pp. 1–6.

[18] R. Wu, J. O. Gonzalez, Z. Davletzhanova, P. A. Mawby, and O. Alatise, "The Potential of SiC Cascode JFETs in Electric Vehicle Traction Inverters," *IEEE Transactions on Transportation Electrification*, vol. 5, no. 4, pp. 1349–1359, 2019. DOI: 10.1109/TTE.2019.2954654.

[19] H. Umegami, T. Harada, and K. Nakahara, "Performance Comparison of Si IGBT and SiC MOSFET Power Module Driving IPMSM or IM under WLTC," *World Electric Vehicle Journal*, vol. 14, no. 4, 2023. DOI: 10.3390/wevj14040112. [Online]. Available: https://www.mdpi.com/2032-6653/14/4/112.

[20] M. Sayed, S. Araujo, F. Carraro, and R. Kennel, "Investigation of Gate Current Shaping for SiC-based Power Modules on Electrical Drive System Power Losses," in *2021 23rd European Conference on Power Electronics and Applications (EPE'21 ECCE Europe)*, 2021, pp. 11–20. DOI: 10.23919/EPE21ECCEEurope50061.2021.9570513.

[21] W. Frank, D. Levett, and Z. Q. Zheng, "Two-level Slew-rate Control Driver to Optimize IGBT Performance," in *PCIM Europe digital days 2020; International Exhibition and Conference for Power Electronics, Intelligent Motion, Renewable Energy and Energy Management*, 2020, pp. 1866–1871.

[22] W. Frank, N. Thon, and M. Ebli, "Two-Level, Slew-Sate Control Reduces Temperature Stress of Semiconductors in Power Modules," in *PCIM Europe 2022; International Exhibition and Conference for Power Electronics, Intelligent Motion, Renewable Energy and Energy Management*, 2022, pp. 1612–1617. DOI: 10.30420/565822223.

[23] Infineon Technologies AG. "EiceDRIVER™ 1ED32xxMC12H Two-level Slew-rate Control (2L-SRC)." accessed: 2024-04-02. (2023), [Online]. Available: https://shorturl.at/gsFJR.

PCIM Europe 2024, 11– 13 June 2024, Nuremberg     DOI: 10.30420/566262093

# Performance Evaluation of TCM-based, Zero-Voltage Switching (ZVS) Three-Phase Inverters for Electric Vehicle Drive Systems

Khizra Abbas ⊚, Hans-Peter Nee

KTH Royal Institute of Technology, Sweden

Corresponding author:     Khizra Abbas, khizra@kth.se
Speaker:                    Khizra Abbas, khizra@kth.se

## Abstract

This paper investigates the performance and applicability of a triangular current mode (TCM)–based zero voltage switching (ZVS) three-phase inverter, incorporated with novel autonomous gate drivers (AGDs), for electric vehicle (EV) drive systems. By integrating the controller into the gate driver, delays are minimized, enabling ZVS during the turn-on and turn-off operations. Sinusoidal output waveforms aid in reducing electromagnetic interference (EMI). This study examines the potential interference of the LC filter with torque control and evaluates the inverter's ability to respond quickly and accurately to torque commands without introducing delays, overshoots, or oscillations. The proposed 10 kW inverter achieved over 99% efficiency without including filter inductor losses. Torque steps were executed significantly faster than required for practical applications. No delays or oscillations were observed, indicating the inverter's suitability for EV drive systems.

## 1   Introduction

The electrification of transportation, particularly through electric vehicles (EVs), has gained significant momentum in addressing sustainability challenges like global warming, fossil fuel depletion, and greenhouse gas emissions [1]. Despite their numerous benefits, EVs face challenges such as cost, size, weight, efficiency, and electromagnetic interference (EMI). Addressing these challenges is critical; however, resolving one issue may unintentionally create conflicts with solutions for others. The core of EV power systems is the traction inverter, a pivotal component that converts DC power from the battery into the essential AC power needed to drive the electric motor. Efficient and reliable inverters not only enhance the overall performance of EVs but also contribute to reducing their environmental impact. Therefore, ongoing research and development efforts focused on improving inverter design and performance are essential for driving the widespread adoption of EVs and accelerating the transition to sustainable transportation.

The EV industry relies on IGBT-based inverters and PWM techniques for power switching. However, achieving a sinusoidal output current requires high switching frequencies, leading to challenges like increased losses, dv/dt, and common-mode voltages [2]. These issues can cause motor failure, bearing, insulation failures, and high EMI [3]. Solutions include new switching materials, resonant converters, and integration of filters.

SiC technology is transforming the EV industry, offering advantages in power electronics [4], promising enhanced performance, extended range, and reduced reliance on battery storage [5]. However, high-speed switching introduces EMI issues [6], calling for soft-switching techniques.

Soft-switching (SSW) power converters minimize switching losses, particularly at high frequencies, enhancing system efficiency and reliability. Zero-voltage switching (ZVS) [7] is a prominent technique achieved by reducing switch voltage to zero through the prior conduction of the MOSFET's body diode and using a capacitive snubber during turn-off. ZVS enables high-frequency operation while reducing noise and switching losses.

The triangular current mode (TCM) [8] of the filter inductor current is necessary for ZVS, where positive and negative peaks aid in achieving ZVS for both switches in a half-bridge configuration.

Unipolar SiC power transistors, such as SiC MOSFETs, are superior to soft-switching converters be-

PCIM Europe 2024, 11– 13 June 2024, Nuremberg          DOI: 10.30420/566262093

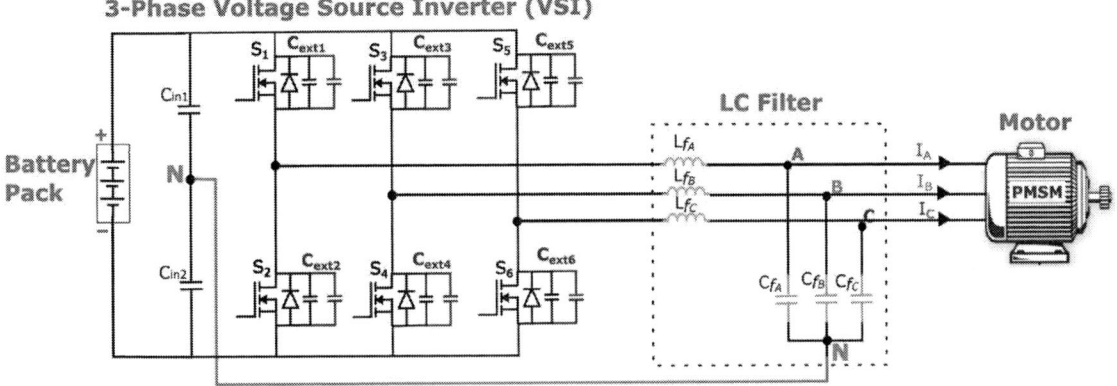

**Fig. 1:** Schematic representation of the two-level three-phase soft-switched inverter for EV drive system.

cause of their absence of conductivity modulation lag and tail currents [9]. In the context of traction converters for EVs, this may enable highly efficient, compact, and potentially cost-effective systems.

One of the most promising concepts in this context is the TCM-based ZVS three-phase inverter, which builds upon the ZVS clamped-voltage concept for flyback converters [10]. This technology has been successfully applied in various domains, such as power factor correction [8] and the Google Little-Box Challenge [11], featuring high efficiency, compactness, and excellent EMI performance. Consequently, the TCM-based ZVS inverter holds promise as a competitive technology for EVs. Additionally, combining a traction inverter employing SiC power transistors with advanced ceramic capacitor technologies [12] may enable operation at elevated temperatures. This capability could potentially eliminate the need for a dedicated cooling system, allowing the inverter to be placed almost anywhere in the vehicle and thermally attached to the chassis of the vehicle.

The TCM-based ZVS inverter [13], incorporating novel autonomous gate drivers (AGDs) [14], offers several advantages for three-phase inverters. These include a simple topology, a current modulation scheme, and high power density facilitated by a high switching frequency and a small filter inductor. Integrating the controller onto the gate driver enhances efficiency by minimizing time delays and achieving ZVS during both turn-on and turn-off. Additionally, the sinusoidal output current and voltage waveforms help in mitigating EMI.

The TCM-based ZVS inverter employs an LC filter. The question arises: will this filter interfere with torque control during transients? Such interfer-

ences could appear as a slow response, excessive overshoots after torque transients, or undamped oscillations following a torque (or load) transient. It is, therefore, important to establish whether the inverter can provide the output current after the LC filter according to the change in reference value ordered by the torque controller. For this purpose, it is not necessary to implement a total vector control system with a detailed model of the motor. This paper investigates whether the proposed inverter can provide the desired output current (required for a torque step) without excessive delays, overshoots, or oscillations.

The key contributions of this paper include:

– Analysis of ZVS at both turn-on and turn-off.

– Discussion on the impact of switching frequency variation on filter design and output waveforms.

– Examination of sinusoidal output waveforms and their impact on EMI.

– Evaluation of the torque step response.

The paper is structured as follows: Section 2 introduces the circuit configuration, current modulation scheme, and operating principle of the TCM-based ZVS inverter. In Section 3, the design considerations of the proposed inverter are discussed. Section 4 presents the results and provides a comprehensive discussion. Finally, Section 5 concludes the research work.

## 2 Proposed Soft-Switched Inverter

### 2.1 Circuit Configuration

The two-level three-phase soft-switched inverter designed for EV drive systems, as depicted in Fig.

1, comprises DC-link capacitors, LC low-pass filters, SiC MOSFET switches, external snubber capacitors, and a permanent magnet synchronous machine (PMSM) serving as the load. In comparison to the conventional two-level three-phase hard-switched inverter, this soft-switched alternative presents the following notable differences:

- Integration of LC low-pass filter: Helps in achieving turn-on ZVS.

- Incorporation of external snubber capacitance: Aids in achieving turn-off ZVS.

- Linking the neutral point of the filter capacitors to the midpoint of the DC-link: Assists in decreasing EMI, leakage current, and bearing currents by controlling the common-mode voltages of the filter capacitor.

## 2.2 Current Modulation Scheme

A key innovation in the design of the inverter lies in the smooth integration of modulation and current control on its gate drivers [14]. This novel approach combines both functions into a single step, significantly improving efficiency and performance. Drawing inspiration from hysteresis control utilized in motors, this method ensures rapid and precise current control. Notably, this design effectively reduces conducted EMI, presenting a significant advantage over the conventional systems.

The autonomous gate drivers (AGDs) [14] enable ZVS during turn-on and turn-off intervals using bidirectional inductor current and external snubber capacitance. AGDs generate turn-on and turn-off signals rapidly, enhancing switching control.

- For Turn-on ZVS, the AGDs sense voltage across the switch. The AGDs generate a turn-on signal when the voltage is less than zero.

- For Turn-off ZVS, the AGDs sense current across the switch. The AGDs generate a turn-off signal when the current exceeds a predefined value of the reference current. The switch turn-off signals are determined by adjustable positive/negative current reference values as shown in Fig. 2.

- The turn-on and turn-off signals are transferred to the RS latch, generating the gate-source voltage ($V_{gs}$) signal for the switch.

Variations in the current references directly influence the output load current as shown in Fig. 3.

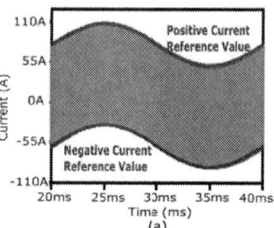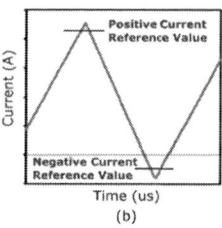

**Fig. 2:** Filter inductor current. (a) Filter inductor current with positive and negative current references. (b) One switching cycle of filter inductor current with reference currents.

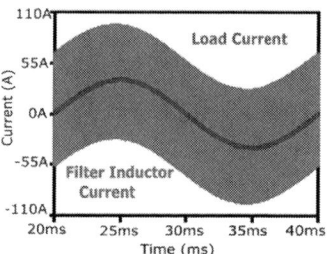

**Fig. 3:** Filter inductor current and average load current for Phase A.

## 2.3 Principle of Operation

ZVS throughout the line cycle is achieved through a systematic strategy that optimizes the utilization of both the MOSFET's output capacitance and its body diode.

A key element of this approach is the incorporation of bidirectional inductor current, which plays a crucial role in discharging the output capacitance across the switch, including both the drain-source capacitance and externally added snubber capacitance. The negative peak of the filter inductor current enables turn-on ZVS for the high-side switch ($S_1$), while the positive current peak achieves the same for the low-side switch ($S_2$). Additionally, the external snubber capacitance aids in the turn-off of ZVS by facilitating gradual charging and discharging. As the voltage across the switch approaches zero, the body diode initiates conduction. This precise timing ensures the smooth initiation of switch activation at zero voltage, thereby guaranteeing ZVS.

To grasp the operating principle of a single-phase half-bridge inverter, it's essential to break down each switching cycle of the inductor current into eight distinct modes, illustrated in Fig. 4. These modes are determined by the direction of the inductor current and the conduction states of the associated switching devices, such as MOSFETs and body diodes. In Fig. 4, the positive and negative signs in each mode represent the direction of the current flow. The direction of the positive current is denoted by the (+) sign, while the direction of the negative current is indicated by the (-) sign. Further detailed analysis of the operating principle of the proposed inverter is available in [13].

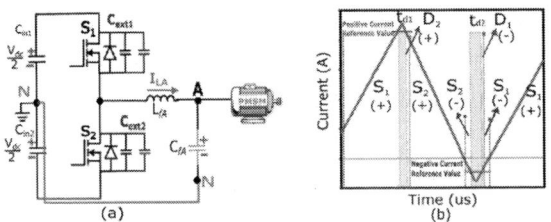

**Fig. 4:** Operating principle of the proposed inverter: (a) Schematic diagram of single-phase half-bridge inverter. (b) Highlighted eight distinct modes of filter inductor current.

## 3 Design Considerations

### 3.1 Snubber Capacitance

Connecting an external capacitance across the drain-source terminals can mitigate the turn-off losses of the MOSFET [15]. This additional capacitance extends the rise time of $V_{ds}$ during the turn-off transient, effectively minimizing the overlap of non-zero $V_{ds}$ and $I_{ds}$ by distributing it.

The symbol $C_{ds, eq}$ represents the effective drain-source capacitance of each MOSFET, calculated as the sum of the external snubber capacitance $C_{ext}$, the intrinsic drain-source capacitance of the MOSFET $C_{oss}$, and any parasitic capacitances of the PCB. The energy stored in the $C_{ds, eq}$ is recovered during the discharging process and is not dissipated in the circuit.

Increasing the drain-source capacitance by adding external snubber capacitance requires careful consideration of two trade-offs:

1. Adding capacitance negatively impacts the turn-off transient switching time. This occurs because the presence of a large $C_{ds, eq}$ requires more time for charging and discharging by the filter inductor current $I_{Lf}$. This limitation becomes more pronounced in scenarios with very high switching frequency requirements.

2. Increasing the $C_{ds, eq}$ of the MOSFET will also raise the turn-on loss experienced during hard-switching. This occurs because the energy stored in the drain-source capacitance dissipates within the MOSFETs during hard-switching. However, this is not a problem for soft switching as it recovers this energy.

The relationship between switching transient time $\Delta T_{sw}$, effective drain-source capacitance of each MOSFET $C_{ds, eq}$, DC link voltage $V_{dc}$, and the instantaneous value of the filter inductor current $I_{Lf, inst}$ can be represented by Eq. (1) [16]:

$$\Delta T_{sw} = \frac{2C_{ds,eq} \cdot V_{dc}}{I_{Lf,inst}} \tag{1}$$

Equation (1) holds for switching instances where the energy stored in the inductor $L_F$ significantly exceeds the energy required for commutating the voltage of $C_{ds, eq}$ [16].

From Eq. (1), the following insights can be derived:

- **Adaptive Blanking Time:** The blanking time needs adjustment to a duration exceeding the turn-off switching time to ensure that the soft-switching process finishes before initiating the complementary switch. The AGD effectively controls the turn-on signal, activating the switch after zero voltage. As the filter inductor current is sinusoidal, an adaptive blanking time is necessary to achieve turn-on and turn-off ZVS.

- **Filter Inductor Current:** The filter inductor current can be assumed to be constant because the snubber capacitance is so small during commutation. Despite the triangular variation of the filter inductor current, the rapid voltage changes during commutation result in relatively small current variations. Thus, for simplicity, the filter inductor current can be approximated as constant during this period. While it's not entirely accurate due to sinusoidal variation, this variation is minimal.

- **Equivalent Snubber Capacitance:** Although the two snubber capacitors may appear to be in series, they are effectively parallel from a circuit perspective. This is because one capacitor charges while the other discharges, with half of the current flowing through each. Therefore, placing these two snubber capacitances in parallel with each other is equivalent to having only one, with no practical difference in operation.

- **Optimum Value of Snubber Capacitance:**
  While there isn't a single universally optimal value of snubber capacitance for an entire line cycle, there may be one for a specific operating point. For instance, when operating as a DC-DC converter with constant current, it's possible to determine an optimal value of snubber capacitance to achieve maximum efficiency [17]. However, in the dynamic setting of a three-phase inverter, such as in an electric vehicle motor drive where currents undergo continuous variations, identifying a straightforward optimal value for snubber capacitance becomes more complex.

## 3.2 Variable Switching Frequency

In the TCM-based ZVS three-phase inverter, the focus is on regulating the current ripple, and the switching frequency isn't directly controlled. Depending on specific operational conditions, this naturally results in a certain instantaneous switching frequency, which varies from cycle to cycle.

In a standard hard-switched inverter, an increase in switching frequency leads to notable switching losses. However, in TCM-based ZVS inverters, higher switching frequencies don't necessarily correspond to excessive losses; soft switching keeps losses low.

To select the LC filter component values for a TCM-based ZVS inverter, a minimum switching frequency is required to estimate the value of filter components. However, when the actual switching frequency is lower than expected, a particular system experiences a higher voltage ripple on the filter capacitor. Meanwhile, the filter inductor current ripple remains constant due to current references.

In such instances, two options are available:

- Attempt to reduce the ripple, although this might not be feasible without compromising the ability to operate at the highest currents.

- Alternatively, accept the higher voltage ripple applied to the motor or increase the size of the filter capacitor to mitigate the voltage ripple on the motor.

## 3.3 LC Low-Pass Filter

The goal of LC filter design is to achieve ZVS and generate sinusoidal output current and voltage waveforms while minimizing total harmonic distortion (THD). This involves thorough consideration of factors such as switching frequency, filter inductance, and filter capacitance. However, achieving these goals often requires making trade-offs to achieve an optimal design.

### 3.3.1 Filter Inductance

In a TCM-based ZVS three-phase inverter, the inductor current has positive and negative peaks for turn-on ZVS. It consists of a high-frequency component ($I_{Lpp}$) and a sinusoidal fundamental component ($I_{Lavg}$). The load primarily carries the sinusoidal component, while the capacitor primarily carries the high-frequency component. The ratio of the peak-to-peak inductor current ripple ($I_{Lpp}$) to the average inductor current ($I_{Lavg}$), is computed as follows:

$$K \cdot I_{Lavg} = I_{Lpp} \tag{2}$$

Here, $K$ represents a constant.

For a TCM-based ZVS single-phase half-bridge inverter, the filter inductance ($L_f$) can be determined as follows:

$$
\begin{aligned}
L_f &= \frac{(V_{\text{in}} - V_{\text{out}})}{K \cdot I_{\text{avg}}} \cdot \frac{1}{f_{\text{sw}}} \\
&= \frac{(V_{\text{dc}}/2) - ((V_{\text{dc}}/2) \times m)}{K \cdot I_{\text{avg}}} \cdot \frac{1}{f_{\text{sw(min)}}}
\end{aligned} \tag{3}
$$

Here, $V_{\text{in}}$ represents the input voltage, $V_{\text{out}}$ denotes the output voltage, $I_{\text{avg}}$ stands for the average inductor current, $V_{\text{dc}}$ signifies the direct voltage, $m$ denotes the modulation index, and $f_{\text{sw(min)}}$ represents the minimum switching frequency.

### 3.3.2 Filter Capacitance

The filter capacitor within an LC low-pass filter of a TCM-based ZVS three-phase inverter is responsible for smoothing voltage fluctuations by absorbing high-frequency harmonic current components. This ensures a stable and reliable voltage output, which is essential for EV applications.

The determination of the filter capacitor size, based on the percentage of output voltage ripple $\Delta V_c$ [18] can be expressed as:

$$C = \frac{\Delta Q}{\Delta V_c} = \frac{\Delta i_L}{8 f_{sw(min)} \times \Delta V_c}, \tag{4}$$

In this context, $\Delta i_L$ signifies the alteration in the inductor current over one switching cycle, while $f_{sw(min)}$ denotes the soft-switched inverter's minimum switching frequency.

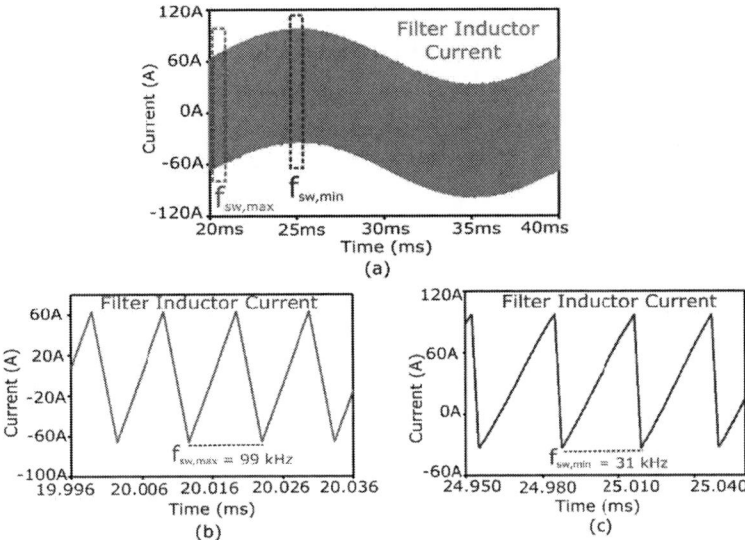

**Fig. 5:** Variation in switching frequency: (a) Filter inductor current of phase A highlighting two distinct areas. (b) Maximum switching frequency observed at zero-crossing. (c) Minimum switching frequency occurs at maximum load current.

## 4 Results and Discussion

The performance of the TCM-based ZVS three-phase inverter with an LC filter was assessed using LTspice software under full-load operational conditions, with a rated power of 10 kW in electric vehicle drive systems.

Table 1 presents the electrical parameters of the inverters, LC filter, and permanent magnet synchronous motor (PMSM) utilized in the LTspice simulation. The load model incorporates an inductor and resistor, which represent the terminal impedance of the PMSM.

**Tab. 1:** Electrical Parameters of Inverter, LC filter and PMSM

| Parameters | Values |
| --- | --- |
| Power Rating | 10 kW |
| Input Voltage | 500 $V_{dc}$ |
| Fundamental Frequency ($f_o$) | 50 Hz |
| Minimum switching frequency | 20 kHz |
| Snubber Capacitance | 2 nF |
| Filter Inductance | 9 $\mu$H |
| Filter Capacitance | 50 $\mu$F |
| Modulation index (m) | 0.9 |
| Power factor | 0.9 |
| R-load | 5.63 $\Omega$ |
| L-load | 8.83 mH |

### 4.1 Variation in Switching Frequency

Although the turn-on signal is controlled by the AGDs, the turn-off process lies within the designer's control. Consequently, the switching frequency isn't directly regulated by the designer, leading to variable switching frequency to achieve ZVS. Figure 5 illustrates this variation over one line cycle. In Fig. 5a, the filter inductor current of phase A is depicted, highlighting two areas: minimum and maximum filter inductor current. At zero-crossing, the switching frequency is at its maximum as illustrated in Fig. 5b, whereas it's at a minimum when the average load current is at its peak as shown in Fig. 5c.

### 4.2 Analysis of Turn-on and Turn-off ZVS

In a TCM-based ZVS inverter, the bidirectional filter inductor current is essential for achieving turn-on ZVS, while external snubber capacitance across each switch facilitates turn-off ZVS. Figure 6 displays waveforms illustrating the turn-on and turn-off ZVS events of the proposed inverter design. These visual representations demonstrate the voltage behavior across the high-side switch ($S_1$) in phase A at peak average load current, enabling ZVS operation.

**Turn-on ZVS:**

- In Fig. 6a, the voltage across the switch becomes zero due to the prior discharge of the drain-source capacitance including external snubber capacitance across the switch, and

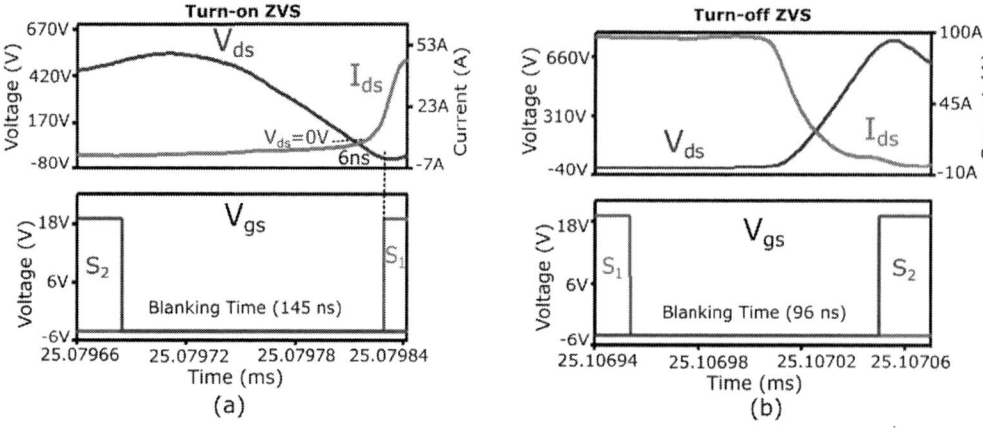

**Fig. 6:** ZVS analysis of the high-side switch $S_1$ of phase A: (a) Turn-on ZVS. (b) Turn-off ZVS.

then the body diode starts conducting. This moment presents the ideal timing to turn on the switch at zero voltage. The AGDs detect the zero voltage across the switch and, after just a 6 nsec delay, initiate the switch turn-on process.

– The blanking time is comparatively high (145 nsec) because the small negative inductor current takes more time to discharge the output capacitance of the switch.

**Turn-off ZVS:**

– In Fig. 6b, during turn-off, there is a small overlapping of $V_{ds}$ and $I_{ds}$. This occurs because the filter inductor current is at its positive peak, quickly charging the output capacitance. However, if the filter inductor current is of a smaller quantity, then there will be no overlapping, achieving turn-off ZVS.

– The blanking time is low (96 nsec) because the large positive inductor current quickly charges the output capacitance of $S_1$ and discharges the output capacitance of $S_2$ switch.

### 4.3 Conduction Losses Reduction

To achieve ZVS in AGD-based inverters, a high ripple inductor current is required. However, this ripple contributes to increased conduction losses in the switches. There are two approaches to addressing conduction losses in TCM-based ZVS inverters:

1. Increasing the die area reduces the resistance of the channel, thereby lowering conduction losses.

2. Employing multiple MOSFETs connected in parallel can also lower the $R_{ds(on)}$, thereby effectively reducing conduction losses.

### 4.4 Sinusoidal Output Waveforms & EMI Mitigation

The TCM-based ZVS inverter generates sinusoidal output waveforms, pivotal in protecting the motor windings and bearings, thus extending the lifespan of the electric motor.

– Figure 7 displays the sinusoidal waveforms of current and voltage at full rated power of 10 kW, while Fig. 8 illustrates the sinusoidal waveforms at low power of 3.5 kW.

– As observed from Fig. 7 and Fig. 8, it's clear that conductive EMI is almost absent due to the sinusoidal nature of these waveforms. Although the fundamental component remains the primary contributor to EMI, its intensity is notably reduced compared to that of a standard two-level converter, largely due to the minimal presence of harmonic components.

– Similarly, radiated EMI is minimized, mainly originating from the fundamental component. Despite its inherent nature, there is a substantial reduction in the magnitude of radiated EMI.

### 4.5 Efficiency

The average input power of a three-phase inverter is calculated as the product of the constant DC-link voltage $V_d$ and the average value of the DC-link current $I_{d(avg)}$ as shown in Eq. (5).

$$P_{\text{in(avg)}} = V_{\text{d}} \cdot I_{\text{d(avg)}} \tag{5}$$

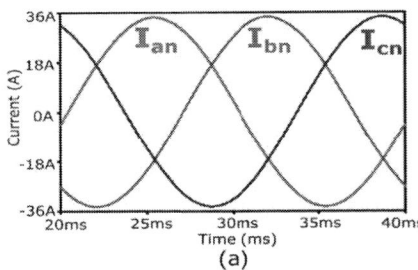 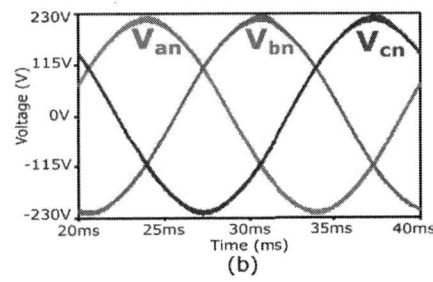

**Fig. 7:** Sinusoidal output waveforms at full rated power 10 kW: (a) Phase currents. (b) Phase voltages.

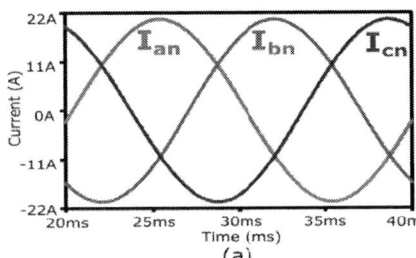 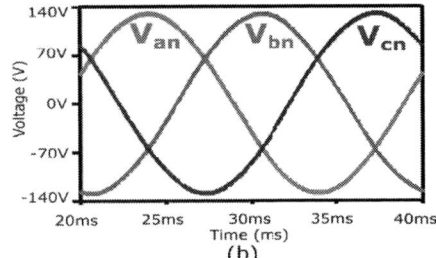

**Fig. 8:** Sinusoidal output waveforms at low power 3.5 kW: (a) Phase currents. (b) Phase voltages.

The output power of the inverter, expressed in Eq. (6), considers the integration of the product of phase voltages $V_a$, $V_b$, and $V_c$ with their respective phase currents $I_a$, $I_b$, and $I_c$ over one switching period $T$.

$$P_{\text{out(avg)}} = \frac{1}{T} \left[ \int V_a(t) I_a(t)\, dt \right.$$
$$\left. + \int V_b(t) I_b(t)\, dt + \int V_c(t) I_c(t)\, dt \right] \tag{6}$$

Efficiency ($\eta$) of the proposed inverter, evaluated at steady-state, is determined by the ratio of the average output power $P_{\text{out(avg)}}$ to the average input power $P_{\text{in(avg)}}$, as outlined in Eq. (7). This efficiency is typically expressed as a percentage.

$$\eta = \frac{P_{\text{out(avg)}}}{P_{\text{in(avg)}}} \times 100 \tag{7}$$

During LTspice simulation, the proposed inverter demonstrated an efficiency of **99%**. However, it's noteworthy that while power device losses were accurately modeled, the inductor losses were not adequately represented due to the assumption of a lossless inductor, which does not align with real-world conditions.

### 4.6 Torque Step Dynamics and Speed Evaluation

In a conventional hard-switched inverter, it's evident that current can be altered rapidly by increasing the current through the motor's inductance. However, achieving a torque step with a rise time as short as 1 millisecond, though feasible, may not be practical for most vehicle applications. Yet, in vehicles, excessively rapid torque steps can adversely affect transmission systems, potentially damaging gearboxes, and related components. Thus, it's unlikely that such rapid torque steps would be employed.

A TCM-based ZVS three-phase inverter with an AGDs concept for EVs has been proposed. Concerns may arise regarding the impact of adding an LC filter, which could potentially slow down torque step dynamics. Therefore, it makes sense to check whether torque steps can be produced sufficiently fast. It would be interesting to simulate and assess whether there are prolonged delays or excessive oscillations.

Two scenarios are considered:

- In the first case, the current rapidly transitions from a low value (20A) to nearly (35A) as illustrated in Fig. 9.

- In the second case, the current rapidly drops from the full value (35A) to approximately 50% (20A) as presented in Fig. 10.

Upon examination, it is demonstrated that excessive oscillations are not observed. Torque steps can be achieved much faster than necessary for practical applications. For example, changing the torque

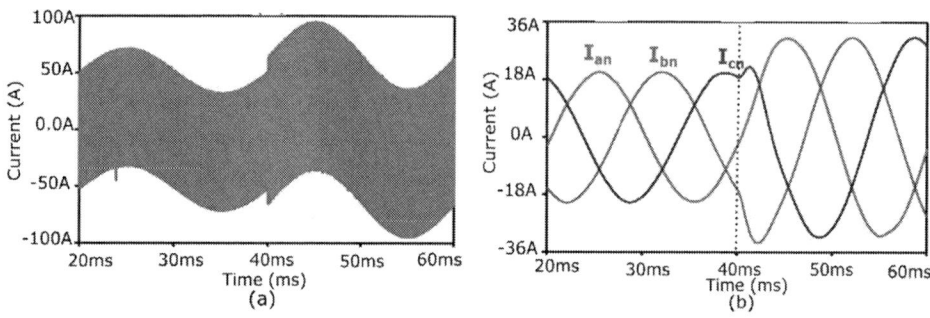

**Fig. 9:** Torque step analysis from 20A to 35A current: (a) Filter inductor current of phase A. (b) Phase currents stepping up from 20A to 35A.

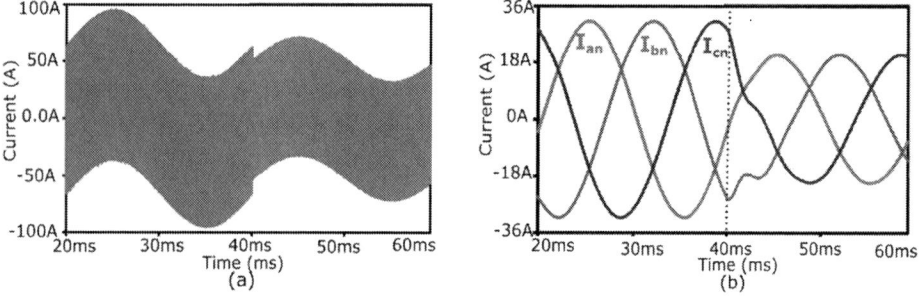

**Fig. 10:** Torque step analysis from 35A to 20A current: (a) Filter inductor current of phase A. (b) Phase currents stepping down from 35A to 20A.

step by 50% of the rated torque within 1 millisecond would be impractical and potentially harmful. A more realistic time frame for torque step adjustments in vehicles would be around 100 milliseconds, which is already approximately rapid. Hence, there is a margin of approximately 100 times faster than required in our case. However, it remains possible to adjust torque almost arbitrarily quickly. One might believe that achieving fast torque steps is challenging. However, it has been shown that torque can be changed almost arbitrarily fast.

## 5 Conclusion

This study evaluated a TCM-based ZVS inverter enhanced with AGDs for electric vehicle drive systems with a rated power of 10 kW. Concerns were raised regarding the potential interference of the LC filter with torque control during transients. The research demonstrated that torque steps can be executed much faster than needed for practical applications, with a margin of approximately 100 times faster than required. Moreover, there were no delays, overshoots, or oscillations observed. Despite this high speed, the system retains the flexibility to adjust torque almost arbitrarily quickly, providing significant advantages in torque control. This

analysis of torque steps recommends the proposed inverter's suitability for EV drive systems.

## References

[1] C. M, "An Overview of the Importance of Power Electronic Converters in Electric Vehicle Technologies," 2023 9th International Conference on Advanced Computing and Communication Systems (ICACCS), Coimbatore, India, 2023, pp. 199-204, doi: 10.1109/ICACCS57279.2023.10113067.

[2] H. Abu-Rub, A. Iqbal, J. Guzinski, "High Performance Control of AC Drives with Matlab/Simulink", Wiley, UK. April 2012, edition-1.

[3] A. A. Al-Abduallah, A. Iqbal, A. A. A. Hamadi, K. Alawi and M. Saleh, "Five-phase induction motor drive system with inverter output LC filter," 2013 7th IEEE GCC Conference and Exhibition (GCC), Doha, Qatar, 2013, pp. 153-158, doi: 10.1109/IEEEGCC.2013.6705767.

[4] M. Su, and C. Chen, "Performance and Reliability Requirements for the Application

of SiC Power MOSFET in Electrified Vehicle Drive Systems," Materials Science Forum ISSN: 1662-9752, Vol. 924, pp 887-890 doi:10.4028/www.scientific.net/MSF.924.887

[5] S. Yu, J. Wang, X. Zhang, Y. Liu, N. Jiang and W. Wang, "The Potential Impact of Using Traction Inverters With SiC MOSFETs for Electric Buses," in IEEE Access, vol. 9, pp. 51561-51572, 2021, doi: 10.1109/ACCESS.2021.3069268.

[6] M. Haider, P. S. Niklaus, M. Madlener, G. Rohner and J. W. Kolar, "Comparative Evaluation of Gate Driver and LC-Filter Based dv/dt-Limitation for SiC-Based Motor-Integrated Variable Speed Drive Inverters," in IEEE Open Journal of Power Electronics, vol. 4, pp. 450-462, 2023, doi: 10.1109/OJPEL.2023.3283052.

[7] J. A. Sabate, V. Vlatkovic, R. B. Ridley, F. Lee, and B. H. Cho, "Design considerations for high-voltage high-power full-bridge zero-voltage-switched PWM converter," in Applied Power Electronics Conference and Exposition, 1990. APEC '90, Conference Proceedings 1990., Fifth Annual, 1990, pp. 275-284.

[8] C. Marxgut, F. Krismer, D. Bortis, and J. W. Kolar, "Ultraflat Interleaved Triangular Current Mode (TCM) Single-Phase PFC Rectifier," IEEE Transactions on Power Electronics, vol. 29, no. 2, pp. 873-882, 2014.

[9] P. Ranstad, H.-P. Nee, J. Linner, and D. Peftitsis, "An Experimental Evaluation of SiC Switches in Soft-Switching Converters," IEEE Trans. Power Electron., vol. 29, no. 5, pp. 2527–2538, May 2014.

[10] C. P. Henze, H. C. Martin, and D. W. Parsley, "Zero-Voltage Switching in High-Frequency Power Converters Using Pulse Width Modulation," Proc. of APEC'88, 1988, pp.33–40.

[11] D. Neumayr, D. Bortis, E. Hatipoglu, J. W. Kolar, and G. Deboy, "Novel Efficiency-Optimal Frequency Modulation for High Power Density DC/AC Converter Systems," Proc. of the IEEE Third International Future Energy Electronics Conference and ECCE Asia (IFEEC 2017–ECCE Asia), Kaohsiung, Taiwan, 3-7 Jun., 2017.

[12] D. Neumayr, D. Bortis, J. W. Kolar, M. Koini, and J. Konrad, "Comprehensive Large-Signal Performance Analysis of Ceramic Capacitors for Power Pulsation Buffers," Proc. of the 17th IEEE Workshop on Control and Modeling of Power Electronics (COMPEL), Trondheim, Norway, 27-30 Jun. 2016.

[13] K. Abbas, H. -P. Nee and K. Kostov, "Comprehensive Insight into the Operational Dynamics of TCM-Based Zero-Voltage Switching (ZVS) Two-Level Three-Phase Inverters for Electric Vehicle (EV) Motor-Drive Applications," 2024 IEEE Texas Power and Energy Conference (TPEC), College Station, TX, USA, 2024, pp. 1-6, doi: 10.1109/TPEC60005.2024.10472258

[14] K. Abbas and H.-P. Nee, "Autonomous Gate Drivers Tailored for Triangular Current Mode-Based Zero-Voltage Switching Two-Level Three-Phase Inverters for Electric Vehicle Drive Systems," Energies 2024, 17, 1060. https://doi.org/10.3390/en17051060

[15] B. Agrawal, M. Preindl and A. Emadi, "Turn-off energy minimization for soft-switching power converters with wide bandgap devices," 2017 IEEE International Conference on Industrial Technology (ICIT), Toronto, ON, Canada, 2017, pp. 236-241, doi: 10.1109/ICIT.2017.7913089.

[16] M. Jahnes, L. Zhou, Y. Fahmy and M. Preindl, "A Peak 1.2MHz, >99.5% Efficiency, and >10kW/L Power Density Soft-Switched Inverter for EV Fast Charging Applications," in IEEE Transactions on Transportation Electrification, doi: 10.1109/TTE.2023.3330997.

[17] K. Abbas, H. -P. Nee and K. Kostov, "Autonomously Modulating Gate Drivers For Triangular-Current Mode (TCM) Zero-Voltage Switching (ZVS) Buck Converter," 2023 22nd International Symposium on Power Electronics (Ee), Novi Sad, Serbia, 2023, pp. 1-6, doi: 10.1109/Ee59906.2023.10346144.

[18] R. H. Ashique and Z. Salam, "A Family of True Zero Voltage Zero Current Switching (ZVZCS) Nonisolated Bidirectional DC–DC Converter With Wide Soft Switching Range," in IEEE Transactions on Industrial Electronics, vol. 64, no. 7, pp. 5416-5427, July 2017, doi: 10.1109/TIE.2017.2669884

PCIM Europe 2024, 11– 13 June 2024, Nuremberg        DOI: 10.30420/566262094

# A Partial Load Three-Phase Triangular Current Mode Modulation Concept with an Optimized Filter Inductor for High Efficiency Traction Drives

Bhaskar Chatterjee[1,2], Jan Allgeier[1], Thomas Plum[1], Marc Hiller[2]

[1] Robert Bosch GmbH, Germany
[2] Karlsruhe Institute of Technology, Germany

Corresponding author:     Bhaskar Chatterjee, bhaskar.chatterjee@de.bosch.com
Speaker:                  Bhaskar Chatterjee, bhaskar.chatterjee@de.bosch.com

## Abstract

This paper introduces a partial-load Triangular Current Mode (TCM) modulation concept for high efficiency traction drive applications without compromising peak performance. This concept relies on a disconnectable AC filter which allows TCM modulation at low current loads and SVPWM modulation at higher current loads. Here, a low power-loss and high power-density saturable filter inductor is designed and optimized which is a key component for the concept implementation and its high efficiency operation. A total power-loss saving of up to 14% is expected with this concept for a 300 kW, 800 V e-vehicle traction drive over the WLTP cycle.

## 1  Introduction

### 1.1  Partial Load Operation

Modern passenger cars increasingly demand higher energy efficiency for covering a higher range or lowering the battery cost, while still constrained to fulfill the performance and speed requirements. However, when observing real driving data [1] low & partial load operation form a significant portion of driving duration and cumulative energy consumption, as seen in the driving energy distribution histogram in Fig.1. This driving distribution trend can be seen in various passenger vehicle classes [1] and also observed in the WLTP [2] operation, simulated for the same vehicle (Fig.1). High power efficiency during low & partial loads is therefore a significant factor for the overall energy efficiency of the traction drive.

### 1.2  Harmonic Losses in Machine

Additionally, in the last decade the trend for traction drive DC-Link voltage has shifted from 400 V to 800 V especially in 100 to 300 kW systems [3], largely due to push for fast & efficient battery charging, higher power density demand and availability & adoption of wide-bandgap semiconductor switches. The increased DC-Link voltage also relatively reduces the fundamental frequency current and conduction losses. However, when this voltage is applied on the machine using standard PWM modulation strategies like SVPWM, it leads to significant eddy current & hysteresis losses in the machine iron [4] and magnets [5] in case of PMSM.
As the PWM harmonic losses in the machine are dependent on the DC-link voltage, the switching frequency and the modulation index, their contribution is relatively

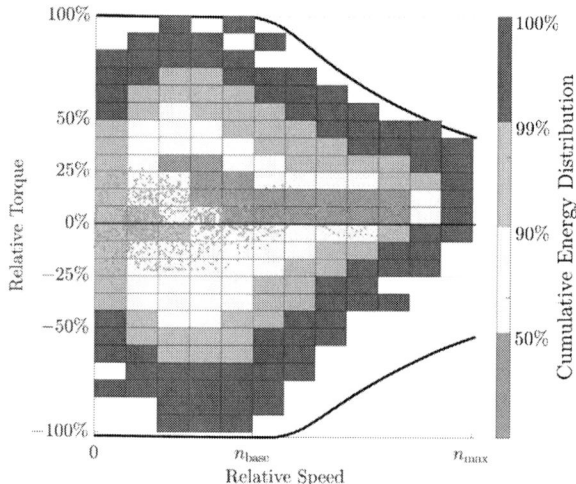

**Fig. 1:** Histogram shows cumulative energy distribution over speed & torque for the e-vehicle *NISSAN Leaf* (First Generation) from the *2012 California Household Travel Survey* [1] (Approximately 300 hours of real speed data and simulated torque for an assumed 1500 kg, 80 kW electric vehicle). Histogram bins differentiating the total 'transacted' energy (absolute value of motoric & generatoric) at 50%, 90%, 99% & 100%. Superimposed on top (*blue* scattered) is the WLTP [2] cycle sampled at 1 s intervals, simulated for the same vehicle.

higher for partial loads. This can be seen in Fig.2 which shows relative energy loss contributions of a simulated 800 V, 300 kW electric drive (inverter & machine) in the WLTP cycle. This simulation meant to gauge the efficiency improvement possibilities in a high power drive is modelled for a 2500 kg vehicle. The drive consists

PCIM Europe 2024, 11– 13 June 2024, Nuremberg        DOI: 10.30420/566262094

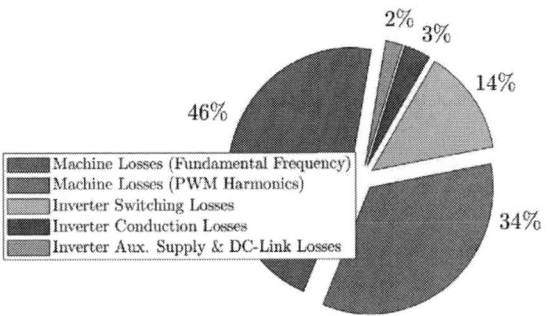

**Fig. 2:** Relative distribution of energy losses in a 800 V, 600 $A_{rms}$ 300 kW electric drive during WLTP at constant switching frequency of 10 kHz for 2500 kg vehicle powered by a 4 pole-pair Permanent Magnet Synchronous Machine & 1200 V SiC switches

of a 800 V, 300 kW 4 pole-pair Permanent Magnet Synchronous Machine [6] and a 1200 V, 600 $A_{rms}$ SiC MOSFET inverter. The inverter is switched at 20 V/ns in SVPWM schema at constant switching frequency of 10 kHz. PWM voltage harmonic losses in the machine contribute to more than a third of the overall WLTP losses in the drive and this work is aimed to lower these losses. These PWM harmonic losses, apart from affecting the overall efficiency also lead to higher machine rotor temperatures affecting the machine performance in PMSM [7]. Additionally, the high slope of the pulsed voltage wavefront puts a constraint on the machine insulation [8] and leads to high transient bearings currents causing damage to the shaft and the race [9].

### 1.3 Existing Solutions

To mitigate or attenuate the higher order voltage harmonics impressed on the machine, a wide range of inverter topologies are proposed in literature. Multilevel inverter topologies like NPC [10], ANPC [11], TNPC [12] etc. do reduce the PWM harmonic amplitudes by a factor of voltage level introduced, but have the drawback of introducing additional semiconductor switches or passive elements [13] in series which cannot be decoupled easily, leading to higher costs and more complex power module and driver design.

Other topologies propose to create a higher fidelity voltage PWM signal at the machine side by increasing the switching frequency either directly incurring high inverter switching losses or using an auxiliary circuit [14] for low loss switching, however the overall efficiency gains are generally lower compared to increasing an additional voltage level [15].

Additionally, low-pass filter elements between the inverter-machine interface are also used to attenuate the PWM voltage harmonics [16]. Other filtering solutions include an active boost-stage like the Y-Inverter [17] or with Zero Voltage Switching in a TCM inverter [18][19] to have high frequency low loss switching. However, in traction application there is a high variation in the load setpoint and low & partial loads have a greater

significance. This leads to a scaling problem as passive components are not easily decoupled between different operating points, leading to over-design and suboptimal power efficiencies.

This paper introduces a partial load triangular current mode (TCM) modulation concept for the high-power traction drives with the feature to decouple low & high operational load points. It adds a custom designed low pass filter with a load dependent switch and a saturable inductor to deploy TCM modulation. This allows for two separate modulation schemas to be operated based on load demand. Partial loads are operated under TCM, providing near sinusoidal voltages at the machine, mitigating PWM harmonic losses. For higher loads, the AC filter is effectively deactivated and the drive is operated under SVPWM leading to no higher current stress of the power module. This concept allows for minimal changes to an existing 2-level power-module and DC-Link design in terms of semiconductor chip area, DC-Link capacitor and busbar designs. This work elaborates on the design principles of a partial load disconnectable filter and highlights the key constraints in dimensioning and operation.

## 2 Partial-Load TCM Modulation Concept

Figure 3 shows the schematic of a disconnectable AC filter between a three-phase inverter and machine. The inverter is operated under TCM modulation at partial loads with the AC filter Fig.3(a) and SVPWM modulation at higher loads with the filter disconnected Fig.3(b). The auxiliary switches $A_{1,2,3}$ disconnect the filter capacitors $C_{1,2,3}$ at load currents higher than the designated boundary current $i_B$. The filter inductors $L_{1,2,3}$ which are still in the load path can be effectively deactivated by the magnetic saturation of the core material at boundary load $i_B$.

### 2.1 TCM Operation

Under TCM Modulation an AC-side filter (as seen in Fig.3(a) with inductors $L_{1,2,3}$ and capacitors $C_{1,2,3}$ ) is doubly utilized, firstly to enable ZVS (zero voltage switching) by subsequent reversal of inverter current direction (or atleast current nullification) between every switching event and secondly to filter out higher frequency component for a near sinusoidal voltage at the machine terminals.

For small time duration $\Delta t$ ($<< 1/2\pi\sqrt{LC}$, characteristic period of the AC filter), the rise and fall times of inverter node current $i_{Node}$ can be described as [20]

$$\Delta t_{rise}(t) = L\frac{\Delta i_{Node}(t)}{U_{DC} - v_{Pha,T_n}(t)} \tag{1}$$

$$\Delta t_{fall}(t) = L\frac{\Delta i_{Node}(t)}{v_{Pha,T_n}(t)} \tag{2}$$

where $U_{DC}$ is the DC-link voltage & $v_{Pha,T_n}$ is the voltage between the machine terminal to the negative DC-link

775

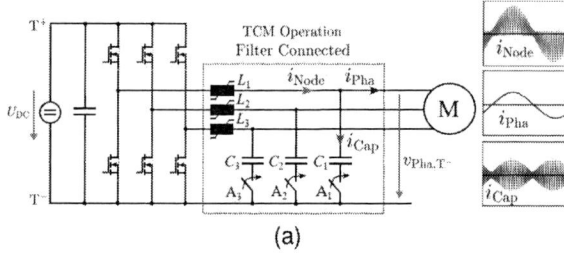

(a)

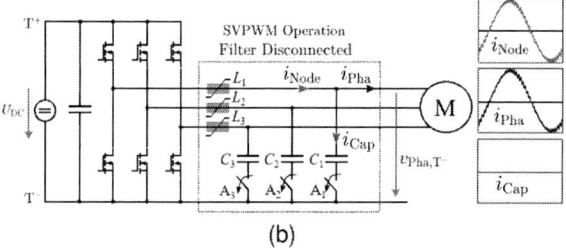

(b)

**Fig. 3:** Partial-Load TCM Modulation Concept with a 'disconnectable AC filter' between the traction inverter and machine operated under (a) TCM modulation scheme at partial-loads with load current under the designed boundary current $i_{\mathrm{B}}$ with auxiliary switches $A_{1,2,3}$ closed and filter inductors $L_{1,2,3}$ unsaturated ; (b) SVPWM scheme at higher loads with load current greater than $i_{\mathrm{B}}$ with $A_{1,2,3}$ open and $L_{1,2,3}$ saturated.

rail. To simplify the above expression $v_{\mathrm{Pha,T_n}}$ can be rewritten in terms of normalized voltage $m(t)$

$$v_{\mathrm{Pha,T_n}}(t) = \frac{U_{\mathrm{DC}}}{2}\big(1 + m(t)\big) \tag{3}$$

Here $m(t)$ incorporates a generalized voltage waveform including common-mode components for higher voltage utilization. For a 3-phase system with a 3$^{\mathrm{rd}}$ order component included, $m(t)$ can be written as

$$m(t) = m_1 \sin(\omega t + \phi) + m_3 \sin(3\omega t + 3\phi + 3\theta_0) \tag{4}$$

Here $m_1$, $\omega$ & $\phi$ are the machine operating point dependent modulation index, fundamental frequency & power factor angle respectively. $m_3$ is the normalized 3$^{\mathrm{rd}}$ order amplitude and can be varied depending on the machine operating point and $\theta_0$ is defined as common mode off-centering angle required for better voltage utlization due to the added inductor voltage drop as can be seen in Fig.4 & 5 for the set of high-speed operating points {3,3'}.

Neglecting the commutation time of the ZVS resonant circuit, the total switching time period and the time varying switching frequency can be formulated as

$$\Delta t_{\mathrm{Swt}}(t) = \Delta t_{\mathrm{rise}}(t) + \Delta t_{\mathrm{fall}}(t) \tag{5}$$

$$f_{\mathrm{Swt}}(t) = \frac{U_{\mathrm{DC}}}{4L} \cdot \frac{1 - m(t)^2}{\Delta i_{\mathrm{Node}}(t)} \tag{6}$$

The upper bound of the switching frequency $f_{\mathrm{Swt}}^{\mathrm{max}}$ can be incorporated into a voltage dependent minimum current

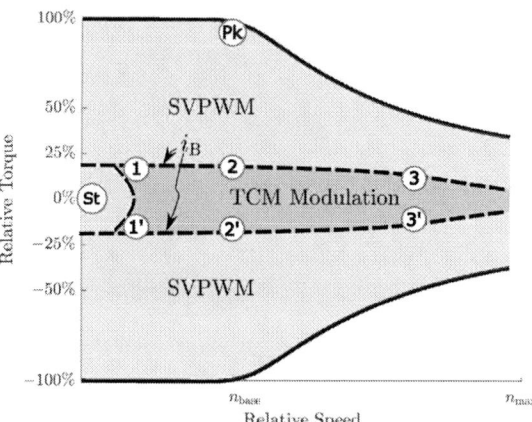

**Fig. 4:** Overview of the traction machine operating region with a bifurcation around 'boundary current' $i_{\mathrm{B}}$ between partial and higher loads. Labelled are sets of points {1,1'}, {2,2'}, {3,3'} in the TCM motoric and generatoric region respectively for increasing speeds and the peak operating point 'Pk' & the no-load stationary point 'St'.

band $\Delta i_{\mathrm{Node}}^{\mathrm{min}}$

$$\Delta i_{\mathrm{Node}}^{\mathrm{min}}(t) = \frac{U_{\mathrm{DC}}}{4L f_{\mathrm{Swt}}^{\mathrm{max}}}\big(1 - m(t)^2\big) \tag{7}$$

For the partial load TCM case, as the filter inductor is designed to saturate at a load current of $i_{\mathrm{B}}$ (absolute value of $2i_{\mathrm{B}}$), the minimum node envelop current $\Delta i_{\mathrm{Node}}^{\mathrm{min}}$ cannot exceed this value in the electrical period, which leads to the lower bound for the TCM filter inductor

$$2i_{\mathrm{B}} \geq \Delta i_{\mathrm{Node}}^{\mathrm{min}}(t) \tag{8}$$

$$L \geq \frac{U_{\mathrm{DC}}}{8 f_{\mathrm{Swt}}^{\mathrm{max}} i_{\mathrm{B}}} \tag{9}$$

Additionally in traction applications where the varied range of voltage is intended to be utilized, a lower bound for the switching frequency $f_{\mathrm{Swt}}^{\mathrm{min}}$ also needs to be defined. This can be incorporated into a current dependent maximum voltage $m^{\mathrm{max}}$

$$m^{\mathrm{max}}(t) = \pm\sqrt{1 - \frac{4L f_{\mathrm{Swt}}^{\mathrm{min}}}{U_{\mathrm{DC}}}\Delta i_{\mathrm{Node}}(t)} \tag{10}$$

It can be seen from Eqn.10 that unlike SVPWM, in TCM, full DC-link voltage cannot be impressed over the machine. This minimum inductor voltage $v_L^{\mathrm{min}}$ needed to drive the triangular current can be expressed as

$$v_L^{\mathrm{min}}(t) = \frac{U_{\mathrm{DC}}}{2}\big(1 - m^{\mathrm{max}}(t)\big) \tag{11}$$

and for a sufficiently small $L$, $v_L^{\mathrm{min}}$ can be linearized as,

$$v_L^{\mathrm{min}}(t) \approx L f_{\mathrm{Swt}}^{\mathrm{min}} \Delta i_{\mathrm{Node}}(t) \tag{12}$$

The inductor voltage drop at the fundamental frequency current waveform $i_{\mathrm{Pha}}$ is neglected here as the fundamental frequency is more than an order of magnitude

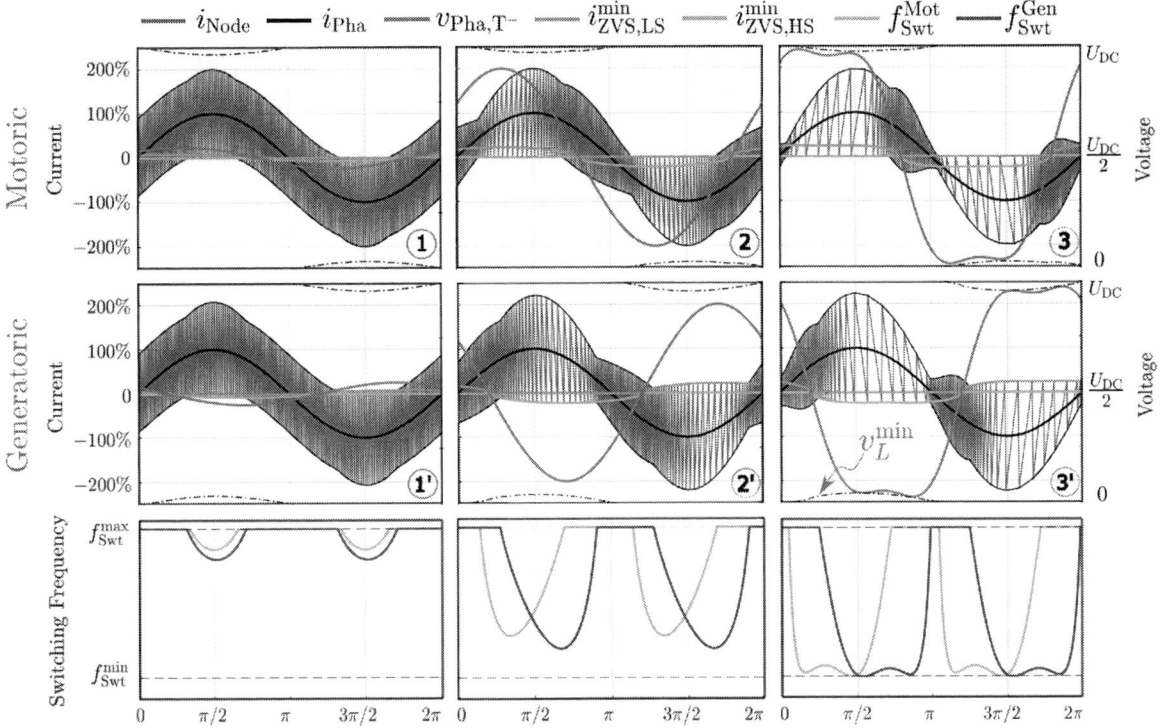

**Fig. 5:** Current, voltage & switching frequency over an electrical period under TCM operation in motoric and generatic modes at operating points {1,1'},{2,2'} & {3,3'} as defined in Fig.4.

lower than $f_{\mathrm{Swt}}^{\min}$ by design. For a PMSM traction machine this loss of voltage utilization will lead to an earlier field-weakening & lost torque and has to be compensated with additional D-axis & Q-axis currents respectively. Balancing the inverter side voltage deficit amplitude to machine side inductive voltage drop leads to

$$\left(\hat{v}_L^{\min}\right)^2 = \omega^2 L_{\mathrm{D}}^2 \Delta i_{\mathrm{D}}^2 + \omega^2 L_{\mathrm{Q}}^2 \Delta i_{\mathrm{Q}}^2 \qquad (13)$$

where $L_{\mathrm{D}}$ & $L_{\mathrm{Q}}$ and $\Delta i_{\mathrm{D}}$ & $\Delta i_{\mathrm{Q}}$ are the D-axis & Q-axis machine inductances and additional currents respectively. With an added assumption that the space-vector direction of $\Delta i_{\mathrm{Node}}$ (or linearized $v_L^{\min}$) is close to phase current $i_{\mathrm{Pha}}$ especially for high speed (high $m$) operating points (see Fig.5) where such a limit arises, one can additionally specify

$$\Delta i_{\mathrm{D}} = -\frac{L}{L_{\mathrm{D}}} \cdot \frac{f_{\mathrm{Swt}}^{\min}}{\omega} \cdot \frac{\Delta \hat{i}_{\mathrm{Node}}}{\hat{i}_{\mathrm{Pha}}} \cdot i_{\mathrm{Q}} \qquad (14)$$

$$\Delta i_{\mathrm{Q}} = -\frac{L}{L_{\mathrm{Q}}} \cdot \frac{f_{\mathrm{Swt}}^{\min}}{\omega} \cdot \frac{\Delta \hat{i}_{\mathrm{Node}}}{\hat{i}_{\mathrm{Pha}}} \cdot i_{\mathrm{D}} \qquad (15)$$

The key takeaway here is that, the ratio of current amplitudes ($\Delta \hat{i}_{\mathrm{Node}}/\hat{i}_{\mathrm{Pha}}$) is roughly a little larger than '2' due to the nature of TCM modulation. The ratio of frequencies ($f_{\mathrm{Swt}}^{\min}/\omega$) is roughly one order of magnitude due to the requirement of a high fidelity filter design between stop & pass frequencies. Which means that, to have sufficiently low increase in additional fundamental frequency

current and conduction losses in field-weakening region of PMSM, the TCM filter inductor $L$ must be roughly two orders of magnitude smaller than the set of machine inductances $\{L_{\mathrm{D}}, L_{\mathrm{Q}}\}$. For 100 kW to 300 kW traction machines the $\{L_{\mathrm{D}}, L_{\mathrm{Q}}\}$ can range between 250 μH to 650 μH leading to the upper bounds of $L$ as roughly <10 μH.

For a complete ZVS Turn-On operation & therefore a full voltage commutation in both Low & High side switches in a bridge leg, a minimum sufficient energy is required in the filter inductor [21] requiring a minimum reverse inverter node current $i_{\mathrm{ZVS}}^{\min}$ [19][22]. For the Low & High side switches (LS & HS) these can be respectively defined as

$$i_{\mathrm{ZVS,LS}}^{\min}(t) = \begin{cases} 0 & m(t) < 0 \\ \frac{U_{\mathrm{DC}}}{Z_c} \sqrt{m(t)} & m(t) \geq 0 \end{cases} \qquad (16)$$

$$i_{\mathrm{ZVS,HS}}^{\min}(t) = \begin{cases} -\frac{U_{\mathrm{DC}}}{Z_c} \sqrt{-m(t)} & m(t) \leq 0 \\ 0 & m(t) > 0 \end{cases} \qquad (17)$$

where $Z_c$ is the characteristic impedance of the resonant circuit between switch output capacitances $C_{\mathrm{oss}}$ (assumed constant) and filter inductor $L$

$$Z_c = \sqrt{\frac{L}{2C_{\mathrm{oss}}}} \qquad (18)$$

In the partial load TCM concept, it is prudent to have the lowest possible time-varying ZVS reverse current instead of a constant value, because $Z_c$ is relatively low for high-power traction drives. The semiconductor chip area is dimensioned for peak current loads and hence have a high effective $C_{\mathrm{oss}}$. Additionally, as discussed above the chosen filter inductor is relatively small ($<10\,\mu\mathrm{H}$). The resulting $i_{\mathrm{ZVS}}$ peak amplitude is non-trivial compared to $i_{\mathrm{Pha}}$ especially at high machine speeds (as seen in Fig.5 & Table 1) and can be variably deployed at low power factor or generatoric operating mode.

With this, the requirements of ZVS and maximum switching frequency bounds can be collected for the upper and lower envelops of the triangular node current $i_{\mathrm{Node}}$ as

$$i_{\mathrm{Node}}^{\mathrm{Upper}}(t) = \max \left\{ \begin{array}{c} 2i_{\mathrm{Pha}}(t) - i_{\mathrm{ZVS,HS}}^{\mathrm{min}}(t) \\ i_{\mathrm{ZVS,LS}}^{\mathrm{min}}(t) \\ i_{\mathrm{Pha}}(t) + \frac{\Delta i_{\mathrm{Node}}^{\mathrm{min}}(t)}{2} \end{array} \right\} \quad (19)$$

$$i_{\mathrm{Node}}^{\mathrm{Lower}}(t) = \min \left\{ \begin{array}{c} 2i_{\mathrm{Pha}}(t) - i_{\mathrm{ZVS,LS}}^{\mathrm{min}}(t) \\ i_{\mathrm{ZVS,HS}}^{\mathrm{min}}(t) \\ i_{\mathrm{Pha}}(t) - \frac{\Delta i_{\mathrm{Node}}^{\mathrm{min}}(t)}{2} \end{array} \right\} \quad (20)$$

Here the three parts correspond respectively to the three conditions imposed on the upper & lower envelops. First incorporates the required reversal of current for ZVS of the complimentary switch. Second corresponds to ZVS of the conducting switch especially in a case of a very low current. And third constraint is to avoid the violation of maximum switching frequency.

## 2.2 SVPWM Operation

For operation above the boundary current load $i_{\mathrm{B}}$ (Fig.4), the auxiliary switch $A_{1,2,3}$ are open and the TCM inductor is designed to be in saturated state as seen in Fig.3(b). Here the drive is operated under the SVPWM scheme with pulsed voltages impressed on the machine. The filter capacitor current $i_{\mathrm{Cap}}$ is negligible because the filter leg now only contains the parasitic output capacitance of the auxiliary switch.

The TCM filter inductor saturation is designed to be approximately at 130% of the boundary current $i_{\mathrm{B}}$, to incorporate variation in relative permeability in the ferrite core (discussed in section 3.3) based on temperature and to aid in transition between the TCM and SVPWM operation modes. Therefore in SVPWM operation near $i_{\mathrm{B}}$ there is an inductive voltage drop across TCM inductor $L$. However, the additional current requirement is small due to the series connection of $L$ with machine inductances $\{L_{\mathrm{D}}, L_{\mathrm{Q}}\}$

$$\frac{\Delta i_{\mathrm{D}}}{i_{\mathrm{D}}} = \frac{L}{L_{\mathrm{D}}} \quad (21)$$

$$\frac{\Delta i_{\mathrm{Q}}}{i_{\mathrm{Q}}} = \frac{L}{L_{\mathrm{Q}}} \quad (22)$$

SVPWM can also be operated near the vicinity of the stationary no-load operating point (operating point 'St' in

Fig.4) with near zero voltage and current requirements to avoid TCM conduction & switching losses at $i_{\mathrm{B}}$ and $f_{\mathrm{Swt}}^{\mathrm{max}}$ as seen in Eqn.7. A minimum normalized voltage $m^{\mathrm{min}}$ can be calculated based on the net loss savings at low speeds for a specified drive.

It can be stated that the DC-Link capacitor, the chip area of the main switches in the inverter bridges, the AC & DC side busbars generally do not require a modification. Because, firstly the boundary current $i_{\mathrm{B}}$ is chosen to be sufficiently small (see section 3.2). Such that the maximum worst-case rms current during TCM of $\sqrt{5/6} \cdot i_{\mathrm{B}}$ still does not exceed the continuous operation rating of the inverter and is much lower than the peak current (at operating point 'Pk' Fig.4), which is the limiting constraint for the chip area of main switches. Additionally, the operating switching frequency in TCM is higher than the case with hard switched SVPWM thus unaffecting the DC-link capacitor requirement.

As this paper focusses on the partial load TCM concept and the design of the filter inductor, the full inverter implementation is outside its scope however it can be hinted that the transition between SVPWM & TCM operating modes can be made one phase at a time during the zero crossing of the phase current $i_{\mathrm{Pha}}$. The concept also requires additional logic for the control of the auxiliary switches $A_{1,2,3}$. A hysteresis based controller for the transition can also be implemented. A similar concept as in [23] can be implemented for the TCM zero-current detection.

## 3 TCM Filter Design

### 3.1 Filter Dimensioning

Assuming a low-loss high quality factor TCM filter, the filter cutoff frequency can be determined by defining a distortion gain factor $k_{\mathrm{pass}}$ for the highest pass-frequency. As the filter is leg is connected to the negative DC-link rail, the frequency of the common-mode voltage also has to be accounted. Therefore, the third harmonic of the highest machine fundamental frequency $f_1$ has to be taken into account for the filter frequency calculation

$$f_{\mathrm{Filter}} = 3 \cdot \sqrt{\frac{1 + k_{\mathrm{pass}}}{k_{\mathrm{pass}}}} \cdot f_1^{\mathrm{max}} \quad (23)$$

Defining similarly an attenuation factor $k_{\mathrm{stop}}$ for the lowest stop-frequency $f_{\mathrm{Swt}}^{\mathrm{min}}$ provides

$$f_{\mathrm{Swt}}^{\mathrm{min}} = \sqrt{\frac{1 + k_{\mathrm{stop}}}{k_{\mathrm{stop}}}} \cdot f_{\mathrm{Filter}} \quad (24)$$

From Eqn.10 it can be seen that the minimum switching frequency $f_{\mathrm{Swt}}^{\mathrm{min}}$ is needed to be as low as possible for greater utilization of the DC-link voltage. Additionally considering SiC switches using Eqn.9 $f_{\mathrm{Swt}}^{\mathrm{max}}$ can be constrained as

$$\frac{U_{\mathrm{DC}}}{8 L i_{\mathrm{B}}} \leq f_{\mathrm{Swt}}^{\mathrm{max}} \leq \min \left\{ \frac{U_{\mathrm{DC}}}{4 L i_{\mathrm{Ch,ZTL}}^{\mathrm{max}}}; 150\,\mathrm{kHz} \right\} \quad (25)$$

Where $i_{\text{Ch,ZTL}}^{\max}$ is the maximum channel current for the Zero Turn-Off Losses for a power module with sufficiently low commutation loop parasitic inductances during fast active turn-off events [24][25]. In an ideal power module design for the presented concept, the $i_{\text{Ch,ZTL}}^{\max}$ should approach $2i_{\text{B}}$. Additionally, a maximum limit of 150 kHz is set by the FCC & CISPR norms. Realistically, as the inductor, busbar copper and gate-driver losses increase with frequency, $f_{\text{Swt}}^{\max}$ is chosen close to its lower bound. The filter capacitance value is simply derived as

$$C = \frac{1}{(2\pi f_{\text{Filter}})^2 L} \tag{26}$$

Both the capacitor and the auxiliary switches have to be rated for $U_{\text{DC}}$ and for the worst case rms current value through filter leg

$$i_{\text{Cap}}^{\text{rms,max}} = \frac{i_{\text{B}}}{\sqrt{3}} \tag{27}$$

For the three auxiliary switches $A_{1,2,3}$, smaller chip area devices of similar make as the main inverter switches can be used. Given the peak current value for the drive system $i_{\text{Pk}}$ the auxiliary switch chip area $A^{\text{Swt,Aux}}$ can be formulated in terms of the main switch area $A^{\text{Swt,Main}}$ as

$$\frac{A^{\text{Swt,Aux}}}{A^{\text{Swt,Main}}} = \frac{R_{\text{dsOn}}^{\text{Swt,Main}}}{R_{\text{dsOn}}^{\text{Swt,Aux}}} = \frac{R_{\text{th}}^{\text{Swt,Main}}}{R_{\text{th}}^{\text{Swt,Aux}}} \leq \frac{i_{\text{B}}}{\sqrt{3}i_{\text{Pk}}} \tag{28}$$

This is assuming both devices are allowed the same maximum chip hotspot temperature and that the device electrical resistance $R_{\text{dsOn}}$ & the thermal contact resistance $R_{\text{th}}$ scale inversely with chip area. Unlike the main switches the auxiliary switching devices only switch during SVPWM-TCM transitions and their meaningful loss expenditure is only conduction losses during TCM operation. The resulting total effective semiconductor device area increases thus by a small proportionate factor of less than $\frac{i_{\text{B}}}{2\sqrt{3}i_{\text{Pk}}}$.

## 3.2 Optimal Boundary Current

For a high efficiency filter design the boundary current $i_{\text{B}}$ and the filter inductance $L$ should be derived simultaneously using the additional loss and load-collective information of the drive system for optimum results.

Ideally when given information about the operating load-collective and the traction machine operating maps, the trajectories of the electrical variables for the traction drive operation over entire lifetime duration $t_{\text{life}}$ can be formulated as

$$\left\{ n_{\text{EM}}(t), T_{\text{EM}}(t), U_{\text{DC}}(t) \right\} \Big|_{t=0}^{t_{\text{life}}} \mapsto$$
$$\left\{ f_1(t), i_{\text{Pha}}(t), m(t), \cos(\phi(t)) \dots \right\} \tag{29}$$

with $n_{\text{EM}}$ & $T_{\text{EM}}$ are respectively the speed & torque operating point of the drive and $U_{\text{DC}}$ is the DC-link (battery) voltage. Eqn. 29 also forms the basis for creating the

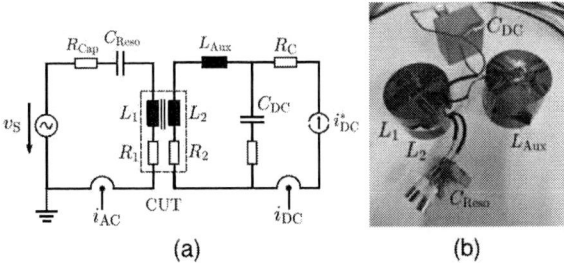

**Fig. 6:** (a) Core material characterization circuit for *Ferroxcube* '3C97' with DC-bias using the (b) 'P66/56' core geometry with 3:1 turns for $L_1$ (35 µH) & $L_2$ (3 µH) and 60 turns for $L_{\text{Aux}}$ (2 mH) with 2 µF $C_{\text{DC}}$.

total traction drive power loss trajectory at SVPWM or TCM operating strategy over the two parameters, boundary current $i_B$ & filter inductance $L$

$$P_{\text{Loss,Drive}}(i_{\text{Pha}}(t), \dots)\Big|_{\{i_B, L\}} =$$
$$\begin{cases} P_{\text{Loss,Drive}}^{\text{TCM}}(i_{\text{Pha}}(t), \dots, L) & i_{\text{Pha}} \leq i_B \\ P_{\text{Loss,Drive}}^{\text{SVPWM}}(i_{\text{Pha}}(t)) & i_{\text{Pha}} > i_B \end{cases} \tag{30}$$

Subsequently, the optimum boundary current & filter inductance values would be the argument at the minima of overall energy loss consumption as

$$\{i_B, L\} = \underset{\{i_B, L\}}{\arg\min} \left\{ \int_{t=0}^{t_{\text{life}}} P_{\text{Loss,Drive}}(t) \cdot dt \right\} \tag{31}$$

For a practical filter design, Eqn. 31 can be solved iteratively with the inductor geometry optimization process (section 3.3.3). Iteratively calculating $i_B$ & $L$ with updated inductor power loss values for the optimum geometric designs should converge to a solution, as the Eqn. 31 is a convex function for both these parameters.

## 3.3 Inductor Design

### 3.3.1 Core Characterization

A P-type core is taken as a starting design for better shielding against stray magnetic flux. For the core material ferrite '3C97' from *Ferroxcube* is chosen due to its low-loss profile and a stable behaviour at high temperature. It also have a relatively high saturation magnetic flux density $B_{\text{Sat}}$ of 300 mT near 140 °C and has a high relative permeabilty $\mu_R \approx 4500$.

To characterize the core material and derive a suitable loss model for non-uniform geometries, initial loss validation tests were made with toroid and P-type cores. The simulation tool FEMM [26] is used create models for magnetic flux density and current density inside the core and winding respectively. The losses are calculated by the volumetric integration of the local power loss density. For specifically the core losses, the simulated magnetic flux density (Fig. 8) is scaled with the fitted loss function from the characteristic measurements as seen in Fig. 7.

The DC-bias characterization of the core is made using the circuit as seen in Fig. 6(a). The CUT ('P66/56' from *Ferroxcube*) has 2 sets of windings for AC & DC excitation. A small air-gap (200 µm) is introduced to avoid uneven contact of the two halves of the P-core. A resonance circuit is used on the AC side with different resonance capacitors $C_{Reso}$ for different test frequencies. The CUT is also characterized for higher temperatures upto 140 °C by heating the top & bottom surfaces of the P-core and making measurements during the cool-down phase with uniform temperatures.

The voltage across the inductor $v_L$ is measured, along with the DC & AC currents. The AC excitation $B_{AC}$ is calcuated from the flux linkage and verified with simulation at no DC-bias. The DC-bias $B_{DC}$ can then be calculated using the two current & the turns ratio of $L_1$ & $L_2$. The AC source supplies the power loss in the core, the winding and the resonant capacitor $C_{Reso}$. Relatively low-loss & stable foil capacitors are used. Their equivalent series resistance can be measured via small signal impedance analyser and its losses excluded. The coil winding resistance can also be measured in a similar manner as core losses scale faster with voltage and diminish at smaller voltages.

Figure 7(a,b) shows the absolute power loss density measured for the core material with respect to the AC excitation $B_{AC}$ and the excitation frequency $f$ and fit well with an exponential function approximation of the Steimetz equation [27]. Figure 7(c,d) shows the pecentage increase in core losses with respect to increasing temperature and DC bias. The relative increase in losses is similar over different AC excitation strengths and frequencies, therefore a loss scaling function can be used to incorporate these effects as in [28]. The constructed loss model over sinusoidal exciation is transformed for triangular excitation using the 'improved generalized Steinmetz equation' (iGSE) as in [29].

### 3.3.2 Winding Selection

The maximum current seen by the filter inductor during TCM is $\approx 2i_B$), however the winding still needs to conduct the entire load including the peak load current $i_{Pk}$. This constraints how small the winding cross section area can be and hence the overall volume of the inductor. However, it should be noted that if the thermal time constant of the inductor (centre) is much larger than the required peak current time duration of the drive, a smaller cross-section is possible. In this case winding cross-section was chosen for the peak current density between 6 to 20 A/mm$^2$. To accommodate the large currents at lower frequency and relatively smaller current at high frequencies solid rectagular wires & litz wires are chosen.

### 3.3.3 Inductor Geometry Optimization

For given values of the inductance $L$, the boundary current $i_B$, maximum magnetic flux density $B_{max}$ & relative permeability $\mu_R$, an optmization function is created for geometric parameters of the inductor. These are namely,

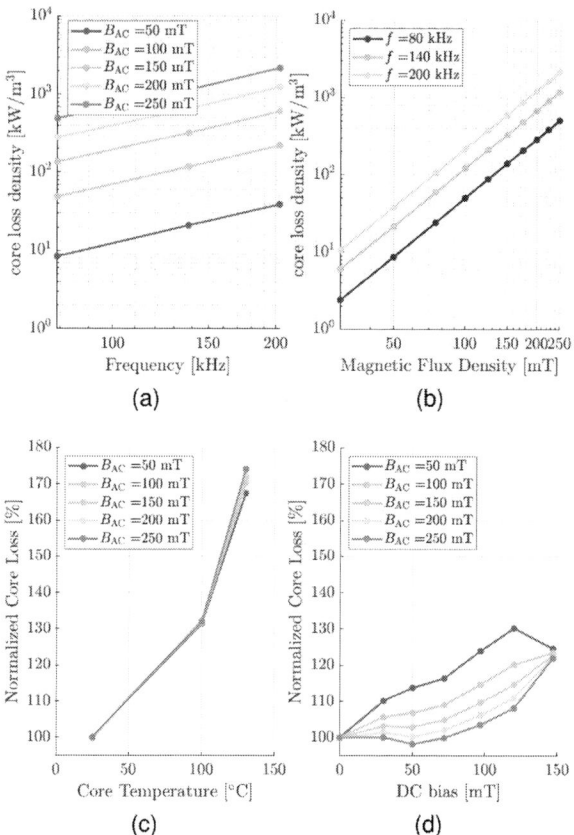

**Fig. 7:** *Top:* Power loss density in ferrite '3C97' measured for P-type 'P66/56' core at 25 °C and 0 mT DC-bias over (a) increasing freqeuncy & (b) increasing magnetic flux density. *Bottom:* Percentange increase in measured core losses at 140 kHz for (c) increasing core temperature at 0 mT DC-bias, (d) increasing DC-bias at 25 °C.

1.) turns number, 2.) core-cross section area, 3.) winding pitch, 4.) air-gap length, 5.) the distance of the winding from the air-gap and 6.) the wire cross-section dimensions. Next the inductor box-volume, total core & winding weight and the inductor power-loss are established as the cost-objective variables. During the optimization run, a tolerance band of 10% is kept for any deviation in the resulting inductance value.

For the loss modelling, the flux density and current density are simulated in FEMM [26] and the powerloss density is calculated using the core characterization function (section 3.3.1) as seen in Fig. 8. The losses are then scaled for the current amplitudes of harmonic components for the TCM node current $i_{node}$. This process is repeated for each machine opearting point to cover the entire operating region (Fig.4). For operating points with load current greater than $i_B$, the simulation is made for an air core.

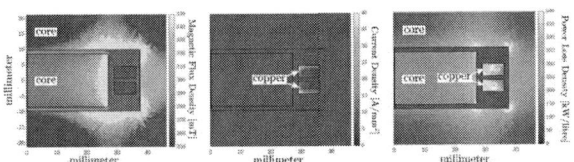

**Fig. 8:** A representative inductor cross-section with simulated magnetic flux density (*left*), the current density (*center*) and the calculated power loss density (*right*).

|  | $L_A$ | $L_B$ | $L_C$ |
|---|---|---|---|
| Inductance | 5.3 µH | 5.1 µH | 5.3 µH |
| Height | 100 mm | 83 mm | 43 mm |
| Diameter | 83 mm | 91 mm | 106 mm |
| Airgap | 9.1 mm | 5.2 mm | 4.5 mm |

**Tab. 2:** Design parameters for the constructed TCM filter inductors as shown in Fig.9. Inductance values are measured for small signals at 140 kHz. Total airgap shown is split in two parts.

## 4 Measurement & Validation

A 300 kW, 600 $A_{rms}$, 800 V drive system is taken in focus with an inverter consisting of 6 parallel 30 mm$^2$ SiC MOSFET devices switched at 20 V/ns during SVPWM operation. For this aforementioned drive and an appropriate load collective, the TCM design and operational parameters are calculated as seen in Table 1. Here the pass & stop gains of 5% & 50% respectively are considered and the maximum rms current values are considered for a motoric operating point. The boundary current $i_B$ comes to 17% of the peak current based on the load collective. The net increase in overall semiconductor device area due additional auxiliary switches $A_{1,2,3}$ comes to approximately 7%.

| $U_{DC}$ | $i_{Pk}^{rms}$ | $f_1^{max}$ |
|---|---|---|
| 800 V | 600 $A_{rms}$ | 1.2 kHz |

| $L_{1,2,3}$ | $C_{1,2,3}$ | $f_{Filter}$ |
|---|---|---|
| 5 µH | 20 µF | 16 kHz |

| $i_B^{rms}$ | $f_{Swt}^{min}$ | $f_{Swt}^{max}$ |
|---|---|---|
| 100 $A_{rms}$ | 40 kHz | 140 kHz |

| $i_{Node}^{rms,max}$ | $i_{Cap}^{rms,max}$ | $i_{ZVS}^{min}$ |
|---|---|---|
| 129 $A_{rms}$ | 81 $A_{rms}$ | −29 A |

**Tab. 1:** Design & operational parameter values obtained for partial load TCM concept for a 300 kW, 800 V drive.

### 4.1 Constructed Inductors

**Fig. 9:** 5 µH custom saturable inductors with saturation at 300 A and rated to conduct 850 A peak current. Constructed using *Ferroxcube* '3C97' core and litz winding (4200 * 142 µm) for $L_A$ (4 turns) & $L_B$ (3 turns) and solid rectangular wire winding (20*2 mm) for $L_C$ (3 turns).

Three different inductor types are constructed with varying degrees of compactness as seen in Fig.9 with design parameters shown in Table 2. Figure 10 shows their simulated and measured winding resistances (at small signal voltage). The more compact solid wire inductance $L_C$ has a higher $R_{AC}$ at lower frequencies due to skin & proximity effect but also because of a lower copper cross-section area. The strand diameter 142 µm of litz wires ($L_A$ & $L_B$) is chosen based on the skin depth of copper at 140 kHz. However as can be seen in Fig.10 due to their large cross section & relatively short lengths these behave more like a plain stranded wire. The modelling of these wires is done by simulating a linear combination of litz and plain wire resistances as in [30]. The largest inductor $L_A$ even though has a higher number of turns, still has a larger core size and a wire length. This is due to the large wire cross section and turn radius of the litz wire. It is constructed to investigate the plain wire phenomenon in thicker litz wires. The larger cross section of litz is chosen based on nearest market availability.

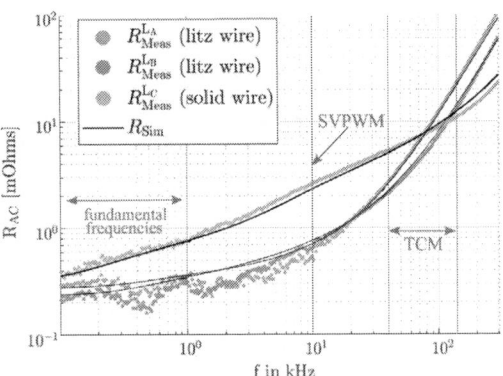

**Fig. 10:** Small signal AC resitance of the three constructed inductors $L_A$, $L_B$ & $L_C$. Shown additionally are the range of frequencies for TCM (40 to 140 kHz), SVPWM (10 kHz) and fundamental frequencies for the machine (under 1 kHz).

### 4.2 Losses during WLTP

Figure 11 shows a comparison of WLTP losses for a 300 kW, 600 $A_{rms}$ drive, simulated at 800 V DC. The inverter & machine harmonic losses for 10 kHz SVPWM operation are calculated to be 1.25 kWh per 100 km.

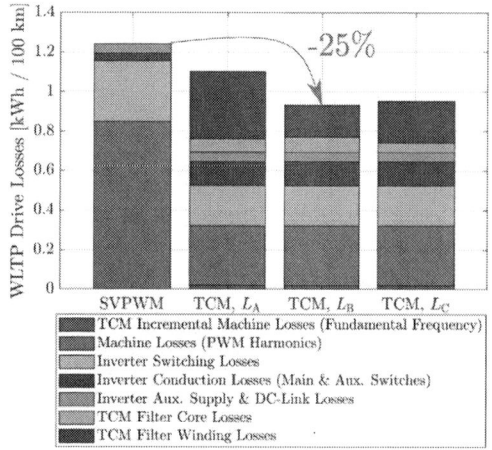

**Fig. 11:** Comparison of energy losses in a WLTP cycle (scaled to 100 km) for a 300 kW, 600 A$_{rms}$, 800 V traction drive for 10 kHz SVPWM and partial load TCM modulation for filter inductors $L_A$, $L_B$ & $L_C$ operated between 40 to 140 kHz with a boundary current of $i_B$ of 100 Arms.

For TCM operation additionally TCM filter losses and fundamental frequency machine losses due to early field-weakening are also included. It can be seen that in the TCM operation, PWM harmonic losses are less than halved but not fully diminished due to the intermittent transition to SVPWM mode in either higher loads or low speeds. Additionally the switching losses are lowered due to only channel turn-off losses at higher frequencies. However, total semiconductor conduction losses (main and auxiliary switches) are increased by a factor of more than three due to higher rms current in TCM but also due to the conduction losses in the auxiliary switches with relatively smaller chip area. The inductor core losses vary in proportion to the core volume of the inductors with lowest losses occuring in the compact $L_C$ inductor. The TCM filter winding losses which also include the relatively small filter capacitor losses correlate with the $R_{AC}$ values at around 120 kHz. The most loss savings are for the $L_B$ design which saves roughly an equivalent of 0.32 kWh for 100 km of driving which is approximately 25% of PWM harmonic & inverter losses & 14% of the total drive losses. It is closely followed by the solid wire inductor $L_C$ with 0.3 kWh per 100 km. $L_A$ is the least performing inductor with reduced savings of 0.15 kWh per 100 km.

# 5 Conclusion

This paper introduced a new modulation concept for high efficiency traction drives which saves machine PWM harmonic losses at partial loads. The concept relies on an additional disconnectable low-pass filter at the inverter-machine interface which allows for TCM modulation at partial load operation and SVPWM at high load. This paper elaborated on the general design principles, constraints and operation for such a filter which effectively

deactivates above a designated partial load boundary current by the active disconnection of filter legs and by the saturation of the filter inductor. Design of an optimum saturable inductor is also discussed, and three designs are constructed and validated as a proof-of-concept. Lastly, expected energy loss savings for a 300 kW, 800 V drive is also calculated for the WLTP cycle, leading upto 25% saving in inverter and machine PWM harmonic losses (overall drive energy loss saving of upto 14%). Future work for this concept can include, design of a low parasitic power module to facilitate low loss (or no-loss) turn-off events. Additionally, further work can be done on the realization of the partial load TCM inverter as a proof-of-concept with modulation & control of the drive over its entire operating region. Lastly, focus can also be given to the size or device are of the auxiliary switches which are currently only ≈7% of the total device area but can be made bigger to have lower on-state conduction losses.

**Acknowledgement** The authors would like to thank Mr. Dennis Bura & Mr. Johannes Rickert for their kind support during measurements in the laboratory.

# References

[1] *2012 California Household Travel Survey, Transportation Secure Data Center*, Accessed Jan. 15, 2017, National Renewable Energy Laboratory, 2017.

[2] *Proposal for a new global technical regulation on the Worldwide harmonized Light vehicles Test Procedure (WLTP)*, World Forum for Harmonization of Vehicle Regulations, Inland Transport Committee, United Nations Economic Commission for Europe, 2014.

[3] I. Husain, B. Ozpineci, M. S. Islam, E. Gurpinar, G.-J. Su, et al., "Electric drive technology trends, challenges, and opportunities for future electric vehicles", *Proceedings of the IEEE*, vol. 109, no. 6, pp. 1039–1059, 2021. DOI: 10.1109/JPROC.2020.3046112.

[4] K. Yamazaki and Y. Seto, "Iron loss analysis of interior permanent-magnet synchronous motors-variation of main loss factors due to driving condition", *IEEE Transactions on Industry Applications*, vol. 42, no. 4, pp. 1045–1052, Jul. 2006. DOI: 10.1109/TIA.2006.876080.

[5] Y. Miyama, M. Hazeyama, S. Hanioka, N. Watanabe, A. Daikoku, and M. Inoue, "PWM Carrier Harmonic Iron Loss Reduction Technique of Permanent-Magnet Motors for Electric Vehicles", *IEEE Transactions on Industry Applications*, vol. 52, no. 4, pp. 2865–2871, Jul. 2016. DOI: 10.1109/TIA.2016.2533598.

[6] M. Reinlein, T. Hubert, A. Hoffmann, and A. Kremser, "Optimization of analytical iron loss approaches for electrical machines", in *2013 3rd International Electric Drives Production Conference (EDPC)*, 2013, pp. 1–7. DOI: 10.1109/EDPC.2013.6689759.

[7] N. Zhao, Z. Q. Zhu, and W. Liu, "Rotor eddy current loss calculation and thermal analysis of permanent magnet motor and generator", *IEEE Transactions on Magnetics*, vol. 47, no. 10, pp. 4199–4202, 2011. DOI: 10.1109/TMAG.2011.2155042.

[8] T. Petri, M. Keller, and N. Parspour, "The influence of voltage form on the insulation resilience of inverter-fed low voltage traction machines with hairpin windings", in *2023 IEEE International Electric Machines & Drives Conference (IEMDC)*, 2023, pp. 1–6. DOI: 10.1109/IEMDC55163.2023.10238888.

[9] M. Asefi and J. Nazarzadeh, "Survey on high frequency models of PWM electric drives for shaft voltage and bearing current analysis", en, *IET Electrical Systems in Transportation*, vol. 7, no. 3, pp. 179–189, Sep. 2017. DOI: 10.1049/iet-est.2016.0051.

[10] A. Nabae, I. Takahashi, and H. Akagi, "A new neutral-point-clamped pwm inverter", *IEEE Transactions on Industry Applications*, vol. IA-17, no. 5, pp. 518–523, 1981. DOI: 10.1109/TIA.1981.4503992.

[11] T. Bruckner and S. Bernet, "Loss balancing in three-level voltage source inverters applying active npc switches", in *2001 IEEE 32nd Annual Power Electronics Specialists Conference (IEEE Cat. No.01CH37230)*, vol. 2, 2001, 1135–1140 vol.2. DOI: 10.1109/PESC.2001.954272.

[12] M. Schweizer and J. W. Kolar, "High efficiency drive system with 3-level t-type inverter", in *Proceedings of the 2011 14th European Conference on Power Electronics and Applications*, 2011, pp. 1–10.

[13] J. Ewanchuk, J. Salmon, and A. Knight, "Performance of a high speed motor drive system using a novel multi-level inverter topology", in *2008 IEEE Industry Applications Society Annual Meeting*, 2008, pp. 1–8. DOI: 10.1109/08IAS.2008.192.

[14] R. De Doncker and J. Lyons, "The auxiliary resonant commutated pole converter", in *Conference Record of the 1990 IEEE Industry Applications Society Annual Meeting*, 1990, 1228–1235 vol.2. DOI: 10.1109/IAS.1990.152341.

[15] M. Schweizer, T. Friedli, and J. W. Kolar, "Comparative evaluation of advanced three-phase three-level inverter/converter topologies against two-level systems", *IEEE Transactions on Industrial Electronics*, vol. 60, no. 12, pp. 5515–5527, 2013. DOI: 10.1109/TIE.2012.2233698.

[16] C. Attaianese, M. Di Monaco, V. Nardi, and G. Tomasso, "Motor side active filter (msaf)", in *2007 IEEE Industry Applications Annual Meeting*, 2007, pp. 1628–1635. DOI: 10.1109/07IAS.2007.251.

[17] M. Antivachis, N. Kleynhans, and J. W. Kolar, "Three-phase sinusoidal output buck-boost gan y-inverter for advanced variable speed ac drives", *IEEE Journal of Emerging and Selected Topics in Power Electronics*, vol. 10, no. 3, pp. 3459–3476, 2022. DOI: 10.1109/JESTPE.2020.3026742.

[18] J. Cho, D. Hu, and G. Cho, "Three phase sine wave voltage source inverter using the soft switched resonant poles", in *15th Annual Conference of IEEE Industrial Electronics Society*, 1989, 48–53 vol.1. DOI: 10.1109/IECON.1989.69610.

[19] C. Marxgut, J. Biela, and J. W. Kolar, "Interleaved triangular current mode (tcm) resonant transition, single phase pfc rectifier with high efficiency and high power density", in *The 2010 International Power Electronics Conference - ECCE ASIA -*, 2010, pp. 1725–1732. DOI: 10.1109/IPEC.2010.5542048.

[20] M. Haider, J. A. Anderson, N. Nain, G. Zulauf, J. W. Kolar, *et al.*, "Analytical Calculation of the Residual ZVS Losses of TCM-Operated Single-Phase PFC Rectifiers", *IEEE Open Journal of Power Electronics*, vol. 2, pp. 250–264, 2021. DOI: 10.1109/OJPEL.2021.3058048.

[21] M. Kasper, R. Burkat, F. Deboy, and J. Kolar, "ZVS of Power MOSFETs Revisited", *IEEE Transactions on Power Electronics*, pp. 1–1, 2016. DOI: 10.1109/TPEL.2016.2574998.

[22] C. Marxgut, F. Krismer, D. Bortis, and J. W. Kolar, "Ultraflat interleaved triangular current mode (tcm) single-phase pfc rectifier", *IEEE Transactions on Power Electronics*, vol. 29, no. 2, pp. 873–882, 2014. DOI: 10.1109/TPEL.2013.2258941.

[23] IEEE Power Electronic Systems Laboratory and D. Neumayr, "The Essence of the Little Box Challenge-Part A: Key Design Challenges & Solutions", *CPSS Transactions on Power Electronics and Applications*, vol. 5, no. 2, pp. 158–179, Jun. 2020. DOI: 10.24295/CPSSTPEA.2020.00014.

[24] X. Li, L. Zhang, S. Guo, Y. Lei, A. Q. Huang, and B. Zhang, "Understanding switching losses in sic mosfet: Toward lossless switching", in *2015 IEEE 3rd Workshop on Wide Bandgap Power Devices and Applications (WiPDA)*, 2015, pp. 257–262. DOI: 10.1109/WiPDA.2015.7369295.

[25] X. Li, X. Li, P. Liu, S. Guo, L. Zhang, *et al.*, "Achieving zero switching loss in silicon carbide mosfet", *IEEE Transactions on Power Electronics*, vol. 34, no. 12, pp. 12 193–12 199, 2019. DOI: 10.1109/TPEL.2019.2906352.

[26] D. Meeker, *Finite Element Method Magnetics*, Version 4.2, 2006.

[27] C. Steinmetz, "On the law of hysteresis", *Proceedings of the IEEE*, vol. 72, no. 2, pp. 197–221, 1984. DOI: 10.1109/PROC.1984.12842.

[28] J. Muhlethaler, J. Biela, J. W. Kolar, and A. Ecklebe, "Core losses under the dc bias condition based on steinmetz parameters", *IEEE Transactions on Power Electronics*, vol. 27, no. 2, pp. 953–963, 2012. DOI: 10.1109/TPEL.2011.2160971.

[29] J. Reinert, A. Brockmeyer, and R. De Doncker, "Calculation of losses in ferro- and ferrimagnetic materials based on the modified Steinmetz equation", *IEEE Transactions on Industry Applications*, vol. 37, no. 4, pp. 1055–1061, Aug. 2001. DOI: 10.1109/28.936396.

[30] H. Rossmanith, M. Doebroenti, M. Albach, and D. Exner, "Measurement and characterization of high frequency losses in nonideal litz wires", *IEEE Transactions on Power Electronics*, vol. 26, no. 11, pp. 3386–3394, 2011. DOI: 10.1109/TPEL.2011.2143729.

PCIM Europe 2024, 11– 13 June 2024, Nuremberg    DOI: 10.30420/566262095

# GaN vs Si Synchronous Rectifier for LLC Converter

Gokhan Sen[1], A. Cem Gungor[2], Milko Paolucci[1], Sriram Jagannath[1] and Serkan Dusmez[3]

[1] Infineon Technologies AG, Austria
[2] RWTH Aachen University, Germany
[3] Huawei Technologies Duesseldorf GmbH, Nuremberg Research Center, Germany

Corresponding author:     Gokhan Sen, goekhan.sen@infineon.com
Speaker:                  Gokhan Sen, goekhan.sen@infineon.com

## Abstract

Adoption of GaN switches on the primary side of soft-switching converters, i.e. LLC, to achieve higher frequencies has become widely accepted in the industry due to the benefits such as lower circulating current requirement for ZVS, lower gate drive losses, and the reduction of duty cycle loss due to faster switch-node voltage transitions. Similarly, using GaN switches in the synchronous rectifier provides clear advantages due to lower $R_{dson}$ per area, lack of $Q_{rr}$, and low output capacitances. This paper analyzes and investigates the benefits of these low parasitics on the system performance. It has been shown that usage of GaN switches on the secondary side, increases ZVS performance, facilitates voltage regulation at low loads, and reduces losses in both below and above-resonance situations.

## 1 Introduction

So far, silicon (Si)-based switches are typically observed on the secondary side as synchronous rectifiers (SR) for LLC converters, despite their higher $Q_g$ related losses compared to GaN [1-3]. On the other hand, studies have shown that the high $C_{oss}$ of the secondary switch increases the equivalent capacitance of the primary switching node like transformer parasitic (intra-winding) capacitance, necessitating a higher peak $i_{Lm}$ for ZVS [4]. Similarly, at no load or very light load conditions, LLC tank gain behavior changes due to the high frequency secondary resonance occurring because of the characteristics of the load, changing from resistive to capacitive caused by the SR $C_{oss}$ as well as the transformer parasitic capacitances [5]. This results in an unexpected increase of LLC tank gain at high switching frequencies, making no load regulation challenging by putting a limit to maximum switching frequency and/or requiring a dummy load.

It is known that one of the key features in LLC topology for high performance is the ZVS of the primary side switches [6]. This requires sufficient magnetic energy to be stored such that the equivalent capacitive energy at the primary switch node is removed completely during the dead-time. This energy equality needs to be satisfied for all operating conditions to utilize ZVS feature of LLC, including input and output voltage corners as well as minimum and maximum load conditions.

Moreover, ZVS condition is also affected by the below or above resonance operations of LLC. One distinction comes from the fact that during above resonance operation, the load current being non-zero at the switching instant helps ZVS by increasing the available inductive energy together with the magnetizing inductance energy. On the other hand, when Si MOSFET is used as SR, $Q_{rr}$ acts as an additional capacitive charge like $Q_{oss}$, which needs to be removed as part of the ZVS process. Also, $Q_{rr}$ value increases with increased load current which counteracts its positive effect.

Another distinction between GaN and Si as SR in LLC is the no-load behavior, which has been studied in detail in the literature. It has been shown that load characteristics change from resistive to capacitive at no-load operation [7]. In [8], time domain analysis has shown that there is a correlation between the switching instant and the oscillation frequency caused by SR $C_{oss}$, which can increase the output voltage at no-load operation. Although it is common to employ burst mode modulation as a remedy to no load regulation problem, audible noise might be an application specific obstacle. Here, the device capacitance related difference between GaN and Si SR needs to be investigated since dummy load requirement depends on this operation mode.

This study provides a comprehensive understanding of the above discussed differences of GaN and Si during the below and above resonance operations of LLC by providing in-depth discussions and modeling. After underlying the fundamental mechanisms, analysis of power loss components and a comparative evaluation between GaN FETs and Si FETs when used as SR FETs in an LLC have been provided. The presented differences of GaN and Si SRs have been verified on a 500W / 24V LLC converter prototype.

## 2   GaN vs Si: ZVS Energy

The ZVS of primary side switches in the below resonance operation is different than the above resonance case because the net energy transfer from primary to secondary is completed naturally before the end of a switching period, after which in-circuit oscillations starts. In the above resonance operation, there is a non-zero current on the secondary side at the primary switching instant in which case, $Q_{rr}$ comes into play if MOSFET SR is used, and the primary and secondary device voltages are clamped. In both operation modes, the equivalent capacitances seen on the primary half bridge switch node is composed of primary and secondary drain-source capacitances as shown in Fig. 1. Component parasitic capacitances and layout related capacitances are also contributing to the ZVS process in practice, which are not considered in this analysis for simplicity. The main difference in below resonance is the random nature of the SR capacitive energy due to the free oscillation between parasitic inductances and capacitances.

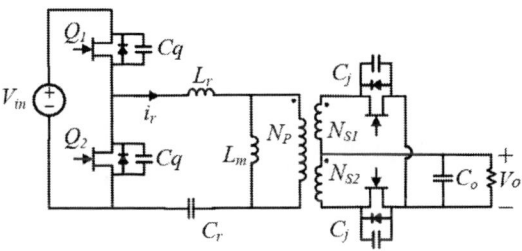

**Fig. 1.** LLC converter with switch parasitics.

To have a better understanding, mathematical derivation of the current is essential to observe this complex relationship. Additionally, precise knowledge of the RMS current after the switch node voltage transition completed within the dead-time period is required to calculate the dead-time losses. An analytical representation of the circuit in Fig. 1 in below resonance free oscillation region is given in Fig. 2, where the resonance current and the magnetizing current become equal after the energy transfer is completed and the next switching cycle has not started yet. The primary current oscillation during this time frame can be observed

in Fig. 3. At the switching instant of primary switches, the primary side current can be anywhere between its peak value and valley point, which determines the amount of capacitive energy stored at the switching instant. This oscillation in $i_{Lr}$ accelerates or decelerates the voltage transition of the primary switch node, depending on when the primary switches are turned off.

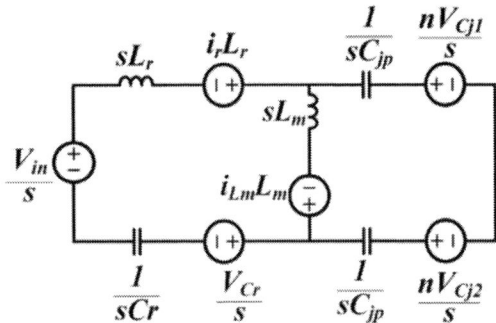

**Fig. 2.** Equivalent circuit of LLC in below resonance operation when $i_{Lr} = i_{Lm}$.

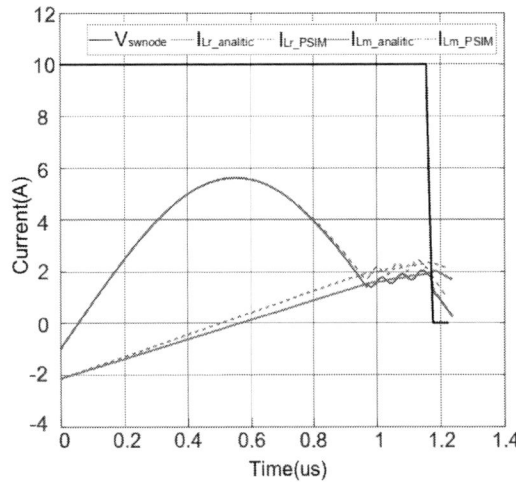

**Fig. 3.** Analytical model and simulation waveforms.

By employing the equivalent circuit models for all the operation modes of below resonance operation and utilizing an iterative algorithm to determine initial conditions, comprehensive circuit information can be derived.

The impact of the SR capacitive characteristics become evident in the required ZVS energy - determination of $L_m$, as well as circulating and dead-time losses. In essence, the magnetizing current used for the ZVS of the primary-side switches circulates in the primary-side of the circuit, leading to additional conduction losses. It can be concluded that the more the respective SR capacitive energy required to be removed, the more magnetizing inductance energy is required, which in turn increases the primary-side RMS currents.

To have a more quantitative understanding, simulations have been conducted based on a 500 W, 24 V output, half-bridge, center-tapped LLC converter with a resonance frequency of 500 kHz. The impact of secondary side SRs on the switch node transition can be observed in Fig. 4, where the switch node voltage transition is plotted, with primary switch capacitances are maintained at 100 pF, and secondary switch $C_{oss}$ values have been swept between 1 nF and 5 nF.

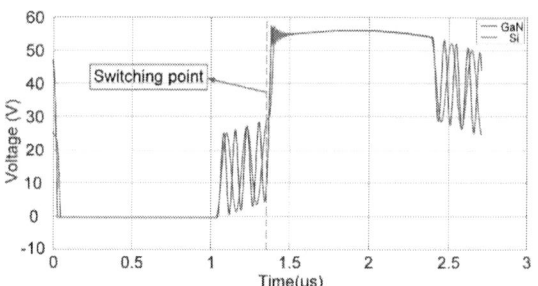

**Fig. 5.** Simulated Vds waveforms for different $C_{oss}$ of GaN and Si SR switches.

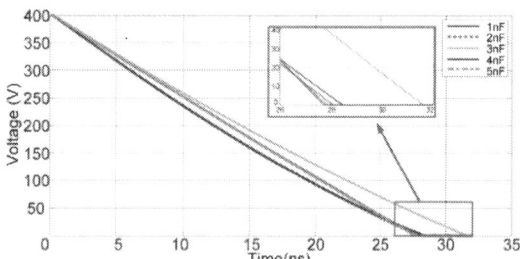

**Fig. 4.** Switch node voltage transition for different secondary side FET parasitic capacitances.

Fig. 4 reveals that there is no proportional relationship between the switch node transition speed and the output capacitance value of the SR switch. This is because different $C_{oss, sec}$ values lead to different oscillation frequencies after $i_{Lr} = i_{Lm}$ as shown in Fig. 5. Depending on the turn-off timing of the primary side FETs, SR $V_{ds}$ can be switched anywhere between its peak and valley points in terms of amplitude. This may either accelerate or decelerate the transition depending on the instantaneous energy stored at the time of switching.

For instance, in this case, $C_{oss, sec}$ = 4 nF has resulted in the fastest slew rate. In a different switching frequency, the effect could be reversed; hence, a larger dead-time would be required to ensure ZVS. The voltage transition time is crucial concerning the power loss occurring between the time switch node voltage reaching zero and the fixed dead-time due to the large body diode or 3rd quadrant voltage drop of primary FETs.

Here it can be understood that, in addition to the instantaneous SR $V_{ds}$ voltage, SR $C_{oss}$ value itself also affects the primary switching transition due to direct proportionality. Based on this relation, one can assume that for the same peak $V_{ds}$ voltage, GaN SR having a lower $C_{oss}$, will store less energy at the time of switching compared to Si SR. This will result in faster transition in below resonance.

However, it should be noted that the below resonance operation is not the worst case for ZVS since the switching frequency is lower than that of above resonance operation which increases the volt-seconds applied to the transformer primary terminals. So, in an LLC converter design if $L_m$ is dimensioned for ZVS at all conditions including above resonance, ZVS in below resonance will also be covered. On the other hand, in case of Si SR, peak capacitive energy is higher which results in wider range of dv/dt and higher dead time losses in a constant dead-time controller, since dead-time will be determined based on worst case condition.

It can be observed in Fig. 6 that until the switching action occurs in the above resonance operation, primary to secondary energy transfer continues which means there is a non-zero load current flowing through the SRs. In this case there will not be any $V_{ds}$ oscillation as in below resonance.

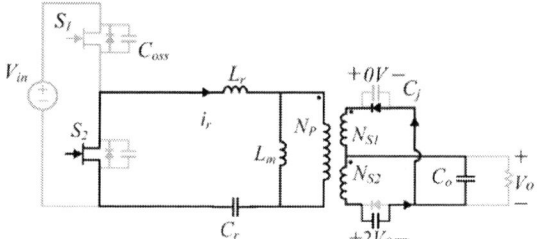

**Fig. 6.** LLC equivalent circuit during SR switching transition in the above resonance operation.

Lowest energy in $L_m$ is stored in maximum operating frequency due to lowest volt-seconds applied to the transformer terminals. In this condition, the equivalent capacitive energy, which is composed of primary and SR $C_{oss}$, transformer intra-winding and layout capacitances, needs to be removed. In case of Si SR, since there is a non-zero current flowing through the body diode, $Q_{rr}$ comes as a contributor to the total capacitive charge, increasing the required inductive energy to be stored in $L_m$.

# 3 Gan vs Si: No Load

At no-load operation, the output voltage of LLC converter exhibits a higher value than the expected value obtained from the first harmonic approximation (FHA) model. This voltage rise requires attention during the design phase for closed loop voltage regulation. The mechanism behind this can be explained by analyzing the primary and secondary switching transition time intervals based on the equivalent circuit given in Fig. 7. Here the left and right-hand sides of the circuit represent the primary and secondary switch capacitances with initial conditions, respectively, where LLC tank acts as an inductive buffer in between. It can be predicted from Fig. 7 that there is a resonance between SR capacitance and transformer leakage inductance where primary switch voltage transitions act as a ramp input voltage.

As switch capacitance is a key parameter for the circuit dynamics during the voltage transition, the selection between GaN or Si SR can significantly impact the performance. Figure 8 shows a no-load simulation comparing the currents flowing to the output, where only SR switches are different in the simulated circuits. One intriguing observation is that the output currents in both cases are non-zero with a triangular shape. However, this seems to be controversial since the load is completely removed in the simulation. This implies an uncontrolled increase in the output voltage there is no load connected to the output that would drain current.

Contrary to Fig. 8, in Fig. 9 where input ramp timing and SR $C_{oss}$ - $L_r$ oscillation period are well matched, the net output current becomes zero. This implies the capacitive resonance current trapped in the former case is completely removed in the latter scenario. Moreover, for the same timing condition, Fig. 8 showed that GaN SR produces less net output current compared to Si SR in a time mismatched condition.

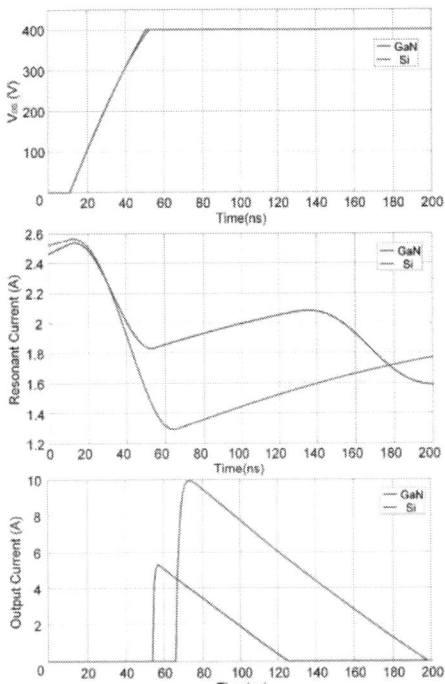

**Fig. 8.** No load simulation results showing primary side voltage transition and circuit currents for GaN vs Si SRs in the worst-case scenario.

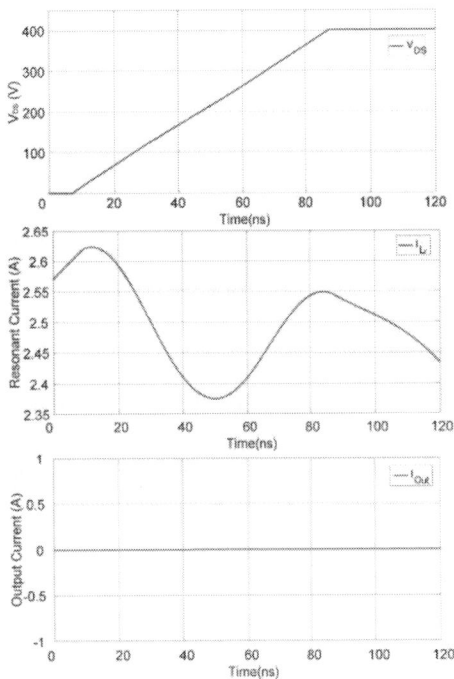

**Fig. 9.** No load simulation results showing primary side voltage transition and circuit currents for GaN vs Si SRs in a well-matched timing scenario.

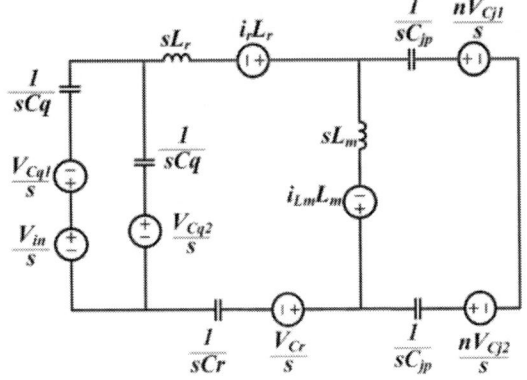

**Fig. 7.** LLC equivalent circuit during primary-side switching at no load condition.

## 4    GaN vs Si: Experiments

Figure 10 shows a 500 W LLC prototype designed for 500 kHz resonance frequency, utilizing Infineon IGLD65R140D2 as primary GaN FET and IGC033S101 as SR GaN FET. For comparison purposes, the tests have been repeated using BSC034N10LS5 Si FET as SR. The primary power stage, the LLC tank, and the transformer are kept the same for both cases. The primary half-bridge dead-time for both Si SR and GaN SR experiments have been adjusted for minimum body diode conduction so that the primary circulating current needed for ZVS is kept inside the primary device channels as much as possible to minimize the losses for both cases. The output voltage is fixed to 24 VDC. The input voltage has been swept between 365 and 405 VDC to observe below-resonance and above-resonance performances and waveforms.

As depicted in Fig. 11, the $i_{Lr}$ current, which is practically equal to $i_{Lm}$ at no-load, is dragged down more compared to GaN SR case at the time of switching when Si FETs are used due to the larger $C_{oss,\,sec}$. This issue becomes more pronounced at high current levels in the above resonance region, as $Q_{rr}$ increases with load current, which results in a drop in primary ZVS energy.

Figure 12 shows experimental waveforms with GaN and Si SR. The waveforms include the primary resonant current (purple), primary half-bridge switching point voltage (blue), and SR drain-to-source voltages for both secondary center-tap sections (red, green). The operating conditions for this measurement are Vin = 405 VDC, $V_{out}$ = 24 VDC. The operation is in the above resonance region, which can also be observed from the waveshape of the resonant current (purple). This waveshape translates into a non-ZCS turn-off for the secondary side SR switches which brings $Q_{rr}$-related effects into play.

From these results any increase in any of the switch capacitances will necessitate a higher ZVS energy. This can be obtained by lowering the transformer magnetizing inductance to increase the peak current, which in turn increases the circulating losses on the primary-side. GaN, having a lower capacitance both on the primary-side and the secondary-side, is a high-efficiency enabler from this perspective. In addition, the lower magnetizing energy requirement implies a higher magnetizing inductance, resulting in a lower air gap for the same number of turns. This results in a lower fringing flux, which is a significant contributor to the transformer winding AC resistance.

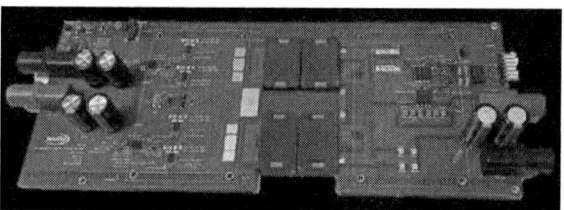

**Fig. 10.** Low-profile GaN LLC converter prototype.

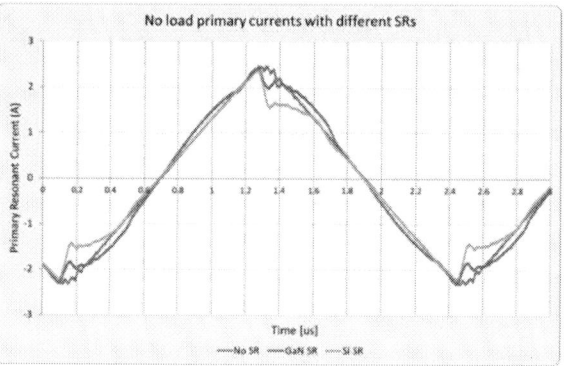

**Fig. 11.** Primary side current for different SR FETs.

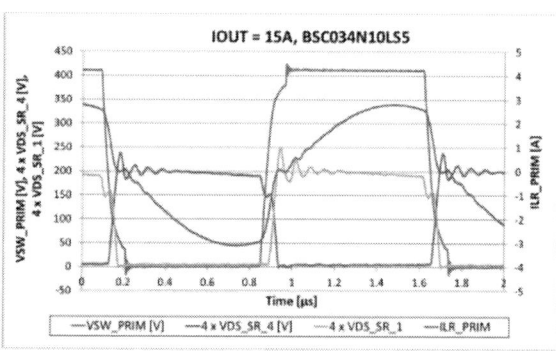

(a)

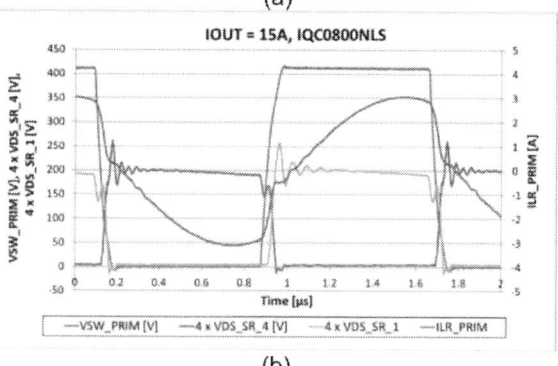

(b)

**Fig. 12.** Switching waveforms at Iout=15A in the case of using (a) Si SR, (b) GaN SR.

Moreover, the measured no-load output voltage is illustrated in Fig. 13, where Si SR has an earlier increase compared to GaN SR, which affects the dummy load sizing. As discussed earlier, if a resonant current is trapped on the secondary side during switching in no load, the output voltage will increase unless there is an equivalent dummy load draining that current, which is frequently used in practice.

The distinction in power losses resulting from the use of GaN and Si SRs can be summarized below as in Eq. (1).

$$\Delta P_s = \Delta P_g + \Delta P_{cir}^{'} + \Delta P_{dead} + \Delta P Q_{rr} + \Delta P_{dum} \qquad (1)$$

Here $P_g$ stands for gate drive, $P_{cir}$ for circulating, $P_{dead}$ for dead-time, $P_{Qrr}$ for reverse recovery and $P_{dum}$ for dummy load losses, respectively. Using the analytically derived models and Eq. (1), the power losses for using GaN and Si FETs as SRs were calculated. The power loss differences for no-load condition are around 0.94 W, and 0.8 W in the below and above resonance regions, respectively. Under full load (500 W), the power loss difference widens to 1.08 W and 1.36 W in the below and above resonance regions, respectively.

Figure 14 shows efficiency measurement results for Vin = 365 VDC and 405 VDC conditions, which are corresponding to a) below-resonance and b) above-resonance operations, respectively. The graphs compare GaN vs Si SR for various load conditions in each case. The GaN SR case shows a clear superiority for both cases for all load conditions. The difference between the GaN SR and the Si SR performance can also be observed in the thermal pictures shown in Fig. 15. SR switch temperatures for both cases differ from each other up to 4°C.

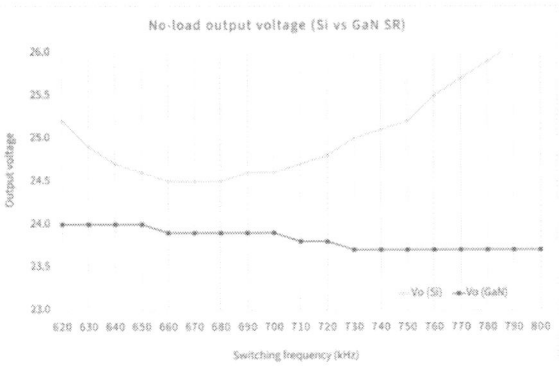

**Fig. 13.** Measured LLC output voltage vs switching frequency at no-load for Si and GaN SR.

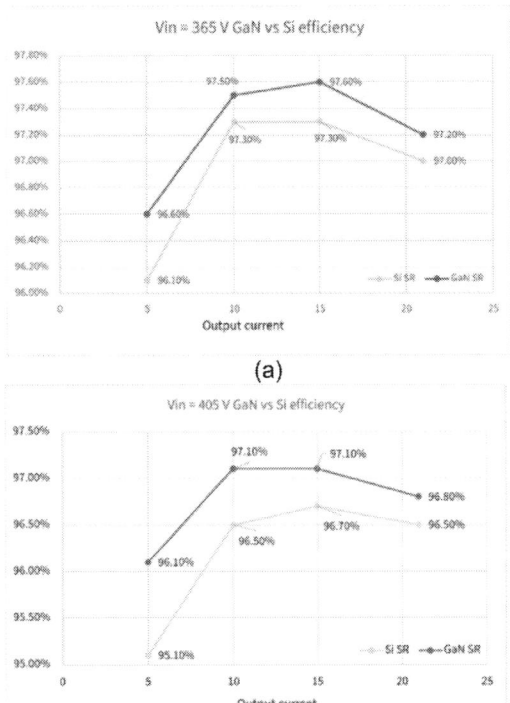

**Fig. 14.** Efficiency comparison of the 500 W LLC.

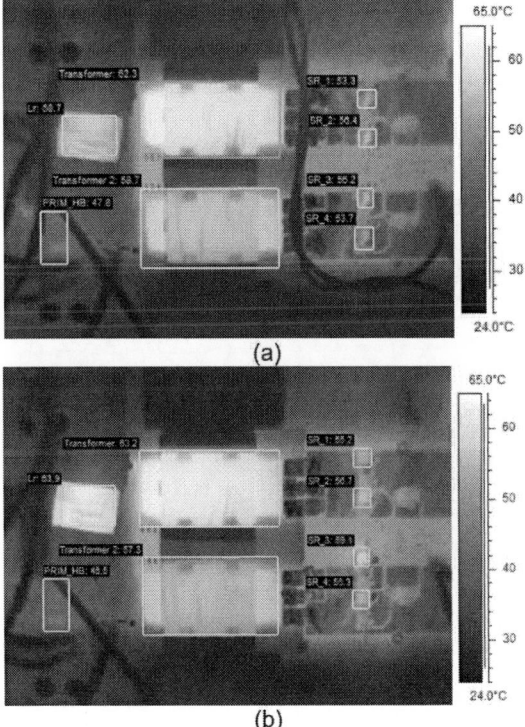

**Fig. 15** Thermal images at rated power operation; (a) GaN SR, (b) Si SR.

## Conclusions

This investigation underscores the suitability of Gallium Nitride (GaN) switches, particularly when employed as synchronous rectifiers (SRs) on the secondary side of LLC converters. The study meticulously quantifies several key advantages associated with GaN SRs:

A. Low Gate Driver Losses: GaN SRs exhibit minimal gate driver losses, contributing to overall system efficiency.

B. Reduced Energy Demand for Primary ZVS: GaN SRs facilitate reduced energy requirements during primary zero-voltage switching (ZVS) due to their inherently lower reflected output capacitance (Coss).

C. Enhanced Voltage Regulation Under No-Load Conditions: The incorporation of GaN SRs enhances the LLC converter's ability to maintain stable voltage levels even during no-load scenarios.

Experimental validation using a 500W / 24V LLC prototype substantiates the practical significance of these findings. Furthermore, a comprehensive power loss analysis, encompassing gate charge losses, circulating losses, dead-time losses, and reverse recovery-related losses, provides conclusive evidence of GaN SRs outperforming traditional silicon (Si) switches across various operational scenarios.

## References

[1] D. Fu, B. Lu and F. C. Lee, "1MHz High Efficiency LLC Resonant Converters with Synchronous Rectifier," 2007 IEEE Power Electronics Specialists Conference, Orlando, FL, USA, 2007

[2] X. Zhou et al., "Analysis and Design of SR Driver Circuit for LLC DC-DC Converter Under High Load Current Application," 2019 IEEE Energy Conversion Congress and Exposition (ECCE), Baltimore, MD, USA, 2019

[3] R. Yu, G. K. Y. Ho, B. M. H. Pong, B. W. -K. Ling and J. Lam, "Computer-Aided Design and Optimization of High-Efficiency LLC Series Resonant Converter," in IEEE Transactions on Power Electronics, vol. 27, no. 7, pp. 3243-3256, July 2012

[4] C. -W. Chen, X. Zhao, C. -S. Yeh and J. -S. Lai, "Analysis of the Zero-Voltage Switching Condition in LLC Series Resonant Converter with Secondary Parasitic Capacitors," 2019 IEEE Applied Power Electronics Conference and Exposition (APEC), Anaheim, CA, USA, 2019

[5] C. -O. Yeon, J. -W. Kim, M. -H. Park, I. -O. Lee and G. -W. Moon, "Improving the Light-Load Regulation Capability of LLC Series Resonant Converter Using Impedance Analysis," in IEEE Transactions on Power Electronics, vol. 32, no. 9, pp. 7056-7067, Sept. 2017

[6] Bo Yang, F. C. Lee, A. J. Zhang and Guisong Huang, "LLC resonant converter for front end DC/DC conversion," APEC. Seventeenth Annual IEEE Applied Power Electronics Conference and Exposition, Dallas, TX, USA, 2002

[7] M. -J. Kwon and W. -C. Lee, "A Study on the Analysis and Control of No-load Characteristics of LLC Resonant Converter for Plasma Process," 2018 International Power Electronics Conference (IPEC-Niigata 2018 -ECCE Asia), Niigata, Japan, 2018

[8] J. -W. Kim, M. -H. Park, B. -H. Lee and J. -S. Lai, "Analysis and Design of LLC Converter Considering Output Voltage Regulation Under No-Load Condition," in IEEE Transactions on Power Electronics, vol. 35, no. 1, pp. 522-534, Jan. 2020

[9] H. Wen at all, "Analysis of Diode Reverse Recovery Effect on ZVS Condition for GaN-Based LLC Resonant Converter," in IEEE Transactions on Power Electronics, vol. 34, no. 12, pp. 11952-11963, Dec. 2019

PCIM Europe 2024, 11– 13 June 2024, Nuremberg    DOI: 10.30420/566262096

# Co-Simulation Design of A GaN-Based Three-Phase LLC Converter with Integrated Three-Phase Magnetics

Jhih-Cheng Hu[1], Shih-Cyuan Kuo[1], Hong-Xuan Liao[1], Ming-Shi Huang[1]

[1] National Taipei University of Technology, Taipei, Taiwan

Corresponding author:  Jhih-Cheng Hu, t111319010@ntut.org.tw
Speaker:                      Jhih-Cheng Hu, t111319010@ntut.org.tw

## Abstract

This paper proposes a co-simulation method using Ansys software to design a GaN-based three-phase interleaved LLC converter with integrated three-phase magnetics, with the aim of reducing circuit design costs and time. Furthermore, to minimize the size of integrated magnetics, the switching frequency is increased to above 800 kHz.  Additionally, three-phase five-leg ferrite core is used to replace traditional three-leg topology, reducing the height through flux sharing in the side-leg. Finally, a 4.8 kW test platform with a 400 V input voltage is constructed to verify the effectiveness of the proposed method, achieving a maximum efficiency of 96.4 % at 400 V output voltage and 3 kW.

## 1 Introduction

Electric vehicles (EVs) have become increasingly popular in recent years due to energy transition concerns, leading to a rising demand for EV battery chargers. The power supply mode of chargers can be divided into AC charging and DC charging [1]. DC charging offers the advantage of providing high power levels and shorter charging durations.

The LLC converter is commonly used to ensure galvanic isolation due to its high efficiency and soft-switching characteristics [2]-[3]. With the development of wide-bandgap semiconductors, Gallium nitride (GaN) offers fast switching speeds and no reverse recovery losses. This allows for an increase in the switching frequency of LLC converters above several hundred kHz, even reaching MHz levels, which can lead to a reduction in the volume of passive components. Furthermore, the use of an integrated magnetic approach can further enhance power density [4].

To increase the output power, the half-bridge structure is usually replaced with a full-bridge structure [5]. However, at higher power levels, the output current ripples increase, resulting in a larger size and higher cost for the required filtering capacitors. Two-phase interleaved LLC converter is proposed to reduce the current ripple [6]-[7]. Nonetheless, this configuration typically requires additional components or control methods to achieve effective current sharing. Another approach studied in the literature is the three-phase interleaved LLC converter with star-connected configuration [8]-[10]. This topology can automatically achieve current sharing through three transformers are star-connected in the primary with a floating star point. Additionally, it can also help reduce the root mean square (RMS) current. Therefore, the three-phase interleaved LLC converter emerges as a superior candidate due to these aforementioned benefits.

This paper proposes a co-simulation design approach for a GaN-based three-phase interleaved LLC converter with integrated three-phase magnetics. The proposed co-simulation design flowchart is shown in Fig. 1. Additionally, a comparison between the three-phase three-leg ferrite and five-leg ferrite is analyzed using Ansys Maxwell 3D. The designed values of magnetics are verified using the impedance analyzer. Subsequently, a co-simulation circuit model is constructed in Ansys Twin-Builder. Finally, a digitally controlled 4.8 kW test platform with an input voltage $V_{in}$ of 400 V is established. The maximum efficiency is 96.4 % at 3 kW. Furthermore, the converter can achieve an output voltage $V_o$ of 400 V when operating at a switching frequency $f_s$ of 800 kHz. The experimental results can verify the effectiveness of the proposed co-simulation design approach.

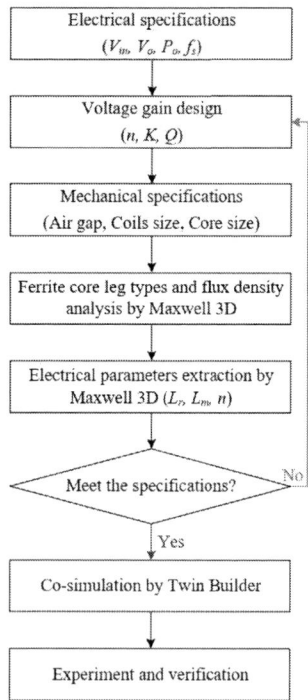

**Fig. 1:** Proposed co-simulation design flowchart.

| Rated Power ($P_o$) | 4.8 kW |
|---|---|
| Input Voltage ($V_{in}$) | 400 V |
| Output Voltage ($V_o$) | 400 V |
| Turn Ratio ($n$) | 1 |

**Table 1:** Specifications of a three-phase LLC converter.

# 2 Three-Phase Interleaved LLC Converter Design

## 2.1 Topology and Specifications

Fig. 2 displays the three-phase LLC converter with star-connection, and its specifications are presented in Table 1. At the primary side, there are three half-bridges based on GaN, and three identical sets of resonant tanks. The three-phase legs are controlled in an interleaved method with a 120° phase shift. Each resonant tank includes a resonant capacitor $C_r$, a resonant inductor $L_r$, and a magnetizing inductor $L_m$. To reduce the volume, $L_r$ and transformer use integrated three-phase magnetics. At the secondary side, there are six diodes that comprise a three-phase full-wave rectifier.

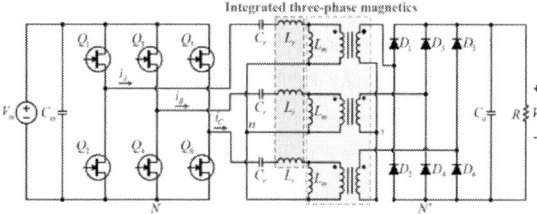

**Fig. 2:** Proposed three-phase interleaved LLC converter.

## 2.2 Power Semiconductor Selection

When the LLC converter operates at the resonant frequency $f_r$ or above it, the primary side switches can achieve zero-voltage switching (ZVS). However, the switches still experience hard-switching at turn-off. Therefore, having a small output capacitance $C_{oss}$ is crucial for achieving ZVS with a lower magnetizing current, which contributes to reducing conduction losses and turn-off switching losses. In addition, this paper uses GaN power switches to increase $f_s$ and improve power density. In this paper, the primary-side GaNs and the secondary-side SiC diodes use Panasonic PGA26E026 and Hestia Power H3D065E030, respectively.

## 2.3 Resonant Tanks Parameters Design

Table 2 illustrated the key design parameters in this paper. To achieve high power density, the $f_r$ of 800 kHz is selected. Based on the $C_{oss}$ of GaN, $L_m$ is designed to be 10 µH with a dead time of 62.5 ns. Then, $L_r$ is selected to be 1.6 µH based on the required inductor ratio $K$, which is 6.2. Therefore, $C_r$ equals to 24 nF using the formula for $f_r$.

| Resonant Frequency ($f_r$) | 800 kHz |
|---|---|
| Switching Frequency ($f_s$) | 800 kHz |
| Resonant Inductor ($L_r$) | 1.6 µH |
| Magnetizing Inductor ($L_m$) | 10.0 µH |
| Resonant Capacitor ($C_r$) | 24.0 nF |

**Table 2:** Key design parameters.

## 2.4 Three-Phase Magnetics Analysis

The common three-phase transformer includes both a three-leg and a five-leg structures. The five-leg structure is advantageous as it can reduce the height of ferrite core through flux sharing in the side-leg, leading to a lower flux density between phases compared to the three-leg topology.

Simulation models and the results of the flux density at 800kHz in Maxwell 3D are presented in Fig. 3 and Fig.4, respectively. When phase A reaches its peak value ($\Phi_A=\Phi_{max}$, $\Phi_B=\Phi_C=-0.5\Phi_{max}$), flux density in both structures remains below the 50mT saturation flux density at the same height. The five-leg transformer has a lower flux density between phases. Additionally, the five-leg structure can prevent the magnetic leakage at the side windings, as shown in the red circles of Fig. 4(a).

Therefore, this paper employs a five-leg transformer to design integrated three-phase magnetics using Ferroxcube 3F46 ferrite brick with ten-layer PCB windings, as displayed in Fig. 5 and Fig.6, respectively. In addition, to further decrease the volume, the flux density after height reduction is depicted in Fig. 7, confirming the effectiveness of the height reduction.

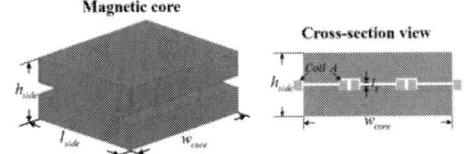

(a) Three-leg (51.2mm*42.4mm*21.2mm)

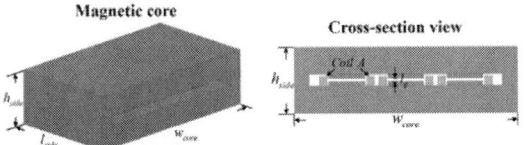

(b) Five-leg (72.8mm*42.4mm*21.2mm)

**Fig. 3:** Three-phase integrated transformer.

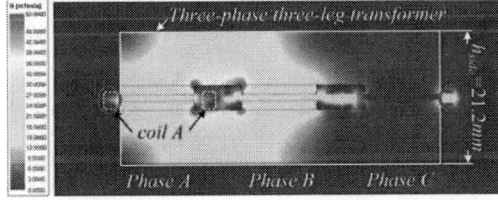

(a) Three-leg (51.2mm*42.4mm*21.2mm)

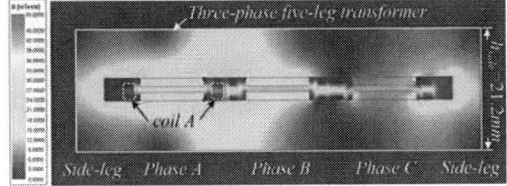

(b) Five-leg (72.8mm*42.4mm*21.2mm)

**Fig. 4:** Magnetic flux density of the three-phase integrated transformer ($\Phi_A=\Phi_{max}$, $\Phi_B=\Phi_C=-0.5\Phi_{max}$).

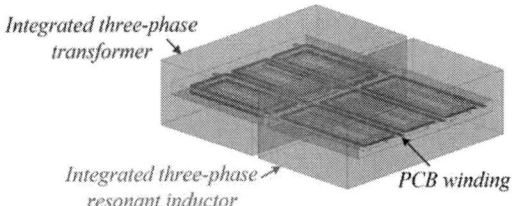

**Fig. 5:** Proposed three-phase five-leg integrated magnetics.

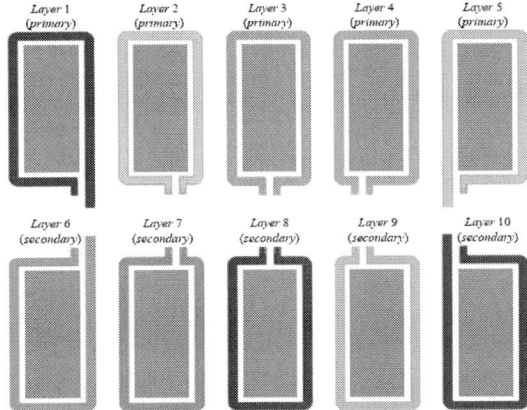

**Fig. 6:** PCB winding implementation for one phase of the proposed integrated three-phase transformer.

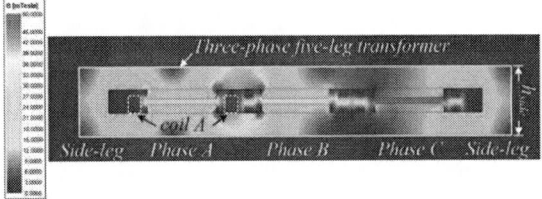

**Fig. 7:** Flux density of the five-leg transformer after height reduction ($h_{side}$=11mm, $\Phi_A=\Phi_{max}$, $\Phi_B=\Phi_C=-0.5\Phi_{max}$).

# 3 Co-Simulation Results and Experimental Verification

## 3.1 Measurement of the Integrated Three-Phase Magnetics

Fig. 8 displays the measurement setup for the integrated three-phase magnetic. The $L_r$, $L_m$, and leakage inductance $L_{lk}$ of each phase are measured using an impdance analyzer (WK 6500B). Both open-circuit and short-circuit methods are employed, as depicted in Fig. 9. The measurement procedure details are described as follows:

(i) Fig. 9 (a) displays the open-circuit test when the secondary-side is open-circuited. $L_{1o}$ can be expressed as follows:

$$L_{1o} = L_{lk1} + L_m \qquad (1)$$

(ii) Fig. 9 (b) displays the open-circuit test when the primary-side is open-circuited. $L_{2o}$ can be expressed as follows:

$$L_{2o} = L_{lk2} + \frac{L_m}{n^2} \qquad (2)$$

(iii) Fig. 9 (c) displays the short-circuit test method when the secondary-side is short-circuited. $L_{1s}$ can be expressed as follows:

$$L_{1s} = L_{lk1} + \frac{n^2 L_m L_{lk2}}{L_m + n^2 L_{lk2}} \qquad (3)$$

(iv) The turn ratio $n$ can be derived from Eq. (1) to Eq. (3).

The measurement results and simulation results using Maxwell 3D are listed in Table 3. It is worth noting that the errors between simulation and measurement are less than 10%, confirming the effectiveness of the simulation.

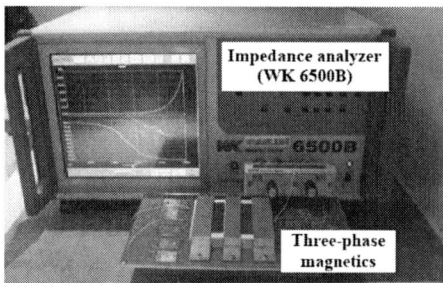

**Fig. 8:** Measurement setup for key parameters of integrated three-phase magnetic.

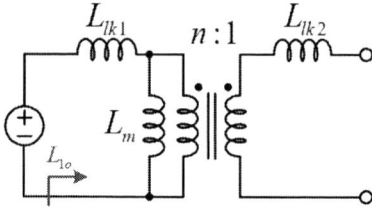

(a) Open-circuit test (primary-side)

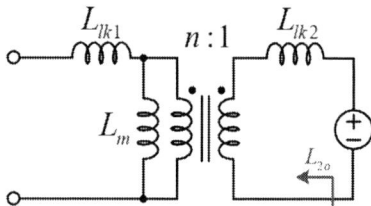

(b) Open-circuit test (secondary-side)

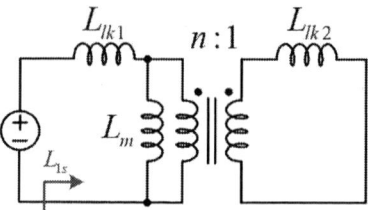

(c) Short-circuit test (secondary-side)

**Fig. 9:** Measurement methods for parameters of the transformer.

| | Maxwell 3D | WK 6500B | Error |
|---|---|---|---|
| $L_{m,a}$ ($\mu$H) | 11.10 | 10.86 | 2.2% |
| $L_{m,b}$ ($\mu$H) | 10.98 | 10.83 | 1.4% |
| $L_{m,c}$ ($\mu$H) | 11.10 | 10.86 | 2.2% |
| $L_{lk,a}$ ($\mu$H) | 0.37 | 0.39 | 5.1% |
| $L_{lk,b}$ ($\mu$H) | 0.37 | 0.39 | 5.1% |
| $L_{lk,c}$ ($\mu$H) | 0.37 | 0.39 | 5.1% |
| $L_{r,a}$ ($\mu$H) | 1.38 | 1.43 | 3.6% |
| $L_{r,b}$ ($\mu$H) | 1.36 | 1.39 | 2.2% |
| $L_{r,c}$ ($\mu$H) | 1.38 | 1.42 | 2.8% |

**Table 2:** The simulation and measurement results of three-phase integrated magnetics.

## 3.2 Co-Simulation and Experimental Results

The simulation model and experimental setup for the proposed three-phase LLC converter are shown in Fig. 10. In addition, the three-phase integrated magnetics model in Ansys Maxwell 3D are imported into Ansys Twin-Builder for the co-simulation of electrical and magnetic fields.

The co-simulation and experimental waveforms at 3.0 kW and 4.8 kW, operating at 800 kHz, are depicted in Fig. 11 and Fig. 12, respectively. It is evident that the co-simulation waveforms in Ansys Twin-Builder closely resemble the experimental results, verifying the effectiveness of the co-simulation model.

Additionally, it is worth noting that the three-phase currents achieve good current sharing due to the star-connected structure of the three-phase transformer.

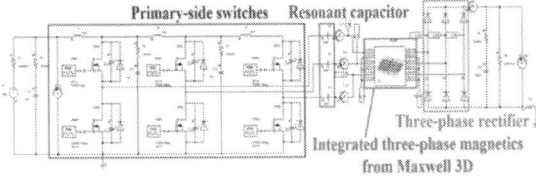

(a) Co-simulation model

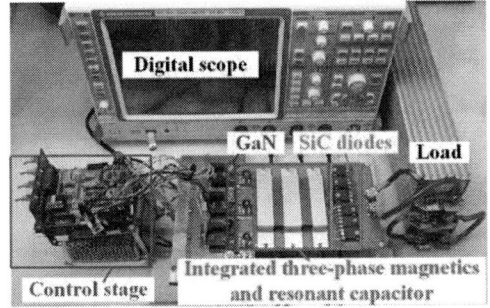

(b) Test platform

**Fig. 10:** Co-simulation model and test platform of three-phase interleaved LLC converter.

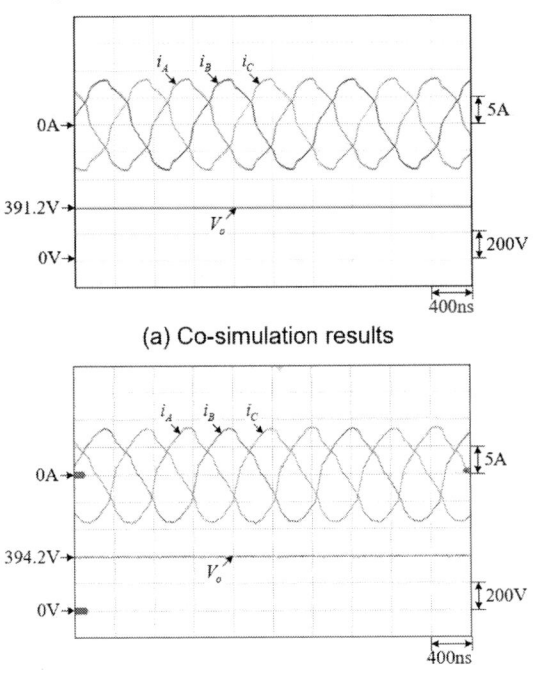

**Fig. 11:** The key waveform of three-phase LLC converter at 3.0 kW and 800 kHz.

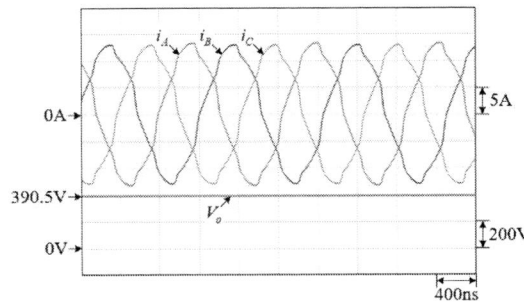

(a) Co-simulation results

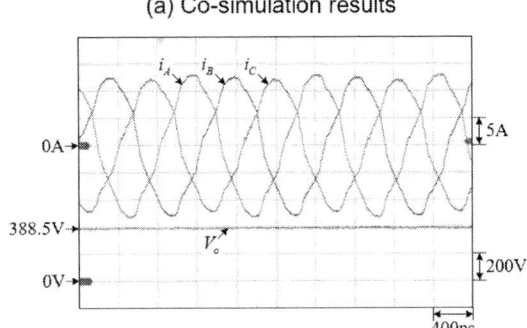

(b) Experimental results

**Fig. 12:** The key waveform of three-phase LLC converter at 4.8 kW and 800 kHz.

### 3.3 Efficiency Measurement

In this paper, the efficiency is measured using a power analyzer (Keysight PA2201A). The efficiency curve of the three-phase interleaved LLC converter under different load conditions is displayed in Fig. 13. The converter operates at 800 kHz, resulting in an output voltage of 400 V. The maximum and full-load efficiencies are 96.4 % and 95.8 % at 3.0 kW and 4.8 kW, respectively.

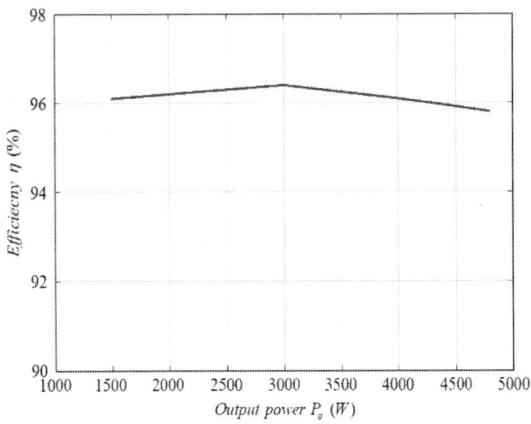

**Fig. 13:** The efficiency curve of the three-phase interleaved LLC converter at 800 kHz.

## 4 Conclusions

This paper utilizes a co-simulation approach to design a 4.8 kW GaN-based three-phase interleaved LLC converter operating at 800 kHz. Additionally, integrated three-phase five-leg magnetics are assist-designed using Maxwell 3D, resulting in height reduction through side-leg flux sharing. The measured electrical parameters are consistent with the simulation results. Finally, the effectiveness of the proposed method is verified using the platform, achieving a maximum efficiency of 96.4 % at an output voltage of 400 V and 3 kW.

## 5 Reference

[1] Mornsun, "AC/DC Power supply Design Requirements for EV Charging Stations." [Online]. Available: https://www.mornsun-power.com/html/support-detail/555.html

[2] B. Yang, F. C. Lee, A. J. Zhang, and G. Huang, "LLC Resonant Converter for Front End DC/DC Conversion," in *Proc. IEEE Appl. Power Electron. Conf. (APEC)*, Dallas, TX, USA, Mar. 2002, pp. 1108-1112.

[3] B. Lu, W. Liu, Y. Liang, F. C. Lee, and J. D. van Wyk, "Optimal Design Methodology for LLC Resonant Converter," in *Proc. IEEE Appl. Power Electron. Conf. (APEC)*, Dallas, TX, USA, Mar. 2006, pp. 533-538.

[4] C. Fei, R. Gadelrab, Q. Li, and F. C. Lee, "High-Frequency Three-Phase Interleaved LLC Resonant Converter with GaN Devices and Integrated Planar Magnetics," *IEEE J. Emerging Sel. Top. Power Electron.*, vol. 7, no. 2, pp. 653-663, Jan. 2019.

[5] J. Biela, U. Badstuebne, and J. W. Kolar, "Design of a 5-kW, 1-U, 10-kW/dm$^3$ Resonant DC–DC Converter for Telecom Applications," *IEEE Trans. Power Electron.*, vol. 24, no. 7, pp. 1701–1710, Jul. 2009.

[6] K. H. Yi and G. W. Moon, "Novel Two-Phase Interleaved LLC Series-Resonant Converter Using a Phase of the Resonant Capacitor," *IEEE Trans. Ind. Electron.*, vol. 56, no. 5, pp. 1815–1819, May 2009.

[7] H. Wu, X. Zhan, and Y. Xing, "Interleaved LLC resonant converter with hybrid rectifier and variable-frequency plus phase-shift control for wide output voltage range applications," *IEEE Trans. Power Electron.*, vol. 32, no. 6, pp. 4246–4257, Jun. 2017.

[8] E. Orietti, P. Mattavelli, G. Spiazzi, C. Adragna, and G. Gattavari, "Current Sharing in Three-Phase LLC Interleaved Resonant Converter," in *Proc. IEEE Energy Convers. Congr. Expo. (ECCE)*, San Jose, CA, USA, Sep. 2009, pp. 1145–1152.

[9] Y. Nakakohara, H. Otake, T. M. Evans, T. Yoshida, M. Tsuruya, and K. Nakahara, "Three-Phase LLC Series Resonant DC/DC Converter Using SiC MOSFETs to Realize High-Voltage and High-Frequency Operation," *IEEE Trans. Ind. Electron.*, vol. 63, no. 4, pp. 2103-2110, Apr. 2016.

[10] C. Wei, Z. Hu, J. Chen, F. Zhang, H. Zhan, and A. Narain, "A SiC Based 60kW Three Phases Interleaved LLC Converter for EV Fast Charger," in *Proc. PCIM Europe*, Nuremberg, Germany, 2023, pp. 1-6.

PCIM Europe 2024, 11– 13 June 2024, Nuremberg    DOI: 10.30420/566262097

# Switching Assisting Circuit Improving the Efficiency of DC-DC Converters Based on Piezoelectric Resonators

Valentin Breton[1], Emile Bigot[1], François Costa[2], Ghislain Despesse[1]

[1] Univ. Grenoble Alpes, CEA, Leti, France

[2] Ecole Supérieure du Professorat et de l'Education, Université Paris-Est Créteil, France

Corresponding author:    Ghislain Despesse, ghislain.despesse@cea.fr
Speaker:                 Ghislain Despesse, ghislain.despesse@cea.fr

## Abstract

A new type of converter based on piezoelectric resonators has been developed in recent years. One limitation of these converters is the parallel capacitance in the equivalent circuit of the resonator. This one needs to be charged and discharged, inducing a circulating current, which induces losses. To overcome that limitation, this paper presents an assisting circuit to reduce the piezoelectric voltage swing and then the circulating current. This one is applied to an isolated dual bridge dc-dc converter based on two piezoelectric resonators (PR). The principle is validated experimentally, showing improved efficiency of up to 11% and 5% on average.

## 1    Introduction

The magnetic-less dc-dc converters based on piezoelectric resonators are making increasing contribution to the field of power conversion. Recent progress in the dc-dc piezoelectric converters area [1], [2], [3], [4] has led to new structures [5], [6], control improvements [7], [8], achieving higher power and power density levels [9], [10], [11]. All these converters demonstrate high-performance capabilities thanks to the attractive characteristics of piezoelectric resonators (PRs) such as high quality factor and frequency stability. However, these recently developed converters still face some limits such as high voltage issues or materials challenges with specific losses mechanisms [12].

The recent development we are interested in is the isolated topology [6] based on two serial-working piezoelectric resonators. This particular topology has to be controlled by a dedicated and unique switching sequence to ensure the isolation principle. The converter faces limits at higher voltage operation using thin piezoelectric discs due to the circulating current of the resonators, directly related to the voltage range of the implemented conversion cycle.

This paper proposes a method allowing the isolated dc-dc converter based on piezoelectric resonators to work with a lower amplitude conversion cycle while keeping the isolation and ZVS operation. Two assisting circuits are designed, implemented and experimentally tested to improve the efficiency and power density of the converter.

## 2    Theoretical analysis

### 2.1    Topology and switching sequence

The topology of Fig. 1 is the dual-bridge isolated piezoelectric resonator based converter (DB-IPRC) presented in [6].

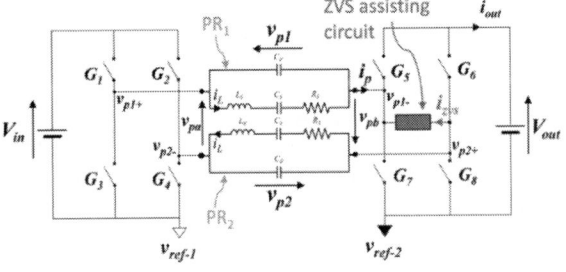

**Fig. 1**    DB-IPRC structure including assisting circuit. PR representation based on Van-dyke model [12].

This topology consists of two dual-bridge physically separated by two PRs and allows bidirectional power transfer. Voltage across $PR_1$ and $PR_2$ are, respectively, $v_{p1}$ and $v_{p2}$. Voltages between

PRs terminals on the primary and secondary side are respectively $v_{pa}$ and $v_{pb}$, with an equivalent global piezoelectric voltage given by:

$$v_p = v_{p1} + v_{p2} \qquad (1)$$

It is shown in [6] that the $\{-V_{in}-V_{out}; V_{out} - V_{in}; V_{in} - V_{out}\}$ conversion cycle with ZVS operation required a voltage swing up to $V_{in} + V_{out}$ during $\Phi_3$ (see Fig. 2). This voltage excursion is needed to ensure zero-voltage switching, while avoiding common-mode injection. This is used to clamp $v_{pa}$ to $V_{in}$ at $v_p = V_{in} + V_{out}$ and then to reverse the $v_{pb}$ potential after the current has become positive, ensuring ZVS operation. This cycle is plotted in Fig. 2 as $v_{p-ZVS}$. Between the voltage stages $V_b = V_{out} - V_{in}$ and $V_a = V_{in} - V_{out}$ the polarity of both $v_{pa}$ and $v_{pb}$ are reversed, $v_{pa}$ going from $-V_{in}$ at $\alpha_2$ to $+V_{in}$ at $\alpha_{3a}$ and $v_{pb}$ going from $+V_{out}$ at $\alpha_{3a}$ to $-V_{out}$ at $\alpha_{3b}$.

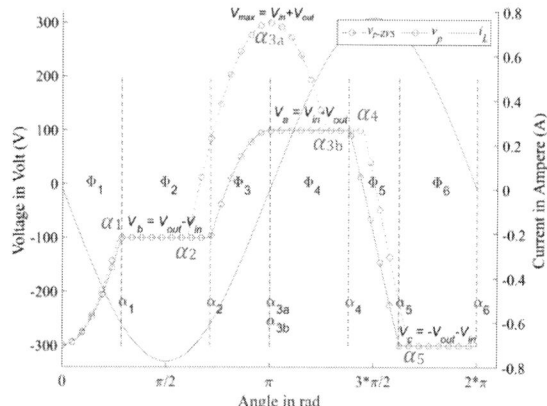

**Fig. 2** Three-stage cycle ($v_p$), the internal $i_L$ current for the DB-IPRC with ZVS assisting circuit, and the previously used three-stage cycle ($v_{p-ZVS}$) with the $V_{max} = V_{in} + V_{out}$ voltage swing. The switching angles $\alpha_{1-6}$ and the phases $\Phi_{1-6}$ are related to the $v_p$ cycle.

The new design uses the same $\{-V_{in}-V_{out}; V_{out} - V_{in}; V_{in} - V_{out}\}$ conversion cycle, but without the $V_{in} + V_{out}$ voltage swing. The corresponding cycle of this case study is plotted in Fig. 2 as $v_p$. The control is simplified assuming $\alpha_{3b}$ angle equal to $\alpha_{3a}$. To keep the advantages of this topology such as the isolation and to achieve ZVS operating, we add a switching assisting circuit (ZVS assisting circuit on Fig. 1) between the two middle-point ($v_{p1-}, v_{p2+}$) of the output half-bridges ($v_{pb}$). This assisting circuit reverse the polarity of $v_{pb}$ between $\alpha_2$ and $\alpha_{3a}$ (see Fig. 4) removing the need of the $\alpha_{3a}$ to $\alpha_{3b}$ phase and the need for $v_p$ to go up

to $V_{in} + V_{out}$. Therefore, a bigger part of the period ensures power transfer on the constant voltage phases and a lower circulating current is required to charge/discharge the piezoelectric parallel capacitor $C_p$ to voltage extrema. We chose to implement two different assisting circuits: a piezoelectric resonator and an inductor with a serial diode in the $i_{zvs}$ current direction (see Fig. 3).

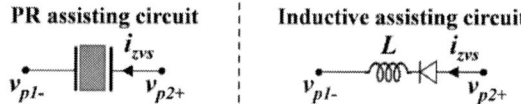

**Fig. 3** Piezoelectric (left) and inductive switching assisting circuit (right)

The equivalent Van-Dyke model [13] parameters of the selected PRs are given in Table 1.

| Equivalent parameters | Values 1*25 mm (PR₁ and PR₂) | Values 2*25 mm (PRzvs) |
|---|---|---|
| $C_p$ | 4.24 nF | 2.12 nF |
| $R_s$ | 0.5 Ω | 3.3 Ω |
| $L_s$ | 2.21 mH | 2.67 mH |
| $C_s$ | 1.43 nF | 1.20 nF |
| $[f_r, f_{ar}]$ | [89, 104] kHz | [89, 104] kHz |

**Table 1** Equivalent parameters used in the Van-Dyke model.

A simulation of currents and voltages in the converter by using an inductive assisting circuit is shown Fig. 3, $i_{ZVS}$ being the current crossing the assisting circuit.

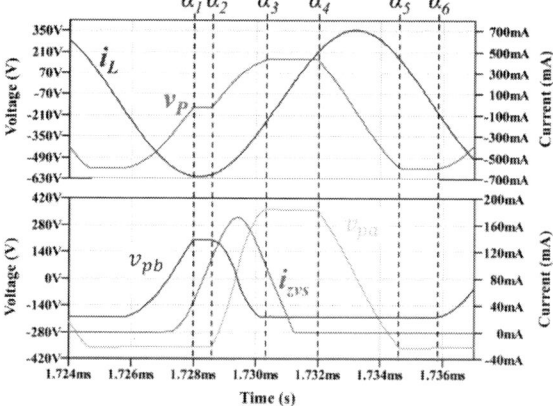

**Fig. 4** Simulated waveforms with an inductive (1.8 mH) + diode assisting circuit and two 25D x 1th mm PRs. $V_{in}$ = 360 V, $V_{out}$ = 200 V, $P_{out}$ = 20 W.

The inductor is loaded during phase $\Phi2$, therefore its energy is restored during phase $\Phi3$ reversing $v_{pb}$. The assisting circuit needs to reverse $v_{pb}$

only in one direction, so a diode is used to bloc the reverse current.

## 2.2 Internal piezoelectric current

The use of the $v_{p-ZVS}$ cycle of Fig. 2 limits the common mode voltage or current still working with ZVS operation. The major inconvenient is that it induces a voltage swing from $-V_{in} - V_{out}$ to $V_{max} = +V_{in} + V_{out}$, i.e. a maximal voltage range $\Delta V = 2 * (V_{in} + V_{out})$.

$$I_L = I_{circul} + I_{useful} = \frac{\pi}{2} * I_{out} + C_p \omega * \frac{\Delta V}{2} \quad (2)$$

The literature [6] gives us (2) and shows that $I_{useful}$ is proportional to the $I_{out}$ current by the ratio π/2 while $I_{circul}$ is constant whatever the output current, but directly linked with the $\Delta V$ voltage range. Thus, the maximum voltage range $\Delta V$ between the lower and the upper PR voltage $v_{p-zvs}$ during the cycle consumes a part ($I_{circul}$) of the $I_L$ (2) current to charge and discharge the parallel capacitance $C_P$ of the piezoelectric resonator. The PR physical limits (either thermal or mechanical) induce a maximum $I_L$ of operation to not risk its depolarization. Thus, the higher the $\Delta V$ voltage, the higher the circulating current, the lower the output current capability. Previous work [6] also shows that it exists a voltage limit with 1 mm thick piezoelectric resonators and a large circulating current due to big voltage range $\Delta V$ from $(-V_{in} - V_{out})$ to $(+V_{in} + V_{out})$. By decreasing the internal current, the total mechanical losses $R_s * I_L^2$ in the material and the conduction losses in the switches are also decreased.

Consequently, the aim of the implemented assisting circuits (see Fig. 3) is to avoid the $V_{in} + V_{out}$ voltage swing required to ensure ZVS operation. With the assisting circuit, the voltage range $\Delta V$ between the lower and the upper PR voltage $v_p$ falls to $(-V_{in} - V_{out}) - (+V_{in} - V_{out}) = 2 * V_{in}$, reducing the circulating current $I_{circul}$.

## 2.3 Inductive assisting circuit principle

The idea is to reverse $v_{pa}$ and $v_{pb}$ voltages without using the natural
$$v_p$$
voltage change in open circuit (1), but by using an inductive component placed across $v_{pa}$ and/or $v_{pb}$ voltages. In the case of the step-down converter, we choose to place it across $v_{pb}$, the lowest voltage between $v_{pa}$ and $v_{pb}$ to reverse. The principle is to load the inductor between $\alpha_1$ and $\alpha_2$ under $V_{pb} = V_{out}$. At $\alpha_2$ all the switches

are open to let the PR goes to the next constant voltage level. The $i_{ZVS}$ pre-loaded current charges the switches parasitic capacitances. The voltage $v_{pb}$ goes from $V_{out}$ to $-V_{out}$ and $v_{pa}$ goes from $-V_{in}$ to $-V_{in} + 2V_{out}$ thanks to the assisting circuit. The remaining $2V_{in} - 2V_{out}$ voltage swing required to reach $+V_{in}$ is done thanks to the open circuit operation of $v_p$. The inductor is only loaded between $\alpha_1$ and $\alpha_2$ with a positive polarity. It is useless to charge the inductor with a negative current between $\alpha_3$ and $\alpha_6$. To do so, a high-voltage diode (ZVS assisting diode in Table 2) is placed in series with the inductor, loading the inductor only during the first half-period with a positive $v_{pb}$ polarity.

The use of the dual-active bridge topology implies the charge of the parasitic capacity of the half of the switches and the discharge of the one of the other half. We note $Q_{switches}$ the sum of the charges of four switches (GaN+diode). The $i_{zvs}$ current is the integral of the output voltage over the inductor value. The $v_{pb}$ voltage being alternately equal to $V_{out}$ and $-V_{out}$, $i_{zvs}$ rise in $\frac{t*V_{out}}{L}$ and decrease in $\frac{-t*V_{out}}{L}$. Assuming that the voltage inversion is completely achieved between $\alpha_2$ and $\alpha_3$, we calculate the $i_{zvs}$ current (3) needed to charge the parasitic capacitances in that duration. The value of the inductor is calculated in (4) considering that the $L$ is charged between the constant voltage stage between $\alpha_1$ and $\alpha_2$. The $t_x$ notation corresponds to the temporal switching times of the conversion cycle with $t_x = \frac{\alpha_x}{2*\pi*f}$ ($f$ being the working frequency).

$$i_{zvs} = \frac{Q_{switches}}{(t_3 - t_2)} \quad (3)$$

$$L = \frac{V_{out} * (t_2 - t_1)}{i_{zvs}} \quad (4)$$

A numeric application using $Q_{switches}$=280 nC, $V_{out}$=200 V, $T$=10.02 μs, $(t_2 - t_1)$=1.7 μs and $(t_3 - t_2)$=1.86 μs gives us: $i_{zvs}$=150.5 mA, and $L_{max}$=2.26 mH. In fact, a lower inductor value work either, but with a higher current and then higher losses. A higher inductor value results in an incomplete voltage reverse, but still better than no voltage reverse.

## 2.4 Piezoelectric assisting circuit

Another solution is to excite a same-working-frequency piezoelectric resonator (PR$_{ZVS}$) to use his internal current to charge and discharge parasitic capacitances. To do so, the control frequency needs to be in the range between the resonance frequency and the anti-resonance of the $PR_{ZVS}$. The $i_{zvs}$ current is then in quadrature with the $v_{pb}$ voltage with an extrema around $\alpha_3$, perfect to reverse the $v_{pb}$ voltage between $\alpha_2$ and $\alpha_3$. This circuit, non-sensitive to dc-voltage can be placed both on $v_{pa}$ and $v_{pb}$ terminals. Working in radial mode, the $PR_{ZVS}$ should be same diameter (25 mm) as $PR_1$ and $PR_2$ to work at the same frequency. We choose to work with a two-time-thicker PR (2 mm) to reduce by half the parallel capacitance $C_P$ of the $PR_{ZVS}$ to obtain the just sufficient $i_{zvs}$ current. Thus, a 25 D x 2th mm PR with equivalent parameters of Van-Dyke model [13] presented in Table 1 is placed between $v_{p1-}$ and $v_{p2+}$ across the $v_{pb}$ voltage for the experimental measurements of Fig. 7.

## 3 Experimental measurements

### 3.1 Converter Prototype

We developed an isolated dual bridge dc-dc converter prototype of Fig. 5 for tests and validation. On the secondary side, between $v_{p1-}$ and $v_{p2+}$ of the secondary bridge, a 2mm thick piezoelectric assisting circuit is inserted (Fig. 5) or an inductive assisting circuit (Fig. 6).

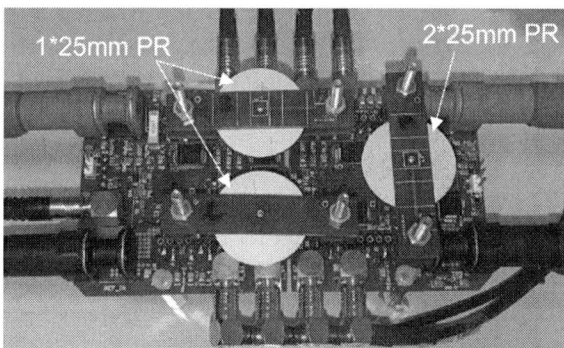

**Fig. 5** Prototypes of the dc-dc converter with piezoelectric assisting circuit (right).

**Fig. 6** Prototypes of the dc-dc converter with inductive assisting circuit (right).

The prototypes uses two disc-shaped 25D x 1th mm PRs working around 95 kHz operating frequency. The chosen transistors are system-in-package 600 V integrated half-bridge with dedicated driver simplifying the design and preventing the short-circuit risks. Schottky power diodes are placed in parallel of each GaN transistors, making up the switches. A custom command from a field programmable gate array (FPGA) board was developed in VHDL language to generate control signals.

The component references are listed in Table 2.

| Component | Reference |
|---|---|
| Piezoelectric resonators | 25 D x 1th mm, PZT C213, *Fuji Ceramics* |
| Schottky diode (and ZVS assisting diode) | 3 A, 400 V ES3G-E3/57T, Vishay |
| Active switch | MasterGaN 600 V SIP integrating half-bridge gate driver and high-voltage power GaN transistor, *ST Microelectronics* |
| ZVS assisting inductor | 1.8 mH, RFC1010B-185KE, *Coilcraft* |
| ZVS assisting PR | 25 D x 2th mm, C213, *Fuji Ceramics* |
| FPGA Board | Dev-board Arty-A7, *Digilent* |

**Table 2** Prototype Netlist

The switching frequency is between the resonant and anti-resonant frequency [89–104 kHz] of the PRs operating in radial mode. Gate signals are generated according to an open-loop control with a constant-voltage load. Switching angles and frequency are precomputed and fine-tuned manually to get the desired output power and to ensure perfect ZVS operation.

### 3.2 Efficiency and Output Power

We made a first characterization work for the converter without ZVS assisting circuit, using $v_{p-zvs}$ cycle of Fig. 2. Then we made a characterization using $v_p$ cycle of Fig. 2 with a 1.8 mH inductive assisting circuit with diode, and a last one with a disc-shaped 25D x 2th mm PR assisting circuit mounted across the $v_{pb}$ voltage on the secondary part of the converter (see Fig. 5). As the $PR_{ZVS}$ assisting circuit is working at the same frequency as $PR_1$ and $PR_2$, it directly depends on $I_{out}$. A low output current implies an operating frequency close to the anti-resonant frequency with a low internal current. The current ($i_{zvs}$) is then insufficient to reverse the $v_{pb}$ voltage between $\alpha_2$ and $\alpha_3$. When the output current becomes significant, the inside current of the $PR_{ZVS}$ becomes sufficient for the ZVS assistance. This case should be reserved for dedicated operating points.

PCIM Europe 2024, 11– 13 June 2024, Nuremberg        DOI: 10.30420/566262097

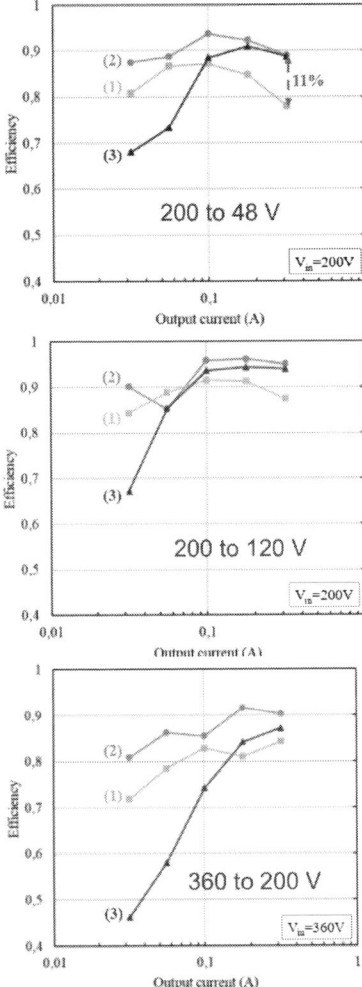

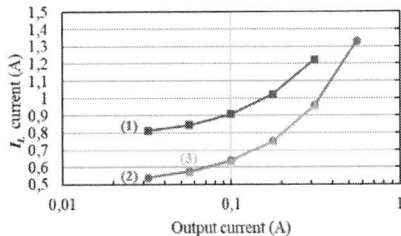

**Fig. 9**  Internal PR current ($I_L$) depending on the output current ($I_{out}$) without assisting circuit (1) ($v_{p-zvs}$ on Fig. 2), with inductive assisting circuit (2), and with 25 D× 2th mm PR assisting circuit (3) for $V_{in}$=360 V and $V_{out}$=200 V.

Fig.7 shows a significantly improved efficiency with the use of an inductive assisting circuit by 5 % on average and up to 11% compared with the converter not using ZVS assisting circuit. Fig. 9 shows the PR internal current $I_L$ in function of $I_{out}$. By reducing $\Delta V$ voltage range thank to the assisting circuit, it significantly decreases the current $I_L$, by 0.25 A on average. As $I_{circul}$ (2) decreases, the PR can provide a higher $I_{useful}$ current, so a higher output current ($I_{out}$).

## 4   Conclusion

This new design improves the dual-bridge isolated piezoelectric resonator converter (DB-IPRC) previously presented in [6] improving the efficiency, while keeping the isolation principle of the converter. For a 200 V to 120 V conversion, the converter shows an efficiency of 96.2% with the inductive assisting circuit, 94.3% with the piezoelectric one and 87.4% without any assisting circuit. The PR assisting circuit offers a gain in efficiency over a smaller operating range than the inductance, but leads to a flatter converter.

## References

[1] B. Pollet, F. Costa and G. Despesse, « A new inductorless DC-DC piezoelectric flyback converter, » 2018 IEEE International Conference on Industrial Technology (ICIT), Lyon, France, 2018, pp. 585-590, doi: 10.1109/ICIT.2018.8352243.

[2] J. D. Boles, J. J. Piel, N. Elaine, J. E. Bonavia, J. H. Lang, et D. J. Perreault, « Piezoelectric-Based Power Conversion: Recent Progress, Opportunities, and Challenges », in 2022 IEEE Custom Integrated Circuits Conference (CICC), avr. 2022, p. 1-8. doi: 10.1109/CICC53496.2022.9772801.

[3] B. Pollet, M. Touhami, G. Despesse and F. Costa, "Effects of Disc-Shaped Piezoelectric Size Reduction on Resonant Inductorless DC-

**Fig. 7**  Efficiency in function of the output current ($I_{out}$) for various conversion ratios using 25D x 1th mm PRs. Each graph presents the efficiency without any assisting circuit (1), with inductive assisting circuit (2), and with disc-shaped PR 25D x 2th mm assisting circuit (3). The case without assisting circuit use the $v_{p-zvs}$ conversion cycle of Fig. 2.

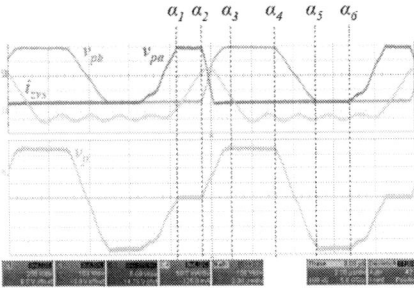

**Fig. 8**  Experimental three-stage cycle $v_p$, full-bridge voltages $v_{pa}$ and $v_{pb}$, and the $i_{zvs}$ current in the inductive assisting circuit at $V_{in}$=360 V, $V_{out}$=200 V, and 20 W output power,  $i_{zvs\ p-p}$= 192 mA.

DC Converter," 2019 20th International Conference on Solid-State Sensors, Actuators and Microsystems & Eurosensors XXXIII, Berlin, Germany, 2019, pp. 1423-1426, doi: 10.1109/TRANSDUCERS.2019.8808328.

[4] W. D. Braun et al., « Optimized Resonators for Piezoelectric Power Conversion », IEEE Open J. Power Electron., vol. 2, p. 212-224, 2021, doi: 10.1109/OJPEL.2021.3067020.

[5] M. Touhami, G. Despesse, et F. Costa, « A New Topology of DC-DC Converter Based On Piezoelectric Resonator », in 2020 IEEE 21st Workshop on Control and Modeling for Power Electronics (COMPEL), nov. 2020, p. 1-7. doi: 10.1109/COMPEL49091.2020.9265767.

[6] V. Breton, E. Bigot, G. Despesse, et F. Costa, « A New Isolated Topology of DC–DC Converter Based on Piezoelectric Resonators », IEEE Trans. Power Electron., vol. 38, no 8, p. 10012-10025, août 2023, doi: 10.1109/TPEL.2023.3276478.

[7] E. A. Stolt, W. D. Braun, et J. M. Rivas-Davila, « Forward-Zero Cycle Closed-Loop Control of Piezoelectric Resonator DC-DC Converters », in 2022 IEEE 23rd Workshop on Control and Modeling for Power Electronics (COMPEL), juin 2022, p. 1-6. doi: 10.1109/COMPEL53829.2022.9829965.

[8] J. Forrester, M. P. Foster, et J. N. Davidson, « Resonant current estimation and phase-locked loop control system for inductorless step-up single piezo element-based (SUPRC) DC-DC converter », in IECON 2022 – 48th Annual Conference of the IEEE Industrial Electronics Society, oct. 2022, p. 1-6. doi: 10.1109/IECON49645.2022.9969043.

[9] J. D. Boles, J. E. Bonavia, J. H. Lang, et D. J. Perreault, « A Piezoelectric-Resonator-Based DC–DC Converter Demonstrating 1 kW/cm Resonator Power Density », IEEE Trans. Power Electron., vol. 38, no 3, p. 2811-2815, mars 2023, doi: 10.1109/TPEL.2022.3217773.

[10] M. Touhami et al., « Piezoelectric Materials for the DC-DC Converters Based on Piezoelectric Resonators », in 2021 IEEE 22nd Workshop on Control and Modelling of Power Electronics (COMPEL), nov. 2021, p. 1-8. doi: 10.1109/COMPEL52922.2021.9645999.

[11] E. Stolt, W. Braun, K. Nguyen, V. Chulukhadze, R. Lu and J. Rivas-Davila, « A Spurious-Free Piezoelectric Resonator Based 3.2 kW DC–DC Converter for EV On-Board Chargers, » in IEEE Transactions on Power Electronics, vol. 39, no. 2, pp. 2478-2488, Feb. 2024, doi: 10.1109/TPEL.2023.3334211.

[12] K. Uchino, « High-Power Piezoelectrics and Loss Mechanisms. » Boca Raton: CRC Press, 2020. doi: 10.1201/9781003087519.

[13] K. S. Van Dyke, « The Piezo-Electric Resonator and Its Equivalent Network, » in Proceedings of the Institute of Radio Engineers, vol. 16, no. 6, pp. 742-764, June 1928, doi: 10.1109/JRPROC.1928.221466.

PCIM Europe 2024, 11– 13 June 2024, Nuremberg    DOI: 10.30420/566262098

# Comparative Evaluation of Transformer-based Fixed-ratio DC-DC Converters for 48V Data Centers

Xufu Ren[1], Jinfeng Zhang[1], Pengcheng Xu[2], Jibin Song[2], Teng Long[1]

[1] University of Cambridge, United Kingdom
[2] EPIC Tech, China

Corresponding author:    Teng Long, tl322@cam.ac.uk
Speaker:                          Xufu Ren, xr222@cam.ac.uk

## Abstract

Transformer-based fixed-ratio DC-DC converters are widely used in 48V data centers to perform voltage step-down. This paper aims to conduct a comparative evaluation of various topologies in terms of operation principles, voltage conversion ratios, current and voltage stress, and practical design considerations. Based on 4:1 intermediate bus architecture, we selected an optimal topology and developed a 1kW 4:1 bus converter with superior 98.2% efficiency and 5.4kW/in$^3$ power density. This prototype is configured into a system board equipped with multi-phase voltage regulators (VRs) capable of delivering kilo-amperes current and providing fast transient response for AI computing cores.

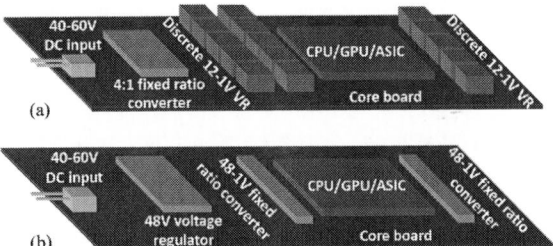

**Fig. 1:** Examples of commonly used VRM architectures. (a) IBA. (b) FPA.

## 1 Introduction

Acting as the brain of AI computing, high-performance XPUs (CPUs, GPUs) used in modern data centers integrate billions of transistors and ask the power demands for hundreds of amperes of current at a low voltage (e.g., ≤1V). Highly efficient and compact power conversion system is urgently required to accommodate the increasing power supply requirements for processors [1][2].
48V voltage bus is becoming a common trend in XPU power architecture [3]. The onboard input voltage ranges from 40V to 60V (typical 54V), subsequently transformed into a lower voltage to supply power to the processors. High voltage transfer ratio, large output currents, and fast transient response are main challenges for those onboard voltage regulator modules (VRMs). Figure 1 depicts two commonly adopted solutions including intermediate bus architecture (IBA) and factorized power architecture (FPA). The IBA consists of a first-stage DC-DC converter performing a fixed-ratio voltage conversion (e.g., 4:1, 8:1, etc.), and the variable low voltage bus is regulated by the second-stage voltage regulators (VRs) like multi-phase buck converters [3][4][5]. For FPA, the voltage regulation is set on the first stage to regulate the input voltage to a constant high voltage bus (e.g., 48V, 36V,

etc.), followed by a DC transformer (DCX) with high voltage transfer ratio [6]. For both architectures, unregulated DC-DC converters are essential components to perform fixed-ratio voltage step-down. To satisfy the heat dissipation in strongly limited space on the motherboard, it's challenging to design the converters with both high efficiency and high power density.
Fixed-ratio DC-DC converters utilized in 48V data centers can be generally categorized as switched capacitor based converters and transformer-based converters. Employing ceramic capacitors with high energy density, switched capacitor converters can realize high power density without the need for transformers [7][8]. However, floating gate drivers and start-up control might cause practical issues in power integration and reliability for industrial applications. Transformer-based converters are more attractive in nowadays 48V data center products.

PCIM Europe 2024, 11– 13 June 2024, Nuremberg    DOI: 10.30420/566262098

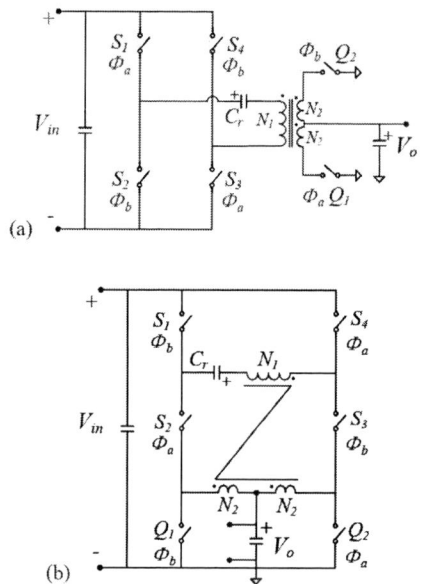

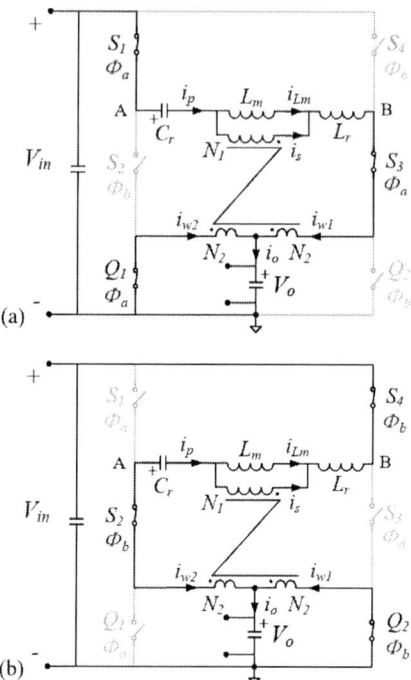

**Fig. 2:** Full-bridge resonant topologies. (a) Isolated converter. (b) Non-isolated converter.

**Fig. 3:** The equivalent circuit of non-isolated LLC converter shown in Fig. 2 (b). (a) Positive half. (b) Negative half.

Besides the conventional isolated resonant converters, non-isolated resonant converters have been investigated due to their improved voltage transfer ratio and reduced power losses [9][10][11].

This paper will perform a comparative evaluation of various transformer-based DC-DC converters for 48V data centers. Section 2 will outline topologies and introduce their operation principles. In Section 3, a comparative assessment of various topologies is carried out. Section 4 will present experiments conducted with highly efficient and compact 4: 1 bus converters. Finally, conclusions are given in Section 5.

## 2 Topologies and Operation Principles

Transformer-based topologies used in 48V data centers primarily include conventional isolated resonant converters and non-isolated resonant converters. In this Section, we will firstly introduce the operation principles of auto-transformer-based non-isolated resonant converters. Subsequently we will investigate other topology variants.

### 2.1 Non-isolated resonant converters

Isolated resonant converters are commonly applied in 48V data centers. Operating at resonant frequency can enable soft-switching and ensure high efficiency performance. Fig. 2 (a) shows an isolated full-bridge (FB) LLC converter. The indepen-

dent output rectifiers offer flexibility in transformer design, facilitating easy paralleling through magnetic integration. This allows for adjustable output power and voltage conversion ratio. Figure 2 (b) shows a non-isolated LLC converter derived from the isolated LLC converter depicted in Fig. 2 (a) [11]. A cross-link auto-transformer is implemented in this topology. The transformer winding voltage is switched in series and parallel alternatively to realize voltage conversion. Utilizing the topology in Fig. 2 (b) as an example, we will explain the operation principles of non-isolated resonant converters. The equivalent circuit and operation waveform are depicted in Fig. 3 and Fig. 4 respectively. At time $t_0$, power switches $S_1$, $S_3$, and $Q_1$ are turned on, and the converter operates in the positive half. During this period, $L_r$ resonates with $C_r$ at the resonant frequency $f_r = \frac{1}{2\pi\sqrt{L_rC_r}}$. Using Kirchhoff's voltage law (KVL), the voltage transfer ratio can be calculated as:

$$V_{in} = (n + 2)V_o \tag{1}$$

where $n$ is defined as the transformer turn ratio $N_1/N_2$, $N_1$ and $N_2$ are turn numbers of primary winding and secondary winding respectively. For non-isolated LLC converter, reduced transformer

804

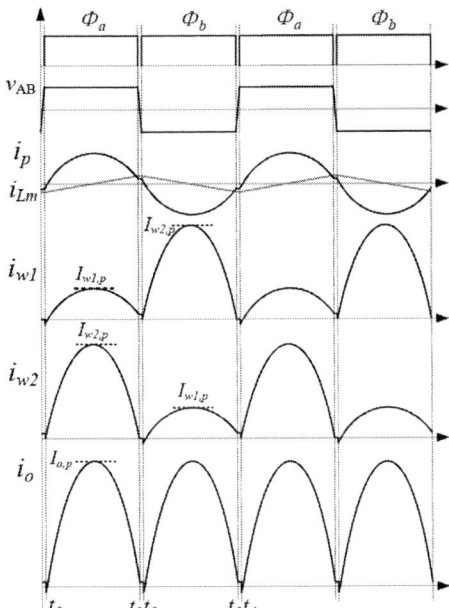

**Fig. 4:** The operation waveform of non-isolated LLC converter depicted in Fig. 2 (b).

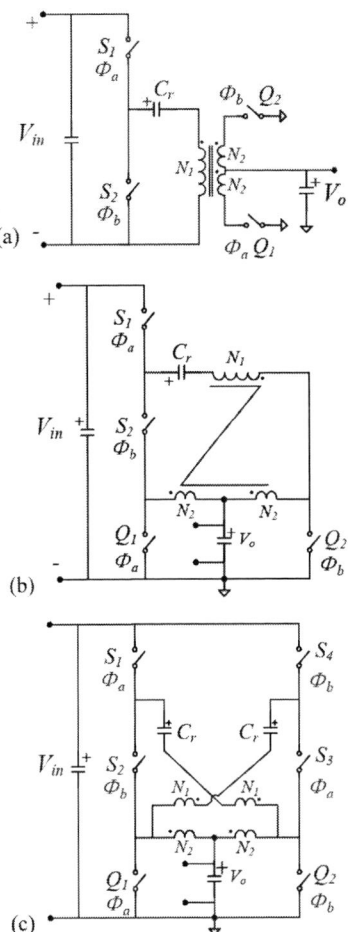

**Fig. 5:** Half-bridge resonant topologies. (a) Isolated converter. (b) Non-isolated converter. (c) Interleaved non-isolated converter (i.e., HSC noted in [10]).

turn ratio can realize the same voltage conversion ratio according to Eq. (1).

During the positive half, the primary winding is connected in series with one secondary winding, allowing for $i_p = i_{w1}$. The magnetizing inductance $L_m$ is clamped by the voltage $nV_o$, causing the magnetizing current $i_{Lm}$ to increase linearly. After half switching period, the secondary winding current $i_{w2}$ resonates to zero at time $t_1$, and the SR switch $Q_2$ achieves ZCS turn-off. The duration $t_1$ to $t_2$ is the dead time, which is much smaller than the switching period. The magnetizing current in dead time, $I_{Lm}$, is calculated as: $\frac{nV_o}{4f_sL_m}$, which is the ZVS current to ensure the ZVS turn-on operation of power switches. As shown in Fig. 3 (a), during the dead time, the ZVS current discharges the output capacitance of $S_2$, $S_4$, and $Q_2$, and charges the output capacitance of $S_1$, $S_3$, and $Q_1$, creating the ZVS turn-on condition of $S_2$, $S_4$, and $Q_2$ at time $t_2$.

Figure 4 depicts the current waveform in the transformer windings. The magnetizing current $i_{Lm}$ is designed to ensure ZVS, typically remaining significantly lower than resonant current $i_p$. In this case, resonant current $i_p$ can be approximated as an ideal sinusoidal waveform if neglecting $i_{Lm}$. By applying the flux conservation law to the ideal transformer, we can derive the following equation:

$$i_{w2} = (n+1)i_{w1} \tag{2}$$

The output current $i_o$ equals the sum of winding current $i_{w1}$ and $i_{w2}$:

$$i_o = i_{w1} + i_{w2} \tag{3}$$

Assuming the average output current is $I_o$, then the peak current of the output current $I_{o,p}$ shown in Fig. 4 is $\frac{\pi}{2}I_o$. Combing Eq. (2) and Eq. (3), the peak current of the secondary winding shown in Fig. 4 is calculated as:

$$\begin{cases} I_{w1,p} = \frac{1}{n+2}I_{o,p} = \frac{\pi I_o}{2(n+2)} \\[2mm] I_{w2,p} = \frac{n+1}{n+2}I_{o,p} = \frac{\pi(n+1)I_o}{2(n+2)} \end{cases} \tag{4}$$

The operation in the negative half ($t_2$ to $t_4$) can be investigated using the same analysis. As illustrated in Fig. 4, the secondary winding currents

PCIM Europe 2024, 11– 13 June 2024, Nuremberg   DOI: 10.30420/566262098

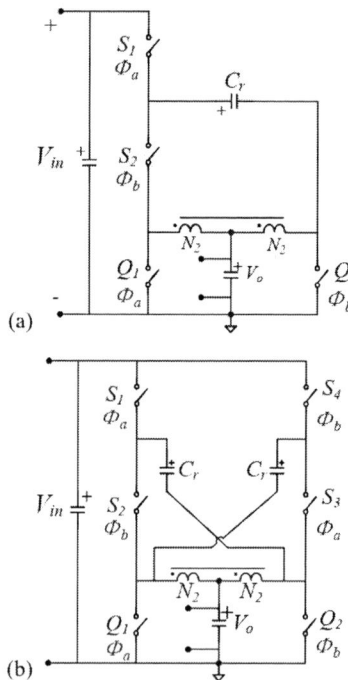

(a)

(b)

**Fig. 6:** Half-bridge non-isolated resonant topologies with fixed 4: 1 voltage conversion ratio. (a) Without interleaving. (c) With interleaving.

$i_{w1}$ and $i_{w2}$ have the symmetric current distribution compared with that in the positive half, keeping the same output current.

## 2.2 Topology variants

In Section 2.1, we analyzed the operation principles of an FB non-isolated resonant converter, which is derived from a conventional isolated resonant converter. The same derivation can be implemented on half-bridge (HB) configuration. Figure 5 shows a isolated HB resonant converter and its non-isolated variants [10][11]. The topology illustrated in Fig. 5 (b) has same operation principles with the converter introduced in Section 2.1. There is an auto-transformer configured between power switches. Different from the FB converter depicted in Fig. 2, the resonant capacitors in HB circuit withstand a DC bias voltage ($V_{in}/2$).

The topology illustrated in Fig. 5 (c) is proposed in [10] as the hybrid switched capacitor converter (HSC). This configuration can be seen as an interleaving of the resonant branch from the topology shown in Fig. 5 (b). Therefore, additional resonant capacitors and transformer winding are required. In the HSC, the two resonant branches operate in parallel, carrying the same current but in opposite

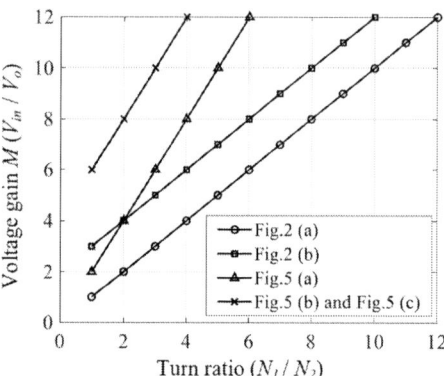

**Fig. 7:** Voltage conversion ratio of different topologies.

directions. Similarly, the DC bias voltage ($V_{in}/2$) is distributed across the resonant capacitors. HSC serves the same voltage gain function as the topology shown in Fig. 5 (b).

The last two topologies are depicted in Fig. 6, both capable of providing a fixed 4: 1 voltage conversion ratio. From a topological perspective, Fig. 6 (a) has a similar configuration with series capacitor buck converter [12]. However, this topology operates on resonance, resulting in a two-winding transformer output rather than a coupled / uncoupled inductor output as reported in [12]. The resonant branch in Fig. 6 (a) can also be interleaved by configuring a symmetry circuit, forming a topology shown in Fig. 6 (b). These two topologies have the same voltage gain equation and both have DC bias voltage ($V_{in}/2$) across the resonant capacitors.

## 3 Comparative Evaluation of Different Topologies

In this section, we will carry out a comparative evaluation of various topologies introduced in Section 2. The same analysis of FB non-isolated LLC converter in Section 2.1 can be applied on other non-isolated LLC topologies. Table 1 gives a comprehensive comparison of different topologies in terms of their voltage gain, voltage and current stress of power switches, and current stress of transformer windings. $M$ is the voltage conversion ratio, which equals to $V_{in}/V_o$, and $I_o$ is the average output current.

The non-isolated converters offer several advantages over isolated converters:

1) Lower transformer turn ratio. Figure 7 shows the voltage gain with respect to transformer turn ratio of different topologies. It indicates that the non-isolated converters need lower transformer turn

806

**Tab. 1:** The performance comparison of different topologies.

| Topology | | Fig. 2(a) | Fig. 2(b) | Fig. 5(a) | Fig. 5(b) | Fig. 5(c) | Fig. 6(a) | Fig. 6(b) |
|---|---|---|---|---|---|---|---|---|
| Voltage conversion ratio $V_{in}/V_o$ | | $M$ | $M$ | $M$ | $M$ | $M$ | $4$ | $4$ |
| Turn ratio $N_1/N_2$ | | $M$ | $M-2$ | $\frac{M}{2}$ | $\frac{M}{2}-2$ | $\frac{M}{2}-2$ | $\diagdown$ | $\diagdown$ |
| Voltage Stress | $S_1,(S_4)$ | $MV_o$ | $(M-2)V_o$ | $MV_o$ | $(M-2)V_o$ | $(M-2)V_o$ | $2V_o$ | $2V_o$ |
| | $S_2,(S_3)$ | $MV_o$ | $MV_o$ | $MV_o$ | $MV_o$ | $MV_o$ | $4V_o$ | $4V_o$ |
| | $Q_1,Q_2$ | $2V_o$ | $2V_o$ | $2V_o$ | $2V_o$ | $2V_o$ | $2V_o$ | $2V_o$ |
| Current Stress (RMS) | $S_1,S_2,(S_3,S_4)$ | $\frac{\pi I_o}{4M}$ | $\frac{\pi I_o}{4M}$ | $\frac{\pi I_o}{2M}$ | $\frac{\pi I_o}{2M}$ | $\frac{\pi I_o}{4M}$ | $\frac{\pi I_o}{8}$ | $\frac{\pi I_o}{16}$ |
| | $Q_1$ | $\frac{\pi I_o}{4}$ | $\frac{\pi (M-1) I_o}{4M}$ | $\frac{\pi I_o}{4}$ | $\frac{\pi (M-2) I_o}{4M}$ | $\frac{\pi (M-1) I_o}{4M}$ | $\frac{\pi I_o}{8}$ | $\frac{3\pi I_o}{16}$ |
| | $Q_2$ | $\frac{\pi I_o}{4}$ | $\frac{\pi (M-1) I_o}{4M}$ | $\frac{\pi I_o}{4}$ | $\frac{\pi I_o}{4}$ | $\frac{\pi (M-2) I_o}{4M}$ | $\frac{\pi I_o}{4}$ | $\frac{3\pi I_o}{16}$ |
| Transformer Winding Current (RMS) | Primary | $\frac{\sqrt{2}\pi I_o}{4M}$ | $\frac{\sqrt{2}\pi I_o}{4M}$ | $\frac{\sqrt{2}\pi I_o}{2M}$ | $\frac{\sqrt{2}\pi I_o}{2M}$ | $\frac{\sqrt{2}\pi I_o}{4M}$ | $\diagdown$ | $\diagdown$ |
| | Secondary | $\frac{\pi I_o}{4}$ | $\frac{\pi I_o \sqrt{(M-1)^2+1}}{4M}$ | $\frac{\pi I_o}{4}$ | $\frac{\pi I_o \sqrt{(M-2)^2+4}}{4M}$ | $\frac{\pi I_o \sqrt{(M-2)^2+4}}{4M}$ | $\frac{\sqrt{2}\pi I_o}{8}$ | $\frac{\sqrt{2}\pi I_o}{8}$ |

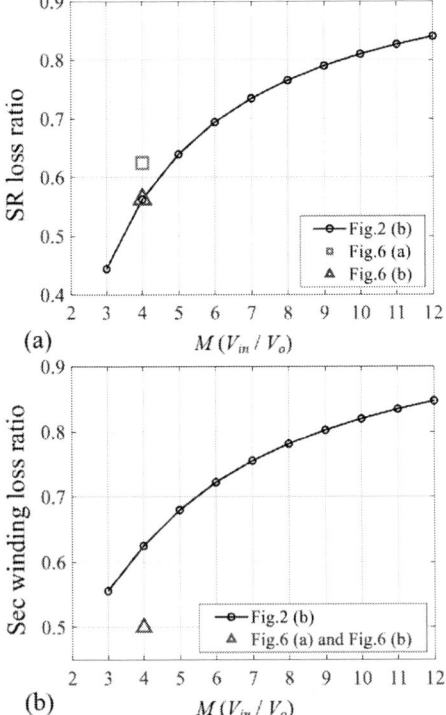

**Fig. 8:** Loss ratio of different topologies in FB group. (a) SR loss ratio. (b) Secondary winding loss ratio.

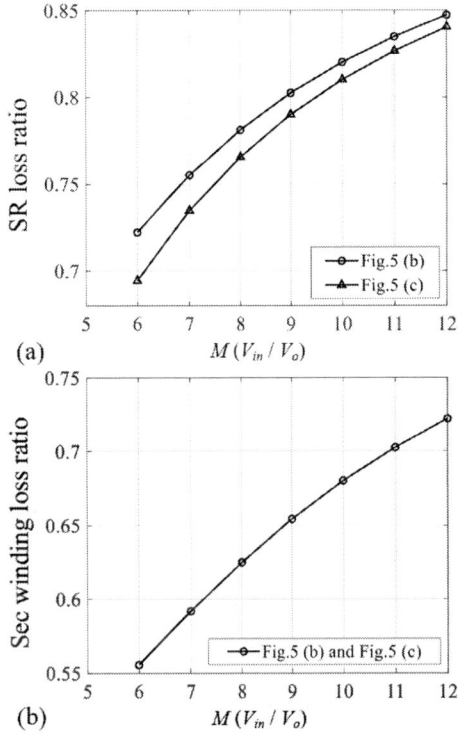

**Fig. 9:** Loss ratio of different topologies in HB group. (a) SR loss ratio. (b) Secondary winding loss ratio.

ratio to achieve the same voltage gain, offering benefits for transformer winding design and loss reduction. In addition, the non-isolated HB converters are more suitable for higher voltage transfer ratio applications. For example, a 2: 1 turn ratio transformer can realize 8: 1 voltage conversion by employing topologies shown in Fig. 5 (b) and (c).

2) Decreased voltage stress. According to table 1, the non-isolated converters can effectively reduce the voltage stress on upper side power switches $S_1$ (and $S_4$), allowing using devices with lower blocking voltage and better on-state resistance.

3) Reduced conduction losses. The topologies can be divided into two groups: the FB group (Fig. 2(a) and (b)) and the HB group (Fig. 5(a), (b), and (c)). Those shown in Fig. 6 (a) and (b) are categorized

into FB group since their volatge conversion ratio ($M = 4$) is less than the minimum voltage conversion ratio of Fig. 5 (b) and (c), as indicated in Fig. 7. Using FB group as an example, we take the isolated FB converter's loss shown in Fig. 2 (a) as the benchmark to compare against other topologies, thereby obtaining the loss ratio. For high-current and low-voltage applications, our analysis primarily focuses on conduction losses related to synchronous rectification devices (SR) and secondary transformer winding. Given that conduction loss is directly proportional to the square of RMS current, we can derive the SR loss ratio between the topology shown in Fig. 2 (b) and that in Fig. 2 (a) based on the RMS current values provided in table 1:

$$\frac{P_{SR,Fig.2(b)}}{P_{SR,Fig.2(a)}} = (\frac{M-1}{M})^2 \qquad (5)$$

Accordingly, the loss ratio of secondary winding can be calculated as:

$$\frac{P_{w,Fig.2(b)}}{P_{w,Fig.2(a)}} = \frac{(M-1)^2 + 1}{M^2} \qquad (6)$$

Figure 8 (a) illustrates the loss ratio of topologies in FB group. The loss ratios of topologies depicted in Fig. 6 remain constant since they can only perform a fixed 4: 1 voltage conversion. As indicated in Fig. 8 (a), when $M$ is low, the conduction loss of SR switches in the non-isolated converter demonstrates a significant reduction compared to that in the isolated converter. As $M$ increases, the loss in the non-isolated converter converges towards that observed in the isolated converter. This trend highlights the suitability of non-isolated FB converters for applications with lower voltage transfer ratios, such as the 4:1 ratio.

Non-isolated converters also exhibit reduced secondary winding losses compared to isolated converters, as illustrated in Fig. 8 (b). According to the operation waveform shown in Fig. 4, in every half switching period, there are currents flowing through both secondary windings to the output, resulting in diminished conduction losses compared with isolated LLC converter. For specific 4: 1 voltage conversion ratio, the current distribution ratio in two secondary windings is 1: 3 for Fig. 2 (b) topology, while which is 1: 1 for Fig. 6 topologies. Consequently, the topologies depicted in Fig. 6 experience lowest secondary winding losses.

Figure 9 shows the loss ratio of topologies within the HB group. Similarly, the conduction losses of non-isolated converters exhibit a significant decrease compared with isolated converters when $M$ is small. However, as $M$ increases, their losses become close to each other. Thereby the non-isolated HB converters are suitable for applications with voltage transfer ratios such as the 6: 1 and 8: 1. It should be mentioned that the topology shown in Fig. 5 (b) has an asymmetric current loop, resulting in unbalanced current stress of SR switches $Q_1$ and $Q_2$. Therefore, the SR conduction losses of topology shown in Fig. 5 (b) are higher than that of topology shown in Fig. 5 (c).

Although the non-isolated converters offer several advantages, it does not imply that isolated converters are unsuitable for 48V data centers. One drawback of non-isolated converter is that the output full-wave rectifier cannot be easily paralleled to increase output current. Additionally, the auto-transformer presents a more complex current flow path, leading to challenges in magnetic integration within multi-winding transformer designs. As previously discussed, for applications requiring large voltage conversion ratios, the reduction in losses with non-isolated topologies may not be significant. In our preliminary work, we have developed a 48-to-1V DCX using isolated LLC converters. By implementing series connected transformer windings and parallel connected rectifiers, the proposed LLC DC transformer has achieved high voltage conversion ratio and high output current ($>$300A) within a 58 by 15 by 5.6 mm volume (1kW/in$^3$ power density) [13]. Such a design case is difficult to achieve using non-isolated converters.

## 4 Experiments

The IBA with 4:1 bus converters is a widely adopted solution in the data center industry [4][8][14]. Based on the topology analysis, this paper aims to develop a highly efficient and compact 4: 1 bus converter. According to the analysis in Section 3, four topologies shown in Fig. 2 (a), (b) and Fig. 6 (a), (b) are selected as candidates. Analyzing the loss ratio plotted in Fig. 8, it is observed that the topology depicted in Fig. 6 (b) achieves the lowest conduction loss. However, both topologies shown in Fig. 6 endure a DC bias voltage ($V_{in}/2$) on their resonant capacitors. For high density bus converters, Class II capacitors are frequently utilized due to their large energy density. However, this kind of capacitors exhibit decreased capacitance and increased equivalent resistance (ESR) under voltage

**Fig. 10:** The 1kW 4: 1 bus converter prototype.

**Fig. 11:** System board equipped with 4: 1 bus converters and multi-phase VRs.

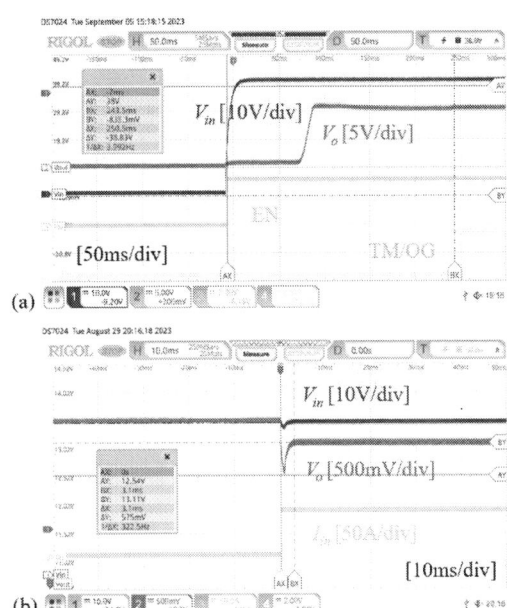

**Fig. 12:** Experimental results of the 1kW 4: 1 bus converter. (a)Soft-start. (b)Transient performance (no-load to full-load).

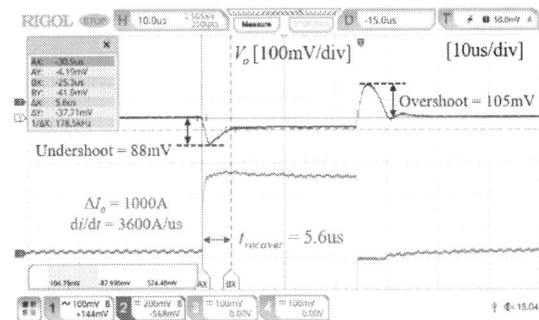

**Fig. 13:** Fast transient response of the VRM system. (0-1000A current step with 3600A/us slew rate)

bias, affecting the resonant process and leading to larger losses [15]. Therefore, we selected the topology depicted in Fig. 2 (b) as our design.

According to the parameters listed in table 1, the current and voltage stress of power switches can be calculated. The 80V Infineon MOSFETs are selected for the power switches $S_1 - S_4$, and 40V Infineon MOSFETs are selected for the SR switches $Q_1$ and $Q_2$. The constructed prototype is shown in Fig. 10, which has a dimension of 23.0 by 17.0 by 7.7mm, reaching a power density of 5.4kW/in$^3$. All controller, gate driver, and auxiliary power are included in this prototype, featuring pin-to-pin compatibility with the state-of-the-art products.

This prototype has been comprehensively tested under industry standard. Figure 12 displays the soft-start and transient performance. To further validate its reliability, the converter is integrated into a system board containing multi-phase VRs and dynamic loads, as shown in Fig. 11. The entire VRM system can deliver 1000A continuous current with less than 1V regulated output voltage. The fast transient response performance is evaluated, as shown in Fig.13, indicating a 88mV undershoot voltage for a 1000A current step with a slew rate of 3600A/us.

The power consumption of Nvidia's H100 Tensor Core GPU has reached up to 700W. To satisfy the increasing power demands of modern processors, we are developing a 1.5kW 4: 1 bus converter for next-generation XPUs, still within a compact volume of 23.0 by 17.0 by 7.7mm. The measured efficiency of 1kW and 1.5kW 4: 1 bus converters are shown in Fig. 14 and Fig. 15 respectively. For the typical 54V input voltage, the 1kW prototype reaches a peak efficiency of 98.2% and full-load efficiency of 97.3%, and the 1.5kW prototype reaches a peak efficiency of 98.0% and full-load efficiency of 97.0%, both including the auxiliary powers for controller and gate driving. The converters have been commercialized and currently being promoted

**Tab. 2:** The performance comparison with the state-of-the-art 4: 1 bus converters. ($V_{in} = 54$V)

| | Power | Peak effi. | Full-load effi. | Volume | Note |
|---|---|---|---|---|---|
| This work | 1kW | 98.2% | 97.3% | 23/17/7.7mm | Product |
| This work | 1.5kW | 98.0% | 97.0% | 23/17/7.7mm | Product |
| Vicor NBM2317 [16] | 1kW | 97.9% | 97.0% | 23/17/5.2mm | Product |
| FPM BMR313 [14] | 1kW | 97.2% | 96.4% | 23/17/7.6mm | Product |
| MPS MPC12106 [17] | 800W | 98.1% | 96.5% | 24/18/8.6mm | Product |
| Berkeley [7] | 800W | 99.0% | 98.1% | 23/17/6.2mm | Laboratory prototype |
| CPES [5] | 900W | 98.4% | 98.0% | 36/37/7.0mm | Laboratory prototype |

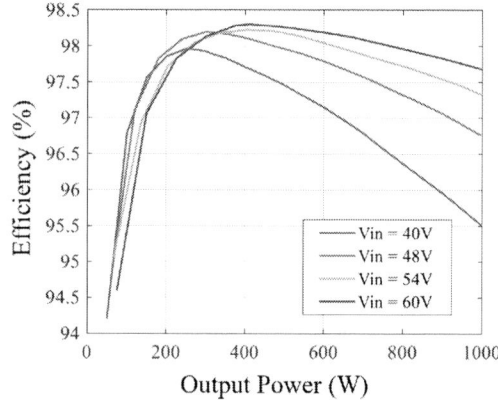

**Fig. 14:** Measured efficiency of the 1kW 4: 1 bus converter.

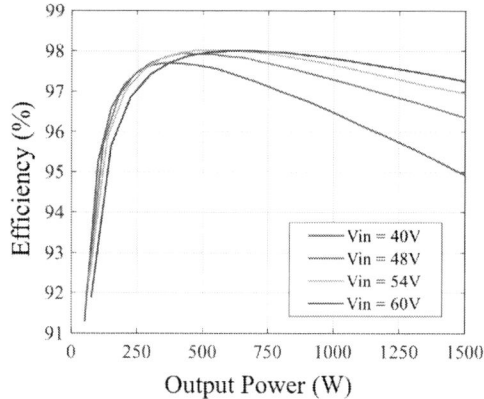

**Fig. 15:** Measured efficiency of the 1.5kW 4: 1 bus converter.

to 48V data center markets.

Table 2 lists the performance comparison with the state-of-the-art 4: 1 bus converters, demonstrating that this work achieves superior efficiency and power density. In addition to industry products, two academic research studies are compared. Utilizing resonant switched capacitor converter, the work in [7] achieved a highly effcient and compact design. However, this study focused on the power stage, omitting auxiliary power, controller, and other es-

sential circuits such as extra buck converter for soft-start, which also affect efficiency and power density performance. The research conducted by CPES [5] employed a conventional isolated LLC topology with a matrix transformer, exhibiting high efficiency but limited power density.

## 5 Conclusions

This paper conducts a comparative analysis of various transformer-based resonant converters for 48V data centers. The performance comparison including voltage conversion ratio, current and voltage stress, and practical design considerations of different topologies are carried out. The suitability of these topologies is discussed in detail. Based on the topology analysis, we have developed high power 4: 1 bus converters, demonstrating superior efficiency and power density compared to existing state-of-the-art solutions.

## References

[1] K. Radhakrishnan, M. Swaminathan, and B. K. Bhattacharyya, "Power delivery for high-performance microprocessors—challenges, solutions, and future trends," *IEEE Transactions on Components, Packaging and Manufacturing Technology*, vol. 11, no. 4, pp. 655–671, 2021. DOI: 10.1109/TCPMT.2021.3065690.

[2] M. Chen, S. Jiang, J. A. Cobos, and B. Lehman, "Design considerations for 48-v vrm: Architecture, magnetics, and performance tradeoffs," in *2023 Fourth International Symposium on 3D Power Electronics Integration and Manufacturing (3D-PEIM)*, 2023, pp. 1–9. DOI: 10.1109/3D-PEIM55914.2023.10052608.

[3] S. McCauley and S. Jiang, *Google 48V Update: Flatbed and STC*, https://www.opencompute.org/files/External-2018-OCP-Summit-Google-48V-Update-Flatbed-and-STC-20180321.pdf.

[4] MPS, *MPS 48V Data Center Solutions*, https://www.monolithicpower.com/en/products/power-management/48v-data-center.html.

[5] M. H. Ahmed, F. C. Lee, and Q. Li, "Two-stage 48-v vrm with intermediate bus voltage optimization for data centers," *IEEE Journal of Emerging and Selected Topics in Power Electronics*, vol. 9, no. 1, pp. 702–715, 2021. DOI: 10.1109/JESTPE.2020.2976107.

[6] T. Curatolo, *Factorized Power Architecture: Achieving high density and efficiency in board mounted power*, https://www.vicorpower.com/documents/whitepapers/wp-FPA-Achieving-high-density-efficiency-VICOR.pdf.

[7] T. Ge, Z. Ye, and R. C. Pilawa-Podgurski, "A 48-to-12 v cascaded multi-resonant switched capacitor converter with 4700 w/in3 power density and 98.9% efficiency," in *2021 IEEE Energy Conversion Congress and Exposition (ECCE)*, 2021, pp. 1959–1965. DOI: 10.1109/ECCE47101.2021.9595943.

[8] S. Jiang, S. Saggini, C. Nan, X. Li, C. Chung, and M. Yazdani, "Switched tank converters," *IEEE Transactions on Power Electronics*, vol. 34, no. 6, pp. 5048–5062, 2019. DOI: 10.1109/TPEL.2018.2868447.

[9] X. Ren, J. Zhang, P. Xu, and T. Long, "A non-isolated fixed-ratio dc-dc converter using switched auto-transformer (satx) for data center applications," in *2023 IEEE Energy Conversion Congress and Exposition (ECCE)*, 2023, pp. 3321–3324. DOI: 10.1109/ECCE53617.2023.10362385.

[10] R. Rizzolatti, C. Rainer, S. Saggini, and M. Ursino, "High density hybrid switched capacitor converter for data-center application," in *2021 IEEE Applied Power Electronics Conference and Exposition (APEC)*, 2021, pp. 1288–1293. DOI: 10.1109/APEC42165.2021.9487136.

[11] D. Huang, X. Wu, and F. C. Lee, "Novel non-isolated llc resonant converters," in *2012 Twenty-Seventh Annual IEEE Applied Power Electronics Conference and Exposition (APEC)*, 2012, pp. 1373–1380. DOI: 10.1109/APEC.2012.6165999.

[12] Y. Jang, M. Jovanovic, and Y. Panov, "Multiphase buck converters with extended duty cycle," in *Twenty-First Annual IEEE Applied Power Electronics Conference and Exposition, 2006. APEC '06.*, 2006, 7 pp.-. DOI: 10.1109/APEC.2006.1620513.

[13] X. Ren, J. Zhang, Y. Jiang, X. Li, and T. Long, "A 48-to-1v llc dc transformer," in *2023 IEEE 24th Workshop on Control and Modeling for Power Electronics (COMPEL)*, 2023, pp. 1–5. DOI: 10.1109/COMPEL52896.2023.10221067.

[14] FPM, *BMR3131011/001*, https://flexpowermodules.com/resources/fpm-datasheet-bmr313.

[15] Y. Jiang, B. Hu, Y. Shen, X. Ren, S. Sandler, et al., "Loss characterization and modeling of class ii multilayer ceramic capacitors: A synergistic material-microstructure-device approach," *IEEE Transactions on Power Electronics*, vol. 38, no. 11, pp. 13 535–13 554, 2023. DOI: 10.1109/TPEL.2023.3286818.

[16] Vicor, *NBM2317S60D1580T0R*, https://www.vicorpower.com/documents/datasheets/ds-NBM2317S60D1580T0R-VICOR.pdf.

[17] MPS, *MPC12106-54-0750*, https://www.monolithicpower.com/en/products/power-management/48v-data-center.html.

PCIM Europe 2024, 11– 13 June 2024, Nuremberg      DOI: 10.30420/566262099

# High-Density 3.3 kW GaN Rectifier for Server Applications Comprising a 130 kHz Totem-Pole PFC and a 500 kHz LLC

Antonello Laneve[1], Matteo-Alessandro Kutschak[1], David Meneses Herrera[2], Manuel Escudero Rodriguez[1]

[1] Infineon Technologies AG, Austria
[2] Infineon Technologies AG, Finland

Corresponding author:   Manuel Escudero Rodriguez, Manuel.EscuderoRodriguez@infineon.com
Speaker:          Manuel Escudero Rodriguez, Manuel.EscuderoRodriguez@infineon.com

## Abstract

In this paper, a complete 3.3 kW full power supply unit (PSU) for server applications and the related design challenges are discussed. The proposed unit can target high power in very small form factor by means of the use of wide-bandgap (WBG) GaN and SiC switches, and features hold-up time extension as per server requirements. The designed unit is capable of 3.3 kW maximum steady state output power at 230 $V_{AC}$ down to 176 $V_{AC}$ input, with 50 $V_{DC}$ output voltage. It fits 72 mm x 40 mm x 192 mm dimensions, total peak efficiency of 97.4 % at 230 $V_{AC}$ input, resulting in a total power density of 96 W / in³. The PSU will also incorporate a baby-boost converter to enable 10 ms hold-up during AC line dropout conditions.

## 1  Introduction

The trend in server and telecom applications is aiming for continuous increase in power density requirements. As an example the Open Compute Project (OCP) specification for server Power Supply Unit (PSU) has increased the power by 80 % while only increasing the overall volume by 20 % between 2022 and 2023 (i.e. 1.5 times higher power density) [1]-[2]. To accommodate for these new power density requirements, it is needed to shrink the size of all the components and especially the passive elements, mostly bulky magnetic components (i.e. inductors and transformers). One of the possible approaches to shrink the magnetics is reducing the applied voltage-time-area by increasing the switching frequency of the commutated converter. However, other than reducing the volume, efficiency must still be high enough to overcome the cooling constraints and address the challenge of extracting the same or more dissipated power within a reduced space. A key enabler of high-switching frequency and high-efficiency in power converters is wide-bandgap (WBG) semiconductors such as SiC MOSFETs and GaN HEMTs, that feature fast commutations with low switching and driving loss.

The other main passive elements, the capacitors, occupy anther large fraction of the space. Indeed, in the standard server architecture, a bulky high voltage capacitance is required to comply the hold-

up time requirements. The PSU must be able to maintaining the nominal output voltage at full load even if the input AC grid is missing for up to 10 ms at full power (3.3 kW). In state-of-the-art converters this is achieved by oversizing the bulk capacitance to minimize the voltage drop of the intermediate bus during an AC line cycle drop-out (LCDO) [3]. This problem can be addressed with proper modifications of the converter structure.

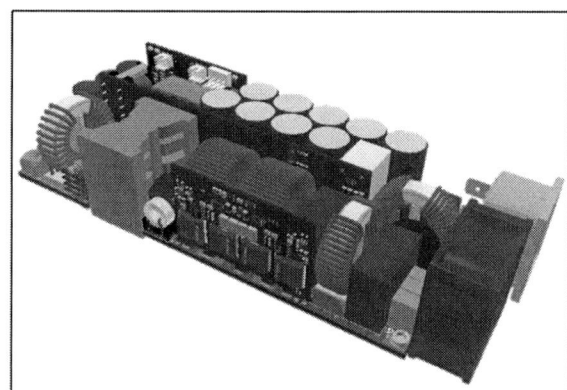

**Fig. 1**   Image of the proposed 3.3 kW GaN rectifier for server and datacenter applications.

In this work, a complete 3.3 kW power supply has been built and tested under steady state and dynamic load conditions, achieving a peak efficiency of 97.4 %, and a power density of 96 W/in³ without

PCIM Europe 2024, 11– 13 June 2024, Nuremberg    DOI: 10.30420/566262099

compromising hold-up time. The overall dimensions of the converter being 72 mm x 40 mm x 192 mm. On the other hand, the OCP specifications require much lower power density: 73.5 mm x 40 mm x 524.5 mm (about 31 W/in$^3$) for the 3 kW and 73.5 mm x 40 mm x 640 mm (about 47 W/in$^3$) for the 5.5 kW power converters.

In this paper, Section 2 presents the PSU architecture, Section 3 discusses the Interleaved totem pole (ITTP) PFC stage, Section 4 the DC-DC LLC converter, and finally Section 5 presents the experimental results. Conclusions are eventually reported in Section 6.

## 2    PSU architecture

The most dominant converter architecture in server applications is a two-stage system, comprising a front-end non-isolated AC-DC converter cascaded with a back-end isolated DC-DC converter. The non-isolated AC-DC stage provides power factor correction (PFC), reduces input current total harmonic distortion (THD) at the AC input, and regulates an intermediate bus, commonly around 400 $V_{DC}$. The intermediate bus feeds the isolated DC to DC converter, which provides a tightly regulated 50 $V_{DC}$ voltage at the output under all load conditions.

For the front-end PFC block, a bridgeless totem-pole active rectifier can achieve a very high efficiency, although it is well acknowledged that it requires WBG devices when operating in Continuous Conduction Mode (CCM) to have low or zero

reverse recovery charge and a reasonable low switching loss. For the DCDC stage, the series-parallel resonant converter (LLC) is commonly the most preferred solution, being capable of achieving very high efficiencies with a reduced number of components [3].

Therefore, the converter in this work follows a two-stage approach. Fig. 2 shows a simplified schematic of the proposed 3.3 kW PSU.

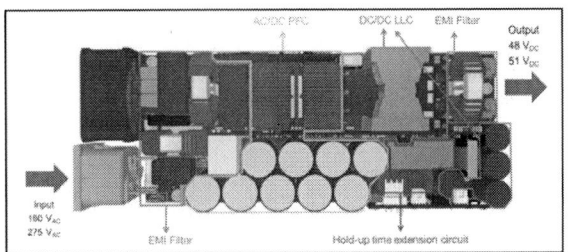

**Fig. 3**    Overview of the hardware blocks of the proposed 3.3 kW GaN rectifier supply.

For the front-end converter a two times interleaved bridgeless totem-pole (ITTP) PFC operating in CCM has been selected. The two so-called fast-switching legs commutate at 65 kHz, with an effective 130 kHz ripple at the AC side. The two legs use four CoolSiC™ IMT65R057M1H 650 V Silicon Carbide (SiC) MOSFETs [4] from Infineon. Whereas active rectification half-bridge (HB) leg of the PFC converter commutate at the grid frequency (50 Hz – 60 Hz) and uses four (two times in parallel) CoolMOS™ IPT60R016CM8 Silicon (Si) super-junction MOSFETs from Infineon. The

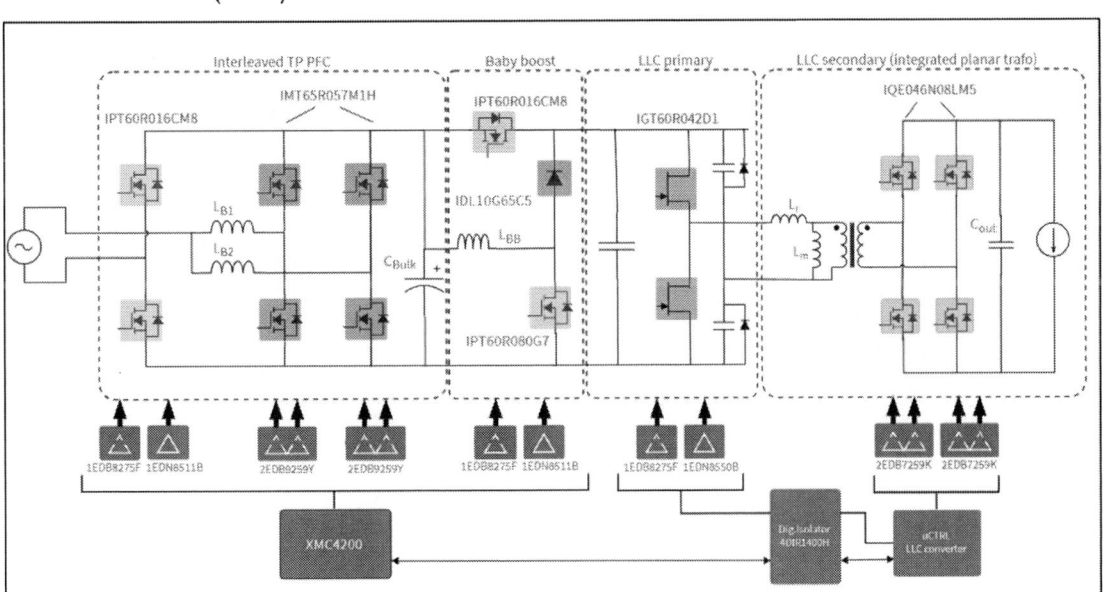

**Fig. 2**    Simplified schematic of the prototype PSU. Front-end totem-pole PFC followed by a half-bridge LLC resonant converter.

813

paralleling at device level and at stage level (interleaving) allows a better heat-spreading and therefore a reduced conduction loss.

In the back-end DC-DC converter, the inverter stage is realized with a half-bridge made of GaN HEMTs that switch near the series resonant frequency of the LLC tank (500 kHz), therefore enabling further shrinking of the passive components. Moreover, the synchronous rectification (SR) stage of the LLC converter is fully embedded in an integrated planar magnetic structure, which comprises the primary and secondary PCB windings, the main transformer core, and the series and parallel resonant inductors of the LLC converter.

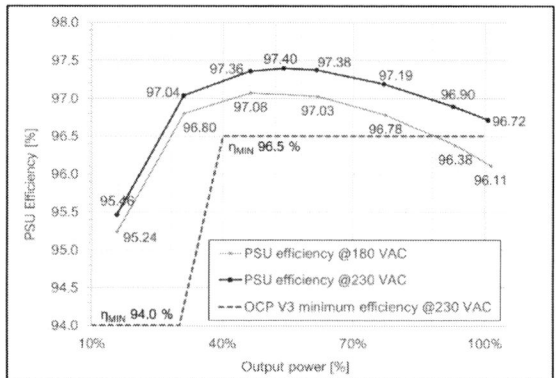

**Fig. 4** Final measured efficiency of the PSU.

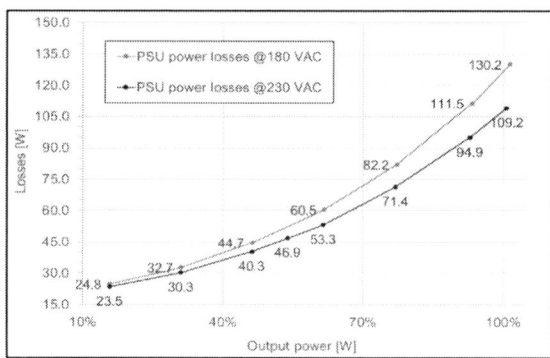

**Fig. 5** Measured losses of the PSU.

All the choices above allow a shrink of all the magnetic components, but not of the total bulk capacitance which should still comply with the hold-up time requirements. To further increase the power density, the total bulk capacitance connected to the DC-link has been reduced thanks to an auxiliary boost converter, so-called "baby-boost", between the LLC and the PFC, which decouples the LLC input voltage from the PFC during an AC LCDO event [5], allowing larger voltage drop on the bulk to power the load, while the LLC has a less stringent minimum input voltage requirement. Moreover, this approach allows to design the LLC

tank to work more efficiently close to the resonant frequency and to keep high overall efficiency in the PSU in steady state.

An overview of the hardware implementation of the proposed power supply unit is shown in Fig. 3. A summary of the overall efficiency results and the measured losses is presented in Fig. 4 and Fig. 5 respectively.

## 3 Interleaved Totem-Pole PFC

### 3.1 Hardware implementation

The whole PSU has been designed to follow a modular daughter card approach to exploit all the available space within the specified dimensions. For this reason, the PFC stage is split among two different power daughter cards. The synchronous rectification HB leg is on the power daughter card which hosts also the LLC primary HB and the flyback auxiliary supply, while the two high frequency (HF) legs of the ITTP PFC are on a second dedicated power daughter card, as shown in Fig. 6.

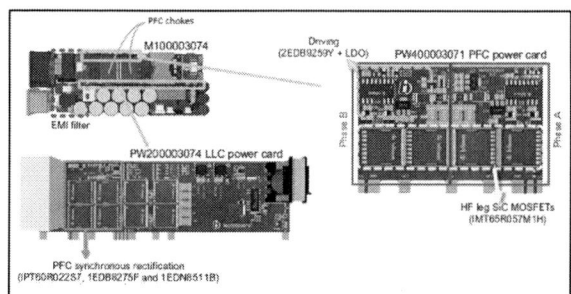

**Fig. 6** Hardware implementation of the PFC stage.

| Parameter | Value |
|---|---|
| Nominal $V_{IN}$ | 176 $V_{AC}$ – 277 $V_{AC}$ |
| Nominal $V_{OUT}$ | 410 $V_{DC}$ |
| Rated power | 3300 W |
| Switching freq. | 65 kHz |
| Boost inductors | 385 µH (x2 phases) nominal |
| Core material | CH33060GT High Flux GT 60 µ |
| Inductor turns | 64 (AWG 15) |
| SR MOSFETs | 4x IPT60R016CM8 (x2 parallel) |
| HF MOSFETs | 4x IMT65R057M1H (x2 legs) |

**Table 1** Key specifications of the PFC prototype.

Table 1 shows the key parameters of the ITTP PFC stage, including the PFC chokes (cores and windings). SiC MOSFETs have been chosen for the two HF legs due to the benefit out of the fast switching of SiC, and the lower increase of the $R_{DS,ON}$ with temperature.

Final losses and efficiency of the stand-alone ITTP PFC stage are reported in Fig. 7. Overall, peak efficiency of 98.9 % at 70 % of load is achieved, most of the losses coming from the switching of the SiC devices, conduction losses of the MOSFETs and copper losses of the PFC chokes (Fig. 8). A comparison of performance between 57 mΩ and 72 mΩ SiC MOSFETs in the HF leg will be discussed extensively in Section 5.1.

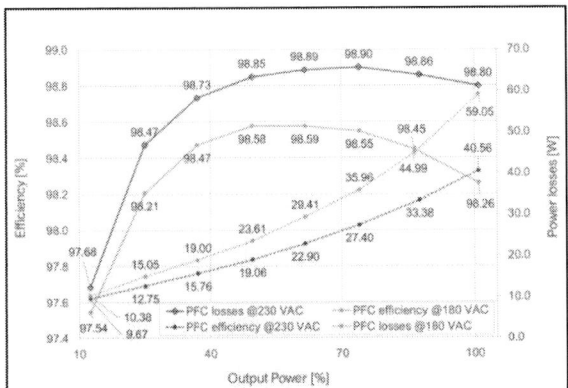

**Fig. 7** Final PFC efficiency and losses.

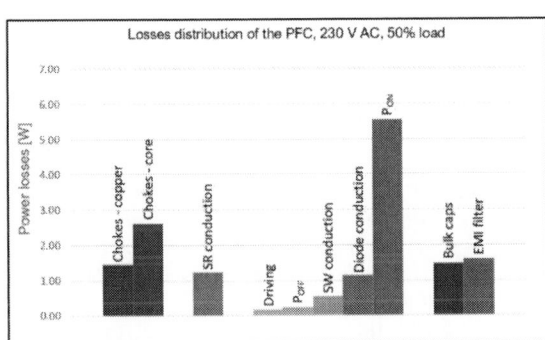

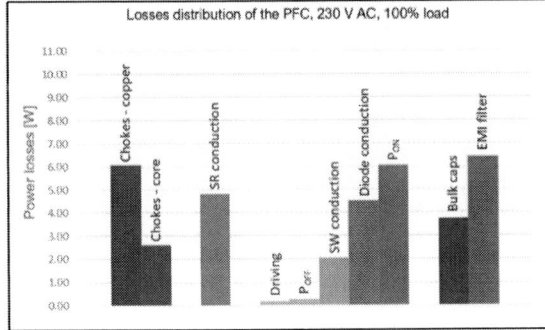

**Fig. 8** PFC losses distribution at 50 % and 100 % of load.

# 4 The LLC DCDC stage

The DC-DC LLC converter HB primary switches are on the LLC power daughter card together with the PFC SR stage and the auxiliary power supply, while the split resonant capacitors are on the main board (Fig. 9).

The secondary side of the LLC has been integrated within the magnetic transformer structure, which comprises the main transformer, the secondary side synchronous rectifier, the series resonant and the parallel inductors (Fig. 9). Non-isolated hybrid driving is chosen for the primary half-bridge [6] as the isolation boundary is implemented via isolators on the control card.

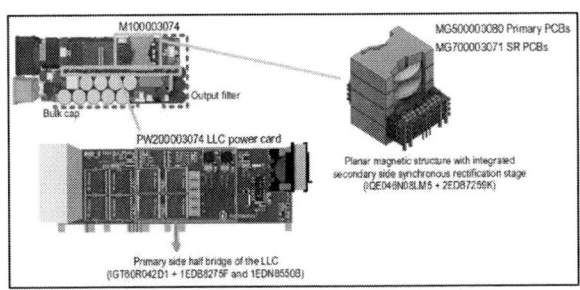

**Fig. 9** Hardware implementation of the LLC stage.

Table 2 reports key parameters of the LLC stage, including main specification of the LLC integrated magnetic structure and the MOSFETs. A detailed description of the integrated magnetic structure will follow in Section 0.

| Parameter | Value |
| --- | --- |
| Nominal $V_{IN}$ | 380 - 430 $V_{DC}$ |
| Nominal $V_{OUT}$ | 50 $V_{DC}$ |
| Rated power | 3300 W |
| Resonant freq. | 500 kHz |
| Series inductor | 1.65 µH nominal |
| Parallel inductor | 10.0 µH nominal |
| Resonant cap. | 60 nF |
| Turns ratio | 8:2 |
| HB MOSFETs | 4x IGT60R042D1 (2x parallel) |
| SR MOSFETs | 32x IQE046N08LM5 (8x parallel) |

**Table 2** Key specifications of the LLC prototype.

The measured efficiency of the half-bridge LLC converter is plotted for 400 $V_{DC}$ nominal input voltage in Fig. 10. Efficiency is near 98.45 % at 50 % of the rated load and still near 98 % at full load.

Fig. 11 show an estimation of the power losses breakdown for the LLC converter only. Main contributors to power losses are conduction losses of the primary side, of the synchronous rectifiers, and total copper losses of the series and parallel inductance, and of the main transformer itself.

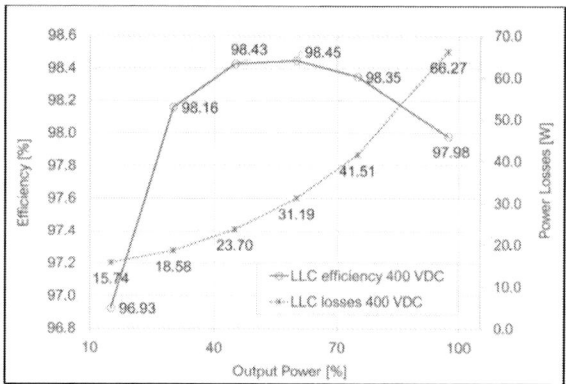

**Fig. 10** Final PFC efficiency and losses.

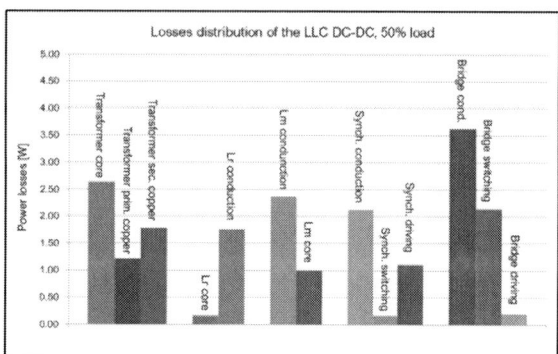

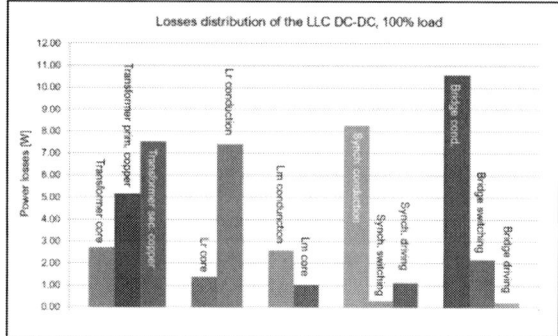

**Fig. 11** LLC losses distribution at 50 % and 100 % of load.

## 4.1 Integrated planar transformer

To have better control on the series resonant and the parallel inductors values, and avoid compromising efficiency, the implementation of the series and parallel inductors does not rely on the leakage and magnetizing inductance of the main trans-

former. On the contrary, they are discrete components but, to not compromise the power density, integrated within the magnetic structure on top of the main transformer and the LLC synchronous rectification stage. Fig. 12 shows the cross section of the transformer structure. This choice is one of the key design aspects to achieve the target efficiency of 98.5 % (LLC only), and the required ~100 W/in³ power density.

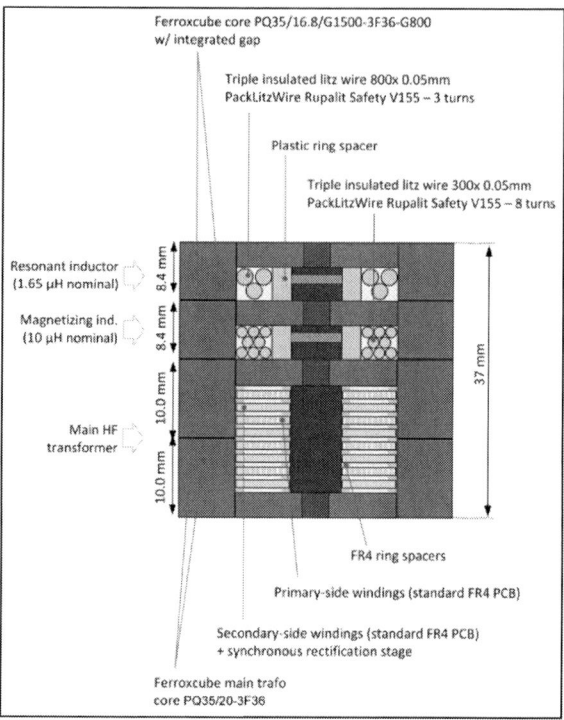

**Fig. 12** Planar transformer assembly with integrated synchronous rectification stage.

The full magnetic structure including series and parallel inductors of the LLC converter has an overall dimensions of 35 x 37 x 47.5 mm, which allows to fit the full power supply unit in a 1U form factor (40 mm maximum height) as per OCP requirements.

For the resonant series inductor and the parallel inductor of the resonant tank, two customized PQ35/16.8/G1500-3F36-G800 cores with integrated gap from Ferroxcube have been used, with respectively 3 turns of 1000 x 50 μm and 8 turns of 1000 x 50 μm triple insulated litz wire from PackLitzWire.

The arrangement of the two wire windings around the central leg of the ferrite core has been kept away from the gap with a dielectric ring, avoiding additional copper losses due to the intense fringing fields.

For the main 8:2 transformer, PQ35/20-3F36 cores from Ferroxcube have been used. The main transformer stack uses four primary and four secondary PCBs (Fig. 13 and Fig. 14) with interleaving to reduce high frequency copper losses. The four primary PCBs have four turns each, tow times in parallel and two times in series to achieve a total of 8 turns on the primary side.

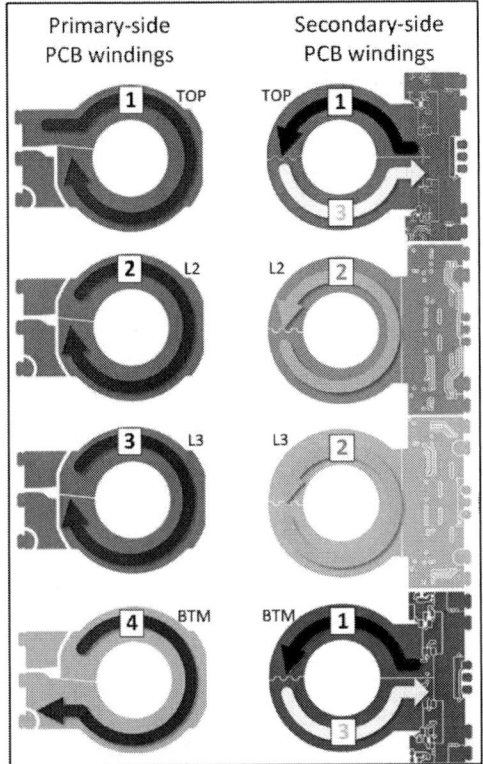

**Fig. 13** Primary and secondary side PCBs of the integrated main transformer.

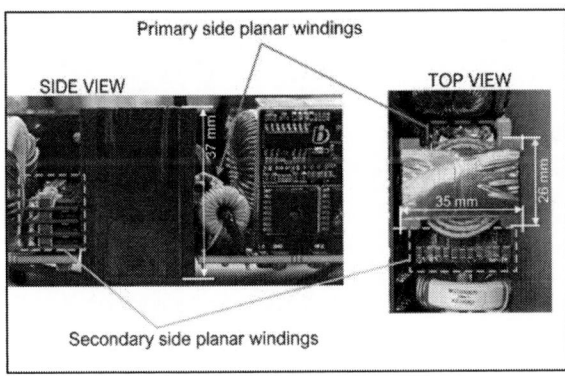

**Fig. 14** Side and top view of the magnetic structure integrating the main transformer, the series and parallel resonant inductors, and the synchronous rectification stage of the LLC converter.

The four secondary side PCBs have two times two turns each, and they are all in parallel, resulting in overall in 2 turns. In between each interleaved primary and secondary PCBs, a FR4 spacer is also inserted to increase the distance between each primary and secondary winding, and consequently keep a relatively low inter-winding capacitance.

The inter-winding capacitance has been measured with and without the ring spacers between the PCBs. A decrement from circa 140 pF (no spacers) to near 60 pF (with FR4 spacers) has been measured with the proposed FR4 spacing technique.

## 4.2 Main transformer conduction losses at full load

Distribution of conduction losses of the primary and secondary planar windings of the main transformer have been estimated for this converter through finite element modeling (FEM) analysis before testing. The conduction losses on primary and secondary PCB windings at resonance and full load were estimated to be 6.0 W for the primary side and 7.8 W for the secondary side. The estimations are in line with the overall loss forecasts (see "transformer sec. copper" in Fig. 11, full load). Therefore, this construction was adopted for the final main transformer assembly.

A sketch of the ohmic losses spatial distribution in the windings is reported in Fig. 15

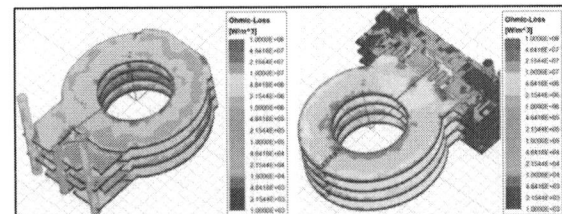

**Fig. 15** Overview of the FEM results for conduction losses and their distribution.

# 5 Experimental results

## 5.1 Interleaved totem-pole PFC

The ITTP PFC and the LLC DC-DC converters have been tested standalone first, and then together for the full PSU performance. In Fig. 16 the steady state waveforms at full load and 230 $V_{AC}$ voltage are reported.

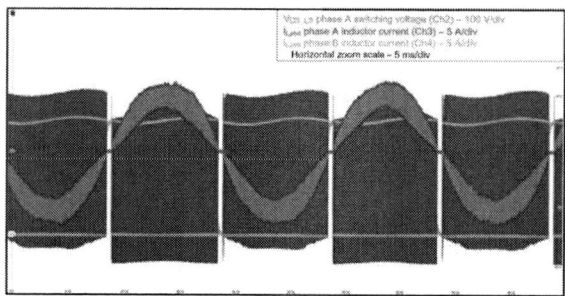

**Fig. 16** PFC drain voltage (phase A) overshoot and currents (phase A and phase B) at full load and 230 V$_{AC}$ line input.

Meanwhile, Fig. 17 and Fig. 18 report a detailed view of the PFC zero crossing during both grid polarity transitions. A 400 µs zero-crossing time has been chosen to avoid affecting the THD.

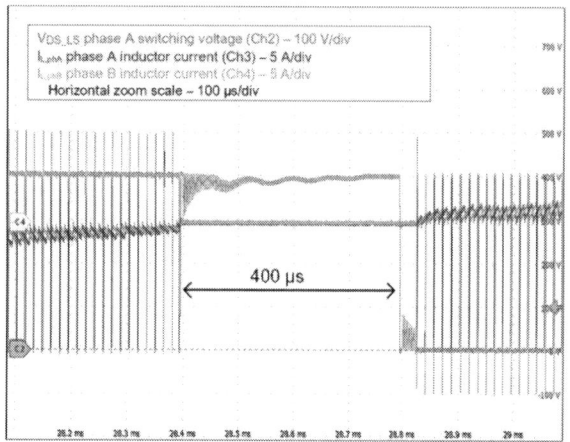

**Fig. 17** Detailed view of the positive V$_{AC}$ zero crossing in the PFC totem-pole at 230 V$_{AC}$.

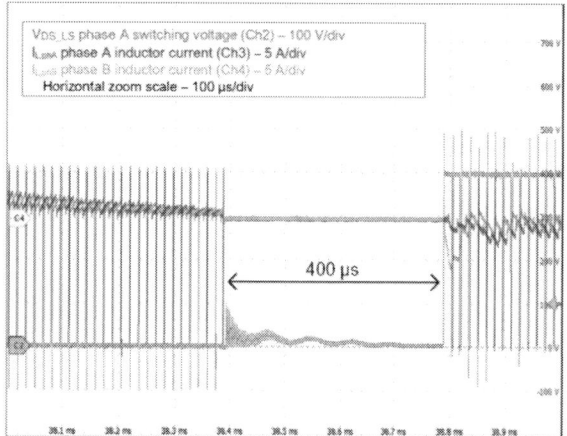

**Fig. 18** Detailed view of the negative V$_{AC}$ zero crossing in the PFC totem-pole at 230 V$_{AC}$.

### 5.1.1 Performance with 57 mΩ vs 72 mΩ CoolSiC™

During the design process, PFC performance have been compared between CoolSiC™ IMT65R072M1H and IMT65R057M1H with the full PSU enclosed. Main reason for this evaluation was an excessive temperature on the PFC fast leg when using IMT65R072M1H. Indeed, at 180 V$_{AC}$ the fast leg of the ITTP PFC becomes the hotspot of the converter due to the higher input current, and the power card placement (reduced airflow). It has been observed that IMT65R072M1H devices were experiencing T$_{CASE}$ ~120 °C at full load, 180 V$_{AC}$ input line voltage, and 25 °C ambient temperature. While this is within acceptable boundaries, it does not give adequate margin to operate at the maximum ambient temperature of 45 °C.

By using CoolSiC™ IMT65R057M1H, a total reduction of 6.3 W power dissipation at full load and 180 V$_{AC}$ line voltage has been observed for the HF switches which resulted in a significantly lower PFC temperature.

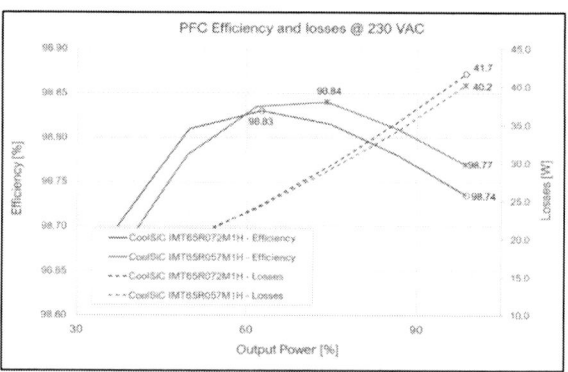

**Fig. 19** Comparison between IMT65R057M1H and IMT65R072M1H in PFC high frequency leg for 230 V$_{AC}$.

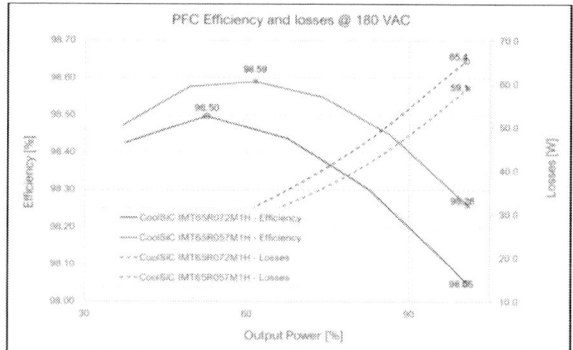

**Fig. 20** Comparison between IMT65R057M1H and IMT65R072M1H in PFC high frequency leg for 180 V$_{AC}$.

In Fig. 19 and Fig. 20 efficiency and losses of the PFC are reported for 230 V$_{AC}$ and 180 V$_{AC}$ input. At 230 V$_{AC}$ a crossing point can be observed. This is related to the change in the $R_{dsON}$ and $Q_g$ of the MOSFETs ($R_{dsON}$ @25°C and $Q_g$ @18 $V_{DC}$ driving are 72 mΩ, 22 nC and 57 mΩ, 28 nC respectively). At 230 $V_{AC}$ no major benefit is observed. As expected however, the use of the IMT65R057M1H is particularly beneficial at the lowest input line voltage, as shown in Fig. 20.

With CoolSiC™ IMT65R057M1H the peak efficiency becomes +0.1 % higher at mid-load, and +0.2 % at full load. The temperature profiles with IMT65R057M1H are also reported in Fig. (PSU fully enclosed and fan at full speed). The final resulting PFC efficiency (other modifications included) with CoolSiC™ IMT65R057M1H is reported in Fig. 7.

## 5.2  Resonant LLC converter

The resonant LLC converter is fed by the bulk voltage of the PFC, with an average 410 $V_{DC}$ and a peak to peak ripple voltage of 30 V at full load (Fig. 21), because of the reduced bulk capacitance (900 µF) to meet the power density target. Under these conditions, the LLC converter frequency spans from 570 kHz to 460 kHz, which has been only possible by using Infineon CoolGaN™ IGT60R042D1 GaN HEMTs at the HB primary side of the LLC. A detailed view of the LLC frequency is reported in Fig. 22 and Fig. 23 for the minimum and maximum bulk voltage conditions respectively.

Fig. 24 reports the primary and secondary voltages, and resonant current at full load, while Fig. 25 shows the output voltage and bulk voltage ripple at 10 % load.

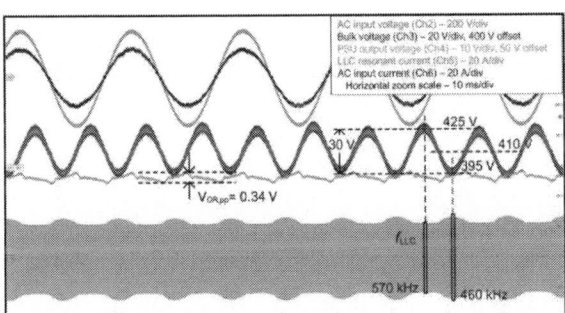

**Fig. 21**  Output voltage ripple for 100 % load conditions.

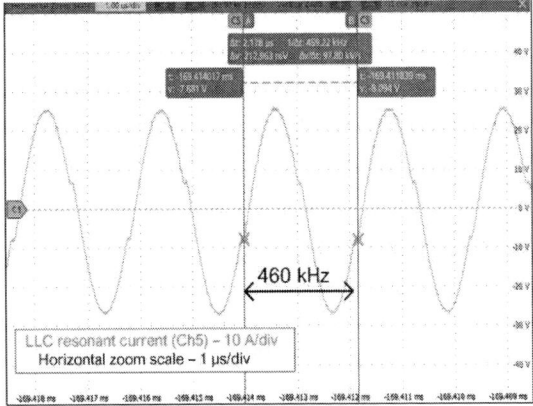

**Fig. 22**  LLC resonant current at 100 % load conditions and lowest bulk voltage (395 V) in steady state operation

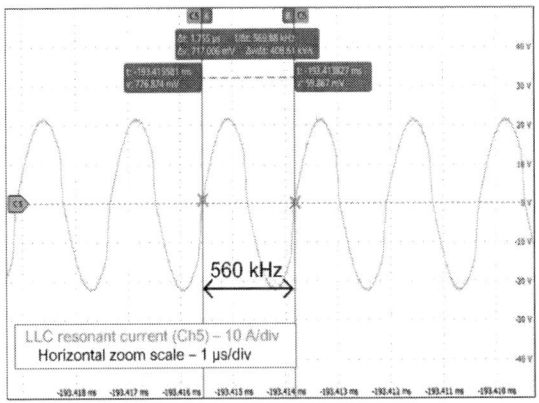

**Fig. 23**  LLC resonant current at 100 % load conditions and highest bulk voltage (425 V) in steady state operation

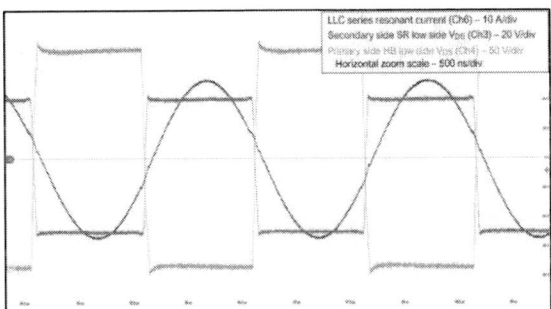

**Fig. 24**  LLC resonant current, primary, and secondary side drain-source voltages (low-side switch at full load

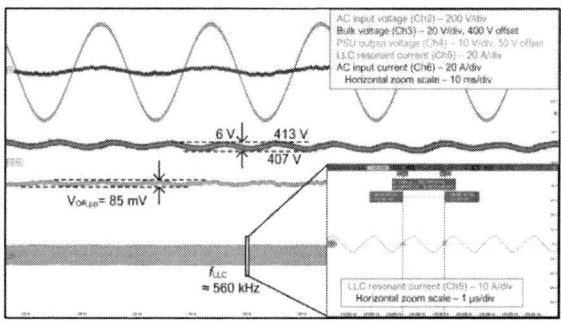

**Fig. 25** Output voltage ripple for 10 % load conditions.

## 5.3 Load transients

Load transients has been tested at 10% - 90% and 10 % - 50 % of full load, 20 Hz repetition rate as per OCP requirements. Because of the reduced bulk capacitance to meet the power density constraint, load transients are more critical compared to a state-of-art PSU, i.e. same load jumps of the output current would lead to faster discharge or charge of the bulk voltage during transients. To make possible 10%-90% jumps with a 20 Hz high frequency repetition rate with only 900 μF bulk capacitance, feedforward (FF) of the output current to the PFC voltage control loop has been implemented.

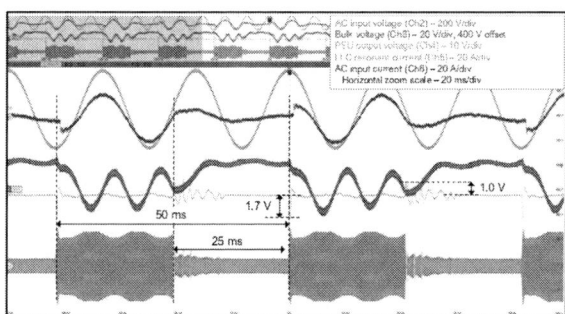

**Fig. 26** 10 % - 90 % load transients of the full PSU, 20 Hz repetition rate.

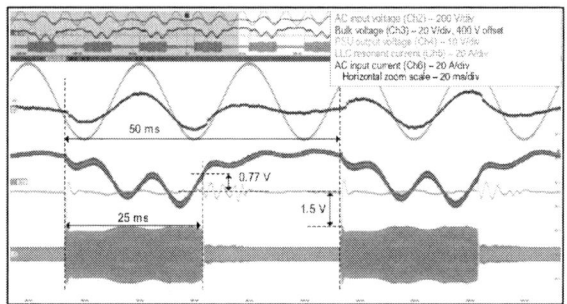

**Fig. 27** 10 % - 50 % load transients of the full PSU, 20 Hz repetition rate.

The output current information is sent from the LLC to the PFC controller via UART interface, and

this additional FF information is used to regulate the bulk voltage beyond the control bandwidth of the voltage loop. Results obtained for the proposed PSU are shown in Fig. 26 (10% - 90 % at 20 Hz) and Fig. 27 (10 % - 50 % at 20 Hz).

## 5.4 Power supply thermals

Thermal performance of the full rectifier has been measured with Type J thermocouples, fan supplied externally and at 25°C ambient temperature. It must be remarked that the PSU temperature has been taken with full enclosure to provide proper cooling. Indeed, the enclosure is designed to convey the airflow through the high temperature components on the right-hand side of the converter (Fig. 28). Critical hotspots such as PFC high frequency leg, LLC primary side and SR MOSFETs and drivers have been reported in Fig. 29.

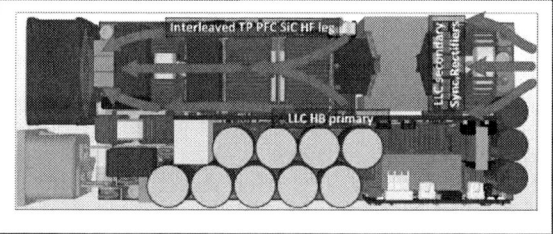

**Fig. 28** Main airflow "pipe" of the PSU when assembled with full enclosure.

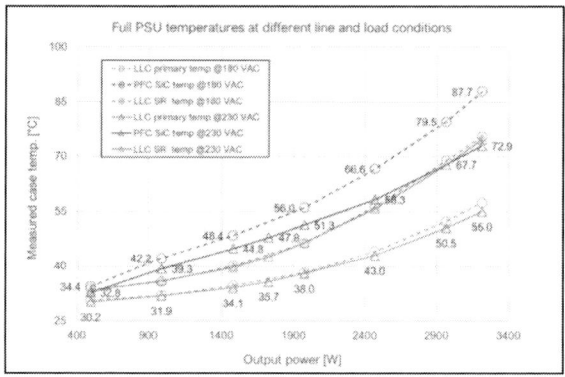

**Fig. 29** Full PSU temperature profile with enclosure, fan at full speed, and IMT65R057M1H.

A maximum temperature of 88 °C has been observed in the PFC SiC MOSFETs with the design choices discussed in Section 5.1.1. This also provides enough margin with respect to the maximum ambient temperature of 45°C for server applications. It is remarkable that for the ITTP PFC a L-shaped heatsink has been required to dissipate losses, while the LLC HB and SR are cooled without any heatsink.

# 6 Conclusions

As per the latest OCP standard updates, power demand at rack level in server and datacenter keeps growing substantially due to higher computing workload that must be accommodated in less floor space. A power supply unit based on Infineon components has been proposed in this paper, targeting 3.3 kW within 72 mm x 40 mm x 192 mm and reaching 96 W/in$^3$, which is 3x higher compared to the current OCP V3 that requires "only" 3 kW within 73.5 mm x 40 mm x 524.5 mm (31 W/in$^3$). Moreover, it is 2x higher compared to the most recent 5.5 kW OCP specification that recommends a 73.5 mm x 40 mm x 640 mm form factor (47 W/in$^3$). Therefore, the design described in this paper could be potentially scaled to higher power ratings withing the specified dimensions.

To achieve the outstanding power density, an integrated magnetic structure has been proposed. It comprises the main transformer, the series and parallel resonant inductors, and the synchronous rectification stages. Simulation, testing and implementation results within the PSU have been presented.

The PSU achieves an overall peak efficiency of 97.4 % (fan consumption not included), which could be further improved by slightly reducing conduction losses on the PFC SR stage and the so-called "static switch", in order to meet the 97.5 % peak efficiency of the OCP requirements. On the other hand, the PSU also already exceeds the OCP V3 minimum efficiency, which makes this design a suitable starting point for future server applications.

The analysis of the results and the thermal captures show that the main contributions to the overall power loss are the integrated magnetic structure of the LLC, the switching and conduction losses of the SiC MOSFETs in the PFC, copper losses in the PFC chokes and input filter (especially at 180 $V_{AC}$ input) and the conduction and driving losses of the secondary side SRs in the LLC.

A comparison between 57 mΩ and 72 mΩ SiC device in the PFC stage has been also reported, which shows the 57 mΩ better performing at the higher temperatures and lower input voltages, e.g. at 180 $V_{AC}$ input.

Finally, as next step the 10 ms hold up time will be implemented within the same hardware by means of an auxiliary baby-boost converter, while maintaining 96 W/in$^3$ power density.

# References

[1] Open Compute Project (OCP), Open Rack V3 48V PSU Specification Rev: 1.0. Available online: http://www.opencompute.org (accessed on 8 August 2023).

[2] Open Compute Project (OCP), Open Rack V3 48V 5.5kW PSU Specification Rev: 0.2. Available online: http://www.opencompute.org (accessed on 27 July 2023).

[3] M. Escudero, M. -A. Kutschak, D. Meneses, N. Rodriguez and D. P. Morales, "High Efficiency, Narrow Output Range and Extended Hold-Up Time Power Supply with Planar and Integrated Magnetics for Server Applications," PCIM Europe digital days 2021; International Exhibition and Conference for Power Electronics, Intelligent Motion, Renewable Energy and Energy Management, Online, 2021, pp. 1-8.

[4] M.-A. Kutschak, D. Meneses, and F. Pulsinelli, "3300 W CCM totem pole with 650 V Cool-SiC™ in TOLL package and XMC™," 2024 Infineon Technologies AG, Application Note AN_2307_PL52_2308_140931.

[5] Yan Xing, Lipei Huang, Xuansan Cai and Stan Sun, "A combined front end DC/DC converter," Eighteenth Annual IEEE Applied Power Electronics Conference and Exposition, 2003. APEC '03., Miami Beach, FL, USA, 2003, pp. 1095-1099 vol.2, doi: 10.1109/APEC.2003.1179353.

[6] A. Laneve, A. Rossi, D. Varajao, "A Non-isolated and Cost-Effective hybrid driving solution for High Voltage GaN HEMTs", APEC '24

[7] F. Pulsinelli, D. Meneses, S. Abdel-Rahman, "Hybrid SiC and GaN implementation of a Totem-Pole PFC", 2023 IEEE Applied Power Electronics Conference and Exposition (APEC), Long Beach, California, USA, 2023, Industry Session.

[8] Yan Xing, Lipei Huang, Xuansan Cai and Stan Sun, "A combined front end DC/DC converter," Eighteenth Annual IEEE Applied Power Electronics Conference and Exposition, 2003. APEC '03., Miami Beach, FL, USA, 2003, pp. 1095-1099 vol.2.

[9] P. R. Prakash, A. Nabih and Q. Li, "Investigation and Solutions for High Termination Losses in Planar Matrix Transformers with Full-Bridge Rectifiers," 2022 IEEE Applied Power Electronics Conference and Exposition (APEC), Houston, TX, USA, 2022, pp. 27-34.

PCIM Europe 2024, 11– 13 June 2024, Nuremberg    DOI: 10.30420/566262100

# Addressing Power Switch Technology Selection Si/SiC/GaN in High Efficiency ZVS-PFC Resonant Converters

Marco Torrisi[1], Sebastiano Messina[1], Daniele Giovanni Sfilio[1], Angelo Giordano[1] and Mario Cacciato[2]

[1] ST Microelectronics, Italy
[2] University of Catania, Italy

Corresponding author:    Marco Torrisi[1], marco.torrisi-sl@st.com
Speaker:                 Marco Torrisi[1], marco.torrisi-sl@st.com

## Abstract

The use of Gallium Nitride (GaN) and Silicon Carbide (SiC) in power electronics has increased due to their advantages at the system level. They are now being used in various applications, even in fields where Silicon (Si) MOSFETs were preferred. However, while GaN and SiC offer better switching performance, in high efficiency applications all the losses contributions must be evaluated. For example, in resonant circuits that typically operate with zero voltage switching (ZVS), GaN or SiC devices may lead to a different switching frequency compared to Si, consequently a different losses distribution could be observed. This paper compares the overall performance of Si, SiC, and GaN MOSFETs in a resonant ZVS-PFC topology, with 99.2% efficiency, to provide general guidance on technology selection for a high efficiency design.

## 1 Introduction

The future of power electronics is oriented towards very high power-density, which higher switching frequencies facilitate, but it is also necessary to increase efficiency surrounding the thermal aspects. Possible solutions are improving the topology (bridgeless, interleaving, etc.), using superior devices, and improving control techniques. Due to their excellent physical properties, wide-bandgap (WBG) semiconductors like SiC and GaN are the perfect candidates for future power electronics. Their higher bandgap allows the material to withstand higher electrical fields, meaning higher breakdown voltages (BV) and lower on-state resistance ($R_{ds(on)}$) on a much smaller die [1], allowing significant reduction in final package dimensions compared to Si devices. Furthermore, their higher electron mobility and saturation velocity allow faster switching [2], rendering these new technologies ideal for addressing the previously mentioned challenges. Figure 1 shows the switch technology trends for ZVS-PFC: while SiC is generally preferred for high current applications due to the excellent $R_{ds(on)}$ temperature stability, its switching cannot typically exceed certain limits. GaN, on the other hand, presents excellent performance at very high frequency, but its $R_{ds(on)}$ is not as stable in temperature as SiC [3] and reverse conduction losses could become relevant at higher rms currents.

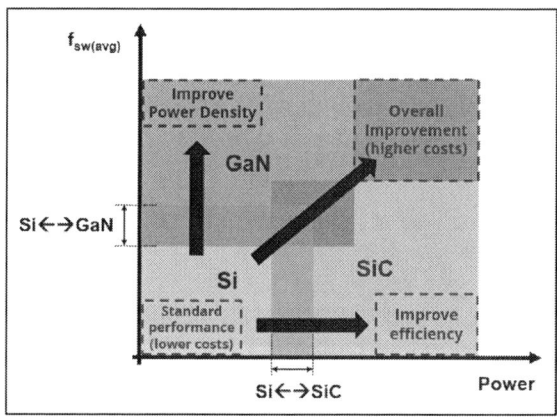

Fig. 1    Trend of switch technology selection areas in ZVS-PFC.

On the other hand, Si offers a good compromise between conduction and switching performance, in a relatively lower frequency range. Also considering its lower cost Si is very attractive. This paper compares the three different technologies in a high-efficiency, 3-channel, totem pole ZVS-PFC. In addition to the benefits of bridgeless and interleaved topologies, highlighted in [4] and [5], the ZVS control technique eliminates device turn-on losses by exploiting the resonance between the output capacitance of the MOSFET ($C_{oss}$) and the boost inductance L, as shown in [6]. Typically, in ZVS-PFC, the device power losses are due to conduction, turn-off, reverse and driving. However, it is not easy to determine the most suitable technology due to the numerous parameters involved, including rms current, switching frequency, $R_{ds(on)}$ vs temperature, etc.

## 2 ZVS Totem Pole PFC

### 2.1 Adopted control scheme

Figure 2 shows the proposed control scheme for a 3-channel, interleaved, totem pole, ZVS-PFC with hysteresis current control.

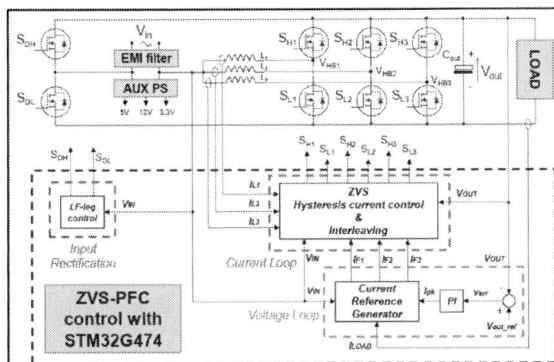

**Fig. 2** Control scheme of 3-channel interleaved totem pole ZVS-PFC converter with hysteresis current control.

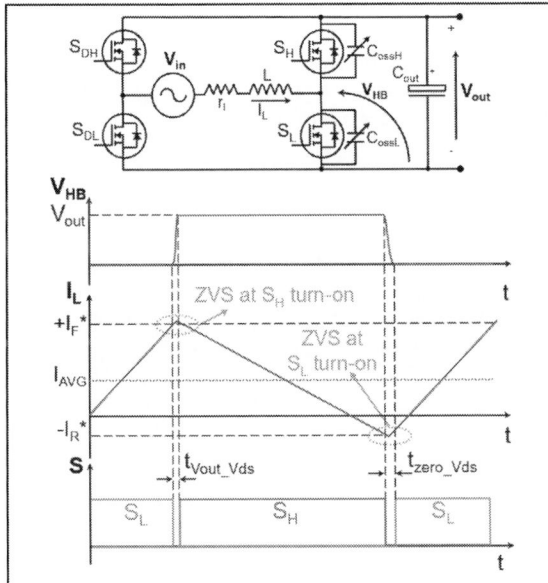

**Fig. 3** Single channel totem pole ZVS-PFC circuit and high frequency switching waveforms.

The outer voltage loop is implemented with a standard PI controller, which gives the PFC input peak current value $I_{pk}$, converted into three current references ($I_{F1}$, $I_{F2}$, $I_{F3}$) for the three channels, and the inner hysteresis current control must guarantee the ZVS condition for all the input and output conditions by satisfying the resonance equation:

$$\begin{cases} s^2 + 2\alpha s + \omega_0^2 = 0 \\ v_C(t) = C_1 e^{-\alpha t} \cos(\beta t + \varphi_1) + V_C(\infty) \\ i_L(t) = C_2 e^{-\alpha t} \cos(\beta t + \varphi_2) + I_L(\infty) \end{cases} \quad (1)$$

where $\alpha = r_l/2L$, $\beta = \sqrt{\omega_0^2 - \alpha^2}$, $\omega_0^2 = 1/LC$ $V_C(\infty) = V_{IN}$ and $I_L(\infty) = 0$. Due to the complexity of the analytical model, two adaptive internal current threshold references ($I_F^*$, $I_R^*$) and resonance times ($t_{zero\_Vds}$, $t_{Vout\_Vds}$; i.e., dead times) are selected thanks to pre-calculated and stored lookup tables for a finite set of input voltages $V_{in}$. The entire control scheme is implemented in an STM32G474 MCU of STMicroelectronics.
The theoretical waveforms of ZVS-PFC are shown in Fig. 3 (at the switching frequency) and Fig. 4 (at the line frequency).

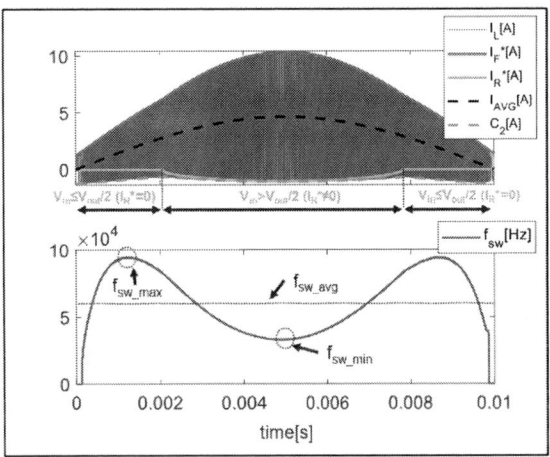

**Fig. 4** Half-line cycle waveforms of ZVS-PFC [6].

### 2.2 Device losses contribution

Unlike the standard CCM fixed frequency, the ZVS-PFC operates at variable switching frequency (see Fig. 4), and the boost inductance, L, together with the $C_{oss}$ value of the switch (inversely proportional to its $R_{ds(on)}$), set the average switching frequency of the converter, as qualitatively shown in Fig. 5 by simulation results.
A switch with a lower $R_{ds(on)}$ leads to lower conduction losses, while its higher $C_{oss}$ implies both a

higher switch-off energy and a lower average switching frequency, which equates to less switch-off events per second. It is therefore difficult to predict the switch-off loss behavior because a higher switch-off energy will always be counterbalanced by a lower average switching frequency; the same observation is also true for gate driving losses.

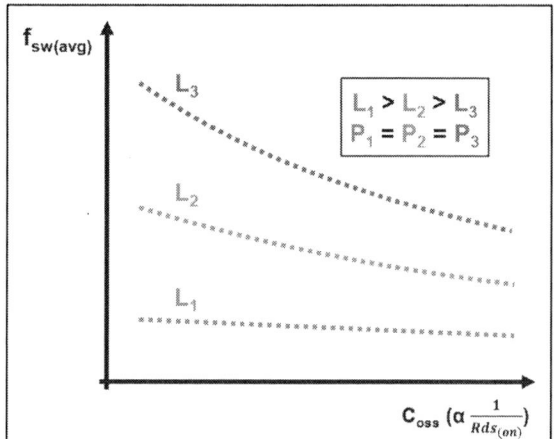

**Fig. 5** Behaviour of $f_{sw(avg)}$ vs $C_{oss}$ for different values of boost inductance L.

Thanks to the ZVS operation, the turn-on losses are eliminated, so the contributions of the device losses are given by conduction, driving, turn-off, and reverse conduction.

- *Conduction losses:*

$$P_{cond} = R_{DS(on)} \cdot I_{SW(rms)}^2 \qquad (2)$$

where $R_{DS(on)}$ is the drain-source resistance evaluated at the operating temperature of the device and $I_{SW(rms)}$ is the rms current of the switch.

- *Driving losses:*

$$P_g = V_{drv} \cdot Q_g \cdot f_{SW(avg)} \qquad (3)$$

where $V_{drv}$ is the gate driving voltage, $Q_g$ is the total gate charge of the switch and $f_{SW(avg)}$ is the average switching frequency of the converter over one half-line cycle.

- *Turn-off losses:*

$$P_{off} = 0.5 \cdot I_{Lpk(avg)} \cdot V_{out} \cdot t_{fall} \cdot f_{sw(avg)} \qquad (4)$$

where $I_{Lpk(avg)} = I_{Lpk}/(\pi/2)$ is the average peak value of inductor current over the half-line cycle and $t_{fall}$ is the fall time of the switch. Due to the uncertainty in the $t_{fall}$ estimation, Eq. 4 may deliver inaccurate results (typically overestimation) with respect to actual losses, especially for fast-switching WBG

devices. For this reason, the use of switching energy, instead of $t_{fall}$, may be preferable, but it implies the use of the same test conditions as given in the datasheet of the device (not always possible). Hence, to address this issue, the turn-off losses are obtained indirectly from thermal analysis after subtracting all the other contributions.

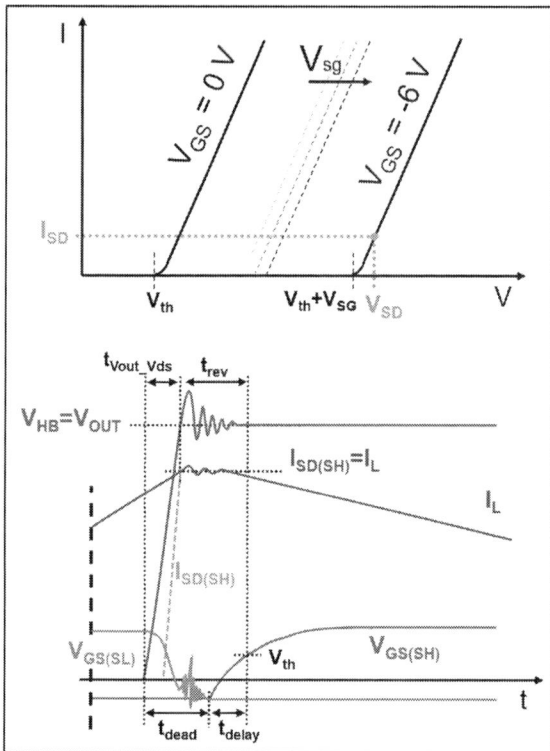

**Fig. 6** Third quadrant operation of the switch and reverse conduction time, $t_{rev}$, detection in the switching waveforms.

- *Reverse conduction losses:*

$$P_{rev} = V_{SD} \cdot I_{SD} \cdot t_{rev} \cdot f_{SW(avg)} \qquad (5)$$

This contribution relates to the reverse (source to drain) voltage $V_{SD}$, generated by a reverse conduction of the peak inductor current ($I_{SD(SH)}=I_L$) in the rectifier switch $S_H$ (after the turn-off of the boost switch $S_L$). This behavior is observed during the time interval, $t_{rev}$, between the rising of the midpoint voltage, $V_{HB}=V_{OUT}$, and the turn-on of $S_H$ ($V_{GS(SH)}=V_{th}$), as indicated in Fig. 6.

In ZVS-PFC, the reverse conduction loss of the boost switch $S_L$ can be neglected as its reverse current is always around zero (or slightly negative, see Fig. 8). The reverse conduction time, $t_{rev}$, typically includes part of the dead time and a driving delay, which are normally present to ensure adequate safety margin in preventing cross conduction on the half bridge.

The reverse conduction losses could be relevant for WBG devices, especially when a negative gate source voltage (e.g. $V_{GS}$ = -6 V) is applied during the turn-off. Figure 6 illustrates the typical voltage-current diagram of third quadrant operation for the WBG devices, which is used to evaluate the reverse voltage. Hence, for a given reverse current (different from zero), the following relationship is always valid:

$$V_{SD} > V_{th} + V_{SG} \qquad (6)$$

Higher values of the reverse voltage $V_{SD}$ are typically obtained for WGB devices (especially for GaN), while for silicon devices, the typical reverse diode forward characteristics (typically < 1 V) is considered at the peak inductor current.

# 3 Experimental results and technology comparison

## 3.1 Board specs and device parameters

The technology comparison is performed on the STEVAL-TTPPFC01 evaluation board [7].

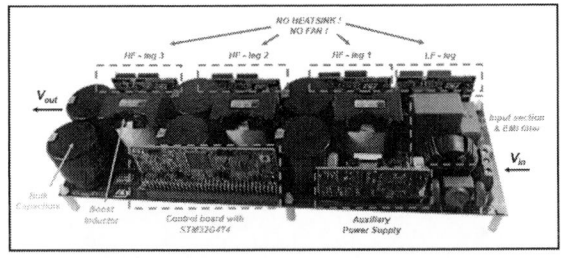

**Fig. 7** Photo of the STEVAL-TTPPFC01 evaluation board.

| Parameter | Value |
|---|---|
| Output Power | 2 kW |
| Output Voltage | 390 Vdc |
| Input Voltage | 230 Vac ± 10% |
| Variable Switching Frequency | (25 – 300) kHz |
| Boost Inductor | 160 µH |
| Bulk Capacitors | 6 x 330 µF |
| LF-leg MOSFET | STY145N65M5 (12mΩ typ.) [8] |
| Microcontroller | STM32G474QET [9] |

**Table 1** Specs of the STEVAL-TTPPFC01 evaluation board.

The STEVAL-TTPPFC01, illustrated in Fig. 7, is a 2 kW, three channel interleaved, totem pole ZVS-PFC with the main specs in table 1. The performance of Si, SiC, and GaN MOSFETs are evaluated in the three high-frequency legs, selecting devices with similar $R_{ds(on)}$ and 650V breakdown voltage as indicated in table 2.

| Compared technology | $Q_g$ [nC] | $R_{ds(on)}$ [mΩ] (Typ@25°C) | $V_{DS}$ [V] | $C_{oss(eq\_tr)}$ [pF] |
|---|---|---|---|---|
| STWA75N65DM6 Si [10] | 118 | 33 | 650 | 960 |
| SCT027W65G3 SiC[a] | 51 | 29 | 650 | 241 |
| SGT40R65ALD GaN[a] | 8.6 | 30 | 650 | 297 |

[a] Tests done with engineering samples.

**Table 2** Main parameters of the devices used for the technology comparison.

For the sake of simplicity, in the following sections the notation Si, SiC, and GaN will refer to the corresponding part numbers indicated in table 2.

## 3.2 Switching waveforms

Figure 8 shows the waveforms of one line-cycle together with a zoom of one high frequency switching cycle at the peak of the input voltage $V_{in}$=230$V_{ac}$ and full load.

The switching frequency behavior and the inductor current envelope precisely match the theoretical simulation of Fig. 4. The correct operation of the single high frequency leg is shown for all the technologies, since the ZVS condition is achieved at the turn-on of both low-side and high-side switches (i.e., $S_L$ and $S_H$).

The midpoint voltage variation $V_{HB}$ during the resonant transitions highlights the more linear behavior of WBG devices in terms of $C_{oss}$. Moreover, the lower $C_{oss}$ of WBG devices leads to the reduction of the ZVS parameter used in the adopted control technique: lower values of minimum reverse current (refer to -$I_R{}^*$ in Fig. 3), which means narrower hysteresis band, combined with lower resonance times. Consequently, a higher average switching frequency of about 70 kHz is observed for SiC and GaN devices, over the 55 kHz reached with the Silicon one.

The reverse conduction of the rectifier switch, $S_H$, can be seen from the overshoot of the mid-point voltage, $V_{HB}$, after the turn-off of the boost switch $S_L$. As expected, this behavior is more evident in the WBG devices due to the higher reverse source-drain voltage $V_{SD}$.

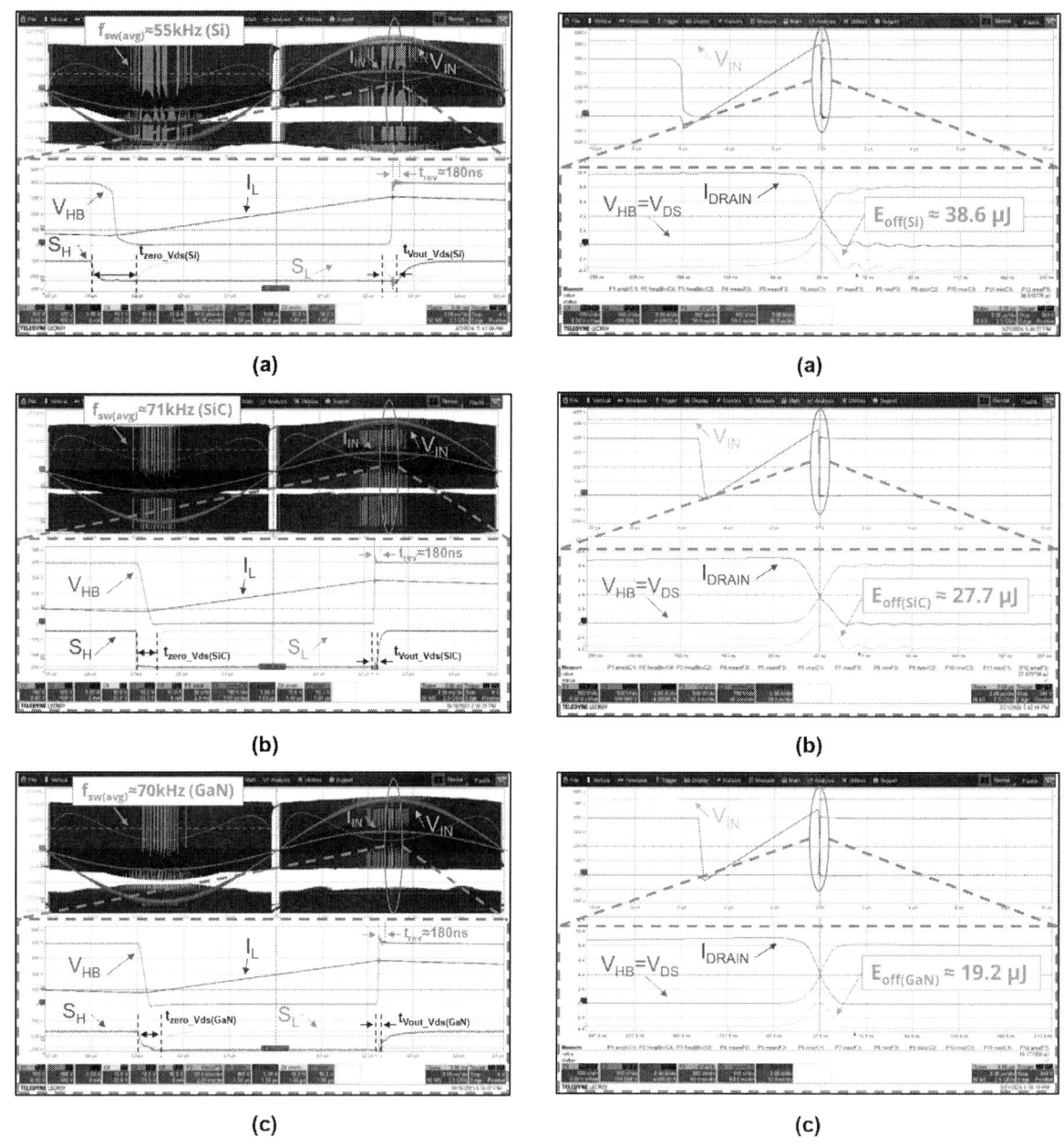

**Fig. 8** Switching waveforms of Si (a), SiC (b), and GaN (c) devices at the line frequency with a zoomed switching cycle at the peak of the input nominal voltage $V_{in}$=230$V_{ac}$ and full load.

**Fig. 9** Switch-off energy measurement and waveforms of Si (a), SiC (b), and GaN (c) devices at the peak of the input nominal voltage $V_{in}$=230$V_{ac}$ and full load.

The duration of the reverse conduction time is around 180 ns, which can also be considered as the average value over the half-line cycle $t_{rev(avg)}$≈180 ns to evaluate its impact in loss calculations.

Figure 9 shows the switch-off energy measurement at the peak of the input line voltage and full load, confirming the best switching performance of the GaN device followed by SiC and Si technologies ($E_{off(GaN)}$<$E_{off(SiC)}$<$E_{off(Si)}$).

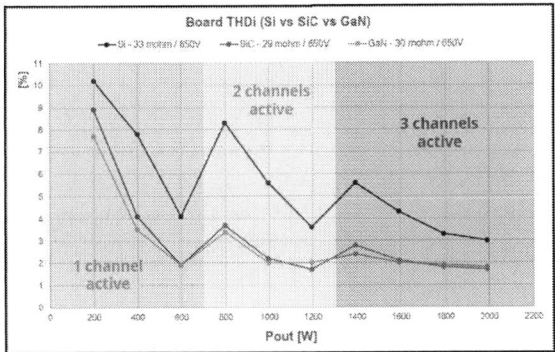

**(a)**

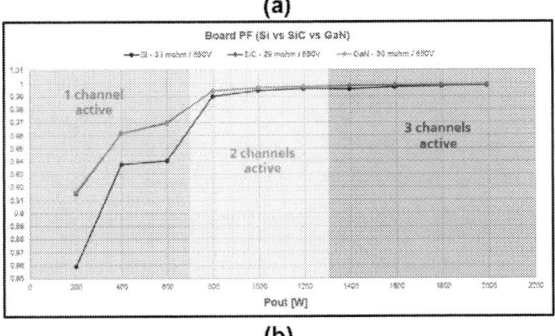

**(b)**

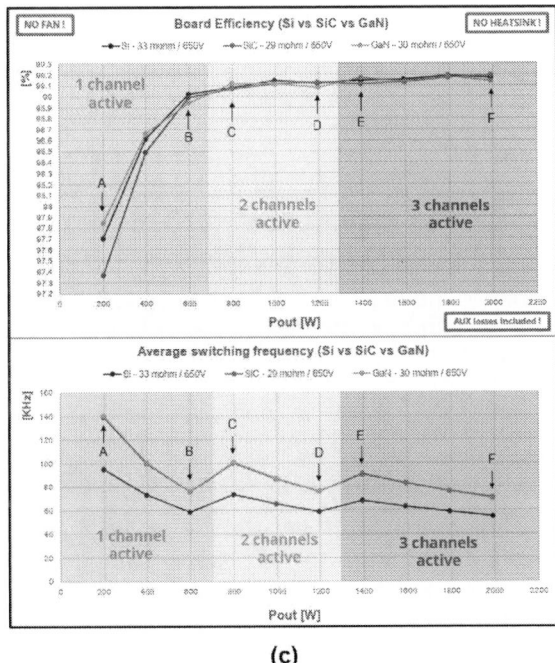

**(c)**

**Fig. 10** THDi (a), power factor PF (b), efficiency and average switching frequency (c) at nominal $V_{in}=230V_{ac}$.

## 3.3 Steady-state performance and thermal measurements

A precision power analyzer, YOKOGAWA WT1804E, was used to measure the steady-state performance for all the technologies.

Thanks to the lower resonance times (i.e. lower dead-times) the WBG devices show, in Fig. 10(a), an overall improvement of the total harmonic distortion of the input current (THDi), as expected from [11] and [12].

A better power factor is also evident below 50% of load, as highlighted in Fig. 10(b).

An outstanding converter efficiency, around to 99.2%, is visible in Fig.10(c) for all the technologies with loads higher than 50%, but the WBG devices operate at 30% higher switching frequency than the Si device, confirming the higher switching performance of WBG devices vs Si. For light loads, the lowest switch-off energy of GaN completely counterbalances its higher switching frequency, allowing a slightly better efficiency than Si and SiC. SiC showed the best efficiency at maximum load, confirming its superior $R_{DS(on)}$ behavior over temperature.

Regarding efficiency, it is interesting to compare GaN and SiC, not only at light or full load, but also across the phase shedding transitions (i.e., from 1- to 2-channel operation and from 2- to 3-channel operation). In fact, when an additional channel is enabled, the average switching frequency of each channel increases, while its power level (hence rms current) falls. Six check points (A-F) are placed in Fig. 10(c) to provide more clarity:

- **A**: 200 W power (10% load) with high average switching frequency (≈140 kHz). The GaN device offers the highest efficiency thanks to its superior switching performance (lower $Q_g$ and $E_{off}$).
- **B**: higher power (600 W) and lower switching frequency (≈75 kHz), compared to point A. The better $R_{DS(on)}$ and reverse behaviors of the SiC are dominant versus the switching losses allowing better efficiency.
- **C**: the 2nd channel is enabled, leading to lower power (400W) and higher switching frequency (≈100 kHz) for each channel. Here GaN outperforms SiC.
- **D**: same behavior as in point B.
- **E**: the 3rd channel is enabled and the same behavior as in point C is observed.
- **F**: full load operation with highest power (667 W) and lowest average switching frequency per channel (70 kHz). As in point B and D, SiC device offers slightly higher efficiency.

827

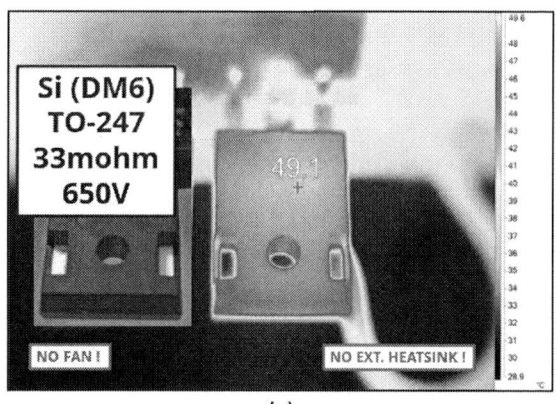

(a)

(b)

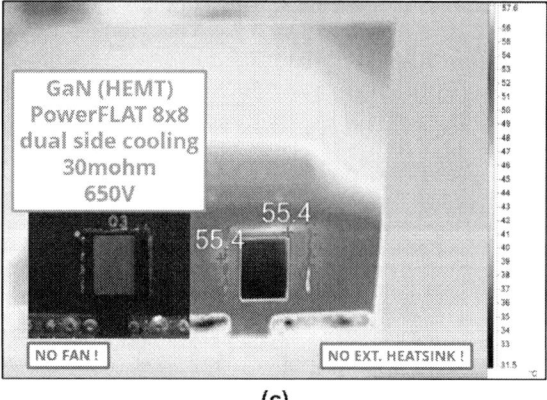

(c)

**Fig. 11** Thermal measurements of Si (a), SiC (b), and GaN (c) devices at the nominal input voltage $V_{in}$=230$V_{ac}$ and full load.

The efficiency of the Si device is somewhere in the middle range between SiC and GaN for almost all the load levels. The previous analysis might be useful to offer an initial appreciation of the general performance trends offered by each technology in a given design, once the power and switching frequency are set. However, for a comprehensive

analysis and final losses distribution, the thermal measurements (open frame, no heatsink, no fan) for all the technologies were conducted using a thermal imager Fluke Ti480 PRO.

As stated in the previous section, the thermal measurement is used to indirectly derive turn-off losses from the single device power loss, after subtracting all the other contributions.

The total thermal power loss for the single switch can be obtained from the typical thermal relationship:

$$P_{term} = (T_{dev} - T_{amb})/R_{th-JA} \qquad (7)$$

where $T_{dev}$ is the switch temperature, $T_{amb}$ is the ambient temperature, and $R_{th-JA}$ is the junction-to-ambient thermal resistance of the device. In this case, since the driving loss is not responsible for device heating, it is not included in the calculated thermal power loss, $P_{term}$. Table 3 contains the results of (7) in the specified conditions.

| Device | $R_{th-JA}$ [°C/W] | $T_{amb}$ [°C] | $T_{dev}$ [°C] | Temp. coeff. for $R_{ds(on)}$ | $P_{term}$ [W] |
|---|---|---|---|---|---|
| Si | 50 | 27 | 49.1 | 1.2 | 0.442 |
| SiC | 40 | 28 | 43.6 | 0.95 | 0.390 |
| GaN | 47[a] | 26.8 | 55.4 | 1.25 | 0.609 |

[a] Measured experimentally.

**Table 3** Devices temperature and estimated thermal loss at nominal input voltage $V_{in}$=230$V_{ac}$ and full load.

## 3.4 Losses distribution and considerations on technology selection trend

Thanks to the thermal results obtained in the previous section, it is possible to estimate each contribution to the device losses from equations (2), (3), and (5) for the same average reverse conduction time, $t_{rev(avg)} \approx 180$ ns, and average peak current, $I_{Lpk(avg)} = I_{SD(avg)} \approx 5.7$ A, for all the cases.

| Device | $I_{SW(rms)}$ [A] | $V_{drv(on)/(off)}$ [V] | $V_{th(min)}$ [V] | $V_{SD}$ [V] | $f_{SW(avg)}$ [kHz] |
|---|---|---|---|---|---|
| Si | 2.755 | 12 / 0 | 3.25 | 0.75 | 55 |
| SiC | 2.548 | 18 / -5 | 1.8 | 2.6 | 71 |
| GaN | 2.559 | 6 / -6 | 1.4 | 8.2 | 70 |

**Table 4** Devices parameters used to estimate the losses contributions (Vin=230Vac, full load).

Finally, the turn-off loss is indirectly calculated for all the devices:

$$P_{off(avg)} = P_{term} - P_{cond} - P_{rev(avg)} \qquad (8)$$

where $P_{off(avg)}=P_{off}/2$ and $P_{rev(avg)}=P_{rev}/2$ are the average values in one line cycle of the turn-off and reverse losses respectively, while $P_{cond}$ in (8) is already averaged, since the rms currents $I_{SW(rms)}$ in table 4 are measured in the entire line cycle. This approach bypasses the previously described issue of uncertainty in $P_{off}$ estimation.

Even if the three devices show comparable loss values, the results in Fig. 12 reveal significantly different loss distributions among the three technologies.

For the Si device, the conduction losses are dominant, while a comparable value of $P_{off}$ and $P_g$ is evident for Si and SiC, since the higher $Q_g$ and $E_{off}$ of the Si are counterbalanced by its lower switching frequency (55 kHz vs the 71kHz of the SiC). The Si shows the lowest reverse conduction loss, while the SiC offers the lowest conduction loss.

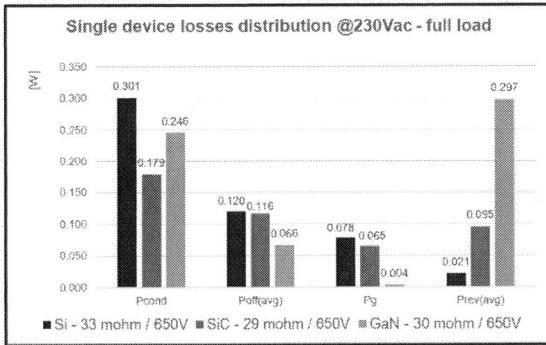

**Fig. 12** Losses distribution for Si, SiC, and GaN devices at the nominal input voltage $V_{in}=230V_{ac}$ and full load.

The GaN device showed the best switching performance thanks to the lowest $Q_g$ and $E_{off}$, resulting in approximately one-half the $P_{off}$ of Si and SiC, combined with an almost zero $P_g$. However, this result was partially nullified by the highest reverse conduction losses $P_{rev}$, which is a well-known issue in GaN technologies. To reduce this contribution and further drive the GaN at higher rms currents, certain adjustments can be adopted: a value closer to zero for $V_{drv(off)}$ can decrease the $V_{SD}$, and a lower dead-time may reduce the reverse conduction time $t_{rev}$. These adjustments, however, must still take into account the higher safety margins required for high power levels, such as more negative $V_{drv(off)}$ and higher dead-times to avoid the risk of cross-conduction in the high-frequency legs. Hence a compromise between performance and robustness must always be sought.

According to Fig. 10 and Fig. 12, all three technologies offer design and performance benefits. Then Fig. 13 offers, finally, clear general guidance on

technology selection, even at different power levels and/or switching frequencies, to provide high efficiency ZVS-PFC designs.

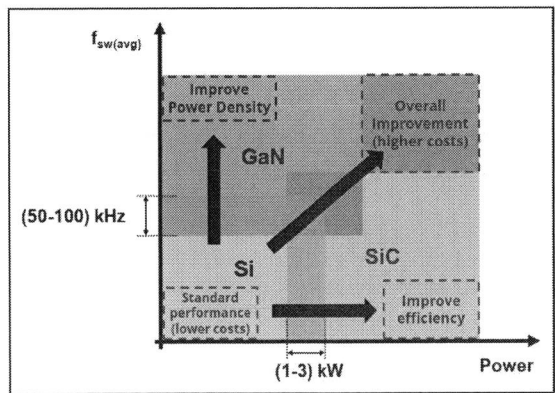

**Fig. 13** Trend of switch technology selection areas in ZVS-PFC for high efficiency designs.

# 4 Conclusion

In this paper a performance comparison among Si, SiC, and GaN MOSFET technologies has been provided in a 2 kW, 3-channel totem-pole resonant ZVS-PFC, to suggest the selection trend in high efficiency designs. An improvement is evident for WBG devices in terms of THDi and power factor compared to Si. Regarding efficiency, even if the three devices showed comparable loss values, the loss distributions are totally dissimilar.

The GaN device showed the best performance at lower loads, confirming its superiority at higher switching frequencies (>100 kHz) and lower rms currents, where the reverse conduction losses are not as impactful. Regarding SiC device, its better $R_{ds(on)}$ temperature stability renders it the best choice for higher rms current combined with switching frequencies that are not exceptionally high. The Si device works well in the intermediate range between high current and high switching frequency. Hence, if we also consider the costs, the Si offers the best compromise between performance and cost.

All the three technologies are essentially valid, high-performance alternatives for the given design. The results of this paper provide clear indications for selecting the most suitable technology, including at different power levels and/or switching frequencies, to provide ZVS-PFC design with high conversion efficiency.

# References

[1] J. Weimer; D. Koch; M. Nitzsche; J. Haarer; J. Roth-Stielow; I. Kallfass "Miniaturization and Thermal Design of a 170 W AC/DC Battery Charger Utilizing GaN Power Devices" IEEE Open Journal of Power Electronic, DOI 10.1109/OJPEL.2021.3137093.

[2] Fabrizio Roccaforte, Giuseppe Greco, Patrick Fiorenza 1 and Ferdinando Iucolano, "An Overview of Normally-Off GaN-Based High Electron Mobility transistors" in www.mdpi.com/journal/materials, Materials 2019, 12, 1599; doi:10.3390/ma12101599.

[3] N. Kaminski and O. Hilt, "Sic and gan devices-competition or coexistence?" In 2012 7th International Conference on Integrated Power Electronics Systems (CIPS), IEEE, 2012, pp. 1–11.

[4] L. Huber, Y. Jang and M. M. Jovanovic, "Performance Evaluation of Bridgeless PFC Boost Rectifiers," in IEEE Transactions on Power Electronics, vol. 23, no. 3, pp. 1381-1390, May 2008.

[5] M. Ancuti, M. Svoboda, S. Musuroi, A. Hedes, N. Olarescu and M. Wienmann, "Boost interleaved PFC versus bridgeless boost interleaved PFC converter performance/efficiency analysis," 2014 International Conference on Applied and Theoretical Electricity (ICATE), Craiova, 2014, pp. 1-6.

[6] M. Torrisi, S. Messina and M. Cacciato, "Hysteresis Current Control for a High Efficiency Totem Pole PFC in Zero Voltage Switching," 2021 23rd European Conference on Power Electronics and Applications (EPE'21 ECCE Europe), 2021, pp. P.1-P.10, doi: 10.23919/EPE21ECCEEurope50061.2021.9570441.

[7] AN5964, "99.3% efficiency, 2 kW, 3-channel interleaved totem pole PFC with resonant ZVS digital control - Application Note.

[8] STMicroelectronics STY145N65M5 N-channel 650V, 0.012 Ω typ., 138 A MDmesh™ M5 - Datasheet.

[9] STMicroelectronics STM32G474xB/C/E Arm® Cortex®-M4 32-bit MCU+FPU, 170MHz / 213 DMIPS, 128 KB SRAM, rich analog, math acc, 184ps 12 chan Hi-res time, Datasheet - production data.

[10] STMicroelectronics STWA75N65DM6 N-channel 650V, 33 mΩ typ., 75 A MDmesh™ DM6 Power MOSFET - Datasheet.

[11] N. Jiao, S. Wang, T. Liu, Y. Wang and Z. Chen, "Harmonic Quantitative Analysis for Dead-Time Effects in SPWM Inverters," in IEEE Access, vol. 7, pp. 43143-43152, 2019, doi: 10.1109/ACCESS.2019.2907176.

[12] Y. Yang, K. Zhou, H. Wang and F. Blaabjerg, "Harmonics mitigation of dead time effects in PWM converters using a repetitive controller," 2015 IEEE Applied Power Electronics Conference and Exposition (APEC), 2015, pp. 1479-1486, doi: 10.1109/APEC.2015.7104543.

PCIM Europe 2024, 11– 13 June 2024, Nuremberg          DOI: 10.30420/566262101

# Buck-Type Current Unfolding Converter With Discontinuous Conduction Mode in Ultra-Low Power-Factor Operation

Tomoyuki Mannen [1], Boseung Seo[1], Takanori Isobe[1], Ha Pham N.[2]

[1] University of Tsukuba, Japan
[2] University of Technology, Sydney, Australia

Corresponding author:     Tomoyuki Mannen, mannen@ieee.org
Speaker:                   Tomoyuki Mannen, mannen@ieee.org

## Abstract

This paper proposes a control method for a buck-type current unfolding converter, especially operating under non-unity power factor. The proposed method utilizes discontinuous conduction mode (DCM) to control the inductor current. DCM enables the converter to drastically reduce its inductors and allows the current controller to handle step changes in current due to non-unity power factor operations. Furthermore, this method enables the application of a feedforward-based control strategy by utilizing DCM, thereby eliminating the need for current sensors in the converter. Experimental results demonstrate sinusoidal output current waveforms at both unity and zero power factors, achieved without any current sensors in the converter prototype. These results confirm the effectiveness of the proposed method, particularly its robust current control capabilities. The proposed method is expected to increase the switching frequency and reduce the size of the passive components.

## 1   Introduction

Growing demand for renewable energy accelerates the requirement for improved efficiency and miniaturization of power converters, leading to a wide variety of research efforts. Utilizing new power device structures and wide-band-gap semiconductors can enhance converter performance due to their fast switching and low on-resistance, effectively reducing losses and allowing for smaller cooling components and passive elements [1,2].
Moreover, various circuit topologies, such as multilevel topologies, neutral-point-clamped (NPC) converters, and T-type configurations, have attracted attention for their ability to achieve higher efficiency and smaller sizes [3–5]. Discontinuous PWM, which is a well known-method, reduces the number of switching to 2/3 by using common-mode voltage in the three-phase converter. Reference [6] has proposed one-phase PWM for three-phase converters which can reduces the number of the switching to 1/3 and improve its efficiency.
Current unfolding converters separate a current control unit and a current unfolding unit, enabling them to reduce the number of switches operating with high-frequency PWM. As a result, the current unfolding converters can reduce the total number

of switching and inductors compared to conventional three-phase converters [7,8]. However, it may be difficult for the current unfolding converters to regulate the output currents during leading or lagging power factor operations due to sudden current changes [9]. This is because popular converters, includes the current unfolding converters, operate in continuous conduction mode (CCM), leading to delayed and restricted response from the current controller.
In contrast, the discontinuous conduction mode (DCM) offers faster current control, zero current switching, and increased stability by implementing feedforward-based control without accumulating control errors [10]. The DCM can increase the switching frequency and contribute to reducing the size of the inductor [11,12].
This paper proposes a control method utilizing DCM for the current unfolding converter in order to enable ultra-low power factor operation. The current control using the DCM makes it possible to follow step change references caused by the non-unity power factor operations. Since the proposed control method is based on feedforward control, the current unfolding converter can eliminate current sensors for its feedback control and instability

problem due to control delay. Experimental verification using a prototype of the current unfolding converter clarifies a good current control capability of the proposed method. The experimental results exhibit sinusoidal current waveforms at both unity and zero power factor, even though there is no current sensor in the converter. The proposed method is useful for increasing the switching frequency and reducing passive components.

## 2 Circuit Configuration and Operating Principle

Fig. 1 shows the circuit diagram of a buck configuration for a current unfolding converter. This converter is divided into two units: a current regulation unit and a current unfolding unit. The current regulation unit consists of four switches and two inductors, upper and lower choppers, which are connected in a vertically symmetrical manner. This unit operates with high-frequency PWM, and the two choppers regulate two intermediate currents, $i_+$ and $i_-$. Since the sum of the three-phase currents should be zero, the remaining intermediate current $i_0$ is regulated without further modulation once $i_+$ and $i_-$ have been determined. Therefore, the current regulation unit forms the intermediate currents $i_+$, $i_0$, and $i_-$ as part of the three-phase currents.

The current unfolding unit consists of a three-phase T-type NPC converter configuration. However, this unit operates at the line frequency, and unfolds and rearranges the three intermediate currents $i_+$, $i_0$, and $i_-$. Fig. 2 shows the switching patterns of the current unfolding unit. The operation of the current unfolding unit depends on the magnitude order of the phase voltages of the ac source.

The phase with the highest voltage is connected to the upper-side intermediate terminal through one of the upper-side switches, $S_{u+}$, $S_{v+}$, or $S_{w+}$. Similarly, the phase with the lowest voltage is connected to the lower-side intermediate terminal through one of the lower side switches, $S_{u-}$, $S_{v-}$, or $S_{w-}$. The other phase, with the middle voltage, connects to the common terminal through one of the common-side bidirectional switches, $S_{u0}$, $S_{v0}$, or $S_{v0}$. When the output of the converter is connected to a balanced three-phase ac line, each switching device operates at the line frequency or twice the line frequency. The rearranged currents form the three-phase ac currents $i_u$, $i_v$, and $i_w$.

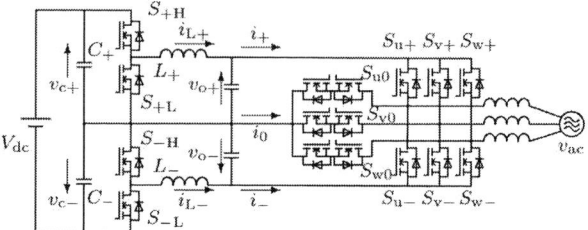

**Fig. 1:** A circuit diagram of the current unfolding converter.

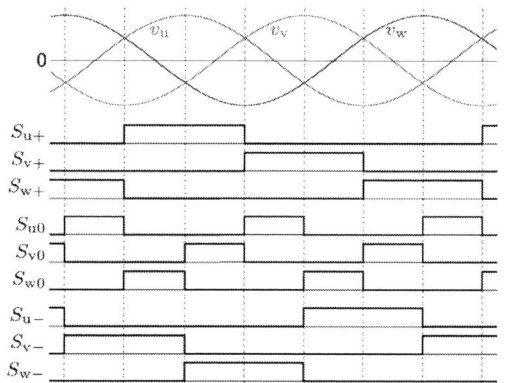

**Fig. 2:** Operating principles of the current unfolding unit.

## 3 Non-unity Power Factor Operation

Fig. 3 shows schematic waveforms of the current unfolding converters. Fig. 3a represents the operation under the unity power factor condition. Each intermediate current exhibits a continuous waveform because the crossing point of the output current coincides with the commutation timing of the current unfolding unit.

On the other hand, Fig. 3b shows the waveforms of non-unity power factor operation. The output current is shifted according to the power factor, and the intermediate current waveforms exhibit jumps at the point of commutation in the unfolding unit. In this case, the current unfolding unit swaps two current paths that have different amplitudes, due to the phase difference between the voltage and current.

### 3.1 Operating with Continuous Conduction Mode (CCM)

Since the inductor current in CCM requires feedback for current control, the current regulation unit employs proportional control with a feedback gain $K_C$. Additionally, integral or resonant control may

PCIM Europe 2024, 11– 13 June 2024, Nuremberg    DOI: 10.30420/566262101

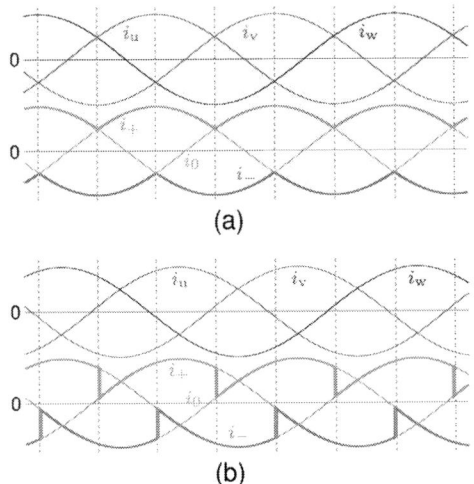

**Fig. 3:** A schematic waveforms of intermediate currents in (a) the unity power factor and (b) the low power factor.

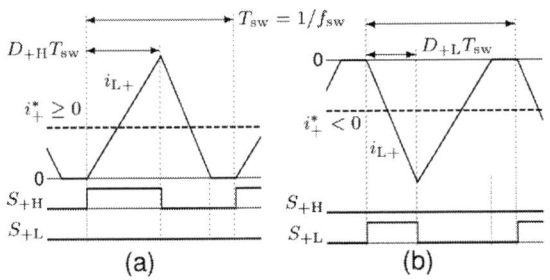

**Fig. 4:** Operation principles of current control in DCM when (a) $i_+^* \geq 0$ and (b) $i_+^* < 0$.

sometimes enhance the feedback characteristics in steady states. In non-unity power factor operation, the current controller must be able to track step changes in the current waveforms. Increasing the feedback gain extends the control bandwidth; however, it can also induce stability problems. A higher switching frequency and/or a smaller ac inductor have the potential to achieve a wider control bandwidth. Nevertheless, the actual system cannot reduce the control delay, which is due to the sensors used for feedback and the calculations in DSPs. Since control delay is the most critical factor inducing instability in the feedback control, the current control bandwidth is limited, making it impossible for CCM operation to exceed these limits.

### 3.2 Operating with Discontinuous Conduction Mode (DCM)

The current regulation unit operating in discontinuous conduction mode (DCM) may improve cur-

rent control performance because DCM does not require feedback control and can operate using feedforward-based control. Fig. 4 shows the operation principles of the current regulation unit with DCM. When the operating power factor is close to unity, the upper- and lower-side chopper operate in uni-direction, with $i_{L+} > 0$ and $i_{L-} > 0$, due to the operation in the current unfolding unit. Conversely, when the output power factor is low, both the upper- and lower-side choppers require bi-directional operation in the fundamental cycle of ac mains.

Focusing on the upper-side chopper, which regulates the upper-side intermediate current $i_+$, when the direction of the intermediate current is $i_+ > 0$, only the switch $S_{+H}$ operates with PWM and the other one $S_{+L}$ remains turned off. Assuming that the average of the upper-side inductor current $i_{L+}$ corresponds to the intermediate current $i_+$, the duty ratio of the switch $S_{+H}$ is given by

$$D_{+H} = \sqrt{\frac{2f_{sw}L_+I_{L+}v_{o+}}{v_{c+}(v_{c+} - v_{o+})}} \tag{1}$$

where $v_{o+}$ is the output voltage of the chopper and $v_{c+}$ is the voltage of the upper-side capacitor $C_+$, which serves as the input voltage of the chopper, and $f_{sw}$ is the switching frequency of the chopper. On the other hand, when the intermediate current has the opposite direction, $i_+ < 0$, only the switch $S_{+L}$ operates with PWM, and $S_{+H}$ remains turned off. Similarly, the duty ratio for $S_{+L}$ is given by

$$D_{+L} = \sqrt{\frac{2f_{sw}L_+I_{L+}(v_{o+} - v_{c+})}{v_{o+}v_{c+}}}. \tag{2}$$

In the same manner, the lower-side chopper regulates the lower-side intermediate current $i_-$. The direction of $i_-$ also determines which switch operates with PWM: $S_{-H}$ when $i_- > 0$ and $S_{-L}$ when $i_- > 0$. Its duty ratios $D_{-H}$ and $D_{-L}$ are also calculated by

$$D_{-H} = \sqrt{\frac{2f_{sw}L_-I_{L-}(v_{o-} - v_{c-})}{v_{o}-v_{c-}}}, \tag{3}$$

$$D_{-L} = \sqrt{\frac{2f_{sw}L_-I_{L-}v_{o-}}{v_{c-}(v_{c-} - v_{o-})}} \tag{4}$$

where $v_{o-}$ is the output voltage of the chopper and $v_{c-}$ is the voltage of the lower-side capacitor $C_-$. Here, the phase difference between the output current and ac line voltage is defined as $\phi$. When $|\phi| \geq \pi/6$, the choppers switch the operating mode with PWM according to the phase angle.

PCIM Europe 2024, 11– 13 June 2024, Nuremberg          DOI: 10.30420/566262101

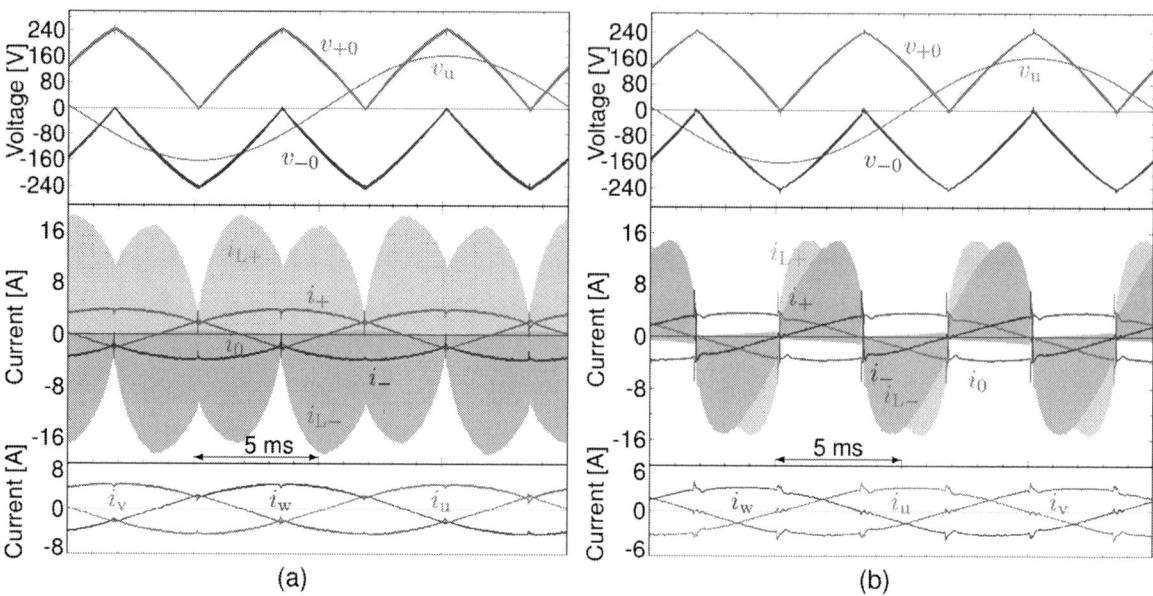

**Fig. 5:** Experimental waveforms of the proposed current unfolding converter operating under (a) unity power factor $\cos\phi = 1$ and (b) ultra-low power factor $\cos\phi = 0$.

**Tab. 1:** Circuit parameters used for experiments

| DC source voltage | $V_{dc}$ | 550 V |
|---|---|---|
| AC line-to-line voltage | $V_{ac}$ | 200 V |
| AC Line frequency | $f_{ac}$ | 50 Hz |
| Switching frequency | $f_{sw}$ | 50 kHz |
| DCM inductor | $L_+, L_-$ | 30 $\mu$H |
| Filter capacitor | $C_f$ | 4.7 $\mu$F |
| AC Filter inductor | $L_{ac}$ | 150 $\mu$H |

## 4 Experimental Results

Fig. 5 shows experimental waveforms of the proposed current unfolding converter operating with DCM. A fabricated prototype of the proposed converter employs SiC-MOSFETs for both the current regulation unit and the current unfolding unit. Since the switching frequency of the current regulation unit was fixed at $f_{sw} = 50$ kHz, the switching inductors and the filter capacitors for DCM were designed as $L_+ = L_- = 30\,\mu$H and $C_f = 4.7\,\mu$F, respectively. The prototype was connected with a dc voltage source, whose set voltage was $V_{dc} = 550$ V, on its high-voltage side of the current regulation unit though series-connected dc capacitors. The ac-side of the prototype was also connected to a three-phase ac voltage source through an additional three-phase filter inductor. The set voltage

of the ac source was $V_{ac} = 200$ V and its frequency was $f_{ac} = 50$ Hz. In the following experiments, the ac current reference was set to $I_{ac}^* = 2.8$ A, resulting in an output of 1 kVA.

Fig. 5a shows measured waveforms operating under a unity power factor $\cos\phi = 1$. The intermediate voltages $v_{o+}$ and $v_{o-}$ followed the difference between the highest and middle ac voltages $v_+ - v_0$ and the lowest and middle ac voltages $v_- - v_0$, respectively. The inductor currents $i_{L+}$ and $i_{L-}$ included large current ripples due to DCM operation; however, the intermediate currents $i_+$, $i_0$, and $i_-$ were well regulated as rearranged three-phase sinusoidal currents. As a result, the ac currents $i_u$, $i_v$, and $i_w$ had a sinusoidal current shape. In unity power factor operation, the peak of inductor current was 18 A which is 4.5 times higher than the ac current.

Fig. 5b is the waveforms when the converter operates under a power factor $\cos\phi = 0$. Its voltage waveforms were the same shape as those in Fig. 5a. On the other hand, both inductor currents $i_{L+}$ and $i_{L-}$ exhibited bidirectional output due to ultra-low power factor operation. Even though the intermediate currents $i_+$, $i_0$, and $i_-$ included step changes when the intermediate voltage $v_{o+}$ or $v_{o-}$ became to zero, the currents followed the step changes well. The ac output currents were sometimes distorted around $v_{o+} = 0$ or $v_{o-} = 0$, but the

834

PCIM Europe 2024, 11– 13 June 2024, Nuremberg          DOI: 10.30420/566262101

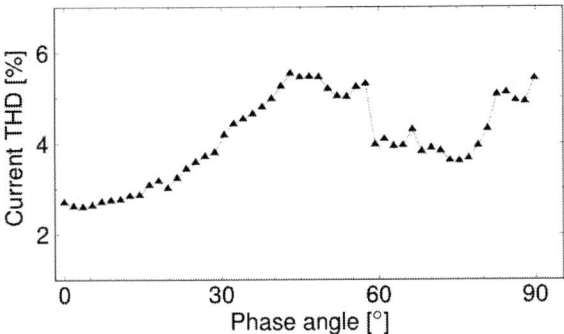

**Fig. 6:** Measured current THDs of the proposed current unfolding converter in various operating power factor.

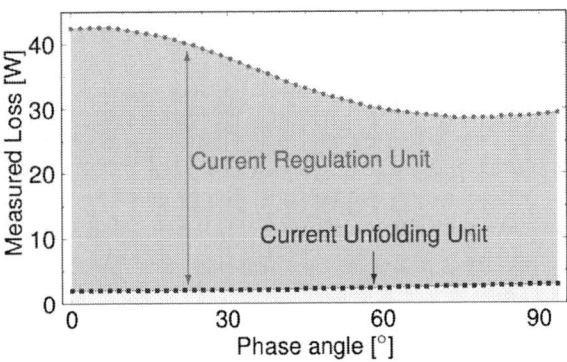

**Fig. 7:** Measured loss of the proposed current unfolding converter in various operating power factor.

current waveforms maintained a sinusoidal shape. Therefore, the DCM enables the proposed converter possible to operate under ultra-low power factor conditions, even though the current reference includes step changes.

Fig. 6 shows measured total harmonic distortion (THD) of the ac current in the proposed converter under various operating power factors. The current phase angle $\phi$ was shifted from 0 to $\pi/2$. When the operating point was close to the unity power factor, the THD reached a minimum value of 2.6%. On the other hand, the maximum THD was 5.4% around $\phi = 45°$ and $90°$ due to large steps in the intermediate currents.

Fig. 7 shows measured conversion losses of the proposed converter across various operating power factors. The loss in the current unfolding unit slightly increased with the increase of phase angle $\phi$, however, the loss was less than 10% of the total loss. Since the current unfolding unit operates at the ac-line frequency or double line frequency, there is almost no switching loss and the dominant loss is due to conduction. Conversely, since the current regulation unit operates with high-frequency PWM at 50 kHz, the predominant losses of the proposed converter occur in this unit. As the peak of the inductor current decreased with the increase of phase angle $\phi$, the loss in the current regulation unit also decreased.

## 5   Conclusion

This paper proposed a control method utilizing DCM for the buck-type current unfolding converter. The DCM operation enables the current regulator of the unfolding converter to follow step change references required by non-unity power factor op-

erations. In ultra-low power factor operations, where $|\phi| \geq \pi/6$, the proposed control method selects two of the four switches operating at the high-frequency PWM in the current regulation unit based on the current direction. Since the proposed control method is based on feedforward control, it allows the current unfolding converter to eliminate the need for current sensors for its feedback control, thus freeing it from instability problems associated with higher switching frequencies and/or faster current regulation.

This paper verified the validity of the proposed method through experiments using a prototype of the current unfolding converter. The experimental results demonstrated that the DCM operation successfully followed step changes in the intermediate currents. As a result, the ac current THD of the proposed method achieved 2.6% at the unity power factor and 5.4% at the low power factor, even though the converter operated solely with feedforward control and without current sensors.

The proposed method is useful for increasing the switching frequency and reducing passive components.

## References

[1] J.W. Kolar, D. Neumayr, D. Bortis, "Google Little Box Reloaded: How to Achieve 200W/in$^3$ & Beyond? Concepts - Evaluation - Barriers - Future," *IEEE APEC 2017*, 2017.

[2] A. Hariya, T. Koga, K. Matsuura, H. Yanagi, S. Tomioka, Y. Ishizuka, T. Ninomiya, "Circuit Design Techniques for Reducing the Effects of Magnetic Flux on GaN-HEMTs in 5-MHz 100-W High Power-Density LLC Resonant DC–DC Converters," *IEEE Transactions*

on *Power Electronics*, vol. 32, no. 8, pp. 5953-5963, 2017.

[3] D. M. Baker, V. G. Agelidis and J. Y. Chen, "A five-level zero average current error controlled single-phase grid-interactive inverter," *International Conference on Power Electronic Drives and Energy Systems for Industrial Growth*, vol. 1, no. 6, pp. 50–55, 1998.

[4] L. B. G. Campanhol, S. A. O. da Silva, A. A. de Oliveira and V. D. Bacon, "Dynamic Performance Improvement of a Grid-Tied PV System Using a Feed-Forward Control Loop Acting on the NPC Inverter Currents," *IEEE Transactions on Industrial Electronics*, vol. 64, no. 3, pp. 2092–2101, 2017.

[5] A. Anthon, Z. Zhang, M. A. E. Andersen, D. G. Holmes, B. McGrath and C. A. Teixeira, "The Benefits of SiC mosfets in a T-Type Inverter for Grid-Tie Applications," *IEEE Transactions on Power Electronics*, vol. 32, no. 4, pp. 2808–2821, 2017.

[6] E. Serban, F. Paz and M. Ordonez, "Improved PV Inverter Operating Range Using a Mini-boost," *IEEE Transactions on Power Electronics*, vol. 32, no. 11, pp. 8470–8485, 2017.

[7] T. B. Soeiro, T. Friedli and J. W. Kolar, "Swiss rectifier – A novel three-phase buck-type PFC topology for Electric Vehicle battery charging," 2012 Twenty-Seventh Annual IEEE Applied Power Electronics Conference and Exposition (APEC), pp. 2617–2624, 2012.

[8] B. Seo, T. Mannen and T. Isobe, "Suppression Method of DC Capacitor Currents in a Three-Phase Current Unfolding Inverter Equipped With Ultra-Small DC Capacitors," *2022 IEEE 31st International Symposium on Industrial Electronics (ISIE).*, pp. 939–942, 2022.

[9] T. Mannen, P. N. Ha and K. Wada, "Performance Evaluation of a Boost Integrated Three-Phase PV Inverter Operating With Current Unfolding Principle," *2019 21st European Conference on Power Electronics and Applications (EPE'19 ECCE Europe).*, pp. 1–8, 2019.

[10] D. Murillo-Yarce, C. Restrepo, D. G. Lamar and J. Sebasti´an, "A General Method to Study Multiple Discontinuous Conduction Modes in DC-DC Converters With One Transistor and Its Application to the Versatile Buck-Boost Converter," *IEEE Transactions on Power Electronics*, vol. 37, no. 11, pp. 13030–13046, 2022.

[11] J. Roy, A. Gupta and R. Ayyanar, "Discontinuous Conduction Mode Analysis of High Gain Extended-Duty-Ratio Boost Converter," *IEEE Open Journal of the Industrial Electronics Society*, vol. 2, pp. 372–387, 2021.

[12] V. Leonavicius, M. Duffy, U. Boeke and S. C. O. Mathuna, "Comparison of realization techniques for PFC inductor operating in discontinuous conduction mode," *IEEE Transactions on Power Electronics*, vol. 19, no. 2, pp. 531–541, 2004.

PCIM Europe 2024, 11– 13 June 2024, Nuremberg          DOI: 10.30420/566262102

# GaN Based Bi-Directional 6.6kW Interleaved Totem-Pole PFC with 13kW/L Power Density and High Efficiency

Esmaeil Jalalabadi[1,2], Yang Jiao[1], Juncheng (Lucas) Lu[1], Xiaoyu Wang[2]

[1] Infineon Technologies AG., Ottawa, Canada.
[2] Carleton University, Department of Electronics, Ottawa, Canada

Corresponding author:     *Esmaeil Jalalabadi, Esmaeil.Jalalabadi@infineon.com*
Speaker:                  *Juncheng (Lucas) Lu, Lucas.Lu@infineon.com*

## Abstract

This paper presents a high-power density bidirectional interleaved 6.6kW Power Factor Correction (PFC) to be used for AC/DC stage of a two-stage isolated bidirectional on-board charger (OBC). Gallium Nitride (GaN) technology is employed to achieve higher switching frequency leading to more compact passive components solutions. The detailed design process for two-phase interleaved inductors value, EMI filter, controller design, and the system level loss breakdown are provided. The 6.6kW PFC prototype has been implemented and tested in both inverter mode and rectifier mode. Implemented PFC achieves 13 kW/L power density and 98.5% measured peak efficiency using phase shedding at 2.2kW rectifier mode.

## 1. Introduction

It is commonly acknowledged that battery chargers play a crucial role in Electrical Vehicles (EV), and the charging time and battery lifetime are directly tied to charger parameters. A battery charger must be efficient and dependable, with low cost, low volume, and light weight. OBCs designed with GaN device can have smaller volume, higher efficiency and higher power density, [1-3] due to the superior device performance. It also enables bidirectional topology to be used in AC/DC stage, so that the OBC can also be used to supply AC load or to be connected to utility grid to support grid demand, expanding their application to reach much beyond simply charging the batteries. Lightweight charging solutions are especially significant in the automobile sector since OBC must fulfil tight dimensions constraints to fit inside an EV [4, 5].
To stay ahead of the game, designers are now pursuing ambitious targets, such as increasing the power–density of OBCs. If the state-of-the-art density was less than 2 kW/L yesterday, current designs are going toward 4 kW/L and suppose to increase to more than 6 kW/L by the end of the decade. Charting a course toward achieving this figure over the longer term will be multi-faceted, requiring wide-bandgap(WBG) semiconductors in novel circuit topologies and innovations in packaging on-board charge assemblies [6, 7]. Five main challenges to design an OBC with their suggested solutions are known as power rating, power density, efficiency, bidirectionality and voltage class (400V/800V). WBG devices have a

critical role to address these challenges to reach the best design solutions [8, 9].
This paper introduces a high-power density interleaved PFC design for 400V EVs with GaN device that is part of a two-stage bidirectional OBC. The proposed design can support 7.2kW in both inverter and rectifier modes based on thermal tests, achieving 14 kW/L power density. The design also includes autonomous output capacitors discharge circuit, latched analogue voltage and current protections, and employs high resolution PWMs and zero crossing detection.

## 2. PFC Design

The OBC structure is shown in figure 1, including interleaved totem-pole PFC stage and isolated DC/DC converter. Fig 2 illustrates the sensing and control circuitry of the interleaved PFC. Each phase current is measured through a fast Hall-effect current sensor with 1MHz bandwidth. Input AC voltage and output DC voltage are measured for control and protection purposes as well.

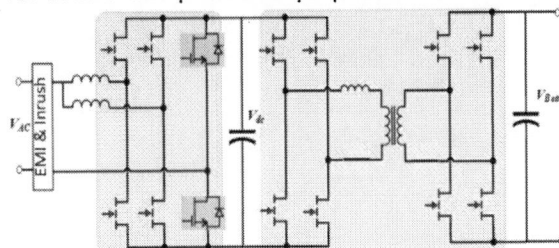

**Fig 1.** simplified structure of proposed single-phase bidirectional onboard charger (OBC)

MCU board is processing the inputs and send the control signals to a driver card that contains three phase-legs with corresponding IMS boards. The PFC benefits from inrush current circuit to provide smooth and safe pre-charging of output caps at the start-up. EMI filter is considered in the AC side to ensure EMI/EMC compliance.

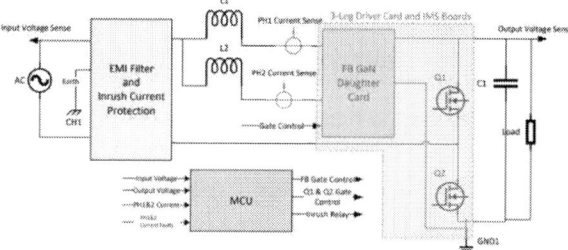

**Fig 2** Control and Sensing Structure of Interleaved PFC

The current waveform of interleaved PFC is shown in Fig 3. For a two-phase interleaved PFC, the switching of phase A is 180 degrees shifted with respect to phase B. This shifted switching will reduce the ripple current significantly, leading to a much smaller EMI filter and reduced inductor sizes that are investigated in the next sections.

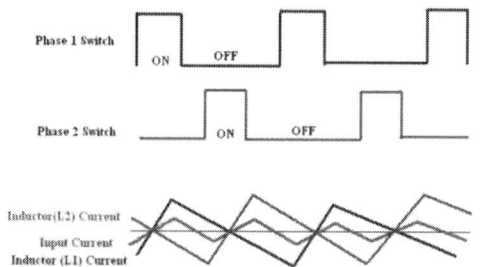

**Fig 3** Ripple reduction through interleaved approach.

## 2.1 Interleaved Inductor Design

The first step to design inductors for a boost PFC is to limit the maximum current ripple. The lower the ripple current the less current harmonics will be injected from the grid. However, there is a trade-off among ripple current, inductance value, and switching frequency.

For a two-phase interleaved PFC, the derivation of inductance value is based on inductor ripple current as follows:

$$Rpl_{int}(t) = \frac{\left(1 - \frac{V_{in}(t)}{V_{out}}\right) V_{in}(t)}{L_{pfc_{int}} F_{sw} \times \frac{I_{inp}}{2}} \times \rho_{int}(t) \quad (1)$$

Here, the peak current is half of the input current as well as having an interleaved coefficient

$0 < \rho_{int}(t) < 1$ that is obtained based on 180-degree shift of switching as follows:

$$\rho_{int}(t) = \begin{cases} \frac{1-2D(t)}{1-D(t)} & 0 < D(t) < 0.5 \\ \frac{2D(t)-1}{D(t)} & 0.5 < D(t) < 1 \end{cases} \quad (2)$$

The maximum ripple occurs at two different values in the AC input voltage as follows:

$$\frac{\partial Rpl_{int}(t)}{\partial V_{in}(t)} = 0 \rightarrow V_{in}(t) = \begin{cases} \frac{V_{out}}{4} \\ \frac{3V_{out}}{4} \end{cases} \rightarrow$$

$$Rpl_{int_{Max}} = \frac{V_{out}}{4 \times L_{pfc_{int}} F_{sw} I_{inp}} \quad (3)$$

Therefore, the inductance value is obtained as follows:

$$L_{pfc_{int}} = \frac{V_{out}}{4 \times Rpl_{int_{Max}} \times F_{sw} \times I_{inp}} \quad (4)$$

The 6.6kW PFC is designed to work at 120kHz Continous Conduction Mode (CCM) with maximum total ripple current of 20%, output voltage of 385v and input RMS AC voltage range of 85v to 265v. The inductor designed values are 100 μH and they are designed based on EELP ferrite cores to meet the required high power density. Figure 4 shows the PFC inductance value for a range of switching frequency. GaN device with better switching performance is more suitable for high switching frequency applications with reduced PFC inductor size.

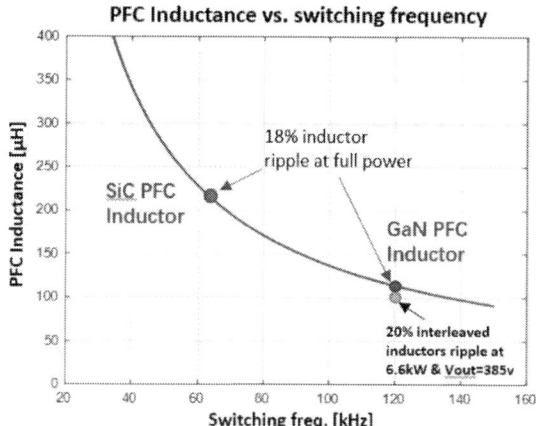

**Fig 4** PFC Inductance vs. Switching Frequency

Table 1 shows a comparison of various inductor designs and shapes with corresponding specifications for PFC applications including proposed design for 6.6kW interleaved PFC

PCIM Europe 2024, 11– 13 June 2024, Nuremberg    DOI: 10.30420/566262102

inductors. Proposed inductor design occupies 40% smaller volumetric space comparing to the state-of-the-art coupled inductor design.

**Table 1** PFC inductors of different projects/technologies

| Spec | Design 1 | Design 2 | Design 3 | Proposed design |
|---|---|---|---|---|
| PFC Inductor | 230 µH (220 µH Min) | 2*514 µH (interleaved) | 2*77 µH (coupled inductors) | 2*100 µH (Interleaved) |
| Freq. | 67 kHz | 70 kHz | 120 kHz | 120 kHz |
| Inductor ripple | --- | --- | --- | 20% |
| DCR | 15 mΩ Max | --- | 12 mΩ /winding. | 18 mΩ /winding. |
| Size | 60*60* 45 | 70*70* 56 | 50*47*47 (0.12 L) | 80*56*16 (0.07 L) |
| Picture | | | | |

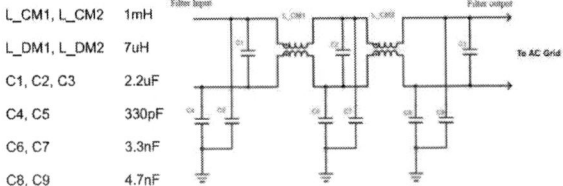

## 2.2 EMI Filter Design

Based on switching node voltages and inductor currents waveforms, the following EMI filter with corresponding common-mode and differential mode impedance has been designed to pass the EMI/EMC compliance.

| | |
|---|---|
| L_CM1, L_CM2 | 1mH |
| L_DM1, L_DM2 | 7uH |
| C1, C2, C3 | 2.2uF |
| C4, C5 | 330pF |
| C6, C7 | 3.3nF |
| C8, C9 | 4.7nF |

**Fig 5** EMI filter and corresponding values.

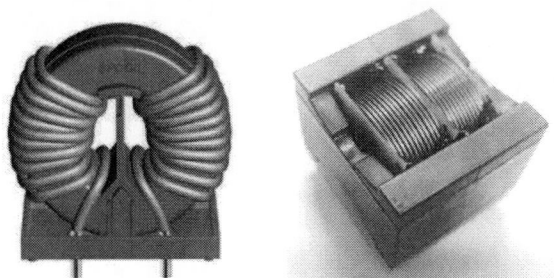

**Fig 6.** Conventional (left 0.1L) and Designed (right 0.025L) EMI Choke

the proposed design takes the advantage of GaN device performance and switches at high frequency to reduce the EMI filter size. Common mode EMI chocks with the required magnetizing inductance and leakage inductance have been designed to considerably improve the power density. Flat wire and PQ cores have been incorporated to achieve 0.025L size for each EMI choke that is about 4 times smaller than available Toroidal EMI choke with the same characteristics in the market (figure 6).

## 2.3 Driver Board and IMS Boards

The reference design uses automotive grade Enhanced mode GaN device GS-065-060-5-B-A and IPT60R028G7 Si CoolMOS from Infineon for the two interleaved high frequency legs and the grid frequency leg, respectively. Figure 7 shows the 3D view of designed IMS boards and Driver Board that includes two fast switching GaN legs and one low frequency Si MOSFET.

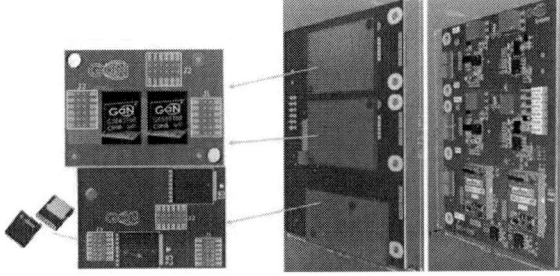

**Fig 7** Three-leg driver card and IMS boards.

The simulated system-level design loss breakdown and efficiency results are shown in figures 8 and 9. Moreover, phase shedding is considered to improve the light load efficiency.

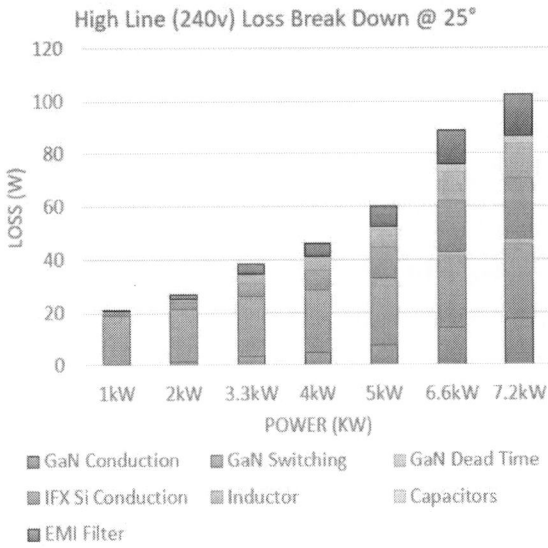

**Fig 8.** PFC loss break down at 25°

839

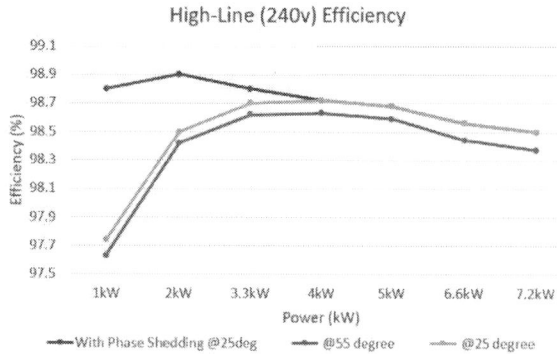

**Fig 9.** High line (240v) efficiency analysis

### 2.4 PFC Controller Design

The closed loop digital control design is based on 120 kHz current loop (fast loop) sampling and 12 kHz voltage loop (slow loop) sampling frequency. The controller is designed to achieve 8 Hz bandwidth for slow loop (voltage feedback loop) with 60-degree phase margin, and to achieve 8 kHz bandwidth for fast loop (each current feedback loop) with 50-degree phase margin (figure 10). Blue line represents the plant model, the red line represents the compensator model, and the green line represents the compensated model after applying the designed controller.

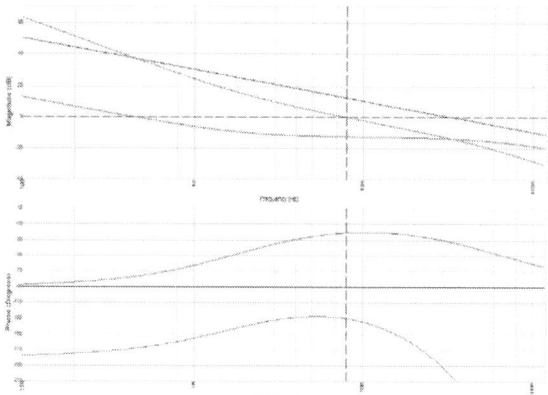

**Fig 10.** Current loop controller design

## 3. PFC Prototype Test Results

The final prototype of interleaved 6.6kW Totem Pole PFC is shown in figure 11. The measured dimensions are 135mm*88mm*43mm, resulting in a 13 kW/L power density. The achieved power density shows 100% improvement comparing to the State-of-the-art design with similar specifications.

Figure 12 shows the inverter mode test results of PFC at nominal input and output voltage condition for 4 kW AC resistive load. The presented waveforms include interleaved inductor current, AC load current, AC voltage, and high frequency switching point voltage.

Soft start implementation is critical that is being realized through gradually increasing the reference voltage to the desired voltage, and the soft-start result is as figure 13.

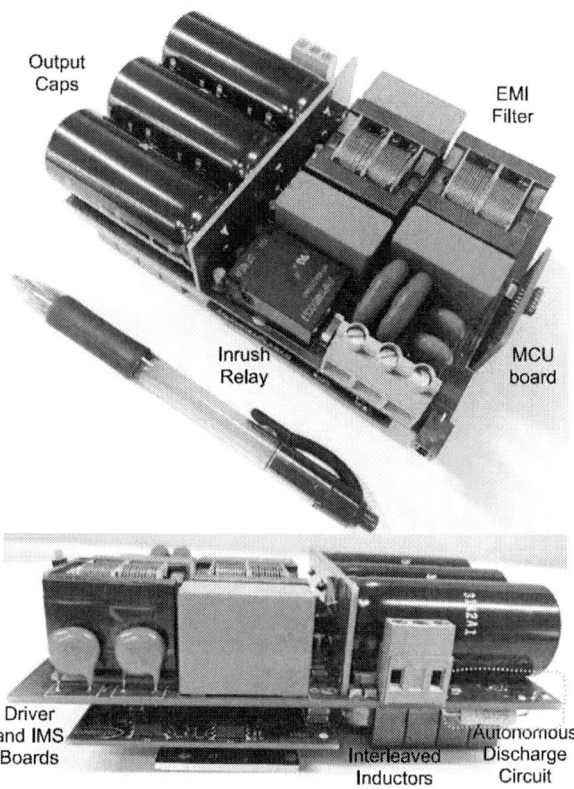

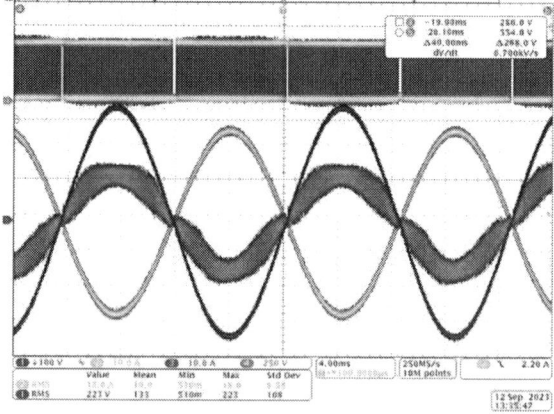

**Fig 11.** Top and side view of the designed PFC

**Fig 12.** Inverter mode (V2L) test waveforms.

Figure 14 shows the test results for rated power in rectifier mode under nominal input and output voltage conditions.

**Table 2.** Steady-state temperature of PFC hot components at 7.2kW rectifier mode

| | GaN IMS | Si MOS IMS | Interleaved Inductor | EMI Chokes |
|---|---|---|---|---|
| Temperature Rise | 69 °C | 65 °C | 45 °C | 58 °C |

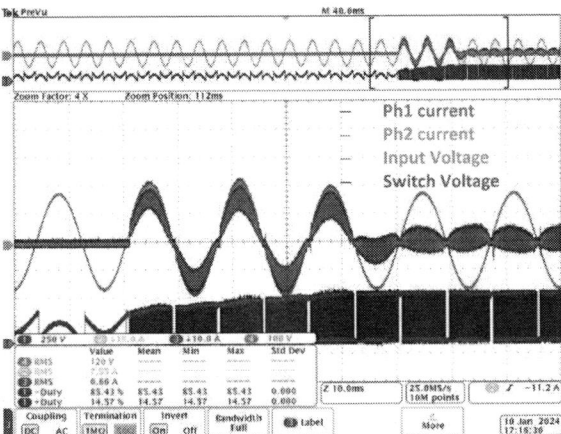

**Fig 13** Rectifier mode soft start performance

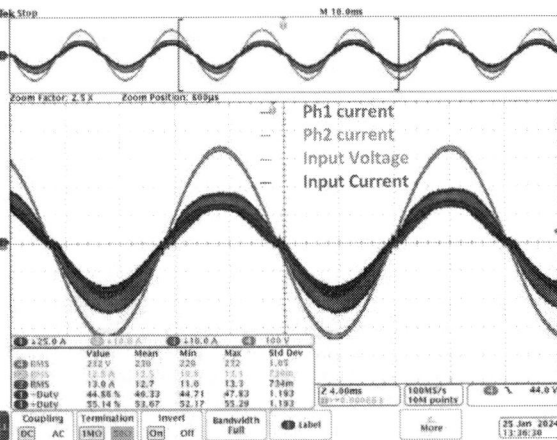

**Fig 14** Rated power rectifier mode test results.

30 minutes continuous running under full load has been conducted to verify thermal performance of designed PFC, and steady state thermal results are shown in table 2.

### 3.1 PFC Dynamic responses

The transient response of the designed PFC has been evaluated by applying negative and positive step changes to output power and input AC voltage. The results provided in figures 15 and 16 verified stable and smooth dynamic response under load step changes. The step changes in output load are regulated by designed voltage loop controller and the step changes in AC input voltage are regulated mainly by feedforward

control. The RMS value of input AC voltage is measured in real-time, and the changes are immediately applied to the regulated AC reference current to ensure a stable, robust, and smooth transient response.

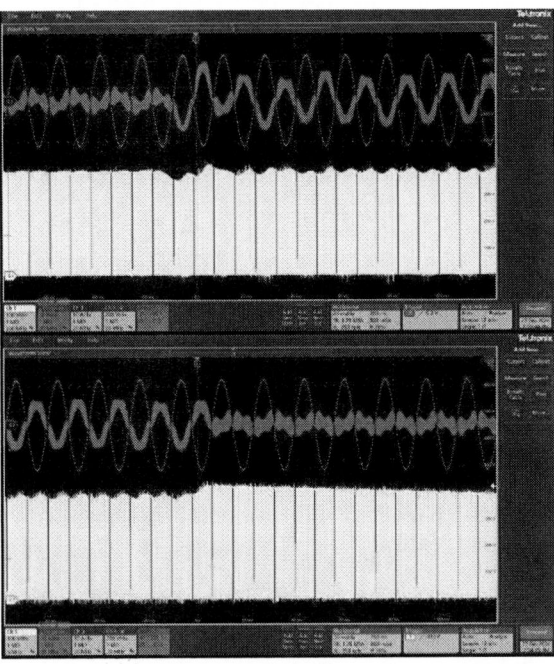

**Fig 15.** Rectifier dynamic response of PFC controller for load step changes (Top: increase, Bottom: decrease)

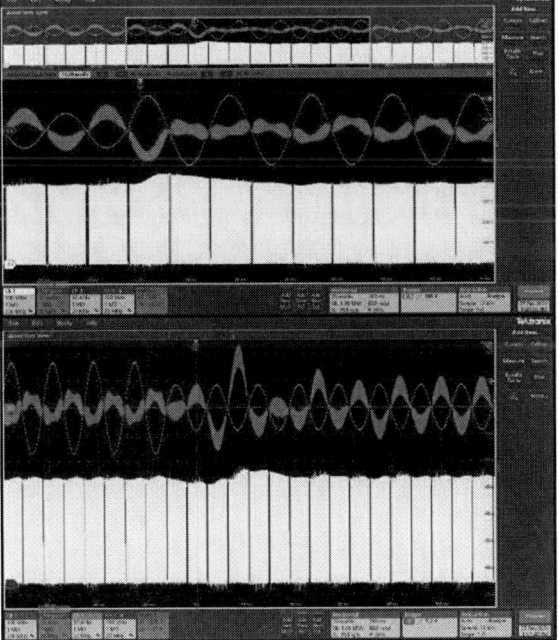

**Fig 16.** Feedforward controller dynamic response for input voltage step change (120v to 240v and vice versa)

The final PFC design has been compared to the existing big three reference designs for 6.6kW PFCs/OBCs in table 3 [1, 10]. The total power density improves by about 100% compared to the highest power density reference design. Moreover, the total peak efficiency is also proved to be higher than the other reference designs for the proposed design with GaN devices as shown in figure 17.

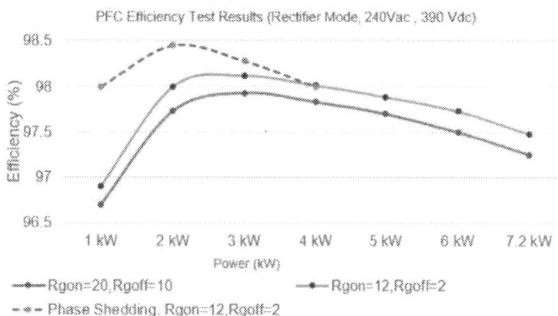

**Fig 17.** Measured efficiency results (rectifier mode)

**Table 3.** 6.6kW PFC reference design comparison

| Spec. | Design 1 | Design 2 | Design 3 | Proposed Design |
|---|---|---|---|---|
| Switching Frequency | AC/DC: 67kHz | AC/DC: 70kHz | AC/DC: 120kHz (CCM) | AC/DC: 120kHz (CCM) |
| Peak Efficiency | 98.1% @230V | >97% @240V | 98.5% @240V | 98.5% @240V |
| Power Density | 5.8 kW/L | 2.1 kW/L | 6.5 kW/L | **13 kW/L** 135*88*43mm |

## 4. Conclusion

Employing GaN devices enables more compact passive components solutions with higher switching frequency. This paper introduced a high power density interleaved PFC design for 400V OBCs while ensuring high efficiency using GaN devices from Infineon. Two loop feedback controller and feedforward controllers have been implemented to achieve high performance dynamic responses. The test results of the early prototype verified the proposed PFC design. 13 kW/L power density for 6.6kW power rating (14kW/L for 7.2kW rating) and 98.5% measured peak efficiency using phase shedding at light loads is achieved.

## 5. References

[1] J. Yuan, L. Dorn-Gomba, A. D. Callegaro, J. Reimers and A. Emadi, "A Review of Bidirectional On-Board Chargers for Electric Vehicles," in *IEEE Access*, vol. 9, pp. 51501-51518, 2021, doi: 10.1109/ACCESS.2021.3069448.

[2] U. R. Prasanna, A. K. Singh and K. Rajashekara, "Novel Bidirectional Single-phase Single-Stage Isolated AC–DC Converter with PFC for Charging of Electric Vehicles," in IEEE Transactions on Transportation Electrification, vol. 3, no. 3, pp. 536-544, Sept. 2017, doi: 10.1109/TTE.2017.2691327.

[3] Y. Hu, I. Liu, K. Xie, T. Bhatia and A. Narain, "High Efficiency and High Power Density Totem-Pole PFC with SiC MOSFETs," PCIM Europe 2023; International Exhibition and Conference for Power Electronics, Intelligent Motion, Renewable Energy and Energy Management, Nuremberg, Germany, 2023, pp. 1-6, doi: 10.30420/566091091.

[4] D. Nardo, G. Aiello and F. Gennaro, "Analysis of WBG Based Hybrid Semiconductors Approach for Bidirectional PFC in On-board Charger Applications," PCIM Europe 2023; International Exhibition and Conference for Power Electronics, Intelligent Motion, Renewable Energy and Energy Management, Nuremberg, Germany, 2023, pp. 1-7, doi: 10.30420/566091240.

[5] Z. Chen, B. Liu, Y. Yang, P. Davari and H. Wang, "Bridgeless pfc topology simplification and design for performance benchmarking", IEEE Trans. Power Electron, vol. 36, no. 5, pp. 5398-5414, May. 2021.

[6] D. Zhang, H. Lin, Q. Zhang, S. Kang and Z. Lv, "Analysis design and implementation of a single-stage multi-pulse flexible-topology thyristor rectifier for battery charging in electric vehicles", IEEE Trans. Energy Convers., vol. 34, no. 1, pp. 47-57, Mar. 2019.

[7] B. Su, and Z. Lu, " An interleaved totem-pole boost bridgeless rectifier with reduced reverse-recovery problems for power factor correction," IEEE Trans. on Power Electron., vol. 25, no. 6, pp. 1406-1415, Jun 2010.

[8] J. W.-T. Fan, R. S.-C. Yeung, and H. S.-H. Chung, " Optimized hybrid PWM scheme for mitigating zero-crossing distortion in totem-pole bridgeless pfc, " IEEE Trans. on Power Electron., vol. 34, no. 1, pp. 928-942, Jan 2019.

[9] Q. Huang, "Review of SiC yotem-pole bridgeless pfc, " CPSS Trans. on Power Electron. and App., vol. 2, no. 3, pp. 187-196, Sept 2017.

[10] C. Wei, D. Zhu, H. Xie and J. Shao, "A 6.6kW high power density bi-directional EV on-board charger based on SiC MOSFETs," PCIM Europe 2019; International Exhibition and Conference for Power Electronics, Intelligent Motion, Renewable Energy and Energy Management, Nuremberg, Germany, 2019, pp. 246-252.

PCIM Europe 2024, 11– 13 June 2024, Nuremberg          DOI: 10.30420/566262103

# The Design of a 2kV 1700A SiC MOSFET Dual Module

Jorge Mari [1], Tobias Schuetz [1], Xiaoting Dong[1], Yanfeng Shen [1], Michael Kirner [1], Luigi Findanno [1]
[1] Semikron Danfoss, Technology & Research, Ismaning, Germany

Corresponding author:    Jorge Mari, jorge.mari@semikron-danfoss.com
Speaker:                 Jorge Mari, jorge.mari@semikron-danfoss.com

## Abstract

In this paper we review some aspects of modern high power module design. We focus on the electrical design methodology applied to develop a multi-chip module built using 2kV silicon carbide (SiC) MOSFETs to reach an astounding 1700A rating. Switching waveforms of the module in its full safe operating area are shown validating the design. The challenges that had to be overcome included current balancing across 28 chips in parallel at each of the two topological switches in the module, and the elimination of high frequency internal ringing across different groups of chips. For this purpose, advanced full time and frequency domain simulation models were used.

## 1    Background

Power electronics is one of the key technology areas playing a decisive role in the creation of solutions to combat the climate change. Among the driving forces behind the flourishing of power electronics we mention a) the development of new electrification applications and b) the strive for less wasted power. These two forces converge and are met by the power semiconductor industry through the creation of modules which can operate with DC-link voltages of up to 1500V, and which employ materials like SiC. It has been proven many times now [1], [2], that in order to fully exploit the benefits of SiC in voltage source converters, a very low commutation inductance is required. Besides this requirement, the market favors standard package solutions which can be offered by multiple manufacturers [3]. One module that was conceived to enable both these requirements is the Semikron Danfoss ST20 which was initially deployed using Si IGBTs. Nevertheless, the use of these type of modules promptly revealed that the electrical design required to exhibit reasonable switching waveforms was far from trivial [4], [5], [6].

## 2    Module description

The module contains a half-bridge using 2kV MOSFET chips. In addition an NTC is used as a temperature sensor.

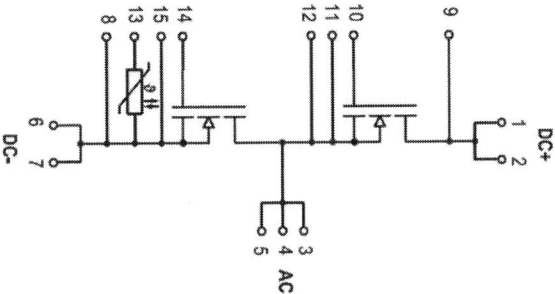

**Fig. 1** Circuit diagram of the half-bridge module

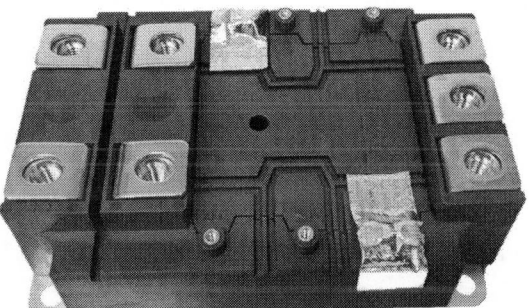

**Fig. 2** Picture of the ST20 module produced.

## 3    Design methodology

The design methodology for power modules is iterative.

843

PCIM Europe 2024, 11– 13 June 2024, Nuremberg          DOI: 10.30420/566262103

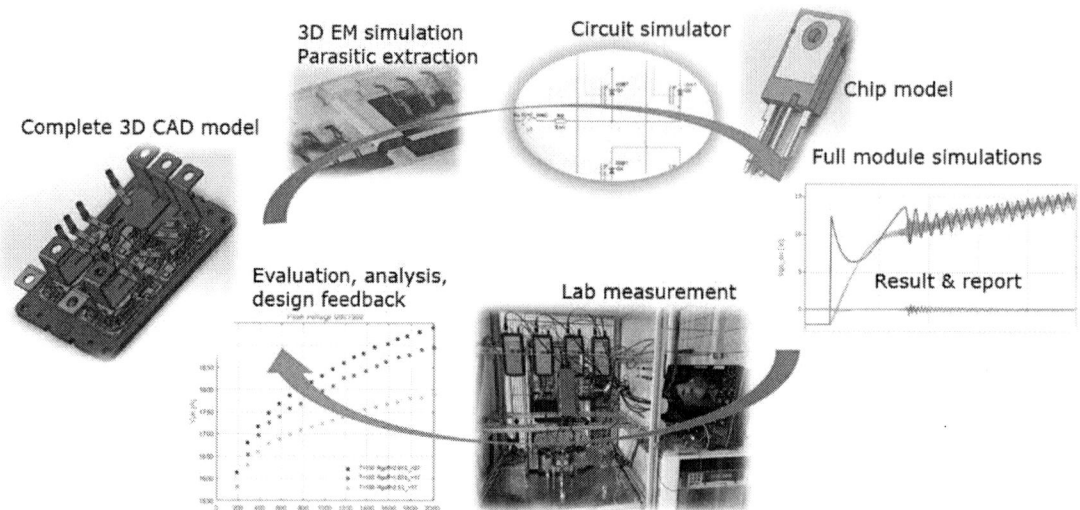

**Fig. 3** The picture illustrates the circular design iteration loop in Semikron Danfoss. An open ST20 module design can be seen on the left and a picture of the measurement testbench can be seen below.

The first step is the creation of a chip layout disposition to fit each of the substrates. The dual module has four substrates, two per side. The substrates are identical for each of the sides to reduce the number of single parts and minimize the costs through economy of scale. The position of the chips is constrained by several factors. There are keep out zones close to the places where ultrasonic welding of the power terminals is done. There are strict constraints on the distance to the edges which include etching tolerances of the copper plated AMB substrates. The size of the gaps that must be respected is derived from company proprietary design rules. Thermo-mechanical simulations of the stacked substrates are performed to result in acceptable pre- and post-sintering deflections, before the thickness of each layer and the final chip position is selected. Thermal simulations with the fully populated baseplates on top of different heat sinks are done to understand the thermal resistance. This allows controlling the heat spreading from chips to baseplate as well as the thermal cross talk. This is only one big set of constraints.

Next the current sharing among the large number of SiC chips in parallel must be ensured. After experimenting with several parallel array dispositions which did not meet our expectations, we opted for a novel semi-circular design. An example of the unsatisfactory performance during the design process is shown in Fig. 4.

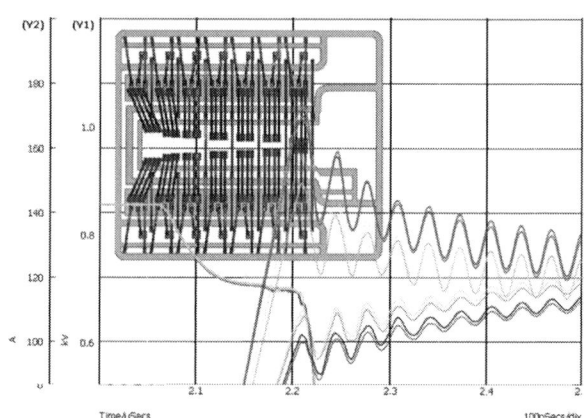

**Fig. 4** Early design with a two-row chip disposition.

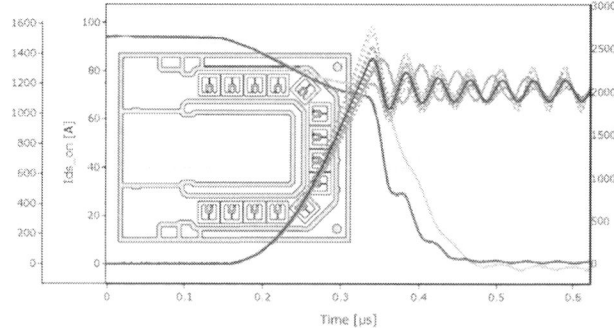

**Fig. 5** Final layout together with a turn-on simulation (blue, violet) at 1.5kV link and 2kA commutation with superposed measurement results (orange, green).

844

In the initial layout a large unbalance across the chips' currents is clearly seen (100A to 175A initial peaks, +/- 30%). The final chip disposition is shown in Fig. 5. The right axis show module current, and the inner left axis is current for chips drawn with dotted lines. Dispersion is now 70A to 100A. Including the reverse recovery peaks IRM of ~400A, only a +/- 16% unbalance stays. This is a very good result for a module with so many chips in parallel using classical wire-bonding technology.

In general, once the complete CAD drawings are available, including the bonding structure, the first step is an electromagnetic simulation to obtain the whole system inductance matrix. See Fig. 6 and Fig. 7 for an example of some readily understandable output values.

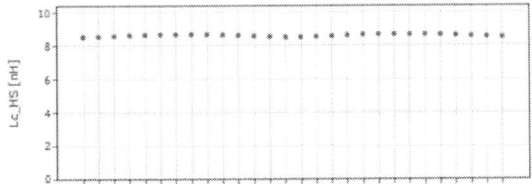

**Fig. 6** Chip individual power loop inductances of the high side of the module.

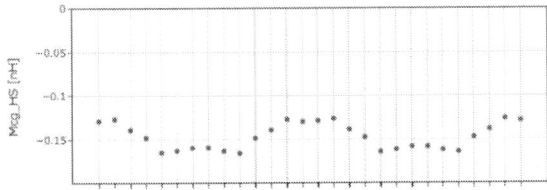

**Fig. 7** Chip individual coupling coefficients from the power loop current paths to each gate-source loop.

Note that there is no use of a Kelvin source (as customary with IGBTs) and that there is more negative coupling in short circuits than in the normal commutation (-0.8 nH/chip, not shown in Fig. 7, but used to decrease the level of short circuit current).

This inductance matrix is combined with single chip Spice models into a big simulation model. When we source the chips from external suppliers, we ask the suppliers to provide chip models. These are often done based on characterized TO247 packages. Many times, the models provided are not accurate enough. When that is the case, we typically construct other simpler modules on previously studied platforms and proceed to ex-

tract the model parameters which were judged unsatisfactory. This situation is seen in Fig. 8 and Fig. 9.

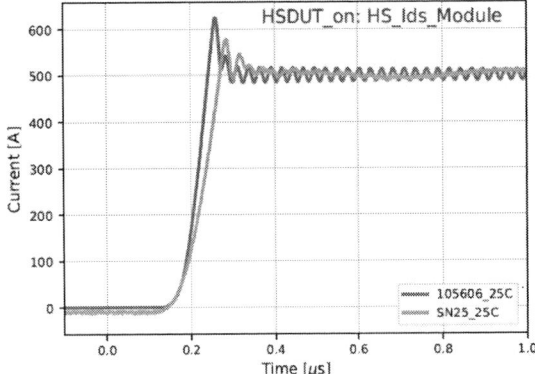

**Fig. 8** The graphs show the mismatch in current slope at turn-on of an auxiliary module using the supplier chip model compared to an actual measurement.

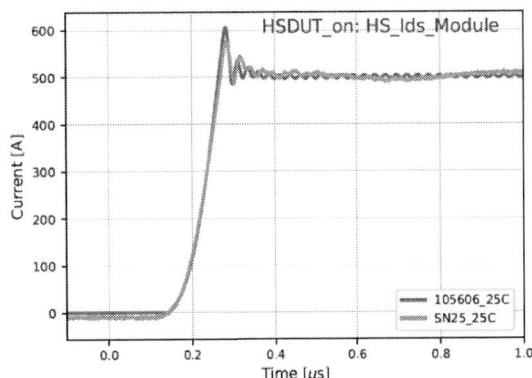

**Fig. 9** These graphs show the level of agreement between measurement and the own developed model.

Afterwards different driver settings are selected, and multiple operating points are simulated. These are transient time domain simulations which are used to analyze the switching waveforms and extract the main characteristics. At this stage we try to answer the following questions:

a) are all gate-source voltages of both DUT and AUX chips tightly bundled together?
b) is the maximum spreading between each of the chip currents at each switch event small enough?
c) are there any high frequency oscillations appearing in each of the gate-source or power loop signals?
d) are the drain-source voltage overshoots below the blocking voltage?

e) are the gate-source undershoots acceptable with respect to the chip supplier specifications?

f) is the module immune to parasitic turn-on?

Key extracted features include current gradients, voltage gradients, current and voltage spikes and switching losses. The first objective is to create a module where all these magnitudes are approximately the same across the whole set of chips.

Later, sensitivity analyses are done, where we vary the values of some key chip parameters as temperature, threshold, transconductances, and derive bounds for the acceptable tolerance required from the suppliers, while keeping the performance specifications within range. As an example, consider the threshold voltage variation.

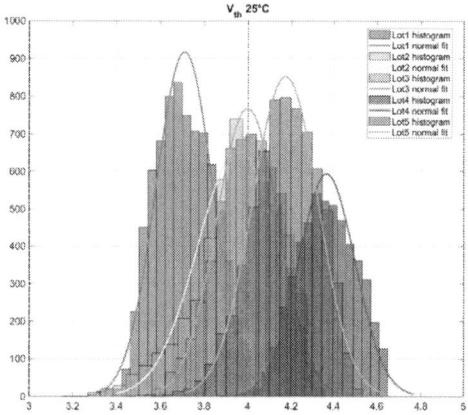

**Fig. 10** Histograms of threshold voltage for 5 different lots and their gaussian fittings.

Ultimately, modules must be constructed with chips coming from any of those lots.

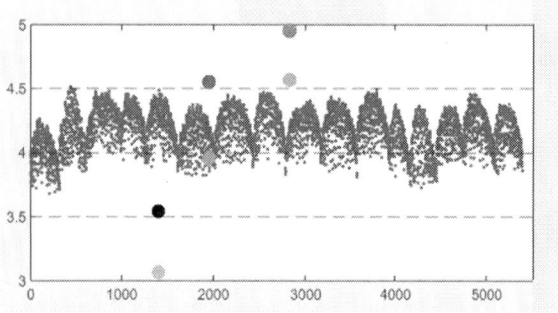

**Fig. 11** Example of $V_{th}$ in consecutive chips arranged per wafer of one lot and choice for simulations.

Although the spreading intra-lot is moderate the spreading across lots is somewhat large and this compromises the switching characteristics of the modules, especially at high temperatures. To counteract this effect, it was decided to classify the chips into 4 intervals and construct modules coming only from those intervals. To assess if this is satisfactory, simulations with chips from the extremes of the binning classes located at each of the four substrates were performed.

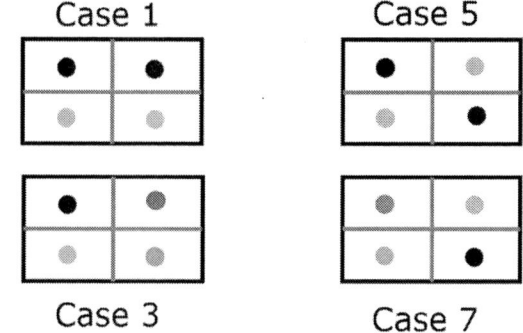

**Fig. 12** Modules built out of substrates with chips in several extremes of the binning classes.

Simulations are done both at 25°C and 150°C considering also the temperature dependence of several other chip parameters. The results were judged satisfactory.

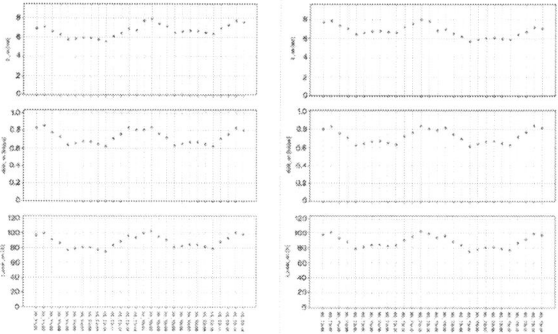

**Fig. 13** From top down, $E_{on}$, di/dt on and $I_{pk}$ for each of the 28 chips at the high side. Left is Case 1 at 25°C and right is Case 7 at 25°C. Switching condition is turn on 1.5kV, 2kA commuting from body diode conduction.

## 4 Joining technology

For the module it was decided to use a Si3N4 ceramic which has a far better thermal conductivity than classical Al2O3 or HPS. The AMB stack has the ceramic sandwiched between asymmetric layers of copper chosen for thermomechanical endurance for sintering as well as ampacity. To get robust solder joints we chose SnSb5 to solder to a thick Cu baseplate. Temperature shock tests -

PCIM Europe 2024, 11– 13 June 2024, Nuremberg          DOI: 10.30420/566262103

40°C to +125°C up to 500 times were successful. AlSiC baseplates are also possible and used in traction modules albeit at a higher cost. The chosen ceramic has a better robustness in friction welding and is demonstrably better than Al2O3. It is also best to reduce the mechanical stress between chip and substrate at the sinter layer, due to the very close coefficients of thermal expansion. This is of high importance since SiC can reach very high temperatures under certain transients. The vertical stack layout of the substrates is shown in Fig. 14.

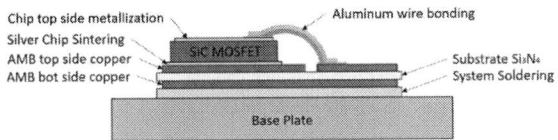

**Fig. 14** Vertical stack layout of the substrates

The chips were interconnected and connected to the copper traces with thick Aluminum wire bonds. The length of the bonds was kept short, and the diameter was chosen wide enough to ensure that the maximum bond wire temperature stays below the maximum tolerable temperature of the silicone mold during the targeted 25 years of lifetime. To simulate the maximum bond wire temperature, the maximum chip junction temperature of 150°C and a copper layer temperature of 125°C at the second foot of the bond wire were considered.

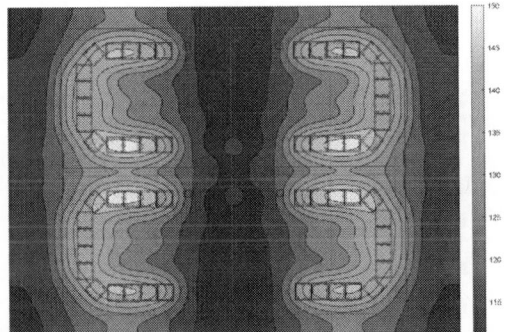

**Fig. 15** Temperature field at the top Cu layer

Fig. 15 shows the temperature field of the module mounted with thermal grease on top of an air-cooled heat sink with air at 60°C with uniform heat transfer coefficient at the bottom side. The total module power dissipation simulated was 2kW.

# 5    Module performance

The most difficult challenge to meet with multichip SiC MOSFET based modules is to reach a clean switching performance devoid as much as possible of internal oscillations and still fast enough to reach very low switching losses. This, combined

with the low forward conduction losses which are natural to SiC, results in a module very attractive for multiple applications. The switching performance was measured in extensive DOEs covering all temperature ranges from -20°C to +150°C, all voltage levels from 500V to 1500V and all current levels from 20A up to 2600A. Apart from favorable low commutation inductance $L_{com}$ environments (needed to exploit to its maximum the module entitlement), less propitious conditions with higher $L_{com}$ (we tested up to 80nH) and higher turn-on and turn-off driver resistances were exercised. This allows prospective users in IGBT retrofit scenarios or with less degrees of freedom in the design of their converters to profit from some of the benefits of SiC modules, for example the low forward voltage drop. Here we only illustrate a couple of switch events.

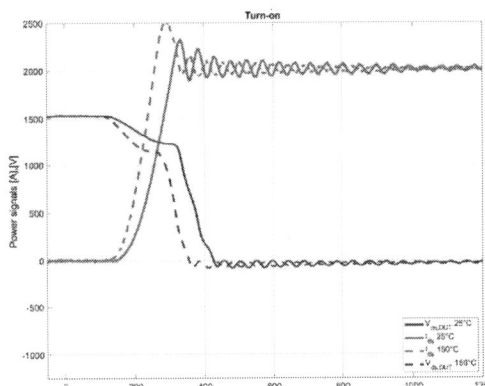

**Fig. 16** Turn-on waveform of the module as measured in a double pulse test. Solid lines are 25°C and dashed lines are 150°C. Switched current from body diode to MOSFET DUT was 2000A at 1500V link voltage. Commutation inductance 17nH

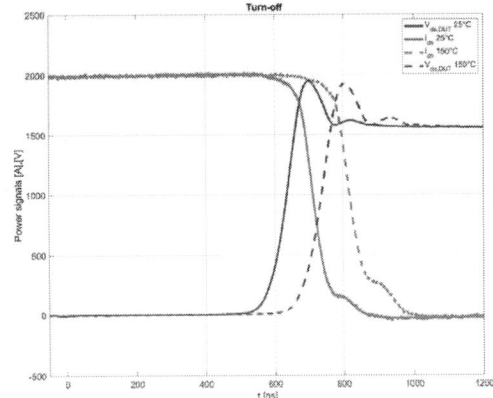

**Fig. 17** Turn-off under same conditions as in Fig. 16

The interplay between chip and layout plays a significant role in the overall module performance. Semikron Danfoss practices a multi-source approach to chips. The results shown here were created with only one brand of chips, but the approach used is equally applicable to other chip types and equivalent results have been obtained. Particular attention was paid to the avoidance of parasitic turn-on (also known as crosstalk). Without a clear access to individual chips for measurement reasons, the methodology explained in [7] was applied. The key aspect of this method is to compute the net bipolar charge (at all temperatures, voltages, currents and AUX gate biases) which is the charge associated to the AUX device reduced by the charge associated to the output capacitance. The resulting curves can be seen in Fig. 18.

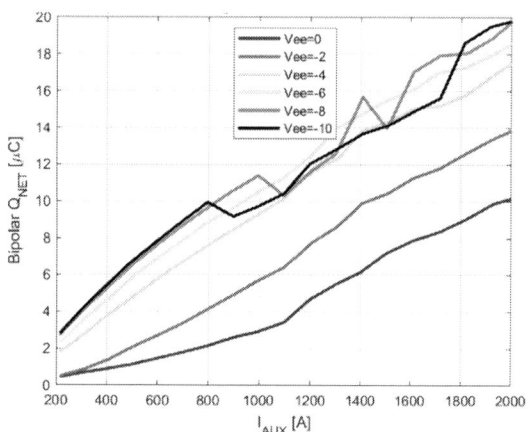

**Fig. 18** Net bipolar and potential PTO charge estimate at turn-on, 150°C, 1500V.

For all current levels tested, for any given current as we move up in Vee (the bias of the AUX device), the net charge decreases. The reason for the decrease is the progressively less forward current in the body diode prior to turning it off. No PTO happens, because if it did at any vertical line the $Q_{NET}$ would start to increase as Vee increases towards 0.

## 6 Conclusions

This paper presents the design and performance evaluation of a 2-kV 1700-A SiC MOSFET dual module product. The challenges associated with paralleling and packaging 28 SiC MOSFET chips have been meticulously addressed. We have developed comprehensive time- and frequency-domain simulation models to facilitate design iteration and optimization. Additionally, our analysis of chip parameter sensitivity demonstrates the power module's robustness in the face of variations. Leveraging advanced joining technologies, we can achieve higher thermomechanical reliability. Notably, the developed module exhibits outstanding electrothermal performance across a broad range of operating conditions: from -20°C to +150°C, voltages spanning 0.5 kV to 1.5 kV, and currents ranging from 20 A to 2.6 kA.

## References

[1] J. Lutz, R. Baburske "Some aspects on ruggedness of SiC power devices", Microelectronics Reliability, 54 (2014) 49-56

[2] D. Kawase, et al, "High voltage module with low internal inductance for next chip generation - next High-Power Density Dual (nHPD2)," Proceedings of PCIM Europe 2015; pp. 1-7.

[3] "New generation power semiconductor". October 2016. http://www.roll2rail.eu/home.aspx

[4] H. Li, et al, "Circuit mismatch and current coupling effect influence on paralleling SiC MOSFETs in multichip power modules," Proceedings of PCIM Europe 2015; pp. 1-8.

[5] M. Spang, et al, "Differential-mode oscillations between parallel IGBTs in power modules," EPE'15 ECCE-Europe, Geneva, Switzerland, 2015, pp. 1-10, 10.1109/EPE.2015.7309089.

[6] U. Drofenik et al "Modelling the Thermal Coupling between Internal Power Semiconductor Dies of a Water-Cooled 3300V/1200A HiPak IGBT Module", PCIM Europe 2007

[7] J. Mari, F. Carastro and M. -J. Kell, "Assessing the Presence of Parasitic Turn On in SiC Mosfet Power Modules," 2021 23rd European Conference on Power Electronics and Applications (EPE'21 ECCE Europe), Ghent, Belgium, 2021

# Technological Approaches to High-Power Density SiC Power Module for Automotive

Takeshi Tokorozuki[1], Hideo Komo[1], Kazuhiro Nishimura[2], Rei Yoneyama[1], Gourab Majumdar[1]

[1] MITSUBISHI ELECTRIC Corp, POWER DEVICE WORKS
[2] MELCO SEMICONDUCTOR ENGINEERING Corp.

Corresponding author: Takeshi Tokorozuki, tokorozuki.takeshi@dn.mitsubishielectric.co.jp

## Abstract

In recent years, environmental issues such as global warming have become increasingly serious. To address these problems, there is a global demand for the electrification of automobiles. One important aspect of electrification is the reduction of power module losses. We are developing a new series of power modules using SiC-MOSFET, which significantly reduce losses and improve power density compared to conventional Si-IGBT power modules. In this paper, we will describe the details of the new module series, including new heat dissipation technologies, functional integration for downsizing, and an extremely high-speed overcurrent protection method. Finally, we will investigate the application benefits provided from the perspective of loss reduction through power modules.

## 1 Introduction

Currently, the promotion of xEV (electric vehicles) is being globally advocated as an energy-saving and decarbonization policy. To achieve the widespread adoption of xEVs, it is necessary to have long driving ranges and reduce battery costs. To accomplish this, it is essential to reduce losses and downsize power modules used within inverters. Our company has been contributing to the development of xEVs through power modules since 1997 when we introduced the IPM (Intelligent Power Module), which integrates control circuits into power modules, and in 2001, we launched the T-PM (Transfer-molded Power Module) with small size and high reliability. While Si (silicon) has been commonly used as the semiconductor in conventional power modules, SiC (silicon carbide), which has significantly lower energy losses, has gained attention in recent years. SiC power module development is rapidly progressing in various countries. We have been developing SiC since the 1990s and have been expanding its use in a wide range of applications, including railway, consumer, and automotive. Currently, we are developing the latest trench-type structure. To harness the advantages of SiC, low inductance to minimize surges, low thermal resistance to withstand high-temperature operation, and high reliability are essential technologies [1]. Our company possesses transfer-mold technology that greatly improves module reliability and has been developing modules that combine accumulated SiC chip technology to maximize the strengths of SiC.

## 2 Features of new power module series

### 2.1 Overview

Fig. 1 illustrates the specifications and lineup of the new module. This configuration is a half-bridge compact module with a package size of 26.5mm x 53.9mm x 6.92mm (molded resin size). In addition to the module core, we are planning to offer models with single mounting to the cooler and models with 2 parallel mounting. We are currently developing both Cu and Al materials for the cooler. The performance characteristics of this product are as follows:

(1) Minimization of inverter power losses through innovative SiC devices and packaging technology.
(2) High-power density and inductance (5nH in the case of the 2 parallel specification).
(3) Excellent heat dissipation structure.
(4) Adoption of multi-functional chips, which contribute to the miniaturization of the package while incorporating temperature sensing, balance resistors, and DESAT diodes, simplifying customer circuit design.
(5) High-speed short-circuit protection using our unique short-circuit protection.

As a result, this module boasts high functionality and industry-leading power density.

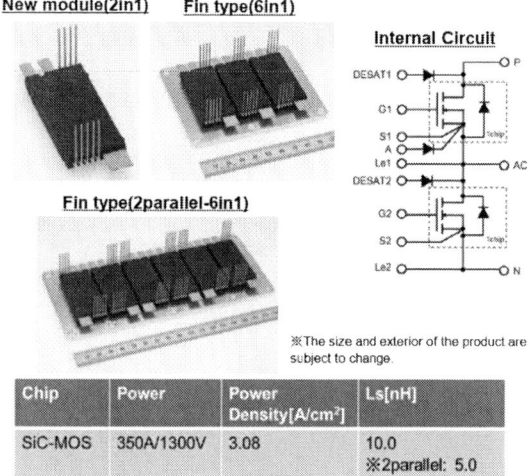

| Chip | Power | Power Density[A/cm²] | Ls[nH] |
|------|-------|---------------------|--------|
| SiC-MOS | 350A/1300V | 3.08 | 10.0 ※2parallel: 5.0 |

**Fig. 1** New power module specification and lineup

## 2.2 Required technology

There are several challenges in achieving high power density and minimizing Ls [2] through package miniaturization. On the chip side, it is necessary to develop a low on-resistance device structure to reduce heat generation and enhance the performance of SiC. On the package side, low thermal resistance is essential. This requires the use of high thermal conductivity bonding materials and cooling fins to improve heat dissipation. Additionally, SiC has a high chip cost, and to ensure effective area allocation for heat dissipation, it is common to externally mount current and temperature sensing functions for protection applications. However, incorporating these detection functions, such as thermistors and DESAT diodes, requires pattern division and area allocation, which poses challenges for miniaturization. Furthermore, there are challenges in terms of short-circuit withstand capability. SiC has low on-resistance and a large saturation current, necessitating an increase in chip volume to improve short-circuit withstand capability. As a result, chip costs increase, and the package size also increases, leading to an increase in Ls. In the next chapter, we will provide a detailed explanation of our applied technologies to address these challenges.

# 3 Technologies applied to new series

## 3.1 SiC-MOSFET technology

In recent years, SiC has gained global attention as a power semiconductor due to its advantages such as low switching losses, high-temperature operation capability, and low on-resistance, despite Si being commonly used in the past [3][4]. Mitsubishi Electric started developing SiC power devices in the 1990s and released power modules for electric railways equipped with the first-generation SiC power MOSFET (Metal-Oxide-Semiconductor Field-Effect Transistor) in 2010. In 2013, mass production of second-generation devices with optimized cell size and carrier injection mechanism began, and currently, we are developing an unique SiC-MOSFET with a new gate trench structure, which will be incorporated into this module. The structure of the conventional design and the trench gate structure are shown in Fig. 2. Compared to the conventional planar gate, the trench gate structure allows for downsizing of the unit cell and enables low-loss operation through high integration. The challenges of the trench structure include electric field concentration at the trench bottom, leading to device breakdown and degradation of the gate insulating film [5]. To address this, we have mitigated field concentration by forming a Bottom P Well (BPW) at the trench bottom. However, the BPW has the following issues:

(1) Increasing switching losses when the potential becomes unstable.
(2) Narrowing of the current path due to the depletion layer extended from the BPW.

To stabilize the potential of the BPW, we applied Sidewall Connection (SC) for (1), and to suppress the extension of the depletion layer, we formed JFET Doping (JD) for (2). As a result of these improvements, the trench structure SiC-MOSFET achieved a 50% reduction in on-resistance at Vth=4.1V and a rated voltage of 1.2kV at room temperature, compared to the conventional planar type (shown in Fig. 3) [6].

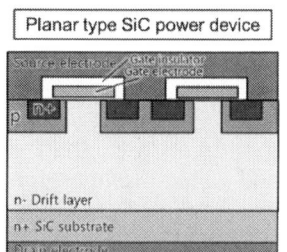
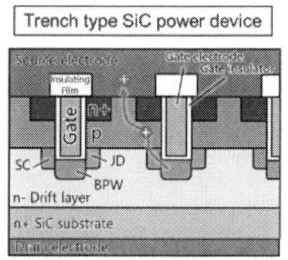

**Fig.2** Planar type and trench type structure.

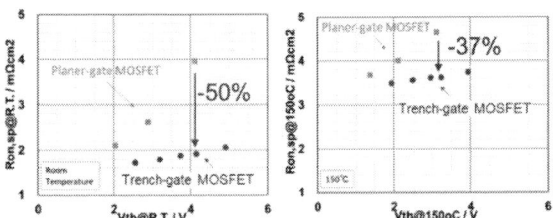

**Fig. 3** Comparison of on-resistance between planar-gate MOSFET and trench-gate MOSFET.

## 3.2 Heat dissipation technology

To achieve industry-leading miniaturization and high-power density, reducing thermal resistance is essential. Previously, our company used a structure in power modules that consisted of a thick copper heat spreader with a high thermal conductivity insulation sheet underneath, which was higher than that of mold resin. However, there were drawbacks to this design, such as the need for secure fixation due to using grease for cooling and the lower heat dissipation performance [7]. Fig. 4 shows the structure comparison between conventional and new structure. In this development, we have revamped the heat dissipation structure from the conventional design. To achieve excellent heat dissipation, we have adopted an insulator with high thermal conductivity in the module, replacing the insulation sheet that was previously used for insulation. For the chip bonding material, we used a material with high thermal conductivity and high thermal reliability compared to conventional lead-free solder. The most significant contribution to reducing thermal resistance in this structural change is the replacement of the grease with solder for the bonding material at the bottom of the module. This not only improves heat dissipation but also eliminates the need to consider springs and retaining plates for module mounting, which is a welcome point for users. We are planning to offer a lineup of Cu and Al coolers, and for Al, we are considering adopting a newly developed pin-fin that exhibits higher thermal conductivity than the conventional cylindrical fins. Through these combinations, we expect to successfully achieve a reduction in thermal resistance of over 30% compared to the conventional design in Al (shown in Fig. 5).

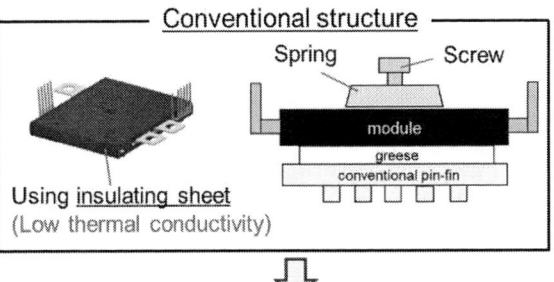

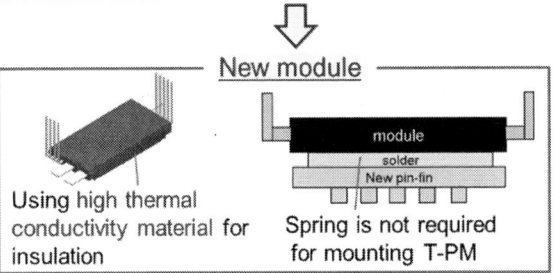

**Fig. 4** Structure comparison between conventional and new structure

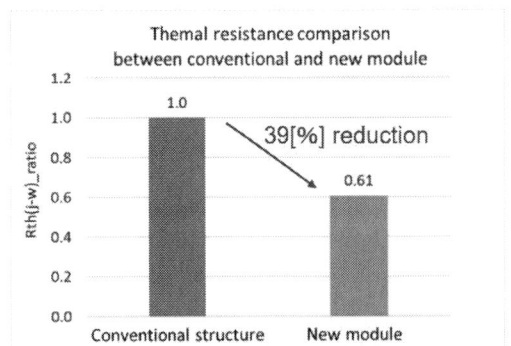

**Fig. 5** Structure and Thermal resistance Comparison

## 3.3 Consolidation of peripheral functions

SiC is known for its high cost per area, and it is common to provide detection functions such as temperature sensing and balance resistance outside the chip. Integrating balance resistance into the chip can lead to thermal interference, causing risks such as changes in switching speed and oscillation. However, the decrease in temperature sensing speed and the need to secure mounting space become barriers to module miniaturization. In our company, we have successfully developed a new Si chip that integrates these functions, reducing the space required for device mounting. Fig. 6 shows a schematic diagram of a multi-functional chip. It can eliminate the need for pattern cutting so increased the heat dissipation efficiency of SiC, contributing significantly to the miniaturization of the package. With this miniaturization, the package size has been reduced by 16% compared to using a multi-functional chip, and Fig. 7 shows that Ls has been reduced by 20%. Furthermore, by incorporating the DESAT-Di, which is typically installed externally, into the multi-function chip and inserting it into the package, there is no need to separate the high-voltage and low-voltage lines for DESAT within the PCB substrate. In other words, there is no need to consider the mounting position and insulation layout of the device within the control board, significantly reducing the circuit design burden for users.

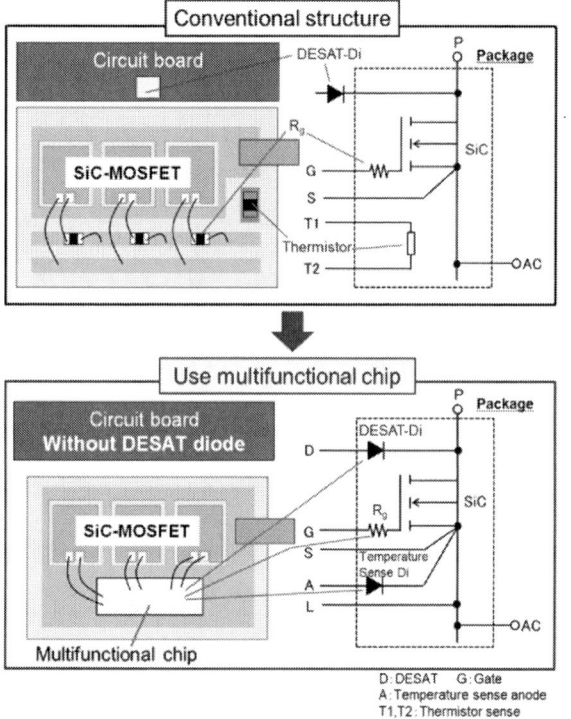

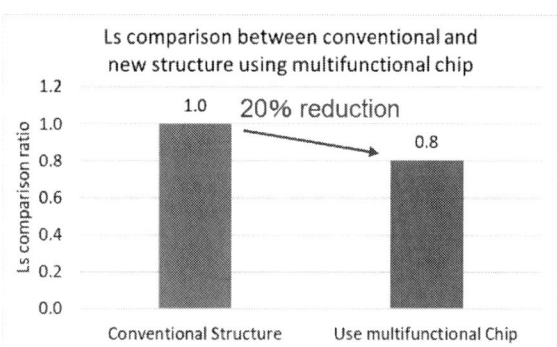

Fig. 6 Package miniaturization with multifunctional chip

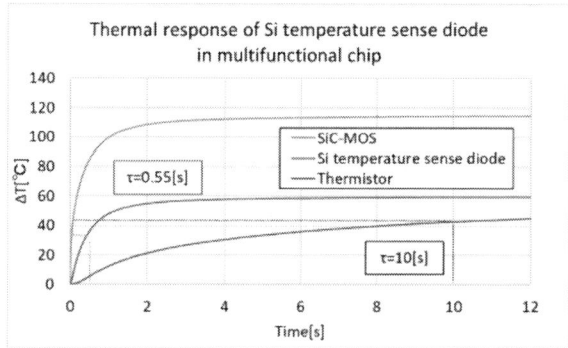

Fig. 7 Ls reduction effect with multifunctional chip

This device has bonding pads on the surface, allowing direct wire bonding to the chip and control terminals without the need for a pattern. As a result, it can be placed close to SiC. Additionally, unlike thermistors, it does not use resin for insulation but

instead utilizes temperature sensing with Si diodes, enabling fast thermal response. Fig. 8 presents the simulation results of the temperature sensing Ts of the multi-function chip when SiC is heated. It can be seen that it can rapidly detect temperature with a time constant of 0.55s against a temperature rise of approximately 120°C in SiC, indicating excellent thermal response characteristics. When compared to a thermistor placed in the same position as the multi-functional chip, Si exhibits a significantly higher thermal response speed due to its high thermal conductivity. (In the case of a thermistor, actual implementation is required at a position further away from SiC due to wire bonding issues, and pattern division is necessary, further reducing the thermal response speed). With this multi-function chip, we have achieved module miniaturization, reduced Ls, and increased module functionality.

Fig. 8 Built-in temperature sensing of multi-function chip and thermal response of thermistors.

## 3.4 About Short-Circuit Protection

SiC has significantly lower switching losses compared to Si, making it high-performance. However, due to its large saturation current, it has challenges in short-circuit protection [8]. One way to improve the short-circuit withstand capability of the chip is to increase its size, but this comes with the drawback of increased cost and larger module size. To solve this problem, we have successfully used a new short-circuit protection system (SCM terminal system) to use SiC without changing the chip size. The circuit and module SCM terminal positions are shown in Fig. 9, and the operating principle is illustrated in Fig. 10.

The purpose of this SCM terminal method is to reduce the short-circuit energy during the period until the short-circuit current is completely interrupted. The principle is as follows:

when a short circuit occurs, a reverse electromotive force is generated by the busbar inductance. This surge voltage is detected, and the gate voltage is reduced to suppress the short-circuit current before DESAT protection is activated. Finally, the DESAT protection is activated, and the interruption is completed. Fig. 11 compares the energy during a short circuit using this system. We have successfully achieved a significant reduction in short-circuit energy compared to the case of DESAT protection alone. (@Vgs=21.2V, Tj=175°C, Vcc=500V).

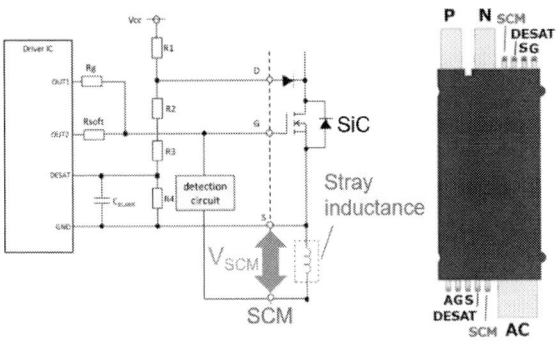

**Fig. 9** Short-circuit protection circuit and SCM terminal

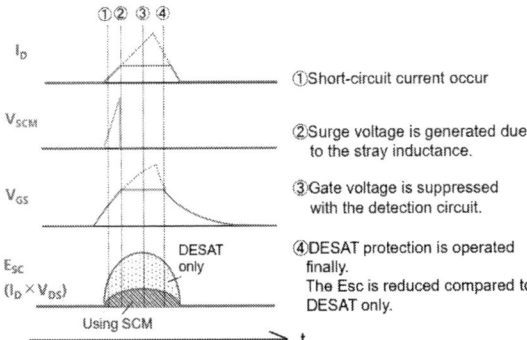

**Fig. 10** Operation principle of SCM

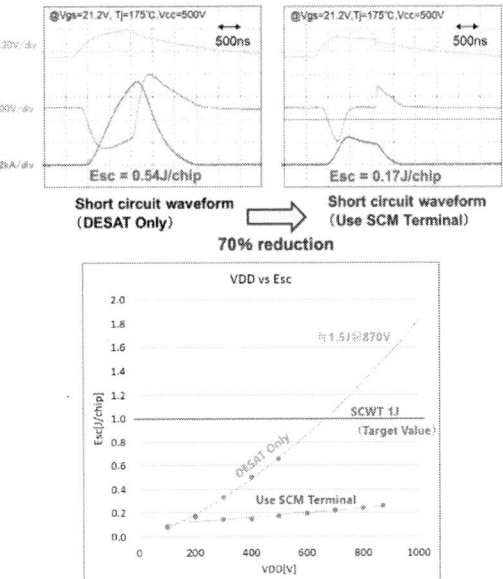

**Fig. 11** Esc reduction waveform with SCM

## 3.5 Output current

Fig. 12 shows the calculated inverter losses for each frequency at an operating temperature of 175°C, with a main battery voltage of 800V, coolant temperature ($T_W$) of 65°C, and coolant flow rate of 10 l/min (calculated for the case of Al fins). Under these conditions, it is possible to achieve an output current exceeding 500Arms at a frequency of 10kHz for a configuration with 2 parallel modules. This corresponds to an output power of over 250kW. In the single module configuration, an output of 250Arms is possible. With this lineup, a wide range of output power from 50kW to 250kW can be covered. We anticipate that the high-performance SiC chip, the revamped heat dissipation structure, the multi-function chip, and the short-circuit protection technology previously mentioned will achieve high output density and enhanced functionality in the module.

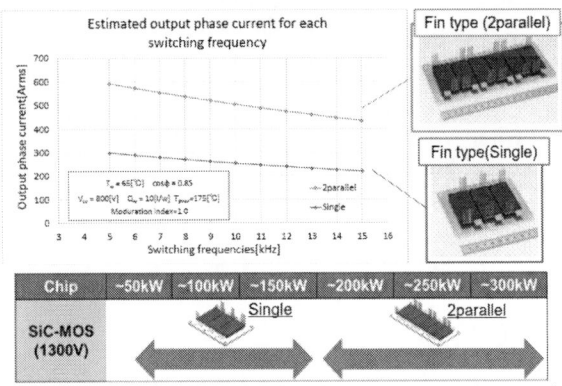

**Fig. 12** Output achievable phase current for each frequency.

# 4 Conclusion

In this paper, Mitsubishi Electric Corporation discussed its newly developed SiC power module. SiC chips offer low losses, but achieving the best performance requires addressing various challenges. The identified challenges are being addressed with countermeasure technologies to achieve low Ls and high-power density. Furthermore, the newly developed multifunctional chips allowed for incorporating protection and other applications into the module, resulting in compact and high-functionality modules. Through these developments, Mitsubishi Electric continues to contribute to accelerating the electrification of automobiles.

# Reference

[1] Biwei Zhang, Antoni Ruiz, Milad Maleki: On Superior Power Cycling capability of a High Power Density SiC Power Module for e-Mobility Application, PCIM Europe 2022, 10–12 (2022)

[2] Luciano Salvo, Fabio Occhipinti, Mario Pulvirenti, Alessandra Raffa, Edoardo Zanetti, Angelo G. Sciacca, Massimo Nania, Gionatan Montoro: Performance Evaluation of SiC MOSFET-Based Half-Bridge Converters Under Dynamic Voltage Clamp Limits, PCIM Europe 2022, 10 – 12 May 2022, Nuremberg

[3] A. Nisch, C. Klöffer, J. Weigold, W. Wondrak: Effects of a SiC TMOSFET tractions inverters on the electric vehicle drivetrain, PCIM Europe 2018, 95–102 (2018)

[4] Mesemanolis, Athanasios, Milad Maleki, Samuel Hartmann, Antoni Ruiz, David Weiss, Gontran Paques, and Tobias Keller. "Fast and reliable switching of parallel SiC MOSFET chips in a Half-bridge module." In PCIM Europe digital days 2020; International Exhibition and Conference for Power Electronics, Intelligent Motion, Renewable Energy and Energy Management, pp. 1-8. VDE, 2020.

[5] K. Sugawara, Y. Fukui, R. Tanaka, K, Adachi, Y. Kagawa, S. Tomohisa, N. Miura, E. Suekawa, Y. Terasaki: A Novel Trench SiC-MOSFETs Fabricated by Multiple-Ion-Implantation into Tilted Trench Side Walls (MIT2-MOS), PCIM Europe digital days 2021 (2021)

[6] K. Sugawara, Y. Fukui: Development of SiC Trench MOSFET with Novel Structure Enabling Lower Losses, Mitsubishi Electric ADVANCE September 2022, 19-21 (2022)

[7] Hideo Komo, Rei Yoneyama, Shoichi Orita, Gourab Majumdar: An advanced SiC power module designated for automotive, PCIM Asia 2023, 29 – 31 August 2023, Shanghai, China

[8] Vikneswaran, Thayumanasamy, Carlos Fuentes, Kevin Lenz, Ingo Rabl, Jürgen Engstler: hort Circuit Protection of a Power Module with Trench-SiC-MOSFET. Can DESAT be Fast Enough?, PCIM Europe 2022, 10 – 12 May 2022, Nuremberg

PCIM Europe 2024, 11– 13 June 2024, Nuremberg    DOI: 10.30420/566262105

# Extremely Compact SiC Power Module for EV Traction Inverters in the 250 kW Class

Raffael Schnell[1] , Rémi Guillemin [1], Roger Stark[1] , Sven Matthias[1], Coris Li[2], Samuel Hartmann[3]

[1] SwissSEM Technologies AG, Switzerland

[2] Sunking Pacific Semiconductors Technology Ltd., China

[3] MFIS GmbH, Switzerland

Corresponding author:   Raffael Schnell, raffael.schnell@swiss-sem.com
Speaker:                 Raffael Schnell, raffael.schnell@swiss-sem.com

## Abstract

This paper describes a novel phase-leg SiC module for electric vehicle (EV) applications which is designed to achieve a record power density and features exceptional homogenous internal current sharing, low connection resistance and inductance as well as very efficient direct water cooling without the need of a copper baseplate and corresponding joint to the substrate vulnerable to power cycling. We will highlight results of our electromagnetic and spice simulations and compare them to real measurements showcasing the excellent switching performance. We also will touch thermal simulations and show results of measurements.

## 1    Introduction

Electric Vehicles are well known for their superb energy efficiency compared to combustion engine powered counterparts. This is thanks to the unrivaled energy conversion efficiency from electricity to mechanical motion, but as well due to the fact, that breaking energy can be harvested to recharge the battery.

Converters in the range of 200 to 250 kW power are popular for passenger cars but can also be used as single wheel or axle drives for smaller size bus and truck applications.

The new "highly efficient module for electric vehicles" (HEEV) is designed to offer the world's most compact design of SiC power modules for 200 to 250 kW 800 V battery traction inverters, with a footprint of only 58 x 45 mm² it is considerably smaller than similar existing modules [1]. Its design features a very homogenous internal current sharing and very low stray inductance of only 4.5 nH enabling fast switching of SiC MOSFETs without the need of an additional gate resistor, and heavily reduces the SiC MOSFET tendency to generate unwanted oscillations. Also, the module internal resistance is reduced to less than 0.25 mΩ to make full use of the low $r_{DSon}$ MOSFET of 2.1 mΩ at 25°C for the full module rated at 1200 V $V_{DS}$.

To eliminate the large area joint between substrate and (pin-fin) baseplate, which is the bottleneck for power cycling capability, a novel cooling concept has been developed. The cooling structure is directly attached to the bottom substrate metallization. This enables higher power cycling performance and a more compact and lightweight module design. Despite the compact design, low thermal resistance values and no cross heating to neighboring modules has been achieved thanks to an optimized cooling fin arrangement. The module can be easily installed onto a cooling element with a clamp bridge. The sealing between the module and the cooler is realized with well-established O-rings.

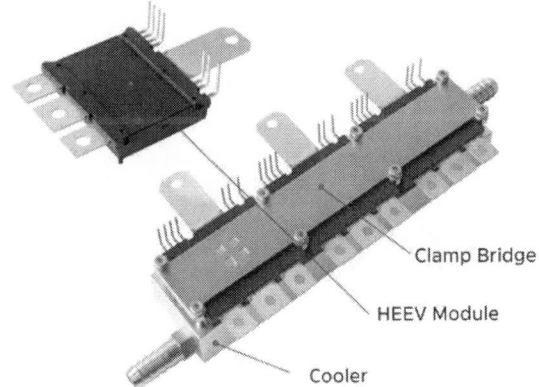

**Fig. 1** HEEV Module Assembly on a Cooler

## 2 Electrical Design

### 2.1 Design Goals and Simulations

SiC MOSFETs offer much faster switching speed compared to its silicon IGBT counterparts. Although, the manufacturing technology and process control is still not as mature as for established silicon processing leading to significantly higher parameter variations, especially for the threshold voltage. Combined with lower capacitance values, usually lower internal gate resistance and a more sophisticated module layout due to the need of paralleling many chips, there is a high risk of severe device oscillations rather than proper switching. So special emphasis has been taken on the HEEV module design to ensure homogenous internal current sharing with slight gate-counter coupling to damp unwanted oscillations [2]. The six paralleled chips have been arranged to have very similar source impedance and gate-coupling (MG), see Figure 2. Special attention has also been paid to chip-to-chip impedance to reduce the risk for internal gate-oscillations. The design also must be optimized for low internal stray inductance. For the HEEV module a very low internal stray inductance of 4.5 nH has been achieved.

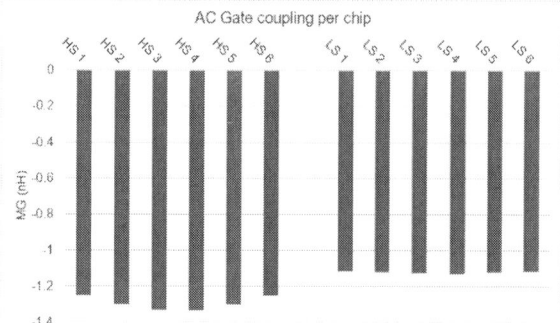

**Fig. 2** Gate Coupling per Chip

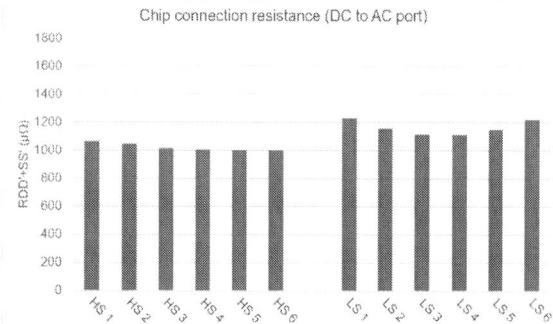

**Fig. 3** Connection Resistance per Chip

For static current sharing low and homogeneous

connection resistance was achieved for all paralleled chips (Fig. 3). The total connection resistance of the module is with $R_{DD'}+R_{SS'}$ of 195 µΩ for the LS and 170 µΩ for the HS switch very low. To further judge the module performance, the most promising variant of the electromagnetic simulation with Q3D was exported as a netlist containing all direct and mutual impedances and capacitances [3]. This netlist was then imported into a SPICE simulator to make a first check of switching performance with SiC MOSFET models of the used chips. Figure 4 shows the simulated turn-on behavior with VGS and ID of each chip shown separately.

The switching speed (rise time of around 45 ns) of all 6 paralleled MOSFETs is very similar and the current sharing ideal, so hardly any difference can be observed. The graph also shows the external VGS (black) and the slight counter coupling can be seen by the difference of the external gate-voltage to the individual chip gate-voltages. Notably, the slight counter coupling helps to prevent the opposite switch (HS) from suffering parasitic turn-on as the individual VGS are pulled down from the external VGS. Hence no parasitic turn-on of the opposite switch could be observed at a current rise time of 45 ns, despite the fast $dV_{DS}/dt$ occurring.

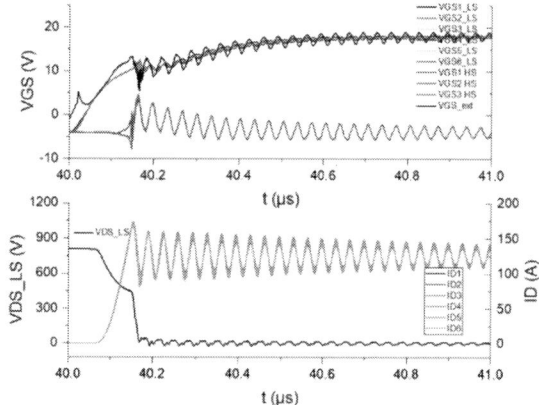

**Fig. 4** Simulated LS-switch turn-on @ 800 V, 750 A, $R_G$=3 R, $V_{GS}$= +18 /- 4 V, $L_{s,ext}$=15 nH

The simulated turn-off of the HEEV module for both LS and HS switch (Fig. 5 /Fig. 6) shows excellent current sharing and no dangerous oscillations. Thanks to very similar gate coupling the low-side (LS) and high side (HS) switch have a closely matched switching speed and hence an identical peak voltage. Another important aspect to look at for SiC MOSFETs is negative gate-voltage peaks of the opposite switch due to gate coupling and dv/dt displacement current through the miller capacitance. Too high negative peak voltages can negatively impact long-term reliability. The simulation of the opposite switch VGS directly at the chip shows no dangerous negative values beyond the datasheet maximum limits from the chip manufacturer. Since a direct measurement of the on-chip VGS is hardly possible, simulations are the only practical way to verify such effects.

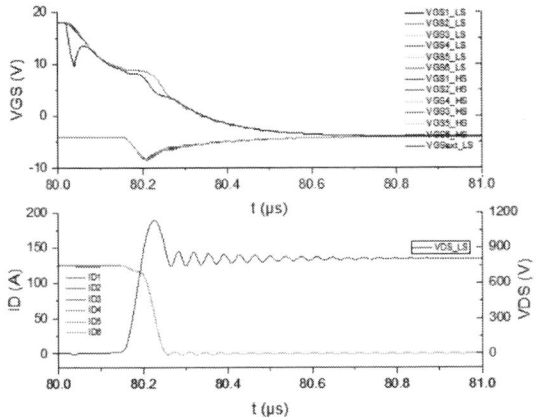

**Fig. 5** Simulated LS-switch turn-off @ 800 V, 750 A, $R_G$=4 R, $V_{GS}$= +18 /- 4 V, $L_{s,ext}$=15 nH

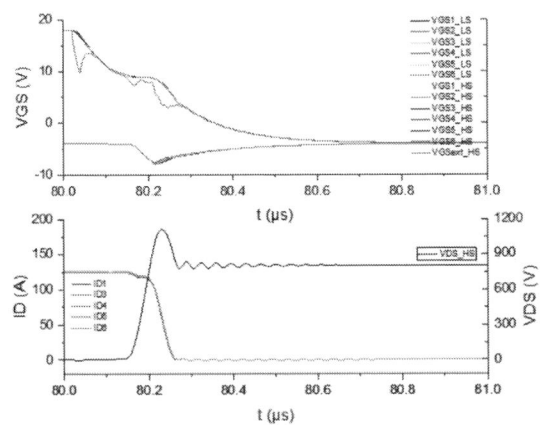

**Fig. 6** Simulated HS-switch turn-off @ 800 V, 750 A, $R_G$=4 R, $V_{GS}$= +18 /- 4 V, $L_{s,ext}$=15 nH

## 2.2 Electric Characterization

To verify the simulation results, find optimal driving conditions and characterize the module, tests in a double pulse tester have been carried out with an external stray inductance of around 15 nH which leads to around 20 nH total when taking the module internal inductance into account. Figure 7 shows the measured turn-on and figure 8 the turn-off of the HEEV module with similar conditions than in the simulation, the measurements are in

good accordance with the simulation and proof the excellent switching performance of the HEEV module.

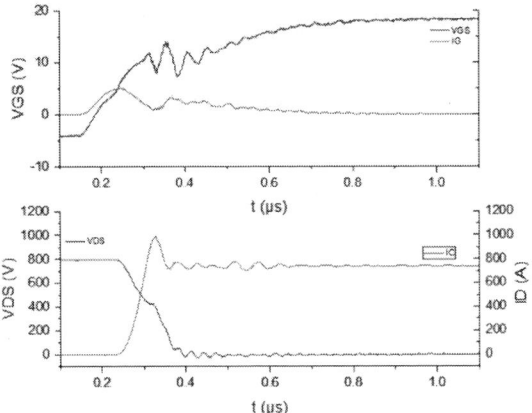

**Fig. 7** Measured LS-switch turn-on @ 800 V, 750 A, $R_G$=3 R, $V_{GS}$= +18 /- 4 V, $L_s$=20 nH, 25°C

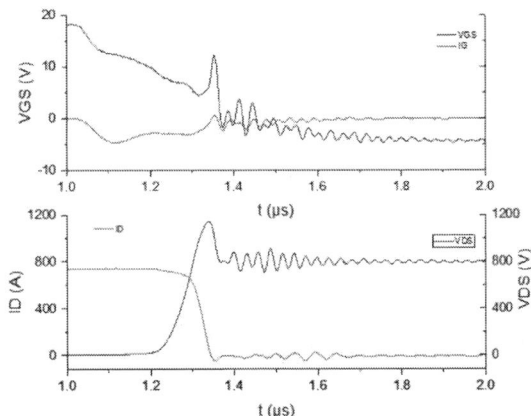

**Fig. 8** Measured LS-switch turn-off @ 800 V, 750 A, $R_G$=4 R, $V_{GS}$= +18 /- 4 V, $L_s$=20 nH, 25°C

To test the module SOA capability, we've tested it up to a switching current of 1500 A. Figure 9 shows the turn-off at 1500 A and 800 V DC-link voltage at 25°C. The same test was repeated at the maximum junction temperature of 175°C (Fig. 10). Even with these harsh conditions, the HEEV module exhibits a smooth switching behavior, and the peak voltage remains in the range of 1200 V. Both tests were carried out with a total stray inductance of about 20 nH on a test bench with all the metering equipped. If we consider that the stray inductance can potentially be reduced as low as 10 nH with a dedicated direct mount capacitor, the peak voltage would nearly be halved, and the switching behavior would further improve.

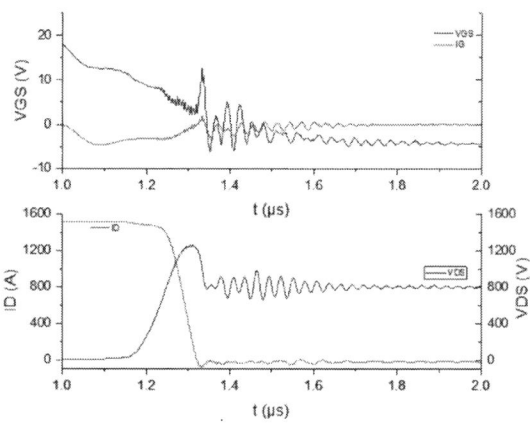

**Fig. 9** Measured LS-switch turn-off @ 800 V, 1500 A, $R_G$=4 R, $V_{GS}$= +18 /- 4 V, $L_s$=20 nH, 25°C

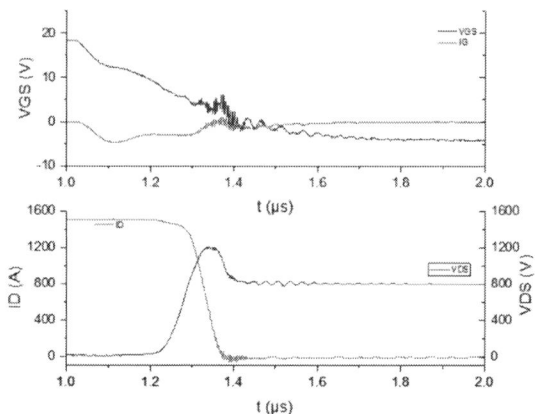

**Fig. 10** Measured LS-switch turn-off @ 800 V, 1500 A, $R_G$=4 R, $V_{GS}$= +18 /- 4 V, $L_s$=20 nH, 175°C

SiC MOSFETs are known for their low switching losses, which come with a price to pay, fast switching speed. This needs to be balanced for real life applications, for instance switching-on with a lower gate resistor RG increases the switching speed and potentially reduces the switching losses Eon, but it also increases the recovery losses of the opposite switch intrinsic Diode. In addition to this, fast dv/dt of the intrinsic Diode during recovery (Fig. 11) causes a depletion current (1) through the MOSFET $C_{DG}$ pumping-up its $V_{GD}$.

$$I_{DG} = \frac{dV_{DS}}{dt} \times C_{DG} \qquad (1)$$

Due to the lower threshold voltage and less allowed negative VGD during off-state, there is a risk of parasitic partial turn-on of the opposite switch MOSFET leading to higher peak turn-on current, which can increase the losses in the MOSFET and the Diode.

Figure 11 shows the intrinsic diode recovery at 175°C and 500 A. The dv/dt becomes higher with temperature and reaches up to 24 kV/us at 175°C and with RG=3 Ohms.

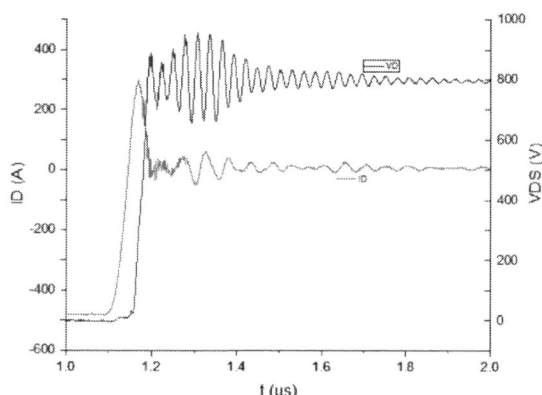

**Fig. 11** Measured HS-switch recovery @ 800 V, 500 A, $R_G$=3 R, $V_{GS}$= +18 /- 4 V, $L_s$=20 nH, 175°C

To define an optimal $R_G$ for MOSFET turn-on we have hence to look at the total losses Eon+Err of the diode and consider the switching waveforms.

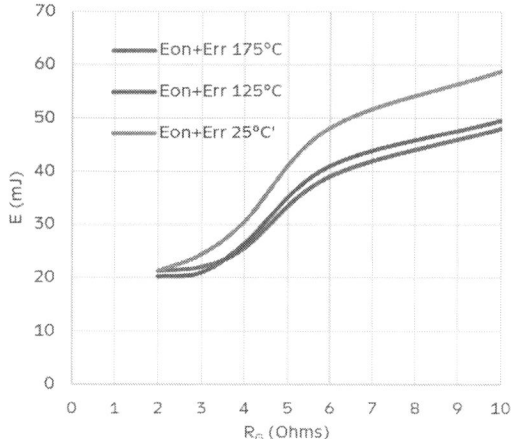

**Fig. 12** $E_{on}$+$E_{rr}$ versus $R_G$ at 800 V, 500 A, $L_s$=20 nH

Figure 12 shows that in general the switching losses are reduced with lower $R_{Gon}$, but at elevated temperatures there is a saturation at around 3 Ohms. Hence, technically it does not make sense to go lower than this, also taking the negative aspects like the risk for higher recovery peak voltage and the tendency for more oscillations into account that you buy with faster switching.
Thanks to the balanced electrical design and low internal stray inductance the HEEV module can be

switched at fast switching speed and hence low switching losses result, see Figure 13 for the switching losses characteristics as a function of current:

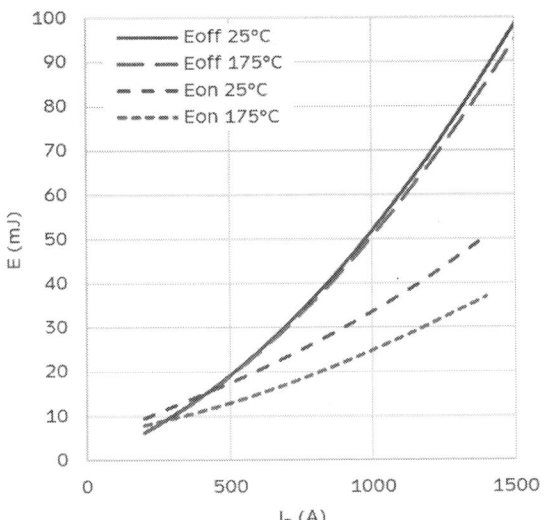

**Fig. 13** Switching losses versus $I_D$ @ 800 VDC, $R_{Gon}$=3 R, $R_{Goff}$=4 R, $V_{GS}$= +18 /-4 V, $L_s$=20 nH

The corresponding switching time characteristics can be found in Figure 14.

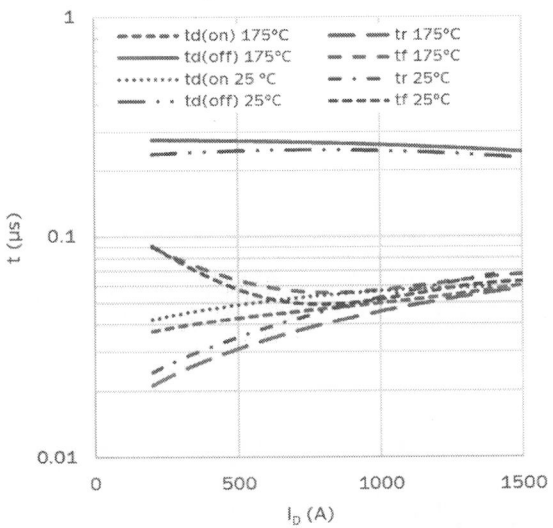

**Fig. 14** Switching times versus $I_D$ @ 800 VDC, $R_{Gon}$=3 R, $R_{Goff}$=4 R, $V_{GS}$= +18 /-4 V, $L_s$=20 nH

## 3 Thermal and Mechanical Design

### 3.1 Design Principles

The main design target for the HEEV module was to reach a record high current density and improve the module reliability. Transfer molded modules are well known for improved protection against environmental impact and if the right mold is chosen, can also support somewhat higher operation temperatures compared to classic silicone gel encapsulated modules. We also employ silver sintering to attach the SiC chips to improve the power cycling performance and use a die top system and copper wire bonds to reduce the connection resistance. Nevertheless, the Achilles heel for power cycling in traction application is for both classic gel-filled and transfer molded modules, the large area solder (or sinter) joint between the insulating substrate (DBC or AMB) and the classic copper baseplate or pin-fin base.

Our goal was to get rid of this large area joint and the pin-fin base for power cycling, but as well size reasons, but still enable efficient direct water cooling.

As a solution to this we have chosen a Silicon-Nitride AMB substrate with thicker copper metallization and Nickel plating on the cooling side to protect from erosion effects of the cooling liquid. The cooling-fins to enhance the heat-transfer to the cooling water by generating the required water turbulence are welded directly to the substrate backside. The shape and arrangement of the cooling-fins was carefully optimized with CFD simulations together with the cooler enclosure on which the modules are mounted with a clamp bridge and O-ring seal (Figure 1).

The cooling structure of the HEEV module is shown in Figure 15. The fin-structure and alignment to the water flow direction is optimized to generate maximum turbulence and hence water mixture to enhance the heat take-up of the cooling liquid, whilst avoiding too high surface velocity on the cooling structure that could promote corrosion and hence reduce the lifetime.

**Fig. 15** HEEV cooling structure directly welded to the AMB substrate

## 3.2 Cooler Arrangement and Simulation

Beneath the general module design, the integration onto a cooler plays an important aspect for overall cooling efficiency. We have studied several concepts and compared the results of CFD simulations. For a traction inverter the usual arrangement is to place 3 half bridge modules on a heatsink. The simplest approach is to cool the modules in series (daisy chain) see Figure 16.

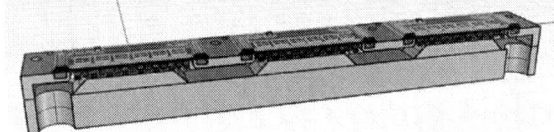

**Fig. 16** Series cooling of modules

However, this approach has the disadvantage, that the temperature increases significantly from module to module in flow direction. Also, within the module the chip rows being arranged in flow direction leads to a strong temperature difference in flow direction (Fig. 18). Hence, we've also investigated a parallel cooling approach in which each module gets cooling water with the same inlet temperature (Fig. 17).

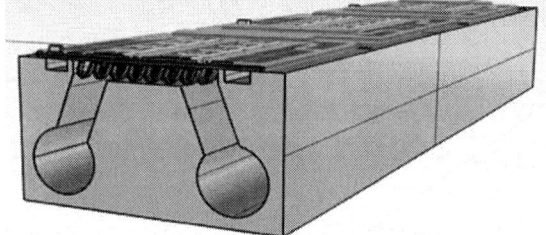

**Fig. 17** Parallel cooling of modules

Figure 18 shows the temperature distribution of the chips at arbitrary scale inside a single module with water flow from left to right. Obviously, chips get heated along the water flow and the hottest chip has 10% more Rth than the average or even close to 20% more than the coolest chip.

With the parallel cooling approach, the situation is considerably improved. We still have an unavoidable temperature difference, but it is significantly lower compared to the serial arrangement and chip temperatures within one switch are more homogeneous, which is advantageous for the electrical current sharing. Figure 19 shows the parallel cooling arrangement with water flow from bottom to top.

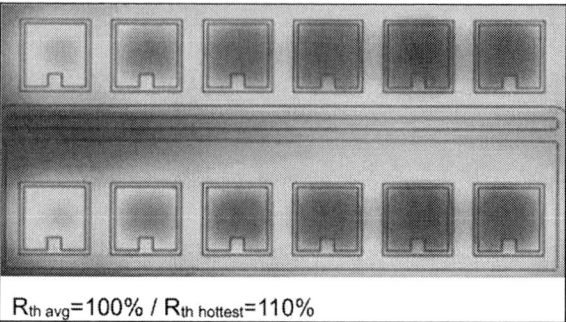

$R_{th\ avg}$=100% / $R_{th\ hottest}$=110%

**Fig. 18** Temperature inside a serial cooling arrangement

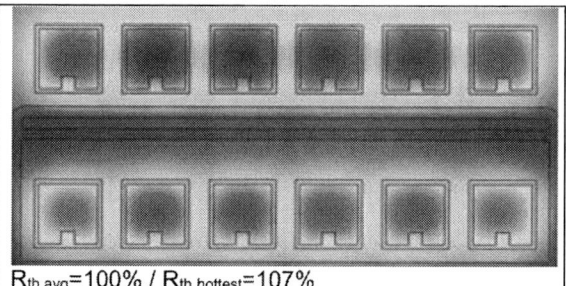

$R_{th\ avg}$=100% / $R_{th\ hottest}$=107%

**Fig. 19** Temperature inside a parallel cooling arrangement

To further improve the cooling, we took a detailed look at the shape and arrangement of the cooling fins. In classic arrangements (like serial) the cooling fluid flows through the openings of the cooling fins – but this approach causes low turbulence/water mixture, and the thermal performance is not optimal. Figure 20 shows the simulated cooling liquid velocity, and we can see that there are ineffective channels where the water flows at high velocity (red stream) but does not take-up a lot of heat and the surface velocity at the module side, which has the largest most direct cooling area, is low.

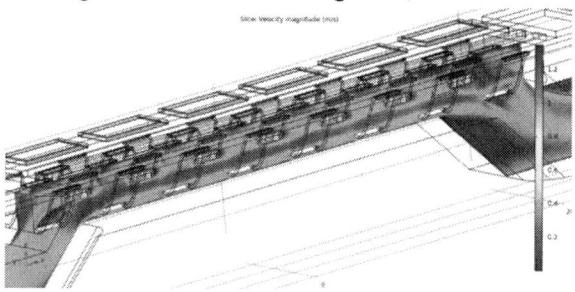

**Fig. 20** Velocity in a slice for "classic" design

To increase the turbulence and increase the cooling efficiency we rotated the cooling fins by 90° and optimized its shape and spacing to get a similar maximum surface velocity as in the classic approach but with much more turbulence. We have also investigated the shear stress due to the coolant surface velocity and keep it within reasonable levels reported in [4] to avoid erosion in the long run.

Figure 21 shows the streamlines for the improved and 90° rotated fin design. The fins cause eddies that mix the cooling fluid very nicely.

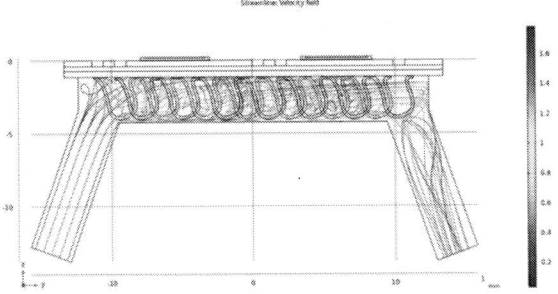

**Fig. 21** Streamlines for the improved design

### 3.3 Thermal measurements

Before going into production, the cooling design of the module was optimized completely using simulation methods, considering processing of the cooling fin welding, and shaping. The most promising design was then assembled, and the thermal performance was measured in a test bench and compared to simulation results. The test method for Rth is based on an indirect measurement of the junction temperature. For this we use the body diode forward voltage characteristics as a temperature dependent parameter. Depending on the MOSFET design, the body diode forward voltage needs to be measured at sufficiently large negative gate voltage to avoid errors due to gate-threshold drift and hysteresis effects. We have measured the Rth(j-a) with 50% glycol water mixture as this is used in many EV applications. The Rth(j-a) is defined as in (2):

$$R_{th(j-a)} = \frac{T_j - 0.5 \times (T_{outlet} + T_{inlet})}{V_{DSon} \times I_D} \quad (2)$$

The measurement results compare nicely with the simulation results, whereas the high-side (HS) switch, which was closer to the inlet in the measurement, shows as Rth. The same effect was visible in the simulation (Fig. 19), where the lower row of chips close to the inlet showed lower chip temperatures. Also, we observe that the Rth improves slightly with a higher coolant temperature. The simulated Rth represents an average of LS and HS switches with both switches simultaneously

heated. For the measurement also both switches are heated, but Rth is evaluated individually for the LS and HS switch. The measured and simulated Rth(j-a) versus flow rate is shown in Figure 22. At 10 l/min, 50% glycol and 50°C coolant temperature a Rth(j-a) of 110 K/kW can be reached.

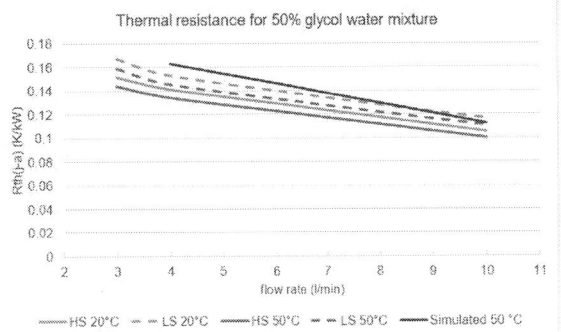

**Fig. 22** $R_{th(j-a)}$ versus flow rate

## 4 Application Performance

Taking the measured electrical losses and thermal performance into account we can make a first estimation on the achievable performance in a 3-phase 2-level inverter such as typically used in EV Traction chains.

Using calculation methods [5] we can simulate the achievable output current of the HEEV module versus switching frequency with $T_{water}$ set to 50°C and the $T_j$ fixed to 175°C. We assume a water flow rate of 10 l/min and 50% water glycol mixture.

The achievable phase-output current with this method and conditions is shown in figure 23: Up to a switching frequency of 13 kHz we achieve an output current of above 600 Arms or above.

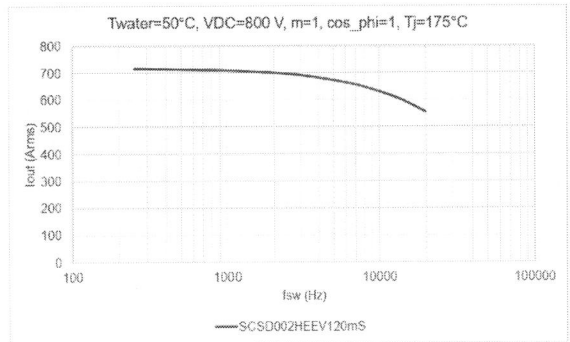

**Fig. 23** HEEV output current versus switching frequency

# 5   Conclusions & Outlook

In this paper we have presented the design and simulation methodology of the HEEV module as well as novel module concepts to achieve record high power power-density. With a module footprint area of just 58 x 45 mm$^2$ for a single phase-leg module more than 250 kW output power can be achieved in a 3-phase inverter.

Despite of the very small module footprint, we achieved a record low inductive and resistive module and very smooth switching of SiC MOSFETs thanks to a homogeneous electrical design.

The base-less module architecture without large area interconnects promises an excellent power cycling performance especially at longer cycles and thermal shock test (TST) capability. The tests for this are still on-going.

# References

Please follow international scientific citation rules.

[1] F. Carastro, Z. Chen, A. Streibel and O. Muehlfeld, "DCM™1000X – Automotive Power Module Technology Platform Optimized for SiC Traction Inverters," 2021 IEEE Applied Power Electronics Conference and Exposition (APEC), Phoenix, AZ, USA, 2021

[2] Schnell, R., Hartmann, S., Truessel, D., Fischer, F., Baschnagel, A., & Rahimo, M.T., "LinPak, a new low inductive phase-leg IGBT module with easy paralleling for high power density converter designs", Proc. PCIM 2015, Nuremberg, Germany, 2015

[3] R. Schnell, S. Hartmann, "3D bond wire modelling and electro-magnetic simulation accelerates IGBT module development", Bodos Power Systems Magazine, January 2021

[4] R. Remsburg, J. Gilmore, "Analytical and Experimental Characterization of Erosion Effects According to Pin-Fin Shape in Electronics Cooling Loops", Proc. PCIM 2012, Nuremberg, Germany, 2012

[5] R. Schnell, U. Schlapbach, "Realistic benchmarking of IGBT-modules with the help of a fast and easy to use simulation-tool", Proc. PCIM'04, Nuremberg, Germany, 2004

PCIM Europe 2024, 11– 13 June 2024, Nuremberg    DOI: 10.30420/566262106

# Benefits of .XT Interconnection Technology for 3.3 kV XHP 2 Module with 3.3 kV CoolSiC MOSFET

Matthias Bürger, Tobias N. Wassermann, Henry Foester

Infineon Technologies AG, Germany

Corresponding author: Matthias Bürger, matthias.buerger@infineon.com

## Abstract

CoolSiC™ MOSFETs enable operation at higher frequencies for high-voltage applications in the 3.3 kV voltage class. Equipped with CoolSiC™ MOSFETs, the Infineon 3.3 kV XHP™ 2 module enables high power densities in applications such as traction. With the trend towards higher power densities, robustness against failure such as surge current and short circuits is becoming increasingly important. Furthermore, traction applications demand high power cycling capabilities to ensure an ample service life of the power module. Combined with the Infineon .XT interconnect technology in an XHP™ 2 package, CoolSiC™ MOSFETs can address these requirements for traction applications. This paper describes the performance of the XHP™ 2 3.3 kV CoolSiC™ MOSFET .XT module with regard to surge current, short circuit, and power cycling.

## 1   Introduction

Today silicon-based IGBT power semiconductors are state-of-the-art in high-power switching systems. However, reducing energy consumption is gaining more importance in new railway traction systems. Silicon carbide (SiC)-based devices support the trend towards more energy efficiency and power density in traction applications. Introduction of SiC-based power modules enables high frequency operation. Thus, reducing transformer losses and leakage inductance. It also reduces losses in the electric motor. Thus, SiC-based power modules optimize the efficiency of the overall traction system [1].

In traction applications, the handling of failure events, such as surge current and short circuit, is crucial for high operational reliability. The trend towards higher power density and smaller, more energy-efficient transformers create additional pressure on the power modules in such cases. Furthermore, higher current density leads to higher temperature ripple amplitudes under cyclic load conditions. Handling this requires greater power cycling capabilities of the semiconductor modules [2].

Infineon's trench MOSFET technology (CoolSiC™) is well suited for various applications and has a proven track record in the 1200-2000 V

voltage class modules [3]. This technology has now been adjusted to cover the 3.3 kV voltage class to enable optimum performance of applications such as traction and medium voltage drives (MVD) [4]. The low inductive XHP™ 2 package is equipped with 3.3 kV CoolSiC™ MOSFET technology [5]. The XHP™ 2 package (FF2000UXTR33T2M1) with its half-bridge topology is shown in Fig. 1.

To provide optimum reliability and temperature stability for cyclic mission profiles in traction applications, Infineon's .XT technology is applied to the 3.3 kV XHP™ 2 package with CoolSiC™ MOSFETs [6].

The internal MOSFET body diode is used for current commutation within the module instead of a Schottky diode. This reduces the complexity in production and frees up space for the integration of additional MOSFET chip area.

As a result, 1000 A nominal current with low drain-source on-resistance $R_{DS(on)} = 1.9 \text{ m}\Omega$ is ensured at T = 25°C. The tailored 3.3 kV chip design in combination with a low inductive module enables fast switching with low dynamic losses and fewer oscillations [7]. This paper describes the .XT technology for the 3.3 kV CoolSiC™ MOSFETs and its benefits with regard to surge current, short circuit, and power cycling capability.

PCIM Europe 2024, 11– 13 June 2024, Nuremberg          DOI: 10.30420/566262106

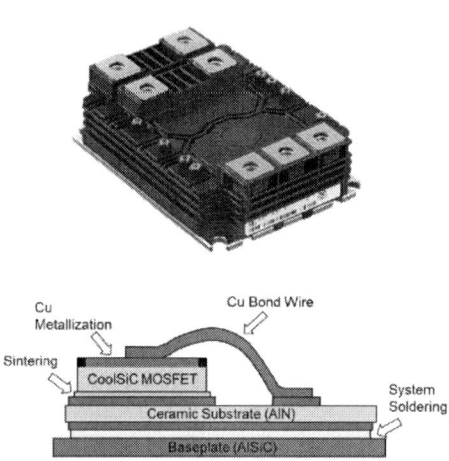

**Fig. 1:** Picture of the XHP™ 2 half-bridge module equipped with 3.3 kV CoolSiC™ MOSFET chips and schematic vertical structure with .XT technology.

## 2 The 3.3 kV XHP™ 2 CoolSiC™ MOSFET module

The standard joining technology of aluminum (Al) wire bonding and soft soldering is applied to most of the existing high-power modules. In the 3.3 kV XHP™ 2 package with CoolSiC™ MOSFETs, the high current density, if combined with the standard Al joining technology, would reduce its performance in the event of a surge current, short circuit, and for power cycling. This is because the standard Al joining technology suffers from low inherent mechanical strength and low melting temperature. To overcome these problems, for the first time, the .XT technology is applied to 3.3 kV CoolSiC™ MOSFETs in the XHP™ 2 package.

Figure 1 shows the different components of the .XT technology in the vertical structure of the module. The front side copper (Cu) chip material is connected via Cu bonds to the substrate metallization tracks. The chip's back-side is sintered with silver (Ag) particles on an aluminum nitride (AlN) ceramic substrate. A newly developed process for system soldering is used to interconnect the substrate's back side with an aluminum silicon carbide (AlSiC) composite baseplate.

Figure 2 shows the simulated transient thermal resistances of SiC MOSFETs, in the XHP™ 2

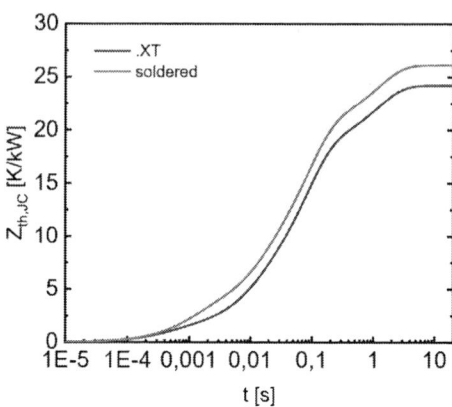

(a) $Z_{th,JC}$ characteristics of standard and .XT joining technology.

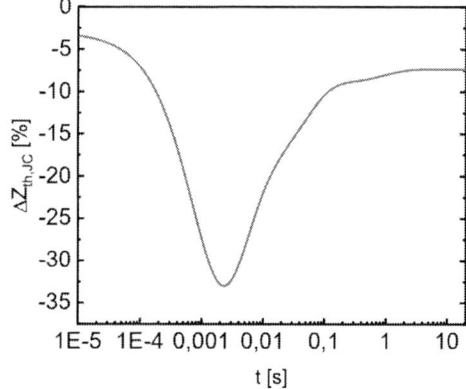

(b) Relative $Z_{th,JC}$ improvements at different time values.

**Fig. 2:** Simulation of transient thermal resistances of .XHP™ 2 equipped with 3.3 kV CoolSiC™ MOSFETs with .XT compared to standard soft soldered joining technology.

package, with standard soft soldered and with .XT joining technologies. As can be seen, a significant reduction of more than 30% in $Z_{th,jc}$ can be realized in a short time (t = 2 ms). This improvement in $Z_{th}$ is especially helpful during fault events such as surge currents that typically occur in these time regimes. The impact in the longer term i.e., t > 10 s is ~7%.

This lower thermal resistance, combined with the higher mechanical strength of the .XT technology interconnection, is beneficial for power cycling. Furthermore, the maximum operating temperature of $T_{vj,max} = 175°C$ is small compared to the high melting temperature of the Cu metallization and Ag

PCIM Europe 2024, 11– 13 June 2024, Nuremberg        DOI: 10.30420/566262106

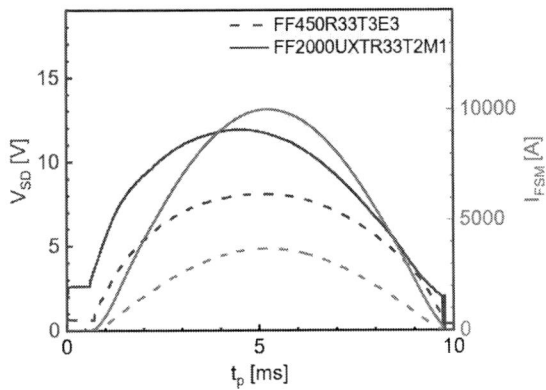

**Fig. 3:** Comparison of surge current waveforms for the XHP™ 3 (FF450R33T3E3, T=150°C) module and the XHP™ 2 (FF2000UXTR33T2M1, T=175°C, $V_{GS}$ = -5 V) module package with 3.3 kV CoolSiC™ MOSFETs and .XT technology.

sinter interconnection, which minimizes material creeping [8]. This shows that the .XT technology supports the robustness of the CoolSiC™ MOSFET XHP™ 2 modules against surge current and short circuit events.

## 3   Benefits of the .XT Technology

### 3.1 Surge Current

For traction applications, transformers are key components for high system efficiency. Enabled by CoolSiC™ MOSFETs, operating at high frequencies allows for smaller transformers with fewer windings accompanied by low ohmic resistances and stray inductances [9].

Usually, the parasitic stray inductance of a transformer helps limit surge currents in the application. The use of small and energy-efficient transformers makes handling such failure events more challenging and requires high surge current capabilities of the power modules. The CoolSiC™ MOSFET power modules provide high surge current capabilities to ensure that simplified safety measures can be installed in the inverter system.

Silicon (Si) diodes with standard, soft-soldered joining technology usually have an Al front-side metallization. The Al diodes are limited in their surge current capabilities due to the low melting point of the front-side metal or the chip solder. The combination of Cu front-side metallization and

sintered die attach, introduced as the .XT technology, overcomes this obstacle, and results in increased surge current capabilities.

In power modules using SiC MOSFETs with paralleled Schottky barrier diodes (SBD), surge currents could be further limited. High surge currents can lead to a critical rise in chip temperatures that can damage the internal Schottky contact of the diode, which is indicated by increased leakage currents. The new optimized chip process technology enables the use of the internal body diode in the 3.3 kV CoolSiC™ MOSFET module. The elimination of SBDs combined with the .XT technology for SiC MOSFETs extends the surge current limit.

In Fig. 3, the waveforms of specified surge current values for the 3.3 kV XHP™ 2 CoolSiC™ MOSFET module (FF2000UXTR33T2M1, solid lines) are compared to the waveforms of the XHP™ 3 module (FF450R33T3E3, dashed lines) using the state-of-the-art Si IGBT3 technology with the same footprint. The currents are indicated by red lines while the corresponding voltages are marked in blue. The maximum surge current, $I_{FSM}$, for the XHP™ 3 is 3700 A, resulting in a $I^2t$ value of 68 kA²s for a pulse width of $t_p$ = 10 ms at T = 150°C. For the XHP™ 2 CoolSiC™ MOSFET module, the peak $I_{FSM}$ reaches 10000 A, resulting in a surge current limit, which is increased by more than a factor of 7 to $I^2t$ = 500 kA²s (T = 175°C, $V_{GS}$ = -5 V).

Additionally, the CoolSiC™ MOSFET XHP™ 2 modules do not require a derating of surge currents due to very low leakage currents.

The diode forward voltage shown in Fig. 3 indicate a fully bipolar operation up to the maximum current of both modules. The use of the internal body diode at $V_{GS}$ = -5 V results in a homogeneous current distribution inside the chip. A symmetric module layout avoids an overload of individual chips and enables a high module surge current capability.

The Ag sintered die attach, the Cu front side, and Cu bonds have higher melting points (T = ~1000°C) compared to a standard soldered die attach with Al wire bond technology. The physical limit of the 3.3 kV CoolSiC™ MOSFET is defined by a collapse of the gate oxide due to thermal stress induced by different thermal expansion coefficients of the chip's front side materials.

Fully bipolar operation is ensured at a negative gate voltage of $V_{GS}$ = -5 V. Surge current with open

865

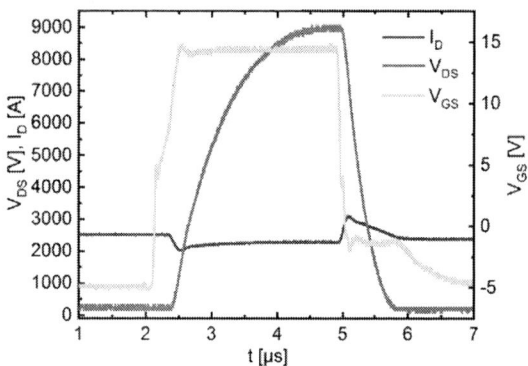

**Fig. 4:** Short circuit waveform of the XHP™ 2 CoolSiC™ MOSFET module at $V_{DS} = 2400$ V at $T = 175$ °C and $t_p = 3$ µs.

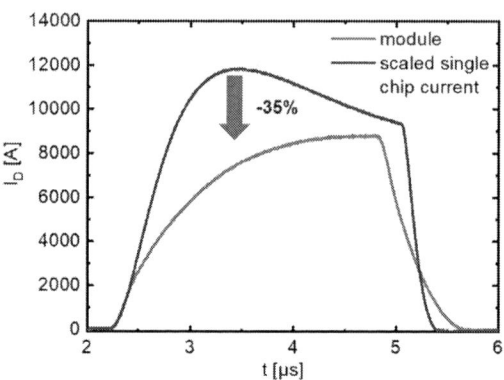

**Fig. 5:** Comparison of module and scaled single chip short circuit current

MOSFET channels ($V_{GS} = 15$ V) can cause unipolar and bipolar operations in parallel. This can lead to current crowding in single chips and thus, to a reduction in the surge current ruggedness. Therefore, surge currents beyond the Safe Operating Area (SOA) limits with open MOSFET channels in the reverse direction must be avoided.

## 3.2 Short Circuit

An important requirement for many customers of traction applications is a sufficient short circuit withstand time of the power module. In the case of an unforeseeable failure, the gate driver should have a certain time to react and turn off the short circuit current.

During the development of the 3.3 kV CoolSiC™ MOSFET chip, a cell design study was carried out to find a suitable rugged chip variant for short circuits. The trade-off between the saturation current and $R_{DS(on)}$ was investigated to tailor the chip to meet the requirements of high-voltage applications [4].

A typical short circuit waveform is shown in Fig. 4 at $V_{DS} = 2400$ V (blue line), $V_{GS} = -5/15$ V (green line), and T = 175°C. The short circuit is turned on and off under the gate resistor values given in the product datasheet – $R_{g,on} = 0.75$ Ω and $R_{g,off} = 1.2$ Ω. The channel of the passive switch opens at a higher gate voltage, while $V_{GS} = -5/15$ V is applied to the active switch. The 3.3 kV XHP™ 2 CoolSiC™ MOSFET module provides a short

circuit withstand time of $t_p = 3$ µs. The short circuit withstand time is determined by the ECPE guideline AQG 324 [10]. The AQG guideline defines the short circuit withstand time for power modules and defines $t_p$ between 10% of the rising and falling edges of the maximum short circuit current, $I_{D,Peak}$ (red line in Fig. 4).

After the short circuit turn-on, the current rises quickly to $I_D = 2500$ A. Then a bend can be observed in the $I_D$ transient followed by a decreasing di/dt phase. The maximum short circuit current appears at the turn-off. This behavior strongly differs from the typical short circuit event of single CoolSiC™ MOSFET chips, where the maximum short circuit current after turn-on is followed by a pattern of decreasing currents due to the heating of the MOSFETs. To limit the dissipated short circuit energy, it is important to reduce the rise in current after short circuit turn-on.

On the one hand, the maximum short circuit current can be limited by the saturation current of the chip, as discussed in [4], and on the other, the design of the power module itself can be used to suppress the current rise during a short circuit event.

A comparison between the short circuit current transients for an XHP™ 2 CoolSiC™ module (red line) and a single chip (multiplied by the number of paralleled chips, blue line) is shown in Fig. 5. The single chips are measured in a TO247 package with a stray inductance, and gate resistances scaled to the module. The splitting of both currents starts at $I_D = 2500$ A without influencing the SOA region ($2I_{D,nom} = 2000$ A) of the module. The scaled single-chip current transient reaches

Figure 6 compares short circuit transients for lower (low-side, LS) switch (blue) and upper (high-side, HS) switch (red) at $V_{DS} = 2400$ V and $T = 175°C$ ($R_{g,on} = 0.75$ $\Omega$ and $R_{g,off} = 1.2$ $\Omega$). As can be seen in the image, the current rise starts symmetrical for both switches and splits up after $I_D = $ ~1500 A, which is above the typical application currents. For dynamic switching up to nominal conditions, a high degree of symmetry between LS and HS switch is ensured [7]. Then the HS current rises steeper compared to the LS switch and reaches 15% of the higher maximum short circuit current before turn-off. Despite a higher short circuit current on the HS, $V_{DSmax}$ is still below the maximum breakdown voltage of 3.3 kV. However, the higher short circuit current for the HS correlates to a higher energy dissipation. Considering the full half-bridge module, the HS switch defines the short circuit ruggedness of the XHP™ 2 CoolSiC™ MOSFET module.

Both short circuit transients for LS and HS are very smooth for the recorded $V_{DS}$, $I_D$, and $V_{GS}$ signals, which can be helpful for external short circuit detection methods.

While the gate impedance can be adjusted to balance the LS and HS switching, the short circuit current rise can also be limited. Furthermore, implementing the .XT technology combined with a short circuit rugged chip design results in a short circuit withstand time of $t_p = 3$ µs, which is well suited for traction applications.

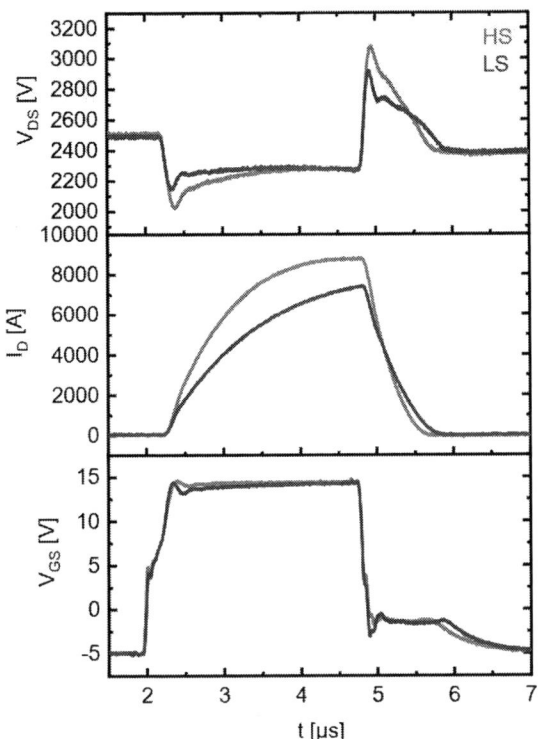

**Fig. 6:** Comparison of short circuit transients for LS and HS at $V_{DS} = 2400$ V and $T = 175°C$

$I_{D,Peak} = 12000$ A. Afterwards the scaled chip current, $I_{D,Chip}$, decreases to a comparable value of the module, $I_{D,module}$, at turn-off.

The XHP™ 2 CoolSiC™ module is designed with inherent negative feedback into the gate loop, which slows down the current rise during a short circuit event. Consequently, the maximum short circuit current reduces by 35% and lesser short circuit energy is dissipated in the same time frame compared to the scaled single-chip transient.

Typical failure signatures of SiC MOSFETs with Al metallization are melting signatures after short circuit tests. The implementation of the .XT technology with Cu front-side metallization and Cu bonds overcomes this limitation. Repetitive short circuit pulses do not show any degradation of the front-side metal. At the intrinsic destruction limit, however, mechanisms such as gate oxide breakdown or thermal runaway can lead to failures.

## 3.3 Power Cycling

For power cycling, it is typically assumed that the SiC MOSFET power modules exhibit a reduced performance compared to their Si IGBT counterparts. This has been attributed to the degradation of the soft-solder die attach, mainly caused by the greater rigidity of the SiC chips [11]. Recent findings, however, indicate that their performance can be significantly improved by switching to a more robust die attach, such as silver sintering. This, then, shifts the dominant failure mechanism from the die-attach layer to the bond wires, which would be lifted up from the chip's front side [12]. In this way, the power-cycling curves of the SiC modules can be pushed to the same limits as of typical Si IGBT modules with the same failure mechanism at the wire bonds [13].

Further improvements, thus, require a change in the wire bonding technology. Here, CoolSiC™ MOSFETs combined with the .XT technology can

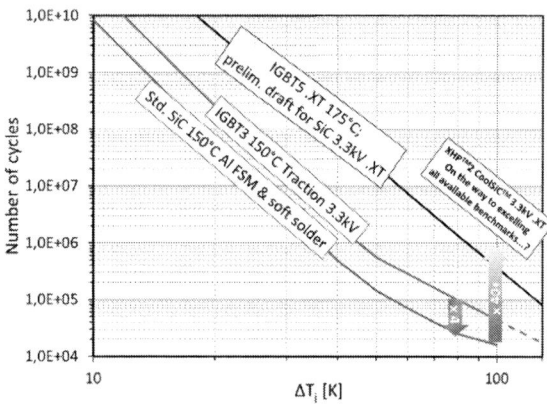

**Fig. 7:** Benchmark power cycling performance curves for XHP™ 2 CoolSiC™ 3.3 kV .XT. The diagram also contains indications from power-cycling results in first tests, pointing at a further improvement beyond the benchmark performance levels.

offer another advantage. The more robust Cu bond wire connection on the chip's Cu front-side metallization (FSM) is expected to further increase the power cycling performance of SiC modules, as already demonstrated for Si IGBT power modules [14, 15].

We tested the XHP™ 2 CoolSiC™ module at $T_{vjmax} = 175°C$ and $\Delta T = 100$ K. The results indicate that this combination can overcome even the benchmark performance of Si IGBT power modules with .XT joining technology [14, 15]. Compared to standard Si IGBT traction modules, this constitutes an improvement by an order of magnitude, and a quantum leap when compared to the reference performances of standard, soft-soldered SiC MOSFET power modules (see Fig. 7).

The improvement over Si IGBT power modules with .XT joining technology can, among other subtle influences of the substrate layout, be attributed mainly to the improved match between the coefficients of thermal expansion (CTE) of the different materials that comprise the XHP™ 2 CoolSiC™ module (see Fig. 1). With approximately 4.5 ppm/K, the AlN ceramic substrates provide an almost perfect match to the SiC chip and a good link to the AlSiC base plate (ca. 7.5 ppm/K) [16], so that all differences along the stack remain well below a factor of 2. Such good matches restrict the relative movement in the interconnecting layers and, thus, reduce the stress on the interconnections. For a Si IGBT power module with $Al_2O_3$ ceramic substrate and a Cu base plate (CTE > 17 ppm/K), the spread of CTE along the stack is much larger, causing stronger relative movements during temperature swings.

In summary, this combination of Infineon's CoolSiC™ and .XT interconnection technology has the capability to successfully overcome the power cycling limitations of the SiC MOSFET power modules. The chosen material combination in the XHP™ 2 CoolSiC™ 3.3 kV module even has the potential to excel all benchmarks, including that of the Si IGBT .XT power modules currently available.

# 4 Summary

This paper describes the key benefits of combining 3.3 kV CoolSiC™ MOSFET chips with Infineon's XT joining technology in the low-inductive XHP™ 2 package. The .XT technology pushes the power cycling limit to a new level for CoolSiC™ MOSFETs – a crucial door-opener for long-running and demanding applications such as traction. The elimination of the Schottky diode combined with the usage of .XT technology provides the basis for a high level of surge current ruggedness. The significant rise in surge current robustness and the ample short circuit withstand capability are additional features that can make Infineon's XHP™ 2 CoolSiC™ MOSFET 3.3 kV .XT module the new benchmark in traction.

# 5 Acknowledgments

This work was supported by the Federal Ministry for Economic Affairs and Climate Action on the basis of a decision by the German Bundestag.

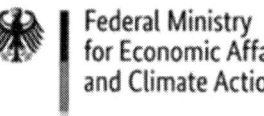

Federal Ministry for Economic Affairs and Climate Action

# 6 References

[1] B. Laska, J. Weigel, S. Buchholz, W. Brekel, M. Wissen, T. Gutt, R. Schrader, P. Münster, T-M. Plötz, I. Kirchner, and H-G. Eckel, "New Traction Converter with low Inductive High-Voltage Half Bridge IGBT Module", PCIM Europe 2018, pp. 1139-1145.

[2] V. Jadhav, S. Mansueto, M. Bürger, U. Schwarzer, D. Car, H.P. Felsl, T. Soellradl, and T. Kurzmann, "Latest IGBT4 chip technology enables the first 2000 A 3300 V module in IHV package", PCIM Europe 2020, pp. 237–242.

[3] D. Peters, T. Basler, B. Zippelius, T. Aichinger, W. Bergner, R. Esteve, D. Kueck, and R. Siemieniec, "The new CoolSiC™ Trench MOSFET Technology for Low Gate Oxide Stress and High Performance", PCIM Europe, 2017, pp. 168–174.

[4] C. Leendertz, M. Hell, G. Zeng, T. Ganner, T. Söllradl, P. Sochor, R. Elpelt, K. Schraml, and D. Peters, "CoolSiC™ Trench MOSFET Chip Design for the 3.3 kV Class", PCIM Europe, 2023, pp. 589–594.

[5] W. Brekel, W. Rusche, A. Höhn and W. Bücker, "XHP 2 – The Low Inductive, Multi-Package Housing for the Next Generation of High-Power Applications", PCIM Europe, 2020, pp. 222–228.

[6] V. Jadhav, U. Schwarzer, S. Buchholz, and W. Brekel, "Novel 550A/3300V module with IGBT4.XT Technology in XHP™ 3 package to enhance power density and lifetime for next generation power converters", PCIM Europe, 2019, pp. 539–543.

[7] M. Buerger, K.-H. Hoppe, K. Schraml, and A. Wedi, "The New XHP™ 2 Module using 3.3kV CoolSiC™ MOSFET and .XT Technology", PCIM Europe, 2023, pp. 832–838.

[8] W. Rusche and N. Heuck, "Lifetime Analysis of PrimePACK™ Modules with IGBT5 and .XT, Bodo´s Power Systems", July 2016, pp. 18–21.

[9] J. Vishal, J. Czichon, and M. Bürger, "Efficient and Optimized Traction Converter Systems Enabled by the New 3.3 kV CoolSiC™ MOSFET and .XT in an XHP™ 2 Package", PCIM 2023, pp. 859–864.

[10] "ECPE Guideline AQG 324: Qualification of Power Modules for Use in Power Electronics.

[11] P. Salmen, T. Methfessel, C. Kuenzel, and O. Schilling, "Impact of load-pulse duration on power-cycling capability of SiC devices", CIPS 2020, pp. 1–4.

[12] M. Hanf, F. Hoffmann, J.-H. Peters, S. Clausner, and N. Kaminski, "Power Cycling Performance of 3.3 kV SiC-MOSFETs and the Impact of the Thermo-Mechanical Stress on Humidity Induced Degradation", ICSCRM 2023.

[13] E. Mengotti, I. Kovacevic-Badstuebner, E. Bianda, D. Baumann, C. Kenel, P. Natzke, S. Race, and U. Grossner, "Power cycling of SiC MOSFETs packaged in different module's solutions", ICSCRM 2023.

[14] T. Methfessel and H. Jähme, "End-of-life mechanism due to cyclic thermomechanical loading of power modules with .XT joining technology", CIPS 2020, pp. 185–89.

[15] T. Methfessel, F. Sauerland, K. Mainka, and O. Schilling, "Enhanced lifetime and power-cycling modelling for PrimePACK™ .XT power modules", PCIM 2020, pp. 583–590.

[16] M. A. Occhionero, R. A. Hay, R. W. Adams, and K. P. Fennessy, "Aluminum Silicon Carbide (AlSiC) for Thermal Management Solutions and Functional Packaging Designs", Proceedings of SPIE - The International Society for Optical Engineering 1998.

PCIM Europe 2024, 11– 13 June 2024, Nuremberg    DOI: 10.30420/566262107

# Large-area Bonding with LMEE: Suppression of the Degradation of the Junction-to-Water Thermal Resistance in Power Modules

Yo Mochizuki[1], Takukazu Otsuka[1], Navid Kazem[2], Dylan Shah[2] and Ken Nakahara[1]

[1]ROHM Co., Ltd., Japan
[2]Arieca Inc., USA

Corresponding author:    Yo Mochizuki, yo.mochizuki@rohm.co.jp
Speaker:                 Yo Mochizuki, yo.mochizuki@rohm.co.jp

## Abstract

This paper investigates the junction-to-water thermal resistance ($R_{th}$) durability of power modules (PMs) during the thermal shock test (TST), in which the PMs were subjected to the thermal cycle from −40°C to 125°C with dwell time 20 min for up to 1000 cycles. The PMs contained silicon carbide power devices, and were bonded to a water cooler by using liquid metal embedded elastomers (LMEE). The bonding was performed at 150°C for 30 minutes under a pressure of 0.1 MPa. The $R_{th}$ showed no significant degradation during the TST. This indicates that LMEE is a good candidate as a bonding material for PM-and-cooler assembly for enhancing its durability.

## 1    Introduction

Power modules (PMs) with high power density are highly anticipated for use in power conversion systems, such as those installed in electric vehicles (EVs). Power loss in the conversion inevitably generates heat and then the thermal resistance ($R_{th}$) plays a key role in maintaining system operations. The heat generated in such PMs is typically dissipated by a water cooler bonded to the PMs. Thermal interface material (TIM) is widely used for this bond, and thermal grease has typified such material [1]. However, the thermal conductivity ($k$) of commercial thermal grease is typically about 4 W/mK at maximum, which worsens the total $R_{th}$ of PMs. In addition, a problem has been reported that grease is easily pumped out during the operation of PMs [2]. Thus, sintered silver (s-Ag) bonding has been actively studied due to its high $k$ and high melting point [3,4]. However, the maximum bonding area is limited to approximately 10×10 mm because many cracks are easily induced by thermal stress [4].

The authors propose liquid metal embedded elastomers (LMEE) as a candidate for TIM to solve the above problems. LMEE consists of a silicone elastomer embedded with liquid phase eutectic gallium indium. LMEE typically exhibits a fracture strain of 300%, $k$ = 10 W/mK, and a Young's modulus of 0.4 MPa. In particular, this higher $k$ than that of grease is likely to be attractive for achieving low $R_{th}$ [5].

In this study, the $R_{th}$ of a LMEE-bonded PM and water cooler assembly was observed during a thermal shock test (TST). The next section describes the experimental setup in detail. In the last section, we show the effectiveness of LMEE on the $R_{th}$ durability.

## 2    Experimental methodology

Fig. 1 shows a cross-sectional sketch of the structure used in this study. "SiC" denotes silicon carbide, and "AMB" denotes a silicon nitride ceramic plate sandwiched with copper plates. SiC chips were bonded onto the AMB at 300 °C by using s-Ag (average diameter: 18 nm) for 10 minutes under a pressure of 30 MPa (CYPM-200, manufactured by SHINTOKOGIO, LTD, Aichi, Japan). Lead flames were bonded onto the top surface of the AMB by soldering with Sn-Ag3%-Cu0.5% system at 230 °C in the formic acid reduction atmosphere. Then, wedge Al wire bonding was performed between the flame and the SiC chips surfaces in an air atmosphere. These bonded bodies were molded with epoxy resin, and finally the molded bodies, each representing a PM in actual use, were bonded to water coolers with s-Ag and LMEE. For s-Ag

870

PCIM Europe 2024, 11– 13 June 2024, Nuremberg          DOI: 10.30420/566262107

bonding, 210°C for 10 minutes under a pressure of 20 MPa was used, while for LMEE bonding, 150°C for 30 minutes under a pressure of 0.1 MPa was used. The dimensions of the bodies are listed in Table 1.

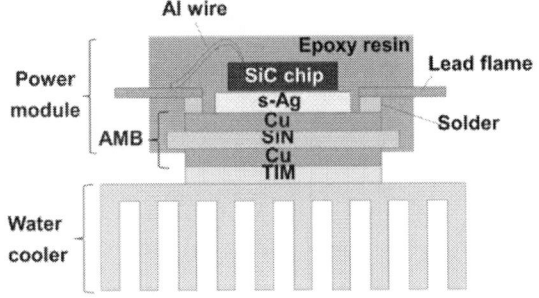

**Fig. 1**   Cross-sectional schematic view of an evaluated assemble.

|              | Width | Length | Thickness            |
|--------------|-------|--------|----------------------|
| SiC chip     | 4.8   | 4.8    | 0.35                 |
| s-Ag         | 7     | 7      | 0.05                 |
| AMB          | 30    | 42     | 1.12                 |
| TIM          | 30    | 42     | 0.05(LMEE) 0.15(s-Ag)|
| Water cooler | 142   | 59     | 13                   |

**Table 1**   The dimensions of each part （Unit: mm）

Figs. 2 show the cross-sectional images of the material interfaces in an assembled structure. Figs. 2 (a) and (b) are the scanning electron micrography (SEM) images of a s-Ag layer, while Figs (c) and (d) are optical microscope images. SEM could not be used to charge up the LMEE. The s-Ag layer shows a porosity of $\rho$ = 16.5%, determined from the binarized image of Fig. 2 (b).

The assemblies were placed in a commercial chamber (TSA41-EL-A, manufactured by Espec, Osaka, Japan), and the TST cycling was from −40°C to 125°C maintaining 30 minutes at each temperature for up to 1000 cycles. $R_{th}$ was evaluated at TST 100-cycle intervals by transition thermal resistance measurement (A07M355, manufactured by Coper Electronics Co., Ltd., Kanagawa, Japan).

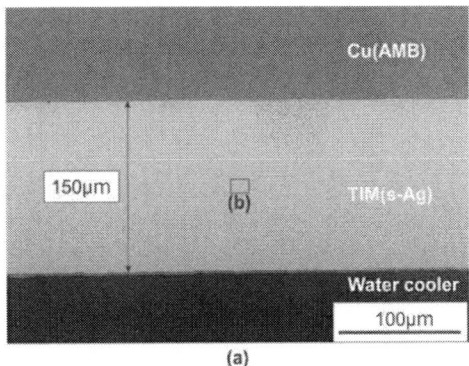

(a)

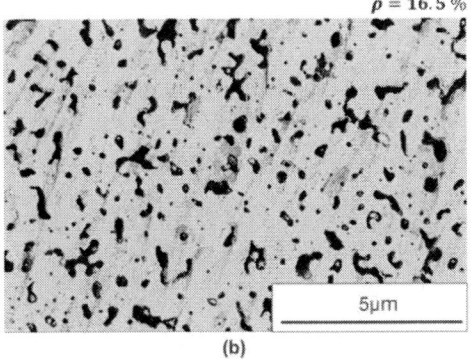

(b)

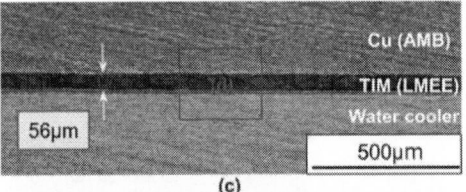

(c)

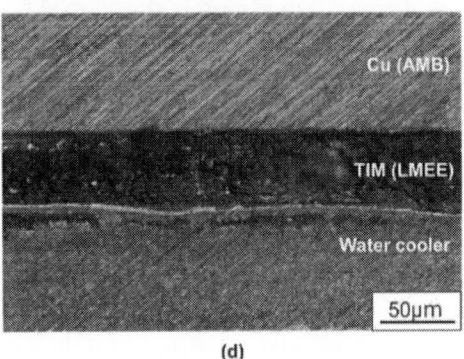

(d)

**Fig. 2**   Typical cross-sectional images of an assembled structure. (a): Overall picture and (b): microstructure of s-Ag observed with SEM. (c):Overall picture and (d): magnified picture of LMEE with an optical microscope.

# 3    Results & Discussion

Fig. 3 (a) shows the $R_{th}$ measurement results at every 100 cycles of the TST. The vertical axis represents the increase rate of $R_{th}$, $r_{th}$, which is defined by the measured increase of $R_{th}$ divided by the initial $R_{th}$. The initial $R_{th}$ value of s-Ag and LMEE was almost identical, suggesting that LMEE has almost the same heat dissipation capacity as s-Ag, but $r_{th}$ completely differed between these two materials. For s-Ag, as shown in Fig. 3 (a), $r_{th}$ jumps up by about 10% just after 100 cycles of TST, and reaches 35% after 300 cycles of TST. This durability is not satisfied at all with the requirement on the $R_{th}$ failure requirement in the EV market, i.e. 20% [6]. In stark contrast, however, the $r_{th}$ for LMEE is approximately 4.1% even after 1000 cycles of TST. This means that LMEE has the excellent capacity as TIM to meet the requirement on the $R_{th}$ failure requirement in the EV market.

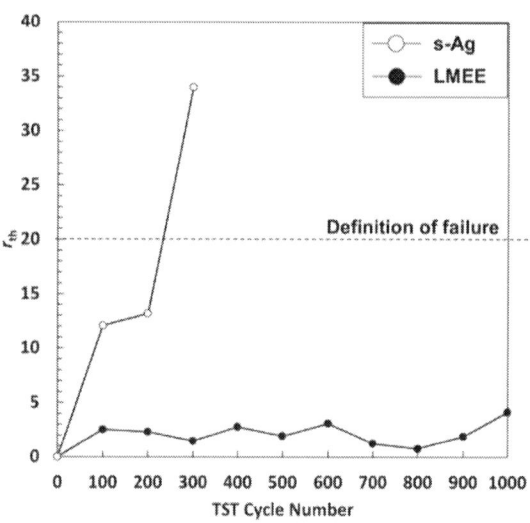

**Fig. 3 (a)** $R_{th}$ measurement results during the TST.

Fig. 3 (b) shows the scanning acoustic tomography (SAT) image of s-Ag detected at bonded area after 300 cycles of TST (IS-350, manufactured by Insight k.k. Co., Ltd., Tokyo, Japan). The SAT image was obtained using an ultrasonic probe with 25 MHz. In the SAT image, a black-colored area shows bonded, while white and pale gray color means delamination. Therefore, the delamination was considered to increase $r_{th}$ of s-Ag shown in Fig. 3 (a).

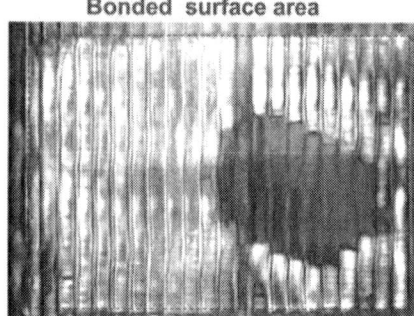

**Fig. 3 (b)** SAT image of s-Ag at bonded area after 300 cycles of TST. Red dash-square is bonded surface area.

Finite element analysis (FEA) was conducted to understand the out-of-plane deformation distribution, using software, FEMTET, in which simulation a cyclic temperature between -40 °C and 125 °C was set to the ambient atmosphere as a driving force of induced stress. The material characteristic values used for the FEA are listed in Table 2. In LMEE, elastic behavior was considered [5], In s-Ag the elastic-plastic behavior, as shown Fig. 4 [7].

| Material | Young's Modulus (GPa) | Poison's Ratio | CTE ($\times 10^{-6}$) |
|---|---|---|---|
| Epoxy | 15 | 0.3 | 12 |
| SiC | 412 | 0.17 | 3.0 |
| Cu | 118 | 0.35 | 16.8 |
| $Si_3N_4$ | 300 | 0.3 | 2.7 |
| s-Ag |  | 0.35 | 19.5 |
| LMEE | 0.4 | 0.5 | 33 |
| Al | 68.5 | 0.34 | 23.1 |

**Table 2** Material mechanical properties used in FEA. Young's Modulus of the s-Ag shown in Fig. 4.

PCIM Europe 2024, 11– 13 June 2024, Nuremberg          DOI: 10.30420/566262107

**High temperature (150 °C)**

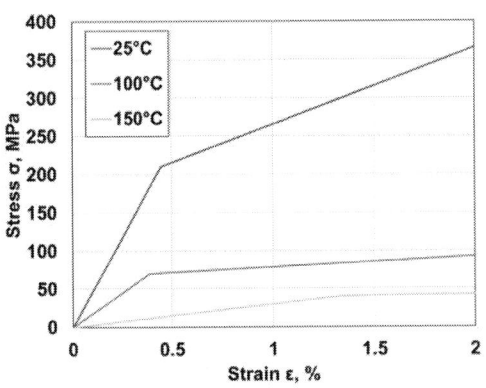

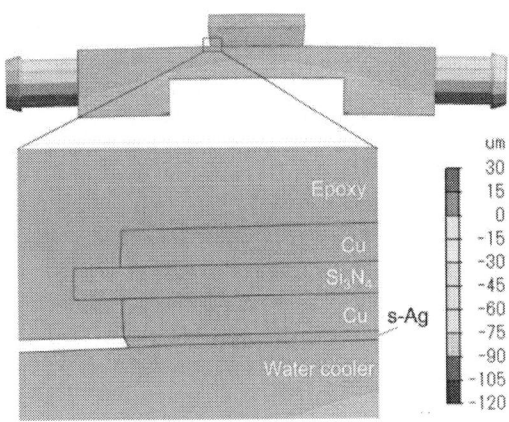

**Fig. 4** *S-S* curves for each temperature with s-Ag [7] in FEA.

**Low temperature (-40 °C)**

Figs. 5 depicts the FEA results for high (150 °C) and low (-40 °C) temperatures. The displacement diagram in FEA is the vertical axis component, shown as figures multiplied by 30. The s-Ag deforms more than 100 µm with temperature change, as shown Fig. 5 (b). In contrast, The LMEE only deforms less than 10 µm, as shown Fig 5 (c). Therefore, the s-Ag is considered to have degradation because of its greater displacement than LMEE.

**FEA model**

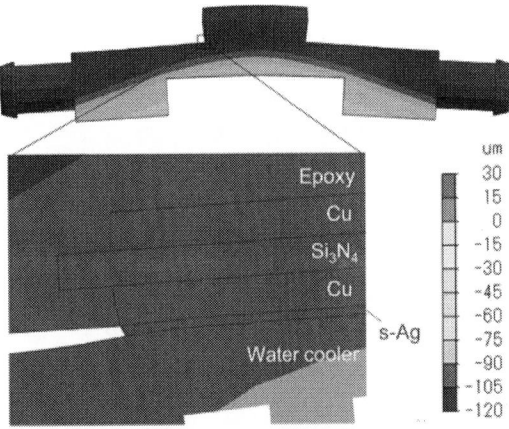

**Fig. 5 (b)** The out-of-plane deformation distribution with s-Ag in FEA.

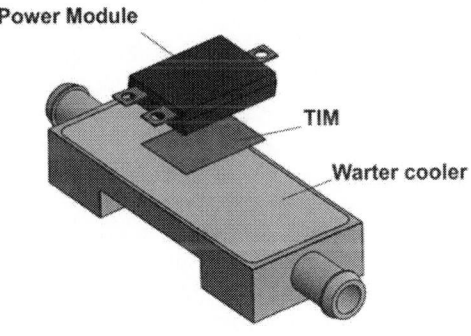

**Fig. 5 (a)** simulation model in FEA.

873

**High temperature (150 °C)**

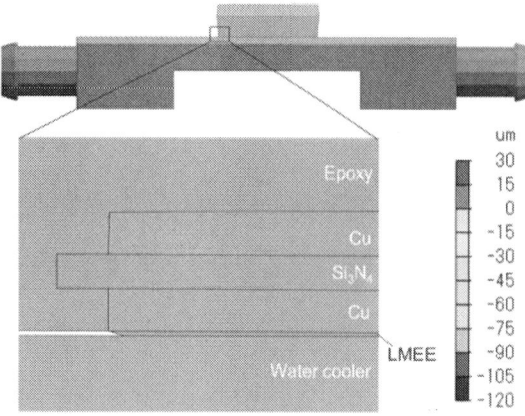

**Low temperature (-40 °C)**

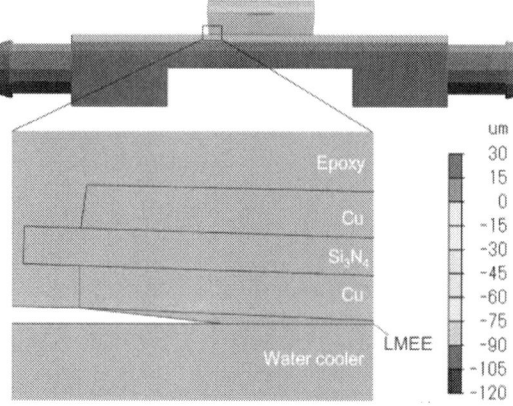

**Fig. 5 (c)** The out-of-plane deformation distribution with LMEE in FEA.

As for our future work, TST is not the only reliability test in the EV market, and accordingly we will conduct vibration tests, high-temperature storage tests, and humidity tests to verify LMEE's reliability potential.

# References

[1] N. S. Goel et al., "Technical review of characterization methods for thermal interface Materials (TIM)," Intersociety Conference on Thermal and Thermomechanical Phenomena in Electronic Systems, May 2008.

[2] F. Sarvar, D. C. Whalley, and P. P. Conway, "Thermal Interface Materials - A Review of the State of the Art," Electronic Systemintegration Technology Conference, Sep, 2006, doi: 10.1109/ESTC.2006.280178.

[3] K. Wakamoto, T. Otsuka, K. Nakahara, V. Mugilgeethan, R. Matsumoto, and T. Namazu, "Degradation Mechanism of Silver Sintering Die Attach Based on Thermal and Mechanical Reliability Testing," IEEE Transactions on Components, Packaging and Manufacturing Technology, vol. 13, pp.197-210, Feb, 2023, doi: 10.1109/TCPMT.2023.3242423.

[4] P. Peng, A. Hu, A. P. Gerlich, G. Zou, L. Liu, and Y. N. Zhou, "Joining of Silver Nanomaterials at Low Temperatures: Processes, Properties, and Applications," ACS Appl Mater Interfaces, vol. 7, no. 23, pp. 1259712618, Jun. 2015

[5] W. Li, S. Liu, C. Lou, M. Sang, X. Gong, K. Leung, and S. Xuan "Magneto-induced self-stratifying liquid metal-elastomer composites with high thermal conductivity for soft actuator" Li et al., Cell Reports Physical Science, vol. 4, no. 1, p. 101209, Jan. 2023, doi: https://doi.org/10.1016/j.xcrp.2022.101209.

[6] "Qualification of Power Modules for Use in Power Electronics Converter Units in Motor Vehicles" EPCE Guideline AQG 324 (2021), Chap. 8.2, p. 37.

[7] Keisuke Wakamoto, Yo Mochizuki, Takukazu Otsuka, Ken Nakahara and Takahiro Namazu "Tensile mechanical properties of sintered porous silver films and their dependence on porosity," Japanese journal of Applied Physics, vol. 58, no. SD, pp. SDDL08-1-5,2019,doi: 10.7567/1347-4065/ab0491.

PCIM Europe 2024, 11– 13 June 2024, Nuremberg          DOI: 10.30420/566262108

# Active Thermal Control of SiC MOSFETs Utilizing Transient Thermal Characterization

Varaha Satya Bharath Kurukuru [1], Md. Nazmul Hassan [1], Roberto Petrella [1,2]

[1] Research Division Power Electronics, Silicon Austria Labs GmbH,
Austria
[2] PEMD Laboratory, Polytechnic Department of Engineering and Architecture,
Italy

Corresponding author: Varaha Satya Bharath Kurukuru, varaha.kurukuru@silicon-austria.com
Speaker: Varaha Satya Bharath Kurukuru, varaha.kurukuru@silicon-austria.com

## Abstract

Effectively managing the temperature of Silicon Carbide (SiC) MOSFETs is essential for optimizing performance, especially in applications demanding high reliability and energy efficiency. Traditional thermal management techniques often struggle with the rapid thermal transients characteristic of these semiconductors, leading to potential overheating and reduced operational lifespan. Addressing this challenge, this research introduces an advanced active thermal control (ATC) strategy for a 3-phase dual converter system, employing field-oriented control (FOC) with discontinuous pulse-width modulation (DPWM) to drive permanent magnet synchronous machines (PMSM). By integrating transient thermal characterization into a support vector data descriptor (SVDD), this research achieves accurate estimation of virtual junction temperatures $(T_{vj})$, enabling more responsive thermal management. Simulations demonstrate the ATC system's effectiveness, revealing a significant reduction in power loss. The proposed approach not only mitigates the risk of overheating but also promotes improved efficiency and longevity in SiC MOSFET-based power systems.

## 1    Introduction

In the dynamic landscape of power electronics, active thermal control (ATC) has become increasingly critical, especially when it comes to optimizing the performance of silicon carbide (SiC) MOSFETs [1]. These semiconductors are integral to a wide array of applications, with a growing emphasis on energy efficiency and reliability. As the demand for high-performance systems continues to increase, effective thermal management becomes a mandatory specification. Current trends in ATC have shown promising improvements, with various methods employed to monitor and control temperature in SiC MOSFETs [2]. However, these trends faced significant challenges due to inaccurate temperature estimation. Besides, many current approaches lack the capability to address the rapid thermal dynamics of SiC MOSFETs, resulting in a lack of real-time insights into temperature fluctuations [3].
Traditional thermal management strategies, such as passive cooling (heat sinks) and active cooling (forced air or liquid), often fail to keep pace with

the quick transient changes in temperature associated with high-power applications involving SiC MOSFETs [4]. Moreover, methods like direct temperature measurement using thermocouples or infrared cameras can provide accurate data but are intrusive and may not offer the necessary response speed or integration simplicity for real-time control [5]. On the other hand, indirect temperature estimation methods, such as those based on electrical parameters (on-state resistance measurement), although non-invasive, are hindered by significant estimation errors under dynamic operating conditions, where power and environmental fluctuations are common [6], [7].
This is where transient thermal characterization-based ATC steps in. It offers a dynamic approach to temperature monitoring by considering the fast-changing thermal dynamics of SiC MOSFETs. The transient thermal characterization provides a more accurate and real-time understanding of temperature fluctuations, enabling proactive thermal management [8], [9]. Such a solution holds the potential to revolutionize power electronics by mitigating the

issues associated with overheating, ultimately enhancing efficiency and extending the operational lifespan of critical systems. Furthermore, integrating machine learning (ML) into transient thermal characterization enables the effective identification of abnormal temperature spikes, helping to predict potential overheating incidents before they occur. Considering these advancements, the proposed approach overcomes the constraints of current thermal management methods by harnessing the fast, non-intrusive estimation features of the Support Vector Data Descriptor (SVDD) [10] model, combined with transient thermal characterization. This integration offers several advantages:

- Unlike traditional methods that may lag in dynamic environments, the proposed approach utilizes real-time data processing capabilities of SVDD, providing faster and more accurate temperature estimations.

- By using electrical characteristics already available in the system for SVDD training, the approach avoids the need for additional sensors, reducing system complexity and potential points of failure.

- The system not only monitors but anticipates thermal issues before they reach critical thresholds, enabling preemptive actions that enhance system reliability and prevent downtime.

In view of these developments, this paper presents a method to attain ATC for SiC MOSFETs employing transient thermal characterization alongside SVDD. The methodological details are explored, and a comprehensive explanation of how SVDD aids in improving the efficiency of virtual junction temperature ($T_{vj}$) estimation is provided. The major contributions of this article are as follows:

- Deriving a temperature-dependent RC network for the transient thermal characterization of SiC MOSFETs.

- Using SVDD to accurately estimate the $T_{vj}$ from the power loss information of each switch and the temperature-dependent RC network.

- Introducing a more robust ATC system designed to act proactively based on accurately estimated $T_{vj}$.

The subsequent sections of the paper are organized as follows: Section I introduces the background and motivation for this research, followed by a detailed methodology in Section II that explains the converter model, and the transient thermal characterization approach. Section III discusses the development of SVDD for estimating $T_{vj}$, and the modified control approach based on the estimated temperature using SVDD for ATC. The results and analysis, along with the discussion on findings, are provided in Section IV, and the research is concluded in Section V.

## 2 Model Specifications and Thermal Characterization

### 2.1 Converter Topology

The converter topology utilized in this research employs a 3-phase dual inverter configuration as shown in Fig. 1, designed to handle high power outputs while optimizing the converter's physical layout. This configuration is structured with two interconnected half-bridges per phase, effectively doubling the current carrying capacity per phase and thereby minimizing voltage drops and power dissipation across the switches. The detailed technical specifications of the converter are shown in Tab. 1.

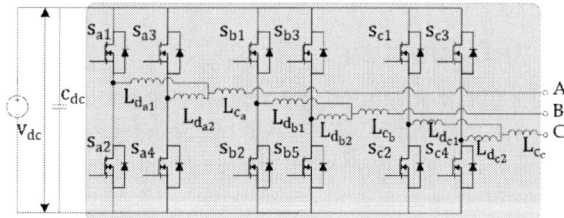

**Fig. 1:** Schematic of a 3-Phase Dual Inverter Configuration.

| Parameter | Value |
|---|---|
| DC Voltage Range | 40-900 V |
| DC Link Capacitor Value | C = 370 µF per Stack |
| Switching Frequency Range | 2-100 kHz |
| Differential Mode Inductances | $L_d$ = 300 µH per phase |
| Common Mode Inductances | $L_c$ = 500 nH per phase |
| Nominal Current (RMS) | 315 A per half-bridge |
| Nominal Power Ratings | 440 kVA per Stack |

**Tab. 1:** Technical Specifications for the 3-Phase Dual Inverter Configuration

The converter drives a Permanent Magnet Synchronous Motor (PMSM) using Field-Oriented Control (FOC) . It operates over a DC Voltage Range of $40-900 V_{DC}$. The DC Link Capacitor is rated at $370 \mu F$ per stack, crucial for smoothing voltage fluctuations on the DC link, particularly important for dynamic load changes. The Switching Frequency Range of the converter spans from $2-100 kHz$. This wide range allows for strategic adjustments to optimize the trade-off between electromagnetic interference and switching losses. Utilizing higher frequencies reduces the size of the required passive components, while lower frequencies enhance operational efficiency by minimizing losses. The converter is equipped with differential and common mode inductances of $300 \mu H$ and less than $500 nH$, respectively. These values are optimized to minimize energy loss during the filtering of PWM signals and to support rapid changes in current without significant voltage drops. Additionally, the Nominal Current and Power Ratings are set at 315 A per half-bridge and 440 kVA per stack, respectively, highlighting the converter's capability to manage high power outputs.

## 2.2 Transient Thermal Characterization Approach

A 1200 V, 480 A Silicon Carbide (SiC) module used in the converter discussed above is subjected to transient thermal characterization to model and predict thermal behaviors under operational stress. SiC MOSFETs are chosen for their high thermal conductivity and elevated breakdown voltages, which are crucial for handling high power densities and improving reliability in high-power applications. However, managing the heat in such devices is challenging due to their potential for rapid temperature increases when subjected to high current densities. Accurate thermal management is essential to prevent thermal runaway and ensure device longevity.

### 2.2.1 Heat Dissipation Path and Thermal Crosstalk:

The analysis involves a detailed investigation of heat dissipation paths within the SiC MOSFET module as given in Fig. 2. Recognizing these paths allows for pinpointing critical zones where thermal crosstalk—heat transfer from one device affecting the temperature of an adjacent device—might impair device performance or lead to failure. The model simulates different operational scenarios as discussed in [11] to accurately forecast how heat is distributed across the module, enabling targeted

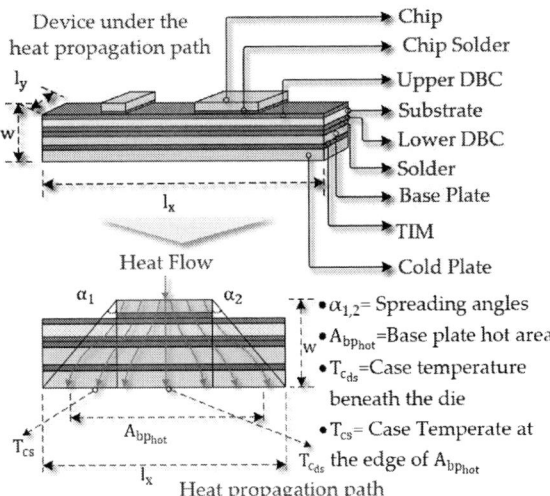

**Fig. 2:** SiC MOSFET Layer Configuration and Heat Propagation Analysis.

cooling interventions and design optimizations to mitigate adverse thermal effects.

### 2.2.2 Dynamic Cauer Network Model:

A dynamic Cauer network model forms an integral part of this characterization [11], portraying the module's thermal resistances and capacitances as interconnected RC networks that respond to temperature changes. This model is adept at representing complex thermal dynamics within power electronic modules due to its ability to independently analyze each thermal node. Such detailed and responsive modeling is critical for crafting effective cooling strategies and reliably predicting the operational lifespan of the components under varying conditions. The adaptability of the Cauer model also aids in refining thermal management systems to preemptively address potential overheating issues before they compromise device functionality.

## 3 Advanced Thermal Management Strategy

### 3.1 Thermal Observer

Data from transient thermal characterization is employed to inform the training of an SVDD model, which is designed for precise temperature estimation within the power module. The training process constructs a hypersphere characterized by radius $R$ and center $a$, which encompasses all relevant data points. The objective is to reduce the hypersphere's volume by minimizing $R^2$ while an error function $F(R, a)$ is introduced. This function includes a penalty term involving slack vari-

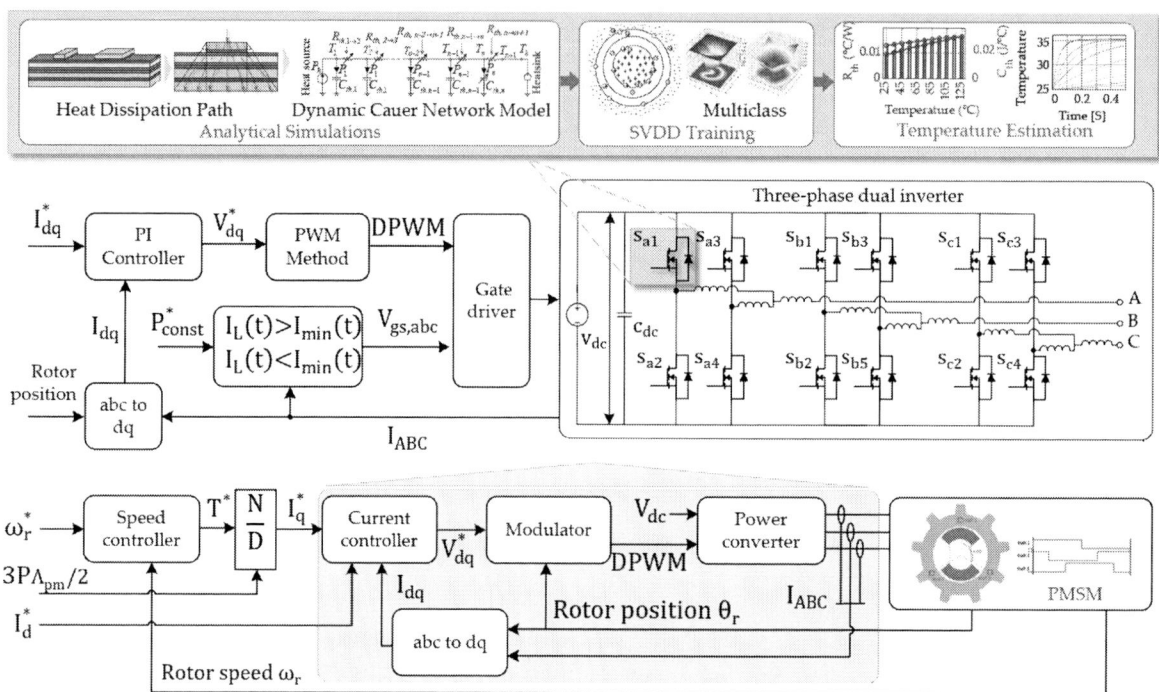

**Fig. 3:** Integrated Control and Thermal Management System for SiC MOSFETs in PMSM Drive Applications.

ables $\xi_i$ to effectively handle data points that fall outside the hypersphere, thereby maintaining data integrity and managing anomalies. This method parallels the optimization problem seen in Support Vector Machines (SVM), wherein the goal is to minimize error subject to a user-specified parameter that dictates the extent of data isolation. Kernelization further refines the approach, providing system flexibility by substituting inner products with kernel functions. Consequently, the optimization problem mirrors that of SVMs, showcasing the method's adaptability. The technique is detailed further in [12]. The SVDD model is trained using labeled temperature data indicative of normal and excessive thermal conditions. Once the model encapsulates the "normal" operating data within the minimum hypersphere, maximizing the boundary for anomaly detection, it validates itself by estimating junction temperatures against power loss data. When estimated temperatures exceed a predetermined safety threshold, the model signals the potential for overheating, prompting preemptive measures. This training and validation process, is iteratively refined to accommodate changing operational scenarios, proving invaluable for real-time thermal regulation and MOSFET protection. Integrated within the thermal control system, the model

delivers MOSFET temperature estimates in a feedback loop. Consequently, the control system is able to exert subtle thermal adjustments, ensuring efficient and non-intrusive temperature management. Fig. 3 outlines the complete control and thermal management workflow for a SiC MOSFET-based 3-phase dual inverter driving a PMSM. It encapsulates the process from thermal characterization to active thermal control, illustrating the dynamic Cauer network model used for analytical simulations of the heat dissipation path, the multiclass SVDD training for temperature estimation, and the subsequent implementation of active loss manipulation method which will be discussed in further sections. The architecture highlights how estimated junction temperatures influence the inverter control strategies, demonstrating the system's ability to maintain optimal performance and thermal conditions in real-time operational scenarios.

## 3.2 Methods for active loss manipulation

The method for active loss manipulation aiming to achieve ATC utilizes the thermal observer discussed above. The control scheme is developed for the 3-phase dual inverter system, driving a PMSM through FOC. Typically, space-vector modulation is employed, which is replaced with discontinuous pulse-width modulation (DPWM) for active loss manipulation [13], [14]. DPWM confines the phase leg to either the positive or negative terminals of the DC bus during specific intervals by introducing a zero-sequence component $u_z^*$ to the original references $u_{jo}^*$. The generation of the zero-sequence component is based on the output voltage references.

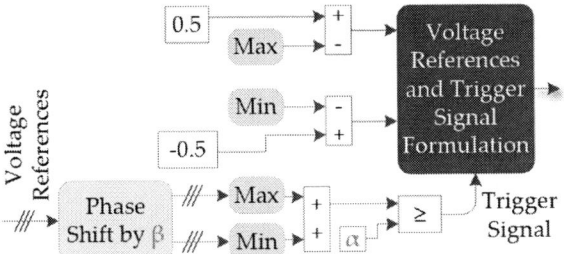

**Fig. 4:** Zero-Sequence Component Generation for Discontinuous Pulse Width Modulation.

For a three-phase converter, the normalized output voltage references can be expressed as follows:

$$
\begin{aligned}
u_{a0}^* &= \frac{1}{2}m\sin(\omega_0 t), \\
u_{b0}^* &= \frac{1}{2}m\sin(\omega_0 t - 120°), \\
u_{c0}^* &= \frac{1}{2}m\sin(\omega_0 t + 120°),
\end{aligned}
\tag{1}
$$

where $m$ represents the modulation index and $\omega_0$ denotes the angular frequency of the AC voltage. The maximum ($u_{max}$) and minimum ($u_{min}$) values of the normalized three-phase output voltage references are defined as:

$$
\begin{aligned}
u_{max} &= \max(u_{a0}^*, u_{b0}^*, u_{c0}^*), \\
u_{min} &= \min(u_{a0}^*, u_{b0}^*, u_{c0}^*).
\end{aligned}
\tag{2}
$$

In discontinuous modulation, the maximum clamping interval within one fundamental period remains constant at 120 deg for one phase, irrespective of whether it is clamped to the positive or negative terminals of the DC bus. The DPWM introduces a new variable, $\alpha$, to determine the positive ($\theta_p$) and negative ($\theta_n$) intervals, representing the clamping interval over one fundamental period:

$$
u_{z1}^* = \begin{cases} \frac{1}{2} - u_{max}, & \text{if } u_{max} + u_{min} \geq \alpha, \\ -\frac{1}{2} - u_{min}, & \text{if } u_{max} + u_{min} < \alpha. \end{cases}
\tag{3}
$$

$\alpha$ is restricted within the range $[-0.25m, 0.25m]$. A phase angle control variable, $\beta$, is introduced to generate a trigger signal for the zero-sequence component. The modification of normalized three-phase output voltage references to obtain the trigger signal is as follows:

$$
\begin{aligned}
u_{a0t}^* &= \frac{1}{2}m\sin(\omega_0 t - \beta), \\
u_{b0t}^* &= \frac{1}{2}m\sin(\omega_0 t - 120° - \beta), \\
u_{c0t}^* &= \frac{1}{2}m\sin(\omega_0 t + 120° - \beta),
\end{aligned}
\tag{4}
$$

where $u_{maxt}$ and $u_{mint}$ values are determined by the modified normalized three-phase output voltage references.

The schematic in Fig. 4 offers an overview of the zero-sequence component generator within the DPWM. Furthermore, the selection of $\alpha$ and $\beta$ is dynamically adjusted based on real-time temperature feedback, optimizing the modulation strategy to ensure MOSFETs operate within safe temperature limits. This complex mapping requires continuous monitoring and an adaptive control strategy to dynamically adjust $\alpha$ and $\beta$, balancing power losses and thermal conditions effectively.

## 4 Results and Discussion

This section presents the outcomes of the simulations conducted to evaluate the effectiveness of the SVDD integrated ATC system. The key focus is on assessing the responsiveness of the control strategies (FOC and DPWM), and the overall impact on power loss of the converter. The dual inverter topology was modeled in Simulink using the parameters given in Tab. 1. FOC was employed to manage the torque and speed of the PMSM, while DPWM was used to optimize switching events and minimize thermal load on the MOSFETs. The simulation involved dynamic load scenarios where the motor was subjected to varying torque demands. The switching frequency of the converter is set at 20 kHz. Using MATLAB, transient thermal behavior of the SiC MOSFETs was modeled based on a temperature-dependent Cauer network. The model provided a detailed map of thermal propagation and crosstalk within the module, crucial for identifying potential failure points due to overheating. The SVDD model was trained with data generated from the thermal simulations, which included a range of power loss scenarios and their corresponding temperatures. The model's accuracy in encapsulating the normal operation data within a defined hypersphere was validated against a set of unseen data, ensuring robustness in anomaly detection. The trained SVDD model was integrated into the Simulink environment as a custom MATLAB Function Block. This block received real-time power loss measurements from each MOSFET and provided temperature estimates which were fed into the control system. Adjustments to the DPWM control signals based on these temperature estimates allowed for real-time thermal management. The simulation results offer a comprehensive insight into the performance of the ATC system .

command; however, the system demonstrates robustness by promptly returning to steady operation without any oscillations, implying excellent damping characteristics and control loop stability. The rotor speed remains steady at approximately 3000 rpm with minor deviations that correspond to the transient events observed in the torque profile. The swift recovery to the target speed post-transient suggests a highly responsive and accurate speed control mechanism within the drive system.

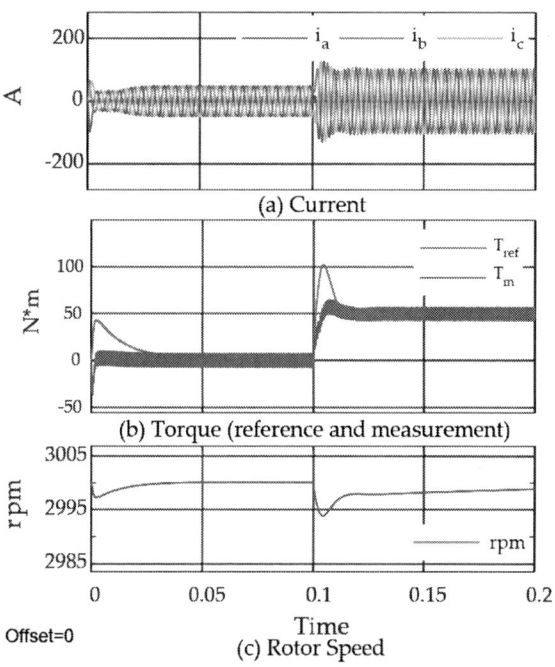

**Fig. 5:** Current, Torque, and Speed Parameters of the System Operating with Active Thermal Control.

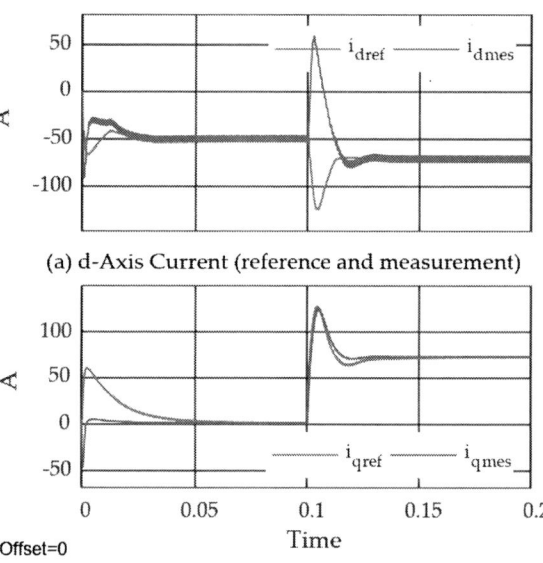

**Fig. 6:** D-axis and Q-axis Current Profiles with Active Thermal Control.

Fig. 5 illustrates the performance metrics of a 3-phase dual inverter system driving a PMSM, showcasing phase currents, torque behavior, and rotor speed over time. Starting with phase currents $(i_a, i_b, i_c)$, the waveforms maintain a balanced and consistent amplitude throughout the observation period. This indicates a well-regulated inverter operation with minimal current distortion, which is essential for maintaining motor efficiency and performance. The torque graph displays the reference torque $(T_{ref})$ and measured torque $(T_m)$. Initial torque response follows the reference closely, suggesting the control system's precision in tracking demanded torque. A sharp transient spike, observed in both the reference and measured values, is indicative of a rapid load change or a control

The Fig. 6 presents a detailed view of the current components in a FOC scenario, with the d-axis and q-axis currents representing the flux and torque-producing components, respectively. By examining these responses reveals that the system can maintain current control despite transient events, with both the reference and measured values converging smoothly after an initial overshoot. This is essential for effective vector control in FOC systems, where decoupling of torque and flux-producing currents is pivotal.

The Fig. 7 contrasts the inverter power loss between a standard FOC-SVPWM control strategy and the developed ATC system. With FOC-SVPWM, the power loss peaks at approximately 450W, whereas the developed ATC system contains the power loss to under 150W during similar transient conditions. This significant reduction in power loss not only implies enhanced thermal man-

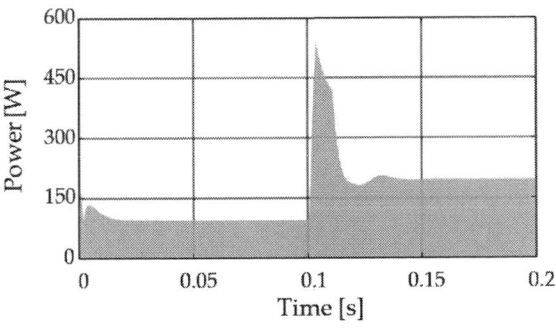

(a) Inverter power loss with FOC-SVPWM

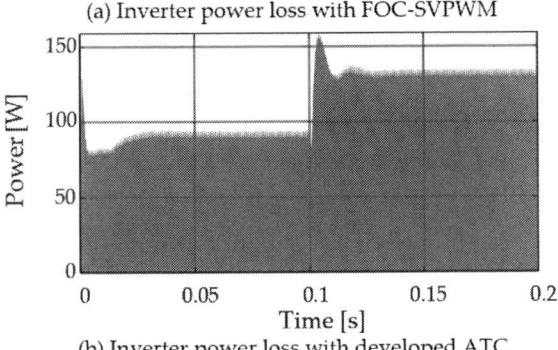

(b) Inverter power loss with developed ATC

**Fig. 7:** Comparison of Inverter Power Loss Between Regular FOC-SVPWM and the Developed Active Thermal Control.

agement but also points to improved overall efficiency of the inverter system. From the simulations and results, the ATC system, supported by transient thermal characterization and optimized by SVDD, enhances the SiC MOSFETs' performance, as depicted by the simulation data. The results highlight the system's capability to minimize the power loss and maintain operational integrity under dynamic conditions. Such advancements establish the developed ATC method as a highly effective approach for modern power electronic systems, addressing the critical challenge of thermal management with demonstrable superiority.

## 5 Conclusion

In conclusion, this results successfully demonstrated the advancement in the management of SiC MOSFET temperatures within a 3-phase dual inverter system by integrating transient thermal characterization with SVDD. The approach has led to a responsive ATC system that accurately estimates junction temperatures, a critical factor for reliable and efficient performance in power electronic applications. Simulations confirm that this method substantially lowers power losses with the devel-

oped active thermal control approach.

## Acknowledgment

This project has received funding from the ECSEL Joint Undertaking (JU) under grant agreement No 101007281. The JU receives support from the European Union's Horizon 2020 research and innovation programme and Austria, Germany, Slovenia, Netherlands, Belgium, Slovakia, France, Italy, Turkey.

## References

[1] A. Moghassemi, S. M. I. Rahman, G. Ozkan, C. S. Edrington, Z. Zhang, and P. K. Chamarthi, "Power converters coolant: Past, present, future, and a path toward active thermal control in electrified ship power systems," *IEEE Access*, vol. 11, pp. 91 620–91 659, 2023. DOI: 10.1109/ACCESS. 2023.3308523.

[2] G. L. Rødal and D. Peftitsis, "An adaptive current-source gate driver for high-voltage sic mosfets," *IEEE Transactions on Power Electronics*, vol. 38, no. 2, pp. 1732–1746, 2023. DOI: 10.1109/TPEL. 2022.3208827.

[3] M. Farhadi, F. Yang, S. Pu, B. T. Vankayalapati, and B. Akin, "Temperature-independent gate-oxide degradation monitoring of sic mosfets based on junction capacitances," *IEEE Transactions on Power Electronics*, vol. 36, no. 7, pp. 8308–8324, 2021. DOI: 10.1109/TPEL.2021.3049394.

[4] S. Jones-Jackson, R. Rodriguez, Y. Yang, L. Lopera, and A. Emadi, "Overview of current thermal management of automotive power electronics for traction purposes and future directions," *IEEE Transactions on Transportation Electrification*, vol. 8, no. 2, pp. 2412–2428, 2022. DOI: 10.1109/TTE.2022.3147976.

[5] S. Kalker, C. H. van der Broeck, and R. W. De Doncker, "Online junction-temperature sensing of sic mosfets with minimal calibration effort," in *PCIM Europe digital days 2020; International Exhibition and Conference for Power Electronics, Intelligent Motion, Renewable Energy and Energy Management*, 2020, pp. 1–7.

[6] F. Erturk, E. Ugur, J. Olson, and B. Akin, "Real-time aging detection of sic mosfets," *IEEE Transactions on Industry Applications*, vol. 55, no. 1, pp. 600–609, 2019. DOI: 10.1109/TIA.2018. 2867820.

[7] F. Karakaya, A. Maheshwari, A. Banerjee, and J. S. Donnal, "An approach for online estimation of on-state resistance in sic mosfets without current measurement," *IEEE Transactions on Power Electronics*, vol. 38, no. 9, pp. 11 463–11 473, 2023. DOI: 10.1109/TPEL.2023.3282585.

[8] Y. Zhang, Y. Zhang, Z. Xu, Z. Wang, V. H. Wong, et al., "Guideline for reproducible sic mosfet thermal characterization based on source-drain voltage," *IEEE Transactions on Industry Applications*, pp. 1–10, 2024. DOI: 10.1109/TIA.2024.3352548.

[9] Z. Ni, S. Zheng, M. S. Chinthavali, and D. Cao, "Investigation of dynamic temperature-sensitive electrical parameters for medium-voltage sic and si devices," *IEEE Journal of Emerging and Selected Topics in Power Electronics*, vol. 9, no. 5, pp. 6408–6423, 2021. DOI: 10.1109/JESTPE. 2021.3054018.

[10] D. M. Tax and R. P. Duin, "Support vector data description," *Machine Learning*, vol. 54, no. 1, pp. 45–66, Jan. 2004. DOI: 10.1023/B:MACH. 0000008084.60811.49.

[11] V. S. Bharath Kurukuru, M. N. Hasan, and R. Petrella, "Thermal characterization of packaged sic devices for high-temperature applications," in *2023 25th European Conference on Power Electronics and Applications (EPE'23 ECCE Europe)*, 2023, pp. 1–11. DOI: 10.23919/ EPE23ECCEEurope58414.2023.10264637.

[12] V. S. B. Kurukuru, M. A. Khan, and R. Singh, "Health monitoring framework for electric vehicle drive train in digital twin," in *2023 25th European Conference on Power Electronics and Applications (EPE'23 ECCE Europe)*, 2023, pp. 1–10. DOI: 10.23919/EPE23ECCEEurope58414.2023. 10264291.

[13] F. Liu, K. Xin, and Y. Liu, "An adaptive discontinuous pulse width modulation (dpwm) method for three phase inverter," in *2017 IEEE Applied Power Electronics Conference and Exposition (APEC)*, 2017, pp. 1467–1472. DOI: 10.1109/APEC.2017. 7930892.

[14] D. Nayak and Y. R. Kumar, "Discontinuous modulation based active thermal control for enhancing the kva limit of three-phase inverters," pp. 0–9, 2023. DOI: 10.36227/techrxiv.24591477.v1.

PCIM Europe 2024, 11– 13 June 2024, Nuremberg          DOI: 10.30420/566262109

# Thermal Management Solutions by Additive Manufacturing – Powder Bed Fusion and Diffusion Bonding

Simon Jahn[1], Felix Gemse[1], Steffen Dahms[1], Maximilian Streinz[1]

[1] ifw Jena – Guenter-Koehler-Institut fuer Werkstoffpruefung und Fuegetechnik GmbH, Germany

Corresponding author:     Simon Jahn, sjahn@ifw-jena.de
Speaker:                          Simon Jahn, sjahn@ifw-jena.de

## Abstract

The market share of products made by Additive Manufacturing increased significantly in the last years. Due to the freedom of design, parts can have very complex internal structures for heat transfer or cooling. The heat transfer can be further promoted by using a multi material approach. With several additive manufacturing processes enabling the use of two or more materials within one component, the interest grows. In the contribution examples made by powder bed fusion and sheet lamination combined with diffusion bonding were presented.

## 1  Introduction

### 1.1  Additive Manufacturing

The basic idea of additive manufacturing is to reduce a three-dimensional manufacturing problem to several simple two-dimensional tasks. In a first step, the 3D volume model of the component is sliced into a large number of thin layers of equal thickness (slicing). The resulting 2D and 2½D layer data forms the manufacturing information for the processes that build the component layer by layer. This approach is not new in itself. Even the ancient Egyptian master builders used the idea of building in layers. Further examples are the production of topographic relief maps, for which the layer principle has been used since the end of the 19th century, or today's 3D puzzles, which create a model layer by layer.

Additive manufacturing processes automate the layered build principle and transfer it to component production. The desired geometry is created by joining small volume elements together. This leads to a completely new approach in design and manufacturing, allows significantly greater geometrical freedom and offers the opportunity to reduce the use of complex, product-specific tools and molds [1].

In the past, additive processes were summarized in a group of generative processes (from the Latin "generare": generating, primary forming). In the international area, all processes that are additive are currently summarized under the term "3D printing".

This trend is noticeable especially in the non-technical environment.

A closer look at the processes reveals a number of conceivable ordering principles that cannot all be described in detail here. An example is the form of material to be used, e.g. powder or wire/filament or sheet. According to the standard ISO/ASTM 52900:2018-06, additive processes are divided into seven so-called "process categories":

1.     vat photopolymerisation, VPP
2.     material jetting, MJT
3.     binder jetting, BJT
4.     powder bed fusion, PBF
5.     material extrusion, MEX
6.     directed energy deposition, DED
7.     sheet lamination, SHL

The market share of products made by Additive Manufacturing (AM) increased significantly in the last years. This is driven by several applications and thermal management is one of the important ones. Besides conformal cooling in molding, casting and forging industry, thermal management in systems of the semiconductor sector as well as cooling of electronic devices are important markets. Due to the freedom of design, parts can have very complex inside structures for heat transfer or cooling. The heat transfer can be promoted further by using different materials, e.g. by a multi material approach. With several additive manufacturing processes enabling the use of two or more materials within one component, the interest grows. Properties, which are marked as disadvantage of

883

some AM processes can contribute to the performance of thermal management components. In this contribution, the technology and mainly examples, successfully realized by powder bed fusion and sheet lamination combined with diffusion bonding, are presented in the following.

## 1.2 Sheet lamination combined with diffusion bonding

Diffusion bonding is a solid state joining principle without liquid phase. The process inventor Kazakov descripted the basic principle and the state of the art regarding systems and application in 1985 [2]. One main advantage of diffusion bonding is the joining of areas without any additional material. This is of interest especially for harsh environment applications with demands on corrosion resistance. Another advantage to remark is the possibility to join material combinations, which cannot be joined in a common liquid phase due to intermetallic formation. In the last years, there were significant development steps with regard to the diffusion bonding systems.

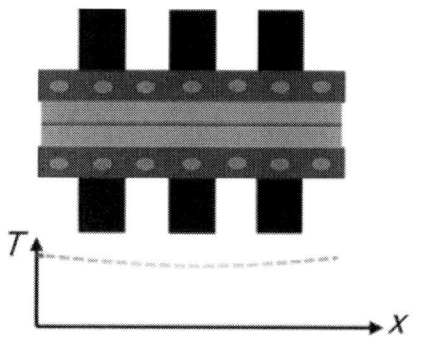

**Fig. 1** Diffusion bonding system with heated pressing plates (top) and principle (below)

Materials for thermal management solutions, mostly Al- or Cu-based, have to be diffusion bonded in vacuum. The heat transfer by radiation is low, especially below 500 °C. To overcome this disadvantage, a system with heated pressing plates was developed (Fig. 1).

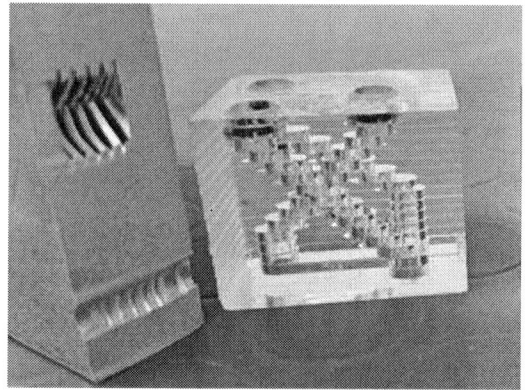

**Fig. 2** Al-made cold plate (top left), glass made cube with inside structures (top right), multi material heat transfer element (bottom left) and heat exchanger (bottom right)

Since a few years, the joining technology diffusion bonding grows significantly each year. Reasons for the intensified use nowadays are the capability to join large areas without any additional filler material and the high resulting bond strength. The bond strength can be as high as the strength of the base material for similar material joints. For corrosion resistance it is important to have the same material everywhere. The corrosion resistance is not reduced locally (compared to brazing). Diffusion bonded parts are applied in lithography systems for semiconductor production. By staking laser or waterjet cut layers and subsequent diffusion bonding, parts with inside structures can be realized. This additive manufacturing principle is called sheet lamination (SHL). Temperature management is the main function of diffusion bonded parts today, e.g. in molds, dies, heat exchangers or heat sinks. One important fact for controlling temperature is the scale effect. The surface size matters, because this area interacts with fluids (Fig. 2). For smaller cross sectional areas of fluid channels, the heat transfer rate raises. The reasons for this behavior are on one side the shorter distance within the fluid to get in contact with the channel surface. On the other side, and of higher importance, is the ratio between cross sectional area and cross sectional area perimeter. For a quadratic channel, the edge length correlates linear with the perimeter length (4 times the edge length). The ratio between perimeter and cross sectional area scales inversely proportional with regard to the edge length. For small channels, this ratio increases and thus also the heat transfer.

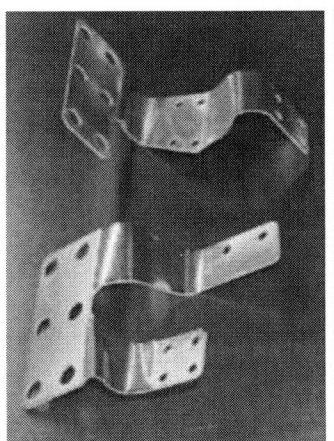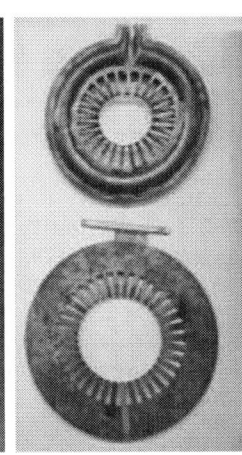

**Fig. 3** Flexible heat transfer elements made of Al or Ag (left) and Cold plates made form Cu (right)

If heat transfer should be applied without transfer of vibration, flexible elements are in use. In Fig. 3 on the left side such elements, which can be made of Al or Ag, are displayed. One advantage in comparison to traditional elements is, that there are more than two bonded areas and the elements were formed into the necessary shape during bonding.

Cold plates can be bonded without additional material. In Fig. 3 (right) two Cu-made cold plates are shown in the left with diameter of 350 mm and 500 mm. The smaller one passed a burst test at 10-times of the application pressure without failing. Such heat sinks, made of Al, are part for E-motors. Because of the heat "capture" directly at the source, the system volume can be shrinked to a quarter of comparable systems with traditional water jackets for cooling (Fig. 4). The design is patended and such systems are in use for test benches, where short time overloading does not cause any damage due to the better cooling.

**Fig. 4** Al-made heat sinks in E-motor

## 1.3 Powder bed fusion

Powder bed fusion (PBF) is the most widely used and investigated process in AM. Each layer of the powder bed is created by a manufacturer-specific coating system after lowering the building base with regard to the used layer thickness. An energy source, a laser- or electron beam, is used to fuse powder particles of selected regions in the powder bed and to join them with the layer below. The process cycle is repeated to create a solid three-dimensional component.

**Fig. 5** Nozzle shielding gas distribution channels for welding

Depending on part design and building orientation, it is necessary to add support structures to the part before building. In most cases this is a drawback, but not always. As shown in Figure 5, such support structures can be used for adjusting a gas flow and get better flow covering.

With additive manufacturing, it is possible to realize large areas for heat exchange in small parts. Different examples are shown in Figure 6. Even 3D lattice structures can be integrated. With the design of the structures inside, the flow properties and pressure loss can be adjusted to the applications need.

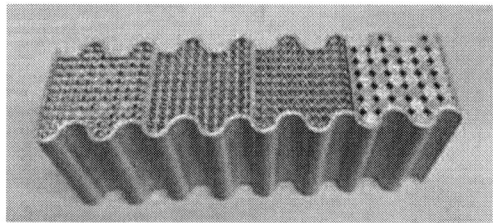

**Fig. 6**  Different structures for heat exchange in air cooling applications.

Typically, PBF parts show a rougher surface. If such a surface is used in a heat sink for air cooling it is a feature, not a disadvantage. Accordingly, the flow behavior of air along the fins changes and is more efficient (Fig. 7).

**Fig. 7**   Al made heat sink with rough surface area

## 1.4   Multi-material parts

By combining different materials, the thermal behavior of a part can be designed, too. For example, a corrosion resistant material (e.g. steel) can be used in the area with media contact and directly behind it a material with high thermal conductivity for faster heat transfer like Cu- or Al-can be applied. For some applications silver foil material is used for the flexible part of the heat transfer elements. Figure 8 contains such a multi-material heat transfer element, where the connection plate was made out of Cu and the foil stripes out of silver.

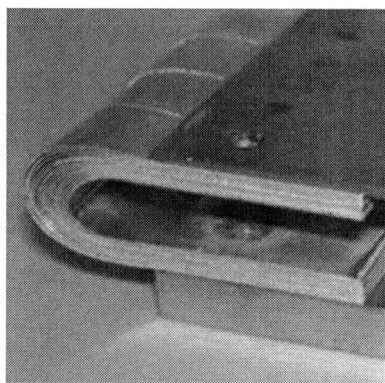

**Fig. 8**   Heat transfer element made of silver and copper

There are several approaches to manufacture multi-material parts with the PBF principle. Further, a combination with other AM principles is possible too, e.g. material extrusion of metals (Fig. 9).

**Fig. 9** Multi-material parts made of Cu and steel

## 1.5   Flow optimisation

The new freedom for 3D-design by additive manufacturing combined with finite element analysis (FEA) respectively computational fluid dynamics (CFD) unlocks the full potential and reduces the experimental effort.

Besides the heat sink itself, the flow over or towards the heat sink influences the efficiency and sound level for air cooling as well. With additive manufacturing, complex shaped nozzles can be realized to get an optimized airflow field. Further, the sound level can be reduced without a decrease of cooling efficiency.

In Figure 10 such a part, a nozzle, is shown. The gas exit is designed as Laval type with a slot of 0,25 mm (Fig. 10 below left).

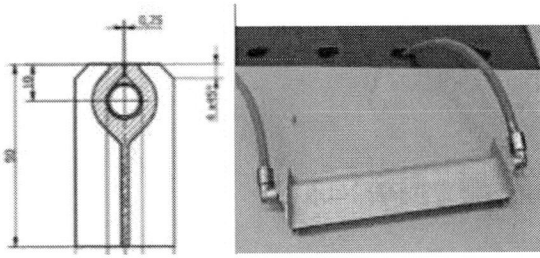

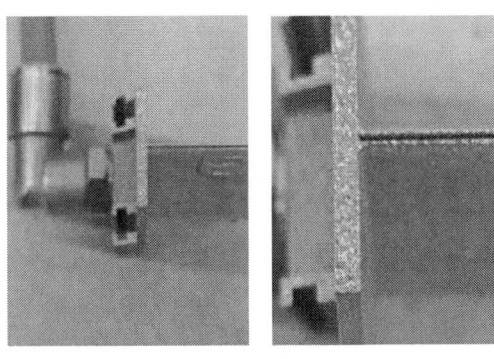

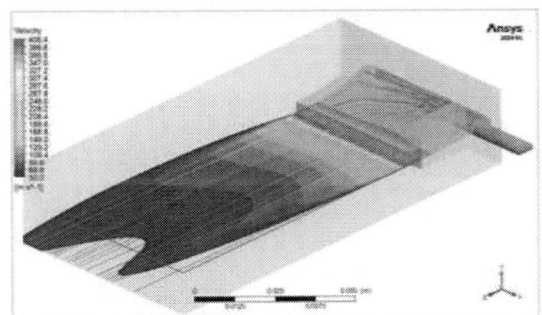

**Fig. 10**  Demonstrator for optimized flow

In our investigations, several designs were tested, starting with FEA/CFD calculations (Fig. 11 top). After this analysis, numerous nozzles were built by PBF in order to validate the FEA results and to test the nozzles in real applications. Particle Image Velocimetry (PIV) imaging technology was used for the validation (Fig. 11 below).

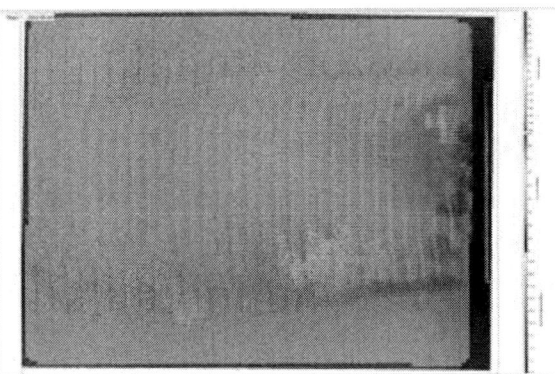

**Fig. 11**  FEA/CFD result of flow velocity

The PIV results confirmed the results of the design optimization as well as the application tests. The sound level was reduced significantly as well as the necessary air flow rate. Only the area, which has to be covered by the air, is "flooded".

[PIV measurement image]

**Fig. 12**  PIV measurement result of flow velocity

## 2  Summary

With additive manufacturing processes, the thermal management of component temperatures can be adjusted more effectively and efficiently.

The most important reason is the freedom of geometrical design. Flow channels can be designed in individual shapes and length.

Second, the utilization of scale effects (increase of surface in smaller channels) has to be mentioned. In hand sized cubes, inner surfaces of several square meters can realized.

Third point to consider is the multimaterial approach. Due to the process development for several AM principles, components made of two or three materials, can be manufactured in one process. The advantage is, that the right material is in the location where it is necessary, but close to another with different properties.

With flow optimization, the efficiency of heat sinks can be increased further.

## References

[1] Gebhardt, A.: Generative Fertigungsverfahren. 3rd edition, Carl Hanser Verlag, Munich, 2007.

[2] Kazakov, N.F. (Edt.): Diffusion Bonding of Materials, Mir Publishers, Moscow, 1985.

# Advanced Pumped Two-Phase Cold Plate for Cooling Power Electronics.

Elizabeth K. Seber[1], Michael C. Ellis[1]

[1] Advanced Cooling Technologies, Inc., United States

Corresponding author: Elizabeth Seber, elizabeth.seber@1-act.com
Speaker: Elizabeth Seber, elizabeth.seber@1-act.com

## Abstract

Advanced Cooling Technologies, Inc. (ACT) has developed an advanced pumped two-phase (P2P) system that outperforms typical P2P systems through increased cooling performance and reduced size, weight, and power (SWaP) requirements of the thermal management system. This advanced system uses a hybrid evaporator that uses thin-film evaporation as the mechanism of cooling, making the system capable of removing 1000+ W/cm² of heat density, up to multiple kW of power. A case study is performed to demonstrate the improvements in the thermal management system made by the hybrid evaporator, starting at the component level and expanding through the system design.

## 1 Introduction

### 1.1 High Heat Flux Electronics Cooling

Power electronics have been experiencing a trend of increasing power densities, while at the same time experiencing a desired trend in decreasing form factor (size and weight) [1]. Due to unavoidable inefficiencies, as the power density increases, the heat dissipated by such components also increases. As such, improved thermal management systems are required to cool the components to an acceptable operating temperature, and must do so with reduced Size, Weight, and Power (SWaP).

Pumped Two-Phase (P2P) systems are known for their ability to handle high heat loads by providing large cooling flow rates to the heat source. Due to the nature of two-phase heat transfer, P2P systems demonstrate superior cooling capacity compared to single-phase cooling flows. However, P2P systems rely on flow boiling in the evaporator, also referred to as a cold plate. Flow boiling exhibits a heat transfer coefficient (HTC) typically less than 10,000 W/m²·K, and the maximum heat flux that these systems can support is often limited by the Critical Heat Flux (CHF) associated with the evaporator design.

Capillary systems, another thermal architecture, rely on thin-film evaporation from a wicked structure, which can have an HTC as high as 120,000 W/m²·K [2]. In devices such as vapor chambers, the wick can manage heat fluxes as high as 1000 W/cm². The main drawback of capillary systems is the liquid transport mechanism of the wick structure, which is driven by capillary force, and limits the maximum heat load they can handle before dry-out occurs in the wick.

To overcome the limits of these two architectures, Advanced Cooling Technologies, Inc. (ACT) has developed a hybrid, two-phase cooling system (HTPCS) that merges the benefits of mechanically pumped two-phase cooling with those of thin-film evaporation from wicked structures [3]. This novel evaporator eliminates the weaknesses of the two described architectures and results in a thermal management system capable of addressing high heat fluxes and transporting high heat loads, thereby outperforming existing two-phase systems.

Through a case study comparison, ACT demonstrated the improved performance of the novel cold plate at both a component level and at a system level when compared to a State-of-the-Art (SOTA) P2P System. The funding for this research was provided by the US Air Force and the Department of Defense under a Phase I SBIR effort. As such, specific information that defined the cooling application is withheld from the published results herein. However, a qualitative and quantitative description of the performance of each cold plate is given.

### 1.2 Cold Plate Architectures

To compare cold plate architectures, ACT established an array of high-heat flux heat sources to cool. This three-by-five array of heat sources was chosen to mimic a potential laser diode array.

The SOTA cold plates used in P2P systems are comprised of an array of macro-channels, in which a working fluid flows in from an inlet manifold. Heat sources mounted directly to the cold plate evaporate the liquid within these channels, and the working fluid exits the cold plate as a two-phase mixture, typically around 0.80 quality. An example of this type of cold plate architecture is shown in Fig. 1. In this figure, the red boxes denote the array of high-density heat sources.

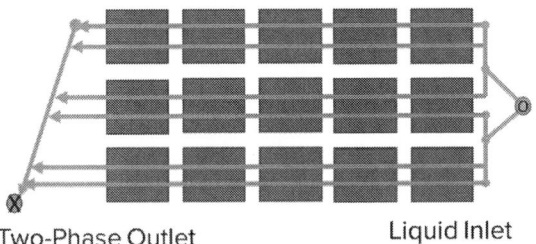

Two-Phase Outlet           Liquid Inlet

**Fig. 1** Schematic of a State-of-the-Art Channel Cooled Pumped Two-Phase Cold Plate.

A schematic of the novel, Hybrid cold plate is shown in Fig. 2, where incoming subcooled liquid enters the cold plate, flows through, and then exits the cold plate through an excess liquid flow outlet. The working fluid is drawn in through capillary action in the wicks to a converging wick structure and monolayer wick. The monolayer wick covers the heat sources denoted by the red boxes in the figure, and thin-film evaporation occurs over the wick structure. This thin film evaporation results in reduced thermal resistance between the heat sources and the working fluid, allowing for improved heat transfer for the high-heat flux sources. The vaporized working fluid then exits the system through a vapor manifold. Vapor generated in this way is kept separate from the liquid working fluid by a non-permeable barrier and exits at a vapor quality of 1.00.

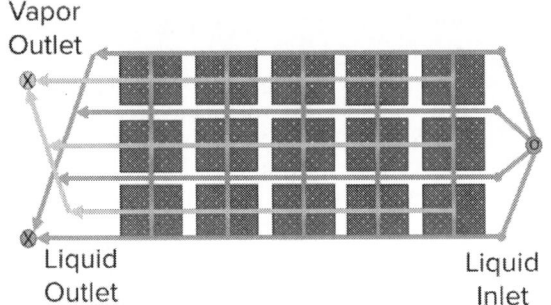

Vapor
Outlet

Liquid                      Liquid
Outlet                     Inlet

**Fig. 2** Schematic of ACT's Novel, Hybrid Evaporator.

## 2 Experimental Setup

### 2.1 Testing Apparatus

ACT used an in-house P2P testing apparatus, as depicted in Fig. 3., to characterize the thermal performance of each cold plate. R134a, a common refrigerant used in P2P systems, was used as the working fluid.

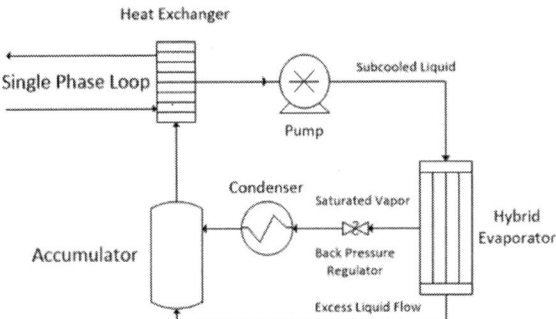

**Fig. 3** System Schematic of Hybrid Evaporator in Pumped Two-Phase Test Loop.

Subcooled liquid was pumped from an accumulator, through a heat exchanger, and into the prototype cold plate. Excess liquid flow from the cold plate was directed back to the accumulator. The saturated vapor that exited the Hybrid Evaporator went through a condenser before returning to the accumulator. A back-pressure regulator on the saturated vapor line enabled control of the vapor pressure. This pressure control was found to be vital for the performance of the hybrid cold plate, as a balance is needed against the capillary pressure of the saturated liquid within the wicked structure to prevent flooding of the evaporator, which would have led to flow boiling and increased thermal resistance.

The SOTA cold plate was also tested in the P2P apparatus, though instead of having two outlet ports, a single outlet port was used. The exiting two-phase mixture would pass through a condenser before returning to the accumulator.

Additionally, thermocouples were placed throughout the test loop, as well as pressure transducers to characterize the quality of the fluid throughout the loop. Resistive heaters were used as the heat sources, as they could be easily powered and controlled, and were mounted to the tops of each cold plate. A thermocouple was placed between each heater and the surface of the cold plate to measure the effective heat transfer coefficient of the cold plates. Thermocouples located at the outlets of each cold plate were also for measuring the effective HTC, as the exiting fluids would be at saturation and therefore would be closely matched in temperature to the fluid at the location of evaporation.

## 2.2 Testing Procedure

Each cold plate was tested on the P2P testing apparatus in the same fashion. At a steady flow rate, the power supplied to each heat source was held constant until a steady-state condition was reached. Then, the power supplied was increased incrementally until a new steady-state condition was reached. This was repeated for a total power ranging from approximately 100 to 1000 W. This was repeated for each cold plate, at three separate liquid inlet flow rates: 0.27, 0.5, and 1.0 liters per minute (LPM).

## 3 Experimental Data

### 3.1 Maximum Temperature

Figure 4 depicts the maximum temperature experienced by the thermocouples mounted between the heat sources and the cold plate versus the average power supplied to each heat source. Both the SOTA macro-channel cooled cold plate and the Hybrid Evaporator are represented in this graph, with each line indicating a different test with different flow rates.

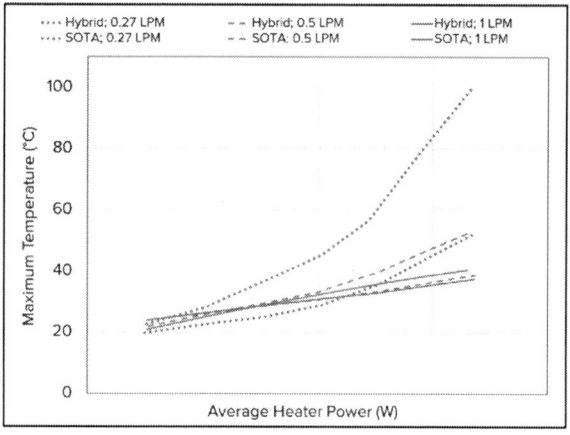

**Fig. 4** Maximum Temperature vs. Average Heater Power for Hybrid and SOTA cold plates.

At lower powers, each cold plate exhibits similar maximum temperatures. However, as the total power supplied increases, the maximum temperature for each test case also increases. For some tests, such as the SOTA cold plate with an inlet liquid flow rate of 0.27 LPM, the maximum temperature greatly increases to 100°C at the highest power. This is a 50°C increase in maximum temperature compared to the Hybrid cold plate test case with a 0.27 LPM liquid inlet flow rate. This jump in temperature within the SOTA cold plate is

likely due to dry out occurring within one of the channels, as this temperature was recorded by a heat source close to the outlet manifold of the cold plate.

When looking at the test cases with higher flow rates, in all cases the Hybrid cold plate demonstrates lower maximum temperatures compared to the SOTA cold plate at the same flow rate.

Notably, the Hybrid cold plate tested at a flow rate of 0.5 LPM demonstrates just slightly lower temperatures compared to the SOTA cold plate tested at a flow rate of 1.0 LPM. This 50% decrease in flow rate for the same level of performance suggests that a lower pumping requirement can be achieved by using a Hybrid cold plate.

### 3.2 Temperature Spread

Figure 5 depicts the maximum temperature difference for each cold plate for the range of coolant flow rates, versus the average heater power. This temperature difference was determined by subtracting the minimum temperature recorded from the maximum temperature recorded for each heat source by each heat source.

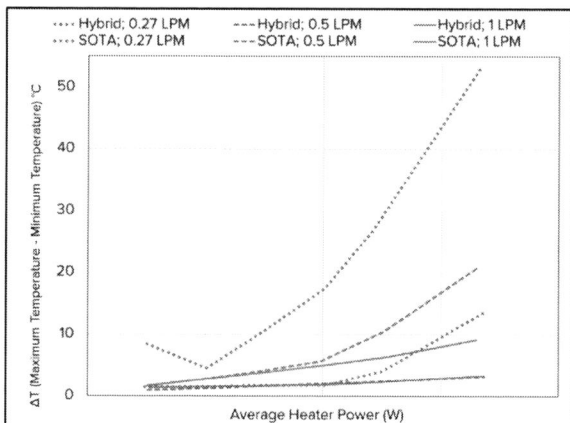

**Fig. 5** Temperature Difference (Maximum – Minimum Temperature) vs. Average Heater Power for Hybrid and SOTA cold plates.

As anticipated, as the average heater power increases, the temperature spread across the cold plate also increases. Notably, for the SOTA channel-cooled cold plate, the minimum temperatures recorded were all closest to the subcooled liquid inlet manifold. The maximum temperatures recorded were all closest to the two-phase mixture outlet manifold. This suggests that by the time the working fluid passed through the channel and reached the heat sources located at the end of the flow path, the liquid was either no longer subcooled or may have evaporated entirely, leading to drying out of the channel.

For the Hybrid cold plate, the temperature difference experienced at coolant flow rates of 0.5 and 1.0 LPM are nearly identical, suggesting a high degree of temperature uniformity. Notably, the test case with a coolant flow rate of 0.27 LPM for the Hybrid cold plate also remains identical to the other flow rates until a certain heat load is reached. At this point, dry-out is starting to occur in the wicks located closest to the outlet manifolds, resulting in an increased temperature difference across the plate.

The recorded temperature difference between the maximum and minimum temperatures of the cold plate describes only part of the temperature uniformity of the cold plates. To better visualize the temperature uniformity, the temperatures for each heat source, and the total power supplied are depicted in Fig. 6 and Fig. 7 for the SOTA and Hybrid cold plates, respectively.

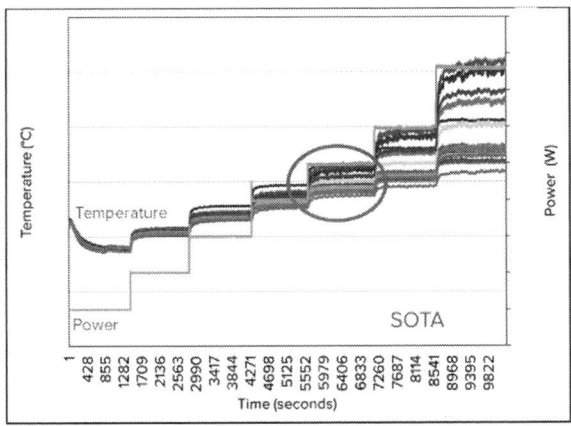

**Fig. 6** Temperature of Heat Sources and Total Power versus Time for the SOTA Cold Plate

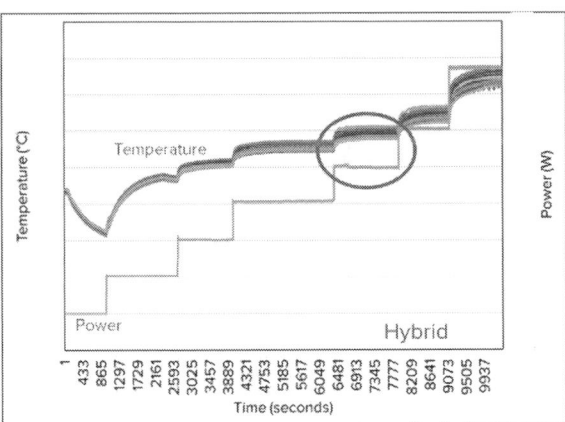

**Fig. 7** Temperature of Heat Sources and Total Power versus Time for the Hybrid Cold Plate

Initially in these tests, when the total power level is low, both cold plates exhibit a tight temperature control for each heat source. As time goes on, and as the power load increases, the temperature spread for the SOTA cold plate increases significantly, while the Hybrid cold plate temperature spread remains tight until the highest power load. Circled in red are the target power, and corresponding temperature data, for the program application.

## 3.3 Average Heat Transfer Coefficient

The effective HTC was calculated using Eqn. 1 for each heat source, where Q is the power load for that heater in watts, A is the surface area of the heater, in meters squared, $T_{Heater}$ is the temperature recorded by the thermocouple under the heat source in Kelvin, and $T_{Outlet}$ is the temperature recorded at the outlet of the cold plate in Kelvin. Due to the nature of two-phase flow, the temperature at the outlet is a good approximation for the temperature at the site of evaporation. The average was taken for each HTC and is depicted in Fig. 8.

$$HTC = \frac{Q}{A \times (T_{Heater} - T_{Outlet})} \qquad (1)$$

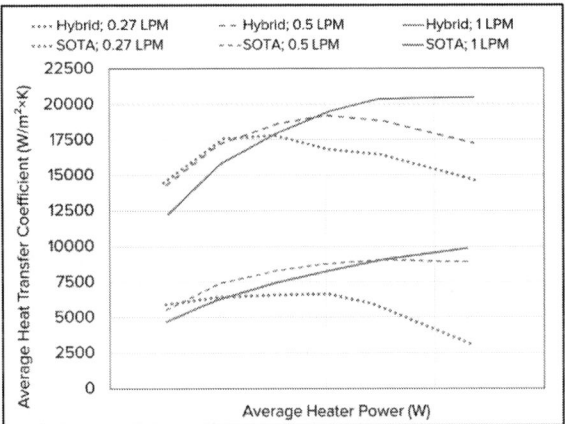

**Fig. 8** Average Heat Transfer Coefficient vs. Average Heater Power for Hybrid and SOTA cold plates.

Notably, regardless of the flow rate or heat load, the Hybrid cold plate demonstrates an effective average heat transfer coefficient nearly twice that of the SOTA cold plate. This is an indication that the thermal resistance of the Hybrid cold plate is significantly lower than the SOTA cold plate.

The trendline for each test case suggests that there is a limit to the HTC where the curve flattens before decreasing.

# 4 Conclusions

ACT designed, fabricated, and tested prototype cold plates to compare the performance between two different pumped two-phase cold plate architectures. The first cold plate architecture, representative of the state-of-the-art, relied on flow boiling through macro-channels. The second cold plate architecture was a novel design developed by ACT, referred to as the Hybrid cold plate as the architecture combined the benefits of pumped two-phase with those of thin-film evaporation off wicked structures.

The Hybrid cold plate achieved overall better thermal performance than the SOTA channel-cooled cold plate. The Hybrid cold plate exhibited overall lower temperatures when testing at higher powers. At a coolant flow rate of 0.5 LPM, the Hybrid cold plate outperformed the SOTA cold plate by 13°C. At a coolant flow rate of 0.27 LPM, the difference was closer to 50°C.

In terms of temperature uniformity, at a coolant flow rate of 0.5 LPM, the Hybrid cold plate exhibited a temperature uniformity improvement of 250% compared to the SOTA.

Additionally, the Hybrid cold plate's effective, average heat transfer coefficient was consistently nearly twice that of the SOTA cold plate.

The thermal performance benefits become increasingly apparent when operating and higher heat loads, and at lower flow rates. These performance benefits result in reduced pumping power requirements. This reduced requirement equates to reduced size, weight, and power properties of the entire thermal management system.

# 5 Acknowledgements

The authors would like to thank Megan Gettle and Max Demydovych for their work in fabricating and testing the prototype cold plates. The authors would like to acknowledge and thank the U.S. Air Force SBIR Program (Contract No. FA945122PA018) for financially supporting the project.

# References

[1] S. Jagdale, "What Are the Tradeoffs to Achieving High Power Density?," Power Electronics News, 24 January 2023. [Online]. Available: https://www.powerelectronicsnews.com/what-are-the-tradeoffs-to-achieving-high-power-density/. [Accessed 2024].

[2] T. W. Davis and S. V. Garimella, "THERMAL RESISTANCE MEASUREMENT ACROSS A WICK STRUCTURE USING A NOVLE THERMOSYPHON TEST CHAMBER," CTRC Research Publications, West Lafayette, 2008.

[3] M. R. Shaeri, R. W. Bonner, M. C. Ellis, E. K. Seber, and M. V. Demydovych, "Heat Transfer Device Having an Enclosure and a Non-Permeable Barrer Inside the Enclosure". United States Patent 11,408,683, 26 August 2022.

PCIM Europe 2024, 11– 13 June 2024, Nuremberg    DOI: 10.30420/566262111

# Feasibility Study of High-Power Density of Modified Isolated CLLC dc-dc Interface with Wide Range of Voltage/Current Regulation

Oleksandr Husev[1], Oleksandr Matiushkin[1], Parham Mohseni[1], Francisco Canales[2], Dmitri Vinnikov[1]

1 Tallinn University of Technology (TalTech), Estonia

[2] ABB, Switzerland

Corresponding author:   Oleksandr Husev, oleksandr.husev@taltech.ee

## Abstract

Isolated dc-dc converters are widely used in power electronics. In particular for battery chargers where isolation between grid and battery is required along with wide range of battery voltage regulation. In this context, we propose the high-power density asymmetrical CLLC isolated dc-dc GaN/SiC based interface with capability to control output voltage/current in the wide range. This paper discloses the operation principle along with components selection guidelines. Also, simulation results disclose all potential of the proposed approach. Compact simplified 5 kW demonstrator working at 500 kHz was designed in order to demonstrate feasibility of the proposed approach.

## 1   Introduction

One of the greatest problems the world is facing today are the environmental pollution, the depletion of natural resources and constant grow of electric energy consumption [1]. People are trying to improve the situation. Efforts are underway to develop cheaper and more sustainable energy, such as solar Photovoltaic (PV) power, wind power and other forms of Renewable Energy Sources (RESs) that can replace oil and fossil fuels [2]. Another direction lies in the reduction of emissions by using new green technologies in the massive industry, Electrical vehicles (EVs) and safe storage technologies [3]-[7].

At the same time, the plan of total vehicle electrification exacerbates the problem of electric energy demand. In advance to this, the current ac grid infrastructure is not very compatible with renewable energy integration due to its inherent unpredictable generation capability and its inertialess nature. Production of green hydrogen may partially help to mitigate this problem; at the same time, many researchers and technical experts are skeptical about its global implementation because of its high cost and low efficiency [8].

The Low Voltage dc (LVdc) is naturally applicable in a scenario with a high penetration level of RESs, Battery-Based Energy Storage Systems (BESSs), home appliances, and EVs. It may lead to reduced power electronics stages, higher efficiency and resilience, cost reduction in energy distribution, space and weight savings, with a flexible placement of electrical equipment [9]-[11]. Furthermore, this modern system will no longer operate at a specific frequency, facilitating its control and definition of standards. The implementation and spreading of the LVdc grid can solve problems with a further increase in renewable energy generation, vehicle electrification and reduction of environmental pollution.

From the research point of view, LVdc has no novelty. The main question is that despite on well awareness of the LVdc concept and benefits why its implementation can be described as nearly next to nothing [12]. There are several main problems towards its implementation. It is evident that has high "saturation" levels of bulky power electronics devices including fast Solid State Circuit Breakers (SSCB), different dc-dc and isolated interlink (dc-ac) converters. The very first and main problem is non mature level of power electronics. In this regard, dc-dc transformer (DCX) which is considered like a key enabling technology for dc microgrids between medium and low voltage. Size, cost and efficiency of DCX are factors for further dc microgrid implementation [13]-[16].

Also, isolated dc-dc converters are widely used in many other applications of power electronics. For battery chargers where isolation between ac grid and battery is required due to the common mode leakage current limitation. It can be onboard

893

Electric Vehicle (EV) chargers or fast EV dc chargers, but requirements for power density and efficiency becomes more and more demanded [17]. Especially, taking into account versatility of different battery packs and their voltage range [4]. This infrastructure is under development and effective power electronics solutions play a key role.

At the same time, it is well-known that technological development of novel Wide Band Gap (WBG) semiconductors is a motive force for power electronics development, in particular in terms of size and efficiency [18], [19]. From another side, it is also obvious that "Ideal switch is not enough" [20] especially at switching frequency increasing. It is already reported in many papers, that magnetic components become a dominative losses component in the new generation of power electronics devices based on WBG semiconductors.

In this context, this work proposes the approach how to realize high-power density isolated dc-dc WBG based converter with air transformer and reasonable voltage regulation range. The paper is organized as follows. Section II describes the proposed concept. Section III is devoted to the Zero Voltage Switching (ZVS) capability analysis and the output voltage/current regulation range. Section IV devoted to the simulation and experimental verification, while conclusions are presented and discussed in Section V.

## 2 Concept of the modified asymmetrical LCCL converter

The proposed solution is shown in Fig. 1a. Instead of DAB or LLC converter, typically used in similar applications, it has CLLC isolation stage configuration.

### 2.1 Proposed modifications

It should be noted that considered CLLC configuration is already tested in several applications including DCX [13], [21]-[27]. At the same time, there are no reports about very high switching frequency applicability and wide range of regulation at the same time. Mostly it is explained by required complex control and limited capability of regulation maintaining Zero Current Switching (ZCS)/ZVS conditions. ZVS condition is mostly achieved with frequency control, but voltage gain also depends on frequency control which creates limitations. Works [23], [24] disclose this issue. Also, it should be noticed that despite on the smallest value of magnetizing inductance, some of the solutions consider utilizing transformer even at high switching frequency [21], [22].

Also, it is important that all above mentioned solutions utilize leakage inductance compensation approach (Fig. 1b), which means that relatively large magnetizing inductance is required. In opposite to above mentioned solutions we propose to utilize approach which is similar to Inductive Power Transfer (IPT) approach [28] with very low value of inductances and ZCS at nominal power and resonance switching frequency (Fig. 1c).

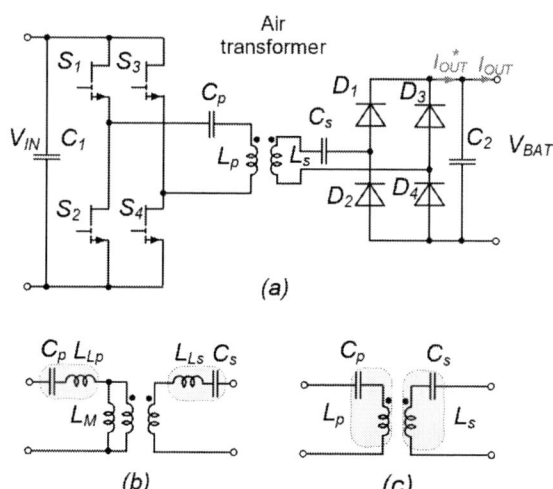

**Fig. 1** Proposed isolated dc-dc interface which is based on high switching frequency WBG semiconductors along with air transformer with full compensation (CLLC configuration) (a) with classical resonant circuit (b) and proposed resonant circuit (c).

ZCS feature along with WBG transistor utilization potentially allow to realize very high switching frequency which in turn means very low inductance that can be realized without ferrite core. This is the main difference to the approach proposed in [29] or many others with DAB or LLC converter. DAB and LLC circuits require high magnetizing inductance which is not possible to achieve without ferrite core. It is well-known that magnetics (and ferrite core in particular) is the main source of losses and obstacle toward further switching frequency increasing. By this approach very high-power density with decent efficiency can be achieved. At the same time, due to the fixed coupling between primary and secondary coils there are not any problems with tuning of compensation circuits. Design methodology is very similar to SS compensation case with fixed coupling. In [28] it is mentioned that Load Matching Factor (LMF) for SS topology should be selected according to the nominal coupling $k$:

$$\gamma \approx k, \qquad (1)$$

and defined as:

$$\gamma = \frac{R_L}{\omega_{res} \cdot L_S}, \qquad (2)$$

where $\gamma$ – LMF, $\omega_{res}$ – resonance frequency, $L_S$ – secondary side self-inductance, $R_L$ – equivalent load, which in turn calculated as:

$$R_L = \frac{8}{\pi^2} \cdot \frac{V_{SEC}^2}{P}. \qquad (3)$$

From (2), the secondary and primary side self-inductances can be expressed:

$$L_S = \frac{R_L}{\omega_{res} \cdot \gamma}, \ L_P = L_S \cdot \left(\frac{V_{PR}}{V_{SEC}}\right)^2. \qquad (4)$$

Compensation elements should be tuned to the resonance frequency that can be found by the equation:

$$C_P = \frac{1}{L_P \cdot \omega_{res}^2}, \qquad C_S = \frac{1}{\omega_{res}^2 \cdot L_S}. \qquad (5)$$

As a conclusion from well-known design approach, the mutual inductance is defined by opposite value of the power and switching frequency. As higher power, as smaller value of mutual inductance is required. In case of low mutual coupling, the mutual inductance, which is defined by power, remains the same, but self-inductance of the primary and secondary side should be increased with correspondent decreasing of resonance capacitors. Another conclusion, from this design approach is that air core transformer looks like optimal solution at high switching frequency as big magnetizing current is not required.

## 2.2 Cycle skipping control approach for current/voltage regulation

The cycle skipping control approach is an integral component of the proposed solution. Fig. 2 illustrates the main features (including control) of the proposed dc-dc stage. In order to provide desired operation 4 conventional equivalent circuits are utilized (3 of them are demonstrated). At the same time, to drop the output power or output voltage the modulation index of primary side voltage can be decreased by introducing zero state, which in turn deteriorates ZCS condition and decrease efficiency. Similar effects can be achieved by switching frequency variation which is well-known in inductive power transfer [29]. But due to the high switching frequency of the dc-dc stage, we consider cycle skipping approach as the most suitable for this case.

The first and second circuits generate positive and negative waves while the circuit 3 corresponds to shirt circuit of resonance circuit (zero state). It allows control the amplitude of output current and staying in ZCS condition. The sequence of possible circuits utilization is demonstrated in Fig. 2. As a result, the resonance link is working with very high constant switching frequency, while the utilization of cycle skipping control can allow regulate output voltage or current (in case of fixed secondary side voltage).

# 3 ZVS/ZCS condition and regulation range

It is evident that minimization of magnetic components is not enough and soft switching along with wide range of current/voltage regulation are very important features to be addressed.

## 3.1 ZCS vs ZVS

In case of GaN transistors utilizing in ZCS condition it is well-known that power losses are defined by Eoss and Eqoss losses where the Eoss loss is introduced by the capacitance self-discharging current of the switch device itself and Eqoss loss is introduced by the capacitance charging current from the opposite switch device [30].

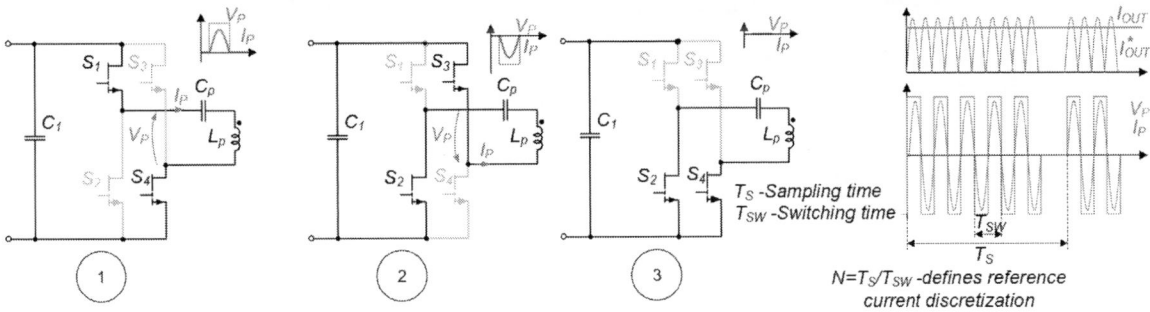

**Fig. 2** Equivalent circuits of the proposed solution with ZCS along with Illustration of cycle skipping approach which allows regulating output voltage or current.

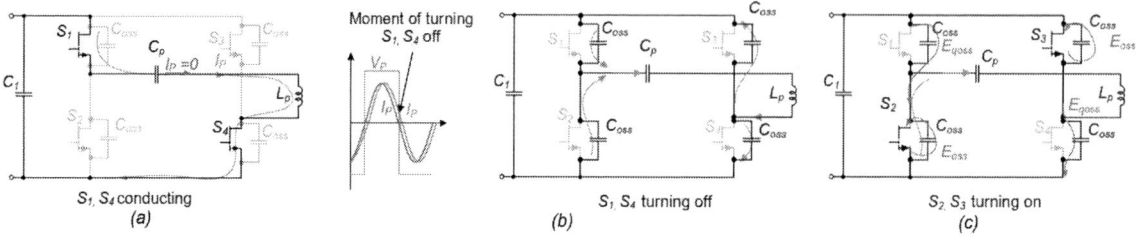

**Fig. 3** Equivalent Proposed isolated dc-dc interface which is based on high switching frequency WBG semiconductors along with air transformer with full compensation (CLLC configuration).

It is well-known that even in the case of GaN semiconductors, the ZCS is not enough and at high frequency (500+ kHz) switching losses dominates over conduction losses. It means that ZVS condition becomes more important compared to ZCS in case of very high switching frequency. In [31] it was clearly shown that for a design and optimization of a converter system with ZVS, it is crucial to identify the conditions under which soft switching can be achieved. A basic requirement for ZVS is a semiconductor half-bridge with an inductive element connected to the midpoint, which is also common to many topologies.

The inductance present is not enough as ZVS condition requires the presence of an impressed current of an inductive component which charges/discharges the output capacitances of the MOSFETs within a bridge leg during the interlocking time of the associated gate signals.

As it was shown in [31], the condition of ZVS in case of half-bridge converter with an inductive element connected to the midpoint is following:

$$\frac{1}{2}L \cdot I_I^2 = Q_{oss}(V_{DC}) \cdot V_{DC}, \qquad (6)$$

where $I_I$ is the initial current in inductor $L$, $V_{DC}$ – dc-link voltage and $Q_{oss}(V_{DC})$ the stored charge of the MOSFETs [32].

However, in our particular case, the condition of ZVS/ZCS is illustrated in Fig. 3. It shows that in case of ZCS (red color), in the moment before commutation (Fig. 3a), the current is close to zero and straight after transistors $S_1$, $S_4$ off there is no enough energy to recharge the $C_{OSS}$ capacitors (Fig. 3b) and as a result all energy accumulated in $C_{OSS}$ is discharging through the transistor body when transistors $S_2$, $S_3$ are ON (Fig.3c).

In case of current conducting during the switching transistors OFF, the situation is very similar to the described case in [31]. At the same time, condition described in eq. (7) cannot be used directly due to the presence of series compensator. At the same time, from IPT application it is known that SS compensation approach in case of symmetrical full compensation may have only capacitive condition [33]. It means that current phase depicted by green color in Fig. 3 is not achievable in symmetrical circuit. It is reasonable to analyze asymmetrical case.

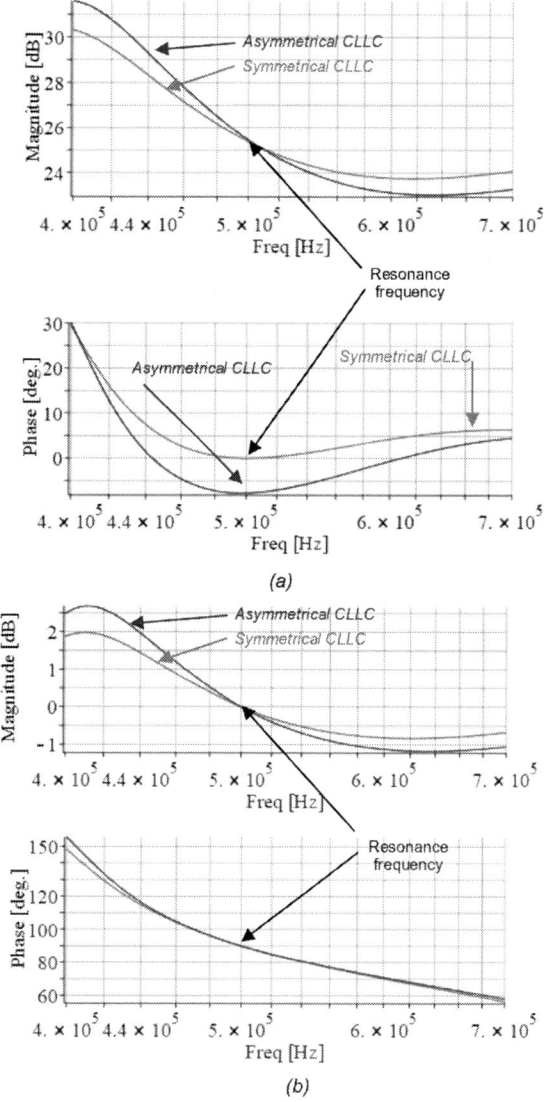

**Fig. 4** Bode diagram of input impedance (a) and gain factor (b).

Fig. 4a shows Bode diagram of the initial symmetrical and asymmetrical case. The asymmetrical case is provided by means of reduced capacitor on the secondary side. Due to the reduced compensation capacitor on the secondary side, the input current has a slight inductive feature which provides ZVS during commutation during turn on.

At the same time, the current value at the moment of commutation has to be defined. Eq. (7) is not appropriate in this case as we have different configuration with series capacitor and due to the phase shift between primary and secondary side, energy is not accumulated only in the leakage inductance. All together it leads to a set of differential equations with simple, but bulky solutions which takes into account all above mentioned parameters.

## 3.2 Voltage/current regulation capability

Considering that cycle skipping control can provide any level of buck operation it is reasonable to study of the boosting capability. Fig. 4b shows Bode diagram of the Gain factor, where resonant frequency is equal to 500 kHz. Gain factor is equal to 1 unit in case of ideal matching of switching and resonant frequencies. It can be seen, that not significant switching frequency changes leads to significant gain factor tuning. It gives the opportunity to regulate the output voltage in the wide range. Also, it should be noticed, that asymmetrical solution has higher maximum gain factor.

# 4 Experimental verification

Fig. 5 shows the realized experimental prototype of the converter rated for 5 kW power. Table I summarizes the passive components along with semiconductors that are selected to provide reference converter. GaN semiconductors were selected as high-frequency switching transistors S1-S4. At the same time, SiC diodes SCS230AE2HRC11 were used for rectification stage.

Fig. 5a shows the general view of the full experimental setup. It has split primary and secondary side PCBs with heatsinks, air transformer and control board. The control board is used in order to provide open loop switching signals and is realized on TMS32028379DPTPT. This is a µC with two cores and two sub-cores specially designed for power electronics application and allows realizing any complex control algorithm with high switching frequency.

The primary side setup is shown in Fig. 5b. It has common heatsink for all GaN semiconductors, which helps to dissipate the power losses and equalize the temperature of semiconductors. It has drivers, auxiliary isolated dc-dc supplies for drivers and series ceramic capacitors along dc-link capacitors. The secondary side setup is shown in Fig. 5c.

It has the same common heatsink for all rectification diodes. Such an approach helps to measure and compare of power losses in primary and secondary side converters. Also, it has series secondary side ceramic capacitors along with output filtering capacitors.

**Table I** Experimental prototype specifications.

| Parameter | Value |
|---|---|
| Capacitors $C_1$, $C_2$ | 7 µF |
| Capacitor $C_P$ | 17 nF |
| Capacitor $C_P$ | 17nF/9 nF |
| Air transformer ($L_P$, $L_S$, M) | 6.4 µH/6.1 µH/5 µH |
| Switching frequency | 500 kHz |
| High switching frequency transistors $S_1$-$S_4$ | GS66516T-TR |
| Primary side heatsink | 1.5 K/W |
| High switching frequency di- | SCS230AE2HRC11 |
| Secondary side heatsink | 1.5 K/W |
| Thermal film for GaN | 1.4 K/W |

## 4.1 Experimental tests

Further, we describe the experimental verification of the proposed solution. All the measurement results were obtained by voltage probes Tektronix TPA-BNC, current probes Tektronix TCP0150 and by the digital oscilloscope Tektronix MDO4034B-3. Chroma 62150H-1000S power supply was used as input voltage source. Bidirectional dc voltage source along with the dc electronic load were used to emulate the dc grid and were used as input and output voltage sources. Precision power analyzer YOKOGAWA WT1800 was used to estimate the efficiency of the converter.

The experimental study was targeted to achieve several aims. The first aim was to evaluate the operation of the proposed solution. Detailed efficiency study of the solutions in different operating points was a second goal. The first experimental tests were devoted to the study of impact of asymmetrical approach. The very first test is devoted to symmetrical case when primary and secondary side series capacitors were the same 17 nF. The oscillograms are shown in Fig. 6a. In this case the efficiency was about 91.8% and the temperature of transistors was about 90°C, which indicates significant losses. Due to this, the input and output voltage was at 120 V and power only 500 W. Straight after transistor $S_1$ is turned off, the energy in equivalent series inductance is not enough to recharge $C_{OSS}$ capacitors. Next test was conducted in slightly asymmetrical case, when secondary side series capacitor is reduced to 14 nF and is shown in Fig. 6b. The efficiency is better (92.9%) and temperature dropped to 75°C. The final test at the same voltage level was conducted in case of secondary side series capacitor is reduced to 9 nF.

The oscillograms are shown in Fig. 6c. In this case, the efficiency is significantly better (93.7%) and temperature dropped to 33°C. The current on the primary side in the moment of commutation was equal to 10 A which corresponds to the analytically predicted value. It means that switching losses in GaN semiconductors related to the $C_{OSS}$ capacitors are eliminated and further voltage and power rising is possible.

Fig. 7 is devoted to the demonstration of steady-state operation in the nominal operation point with 350 V input output and 4.2 kW maximum power. In this case, efficiency was around 95 %. Three figures cover all main diagrams that disclose the operation. In particular, Fig. 7a shows voltage across transistors along with primary and secondary currents. Fig. 7b shows the transient process of the voltage across transistors along with primary and secondary currents. Eventually, Fig. 7c shows voltage across transistors along with input and output currents to demonstrate non-significant ripple in the current.

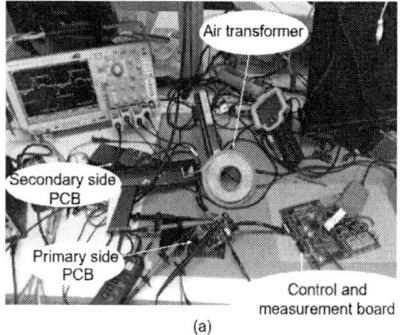 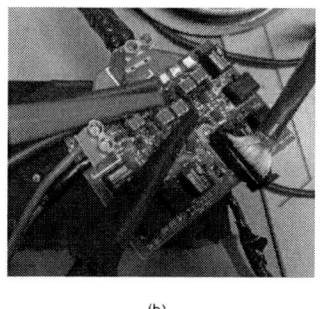 

(a) (b) (c)

**Fig. 5**  General view of the laboratory experimental setup (a), primary side PCB with GaN semiconductors (b) and secondary side PCB with SiC semiconductors (c).

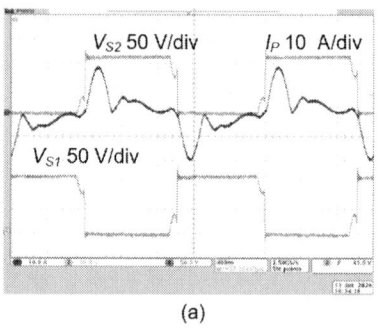 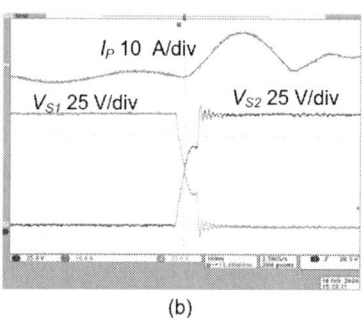 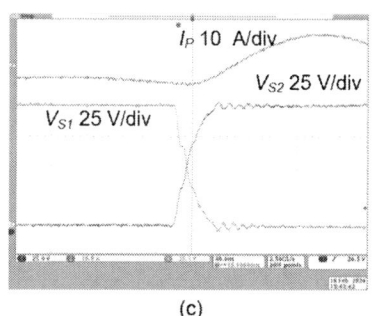

(a) (b) (c)

**Fig. 6**  Experimental results of the steady-state operation in symmetrical and asymmetrical mode: symmetrical mode with 500 W and 120 V input/output voltage (a), asymmetrical mode with 500 W and 120 V input/output voltage but without ZVS (b), asymmetrical mode with 500 W and 120 V input/output voltage with ZVS (c).

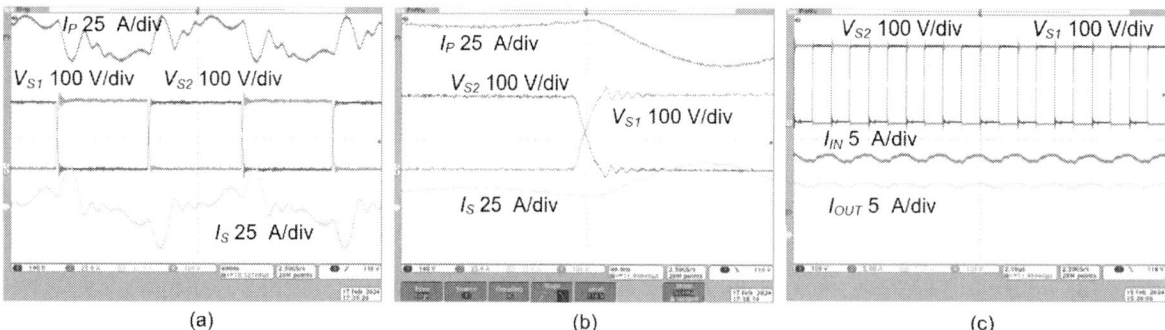

(a) (b) (c)

**Fig. 7**  Experimental results of the steady-state operation in nominal power point with 4.2 kW: voltage across transistors along with primary and secondary currents (a), transient process of the voltage across transistors along with primary and secondary currents (b), voltage across transistors along with input and output currents (c).

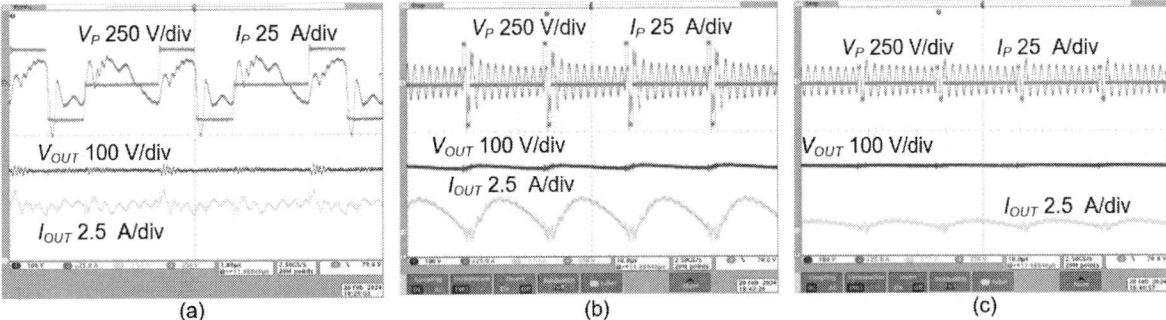

(a)                    (b)                    (c)

**Fig. 8** Experimental results in case of output current/voltage control by means of cycle skipping approach: waveforms of the primary side voltage and current along with output current and voltage in case of 50% skipping and 350 V input voltage (a), waveforms of the primary side voltage and current along with output current and voltage in case of 90% skipping and 350 V input voltage (b), waveforms of the primary side voltage and current along with output current and voltage in case of 90% skipping and 150 V input voltage (c).

The main conclusion from Fig. 7 is that converter is working as expected according to the theoretical expectation and simulation verification. This mode is similar to any DCX operation.

At the same time, the main goal of this solution is to realize capability to work in case of the wide range of current and voltage regulation. In theoretical analysis and simulation verification it was clearly demonstrated that capabilities. Fig. 8 is devoted to the experimental results of cycle skipping approach in different conditions. It is evident that no problems expected when the duration of the active switching is close to maximum. Fig. 8a shows the waveforms of the primary side voltage and current along with output current and voltage in case of 50% skipping and 350 V input and output voltage. In this case, half of the nominal power is delivered to the output side. It can be seen, that the current ripple is not significantly increased compared to the nominal power. Fig. 8b shows extreme case when only 10% of active pulses are given. It shows the same waveforms of the primary side voltage and current along with output current and output voltage. The input voltage was the same 350 V. In this case the output current ripple is significant and almost equal to average component, but it is acceptable for only 10% of nominal power. Eventually, the last figure (Fig. 8c) shows

the case when the input voltage was dropped to 150 V while output voltage was remaining the same 350 V. Due to the current source feature of the proposed solution, it still can provide power delivery to the secondary side. It shows the same waveforms of the primary side voltage and current along with output current and voltage.

## 4.2 Efficiency study

After proving the general concept operation capability, the second goal was to analyze whether it's feasible from efficiency point of view. The very precise four wire measurement method by means of Yokogawa WT1800 was used. The power was investigated in a range from 400 W to 4.2 kW. In order to evaluate the losses in semiconductors and inductor, the thermal camera was used. Fig. 9 is dedicated to the thermal picture of the operated converter at maximum power 1 hour after launch. It shows primary side PCB with GaN transistors (Fig. 9a), secondary side PCB with SiC diodes (Fig. 9b), and air transformer (Fig. 9c). The main conclusion of this figure is that semiconductors themselves were far from their maximum temperature condition which means that there is still potential for further power/switching frequency increasing.

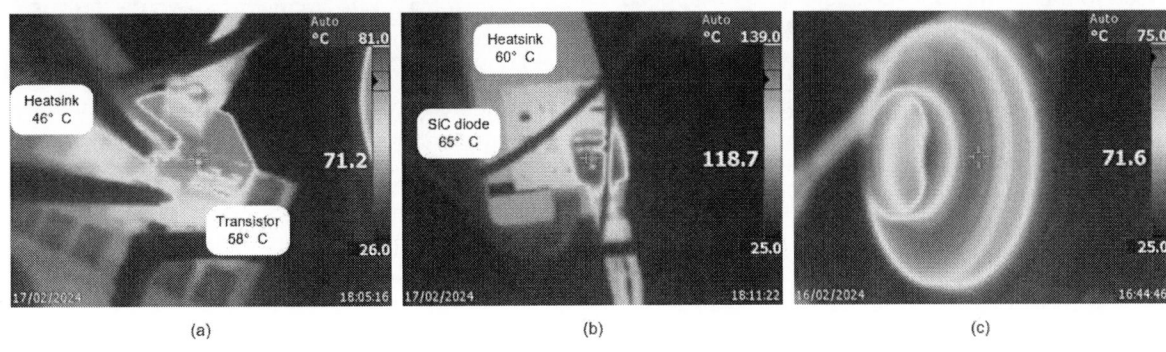

(a)                    (b)                    (c)

**Fig. 9** Thermal pictures: primary side PCB with GaN transistors (a), secondary side PCB with SiC diodes (b), and air transformer (c).

For example, in case of primary side PCB the hottest point was corresponding to the terminal connection, while the heatsink temperature was 46°C and GaN transistors only 65°C.

However, it should be mentioned that thermal design is one of the main challenges that have to be addressed during converter design with GaN semiconductors. The preliminary tests demonstrated that top-cooled semiconductors have better potential to optimize switching cells along with thermal management [17]. At the same time, thermal film conductivity is a "bottle neck" and the best one was used in this case (Table I).

Fig. 10 shows losses breakdown analysis in the nominal power point and a half of it. It can be seen that at nominal power point the GaN transistors contribute by 34 W which is less than 1 % of total power. The major losses come from air transformer. Fig. 9c shows that despite the neglecting skin effect, the proximity effect is very significant and current distribution in the wires are not equal, which in turns cause the extra heating and losses. Another point that should be emphasized is that other losses mostly related to PCB and wires are also significant and can be optimized.

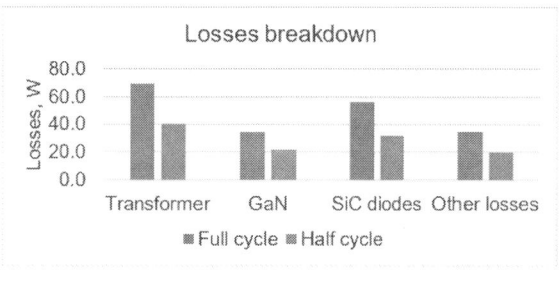

**Fig. 10** Losses breakdown analysis for nominal and a half of power.

At the same time, with further input voltage rising, the power will rise as well, which is not a problem from semiconductor point of view.

The efficiency profile as the function of the input power is illustrated in Fig. 11. This concept provides maximum efficiency at maximum power level, which in turns makes it very attractive for applications where maximum power is the most expected operating power. In case of cycle skipping appearing, the efficiency is declined due to the current oscillation on the primary side. In fact, except energy losses, it does not create any limitations to the thermal design. Also, the effect of efficiency decline at low power is very typical for any type of converter. However, in this case, the declining ramp is not high, and the efficiency is over 90 % at only 10 % of nominal power.

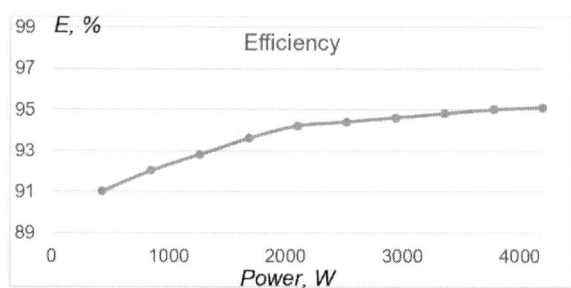

**Fig. 11** Efficiency curve versus input power in case of equal input and output voltage equal to 350 V.

## 5   Conclusions

The proposed modified asymmetrical CLLC isolated dc-dc GaN/SiC based interface with capability to control output voltage/current in the wide range is disclosed in this paper. In the case of classical CLLC with leakage inductance compensation and symmetrical circuits high value of magnetizing inductance is expected which results in correspondent losses at very high switching frequencies.

In case of the proposed solution air core transformer can be utilized. Theoretical analysis revealed that to provide full ZVS operation, an asymmetrical approach is required. Compact simplified 5 kW demonstrator was designed in order to demonstrate feasibility of the proposed approach. Despite that not extraordinary efficiency is achieved, it shows good utilization potential as GaN semiconductors introduced less than 1 % of losses. In the case of further air transformer and general prototyping optimization very high efficiency and power density can be achieved.

Also, this solution has maximum efficiency in the point with maximum power, which makes it suitable for many applications where maximum power is a most expected operation point like in EV charging systems.

## Acknowledgments

This research was supported by the Estonian Research Council grants PRG675 and PUTJD1209.

## References

[1] https://yearbook.enerdata.net/total-energy/world-consumption-statistics.html

[2] "Renewables 2022 Global Status Report," Renewable Energy Policy Network for the 21st Century, Paris, France, June 2023.

[3] J. Jaworski, N. Zheng, M. Preindl and B. Xu, "Vehicle-to-Grid Fleet Service Provision considering Nonlinear Battery Behaviors," in IEEE Transactions on Transportation Electrification., early access.

[4] P. Mohseni, O. Husev, D. Vinnikov, R. Strzelecki, E. Romero-Cadaval and I. Tokarski, "Battery Technologies in Electric Vehicles: Improvements in Electric Battery Packs," in IEEE Industrial Electronics Magazine, vol. 17, no. 4, pp. 55-65, Dec. 2023.

[5] J. Sabata, S. Shom, A. Almaghrebi, A. McCollister and M. Alahmad, "Incentivizing Electric Vehicle Adoption Through State and Federal Policies: Reviewing influential policies," in IEEE Electrification Magazine, vol. 11, no. 2, pp. 12-23, June 2023.

[6] S. Qazi et al., "Powering Maritime: Challenges and prospects in ship electrification," in IEEE Electrification Magazine, vol. 11, no. 2, pp. 74-87, June 2023.

[7] A. A. Pesaran, "Lithium-Ion Battery Technologies for Electric Vehicles: Progress and challenges," in IEEE Electrification Magazine, vol. 11, no. 2, pp. 35-43, June 2023.

[8] https://www.ricardo.com/en/news-and-insights/insights/the-challenges-and-opportunities-of-electrification-in-the-automotive-sector#:~:text=However%2C%20there%20are%20multiple%20challenges,is%20to%20source%20these%20materials.

[9] T. Dragičević, et al, "DC Microgrids-Part II: A Review of Power Architectures, Applications, and Standardization Issues," in IEEE Trans. Power Electron., vol. 31, no. 5, pp. 3528–3549, 2016.

[10] P. Purgat, A. Shekhar, Z. Qin and P. Bauer, "Low-Voltage dc System Building Blocks: Integrated Power Flow Control and Short Circuit Protection," in IEEE Ind. Electron. Magazine, vol. 17, no. 1, pp. 6-20, March 2023.

[11] Binbin Li, et al, "DC/DC Converter for Bipolar LVdc System with Integrated Voltage Balance Capability," in IEEE Trans. Power Electron., vol. 36, no. 5, pp. 54114–5424, 2021.

[12] O. Husev, D. Vinnikov, S. Kouro, F. Blaabjerg and C. Roncero-Clemente, "Dual-Purpose Converters for DC or AC Grid as Energy Transition Solution: Perspectives and Challenges," in IEEE Industrial Electronics Magazine, early access.

[13] P. Czyz et al., "Analysis of the Performance Limits of 166 kW/7 kV Air- and Magnetic-Core Medium-Voltage Medium-Frequency Transformers for 1:1-DCX Applications," in IEEE Journal of Emerging and Selected Topics in Power Electronics, vol. 10, no. 3, pp. 2989-3012, June 2022.

[14] D. Atkar, P. Chaturvedi, H. M. Suryawanshi, M. Liserre and P. Nachankar, "Adaptive Control Algorithm for Two-Stage Integrated DC Transformer in DC Microgrid Applications," in IEEE Trans. on Ind. Electron., vol. 70, no. 6, pp. 5771-5783, June 2023.

[15] D. Menzi, A. Yang, S. Chhawchharia, S. Coday and J. W. Kolar, "Novel Three-Phase Electronic Transformer," in IEEE Transactions on Power Electronics, Early access.

[16] Y. Cao, K. Ngo and D. Dong, "A Scalable Electronic-Embedded Transformer, a New Concept Toward Ultra-High-Frequency High-Power Transformer in DC–DC Converters," in IEEE Transactions on Power Electronics, vol. 38, no. 8, pp. 9278-9293, Aug. 2023.

[17] D. B. Yelaverthi, R. Hatch, M. Mansour, H. Wang and R. Zane, "3-Level Asymmetric Full-Bridge Soft-Switched PWM Converter for 3-Phase Unfolding Based Battery Charger Topology," 2019 IEEE Energy Conversion Congress and Exposition (ECCE), 2019, pp. 2737-2743.

[18] O. Husev, T. Jalakas, D. Vinnikov, N. Vosoughi and E. Persson, "PCB Design Impact on GaN-Based Converter Operation," 2023 IEEE Applied Power Electronics Conference and Exposition (APEC), Orlando, FL, USA, 2023, pp. 1-6.

[19] E. Persson and D. Wilhelm, "Gate Drive Concept for dv/dt Control of GaN GIT-Based Motor Drive Inverters," 2020 IEEE International Electron Devices Meeting (IEDM), San Francisco, CA, USA, 2020, pp. 27.6.1-27.6.4.

[20] J. W. Kolar, D. Bortis and D. Neumayr, "The ideal switch is not enough," 2016 28th International Symposium on Power Semiconductor Devices and ICs (ISPSD), Prague, Czech Republic, 2016, pp. 15-22.

[21] C. -H. Jo, G. Li and D. -H. Kim, "Design Methodology for Bidirectional Resonant Converter With Dual Resonant Frequencies for Wide Voltage Range," in IEEE Transactions on Power Electronics, vol. 39, no. 2, pp. 2372-2384, Feb. 2024.

[22] Z. Zhang, C. Liu, M. Wang, Y. Si, Y. Liu and Q. Lei, "High-Efficiency High-Power-Density CLLC Resonant Converter With Low-Stray-Capacitance and Well-Heat-Dissipated Planar Transformer for EV On-Board Charger," in IEEE Transactions on Power Electronics, vol. 35, no. 10, pp. 10831-10851, Oct. 2020.

[23] L. Zhao, Y. Pei, L. Wang, L. Pei, W. Cao and Y. Gan, "Design Methodology of Bidirectional Resonant CLLC Charger for Wide Voltage Range Based on Parameter Equivalent and Time Domain Model," in IEEE Transactions on Power Electronics, vol. 37, no. 10, pp. 12041-12064, Oct. 2022, doi: 10.1109/TPEL.2022.3170101.

[24] H. Chen, L. Wang, K. Sun and L. Lu, "A Switching Delay Strategy for Sensorless Synchronous Rectification in CLLC Converters," in IEEE Transactions on Power Electronics, vol. 39, no. 1, pp. 280-293, Jan. 2024

[25] S. V. Anbuselvi, R. Brinda, B. Sripriya and R. P. K. Devi, "Performance Analysis of 2.4KW CLLC Resonant Dual Active Bridge Converter with Different Phase Shift Modulation Techniques for EV Charging Applications," 2023 IEEE Transportation Electrification Conference and Expo, Asia-Pacific (ITEC Asia-Pacific), Chiang Mai, Thailand, 2023, pp. 1-8

[26] L. Du, X. Du, H. Cao, H. Yang and H. A. Mantooth, "An Online High-Frequency Resonant Current Digitalization Method for CLLC Converters," 2023 IEEE Energy Conversion Congress and Exposition (ECCE), Nashville, TN, USA, 2023, pp. 6317-6322

[27] H. Zhu, S. Hu, M. Tahir, Y. Bai and X. Wu, "CLLC Modeling and Control in V2G Mode to Mitigate Double Line Frequency Current for High Power Density On-board Charger," in IEEE Journal of Emerging and Selected Topics in Power Electronics, early access.

[28] R. Bosshard, J. W. Kolar, J. Mühlethaler, I. Stevanović, B. Wunsch and F. Canales, "Modeling and η-α-Pareto Optimization of Inductive Power Transfer Coils for Electric Vehicles," in IEEE Journal of Emerging and Selected Topics in Power Electronics, vol. 3, no. 1, pp. 50-64, March 2015.

[29] J. A. Anderson, M. Haider, D. Bortis, J. W. Kolar, M. Kasper and G. Deboy, "New Synergetic Control of a 20kW Isolated VIENNA Rectifier Front-End EV Battery Charger," 2019 20th Workshop on Control and Modeling for Power Electronics (COMPEL), Toronto, ON, Canada, 2019, pp. 1-8.

[30] R. Hou, J. Lu and D. Chen, "Parasitic capacitance Eqoss loss mechanism, calculation, and measurement in hard-switching for GaN HEMTs," 2018 IEEE Applied Power Electronics Conference and Exposition (APEC), San Antonio, TX, USA, 2018, pp. 919-924.

[31] M. Kasper, R. M. Burkart, G. Deboy and J. W. Kolar, "ZVS of Power MOSFETs Revisited," in IEEE Transactions on Power Electronics, vol. 31, no. 12, pp. 8063-8067, Dec. 2016.

[32] R. Miftakhutdinov, "New aspects on analyzing ZVS conditions for converters using superjunction Si and wide bandgap SiC and GaN power FETs," 2014 16th European Conference on Power Electronics and Applications, Lappeenranta, Finland, 2014, pp. 1-9.

[33] V. Shevchenko, O. Husev, R. Strzelecki, B. Pakhaliuk, N. Poliakov and N. Strzelecka, "Compensation Topologies in IPT Systems: Standards, Requirements, Classification, Analysis, Comparison and Application," in IEEE Access, vol. 7, pp. 120559-120580, 2019

PCIM Europe 2024, 11– 13 June 2024, Nuremberg     DOI: 10.30420/566262112

# DC-Bias Reduction in High-Frequency Dual Active Bridge DC-DC Converters through Slow DC Measurements

Patrick Lenzen[1], Martin Pfost[1]

[1] Chair of Energy Conversion, TU Dortmund University, Germany

Corresponding author:     Patrick Lenzen, patrick.lenzen@tu-dortmund.de
Speaker:     Patrick Lenzen, patrick.lenzen@tu-dortmund.de

## Abstract

This paper addresses the issue of DC bias in high-frequency Dual Active Bridge DC/DC converters. A new method to identify and reduce the DC bias by measuring only the DC-link voltage at each switching event is proposed. Experimental results confirm that the DC bias causes DC-link voltage ripple between both switching events during one period. This ripple is extracted by an analog circuit that allows a slow DC measurement as possible with many uC-interpreted ADCs. This DC value corresponds to the DC bias and can be eliminated by controlling the Dual Active Bridge with duty cycle control in addition to the phase shift control.

## 1 Introduction

High-frequency transformers in Dual Active Bridge (DAB) converters have the disadvantage that the transformer core can saturate. The current through the transformer increases when the integral of the voltage across the magnetizing inductance is not zero during a switching period. Because of the higher current, the transformer saturates, which can lead to overheating of the whole converter and semiconductor overload. The saturation can be caused by the controller [1], asymmetrical setup, input/output voltage changes, or temperature variations. Especially for fast-switching converters using wide bandgap semiconductors, the DC bias can be higher than for IGBT devices with lower switching frequencies [2]. A short propagation delay generates with SiC MOSFETs and a switching frequency of 100 kHz a DC bias that is 20 times higher than with comparable IGBT devices and a switching frequency of 8 kHz.

Previously, saturation is prevented by adding DC blocking capacitors in series [3] or by removing the DC bias through measurements and controller adjustments. For the identification, a flux density transducer called "Magnetic Ear", as proposed in [4], or a flux-gate sensor [5] can be used. However, these solutions have their drawbacks, passive solutions cause additional losses, while the indirect solutions with identification and measure-

ment require many additional components. A more streamlined approach, as proposed in [6], eliminates the need for additional measurement equipment. This method detects the presence of DC bias by analyzing the asymmetrical voltage ripple of the input and output.

For practical applications, DAB converters require galvanic isolation. Typically, the microcontroller of the converter can only be referenced to either the primary or secondary ground potential. Consequently, voltage/current measurements from the opposite side are usually performed with isolated amplifiers to avoid system complexity of integrating a second microcontroller. The technique presented in [6] cannot be applied for high-frequency converters, here 500 kHz, due to the low cutoff frequency of the commercially available isolated amplifiers (e.g. Texas Instruments AMC3311: 275 kHz). Furthermore, the sample & hold time of the ADC should be as short as possible to achieve a measurement cycle of two measurements per period for the input and output voltage. For a switching frequency of 500 kHz, the measurement cycle must be 1 MHz. Typical converters require additional measurements such as input current and temperature measurements which may exceed the capability of the ADC. This paper presents a novel approach to eliminate the DC bias by extracting the DC bias in analog and providing a DC signal to the controller for control.

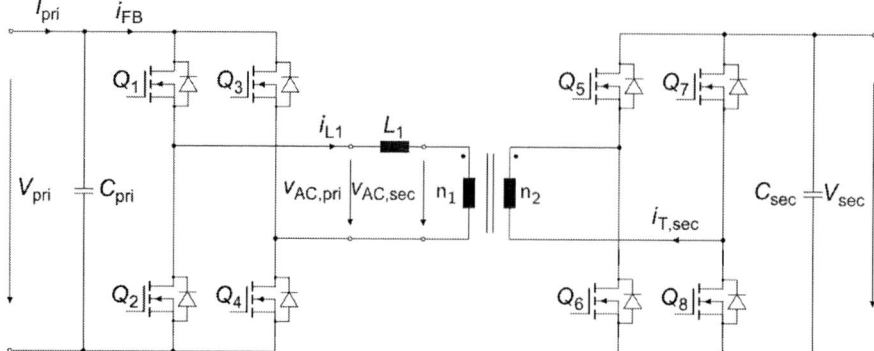

**Fig. 1:** Schematic of the converter with one primary and one secondary full bridge. The input and output filters are not shown.

## 2 Proposed System

This paper investigates a DC-bias elimination method for a DAB, as depicted in Fig. 1, with high switching frequency. The converter uses phase shift control. An input filter is required to allow voltage variation across the DC-link capacitors. The DC-link capacitor should be designed that the voltage variation is between 1-5%.

## 3 Proposed Method

The identification of the DC bias with the DC-link voltage is based on the following physical relationship: A DC bias in the primary windings, see $i_{L1}$ in Fig. 2, results in an asymmetrical current $i_{FB}$ from the input. Due to the input filter, a constant input current $I_{pri}$ is assumed. Therefore, the AC component of the asymmetrical current is buffered by the DC-link capacitor $C_{pri}$. Because of this, the DC-link voltage $V_{pri}$ is different for the two half-cycles of the primary side, see Fig. 2. If using two separate DC-link capacitors for each half bridge ($Q_1$, $Q_2$ and $Q_3$, $Q_4$), the measurement on the PCB should be geometrically exactly between the two half bridges. Otherwise, the impact of a positive DC bias on the DC-link voltage asymmetry is smaller than for a negative DC bias or vice versa. If an exact placement on the PCB is not possible, this is usually not a problem because the controller tries to reduce the asymmetry and the gain has no direct influence. The gain is only important for the stability of the controller which should be checked first.

For a DC bias in the secondary windings, the procedure is the same. The influence of a primary DC bias on the secondary DC link voltage and vice

versa is small, see [7]. Therefore, a slight asymmetry in the DC-link voltage is tolerated and not eliminated.

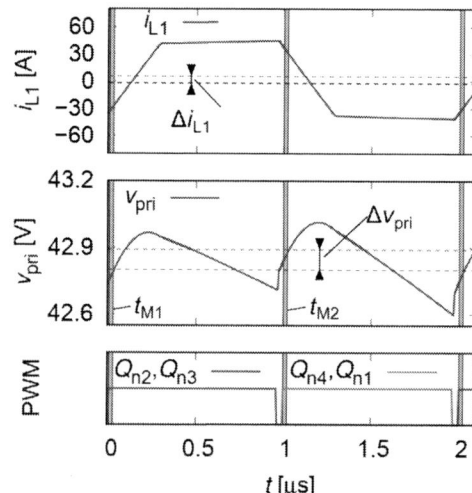

**Fig. 2:** From top to bottom, the simulation results show the inductor current $i_{L1}$ including a 5 A offset, voltage across the primary DC-link capacitor, and the PWM signal utilized to control the primary bridge.

To identify the DC bias with the controller, the DC-link voltage cannot be measured directly for converters with high switching frequencies because in our case the ADC sampling rate is too slow and the cutoff frequency of the isolated amplifier is too small, as described before. Therefore, the values at each switching event are extracted in analog and a DC signal corresponding to the DC bias is sent to the ADC of the controller.

In Fig. 3 the different blocks of the analog DC bias detection circuit are shown. To keep the circuit as cheap as possible, the whole circuit is supplied with solely 5 V. In the first stage, the DC-link voltage is prepared for the following steps. Two different peak detection modules are needed to generate a constant DC value of the input voltage at two points in time. The first measurement is at the rising edge of the PWM signal $Q_{n2}$, $Q_{n3}$ ($t_{M1}$) and the second measurement at the positive edge of the PWM signal of $Q_{n1}$, $Q_{n4}$ ($t_{M2}$), see Fig. 2. Afterwards, the difference of both signals is calculated, resulting in an equivalent value for the amount of DC bias which can be measured by the microcontroller.

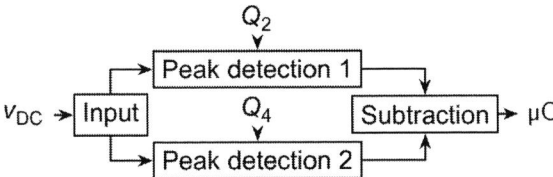

**Fig. 3:** Overview of the analog circuit with the DC-link voltage as input and the microcontroller input as output. The peak detection is triggered by the PWM signals of $Q_2$ and $Q_4$, respectively.

In Fig. 4 the detailed circuit of the input stage from Fig. 3 is shown. The DC-link signal is high-pass filtered by $C_1$ and $R_1 \| (R_2 + R_3)$ to decouple the AC signal from the DC input voltage and achieve a higher resolution. The cutoff frequency is 160 kHz. The resistors $R_2$ and $R_3$ divide the input voltage to a suitable voltage for the amplifier. The additional Zener diode parallel to $R_3$ protects the input of the amplifier. $R_4 - R_7$ form with $OP1$ a non-inverting amplifier with unity gain and add 1.65 V. Due to the voltage divider of $R_3 - R_5$ an additional voltage shift is added to the input signal. However, this is not a problem because in the last stage the signals of both timings $t_{M1}$ and $t_{M2}$ are subtracted from each other. The output resistor $R_8$ generates a minimum load to reduce the ringing when the load is changed by the following stage.

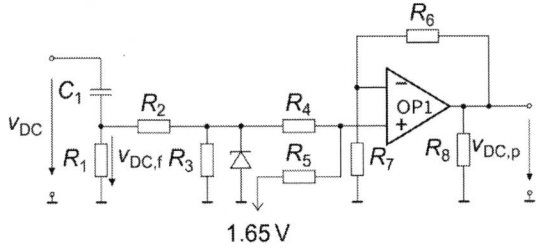

**Fig. 4:** Analog circuit decouples the DC signal and shifts the signal by 1.65 V. The input signal of the circuit is the DC link voltage.

The peak detection circuit from Fig. 3 is shown in detail in Fig. 5. The circuit extracts the voltage at one specific time from the prepared DC-link voltage $v_{DC,p}$. The DC-link voltage is prepared by the previous circuit as described in the previous paragraph, see Fig. 4. As shown in Fig. 3, the circuit is realized twice, the peak detection circuit 1 for $t_{M1}$ and the peak detection circuit 2 for $t_{M2}$. The output voltage $v_{DC,p}$ from the input stage is only allowed to pass to the precision rectifier for a 100 ns long interval after a positive edge of the PWM signal, as indicated in gray in Fig. 2. During the periods in between, the voltage is blocked by the MOSFET $Q_{sw}$. The signal for $Q_{sw}$ is generated using the PWM signal $Q_2$ for $t_{M1}$ and $Q_4$ for $t_{M2}$, respectively. The PWM signal is fed into a 74AHC123 monostable multivibrator to generate a low signal after a rising edge of the PWM signal for 100 ns.

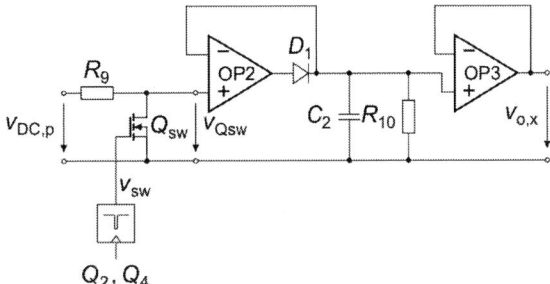

**Fig. 5:** Precision rectifier with the input signal processing is shown. The switch will be closed for a small time after the switching event of $Q_2$ or $Q_4$.

In the peak detection circuit 1, see Fig. 3, the precision rectifier with the amplifier $OP2$ (in detail in Fig. 5) outputs a DC signal corresponding to the maximum value of the DC-link voltage at the time $t_{M1}$. Subsequently, in peak detection circuit 2, this signal is instead generated during the time $t_{M2}$. The precision amplifier $OP2$ includes the diode $D_1$ and capacitor $C_2$. The resistor $R_{10}$ is needed to avoid saturation. Therefore, the output voltage of $OP2$ is slightly lower than the peak voltage of $V_{Qsw}$. After the precision rectifier an impedance amplifier $OP3$ is used to avoid load dependence. This is necessary because the input impedance of the following stage is different for both stages.

Both $v_{o,1}$, $v_{o,2}$ corresponding to $t_{M1}$ and $t_{M2}$ are subtracted by a differential unity gain amplifier $OP4$ to obtain an equivalent value representing the DC bias, see Fig. 6. Additionally, the signals are shifted by 1.65 V to perform a single-ended ADC measurement. A value above 1.65 V represents a positive DC bias and value below 1.65 V represents a negative DC bias. For the primary side, the

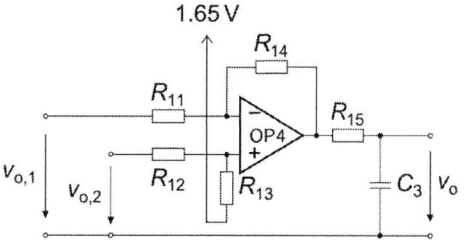

**Fig. 6:** Differential amplifier with a voltage addition of the 1.65 V reference signal.

output is referenced to the controller ground; the DC signal can be measured directly by the ADC of the microcontroller. For the secondary side, the DC signal is passed through an AMC1311 isolated amplifier to the ADC of the controller.

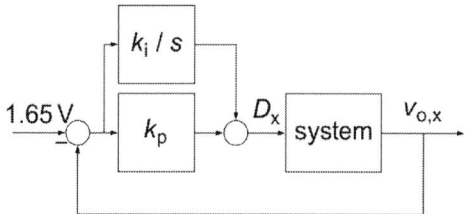

**Fig. 7:** Control loop with the P($k_p$) and I($k_i/s$) controller. The output of the controller is the duty cycle and the output of the system is the output of the proposed detection circuit.

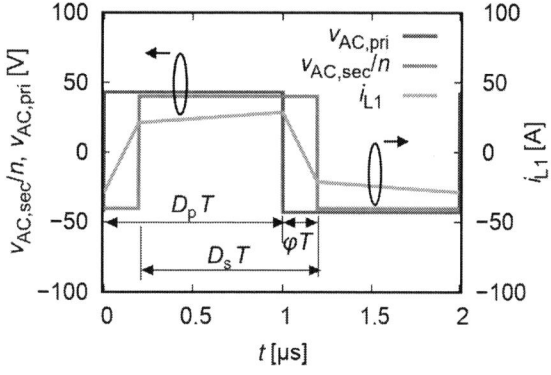

**Fig. 8:** The control method is shown. An outer phase shift $\varphi T$ is implemented between the primary voltage $v_{AC,pri}$ and the secondary voltage $v_{AC,sec}$. The duty cycle of the primary side $v_{AC,pri}$ is $D_p T$ and for $v_{AC,sec}$ $D_s T$, respectively. The current $i_{L1}$ flows through the inductor $L_1$.

The control loop is shown in Fig. 7. The controller tries to approximate the voltage $v_o$ to the reference signal. Here, the reference signal is 1.65 V which was added by the differential amplifier *OP*4. The reference signal can have a small offset; therefore, the circuit should be calibrated first to update

the reference voltage. This can be done by measuring the ADC input voltage before starting the converter. The output of the controller is the inner phase shift $D_x$ of the corresponding side (primary or secondary). By adjusting the inner phase shift $D_p$ for the primary side or $D_s$ for the secondary side, respectively, the volt-seconds on the primary or secondary side of the transformer can be adjusted, see Fig. 8.

# 4 Experimental Results

The measurements were performed with a transformer transfer ratio of $n = 15$, an input primary voltage (LV) of 43 V, an output secondary voltage (HV) of 650 V, and a switching frequency of 500 kHz.

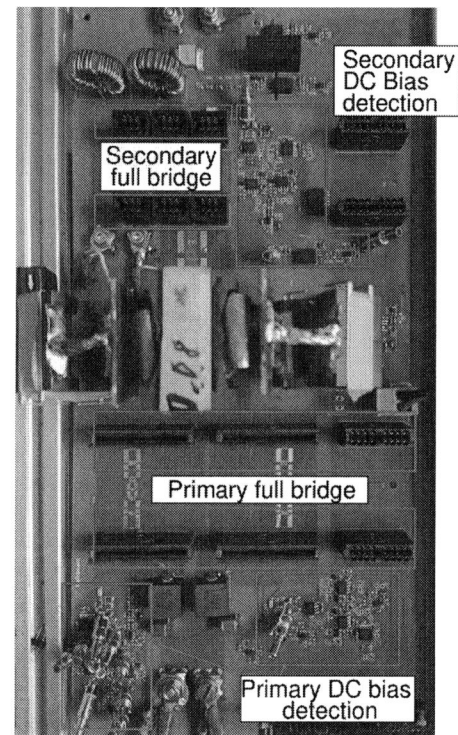

**Fig. 9:** Assembly of the DAB with two primary and one secondary full bridges. The pluggable boards with the full bridges are not shown. A DC bias detection circuit is mounted for each full bridge.

The assembly of the DAB with the DC bias detection is shown in Fig. 9. The motherboard allows the use of two identical primary full bridges. Therefore, two primary detection circuits are implemented. The transformer module on the motherboards contains the series inductors. The connection between the inductor and transformer is set up so that the transformer DC bias can be measured with the current probe. GaN HEMTs of type EPC2021 from EPC are used on the primary side and SiC MOSFETS of type C3M0065100J from Wolfspeed on the secondary side. The DC-link capacitors are designed to achieve the 5% ripple voltage. There are $2 \cdot 33\,\text{nF}$ on the HV side and $2 \cdot 6\,\mu\text{F}$ per full bridge on the LV side.

The experimental results are shown in Fig. 10 for a positive DC bias of 7 A on the LV side and in Fig. 11 for a negative DC bias of 10 A on the LV side. The impact of the DC bias on the DC link voltage $v_{\text{pri}}$ becomes visible. Therefore, the high-pass filtered signal $v_{\text{DC,f}}$ is also asymmetrical, see second plot in Fig. 10, Fig. 11. This signal is not around 0 V because of the voltage divider $R_1 - R_5$ and the reference voltage 1.65 V (Fig. 4). This offset is not a problem because of the differentiation in the last stage (Fig. 6).

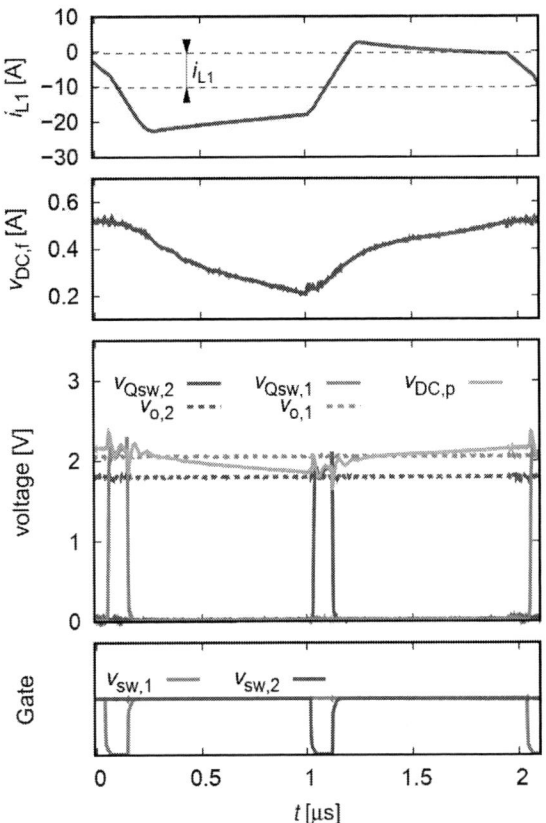

**Fig. 11:** As Fig. 10, but now for negative DC bias of 10 A.

A signal with a pulse width of 100 ns $v_{\text{sw,1}}, v_{\text{sw,2}}$ is generated by the retriggerable monostable multivibrator to drive the MOSFET $Q_{\text{sw}}$, as described in the previous section, see lower plot in Fig. 10, Fig. 11. The output signals of the MOSFETs $v_{\text{Qsw,1}}$ and $v_{\text{Qsw,2}}$ are shown in the third plot in Fig. 10 and Fig. 11. Due to the switching event small ringing is generated. This high frequency ringing is too fast for the following peak amplifier and is therefore ignored.

The output of the first peak amplifier is the voltage $v_{\text{o,1}}$ and for the second amplifier $v_{\text{o,2}}$. In the last stage the difference of $v_{\text{o,2}} - v_{\text{o,1}}$ is calculated by a differential amplifier. The positive DC bias of 7 A results in a voltage at the ADC of 300 mV. The negative DC bias of 10 A results in a voltage at the ADC of 270 mV. This is not symmetric because of an offset which can also be seen in Fig. 12.

Fig. 12 shows the relationship between the primary DC bias and the ADC input voltage. The calculation can be done with $v_{\text{o,x}} = 0.033 i_{\text{L1}} + 1.7$. The circuit is designed with a reference value of 1.65 V.

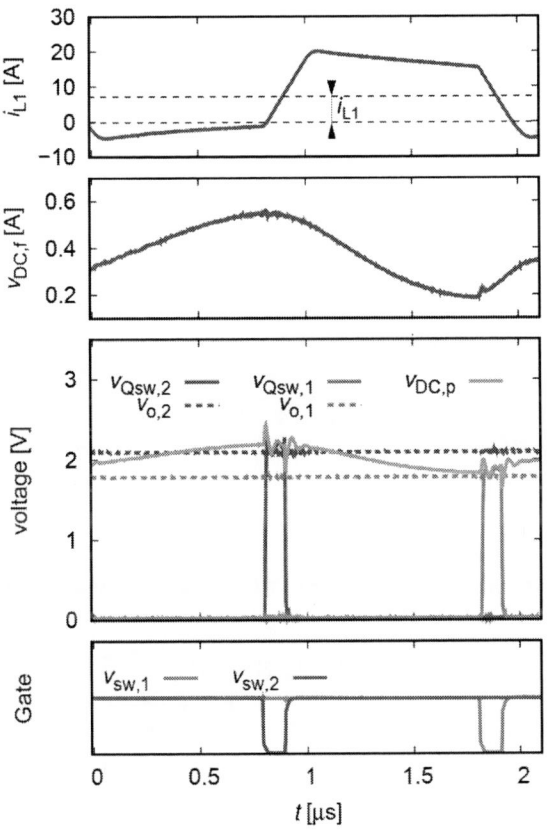

**Fig. 10:** The experimental results for the DAB with a $\Delta i_{\text{L1}} = 7$ A DC bias are presented from top to bottom: Inductor current $i_{\text{L1}}$, High-pass filtered DC-link voltage $v_{\text{in,C}}$, DC-link voltage is measured at $\Delta t_{\text{M1}}$ and $\Delta t_{\text{M2}}$ using the precision amplifier circuit, shown in Fig. 2.

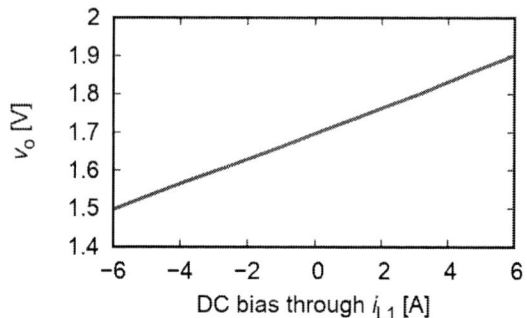

**Fig. 12:** Measured relationship between the primary DC bias and the ADC input voltage.

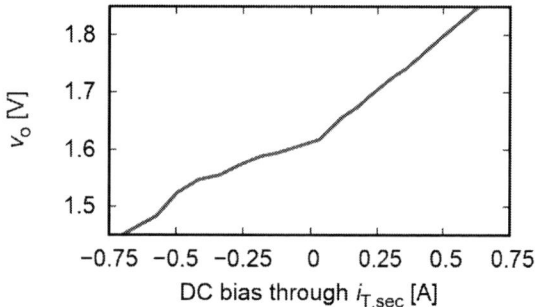

**Fig. 13:** Measured relationship between secondary DC bias and ADC input voltage.

After calibration, it can be seen that this is about 1.7 V. Therefore, calibration should be done before use, because this value is needed in the control loop, see Fig. 7. As described in the previous section, the ADC voltage can be measured before starting the converter.

The relationship for a secondary DC bias is shown in Fig. 13. The reference value of the ADC input is 1.62 V. The slope for the secondary side is different for positive and negative DC bias. This is because the measurement is not exactly in the middle between the two half bridges. In addition, the components have tolerances that must be considered. For the controller this is not a problem because the controller tries to regulate the input signal to the reference value. Small differences in the slope for positive and negative DC bias are tolerated.

## 5 Conclusion

This paper successfully demonstrates a method for detecting and eliminating the DC bias for DAB converters with galvanically isolated primary and secondary sides. The approach offers advantages over previous methods by requiring only two addi-

tional voltage measurements to detect primary and secondary DC bias. High-speed ADC measurements are not required because only a DC value needs to be measured. Therefore, this approach is applicable to converters with high switching frequencies with uC-interpreted ADCs.

## References

[1] M. Wattenberg, U. Schwalbe, and M. Pfost, "Impact of DC-bias on dual active bridge control and how to avoid it," in *2019 21st European Conference on Power Electronics and Applications (EPE '19 ECCE Europe)*, 2019, pp. P.1–P.8.

[2] N. Badenhop, L. Fräger, D. Kampen, S. Langfermann, and M. Owzareck, "DC bias currents in full-bridge DC-DCconverters in context of WBG semiconductors and high switching frequencies," in *2022 24th European Conference on Power Electronics and Applications (EPE'22 ECCE Europe)*, 2022, pp. 1–7.

[3] G. Chryssis, *High-frequency switching power supplies: theory and design*. Mc-Graw-Hill, 1989.

[4] G. Ortiz, L. Fässler, J. W. Kolar, and O. Apeldoorn, "Flux balancing of isolation transformers and application of "the magnetic ear" for closed-loop volt–second compensation," *IEEE Transactions on Power Electronics*, vol. 29, no. 8, pp. 4078–4090, 2014.

[5] G. Qiu, L. Ran, H. Feng, H. Jiang, T. Long, A. J. Forsyth, W. Shao, and X. Hou, "A fluxgate-based current sensor for DC bias elimination in a dual active bridge converter," *IEEE Transactions on Power Electronics*, vol. 37, no. 3, pp. 3233–3246, 2022.

[6] S. M. Kaviri, H. Hajebrahimi, S. Eren, and M. Pahlevani, "A digital active DC-eliminating method for DC/DC converters," *IEEE Transactions on Power Electronics*, vol. 34, no. 4, pp. 3014–3019, 2019.

[7] P. Lenzen and M. Pfost, "DC-bias elimination in high-frequency dual active bridge DC/DC converters through single-sided measurements," in *2024 IEEE Applied Power Electronics Conference and Exposition (APEC)*, 2024.

PCIM Europe 2024, 11– 13 June 2024, Nuremberg    DOI: 10.30420/566262113

# Optimized Current Sharing Technique for Interleaved CLLC Converters for Minimal Output Current Distortion

Martin Gendrin[1], Akshay Mahajan[1]

[1] Fraunhofer Institute for Solar Energy Systems ISE, Freiburg, Germany

Corresponding author:    Martin Gendrin, martin.gendrin@ise.fraunhofer.de
Speaker:    Martin Gendrin, martin.gendrin@ise.fraunhofer.de

## Abstract

An optimized current sharing technique (OCST) is proposed in this paper for a converter with 3 CLLC interleaved converters with slight differences in their respective resonance frequencies due to tolerance in the high-frequency transformers. The underlying optimization problem aims at minimizing the output current ripple and is solved off-line, resulting in an optimization vector that defines the output DC current sharing and the phase lags between the 3 converters. The resulting technique demonstrates a significant reduction both in the DC current and DC voltage ripples.

## 1   Introduction

Hydrogen, as an energy vector, will play several key roles on the path to decarbonisation, including as an electrical storage medium and in the steel industry, thanks to the possibility of producing it locally with renewable energy and its ability to be transported and stored. Many countries therefore consider it strategic to massively increase their production capacity. Germany, for example, plans to increase its green hydrogen production capacity by 5GW until 2030 and by further 5GW until 2040 at the latest. [1].

Bi-directional isolated fast switching DCDC converters, widely employed in the energy distribution sector for EV chargers, are increasingly being used for the green hydrogen generation. Among the existing topologies, the CLLC converter tends to show higher performance and efficiency thanks to the Zero Voltage Switching (ZVS) on the primary side and to the Zero Current Switching (ZCS) on the secondary side. In addition to its compactness and galvanic isolation, the turns ratio of its integrated high-frequency transformer allows a high voltage transformation ratio while maintaining a high efficiency. However, the CLLC alone tends to be unsuitable for applications that requires a wide output voltage range, such as electrolysers. There are several solutions in the literature to overcome this problem, including those listed in [2]. In the presented application, a multi-stage isolated DCDC converter is used, integrating a current controlled

buck converter, which supplies the primary side of a CLLC converter operating as a DC transformer (DCX). In order to achieve the required high output current, three such converters are connected in input-parallel output-parallel (IPOP). Unfortunately, as the output current of a CLLC converter is at best a rectified sine-wave, the output capacitor should be quite important to ensure an acceptable voltage ripple. Interleaving the outputs of the three converters is a common method to increase the output current of a converter while keeping the current ripple low. It has been also acknowledged as a solution to reduce the size of the output capacitor in [3]. The resulting converter is referred in the rest of this paper as interleaved two-stage three-phase DCDC converter and the buck and CLLC association is referred as a phase.

Component tolerance, particularly of the high frequency transformer, leads to discrepancies in the resonant frequencies of interleaved CLLC converters. This can lead to unbalanced current sharing between the parallel converters [4],[5],[6]. To overcome this problem, in [4] the turn-on time of the bottom MOSFETs of the secondary side is shifted to induce boost operation in the converter. In [5], an auxiliary converter is connected in series with the input capacitors of each parallel converter, resulting in a power flow balancing the current. Several papers solve the unbalance based on magnetic-based strategies [7]. In [6], a model for quantifying the current unbalance is proposed based on a two-stage converter using an LLC as DCX, which is quite similar to the phase topology presented phase.

All the aforementioned papers aim to correct the current unbalance between the converters. However, the differences in the resonant frequencies also tend to induce different output current shapes, leading to an increased total output current and thus output voltage ripples when the conventional interleaved design with equal current sharing between the converters and equal distribution over a PWM period is used. Therefore, this paper proposes an optimised current sharing technique (OCST) for the considered interleaved two-stage three-phase converter, which optimises the current repartition and the phase lag between each phases to reduce the output current and hence the voltage distortion. The rest of this article is structured as follows. In section 2, a brief description of the proposed converter is given. In section 3, its simplified model is presented. In section 4, the OCST is developed. Simulation results supporting the design are shown in section 5 .

## 2 The interleaved two-stage three-phase isolated DCDC converters

The considered interleaved two-stage three-phase converter is shown in Fig. 1. As highlighted, it is composed of 3 interleaved phases, each consisting of a buck and a CLLC. All the phases are connected in parallel on both the input and output DC buses. The switches of the buck and of the CLLC primary side are SiC MOSFETs. The secondary side of the CLLC is equipped with Si MOSFETs. In each CLLC, the leakage inductance of the high frequency transformer is used as resonant inductance. This hardware topology is bidirectional, which was required in this project. However for the purposes of this article, we will only consider a unidirectional power flow from input to output, i.e. from primary to secondary side for the CLLC. The table 1 summarises the physical parameters of each phase.

In table 1, the buck inductors are all considered equal, since the maximum deviation of their real inductance doesn't exceed $1.5\,\%$ with respect to the expected value. For the high frequency transformers, both the equivalent resonant inductance $L_{r\,eq}$ and the magnetizing inductance $L_m$ deviate significantly from the expected values due to tolerances and should therefore be considered individually for each phase. This results in three different CLLC resonant frequencies, where $f_{r1} > f_{r2} > f_{r3}$. The

**Tab. 1:** Main parameters of the two-stage converters

|  | *Phase 1* | *Phase 2* | *Phase 3* |
|---|---|---|---|
| $V_{DC\,in}$ | $600\,V\ -\ 800\,V$ | | |
| $L_b$ | $125.81\,\mu H$ | | |
| $C_{DC\,int}$ | $18\,\mu F$ | | |
| $C_{r\,eq}$ | $356\,nF$ | | |
| $L_{r\,eq}$ | $5.96\,\mu H$ | $6.93\,\mu H$ | $7.95\,\mu H$ |
| $L_m$ | $417.9\,\mu H$ | $364.2\,\mu H$ | $431.2\,\mu H$ |
| $f_{rx}$ | $109.3\,kHz$ | $101.33\,kHz$ | $94.6\,kHz$ |
| *turns ratio* | $4:1$ | | |
| $V_{DC\,out}$ | $70\,V\ -\ 150\,V$ | | |
| $I_{DC\,out}$ | $0\,A\ -\ 250\,A$ | | |

expression of $L_{r\,eq}$ and $C_{r\,eq}$ is given by Eq. (1).

$$L_{r\,eq} = L_{rp} + n^2 \cdot L_{rs}$$
$$C_{r\,eq} = \frac{C_{rp} \cdot C_{rs}}{n^2 \cdot C_{rp} + C_{rs}} \tag{1}$$

The control architecture of the interleaved two-stage three-phase DCDC converter is shown in Fig. 2. A common voltage PI controller provides a global output current reference $I_{DC\,out}^*$ which, thanks to the independent internal DC buses of each phase allowing full controllability of the internal DC bus current, can be freely shared between the CLLC converters by adapting the factors $g_{ix}$. No further adjustment than setting the $g_{ix}$ to $1/3$ is required to achieve current balancing. Furthermore, as the three phases are controlled by the same DSP, it is easy to control the phase lag.

With regard to the CLLC in each phase, the first half-bridge on the primary side $(S_1, S_2)$ is controlled by a pair of complementary pulse signals with a duty cycle of $50\,\%$ and integrating a dead time of $200\,ns$. The second half bridge of the primary side $(S_3, S_4)$ is controlled with the same pair of signals shifted by $180°$. Considering the unidirectional power flow mentioned earlier, the secondary side could work as a synchronous rectifier. However, to take advantage of the superior conduction performance of the Si MOSFETs and to avoid the losses induced by the forward voltage of the recovery diodes, the secondary side is also controlled using for its first $(S_5, S_6)$ and second $(S_7, S_8)$ half-bridges respectively the same pairs of signals as the second $(S_3, S_4)$ and first $(S_1, S_2)$ half-bridges of the primary side.

Because the phase converters are interleaved, they

PCIM Europe 2024, 11– 13 June 2024, Nuremberg          DOI: 10.30420/566262113

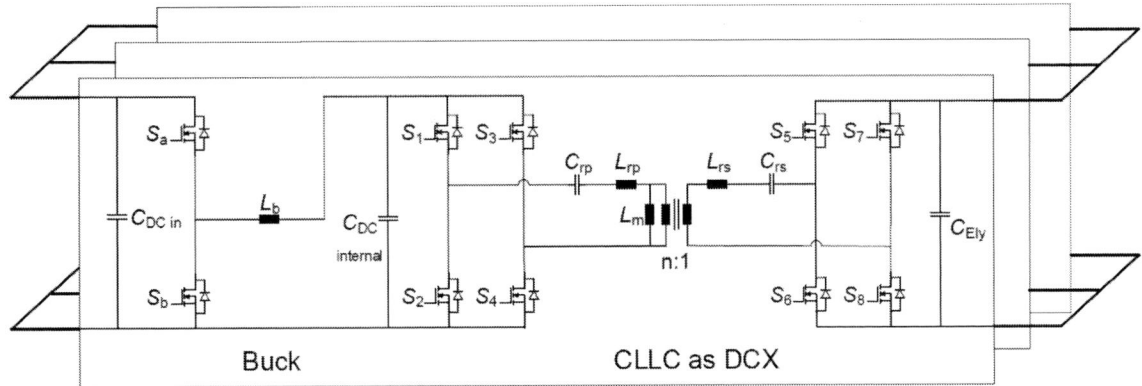

**Fig. 1:** Converter with 3 interleaved units including one buck and one CLLC converter each

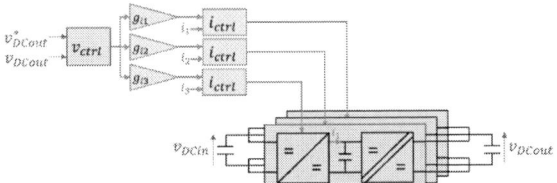

**Fig. 2:** Control architecture of the interleaved two-stage three-phase DCDC converter

share the same switching frequency $f_s$, which in this paper is set to $f_{r1}$, leading to $f_{r2}, f_{r3} < f_s$. The alternative of working with $f_s = f_{r3}$, leading to $f_{r1}, f_{r2} > f_s$ has several disadvantages. The secondary side being controlled, the secondary current is not restrained into one direction during the P-mode [8] [9] [10]. Therefore, regardless of the chosen $f_s$, ZCS is lost in the phases where the resonant frequencies differ from the switching frequency. In addition, the presence of a negative current forces the amplitude of the secondary current to increase in the phase 1 and phase 2 in order to deliver the same DC output current, resulting in increased conduction losses. Finally, the unbalance in the current sharing required to minimise the current ripple is much greater when $f_s = f_{r3}$, which would penalise one phase in particular.

## 3  Model of the interleaved two-stage three-phase DCDC converter

The proposed model will serve as the basis for the optimisation problem solved in our OCST. Therefore, it should provide sufficient accuracy in representing the output current waveform $i_{outx}(t)$ of each phase while remaining simple. In this sense,

the buck converter associated with the internal DC capacitor in each phase is modeled as an ideal voltage source converter supplying the primary side of the CLLC with $V_{DC\,intx}$.

As mentioned before $f_s = f_{r1}$ and $f_{r2}, f_{r3} < f_s$. The CLLC of phase 1 therefore operates at its resonant frequency. As for the other two phases, their respective resonant frequencies cause them to operate almost exclusively in a combination of a P-mode and a N-mode. Those modes are defined in relation with the polarities of the full-bridge voltages on the primary and on the secondary side. On the first PWM half cycle when $S_1, S_3$ turn ON (see Fig. 1), the secondary full-bridge voltage can remain clamped to $-V_{DC\,out}$ (N-mode) or switch to $V_{DC\,out}$ (P-mode) [9]. The second PWM half-cycle is then symmetrical to the first PWM half cycle in reverse direction and polarity. Each CLLC output current $i_{outx}(t)$ can be expressed by the following equation [8]. This equation was developed for a CLLC with a secondary side acting as a synchronous rectifier, which isn't the case here. It will be shown at the end of this section that this equation can also be used in our application.

$$i_{outx}(t) =$$
$$\begin{cases} I_{sx} \cdot sin(2\pi f_{rx}t) \text{ if } t \in (0, t_{px}) \\ I_{sx} \cdot sin(2\pi f_{rx}t) \\ \quad - \dfrac{2nV_{DC\,intx}}{Z_{r\,eqx}} \cdot sin(2\pi f_{rx}(t - t_{px})) \text{ if } t \in \left(t_{px}, \dfrac{T_s}{2}\right) \end{cases}$$
$$(2)$$

In Eq. (2), $T_s$ is the switching period, $Z_{r\,eqx}$ the respective equivalent resonant tank impedance expressed as follows and $n$ the transformer ratio.

911

$$Z_{r\,eqx} = \sqrt{\frac{L_{r\,eqx}}{C_{r\,eqx}}} \qquad (3)$$

Eq. (2) includes the time $t_{px}$ and the output current amplitude $I_{sx}$ that need to be expressed. Considering the mean value of Eq. (2) as the output DC current of the associated CLLC converter $I_{DC\,outx}$ and the boundary condition $i_{outx}\left(\frac{T_s}{2}\right) = 0$, an analytical function can be derived that relates $I_{sx}$ to $I_{DC\,outx}$ and $t_{px}$ to $I_{DC\,outx}$, which will be specific to each CLLC converter. In particular, $t_{px}$ can be expressed as :

$$t_{px} = \frac{T_s}{2} - \frac{asin\left(\dfrac{Z_{r\,eqx} \cdot sin\left(2 \cdot \pi \cdot f_{rx} \cdot \dfrac{T_s}{2}\right)}{2 \cdot V_{DC\,intx}} \cdot \dfrac{I_{sx}}{n}\right)}{2 \cdot \pi \cdot f_{rx}}$$

$$(4)$$

As mentioned earlier, the CLLC of phases 2 and 3 will operate in N-mode directly from the start of the dead time and for a duration of $T_s/2 - t_{px}$. According to Eq. (4), this duration will be in the worst case, i.e. for the phase 3 with maximum output load and minimum input voltage, about $80\,ns$, so within the dead time. During this time, the secondary side works as a synchronous rectifier, justifying the use of Eq. (2).

## 4 The proposed optimised current sharing technique

The proposed OCST foresees an offline resolution of an optimisation problem, based on the model presented in the previous section and aiming at minimising the output current ripple, in order to quantify the current factors $g_{ix}$ and the phase lag. The DC output currents $I_{DC\,out2}$ and $I_{DC\,out3}$ flowing through phase 2 and phase 3, the time delay $\delta t_2$ between phase 1 and phase 2 and the time delay $\delta t_3$ between phase 1 and phase 3 are considered as optimisation variables. $\delta t_2$ and $\delta t_3$ correspond to the time in the PWM period when the associated phase starts to operate in P-mode, corresponding to the zero crossing of $i_{o2}(t)$ and $i_{o3}(t)$ respectively. Note that the current $I_{DC\,out1}$ is constrained by $I_{DC\,out2}$, $I_{DC\,out3}$ and the current reference $I^*_{DC\,out}$.

Given the optimisation vector $x = [I_{DC\,out2}, I_{DC\,out3}, \delta t_2, \delta t_3]$, the current in each

CLLC converter $i_{outx}(t)$ can be derived with Eq. (2), Eq. (4) and the function linking $I_{sx}$ to $I_{DC\,outx}$. The three resulting currents can be used in the calculation of the RMS value of the total output current distortion $\Delta I_{DC\,outRMS}$ expressed below, which is the value the optimisation seeks to minimise.

$$\Delta I_{DC\,outRMS} =$$
$$\sqrt{\frac{2}{T_s} \cdot \int_0^{\frac{T_s}{2}} i_{out1}(t) + i_{out2}(t) + i_{out3}(t) - I^*_{DC\,out}} \qquad (5)$$

To derive the cost function for the optimisation problem, a time-independent function should be expressed by evaluating the integral of Eq. (5). To simplify this evaluation, the duration of the N-mode $T_s/2 - t_{px}$ is neglected ans the integral is simplified to a piecewise integral defined on three time intervals $[0 : \delta t_2]$, $[\delta t_2 : \delta t_3]$ and $[\delta t_3 : T_s/2]$. The result of this evaluation is the cost function $\Delta I_{DC\,outRMS}(x)$, obtained in this paper with Mathcad Prime.

Fig. 3 shows the RMS value of the output current ripple calculated with Eq. (5) for $\delta t_2$ and $\delta t_3$ varying over their entire range $[0 : T_s/2]$ and considering an ideal current balance between the phases. It is clear from this figure that there are two local minima corresponding approximately to an equal distribution over $T_s/2$ of the phase lag with the two possible arrangements phase 1 - phase 2 - phase 3 or phase 1 - phase 3 - phase 2. As labelled in this figure, the first arrangement has a lower minimum than the second independently of the output load.

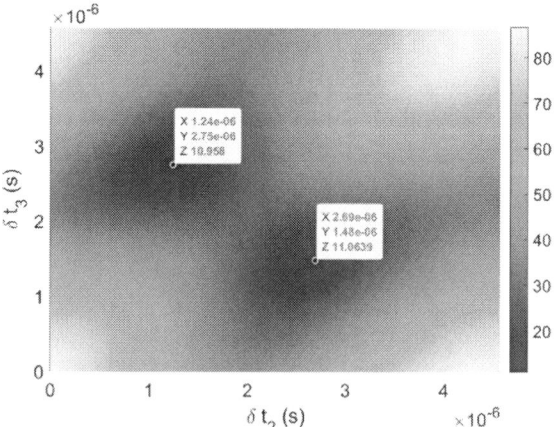

**Fig. 3:** RMS value of the output current ripple calculated with Eq. (5) with respect to $\delta t_2$ and $\delta t_3$ for $I_{DC\,outx} = 66,67\,A$ and $V_{DC\,intx} = 360\,V$

This could be problematic because the Matlab function *fmeansearch* used to solve the optimisation problem in this paper deals with unconstrained non-linear optimisation problems and could therefore converge to the wrong local minima. To avoid this, $\delta t_2$ and $\delta t_3$ are initialised in order to correspond to an equal distribution over $T_s/2$ of the current maxima in the three phases in the correct order and with respect to the phase rise times.

$$\delta t_{20} = \frac{T_s}{6} - \left( \frac{1}{4 \cdot f_{r2}} - \frac{1}{4 \cdot f_{r1}} \right)$$
$$\delta t_{30} = \frac{2 \cdot T_s}{6} - \left( \frac{1}{4 \cdot f_{r3}} - \frac{1}{4 \cdot f_{r1}} \right) \quad (6)$$

Considering Eq. (2) and the formulation of the optimisation problem, it appears clearly that the optimal solution for $x$ depends on the working point through $V_{DC\,intx}$ and $I^*_{DC\,out}$. Fig. 4 shows the evolution of the factors $g_{ix}$ and of the time delays $\delta t_x$ normalised by their respective mean values over the voltage range of the electrolyser $[70\,V : 150\,V]$. This figure shows a negligible evolution of the optimal solution with respect to the working point. The optimisation problem can therefore be solved only for the nominal working point and the results used over the entire range.

## 5 The Simulation results

The proposed OCST has been validated in simulation using a PLECS model. This detailed model accounts for the interleaved two-stage three-phase

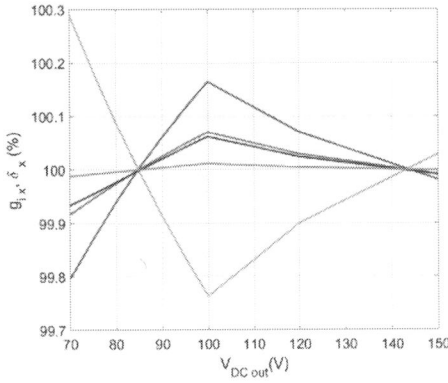

**Fig. 4:** Evolution of the factors $g_{ix}$ and of the time delay $\delta t_x$ normalised with their respective mean values expressed in % over the voltage range $[70\,V : 150\,V]$

DCDC converter shown in Fig. 1, including its control architecture (Fig. 2). Great care has been taken to model the discrete characteristics of the control. Thermal models in form of look-up tables have been added to the PLECS semiconductor models to accurately simulate both the conduction and switching losses of the MOSFETs and diodes. For the SiC MOSFETs and body diodes used in the buck and primary side of the CLLC, a thermal model is supplied by the manufacturer, Wolfspeed, and is used directly. Unfortunately, no thermal models were available for the secondary Si MOSFETs and body diodes. An application specific thermal model has been derived from a comparable SiC MOSFET. Most of the equivalent series resistances of the passives are also included for each phase in both the buck and the CLLC. This high level of detail allows an accurate simulation of the physical converter. It allows the losses of the stages in each phase to be estimated. The converter parameters are taken from table 1. The part of the model representing one phase of the interleaved two-stage three-phase DCDC converter is shown in Fig. 5. The switches of the CLLC secondary side, shown as single switches in Fig. 1, are actually two Si MOSFETs in parallel sharing the same command signals as presented in Fig. 5.

The optimisation problem is solved using the *fminsearch* function of Matlab. The resulting optimised vector $x$ is then fed to the PLECS model. In order to assert the performance of the OCST, the simulation results are compared to the results obtained with :

- the considered interleaved two-stage three-phase converter working with a perfect current balance and the classical interleaved design,

- an ideal interleaved two-stage three-phase converter, where all the resonant frequencies and the switching frequency are equal to $f_{r1}$.

The simulation results are shown in Fig. 6 in steady state with $I_{DC\,out} = 240\,A$ and $V_{DC\,out} = 120\,V$. As expected, the OCST adapts the two delays $\delta t_2$ and $\delta t_3$, which results in a shift of the local minima of the total output current with optimisation compared to the two other total output currents. This modification, together with the optimised current sharing between the phases, allows on one side a reduction of the output current ripple and on the other side the reduction of the output voltage ripple, as clearly shown in Fig. 6.

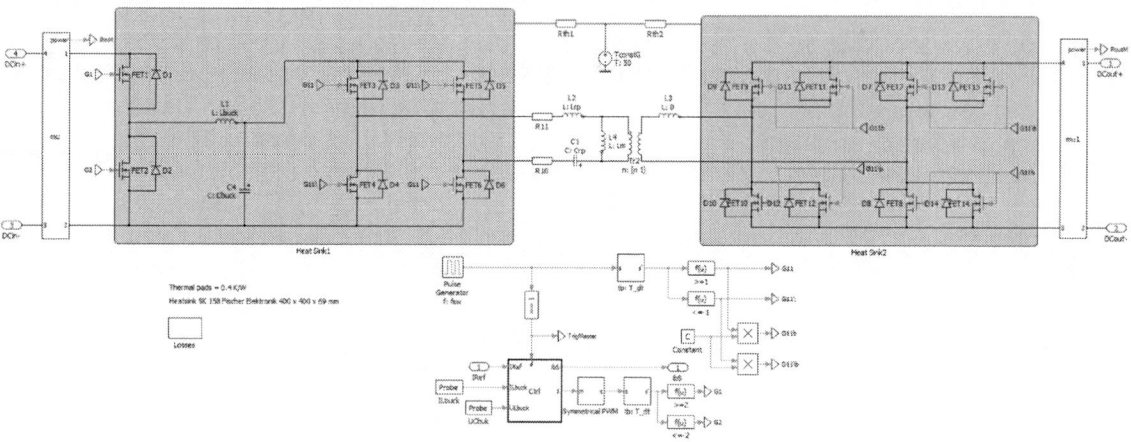

**Fig. 5:** PLECS model representing one phase of the interleaved two-stage three-phase DCDC converter

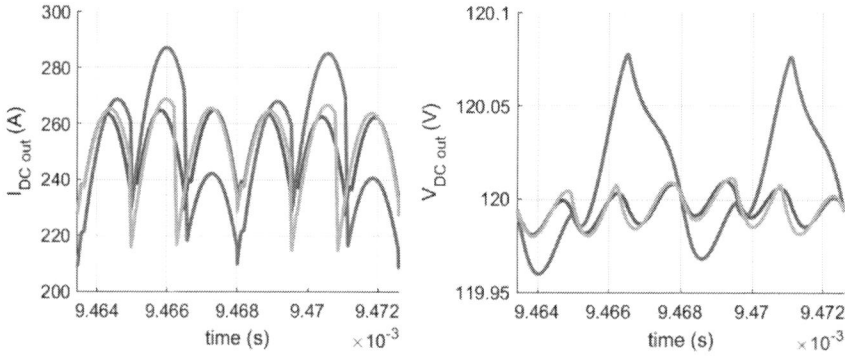

**Fig. 6:** Simulation results with left the overall output current $I_{DC\,out}$ and right the resulting output DC bus voltage $V_{DC\,out}$. The results obtained with the ideal converter are depicted in blue, with optimisation in yellow and without optimisation in red.

**Tab. 2:** RMS values of the output current/voltage ripple

|  | *ideal* | *OCST* | *no OCST* |
|---|---|---|---|
| $\Delta I_{DC\,outRMS}$ | 10.03 A | 13.38 A | 21.25 A |
| $\Delta V_{DC\,outRMS}$ | 7.6 mV | 8.8 mV | 33.9 mV |

The RMS values of the output current and voltage ripples are presented in table 2. Those results show an improvement in the current ripple and, more significantly, in the output DC bus voltage ripple. $\Delta I_{DC\,outRMS}$ with optimisation is only 133.4% that of the ideal converter, while without optimisation this factor rises to 211.86%. For the output voltage ripple, these ratios are 116.08% and 446.27% respectively. The output capacitor could be reduced accordingly while keeping the output voltage ripple acceptable.

The OCST doesn't target the repartition or the minimisation of the phase losses. However, it should be proven that it doesn't cause a significant reduction of the converter efficiency. The table 3 summarises the phase losses obtained for the three simulated cases. For each phase, the losses include the switching and conduction losses of the buck and of the primary and secondary sides of the CLLC.

**Tab. 3:** Phase and total losses of the two-stage three-phase DCDC converter

|  | *ideal* | *OCST* | *no OCST* |
|---|---|---|---|
| $P_{l1}$ | 328.2 W | 293.5 W | 323.7 W |
| $P_{l2}$ | 328.2 W | 380.1 W | 334.7 W |
| $P_{l3}$ | 328.2 W | 348.6 W | 360.9 W |
| $P_{l\,total}$ | 984.6 W | 1022.2 W | 1019.3 W |

As expected, in table 3 the losses are the same in all phases in the ideal case. As for the results obtained without the OCST on the converter with tolerance, the two phases 2 and 3 operating out of the resonant frequency show higher losses compared to the ideal case, with the highest losses for phase 3 operating the farthest from its resonant frequency. According to the simulation, the increased current flow in the body diodes on both the primary and secondary sides during the N-mode associated with their poor conductivity tends to dominate in the increased losses in phases 2 and 3. Regarding the results obtained with our OCST, it should be first noted that the total losses remain at the same level when it is activated, resulting in an unchanged total converter efficiency. However, the OCST tends to increase the loss discrepancies between the phases. Indeed, the optimal point that minimizes the output ripple induces the following factors : $g_{i1} = 0.9446/3$, $g_{i2} = 1.0766/3$ and $g_{i3} = 0.9788/3$. By reducing the current in phases 1 and 3, the associated losses are reduced, while the losses of phase 2 are increased. This results in a standard deviation of $35.78\,W$ with OCST and of $15.6\,W$ without. This effect should be verified experimentally and its consequences, such as thermal disparities, asserted.

## 6  Conclusion

This paper has presented an optimized current sharing technique aimed at minimizing the output current ripple of an interleaved two-stage three-phase DCDC converter with different CLLC resonant frequencies. By adjusting both the current sharing and the phase lag between the phases, this technique manages to reduce the RMS value of the current ripple by about $37\,\%$, which leads to a reduction in the RMS value of the voltage ripple by about $74\,\%$, according to simulation results. This improvement doesn't lead to a reduction of the overall efficiency of the converter, although the evolution of the phase efficiencies must be carefully considered in further work.

## References

[1] B. fuer Wirtschaft und Klimaschutz, *Die Nationale Wasserstoffstrategie*, 1st ed. Bundesministerium fuer Wirtschaft und Energie, 2020.

[2] N. Zanatta, T. Caldognetto, D. Biadene, G. Spiazzi, and P. Mattavelli, "A two-stage dc-dc isolated converter for battery-charging applications," *IEEE Open Journal of Power Electronics*, vol. 4, pp. 343–356, 2023. DOI: 10.1109/OJPEL.2023. 3271227.

[3] K.-H. Yi and G.-W. Moon, "Novel two-phase interleaved llc series-resonant converter using a phase of the resonant capacitor," *IEEE Transactions on Industrial Electronics*, vol. 56, no. 5, pp. 1815–1819, 2009. DOI: 10.1109/TIE.2008.2011310.

[4] M. Sato, R. Takiguchi, J. Imaoka, and M. Shoyama, "A novel secondary pwm-controlled interleaved llc resonant converter for load current sharing," in *2016 IEEE 8th International Power Electronics and Motion Control Conference (IPEMC-ECCE Asia)*, 2016, pp. 2276–2280. DOI: 10.1109/IPEMC.2016.7512652.

[5] M. Wang, X. Zha, S. Pan, J. Gong, and W. Lin, "A current-sharing method for interleaved high-frequency llc converter with partial energy phase shift regulation," *IEEE Journal of Emerging and Selected Topics in Power Electronics*, vol. 10, no. 1, pp. 760–772, 2022. DOI: 10.1109/JESTPE. 2021.3060445.

[6] F. Wang, X. Wang, and X. Ruan, "Model building and current sharing analysis for ipop converter system based on two-stage llc resonant converter," in *IECON 2023- 49th Annual Conference of the IEEE Industrial Electronics Society*, 2023, pp. 1–6. DOI: 10.1109/IECON51785.2023. 10311912.

[7] Y. Nakakohara, H. Otake, T. M. Evans, T. Yoshida, M. Tsuruya, and K. Nakahara, "Three-phase llc series resonant dc/dc converter using sic mosfets to realize high-voltage and high-frequency operation," *IEEE Transactions on Industrial Electronics*, vol. 63, no. 4, pp. 2103–2110, 2016. DOI: 10.1109/TIE.2015.2499721.

[8] Y. Cao, M. Ngo, D. Dong, and R. Burgos, "A simplified time-domain gain model for cllc resonant converter," in *2021 IEEE Energy Conversion Congress and Exposition (ECCE)*, 2021, pp. 3079–3086. DOI: 10.1109/ECCE47101.2021. 9596002.

[9] X. Fang, H. Hu, Z. J. Shen, and I. Batarseh, "Operation mode analysis and peak gain approximation of the llc resonant converter," *IEEE Transactions on Power Electronics*, vol. 27, no. 4, pp. 1985–1995, 2012. DOI: 10.1109/TPEL.2011.2168545.

[10] Z. Lv, X. Yan, Y. Fang, and L. Sun, "Mode analysis and optimum design of bidirectional cllc resonant converter for high-frequency isolation of dc distribution systems," in *2015 IEEE Energy Conversion Congress and Exposition (ECCE)*, 2015, pp. 1513–1520. DOI: 10.1109/ECCE.2015. 7309873.

PCIM Europe 2024, 11– 13 June 2024, Nuremberg     DOI: 10.30420/566262114

# Primary-side Output Regulation Principles in Dynamic Multi-MHz Inductive Power Transfer Systems and Isolated DC/DC Converters

Ioannis Nikiforidis[1,2], Roberto La Rosa[3], Prateek Wangle[1], Paul. D. Mitcheson[1,2]

[1] Imperial College London, United Kingdom
[2] Bumblebee Power, United Kingdom
[3] STMicroelectronics, Italy

Corresponding author:    Ioannis Nikiforidis, i.nikiforidis@imperial.ac.uk
Speaker:    Ioannis Nikiforidis, i.nikiforidis@imperial.ac.uk

## Abstract

Typically, output regulation of inductive power transfer systems is implemented through an additional DC/DC power conversion stage between the high frequency rectifier and the load. However, output regulation can also be achieved directly from the transmitter side, which could reduce cost and complexity and increase system efficiency. High frequency Class E-type inverters are traditionally operated in open-loop, hence both output and efficiency are strongly load dependent. Here, these limitations are overcome by introducing various control schemes, which achieve constant output while maintaining soft-switching. In our experiments, the output is regulated, and soft-switching is maintained across the entire load range, even after introducing significant misalignment.

## 1 Introduction

Unlike the Class D primary drivers, the Class E-type inverters are low bandwidth, operating at a specific frequency, and maintain their optimal operating conditions and soft-switching only for a specific load [1]. Even the output current load-independent Class EF [2] and Class E [3] inverters, cannot maintain stable secondary outputs with changes in coupling, and are still negatively affected by reflected reactance. In addition, changing the frequency in a Class E-type inverter for variations in load, as is the case in the voltage source bridge inverters, would detune the circuit and negatively affect performance. A frequency modulation scheme for maintaining output voltage on a Class E voltage independent inverter is presented in [4]. However, this type of load independence is not ideal for wireless power transfer (WPT), since it cannot accommodate the unloaded primary coil operation.

Also, Class E type circuits are not controllable through pulse width modulation (PWM) because changes in duty cycle do not translate to control of the output, as explained in [1]. Instead, the duty cycle primarily affects the zero voltage switching (ZVS) condition [5] and hence a control scheme on

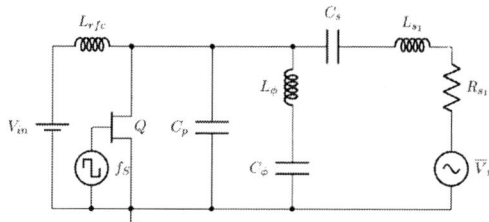

**Fig. 1:** Class EF inverter circuit diagram with the reflected load modelled as an AC voltage source $\overline{V}_r$ at the switching frequency $f_S$ in series with the primary coil $L_{s_1}$.

duty cycle can be used to ensure high efficiency operation.

Another way to adjust the behaviour of the inverter in real-time is by using variable reactive components. However, mechanically variable inductors and capacitors or component arrays [6] are not suitable for high frequency (HF) applications because of their inherently low speed of change and additional losses introduced by the relays and switches required. Electrically controlled capacitors such as varactors [7] or inductors such as saturable reactors [8]–[12] have typically lower $Q$ factors, series resonance frequencies (SRF), and power handling capabilities and thus reduce efficiency.

Here, therefore, we investigate the feasibility of an

PCIM Europe 2024, 11– 13 June 2024, Nuremberg          DOI: 10.30420/566262114

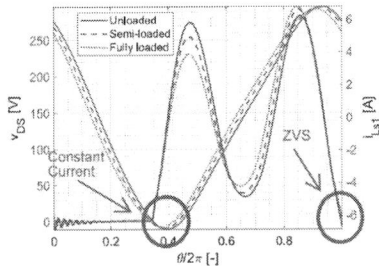

**Fig. 2:** Drain voltage (blue) and output current (orange) of the load-independent Class EF inverter with different load resistances during optimal conditions.

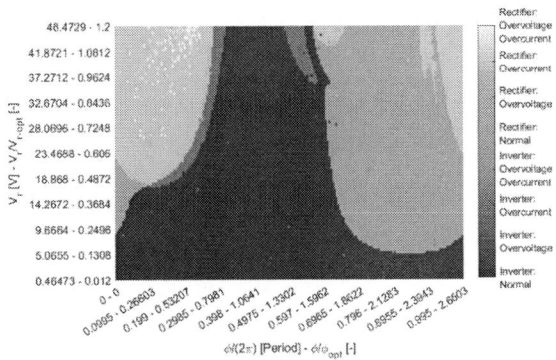

**Fig. 3:** State of operation map for the infinite choke load-independent inverter based on the reflected load. The left values on the horizontal and vertical axes are the absolute and the right ones are the relative to the ideal load. The overvoltage and overcurrent states are based on the limits of the GS66504B transistor from its datasheet.

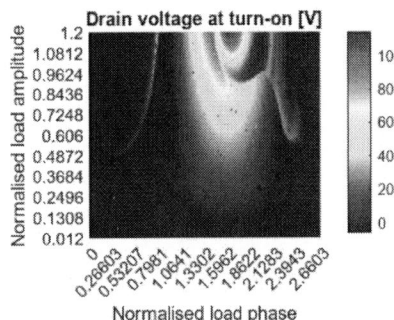

**Fig. 4:** Map of the drain voltage at turn-on of the load-independent Class EF inverter based on the amplitude and phase of the load.

active control scheme for Class E-type inverters by controlling various combinations of input voltage, duty cycle and frequency. As explained in [1], the inverter typically follows a prior power conversion stage to produce the required supply voltage, and hence the inverter input voltage can be controlled through the duty cycle of that power supply. We perform a variety of simulations and experiments to prove the proposed concept of active control to regulate the output of the system while constantly achieving soft-switching across the entire operating range.

## 2  Limitations of Open-loop Operation

In this section we perform an analysis on the operation of the passive Class EF inverter over various loading conditions. We run a comprehensive simulation using the method in [13], where the reflected load is modelled as an AC voltage source at the switching frequency in series with the primary coil, as shown in Fig. 1. Based on the relevant literature, one of the most suitable design candidates for HF inductive power transfer (IPT) systems with the best performance on varying load is the circuit presented in [2], hence the study that follows is based on data from the load-independent Class EF operation. The waveforms of the load-independent Class EF inverter under optimal conditions. i.e. only resistive load is reflected to the primary, are shown in Fig. 2, where ZVS and constant output current is achieved from short circuit to full load.

However, the equivalent input impedance of a high frequency passive rectifier is dependent on the DC load, due to a combination of circuit design and device non-linearities [14]. We simulate the inverter across a large reflected complex load range, where the complex load is modelled as shown in Figure 1

with amplitude from 0 to 1.2 times the equivalent effect of the ideal resistive load and phase from 0 to $2\pi$. The ideal load of 6 Ω corresponds to amplitude of 38.6 V with phase of 2.35 rad. Since such analysis involves the computation of a large amount of data points, we simulated the device with the effects of the non-linear capacitances of the device included but the stray inductances omitted. This decreases the time required for the computations significantly, but reduces the accuracy, as resonant ringing effects are not captured.

The map of the state of operation of the inverter based on the reflected load is presented in Fig. 3. The circuit operates as an inverter when the supply produces power and as a rectifier when the supply consumes power. The circuit enters the overvoltage operation when the peak of the drain voltage exceeds 650 V and overcurrent operation

917

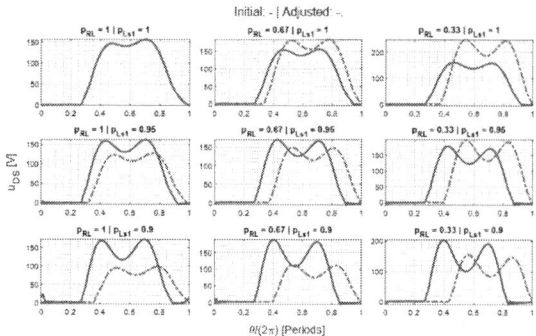

**Fig. 5:** Drain voltage waveforms of the Class EF$_2$ inverter before and after the input voltage and duty cycle adjustment across a range of reflected impedance values. The $p$ parameters represent the fraction of the corresponding reflected load compared to the nominal value.

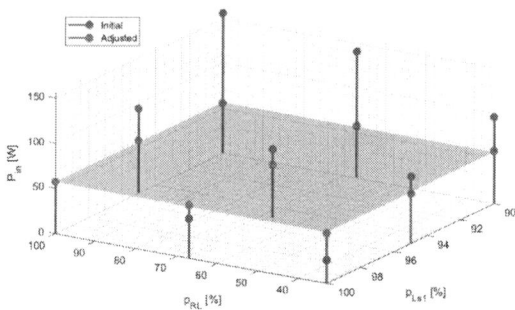

**Fig. 6:** Input power of the Class EF$_2$ inverter before and after the input voltage and duty cycle adjustment across a range of reflected impedance values. The grey plane represents the nominal power level operation.

when the absolute peak current on the device exceeds 15 A. However, normal operation does not necessarily mean safe operation, as the device could be hard switching or reverse conducting beyond its heat dissipating capabilities. As expected, the load-independent Class EF operates as an inverter without exceeding the device rating at any reflected voltage amplitude as long as the phase does not deviate much from the ideal. Non-ideal phase loads can be accommodated as long as their amplitude is very low.

An important metric indicating efficient operation and even possibility of failure is the drain voltage at turn-on, and such information is not shown in Fig. 3. Ideally, the goal is to maintain ZVS (and even zero voltgae derivative switching (ZVDS) if applicable) conditions in a load range as broad as possible. The switching voltage condition map for the inverter example in this section is given in Fig. 4, where the data points declining from dark blue indicate unacceptable operation. When taking into account the information from Fig. 4, the safe operation range of the Fig. 3 is even further restricted.

## 3 Control Scheme Concept

In this section we present examples from both simulations and experiments that prove the concept of the control scheme on Class E-type resonant converters for output power regulation while maintaining high efficiency at various reflected impedance values. The control variables in our examples are the duty cycle, the input voltage, and the switching frequency. The duty cycle has minimal effect on the output characteristics, especially for the case

of the Class EF topology. Changing the duty cycle mainly affect the ZVS and ZVDS conditions, which translates to high drain efficiency. Primarily, hard switching must be avoided, but reverse conduction is also less efficient that normal on-state.

The supply voltage controls the output current, for a given load profile. However, depending on the ratio of output capacitance of the transistor (combined with the reverse external diode if present) to the remaining shunt capacitance, the supply voltage can have a substantial effect to the soft-switching conditions. This effect becomes more prevalent as the switching frequency rises. Supply voltage lower than the nominal value leads to hard switching and supply voltage higher than nominal results in reverse conduction, when the load is ideal.

Altering the switching frequency effectively changes the residual reactance of the output branch. Secondarily, it also affects the shunt and φ-branch (if applicable) impedance, which in turn modifies the soft-switching characteristics. Since the operation of the Class E-type circuits is very sensitive to reactance changes in the output branch, frequency modulation can be used to control the output power of the inverter while remaining within the industrial, scientific, and medical (ISM) band limits, which can be beneficial for HF IPT systems to remain within electro-magnetic interference (EMI) regulations.

For all the above reasons, a control scheme involving the electrical signals present in a Class E-type inverter has the potential to maintain power across a broad range of coupling while at the same time maintaining ZVS and close to ZVDS conditions, and maximising efficiency. A load-independent design is not necessary when such control scheme is

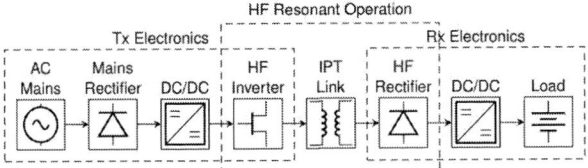

**Fig. 7:** High level block diagram of the various power conversion stages in a typical IPT system.

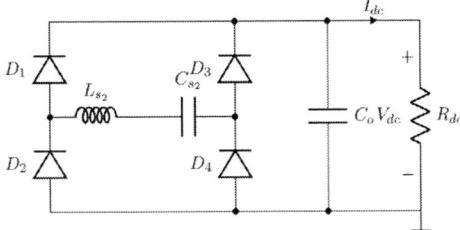

**Fig. 8:** Full bridge Class D current driver rectifier at the Rx side of an IPT system.

applied, and a Class $EF_2$ design with soft current switching can be used instead, providing higher efficiency with lower EMI.

# 4 Simulation examples

In the demonstration examples that follow the adjustment of each cycle of the control variables is carried out hierarchically, with the frequency being the first to be altered, the input voltage following and the duty cycle last. This choice of operation was based on the sensitivity of the inverter to each signal and its overall corresponding effect. The control cycles continue in this manner until all operating conditions are met (mainly output power and soft switching).

## 4.1 Input Voltage - Duty Cycle Control

In this example we use the Class $EF_2$ inverter design used in [13]. The resistive reflected load varies from critically low coupling (33 % of optimal load) to fully loaded and the reactive part ranged from zero to 10 % of the primary coil reactance in the form of capacitive loading. In this specific case we only adjust the input voltage and the duty cycle of the transistor, as there are applications where the switching frequency has to be fixed according to regulations (e.g. WPT at ISM bands). We regulate the input power and while maintaining ZVS.

The drain voltage waveform as well as the input power level at various reflected impedance values across the range for the Class $EF_2$, before and after the adjustment, are shown in Fig. 5 and Fig. 6,

**Tab. 1:** Component values of the HF IPT part of the charger example.

| Parameter | Value | Unit |
|---|---|---|
| $V_{in}$ | 234 | V |
| $f_S$ | 13.56 | MHz |
| $D$ | 24.8 | % |
| $L_{rfc}$ | 72 | µH |
| $C_p$ | 39.4 | pF |
| $L_\phi$ | 910 | nH |
| $C_\phi$ | 38 | pF |
| $C_{s_1}$ | 380 | pF |
| $V_{dd}$ | 5 | V |
| $G$ | LMG1020 | – |
| $Q$ | GS66504B | – |
| $R_g^{low}$ | 1 | Ω |
| $R_g^{high}$ | 0 | Ω |
| $L_{s_1}$ | 1 | µH |
| $L_{s_2}$ | 250 | nH |
| $k$ | 70 | % |
| $C_{s_2}$ | 496 | pF |
| $D_r$ | C3D06060A | – |
| $R_L$ | 62 | Ω |

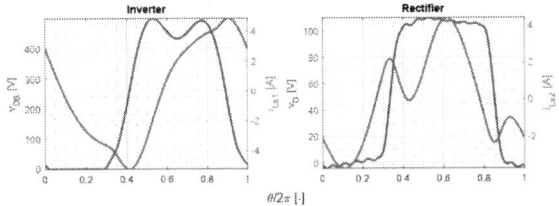

**Fig. 9:** Waveforms at nominal load of the drain voltage, primary coil current, secondary coil current and voltage across a diode of the rectifier in the power supply simulation.

respectively. The change in the duty cycle after the adjustment can be seen from the shift of the turn-off point in Fig. 5. The change of the input voltage is directly connected to the peak of the drain voltage, as its DC component is set by the supply.

Reverse conduction in our study can reach up to almost 20 % of the period, as shown in Figure 5, which would be detrimental to the the drain efficiency as the transistor operates at higher on-resistance for quite substantial amount of time [15]. In addition, there are some cases of hard switching. The input voltage in Figure 6 can swing from half to over three times the expected value, if left unregulated. In all of the instances in Figure 5 and Figure 6, the adaptation of the input voltage and duty cycle leads to ZVS and operation at nominal power.

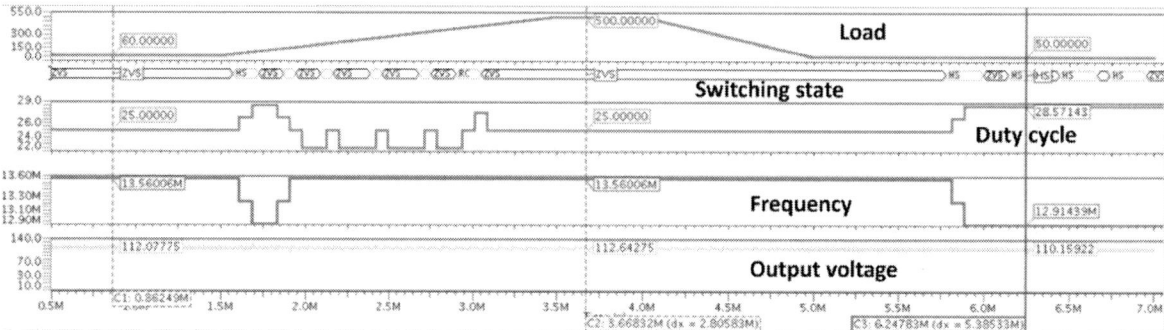

**Fig. 10:** Mixed signal simulation of the isolated air-core power supply. The first row is the load, the second row is the switching state, the third is the duty cycle, the fourth is the frequency, and the bottom is the output voltage.

## 4.2 Isolated Power Supply

Here we demonstrate the operation of the full control scheme $(D, f_S)$ in an isolated power supply. The galvanic isolation is provided by two coupled air core coils in close proximity to achieve the highest possible coupling in the smallest volume, as for example in the case of a PCB core-less transformer [16], [17]. The primary side is driven with an Class EF inverter and the rectifier at the secondary is a full bridge current driven Class D. The rectified mains, after the input line filtering, is stepped down by a buck converter, which can also operate as a power factor correction (PFC) stage [18], in a system block configuration as shown in Fig. 7. The output DC voltage of the charger is regulated from the primary side, and as a result no final voltage regulation power stage is required.

The circuit diagram of a full bridge current driven rectifier along with the secondary coil and load is shown in Fig. 8. It is designed to deliver 100 V at 180 W at nominal operation, according to the equations in [19], and the value of the series resonant capacitance was tuned to absorb the parasitic reactance caused by the C3D06060A Schottky diodes from Wolfspeed. The Class D topology was chosen for this application because of its simplicity and high power density, since no inductors are involved in the design. The high operating frequency (13.56 MHz) also helps keeping the size of the charger small.

The next step is to design the IPT link along with the inverter. A primary coil of 1 μH was chosen to keep the loaded quality factor of the output branch

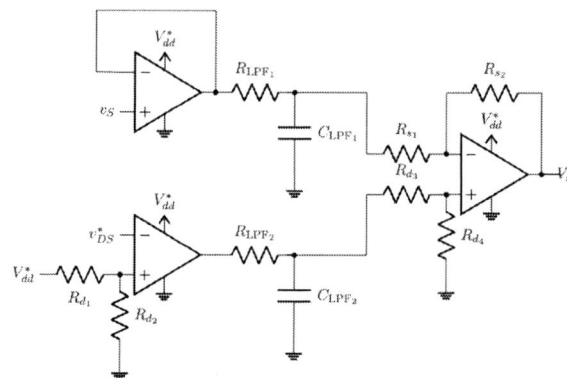

**Fig. 11:** Analog measurement technique of the switching condition of a Class E-type inverter. The supply voltage $V_{dd}^*$ comes from the output of a linear regulator to minimise noise. The instance of the drain voltage $v_{DS}^*$ is measured across a capacitor divider in series with $C_p$. The output DC reference containing the information of the switching condition is the signal $V_s$. The compactors on the left are the LTC6752 and the subtractor on the right is based on the AD8067.

of the inverter reasonably high. The secondary coil was chosen at a quarter of the primary inductance, to reduce the reflected load while also minimise size on the PCB. The primary driver is a Class $EF_2$ single ended inverter, to achieve the highest possible input voltage at the minimal power level. This reduces the range required for the buck converter before the inverter, which simplifies the overall operation and feedback control of the DC/DC converter.

PCIM Europe 2024, 11– 13 June 2024, Nuremberg     DOI: 10.30420/566262114

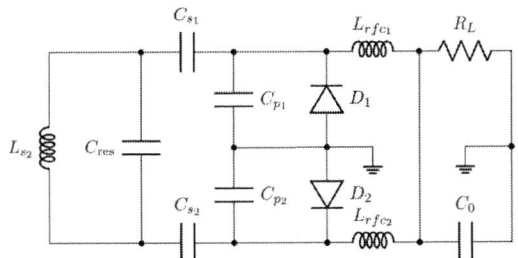

**Fig. 12:** Circuit diagram of a hybrid Class E rectifier. The diode model used is the C3D10065 from Cree.

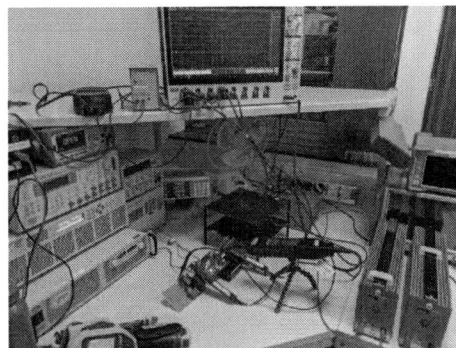

**Fig. 13:** Experimental setup of the IPT DC/DC system for demonstration of the inverter input voltage and duty cycle control. The two dark blue PCBs are the coupled coils, the board at the front is the inverter running from mains (left) through a capacitor divider and a rectifier (strip-board at front), the board at the back is the rectifier, and the load consists if the rheostats on the right.

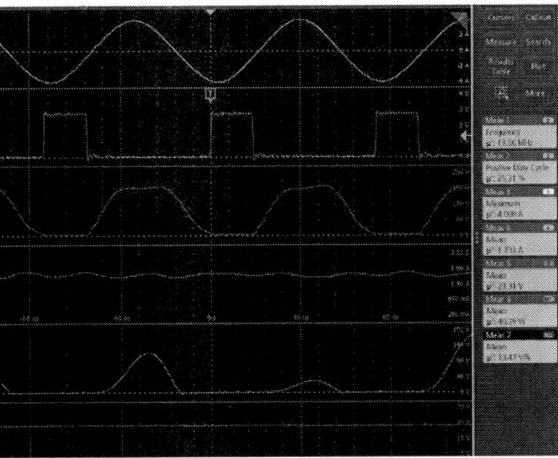

**Fig. 14:** Measured waveforms at nominal operation of the setup in Fig. 13. From top to bottom, in yellow is the primary coil current, in cyan is the switching signal. in red is the drain waveform, in green is the output current, in orange is the voltage across one of the diodes of the rectifier, and in purple is the output voltage. The two measurements of the "Math" functions are the output power ("Meas 6") and load resistance ("Meas 7").

The component values of the HF IPT part of the power supply unit example are given in Tab. 1 and the switching voltages and link current waveforms are shown in Fig. 9. The current on the IPT link are no longer sinusoidal due to the high loaded primary $Q$ factor. This however is not an issue in this particular application, since the whole charger is going to be shielded. Also, it must be mentioned that such non-sinusoidal output current inverter designs are possible because of the tuning method proposed in [13].

The results of the mixed signal simulation in Cadence of the charger across a range of output loads is shown in Fig. 10. The buck converter can run at a much lower frequency compared to the IPT inverter. This property, combined with the 50 Hz input mains signal and DC output severely slows down the simulation speed. For this reason, the value of the stabilising capacitors wherever necessary has

been reduced, hence the relatively large ripple is the DC signals in Fig. 10. The load varies from nominal (maximum power) to practically open and the system manages to keep the output voltage stable, while maintaining soft switching. The points where the algorithm changes the inverter frequency and duty cycle can be spotted by the changes in the corresponding signals.

The switching condition of the inverter is determined by comparing its duty cycle with the drain voltage. Since there is an inherit time delay between the gate signal and the drain waveform, due to the propagation delay of the gate driver and the switching delays of the transistor at this frequency, we transform the information of the duty cycles into DC voltage level, before carrying out the comparison. The circuit diagram of the above analog comparison method is shown in Fig. 11. The comparator on the top left acts as a buffer for the gate signal and the bottom one extracts the actual turn-on time of the transistor, including potential reverse conduction. Hard switching is also detected because of the non trivial transition time in this HF operation. The PWM outputs of the two comparators are filtered and passed to a subtractor, to get the switching information in DC. The values of the

921

(a) Nominal coupling      (b) Low coupling

**Fig. 15:** Photos of the IPT link in Fig. 13 at maximum coupling (left) and maximum gap and misalignment (right). The low coupling configuration has 20.2 % larger gap distance compared to nominal value, and 33.3 % misalignment relative to the coil outer radius.

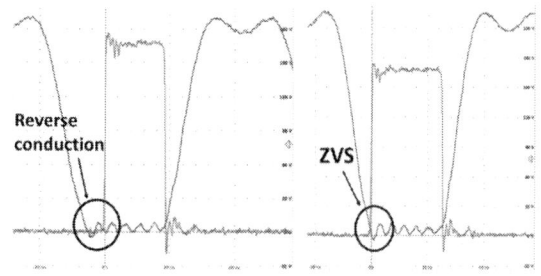

(a) Passive operation      (b) Adjusted operation

**Fig. 16:** Gate signal (cyan) and drain voltage (red) of the inverter in Fig. 13 at low coupling during passive, unadjusted operation (left), and readjusted operation (right). The measurement on the left of the screen are inline with the description in Fig. 14.

potential dividers are calibrated to match the required voltage reference at optimal operation. The duty cycle and frequency are then produced by the AD9833 programmable waveform generator.

## 5 Experimental Verification

In this section we present experimental verification of the concept of two versions of power and efficiency control in an IPT DC/DC system for coupling factor variation. The first example modulates the input voltage and the second the switching frequency to maintain constant output power. The soft switching conditions in both instances are maintained through duty cycle adjustments. The two examples incorporate different inverter and rectifier topologies to demonstrate the robustness of the proposed control approach.

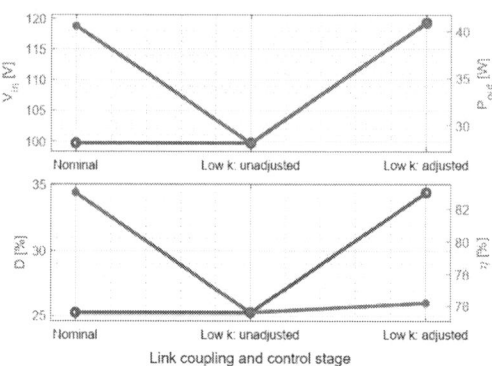

**Fig. 17:** Progression of the control variables and regulated quantities of the IPT system in Fig. 13 at nominal operation, low coupling without regulation, and low coupling with active control.

### 5.1 Input Voltage - Duty Cycle Control

In this setup we use the Class $EF_2$ inverter design discussed in [13] as the primary coil driver and a hybrid push-pull Class E rectifier, shown in Fig. 12, according to the design guidelines in [20]. The nominal input voltage of the inverter is adjusted to 100 V in order to accommodate the equivalent load reflected by the rectifier, and the output power is set to 40 W to a 13.46 $\Omega$ DC load at nominal coupling. A photo of the DC/DC IPT experimental bench setup is shown if Fig. 13 and the corresponding waveforms at nominal operation are displayed in Fig. 14. There is substantial instability in the operation of the rectifier, based on the behaviour of the voltage across the diode, which is primarily caused because of miss-match between the two custom cored output inductors, since the hybrid Class E design uses finite chokes. However, this effect does not propagate neither to the operation of the inverter, nor to the output load. The inverter is operating at both ZVS and ZVDS and the output voltage on the load is stable.

At nominal conditions, the two coils have a gap of 8.4 cm and no misalignment, and the most extreme separation at our example is at 10.1 cm separation and 3.5 cm misalignment. The two instances of coil placements are shown in Fig. 15. If the inverter operates passively, the case of extreme misalignment causes a 45.6 % drop in output power for the same DC load and the overall efficiency drops by 9 % due to primarily decrease in link efficiency but also sub-optimal drain switching conditions.

The measurements before and after the input voltage and duty cycle adjustment with the link and low

PCIM Europe 2024, 11– 13 June 2024, Nuremberg    DOI: 10.30420/566262114

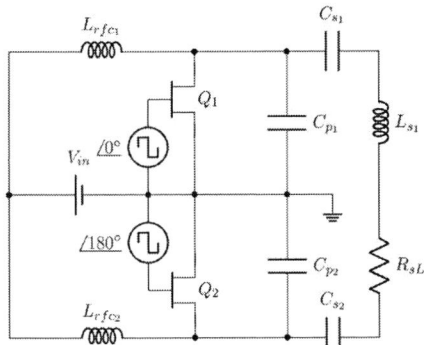

(a) Class E push-pull inverter

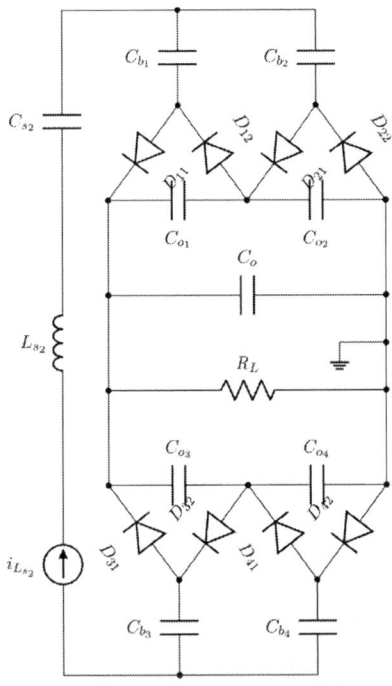

(b) Full-bridge doubler rectifier

**Fig. 18:** Circuit diagrams of the inverter and rectifier topologies used for the experimental test of the switching frequency and duty cycle control.

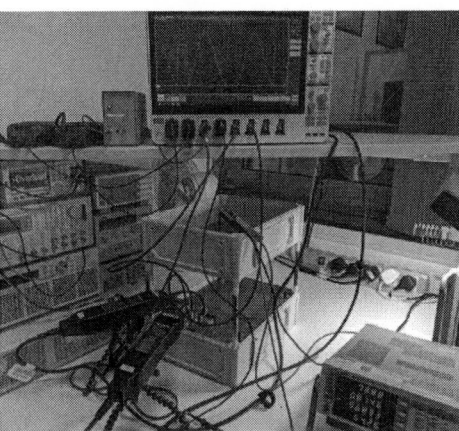

**Fig. 19:** Photo of the experimental setup of the frequency and duty cycle control at nominal operation.

operation. The seemingly low overall DC/DC efficiency increase (1 %) of the system after adjustment is because the majority of the losses are attributed to the link, as shown in [13]. Consequently, all of the efficiency increase is mainly directed to the inverter, and specifically the transistor. This 1 % system efficiency increase is translated to quite substantial gain in drain efficiency, which in turn results in meaningfully lower temperature on the device.

## 5.2 Switching Frequency - Duty Cycle Control

In this subsection we demonstrate an experimental example of a DC/DC IPT system with output voltage and efficiency regulation by controlling the duty cycle and switching frequency of the primary side. The primary driver is a push pull Class E running at 6.78 MHz and the secondary is a series tuned Class D full-wave voltage doubler based on the Cockroft-Walton voltage multiplier [21]–[23]. The circuit diagrams of the primary and secondary are shown in Fig. 18.

A photo of the experimental setup is shown in Fig. 19. The gate an drain and gate waveforms of the inverter at nominal load are shown at the oscilloscope, where both ZVS and ZVDS conditions are met. The supply voltage is kept constant at 50 V, the duty cycle is 29 %, and the nominal output voltage is 70 V at 80 % DC/DC efficiency.

After the tuning of the system, lateral misalignment was introduced to the link until the voltage drop on the fixed load was equivalent of around 20 % of power loss, where the efficiency dropped to 74.8 %

coupling are shown in Fig. 16. Without any adjustment on the inverter, the output voltage and in turn the output power on the load is below the required value and the transistor is reverse conducting, reducing overall efficiency. After the adjustment of the input voltage and duty cycle the output power as well as ZVS are restored.

The control variables $(V_{in}, D)$ as well as the regulated quantities $(P_{\mathrm{out}}, \eta)$ at nominal operation and at the extreme low coupling case, are shown in Fig. 17, both for the passive and active inverter

923

with the transistors at reverse conduction. Decreasing the switching frequency to 6.61 MHz with a duty cycle of 38 %, without changes in the supply voltage, restored ZVS conditions to the Tx driver and the output voltage to nominal value, while increasing the overall efficiency to 76.9 %.

# 6 Conclusion

In this paper we investigated the limitations of open-loop Class E type inverters for IPT applications under various reflected impedance combinations. Specifically, a load-independent Class EF inverter was designed, since it represents the best known configuration of these circuit types for open-loop operation in dynamic conditions. The robustness of the soft switching conditions and output power regulation, especially in the case of reflected reactance, were examined. It was shown that the robustness can be significantly improved by using active primary control scheme.

Then, four examples of control scheme applications were presented, two in simulation and two based on experiment. The goal was output power regulation while maintaining soft switching under dynamic load conditions, with the only control freedom being frequency, duty cycle and input voltage of the inverter. We successfully demonstrated that the operation of the inverter is significantly improved through the use of active control adjustments compared to passive open-loop operation. The improvements are most evident at the extremes of load and coupling factors.

# 7 Acknowledgements

This work was supported in part by EPSRC Quietening Ultralow-loss SiC & GaN Waveforms, under Grant EP/R029504/, in part by EPSRC Safe Power Delivery Using a Reconfigurable Mesh of Inductive Transceivers, under Grant EP/X020606/1, and in part by STMicroelectronics.

# References

[1] K. Van Schuylenbergh and R. Puers, *Inductive powering: basic theory and application to biomedical systems*. Springer Science & Business Media, 2009.

[2] S. Aldhaher, D. C. Yates, and P. D. Mitcheson, "Load-independent Class E/EF inverters and rectifiers for MHz-switching applications," *IEEE Transactions on Power Electronics*, vol. 33, no. 10, pp. 8270–8287, 2018. DOI: 10.1109/TPEL.2018. 2813760.

[3] T. Sensui and H. Koizumi, "Load-independent class e zero-voltage-switching parallel resonant inverter," *IEEE Transactions on Power Electronics*, vol. 36, no. 11, pp. 12 805–12 818, 2021. DOI: 10.1109/TPEL.2021.3077077.

[4] Y. Komiyama, S. Matsuhashi, W. Zhu, T. Mishima, Y. Ito, *et al.*, "Frequency-modulation controlled load-independent class-e inverter," *IEEE Access*, vol. 9, pp. 144 600–144 613, 2021. DOI: 10.1109/ ACCESS.2021.3121781.

[5] Y.-F. Li, "Auto-tuning controller design of class e inverter with resonant components varying," in *2012 IEEE International Symposium on Industrial Electronics*, 2012, pp. 217–221. DOI: 10.1109/ ISIE.2012.6237087.

[6] F. Issi and O. Kaplan, "Design and application of wireless power transfer using class-e inverter based on adaptive impedance-matching network," *ISA Transactions*, vol. 126, pp. 415–427, 2022. DOI: https://doi.org/10.1016/j.isatra.2021.07.050.

[7] A. Reatti, F. Corti, L. Pugi, M. K. Kazimierczuk, G. Migliazza, and E. Lorenzani, "Control strategies for class-e resonant inverter with wide load variation," in *2018 IEEE International Conference on Environment and Electrical Engineering and 2018 IEEE Industrial and Commercial Power Systems Europe (EEEIC / I&CPS Europe)*, 2018, pp. 1–6. DOI: 10.1109/EEEIC.2018.8494505.

[8] S. Aldhaher, P. C.-K. Luk, and J. F. Whidborne, "Electronic tuning of misaligned coils in wireless power transfer systems," *IEEE Transactions on Power Electronics*, vol. 29, no. 11, pp. 5975–5982, 2014. DOI: 10.1109/TPEL.2014.2297993.

[9] S. Aldhaher, P. C.-K. Luk, and J. F. Whidborne, "Tuning class e inverters applied in inductive links using saturable reactors," *IEEE Transactions on Power Electronics*, vol. 29, no. 6, pp. 2969–2978, 2014. DOI: 10.1109/TPEL.2013.2272764.

[10] S. Aldhaher, P. C.-K. Luk, A. Bati, and J. F. Whidborne, "Wireless power transfer using class e inverter with saturable dc-feed inductor," *IEEE Transactions on Industry Applications*, vol. 50, no. 4, pp. 2710–2718, 2014. DOI: 10.1109/TIA. 2014.2300200.

[11] D. Medini and S. Ben-Yaakov, "A current-controlled variable-inductor for high frequency resonant power circuits," in *Proceedings of 1994 IEEE Applied Power Electronics Conference and Exposition - ASPEC'94*, 1994, 219–225 vol.1. DOI: 10.1109/APEC.1994.316396.

[12] Y. Wei, Q. Luo, and A. Mantooth, "Comprehensive analysis and design of llc resonant converter with magnetic control," *CPSS Transactions on Power Electronics and Applications*, vol. 4, no. 4,

pp. 265–275, 2019. DOI: 10.24295/CPSSTPEA. 2019.00025.

[13] I. Nikiforidis, J. M. Arteaga, C. H. Kwan, N. Pucci, D. C. Yates, and P. D. Mitcheson, "Generalized multistage modeling and tuning algorithm for class ef and class $\Phi$ inverters to eliminate iterative re-tuning," *IEEE Transactions on Power Electronics*, vol. 37, no. 10, pp. 12877–12900, 2022. DOI: 10.1109/TPEL.2022.3176391.

[14] H. Sekiya, X. Wei, T. Nagashima, and M. K. Kazimierczuk, "Steady-state analysis and design of class-de inverter at any duty ratio," *IEEE Transactions on Power Electronics*, vol. 30, no. 7, pp. 3685–3694, 2015. DOI: 10.1109/TPEL.2014. 2339355.

[15] *Gs66504b bottom-side cooled 650 v e-mode gan transistor preliminary datasheet*, GaN Systems Inc., 2020.

[16] W. Liang, J. Glaser, and J. Rivas, "13.56 mhz high density dc–dc converter with pcb inductors," *IEEE Transactions on Power Electronics*, vol. 30, no. 8, pp. 4291–4301, 2014.

[17] G. Meneghesso, M. Meneghini, and E. Zanoni, *Gallium nitride-enabled high frequency and high efficiency power conversion*. Springer, 2018.

[18] B. Keogh, "Power factor correction using the buck topology-efficiency benefits and practical design considerations," in *Texas Instruments Power Supply Design Seminar*, vol. 4, 2010, pp. 1–37.

[19] G. Kkelis, "Rectifier design for optimisation of multi-mhz inductive power transfer systems," Ph.D. dissertation, Imperial College London, 2017.

[20] G. Kkelis, D. C. Yates, and P. D. Mitcheson, "Class-e half-wave zero dv/dt rectifiers for inductive power transfer," *IEEE Transactions on Power Electronics*, vol. 32, no. 11, pp. 8322–8337, 2017. DOI: 10.1109/TPEL.2016.2641260.

[21] J. M. Arteaga Saenz, "Design of high frequency inductive power transfer systems for integration into applications," Ph.D. dissertation, Imperial College London, 2019.

[22] C. G. Maennel, "Improvement in the modelling of a half-wave cockroft-walton voltage multiplier," *Review of Scientific Instruments*, vol. 84, no. 6, p. 064701, 2013.

[23] I. C. Kobougias and E. C. Tatakis, "Optimal design of a half wave cockroft-walton voltage multiplier with different capacitances per stage," in *2008 13th International Power Electronics and Motion Control Conference*, IEEE, 2008, pp. 1274–1279.

PCIM Europe 2024, 11– 13 June 2024, Nuremberg    DOI: 10.30420/566262115

# Low Voltage DC-Grids with Galvanic Isolation: System Discussion, Efficiency and Performance Comparison to AC-Feeding

Lukas Fräger[1], Andreas Schnieder[2], Sascha Langfermann[1], Michael Owzareck[1], Dennis Kampen[3], Jens Friebe[4]

[1] BLOCK Transformatoren-Elektronik GmbH, Germany
[2] Herbert Kannegießer GmbH, Germany
[3] Hochschule Bremen, Germany
[4] Universität Kassel, Elektrische Energieversorgungssysteme, Germany

Corresponding author: Lukas Fräger, Lukas.fraeger@block.eu

## Abstract

Low voltage DC-grids have shown to increase efficiency and performance compared to AC-grids. This paper provides an overview of different feeding configurations for low voltage DC-grids. Their advantages and disadvantages are compared in terms of both efficiency and performance. To prove the considerations, a test machine is evaluated under both DC and AC feeding. It is shown that losses of an exemplary industrial machine are reduced by up to 18.6 % when switching to bidirectional DC feeding with galvanic isolation and by up to 9.6 % without galvanic isolation.

## 1  Introduction

Current industrial machines are usually fed by a low-voltage AC grid. During the last decades, the share of electronic loads, that internally work with a DC voltage (i.e. switched mode power supplies, inverters, LED-lighting) has steadily increased. This has led to the idea of DC- instead of AC-feeding of the machines. Much research and standardization in the field of DC-grids was conducted in recent years [1] [2] [3] [4] [5] [6].

This paper provides an insight into the advantages and disadvantages in terms of both efficiency and performance when replacing an AC-grid inside an industrial machine by a DC-grid. Compared to formerly published results, the DC-grid is only kept locally inside one cabinet. The total power of the DC-grid is thus rather low (6 kW). Additionally, the feeding power supply provides a galvanic isolation, which can provide further advantages.

## 2  System Overview

Figure 1 shows an example of the low voltage feeding grid inside an industrial machine. For the examined machine, all loads are electronic loads, that internally work with a DC-voltage.

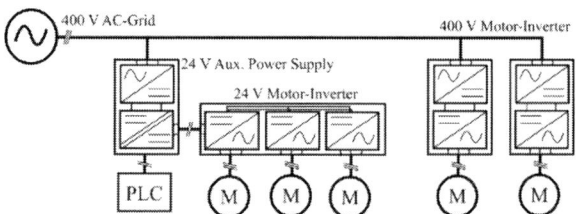

Figure 1: Block diagram of the electrical loads and their feeding grid in the industrial machine in conventional AC-feeding configuration.

The industrial machine, that is described by Figure 1 is shown in Figure 2. On the right side, the cabinet can be seen. In the back, behind the mesh for protection, the dummy mechanical loads are shown. While some of the smaller drives are powered by small 24 V servo inverters, a big blower and a larger servo drive are driven by three-phase, 400 V fed inverters.

Figure 2: Industrial machine with mechanical load dummies for test purposes.

# 3 Proposed System

The alternative to the standard AC-feeding is a DC-bus inside the industrial machine. As all loads already contain a DC-link, the DC-bus can be connected to the DC-link without a change in components.

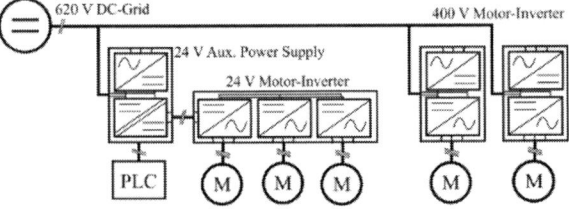

Figure 3: Block diagram of the proposed low voltage DC-grid inside the machine without change of components in comparison to the AC-grid.

The proposed wiring of the exemplary machine from chapter 2 for a DC-supply is presented in Figure 3. It can be seen, that all DC-links, that were formerly supplied by individual rectifiers, are now connected to form one DC-grid.

The main development goals of switching to a DC-grid are:

- Efficiency improvements
- Performance enhancement
- Elimination of braking resistors
- Cost optimization

Efficiency improvements by changing to a DC-feeding were amongst others shown in [7] [8].

When the supplying grid stays an AC grid, the main reason for lower overall losses is the braking energy that is kept inside the system or fed back to the grid instead of being burned in brake-resistors. Further improvements can be realized when renewable energy e.g. from a photovoltaic (PV) system is directly fed to the AC grid instead of being converted back and forth [5]. When the DC-grid is only limited to one cabinet or small machine and the overall power is thus low, direct feeding from a PV system is not useful. Device losses will also decrease, as no rectification is needed inside the load devices. However, forming the DC-grid from an AC-grid still requires a centralized rectifier.

Performance can be enhanced by DC grids in two ways: One is the ability to quickly brake down drives without the need for additional braking resistors. This can increase the machines dynamics. The other is peak power reduction, as the peak power may be supplied from the connected DC-link of the connected DC-loads [8].

Switching to a DC-grid enables an elimination or reduction of braking resistors: No braking resistor is required when the DC supply is bidirectional, the DC link capacitance is large enough to handle the braking energy, or there is sufficient load to compensate for the additional braking energy. If braking energy is still too high, the braking resistors may be centralized and combined to one large braking resistor. This can reduce the number of required resistors and brake choppers as well as their total power.

Cost analysis must be split into two phases: One is the capital expenses that include material costs for the machine and the other are operating expenses that include the power consumption of the machine. Potential to reduce the material cost besides reducing braking resistors exists by eliminating the rectification stage in the DC loads. However, today only few pure DC devices are available. The reduction in energy consumption and peak power on the other hand can lead to reduced operating expenses today.

## 3.1 DC-Grid Supply

As currently existing low voltage distribution grids are AC-grids [9] [10], a conversion to DC inside the industrial machine is necessary. Figure 4 shows different configurations for DC-grids, of which I…III were proposed and evaluated in [1]. The usage of a power supply as connection to the AC-grid (IV) is proposed as use case in [11].

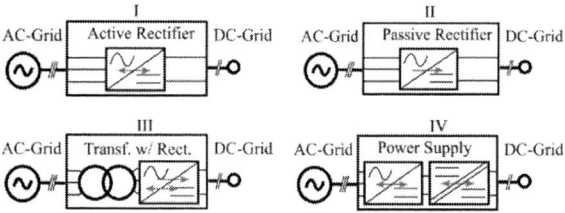

Figure 4: Configurations for the supply of DC-grids with indication of possible power-flow directions.

I: Conversion between AC- and DC-grid with an active rectifier as in [12]. Active rectifiers are well established and can be obtained as off-the-shelf components. However, most available converters are limited to the use with specific loads, especially inverters for servo-drive applications (c.f. [13] [14]). Using an active rectifier for the formation of a DC-grid enables a bidirectional power flow between AC- and DC-grid with a sinusoidal input current and controllable power factor. One very common configuration is the active six switch boost rectifier [15], which is also used in [13] [14]. In this topology, the DC-voltage must always be larger than the rectified AC-voltage $(V_{DC} \geq \sqrt{6} \cdot V_{AC,RMS})$. These active rectifiers also contain a passive rectifier in form of their free-wheeling diodes. In turns, additional effort must be made to protect the rectifier circuit from too high inrush currents due to capacitive DC-loads. A further disadvantage is the common mode voltage, that will lead to an AC-voltage between DC-grid and AC neutral / ground.

II: Conversion between AC- and DC-grid with a passive rectifier (B6U). The passive rectifier will provide a DC output voltage of $V_{DC} = \sqrt{6} \cdot V_{AC,RMS}$ with a ripple voltage of up to $13.4\,\%$ [16]. The power flow is only unidirectional from the AC- to the DC-grid. The current spikes from the rectification will lead to high current harmonics and a low power factor. Furthermore, the DC voltage will contain a common mode voltage as with configuration I. An advantage are very high peak DC currents that can be realized at affordable cost. Those may be helpful to trip load breakers and fuses. However, they still depend on the AC-grid impedance.

III: While both configurations I and II lack a galvanic isolation, configuration III provides a galvanic isolation by a grid-frequency power transformer. This enables an elimination of the common mode voltage and a reduction of the overvoltage category. Also, the transformer can be used to adopt the (minimum) output voltage of the (active) rectifier. However, voltage adjustment must be done manually during installation.

IV: The fourth option is to use a bidirectional power supply with a galvanically isolating DC/DC converter. This option circumvents many disadvantages and combines most advantages of configurations I-III (controlled input current, galvanic isolation, controlled inrush current, controlled output voltage / current). Especially the controlled output voltage may be used to control the start-up of the DC-grid and to realize fuse modes in case of overload or isolation errors. The large output voltage range can be further used to exploit the full potential of the stored energy in the DC-link, as the power from the grid can be controlled even if the voltage in the DC-link drops below the rectification voltage. Disadvantageous can be the higher complexity, reduced energy efficiency compared to I and II on a device level and the reduced maximum peak current compared to II.

An overview of the discussed advantages and disadvantages for the different feeding configurations is provided in Table I.

Table I: Overview of advantages / disadvantages of different supply configurations of DC-grids.

| Feeding Configuration | I | II | III | IV |
|---|---|---|---|---|
| Power Factor | + | - | I/II | + |
| Current Harmonics | + | - | I/II | + |
| Overload capability | o | + | I/II | o |
| Overload control | - | - | - | + |
| Galvanic Isolation | x | x | ✓ | ✓ |
| Common mode voltage | o | - | + | + |
| Supply efficiency | + | + | - | o |
| Inrush current control | - | - | - | + |
| Output voltage adjustment | o | - | I/II | + |
| Utilization of DC capacitance | o | - | I/II | + |

The proposed system (Figure 3) is preferably supplied by a bidirectional, galvanically isolated power supply with the discussed advantages:

- Current controlled precharge
- Independent earthing concept of DC and AC grid
- Constant (controlled) DC-voltage and current (Defined overcurrent behavior)
- Electronic identification of isolation errors possible
- Utilization of DC-capacitance as storage for peak power reduction

# 4 Power Supply

To realize the proposed system, a power supply is necessary. It needs to support universal grid voltages, provide a bidirectional power flow and a current- and voltage-controlled power output with high efficiency and high power-density. The resulting specifications are shown in Table II.

Table II: Power supply specifications.

| Specified Value | Value |
|---|---|
| Input voltage range ($V_{AC,RMS}$) | $200\,V \dots 480\,V$ |
| Output voltage ($V_{DC,Out}$) | $600\,V$ ($550\,V \dots 700\,V$) |
| Power | $6\,kW$ |
| Isolation type | *Reinforced acc.* [17] |

The built-up power supply is shown in Figure 5. Active rectifier and galvanically isolated DC/DC converter as well as LC-filter, EMC-filter and auxiliary supply are integrated in one housing. Cooling is done by forced air cooling and the supply has an overall power density of 0.6 kW/l.

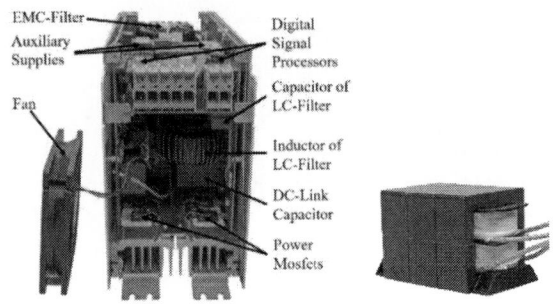

Figure 5:Built-up power supply without housing and removed fan (left) and MF-transformer (right), which is located inside the housing behind the inductors.

Detailed measurement results of the exemplary power supply are provided in [18] and [11].

## 4.1 Efficiency Comparison to LF-Transformer

To compare the efficiency, a simple model was used that assumes division of losses into a constant component and a component proportional the square of the power:

$$P_{loss} = \left( a + (1-a) \left( \frac{P}{P_{nom}} \right)^2 \right) \cdot P_{loss,nom} \quad (1)$$

$a$ is the share of constant losses, $P$ is the transfer power, $P_{nom}$ the nominal transfer power and $P_{loss,nom}$ the power loss at nominal power. For the low frequency transformer $a$ can often be assumed to be $a = 1/2$, as half of the losses are core losses and thus constant [19]. For a specific transformer, losses can be found in datasheets. For the exemplary 6.3 kW transformer [20] the actual value is $a = 0.31$. For the power supply, fitting to the results of [18] leads to approximately $a = 1/5$. This model is extremely simplified and neglects any details. An accurate model can be found in [11]. Still, this model can give a first estimate of the losses in dependence of the load.

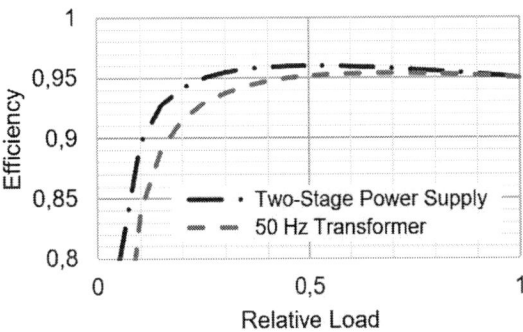

Figure 6: Efficiency model of 50 Hz transformer and 600 V DC Power-Supply.

Figure 6 gives the results of the efficiency model in dependence of the relative output power. Though nominal load efficiency is equal for both the power supply and the transformer, the power supply has the higher efficiency compared to the 50 Hz transformer especially at low and medium loads. Furthermore, the power supply already provides a DC-grid, while the AC output of the transformer yet has to be rectified.

# 5 System Measurements

To evaluate the efficiency advantages of the DC supplied machine, tests were conducted at the built-up industrial test machine described in chapter 2. A picture of the machine is given in Figure 1, the different configurations for AC- and DC-feeding shown in Figure 1 and Figure 3, respectively.

## 5.1 Efficiency Measurements

An overview of the evaluated test setups is given in Figure 7. To provide a complete picture, measurements without a galvanic isolation were also conducted despite the disadvantages described.

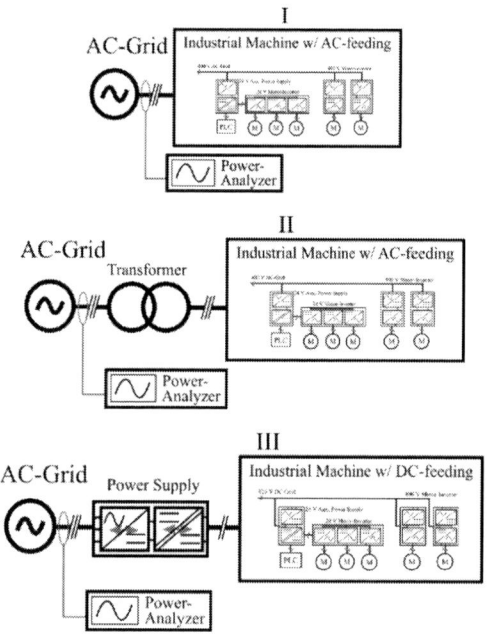

Figure 7: Test configurations for efficiency measurements: I: without galvanic isolation, II: with galvanic isolation by a low frequency isolation transformer, III: with DC grid and galvanic isolation by a power supply.

The test machine works in a periodic process. Measurements of the total energy consumption at the AC-grid coupling point (comp. Figure 7) were performed over 6 periods and the mean power consumption used as result. As measurement equipment a ZES Zimmer LMG670 power analyzer is used. The resulting power consumption in dependence of the feeding configuration for full-load and stand-by operation is given in Figure 8 and Figure 9, respectively. With the measured losses of the AFE and DC/DC converter from [11], the power consumption when supplying the DC-grid with a bidirectional rectifier (AFE) without galvanic isolation were estimated.

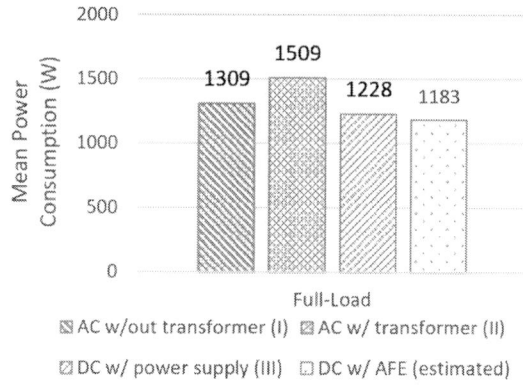

Figure 8: Measurement results of the machine's average power consumption in dependence of the feeding configuration for full-load operation.

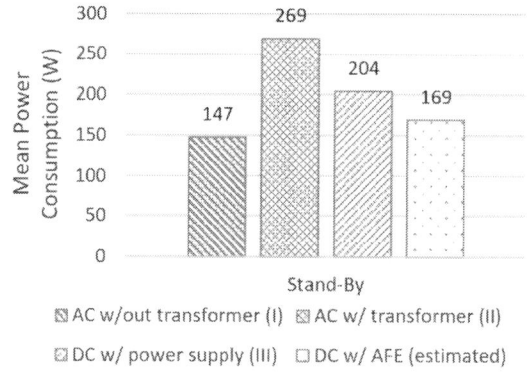

Figure 9: Measurement results of the machine's average power consumption in dependence of the feeding configuration for stand-by operation.

The measurement results reveal that the DC-grid with a highly efficient power supply is more efficient than an AC-grid with a low frequency (50 Hz / 60 Hz) transformer in any measured operating point. While at low power the core losses of the low frequency transformer are responsible for the higher losses, at higher power the DC-fed system benefits from the braking energy, that is fed back to the grid rather than lost in braking resistors or friction.

This also explains the results compared to AC-feeding without galvanic isolation. While the stand-by losses are 57 W higher with the active power supply (losses in semiconductors and magnetics), the efficiency at higher power benefits from feeding back braking energy (81 W lower).

The average power of feeding the machine without a galvanic isolation but when remaining a bidirectional active rectifier is slightly smaller (45 W at full load and 40 W at stand by operation) compared to the feeding with galvanic isolation. However, as discussed in chapter 3.1, additional protection and precharge measures are needed.

## 6 Conclusion and Outlook

The measurement results confirm that DC-grids help to improve the efficiency of industrial machines: Power consumption at full load was reduced by 281 W or 18.6 % when a galvanic isolation is needed and by 126 W or 9.6 % without galvanic isolation. Galvanically isolated DC-grids provide further flexibility and thus may be the more universal approach compared to grids without galvanic isolation. At the same time, they preserve the general advantages of DC-grids.

Future work is needed to provide measurement results for even more use cases. Additionally, power supplies should be developed, that include special modes such as burst mode to further reduce open load power consumption.

## References

[1] B. Sattler and S. Wiesner, "Research Project DC-Industrie - Energiewende meets Industrie 4.0," ZVEI - German Electrical and Eletronic Manufacturers' Association, Frankfurt, 2019.

[2] C. Meyer, M. Hoing, A. Peterson and R. W. D. Doncker, "Control and Design of DC-Grids for Offshore Wind Farms," *Conference Record of the 2006 IEEE Industry Applications Conference Forty-First IAS Annual Meeting,* pp. 1148-1154, 2006.

[3] „NPR 9090:2018 nl".

[4] K. Kim, K. Park, G. Roh and K. Chun, "DC-grid system for ships: a study of benefits and technical considerations," *Journal of*

*International Maritime Safety, Environmental Affairs, and Shipping,* vol. 2, no. 1, pp. 1-12, 2018.

[5] R. Weiss, L. Ott und U. Boeke, „Energy efficient low-voltage DC-grids for commercial buildings," in *2015 IEEE First International Conference on DC Microgrids (ICDCM)*, Atlanta, GA, USA, 2015.

[6] Y. Neyret, "DC Microgrids Principles and Benefits," 2023. [Online]. Available: https://www.dc.systems/assets/public/DC-Systems-White-Paper.pdf.

[7] D. Hölderle, J. Knapp und A. Sauer, „Evaluation of a Production Machine's Energy Efficiency in the Transformation Process From AC To DC Supply," in *2023 IEEE Fifth International Conference on DC Microgrids (ICDCM)*, Auckland, New Zealand, 2023.

[8] S. Warkentin, S. Puls, S. Riethmueller, F. Blank und H. Borcherding, „Measured Advantages of a Production Plant with DC Grid in terms of Energy Efficiency, Peak Power Reduction and Power Quality," in *PCIM Europe 2023*, Nuremberg, Germany, 2023.

[9] "www.worldstandards.eu," 19 12 2021. [Online]. Available: https://www.worldstandards.eu/electricity/three-phase-electric-power/.

[10] "IEC 60038".

[11] L. Fräger und J. Friebe, „Analysis and Optimization of the Loss Distribution for a Two Stage AC/DC Power Supply," *IEEE Open Journal of Power Electronics,* Bd. 5, pp. 272-285, 2024.

[12] L. Fräger, S. Langfermann, M. Owzareck and J. Friebe, "An Analytic Inverter Loss Model for Design and Operation Space Optimization," *2021 23rd European Conference on Power Electronics and Applications,* 2021.

[13] Siemens, "Sinamics S120 / Katalog D21.3," Siemens, 08 2022. [Online]. Available: siemens.com/d21-3. [Accessed 06 04 2024].

[14] B&R, "Acopos Multi Anwenderhandbuch (MAACPM-GER)," 08 2023. [Online].

[Accessed 04 06 2024].

[15] J. W. Kolar und T. Friedli, „The Essence of Three-Phase PFC Rectifier Systems—Part I," *IEEE Transactions on Power Electronics,* Bd. 28, Nr. 1, pp. 176-198, 2013.

[16] J. Specovius, Grundkurs Leistungselektronik - Bauelemente, Schaltungen und Systeme, Wiesbaden, Germany: Springer, 2018.

[17] „IEC 61558".

[18] L. Fräger, S. Langfermann, M. Owzareck, D. Kampen und J. Friebe, „Analysis of the Loss Distribution of a 6 kW two Stage Power Supply for 600 V DC Applications," in *24th European Conference on Power Electronics and Applications (EPE'22 ECCE Europe),* Hannover, Germany, 2022.

[19] W. Hurley und W. Wölfle, Transformers and Inductors for Power Electronics Theory, Design and Applications, Chichester, United Kingdom: John Wiley & Sons Ltd, 2013.

[20] Block Transformatoren-Elektronik GmbH, "Datasheet TT3 6,3-4-4," 23 06 2023. [Online]. Available: https://www.block.eu/fileadmin/2c9ee487874 001a2018747620463566b.Datenblatt_TT3_6 _3_4_4.pdf. [Accessed 06 04 2024].

PCIM Europe 2024, 11– 13 June 2024, Nuremberg          DOI: 10.30420/566262116

# Implementation and Experimental Evaluation of an Adaptive DC Grid Controller for Decentralised Grid Control

Steffen Menzel[1], René Reimann[1], Lorenz Grundhoff[1], Wilfried Holzke[1], Bernd Orlik[2]

[1] Institute for Electrical Drives, Power Electronics and Devices (IALB), University of Bremen, Germany
[2] IALB, University of Bremen, Germany, until 31.03.2023

Corresponding author:    Steffen Menzel, menzel@ialb.uni-bremen.de
Speaker:                 Steffen Menzel, menzel@ialb.uni-bremen.de

## Abstract

Future energy grids will have a high share of power converters to connect renewable power plants or energy storages to AC grids. Moreover, overlaying DC transmission grids in the high voltage sector, connected by additional power converters, are planned in several areas. This work shows the extension of a decentral controller for each power converter for these applications to an adaptive grid control approach. The control approach was implemented on a test bench at IALB and evaluated both in simulation as well as experimentally in different operating scenarios.

## 1    Introduction

The growing impact of fast switching power converters in comparison to huge but slow synchronous generators of conventional power plants on the overall AC grid stability can be seen in various grids and applications [1]. In the worst-case, large-scale power converters can affect each other even over long distances. This becomes increasingly important for the upcoming implementation of overlaying DC transmission grids, which are connected by large-scale power converters to their AC grids. To overcome this issue, a control approach for these power converters based on a fictious machine set has been proposed in previous works [2], [3]. The fundamental working principle has already been shown and proven, but for a fully decentral control in different modes of operation, the controller needs to adapt its working scheme according to the grid health status or power demands.

Therefore, three different control modes are proposed: DC-voltage control, power control and island grid control. Which control modes are possible for each power converter depends on the grid health status of the adjacent grids of each converter as well as on the grid structure. For example, power converters, which are connected to a weak AC grid with deviations in grid frequency or grid voltage due to load changes, are not fully capable to provide a stable DC voltage control on one hand. AC grids without any generators or power plants on the other hand, can only be controlled in island grid control mode, because they cannot provide power in every quadrant and need to prioritise their local grid control over any bidirectional power control. A sample DC grid with four power converters and different AC grid types is shown in Fig. 1.

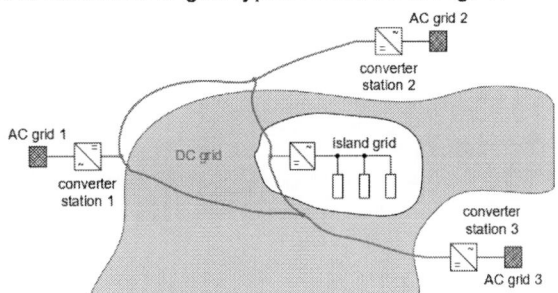

**Fig. 1**    Sample DC grid

The adaptive functionality of the proposed control approach is now implemented by monitoring and evaluating the grids at the power converters and by changing between the control modes if necessary. The following section describes the design of the DC grid controller in detail.

## 2    Design of the DC grid controller

The fundamental control approach is based on the calculation model of a fictitious machine set.

Here, a synchronous machine and a DC machine are rigidly coupled and both machines are separately excited. The synchronous machine represents the coupling to the AC grid, the DC machine stands for the coupling to the DC grid, while each stator winding marks the connection point to the measured local voltage [2], [3]. By using this fictious machine set, the behaviour of a conventional synchronous generator is imprinted on the power converter. Due to the fast-switching ability of the power converter, the du/dt rate could be remarkably high leading to a high di/dt rate in consequence, which is only limited by the inductor on the AC side of the power converter. The imprinted behaviour of the machine set reduces the high dynamic of the power converter on one hand but leads to a much better stability on the other hand [4], [5]. Furthermore, this control approach is working decentralised at each power converter station and no overlaying communication topologies are needed between these stations. This is a key advantage with respect to fail safe operation, which becomes more and more important in the power generation and transmission sector.

In this work, this control approach was upgraded with a monitoring and evaluation block to adapt the operating mode of the controller structure to the grids needs. As mentioned before, there are three different control modes for the overall control structure: DC grid control, power control as well as island grid control. A simplified overview of the control structure is shown in the following Fig. 2.

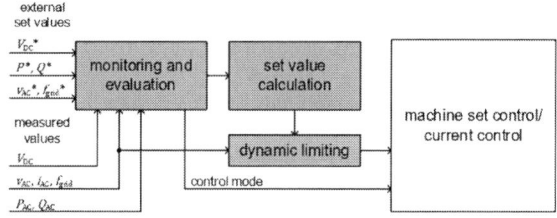

**Fig. 2**  Overview of the control structure

In the beginning each power converter is using in an initial operating mode, which is set by the operator. During operation, the measured values for the DC link voltage, the AC voltages and currents as well as both active and reactive power are constantly monitored. According to the given desired values, the set values for the voltage and current controllers inside the machine set control structure are calculated. These values are then fed into an additional block for dynamic limiting of the values according to the operating mode, the grid health status as well as technical limitations.

## 2.1  Monitoring and evaluation

A change of the control mode is started if the criteria depending on the actual control mode and grid health status are met. In this case, the focus is on the grid health status of the AC grid at the power converter, respectively on its grid voltage and grid frequency. Therefore, the RMS value of the AC grid voltage as well as the grid frequency are monitored according to given boundary values following a norm [6]. The resulting values are shown in the following Table 1.

| $V_{AC}$ | normal operation | 0.96 – 1.04 p.u. |
|---|---|---|
| f | normal operation | 0.996 – 1.004 p.u. |
|  | critical operation (weak grid) | 0.950 – 0.996 p.u. |
|  |  | 1.004 – 1.030 p.u. |
|  | cutoff values | < 0.950 p.u. |
|  |  | > 1.030 p.u. |

**Table 1**  Boundary values for voltage and frequency

The evaluation of the measured values is not only limited to steady state values. During operation, the moving average values are also calculated as well as their rate of change to keep track of dynamic changes within the grid. The resulting control mode is figured out by a state machine, which is shown in Fig. 3.

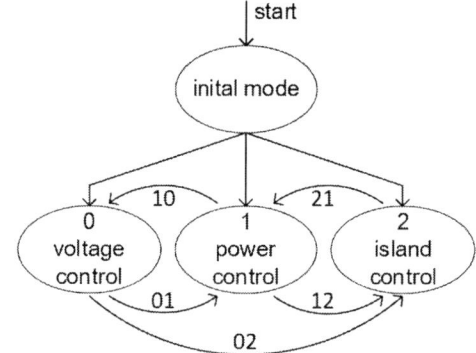

**Fig. 3**  State machine for the control mode

In the beginning, the initial mode of operation is given by the operator. The state machine branches into this first state and, afterwards, the transitions to the other control modes are performed according to the grid health status. The voltage control mode (mode 0) can only be applied if the AC grid is a strong grid.

Therefore, both the AC voltage as well as the grid frequency must be in the range of normal operation, as shown in Table 1. If the grid parameters exceed these values, the grid is a weak grid, so greater deviations in voltage and frequency may occur. There is a variety of reasons for this grid behaviour. The most common reason is a slight imbalance between power generation and consumption, so switching loads on and off in the grid could lead to instabilities. In this case, the controller mode is switched to the power control mode (mode 1), as a reliable DC voltage control cannot be granted in every state of the AC grid. During this operation mode short temporal over and under shootings of the grid parameters are allowed. If the grid voltage or the grid frequency remains lower than its boundary values of 0.96 p.u. respectively 0.95 p.u. for a longer time, it is assumed, that there are no power generation elements in the grid anymore. This happens also if the moving average values of voltage or frequency change quick. In this case, the control mode switches directly to the island grid control (mode 2) for supporting the adjacent AC grid of the power converter. The transitions of the state machine and the conditions are shown in the following Table 2.

| transition | condition |
|---|---|
| 01 | grid gets weak |
| 10 | grid gets strong |
| 02 | power generation lost, |
| 12 | extremely weak grid |
| 21 | power generation restored |

**Table 2** Transitions of the state machine

Furthermore, changing between the sates is also restricted according to the first control mode. If the controller was initially working in the power control mode, it can never switch to the DC voltage control mode. Correspondingly, a controller with initial island grid control can never switch to a higher control mode. This was implemented to ensure grid stability in any case of operation and to avoid undesired and multiple changes in the controlling modes. But of course, power converters with strong AC grids can also operate permanently both in power and island grid control mode. Before each transition, the controller release as well as the set value release are set to 0 and each integrator within the controller structure is also reset.

After the transition is executed, the controller release is activated again, followed by the set value release after a short time delay.

To illustrate the behaviour of the state machine, a voltage and frequency profile was fed into the state machine for a controller with voltage control mode as first mode of operation. The resulting plot is shown in Fig. 4. Here, the voltage and frequency are ramped respectively stepped to rather extreme values for simplification purposes. In the beginning at t = 0, the controller is working in DC voltage control mode with a strong AC grid. Then the AC grid frequency drops below 0.95 p.u. and the controller is switched to mode 2 (island control). Afterwards, the grid frequency recovers, and the grid becomes a weak grid, leading the control mode to change to mode 1 (power control). When the grid frequency reaches the boundary values for the normal operation, the controller switches back to its initial DC voltage control (mode 0). In the following seconds, the grid frequency is ramped up, so the controller switches to power control mode again. In this case, no transition to the state island control is implemented. The island grid control mode is only applicable for grids without any generators and, therefore, it is physically not possible, that the grid frequency rises should only power consumers be connected to the grid.

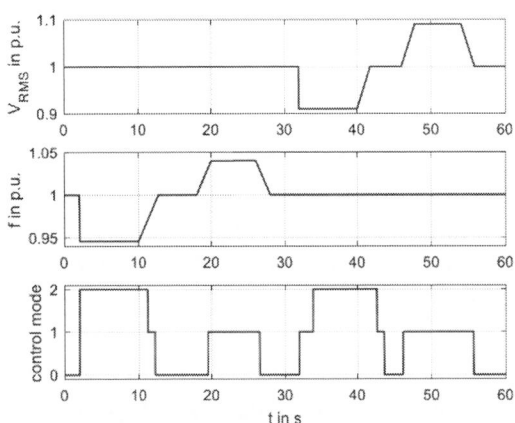

**Fig. 4** Working scheme of the state machine

After the grid frequency went back to its nominal value and the controller is again working in mode 0, the AC grid voltage drops below 0.96 p.u. and the controller switches immediately to mode 1. Because the grid voltage stays low in this case, the control mode switches to mode 2. The voltage recovers and, therefore, the system switches to mode 1 and finally back to mode 0. In the last step, the voltage is ramped up, which leads to the same behaviour as was seen before at the over frequency scenario: the control mode switches to 1.

At the end of this illustration, both voltage and frequency get to their nominal values and the controller is again working in the first control mode 0.

## 2.2 Set value calculation

In the voltage control mode, the desired DC voltage is the only set value, which needs to be fed externally into the controller structure. For controlling the power either a set value for the active or reactive power or eventually for both are needed. To achieve stability – especially during island control – additionally drop charts are added and implemented for both active and reactive power. While the active power is linked to the grid frequency, the reactive power component is connected to the RMS value of the AC grid voltage. In the operating modes 1 and 2 these droop charts are always active, since the main goal of the proposed control structure is stability and to offer grid supporting functionalities. The resulting droop charts with respect to Table 1 are shown in the following Fig. 5 and Fig. 6.

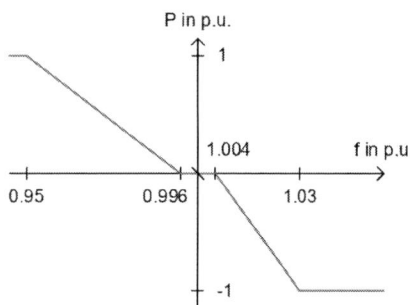

**Fig. 5**    f-P droop chart

The f-P droop chart includes a dead band around the nominal value of the frequency with a width of ±0.004 p.u. respectively ±200 mHz in a 50 Hz grid. Generally, the transmission of active power is increased at a lower than nominal grid frequency to support the AC grid and reduced in case of a higher grid frequency. Normally, the V-Q droop chart does not include a dead band around the nominal voltage value. But it can also be implemented as shown by the dashed line in Fig. 6. The calculated values from the droop charts were added to the eventually given set values for the power controller, respectively to the resulting set currents for the underlaying current controller in the control structure. Since the whole controller structure is calculated within a rotating reference frame, the power flow is also calculated from the measurement values and the amplitude invariant version of the Park and Clarke transformation is used. Under normal operating conditions in a strong grid with a conventional control approach,

the voltage components are calculated by a PLL and, therefore, the q-component is controlled to 0.

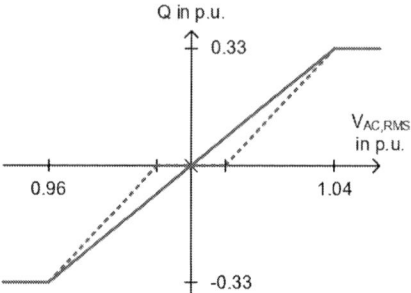

**Fig. 6**    V-Q droop chart

The power calculation in the rotating reference frame and the formulas for the power set values P* and Q* are shown in the following equations.

$$P = \frac{2}{3}\left(V_d I_d + V_q I_q\right) \tag{1}$$

$$P^* = P_{set} + P_{droop} \tag{2}$$

$$Q = \frac{2}{3}\left(V_q I_d + V_d I_q\right) \tag{3}$$

$$Q^* = Q_{set} + Q_{droop} \tag{4}$$

But should the grid could be weak and the grid parameters are not measured at the same location in the converter structure (e.g. current measurement at the semiconductor and voltage measurement before the commutation inductance), the resulting set values need to be corrected by the measured voltage values. The resulting formulas are shown in Eq. 5 and Eq. 6.

$$I_d = \frac{2}{3}\frac{1}{V_d}\frac{P^* + Q^*\frac{V_q}{V_d}}{1 + \frac{V_q^2}{V_d^2}} \tag{5}$$

$$I_q = \frac{2}{3}\frac{1}{V_d}\frac{P^*\frac{V_q}{V_d} - Q^*}{1 + \frac{V_q^2}{V_d^2}} \tag{6}$$

The resulting current values are the set values for the underlaying current controller. In this way, the current values are corrected by the actual voltages at the power converter which leads to a significantly better steady state accuracy. According to the momentary control mode, the set values need to be dynamically limited according to the overview in Fig. 2.

## 2.3 Dynamic limiting

Since the maximum of the absolute value of the current is 1 according to Eq. 7, either the d-component of the current is set, and the maximum value of the q-component is limited or vice versa.

$$I = \sqrt{I_d^2 + I_q^2} = 1 \, \text{p.u.} \qquad (7)$$

Hence, both components cannot be chosen independently from each other. While the component $I_d$ mainly corresponds to the active power, the q-component $I_q$ is linked to the reactive power. The maximum values $I_{d,max}$ and $I_{q,max}$ for the current are given by the rated power, respectively the rated current of the power converter. According to technical application rules, the power converter needs to provide an additional reactive power of up to 0.33 p.u. while feeding an active power of 1.0 p.u. into the grid at a power factor of 0.95, if needed [6]. The simplest way to fulfil this requirement leads to a strict limitation of the d- and q-values as presented in Fig. 7 by the boundary values $I_{d,lim}$ and $I_{q,lim}$. These boundary values represent the values of 1.0 p.u. for $I_{d,lim}$ and 0.33 p.u. for $I_{q,lim}$.

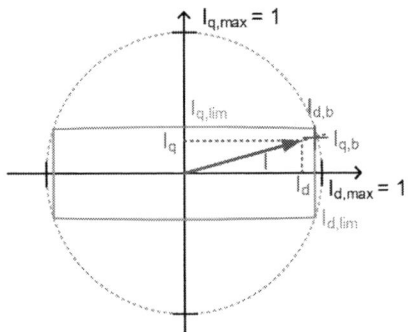

**Fig. 7** Strict limitation of the d- and q-values

Each pair of values, either calculated by Eq. 5 and 6 or manually set, is checked to be inside the limits marked by the green rectangle in Fig. 7. By stretching the current vector **I** to the rectangle, the resulting limit values $I_{d,b}$ and $I_{q,b}$ can be determined. Hence, the resulting current vector **I** can move within this rectangle during operation. However, if its length is close to the border, this could lead to step-shaped changes of the actuating values. This especially occurs at the edges of the rectangle, which poses the risk of current or voltage oscillations. To avoid this, an ellipsoidal limiting is used, as shown in Fig. 8. Here, the resulting limit values along the ellipse can easily be calculated with respect to the given values $I_{d,lim}$ and $I_{q,lim}$, which mark the semi-major and semi-minor.

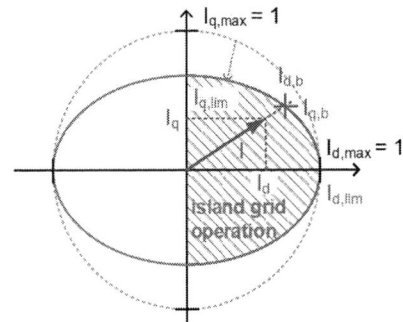

**Fig. 8** Dynamic limitation of d- and q-values

This also leads to a simpler implementation on the controller hardware later. The equation for the ellipse is given by Eq. 8.

$$\frac{I_{d,b}^2}{I_{d,lim}^2} + \frac{I_{q,b}^2}{I_{q,lim}^2} = 1 \qquad (8)$$

The current vector **I** is stretched to the border of the ellipse by the factor k and equals the current $I_b$.

$$I_b = kI \qquad (9)$$

Due to the AAA similarity theorem the two triangles $I_d\,I_q\,I$ and $I_{d,b}\,I_{q,b}\,I_b$ are similar, and the following equations can be derived.

$$I_{d,b} = kI_d \qquad (10)$$

$$I_{q,b} = kI_q \qquad (11)$$

By inserting Eq. 10 and Eq. 11 into Eq. 8 and after rearranging the equation, the factor k is defined as follows:

$$k = I_{d,lim}I_{q,lim}\sqrt{\frac{1}{I_d^2 I_{q,lim}^2 + I_q^2 I_{d,lim}^2}} \qquad (12)$$

Before the factor k is calculated, the values $I_d$ and $I_q$ are checked to be smaller than the absolute limit values $I_{d,lim}$ and $I_{q,lim}$. Finally, the current values for the dynamic limiting are calculated by Eq. 10 and Eq. 11 and the current values $I_d$ as well as $I_q$ are limited to these values if necessary. As shown in Fig. 8, during island grid operation the limitation of the values is only possible in the right-hand plane, as the grid only consists of power consumers respectively loads. Therefore, only positive values for $I_d$ are possible.

## 3 Evaluation in simulation

The evaluation of the proposed controller was done in two steps. First, a sample DC grid with four power converters was set up in MATLAB/Simulink using the PLECS toolbox to model the power converters and the electric components. The controller model was set up in a separate Simulink block, with the desired set values and measurements at the power converters routed via dedicated in- and outputs into this block.

In the second step, this controller block was converted into C code by the MATLAB embedded coder and implemented in the toolchain for programming the test stand. Afterwards, the test stand was configured according to the sample grid and the microcontrollers of the power converters were programmed with the generated code. The structure of the test stand as well as the measurement results are shown in section 4.

## 3.1 Simulation Environment

The sample DC grid with four power converter stations, each connected to a separate AC grid, is shown in Fig. 9. Within the simulation environment, the power converter consists of a two-level IGBT structure, the commutation inductance and the DC link capacitors. For simplification, not every single part for the operation of the test stand was implemented in the simulation environment.

To evaluate the proposed controller structure in multiple operation modes, each AC grid is independently characterised. While AC grids 1 and 3 are rated as strong grids with conventional power plants, AC grid 4 is an island grid with no power generation and AC grid 2 is a weak power grid, with only a small share of generators. For a stable operation both AC grids 2 and 4 need to dissipate power from the DC grid. In the island grid, power is needed to keep up the island grid with respect to the voltage amplitude and the grid frequency. At converter station 2, the power flow in direction of the AC grid is required to offer the grid supporting functions.

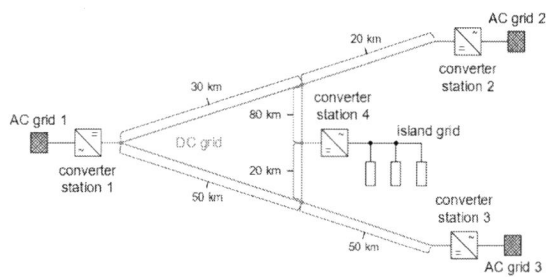

**Fig. 9** Sample grid with four converter stations

All converter stations are connected via transmission line replications to mimic the losses and the dynamic behaviour of the transmission lines.

## 3.2 Simulation results

The following two plots show the simulation results for an operating scenario of the DC grid from Fig. 9 with the proposed control approach in MATLAB/Simulink and PLECS. Initially the controller of station 1 starts in voltage control mode (mode 0) and starts to control the overall DC voltage at $t_1$. The initial control mode of converter station 3 is set to mode 1 (power control mode) and at $t_2$ the converter station 3 starts to feed power into the DC grid. The active power P is calculated by the voltage and the current measured at the AC side of each power converter, with the current counted positive in direction to the AC grid. Therefore, the power flow into the DC grid has the opposite sign. To evaluate the capabilities of the control approach, the initial control mode of station 2 was set to voltage control mode, although AC grid 2 is characterised as a weak grid.

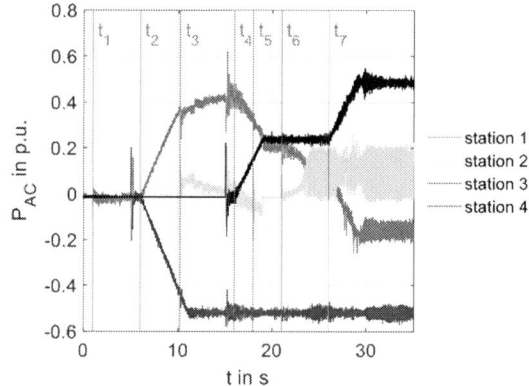

**Fig. 10** Power flows to the AC grids

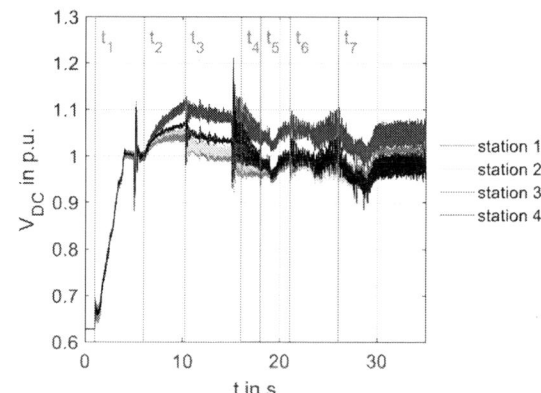

**Fig. 11** Local DC voltages

At $t_3$ the controller of station 2 is started and the station begins to control its local DC link voltage and therefore balances the DC grid. But station 2 is affected by a grid frequency droop starting at $t_5$,

PCIM Europe 2024, 11– 13 June 2024, Nuremberg    DOI: 10.30420/566262116

because the active power demand in AC grid 2 gets bigger than its production. Here, the frequency droop was modelled by a controllable AC grid at converter station 2. The state machine of the controller structure at converter station 2 immediately initiates the transition to the power control mode and restarts the controller at $t_6$. According to the implemented droop charts positive set values for the active power are calculated and fed to the controller to support the local AC grid 2.

At $t_4$ the island grid station 4 demands power from the DC grid. Here, the controller of converter station 2 is still in voltage control mode and balances the DC grid. As a consequence, the power flow at station 2 changes from positive to negative. The power demand of converter station 4 is increased even further at $t_7$. At this point only converter station 1 is controlling the DC voltage of the grid and, therefore, has to balance the power flow to achieve a steady DC voltage. Correspondingly, the power flow at converter station 1 changes from positive to negative shortly after $t_7$. Both plots show, that the overall system remains stable at any point of this evaluation scenario.

# 4    Evaluation on the test stand

## 4.1    Test stand environment

The test stand at IALB consists of up to four two-level IGBT power converters with a rated power of 10 kW each and an isolating transformer at the AC side. To set up the DC grid, the DC links of the converters are connected by using transmission line replications to setup different DC grid structures. Each power converter is individually controlled by its own microcontroller board, which was developed at IALB based on a TI DSP. The software framework consists of two parts. The first part includes the management of hard- and software (e.g. in- and output assignment, data interfaces, state machine) and the second part consists of the controller block, which has been generated from the Simulink block including the controller structure. By following this structure, the second part can be changed easily, and a fast implementation of the proposed control approach is possible.

The test stand is completed by a central test stand control unit, which provides safety functions like the emergency stop, monitoring isolation and can also show a summary of test stand parameters. These are collected by a PLC inside the test stand control cabinet and displayed at a web-based HMI, which also offers the possibility to start and stop the test stand as well as the measurement system.

## 4.2    Measurement results

To evaluate the adaptive DC grid controller, different test stand scenarios were realised by manually applying set values to the control structure of the power converters or by manipulating the grid at the converter. The test stand was set up according to Fig. 9 and the four power converter were connected by transmission line replications, each representing the given length.

In the first scenario, the controller station 1 is operating in mode 1, while the stations 2 and 3 are in power control mode. The scenario is like the one evaluated in simulation before. But in this case, the station 4 is only dissipating a constant and small amount of power from the DC grid. The resulting power flow of each power converter in direction towards the DC grid is shown in Fig. 12 and the resulting DC link voltages are plotted in Fig. 13.

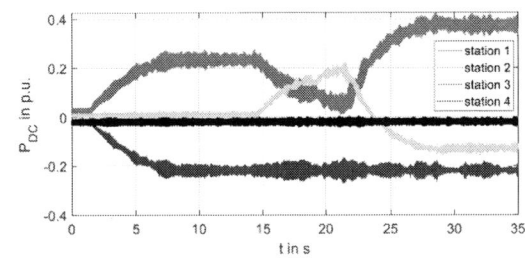

**Fig. 12**   DC power in the first scenario

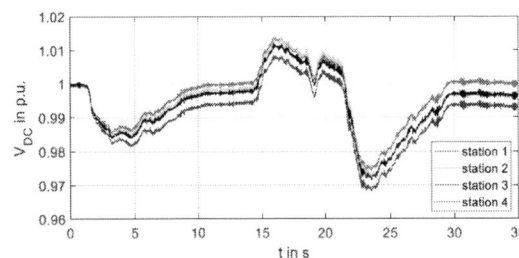

**Fig. 13**   DC voltages in the first scenario

In the beginning, station 3 starts to dissipate power from the DC grid and since station 1 is in voltage control mode, this power is fed into the DC grid by station 1. At around t = 15 s station 2 is started to balance the DC grid in the same manner. But at around t = 22 s, the controller at station 2 switches to also to power control mode and needs to support its local AC grid by drawing power from the grid. Consequently, the power output of station 1 increases. The DC grid is stable during this test, which is also proven by the course of the local DC voltages in Fig. 13.

In the second scenario, the influence of the AC grid voltage on the control approach is evaluated.

939

Therefore, the power converter of station 2 is connected to a variable transformer and the AC voltage and current is measured. Here, station 1 is also in voltage controlling mode, while the other stations are in power controlling mode. The resulting DC link voltages and the power flow are shown in the following Fig. 14 and Fig. 15.

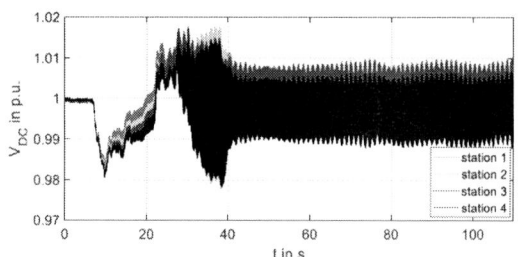

**Fig. 14** DC voltages in the second scenario

First the station 4 starts to draw power from the DC grid. Due to the long transmission line between station 1 and 4, the local DC link voltage drops in the beginning, but is controlled to its nominal value after a few seconds. At t = 22 s station 3 starts to feed power into the DC and, therefore, reduces the power share of station 1. Afterwards the power flow over the DC grid between the power converters remains constant.

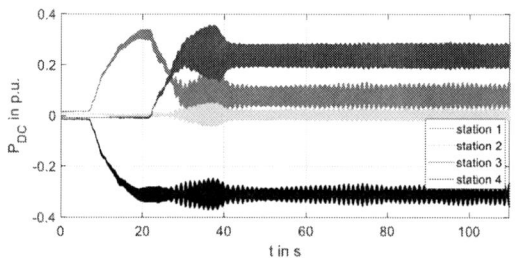

**Fig. 15** DC power in the second scenario

Shortly after t = 50 s the local AC grid voltage at station 2 rises and according to the implemented droop chart (see Fig. 6) the AC current rises to provide reactive power for the AC grid at stations 2.

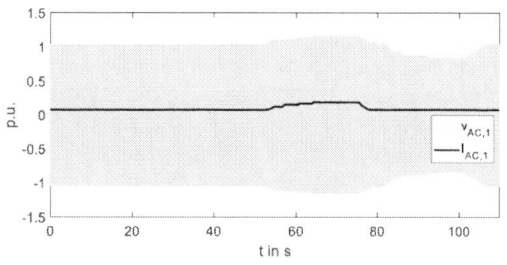

**Fig. 16** AC voltage and RMS value of the AC current at station 2

Due to the fact, that the DC power, respectively the DC current does not change during this process at station 2, it is verified, that only reactive power is provided in this case. Hence, the implemented grid supporting functions are working, which was shown both in simulation and experimental evaluation on the test stand.

In the third and last scenario, dynamic grid situations and are induced by pushing the controller structure to its limits. This test was performed with reduced limit values to avoid damages of the test stand components. The resulting plots for the DC voltages are shown in Fig. 17 and the DC power courses in Fig. 18. The test begins with station 1 controlling the DC voltage and station 4 drawing power from the DC grid. At t = 11 s station 2 is activated and dynamically feeds power into the DC grid. This leads to minor oscillations in the overall DC voltage, but these oscillations are damped, and the grid stability is not directly affected by them.

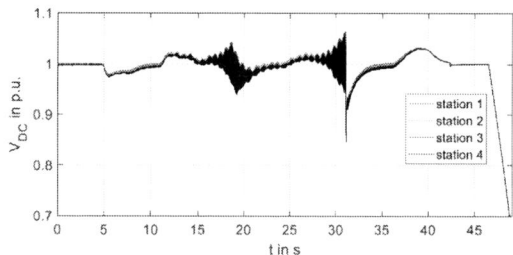

**Fig. 17** DC voltages in the third scenario

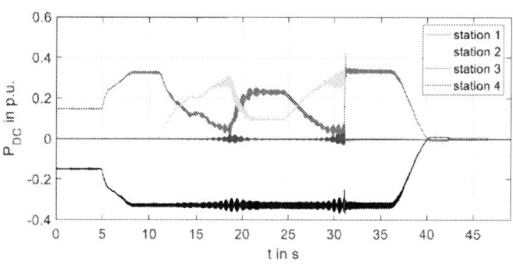

**Fig. 18** DC power in the third scenario

At around t = 31 s the power controller at station 2 is stopped, which leads immediately to a transient and station 1 must quickly ramp up its DC power to avoid a collapse of the DC grid. Here, the DC voltage in Fig. 17 shows a significant dip, but the controller is quick enough to stabilise the grid. At the end of the test, the power demand of the island grid goes back to zero and, afterwards, the voltage controller is switched of.

## 5 Conclusion

In this work the extension of a decentralised control approach to an adaptive DC grid controller was presented. The proposed control approach checks and evaluates the grid health status of its local grids and switches its control mode according to a state machine. Additionally, grid supporting functions were implemented according to a technical application rule with droop charts. The resulting set values for the controller structure are processed by algorithms for dynamic limitation to ensure overall control and grid stability at any time. It has been shown that the proposed control approach can individually control each converter in the grid by ensuring the overall grid stability for different AC grid types in simulation as well as on the test stand. Further improvements are topic of ongoing tasks, and the controller will be evaluated in meshed DC grids respectively grids with an accumulation of switching actions.

## References

[1] Ł. Kocewiak *u. a.*, „Overview, Status and Outline of Stability Analysis in Converter-based Power Systems", in *19th Int'l Wind Integration Workshop*, Nov. 2020.

[2] S. Menzel, R. Reimann, A. Ernst, W. Holzke, H. Raffel, und B. Orlik, „Interconnection of Point-to-point HVDC Links to Form a Multi-Terminal HVDC Grid", in *PCIM Europe 2022; International Exhibition and Conference for Power Electronics, Intelligent Motion, Renewable Energy and Energy Management*, Mai 2022, S. 1–8. doi: 10.30420/565822266.

[3] S. Menzel, R. Reimann, A. Ernst, W. Holzke, H. Raffel, und B. Orlik, „Improving a Machine Set Based Controller for Grid Side Power Converter Applications", in *PCIM Europe 2023; International Exhibition and Conference for Power Electronics, Intelligent Motion, Renewable Energy and Energy Management*, Mai 2023, S. 1–10. doi: 10.30420/566091179.

[4] A. Jain, M. K. Pathak, und N. P. Padhy, „Virtual Synchronous Generator with Fictitious Damper Winding for Compensatory Damping in Renewable Sources Integrated System", in *2022 IEEE International Conference on Power Electronics, Drives and Energy Systems (PEDES)*, Dez. 2022, S. 1–6. doi: 10.1109/PEDES56012.2022.10080307.

[5] J. O. Krah und J. Holtz, „Fast Current Limiting with Virtual Synchronous Generators", in *2024 4th International Conference on Smart Grid and Renewable Energy (SGRE)*, Jan. 2024, S. 1–6. doi: 10.1109/SGRE59715.2024.10428735.

[6] VDE Verlag, „VDE-AR-N 4110 Anwendungsregel: 2023-09 Technische Regeln für den Anschluss von Kundenanlagen an das Mittelspannungsnetz und deren Betrieb (TAR Mittelspannung)". September 2023.

# Demonstrating the Effectiveness of a DC Solid-State Circuit Breaker's Fast Response Time

Ehab Tarmoom[1], Steven Chenetz[1], Dennis Meyer[1], Jason Chiang[1]

[1] Microchip Technology, United States

Corresponding author: Ehab Tarmoom, ehab.tarmoom@microchip.com
Speaker:             Ehab Tarmoom, ehab.tarmoom@microchip.com

## Abstract

The effectiveness of the fast-response time of a DC Solid-State Circuit Breaker (SSCB) over the incumbent solutions, specifically fuses and contactors, is discussed and demonstrated in this paper. Experimental lab measurements followed by simulation models are discussed to show how a solid-state solution improves the safety and reliability of a high-voltage DC distribution system. Specifically, a comparison of a solid-state solution based on Silicon Carbide (SiC) against a DC fuse shows the dramatic decrease in let-through current and energy as well as reduction in arcing and risk of arc-flash.

## 1   Introduction

The electrification of everything has designers pushing DC system voltages higher and higher to achieve system efficiency and power density goals. The conventional circuit protection solutions are not adequate to effectively protect distribution systems at these higher voltages while maintaining both high reliability and high resilience. Solid-state circuit interruption devices, such as a SSCB or an E-Fuse, are solutions that are now being adopted into designs for the many system-level benefits that they offer, primarily the low let-through current and energy. In this paper, the response time of a conventional solution is compared with a SiC-based solid-state solution.

A high-voltage distribution system receives power from at least one electrical power source, such as the electric grid, an Energy Storage System (ESS) or renewable energy source. Many high-voltage DC distribution systems include power electronic converters that efficiently convert the AC or DC input power to a regulated DC bus voltage that feeds power to downstream loads. Figure 1 is a schematic of such a system comprised of converters, loads and circuit protection devices, such as circuit breakers. The electrical wiring in these systems may span from a few meters in length to tens of kilometers. The consequence of this wiring is parasitic line inductance between a power source and the DC bus capacitor, and the parasitic bus inductance between the DC bus capacitor and a load. Both the line inductance and the bus inductance affect the performance of a circuit protection device.

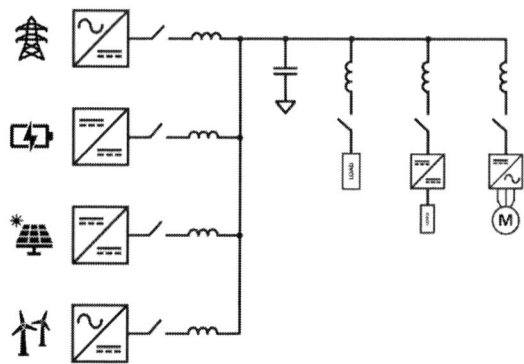

**Fig. 1**    Generalized high-voltage distribution system with lumped parasitic inductance elements.

A solid-state solution's fast response time provides an alternative to contactors and relays, and their associated reliability issues. Disconnecting a load with a DC contactor results in arcing between its contacts due to the system inductance. Arcing degrades the contacts, significantly limiting the life of a contactor. Despite arc-extinguishing features within DC contactors, there remains room for improvement in their reliability. Figure 2 shows the arcing time associated with disconnecting a 400 V DC system under load [1]. As is demonstrated in this paper, a solid-state solution can interrupt high short-circuit current flow in microseconds, orders of magnitude faster than the time needed to develop the small arc shown in the figure at t = 4 ms.

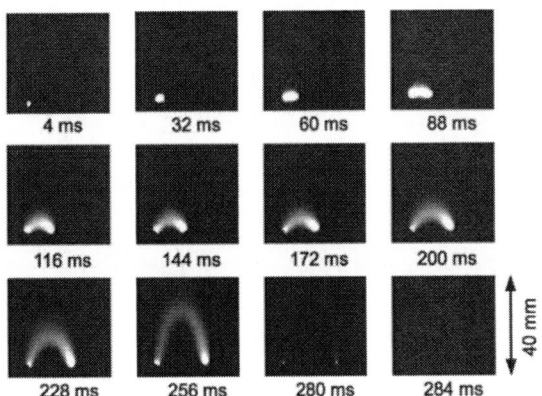

**Fig. 2** Snapshots of arc behavior in 400 V DC disconnection [1].

From a safety perspective, sparks and arcing pose a significant risk in systems that use high-voltage lithium-based batteries, such as in an ESS, EV, and other high-voltage DC distribution system. Electrical discharges can trigger a thermal runaway event, which is a dangerous and potentially catastrophic scenario. High short-circuit currents conducting through improperly-secured connections that may loosen over time may result in arcing. The ability for a circuit interruption device to respond quickly, limiting short-circuit currents to a few hundred Amps compared to thousands or tens of thousands of Amps, can make systems safer by significantly reducing the high arcing current needed to sustain an arc flash event.

## 2 Short-Circuit Test Setup

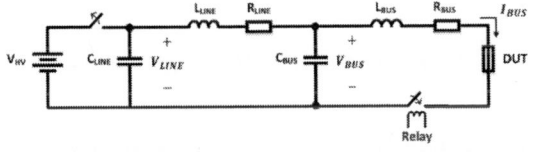

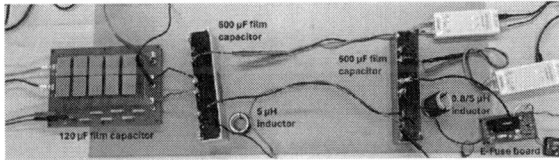

**Fig. 3** Short-circuit test circuit schematic and photo of test setup.

The short-circuit test circuit shown in Fig. 3 consists of a high-voltage DC power supply, $V_{HV}$, line impedance network, bus impedance network, the Device-Under-Test (DUT) and an electromechanical relay. The line impedance consists of a capacitance, $C_{LINE}$, of 120 µF and 500 µF film capacitors

connected in parallel. The line parasitic inductance, $L_{LINE}$, is approx. 5 µH and has a DC resistance represented by $R_{LINE}$. The line inductance and resistance model the impedance between a DC source, such as a high-voltage ESS and a downstream electrical distribution system. The bus network includes a film capacitor, $C_{BUS}$, with a capacitance of 500 µF, an inductance, $L_{BUS}$, of 0.8 µH or 5 µH and a parasitic resistance of $R_{BUS}$. The DUT is a traditional fuse or the Microchip auxiliary E-Fuse demonstrator board. After charging the capacitances, $C_{LINE}$ and $C_{BUS}$, to 450 V at the start of the test, the DC power supply is disconnected from the test circuit. The relay contacts are closed to make a short-circuit connection when the DUT is a traditional fuse. The relay is not used when the DUT is the E-Fuse board because it is turned on into a short-circuit via a serial command over the Local Interconnect Network (LIN).

High-voltage differential voltage probes with 100 MHz bandwidth measure the line voltage, $V_{LINE}$, and the bus voltage, $V_{BUS}$. The points of measurements are at the terminals of the two 500 µF capacitors. The bus current, $I_{BUS}$, measures the short-circuit current at the DUT using a Rogowski coil with a 50 MHz bandwidth. To minimize measurement noise from common-mode currents induced by a fast-changing current in a large loop area, a battery-powered oscilloscope with a 100 MHz was used.

The energy stored in the test circuit calculated in Eq. (1) is 113.4 J.:

$$E_S = \frac{1}{2}(C_{LINE} + C_{BUS}) V_{LINE}^{\,2} = 113.4 J \qquad (1)$$

Much of the energy will be delivered to the DUT. However, there will be energy that is absorbed by the parasitic resistances in the test circuit as well.

The bus inductance, $L_{BUS}$, limits the rate that the short-circuit current rises. In a real-world system the inductance depends on the proximity of a load to the bus capacitance, $C_{BUS}$. The inductance varies from system to system. The values used for $L_{BUS}$ result in an approximate current rate-of-change of 562.5 A/µs and 102.3 A/µs for 0.8 µH and 5 µH bus inductance, respectively.

$$\frac{di}{dt} = \frac{V_{HV}}{L_{BUS}}$$

$$\frac{di}{dt} = \frac{450\,V}{0.8\,\mu H} = 562.5\,\frac{A}{\mu s} \qquad \text{for } L_{BUS} = 0.8\ \mu H \quad (2)$$

$$\frac{di}{dt} = \frac{450\,V}{5\,\mu H} = 90\,\frac{A}{\mu s} \qquad \text{for } L_{BUS} = 5\ \mu H$$

The minimum of a system's short-circuit current capability in Eq. (3) and the product of the DUT's response time and system's short-circuit current rate-of-change in Eq. (4) determines the peak let-through current in Eq. (5).

$$I_{SC} = \frac{V_{HV}}{R_{BUS}} \tag{3}$$

$$I_{DUT} = t_{RES}\left(\frac{V_{HV}}{L_{BUS}}\right) \tag{4}$$

$$I_{PK,LT} = min(I_{SC} + I_{DUT}) \tag{5}$$

# 3 Fuse Test and Simulation

## 3.1 Fuse Short-Circuit Test

Using the test setup with a bus inductance of 5 µH and a traditional high-voltage, fast-acting fuse with a 20 A rating as the DUT, the bus current, line voltage and bus voltage were measured, and the power dissipation and energy were calculated from the bus current and voltage as shown in Fig. 4.

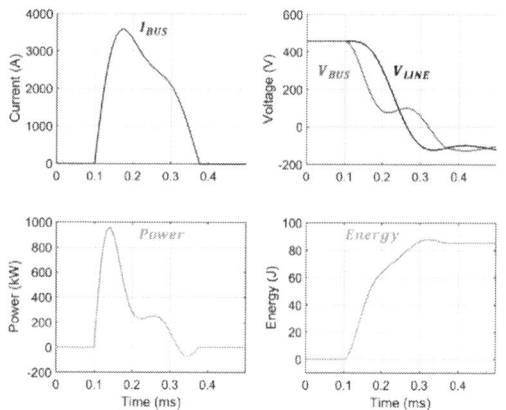

**Fig. 4** Waveforms for short-circuit test with traditional 20 A fuse as the circuit protection device.

The short-circuit is cleared in 276 µs with a peak bus current of 3590 A. The rising edge of current slope of the measured waveform in Eq. (6) is in line with the theoretical slope calculated in Eq. (2):

$$\frac{di}{dt} = \frac{740\,A - 140\,A}{107.6\,\mu s - 100.8\,\mu s} = 88.2\,\frac{A}{\mu s} \tag{6}$$

The capacitances completely discharge and form an LC resonance that causes a negative voltage on the line and bus voltage nodes. This is an artifact of the test setup due to disconnection of the power supply prior to the start of the test and is discussed in the next section. In this test the line voltage drops to -110 V, and the bus voltage initially dips to 78 V at 111 µs into the short-circuit followed by a local peak of 102 V at 163 µs and a final drop to -110 V.

The peak power reached 963 kW and the energy reached 85.4 J.

## 3.2 Fuse Short-Circuit Simulation

Figure 5 shows a SPICE-based model of the test circuit and simulation results using MPLAB® Mindi™ Analog Simulator. The uses the same capacitance and inductance values described in the previous section. The parasitic resistance values for $R_{LINE}$ and $R_{BUS}$ were initially estimated based on the resistance of the wiring, contacts, fuse, fuse holder and the relay. The value for $R_{BUS}$ was slightly adjusted in simulation to match the waveforms close to the test results.

The simulation results show a clearing time for the current to reach down to 0 A of 293 µs. The peak bus current is 3460 A. The bus voltage has a local minimum of 80 V at 120 µs, a local maximum of 103 V at 169 µs followed by a global minimum of -179 V at 298 µs.

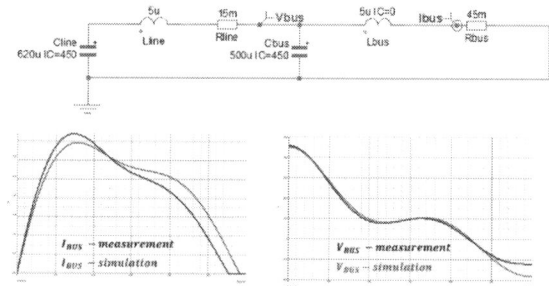

**Fig. 5** SPICE test circuit model and simulation results.

With a circuit model that is a close approximation of the physical system, the impact the line inductance has on the bus voltage can be evaluated, specifically the duration and amount of bus voltage sag. The line capacitance, which could only provide a very limited energy in the lab test, is replaced with an ideal DC source that is representative of a high-energy, low-impedance DC energy storage system as shown in Fig. 6. The dashed-line waveforms are simulation runs using the DC source and line inductances of 5, 15 and 50 µH. The solid-line waveforms, shown as a reference, are with the original circuit model using a line capacitance instead of the DC source.

The simulation results show that the bus voltage drops to 114 V with a 5 µH line inductance. As the inductance increases, its impedance increases, minimizing the DC source's ability to maintain a stable bus voltage. As a result, increasing the inductance to 15 μH and 50 μH causes the bus voltage to further drop to -38 V and -147 V, respectively. The line inductance and bus capacitance form a resonant circuit that causes ringing for several milliseconds before the bus voltage stabilizes. The duration that the bus voltage remains below 300 V during the first dip is 125 μs for a 5 μH inductance and 189 μs for a 15 μH inductance. It takes 1.75 ms to recover the bus voltage above 300 V with a 50 μH inductance. The resonant ringing triggered by the fast-changing short-circuit current not only results in the bus voltage dip but also a high positive voltage exceeding the DC source voltage. In a system, this could stress film capacitors and other components.

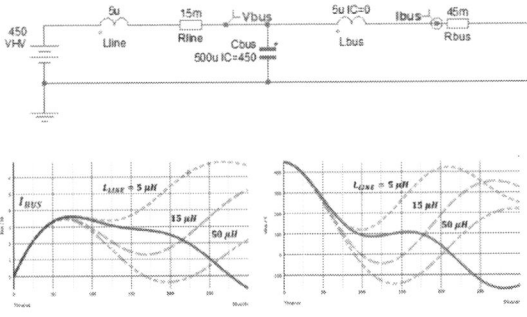

**Fig. 6** Simulation of test circuit model using an ideal DC source and with multiple line inductance values.

# 4 E-Fuse Test and Simulation

## 4.1 E-Fuse Overview

Short-circuit tests were conducted using the Microchip auxiliary E-Fuse demonstrator board in Fig. 7. The board supports different voltage and current options depending on the devices populated and software constants. In this test the 400 V, 30 A variant (p/n MSCDR-EFUSE-003) was used. The specific variant has two 15 mΩ, 700 V SiC MOSFETs connected in parallel to provide unidirectional voltage- and current-blocking in DC systems. The E-Fuse board is resettable after an over-current detection. It is not a one-time use device like a traditional fuse. This provides many benefits including the ability to characterize a single device over several operating points.

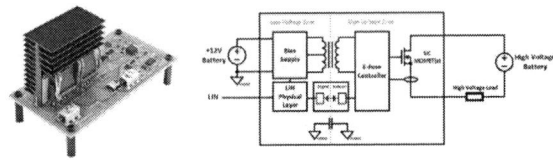

**Fig. 7** Photo and block diagram of Microchip's E-Fuse demonstrator board.

The board consists of a low-voltage zone powered by 12 V that is galvanically-isolated and transfers bias power to the monitor, control and drive circuits in the high-voltage zone. The control circuit includes the capability to control the state of the SiC MOSFETs, configure the over-current trip profile and provide diagnostic information. The trip profile is configurable via LIN and includes a configurable hardware-based, short-circuit monitor with a current threshold resolution of 33 A.

## 4.2 E-Fuse Short-Circuit Test

The first E-Fuse test was with a bus inductance of 0.8 μH and threshold of 33 A. The bus current, bus voltage and line voltage measurements are in the top graphs of Fig. 8. The bus current reached a peak of 216 A and had a clearing time of 672 μs. The rising edge of the bus current, which is based on the bus voltage and bus inductance, had a slope of 586 A/μs which matches the previous calculation in Eq. (2). The theoretical slope of the falling edge in Eq. (7) is based on the bus voltage, inductance and the SiC MOSFET's avalanche-mode breakdown voltage, which is approximately in the range 900 V to 1000 V [2]. Based on the measurement of the falling edge the slope was calculated in to be 615 A/μs. The bus voltage dipped by only 1.5 V to 448.5 V for a brief duration of less than 200 ns.

$$\left(\frac{di}{dt}\right)_{-,theor} = \frac{450\,V - 950\,V}{0.8\,\mu H} = -625\,\frac{A}{\mu s} \quad (7)$$

$$\left(\frac{di}{dt}\right)_{-,meas} = \frac{30\,A - 153\,A}{1.46\,\mu s - 1.26\,\mu s} = -615\,\frac{A}{\mu s}$$

The test was run again with a bus inductance of 5 μH using the same E-Fuse board and a threshold setpoint of 264 A. The results also in Fig. 8 show a clearing time of 6.3 μs and a peak bus current of 287 A. The measured rising-edge current slope of 104.3 A/μs is in line with Eq. (2) and the negative current slope of the theoretical and measured values in Eq. (8) are also close. The bus voltage dipped only by approximately 2 V to 447.8 V and recovered back to the line voltage in 60 μs.

Based on the bus current and bus voltage measurements, the calculated peak power was 129 kW and calculated let-through energy was 406 mJ.

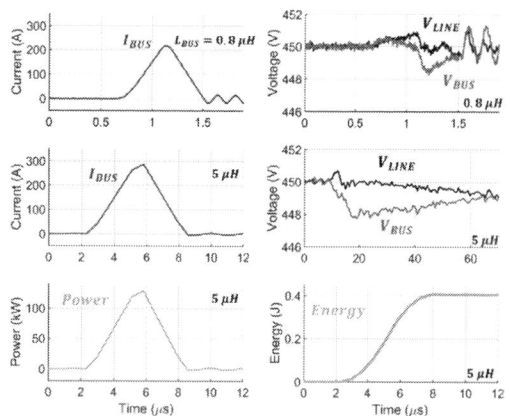

**Fig. 8** E-Fuse measurements with 0.8 µH and 5 µH bus inductance.

$$\left(\frac{di}{dt}\right)_{+,meas} = \frac{264\ A - 45\ A}{5.1\ \mu s - 3\ \mu s} = 104.3\ \frac{A}{\mu s}$$

$$\left(\frac{di}{dt}\right)_{-,theor} = \frac{450\ V - 900\ V}{5\ \mu H} = -90\ \frac{A}{\mu s} \qquad (8)$$

$$\left(\frac{di}{dt}\right)_{-,meas} = \frac{53\ A - 208\ A}{7.9\ \mu s - 6.5\ \mu s} = -110.7\ \frac{A}{\mu s}$$

The E-Fuse board was further tested with a threshold, $I_{SET}$, starting from 33 A to 264 A with 33 A steps and using bus inductance of 5 µH. The waveforms are shown in Fig. 9 and numerical results summarized in Table 1. The range of peak let-through currents was from 45 A to 287 A. The dip in bus voltage was under 1 V for many of the measurements. The range of peak power was from 20 kW to 129 kW. The range of let-through energy was from 28 mJ to 406 mJ. The range of clearing time was 2.1 µs to 6.3 µs.

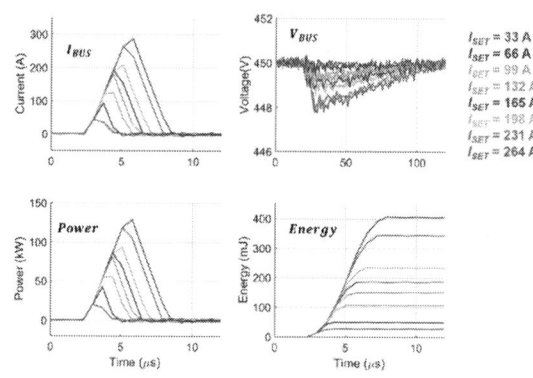

**Fig. 9** E-Fuse short-circuit measurements with increasing over-current threshold.

| # | $I_{SET}$ (A) | $I_{PK,LT}$ (A) | $E_{LT}$ (mJ) | $t_{CLR}$ (µs) |
|---|---|---|---|---|
| 1 | 33 | 45 | 28 | 2.1 |
| 2 | 66 | 95 | 49 | 2.5 |
| 3 | 99 | 126 | 108 | 3.5 |
| 4 | 132 | 181 | 151 | 3.9 |
| 5 | 165 | 193 | 187 | 4.2 |
| 6 | 198 | 208 | 235 | 4.9 |
| 7 | 231 | 264 | 345 | 5.6 |
| 8 | 264 | 287 | 406 | 6.3 |

**Table 1** E-Fuse short-circuit measurements.

## 4.3 E-Fuse Short-Circuit Simulation

A model of the test setup using an E-Fuse was simulated as shown in Fig. 10. The model includes a switch, S1, that represents the SiC MOSFETs channel conduction during a short-circuit. When the switch is turned off, the current then conducts through the diode and 975 V voltage source, $V_{br\_dss}$. The diode and voltage source form a basic model of the SiC MOSFET's avalanche electrical behavior as the bus current decreases to 0 A.

The bus current reaches a peak of 284 A and the clearing time is 6.3 µs. Independent of line inductance the bus voltage momentarily dips to 448.3 V and then begins to increase. The bus voltage increases faster with the lower line inductance.

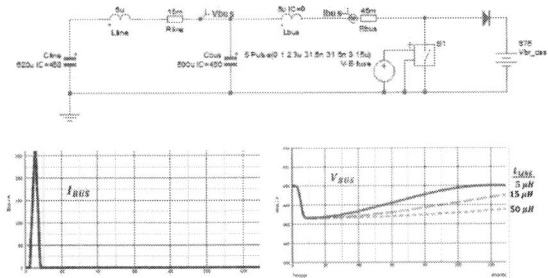

**Fig. 10** Simulation of E-Fuse test circuit model with multiple line inductance values.

The circuit was simulated again but with a DC voltage source instead of a capacitor to simulate practical systems. The circuit model and results are shown in Fig. 11. Independent of the line inductance, the bus voltage sharply dips to 448.3 V. This fast transition triggers the LC resonance between the line inductance and bus capacitance, resulting in several exponentially-decaying AC cycles with a peak voltage of approximately 451.5 V. The resonant frequency varies with the line inductance as shown in Table 2 and matches the theoretical calculations in Eq. (9). The low-amplitude oscillation continues for 4, 10 and 40 ms for the 5, 15 and 50 µH line inductances, respectively.

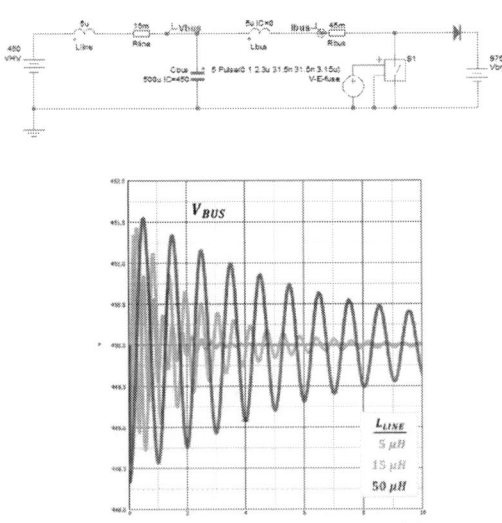

**Fig. 11** Simulation with DC battery source, sweeping multiple line inductance values.

$$f_{RES} = \frac{1}{2\pi\sqrt{L_{LINE}C_{BUS}}}$$

$$f_{5\,\mu H} = \frac{1}{2\pi\sqrt{5\,\mu H * 500\,\mu F}} = 3.18\,kHz \quad (9)$$

$$f_{15\,\mu H} = \frac{1}{2\pi\sqrt{15\,\mu H * 500\,\mu F}} = 1.84\,kHz$$

$$f_{50\,\mu H} = \frac{1}{2\pi\sqrt{50\,\mu H * 500\,\mu F}} = 1.01\,kHz$$

| # | $L_{LINE}$ (µH) | $V_{BUS}$ (V) | | $f_{RES}$ (kHz) |
|---|---|---|---|---|
| | | min | max | |
| 1 | 5 | 448.3 | 451.6 | 3.26 |
| 2 | 15 | 448.3 | 451.5 | 1.84 |
| 3 | 50 | 448.3 | 451.3 | 1.01 |

**Table 2** Circuit simulation measurements.

## 5  Fuse Vs. E-Fuse Performance

Figure 12 captures the measurements of the traditional fuse (thick lines) and the E-Fuse at multiple over-current trip thresholds (thin lines). The measurements include bus current, line voltage and bus voltage as well as calculations of the power and let-through energy based on the bus current and bus voltage. The graphs on the left capture all the measurements over the entire over-current duration in a linear scale measured in units of milliseconds. The graphs on the right, measured in units of microseconds, are zoomed in to the start of the over-current event to see the E-Fuse waveforms. The numerical results are summarized in Table 3.

The E-Fuse solution's fast response to a short-circuit significantly minimizes the peak current and power, and results in 3000 times lower let-through energy in a system over a traditional fuse.

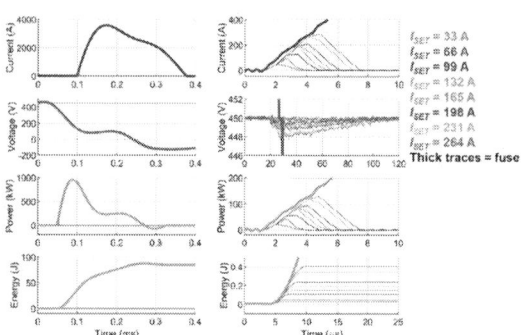

**Fig. 12** Lab measurement waveforms of traditional fuse vs. E-Fuse triggered at multiple over-current thresholds.

| Parameter | Fuse | E-Fuse |
|---|---|---|
| Clearing time | 276 µs | 2.1 to 6.3 µs |
| Peak bus current | 3590 A | 45 to 287 A |
| Min. bus voltage | -110 V | 447.8 to 449.6 V |
| Peak power | 963 kW | 20 to 129 kW |
| Let-through energy | 85.4 J | 28 to 406 mJ |

**Table 3** Summary of fuse and E-Fuse short-circuit test results.

## 6 Conclusion

The work discussed in this paper demonstrates some of the benefits of the fast response time of DC solid-state circuit protection over traditional solutions. It was shown that an E-Fuse solution using silicon carbide reduced the 276 µs clearing time of a traditional fuse down to a configurable range of 2.1 µs to 6.3 µs. Similarly, a peak short-circuit let-through current of 3590 A and let-through energy of 85.4 J of a circuit protected by a fuse were reduced to a peak current range of 45 A to 287 A and energy range of 28 mJ to 406 mJ on a circuit protected by an E-Fuse. Additionally, the E-Fuse solution limited the voltage dip on the 450 V DC bus down to only 447.8 V. Using a traditional fuse under the same test conditions completely discharged the capacitors that were initially charged

to 450 V and even when simulated with a DC voltage source the bus voltage dropped substantially.

The lower let-through current and energy a solid-state solution offers limits the severity of downstream faults and, in some cases, may prevent a fault in a downstream load from becoming a hard failure. The quick response and minimal let-through current demonstrated prevent arc sustainment and reduces the likelihood of arc flash incidents from loose connections that may develop in a system over time. The E-Fuse solution quickly isolates faults and provides for a stable DC bus voltage allowing continued operation of other system components and maximizing uptime, safety and reliability.

## References

Please follow international scientific citation rules.

[1]     K. Tan, X. Song, C. Peng, P. Liu and A. Q. Huang, "Hierarchical protection architecture for 380V DC data center application," *2016 IEEE Energy Conversion Congress and Exposition (ECCE), Milwaukee*, WI, USA, 2016, pp. 1-8, doi: 10.1109/ECCE.2016.7855145.

[2]     E. Tarmoom, S. Chenetz, D. Meyer, X. Zhang, K. Speer and O. Esparza, "SiC-Based Auxiliary E-Fuse Technology Demonstrator for EV Applications," *PCIM Europe 2023; International Exhibition and Conference for Power Electronics, Intelligent Motion, Renewable Energy and Energy Management*, Nuremberg, Germany, 2023, pp. 1-10, doi: 10.30420/566091230.

PCIM Europe 2024, 11– 13 June 2024, Nuremberg          DOI: 10.30420/566262118

# Modelling and Sizing Sensitivity Analysis of a Fully Renewable Energy-Based Electric Vehicle Charging Station Microgrid

Mobin Naderi[1] , Diane Palmer[1], Maria N. Munoz[2] , Yazan Al-Wreikat[3] , Matthew Smith[1], Ewan Fraser[3] , Erica E. F. Ballantyne[2] , David A. Stone[1] , Daniel T. Gladwin[1] , Martin P. Foster[1]

[1] Department of Electronic and Electrical Engineering, University of Sheffield, Sheffield, UK

[2] Sheffield University Business School, University of Sheffield, UK

[3] School of Engineering, University of Southampton, Southampton, UK

Corresponding author:       Mobin Naderi, m.naderi@sheffield.ac.uk
Speaker:                    David A. Stone, d.a.stone@sheffield.ac.uk

## Abstract

This paper presents long-term modelling and second-by-second simulation of an autonomous microgrid (MG), including only renewable energy sources (RESs) and a hybrid energy storage system (HESS) as energy provider, and an electric vehicle (EV) Charge Station as a group load. The model uses forecast data for wind speed and solar radiation to provide wind turbine (WT) and photovoltaic (PV) generated powers, and statistical data for vehicles within a defined car park to model the EV demand. It is flexible and can support varying several planning parameters, e.g. varying sizes of WT and PV generation as well as various capacities of energy storage systems (ESSs). Therefore, in order to examine the impact of variations in RESs and ESS sizes, as well as the impact of EV demand uncertainties on the performance and efficiency of the MG, e.g. EV unmet energy, several sensitivity analyses are provided. Based on sensitivity analysis results, one can find reasonable ranges of MG module sizes, and make a decision for sizing of the overall system. For the case study represented here, results show that at least one WT is required, increasing PV panels is more effective to meet the midday EV load in at the target location, and a lower level of Li-ion ESS capacity is sufficient storage for the charging/discharging of the EVs.

## 1 Introduction

The United Kingdom (UK) government is set to end the sale of new petrol and diesel vehicles by 2030 and only allow the sales of zero-emissions vehicles from 2035 to reduce transport sector emissions [1]. Charging electric vehicles (EVs) from electricity networks powered by renewable energy has the potential to maximise the decarbonisation plan for road transport. As the number of EVs grow in the UK, the rise in charging infrastructure is projected to increase to meet the demand to charge these vehicles, reaching a minimum number of 300,000 public charging points by 2030 [2]. Nevertheless, some challenges, like grid demand increase and supplying EVs by fossil fuels via converting to electrical energy, increases the importance of grid-

isolated renewable energy-based charging stations.

The solution proposed here is a fully grid-independent EV charging station powered exclusively by renewable energy. This solution is developed as part of Future Electric Vehicle Energy networks supporting Renewables (FEVER) project [3]. RES generation, such as by solar and wind, supply EV chargers via an off-vehicle energy store (OVES), containing at least one energy storage technology to provide a cost-effective alternative to a high power, high-cost grid connection. Future publications will describe more hybrid OVESs with multiple ESS technologies.

In design and construction of electrical systems, e.g. the FEVER microgrid (MG), sizing of elements, particularly large elements, has a significant importance. On the other hand, MG planning and element sizing studies require modelling of the systems. Although short-term electrical modelling of MGs is necessary for stability analysis, control design, and short-term performance validation, long-term modelling should be done for MG planning and element sizing to cope with inter-seasonal variations in generation, for example. In order to calculate a long-term model of MGs, their elements/modules should be modelled, i.e. RESs, the load demand, and ESSs.

Renewable power output may be modelled from weather data. Ground-based records at the site of interest are arguably the most accurate resource. The alternative, where these are not available, is measurements from satellite instruments. A full comparison is given by [4]. IEA-PVPS has published a report providing guidance on the choice of data provider and radiation model [5]. The algorithms which describe the conversion of solar radiation data via the photovoltaic process to electrical yield are covered by [6].

An explanation of how to model the power output of a wind turbine and an example calculation is given by [7]. Wind speed is conventionally measured at 10 m height by meteorological stations. Zhou et al detail the power law approach to simulating wind speed at turbine hub height [8]. The correlation between hub-height wind speed and wind turbine yield is modelled by power curves provided by turbine manufacturers. A complete review of power curve generation methods is given in [9].

EV charging/load modelling methods and fundamentals are reviewed in [10]. Although EV load profiles can be modelled using empirical data of EV fleets to study the load behaviour after EV fleet construction [11], it is not possible for planning studies, i.e. before fleet construction. In [12], EV load modelling is done using distribution functions of charging duration, charging start time, and transaction energy for eight different EV models, which is useful for planning and sizing studies.

ESSs can be modelled in several levels of detail and for different applications. Although voltage, current, and state of charge (SOC) need to be modelled via electrical relationships in short-term studies, state of health (SOH), SOC, and calendar/cycle ageing should be modelled through power flow relationships in long-term studies, e.g. ESS sizing and system planning [13]. More challenging topics are calendar and cycle loss modelling, where several theoretical, empirical, and semi-empirical [14] models have been presented in the literature. Theoretical models based on several complicated electro-chemical and physical relationships, and empirical models requiring long periods to obtain results of several experiments are useful for studies [15], where the focus is on ESS modelling details. However, in MG sizing studies, where several MG modules should be modelled, semi-empirical [13, 14], and validated simplified [16] models are much more useful, where less detailed models are used based on empirical parameters.

Sensitivity analysis is a common tool to depict nonlinear and multivariable relationships between important features and changeable parameters of a system, which has been used in various research areas of small and large-scale power systems, e.g. stability analysis [17], controller design, optimal planning [18], and techno-socio-economical sizing of MGs [19]. In this paper, sensitivity analysis is used for MG module sizing in an isolated renewable-based EV charging station. The first contribution of the paper is to provide data-driven long-term annual models for WT and PV energy generation units, and EV charging station load demand as well as validated power-in-power-out model for the HESS. The second contribution is to propose details of sensitivity analysis for making decisions around the MG sizing, and to provide results for a real case study.

Modelling of the EV charging station MG is discussed in Section 2. The methodology of sensitivity analysis for MG module sizing using a long-term model is provided in Section 3. Section 4 presents the results and relevant discussions, and Section 5 concludes the paper.

## 2 EV Charging Station Microgrid Modelling

Figure 1 Shows an energy-based schematic of the studied MG including EV demand, i.e. an EV charge station, wind energy generation using WTs, solar energy generation using PV panels, and a hybrid ESS (HESS). An energy management and measurement system is required to schedule charging/discharging of HESS based on the difference signal of generation and consumption powers. It also uses the data to size the model elements required for planning studies, and provides reports required for sizing and decision making, accordingly. This block needs instantaneous values of EV demand, WT and PV generation, current SOC, maximum allowable C-rate, and some other features of the ESS to function.

As an example, a site for the proposed MG has been established at a general outdoor car park in the UK. To this end, the weather data required for RES generation prediction is obtained at the target

PCIM Europe 2024, 11– 13 June 2024, Nuremberg    DOI: 10.30420/566262118

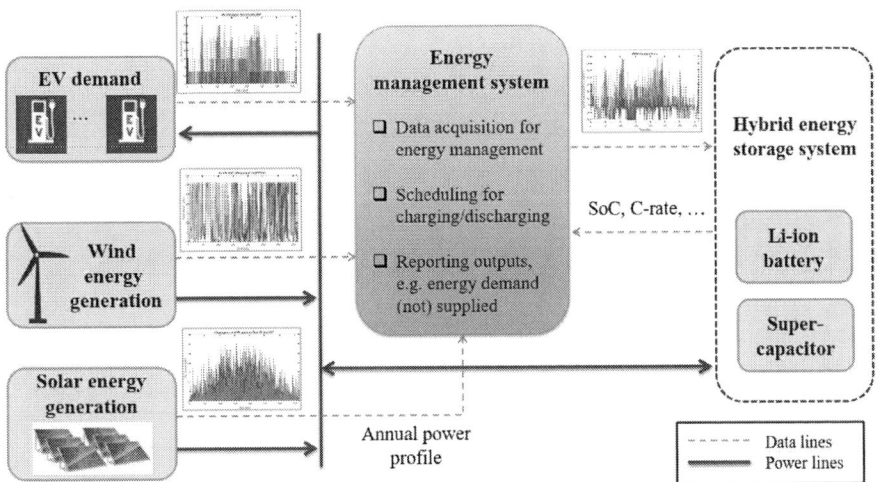

**Fig. 1** An energy-based schematic of the studied MG useful in planning and sizing studies.

location in southern England, with a longitude 1.31° W and a latitude 51° N.

## 2.1  WT generation unit modelling

In order to calculate electrical yield, the input wind speed data is initially scaled to hub height. Since a relatively low wind turbine is to be employed here, the log law is applied because it is more reliable up to 20 m above the ground:

$$U_z = U_{z,ref} \times \ln\left(\frac{z}{z_0}\right) / \ln\left(\frac{z_{ref}}{z_0}\right), \quad (1)$$

where $z$ and $z_{ref}$ are new height (m) and reference height (m), respectively, and $U_z$ and $U_{z,ref}$ are mean wind speed at the new height (m/s) and mean wind speed at the reference height (m/s), respectively. $z_0$ is surface roughness length (m). Surface roughness is a value based on protrusion of land cover e.g. grass, trees, buildings.

Calculation of wind yield i.e. power (kW) is achieved via interpolation of the manufacturer's power curve. The power curve is based on actual measurements. Interpolation supplies values which fall between measurements. Here, the power curve of Aventa AV-7 WT is used, which has a rated power as 6.2 kW [20].

## 2.2  PV generation unit modelling

The PV power output ($PV_{OUT}$) has been obtained by transforming solar irradiance into power as follows:

$$P_{pv} [kW] = G_\beta \times \eta_i \times \eta_p \times p_d \times N_{pv}, \quad (2)$$

where $\eta_i$ is the inverter efficiency, $\eta_p$ is the panel efficiency, $p_d$ is the panel dimension (m$^2$), and $G_\beta$ is the solar irradiance on inclined surfaces (kW/m$^2$).

$G_\beta$ has been estimated by using a solar model developed in MATLAB. A detailed explanation of

the development of the solar model can be found in [21] and in [22]. The input data for the model (i.e., global horizontal solar irradiation) was obtained through the Centre for Environmental Data Analysis (CEDA) archive. The hourly input data was measured at a weather station located in London between 2012 and 2019 in kJ/m$^2$.

## 2.3  EV load demand modelling

To determine the normal EV loading on the system it is possible to examine typical travel profiles for visitors to the selected location. In this case, the car park opens for visitors from 10 am until 5 pm, and this bounds when EVs can plug in and charge at the car park. The system is modelled using 10 uncontrolled AC chargers with a power rating of 7 kW each. The daily energy load demand for charging the EVs was calculated using the chargers' usage profile based on the total number of visitors arriving at the car park each day in 2019 and the hourly visitors' arrival profile. The model calculates the number of EVs arriving at the car park assuming four visitors per car arrive at the car park, with 3% of these cars being EVs.

Visitors typically park their cars for 4 hours while visiting the car park, used in this model as the plug-in period for each EV if any of the 10 chargers are free when arriving at the car park. The model assumes the average efficiency of an EV is 4 miles per kWh and needs to charge to cover a total distance of 30 miles. Figure 2 shows typical EVs behaviour at the car park and EV charging demand for a single day.

## 2.4  ESS modelling

Here, a general long-term model of ESSs is used, which is presented in detail, and is validated in [16]. It is a power-in-power-out model including

951

parameters of SOC, SOH and degradation, a local logic-based ESS management system, charging/discharging losses, and import/export converter losses. Scheduled power of the ESS model, e.g. $P_{b,sch}$ for the Li-ion battery shown in Fig. 3, is obtained from the difference between the vehicle demand and the renewable generation. It can be either positive, meaning a load demand to discharge the ESS, or negative, which means a generation power to charge the ESS. When the SOC of the HESS is between its low and high limits, and the power demand does not lead to exceeding the maximum allowable c-rate of the ESS, the ESS can be charged or discharged successfully for the given scheduled power.

The ESS model can be generalised for different types of batteries, super-capacitors, and some other electrical and electro-chemical ESSs without any structural changes to the model, and only requiring reasonable parameters for each ESS technology. Table 1 shows the parameters for the

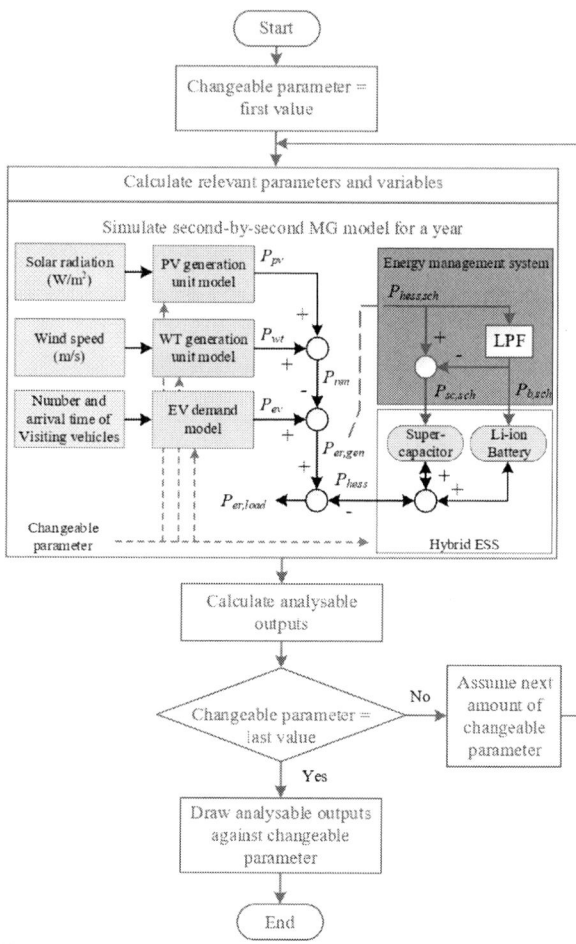

**Fig. 3** Flowchart of the sensitivity analysis method used in MG module sizing.

studied hybrid ESS including the Li-ion battery and the super-capacitor.

A simple charging/discharging schedule is used in the energy management system according to a low-pass filter (LPF) performance such that slow charges/discharges are considered for the high-energy Li-ion battery ESS, and the high-power supercapacitor-based ESS is responsible for fast changes. The simulation part in Fig. 3 shows a schematic of the energy management system and the HESS modules inside the flowchart.

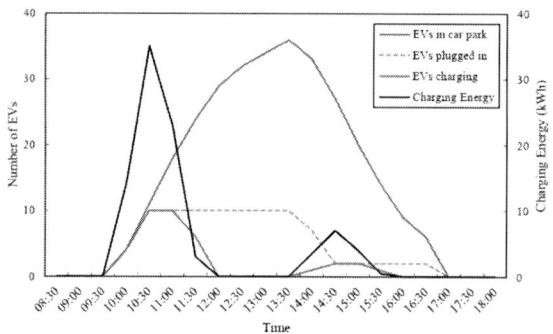

**Fig. 2** EVs behaviour and load demand for an example day (20th of April).

| ESS parameters | Battery value (unit) | Super-capacitor values (unit) |
|---|---|---|
| Nominal capacity | 500 (kWh) | 10 (kWh) |
| Initial capacity | 500 (kWh) | 10 (kWh) |
| Charging/discharging loss | 3 (%) | 3 (%) |
| Converter import/export loss | 3 (%) | 3 (%) |
| Maximum C-rate | 1 | 10 |
| SOC low limit | 3 (%) | 0 (%) |
| SOC high limit | 97 (%) | 100 (%) |
| Initial SOC | 43 (%) | 100 (%) |
| SOH loss per 1000 cycles | 7 (%) | 0.1 (%) |
| SOH calendar loss per month | 0.5 (%) | 0.005 (%) |

**Table 1** Hybrid ESS parameters and their values used in simulations.

## 2.5 Overall Microgrid Modelling

After modelling the MG modules separately, they may be configured into the entire MG system. The simulation part in Fig. 3 shows the connections between different modules of the MG. According to the ESS/HESS model assumptions, the load demand is considered as positive, and the RES generation is denoted with negative values. The

RES generation power, $P_{res}$, is the sum of the WT generation power, $P_{wt}$ and the PV generation power, $P_{pv}$. From this, it is possible to calculate the difference between the RES generation power and the EV demand power, $P_{ev}$, as follows:

$$P_{er,gen}(t) = P_{ev}(t) - P_{ren}(t). \qquad (3)$$

where $P_{er,gen}$ is the generation error power. When its value is positive, it means the generation is not enough, and its negative value implies more generation power than the EV demand power. Assuming charging/discharging power of the HESS, $P_{hess}$, one can write the instantaneous power balance of the MG, as follows:

$$P_{er,load}(t) = P_{er,gen}(t) + P_{hess}(t). \qquad (4)$$

where $P_{er,load}$ is the load error power. A positive value of the $P_{er,load}$ means EV demand power is not met, and a negative value of the $P_{er,load}$ means excess RES power neither consumed by the EV load nor stored in the HESS. Note that $P_{hess}$ in Eq. (4) has positive values during discharging the HESS, and negative values during charging the HESS. The HESS power, $P_{hess}$, is obtained according to energy management system calculations for the HESS scheduled power, $P_{hess,sch}$.

The HESS scheduled power and the generation error power, i.e. $P_{hess.sch}$ and $P_{er.aen}$, are the same values (see Fig. 3). However, the generation error power and its equivalent energy are the actual power and energy signals, but the HESS scheduled power and its equivalent energy are control signals which are used in the energy management system. Note that the HESS scheduled energy and the generation error energy can be easily calculated by integrating $P_{hess,sch}$ and $P_{er,gen}$ with respect to time.

Figure 4 shows the annual power profile of $P_{hess,sch}$, where it is divided into two parts, i.e positive values as discharging power, $P_{hess,sch}^{dis}$, required for supplying remainder of the EV demand, and negative values as the charging power, $P_{hess,sch}^{ch}$, to store the excess RES power. The energy management system can be modelled as follows:

$$P_{b,sch}(t) = G_{lpf}(s) \times P_{hess,sch}(t), \qquad (4a)$$
$$P_{sc,sch}(t) = (1 - G_{lpf}(s)) \times P_{hess,sch}(t), \qquad (4b)$$

where $P_{b,sch}(t)$ and $P_{sc,sch}(t)$ are the scheduled powers of the Li-ion battery and super-capacitor, respectively. $G_{lpf}(s)$ is the LPF transfer function with $\omega_c$ as the cut of frequency as follows:

$$G_{lpf}(s) = \frac{\omega_c}{s + \omega_c}. \qquad (5)$$

Finally, one can calculate the HESS power, according to the output powers of each ESSs as follows:

$$P_{hess,}(t) = P_{b,ex}(t) + P_{sc,ex}(t). \qquad (6)$$

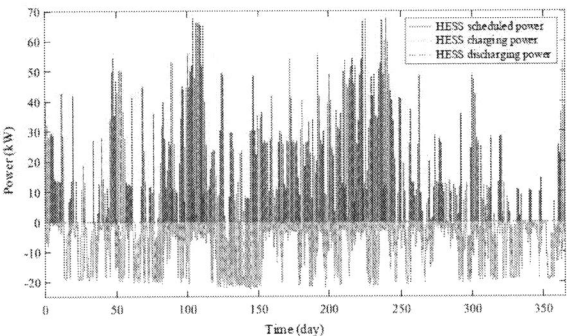

**Fig. 4** Annual power profile of demand and generation difference as the HESS scheduled power (positive amounts for demand after subtracting the RES power and negative amounts for excess power of the RESs.

# 3 Sensitivity Analysis for MG Module Sizing

In the sensitivity analysis approach, there are two important concepts including a changeable parameter and an analysable variable. A changeable parameter is one of the parameters of the model, which is required to be changed to see the impact of these changes on the system performance. The performance is usually assessed using an (a few) output variable(s), which are valuable to be analysed, and can be called analysable variables. In this section, it is explained how the sensitivity analysis method is used to show the relationships between changeable parameters and analysable variables, which are important for module sizing. Since the analysable variables are selected to be different energy concepts in the MG, they are explained before the sensitivity analysis procedure.

## 3.1 Important energy concepts in MG sizing

MG sizing and planning require long-term calculations, where energy signals are more appropriate than power signals due to their cumulative feature with respect to the instantaneous power signals. Energy demand of the EV chargers, available RES energies, and HESS scheduled energy are the most important energy concepts to allow HESS scheduling by the energy management system. These can be obtained by integrating $P_{ev}(t)$, $P_{wt}(t)$, $P_{pv}(t)$, and $P_{hess,in}$ with respect to time.

Figure 5 shows annual energy profiles of the EV demand, one WT, 40 PV panels, and the HESS scheduled energy. The negative amount of the HESS scheduled energy shows that the available energy of the RESs is enough to meet the load demand if the generation and consumption times have enough overlap. However, both the EV demand in the car park and the RES generations have stochastic behaviours. Therefore, a negative HESS scheduled energy indicates enough excess RES energy generated to be able to charge the ESSs after supplying the vehicle charging load.

After scheduling the HESS, $P_{hess}(t)$ and $P_{er,load}(t)$ are important power signals. They include both positive and negative values as explained in Section 2.5, thereby these amounts should be separated before integrating them to calculate meaningful energy concepts as follows:

$$W_{hess,su} = \int P_{hess}^{+}(t)\, dt, \tag{7a}$$
$$W_{hess,st} = \int P_{hess}^{-}(t)\, dt, \tag{7b}$$
$$W_{ev,um} = \int P_{er,load}^{+}(t)\, dt, \tag{7c}$$
$$W_{gen,nst} = \int P_{er,load}^{-}(t)\, dt, \tag{7d}$$

where + and − denote positive and negative amounts, respectively. $W_{hess,su}$ is the energy supplied by the HESS, $W_{hess,st}$ is the stored energy in the HESS, $W_{ev,um}$ is the total EV demand unmet energy, and $W_{gen,nst}$ is the total excess generated energy of the RESs, which is not possible to be stored and/or consumed.

Although all these energy concepts are important in the MG sizing and planning, the HESS scheduled energy and the required energy to be supplied by HESS before scheduling the HESS, and energy supplied by the HESS and the EV demand unmet energy after scheduling the HESS are selected as analysable variables in the sensitivity analysis. Note that the required energy to be supplied by the HESS is calculated by integrating $P_{hess,sch}^{dis}$ with respect to time.

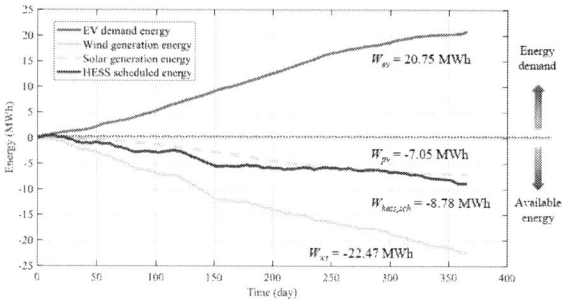

**Fig. 5** Annual energy profiles of the EV demand, the WT, the PV panels, and the HESS scheduled energy.

### 3.2 Sensitivity analysis approach

Since the number of WT, number of PV panels, and ESS capacity need to be determined in the MG sizing, these are selected as changeable parameters. EV proportion is also selected as a changeable parameter to see the impact of EV demand fluctuations on the MG performance.

Furthermore, as mentioned in the previous subsection, the HESS scheduled energy, the required energy to be supplied by the HESS, the energy supplied by the HESS, and the EV unmet energy are selected as analysable variables.

Figure 3 shows a flowchart diagram of the sensitivity analysis method used for the MG module sizing. Each changeable parameter includes a vector of a range of reasonable values. For each value, relevant parameters and variables in the MG model are updated, then the MG model is simulated second by second for the studied year and corresponding input data. The analysable outputs are calculated for each value, and after completing the process for all changeable parameter values, they will be drawn against the changeable parameter values.

## 4 Results and Discussion

The base case parameters used in simulations include 1 WT, 40 PV panels, and characteristics represented in Table 1 for the HESS. The area of the chosen PV panels is 1.67 m², $\eta i$ and $\eta p$ are assumed to be 78% and 12%, respectively. Moreover, EV proportion, number of EV chargers, the charger rated power, average EV park time, and EV average travel distance are assumed to be 3 %, 10, 7 kW, 4 h, and 48 km, respectively.

### 4.1 Sensitivity analysis for WT generation unit sizing

Figure 6 shows sensitivity analysis results for the number of wind turbines, between 0 and 3. As shown in Fig. 6(a), the HESS scheduled energy considerably increases in a negative direction with the increasing number of WTs. However, the energy required to be supplied by the HESS, shown by blue bars in Fig. 6(b), does not show a large decrease, which means a lot of the generated WT power is at different times from EV power demand. The orange bars show the total energy supplied by the HESS, which shows a large increase. Nevertheless, the EV unmet energy, shown by yellow bars, does not become zero even using three WTs. This is because the power profile of the EV demand and WT generation do not match. Therefore, having one WT is a reasonable choice, as the increase in the number of wind turbines from 0 to 1 gives a large decrease on the EV unmet energy, whilst a higher number of WTs may neither be technically reasonable nor cost-effective, as there are diminishing returns when considering the EV unmet energy.

## 4.2 Sensitivity analysis for PV generation unit sizing

Figure 7 shows sensitivity analysis results for PV panel number changeable between 0 and 50. By increasing the PV panel number, the HESS scheduled power, shown in Fig. 7(a), increases in negative values but in smaller steps with respect to the effect of a change in the number of WT. This increases the flexibility of the MG sizing. Figure 7(b) shows 3.35 MWh decrease, 2.95 MWh increase, and 6.3 MWh decrease in the required energy to be supplied by HESS, the HESS energy supplied, and the EV unmet energy, respectively. The EV unmet energy is considerably decreased by both increasing the PV generated power directly used to charge vehicles and increasing the charging/discharging of the HESS indirectly. It may be due to a better match between the solar power generation profile and the EV power demand profile when compared with the WT power profile. Therefore, increasing the PV panel number to improve the MG performance is more reasonable and provides better 'fine tuning' of the characteristics that is seen by increasing WTs.

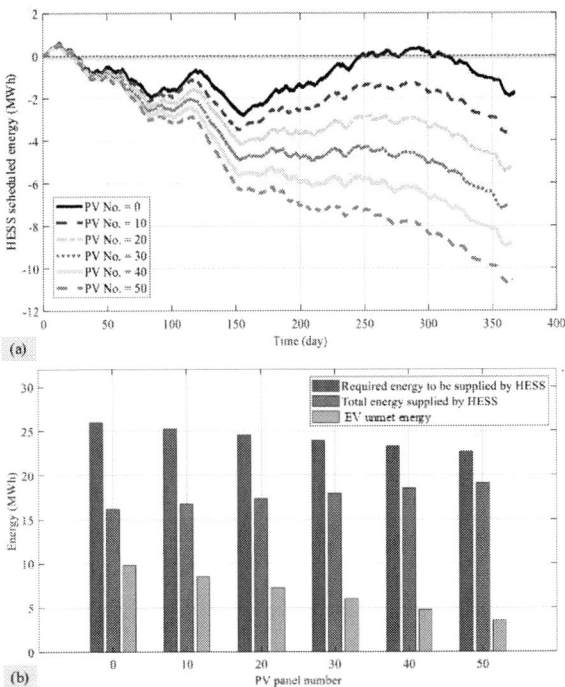

(a)

(b)

**Fig. 7** Sensitivity analysis results of the PV panel number: (a) net input energy to the HESS, (b) required energy to be supplied by HESS, energy supplied by the HESS, and EV unmet energy.

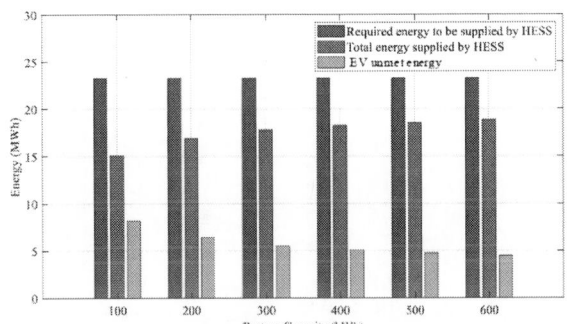

**Fig. 8** Sensitivity analysis results of the ESS nominal capacity including required energy to be supplied by HESS, energy supplied by the HESS, and EV unmet energy.

## 4.3 Sensitivity analysis for ESS sizing

In this section, the sensitivity analysis is done for the Li-ion battery nominal capacity from 100 kWh to 600 kWh when the super-capacitor is disconnected. Since, the changeable parameter makes changes only on the output side of the HESS, the HESS input side energies do not change, e.g., the HESS scheduled energy and the required energy to be supplied by the ESS. Figure 8 shows an exponential behaviour for both the

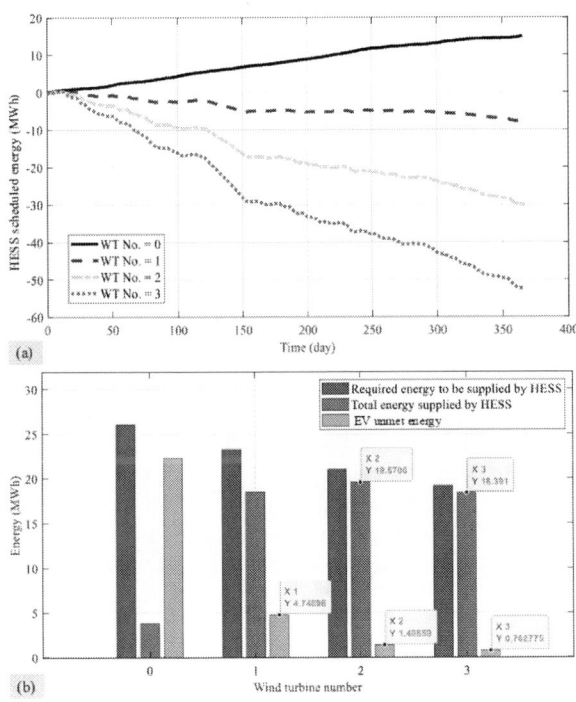

(a)

(b)

**Fig. 6** Sensitivity analysis results of the WT number: (a) net input energy to the HESS, (b) required energy to be supplied by HESS, energy supplied by the HESS, and EV unmet energy.

energy supplied by the HESS and the EV unmet energy.

In fact, they will be saturated for high amounts of the nominal capacity. Increasing the nominal capacity above 400 kWh does not lead to a significant decrease in the EV unmet energy. Therefore, the best size of the ESS should be in the lower band of the selected nominal capacity values.

### 4.4 Sensitivity analysis to study EV uncertainty

Since the number of EVs and their arrival time are stochastic, the load demand causes an uncertainty in long-term studies. Figure 9 shows required energy to be supplied by HESS, energy supplied by the HESS, and the EV unmet energy for different EV proportions as a metric of uncertainties. The MG needs more energy to be provided by the HESS when the EV proportion increases. However, RESs and the HESS sizes are assumed constant, which results in saturation of the total energy supplied by the HESS, and a linear increase in the EV unmet energy. In order to have a robust sizing approach, uncertainties should be included in sizing processes. In addition, a narrower band of uncertainties can be obtained by improving the model through adding details about the uncertain parameters as much as possible.

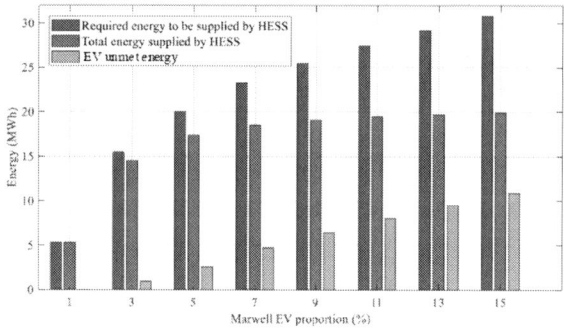

**Fig. 9** Required energy to be supplied by HESS, energy supplied by the HESS, and EV unmet energy for different EV proportions.

## 5 Conclusion

Sensitivity analysis is a powerful, and at the same time simple, tool to inspect linear/nonlinear relationships between important features and effective parameters of a system. Here, it was employed on a long-term model of an electric vehicle charging station microgrid, composed of renewable energy generation resources, and a hybrid energy storage. Although having two or three wind turbines with 6.2 kW rated power results in very low EV unmet energy, the decrease with respect to having one wind turbine is not significant enough to encourage stakeholders to pay two to three times the cost of a single wind turbine. In fact, as a more flexible solution, such small amounts of EV unmet energies can be provided by increasing the number of solar panels. Furthermore, the results show that a nominal capacity in the range 100 kWh-200 kWh is the most suitable choice for the Li-ion battery pack, as increasing the nominal capacity above this level leads to a negligible decrease in the EV unmet energy, which is both technically and economically undesirable. In order to have a robust approach to microgrid module sizing, uncertainties, e.g. electric vehicle number, should be taken into account.

## 6 Acknowledgements

The authors acknowledge the financial support received from the Engineering and Physical Sciences Research Council (EPSRC) through Future Electric Vehicle Energy Networks supporting Renewables (FEVER) grant EP/W005883/1.

## 7 References

[1] 'Transport decarbonisation plan', UK Government, https://www.gov.uk/government/publications/transport-decarbonisation-plan, 2021.

[2] 'UK electric vehicle infrastructure strategy', UK Government,https://www.gov.uk/government/publications/uk-electric-vehicle-infrastructure-strategy, 2022.

[3] 'Future Electric Vehicle Energy netwoks supporting Renewables', https://www.fever-ev.ac.uk/, 2023.

[4] Palmer, D., Koubli, E., Cole, I., et al: 'Satellite or ground-based measurements for production of site specific hourly irradiance data: Which is most accurate and where?', Solar Energy, 2018, 165, pp 240-255.

[5] IEA-PVPS., 'Worldwide Benchmark of Modelled Solar Irradiance Data' (IEA-PVPS T16-05, 2023), pp 1-39.

[6] Koubli, E., Palmer, D. Rowley, P., et al: 'Inference of missing data in photovoltaic monitoring datasets', IET Renewable Power Generation, 2016, 10, (4), pp. 434-439.

[7] Ahammed, S.: 'Optimisation of Floating Horizontal Axis Wind Turbine (HAWT) Blades for Aerodynamic Performance Measurement', International Journal of Innovations in Engineering Research and Technology [IJIERT], 2021, 8, (6), pp 2394-3696.

[8] Zhou, X., Qin, J., Li, H.D., et al.:' A statistical method to construct wind speed at turbine height for study of wind power in China.', Theor Appl Climatol, 2020, 141, pp 419–432.

[9] Lydia, M., Kumar, S.S., Selvakumar, A.I., et al.: 'A comprehensive review on wind turbine power curve modeling techniques', Renewable and Sustainable Energy Reviews, 2014, 30, pp 452-460.

[10] Amara-Ouali, Y., Goude, Y., Massart, P., Poggi, J. M., & Yan, H. (2021). A review of electric vehicle load open data and models. Energies, 14(8), 2233.

[11] Schäuble, J., Kaschub, T., Ensslen, A., Jochem, P., & Fichtner, W. (2017). Generating electric vehicle load profiles from empirical data of three EV fleets in Southwest Germany. Journal of Cleaner Production, 150, 253-266.

[12] Uimonen, S., & Lehtonen, M. (2020). Simulation of electric vehicle charging stations load profiles in office buildings based on occupancy data. Energies, 13(21), 5700.

[13] Mathews, I., et al.: 'Technoeconomic model of second-life batteries for utility-scale solar considering calendar and cycle aging', Applied energy, 2020, 269.

[14] Krupp, A., et al.: Calendar aging model for lithium-ion batteries considering the influence of cell characterization', Journal of Energy Storage, 2022, 45.

[15] Lewerenz, M., et al.: 'Systematic aging of commercial LiFePO4| Graphite cylindrical cells including a theory explaining rise of capacity during aging', Journal of Power Sources, 2017, 345, pp 254-263.

[16] Hutchinson, A.J. and Gladwin, D.T.: 'Verification and analysis of a Battery Energy Storage System model', Energy Reports, 2022, 8, pp 41-47.

[17] Naderi, M., et al.: 'Interconnected autonomous ac microgrids via back-to-back converters— Part II: Stability analysis', IEEE Transa. Power Electron., 2020, 35(11), pp 11801-11812.

[18] Naderi, M., Bahramara, S., Khayat, Y., Bevrani, H.: 'Optimal planning in a developing industrial microgrid with sensitive loads', Energy Reports, 2017, 3, pp 124-134.

[19] Rathore, A., Kumar, A., Patidar, N. P.: 'Techno-socio-economic and sensitivity analysis of standalone micro-grid located in Central India', International Journal of Ambient Energy, 2023, pp 1-22.

[20] 'Aventa AV-7 power curve', https://en.wind-turbine-models.com/turbines/ 1529-aventa-av-7.

[21] Nunez Munoz M, Ballantyne E.E.F, Stone D.A.: 'Development and evaluation of empirical models for the estimation of hourly horizontal diffuse solar irradiance in the United Kingdom', Energy, 2022, 241. https://doi.org/10.1016/j.energy.2021.122820.

[22] Nunez Munoz, M., Ballantyne, E.E.F. and Stone, D.A.: 'Assessing the Economic Impact of Introducing Localised PV Solar Energy Generation and Energy Storage for Fleet Electrification', Energies, 2023, 16(8), p 3570. doi:10.3390/en16083570.

PCIM Europe 2024, 11– 13 June 2024, Nuremberg    DOI: 10.30420/566262119

# LED Powered Rotor Telemetry System

Raphael Beyerle[1], Manfred Schrödl[1]

[1] Technische Universität Wien, Institute of Energy Systems and Electrical Drives

Corresponding author:    Raphael Beyerle, raphael.beyerle@tuwien.ac.at
Speaker:                 Raphael Beyerle, raphael.beyerle@tuwien.ac.at

## Abstract

Measuring physical quantities in rotating shafts is often required in industrial applications and test facilities. Since direct wiring is not possible in most cases, existing measurement systems use either infrared sensors, slip rings, magnetically coupled coils, or a combination of batteries and wireless communication, each of which has its own advantages and disadvantages. This study presents the concept of a rotor telemetry system that relies solely on LEDs and photodiodes for power and data transfer. A prototype is built and measurements of power and data transmission are performed. The feasibility of the presented concept is proven and an exemplary measurement on a speed controlled synchronous motor in a laboratory environment is shown.

## 1 Introduction

During the development of rotating machinery, measurements of physical quantities in rotating shafts can be necessary. Quantities of interest may be temperature, acceleration, pressure or strain which further can be used for system modeling or design validation [1]. Under certain circumstances, rotor temperature can be measured thermographically without contacting the point of interest [2] but most other sensor types need to be installed in specific locations which makes direct cabling on freely rotating systems impossible. A widely applied solution are rotor telemetry systems consisting of a rotating and a stationary unit. The rotating unit is often required to fit inside hollow shafts, which poses a problem especially to small appliances since multiple components are required for signal conditioning and communication. Powering those components is either done through coupled coils [3], [4] or by installing an appropriate energy storage beforehand [5]. Previous studies have shown, that LEDs [6] and photodiodes [7] are capable of powering low power devices.

This work proposes a compact, low-cost rotor telemetry system that uses only infrared LEDs and photodiodes for both power and data transmission. The center of the device's rotation axis is kept free to save space for an optional encoder to measure frequently needed rotation angle. Therefore, the influence of eccentric positioning of the communication elements is studied. A time-division multi-plexing (TDM) method is described to avoid spatial separation of power and data paths due to interference. An example measurement is performed on a permanent-magnet synchronous motor (PMSM) to record its magnet temperature.

## 2 System Design

The developed telemetry system is divided into two parts. A stator unit for power transmission, data reception and user interaction and a rotor unit for performing measurements and transmitting its results. For better visualization, a 3D model of the presented setup is shown in Fig. 1.

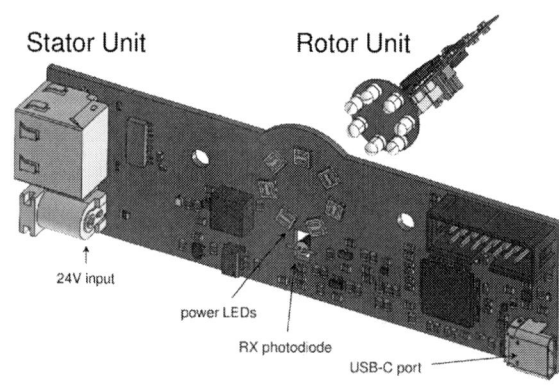

**Fig. 1:** 3D model of the telemetry system prototype.

The rotor unit itself consists of two PCBs (Printed

Circuit Board), called measurement board and LED board, which are connected by a board connector. A detailed view of the rotor unit is shown in Fig. 2. The measurement board has a length of 35 mm, a width of 16 mm and a height of 10 mm. The LED board has a diameter of 16 mm and a height of 7 mm. The combined length of both boards is 42 mm.

**Fig. 2:** Detailed view of the rotor unit divided into measurement and LED boards.

The LED board holds six IR photodiodes and a single IR LED (denoted **TX**). All diodes are aligned on a circle with a radius of 6 mm. The center of the LED board is intentionally left free of diodes so that the magnetic part of an optional AMR (Anisotropic Magnetoresistance) sensor can be mounted to measure the rotation angle $\varphi$ of the rotor. A generalized schematic of the proposed telemetry system is shown in Fig. 3.

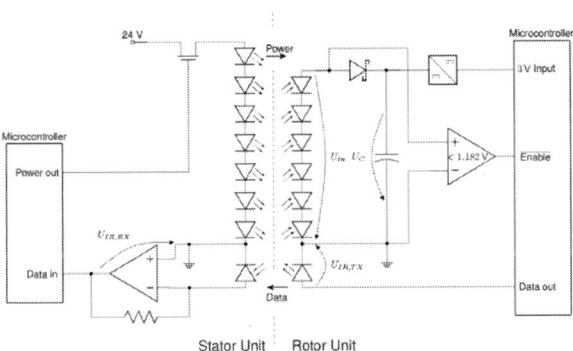

**Fig. 3:** Generalized schematic of the telemetry system design.

The measurement board contains a tantalum capacitor for energy storage, a DC-DC converter for constant voltage supply, a comparator, a microcontroller along with its programming interface and measurement circuitry. For system assembly the entire rotor unit is inserted into the hollow shaft of

the DUT (Device Under Test), where it is connected to previously installed sensors.

The stationary unit is aligned in parallel to the LED board of the rotating unit at a fixed distance $d$. It contains seven high power IR LEDs which are aligned in the same way as the diodes on the LED board of the rotating unit, directly facing each other as shown in Fig. 4. At 14 mm from the center of the diode circle is a single IR photodiode (denoted **RX**). It is slightly tilted by $\gamma = \arctan(d/R)$ so that the center of the cone describing its field of view, shown in gray, is aligned towards the center of the LED board.

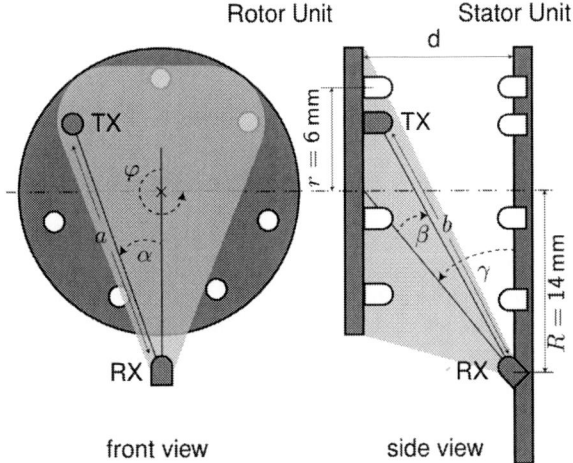

**Fig. 4:** LED arrangement of the rotor unit and the stator unit. The red colored diodes are used for data transmission and the white diodes are used for power supply.

Since the TX and RX diodes are both eccentrically mounted, the distance between them varies with the rotor angle $\varphi$ which affects the received signal strength. In addition to the distance, the signal strength also depends on the half-aperture angles $\theta_{TX}$ and $\theta_{RX}$ of the two cones describing the fields of view of the TX and RX diodes. The distance $D$ between TX and RX and the half-aperture angles $\theta_{TX}$ and $\theta_{RX}$ can be calculated using Eq. (1), Eq. (2) and Eq. (3).

$$D(\varphi) = \sqrt{a(\varphi)^2 + d^2} \tag{1}$$

$$\theta_{TX}(\varphi) = \arctan\left(\frac{\sqrt{[R + \cos(\varphi)]^2 + [r \cdot \sin(\varphi)]^2}}{d}\right) \tag{2}$$

$$\theta_{RX}(\varphi) = \arcsin\left(\frac{\sqrt{[b(\varphi) \cdot \sin(\beta(\varphi))]^2 + [r \cdot \sin(\varphi)]^2}}{D(\varphi)}\right) \tag{3}$$

A worst-case approximation of the input signal strength behavior $E(\varphi)$ can be made using Eq. (4), where $\Phi_{TX}(\theta_{TX})$ and $\Phi_{RX}(\theta_{RX})$ are the relative angle-dependent signal reception intensities of the two diodes as specified by the manufacturer, shown in Fig. 5.

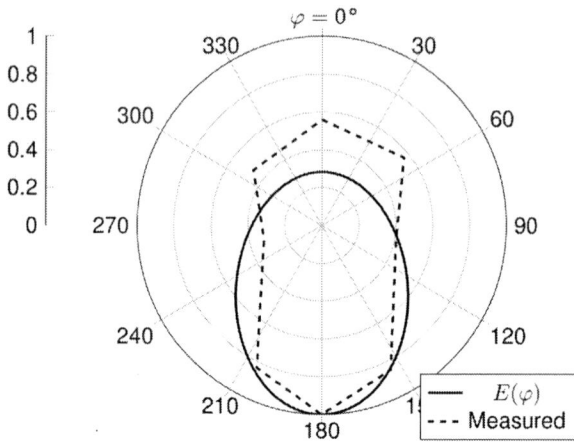

**Fig. 6:** Comparison of estimated and measured normalized signal reception strength as a function of rotor angle $\varphi$ at a distance of $d = 10\,\mathrm{mm}$.

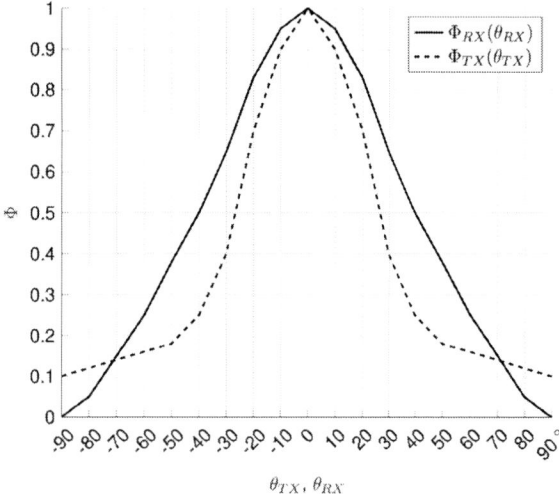

**Fig. 5:** Relative angle-dependent signal reception intensities $\Phi_{TX}(\theta_{TX})$ and $\Phi_{RX}(\theta_{RX})$ of the transmitting IR LED TX and the receiving photodiode RX taken from the corresponding data sheets.

$$E(\varphi) \propto \frac{1}{D(\varphi)^2} \cdot \Phi_{TX}(\theta_{TX}(\varphi)) \cdot \Phi_{RX}(\theta_{RX}(\varphi)) \quad (4)$$

For validation, the signal strength is measured in steps of $\Delta\varphi = 360°/7 \approx 51°$ with an additional measurement at $180°$. The minimum received peak to peak voltage is $0.8\,\mathrm{V}$ which can easily be used for demodulation. A comparison of the polar plot of the normalized function $E(\varphi)$ and the normalized measured values is shown in Fig. 6. The minimum measured signal intensity can be observed at $\varphi \approx 250°$ and not at $\varphi = 0°$ where $D$ reaches its maximum value. The maximum signal strength is received when $\varphi \approx 180°$ where the TX and RX diodes almost face each other.

The receiving RX photodiode is connected to a microcontroller via an amplifier and filter circuit. A USB-C connector allows the user to communicate with the device and receive readings via a serial interface. It is also used to power the controller. The power LEDs are externally supplied with a DC voltage of $24\,\mathrm{V}$ in combination with a current regulator which is controlled by an a optically isolated PWM output of the microcontroller.

## 3 Concept and Implementation

For telemetry operation, the stator unit illuminates the rotor unit with high-power IR LEDs, charging its internal energy storage for a pre-defined period of time. During illumination, the rotor unit is powered down and does not draw significant power from its energy storage. At the end of the illumination phase, onboard amplifiers and a microcontroller are activated in order to measure the desired values. The presented system is equipped with a type K thermocouple to measure the magnet temperature of a PMSM. The values are then sampled by analog-to-digital converters before a bit stream is sent back to the stator unit via the TX IR LED. The received measurement data is amplified and filtered before being sampled by the stator unit's ADC. The sampled data is then demodulated to obtain the bit stream containing the original measurement data. When transmission is complete, the entire cycle begins again.

Since power and data are transmitted over the same physical channel at the same wavelength, both cannot occur simultaneously because illumination would interfere with the data transmission. Therefore, a time-division multiplexing method is used. As described above, data acquisition takes place in periodic cycles. A single measurement cycle can be divided into three continuous phases: I. Charge, II. Compute and III. Transmit. These phases are described in more detail in the following sections.

## I. Charge

The purpose of this phase is to provide the rotor unit with as much energy as it needs to measure one or more values, discretize them and transmit them back to the stator unit. The stator unit is therefore equipped with seven circularly arranged high power IR LEDs having a total radiated flux of $\Phi_{IR,P} = 4.66\,\text{W}$ at $I_F = 1\,\text{A}$ over a broad IR spectrum with its peak wavelength at $860\,\text{nm}$. The forward voltage of a single diode at this current is $V_F = 1.75\,\text{V}$, resulting in an absolute electric power consumption of $12.25\,\text{W}$ during illumination. Since nearly two-thirds of this power is dissipated, excessive heat is generated. Therefore, forced cooling may be necessary if the illumination time exceeds a certain duration.

The rotor unit on the other side is directly facing the high power IR LEDs of the stator unit as shown in Fig. 4 and therefore being illuminated by the radiated flux. Each of the photodiodes has an open circuit voltage of $U_{d,oc} = 0.4\,\text{V}$ and a short circuit current of $I_{d,sc} = 35\,\mu\text{A}$, both specified at an illuminance of $E_V = 1\,\text{mW/cm}^2$ with its peak sensitivity wavelength at $850\,\text{nm}$. As the photodiodes are connected in series, the total input voltage of the rotor unit $U_{in}$ is six times the voltage of a single photodiode $U_d$.

The photodiode's relation between output current $I_d$ and diode voltage $U_d$ at a given photo current $I_{ph}$ can be described by Eq. (5) [8], neglecting internal parasitic components,

$$I_d = I_s \cdot \left[ \exp\left( \frac{q \cdot U_d}{n \cdot k \cdot T} \right) - 1 \right] - I_{ph} \qquad (5)$$

with the Boltzmann constant $k$, absolute temperature of the photodiode $T$, electron charge $q$, ideality factor $n$ and the reverse saturation current $I_s = 5\,\text{nA}$ as specified by the manufacturer. Since the target operating points of the chosen photodiodes are beyond the manufacturer's specifications, $U_{d,oc}$ and $I_{d,sc}$ are measured as a function of the power IR LEDs current $I_F$ in the range $0\,\text{A}$ to $1\,\text{A}$ at a distance of $d = 10\,\text{mm}$, as shown in Fig. 7.

The measured data is used for parameter identification. While measuring the short circuit current so that $U_d = 0\,\text{V}$, Eq. (5) reduces to $I_d = -I_{ph}$. By measuring the open circuit voltage $U_{d,oc}$, $I_d = 0\,\text{A}$ and since the photocurrent is already known from the previous measurement, $n = 1.98$ can be obtained. The ideality factor is calculated using the method of least squares over all measured points.

The result is shown next to the measurement in Fig. 7.

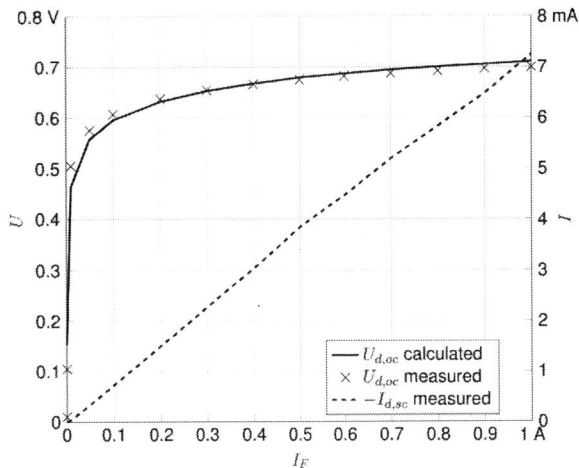

**Fig. 7:** $U_{d,oc}$ and $-I_{d,sc}$ of the photodiodes as a function of the power IR LEDs' current $I_F$ at a distance of $d = 10\,\text{mm}$.

The charging curve of the capacitor $C = 490\,\mu\text{F}$ connected in parallel to six photodiodes as shown in Fig. 3 can be calculated for a given photocurrent. For a numerical solution $I_c(t) = C \cdot \Delta U_c / \Delta t$ is used with $\Delta t = 1\,\mu\text{s}$. Since the capacitor and the photodiodes are connected in parallel, $I_c(t) = -I_d(t)$ and $U_c(t) = 6 \cdot U_d(t)$. The numerical solution besides measured values with $I_{ph}(t) = 7.25\,\text{mA}$ for a duration of $300\,\text{ms}$ is shown in Fig. 8.

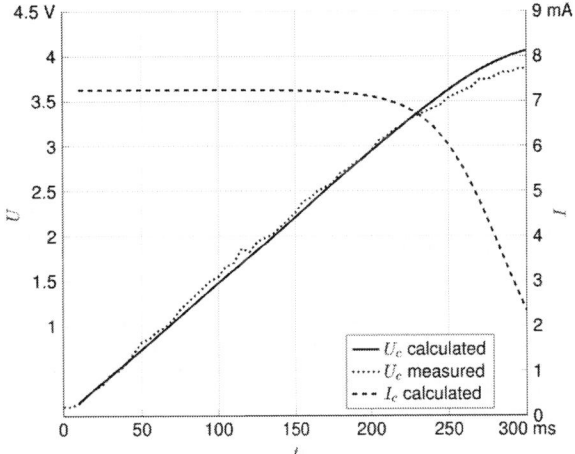

**Fig. 8:** Calculated and measured charging curves of the energy storage capacitor over the duration of $300\,\text{ms}$ with a constant photo current of $I_{ph} = 7.25\,\text{mA}$.

Actual charge time varies because the illumination

depends on the distance between the stator unit and the rotor unit. Illumination may also be affected by housing elements required to secure the rotor unit inside the hollow shaft.

A Schottky barrier diode with less than $10\,\mu A$ of reverse leakage current is placed between the photodiodes and the energy storage to prevent the capacitor from discharging when illumination fades. Since the voltage $U_C$ across the capacitor varies with it's state of charge, a highly integrated DC-DC converter is implemented that produces a constant output voltage of $3\,V$ as long as $U_C > 0.8\,V$. To disable the microcontroller and amplification circuitry during the charge cycle, a comparator compares the input voltage $U_{in}$ to an internal reference voltage of $U_{comp,ref} = 1.182\,V$. As long as $U_{in} > U_{comp,ref}$, the peripherals are disabled.

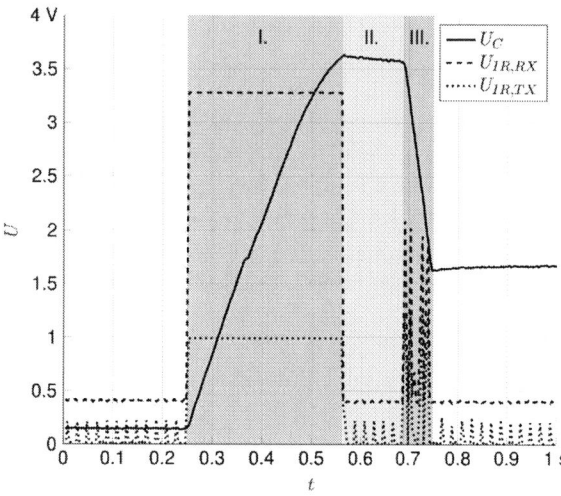

**Fig. 9:** Typical voltage curves of I. Charge, II. Compute and III. Transmit phases at a distance of $d = 10\,mm$ between the rotor unit and the stator unit.

As soon as the power IR LEDs are activated, the capacitor voltage $U_C$ starts to rise as shown in Fig. 9 at $t = 0.25\,s$. When $U_C$ reaches the minimum threshold voltage of the DC-DC converter of about $1.6\,V$, it starts to produce an output voltage of $3\,V$, which supplies the circuit of rotor unit. This can be noticed as a small voltage dip in the charge voltage curve. Measurements on the prototype have shown that $U_C = 3.6\,V$ can be achieved in $300\,ms$ at a distance between the rotor unit and the stator unit of $d = 10\,mm$. During charging, both the photodiode of the stator unit and the IR LED of the rotor unit are fully saturated, so peak voltages are measured. The voltages of the energy storage

capacitor $U_C$, the IR LED of the rotor unit $U_{IR,TX}$ and the output of the transimpedance amplifier $U_{IR,RX}$ are shown in figure Fig. 9. The charging phase ends when illumination stops.

## II. Compute

As soon as the power IR LEDs are turned off and $U_{in}$ falls below the internal comparator reference voltage $U_{comp,ref}$, the microcontroller boots and starts executing code. The controller takes about $150\,ms$ to boot before the first ADC measurement can be taken. During this time the rotor unit has a current consumption of about $250\,\mu A$. The total current consumption of the rotor unit at measurement state $I_{TX} \approx 1\,mA$ can be estimated by summing the active supply currents of the controller running at a frequency of $f_{\mu C,TX} = 600\,kHz$ and two high precision operational amplifiers, according to the manufacturer's data sheets.

Shortly after the core is started, the external peripherals are activated by the microcontroller and the internal 10-bit ADC is started. The ADC is configured to use an internal $1.1\,V$ reference voltage. The rotor unit is equipped with a precision type K thermocouple amplifier with internal cold junction compensation that provides an a output voltage dependent on the measured temperature. The general sensitivity of the amplifier is $5\,mV\,°C^{-1}$, thus limiting the maximum measurable temperature to $220\,°C$ and the achievable accuracy to $0.2\,°C$ with a typical tolerance of $\pm1.5\,°C$, depending on the tolerance class of the sensor used [9].

The measurement data is inserted into a data packet consisting of 4 start bits with alternating high and low states, 10 data bits and 1 stop bit, resulting in a total packet size of 15 bits. Since the communication is asynchronous, the start bits are a necessity for the stator unit to synchronize its clock. The data bits contain the ADC reading. The transmission of raw data allows the system to remain independent of the measured signal, but requires further post processing. The structure of the implemented data packet is shown in Fig. 10.

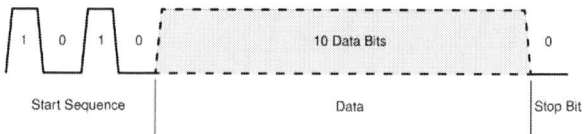

**Fig. 10:** The structure of a single packet of data.

To ensure data integrity, a CRC (cyclic redundancy check) can be calculated and appended to the output data. This allows the stator unit to validate the received data or even perform forward error correction (FEC). Care must be taken because the increased data size requires precise clock synchronization, which is described in the next section. The rotor unit is now ready to send data back to the stator unit.

## III. Transmit

The measurement circuit is deactivated and transmission begins after the measurement data has been calculated. The low power IR LED TX facing the stator unit is connected to the microcontroller via a resistor to limit the drive current. The output voltage signals are OOK (On-Off Keying) modulated at a frequency of 7.5 kHz and sent through the IR LED with a peak forward current of 30 mA driven directly by a digital output pin of the microcontroller. The stator unit is equipped with a photodiode connected to a transimpedance amplifier with a gain of 108 dB and a cutoff frequency of 300 kHz, shown in Fig. 11. An offset voltage of 0.1 V is provided to increase the reverse voltage of the RX diode, reducing its terminal capacitance to 26 pF. This is important to increase receiver bandwidth and amplifier stability.

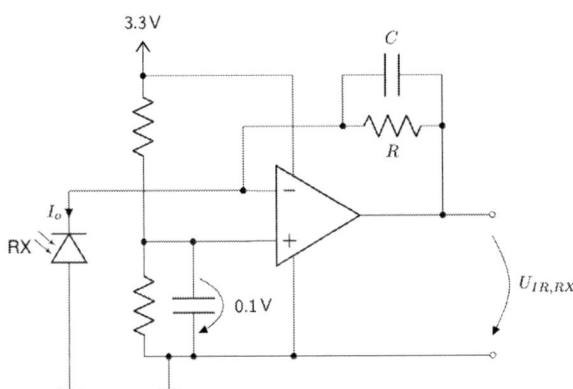

**Fig. 11:** Schematic of the implemented transimpedance amplifier.

The stator unit receives and amplifies the received signal. A sequence of a high-pass filter with a cutoff frequency of about 1 Hz and a low-pass filter with a cutoff frequency of 50 kHz is located between the amplifier and the microcontroller to remove any DC bias caused by ambient IR emission and high-frequency noise. An ADC input on the controller samples the received signal.

For validation purposes, the ADC input data is discarded and 30 data packets with the constant bit sequence $\{1, 0, 1, 0, 1, 0, 1, 0, 1, 0, 1, 0, 1, 0, 0\}$ are sent . The sent and received voltage signals are shown in Fig. 12.

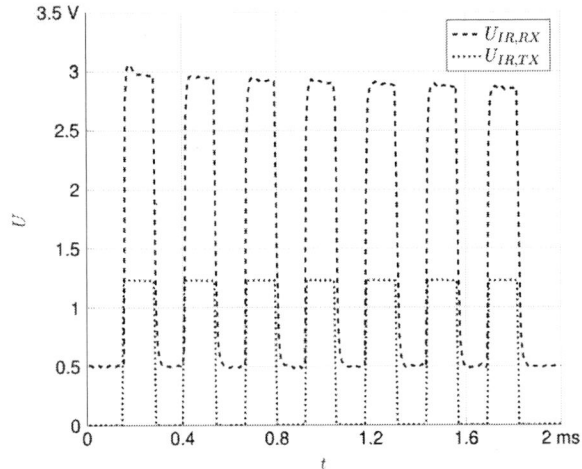

**Fig. 12:** Transmitted and received voltages of the transmitted data at a distance of $d = 7$ mm.

Since no dedicated clock signal is transmitted, a self-synchronization mechanism is implemented using the start sequence $\{1, 0, 1, 0\}$. The demodulation starts after the illumination is stopped by sampling the input signal with a frequency of 200 kHz, which corresponds to a sampling period of 5 µs. Digital on (high) or off (low) input values are distinguished by predefined threshold voltages. Upon detection of the first rising edge, a timer is activated to measure the durations $t_i$, $i \in \{1, 2, 3\}$ between all four edges of the start bit sequence. The actual transmission period is obtained by calculating the mean time $t_s = \mathrm{mean}(t_i)$ between these edges. This allows the signal frequency to be determined under the assumption that the transmission rate is constant over the duration of a single packet transmission. The sampling frequency is set to match the transmission frequency $1/t_s$. Sampling is delayed by $3/2 \cdot t_s$ after the last falling edge of the start sequence to sample at the center of each received bit.

Since the error between the actual and detected sample period can be up to 5 µs/bit either ahead or behind, a sample time offset accumulates with increasing packet transmission time and thus packet length. The transmission of a single bit takes 130 µs. If sampling occurs in the middle of a single bit transmission, the sampling time error margin is half its transmission duration, which is $65\,\mu s$. The maximum sample time error for the last transmitted bit is $5\,\mu s/\text{bit} \cdot 11\,\text{bit} = 55\,\mu s$, which is still within the error margin and therefore allows proper signal demodulation.

The maximum achievable sample rate can be approximated using $I = C \cdot \Delta U/\Delta t$. Charging the capacitor from $U_C = 0\,\text{V}$ to $U_C = 3\,\text{V}$ takes place only during initial charge phase, since it will not be fully discharged later on. Therefore the illumination time is significantly reduced. Assuming a general rotor unit start up time of $t_{start} \approx 150\,\text{ms}$ and a transmission time of $t_{TX} \approx 2\,\text{ms}$ per packet with corresponding currents $I_{start} \approx 250\,\mu A$, $I_{TX} \approx 15\,\text{mA}$ and a charge current of $I_{charge} \approx 5\,\text{mA}$, the total time for one whole measurement cycle $t_{cycle}$ can be calculated. The controller start up reduces $U_C$ from initial $3\,\text{V}$ to $2.92\,\text{V}$. Sending two packets further reduces capacitor voltage to $2.8\,\text{V}$, which must be recharged to its initial state by illuminating it for $t_{charge} \approx 20\,\text{ms}$. Summing over all durations gives $t_{cycle} = t_{start} + 2 \cdot t_{TX} + t_{charge} \approx 174\,\text{ms}$, which results in a maximum sampling frequency of about $f_s = 1/t_{cycle} \approx 5\,\text{Hz}$.

## 4 Example Measurement

A compact $600\,\text{W}$ PMSM having a hollow shaft with an inside diameter of $17\,\text{mm}$ and a length of $41\,\text{mm}$ is equipped with the presented telemetry system to measure the magnet temperature using a previously installed thermocouple. It is driven by a motor controller with a switching frequency of $12.8\,\text{kHz}$ and an output voltage of $24\,\text{V}$. The rotation speed is set to a fixed value of $1625\,\text{min}^{-1}$ under some arbitrary load. The data is recorded during a thermal run under load followed by a cool down period in standstill. A median filter is applied to remove frequent peaks due to EMI (Electromagnetic Interference) caused by the motor controller. The post-processed magnet temperature measured by the proposed system at a sampling rate of $1\,\text{Hz}$ is shown in Fig. 13.

## 5 Conclusion

While existing rotor measurement systems use either thermal cameras, slip rings, coupled coils or battery powered systems for data acquisition, it has been shown that a telemetry system based solely on LEDs and photodiodes for power and data transmission is feasible. By using a time-division multiplexing technique, power and data can be

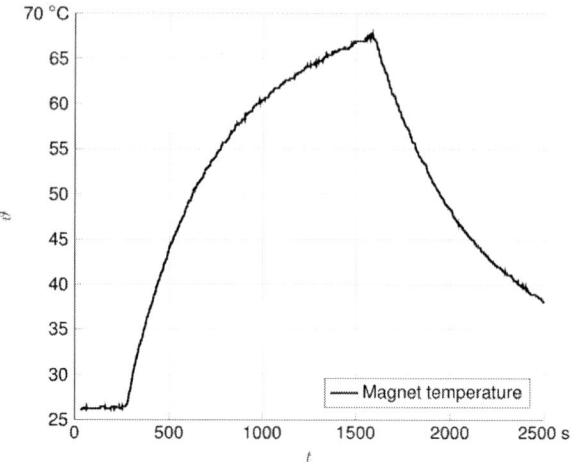

**Fig. 13:** Post-processed example measurement of the magnet temperature of a PMSM while spinning with $1625\,\text{min}^{-1}$ under some arbitrary load.

transmitted without any physical separation, allowing for a compact design. It has been shown that a relatively small energy storage can be charged by the proposed system and that it has sufficient capacity to outlast illumination interruptions. An eccentric arrangement of the diodes for data transmission in order to save space for an AMR sensor proved to be no problem. Despite the lack of EMI immunity of the presented prototype, the system proved to be suitable for measuring the temperature of a rotating system with a sampling rate of $1\,\text{Hz}$ while spinning at $1625\,\text{min}^{-1}$. The cycle time can be significantly reduced by reducing the startup time of the rotor unit. Higher transmission frequency and larger packet size must be accompanied by more accurate clock synchronization which is limited by the maximum sampling frequency of the stator unit. Further efforts are needed to harden the system against EMI, to increase the maximum packet size, and also to reduce the duration of data acquisition.

However, the technology has proven to be suitable for miniaturization and cost reduction. In its current state, the system can be used where high sample rates are not required, such as in many thermal applications.

## References

[1] D. Staton, A. Boglietti, and A. Cavagnino, "Solving the More Difficult Aspects of Electric Motor Thermal Analysis in Small and Medium Size Industrial Induction Motors," en, *IEEE Transactions on En-*

*ergy Conversion*, vol. 20, no. 3, pp. 620–628, Sep. 2005. DOI: 10.1109/TEC.2005.847979.

[2] S. Stipetic, M. Kovacic, Z. Hanic, and M. Vrazic, "Measurement of Excitation Winding Temperature on Synchronous Generator in Rotation Using Infrared Thermography," *IEEE Transactions on Industrial Electronics*, vol. 59, no. 5, pp. 2288–2298, May 2012. DOI: 10.1109/TIE.2011.2158047.

[3] L. Zhang, Z. Tian, Z. Li, S. Cheng, Y. Wang, *et al.*, "Parallel Contactless Transmission of Power and Rotor Temperature of Electrical Machines via Magnetically-Coupled Resonance and Capacitive Radio Frequency," *IEEE Transactions on Industry Applications*, vol. 59, no. 4, pp. 3955–3965, Jul. 2023. DOI: 10.1109/TIA.2023.3262628.

[4] Doležal Ivan, "Simple telemetry system for a rotor temperature measurement," in *Proceedings*, Pilsen, Czech Republic, pp. 1–4.

[5] M. Ganchev, B. Kubicek, and H. Kappeler, "Rotor temperature monitoring system," in *The XIX International Conference on Electrical Machines - ICEM 2010*, Rome, Italy: IEEE, Sep. 2010, pp. 1–5. DOI: 10.1109/ICELMACH.2010.5608051.

[6] I. Haydaroglu and S. Mutlu, "Optical Power Delivery and Data Transmission in a Wireless and Batteryless Microsystem Using a Single Light Emitting Diode," *Journal of Microelectromechanical Systems*, vol. 24, no. 1, pp. 155–165, Feb. 2015. DOI: 10.1109/JMEMS.2014.2323202.

[7] G. M. Pour, M. K. Benyhesan, and W. D. Leon-Salas, "Energy Harvesting Using Substrate Photodiodes," *IEEE Transactions on Circuits and Systems II: Express Briefs*, vol. 61, no. 7, pp. 501–505, Jul. 2014. DOI: 10.1109/TCSII.2014.2327371.

[8] B. E. A. Saleh and M. C. Teich, *Fundamentals of photonics* (Wiley series in pure and applied optics), eng, Third edition. Hoboken, NJ: Wiley, 2019.

[9] ÖVE, *ÖNORM 60584-1: Thermospannungen und Grenzabweichungen*, Vienna, Sep. 2014.

PCIM Europe 2024, 11– 13 June 2024, Nuremberg    DOI: 10.30420/566262120

# 'Infinity Gate Sensor': a Differential Magnetic Field Sensor for Measuring Gate Current of SiC Power Transistors

Yushi Wang[1], Qilei Wang[1], Matthew Appleby[1], Jiaqi Yan[1] , Harry C. P. Dymond[1] ,
Saeed Jahdi[1] , Bernard H. Stark[1]

[1] University of Bristol, United Kingdom

Corresponding author:    Yushi Wang, yushi.wang@bristol.ac.uk
Speaker:                          Yushi Wang, yushi.wang@bristol.ac.uk

## Abstract

For silicon power devices, gate current measurement has been shown to provide a means of inferring temperature and degradation, and it is important for active gate driving. However, in silicon carbide circuits, gate current measurement is challenging due to interference from switching-induced noise and the required low insertion impedance. This paper presents a low-cost, miniature magnetic field current sensor with 500 MHz bandwidth, that has been optimised for high noise immunity, to allow the accurate measurement of gate current for fast-switching SiC devices. Experimental results from 800 V, 10 A and 1200 V, 50 A double-pulsed bridge leg circuits switching at 80-100 V/ns show a high correlation with gate current measurements using current sense resistors and a 1 GHz optically-isolated voltage probe. The sensor's gain is 0.67 V/(A/ns) and its insertion inductance is 3.5 nH at 100 MHz. Magnetic pickup from the adjacent power circuit is seen to contribute less than 1% of the overall measurement, and *dv/dt* susceptibility is quantified through measurement. The theory behind the operation of the sensor, the design principles, the manufacturing detail, and the signal post-processing requirements are presented, providing the user with an alternative to expensive optically isolated probes, and a method of measuring gate current when the addition of a sense resistor is not viable.

## 1    Introduction

Instantaneous gate current measurement of power devices helps in the understanding of device and circuit function [1], provides new options for advanced gate driving [2]-[4], and has been shown to allow the inference of temperature and device degradation [5]-[7]. However, for SiC power transistors, gate current measurement is challenging due to the required low nH-scale insertion impedance, and the high switching-induced noise from high *di/dt* and *dv/dt* in compact layouts.

Current shunts have sufficient bandwidth up to 2 GHz [8], [9], for SiC gate current measurement, however they are not galvanically isolated and coaxial variants with high noise immunity are relatively large, requiring an enlarging of the gate loop layout. Optically-isolated differential voltage probes like IsoVu, DL-ISO, FireFly and SigOFIT offer galvanic isolation and extremely high common mode rejection ratios of over 100 dB, and have been shown to provide a high-bandwidth, low-noise measurement of gate current by probing across a gate resistor [10], [11]. The probes typically cost in excess of 10,000 Euros.

Magnetic field current sensing is a lower-cost alternative that has been reported for measuring source current in SiC and GaN circuits. A variety of PCB-embedded Rogowski coil designs have been reported for power loop current measurement [12], [13] The Bristol Infinity Sensor V2, with its bandwidth of over 1 GHz and an insertion inductance of under 0.2 nH, has been shown to measure the power loop current of GaN devices with low noise [15]. These magnetic sensors, however, have a distinct disadvantage for gate current sensing on wide bandgap devices, since the main aggressor signal, the device's source current, has *di/dt* values that are 2-3 orders higher than those of the gate current to be measured.

This paper makes the following contributions:

In Section 2, it presents a gate sensing concept, named Infinity Gate Sensor, that shows high immunity to the nearby source current transients, allowing it to be used for gate current sensing. A noise-cancelling design is shown, that maintains the advantages of high bandwidth, small size, and low cost.

Section 3 details the implementation of the sensor prototype, how it is integrated into a SiC bridge leg,

966

PCIM Europe 2024, 11– 13 June 2024, Nuremberg    DOI: 10.30420/566262120

and how it is validated against an optically-isolated probe across a shunt resistor.

Section 4 provides an experimental characterisation and validation against an optically isolated probe in two different SiC double-pulse tests. The signal postprocessing method is explained and the sensor's *dv/dt* susceptibility is measured.

## 2 Magnetic gate current sensing

### 2.1 Challenges of magnetic field current sensing in gate current measurement

The main difficulty of magnetic field sensing at the gate of a device is the induced voltage caused by nearby source current transients. Fig. 1 demonstrates the coupling of the source current's magnetic field into a sense coil intended to measure gate current. The source current $i_S$ is aligned on the x-axis and is orthogonal to the much smaller gate current $i_G$ along the y-axis. Their return currents flow closely underneath.

During the turn-on process, a current pulse $i_G$ is applied to the gate, as illustrated in Fig. 2(a). This turns on the SiC device, generating a current rise $i_S$, Fig. 2(b). When a sense coil is placed near the gate, it will be coupled to both currents $i_G$ and $i_S$ and induces a sensor output voltage

$$v_{SENSOR} = M_G(di_G/dt) + M_S(di_S/dt), \quad (1)$$

where $M_G$ and $M_S$ are the sense coils' coupling coefficients to the gate and source currents, respectively. The source coupling $M_S$ is typically smaller, due to the source current being further away from the sensor than the gate current, however the source's *di/dt* is typically at least ten times higher than that of the gate current.

A magnetic sensor's output is integrated using Eq. (2) to obtain the measured current

$$i_M = \frac{1}{M} \int v_{SENSOR} dt. \quad (2)$$

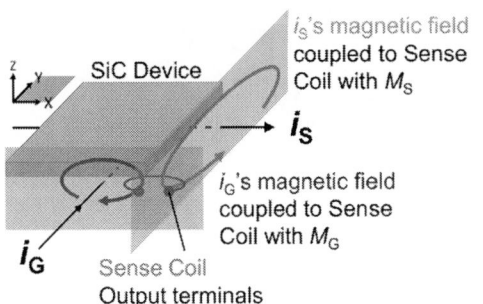

**Fig. 1** Magnetic field coupling into a sense coil, showing that voltages are induced by both gate and source currents.

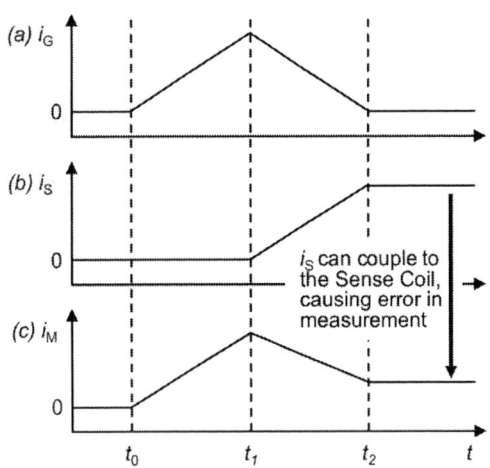

**Fig. 2** Expected magnetic field signal superposition of gate current and source current in the current measured by Sense Coil demonstrated in Fig. 1 during the turn-on process of the SiC device.

If the source current rise couples into a gate sensor, the error shown in Fig. 2(c) results. The error was induced between $t_1$ and $t_2$, thus distorting the measurement. Use of this measurement would result in errors when inferring gate charge or gate capacitance.

Figure 1 shows a simplified scenario, where the source current flows in the X direction, and therefore, a sense coil that lies in a plane that is parallel to the Y-Z plane cannot pick up the source current. However, device packages typically conduct the source current along a path that contains X, Y, and Z components, see Fig. 3, and therefore it would be difficult to find such a plane that would provide immunity. In this case, the source coupling $M_S$ is highly dependent on the package and board layout. This includes the source current trace and that of the return path. An alternative design is therefore required that maximises $M_G$ and minimises $M_S$ from currents in all axes.

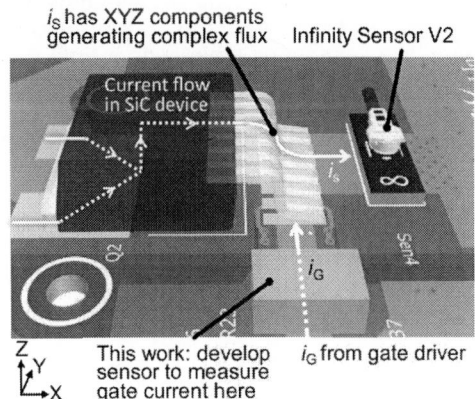

**Fig. 3** Illustration of gate sensor location and immunity requirements, given that source current $i_S$ contains current components in X, Y, and Z directions.

## 2.2 Design concept and theory of the Infinity Gate Sensor

The sensor developed in this paper is optimised for immunity to adjacent source current transients. The Bristol Infinity Sensors [14] and Infinity Sensors V2 [15] are designed for source current measurement, and, as their noise immunity mechanism, use a similar concept to a Rogowski coil. Two coils are wound in opposite directions, so that signals from currents between the coils add up, providing a gain of 0.1 V/(A/ns), whereas signals from currents outside of the sensor body cancel out. This cancellation improves with distance from the sensor and allows relatively large source *di/dt* to be measured in the presence of smaller gate current transients and source current transients in other devices on the same board. However, in the case of gate current sensing, the current to be measured has a significantly lower *di/dt* than that of the same device's source current from which immunity is required, and this aggressor current is significantly closer. Therefore, for gate current sensing, a sensor with higher gain and immunity is required.

To improve gain and immunity, the noise cancelling concept of Fig. 4 has been applied. The concept involves placing a pair of sense coils next to a circular current trace whose current we wish to measure. If the sense coils are close enough together, then they both enclose approximately the same magnetic field, and thus induce the same signal, and therefore subtracting their signals from each other yields zero. If the coils lie a finite distance $\Delta d$ apart, as illustrated in Fig. 4, then the difference between the signals increases as they approach the current trace, i.e. move from right to left in Fig. 4.

This difference is used as the sensor output. If the current trace and sensor coils lie on parallel planes, and are very close, e.g. tenths of a millimetre as they would be in a printed circuit board implementation, then the signal difference is sufficient to provide a useable sensor output.

Conversely, fields that originate from traces whose distance $d_2$ to the sense coils is an order larger than the distance $\Delta d$ between the coils, will provide a degree of signal cancellation. The larger $d_2$ is relative to $\Delta d$, the larger the cancellation, even if the sense coils are perfectly aligned with the field direction as shown in Fig. 4.

Therefore, a three-coil system is used as the sensor, see Fig. 5. The green loop on the left hand represents a circular path in which the gate current is made to flow, and the blue and red loops on the right are two anti-series-connected sense coils, whose voltages $v_A$ and $v_B$ form the output voltage

$$v_{\text{SENSOR}} = v_A - v_B = (M_{GA} - M_{GB})(di_G/dt), \quad (3)$$

where $M_{GA}$ and $M_{GB}$ are the coupling coefficients between the gate current trace, and the sense coils A and B. It is apparent that the smaller the distance between Sense Coil A and gate current trace (high $M_{GA}$), and the larger the distance between the Sense Coil B and the current trace (low $M_{GB}$), the larger the sensor's gain becomes. The gain is also proportional to coil area.

Taking interference from the source current *di/dt* into account, the actual sense coil voltages are

$$\begin{aligned} v_A &= M_{GA}(di_G/dt) \;+\; M_{SA}(di_S/dt) \\ v_B &= M_{GB}(di_G/dt) \;+\; \underbrace{M_{SB}(di_S/dt)}_{\substack{\text{undesired coupling} \\ \text{from source}}} \end{aligned} \quad (4)$$

and therefore, for the influence of the source current to cancel, the coupling from the source current to both sense coils should be approximately equal, i.e. $M_{SA} \cong M_{SB}$. This condition is met if the separation of the source current to the sensor is large relative to the separation of the three coils.

It is worth noting that the sensor is inherently immune to currents in X-direction, or magnetic flux in the Y-Z plane of Fig. 5.

The equivalent circuit of the sensor of Fig. 5, neglecting influence from the source current, is shown in Fig. 6.

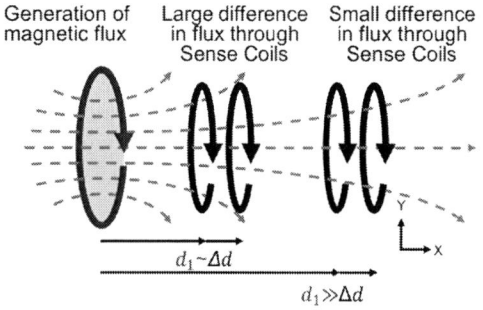

**Fig. 4** Circular trace containing current to be measured, and a pair of sense coils at different distances.

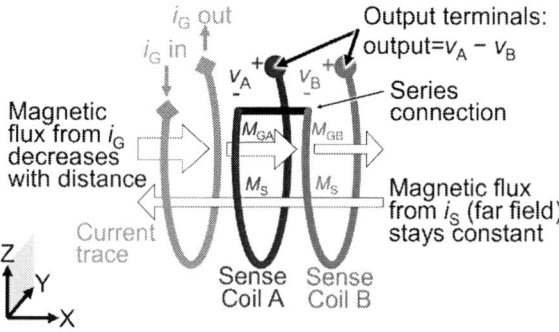

**Fig. 5** Implementation of concept of Fig. 4 as a current sensor.

PCIM Europe 2024, 11– 13 June 2024, Nuremberg          DOI: 10.30420/566262120

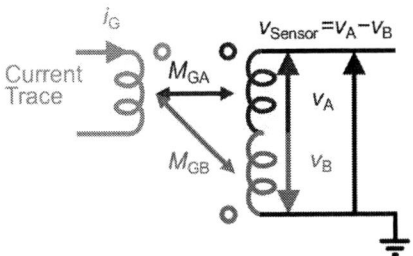

**Fig. 6** Equivalent circuit model of the Infinity Gate Sensor, neglecting coupling between the source current path and the sense coils.

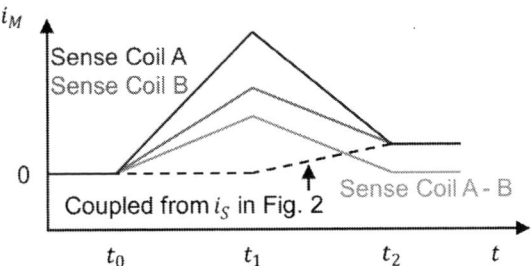

**Fig. 7** Illustration of cancellation of interference from the source current transient.

The model shows the current trace as the primary winding of a transformer, and the sense coils connected in anti-series as the secondary, which provide the differential output voltage.

The cancellation of any pickup from the source current is illustrated in Fig. 7. The diagram illustrates a similar scenario to that of Fig. 2, only that here there are two sense coils. The integrated coil output voltages are shown, where each sense coil provides a different contribution to the gate current measurement due to their different coupling coefficients. Both coils experience the same pick-up from the source current $di/dt$, which cancels out due to the subtraction of $v_A$ and $v_B$.

Note that noise cancellation increases with distance from the noise, and that it is orientation-dependent. In practice, the sensor shown here picks up an error of below 1% at a distance of 5 mm from the source current path (more detail on this in Section 4.3).

# 3 Hardware implementation and validation

## 3.1 Sensor design and fabrication

The Infinity Gate Sensor prototype is implemented as a 4-layer PCB using JLCPCB's 7628 stack up [16]. Figure 8 shows the layout: The gate current trace is placed on the 1st layer, and the sense coils on layers 2 and 4.

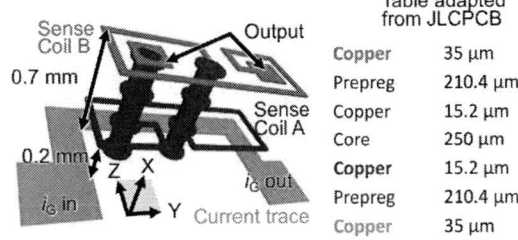

| Table adapted from JLCPCB | |
| --- | --- |
| Copper | 35 µm |
| Prepreg | 210.4 µm |
| Copper | 15.2 µm |
| Core | 250 µm |
| **Copper** | **15.2 µm** |
| Prepreg | 210.4 µm |
| Copper | 35 µm |

**Fig. 8** Infinity Gate Sensor implementation using the 7628 stack-up from JLCPCB [16].

The output is fed into an MHF 4 connector on the 4th layer using vias. The gate current trace follows a partial arc to enclose a similar area to that enclosed by the sense coils. This increases the gain of the sensor, but also its insertion inductance. The advantages of placing Sense Coil B on the 4th layer instead of the 3rd, is increased gain, use of one fewer via, and higher bandwidth, but some immunity is lost. The 0.2 mm dielectric between the gate current trace and the nearest sense coil provides an internal breakdown voltage over 2 kV, however the use of blind vias and a carefully designed external creepage path is needed for high voltage applications. The connection vias in the sensor create a small parasitic loop in the X-Y plane, which can pick up noise. Therefore, the sensor's immunity is the highest against external flux components in the Y-direction, since these couple neither into this loop nor the sense coils. External flux in the X-direction does not couple into this loop either, and its influence is cancelled by the anti-series-connected sense coils. Z-components of flux miss the sense coils, but couple into the parasitic loop, and therefore induce the highest interference.

## 3.2 SiC double pulse test circuit

To validate the Infinity Gate Sensor, a double pulse test is conducted at $V_{DC}$ = 800 V and $I_{LOAD}$ = 10 A, using the discrete C3M0160120J SiC MOSFET shown in Fig. 9. All waveforms are captured on a Tektronix MSO58B 2 GHz 6.25 GSa/s oscilloscope using 64 acquisitions to increase the signal to noise ratio [17]. Signals are de-skewed to ± 0.16 ns. The gate voltage $v_{GS}$ is measured using a 1 GHz Tektronix IsoVu TIVH08 high common-mode rejection differential probe. The switch-node voltage $v_{SW}$ is measured with a PMK HV1000 100:1 400-MHz passive voltage probe via a PCB-mounted coaxial adaptor. The source current $i_S$ is obtained with an Infinity Sensor V2 [15]. The gate current $i_G$ is measured simultaneously with an Infinity Gate Sensor and an IsoVu probe across a 1 Ω shunt resistor, as shown in Fig. 10.

969

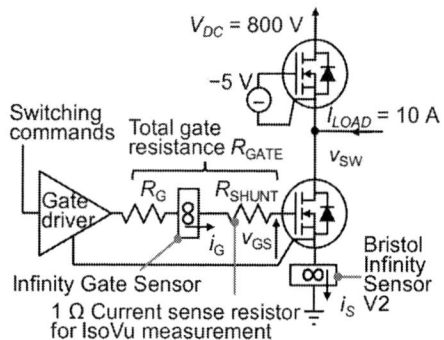

**Fig. 9**  Simplified schematic of test circuit.

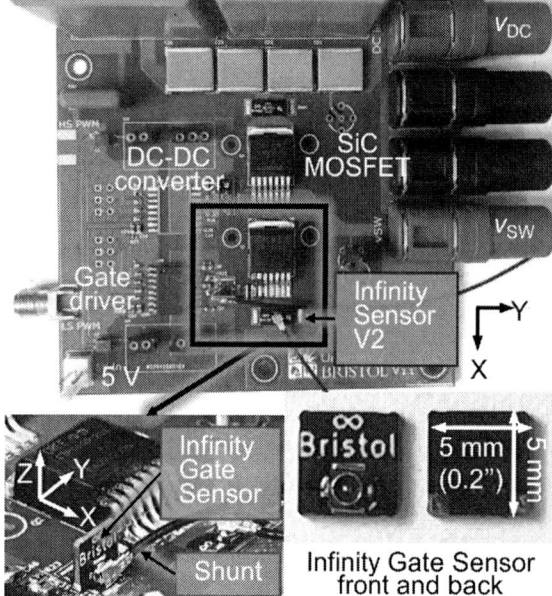

**Fig. 10**  Experimental setup for 800 V 10 A.

The IsoVu probe uses square pins directly located on the shunt resistance, in combination with an MMCX-to-0.1-inch pitch adapter to gain closer access to the circuit board. Both the Infinity Sensor and Infinity Gate Sensor's output voltages are integrated and offset using the method shown in Section 4.2.

In Fig. 10, the coordinate system is aligned with respect to the gate current sensor as it is in Fig. 5 and Fig. 8. The circuit board lies in the X-Y plane and the source current flows mostly in two axes. In the board traces it flows in the X-direction, producing flux components in the Y-Z plane, of which only those in Z couple into the parasitic loop between the vias of the sensor, although these are reduced by raising the sensor above the X-Y plane. In the pins and lead frame of the package, there are current components in the Z-direction, resulting in flux in the X-Y plane, and only those in X require the cancelling concept of the sensor. The Infinity Gate

Sensor is placed approximately 5 mm away from the gate and source pins. On different board layouts, the sensor position that is optimal for reducing interference will vary.

# 4  Experimental results

## 4.1  Sensor bandwidth and insertion impedance

The bandwidth and insertion impedance are two important factors for a current sensor; here they are measured in the gate loop of Fig. 10 and shown in Fig. 11 and Fig. 12.

The sensor gain in Fig. 11 is inferred from an S-parameter measurement using a 9 kHz–6 GHz R&S ZVL vector network analyser (VNA). It shows a flat gain of $M_G = 0.67$ V/(A/ns) across a bandwidth from 1 MHz to 500 MHz. The lower cut-off frequency of the sensor is expected to be lower than 1 MHz, but the measurement is limited by the low $di/dt$ output of the VNA at lower frequencies. The higher cut-off frequency corresponds to the gain change reaching to 3 dB, which is determined by the total length of two sense coils and the length of the current trace. The measurable 10% to 90% current risetime in seconds is limited by [15]:

$$T_{\text{rise}} = 3 \times \frac{0.35}{f_{\text{3dB}}}, \qquad (5)$$

where $f_{\text{3dB}}$ is the sensor bandwidth in Hz. This gives a measurable current risetime range of roughly 2 ns to 1 μs, which meets the bandwidth requirements of typical SiC gate loops.

The insertion inductance at low frequency in Fig. 12 is measured by a 120 MHz Wayne Kerr impedance analyser, while the high frequency above 100 MHz is inferred from an S-parameter measurement from the VNA. The inductance inferred from the VNA measurement matches with the Wayne Kerr at around 100 MHz.

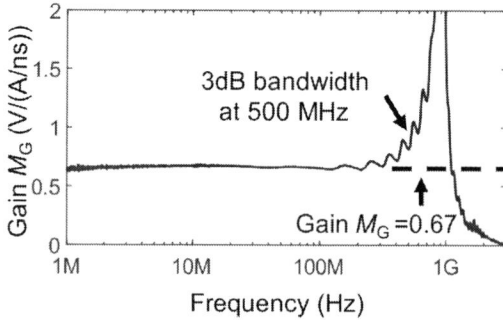

**Fig. 11**  Infinity Gate Sensor frequency response in Fig. 10, inferred from s-parameter measurements.

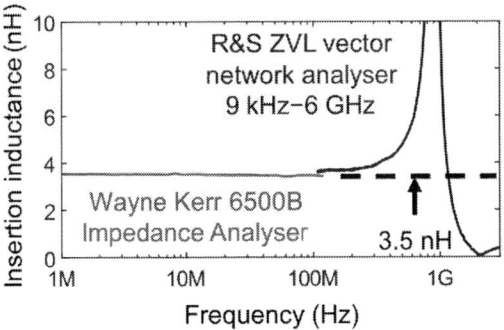

**Fig. 12** Infinity Gate Sensor insertion inductance in Fig. 10, red measured with Wayne Kerr 6500B impedance Analyser, blue inferred from S-parameter measurements of R&S ZVL vector network analyser.

The insertion inductance remains about 3.5 nH in the 120 MHz available bandwidth of the Wayne Kerr Impedance Analyser and rises at higher frequency in the VNA measurement. The insertion inductance is about a quarter of the 14 nH total gate loop inductance, which is acceptable in this circuit. It is larger than the 0.2 nH of the Infinity Sensor V2 due to the curved current trace enclosing a larger loop with the return path.

It should be noted that this insertion inductance is measured in its worse orientation in the setup of Fig. 10. The insertion inductance is measured to be around 0.5 nH if it is placed on the X-Y plane. In circuits where insertion inductance is of concern, the sensor should be placed so that the loop area enclosed by the current trace and its return path is minimised. The gain, immunity and capacitive coupling will also change with the orientation.

## 4.2 Measurement and post-process procedure

The measurement and post-process procedure for the Infinity Gate Sensor, as illustrated in Fig. 13, are:

1. Capture waveform: Obtain the raw sensor measurement data by the oscilloscope with 50 Ω channel termination. It is recommended to take averages to increase the signal to noise ratio.
2. Zoom into transient: Obtain the turn-on or turn-off transient section from data, and <1 µs is recommended, ideally <100 ns.
3. Calculate mean: Calculate the mean of the section before the transient event.
4. Apply offset: Subtract the mean in step 3 from the transient section in step 2.
5. Integrate: Integrate the result of step 4 using Eq. (2).

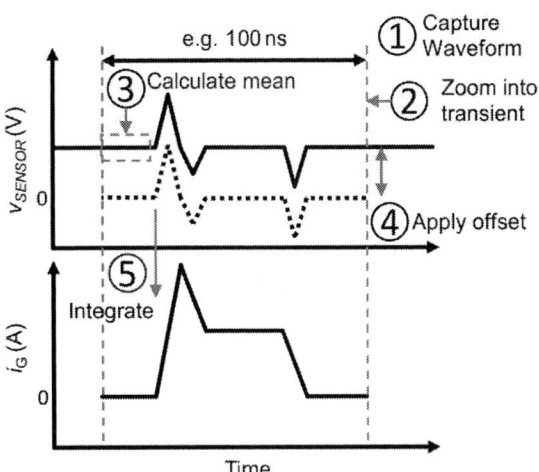

**Fig. 13** Integration procedure to obtain the transient gate current during turn-on. Data sheet and integration code can be found on https://www.infinitysensor.com/

The current measurements shown in this paper are obtained from the raw outputs of Infinity Sensors and Infinity Gate Sensors, post-processed using the above method.

## 4.3 Validation of noise cancelling concept

To validate the noise cancelling concept, a special Infinity Gate Sensor is made with disconnected Sense Coil A and Sense Coil B. The outputs of each sense coil are fed to dedicated connectors. The source current immunity is disrupted, and both gate current and source current can be captured by the sense coils. In the design, the $M_{GA}$ is set as twice that of $M_{GB}$, while $M_{SA}$ and $M_{SB}$ are equal (see Eq. (4)). Figure 14 shows the measured sensor voltage during a complete turn-off and turn-on with the test conditions described in Section 3.2, which are then integrated in Fig. 15 using the method in Section 4.2. The expected gate current can then be obtained by subtracting the current measured by Sense Coil B from that measured by Sense Coil A.

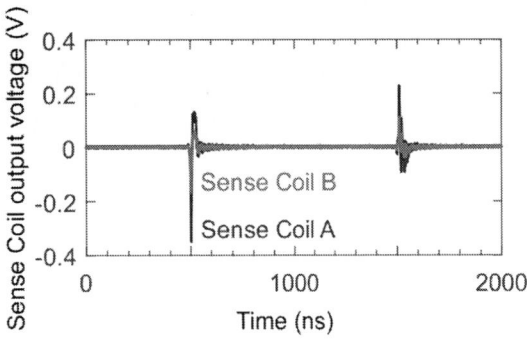

**Fig. 14** Output voltage from Sense Coil A and B.

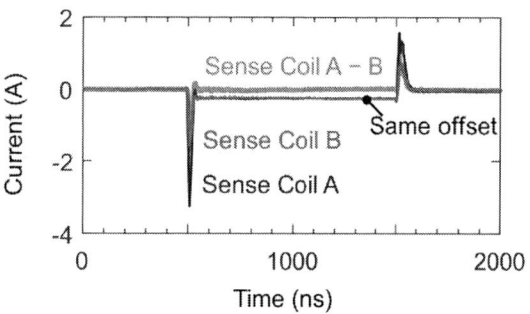

**Fig. 15** Gate current measured by Sense Coil A & B, integrated from signal of Fig. 14.

For Sense Coil A, the first current peak is 3.2 A, and the second is 1.8 A. For Sense Coil B, the first current peak is 1.6 A, and the second is 0.9 A, roughly half of the Sense Coil A. Between the two peaks, both sense coils measure a 250 mA offset, which is coupled from the source current step. By subtracting the signal of Sense Coil B from A, the offset is reduced to less than 1%, supporting the strong source current immunity concept.

### 4.4 Measurement comparisons in 800 V 10 A double-pulse circuit

An 800 V, 10 A double-pulse test is performed using the setup in Section 3.2, and the measured transient waveforms are shown in Fig. 16.

During the turn-on process, the drain voltage drops from 800 V to almost 0 V in 10 ns, and the source current increases from 0 A to a 24 A overshoot in 6 ns. The source current overshoot is caused by the recovery current from the high-side body diode, and the subsequent 180 MHz ringing is caused by the device capacitance resonating with the total power loop inductance.

The gate current waveforms measured by the Infinity Gate Sensor and IsoVu voltage probe across a 1 Ω shunt resistor are shown in the same figure. The two waveforms match well across the whole profile, especially in the rising edge and peaks, which indicates the accurate performance of the Infinity Gate Sensor.

It should be noted that the 180 MHz source current ringing can couple to the gate current through the reverse transfer capacitance of the MOSFET and cause ringing in gate current that can be seen by the IsoVu and the Infinity Gate Sensor. However, the measurement from the sensor shows more oscillations than that from the IsoVu, after the SiC MOSFET completes the turn-on process, which may be induced by the significant source current ringing. This can also be seen on the sensor's output voltage, the amplitude of which reduces after integration. The Infinity Gate Sensor measures

less ringing than the IsoVu during the miller plateau, possibly due to its lower bandwidth.

During the turn-off process, the source current drops slowly at 0.4 A/ns and has much lower oscillations compared to the turn-on process. Meanwhile, there are no oscillations on the gate and source current, and the two waveforms obtained by the Infinity Gate Sensor and IsoVu match well.

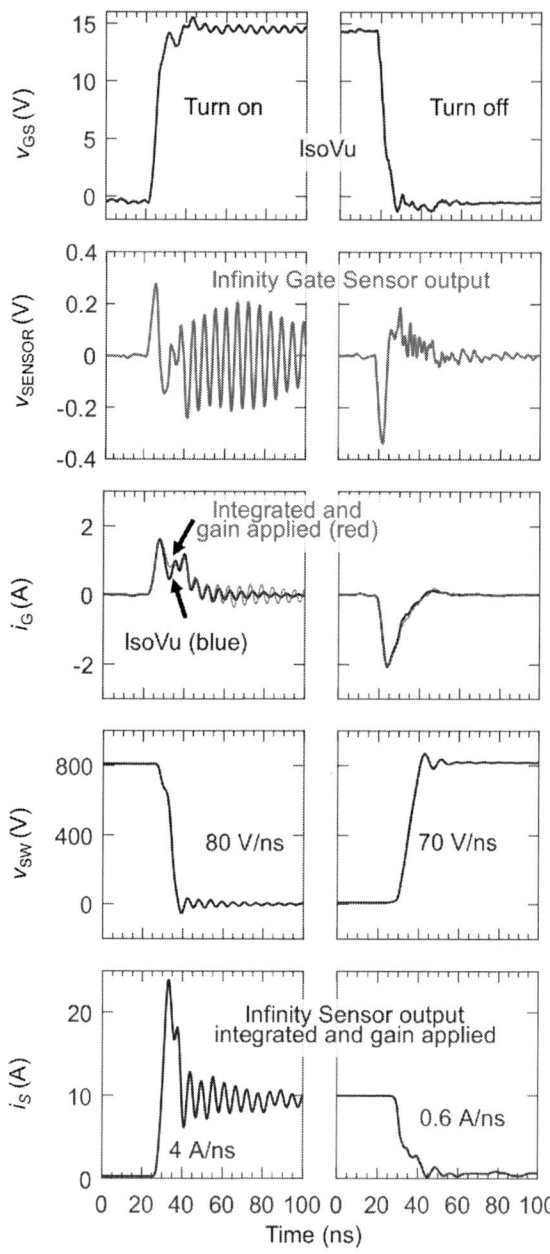

**Fig. 16** Measured gate turn-on and turn off transients of the low-side SiC MOSFET with 1 Ω gate resistor under 800 V, 10 A switching.

PCIM Europe 2024, 11– 13 June 2024, Nuremberg          DOI: 10.30420/566262120

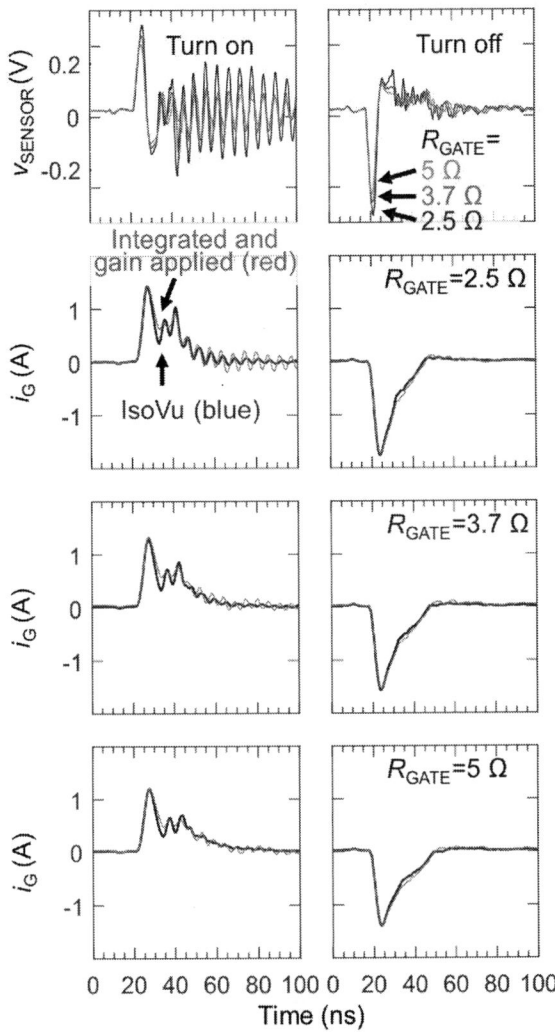

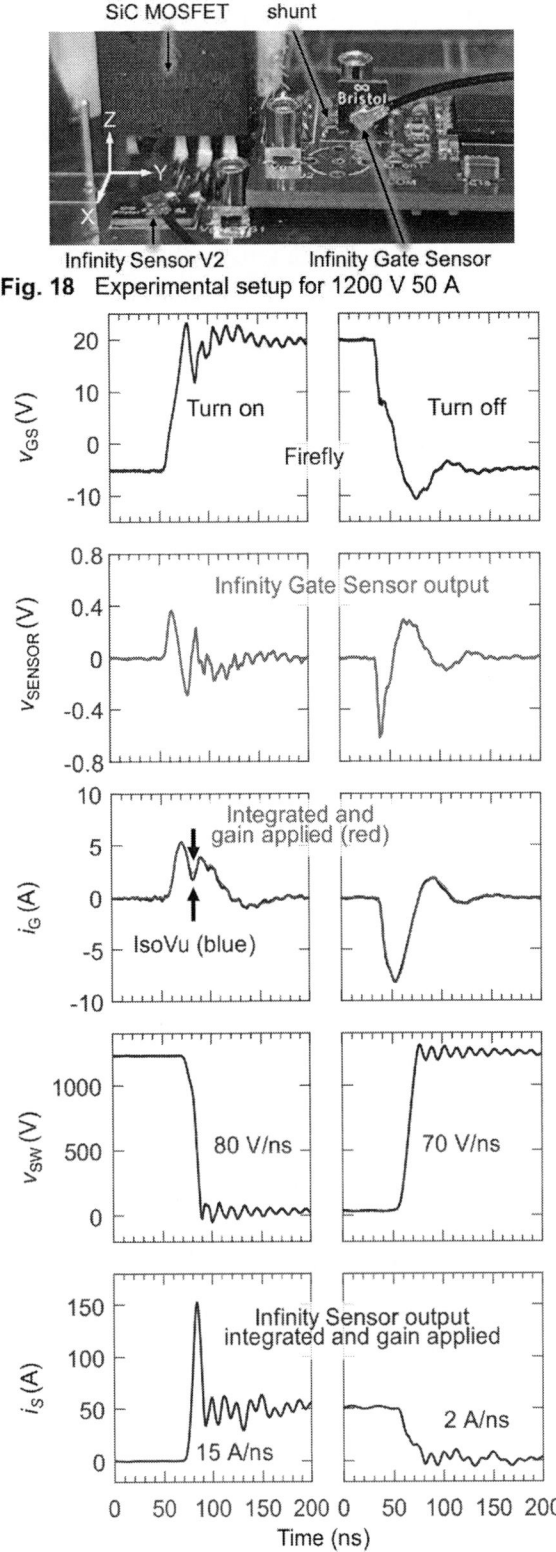

Fig. 18 Experimental setup for 1200 V 50 A

**Fig. 17** Gate current waveforms of the low-side SiC MOSFET with different gate resistances under 800 V 10 A.

The gate current measured by the Infinity Gate Sensor and IsoVu with 2.5 Ω, 3.7 Ω and 5 Ω gate resistance are shown in Fig. 17. During the turn-on process, the gate current ringing amplitude decreases as the gate resistance increases, possibly due to the reduced source current ringing.

## 4.5 Measurement comparisons in 1200 V 50 A double-pulse circuit

To test the Infinity Gate Sensor under higher source current and different layouts, a separate double pulse test is performed with 50 A source current and 1200 V DC voltage using the board shown in Fig. 18. A MSC035SMA170B4 in TO-247-4L package is used vertically on the test PCB, creating a different source current path.

**Fig. 19** Measured gate turn-on and turn off transients of the low-side SiC MOSFET with 0.5 Ω gate resistor under 1200 V, 50 A switching. Please note the time axis is different to Fig. 17 and Fig. 18.

973

The device is driven by an Infineon Eicedriver 1ED3124MU12HXUMA1 with a 0.5 Ω $R_{SHUNT}$ as the gate resistor. The main aggressors are the Y-directional flux from current travelling up and down the device and the flux on Y-Z plane from current flowing along the X-axis direction. The sensor is orientated on the Y-Z plane, similar to Section 3.2, to save space for the compact gate driver design.

Measurements are taken using the equipment listed in Section 3.2 with the exception that the $v_{GS}$ is measured using a PMK FireFly, a >1.5 GHz optically isolated voltage probe. The resulting switching transient is shown in Fig. 19. During the turn-on process, the drain voltage drops from 1200 V to almost 0 V in 15 ns and the source current increases from 0 A to 150 A in 10 ns and settles at 50 A.

This device has a longer risetime than the one used in Section 4.4, and the oscillation frequency is reduced to 80 MHz due to its higher device capacitance and the circuit's higher total loop inductance. Comparing to the previous test, although the source current ringing is higher in the 50 A circuit, it is at lower frequency and further away from the gate sensor. Therefore, the Infinity Gate Sensor is less coupled to the ringing and matches well with the current measured with the IsoVu.

The previous measurements use a gate resistor to allow the simultaneous measurement of gate current by two methods. However, the Infinity Gate Sensor allows elimination of this gate resistor, e.g. to permit faster switching.

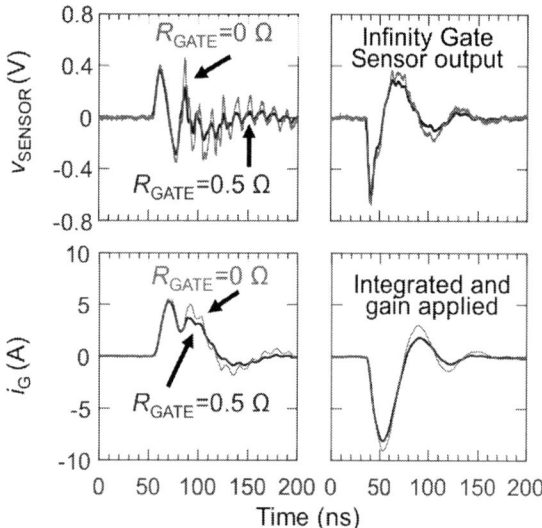

**Fig. 20** Gate current waveforms of the low-side SiC MOSFET with 0 Ω and 0.5 Ω gate resistances under 1200 V, 50 A switching. Left: Turn-on; Right: Turn-off. Please note the time axis is different to Fig. 17 and Fig. 18.

Therefore, the gate currents for a gate resistor of 0 Ω and 0.5 Ω are shown in Fig. 20. It can be seen that the sensor measures more ringing in the gate current at 0 Ω, as may be expected with less damping and faster switching.

## 4.6 Evaluation of *dv/dt* susceptibility

Similar to other magnetic field current sensors, the Infinity Gate Sensor is affected by *dv/dt* from adjacent circuits. The two main sources of *dv/dt* for the bottom device are $v_{SW}$ at the switch node and $v_{GS}$ on the gate current trace. Although the $v_{GS}$ is 2-3 orders smaller than $v_{SW}$, it is more closely coupled to the sense coils, as the gate current trace and the sense coils cover the same area in the Y-Z plane and are only separated by 0.2 mm FR4 in the sensor PCB (Fig. 8), while the switch node is at least 5 mm from the sense coils and is orthogonal to it. Therefore, the sensor output should be affected similarly by both sources.

The capacitive coupling between the sense coils and the current trace is measured to be 0.6 pF for the case in Fig. 10 using the Wayne Kerr 6500B Impedance Analyser. During the low-side turn-on process, the 2 V/ns $v_{GS}$ slew rate can generate a 1.2 mA current flowing to the ground via the lowest impedance in the sensor. At low frequencies, this current flows thorough the sense coil inductance and will not cause a significant change in the sensor output, while at high frequencies, the current flows through the 50 Ω load and generates a signal of around 60 mV. This may contribute to the difference in miller plateau in the gate current measurements of the Infinity Gate Sensor and the IsoVu in Section 4.4 and 4.5.

Therefore, it is expected that the power loop *dv/dt* will not have a significant influence on the sensor when measuring low side gate current. However, it is not recommended for this sensor to be used on the high side due to the high voltage slew rate.

## 5 Conclusion

This paper presents the Infinity Gate Sensor, a low-cost, miniature magnetic field sensor for gate current measurement of power transistors. The raw output from the sensor must be post-processed to obtain current. It has been shown that *dv/dt* immunity is sufficient for the low side, but that this design is not suitable for use in high-side devices.

The measured bandwidth in the presented setup is 1-500 MHz, the insertion inductance at 100 MHz is 3.5 nH, and the gain is 0.67 V/(A/ns), sufficient for SiC devices. These characteristics are dependent on its orientation, which can be optimised for different circuits. A noise current immunity method is proposed and experimentally verified to significantly reduce the effect of nearby much stronger source current. Gate current measurements using the Infinity Gate Sensor were carried out at the low side of 800 V, 10 A and 1200 V, 50 A double pulse circuits with different device packages and current layouts. The measurement results are compared with an optically isolated IsoVu voltage probe across a shunt, showing good agreement in the measurements despite a difference in cost of approximately three orders of magnitude.

# 6 References

[1] N. K. Pilli and S. K. Singh, "Influence of peak gate current and rate of rise of gate current on switching behaviour of SiC MOSFET," 2017 IEEE Transportation Electrification Conference (ITEC-India), Pune, India, 2017, pp. 1-6, doi: 10.1109/ITEC-India.2017.8356960.

[2] Y. Lobsiger and J. W. Kolar, "Closed-loop IGBT gate drive featuring highly dynamic di/dt and dv/dt control," 2012 IEEE Energy Conversion Congress and Exposition (ECCE), Raleigh, NC, USA, 2012, pp. 4754-4761, doi: 10.1109/ECCE.2012.6342173.

[3] A. Schindler, B. Koeppl and B. Wicht, "EMC and switching loss improvement for fast switching power stages by di/dt, dv/dt optimization with 10ns variable current source gate driver," 2015 10th International Workshop on the Electromagnetic Compatibility of Integrated Circuits (EMC Compo), Edinburgh, UK, 2015, pp. 18-23, doi: 10.1109/EMCCompo.2015.7358323.

[4] W. Eberle, Z. Zhang, Y. -F. Liu and P. C. Sen, "A Current Source Gate Driver Achieving Switching Loss Savings and Gate Energy Recovery at 1-MHz," in IEEE Transactions on Power Electronics, vol. 23, no. 2, pp. 678-691, March 2008, doi: 10.1109/TPEL.2007.915769.

[5] N. Baker, S. Munk-Nielsen, F. Iannuzzo and M. Liserre, "Online junction temperature measurement using peak gate current," 2015 IEEE Applied Power Electronics Conference and Exposition (APEC), Charlotte, NC, USA, 2015, pp. 1270-1275, doi: 10.1109/APEC.2015.7104511.

[6] J. Liu, H. Chen, J. Lin, Z. Li, H. Wu and B. Ji, "Investigation of TSEPs based on the feature extraction from the gate current," 11th International Conference on Power Electronics, Machines and Drives (PEMD 2022), Hybrid Conference, Newcastle, UK, 2022, pp. 451-455, doi: 10.1049/icp.2022.1092.

[7] S. Kochoska, J. R. Guitart, L. Richert and B. Vlachakis, " Gate Current Peaks Due to CGD Overcharge in SiC MOSFETs Under Short-Circuit Test," 2023 35th International Symposium on Power Semiconductor Devices and ICs (ISPSD), Hong Kong, 2023, pp. 246-249, doi: 10.1109/ISPSD57135.2023.10147438.

[8] Y. Wang, Z. Zeng, T. Long, P. Sun, L. Wang and M. Zou, "Impedance-Matching Shunt: Current Sensor With Ultra-high Bandwidth and Extremely Low Parasitics for Wide-Bandgap Device," in IEEE Transactions on Power Electronics, vol. 37, no. 10, pp. 11528-11533, Oct. 2022, doi: 10.1109/TPEL.2022.3175973.

[9] T&M Research Products, Inc. SSDN-414-10. [Online]. Available: http://www.tandmresearch.com/

[10] P. D. Judge, R. Mathieson and S. Finney, "A Six Level Gate-Driver Topology with 2.5 ns Resolution for Silicon Carbide MOSFET Active Gate Drive Development," 2021 IEEE 12th Energy Conversion Congress & Exposition - Asia (ECCE-Asia), Singapore, Singapore, 2021, pp. 2133-2138, doi: 10.1109/ECCE-Asia49820.2021.9479081.

[11] K. Klein, E. Hoene and K. Lang, "Power module design for utilizing of WBG switching performance," *PCIM Europe 2019; International Exhibition and Conference for Power Electronics, Intelligent Motion, Renewable Energy and Energy Management*, Nuremberg, Germany, 2019, pp. 1-8.

[12] J. N. Fritz, C. Neeb, and R. W. De Doncker, "A PCB integrated differential Rogowski coil for non-intrusive current measurement featuring high bandwidth and dv/dt immunity," in Proc. Conf. Power Energy Student Summit, Dortmund, Germany, Jan. 2015, pp. 1–6, doi: 10.17877/DE290R-7459.

[13] T. Funk and B. Wicht, "A fully integrated DC to 75 MHz current sensing circuit with on-chip Rogowski coil," *2018 IEEE Custom Integrated Circuits Conference (CICC)*, San Diego, CA, USA, 2018, pp. 1-4, doi: 10.1109/CICC.2018.8357028.

[14] J. Wang et al., "Infinity Sensor: Temperature Sensing in GaN Power Devices using Peak di/dt," *2018 IEEE Energy Conversion Congress and Exposition (ECCE)*, Portland, OR, USA, 2018, pp. 884-890, doi: 10.1109/ECCE.2018.8558287.

[15] H. C. P. Dymond, Y. Wang, S. Jahdi and B. H. Stark, "Probing Techniques for GaN Power Electronics: How to Obtain 400+ MHz Voltage and Current Measurement Bandwidths without Compromising PCB Layout," *PCIM Europe 2022; International Exhibition and Conference for Power Electronics, Intelligent Motion, Renewable Energy and Energy Management*, Nuremberg, Germany, 2022, pp. 1-10, doi: 10.30420/565822010.

[16] JLCPCB, Multilayer high precision PCB's with impedance control [Online]. Available: https://jlcpcb.com/impedance

[17] N. Oswald, B. H. Stark, N. McNeill and D. Holliday, "High-bandwidth, high-fidelity in-circuit measurement of power electronic switching waveforms for EMI generation analysis," *2011 IEEE Energy Conversion Congress and Exposition*, Phoenix, AZ, USA, 2011, pp. 3886-3893, doi: 10.1109/ECCE.2011.6064297.

PCIM Europe 2024, 11– 13 June 2024, Nuremberg    DOI: 10.30420/566262121

# Characterising Wide Bandgap Power Modules: Validating the M-Shunt Concept for High-Power Applications in the Kiloampere Range

Hauke Lutzen[1], Jonas Müller[1], Vladimir Polezhaev[2], Steffen Chemnitz[3], Malte Arndt[1], Lennart Dittmer[1], Till Huesgen[2], Nando Kaminski[1]

[1] University of Bremen, IALB, Germany

[2] University of Applied Science Kempten, EIL, Germany

[3] SubCtech GmbH, Germany

Corresponding author:   Hauke Lutzen    hauke.lutzen@uni-bremen.de
Speaker:                 Hauke Lutzen    hauke.lutzen@uni-bremen.de

## Abstract

In this paper, the viability of the M-Shunt concept [1; 3] for high power levels is presented. The M-Shunt concept is based on a temperature compensated resistive sheet of Manganin™ (MgCuNi), arranged in a multilayer planar setup for best electrical performance regarding current capability, bandwidth, and accuracy. Measurements demonstrate that the M-Shunt is ideally suited for the characterisation of fast current transients of modern power switches and to capture current pulses in the kiloampere range.

The presented M-Shunts are directly screwed onto the contacts of the IGBT modules or soldered in the busbar connections, while the current transients are recorded during double-pulse tests. Preliminary tests with the M-Shunt concept performed in an electromagnetic forming machine have demonstrated, that the M-Shunt concept can measure even much higher current levels without destruction. The M-Shunt concept is confirmed as a priori choice when high bandwidth and high energy are required in conjunction.

## 1   Introduction

In the ever-evolving landscape of power electronics, accurate and robust measurement techniques play a pivotal role in understanding and improving device performance. Among the myriad of current sensing methodologies, M-Shunts have emerged as a promising tool, offering distinct advantages, especially in high power applications. As detailed in [4, 7], M-Shunts exhibit an inherent capability to capture high-current pulses, particularly suited for IGBT, SiC-MOSFET modules or even GaN applications. Unlike conventional solutions such as the FOCS sensor [6], the bandwidth of M-Shunts remains high even for high energy or currents. By low inductive integration of M-Shunts into the measurement circuit, this work aims to push the boundaries of current transient measurement with high bandwidth. Yet, preliminary tests in advanced settings like electromagnetic forming machines suggest that the vast potential of the M-Shunt concept remains untapped. This paper focusses on these details, highlighting the superiority and scalability of M-Shunts in high-energy scenarios, and illustrating their added value compared to slower Rogowski coils or much larger coaxial shunts. Although innovative coil designs and sensor developments (e.g. [5]) achieve the necessary bandwidths, they often do so at the expense of maximum energy, maximum current, or practicality [4].

## 2   Background and Related Work

The architecture of the M-Shunt, initially presented in [8] and further refined in [9], transforms the cylindrical structure of the coaxial shunts into a planar format. As depicted in Figure 1, the M-Shunt is implemented in a six-layer printed circuit board (PCB). The current is conducted through the outermost layers (top and bottom) to the resistor layer, which consists of Manganin or copper. In the centre of the design, the measurement tab is located in an area with reduced electromagnetic fields and does not conduct current, adhering to the fundamental design principles of the coaxial shunt as discussed in references [8] and [9].

To ensure equal current distribution across the shunt's tracks, a distributor structure was introduced as described in [8], with enhancements detailed in this paper (see Figure 1). This study presents various dimensions of the M-Shunt, comparing them to a commercial cylindrical configuration. According to the basic electrical theory, modifications to the length, width, and thickness of the resistive layers affect their resistance, capacitance and inductance. The surface area, calculated as length multiplied by width, is critical for heat dissipation within the shunt. The resistive layer thickness does not only impact the resistance due to material resistivity but also plays a key role in mitigating the skin effect, as elaborated in [2] and [9].

length, or increasing the width. It is crucial to understand that the measurement inductance, which is essential for achieving a large bandwidth, only includes elements between points A and C (Figure 2), whereas the inductance inserted into the circuit encompasses the entire path from input to output, primarily the route from A to B. The same can be said of the introduced resistor and its temperature behaviour in relation to the measuring resistor. While the combination of copper and Manganin in a shunt's load circuit (in to out) has an almost constant resistance due to the negative temperature coefficient of Manganin compared to the positive one of the copper conductors, the temperature behaviour of the pure measuring circuit remains slightly negative as shown below.

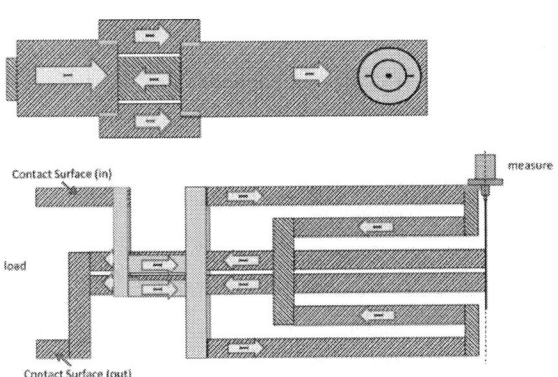

Figure 1 Shunt concept including current distributor

## 2.1 Design Considerations of M-Shunts

Calculating the resistance of an electrical conductor, the formula $R = (\rho l)/A$ is utilised, where $\rho$ represents the material-specific resistivity. In the M-Shunt, the length l corresponds to the length of the resistance layer. The conducting area A is determined by the product of the width and the thickness of the resistance layer.

As depicted in Figure 2, the upper and lower conductors, as well as the resistance from the vias and connections, are primarily composed of copper. Due to copper's low resistivity, these components contribute negligibly to the overall resistance, thus emphasising the resistor portion as the predominant factor in the total resistance calculation. The shunt features two parallel Manganin resistors, each bearing exactly half of the total current.

Increasing the surface area (length and width) decreases the dissipated energy density, while the resistance remains equal. Thereby lowering energy conversion at the shunt, can be achieved by increasing the layer thickness, decreasing the

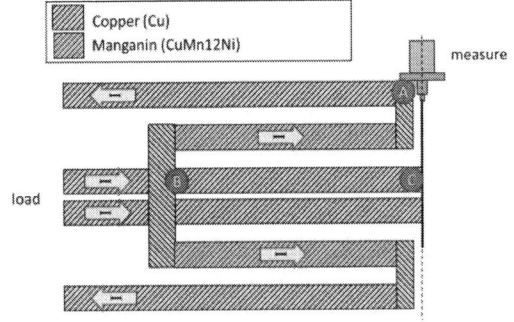

Figure 2 Cross-section of the resistance part of the M-Shunt. Trace A ► C defines the measurement circuit; A ► B is in load and measurement part; B ► C is in measurement circuit only

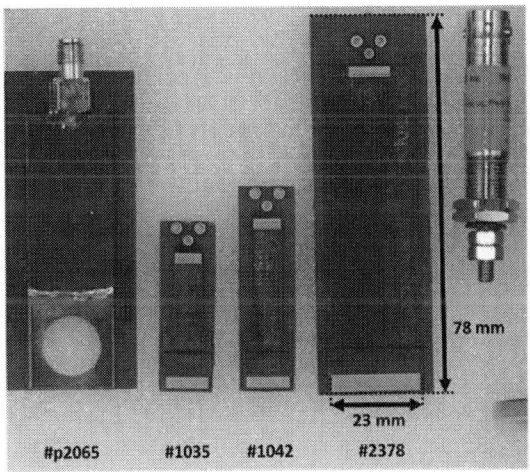

Figure 3 Image of the M-Shunt models #1035, #1042, #2378 in size comparison with T&M shunt SDN015, which has a lower energy level as the #1035 but is double in length and 6 times of the thickness.[7]

In order to ascertain the maximum energy that can be dissipated, an aging experiment was conducted to determine the maximum temperature at which a stable shunt can be reliably guaranteed. The objective of this experiment was to estimate the operational limits and confirm the stability of the M-Shunt. The experiment was handled by connecting the M-Shunt to a DC power source and heating it with a constant load current. The surface temperature of the M-Shunt and the voltage drop were measured using a Picolog TC-08 datalogger with a one-minute sampling rate. The measured current and voltage are used to calculate the resistance ($R_{shunt}$) for each data point. The current was maintained at a constant level for 24 h before the next load step was applied. The load current was increased until 100°C was reached. Finally, the temperature was reduced again to 60°C as a reference.

Table 1 presents the results of the study, averaged over each 24 h period. The average resistance of steps 1 and 6, both at 60°C, is nearly identical. This indicates that the temperature exposure to 100°C for 24 h does not result in a significant degradation of the shunt resistance. However, in subsequent experiments, a slight permanent degradation was observed when the temperature exceeded 120°C. Consequently, 100°C is designated as the upper limit for the operational temperature of the shunt.

Thus, the maximum temperature of Table 1 can be utilised to estimate the maximum energy dissipation using the formula:

$$E_{max} = C_{Mn} \, \Delta T \quad (1)$$

where $C_{Mn}$ represents the heat capacity of the Manganin resistor traces, and $\Delta T$ is the temperature difference between 100°C and ambient. This is an estimation with safety margin, as it neglects heat conduction entirely. Of course, the temperature coefficient of Manganin causes a change in shunt resistance which limits the accuracy of the measurement. The temperature coefficient is less than ±10 ppm/K in the range 20°C to 50°C. Above 50°C the temperature coefficient becomes larger [12]. The average temperature coefficient is -31 ppm/K in the range of 60°C to 100°C, which results in a total resistance change and hence a measurement error of - 0.059%. This systematic error can be reduced by limiting the maximum temperature to 70°C. In this case the maximum energy dissipation is reduced, but the systematic measurement error due the temperature coefficient of resistance remains below ±0.015%.

In the data sheet [11], both a value for $E_{max}$ at 70°C together with a value for $E_{max}$ at 100°C is specified to reduce deviations, which can be classified as ir-

relevant for the overall scattering in practical application. It can be concluded that the 100°C limit represents the maximum energy input, which can be dissipated without the risk of degradation. Even this value can be exceeded for short periods.

*Table 1 Long-term stability through 24-hour time interval temperature stress by loading the M-Shunt with a DC current*

| 24-h step | $T_{shunt}$, [°C] | $I_{load}$, [A] | $R_{shunt}$, [mΩ] |
|---|---|---|---|
| #1 | 60 | 7.97 | 18.123 |
| #2 | 70 | 9.19 | 18.117 |
| #3 | 80 | 10.01 | 18.112 |
| #4 | 90 | 10.94 | 18.105 |
| #5 | 100 | 11.67 | 18.099 |
| #6 | 60 | 7.97 | 18.124 |

The thermal performance of the shunts, which determines their maximum continuous power capacity, is a critical aspect in their design and utilisation. For characterisation purposes, M-Shunts were soldered onto a circuit board and connected to a DC load. The surface temperature of the shunt was monitored by thermocouples and recorded using a datalogger from the moment the DC-load current was activated.

The thermal resistance ($R_{th}$) was calculated by:

$$R_{th} = \frac{T_{s,max} - T_{amb}}{P_{Loss}} \quad (2)$$

where $T_{s,max}$ is the maximum temperature, $T_{amb}$ is the ambient temperature, and $P_{Loss}$ is the power loss determined by the load current and voltage drop across the shunt. It is important to note that the internal thermal resistance from the embedded resistor trace to the copper trace is significantly lower than the convective resistance and is thus negligible. For an exponential approach to the steady state value, the thermal time constant is defined the time required to reach 63.2 % of the maximum temperature rise. Thermal capacitance $C_{th}$ is calculated from this time constant, by dividing it by the thermal resistance $R_{th}$.

The maximum power dissipation is calculated as:

$$P_{max} = \frac{T_{s,max} - T_{amb}}{R_{th}} \quad (3)$$

where $P_{max}$ is the maximum allowable power dissipation, $T_{s,max}$ is the maximum temperature, and $T_{amb}$ is the ambient temperature.

On the one hand, a $C_{th}$ can be determined from the measurement and subsequently an $E_{max}$ can be calculated for the static state. However, this value is problematic in that it is altered by the cooling process over time and is therefore heavily dependent on the test. In order to avoid inaccuracies, the theoretical maximum value is specified as reference. This theoretical $E_{max,theory}$ is significantly lower than any measured value because it is based on an energy input in an infinitesimally short time, without taking into account heat dissipation via the connected copper connections or the surface.

Even the smallest shunt models, #1035 and #1042, exhibit a theoretical maximum energy in the same dimension of comparable coaxial shunts, while it even significantly exceeds it when the heat distribution is taken into account. The coaxial shunt (SDN-015) is shown as a size reference in Figure 3 and has a maximum input of 1.5 J [7], in comparison to the values of 2.56 J for the small versions or even 26 J for the #2378 M-Shunt in comparable dimension. It can be concluded, that it is possible to design M-Shunts, which are even smaller in size but still are suitable to characterise short-term switching below 200 µs.

*Table 2 Overview of $Z_{th}$ measurement results for selected M-Shunt models considered in this study*

| Model | | #2378 | #1035 | #1042 |
|---|---|---|---|---|
| $R_{avg}$ | mΩ | 13.26 | 12,99 | 18,14 |
| $R_{th}$ | K/W | 11,75 | 33,52 | 33,18 |
| $P_{max, continous}$ $\Delta T = 75K$ | W | 6,81 | 2,24 | 2,26 |
| $E_{max, theory}$ $\Delta T = 75K$ | J | 26,99 | 2,56 | 3,87 |
| $I_{max, continous}$ | A | 21,94 | 13,13 | 11,16 |

# 3 Investigation of the measurement quality and its limitations in power applications

The following examines the intricate of shunt resistors in accurately measuring pulsed power applications, with a particular focus on their performance in high-current scenarios exceeding 5 kA. A systematic investigation is conducted into their role in determining switching losses in semiconductor modules, and the underlying failure mechanisms that manifest under such intense electrical stresses are explored.

## 3.1 Single Pulse Tests on Silicon IGBT modules

The fabricated M-Shunts were deployed in a double-pulse test (DPT) setup of a half-bridge to measure the current transients and their superimposed oscillations. Even compared to the

Rogowski coil, a significant advantage of the M-Shunt is its easy integration achieved through the screw-on method. Subsequently, in a series of pulses, the switched energy was incrementally increased, with current pulses reaching the kiloampere range, while still providing excellent results. The measurement setup can be found in [1]. As a reference, a Rogowski with a bandwidth of 23 MHz (T3RC3000-HF [14]) was used.

Figure 5 shows that both, the reference and the M-Shunt record the current transients in a similar way. The calculated energy differs only by small deskew deviations, which are explained in more detail below.

The calculated power dissipated at the shunt is given by:

$$P = I^2 R \quad (4)$$

If this power is integrated over the pulse length, a very low energy is obtained for classic double pulses, far from the thermal limits of the shunts. For example, an ideal double pulse with a maximum current of 1000 A and an average current of 500 A for a 20 mΩ shunt results in a power of 5 kW, based on a pulse length of 100 µs that yields 0.5 J. Therefore, typical double pulse measurements are far away from the thermal limits of the shunts used.

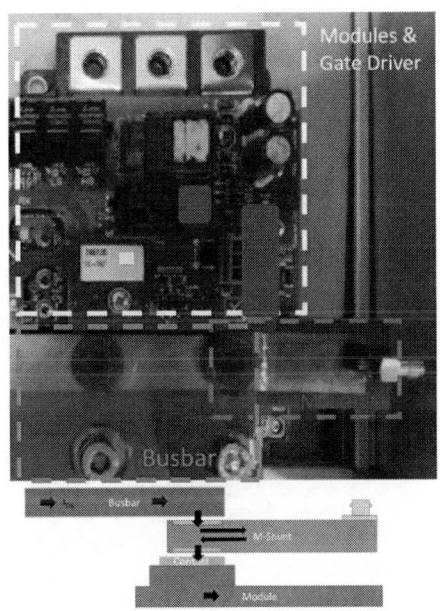

*Figure 4 Photo of a measurement setup with an M-Shunt directly connected to a low inductive silicon module and side view sketch*

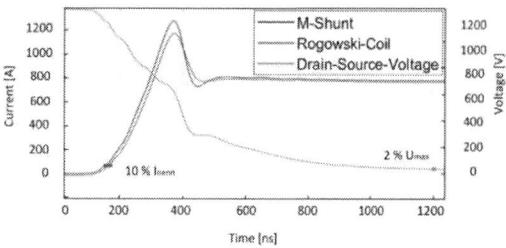

*Figure 5 Current measurements in DPT after delay compensation. Current peaks exceeding 1 kA on a Rogowski-coil (T3RC3000-HF) and M-Shunt (#1035)*

## 3.2 Double Pulse Test on SiC Modules

**Demonstrating the benefits of the current-bandwidth combination of M-Shunts**

In the exploration of optimal current sensing solutions, the integration of bandwidth and current capacity proves to be crucial. Figure 6 demonstrates this by illustrating the switching behaviour of a SiC MOSFET, where measurements using the M-Shunt are compared with those obtained from two different Rogowski coils (model T3RC03000-HF with $I_{max}$ = 3 kA and bandwidth of 23 MHz and model CWTmini50-HF with $I_{max}$ = 600 A and a bandwidth of 50 MHz). The Rogowski coil T3RC3000-HF measurement, characterised by a bandwidth that is insufficient for the application, exhibits a noticeable attenuation of the signal, leading to a less accurate depiction of the MOSFET's true behaviour. In contrast, the second Rogowski coil (CWT50-HF), while providing adequate bandwidth, has a current cut off at approximately 680 A, resulting in the clipping of the signal at this level. These limitations strongly contrast with the performance of the M-Shunt, which captures both the full amplitude and the rapid transient details of the MOSFET's switching without distortion or truncation. This comparison underscores the superiority of the M-Shunt in providing a precise and comprehensive measurement, highlighting its effectiveness in applications demanding high bandwidth and substantial current handling capabilities. This is also depicted in Figure 7 where the mean M-Shunt current was calculated based on a duty time of 50 µs, which is a validation for double pulse tests of roughly 100 µs turn-on time.

Additionally, it is important to acknowledge that more suitable models of the Rogowski coil do exist for specific applications, such as those capable of handling a maximum current of 1200 A. However, an increase in current capacity inevitably leads to a reduction in bandwidth, presenting a trade-off that must be carefully managed in high-perfor-

mance environments, as it can be seen in the Figure 7. The measurements depicted in the previously referenced Figure 5 were conducted using the miniature variant #1035 of the M-Shunt. This context highlights that the M-Shunt, with its capabilities, remains well within safe operational limits, far from the constraints faced by both Rogowski coils. This illustrates the M-Shunt's resilience and adaptability in managing extremes of both, bandwidth (which can be further extended due to [2, 9, 13]) and current.

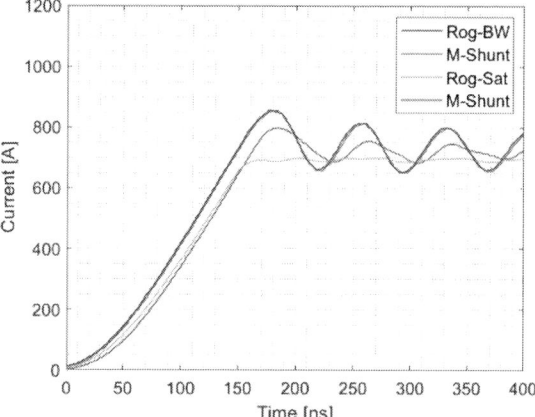

*Figure 6 Turn-on measurement of a 700 A switching measurement with an M-Shunt compared to two Rogowski coils. T3RC3000-HF (too low bandwidth – "BW") and Rogowski coil CWT Mini50HF (maximum current saturation – "Sat")*

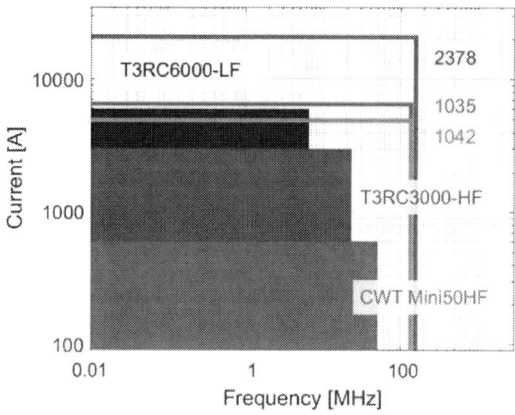

*Figure 7 Comparison of commercially available Rogowski coils considering bandwidth and maximum current in comparison to the M-Shunts calculated according to [4] with a mean duty time of 50 µs. [11]*

## Validation of linearity

The linearity of the Manganin shunts was validated by recording the resistance change against the switched current, as illustrated in Table 3. Initial DC measurements confirmed the linearity of the Manganin-shunts. To further validate this characteristic under dynamic conditions, the resistance value was calibrated against currents measured by a Rogowski coil for each switched current. This approach allowed for a precise examination of potential non-linearities. The results confirmed the initial findings, demonstrating that the shunt exhibited absolute linearity across the range of the tested currents. This consistency proves the shunt is suited for precise current measurements in both static and dynamic settings, supporting its use in high-precision applications where accurate linearity is critical.

*Table 3 Resistance change calculated from the measured switching transients at different turn-on currents. The deviation in the calculated resistance consists of the reference measurement deviation (CWTmini50-HF) combined with the shunt measurement deviation itself.*

| Current | [A] | 200 | 300 | 400 | 500 |
|---------|------|------|------|-------|------|
| #1035 | [mΩ] | 12.7 | 12.5 | 12.5 | 12.6 |
| #1042 | [mΩ] | 17.8 | 17.9 | 18.00 | 17.8 |
| #2378 | [mΩ] | 13.2 | 13.3 | 13.2 | 13.1 |

## Determination of switching losses of SiC modules using small M-shunts

The DPT on SiC modules provides a focused analysis of the switching behaviour of a SiC module, for which the M-Shunt was integrated into the feed busbar. Figure 8 confirms again that the M-Shunt is capable of capturing the current waveforms accurately. Although the Rogowski coil exhibits a slight deficiency in bandwidth, it still fundamentally records the current flow. Notably, the recorded current signal briefly exceeds the 1 kA mark, corresponding to a significantly high-power transfer at over 600 V. This achievement not only confirms the feasibility of using the M-Shunt in high-power applications but also underscores its reliability and precision, even when using the smallest model, #1035. Results obtained with other models yield identical performance, confirming the M-Shunt's consistent and dependable capability across various setups.

For switching loss calculation, the time behaviour of current and voltage is crucial. According to the datasheet of the SiC module in Figure 9, the turn-on characteristic of the nominal operation point of the module is presented. Due to lower bandwidth, the rate of change of the drain current is evidently measured too low by the Rogowski coil

(T3RC3000-HF) compared to the M-shunt (#1035) measurement. Moreover, the amplitude of the ringing ($f_{ringing}$ = 13.5 MHz) during the switching event is measured overdamped by the Rogowski coil as well. It is important to note that internal de-skewing, whether performed within the oscilloscope or as part of the post-processing stage, is of critical importance. This is discussed in more detail below.

The turn-on and turn-off losses depending on current for both the M-Shunt and Rogowski coil are presented in Figure 9. The values from the datasheet are provided as reference. For the calculation of the switching losses according to the datasheet, the double-pulse test is conducted with a stray inductance of 17 nH. However, the comparison of the measured switching losses is conducted using a double pulse setup with at least 38 nH, determined by an LCR meter. The differences in stray inductances within the load loop contribute to the deviation between the calculated energy losses and the datasheet values. In particular, the calculation error of the switching losses can be determined from the behaviour observed at the turn-off. Hence, it is recommended to use a current sensor with high bandwidth, such as the M-Shunt, to ensure precise measurement of the switching losses. In this case, a low bandwidth could result in either lower calculated energy due to the attenuated amplification, or higher values due to the extended calculated switching time. The impact of the smaller bandwidth becomes even more crucial for systems with higher ringing frequencies and higher rates of change of the drain current. So short-term deviations from the data sheet values are obviously no problem.

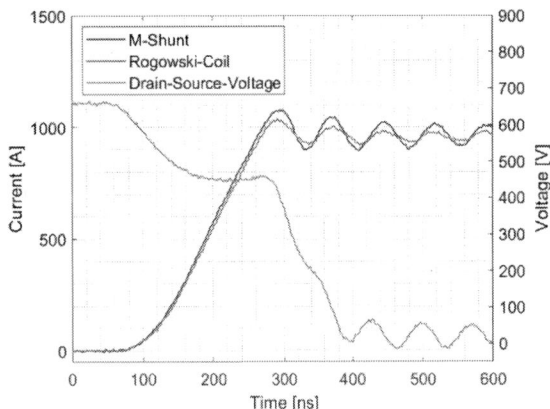

*Figure 8 Turn-on of an SiC Module to determine the switching losses. Measured with M-Shunt #1035 or Rogowski coil (CWTmini50-HF)*

PCIM Europe 2024, 11– 13 June 2024, Nuremberg          DOI: 10.30420/566262121

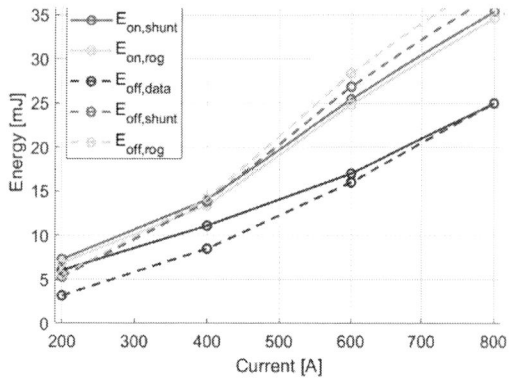

Figure 9 Calculated switching losses due the measurements performed with M-Shunt (#1035) or Rogowski coil (CWTmini50-HF) in comparison with the datasheet values.

**De-skewing for precise switching loss calculation**

As described, the turn on energy losses $E_{on}$ of a MOSFET device are measured according to the IEC standard as the integral of a power waveform computed from 10 % of $I_D$ to 10 % of $V_{DS}$ during turn on, while the turn off energy losses $E_{off}$ are computed by integrating a power waveform from 10 % of $V_{DS}$ to 10 % of $I_D$ during turn off.

Due to its influence on the measured waveform, correct energy loss calculation can only be performed, when the effective inductance of the measurement setup is considered. During a turn on event of a MOSFET device in double pulse measurements for example, the parasitics leads to a $V_{DS}$ transient waveform with a skew between the Drain to Source voltage and Drain current.

This skew is compensated by a de-skew algorithm, where the $I_D$ waveform is chosen as the golden reference, and the de-skew value is applied as a time delay to the probe head of the voltage source of the oscilloscope (here $V_{DS}$).

For the measurements and results presented in this paper, post-processed data or the internal de-skew algorithm of the used Tektronix MSO64 scope has been applied [17], in which the skew of the measurement system is performed as a post-acquisition operation. The algorithm models the $V_{DS}$ waveform as a math function based the acquired $I_D$, $V_{GS}$, and circuit parameters (probe resistance, effective inductance). This modelled $V_{DS}$ waveform is then fitted to the measured $V_{DS}$ waveform, resulting in a time delay. This de-skew value is added to the time offset of the oscilloscope probe head and is then used in subsequent double pulse measurements at same setups.

It is obvious for all energy loss calculations, that the results of the integral of the power waveform are only correct, when the time delay between $I_D$ and $V_{DS}$ are precisely determined. Therefore, it is evident that the process of de-skewing is of paramount importance in order to measure the energy losses.

Performing double pulse measurements on wide band gap devices leads - due to the high switching performance of these devices - to slopes of the Drain current (di/dt) of several amps per nanosecond. This increases the requirements on the measurement setup and especially on the bandwidth of the probe heads and current sensors. The de-skew algorithm covers the influence of the circuit parameters as described above, but cannot compensate the influence of a limited bandwidth of the measurement setup, which modifies the slope and shape of $V_{DS}$ and $I_D$ during turn-on and turn-off. This demonstrates the importance of high bandwidth shunts for the measurement of switching losses.

## 3.3 Investigation of failure mechanisms in pulsed power applications larger than 5 kA

As a glimpse into the limits and possibilities of the M-Shunt concept, currents even higher were measured using M-Shunt prototypes for the characterisation of an electromagnetic pulse forming machine with more than 5 kA. This investigation underlines the robustness and adaptability of the M-Shunt technique in diverse operational conditions.

In high-power electronic applications, the behaviour of PCB based shunts under different switching speeds reveals distinct modes of failure due to electromagnetic and thermal stresses. When subjected to currents, as high as several kiloamperes, PCBs experience significant electromechanical stresses. The high magnetic fields generate forces that can cause immediate physical damage such as trace lifting or delamination, as evidenced by peel tests detailed in [1]. This is due to field generated by the current, which generates the magnetic interactions between conductive traces.

On the other hand, slower switching speeds or DC load shift the primary failure mode to thermal effects caused by joule heating. The resistance of the conductive traces generates heat, leading to thermal expansion, solder fatigue, and degradation of the board's dielectric properties. Such thermal accumulation over time can result in circuit failures from overheating and reduced insulation effectiveness.

Understanding these differing impacts, electromechanical forces at high currents and thermal stresses in long pulsed lower current, is crucial for the M-Shunt design. The measurement in the Electromagnetic forming machine highlights the need for designs that address both, the rapid mechanical stresses induced by high currents and the thermal management challenges posed by prolonged current flow. Enhancing mechanical integrity and incorporating effective thermal management systems are essential strategies to ensure the reliability and longevity of PCBs in high-power settings.

## Measurement setup

The testbench has been established for researching material forming processes and consists of capacitor banks discharged through a tool coil using thyristors (model PPU–SPT–6k–40k–C–001 from Astrol Electronic AG). Each capacitor bank comprises eight capacitors, model E62.S24-503C60 from ELECTRONICON, rated at 3.4 kV and a capacitance of 50 µF. To achieve the necessary voltage rating, two capacitors are connected in series, with three such pairs paralleled to yield an effective capacitance of 75 µF per bank.

The tool coil's inductance forms an oscillatory circuit, with the frequency and pulse width dependent on this inductance. A diode stack (model PPU–SPD–6k–40k–C–001 from Astrol Electronic AG) allows only a single half-wave to pass, creating a defined pulse. A second parallel setup enables the delivery of a subsequent pulse at a variable time interval from the first. This configuration is depicted in Figure 10.

Current flow is measured using a Rogowski coil positioned before the tool coil, serving as a reference measurement. The shunt under test is installed between the tool coil and ground connection, with voltage drop across the shunt measured using a differential probe.

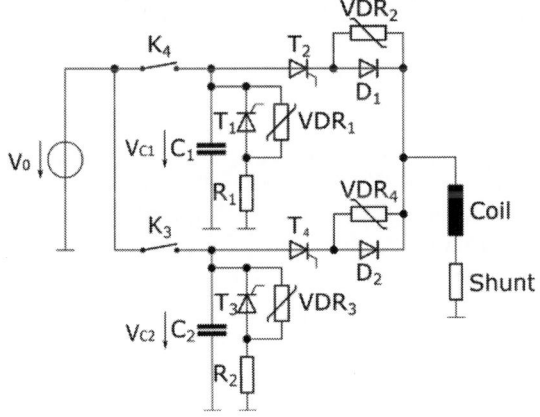

Figure 10 Circuit board of the forming machine

## Measurement results

The results from Manganin shunt models #2378 and #1035 demonstrate that currents exceeding 5 kA do not damage the shunt itself. The energy associated with the destructive pulse (>7 kA) can be referenced from Table 4. In both instances, the point of destruction was in the lead-up wiring, not at the resistor where heat generation occurs. The failures triggered were apparently of mechanical nature, although the resilience limit could be extended by encasing, as subsequent experiments have shown. Overall, the suitability of these shunts for use in high-energy environments, with high currents and sharp transients, has been more than confirmed.

The large overshoot of the theoretical $E_{max,\ theory}$ shows that even for kiloampere pulses, the energy introduced by distribution is significantly higher than calculated.

Using the maximum current of 3 kA specified in the data sheet [11] for the #2378, a pulse of more than 1 ms is still required to reach the theoretical maximum energy. However, this time is so long that the thermal limit in the application is orders of magnitude higher than the theoretical value, this is because the energy is introduced in anything but an infinitesimally short time, so the large cooling surface of the shunt is already at work.

Manganin shunts can be employed in various ways to characterise the switching behaviour of power switches, especially in situations where other sensors fail to operate effectively. While these shunts appear to be over-engineered for the use in standard double-pulse tests, as previously described, it suggests that a reduction in size is be feasible. This finding indicates potential for optimizing shunt design to balance safety and functionality without compromising the ability to handle extreme conditions.

However, in addition to the thermal fault, the electromechanical fault must also be taken into account. While the reduction in layer spacing should be reduced for inductance reasons, a new trade-off arises as this leads to greater Lorentz forces.

Table 4 Measurement results of the High-Energy pulsed measurements

| Shunt-model | $I_{max}$ [kA]* | $E_{max}$ [J]* |
|---|---|---|
| #1035 | 5,040 | 6,75 J |
| #2378 | 7,412 | 15,26 J |

* of the last functional pulse before delamination, shown in Figure 11. The destructive pulse was at 7 kA for #1035 and 8,7 kA for the #2378

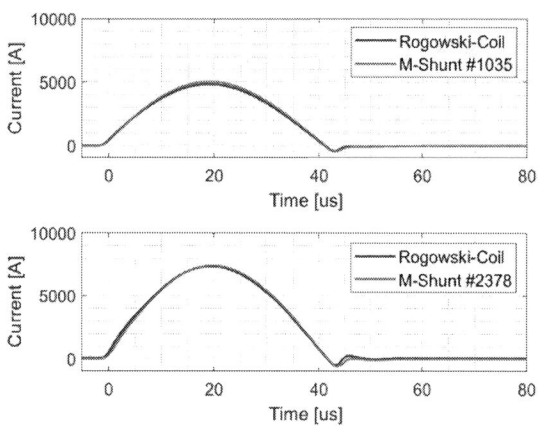

Figure 11 Current measurement of the last functional High-Energy pulse before delamination due to electromagnetic destruction

# 4    Conclusion

In conclusion, the advanced M-Shunt concept offers a promising approach for high fidelity high-current transient measurements, meeting the needs of characterising modern power switches, even in the kiloampere range. Its adaptability and direct integration, as compared to traditional methods such as the Rogowski coil, emphasises its superior measurement precision, especially when observing high frequency effects. Furthermore, preliminary tests on electromagnetic forming machines, suggest the vast potential and broader applicability of the M-Shunt. The straight forward handling and compact integration render it an accessible and space-saving solution for users, although there are only three models available at the moment. [11] As power electronic systems continue to evolve and demand more precise and scalable measurement techniques, the M-Shunt stands out as an excellent tool that combines robustness, accuracy, and versatility.

# Acknowledgement

Parts of this work are based on test benches that were created thanks to the EFRE funding programme. [16]

Europäische Union
Investition in Bremens Zukunft
Europäischer Fonds für
regionale Entwicklung

# References

[[1] H. Lutzen, V. Polezhaev, K. B. Rawal, K. Ahmed, T. Huesgen, and N. Kaminski, "Temperature Compensated M-Shunts for Fast Transient and Low Inductive Current Measurements," in CIPS 2022: 12th International Conference on Integrated Power Electronics Systems, Berlin, Germany, 2022.

[2] H. Lutzen, J. Müller, V. Polezhaev, T. Huesgen, and N. Kaminski, "Reducing the Impact of Skin Effect Induced Measurement Errors in M-Shunts by Deliberate Field Coupling," in 2022 24th European Conference on Power Electronics and Applications (EPE'22 ECCE Europe), Hanover, Germany, 2022.

[3] F. Wilhelmi, A. Schmid, and A. Lindemann, "Assessment of State-of-the-Art Current Sensors for Fast Switching," presented at the PCIM Europe 2022; International Exhibition and Conference for Power Electronics, Intelligent Motion, Renewable Energy and Energy Management, Nuremberg, Germany, 2022.

[4] H. Lutzen, J. Müller, and N. Kaminski, "A Review of Current Sensors in Power Electronics: Fundamentals, Measurement Techniques, and Components to Measure the Fast Transients of Wide Bandgap Devices," in 2023 25th European Conference on Power Electronics and Applications (EPE'23 ECCE Europe), Aalborg, Denmark, 2023.

[5] S. Hain and M.-M. Bakran, "New Rogowski coil design with a high DV/DT immunity and high bandwidth," in 2013 15th European Conference on Power Electronics and Applications (EPE), Lille, France, 2013.

[6] D. Xu, W. Sae-Kok, A. Vujanic, A. Motta, N. Powers, and T. Neo, "Fiber-optic Current Sensor (FOCS): Fully Digital Non-conventional Instrument Transformer," in 2019 IEEE PES GTD Grand International Conference and Exposition Asia (GTD Asia), Bangkok, Thailand, 2019.

[7] M. Billmann, "Coaxialshunt – general information," [Online]. Available: http://www.ib-billmann.de/koax_e.php. [Accessed 30.10.2023].

[8] C. Bödeker, M. Adelmund, and N. Kaminski, "The M-Shunt Structure Applied to Printed Circuit

Boards," in CIPS 2018: 10th International Conference on Integrated Power Electronics Systems, Stuttgart, Germany, 2018.

[9] H. Lutzen, K. Mitsui, D. Silber, K. Wada, and N. Kaminski, "Optimisation and Proof of Concept Studies for the M-Shunt Structure applied to Printed Circuit Boards," in CIPS 2020: 11th International Conference on Integrated Power Electronics Systems, Berlin, Germany, 2020.

[10] K. Schon, "Hochspannungsmesstechnik – Messgeräte und Messverfahren," Springer Verlag, 2010. [Reprint 2016].

[11] "Datasheet M-Shunts of Generation 1," [Online]. Available: https://www.m-shunt.com/gen_one. [Accessed 01.06.2024].

[12] Isabellenhuette, "Datasheets // Manganin," [Online]. Available: https://www.isabellenhuette.de/fileadmin/Daten/Praezisionslegierungen/Datenblaetter_Widerstand/MANGANIN.pdf. [Accessed 10.04.2024].

[13] H. Lutzen, P. Ziegler, J. Müller, J. Roth-Stielow, and N. Kaminski, "Why the 3dB bandwidth of a current sensor does not provide an exact statement about its measurement capability," in CIPS 2024: 13th International Conference on Integrated Power Electronics Systems, Düsseldorf, Germany, 2024.

[14] "Datasheet T3RC3000-HF," [Online]. Available: https://www.teledynelecroy.com/probes/t3rc-current-probes/t3rc3000-hf. [Accessed 10.04.2024].

[15] "Datasheet CWTMini50HF," [Online]. Available: https://www.pemuk.com/Userfiles/CWTmini/CWT_Mini_DS_Feb_2020.pdf. [Accessed 10.04.2024].

[16] "Project report MultiPulse," [Online]. Available: https://www.efre-bremen.de/projekte/multipulse35762?asl=. [Accessed 10.04.2024].

[17] "User Manual Tektronix WGB Double Pulse Test," [Online]. Available: https://download.tek.com/datasheet/WBG-DPT-Datasheet-EN-US-61W-73888-3.pdf. [Accessed 17.04.2024].

PCIM Europe 2024, 11– 13 June 2024, Nuremberg          DOI: 10.30420/566262122

# Characterization of Power-Module Parasitics: Sub-Nanosecond Large Signal Pulsing vs. Double-Pulse Testing

Dennis Helmut[1], Felix Gesele[1], Thomas Brückner[1], Gerhard Groos[1]

[1] University of the Bundeswehr Munich, Germany

Corresponding author:  Dennis Helmut, dennis.helmut@unibw.de
Speaker:                        Gerhard Groos, gerhard.groos@unibw.de

## Abstract

This paper compares the novel sensor gap transmission line pulsing (sgTLP) technique with the established method of double-pulse testing (DPT) for characterizing parasitic inductances in a fast SiC half-bridge power module. SgTLP employs Transmission Line Pulsing with its sub-nanosecond temporal resolution and voltages up to 2 kV to detect parasitic inductances and capacitive effects, which are not typically observable with DPT. While both, sgTLP and established methods, produce corresponding results for the total inductances, sgTLP additionally offers a deeper insight into the structure of the module's internal inductive and even capacitive network.

## 1    Introduction

By the introduction of new materials like Silicon Carbide (SiC) and Gallium Nitride (GaN) [1] in recent years, modern power modules have become more efficient and use smaller packages, particularly allowing higher operating frequencies and temperatures. However, despite these advancements, a major persisting challenge is the parasitic characteristics inherent in the switching device and the module package. Any unwanted parasitic inductance or capacitance can degrade the switching behavior or may result in undesirable resonances, overshoots and electromagnetic interference (EMI) that in turn can impair performance and efficiency [2].

Analysis and characterization of module parasitics help to gain insights into their influences on the switching behavior in order to develop strategies to minimize their negative impact. There are several approaches towards measuring the parasitics [3] including techniques in the frequency [4] and the time domain [5]. All those approaches have the downside that the analysis is performed with small-signal techniques or that the module must be imbedded into complex test circuits, which introduce additional parasitics on their own.

This paper proposes another technique called sensor gap TLP (sgTLP) [6][7] that can extract parasitic parameters by analyzing the step response with adequate voltage levels and time scales [8], without a need for a bandwidth-limiting current sensor. Moreover, sgTLP not only provides current and voltage measurements but is also capable of

extracting the actual excitation present at the ports, enabling the data to be used in simulation tools for in-depth analysis and characterization.

The double-pulse test (DPT), typically used to characterize the switching behavior of power devices, serves as a benchmark in this study. As a well-established method, DPT is also able to determine inductive properties of devices in a setting replicating real-world switching conditions.

The following Section 2 will describe sgTLP, show how it is applied to the power module and how the experimental findings lead to an adequate equivalent circuit model. Section 3 explains the adapted DPT setup and the method used to determine parasitic inductances of the module and show the results gained. Section 4 and 5 discuss the findings and conclude the paper.

An Infineon 750 A 3.3 kV half-bridge SiC module [9] will be used as test object throughout the paper.

## 2    Sensor Gap TLP (sgTLP) Analysis of the Module

This chapter will use sgTLP to study the transient responses at different ports of the SiC half-bridge module on a sub-nanosecond time scale in order to obtain a detailed perspective of the module's parasitic network. The characterization of key inductive parasitics in the power loops, i.e. the loops formed by the main current paths through the module, back to the interface PCB and to the measuring system, involves the analysis of three specific port configurations (c.f. Fig. 1): DC-/AC, AC/DC+,

986

PCIM Europe 2024, 11– 13 June 2024, Nuremberg       DOI: 10.30420/566262122

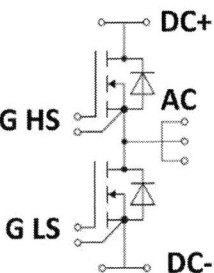

**Fig. 1** General schematic and pins of the half bridge with high side (HS) and low side (LS) switch, reverse diodes and pin nomenclature.

and DC-/DC+. For the gate-circuit analysis, the ports are defined by the gate terminals.

## 2.1 SgTLP as a Characterization Tool

Unlike in electrical port analysis, where signal transients are measured using voltage and current probes as close to the device under test (DUT) as possible, transmission-line theory offers a paradigm that considers the excitation and the DUT reaction as incident and reflected waves and connects them to the current and voltage transients at a specific point along the signal path (i.e. at the port of the DUT) through waveform superposition.

The sgTLP technique [6] is designed to both allow long excitation pulses ($> 1$ µs) and reach very short temporal resolutions ($< 0.1$ ns). Simplified, sgTLP can be seen as a rectangular pulse generator and measuring system with a source impedance of 50 Ω, where current and voltage transients are determined at a certain output reference location. The technique's capability to provide high bandwidth through quick rise times, short cables and absent bandwidth-limiting current probe enables comprehensive characterization not only of the switching dynamics of power devices [10] but also for any examination of fast transient behavior at voltage levels appropriate for power electronics.

When connecting the TLP measurement point to the ports of the DUT, the interface has to be considered critically. Furthermore, specifying an inductance inside the half bridge needs to include the geometry, which closes the current loop between the addressed module pins to properly define the entire inductor loop. The minimal inductor for the considered power loops is the one that completes by a conducting plane right on the top of the module. Therefore, the connection is realized by low inductive PCBs that cover the module (c.f. Fig. 2), mounted on its top, having metal at the bottom where possible and providing an interface to the 50 Ω sgTLP system. Additionally, these (P)CBs can be subject to standalone analysis by

the sgTLP method, allowing their parasitic properties to be extracted and used in simulations.

The usual characterization approach in a small-signal analysis [4] fits a linear model to the response of the DUT to a sinusoidal signal with small amplitudes around single digit volts. In comparison, the sgTLP methodology offers the distinct advantage of applying significantly higher and step like signals of up to some kV. This excitation transient as well as the system response is measured and can be used in a simulation to directly compare them to a model's predictions.

## 2.2 Analysis of the Power Paths

In the following experiments, the excitation has an amplitude of 100 V (power loops) or 15 V (gates), a rise time of 150 ps, and a duration of 500 ns. All sgTLP voltages represent the average values obtained from a sample set of 100 pulses.

The analysis begins by assessing the measurement signals, followed by the application of simple equivalent models that are refined in a step-by-step process to accurately represent the observed phenomena. The developed equivalent-circuit diagrams are shown in Fig. 4.

### 2.2.1 Measured Module Response

Figure 3a, d and e show signal transients for the pulsing configurations AC/DC+ (green / grey), DC-/AC (blue), and DC-/DC+ (orange). The (positive) pulse arrives at the electrode named first and the second electrode is connected to ground. Using the example of AC/DC+, Fig. 3a shows the influence of reversing the pulse polarity by comparing the AC/DC+ with the DC+/AC configuration: The first parts of both signals, featuring clear oscillations, are identical. Only after about 100 ns, in AC/DC+ the diode of the HS switch starts to conduct, while in DC+/AC it keeps on charging in reverse direction. So, the response of the passive parasitic network at the beginning is clearly separated from the later transistor charging or diode

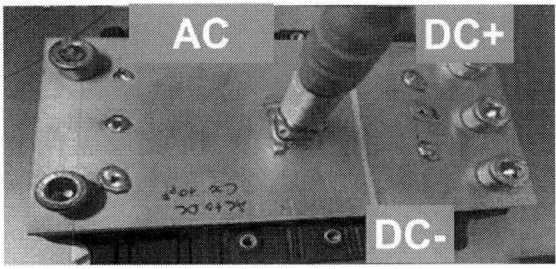

**Fig. 2** Example of low-inductive connection board (CB) for applying the sgTLP pulse to the module: AC/DC+ configuration mounted on the module.

987

conduction, especially no complex transistor switching has to be considered.

Therefore, throughout the study the diode is always operated in forward direction (i.e. reverse direction for the switch), enabling a plain SiC diode model for simulating the switches, which is calibrated using the rear part of the signal transient.

It should be noted, though outside the scope of this paper, that the other polarity, i.e. with blocking diode, could be used to analyze the transistor's nonlinear capacitance with those sgTLP measurements over a wide voltage range [8][11][12].

The signals of all three configurations (Fig. 3d and e) show the passive network's transient response during the first 50 ns. DC-/AC and AC/DC+ exhibit clear damped oscillations with one dominant frequency, although a second, smaller oscillation can be recognized in the AC/DC+ signal (Fig. 3c). The DC-/DC+ oscillation is more complex due to the two switches involved and faster because the higher AC inductance is not included in this path.

At the beginning of the signals, step-like patterns occur in the otherwise smooth transients, as shown in Fig. 3b for AC/DC+. Those steps can also be observed in the response of the connection boards (CBs) alone, measured in an open configuration. Using such experiments, models for the CBs were derived and calibrated, the steps being a consequence of reflections inside the CB. As can be seen later, these CB models achieve a very good agreement between measurement and simulation concerning the steps.

Already those first results highlight the sgTLP's ability to exactly analyze fine details due to its high temporal resolution, leading to independently extracted properties of the CBs in this case. The spectra (Fig. 3f and g) reflect those findings from a different angle. The DC-/AC spectrum is mainly a single low pass, extended by high frequency structures at 2 GHz and above, which belong to the step patterns at the signal onset.

The AC/DC+ signal is similar, but exhibits a clear additional peak at about 180 MHz, corresponding to the oscillation shown in Fig. 3c. As simulation will confirm later, this peak can be understood by the module's layout: The CB for AC/DC+ is mounted directly above the DC- contact (see Fig. 2), giving rise to capacitive coupling from the DC-contact to system ground which is absent during DC-/AC pulsing, but generates the observed oscillation in the AC/DC+ measurements.

Figures 3g and e present the step response and spectrum of the DC-/DC+ configuration, where the significant inductance $L_{AC}$ provided by the AC branch is missing, resulting in faster oscillations. The spectrum generally also shows a low pass

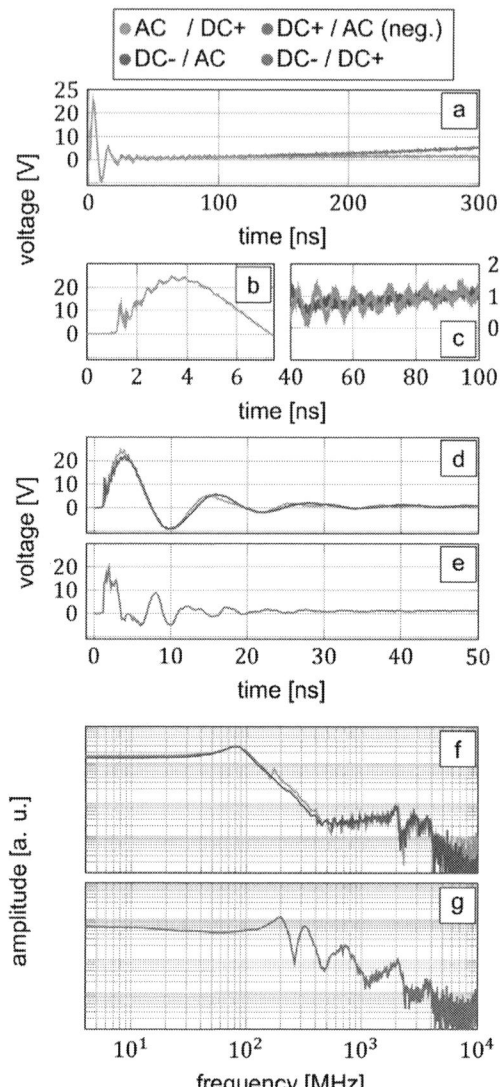

**Fig. 3** Measured signals of the three power loop configurations
a: AC/DC+ and DC+/AC Pulses;
b, c: zoom into beginning and middle of (a);
d, e: transient voltage of all configurations;
f, g: corresponding frequency spectra.

characteristic with a resonance at about 200 MHz, but exhibits two additional broad peaks at about 300 and 700 MHz. The high-frequency features starting at 2 GHz are again caused by reflections in the CBs and correspond to the steps at the pulse onset.

### 2.2.2 Module Circuit Models

The first ansatz for replicating the measured transients is a basic equivalent circuit model (s. Fig 4 I) according to literature [13]. This initial model uses SiC diodes to represent the switches Q1 and Q2

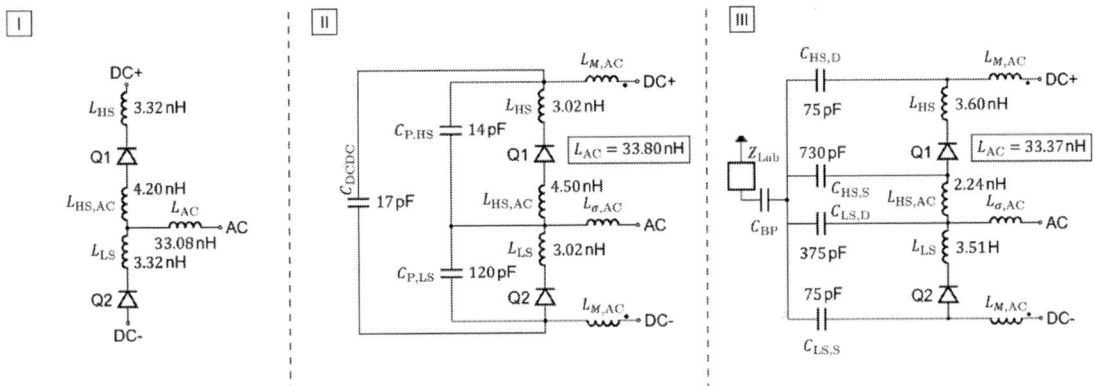

**Fig. 4** Basic structure of the module equivalent circuit models:
I: minimal representation only of the inductive contributions from the three module branches
II: capacitive extension and separation of $L_{\sigma,\text{AC}}$ from $L_{\text{AC}} = L_{\sigma,\text{AC}} + L_{M,\text{AC}}$ to better reproduce the DC-/DC+ signal
III: modelling of the capacitive behavior as coupling to the base plate (BP) and from there to the Lab environment
The equivalent circuits for the CBs and for the laboratory coupling, $Z_{\text{Lab}}$ are not shown (see text).

in the off-state and inductances $L_{\text{HS}} = L_{\text{LS}}$, $L_{\text{HS,AC}}$ and $L_{\text{AC}}$ in the branches of the module. The inductance $L_{\text{HS,AC}}$ includes differences between HS and LS as well as the inductance of the internal connections between the switches.

As shown below, significantly improved results are achieved with model II in Fig. 4 II, which adds separate parallel capacitances to the switches and between DC- and DC+.

The final module representation is model III, which interprets the capacitive behavior as coupling to the base plate (BP) and includes an interaction to the lab environment, indicated by the observations.

The mutual inductance $L_{M,\text{AC}}$ of the closely spaced rails into the DC- and DC+ connection is included in $L_{\text{AC}}$ in model I. But to enable the correct coupling to the capacitive parts of model II and III, those models place $L_{M,\text{AC}}$ directly at the DC- and DC+ pins. The remaining inductance in the AC branch is $L_{\sigma,\text{AC}}$ and the total inductance equivalent of the AC branch: $L_{\text{AC}} = L_{\sigma,\text{AC}} + L_{M,\text{AC}}$ corresponds to $L_{\text{AC}}$ from model I.

Applying the incident pulse transient from the sgTLP measurements to the described models, the resulting DUT voltage $u_{\text{sim}}$ was simulated and compared to the measured transients $u_{\text{meas}}$. The root mean square (RMS) value $\Delta u_{\text{RMS}}$ of the deviation $\Delta u = u_{\text{sim}} - u_{\text{meas}}$ during the first 80 ns was minimized in an optimization process and the reached $\Delta u_{\text{RMS}}$ value used as a measure for the quality of both the model topology and the obtained model parameters.

Figure 5 shows selected data from all three models to enable a focused analysis of the most relevant findings in the following.

*Model I (s. Fig. 4 I)*

According to Fig. 5a and b, model I provides a good fit for DC-/DC+ (not shown) and AC/DC+ only by using the known CB capacitance and adjusting the path inductance. The peak at about 180 MHz caused by a parasitic capacitive path and the CB effects starting at 1 GHz were described above and are also reproduced well. However, model I fails to capture the more complex structure of the DC-/DC+ signal (Fig. 5c), resulting in an overall deviation of $\Delta u_{\text{RMS}} = 1.295$ V.

This deficit can easily be understood by the fact that the total inductance of the DC-/DC+ path only allows one single $LC$ element consisting of the involved module inductances interact with the capacitance of the CB.

*Model II (s. Fig. 4 II)*

To enable the observed multiple resonances in the DC-/DC+ transients, model II (Fig. 4 II) is extended by three capacitances. They should be considered as a minimal approach to reproduce the observed dynamic response, not as physical representation of the module's internal structure.

As Fig. 5e illustrates, the location and overall amplitude of the additional spectral components can be modeled with this approach and the resulting error transient $\Delta u$ (Fig. 5d) is reduced as a consequence (not shown for model I). The overall deviation has decreased to $\Delta u_{\text{RMS}} = 0.629$ V.

The generation of three distinct peaks in the DC-/DC+ configuration marks a significant stride towards an accurate model, though the shape of

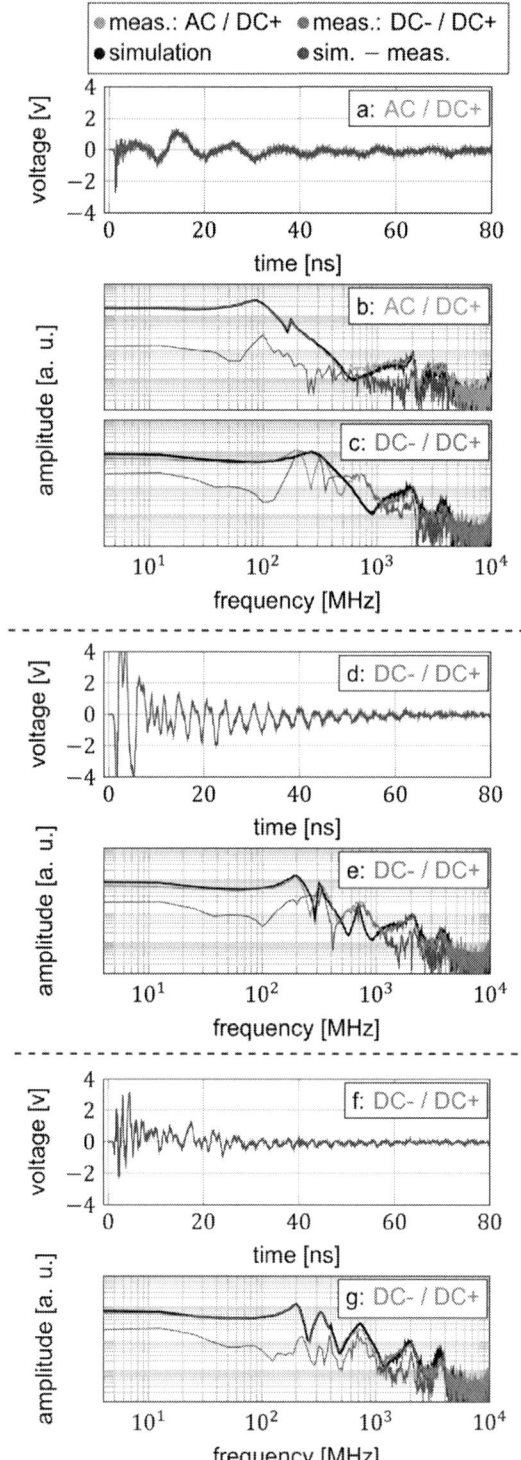

**Fig. 5** Transient voltage deviation $\Delta u = u_{\text{sim}} - u_{\text{meas}}$ and amplitude spectra for selected signals. Color code according to legend at the top.

a, b, c: model I;   d, e: model II;   f, g: model III.

those spectral structures could still do with improvement.

## *Model III (Fig. 4 III)*

The idea for model III is to adequately represent the general capacitive internal structure of the module with only a few parameters, i.e. without building an exact distributed model. The most prominent parasitic capacitive element in such power modules is the base plate, building capacitances towards all metal areas above. Hence, model III (Fig. 4 III) contains four capacitors from the respective nodes to the base plate (BP), which in turn builds a capacitance to the lab environment. As turned out, an additional resonant circuit representing the lab cavity ($Z_{\text{Lab}}$ in the figure) could reduce remaining residuals. A further improvement, motivated by the physics of eddy currents in the base plate, is to separate a certain part of the inductances along the DC-/DC+ path as mutual inductances to a respective counterpart in the base plate, analogous to $L_{M,\text{AC}}$ above. Those mutual parts are not drawn in the figure for clarity.

The achieved results (Fig. 5f and g) show a further significantly reduced deviation transient $\Delta u$ and a nearly perfect agreement of the spectra.

The remaining overall deviation is now only $\Delta u_{\text{RMS}} = 0.317$ V. This means a factor of 2.0 compared to model II and even 4.1 in relation to model I, illustrating the immense improvement achieved.

The extracted parameters are given in Fig. 4 together with those of the previous models. Table 1 lists the main outcomes for model III.

|          | $L_{\text{HS}}$ | $L_{\text{HS,AC}}$ | $L_{\text{LS}}$ | $L_{\text{AC}}$ | $L_{\sigma,\text{AC}}$ | $L_{M,\text{AC}}$ |
|----------|------|------|------|-------|-------|-------|
| $L$      | 3.73 | 2.24 | 3.52 | 33.37 | 17.17 | 16.20 |
| $\sigma_L$ | 0.44 | 0.26 | 0.24 | 0.52  |       |       |

**Tab. 1** Extracted inductive parameters for model III (Fig. 4 III) of the power module (see text). All values in nH.

The methodology is able to determine not only the total inductivity for the main loops, but due to the structure of the spectra, which contain specific resonances that lead to more constraints, can also extract internal components of the inductive network and the capacitive coupling.

Standard deviations $\sigma_L$ in Tab. 1 and Tab. 3 were calculated from the spread of intermediate fits with an RMS error $\Delta u_{\text{RMS}}$ less than 30% above the achieved minimum to give a clue for the accuracy of the results. A more detailed statistical analysis is subject of future work.

In conclusion, the analysis of the power loop by sgTLP and subsequent modelling leads to a schematic of the module's parasitic network which contains not only the main inductive contributions but also details of their allocation like $L_{\mathrm{HS,AC}}$ in Fig. 4 or the mutual fraction of $L_{\mathrm{AC}}$. It also describes the capacitive coupling network to the base plate and even to the lab environment. This highlights the benefits of the sgTLP method not only for the analysis of switching [6], but also for determining the passive parasitics of power modules.

## 2.3 Analysis of the Gate Loops

The analysis of the waveform obtained from the sgTLP method in Fig. 6b and d, identify similar characteristics in both gate loops: initially, the data show an inductive peak indicating leadframe inductance, followed by a rather long gate capacitance's charging phase. This duration is caused by the increased $RC$ constant (50 Ω sgTLP system). A closer examination within the initial 5 ns (Fig. 6c and e) reveals a notable plateau, which has different lengths within the LS and HS gate loops. This plateau may be attributed to a transmission line phenomenon, characterized by an impedance $Z_{\mathrm{TL}}$ and time delay $\tau_{\mathrm{TL}}$, likely resulting from the propagation dynamics within the gate circuits. This transmission line segment can be approximated as a low pass filter [13]: $L_{\mathrm{TL}} = \tau_{\mathrm{TL}} Z_{\mathrm{TL}}$ followed by a parallel $C_{\mathrm{TL}} = \tau_{\mathrm{TL}}/Z_{\mathrm{TL}}$.

The subsequent waveform segments can be modeled by an LRC series circuit model, where $L_{\mathrm{G,int}}$ denotes the inductance attributed to leadframes and bond wires, $R_{\mathrm{G,int}}$ reflects the gate's internal resistance, and $C_{\mathrm{G}}$ encompasses the gate-source capacitance, which dominates the gate charging in this case. The circuit model for the simulation is depicted in Fig. 6a and Tab. 2 shows the results of the parameter fitting outcomes for both gate circuits.

| | $R_{\mathrm{G,int}}$ /mΩ | $C_{\mathrm{G}}$ /nF | $\tau_{\mathrm{TL}}$ /ps | $Z_{\mathrm{TL}}$ /Ω | $L_{\mathrm{G,int}}$ /nH | $L_{\mathrm{G}}$ /nH |
|---|---|---|---|---|---|---|
| LS | 517 | 158 | 344 | 61.4 | 48.8 | 69.8 |
| HS | 519 | 153 | 529 | 61.6 | 45.6 | 77.7 |

**Tab. 2** Parameter fitting results for the gate circuit model (Fig. 6a)

For the overall inductive characteristic, both influences add up to $L_{\mathrm{G}} = L_{\mathrm{G,int}} + \tau_{\mathrm{TL}} Z_{\mathrm{TL}}$. Additionally, the comparison between the model and the actual waveforms (Fig. 6c and e) indicates that the model overestimates the voltage peak. This overestimation is likely due to the use of a lumped element model to approximate the behavior of a distributed network. While this approach offers a simplified representation, it falls short of accurately modeling the network's complexities, resulting in a perceived increase in magnetic flux. This, in turn, suggests that the model inductance is greater by around 1 nH.

## 3 Measurement with Double-Pulse Test (DPT)

### 3.1 Measurement Setup and Principle

The classical approach to determine parasitic inductances, as applied to commutation loops, is to measure the voltage drop across the inductance during the high $\mathrm{d}i/\mathrm{d}t$ of an inductive (natural) commutation process in a double-pulse test (DPT). To measure the different inductances of the half-bridge module, the conventional setup is slightly modified by inserting the DUT as a passive component in the commutation loop as indicated in Fig. 7. There is one specific test setup for each pair of terminals to be tested, e.g. using the AC terminal in the commutation loop. The DUT (marked grey in Fig. 7) is implemented in series to the free-wheeling diode of the active module; the DUT itself remains passive. During the commutation from D2 to S1, the inductance between the terminals of the DUT is determined by

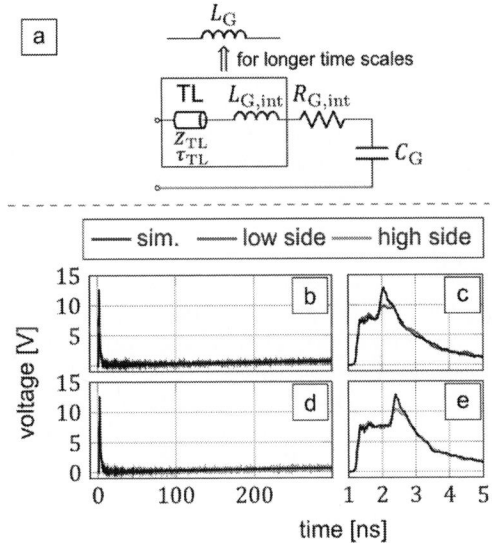

**Fig. 6** Gate model and voltage waveforms
a: Equivalent circuit model for a gate loop
Measured and simulated voltages:
b, c (zoom): at the LS gate
d, e (zoom): at the HS gate

$$u_{\mathrm{M}} = L \cdot \frac{\mathrm{d}i_{\mathrm{BD}}}{\mathrm{d}t} \Rightarrow L = \frac{\int_{t_1}^{t_2} u_{\mathrm{M}}\, \mathrm{d}t}{\Delta i_{\mathrm{BD}}} \qquad (1)$$

where $\Delta i_{\mathrm{BD}} = i_2 - i_1$ is the current difference during the falling edge of the diode current within the evaluation interval and $u_{\mathrm{M}}$ is the voltage across the DUT, integrated over the same interval. An example of a measured waveform is shown in Fig. 9

As already discussed in section 2.1, the measured inductance value is only meaningful if the inductance is defined and the investigated loop is closed. This is achieved by a busbar closely linking the respective module terminals directly on top of the module. Figure 8 shows the physical structures of the three setups, with the current loops of the inductor symbolically indicated (red dashed line). While the DC terminals of the half-bridge module are optimized for low inductance and the loop is almost already closed (Fig. 8b), closing the loop properly with minimum inductance is more crucial for loops incorporating the AC-terminal (Fig. 8a, c), for this measurement and for an application. Even though not the standard case, the AC-terminal inductance becomes critical when assembling e.g. a three-level ANPC inverter of half-bridge modules. The blue arrows in Fig. 8 indicate the position of voltage probes for this measurement, which are connected as close as possible to the small open gap of the inductance loop.

## 3.2 Challenges

The precise determination of small inductances with this approach bears a number of challenges regarding measurement accuracy and resolution that are generally the same as in every DPT:

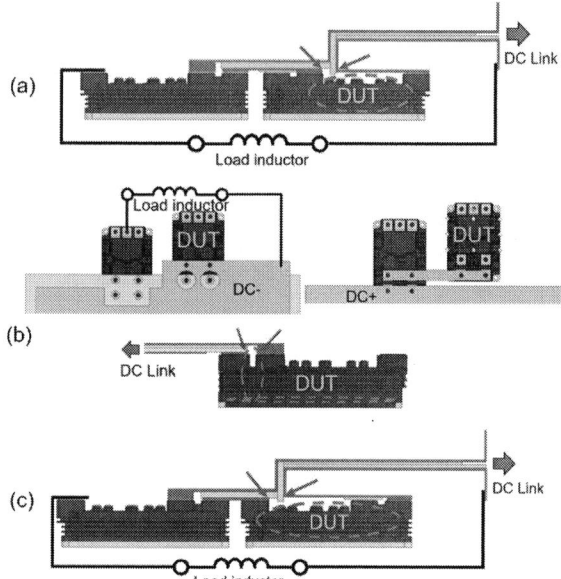

**Fig. 8** Physical structure of all test setups; commutation loop (inductive loop) indicated in red; position of voltage measurement probe pointed in blue.

- limited voltage resolution due to high switched voltages (and the use of an appropriate high-voltage probe),
- high $\mathrm{d}u/\mathrm{d}t$'s causing capacitive coupling currents to ground,
- limited bandwidth of current measurement.

There are some ways to improve the accuracy of the measurement: The MOSFETs of the DUT are constantly gated on such that they do not take any blocking voltage throughout the test; the body diode of the switching module takes the DC-link voltage after the commutation. This leads to the possibility of using probes with a small divider, which improves accuracy.

By separating the oscilloscopes for the DUT and the switching module, the reference for the voltage measurement at the DUT can be freely chosen, preferably at a fix DC-link potential, either DC+ or DC-, compare Fig. 7. This is made possible by an independent power supply of both oscilloscopes via battery and inverter and fiber-optic synchronization, so there is no galvanic connection between the oscilloscopes and versus the ground. Even coupling capacitances in the pF range of high-grade separation transformers could be avoided.

Twelve-bit oscilloscopes LeCroy HDO6104B with 1 GHz bandwidth are used, together with LeCroy PP023 voltage probes (attenuation ratio: 10:1; accuracy: 1%; bandwidth: 500 MHz). As current sensors Rogowski coils CWTMiniHF 15/R/2.5/300/5

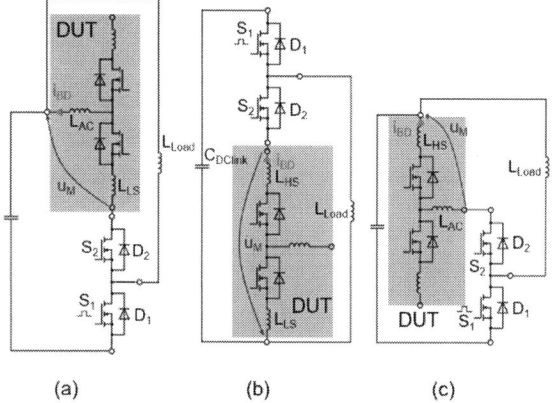

**Fig. 7** Schematics of DPT setup for determining parasitic inductances of half-bridge module: (a) DC-/AC; (b) DC-/DC+; (c) AC/DC+ (The DUT is shaded.)

with a linearity of 0.05 %, an accuracy of 2 % (depending on position of conductor in the coil), a bandwidth of 3 Hz to 18 MHz and a peak $di/dt$ of 80 kA/µs are utilized.

Apparently, the weak point seems to be the current probe with very limited bandwidth and its known sensitivity against voltage steps, even more in comparison to the sgTLP method discussed in the previous section. However, the sole function of the current probe is to deliver a current difference of several hundred amperes over a rather long interval, where the characteristics of the probe prove to be fully sufficient. According to an empirical formula for bandwidth and rise time of flanks [14]

$$t_{\text{rise,min}} \approx \frac{0.44}{f_{\text{BW}}}, \tag{2}$$

rise times down to 24.4 ns can be displayed appropriately, while the evaluated rise times are in the hundreds-ns range.

### 3.2.1 Resistive Voltage Drop

The measured voltage $u_M$ across the DUT includes inductive and ohmic voltage drops in the current path between the terminals. The ohmic part especially includes the $r_{\text{DS,on}}$ of one or two MOSFETs in series (Fig. 7), but also drain- and source-lead resistances. In a first approximation, the equivalent ohmic resistance $R_{\text{corr}}$ can be determined from the voltage at constant current before the commutation. The voltage $u_{L\sigma}$ across the inductance then results as

$$u_L(t) = u_M(t) - R_{\text{corr}} \cdot i_{\text{BD}}(t) \tag{3}$$

for test setups Fig. 7a and c and, due to the opposite polarity of the voltage probe, as

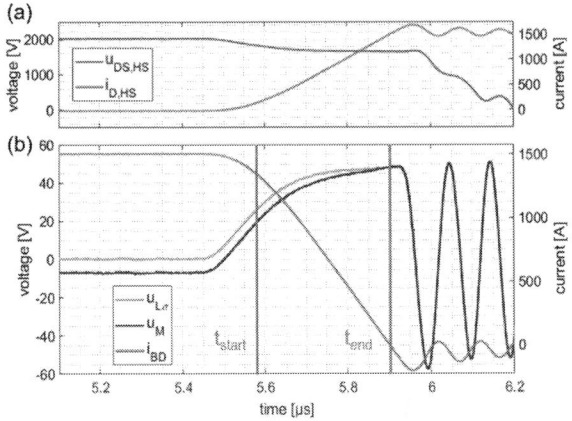

**Fig. 9** Measurement at 2000 V, 1500 A with test setup of Fig. 8b for determining $L_{\text{DC+/DC-}}$;
(a): $u_{\text{DS}}$ and $i_{\text{BD}}$ of switching device;
(b): $i_{\text{BD}}$, $u_M$ and $u_{L\sigma}$

$$u_L(t) = u_M(t) + R_{\text{corr}} \cdot i_{\text{BD}}(t) \tag{4}$$

for the test setup Fig. 7b.

Figure 9 depicts the waveforms of a measurement for $L_{\text{DC-/DC+}}$ (setup: Fig. 7b). Figure 9a shows the voltage (blue) and current (red) of the switched device, while Fig. 9b shows the current $i_{\text{BD}}$ (red) and the measured voltage $u_M$ (dark green) as well as the compensated voltage $u_{L\sigma}$ (bright green) at the DUT. It is clearly visible, that there is a huge mismatch between $u_M$ and the compensated signal $u_{L\sigma}$, so that the compensation is necessary.

### 3.2.2 Value Extraction

For the calculation of the inductance by Eq. (1), the limits of the integration interval need to be defined in a sensible way. On one hand, to limit the impact of noise on the current measurement, a certain value of $\Delta i_{\text{BD}}$ should be evaluated. On the other hand, the integration interval should not comprise the full length of the current commutation to allow the assessment of potential errors and accuracy. Two boundaries of an evaluation window are defined ($t_{\text{start}}$ and $t_{\text{end}}$ in Fig. 9) that proved to provide useful results. $t_{\text{start}}$ is defined at 90 % current level, $t_{\text{end}}$ as the zero crossing of the current. Integration intervals of different length are now shifted within this evaluation window sample by sample, the resulting data are displayed in

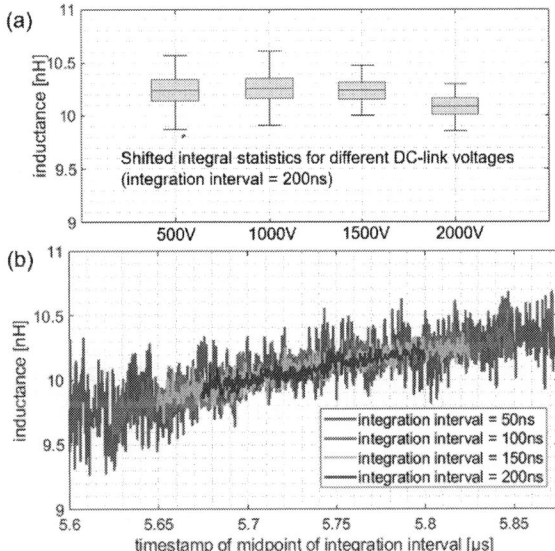

**Fig. 10** Evaluation of inductance values for measurement of Fig. 9:
(a) statistical evaluation for shifted integration interval of 200 ns at different DC-link voltages;
(b) values obtained by integration intervals of different length as a function of the midpoint timestamp according to Fig. 9 ($U_{\text{DC}}$ = 2000 V)

Fig. 10b. The time axis for this plot is aligned with Fig. 9.

The recorded noise stems primarily from the sampled current points; longer integration intervals and consequently larger $\Delta i_{\text{BD}}$ naturally lead to a lower impact of noise. Furthermore, a weak but clear trend towards higher inductance values can be seen, when the integration interval is moved towards $t_{\text{end}}$. Since the compensation by Eqs. (3), (4) is based on the maximum current and does not consider the current dependency of $r_{\text{DS,on}}$, some error will occur for medium currents, while for zero current the compensation is not active. This correlation would need to be quantified, but might contribute to the trend.

Figure 10a shows statistical data for measurements at different DC-link voltages, evaluated by a 200 ns integration interval, shifted between $t_{\text{start}}$ and $t_{\text{end}}$. I.e., all data with 200 ns integration interval from Fig. 10b (in purple) are contained in the right boxplot in Fig. 10a. The results for different voltages and thus different $\text{d}i/\text{d}t$ display a high consistency, proving the good quality of the measurements.

For comparison to the sgTLP method, the mean values of the data obtained by shifted integration intervals of 200ns are used. These values are:

- $L_{\text{DC-/DC+}}$ = 10.20$\pm$0.37 nH
- $L_{\text{AC/DC+}}$ = 37.17$\pm$0.76 nH
- $L_{\text{DC-/AC}}$ = 33.44$\pm$0.40 nH

As uncertainty the standard deviation of the mean values for the different voltages (Fig. 10a for DC-/DC+) was used, added to the mean length of the whiskers in the boxplots to represent the linear trend like in Fig. 10b.

## 4 Gate Path

The inductance in the gate path consists of the inductance of the internal gate and Kelvin-source leads $L_{\text{G,int}}$ and the common-source inductance $L_{\text{CS}}$ (Fig. 11), which are measured in total as gate inductance $L_{\text{G}} = L_{\text{G,int}} + L_{\text{CS}}$. The basic idea to determine this inductance is to apply a voltage step between the gate (G) and Kelvin-source ($S_{\text{aux}}$) terminals using the gate driver and evaluate the first voltage peak across the $G - S_{\text{aux}}$ terminals. Given the assumption that in the first moment there is no gate current, the voltage across $G - S_{\text{aux}}$ is governed by the inductive voltage divider between the module-internal ($L_{\text{G}}$) and the external gate inductance.

To be comparable to the sgTLP measurements, a short coaxial line ($L_{\text{ref}}$) with an equivalent inductance of 126.3 nH (characterized by sgTLP) was

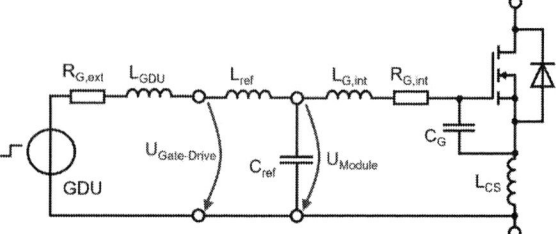

**Fig. 11** Schematic of test setup for determining the inductance in the gate path.

selected as an external reference inductance and introduced between the gate drive and the DUT's $G - S_{\text{aux}}$ terminals. The length of the line was chosen to result in a similar inductance than the gate to improve accuracy. The whole test circuit with some relevant parasitic elements is displayed in Fig. 11.

The voltages for the determination of $L_{\text{G}}$ are measured at the positions shown and the inductance is determined as

$$L_{\text{G}} = \frac{U_{\text{Module}}}{U_{\text{Gate-Drive}} - U_{\text{Module}}} \cdot L_{\text{ref}}. \quad (6)$$

Figure 12 displays one example measurement. Figure 12b shows the complete charging process of the gate capacitance ($C_{\text{G}}$), with a zoom on the first voltage peak as indicated by the red lines. The GDU delivers a voltage step from -6 V to +15 V. The resulting current steps due to the external coaxial cable are not resolved in the DPT measurements, but the line behaves like an inductance. The elements $R_{\text{G,ext}}$, $R_{\text{G,int}}$ and $C_{\text{G}}$ are mainly responsible for the behavior after 250 ns. The first voltage peak is defined by the inductances and their ratio, the oscillation at the first voltage peak

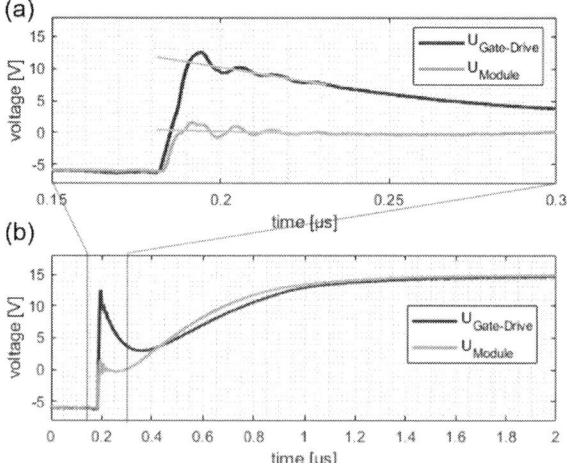

**Fig. 12** Measurement for determining the inductance in the gate path.

is a result of the resonances with the parasitic capacitances of the external circuitry.

This oscillation leads to an overshoot in the first voltage peak, where a direct evaluation of the voltage leads to an error. Therefore, Fig. 12a contains a rough approximation of the voltage curves without these oscillations, indicated in bright green. With Eq. (6) this results to an inductance $L_G$ in the gate path at the LS of 70.6(3.1) nH and at the HS of 79.7(4,7) nH. The given error is the standard deviation of 12 measurements.

# 5  Discussion

The above results assess the main power loops and the gate loops of the considered power module on the one hand with DPT, which has a low source resistance and current rise times of some hundred ns, and on the other hand with sgTLP, offering sub-ns resolution and 50 Ω system impedance.

The most prominent results are compared in Tab. 3. They include the main inductances in the power loop and the gate-loop inductances. An error estimation is given as described above.

|  | Power Loops | | | Gates | |
|---|---|---|---|---|---|
|  | DC− / DC+ | AC / DC+ | DC− / AC | LS | HS |
| sgTLP | 9.48 (.34) | 39.34 (.67) | 36.89 (.66) | 70.15 (.56) | 78.25 (.38) |
| DPT | 10.20 (.37) | 37.17 (.76) | 33.4 (.40) | 70.6 (3.1) | 79.7 (4.7) |
| Diff. | 0.72 7.6% | -2.17 -5.5% | -3.45 -9.3% | 0.45 0.6% | 1.4 1.8% |

**Tab. 3**  Summary of determined inductances. Values in nH, statistical error in brackets.

Both approaches are able to determine the gate inductances and they agree within the limits of uncertainty with deviations of only 0.6 % and 1.8 %. This should partly be a success of the consistent calibration: The reference in the DPT measurements was a transmission line, whose inductance had been characterized very exactly by sgTLP.

The values for the power loops, on the other hand, match to between 5.5 % and 9.3 %, but this is far beyond what would be expected from the statistical analysis of the measurements.

The additional 0.7 nH in the DPT value for the DC-/DC+ inductance could be caused by the connection of the DC+ terminals towards the long side of the module (compare Fig. 8b), which can lead to a constriction of the current in the module-internal terminal connection, and is not present in the corresponding CB for sgTLP. The lower values for AC-/DC+ and DC-/AC in the DPT should be investigated in more detail, they might as well stem from neglected differences of the contact geometries.

Very promising is the accuracy of sgTLP. Even if the remaining discrepancies have to be investigated, sgTLP seems to offer high precision. In the gate analysis, sgTLP achieves a significantly better uncertainty than DPT.

More detailed insights into the internal structure of the module, as its capacitive network, the segmentation of the power-loop inductance, giving values for its parts like $L_{HS,AC}$ and even mutual inductances $L_{M,AC}$ result from the sgTLP analysis of the same measurements.

That capability can be understood from the different source impedances of the methods. Because sgTLP works with a higher system impedance of 50 Ω, inductive response times $\tau_L = L/R$ are significantly smaller, but can be detected with the very fast temporal resolution of the method. On the other hand, capacitive time constants $\tau_C = R \cdot C$ are larger in sgTLP and can be seen, while in DPT, the same capacitances are invisible. This results in the discussed more complex signal transients during sgTLP, which in turn enable the extensive analysis shown.

# 6  Conclusion

This work assesses the potential of sgTLP for the characterization of parasitic inductances within a SiC power module in comparison to DPT.

As the consistent results show, both sgTLP and DPT are very suitable methods to evaluate inductive parasitics in fast power modules and the achieved accuracies are roughly comparable. However, sgTLP can also extract an adequate capacitive network for the module and accesses details of the inductance allocation like $L_{HS,AC}$ or the mutual inductance $L_{M,AC}$, which is at least more difficult and less precise with DPT.

Because of the possible high precision of the sgTLP method, future work should improve the CBs and their calibration as well as extend the statistical analysis to determine the inherent uncertainties more precisely.

## Acknowledgement

Part of this research work (F.G., T.B.) has been carried out within the project DEFINE and is funded by dtec.bw – Digitalization and Technology Research Center of the Bundeswehr, which we gratefully acknowledge. dtec.bw is funded by European Union NextGenerationEU.

# References

[1] L. F. S. Alves *et al.*, "SIC power devices in power electronics: An overview," in *2017 Brazilian Power Electronics Conference (COBEP)*, Juiz de Fora: IEEE, Nov. 2017, pp. 1–8. doi: 10.1109/COBEP.2017.8257396.

[2] M. März, "Parasitics in Power Electronics Avoid them or Turn Enemies into Friends," Zürich, Jan. 27, 2012. Accessed: Mar. 19, 2024. [Online]. Available: https://www.iisb.fraunhofer.de/content/dam/iisb2014/en/Documents/Research-Areas/Energy_Electronics/publications_patents_downloads/Publications/Parasitics_ETHZ_2012_IISB.pdf

[3] S. K. Roy and K. Basu, "Measurement of Circuit Parasitics of SiC MOSFET in a Half-Bridge Configuration," *IEEE Trans. Power Electron.*, vol. 37, no. 10, pp. 11911–11926, Oct. 2022, doi: 10.1109/TPEL.2022.3176114.

[4] Z. Wang, Z. Yuan, and Y. Zhao, "Parasitic Inductances Extraction for SiC Power Modules Using An Enhanced Two-Port S-Parameter Approach," in *2021 IEEE Applied Power Electronics Conference and Exposition (APEC)*, Phoenix, AZ, USA: IEEE, Jun. 2021, pp. 2420–2426. doi: 10.1109/APEC42165.2021.9487325.

[5] Z. Wang, S. Mao, S. Yang, W. Li, Y. Ding, and K. Zeng, "Power Loop Inductance Extraction with High Order Polynomial Fitting Algorithm for SiC MOSFET Power Module Characterization," in *2021 IEEE Workshop on Wide Bandgap Power Devices and Applications in Asia (WiPDA Asia)*, Wuhan, China: IEEE, Aug. 2021, pp. 189–194. doi: 10.1109/WiPDAAsia51810.2021.9656046.

[6] D. Helmut and G. Groos, "Sensor Gap TLP: Expanding Time Resolution and Pulse Duration Beyond Conventional Values of Standard and Very Fast Transmission Line Pulsing," in *2023 45th Annual EOS/ESD Symposium (EOS/ESD)*, Riverside, CA, USA: IEEE, Oct. 2023, pp. 1–11. doi: 10.23919/EOS/ESD58195.2023.10287759.

[7] D. Helmut and G. Groos, "Apparatus and Method for Determining a Response of a Device Under Test to an Electrical Pulse Generated by a Pulse Generator," patent application: EP4226167

[8] D. Helmut, G. Wachutka, and G. Groos, "Extracting Large-Signal Transient Characteristics of Power Electronic Devices Using Nanosecond Pulsing Techniques," in *2018 20th European Conference on Power Electronics and Applications (EPE'18 ECCE Europe)*, 2018, p. P.1-P.7.

[9] "FF2600UXTR33T2M1," Infineon Technologies AG, Preliminary datasheet Rev. 0.20, Nov. 2023.

[10] G. Groos, D. Helmut, G. Wachutka, and G. Schrag, "Sub-Nanosecond Transient Analysis of SiC MOSFET Switching: 'Sensor Gap TLP' as a Versatile Characterization Method with Very High Temporal Resolution," in *CIPS 2022; 12th International Conference on Integrated Power Electronics Systems*, Mar. 2022, pp. 1–5.

[11] D. Helmut, G. Wachutka, and G. Groos, "Transient analysis of latent damage formation in SMD capacitors by Transmission Line Pulsing (TLP)," *Microelectron. Reliab.*, vol. 76–77, pp. 97–101, Sep. 2017, doi: 10.1016/j.microrel.2017.06.076.

[12] G. Groos, D. Helmut, and G. Wachutka, "The Latent Failure Issue Seen from the Other Side: Normal Operation after ESD Induced Degeneration of Devices and Systems," in *2018 40th Electrical Overstress/Electrostatic Discharge Symposium (EOS/ESD)*, 2018, pp. 1–10. doi: 10.23919/EOS/ESD.2018.8509758.

[13] Huibin Zhu, A. R. Hefner, and J.-S. Lai, "Characterization of power electronics system interconnect parasitics using time domain reflectometry," *IEEE Trans. Power Electron.*, vol. 14, no. 4, pp. 622–628, Jul. 1999, doi: 10.1109/63.774198.

[14] F. G. Stremler, *Introduction to communication systems*, 3. ed. in Addison-Wesley series in electrical engineering. Reading u.a: Addison-Wesley Pub. Co, 1990.

PCIM Europe 2024, 11– 13 June 2024, Nuremberg    DOI: 10.30420/566262123

# Statistical Variations in the Parasitic Capacitance of a Coil

Kevin Talits [1,2], Claas Tebruegge [2], Martin Pfost [1]

[1] TU Dortmund, Germany
[2] HELLA GmbH & Co. KGaA

Corresponding author:    Kevin Talits, Kevin.Talits@forvia.com
Speaker:                 Kevin Talits, Kevin.Talits@forvia.com

## Abstract

The used materials and the wire placement have some uncertainties in the production process of an inductor and influence its parasitic capacitance. This study investigates the importance of precise placement of the turns and the deviation in capacitance, dependent on geometrical variations and manufacture tolerances. An ETD 60/30/20 core is wound, and measured 100 times, and compared to simulative results. The results show a direct influence of the shifted wire and a normal distribution of the capacitance. Combined deviations in the production process cause uncertainties of similar magnitude, with variations of up to 13.64 % in extreme cases.

## 1 Introduction

Inductive components with ferrite cores are widely used in power and filter applications and are subject of many studies [1]–[3]. They are included in many electronic devices and are crucial for the functionality of the device. During the winding of a coil, small shifts in the position of the wire can occur. If a wire is not handled with the needed caution, its insulation can be damaged or partially scraped away. Furthermore, displacements of the windings can occur during transport or usage in mobile applications [4].

Because of production tolerances in some parameters of the wire, a certain deviation in the capacitance is to be expected. This can be crucial for Electromagnetic Interference (EMI) suppression, for example, which relies on a precise knowledge of the parasitic capacitance of an inductor [5], [6]. To calculate the EM emission of a device, parasitic capacitance is used and the effect of the parasitics of a coil can reach or exceed the magnitude of the parasitic capacitance of Metal-Oxide-Semiconductor Field-Effect Transistors (MOSFETs) [7], [8].

Finite Element Analysis (FEA) simulations enable precise calculation of inductive components and their electromagnetic properties. By constructing a model of the coil and the core, the parasitic capacitance can be calculated and the influence of the wire placement can be simulated. In this study

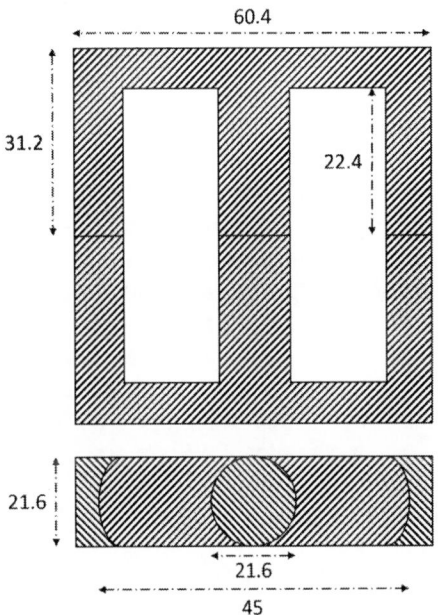

**Fig. 1:** Sketch of the ETD 60/30/20 core with dimensioning in millimeters. Front and top view are shown.

a FEA model for the parasitic capacitance of a coil with a ferrite core is established and verified with measurements. This enables an accurate simulative approach for the variation in wire parameters and to summarize their leverage on the capacitance.

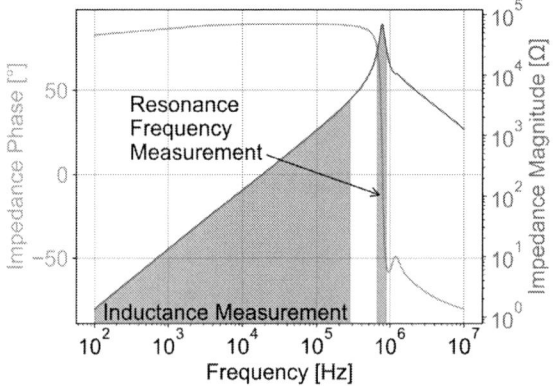

**Fig. 2:** Example of a Bode plot for one measurement of the ETD 60/30/20 core with 20 turns. Measurement areas for inductance in the linear region and resonance frequency within a 3 dB range of the impedance peak are marked.

## 2 Measurement and Simulation Methods

The same inductor with 20 turns is wound 100 times per hand, with an impedance measurement after each assembly. On the core, the height of the winding after the first assembly is measured and used as a reference for the following ones. It is tried to reproduce the same winding height for each coil, but small deviations are to be expected. The core used is an ETD 60/30/20, and can be seen in Fig. 1 with dimensioning in millimeters. For the wires a 24 American Wire Gauge (AWG) CuL wire is used with a diameter of 0.5 mm and an insulation thickness of 0.017 mm.

A Bode 100 impedance analyzer from OMICRON Lab is used for the measurements. Before the first measurement the cables of the impedance analyzer and itself are fixated to a static position for the whole experiment. After this, a short, open and load impedance calibration is performed to eliminate the influence of the setup and to only measure the change in the inductor. Measuring the impedance of the coil, the inductance $L$ is determined in the linear region of the Bode plot. The resonance frequency $f_0$ is calculated from the impedance peak

within a 3 dB measurement range, as shown in Fig. 2. With the formula

$$f_0 = \frac{1}{2\pi\sqrt{L \cdot C_\mathsf{p}}} \tag{1}$$

the parasitic capacitance $C_\mathsf{p}$ can be calculated, with the inductance $L$ and the resonance frequency $f_0$.

$$C_\mathsf{p} = \frac{1}{(2\pi f_0)^2 \cdot L} \tag{2}$$

is used to calculate the parasitic capacitance for the measured data.

The 2D FEA simulation is conducted in ANSYS Maxwell, with the model represented by its vertical cross-section, as seen on the top part of Fig. 1. This representation lacks depth information of the core as seen in the top-view in Fig 1. Simulation results are given as capacitance per meter and is corrected in the post-processing to the original wire length and model depth. Material properties are used from the datasheets of the core and the wire. Relative permittivity of the core is taken from the experimental work in [9]. One volt is used as an excitation of the coil. From one end of the wire to the other, the voltage decreases linear from one to zero volt along the whole wire length [10]. These modifications benefit the accuracy of the models and enable a fast and reliable 2D simulation, which agrees with the measured data shown in section 3. These models are simulated each time with a random shift in every single wire, utilizing the full height of the winding window for the displacement. With the verified FEA model, all parameters with production uncertainties are altered at the same time, and a realistic production batch simulation is created.

## 3 Results

A comparison of the capacitance distribution for measured and simulated data can be seen in Fig. 3 and shows Gaussian distributions with similar shapes. The mean of the measured data is 20.52 pF with an uncertainty interval of $\sigma = 0.94$ pF, with 67 % of the data within one standard deviation, and 96 % within two. The FEA model is validated with the measured data, and the results are in good agreement with the measured data in distribution and magnitude of parasitic capacitance.

This model is used to simulate the influence of manufacturing tolerances on the parasitic capacitance. The diameter of the copper wire can vary in a range of 0.495 mm to 0.505 mm, the insulation

**Tab. 1:** Parameter variations and their influence on the parasitic capacitance with unchanged homogenous wire placement in a FEA simulation. Minimum, average and maximum values of each parameter are used, and the table is split in three separate tables for the relative permittivity $\epsilon_r$.

| @$\epsilon_r$ =2.9 | | Copper Diameter | | |
|---|---|---|---|---|
| | | 0.495 mm | 0.500 mm | 0.505 mm |
| **Insulation** | 0.012 mm | 23.2891 pF | 21.6183 pF | 20.2903 pF |
| **Thickness** | 0.017 mm | 21.5658 pF | 20.2328 pF | 19.1290 pF |
| | 0.022 mm | 20.1877 pF | 19.0925 pF | 18.1639 pF |
| @$\epsilon_r$ =3.05 | | | | |
| **Insulation** | 0.012 mm | 23.6353 pF | 21.8505 pF | 20.4218 pF |
| **Thickness** | 0.017 mm | 21.9285 pF | 20.4939 pF | 19.3185 pF |
| | 0.022 mm | 20.5561 pF | 19.3709 pF | 18.3818 pF |
| @$\epsilon_r$ =3.2 | | | | |
| **Insulation** | 0.012 mm | 23.8668 pF | 21.9976 pF | 20.5430 pF |
| **Thickness** | 0.017 mm | 22.1886 pF | 20.6789 pF | 19.4462 pF |
| | 0.022 mm | 20.8363 pF | 19.5831 pF | 18.5372 pF |

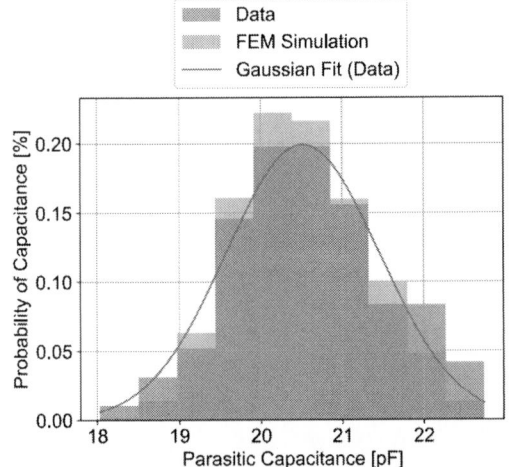

**Fig. 3:** Bar chart of the distribution of parasitic capacitance in measured and simulated data with a Gaussian fit of the measured data. The mean and the uncertainty interval of the Gaussian distribution is given by $x_0$ =20.52 pF and $\sigma$ =0.94 pF, respectively.

thickness from 0.012 mm to 0.022 mm, and the relative permittivity $\epsilon_r$ of the insulation material from 2.9 to 3.2. These variations are taken from the norm IEC 60317 and research in [11]. For a fair comparison, the variations are taken as the boundary of the third uncertainty interval for a Gaussian distribution of the values. A production batch was simulated, where all wire parameters and placement can vary within their limits. The results are shown in table 1 and Fig. 4.

Table 1 shows the combined influences of the manufacturing tolerances on the parasitic capacitance. Combining the maximum deviations of all variables, the variation of the capacitance rises up to 13.64 %, resulting in an uncertainty interval of $\sigma$ =0.95 pF, comparable to the results seen in Fig. 3.

Figure 4 shows the influence of the individual parameters on the parasitic capacitance with a simple linear fit. The impact is converted to the shift of capacitance per uncertainty interval of the parameter. This expression enables a direct comparison of the effect of the different parameters and the severity of their tolerances.

Insulation thickness has the greatest impact of all parameters, with 0.489 pF per uncertainty interval, resulting in a statistical maximum deviation of 1.467 pF. A thicker insulation increases the distance of the exited copper inside the wire to the other wires and the core. The simple equation

$$C = \epsilon_0 \epsilon_r \frac{A}{d} \qquad (3)$$

for the capacitance $C$ of two parallel plates shows the inverse proportion of the distance $d$ between the two surfaces and the capacitance. Thus, a thicker insulation with increased distance between the conducting copper, other wires and the core, decreases the parasitic capacitance.

The copper diameter has a smaller impact, with 0.274 pF per uncertainty interval, resulting in a maximum deviation of 0.822 pF. A larger diameter of the copper increases the surface area of the wire. Equation 3 shows the proportionality of the active surface area and the capacitance. This correlation

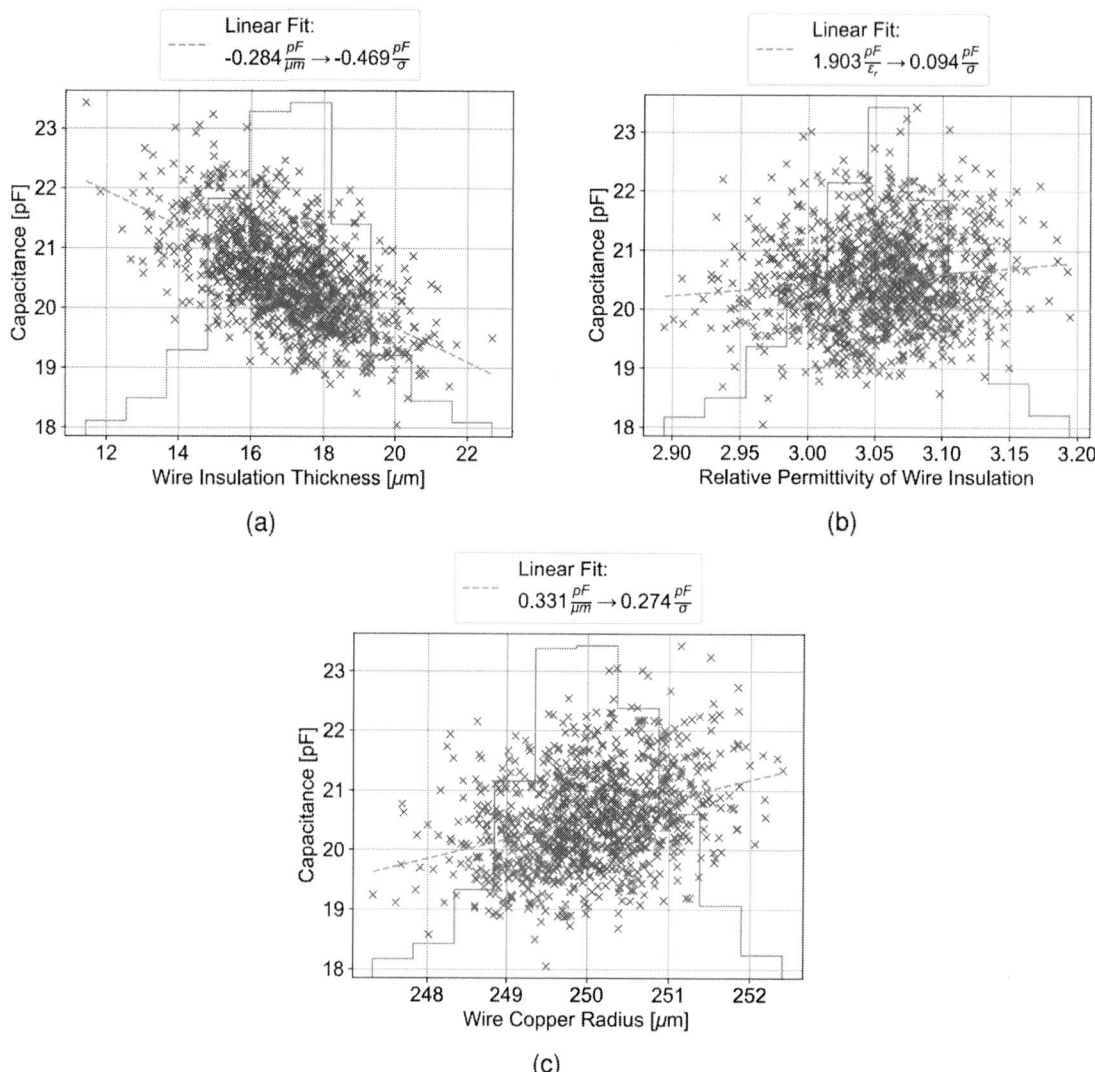

**Fig. 4:** Distribution of CuL wire manufacturing parameter variations and their influence on the parasitic capacitance. A linear fit roughly estimates the change of the parasitic capacitance per uncertainty interval. Parasitic capacitance decreases with the increase of insulation (a) and increases with the increase of relative permittivity (b) and radius (c).

leads to the increase in capacitance with larger copper diameter.

Least impactful is the relative permittivity, with 0.094 pF per uncertainty interval, resulting in a maximum deviation of 0.282 pF. Again Eq. 3 explains the increase in capacitance with higher relative permittivity.

Changing the geometry of the wires with insulation thickness and diameter has an up to five times larger impact on the change of capacitance, than the tolerance in relative permittivity for the used insulation. A single parameter alone changes the

parasitic capacitance in a smaller magnitude than the shift of the wires. Simulating over 1000 models with differing wire and material parameters and changing numbers of turns, a capacitance increase of $(1.056\pm0.009)$ pF per turn is calculated.

## 4  Conclusion and Outlook

This work shows the importance of sophisticated simulative methods to ensure high production quality and safety for varying parameters, like parasitic capacitance. Numerical simulations like this enable a controlled manipulation of multiple influencing factors at once and a high number of testing runs.

Enhancing the deeper understanding of the underlying dependencies of the target value is a key goal of such statistical experiments. Production in a well-controlled environment is crucial, to keep the variance low for high frequency applications in this domain, seen in [5], [6], [12]. The magnitude of the parasitic capacitance found can have a not inconsiderable influence on filter characteristics and can increase in significance as the number of windings or inductive components increases [13]. Together with the parasitic capacitance in MOSFETs, the capacitance of the coil can be a crucial factor for the stability and efficiency of a circuit [7], [8]. These conclusions are not limited to inductors and coils, but can be applied to transformers due to the same physical processes. In the future more wires and core types can be measured and simulated to broaden the understanding of the influence of all presented factors. Established electronic circuits, using parasitic capacitance, can be reevaluated under the question of stability and efficiency with varying capacitance.

# References

[1] J. Kaiser and T. Duerbaum, "An overview of saturable inductors: Applications to power supplies," *IEEE Transactions on Power Electronics*, vol. 36, no. 9, pp. 10 766–10 775, 2021. DOI: 10.1109/TPEL.2021.3063411.

[2] J. Lee, J. Roh, M. Y. Kim, S.-H. Baek, S. Kim, and S.-H. Lee, "A novel solid-state transformer with loosely coupled resonant dual-active-bridge converters," *IEEE Transactions on Industry Applications*, vol. 58, no. 1, pp. 709–719, 2022. DOI: 10.1109/TIA.2021.3119535.

[3] M. A. Bakar, M. F. Alam, A. Majid, and K. Bertilsson, "Dual-mode stable performance phase-shifted full-bridge converter for wide-input and medium-power applications," *IEEE Transactions on Power Electronics*, vol. 36, no. 6, pp. 6375–6388, 2021. DOI: 10.1109/TPEL.2020.3033386.

[4] E. Rahimpour, J. Christian, K. Feser, and H. Mohseni, "Transfer function method to diagnose axial displacement and radial deformation of transformer windings," *IEEE Transactions on power delivery*, vol. 18, no. 2, pp. 493–505, 2003.

[5] S. Wang and F. C. Lee, "Analysis and applications of parasitic capacitance cancellation techniques for EMI suppression," *IEEE Transactions on Industrial Electronics*, vol. 57, no. 9, pp. 3109–3117, 2009.

[6] T. C. Neugebauer and D. J. Perreault, "Parasitic capacitance cancellation in filter inductors," *IEEE Transactions on Power Electronics*, vol. 21, no. 1, pp. 282–288, 2006.

[7] P. Rajeswari and V. Manikandan, "Analysis of effects of mosfet parasitic capacitance on non-synchronous buck converter electromagnetic emission," *Ain shams engineering Journal*, vol. 14, no. 8, p. 102 041, 2023.

[8] S. M. Sharma, A. Singh, S. Dasgupta, and M. V. Kartikeyan, "A review on the compact modeling of parasitic capacitance: From basic to advanced fets," *Journal of Computational Electronics*, vol. 19, no. 3, pp. 1116–1125, 2020.

[9] L. Keuck, "Entwurf eines einstufigen Ladewandlers auf basis eines LLC-Resonanzwandlers," Ph.D. dissertation, University Library Paderborn, 2023.

[10] H. Zhao, S. Luan, Z. Shen, A. J. Hanson, Y. Gao, *et al.*, "Rethinking basic assumptions for modeling parasitic capacitance in inductors," *IEEE Transactions on Power Electronics*, vol. 37, no. 7, pp. 8281–8289, 2021.

[11] B. Riddle, J. Baker-Jarvis, and J. Krupka, "Complex permittivity measurements of common plastics over variable temperatures," *IEEE Transactions on Microwave theory and techniques*, vol. 51, no. 3, pp. 727–733, 2003.

[12] A. J. Drake, K. J. Nowka, T. Y. Nguyen, J. L. Burns, and R. B. Brown, "Resonant clocking using distributed parasitic capacitance," *IEEE Journal of Solid-State Circuits*, vol. 39, no. 9, pp. 1520–1528, 2004.

[13] Z. Wang, H. Li, Z. Chu, C. Zhang, Z. Yang, *et al.*, "A review of EMI research in modular multilevel converter for HVDC applications," *IEEE Transactions on Power Electronics*, vol. 37, no. 12, pp. 14 482–14 498, 2022.

PCIM Europe 2024, 11– 13 June 2024, Nuremberg    DOI: 10.30420/566262124

# A 4.5 kV Fast Recovery Diode Platform for High-Current IGBTs

J. Vobecký[1], U. Vemulapati[2], M. Štencel[1], J. Hylský[1], L. Radvan[1], U. Meier[2], C. Corvasce[2]

[1] Hitachi Energy Czech Republic s.r.o., Novodvorská 1768/138a, Praha 4, Czech Republic
[2] Hitachi Energy Ltd., Fabrikstrasse 3, Lenzburg, Switzerland

Corresponding author: Jan Vobecky, jan.vobecky@hitachienergy.com

## Abstract

A scalable fast recovery diode platform of 4.5 kV class rated to 4 - 5 kA was developed and tested with the aim of increasing the current handling capability of the VSC-HVDC with MMC topology. An increased robustness is achieved using the diode wafers with diameter beyond 100 mm. They are mounted in hermetic housings with pole piece diameters of 110, 119, and 143 mm. The surge current capability then grows from 80 to 120 kA proportionally to device area likewise the maximal average ON-state current. A wide safe operating area (SOA) is demonstrated with turn-off current up to 6 kA, di/dt up to 10 kA/µs, $T_j$ up to 140 °C, and DC link voltage $V_{DC}$ up to 3.6 kV. This SOA can be further expanded depending on testing capability. Safe operation at low current densites at T = 25 °C and $V_{DC} \leq 3.6$ kV is proved as well. The cosmic ray withstanding capability is below 1 FIT up to 3.4 $kV_{DC}$ and 400 FIT at 3.6 $kV_{DC}$.

## 1 Introduction

The recent trend towards the voltage source converter (VSC) HVDC rated up to 8 GW has stimulated the development of 5kA Insulated Gate Bipolar Transistor (IGBT) [1]. To achieve so high rating, the fast recovery diode (FRD) chips are omitted and the IGBT package is fully filled by the IGBT chips. The diode chips are substituted by the new discrete 4.5 kV FRD with enhanced robustness to complement the new 5 kA IGBT StakPak from Hitachi Energy [2]. The advantage is massively increased surge current $I_{FSM}$ and overall robustness up to the DC link voltage of 3.6 kV (requested by the modular multilevel converter (MMC) topology) as presented in this paper.

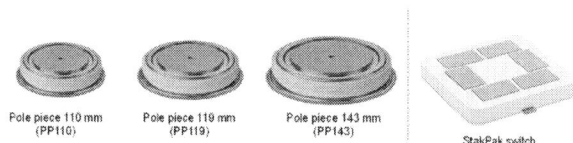

**Fig. 1:** Press-pack FRD platform overview and accompanying press-pack IGBT switch StakPak.

## 2 Large-Area Fast Recovery Diode Platform

As the robustness of a FRD grows with increasing

area of silicon wafer [3], the new FRD platform goes beyond the wafer diameter of 100 mm of our existing portfolio. The corresponding dimensions of housings are summarized in Fig.1 including the abbreviations of individual parts. The diode wafer is classical P-i-N diode with anode buffer for low leakage current and high turn-off robustness [4]. The cathode is equipped with segmented p-n structures for soft reverse recovery at low current densities and high DC link voltage $V_{DC}$ [5].

### 2.1 Static performance

A proper choice of resistivity and thickness of starting silicon is important for high $V_{DC}$ requested by the MMC topology.

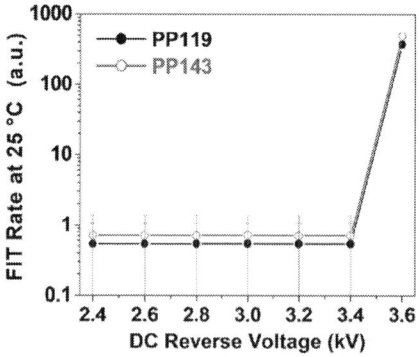

**Fig. 2:** The cosmic ray withstanding capability of diodes PP119 and PP143 tested at room temperature at sea level.

1002

Fig.2 demonstrates the cosmic ray withstanding capability using the FIT rate vs. $V_{DC}$ with diode area as a parameter. For the FIT rate below 1, the presented diodes with $V_{RRM}$ = 4.5 kV can be used up to $V_{DC}$ = 3.4 kV, which has been so far allowed for the diodes of 5.5 kV class.

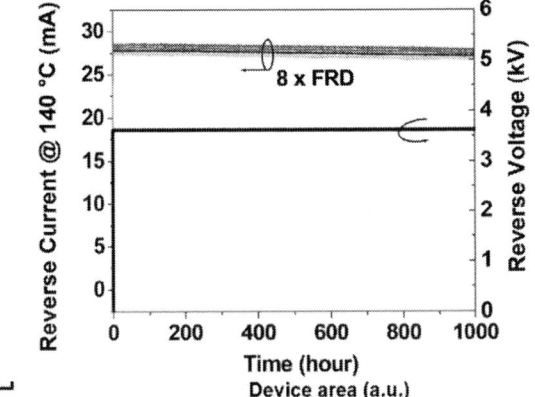

**Fig. 3:** High temperature reverse bias test (HTRB) of eight diodes PP119 at $V_R$ = 3.6 kV at T = 140 °C for the period of six weeks.

As the leakage current increases proportionally to the device area, a special attention has been paid to enable a stable operation up to 140 °C with margin. This has been achieved by work around the junction termination (negative bevel) and optimized dosing of proton and electron irradiation used for carrier lifetime control for soft reverse recovery and wide turn-off safe operation area up to the $V_{DC}$ = 3.6 kV. Fig.3 demonstrates the long-term blocking stability with $I_R$ < 30 mA at $V_{DC}$ = 3.6 kV achieved at $T_j$ = 140 °C.

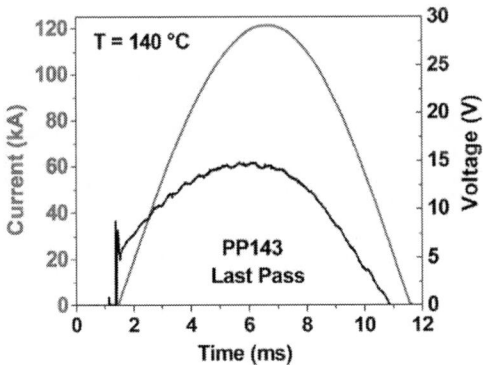

**Fig. 4:** Last pass surge current test for PP143 measured at 10 ms pulse width and starting $T_j$ = 140°C.

Surge current $I_{FSM}$ capability is the key parameter for clearing short circuit fault cases in the VSC-based MMC topology. Compared to the chip diodes in IGBT modules, discrete diodes can take higher surge current ($I_{FSM}$) without destruction and hereby simplify the necessary protection of a MMC cell. Fig.4 shows the measured last pass $I_{FSM}$ test at $I_F$ = 120 kA. Fig.5 illustrates the impact of diode area on the $I_{FSM}$ rating measured for typical ON-state voltage $V_F$ = 1.75 ± 0.05 V at 2.5 kA, $T_j$ = 140 °C and clamping force of $F_M$ = 90 kN, common for all three diodes of the platform. The growing $I_{FSM}$ capability with increasing diode area is obvious. If one needs a diode with $I_{FSM}$ above 100 kA at $T_{jmax}$ = 140 °C, the biggest diode of the presented platform must be chosen.

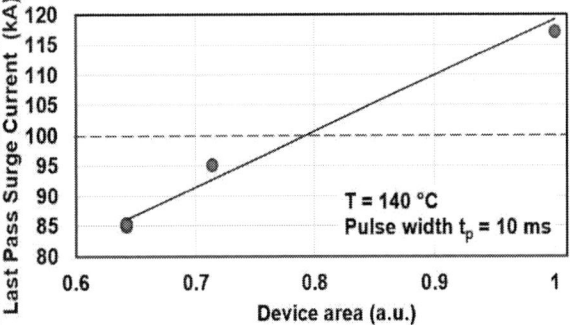

**Fig. 5:** Last pass magnitude of surge current $I_{FSM}$ vs. diode area. $t_p$ = 10 ms, $F_M$ = 90 kN, $T_j$ = 140 °C.

The presented FRD platform is attributed by identical $V_F$ for all three diode sizes. The platform provides scalability of both the load current under normal operation conditions ($I_{FAVmax}$) and the surge current $I_{FSM}$ when exposed to a short circuit event.

## 2.2 Dynamic performance at maximal junction temperature

The switching test under nominal operation conditions, which was performed with 4.5 kV / 5 kA IGBT StakPak module 5SMA 5000L450300 from Hitachi Energy [6, 7] at $T_{jmax}$ = 140 °C, is presented in Fig.6.

The switching test at the edge of the Safe Operation Area (SOA) due to the doubled di/dt at $T_{jmax}$ = 140 °C is shown on Fig.7. The diodes manage the peak power of 20 MW as illustrated for switching of diode PP119 from $V_{DC}$ = 3.4 kV and $I_F$ = 5 kA. The speed of switching at 9.4 kA/µs is at the limit of the tester ($R_g$ = 1 Ω) and is well beyond the requirements laid on a standard device operation.

PCIM Europe 2024, 11– 13 June 2024, Nuremberg   DOI: 10.30420/566262124

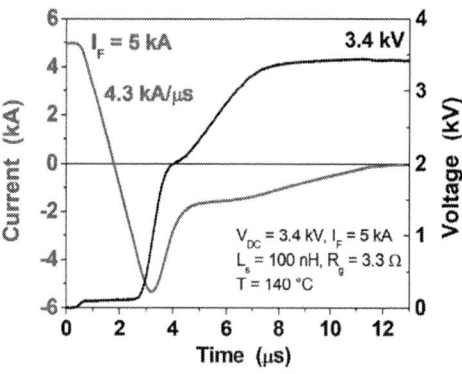

**Fig. 6:** Reverse recovery test of FRD PP119 with IGBT under nominal operation conditions: $V_{DC} = 3.4$ kV, $I_F = 5$ kA, di/dt = 4.3 kA/µs, $R_g = 3.3\ \Omega$, T = 140 °C.

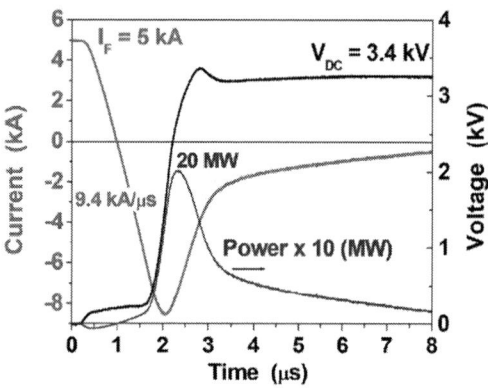

**Fig. 7:** SOA test of FRD PP119 with IGBT: $V_{DC} = 3.4$ kV, $I_F = 5$ kA, di/dt = 9.4 kA/µs, $R_g = 1\ \Omega$, T = 140 °C.

Fast switching with minimal stray inductance from nearly zero current at $V_{DC} = 3.6$ kV is demonstrated in Fig.8. The voltage overshoot is well below the diode rating voltage of 4.5 kV.

The single shot SOA is bounded by the DC link voltage $V_{DC} = 3.4$ kV for which the cosmic ray withstanding capability is below 1 FIT. From electrical viewpoint, the switching at $V_{DC} = 3.6$ kV has been proved safe by single-shot testing. However, one should consider the magnitude of FIT at the level of 400 for such case (see Fig.2).

Regarding the maximal current handling capability, the SOA is defined by $I_F = 6$ kA, which is rather the capability of available tester than that of the diode.

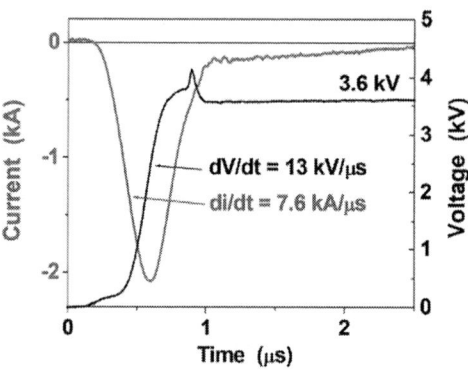

**Fig. 8:** Reverse recovery of diode PP119 under snap-off conditions: $I_F \approx 20$ A, $V_{DC} = 3.6$ kV, $I_F \approx 20$ A, di/dt = 7.6 kA/µs, $R_g = 1\Omega$, T = 140°C.

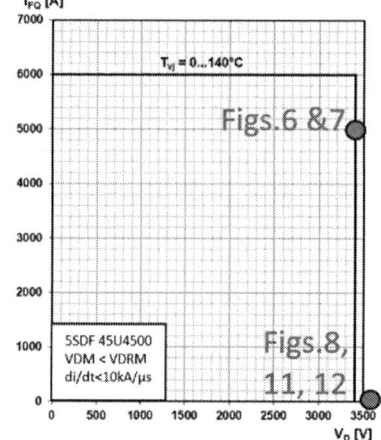

**Fig. 9:** Safe operation area at $T_{jmax} = 140$ °C.

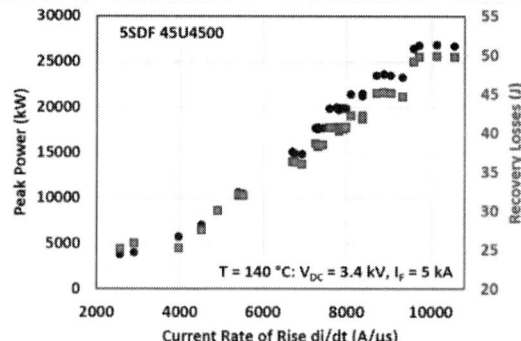

**Fig. 10:** Peak power and reverse recovery losses vs. commutation di/dt of FRD PP119 with 5SMA 5000L450300.

1004

The operation at high currents and at the same time at high di/dt should be realized with caution. The maximal recovery current of the diode, which grows with growing di/dt, can thermally overload the IGBT during turn-on. This is illustrated in Fig.10, which shows that the recovery losses of the diode double from 25 to 50 J after increasing the di/dt from 4 to 10 kA/µs. Under the same conditions, the peak power of the diode increases from 5 to 25 MW. This behavior depends on given IGBT design and may change according to the output characteristics (transconductance) and short circuit current under typical driving voltage $V_{GE}$. In case of the Fig.10, the short circuit current of IGBT was designed to $I_{SC}$ = 18 kA with the goal to minimize the $V_{CE\,sat}$ at $I_C$ = 5 kA for $V_{GE}$ = 15 V. In a hypothetical case of the short circuit current of the IGBT not above 10 kA, the recovery losses of the diode would be significantly reduced with the disadvantage of increased $V_{CE\,sat}$ of the IGBT. This shows the possibility of tuning the distribution of losses between the IGBT and diode to reach an optimum for a given application represented by the ON-state current, $V_{DC}$, and commutation di/dt.

## 2.3 Dynamic performance at room temperature

The reverse recovery at low temperature, high $V_{DC}$ and di/dt can be subject to oscillations of voltage and current (snappiness), when current density in the ON-state prior to turn-off gets close to zero. Resulting overvoltage can be danger for both the IGBT and diode. For the avoidance of the doubt about the diode robustness, and to justify the SOA in Fig. 9 down to $I_{FQ}$ = 0, $V_D$ = 3.6 kV at room temperature, the diodes were tested with the lowest available gate resistor $R_G$ = 1 Ω down to the lowest adjustable ON-state current of 20 A. The voltage and current waveforms for the hardest available test conditions are illustrated in Figs.11a) and b). In the Fig.11a), they were recorded for the pulse width of 150 µs and reveal the overvoltage between the $V_{DC}$ = 3.6 kV and $V_{RRM}$ = 4.5 kV. The magnitude of 4.5 kV is slightly overcome only for the lowest measurable ON-state current of 20 A. This corresponds to the current density in the order of 0.1 A/cm². For the narrow current pulses, represented by $t_p$ = 30 µs from Fig.11b), the overvoltage grows well beyond $V_{RRM}$ = 4.5 kV. Although all diodes and IGBTs passed such test conditions without a failure, it is obvious that the operation with $R_G$ = 1 Ω should be avoided for safety reasons. The overvoltage recorded for all operation conditions is summarized for the ON-state current between 0 and 1 kA in Fig.12.

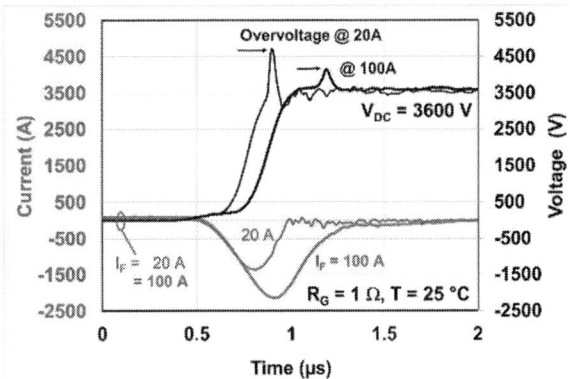

**Fig. 11a:** Reverse recovery of FRD PP119 with IGBT for pulse length of ON-state current $t_p$ = 150 µs. $V_{DC}$ = 3600 V, $R_G$ = 1 Ω, T = 25 °C.

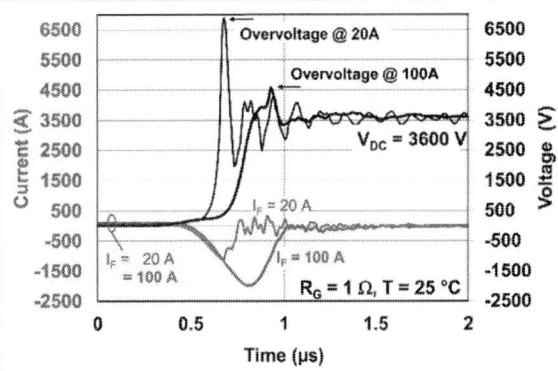

**Fig. 11b:** Reverse recovery of FRD PP119 for $t_p$ = 30 µs, $V_{DC}$ = 3.6 kV, $R_G$ = 1 Ω.

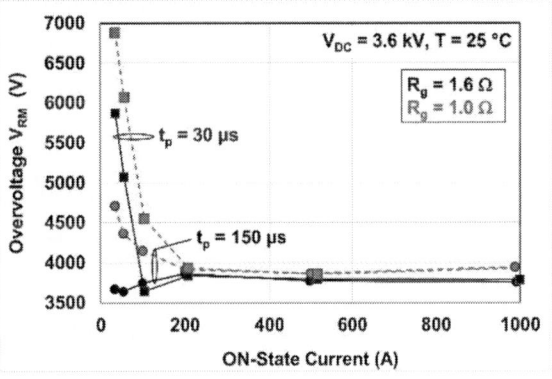

**Fig. 12:** Overvoltage vs. ON-state current during reverse recovery of FRD PP119 for pulse length $t_p$ = 30 and 150 µs, $V_{DC}$ = 3.6 kV, and $R_G$ = 1 and 1.6 Ω at room temperature.

## 2.4 Ratings

The average ON-state current for the sine wave reflecting the achieved ON-state voltage drop $V_F$ and thermal resistance $R_{th(j-c)}$ is summarized in Table 1 below. The presented magnitude of the surge current $I_{FSM}$ is the datasheet value which includes safety margin and is related to the maximal ON-state voltage drop of $V_{F\,max}$ = 1.9 V at $I_F$ = 2.5 kA and $T_{jmax}$ = 140 °C [8].

| Diode | $I_{FSM}$ (A) $T_j$ = 140°C $t_p$ = 10ms Half sine wave | $I_{FAVM}$ (A) | $V_{F\,typical}$ (V) $T_j$ = 140°C $I_F$ = 2.5 / 5.5kA | $R_{th(j-c)}$ (K.kW⁻¹) |
|---|---|---|---|---|
| PP110 | 75 | 4000 | 1.8 / 2.7 | 5.5 |
| PP119 | 80 | 4400 | 1.8 / 2.7 | 4.3 |
| PP143 | 100 | 5000 | 1.8 / 2.7 | 3.3 |

**Table 1**  Key device ratings of the 4.5 kV FRD platform. $T_{case}$ = 55 °C.

## 3 Conclusions

The key numbers and facts of the new 4.5kV large-area fast recovery diode platform for the operation with 5 kA IGBT from Hitachi Energy were presented.

The platform provides scalability in surge current $I_{FSM}$ and maximal average ON-state current $I_{FAVmax}$ while having the same typical ON-state voltage $V_{F\,typ}$ = 1.8 V measured at ON-state current $I_F$ = 2.5 kA and $V_{F\,typ}$ = 2.7 V at $I_F$ = 5.5 kA and at junction temperature of $T_j$ = 140 °C for all three diode sizes.

The excellent cosmic ray withstanding capability below 1 FIT up to the $V_{DC}$ = 3.4 kV and 400 FIT at 3.6 k$V_{DC}$ together with the overall switching robustness verified up to $V_{DC}$ = 3.6 kV is the prerequisite for reliable operation in a 5 kA MMC cell for application in HVDC. The operation at $V_{DC}$ = 3.6 kV is considered for overload conditions, for example when a single stage of the MMC cell is subject to short circuit failure.

The presented diode will be primarily used in the VSC with the MMC topology for 8 GW HVDC, which will benefit from the presented robustness both under surge current conditions as well as fast turn-off. Beside the presented operation with the IGBT StakPak, the diodes passed testing with IEGT and IGCT up to the same DC link voltage and maximal junction temperature as presented for the IGBT.

The production of the 5SDF 45U4500 article, denoted as PP119 in this paper, started in December 2023. The datasheet is available in electronic form at [8].

## Acknowledgment

The Bipolar production line in Lenzburg, Switzerland is acknowledged for processing of devices. Jeremy Jones is acknowledged for the application testing of diodes with the IGBT StakPak.

## References

[1] J. Vobecky, U. Vemulapati, T. Wikström, B. Boksteen, F. Dugal, T. Stiasny, C. Corvasce, "Recent Progress in Silicon Devices for Ultra-High Power Applications," Proc. Int. Electron Devices Meeting 2021, San Francisco, p. IEDM21-773.

[2] E. Tsyplakov, G. Gupta, J. Vobecky. J. Jones, B. Boksteen, L. D. Michelis, C. Winter., M. Chen, "New generation high power semiconductors for 8GW VSC-HVDC applications," Proc. PCIM Asia 2023, Shanghai, 2023, p. 354.

[3] J. Vobecky, "Fast Recovery Diodes for High-Current High-Voltage Insulated Gate Bipolar Transistors," IEEE Electron Device Letters, 2022, Vol. 43, No.8, p.1311.

[4] J. Vobecky, R. Siegrist, M. Arnold, K. Tugan, "Large Area Fast Recovery Diode with very high SOA Capability for IGCT Applications," Proc. PCIM Europe 2011, Nuremberg, p.44.

[5] U. Vemulapati, T. Wikström, M. Lüscher, "An RC-IGCT for application at up to 5.3 kV," Proc. 10th Int. Conference on Power Electronics ICPE 2019-ECCE Asia, Busan, p.787.

[6] 5SMA 5000L450300 StakPak IGBT module, 5SYA 1509-03 09-2022, to be published

[7] J. Jones, G. Gupta, B. Boksteen, E. Tsyplakov, M. Chen, L. De Michelis, G. Paques, Next Generation 4.5 kV IGBT-only StakPak Module with Reduced Losses and High Temperature Capability, to be published at PCIM Europe 2024, Nuremberg.

[8] Fast recovery diodes | Hitachi Energy

PCIM Europe 2024, 11– 13 June 2024, Nuremberg        DOI: 10.30420/566262125

# 6.5 kV Innovative Silicon Power Device (i-Si) Module with High Power Density and Low Loss by Stored Carrier Control

Takashi Hirao ©[1], Tomoyuki Miyoshi[1], Hiroshi Suzuki[1], Yusuke Takada[1], Tomoyasu Furukawa[1], Tsubasa Moritsuka[2], Masaki Shiraishi[2], Isamu Yoshida[2], Takayuki Kushima[2], Yusuke Kanno[3], Yasuhiko Kono[3], Katsumi Ishikawa[3], Mutsuhiro Mori[3]

[1] Research & Development Group, Hitachi, Ltd., Japan
[2] Hitachi Power Semiconductor Device, Ltd., Japan
[3] Railway Systems Business Unit, Hitachi, Ltd., Japan

Corresponding author:    Takashi Hirao, takashi.hirao.nd@hitachi.com
Speaker:                 Takashi Hirao, takashi.hirao.nd@hitachi.com

## Abstract

This paper describes a demonstration of the low loss performance of a 6.5 kV innovative silicon power device (i-Si) module. The i-Si module uses stored carrier control to both the switching device and free wheel diode, which is made entirely from silicon. A 6.5 kV 800 A i-Si module was fabricated at a size of 130 mm × 140 mm. Despite having a package size 2/3 that of a conventional 6.5 kV 750 A IGBT, the fabricated i-Si module has 40% lower loss, similar to that of SiC MOSFETs.

## 1   Introduction

Power semiconductor devices play a crucial role in power conversion. While the application of wide bandgap semiconductors such as SiC and GaN is expanding, silicon is still widely used as a material for power semiconductor devices [1], [2]. Specifically, power modules combining silicon IGBTs and PiN diodes are extensively utilized in high voltage fields. IGBTs and PiN diodes exhibit low conduction loss through conductivity modulation with the stored carriers in the drift layer. We have developed low loss silicon IGBTs and PiN diodes [3]–[6]. However, constraints limit the reduction in switching losses of silicon IGBTs and PiN diodes due to the stored carriers. To reduce switching losses while maintaining low conduction losses through conductivity modulation, some researchers have investigated stored carrier control [7]–[9]. We have proposed an innovative silicon power device (i-Si) to reduce losses further by controlling the stored carriers [10]–[12]. The i-Si module has multiple gate terminals to control the stored carriers. Determining the electrical characteristics of modules controlled by the gate terminals is crucial for accurately estimating the losses.

This paper describes a demonstration of a 6.5 kV

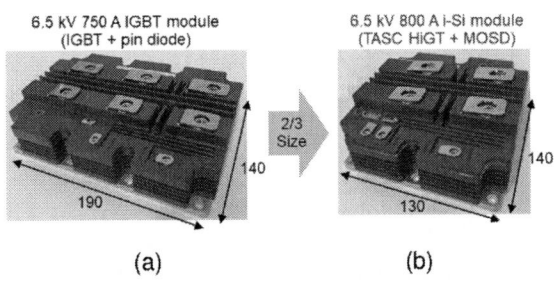

**Fig. 1:** External view of power modules. (a) 6.5 kV 750 A IGBT module, and (b) 6.5 kV 800 A i-Si module.

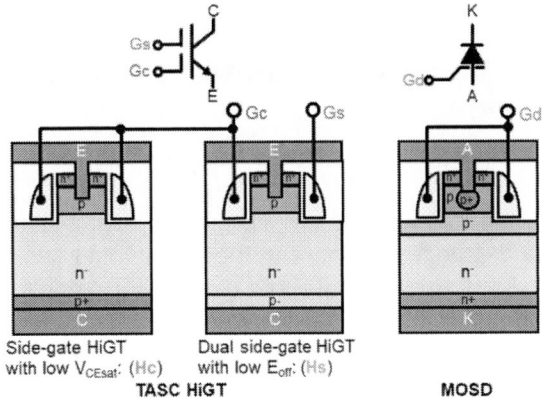

**Fig. 2:** Device configuration and defined circuit symbols of the TASC HiGT and MOSD.

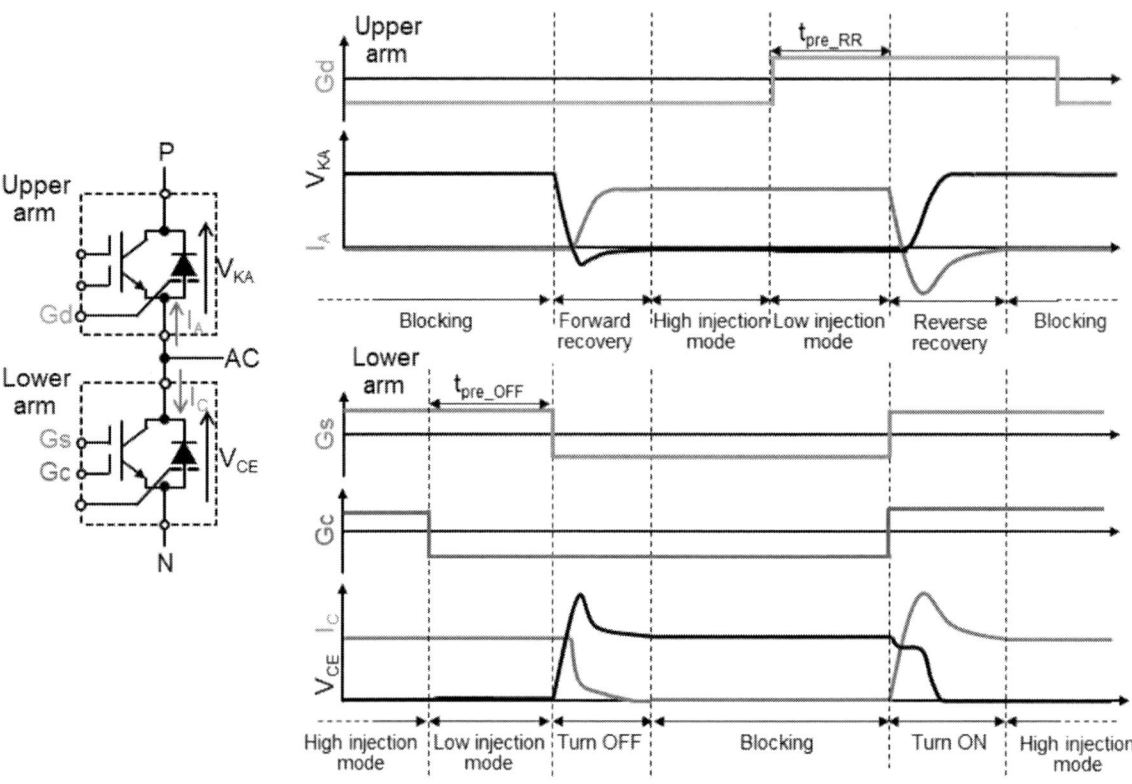

**Fig. 3:** Operation timing chart of the TASC HiGT and MOSD.

800 A i-Si module. First, the operation sequence is explained, focusing on the stored carrier control. Next, the measurement results of the operation are presented. The static characteristics for each voltage pattern of multiple gate terminals are shown. The transient characteristics prior to the preparation period before switching are also displayed. Additionally, the switching waveforms are presented. Finally, a calculation of the inverter losses based on the measurement results is shown.

## 2 Module Configuration and Operation

Figure 1(a) and (b) shows the external views of a 6.5 kV 750 A IGBT module and 6.5 kV 800 A i-Si module, respectively. Despite having the same rated voltage and current, the i-Si module has a package size that is 2/3 that of the IGBT module. Figure 2 illustrates the device configuration in the i-Si module. The switching device in the i-Si module is a dual side-gate high conductivity IGBT through the time and space control of a stored carrier (TASC HiGT). The TASC HiGT has two types of chips: a high conductivity HiGT chip (Hc) and a high speed HiGT chip (Hs) with two gate terminals, Gc and Gs. The free wheel diode (FWD) in the i-Si module is a MOS controllable stored-carrier diode (MOSD) [12]. The MOSD has one gate terminal, Gd. The circuit symbols of the TASC HiGT and MOSD are defined as shown in the top of Fig. 2.

Figure 3 shows the operation timing chart of the TASC HiGT and MOSD. In this chart, the TASC HiGT in the lower arm and the MOSD in the upper arm each conduct. When the IGBT conducts, both the Gs and Gc are ON (high injection mode). The chart has the preparation time $t_{pre\_OFF}$ prior to turning OFF (low injection mode). During $t_{pre\_OFF}$, the Gs is ON, and the Gc is OFF to stop conduction of the Hc chip and to reduce the stored carriers in the Hs chip. As a result, the TASC HiGT turns OFF under conditions of low stored carriers to reduce the turn OFF loss $E_{OFF}$. When the MOSD conducts, the Gd is OFF (high injection mode). The chart has the preparation time $t_{pre\_RR}$ prior to reverse recovery (low injection mode). During $t_{pre\_RR}$, Gd is OFF to reduce the stored carriers of the MOSD. As a result, the reverse recovery current of the MOSD is decreased to reduce the recovery loss $E_{RR}$.

PCIM Europe 2024, 11– 13 June 2024, Nuremberg          DOI: 10.30420/566262125

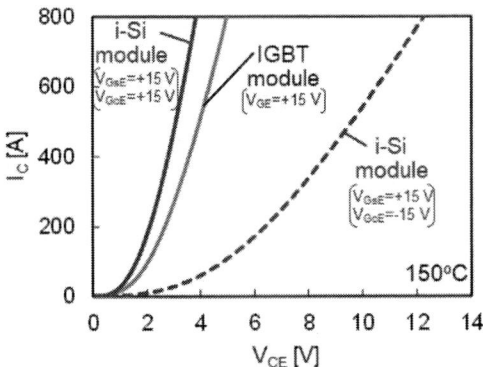

**Fig. 4:** Experimental current-voltage characteristics of switching devices in the IGBT and i-Si modules under the conditions of 150 °C, 800 A, and 3600 V.

# 3 Experimental Results

## 3.1 Static Characteristics

Figure 4 shows the experimental current-voltage characteristics of switching devices in the IGBT and i-Si modules. In Fig. 4, the i-Si module is shown to have two lines. The solid line indicates the TASC HiGT characteristics when both the Gs and Gc were ON ($V_{GsE}$=+15 V, $V_{GcE}$=+15 V). Despite being 2/3 the size, the i-Si module had a lower $V_{CEsat}$ than the IGBT module. The dotted line indicates the TASC HiGT characteristics when the Gs was ON and the Gc was OFF ($V_{GsE}$=+15 V, $V_{GcE}$=−15 V). An increase in $V_{CEsat}$ indicates a decrease in the stored carriers.

Figure 5 shows the experimental current-voltage characteristics of FWD in IGBT and i-Si modules. In Fig. 5, the i-Si module is shown to have three lines. The solid line indicates the MOSD characteristics when the Gd was OFF ($V_{GdA}$=−15 V). Despite being 2/3 the size, the i-Si module had a lower $V_F$ than that of the IGBT module. The dotted lines indicate the MOSD characteristics when $V_{GdA}$=0 V and the Gd was ON ($V_{GdA}$=+15 V). The curve is the same as the solid line when $V_{GdA}$=0 V, but $V_F$ was large when $V_{GdA}$=+15 V, indicating a decrease in the stored carriers.

## 3.2 Dynamic Characteristics

Both the preparation period before switching ($t_{pre\_OFF}$, $t_{pre\_RR}$) and the actual switching period had to be taken into account in the dynamic characteristics of the i-Si module. The loss during the preparation period was small, but it could not be ignored to estimate the loss of the i-Si module ac-

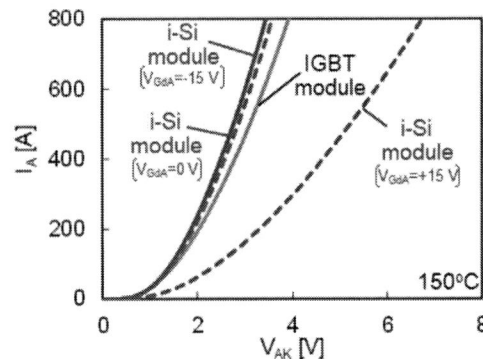

**Fig. 5:** Experimental current-voltage characteristics of the FWD in the IGBT and i-Si modules under the conditions of 150 °C, 800 A, and 3600 V.

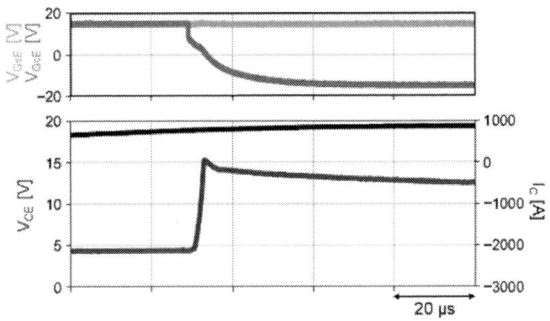

**Fig. 6:** Experimental waveforms of the preparation period before turn OFF under the conditions of 150 °C and 800 A.

curately.

Figure 6 shows the experimental waveforms of the preparation period before turn OFF. When both $V_{GsE}$ and $V_{GcE}$ were +15 V, $V_{CE}$ was the voltage corresponding to the static current-voltage curve (solid line) shown in Fig. 4. When $V_{GcE}$ decreased from +15 V to −15 V, the i-Si module transitioned into a preparation period $t_{pre\_OFF}$. At this time, a small voltage spike occurs in $V_{CE}$ and it gradually converged to the voltage corresponding to the static current-voltage curve (dotted line) shown in Fig. 4. Figure 7 shows the experimental waveforms of the preparation period before reverse recovery. When $V_{GdA}$=−15 V, $V_{AK}$ was the voltage corresponding to the current-voltage curve (solid line) shown in Figure 5. When $V_{GdA}$ increased from −15 V to +15 V, the i-Si module transitioned into a preparation period $t_{pre\_RR}$. At this time, $V_{AK}$ gradually converged to the voltage corresponding to the static current-voltage curve (dotted line) shown in Fig. 5.

Figure 8 shows the experimental $t_{pre\_OFF}$ depen-

1009

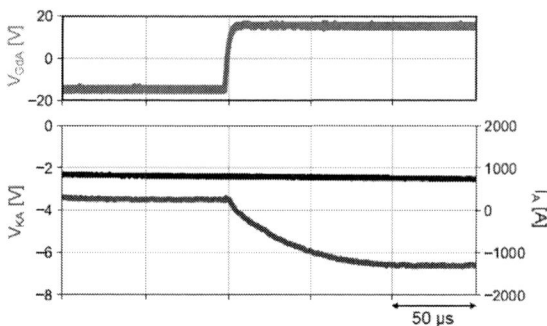

**Fig. 7:** Experimental waveforms of the preparation period before reverse recovery under the conditions of 150 °C and 800 A.

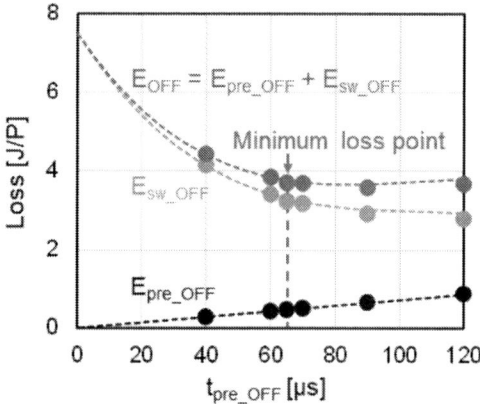

**Fig. 8:** Experimental $t_{pre\_OFF}$ dependence on the switching losses $E_{OFF}$, $E_{pre\_OFF}$ and $E_{SW\_OFF}$ under the conditions of 150 °C, 800 A, and 3600 V.

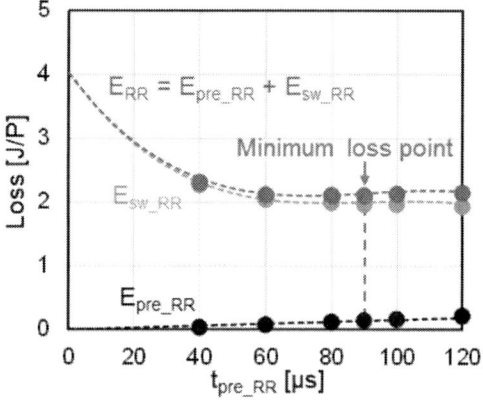

**Fig. 9:** Experimental $t_{pre\_RR}$ dependence on the switching losses $E_{RR}$, $E_{pre\_RR}$ and $E_{SW\_RR}$ under the conditions of 150 °C, 800 A, and 3600 V.

dence on the switching losses $E_{OFF}$, $E_{pre\_OFF}$ and $E_{SW\_OFF}$. $E_{pre\_OFF}$ was the loss during $t_{pre\_OFF}$, and $E_{SW\_OFF}$ was the loss during the actual turn OFF switching.

As shown in Eq. 1, the total turn-off loss $E_{OFF}$ is the sum of the loss during the preparation period and the loss during the actual turn-off period.

$$E_{OFF} = E_{pre\_OFF} + E_{SW\_OFF}, \quad (1)$$

$$\text{where } E_{pre\_OFF} = \int_{t_0}^{t_1} V_{CE}(t)\, I_C(t)\, dt$$
$$- V_{CE}(t_0)\, I_C(t_0)\, t_{pre\_OFF},$$
$$E_{SW\_OFF} = \int_{t_1}^{t_2} V_{CE}(t)\, I_C(t)\, dt,$$
$$t_1 - t_0 = t_{pre\_OFF},\ t_2 - t_1 = t_{SW\_OFF}.$$

$t_{SW\_OFF}$ is the switching time excluding $t_{pre\_OFF}$. Notably, the loss $V_{CE}(t_0)\, I_C(t_0)\, t_{pre\_OFF}$ during this period was eliminated in $E_{pre\_OFF}$ because this was already counted as the conduction loss.

When $t_{pre\_OFF}$ increased, $E_{SW\_OFF}$ decreased, while $E_{pre\_OFF}$ increases. As depicted in Fig. 8, $E_{OFF}$ was minimized when $t_{pre\_OFF}$=65 µs.

Figure 9 shows the experimental $t_{pre\_RR}$ dependence on the switching losses $E_{RR}$, $E_{pre\_RR}$ and $E_{SW\_RR}$. $E_{pre\_RR}$ was the loss during $t_{pre\_RR}$, and $E_{SW\_RR}$ was the loss during the actual reverse recovery.

Similarly, as shown in Eq. 2, recovery loss $E_{RR}$ is the sum of the loss during the preparation period and the loss during the actual reverse recovery period.

$$E_{RR} = E_{pre\_RR} + E_{SW\_RR}, \quad (2)$$

$$\text{where, } E_{pre\_RR} = \int_{t_3}^{t_4} V_{KA}(t)\, I_A(t)\, dt$$
$$- V_{KA}(t_0)\, I_A(t_0)\, t_{pre\_RR},$$
$$E_{SW\_RR} = \int_{t_4}^{t_5} V_{KA}(t)\, I_A(t)\, dt,$$
$$t_4 - t_3 = t_{pre\_RR},\ t_5 - t_4 = t_{SW\_RR}.$$

$t_{SW\_RR}$ is the switching time excluding $t_{pre\_RR}$. Notably, the loss $V_{KA}(t_0)\, I_A(t_0)\, t_{pre\_RR}$ during this period was eliminated in $E_{pre\_RR}$ because this was already counted as the conduction loss.

When $t_{pre\_RR}$ increased, $E_{SW\_RR}$ decreased, while $E_{pre\_RR}$ increased. As depicted in Fig. 9, $E_{RR}$ was minimized when $t_{pre\_RR}$=90 µs.

Figure 10(a)–(c) shows the experimental switching waveforms of the IGBT and i-Si modules. The preparation times $t_{pre\_OFF}$ and $t_{pre\_RR}$ were set to 65 and 90 µs, respectively. In Figure 10(a), the hatched area indicates the tail current due to the stored carriers in the switching devices. The turn OFF waveforms indicate the i-Si module had 35% loss in the tail current period ($E_{tail\_OFF}$) compared with the IGBT module. In total, the turn OFF loss $E_{OFF}$ was reduced by 32%. Notably, the $E_{OFF}$ of the i-Si module included the loss during the $t_{pre\_OFF}$ period, as shown in Eq. 1. In Fig. 10(b), the hatched area indicates the Miller effect period in which the current flowed between the collector and gate. While the IGBT module had a planar gate with a large reverse transfer capacitance Cres, the i-Si module had a side-gate with a small Cres, so the Miller effect was small [6]. The turn ON waveforms indicate the i-Si module had 31% loss during the Miller effect period ($E_{Miller\_ON}$) compared with the IGBT module. Furthermore, by adding the recovery current reduction described next, the turn ON loss $E_{ON}$ was reduced by 41%. In Fig. 10(c), the hatched area indicates the recovery current due to the stored carriers in the FWDs. The recovery waveforms indicate the i-Si module had a 57% loss during the reverse recovery period ($E_{SW\_RR}$). In total, the recovery loss $E_{RR}$ was reduced by 38%, including the loss during the $t_{pre\_RR}$ period, as shown in Eq. 2.

## 4 Inverter Loss Calculation

Figure 11(a) and (b) shows the results of an inverter loss calculation for the IGBT and i-Si modules. For the calculation, we used the the experimental loss data shown in Figs. 4, 5, 8, and 9. The power loss of the i-Si module was 40% lower in the motoring mode and 39% lower in the generating mode than that of the IGBT module. Importantly, the i-Si module had package size 2/3 that of the IGBT module, as shown in Fig. 1(a) and (b).

Figure 12(a) and (b) shows carrier frequency $f_C$ dependency of the inverter loss. The power loss ratio is defined as the ratio to the power loss of IGBT module. The curves of the IGBT and i-Si module were calculated in the same way as those in Fig. 11(a) and (b). The 6.5 kV SiC module has not yet been commercialized, so the curve of SiC MOSFETs was calculated using the simulation characteristics. For the simulation, we chose a DMOS, which is commonly used for high voltage applications. The results indicated that the i-Si

module made entirely from silicon has low loss over a wide $f_C$, similar to a SiC MOSFET.

## 5 Conclusion

This paper described a demonstration of the low loss performance of a 6.5 kV innovative silicon power device (i-Si) module. The i-Si module uses the stored carrier control to both the switching device and free wheel diode, which is made entirely from silicon. A 6.5 kV 800 A i-Si module was fabricated with a size of 130 mm × 140 mm. The preparation period before the actual switching was investigated to estimate the loss accurately. Despite having a package size 2/3 that of a conventional 6.5 kV 750 A IGBT, the fabricated i-Si module has 40% lower loss, similar to that of a SiC MOSFET.

## References

[1] T. Laska, "Progress in Si IGBT technology–as an ongoing competition with WBG power devices," in *2019 IEEE International Electron Devices Meeting (IEDM)*, IEEE, 2019, pp. 12.2.1–4.

[2] N. Iwamuro and T. Laska, "IGBT history, state-of-the-art, and future prospects," *IEEE Transactions on Electron Devices*, vol. 64, no. 3, pp. 741–752, 2017.

[3] M. Mori, H. Kobayashi, and Y. Yasuda, "6.5 kV ultra soft & fast recovery diode (U-SFD) with high reverse recovery capability," in *Proc. 12th International Symposium on Power Semiconductor Devices & ICs (ISPSD)*, IEEE, 2000, pp. 115–118.

[4] M. Mori, K. Oyama, T. Arai, J. Sakano, Y. Nishimura, *et al.*, "A planar-gate high-conductivity IGBT (HiGT) with hole-barrier layer," *IEEE transactions on Electron Devices*, vol. 54, no. 6, pp. 1515–1520, 2007.

[5] M. Mori, K. Oyama, Y. Kohno, J. Sakano, J. Uruno, *et al.*, "A trench-gate high-conductivity IGBT (HiGT) with short-circuit capability," *IEEE transactions on Electron Devices*, vol. 54, no. 8, pp. 2011–2016, 2007.

[6] M. Shiraishi, T. Furukawa, S. Watanabe, T. Arai, and M. Mori, "Side gate HiGT with low dv/dt noise and low loss," in *Proc. 28th International Symposium on Power Semiconductor Devices and ICs (ISPSD)*, IEEE, 2016, pp. 199–202.

[7] M. Sumitomo, H. Sakane, K. Arakawa, Y. Higuchi, and M. Matsui, "Injection control technique for high speed switching with a double gate PNM-IGBT," in *Proc. 25th International Symposium on Power Semiconductor Devices & IC's (ISPSD)*, IEEE, 2013, pp. 33–36.

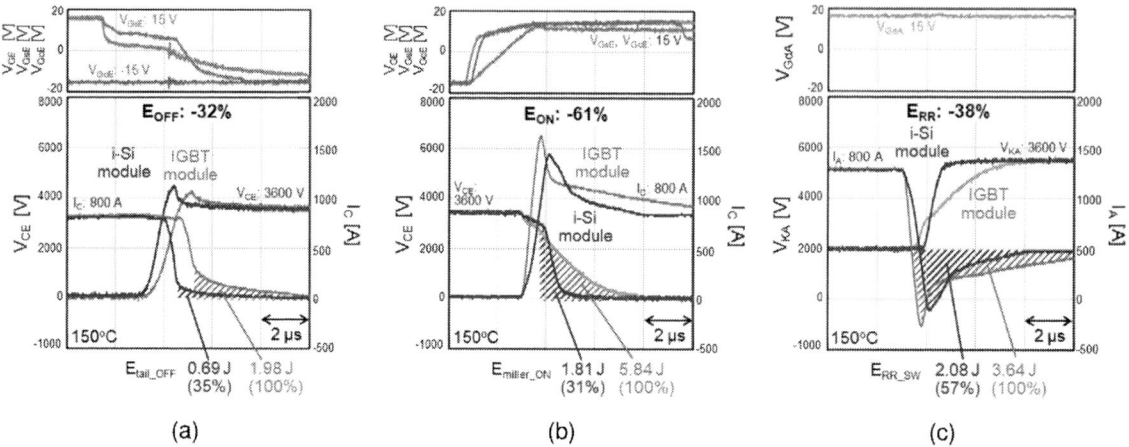

**Fig. 10:** Experimental switching waveforms of IGBT and i-Si modules under the conditions of 150 °C, 800 A, and 3600 V. (a) Turn OFF, (b) turn ON, and (c) reverse recovery.

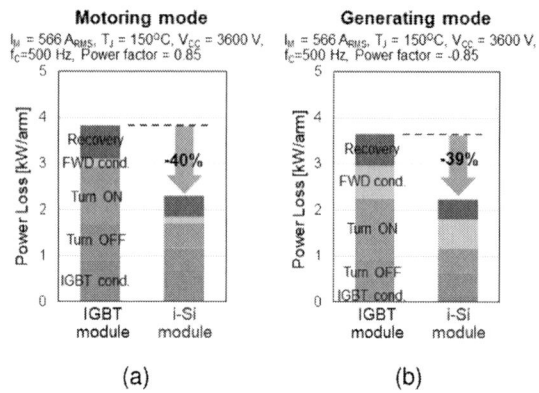

**Fig. 11:** Inverter loss calculation in (a) motoring and (b) generating modes under the conditions of junction temperature $T_J$ of 150 °C, DC bus voltage $V_{CC}$ of 3600 V, motor current $I_M$ of 566 $A_{RMS}$, and carrier frequency $f_C$ of 500 Hz

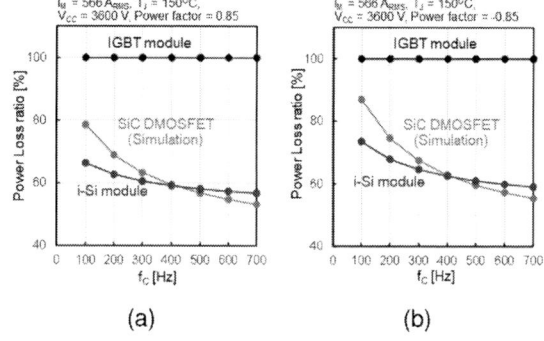

**Fig. 12:** Carrier frequency $f_C$ dependency of inverter loss ratio to IGBT module in (a) motoring and (b) generating modes.

[8] T. Saraya, K. Itou, T. Takakura, M. Fukui, S. Suzuki, *et al.*, "3.3 kV back-gate-controlled IGBT (BC-IGBT) using manufacturable double-side process technology," in *Proc. IEEE International Electron Devices Meeting (IEDM)*, IEEE, 2020, pp. 5.3.1–5.3.4.

[9] Y. Kobayashi, M. Fukui, T. Matsudai, T. Saraya, K. Itou, *et al.*, "Single-back and double-front gate-controlled IGBT for achieving low turn-off loss," in *Proc. 35th International Symposium on Power Semiconductor Devices and ICs (ISPSD)*, IEEE, 2023, pp. 207–210.

[10] M. Mori, T. Miyoshi, T. Furukawa, Y. Takeuchi, Y. Hotta, and M. Shiraishi, "An innovative silicon power device (i-Si) through time and space control of a stored carrier (TASC)," in *Proc. 30th In-*

*ternational Symposium on Power Semiconductor Devices and ICs (ISPSD)*, IEEE, 2018, pp. 520–523.

[11] T. Miyoshi, H. Suzuki, T. Furukawa, S. Watanabe, M. Shiraishi, *et al.*, "A novel 6.5 kV innovative silicon power device (i-Si) with a digital carrier control drive (DCC-drive)," in *Proc. 32nd International Symposium on Power Semiconductor Devices and ICs (ISPSD)*, IEEE, 2020, pp. 46–49.

[12] T. Miyoshi, H. Suzuki, T. Hirao, Y. Takada, T. Furukawa, *et al.*, "A novel MOS controllable stored-carrier diode (MOSD) for an innovative silicon power device (i-Si)," in *2024 36th International Symposium on Power Semiconductor Devices and ICs (ISPSD)*, IEEE, 2024.

PCIM Europe 2024, 11– 13 June 2024, Nuremberg     DOI: 10.30420/566262126

# High Current Density 4.5 kV PressPack IGBTs Push SOA Limits.

Hossein Davoodi[1] ⦿, Peter Waind[1], Paolo Mirone[1], Liutauras Storasta[1], Paul Hailes[2], Julian Pitman[2]

[1] Littelfuse, Germany

[2] Littelfuse, UK

Corresponding author:      Hossein Davoodi, hdavoodi@littelfuse.com
Speaker:                   Hossein Davoodi, hdavoodi@littelfuse.com

## Abstract

Hermetic pressure contact PressPack IGBTs are getting more attention due to their exceptional performance in harsh environmental conditions and long-term reliability. Therefore, demands for devices with higher power rating and improved safe-operating area is ever-increasing. The present work demonstrates a new 4.5 kV PressPack IGBT by Littelfuse which achieves exceptional performance in terms of standard working conditions, SOA, and reliability. Both chip and PressPack developments are discussed in detail and, finally, the experimental results in the PressPack level are analyzed. It will be shown that the chip exhibits a short circuit ruggedness of seven times the nominal current and reverse bias ruggedness of four times the nominal current in a single chip package. Furthermore, when paralleling 34 IGBTs and 18 anti-parallel diodes in a PressPack, the ruggedness is more than twice the nominal current for reverse bias and more than four times the nominal current in short-circuit tests, respectively.

## 1   Introduction

Insulated Gate Bipolar Transistors (IGBTs) have been at the forefront of the electrification in past decades. Demands for more reliable and robust IGBTs for Medium-Voltage Drives (MVD), traction, and High-Voltage DC (HVDC) applications have fueled further developments in both chip and package levels. Thanks to their design, PressPack IGBTs inherit several advantages compared to other counterparts. Among them, hermiticity of the package and therefore robustness to the harsh environmental conditions [1], double sided cooling, stacking possibility, improved symmetry, explosion-free failure mode [2], and lower stray inductance are the most interesting features from the application point of view. Consequently, these features provide more degrees of freedom for the chip designer. Nevertheless, increasing the current rating of the PressPack by increasing the total number of paralleled chips (such as 34 IGBT chips and 18 anti-parallel diodes for 2 kA, 4.5 kV rating in this case) imposes new challenges especially when considering safe operating area [3, 4]. This implies that not only a single chip should be robust enough, but also chip-to-chip variation should be kept to minimum. In other words, the chip design should exhibit a high robustness to process variations. Taking these challenges and opportunities into account, this work presents a newly developed 4.5 kV IGBT and diode chip-set and discuss the most important features of their design enabling an exceptional robustness of HV PressPacks. The discussion starts with a brief review of the chip structures and design. In the next section, the electrical performance of both IGBT and diode are evaluated experimentally and discussed with the support of TCAD simulations. Lastly, we review the PressPack design, and the experimental results of the full chip array designed for 2 kA.

## 2   Chip design and evaluation

### 2.1   IGBT structure

Relying on the well-established Littelfuse XPT™ technology, a new generation of HV-IGBTs (Fig. 1) was developed by adding new features to further improve the cell performance. An enhancement layer [5] has been re-engineered to increase cell avalanche capability during turn-off, reduce the collector-emitter voltage, hence on-state power loss, and precisely control the channel length.

One of the main challenges during the turn-off and in the absence of the channel current is to extract the holes under the termination region. If not carefully mitigated, the ruggedness is limited by filaments around the periphery of the active area. For this device, the periphery of the active area and gate runner, have been redesigned to improve the transition to the termination region. Accordingly, auxiliary paths for draining holes during turn-off is created, which permit a superior RBSOA capability.

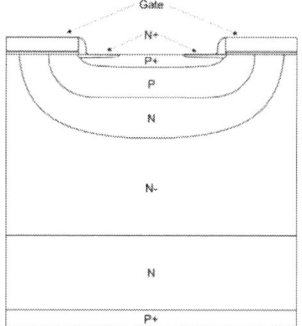

**Fig. 1**    Cross-section of the newly developed HV-IGBT cell.

Latch-up of the parasitic thyristor during turn-off is another failure mechanism which influences the IGBT's safe operating area in reverse bias mode (RBSOA). Therefore, the P+ profile underneath the source has been adjusted to prevent parasitic thyristor latch-up and improve the turn-off capability. The process integration to create and control the P+ profile so that, on one hand, it does not affect the channel properties and, on the other hand, provides maximum protection for the source, was particularly challenging. As shown in Fig. 2, TCAD simulations show the electrons and holes current path in on-state conditions. The images confirm that the newly integrated P+ layer reduces the resistance of the holes' path and improve cell immunity to the latch-up phenomenon. The experimental results in the next session will highlight the increased capability of the SOA.

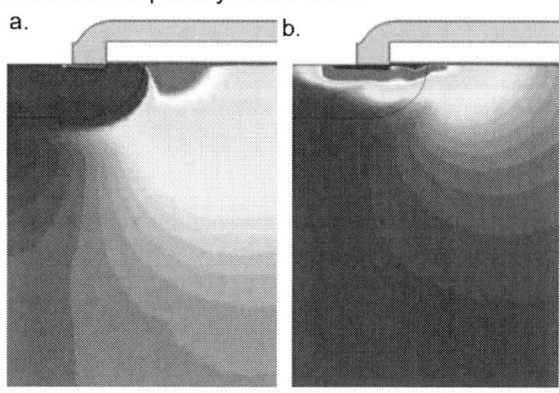

**Fig. 2**    Simulation of the electron (a) and hole (b) current densities in a half-cell during on-state.

In the anode side, the bipolar gain of the IGBT was adjusted by specially designing the collector and the deep-diffused buffer layers. Besides, total channel width was set accordingly to satisfy the power handling of the chip. Optimizing these parameters was a critical step to shape the electric field inside the chip under different working conditions, especially SCSOA [6, 7]. Process and device simulations were conducted using TCAD to properly tailor the design and observe the impact of individual features on the overall performance of the cell. The device simulation results are presented in the next section together with the measurements.

## 2.2 Diode structure

The HV-Sonic diode is designed using a combination of homogeneous and local lifetime control techniques for optimum shaping of the electron-hole plasma profile inside the drift region. In this way, low conduction losses can be achieved without sacrificing the switching performance and smooth current waveform even in the harshest conditions. Lifetime killing is used to control the injection efficiency to optimize the trade-off between reverse recovery energy and softness. Besides, a homogeneous electron irradiation is added to adjust the tail current.

Both IGBT and diode chips benefit from a new compressive Silicon Nitride as the secondary passivation scheme which has reduced the layer stress and enhanced the reliability of the chip.

## 2.3 IGBT characterization

The IGBT chip has been designed with a square footprint, total area of 205 mm$^2$ and active area of 107 mm$^2$, rated for 57 A nominal current. The output characteristic of the device at $V_{ge}$=15 V for both simulation and measurements are plotted in Fig. 3. Simulation and experimental curves are completely fitting up to 120 A where the behavior in the third dimension starts to be significant. Accordingly, the single IGBT chip has a relatively low forward voltage of $V_{cesat}$ = 2.8 V as well as low saturation current of $I_{csat}$ = 180 A. Threshold voltage of the IGBT is $V_{GETH}$ = 6 V.

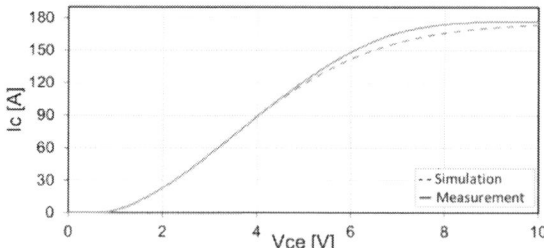

**Fig. 3**    Output characteristic of IGBT chip at room temperature.

In Fig. 4, switching waveforms of the IGBT at 125°C are presented for both simulation and measurements. Simulations were performed in

PCIM Europe 2024, 11– 13 June 2024, Nuremberg          DOI: 10.30420/566262126

mixed mode and enabling inclusion of the electro-thermal boundary conditions. These results confirm improvements achieved by carefully adjusting the bipolar gain which, in turn has multiple benefits. It provides a better control of the current tail and consequently lead to a low turn-off energy of $E_{off} = 170$ mJ Additionally, it helps to form a smooth decay in the tail current which prevents oscillations. Low $I_{csat}$ and controlled di/dt = 200 A/μs lead to a low overshoot current and low turn-on energy of $E_{on} = 125$ mJ.

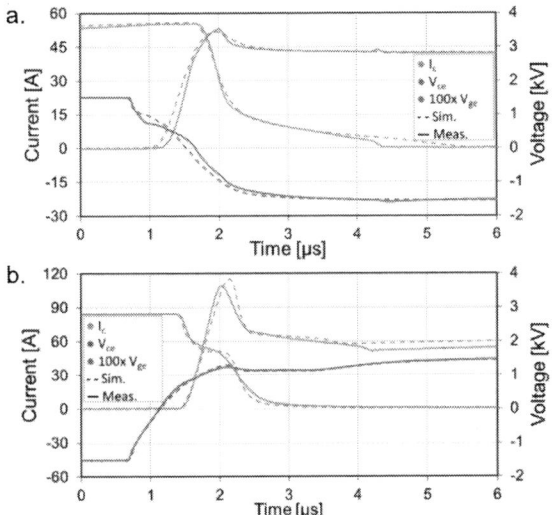

**Fig. 4**    Turn-off (a) and Turn-on (b) waveforms for IGBT chip at 125° C with $R_g = 47\ \Omega$ and $L_s = 5.1$ μH.

RBSOA measurements presented in Fig. 5a were carried out with a gate resistance of $R_g = 47\ \Omega$ and the stray inductance of $L_s = 6$ μH. The waveforms demonstrate a well-controlled dynamic avalanche starting at $V_{CE} = 1500$ V and withstanding almost four times the nominal current. In short circuit safe operating area, the device can easily tolerate a 10 μs pulse with the standard gate bias of $V_{ge} = 15$ V. Nevertheless, the SCSOA capability of the chip is beyond that. As depicted in Fig. 5b, even for a gate bias of $V_{ge} = 20$ V, a gate resistance of $R_g = 47\ \Omega$, and current equal to seven times the nominal current, the device is still robust to the type one SCSOA without any sign of oscillation.

To evaluate the reliability of the new process step and design features introduced, reliability test in overstress conditions were performed. The IGBT dies were tested for High Temperature Reverse Bias test at 125° C and $V_{ce} = 3.6$ kV for 1000 h and High Temperature Gate Bias test at 125° C and $V_{ge} = \pm 30$ V for 1000 h and successfully passed with no observable issue.

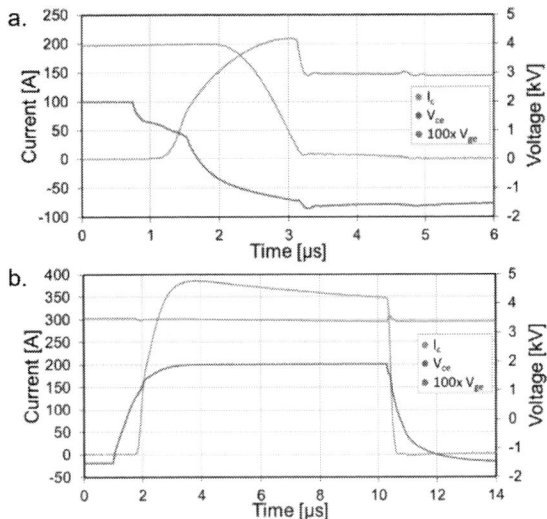

**Fig. 5**    RBSOA (a) and SCSOA (b) waveforms of the IGBT chip at 125° C with $R_g = 47\ \Omega$ and $L_s = 6$ μH (for RBSOA).

## 2.4  Diode characterization

The diode chip has been designed with the same square footprint as the IGBT, rated for 114 A nominal current. Diode forward characteristics are plotted in Fig. 6a, the forward voltage at 114 A is $V_f = 2.9$ V. Diode recovery, when switched with two IGBT chips, is soft over the full range of operating conditions with low reverse recovery current $I_{rm}$ and short tail current resulting in low recovery energy and low IGBT turn-on losses.

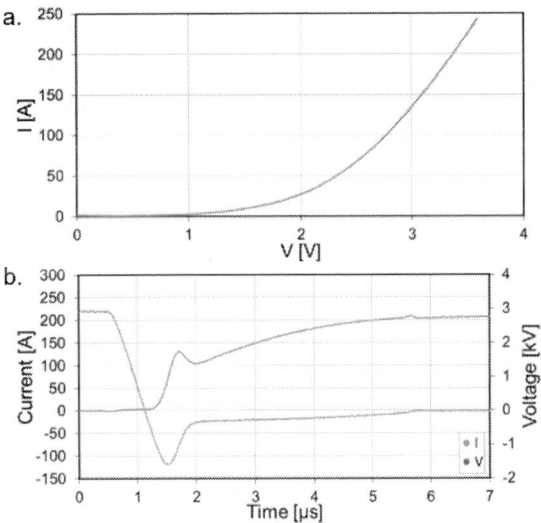

**Fig. 6**    I-V curve at room temperature (a) and RRSOA waveform of the diode chip at 125° C with di/dt = 400 A/μs (b).

1015

The diode dies have also successfully passed High Temperature Reverse Bias test at 125° C and V = 3.6 kV for 1000 h with no observable issue.

# 3 PressPack structure and design

PressPack IGBT technology was introduced by Westcode (now Littelfuse) more than two decades ago. As illustrated in Fig. 7, the packaging construction consists of three steps: individual chips are assembled into their specific cassettes; individual cassettes are then assembled into Press-Pack housing and then sealing the PressPack.

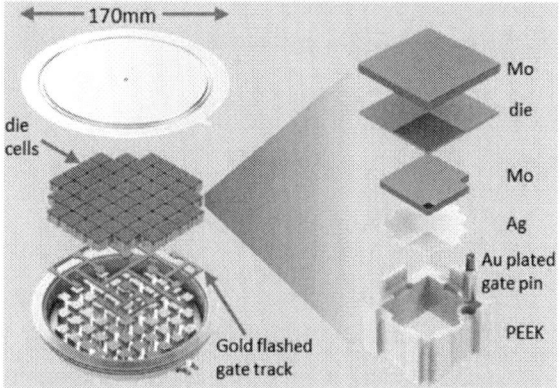

**Fig. 7**    Construction of a PressPack device.

Taking advantage of pressure contacts rather than conventional solder and wirebonds, makes Press-Pack a highly reliable package which can be used in challenging applications and environments [8]. The common cassette structure allows versatility in ratings/package design [9]. Furthermore, Inter-location of cassettes allows for maximum die packing density. With the die being cooled from both sides, thermal resistance is the same for every die. When compared to a conventional module, the internal stray inductance in both emitter and gate connections is much lower. In a robust design, the device fails to a stable short circuit with no package rupture allowing forest-up that offer n+1 redundancy.

## 3.1 PressPack characterization

The IGBT PressPacks with 34 IGBTs and 18 anti-parallel diodes and nominal current rating of 2 kA have been fabricated and tested in both standard switching conditions and SOA tests. The results, depicted in Fig. 8, are in agreement with the single die test results. This means a well-balanced current sharing within the die array, which in turn, indicates the effectiveness of process stability measures to guarantee low chip-to-chip and wafer-to-wafer variations.

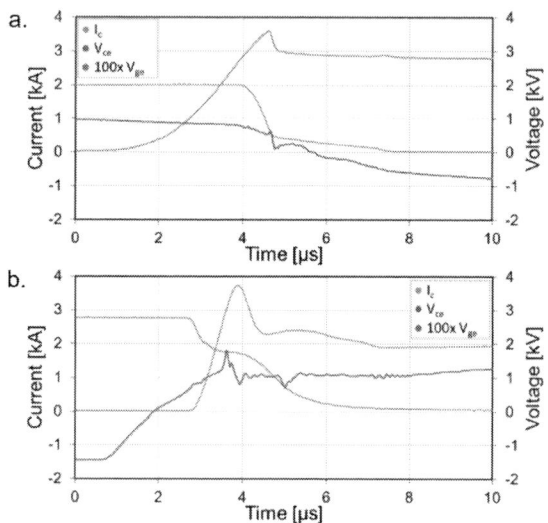

**Fig. 8**    Turn-off (a) and Turn-on (b) waveforms of a PressPack with 34 IGBTs and 18 diodes at 125° C with $R_{g,on}$ = 2.7 Ω, $R_{g,off}$ = 12 Ω, and $L_s$ = 220 nH.

RBSOA capability of more than twice the nominal current was achieved in PressPack level. The waveform also indicates that the PressPack experiences the dynamic avalanche at 2 kV and enters switching self-clamping mode (SSCM) at t = 8 µs.

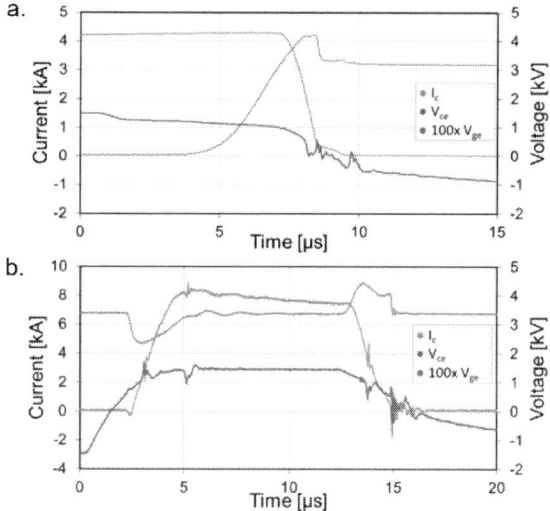

**Fig. 9**    RBSOA (a) and SCSOA (b) waveforms of a PressPack with 34 IGBTs and 18 diodes at 125° C with $R_{g,on}$ = 2.7 Ω, $R_{g,off}$ = 12 Ω, and $L_s$ = 220 nH (for RBSOA).

The SCSOA test in PressPack level shows a capability of more than four times the nominal current With no oscillation. The noises on the SCSOA waveforms are parasitic effects introduced by the test system and are not associated with the Press-Pack performance. It should also be emphasized that the limiting factor for the SCSOA was not the sample under test but the tester itself which could not supply higher gate voltages. SOA results are well aligned with the best in class for this voltage class.

# 4    Conclusion

The improvement introduced in the latest generation of XPT™ IGBT in combination with a properly designed antiparallel diode were presented. These developments have led to a design with high robustness to latch-up and a better handling of the filamentary condition during dynamic avalanche. Accordingly, an RBSOA of $4 \times I_{nom}$ and SCSOA of $7 \times I_{nom}$ were achieved.

The device structure was developed to reduce the losses in turn-on while the anode efficiency was defined to be greatly competitive in the $V_{ce}$ vs $E_{off}$ trade-off.

A reliable and stable fabrication process was also demonstrated to guarantee a balanced current sharing in a full array of 34 IGBTs and 18 anti-parallel diodes, tested in the PressPack. The tests, carried out at nominal current of 2 kA and SOA tests, confirm the good results achieved at single chip level. The new Littelfuse PressPack IGBT exhibits a very high performance in switching and SOA test as well as reliability.

Once the full release process is completed, the new Littelfuse PressPack IGBT will find its position in the market close to the best in class in the field assuring very competitive electrical and reliability performance.

# References

[1]   F. Wakeman, D. Hemmings, W. Findlay, and G. Lockwood, "New high reliability bondless pressure contact IGBTs," in PCIM Europe, 1999.

[2]   F. Wakeman, and G. Lockwood, "Electromechanical evaluation of a bondless pressure contact IGBT," *IEE Proceedings Circuits, Devices and Systems*, vol. 148, no. 2, pp. 89–93, 2001.

[3]   R. Alvarez, S. Bernet, L. Lindenmueller, and F. Filsecker, "Characterization of a new 4.5 kV press pack SPT+ IGBT in Voltage Source Converters with clamp circuit," in 2010 IEEE International Conference on Industrial Technology, Via del Mar, Chile, pp. 702-709, 2010.

[4]   R. Alvarez, F. Filsecker, M. Buschendorf, and S. Bernet, "Characterization of 4.5 kV/2.4 kA press pack IGBT including comparison with IGCT," in 2013 IEEE Energy Conversion Congress and Exposition, Denver, CO, USA, pp. 260-267, 2013.

[5]   M. Rahimo, A. Kopta, and S. Linder, "Novel enhanced-planar IGBT technology rated up to 6.5 kV for lower losses and higher SOA capability," in 2006 IEEE International Symposium on Power Semiconductor Devices and IC's, Naples, Italy, pp. 1-4, 2006.

[6]   A. Kopta, M. Rahimo, U. Schlapbach, N. Kaminski and D. Silber, "Limitation of the short-circuit ruggedness of high-voltage IGBTs," in 2009 21st International Symposium on Power Semiconductor Devices & IC's, Barcelona, Spain, 2009, pp. 33-36, 2009.

[7]   P. D. Reigosa, F. Iannuzzo, M. Rahimo, C. Corvasce and F. Blaabjerg, "Improving the Short-Circuit Reliability in IGBTs: How to Mitigate Oscillations," IEEE Transactions on Power Electronics, vol. 33, no. 7, pp. 5603-5612, 2018.

[8]   M. Sweet, E. S. Narayanan, and S. Steinhoff, "Influence of cassette design upon breakdown performance of a 4.5 kV press-pack IGBT module," in 8th IET International Conference on Power Electronics, Machines and Drives (PEMD 2016), Glasgow, UK, pp. 1-6, 2016.

[9]   F. Wakeman, G. Lockwood, M. Davies, and K. Billett, "Pressure contact IGBT, the ideal switch for high power applications," in IEEE-IAS Conf, Phoenix, USA, 1999.

PCIM Europe 2024, 11– 13 June 2024, Nuremberg          DOI: 10.30420/566262127

# 2.5 kV IGBT Module with High Reliability for Renewable Applications

Akiyoshi Masuda[1], Masaomi Miyazawa[1], Tsuyoshi Uraji[2], Koichi Masuda[3], Narender Lakshmanan[3], Thomas Radke[3]

[1] Mitsubishi Electric Corporation, Power Device Works, Japan
[2] Mitsubishi Electric Corporation, Component Production Engineering Center, Japan
[3] Mitsubishi Electric Europe B.V., Germany

Speaker: Akiyoshi Masuda, Masuda.Akiyoshi@ea.MitsubishiElectric.co.jp

## Abstract

The new 2.5 kV Si-IGBT module has been developed in the LV100 package for renewable applications. This module realizes high reliability, especially power cycling lifetime, by adopting the new structure "SLC+". Available LV100 modules which represent our 7th generation modules achieved a high thermal cycle lifetime by the combination of Insulated Metal Substrate (IMS) and resin encapsulation, which technology was called Solid Cover "SLC". In addition, the newly developed SLC+ structure have mainly two improvements. First item is enhancing the yield strength of bond wire. The newly developed Al-alloy wire is stiff to withstand mechanical stress which aids in robustness against bond wire cracking. Second item is to add hard metallization layer on the surface of chip for preventing the crack in the chip electrode. This function is useful for making full use of wire improvement. Experimental results would show the improvement of power cycling capability with SLC+ structure.

On the other hand, the newly developed 2.5 kV IGBT and diode chip-set is optimized for 1500 $V_{dc}$ usage with consideration of trade-off between losses and LTDS robustness. The 2.5 kV chip-set consisted of 7th generation CSTBT™-IGBTs and RFC-diodes enables power loss reduction and junction temperature reduction. That feature would show the validness for 1500 $V_{dc}$ or 900-1000 $V_{ac}$ operation with simple 2-Level topology. Besides, the suppressed $\Delta T_j$ swing also reduces the thermomechanical stress and contributes to lifetime improvement.

## 1    Introduction

In recent years, 1500 $V_{dc}$ or 900-1000 $V_{ac}$ link voltage has been the strategy toward achieving both high-power density and output current reduction for renewable applications. Due to the corresponding requirements for power modules for renewable energy applications, Mitsubishi Electric has been developing a 2.5 kV Si-IGBT module with high reliability. The 2.5 kV chips are 7th generation CSTBT™-IGBTs and RFC (Relaxed Field of Cathode)-diodes and they are mounted in the LV100 package adopting new technology called Solid Cover structure+ "SLC+".

Achieving a long power cycling lifetime is a primary target for the applications. For example, on a rotor-side application in wind converters, the thermal stress on the diodes could be crucial. It is because junction temperature of the diodes is higher due to the lower output frequency (fo: ~10Hz) of rotor-side application [1]. In addition, the lifetime tends to be limited by the heat dissipation from diodes. Thermomechanical stress with power cycling would cause bond wire lift-off.

The new technology SLC+ will be expected high power cycling lifetime. In the SLC+, the hard resin encapsulation relaxes the mechanical stresses at the interface between the chip and the bond wire, and the new bond wire made of Al-alloy enhances the yield strength too.

Moreover, a new 2.5 kV chip-set consisted of 7th generation CSTBT™-IGBTs and RFC-diodes can realize both low loss and sufficient LTDS robustness for 1500Vdc link voltage with keeping low power losses and junction temperature. Also, the thermal resistance from junction to case of a diode has been improved, so the $\Delta T_j$ swing is suppressed and the thermomechanical stress is relieved. That also contributes to high lifetime.

## 2 Improvement of lifetime

### 2.1 Feature of new package "SLC+"

Solid Cover+ "SLC+" is the new packaging structure which is SLC with mainly two improvements. Table 1 shows the outline of the factors in SLC+ and comparison with conventional structure. These improvements realize the higher lifetime modules toward decades operation for renewable applications.

**Table 1** Comparison of structure outline

| Structure | Gel Cover | Solid Cover (SLC) | Solid Cover + (SLC+) |
|---|---|---|---|
| Representative module types | Std *Image is 6th gen. NX (6th gen.) | NX (7th gen.) LV100 | LV100 (So far) |
| Outline | | | |
| Sealing material | Silicone gel | Resin | |
| Substrate | Ceramic substrate | Insulating metal substrate (IMS) | |
| Bond wire material | Al | | Al-alloy |
| Chip surface | AlSi electrode | | Hard metallization layer |

#### 2.1.1 Solid Cover structure

Our available 7th gen. modules mainly adopt the SLC structure. 6th gen. modules and earlier consist of silicone gel sealing and ceramic substrate. For developing 7th gen. module, Insulating Metal Substrate (IMS) and specific resin for sealing are newly introduced. These materials are expected to improve thermal cycling and power cycling capability.

Ceramic substrate mounted on 6th gen. module has the role of insulation between main current circuit and outside of the module. It is a compound component consisting of the copper patterns and resin for insulation and jointed to copper baseplate and chips with solder in assembly process. It is generally known that the lifetime would be determined by the breakdown of solder layer as thermal cycling capability because the coefficient of thermal expansion (CTE) causes the stress in the solder layer between the copper patterns and the copper baseplate by temperature swings as shown in Fig. 1. In the IMS, the copper patterns and the copper baseplate can be directly bonded to a resin insulation layer, so the solder layer is eliminated. In addition, since the CTE of the resin is close to copper as shown in Table 2 [2], thermo-mechanical stress between the components can be reduced. As a result, thermal cycling capability would be improved.

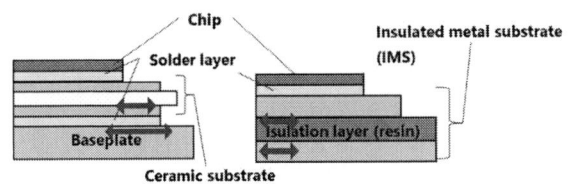

**Fig. 1** Comparison between ceramic substrate and IMS in SLC
↔ : Indicate the degree of thermal expansion

**Table 2** CTE (Coefficient of thermal expansion) of materials [2]

| Material | CTE [ppm/K] |
|---|---|
| Ceramic | 4.5 ~ 7.0 |
| Insulating resin layer | ≈17 |
| Copper | ≈17 |

Besides, resin encapsulation in SLC is expected to realize the higher power cycling capability than gel sealing. The power cycling failure is mainly caused by the breakdown of connecting wire mechanically stressed by junction temperature swings. In SLC, the bond wires are covered by resin, which is done thermal-curing in assembly process. The resin become much harder than silicone gel used for conventional structure and relaxes the mechanical stress in the interface between the wires and surface of the chip homogeneously as shown in Fig. 2 [3]

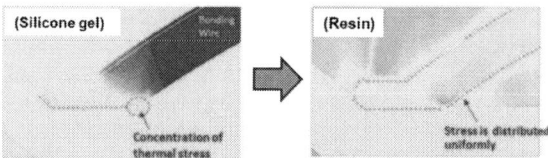

**Fig. 2** Stress concentration to bond wire with gel cover structure and hard resin cover structure

### 2.1.2 Al-alloy bonding wire

In order to provide SLC additional power cycling capability, the new connecting wire made of Al-alloy was adopted. Enhancement of the yield strength of connecting wire itself can suppress breaking wire with power cycling.

Fig. 3 shows the cross section of conventional Al wire and new Al-alloy wire with Electron Backscatter Diffraction (EBSD). The crystal grain size of our Al-alloy is about 5μm or less, which is one tenth finer than the crystal grain size of conventional Al. According to Hall Petch equation (Eq. 1), the yield strength of polycrystalline is thought to be inversely proportional to square root of the crystal grain size. Table 3 shows the yield strength of each bond wire. The yield strength was expected to improve by about three times with Al-alloy. [4]

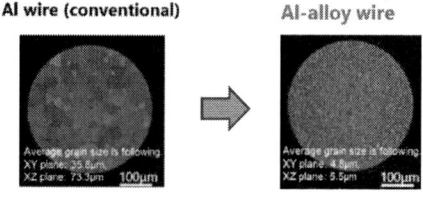

**Fig. 3** Comparison of each wire cross section

$$\sigma = \sigma_0 + \frac{k}{\sqrt{d}}$$

$\sigma$: The yield strength
$\sigma_0$: The yield strength in the case of single crystal
$k$: Constant
$d$: Crystal grain size

**Eq. 1** Hall Petch equation

**Table 3** Yield strength of wire materials

| Wire Material | Yield Str. (MPa) |
|---|---|
| Al | 32 |
| Al alloy | 87 |

### 2.1.3 Hard metallization of chip surface

The chip surface in SLC+ technology is formed by the additional hard metallization layer deposited on the metal electrode. Fig. 4 shows schematic diagrams of crack progress with and without the additional metallization layer on the chip surface. Stress generated by temperature swing of power cycling typically cause crack not only in a bond wire, but also at joint part and in the metal electrode. Our conventional Al bond wire is connected to Al-Si electrode which is formed on the surface of chips. However, in the case of the connection between Al-alloy bond wire and Al-Si electrode, previous research showed the crack progressed inside the electrode. For making full use of the improved yield strength of Al-alloy bond wire, it is necessary to suppress the stress toward chips [4].

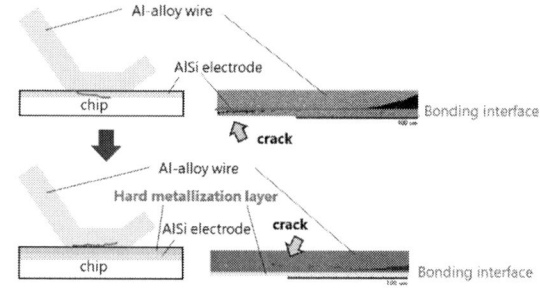

**Fig. 4** Schematic diagram of crack progress

Fig. 5 shows the cross section around joint between Al-alloy wire and hard metallized chip after power cycling tests with the module having SLC+. The grey layer (dashed red line) between the chip and the wire in Fig. 5 indicates the additional metallization layer mainly composed of Ni. The crack was observed inside the Al-alloy wire. This result means that chip metallization could prevent chips being broken physically by thermal stress and the joint part and Al-Si electrode could withstand stress that exceeded the yield strength of Al-alloy. Consequently, power cycling capability is limited by yield strength of Al-alloy.

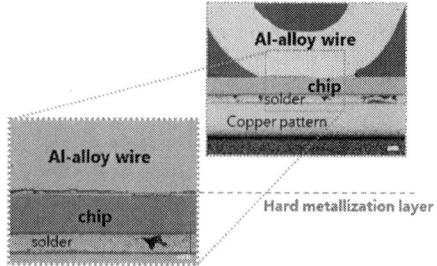

**Fig. 5** The cross section of Al-alloy wire and hard metallized chip after power cycling tests

## 2.2 Power cycling tests

Fig. 6 shows the results of power cycling tests using the SLC+ technology. The main conditions of tests are decided assuming operation in wind applications, as following; $t_{on}$=0.1 s, $T_{jmax}$=150 °C, $\Delta T_j$ =50 K. The number of cycles has been already over 40 million cycles and there have been no cracks inside chips and wires up to 40 million. In this test, the gel encapsulated sample with Al-alloy wire was also examined and already failed at 30 million cycles with wire lift-off. Indicating the impact of the resin encapsulation. Also, it is known that the shorter $t_{on}$ condition, the higher power cycling lifetime in previous research [5]. Compared with reference curve in Fig. 6 (black line, $t_{on}$: 2 s, structure: SLC), it can be expected over 20 times sufficient improvements of lifetime with applying the Al-alloy wire. The $t_{on}$ of about 0.1s is closer to the real application condition. In addition, reduction of $\Delta T_j$ due to power loss improvement would also contribute higher lifetime.

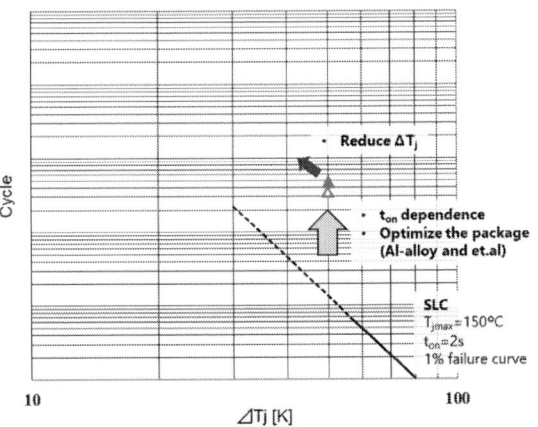

**Fig. 6**   Power cycle curve
   ▲ : SLC+ with Al-alloy wire (no cracks)
   △ : Gel sealing with Al alloy wire (broken)
 --- : Estimated lifetime of SLC, $t_{on}$=2s

## 3    Features of 2.5 kV devices

The 2.5 kV module adopts the 7th generation CSTBT™ IGBTs and RFC diodes. The $V_{CEsat}$-$E_{off}$ trade-off has been improved in the 7th generation IGBTs and diodes versus previous generations. CSTBT™ technology enables power loss reduction by forming a Carrier Stored layer with p-doping and improving conductivity modulation. Furthermore, RFC technology can reduce the thickness of the diodes while maintaining high withstand voltage and tolerance. Finally, the 2.5 kV chip-set has optimized specifications by choosing

points in $V_{CEsat}$-$E_{off}$ and $V_F$-$E_{rr}$ trade-off especially for wind application.

### 3.1  Cosmic ray robustness LTDS

In high DC-link voltage as 1500 $V_{dc}$, the stability against failures induced by cosmic rays needs to be considered, called long term DC stability (LTDS). The failure rate is influenced by the DC voltage applied to the IGBT module in blocking state. Fig. 7 shows the failure rate by cosmic rays, which means the probability of occurring neutron single event burn (SEB). It is estimated that the 2.5 kV IGBT module would have the same FIT rate at $V_{CE}$=1600 V as the 2 kV, LV100 (CM1200DW-40T) module at below 1400V and 1.7 kV, LV100 (CM1200DW-34T) at below 1150V. This means that the 2.5 kV IGBT module has the sufficient LTDS robustness for systems with continuous dc-voltages of 1500 $V_{dc}$ .

These data are results calculated from white neutron irradiation tests. White neutron having similar spectrum to that by cosmic rays enable to accelerate irradiation tests and estimate the failure rate.

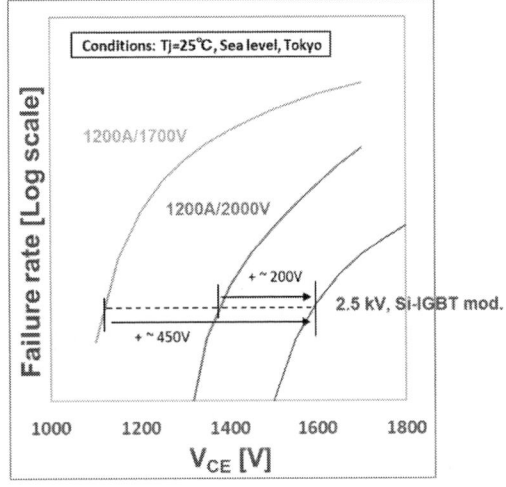

**Fig. 7**   LTDS failure rate of our modules

### 3.2  Power loss and performance

Fig. 8 shows electrical characteristics with comparison of new 2.5 kV device and 7th gen. 1.7 kV device (CM1200DW-34T) and in wind application. The remarkable point is the performance of diode. Assuming the operation as a rectifier converter in wind application, the loss from diode is heavily important. The 2.5 kV IGBT shows about 15 % higher on-state voltage versus 1.7 kV IGBT as a function of $I_c$ at $T_{vj}$=150 °C. Also, the 2.5 kV diode shows only about 5 % higher forward voltage versus 1.7 kV diode as a function of $I_E$ at $T_{vj}$=150 °C.

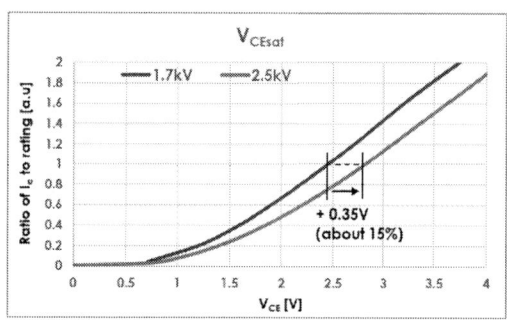

**Fig. 8** Electrical characteristic at $T_{vj}$=150 °C ($V_{CE}$ vs $I_C$/rating)

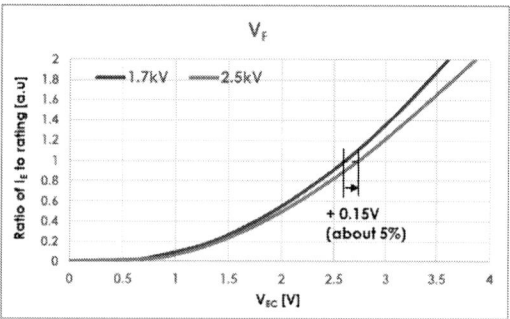

**Fig. 9** Electrical characteristic at $T_{vj}$=150 °C ($V_{EC}$ vs $I_E$ /rating)

Fig. 10-12 show switching wave forms of 2.5 kV module at $T_{vj}$=150 °C. As following the suppressed built-in inductance of the LV100 package, turn-off and recovery surge are low and would not be any problem. Besides, 2.5 kV chips especially diode whose $V_F$-$E_{rr}$ trade-off is optimized could show the smooth and rapid switching.

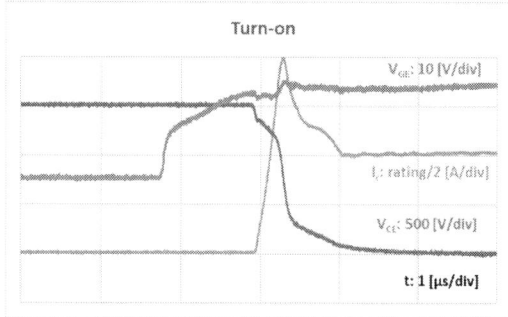

**Fig. 10** Turn-on switching wave form ($T_{vj}$=150 °C, $I_C$= rating, $V_{cc}$= 1500 V, $V_{GE}$=±15 V、$R_G$=1 ohm)

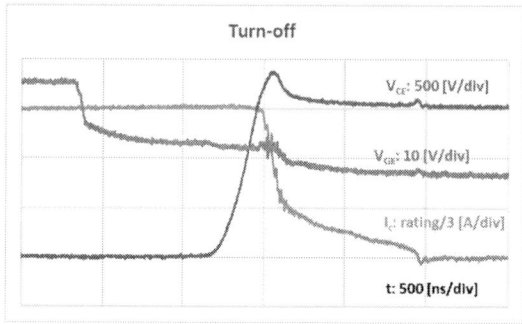

**Fig. 11** Turn-off switching wave form ($T_{vj}$=150 °C, $I_C$=rating, $V_{cc}$=1500 V, $V_{GE}$=±15 V、$R_G$=1 ohm)

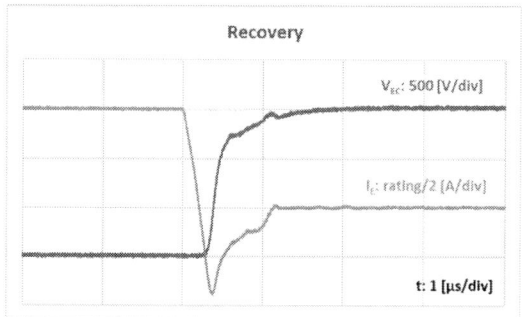

**Fig. 12** Recovery switching wave form ($T_{vj}$=150 °C, $I_E$=rating, $V_{cc}$=1500 V, $R_G$=1 ohm)

Fig. 13 shows the power losses and junction temperatures of the 2.5 kV IGBT module and CM1200DW-34T (1.7 kV, LV100 module) under the operating conditions as Table 4. This comparison indicates that the 2.5 kV IGBT module enables to realize about 20% higher output power ($P_{out}$) versus the 1.7kV module at same junction temperature $T_{j(top)}$: 150 °C. This simulation conditions assume the operation of rotor-side converter of induction generator-based wind power, so $V_{CEsat}$-$E_{off}$ and $V_F$-$E_{rr}$ trade-off of 2.5 kV IGBT module is optimized for regenerative motion in the operation. Besides, the thermal resistance from junction to case has been improved by utilizing a larger diode die. As a result, there are lower 5 K difference of junction temperature between 2.5 kV IGBTs and 2.5 kV diode whereas junction temperature of 1.7 kV diodes is about 25 K higher than 1.7 kV IGBTs. $P_{out}$ of the 1.7 kV module is limited by temperature of diodes, but the 2.5 kV IGBT module takes full advantage of device characteristics by optimizing each heat generation realize high output power.

On the other hand, junction temperature swing ($\Delta T_j$) is significant factor of determining the lifetime of the power module. Fig. 14 briefly shows the $\Delta T_j$ in simulation for the 2.5 kV IGBT module. As per this simulation, it is also expected to have negligible $\Delta T_j$ difference between IGBTs and diodes. As a result, the $\Delta T_j$ of the diode is no longer the main bottleneck and the overall power cycling lifetime can be increased.

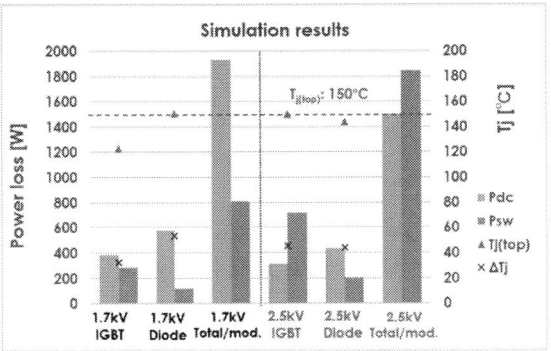

**Fig. 13** Simulation of power loss and junction temperature

**Table 4** Conditions and results of simulation

| | 1.7kV module | 2.5kV module |
|---|---|---|
| Conditions | | |
| Vcc | 1000V | 1500V |
| Vout | 690Vrms | 1000Vrms |
| Topology | 2level | |
| Tj(top) | 150 °C | |
| cos $\phi$ | −0.8 | |
| fc | 1.5 | |
| fout | 6 | |
| Modulation method | 3rd harmonic injection | |
| Modulation index | 0.25 | |
| Results | | |
| Pout | 1.1MW | 1.26MW |
| $\Delta$ Tj (IGBT) | 32.2K | 45.8K |
| $\Delta$ Tj (diode) | 53.7K | 44.2K |

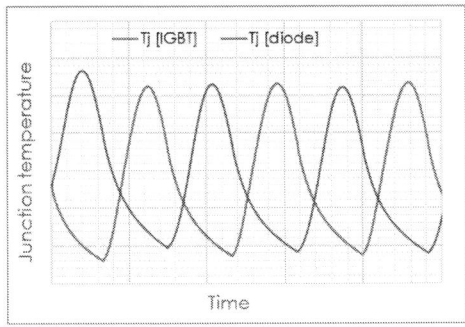

**Fig. 14** The state of junction temperature swing

# 4 Conclusion

The 2.5 kV IGBT module with LV100 package has been developed for renewable applications, mainly wind power application. For meeting this demand, 2.5 kV IGBT module is featured as below.

- **Main targets**
  - High reliability, mainly power cycling capability
  - Suitable for 1500 $V_{dc}$ or 900-1000 $V_{ac}$ operation

- **Specifications for satisfying targets**
  - New SLC+ structure, which consist of IMS, resin encapsulation, Al-alloy bond wire and hard metallized chips
  - Sufficient LTDS robustness
  - Reduce power loss and suppress junction temperature with optimized 7th gen. chipset especially for wind power application

The new packaging technology "SLC+" is expected to have high lifetime. As a feature of our conventional 7th gen. module adopting SLC, IMS enables to improve thermal cycle capability because it is a compound of resin insulating layer and copper pattern, and the difference of their CTE is little. Besides, the resin encapsulation relaxes the thermomechanical stress occurring toward bond wire, so power cycle capability is also improved. As a new factor added with SLC+, power cycling capability has been improved by introducing Al-alloy wire. The grain size of our Al-alloy is about 5µm or less, which is one tenth finer than the crystal grain size of conventional Al. Hall Petch equation shows that the yield strength of connecting wire is improved by about three times with such small grain size. Furthermore, junction temperature swing stress not only the bond wire but also the chip surface, so the crack would progress inside the electrode. Hard metallized chip surfaces are effective to prevent the crack in the electrode. In power cycling tests, there were no cracks through over 40 million cycles at $\Delta T_j$=50 K, $t_{on}$=0.1 s. Compared with reference curve of SLC at $t_{on}$=2 s, it can be expected that over 20 times improvements of lifetime applying-Al alloy wire.

On the other hand, the 2.5 kV IGBT module adopting 7th gen. chip-set shows sufficient LTDS robustness assuming 1500 $V_{dc}$ operation. Despite that, the 2.5 kV diode shows about only 5 % higher forward voltage versus the 1.7 kV diode, because specifications are optimized by choosing points in

$V_{CEsat}$-$E_{off}$ and $V_F$-$E_{rr}$ trade-off. Besides, the thermal resistance of diode from junction to case is also improved due to a larger diode die. Under the simulation assuming the operation of rotor-side converter of induction generator-based wind power, the 2.5 kV IGBT module realizes about 20 % higher output power versus the 1.7 kV module at same junction temperature. This result is showing that the 2.5 kV IGBT module perfectly fit for 1500 $V_{dc}$ or 900-1000 $V_{ac}$ operation with simple 2-Level topology. In addition, junction temperature swing ($\Delta T_j$) of the 2.5 kV IGBT module would also be suppressed and would also contribute to higher power cycling lifetime.

# 5　Reference

[1] LI Hui et. al., "Power Cycling Capabilities Assessment of IGBT Modules in Wind Power Converter Considering the Wind Turbulence Effects" IEEE PEAC 2014.

[2] Takuya Takahashi et. al., "A 1700V-IGBT module and IPM with new insulated metal baseplate (IMB) featuring enhanced isolation properties and thermal conductivity", PCIM Europe 2016.

[3] Thomas Radke et. al., "More Power and Higher Reliability by 7th Gen. IGBT Module with New SLC-Technology", Bodo's Power Systems 2015.

[4] Tsuyoshi Uraji et. al., "Increase of Power Cycling Lifetime in Power Semiconductor Modules Applied Fine-Grained Al Alloy Wire" Mate 2022.

[5] Merouane Ouhab et. al., "Physics-of-Failure Model to Explain the Heating-Time Effect on IGBT Power Modules Lifetime" PCIM Europe 2023.

PCIM Europe 2024, 11– 13 June 2024, Nuremberg          DOI: 10.30420/566262128

# New Generation 4.5kV IGCT and Fast Recovery Diode for Railway Power Supply Applications

Umamaheswara Reddy Vemulapati[1], Remo Baumann[2], Tobias Wikström[1], Thomas Stiasny[1], Chiara Corvasce[1], Christian Winter[1]

[1] Hitachi Energy Ltd., Semiconductors, Lenzburg, Switzerland
[2] Hitachi Energy Ltd., T&SD, Turgi, Switzerland

Corresponding author:   Umamaheswara Vemulapati, umamaheswara.vemulapati@hitachienergy.com
Speaker:                Umamaheswara Vemulapati, umamaheswara.vemulapati@hitachienergy.com

## Abstract

First, we present the device design and experimental electrical performance of our new generation (Gen3) 4.5 kV Asymmetric Integrated Gate Commutated Thyristor (AS-IGCT) and Fast Recovery Diode (FRD), which are optimized for medium-voltage high-power applications. Afterwards, we present the performance improvements such as power level, current control margin (dynamic overloadability), and efficiency using the Gen3 IGCT and FRD product-set compared to previous generation of IGCT and FRD for railway power supply applications. The performance evaluation is done using 3 level-neutral point clamped (3L-NPC) topology simulations both in inverter and rectifier operation modes on power electronic building block (PEBB) level.

## 1    Introduction

Static Frequency Converters (SFCs) offer variable-speed operation of generators and turbines which leads to improved flexibility and efficiency in pumped hydro applications or as 3-phase grid interties where they connect two asynchronous grids [1-4]. SFCs also find applications in railway power supplies where they connect a three-phase 50 or 60 Hz transmission grid with the single-phase railway power grid which operates at 16.7, 25, 50 or 60 Hz. Typically, SFCs use Insulated Gate Bipolar Transistor (IGBT) or IGCT semiconductors at switch positions. The IGCT technology with its major development trends exploits the main advantages of having low conduction losses and hard switching turn-off current capability. New generation (Gen3) IGCTs offer higher power densities, by operating at higher junction temperatures ($T_{vj}$) with smaller thermal resistance as well as with significantly higher maximum turn-off current capability, for a given footprint of the package.

## 2    Device Design & Performance

The design and experimental electrical performance of our new generation (Gen3) 4.5 kV devices (AS-IGCT and FRD) are discussed and compared with that of the Gen2 devices.

## 2.1    Asymmetric IGCT

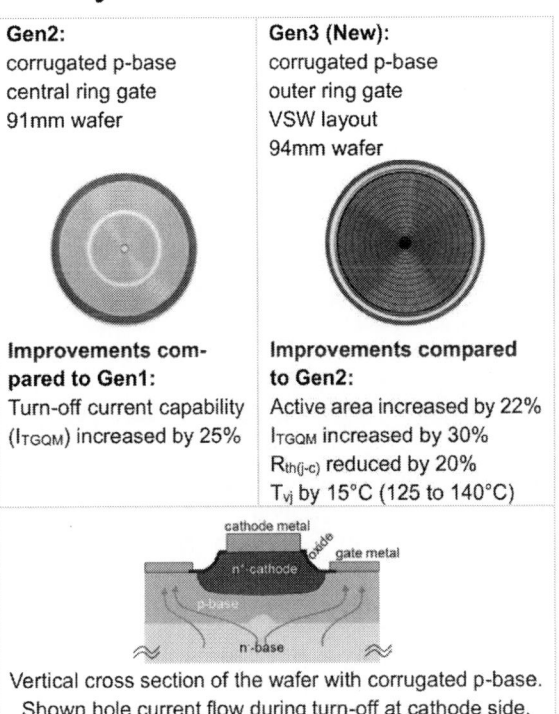

**Gen2:**
corrugated p-base
central ring gate
91mm wafer

**Gen3 (New):**
corrugated p-base
outer ring gate
VSW layout
94mm wafer

**Improvements compared to Gen1:**
Turn-off current capability ($I_{TGQM}$) increased by 25%

**Improvements compared to Gen2:**
Active area increased by 22%
$I_{TGQM}$ increased by 30%
$R_{th(j-c)}$ reduced by 20%
$T_{vj}$ by 15°C (125 to 140°C)

Vertical cross section of the wafer with corrugated p-base. Shown hole current flow during turn-off at cathode side.

**Fig. 1**    Comparison between Gen2 and Gen3 AS-IGCT wafer designs and their performance

Figure 1 illustrates the comparison between different generations of 4.5 kV IGCT and the performance improvements from one generation to another generation. The main limiting factor for Gen1 IGCT was related to the maximum turn-off current capability. The introduction of corrugated p-base, achieved by high power technology "HPT" process [5], in Gen2 IGCT is hailed as a major step for improving the maximum turn-off current capability. For the same footprint of the wafer (91mm), the turn-off current capability increased by 25% compared to Gen1. The corrugated p-base structure as shown in Fig. 1, redirecting the hole current away from cathode segments during turn-off, increases the turn-off current capability of the device.

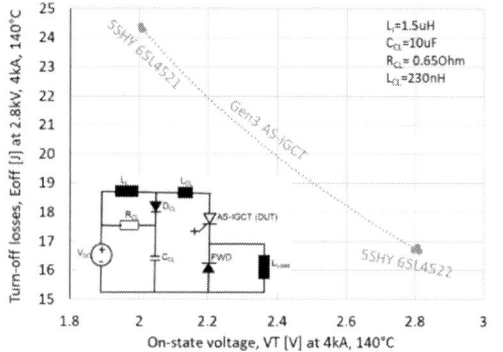

**Fig. 2** Technology trade-off curve of Gen3 AS-IGCT at 4.0 kA, 2.8 kV, 140°C

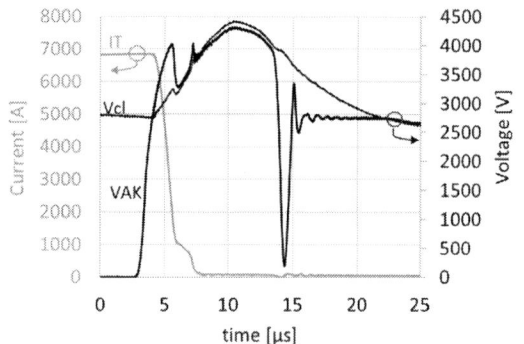

**Fig. 3** Turn-off SOA waveform of Gen3 AS-IGCT (5SHY 65L4521) at 6.8 kA, 2.8 kV, 140°C

Gen3 IGCT design is based on outer ring gate structure which minimizes the stray inductance for the gate signal as well as improves the thermal behavior of the device for a given footprint. Gen3 IGCT employs also corrugated p-base design but achieved from HPT+ process [6]. To further enhance the maximum turn-off current capability, the Gen3 AS-IGCT design utilizes variable segment width (VSW) layout where the segment width increases from inner segment-ring towards outer segment-ring which are farther and closer to the gate contact, respectively [7]. Figure 2 illustrates the technology trade-off curve of Gen3

AS-IGCT at 2.8 kV, 4.0 kA and 140°C. Figure 3 illustrates the turn-off safe operation area (SOA) waveform of Gen3 low on-state variant (5SHY 65L4521) where the device successfully turns-off 6.8 kA at 2.8 kV and 140°C. For the same footprint of the package or housing (85 mm pole-piece diameter), the turn-off current capability of Gen3 AS-IGCT increased by 30% compared to Gen2, while maintaining the technology trade-off at 140°C compared to Gen2 at 125°C (see Fig. 4). Also, it can be seen from Fig. 4 that the Gen3 AS-IGCT offers about 7% reduction in turn-off losses for the same on-state voltage compared to Gen2 under same testing conditions.

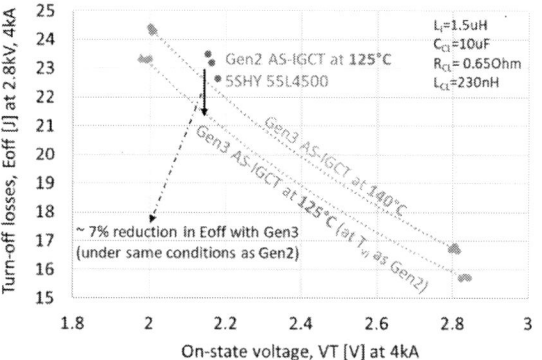

**Fig. 4** Technology trade-off comparison between Gen3 and Gen2 AS-IGCTs at 4.0 kA, 2.8 kV and at different temperatures

## 2.2 Fast Recovery Diode

The new 4.5 kV discrete FRD (5SDF 34L4520) is developed, as an accompanying diode for Gen3 4.5 kV AS-IGCTs, to handle high currents up to 6.5 kA and high reverse recovery di/dt up to 1.7 kA/us at dc-link voltages of up to 2.8 kV. The new generation (Gen3) 4.5 kV FRD wafer diameter is slightly larger than the Gen2 FRD wafer. However, the footprint of the hermetic housing is same as the Gen2 FRD, which is 85 mm polepiece diameter. The Gen3 4.5 kV FRD offers improved technology trade-off performance compared to Gen2 4.5 kV FRD, while maintaining the reverse recovery softness. The Gen3 FRD design is based on cathode segmentation (or structured n⁺-cathode) [8, 9] which offers reverse recovery softness even at lower currents, lower temperatures and at high dc-link voltages. Figure 5 illustrates the schematic structures and wafer pictures of Gen2 and Gen3 FRDs. Thanks to the cathode segmentation as the device can be made on thinner silicon compared to Gen2 FRD hence lower losses and better technology trade-off performance. Figure 6 illustrates the technology trade-off curves comparison between Gen3 and Gen2 4.5 kV FRDs at 2.8 kV, 4.0 kA, 1.7 kA/us and 140°C.

PCIM Europe 2024, 11– 13 June 2024, Nuremberg    DOI: 10.30420/566262128

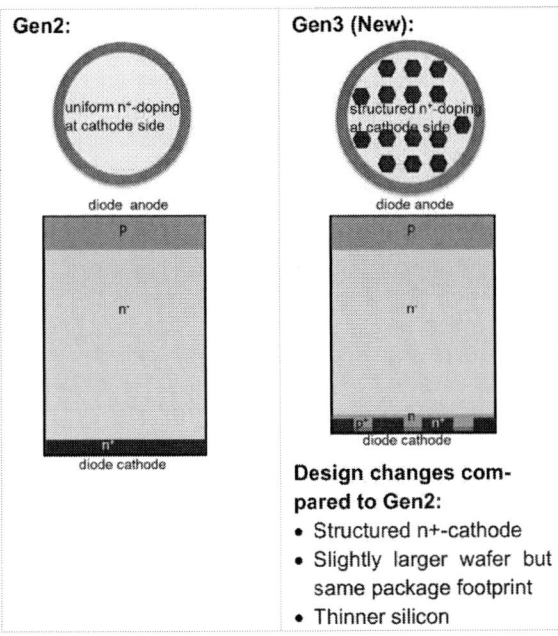

**Design changes compared to Gen2:**
- Structured n+-cathode
- Slightly larger wafer but same package footprint
- Thinner silicon

**Fig. 5** Wafer pictures and schematic structures of Gen2 and Gen3 FRDs

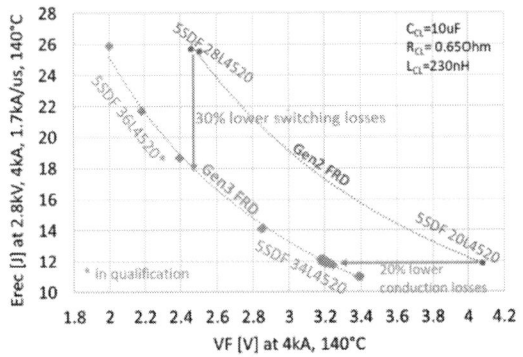

**Fig. 6** Technology trade-off comparison between Gen3 and Gen2 FRDs at 4.0 kA, 2.8 kV, 140°C

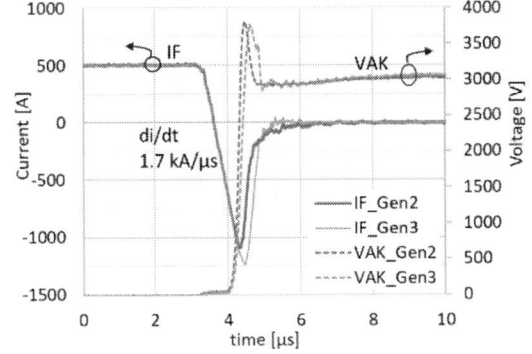

**Fig. 7** Reverse recovery waveforms comparison between Gen3 (5SDF 34L4520) and Gen2 (5SDF 20L4520) FRDs at low currents (at 500 A which is <10% of turn-off SOA current 6500 A of Gen3 FRD) at 2.8 kV, 25°C with di/dt of 1.7 kA/µs

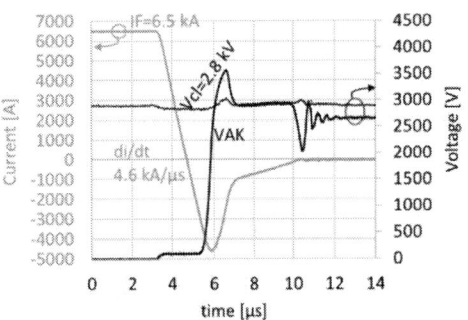

**Fig. 8** Reverse recovery SOA waveform of Gen3 FRD (5SDF 34L4520): Turning-off successfully 6.5 kA at 2.8 kV, 140°C with high di/dt of 4.6 kA/µs

As shown in Fig. 7, the Gen3 4.5 kV FRD maintains reverse recovery softness (the peak voltage is nearly the same) as Gen2 FRD at lower current (500A) and lower temperature (25°C). Figure 8 illustrates the reverse recovery SOA waveform of Gen3 FRD (5SDF 34L4520) at very high di/dt.

# 3 PEBB Level Performance

## 3.1 Application

The performance comparison between Gen3 devices (AS-IGCT & FRD) and Gen2 devices are done using 3L-NPC topology at PEBB level comprising two phases within one mechanical assembly. Two representative inverter and rectifier operation modes were chosen with different power factors. The loss breakdown for rectifier and inverter mode for IGCT and FRD is compared to the actual Gen2 devices at the same set of operation points. The improved turn-off current capability of Gen3 devices is instrumental for an improved dynamic overloadability of the overall converter system. In addition, the better thermal performance of the Gen3 product set can be utilized to achieve a significant improvement in converter power density.

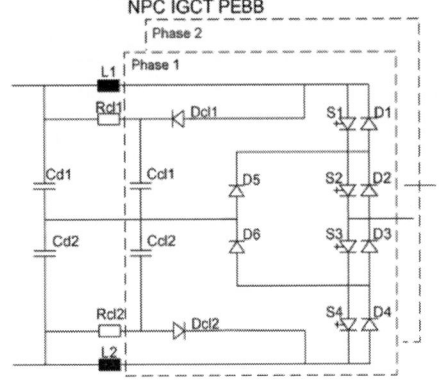

**Fig. 9** Power electronic building block circuit

The PEBB circuit consisting of 2 phases as considered for the comparison is depicted in Fig. 9. A corresponding hardware visualization is shown in Fig. 10.

**Fig. 10** Power electronic building block hardware

The main characteristic values for the simulated performance are summarized in Table 1. The main improvements on the AS-IGCT technology are reflected in a 30% increase of the maximum controllable turn-off current and an increase in junction operation temperature by 15 K.

**Table 1** Device max. parameters for simulation

| Parameter | Symbol | Gen2 AS-IGCT 5SHY 55L4500 | Gen3 AS-IGCT 5SHY 65L4521 |
|---|---|---|---|
| Rep. peak off-state voltage | $V_{DRM}$ | 4500 V | 4500 V |
| Max. controllable turn-off current | $I_{TGQM}$ | 5000 A | 6500 A |
| Threshold voltage | $V_{(T0)}$ | 1.22 V | 1.12 V |
| Slope resistance | $r_T$ | 0.280 mΩ | 0.294 mΩ |
| Turn-off energy ($I_{TGQ}$ = 4000 A) | $E_{off}$ | 31.5 J | 32 J |
| Permanent DC voltage | $V_{DC}$ | 2800 V | 2800 V |
| Junction operating temperature | $T_{vj}$ | 125°C | 140°C |

| Parameter | Symbol | Gen2 FRD 5SDF 28L4520 | Gen3 FRD 5SDF 34L4520 |
|---|---|---|---|
| Repetitive peak reverse voltage | $V_{RRM}$ | 4500 V | 4500 V |
| Maximum on-state current | $I_{FM}$ | 5500 A | 6500 A |
| Threshold voltage | $V_{(F0)}$ | 1.10 V | 1.46 V |
| Slope resistance | $r_F$ | 0.470 mΩ | 0.480 mΩ |
| Turn-off energy ($I_F$ = 4000 A, -1 kA/µs) | $E_{rr}$ | 29.5 J | 11.9 J |
| Permanent DC voltage | $V_{DC}$ | 2800 V | 2800 V |
| Junction operating temperature | $T_{vj}$ | 140°C | 140°C |

The fast recovery diode Gen3 in this comparison has experienced a significant different trade-off improvement between on-state characteristic and turn-off energy. While the Gen3 device threshold voltage is 33% higher compared to the Gen2 FRD, the turn-off energy is reduced by 60%.

## 3.2 Losses Comparison Gen2 to Gen3

The operation conditions revealing highest losses among the comparison on PEBB level is the rectifier operation mode with a power factor of 0.8 using the Gen2 devices. Thus, these losses have been scaled to 1 p.u. serving as a baseline. Losses figures have been generated using the same operation conditions for Gen2 and Gen3 PEBB configurations. The corresponding operation points were based on thermal limitations given by the Gen2 PEBB configuration.

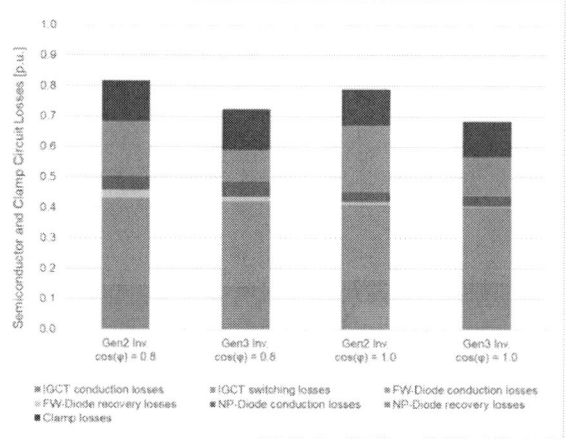

**Fig. 11** Losses breakdown for inverter operation

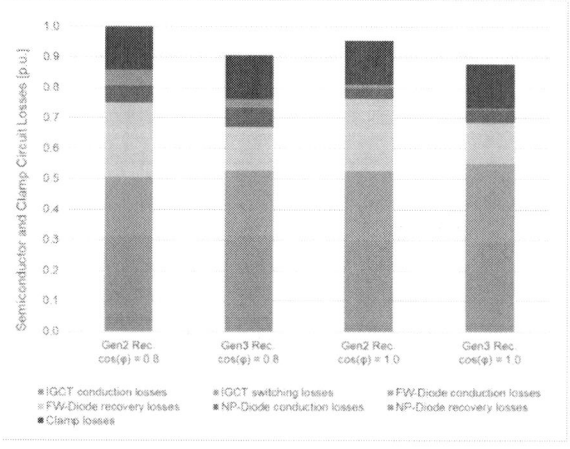

**Fig. 12** Losses breakdown for rectifier operation

Gen3 PEBB configuration outperforms Gen2 PEBB configuration in all considered operation points. The most significant difference in inverter operation (see Fig. 11) are the neutral point diode reverse recovery losses represented by the green bars. A further significant difference is revealed in rectifier operation (see Fig. 12). Freewheeling diode reverse recovery losses (yellow bars) are substantially lower for Gen3 configuration compared to Gen2.

### 3.3 Dynamic Overload Performance

The improved SOA, especially the improved maximum controllable turn-off current of the Gen3 AS-IGCT, is instrumental for a significant improvement in dynamic overload capability (see Fig. 13). While Gen2 devices are limiting the dynamic overload performance of the PEBBs to 1.38 to 1.66 p.u., the new devices are boosting that limit to a range of 1.85 to 2.21 p.u. leading to a better control margin of the overall converter system.

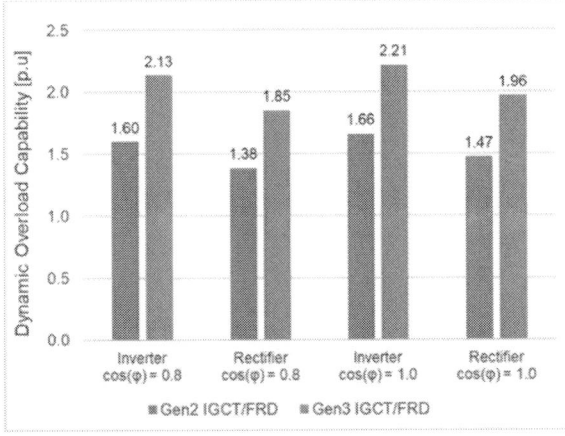

**Fig. 13** Dynamic overload capability, 1 p.u. corresponds to the maximum continuous power rating of the respective operation point with Gen2 devices

### 3.4 Power Density Increase

The last performance comparison of this paper is based on active power capability limited by the corresponding thermal characteristic of the semiconductors. Utilizing the respective thermal performance of the device up to the allowed maximum junction operation temperature, a significant improvement in active power transfer is achieved with Gen3 devices while keeping the conversion efficiency on the same level than PEBBs equipped with Gen2 semiconductors. The highest active power transfer among the investigated operation points is attained in rectifier operation with a power factor of 1.0. The Gen3 devices can boost the PEBB active power by 29% in rectifier operation.

As depicted in Fig. 14 other operation points showing the same trend as well.

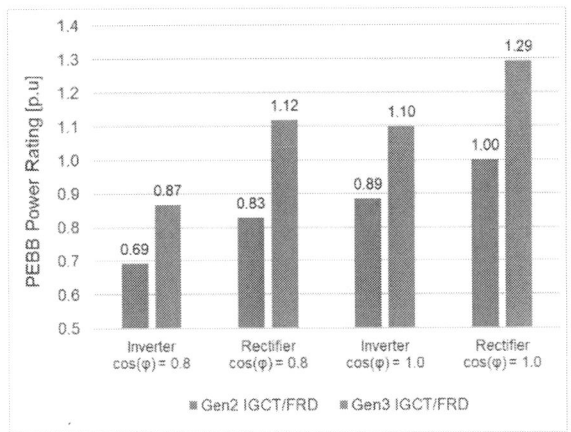

**Fig. 14** Active power rating of PEBBs, 1 p.u. corresponds to the highest power rating achievable among the presented operation points with Gen2 devices

## 4 Conclusion

The device design and experimental electrical performance of new generation (Gen3) 4.5 kV AS-IGCT and FRD product-set are presented and compared with that of the Gen2 devices. The Gen3 4.5 kV AS-IGCT offers 30% higher turn-off current capability, 20% lower thermal resistance $R_{th(j-c)}$ and 15K increase in operating junction temperature compared to Gen2 4.5 kV AS-IGCT, while maintaining the technology trade-off performance. The new (Gen3) 4.5 kV FRD (5SDF 34L4 520) is developed, as an accompanying diode for Gen3 4.5 kV AS-IGCTs, to handle high currents up to 6.5 kA and high reverse recovery di/dt up to 1.7 kA/us at dc-link voltages of up to 2.8 kV. The Gen3 4.5 kV FRD offers improved technology trade-off compared to Gen2 4.5 kV FRD, while maintaining the reverse recovery softness.

The simulation results at the PEBB level (using 3L-NPC topology) reveal that the system with Gen3 devices offers about 10% lower losses compared to system with Gen2 devices, both in rectifier and inverter modes of operation. In addition, the dynamic overload capability of the PEBB with Gen3 devices is increased by 33% compared to Gen2 devices. Furthermore, the Gen3 devices can boost the PEBB active power capability at least by 29% and 23.5% in rectifier and inverter modes of operation, respectively.

# References

[1] P. Steimer, O. Senturk, S. Aubert, and S. Linder, "Converter-fed synchronous machine for pumped hydro storage plants," Proc. IEEE ECCE, 2014, pp. 4561-4567.

[2] C. Haederli, T. Thurnherr, A. Christe, A. Faulstich, and C. Ladreiter-Knauss, "Upper stage of the Malta project: the World's first direct modular multilevel converter for variable-speed pumped storage," Hydro Conference, 2022.

[3] M. Vasiladiotis, R. Baumann, C. Häderli, and J. Steinke, "IGCT-based direct AC/AC modular multilevel converters for pumped hydro storage plants," Proc. IEEE ECCE, 2018, pp. 4837-4844.

[4] B. Ødegård, D. Weiss, T. Wikström, and R. Baumann, "Rugged MMC converter cell for high power applications," Proc. EPE ECCE Europe, 2016.

[5] T. Wikström, T. Stiasny, M. Rahimo, D. Cottet, and P. Streit, "The corrugated p-base IGCT – a new benchmark for large area SOA scaling," Proc. ISPSD, 2007, pp. 29-32.

[6] M. Arnold, T. Wikström, Y. Otani, and T. Stiasny, "High-Temperature operation of HPT+ IGCTs," Proc. PCIM Europe 2011.

[7] T. Wikström, and D. Cottet, "A 6500 A, 4500 V, 94 mm Asymmetric IGCT," Proc. PCIM Europe 2020, pp. 757-761.

[8] U. Vemulapati, T. Wikström, and M. Lüscher, "An RC IGCT for application at up to 5.3 kV," Proc. ICPE-ECCE Asia, 2019.

[9] T. Wikström, U. Vemulapati, and B. Ødegård, "A 4.5kV RC-IGCT with diode segmentation for MMC inverters," PCIM Europe 2022, pp. 150-156.

PCIM Europe 2024, 11– 13 June 2024, Nuremberg      DOI: 10.30420/566262129

# Next Generation 4.5 kV IGBT-only StakPak Module with Reduced Losses and High Temperature Capability

Jeremy Jones[1], Gaurav Gupta[1], Boni Boksteen[1], Evgeny Tsyplakov[1], Makan Chen[1], Luca De Michielis[1], and Gontran Pâques[2]

[1] Hitachi Energy, Fabrikstrasse 3, 5600 Lenzburg, Switzerland

[2] Hitachi Energy Research, Segelhofstrasse 1 A, 5405 Baden-Daettwil, Switzerland

Corresponding author:    Jeremy Jones, jeremy.jones@hitachienergy.com
Speaker:                 Jeremy Jones, jeremy.jones@hitachienergy.com

## Abstract

This paper discusses Hitachi Energy's most recent addition to its StakPak's insulated gate bipolar transistor (IGBT) press-pack platform: the 4.5 kV / 2.5 kA Gen-1 IGBT-only StakPak module. This module uses our improved Gen-1 IGBT technology that has been optimized for losses. With a reduced $V_{CEsat}$ of 600 mV in comparison to its predecessor Gen-0 IGBT technology, current capability is boosted while keeping turn-off losses unaffected. Furthermore, the well demonstrated backside technology design of our Gen-1 IGBT has been proven to reduce collector-emitter leakage current by more than 50%, allowing for best in class high temperature capability $T_{vj}$ = 150°C. Furthermore, at $T_{vj}$=150°C the device has been shown to have reverse bias safe operating area switching capability of $I_C$ > 2x $I_{nom}$ ($V_{CC}$ = 3.6 kV) on module level and short circuit pulse withstand time $t_p$>18 µs ($V_{CC}$ = 3.6 kV) on submodule level. These characteristics make the Gen-1 based IGBT-only StakPak module highly competitive offering major potential for high voltage direct current transmission applications and Hitachi Energy seeks to bring the technologies advantages to current ratings as high as 5 kA.

## 1    Introduction

High voltage direct current transmission (HVDC) based on voltage sourced converter technology (VSC) is taking on an important role in ensuring that the energy of the future is reliable, renewable, and sustainable. VSC-HVDC allows the transfer of large amounts of electricity over long distances with high efficiency, enabling the effective integration of different energy sources into the power grid. Moreover, the technology is also used to efficiently interconnect power grids, adding stability and resilience to this infrastructure. Power semiconductors are critical components of VSC-HVDC technology. The demand for more capable power semiconductors with higher current ratings, lower losses, higher operating temperature, better short circuit performance and improved reliability is therefore ever increasing. Hitachi Energy is a world leader in the design and manufacture of such devices. Our continually evolving StakPak platform, based on insulated gate bipolar transistor (IGBT) and diode press-pack modules as well as our bi-mode IGBTs (BIGT) [1]-[3], is designed to address the unique challenges that HVDC systems face and will continue to push the limits of power ratings and performance. In this work we highlight improvements made to our IGBT technology, with respect to losses and temperature capability.

The paper is structured as follows. Section 2 introduces briefly the StakPak and its underlying IGBT chip technology. Section 3 shows comparative results between Gen-0 (predecessor) and Gen-1 (improved technology) and gives more insight into module operation of Gen-1 technology at high junction temperatures ($T_{vj}$). Finally, section 4 discusses the importance of such results for VSC-HVDC systems.

## 2    The StakPak IGBT-only module

### 2.1    Press-pack module

The StakPak is a press-pack type power module, developed especially for HVDC applications and has extensive field installation receiving excellent feedback. StakPak modules are optimized for assembly in a stack, allowing for more efficient system design, minimizing stray inductance, and enabling uniform current sharing among chips and submodules. In this work we focus on our newly developed StakPak module featuring only IGBTs (IGBT-only) rated at 4.5 kV / 2.5 kA, consisting of three submodules in a sturdy, high temperature capable frame. A submodule consists of IGBT chips soldered to a baseplate on which a housing is mounted. Completing the submodule is a spring package, which is laid into the housing for contact as seen in Fig. 2. Once pressed within the stack, the spring package ensures that each chip is well contacted. The modular nature of the StakPak allows us to scale the current by allowing more submodules within a module frame. We are currently developing an IGBT-only StakPak module with six Gen-1 based submodules using the same footprint as for the module presented in this paper, allowing us to raise the nominal current to $I_C$=5 kA. For use within VSC-HVDC systems the IGBT-only StakPak module can be complemented with one of Hitachi Energy's discrete fast recovery diodes which are designed for high surge current capability [1], [11].

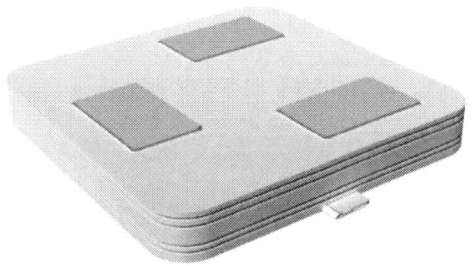

Fig. 1: 4.5 kV / 2.5 kA IGBT-only StakPak module with three submodules.

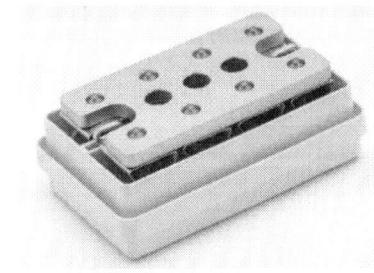

Fig. 2: IGBT-only StakPak submodule

## 2.2 Chip technology

The reference chip technology used in this work is based on Hitachi Energy's enhanced planar cell concept described in [4]. This baseline chip design is then further optimized for lower on-state losses by enhancing the injection efficiency of the backside collector and incorporating higher channel density as required for high current capability [4]. Conventionally, higher injection efficiency would have also resulted in an increase in the collector-emitter leakage current. However, our new chip also features advanced backside technology which not only lowers the sensitivity of collector leakage with respect to its injection efficiency but also helps in reducing the overall leakage current allowing higher temperature operations. The new Gen-1 chip also features improved (shorter) termination design which in turn leads to higher available active area for a given chip size further boosting its current capability. The improved termination design also allowed us to further reduce the thickness of the Si drift region which led to improvement in overall technology trade-off ($E_{off}$ vs $V_{CEsat}$ curve) as compared to the Gen-0 chip.

## 3 Results and discussion

### 3.1 Gen-1 vs. Gen-0 comparison

#### 3.1.1 General remarks

The data presented in section 3.1 serve the purpose of illustrating the improvements achieved through chip optimizations mentioned in section 2.2. The results shown here were obtained through submodule level measurements and are scaled to module level, unless otherwise stated. To allow a fair comparison between the two chip generations, the same test setups and measurement conditions were used for both Gen-0 and Gen-1 and measurements are mostly compared at junction temperature $T_{vj}$=125°C. The testing procedure and test setup used for submodule and module level measurements are comparable. A more detailed description of our testing methodology is found in section 3.2.

#### 3.1.2 On-state and technology trade-off curve

One of the key findings of our measurements is the effect of our chip optimizations on the on-state. Figure 3 shows an on-state comparison performed for Gen-0 and Gen-1 chip technology. The devices were tested at nominal current $I_C$=2500 A, gate voltage $V_{GE}$=15 V and temperatures $T_{vj}$=25°C and 125°C. The Gen-1 chip technology was further tested at $T_{vj}$=135°C and 150°C. As can be seen in the figure, the new generation device shows considerably improved saturation voltage $V_{CEsat}$, reduced by 600 mV compared to its previous generation. Furthermore, the $V_{CEsat}$ at $T_{vj}$=150°C is below 3 V. Turn-off switching losses remain unchanged between the two chip generations, due to thinning of the Si drift region for the Gen-1 chip technology as mentioned in section 2.2. Consequently, the Gen-1 technology curve has shifted to a superior position. Figure 4 shows the improvement of the technology trade-off curve from Gen-0 to Gen-1 at $T_{vj}$=125°C, $V_{CC}$=2800 V and nominal current $I_C$=2500 A.

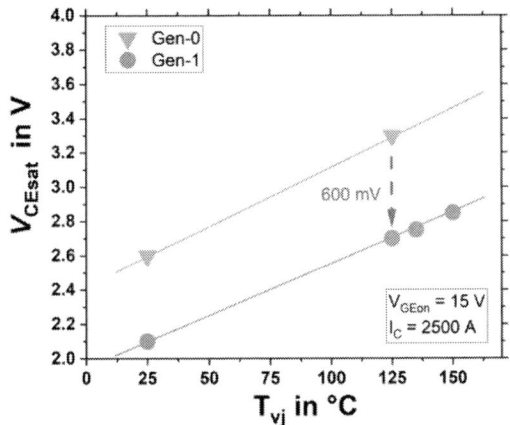

Fig. 3: Saturation voltage ($V_{CEsat}$) of Gen-0 and Gen-1 compared at different temperatures at nominal current $I_C$=2500 A. $V_{CEsat}$ improvement of 600 mV from Gen-0 to Gen-1.

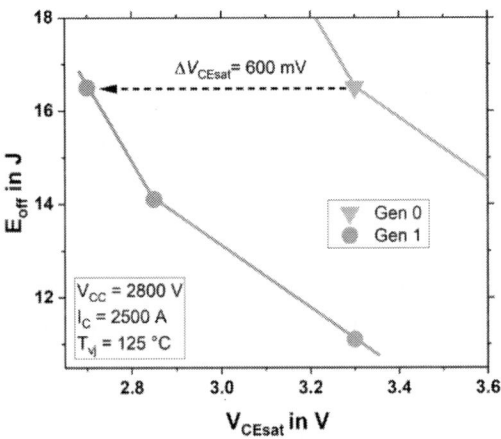

Fig. 4: Technology trade-off curves of Gen-0 and Gen-1 technology at $T_{vj}$=125°C, nominal current $I_C$=2500 A, and $V_{CC}$=2800 V, scaled from submodule to module level. Gen-1 technology shows improved tech-curve.

### 3.1.3 Turn-on switching losses

The effects of the higher channel density design used for Gen-1 technology are partly also seen in the turn-on losses of the new chip generation. Figure 5 highlights the turn-on losses of the old and new chip design at $T_{vj}$=125°C, $V_{CC}$=2800 V and nominal current $I_C$=2500 A, with Gen-1 boasting a reduction of close to 40%.

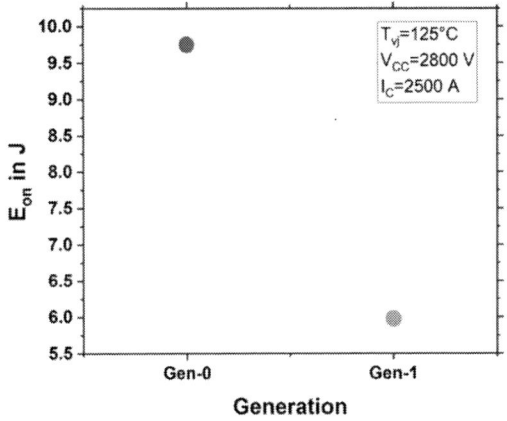

Fig. 5: Turn-on losses scaled from submodule to module level shown for Gen-0 and Gen-1 devices at $T_{vj}$=125°C and nominal conditions $V_{CC}$=2800 V, $I_C$=2500 A.

### 3.1.4 Collector-emitter leakage current

As the need for improved temperature capability of high power IGBTs becomes more important for VSC-HVDC applications [1] we respond with good progress on the new generation chip technology with respect to collector-emitter leakage current $I_{CES}$ as seen in Fig. 6. For the presented collector emitter leakage current measurements, the gate and the emitter of the tested devices were shorted and $V_{CE}$=2800 V, 3400 V and 4500 V were applied for a pulse time $t_p$=100 ms each.

Improvements from one generation to the next at $T_{vj}$=125°C amount to a leakage current reduction well above 50% for all the displayed voltages. An $I_{CES}$ of around 12 mA at the nominal blocking voltage $V_{CE}$=4500 V can be seen for Gen-1. The reduction in the leakage current allows us to raise the operating junction temperature $T_{vj(op)}$ to 150 °C. Figure 7 shows measurements of Gen-1 single chip with and without the new backside process and design at $T_{vj}$=150°C measured at different blocking voltages $V_{CE}$.

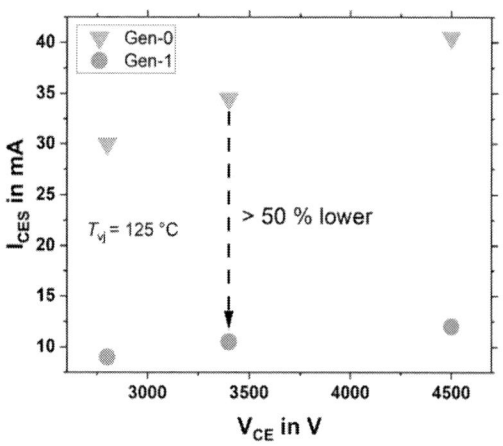

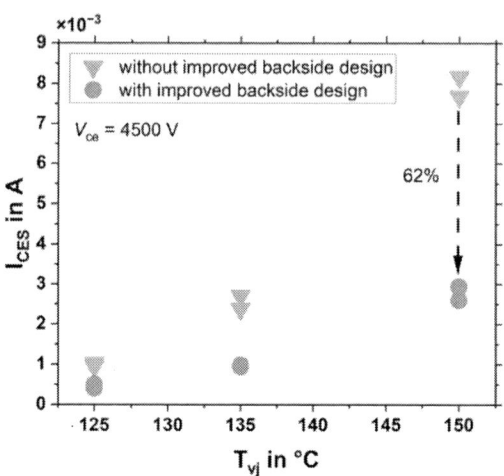

Fig. 6: Collector-emitter leakage current at $T_{vj}$=125°C and different $V_{CE}$ for Gen-0 and Gen-1 IGBT-only technology, scaled to module level.

Fig. 8: Chip level collector-emitter leakage current versus temperature $T_{vj}$ for Gen-1 chip with and without improved backside design at $V_{CE}$=4500 V.

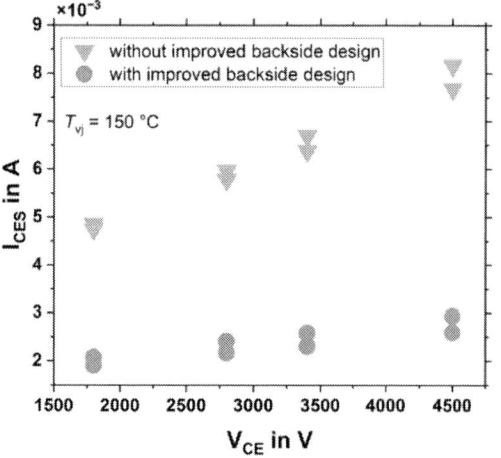

Fig. 7: Chip level collector-emitter leakage current versus different blocking voltages $V_{CE}$ for Gen-1 chip with and without improved backside design at $T_{vj}$=150°C

Figure 8 shows the same chip comparison at $V_{CE}$=4500 V measured at different junction temperatures $T_{vj}$. At $T_{vj}$=150°C and nominal blocking voltage $V_{CE}$=4500 V, the new backside technology reduces the leakage current by over 60%, significantly reducing the risk of thermal runaway.

## 3.2 Gen-1 4.5 kV / 2.5 kA module high temperature operation

### 3.2.1 Measured devices

Data presented in section 3.2 was gathered from 4.5 kV / 2.5 kA Gen-1 IGBT-only StakPak module level measurements with exception of the short-circuit data, where measurements were performed on submodule level and scaled to module level.

### 3.2.2 On-state

On-state measurements were conducted by pushing specified currents from $I_C$=100 A up to 2x $I_{nom}$=5 kA through the device at different temperatures $T_{vj}$=25°C, 135°C and 150°C with $V_{GE}$=15 V.

The on-state characteristics on module level at given conditions is shown in Fig. 9. The characteristics show positive temperature coefficient and for nominal current $I_C$=2500 A, the saturation voltage $V_{CEsat}$ is lower than 3 V. Figure 10 shows the output characteristics of the new Gen-1 based module as usually shown in the data sheet of our devices. On-state was measured for currents up to 2x $I_{nom}$ at $T_{vj}$=150 °C with $V_{GE}$ ranging from 9 V to 19 V.

PCIM Europe 2024, 11– 13 June 2024, Nuremberg        DOI: 10.30420/566262129

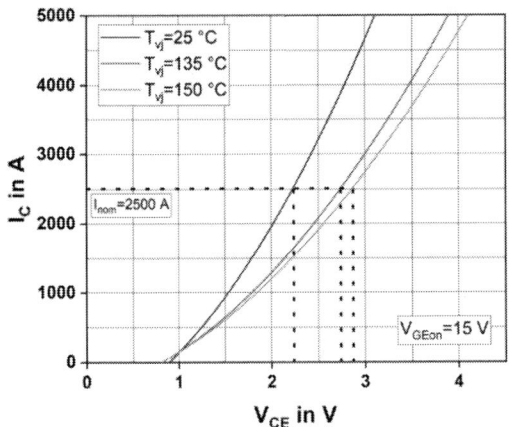

Fig. 9: On-state characteristics for Gen-1 based 4.5 kV / 2.5 kA IGBT-only module at different temperatures and $V_{GEon}$=15 V.

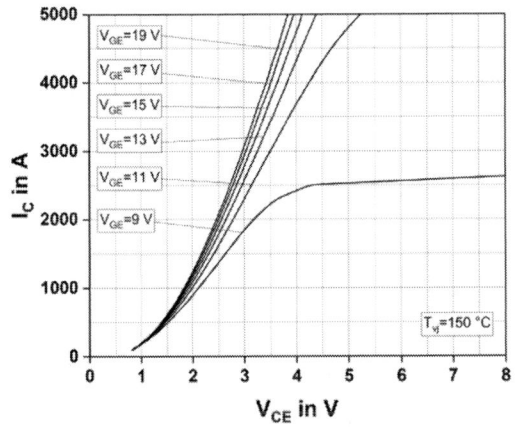

Fig. 10: Output characteristics for Gen-1 based 4.5 kV / 2.5 kA IGBT-only module.

### 3.2.3 Dynamic losses

Dynamic losses were measured using the standard double pulse test with an inductive switching test setup. The test circuit was made up of a phase-leg configuration with a DC-Link capacitor with capacitance $C_{DC\text{-}Link}$=2.1 mF and a set of variable load inductances. The Gen-1 based IGBT-only module was used as the bottom switch and for the top switch we used an adapted version of Hitachi Energies 5SNA 2000K450300 [12] featuring a 1:1 IGBT to diode ratio to safely conduct

the freewheeling current between switching events. The total stray inductance was $L_\sigma$=160 nH. Hitachi Energies custom made programmable gate drive unit was used to control the gate, by varying the gate resistance $R_G$. An external $C_{GE}$=330 nF was added close to the gate of the IGBT-only StakPak module for smoother gate signals.

Figure 11 shows the turn-off switching waveform of our new IGBT-only StakPak module at $T_{vj}$=150°C at nominal current $I_C$=2500 A and $V_{CE}$=2800 V with $R_{Goff}$=14.2 Ω. Similarly, Fig. 12 shows the turn-on switching characteristics of the module at the same conditions with $R_{Gon}$=1.6 Ω.

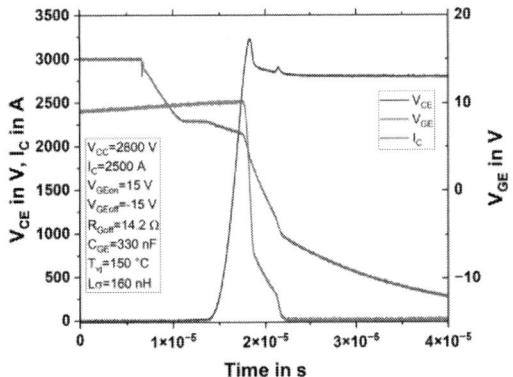

Fig. 11: IGBT turn-off waveform for Gen-1 based 4.5 kV / 2.5 kA IGBT-only module at nominal conditions $V_{CC}$=2800 V, $I_C$=2500 A, $V_{GEoff}$=-15 V, $L_\sigma$=160 nH, $T_{vj}$=150°C.

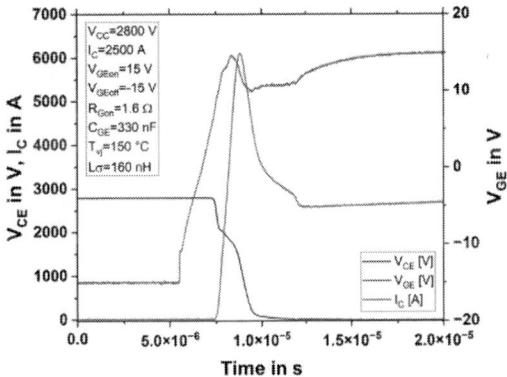

Fig. 12: IGBT turn-on waveform for Gen-1 based 4.5 kV / 2.5 kA IGBT-only module at nominal conditions, $V_{CC}$=2800 V, $I_C$=2500 A, $V_{GEoff}$=-15 V, $L_\sigma$=160 nH, $T_{vj}$=150°C.

The current dependent dynamic switching losses at $T_{vj}$=135°C and 150°C are shown in Fig. 13, where $I_C$ is swept from a few hundred ampere to 2x $I_{nom}$=5 kA, with $V_{GEon}$=15 V and $V_{GEoff}$=-15 V. For switching off the device at $T_{vj}$=150°C two different gate driving conditions are shown. The preferred condition is with an external gate resistor $R_{Goff}$=14.2 Ω. We show the second condition at $R_{Goff}$=7.3 Ω, mainly to compare the turn-off switching losses to operation at $T_{vj}$=135°C where an external gate resistor of $R_{Goff}$=7.3 Ω was used. This change in gate-driving between temperatures is a consequence of a trade-off between losses and reverse bias safe operating area (RBSOA) capability as described in a later section of this paper. The turn-on gate resistor was chosen to be $R_G$=1.6 Ω. Our new IGBT-only StakPak module impresses with low switching losses at nominal conditions $V_{CC}$=2800 V and $I_C$=2500 A. For each switching event IGBT turn-off losses $E_{off}$ fall below 20 J for both temperatures and IGBT turn-on losses $E_{on}$ lie slightly above 10 J. As seen in the figure, at same gate driving conditions, the difference between the two operating temperatures has a minor influence on losses.

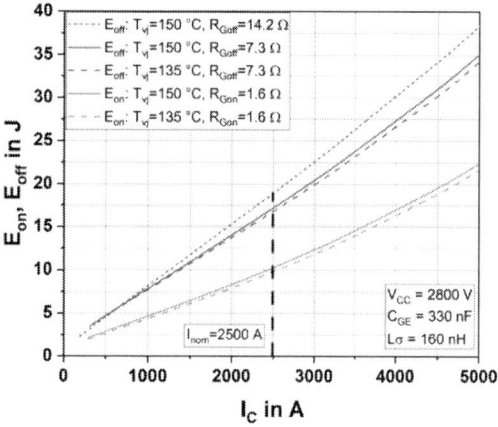

Fig. 13: IGBT turn-on and turn-off switching characteristics versus collector current $I_C$ for Gen-1 based 4.5 kV / 2.5 kA IGBT-only module at different temperatures. $V_{CC}$=2800 V and $L_\sigma$=160 nH.

Fig. 14 presents the switching losses characteristics with respect to gate resistor when the Gen-1 based device operates at $T_{vj}$=135°C and $T_{vj}$=150°C with nominal test conditions $V_{CC}$=2800 V and $I_C$=2500 A. As seen in the graph, the turn-off losses only slowly increase with increasing $R_{Goff}$, as they are strongly influenced by intrinsic factors [6], [7]. In contrast turn-on losses

are strongly dependent on the choice of gate resistor, making a low $R_{Gon}$ the preferred option for low losses. Turn-on switching losses might vary from the values presented here depending on what type of freewheeling diode is used.

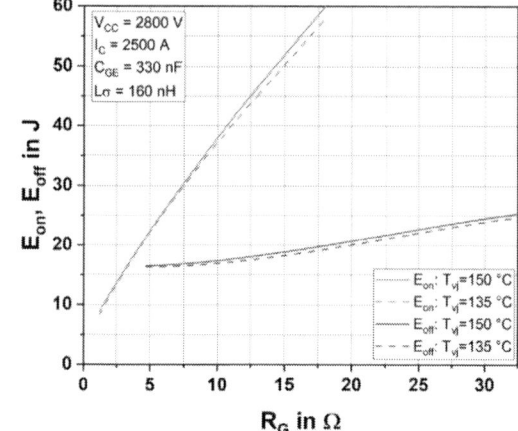

Fig. 14: IGBT turn-on and turn-off switching characteristics versus gate resistor $R_G$ for Gen-1 based 4.5 kV / 2.5 kA IGBT-only module at different temperatures.

### 3.2.4    3.6 kV RBSOA

With our design and process improvements IGBT-only StakPak modules manage to reach impressive RBSOA capability beyond 2x Inom. This capability is demonstrated in Fig. 15, which shows a switching event, where the DC-link voltage $V_{CC}$=3.6 kV, $I_C$=5.5 kA (2.2x $I_{nom}$) and $R_{Goff}$=14.2 Ω at $T_{vj}$=150 °C. At given stray inductance $L_\sigma$=160 nH the device can be safely operated without the risk of reaching critical overvoltage. Another well-known threat to device safety is the occurrence of dynamic avalanche, that can occur in IGBTs and prove limiting to RBSOA capability in high-voltage-high-current IGBTs, particularly at high temperatures [8]-[10]. With our choice of $R_{Goff}$=14.2 Ω we have limited effects of dynamic avalanche and still manage to switch off our device at superior losses as mentioned in section 3.2.3. At $T_{vj}$=135°C losses can be further decreased by choosing the external gate resistor as low as $R_{Goff}$=7.3 Ω with RBSOA capability still good enough to switch off the module at $V_{CC}$=3.6 kV and collector current $I_C$=2.2x Inom.

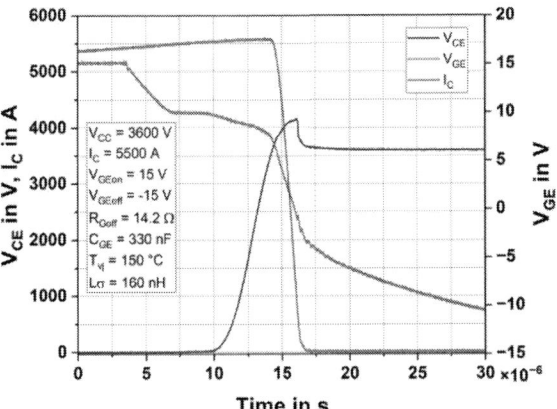

Fig. 15: RBSOA at 2.2x $I_{nom}$ switching current for Gen-1 based 4.5 kV / 2.5 kA IGBT-only module at $T_j$=150°C and $V_{CC}$=3.6 kV.

### 3.2.5  3.6 kV short-circuit SOA

In addition to strong RBSOA capability, the Gen-1 technology also shows impressive short-circuit safe operating area (SCSOA) capability at $T_{vj}$=150°C, withstanding pulses greater than 18 µs on submodule level. This can be seen in Fig. 16 where a short circuit pulse is presented with $V_{CC}$=3.6 kV and $V_{GE}$=15 V. Despite the strong short circuit withstand time of our Gen-1 based device, the magnitude of the maximum short circuit current $I_{SCmax}$=18 kA can be undesirable. Future generations of our IGBT-only StakPak module may feature a reduced short circuit current without compromising on-state characteristics using our novel MOS-cell design principle, reported previously in [5].

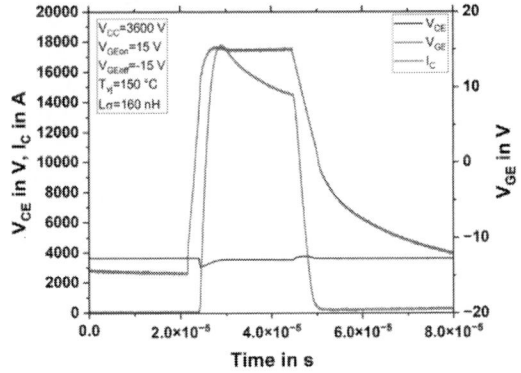

Fig. 16: SCSOA waveform at $V_{CC}$=3.6 kV and $T_{jmax}$=150°C scaled from submodule to module level for Gen-1 IGBT technology.

## 4  GEN-1 IGBT-only in VSC-HVDC applications

VSC-HVDC applications predominantly utilize a half-bridge (HB) cell based modular multi-level converter (MMC) as shown in Fig. 17. Each cell may consist of two in series connected IGBT switches S1 and S2 with an anti-parallel diode each, connected in parallel with a capacitor. Parallel to S1, a mechanical bypass to carry the phase current when S1 fails is installed. Discrete fast recovery diodes of Hitachi Energy are designed to meet the surge current ($I_{fsm}$) demands of the converter and in case of failure can securely carry the current. In a point-to-point HVDC system, typically one converter operates as rectifier and the other converter operates as inverter. In wind power integration, the offshore converter is a rectifier, and the onshore converter is an inverter. In HVDC interconnectors, the power direction may be changed at any time.

To control the active and reactive power, the desired AC voltage is created at the AC terminal of the converter by controlling the cell insertion (S1 off, S2 on) or bypass (S1 on, S2 off). Due to the high number of cells in an arm of the MMC, the switching frequency of the semiconductors in a cell is, typically between 1~3 times the fundamental frequency.

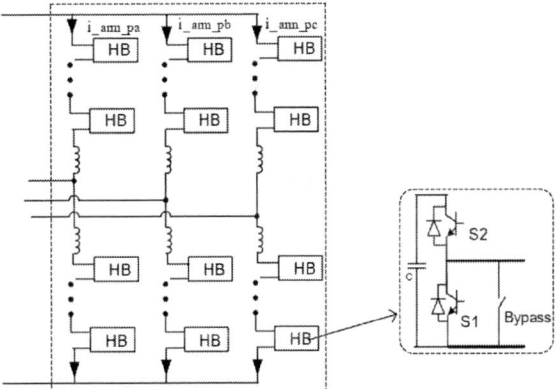

Fig. 17: Schematic of modular multilevel converter predominantly used for VSC-HVDC applications.

This means that most of the losses are comprised by on-state losses, and that's why our Gen-1 based IGBT-only StakPak device with its low $V_{CEsat}$ will significantly contribute to lowering converter loss. By increasing temperature capability of our devices to $T_{vj}$=150°C the output power of HVDC systems can be increased, due to reduction in cooling power needed. Additionally, the overall safety of the systems is increased through high temperature RBSOA and SCSOA capability.

Coupled with the strong $I_{fsm}$ performance of one of our discrete diodes [11], our IGBT-only Gen-1 module makes for a highly effective choice of switch for VSC-HVDC systems.

## 5   Conclusion

In this paper we have presented our advanced Gen1 IGBT chip technology with an improved backside design for reduced leakage currents enabling high temperature operation $T_{vj(op)}$=150°C and a reduced risk of thermal runaway. We have shown that our new generation IGBT-only StakPak is well suited for VSC-HVDC applications thanks to its excellent saturation voltage $V_{CEsat}$ below 3 V at $T_{vj}$=150°C ($I_C$=2500 A), without negatively affecting turn-off switching losses and thus improving the technology trade-off curve. The significantly improved IGBT turn-on losses undoubtfully further improve the overall HVDC system losses, making our Gen-1 IGBT technology powerful. We have shown that our Gen-1 devices can withstand challenging RBSOA conditions at high temperatures ($T_{vj}$=150°C), switching currents beyond $I_C$=2x Inom at $V_{CC}$=3.6 kV adding to overall system safety. Adding to the robustness of this device, they have been shown to withstand long short-circuit pulses of more than $t_p$=18 μs at $T_{vj}$=150°C and $V_{CC}$=3.6 kV as seen on submodule level. The new IGBT-only StakPak module is an important and effective addition to our evolving StakPak platform. Such advancements will enable further increase of current rating e.g. to $I_C$=5 kA with identical module footprint and similarly impressive losses and safe operating area capability. Combined with one of Hitachi Energy's discrete fast recovery diodes, the IGBT-based module is well suited for VSC-HVDC applications, contributing greatly to a sustainable energy future.

## References

[1] E. Tsyplakov et al., New generation high power semiconductors for 8GW VSC-HVDC applications, PCIM Asia 2022

[2] Hitachi Energy, 5SMA 3000L450300, Data Sheet, Doc. No. 5SYA 1430-02, 11-2017

[3] B. Boksteen et al. "Second Generation BIGT Chip Advancing the STakPak Platform" PCIM Europe 2021

[4] M. Andenna et al., "Rugged 4500V HiPak Module with 1500A Current Rating and 150°C Capability for Traction Application", PCIM 2020

[5] G. Gupta et al., "Novel MOS-cell Engineered 4.5 kV Enhanced-planar IGBT Device for Improved Short-Circuit Capability", PCIM Europe 2023

[6] K. Konishi, T. Nitta, T. Tamaki and S. Soneda, "Separate-Bottom Player CSTBT™ for Approaching Turn-off Switching Loss Reduction Limit," 2023 35th International Symposium on Power Semiconductor Devices and ICs (ISPSD), Hong Kong, 2023

[7] S. Machida, K. Ito and Y. Yamashita, "Approaching the limit of switching loss reduction in Si-IGBTs," 2014 IEEE 26th International Symposium on Power Semiconductor Devices & IC's (ISPSD), Waikoloa, HI, USA, 2014

[8] P. Muenster, D. Wigger and H. Eckel, "Impact of the dynamic avalanche on the electrical behavior of HV-IGBTs," Proceedings of PCIM Europe 2015; International Exhibition and Conference for Power Electronics, Intelligent Motion, Renewable Energy and Energy Management, Nuremberg, Germany, 2015

[9] P. Rose, D. Silber, A. Porst and F. Pfirsch, "Investigations on the stability of dynamic avalanche in IGBTs," Proceedings of the 14th International Symposium on Power Semiconductor Devices and Ics, Sante Fe, NM, USA, 2002

[10] A. Bryant et al., "Investigation Into IGBT dV/dt During Turn-Off and Its Temperature Dependence," in IEEE Transactions on Power Electronics, vol. 26, no. 10, pp. 3019-3031, Oct. 2011

[11] J. Vobecky et. al, "A 4.5 kV Fast Recovery Diode Platform for High-Current IGBTs", PCIM Europe 2024

[12] Hitachi Energy, 5SNA 2000K450300, Data Sheet, Doc No. 5SYA 1431-02, 01-2018

PCIM Europe 2024, 11– 13 June 2024, Nuremberg        DOI: 10.30420/566262130

# Finite Element Analysis of the upscaling of Warpage and Bifurcation Hysteresis Loops: from Cu/Si Die to Large Wafers

Vincenzo Vinciguerra[1], Giuseppe Luigi Malgioglio[1], Marco Renna[1]

[1]Quality, Manufacturing and Technology (QMT) – Power & Discrete Technologies R&D Department,

STMicroelectronics, Stradale Primosole 50, 95121 Catania, Italy.

Vincenzo Vinciguerra: https://orcid.org/0000-0003-2188-4178
Corresponding author: Vincenzo Vinciguerra, vincenzo.vinciguerra@st.com
Speaker: Vincenzo Vinciguerra, vincenzo.vinciguerra@st.com

## Abstract

Large semiconductor wafers with thick electrochemical deposited copper (Cu_ECD) layers suffer from severe warpage during the final thinning process. This can lead to an asymmetric warpage or bifurcation of the wafer. Finite element analysis software can predict the phenomenon of warpage and bifurcation, but an accurate prediction requires considering the plastic behavior of the metal layer. The study investigated the extension of a finite element analysis multilinear kinematic hardening model of plastic Cu_ECD from a die level to the wafer level to predict the warpage hysteresis loop and deduce the phenomenon of bifurcation during the thinning process of a 200 mm standard wafer.

## 1    Introduction

Large semiconductor wafers (e.g. 200 mm nominal wafers) metalized with thick electrochemical deposited copper (Cu_ECD) layers suffer severe warpage as they undergo the final thinning process. Moreover, as their thickness reaches a critical value the warpage can degenerate into the bifurcation phenomenon, which determines an asymmetric warpage of the wafer. Although, it is possible to describe the increase in curvature by recurring to finite element analysis software [1-4], an accurate and more precise prediction of the phenomenon of warpage [5-13] and bifurcation [14-21], occurring also in large Cu_ECD metalized wafers, must consider the plastic behavior [22] of the metal layer. Indeed, the plastic behavior of thick Cu_ECD emerges in the investigation of warpage during thermal cycling as a warpage hysteresis loop [23]. In this work, we have investigated the extension of a finite element analysis multilinear kinematic hardening (MKH) model of plastic Cu_ECD, from a die level to the wafer level in order to predict the warpage hysteresis loop at wafer level and deduce the phenomenon of bifurcation as the wafer undergoes the thinning process. First

the bounty of the MKH model has been tested by means of experimental data also reported in the literature [22], that is by simulating the warpage hysteresis loop at die level of 20 µm Cu_ECD thick deposited on a Si (001) die having standard thickness. Then the investigation of the warpage model has been extended to the use-case of a 200 mm standard wafer Cu_ECD metalized, and the results compared in terms of resulting curvatures. As expected, and reasonably, the curvatures were comparable. Given these results we extended the investigation on the warpage hysteresis loop to the case of thinned silicon wafer substrates and investigated the emergence of bifurcation.

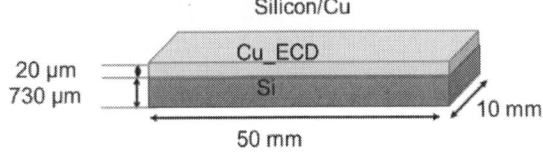

**Fig. 1** Schematic of a silicon (100) 5cm x 1cm die sample, 730 µm thick metalized with a 20 µm thick copper electrochemically deposited (ECD) layer.

# 2 Materials & Methods

A finite element elastic-plastic model of thick e lectrochemical deposited (ECD) copper Cu_EC D layer has been developed, such that it can be of use to simulate the warpage evolution i n Cu redistribution layers [23]. In particular, th e warpage hysteresis loop of Cu_ECD layer d eposited on silicon (100) has been reproduced based on available experimental data [22] an d approximated firstly by an isotropic hardenin g model and hence by a multilinear kinematic hardening (MKH) model.

Finite Element Analysis (FEA) static structural investigation of the warpage have been develo ped by using ANSYS® Mechanical package 2 023/R2.

The simulation was set as follows. The whole system underwent a thermal cycle having 14 steps in one case and 23 steps in a second case. To simulate the warpage hysteresis loop , the starting environment temperature was set at 30° C. The base of the die was fixed with the elastic support set at 0.5 N/mm^3. The e volution of non-linearities were included in the simulation by including the large deflections.

The stress-free condition was considered at t he starting temperature of 30°C.

## 2.1 Bilinear isotropic hardening model

By heating a Cu_ECD/Si(100) die (e,g, as the one sketched in figure 1) the thick copper lay er develops a thermo-mechanical stress, which determines a warpage of the whole structure.
As the temperature increases, along with the non-linearities, deviations from the elastic beha viour emerge. Because of the plastic behaviou r of the copper layer the warpage vs. tempera ture plot shows an hysteresis which results in a warpage hysteresis loop (WHL). A first purp ose of this work is to describe the emergence of the WHL in a Cu_ECD/Si(100) die.
As a first approximation the plastic behaviour of Cu_ECD and the WHL can be described b y setting up a bilinear isotropic hardening mod el.

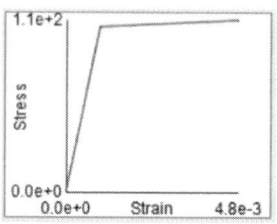

**Fig. 2** Bilinear isotropic hardening model of cop per having a yield strength of 110 MPa and a tan gent modulus of 1150 MPa.

In figure 2, an example of the bilinear isotropi c hardening model has been considered. The model requires to specify only two parameters, the yield strength, which in the specific case of the model of figure 3 is set to a value of 110 MPa, and the tangent modulus which is s et equal to 1150 MPa. In the graph stress-str ain of the bilinear isotropic hardening model, t he yield strength is the point at which deviatio ns from the elastic behaviour emerges.

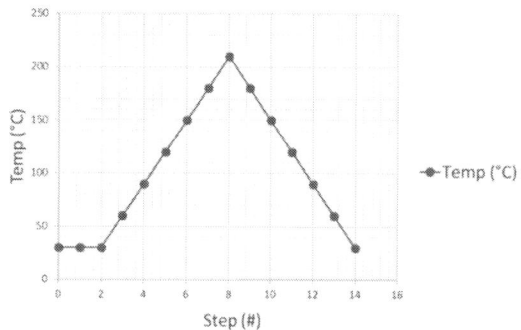

**Fig. 3** Thermal Cycle exploited to simu late the plastic behavior of Copper ECD depos ited on silicon.

The bilinear isotropic hardening model has be en tested against the experimental data of ref [22]. The system Cu_ECD/Si(100) underwent t he thermal cycle reproduced in figure 3. The soak temperature, or maximum temperature, w as regulated to 210°C and the warpage of the 1 cmx 5 cm die monitored.

In figure 4, the warpage of the bilinear isotrop ic hardening model has been reported as a fu nction of the temperature and compared with t he measured values of warpages reported in r ef [22]. To better adapt the model to the expe rimental data the maximum soak temperature of the thermal cycle has been increased to 25 0°C. Moreover, the yield strength has been re duced to 90 MPa. In spite of these changes, i t is clear that a bilinear isotropic hardening m odel can only provide a qualitative and rough description of a WHL. An issue is determined by the abrupt changes which result in the WH L graph.

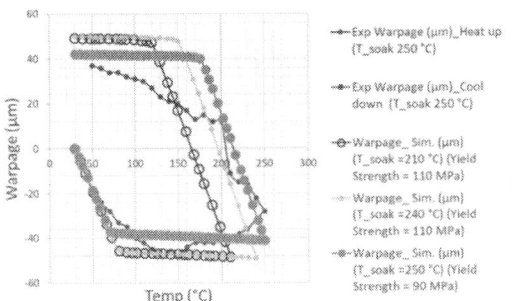

**Fig. 4** Warpage Hysteresis Loop result- ing from the bilinear isotropic harden- ing model of Copper ECD deposited on Silicon.

## 2.2 Designing a Multilinear Kinematic Hardening model of Cu_ECD

A smoother and more accurate approximation of the WHL graph can be obtained by develo ping opportunely a multilinear kinematic harde ning (MKH) model of the Cu_ECD.

In fact, a given thermal budget induces in the film a thermal strain $\varepsilon_{thermal}$ which results fro m the difference of the CTEs (coefficients of t hermal expansion) between the substrate and the thin metal film. Because of the adhesion b etween the film and the substrate, the thin me tal film is constrained to deform accordingly, a nd the thermal strain induces an opposite total strain. Since in a thin metal film, the total st rain has two contributions, the elastic strain $\varepsilon_{elastic}$ and the plastic strain $\varepsilon_{plastic}$, it follows that these quantities are related according to t he equation:

$$\varepsilon_{thermal} = -\varepsilon_{elastic} - \varepsilon_{plastic}. \quad (1)$$

More generally, it is assumed that the total eq ui-biaxial strain $\varepsilon_{equi-biaxial}$ of a thin film depo sited on a thick substrate doesn't change muc h when the substrate deforms,

$$\varepsilon_{thermal} + \varepsilon_{elastic} + \varepsilon_{plastic} = \varepsilon_{total\ equi-biaxial} = const. \quad (2)$$

From this equation, it follows that the time rat e of change of $\varepsilon_{total\ equi-biaxial}$ is zero. Hence

, by considering that the elastic strain is equal to $\sigma(t)/M_f$, the thermal strain $(\alpha_f - \alpha_{Si})\Delta T$, t he rate equation governing the change of the stress of the thin film is

$$\frac{\dot{\sigma}}{M_f} + (\alpha_f - \alpha_{Si})\frac{dT}{dt} + \dot{\varepsilon}_{plastic} = 0 \quad (3).$$

By trials and errors, it is possible to define a stress strain curve, which can reproduce a sm ooth WHL of the Cu_ECD/Si(100) system. Ind eed, the stress has been obtained from the c urvature determined by warpage according to t he Stoney equation [24-25], whereas the strai n was obtained according to the following equ ation:

$$\varepsilon_{Total} = \varepsilon_{Elastic} + \varepsilon_{Plastic} = -\varepsilon_{Thermal} = (\alpha_{Cu} - \alpha_{Si})\Delta T \quad (4)$$

Where $\Delta T$ is the thermal load, $\alpha_{Cu}$ the coefficient of thermal expansion (CTE) of copper and $\alpha_{Si}$ the CTE of silicon.

In figure 5 the resulting MKH model, consistin g in a stress-strain curve has been reproduce d. A best fit of the curve according to the tre nd law:

$$\sigma_{MKH} = \frac{C}{\gamma}\tanh(\gamma\varepsilon_{Total}) \quad (5)$$

has been also reported. From the best fit the values of the parameters C and γ are the fol lowing:

| C (MPa) | γ |
|---------|---|
| 101316.8 | 1133.821 |

**Table 1.** Values of constants C and γ accordi ng to best fitted values.

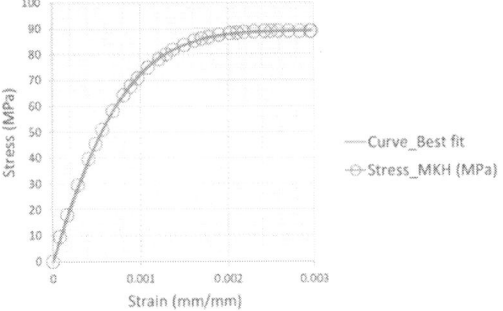

**Fig. 5** Best fit curve of the stress strain curve for the Multilinear Kinematic Model

# 3    Results

The designed multilinear kinematic hardening model has been tested firstly on the 1 cmx 5 cm Cu_ECD/Si(100) die model. In fig.6 the comparison of the warpage resulting from the simulations and with the experimental data of ref [22] has been reported. It is apparent the smooth trend of the MKH model and the good agreement with the experimental data. In order to go further in our investigation, the MKH model has been extended to the case of a 200 mm Si (100) wafer as sketched in figure 7. Moreover, to complete the WHL, the thermal cycle has been extended to -50 °C according to what was reported in figure 8.

Hence, we compared the simulation of the WHL of the Cu_ECD/Si(100) die with that of the wafer system.

In figure 9, the result of such investigation has been reported. Indeed, it can be observed as by upscaling the system, the curvatures resulting from the warpage hysteresis loop, gained for the die with those gained from a 200mm wafer having both the same standard thicknesses of 730 µm, are comparable, consistent and overlapping. This result is what is expected and allowed us extending the investigation to the case of thinned wafers.

In figure 10 the evolution of the WHL for the case of thinned wafers has been reported for the cases of 730 µm, 600 µm, 500 µm and 400 µm. Indeed, as the thickness of the wafer decreases it results that the curvature increases. Moreover, by probing the resulting directional deformation at +250°C and -50°C, we can observe as the wafer bifurcates with a negative and positive curvature, respectively.

In particular, in figure 11. we report the distribution of the directional deformation along the z-direction for the case of a wafer having a thickness of 400 µm when the warpage hysteresis loop reaches the temperature of +250 °C. In fig. 12 we report the distribution of the directional deformation along the z-direction for the case of the same wafer when the warpage hysteresis loop reaches the temperature of -50 °C.

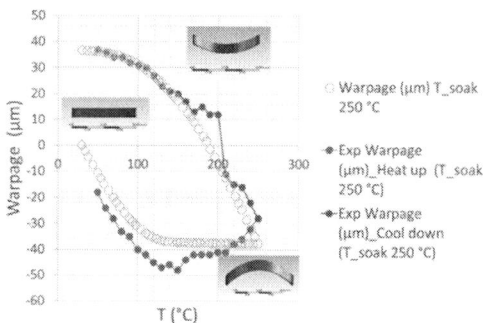

**Fig. 6.** Graph of the warpage hysteresis loop of a 5 cm x 1cm die Cu_ECD/Si (001) reporting the experimental data where the soak temperature was of 250 °C collected from ref [22] and the comparison with a multilinear kinematic hardening (MKH) model set up with a Ansys mechanical enterprise R2/2023.

**Fig.7.** Cu_ECD Layer deposited on a 200 mm Si(100) wafer

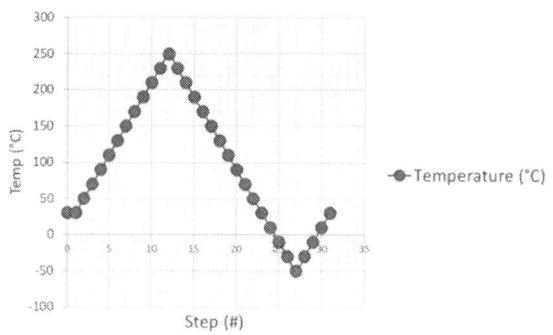

**Fig. 8** Thermal cycle exploited to complete the warpage hysteresis loop.

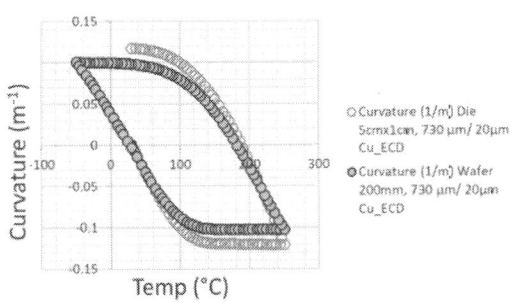

**Fig. 9.** Comparison of the curvatures gained from the simulated warpage hysteresis loop gained of the die of fig. 3 and that simulated for a 200 mm silicon (001) wafer having a thickness of 730 µm, according to the same MKH model.

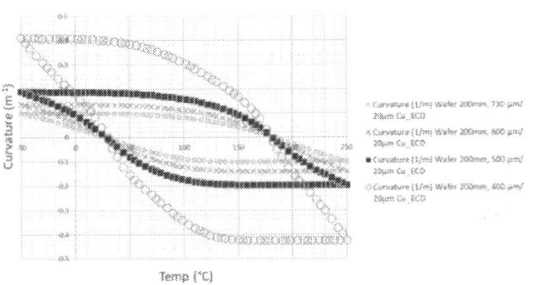

**Fig. 10.** Increase of the curvature of the warpage hysteresis loop of a 200 mm Si (001) wafer metallized with Cu_ECD 20 µm, modelled according to a MKH model, as the thickness of the wafer decreases from 730 µm to 400 µm.

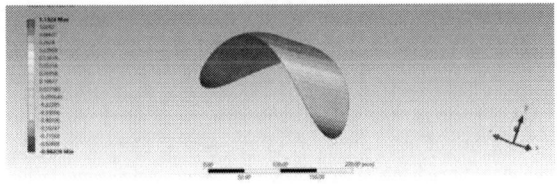

**Fig. 11.** Bifurcation observed at 250 °C in the warpage hysteresis loop of a 400 µm Si (001) 200 mm wafer metalized with a 20 µm Cu_ECD layer.

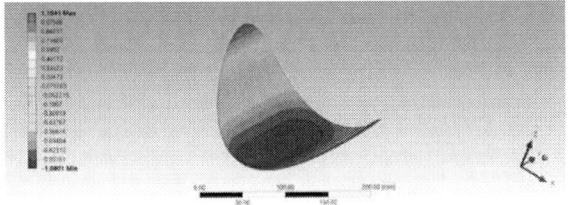

**Fig. 12.** Bifurcation observed at -50 °C in the warpage hysteresis loop of a 400 µm Si (001) 200 mm wafer metalized with a 20 µm Cu_ECD layer.

## 4 Conclusions

The elastic-plastic behavior of Cu_ECD on Si has been successfully modeled, using a multilinear kinematic hardening model. Our model accurately reproduced the experimental warpage hysteresis loop observed at the die level and was consistent with the resulting curvature at the wafer level. An increase in curvature values with the wafer thinning has been observed and the occurrence of bifurcation in the Warpage Hysteresis Loop has been demonstrated. These investigations provide valuable insights into the behavior of Cu_ECD on Si and can help for the design and manufacturing of microelectronics devices.

## 5 Acknowledgments

It is acknowledged the Italian Ministry of Enterprises and Made in Italy (Ministero delle Imprese e del Made in Italy MIMIT) in the frame of the Important Project of Common European Interest (IPCEI) on Microelectronics and Communication Technologies.

## 6 Contributions

V.V. Conceptualization of the work, analytical model elaboration, conceptualization of the ANSYS® simulation experiments, data elaboration, paper drafting, editing, and writing. G.L.M. ANSYS® system design, optimization and simulations. M.R. sponsorship, funding, validation.

## References

[1] A. H. Abdelnaby, G. P. Potirniche, F. Barlow, A. Elshabini, S. Groothuis and R. Parker, "Numerical simulation of silicon wafer warpage due to thin film residual stresses," *2013 IEEE Workshop on Microelectronics and Electron Devices (WMED)*, Boise, ID, USA, 2013, pp. 9-12, http://dx.doi.org/10.1109/WMED.2013.6544506

[2]. A. Mallik and R. Stout, ((2010)). Simulation of process-stress induced warpage of silicon wafers using ANSYS® finite element analysis. in 43rd International Symposium on Microelectronics 2010 IMAPS, (2010).

http://dx.doi.org/10.4071/isom-2010-WA1-Paper3

[3] A. Mallik, R. Stout, J. & Ackaert, (2014). Finite element simulation of different kinds of wafer warpages: Spherical, cylindrical, and saddle. IEEE Transactions on Components, Packaging and Manufacturing Technology, 4(2), 240 - 247. doi: https://doi.org/10.1109/TCPMT.2013.2293873.

[4] J. Schicker, W. Khan, T. Arnold, & C. Hirschl, (2016). Simulating the warping of thin coated Si wafers using Ansys layered shell elements. Composite Structures, 140. http://dx.doi.org/10.1016/j.compstruct.2015.12.062

[5] L. Freund, ((2000).). Substrate curvature due to thin film mismatch strain in the nonlinear deformation range,. Journal of the Mechanics and Physics of Solids, vol. 48, n. 6.

[6]. L. Freund, J. Flora, & E. Chason. (1999). Extensions of the Stoney formula for substrate curvature to configurations with thin substrates or large deformations. Applied Physics Letters, 74(14),

[7]. C.B Masters, N. S. (1990). Deflection shapes due to intrinsic stress in thin films. Mat. Res. Soc. Symp. Proc. Vol. 188, 1990, Materials Research Society.

[8]. C.B Masters, N. S. (1993). Geometrically Nonlinear Stress Deflection Relations for Thin Film/Substrate Systems. Int. J. Engng Sci, Vol 31, No. 6, pp. 915-925.

[9]. D. Shin, J. L. (February 2014.). Analysis of Asymmetric Warpage of Thin Wafers on Flat Plate Considering Bifurcation and Gravitational Force. IEEE Transactions on Components, Packaging and Manufacturing Technology, Vol. 4, No 2.

[10] V. Vinciguerra, G. L. Malgioglio, A. Landi, M. Renna. Determination of the Equivalent Thickness of a Taiko Wafer Using ANSYS Finite Element Analysis. Appl. Sci. 2023, 13, 8528. https://doi.org/10.3390/app13148528.

[11] V. Vinciguerra, G. L. Malgioglio and M. Renna, "A Comparison of Analytical and Finite Element Analysis Methods for Determining the Equivalent Thickness of Large 4H-SiC Taiko Wafers," 2024 25th International Conference on Thermal, Mechanical and Multi-Physics Simulation and Experiments in Microelectronics and Microsystems (EuroSimE), Catania, Italy, 2024, pp. 1-4, http://dx.doi.org/10.1109/EuroSimE60745.2024.10491555

[12] V. Vinciguerra and A. Landi, "On the Way to understand the Warpage in 8" Taiko Semiconductor Wafers for Power Electronics Applications (Si and SiC)," 2021 22nd International Conference on Thermal, Mechanical and Multi-Physics Simulation and Experiments in Microelectronics and Microsystems (EuroSimE), St. Julian, Malta, 2021, pp. 1-14, doi: http://dx.doi.org/10.1109/EuroSimE52062.2021.9410844.

[13]. V. Vinciguerra, G. L. Malgioglio, A. Landi, S. Rascunà, M. Renna, A Comparative Study of Analytical and Finite Element Analysis Numerical Approaches of the Equivalent Thickness of Large 4H-SiC Taiko Wafers, Research Square preprint https://doi.org/10.21203/rs.3.rs-3344668/v1

[14] V. Vinciguerra, G. L. Malgioglio, A. Landi, Modelling the Elastic Energy of a Bifurcated Wafer: A Benchmark of the Analytical Solution vs. The ANSYS Finite Element Analysis. Compos. Struct. 2022, 281, 114996. http://dx.doi.org/10.1016/j.compstruct.2021.114996

[15] V. Vinciguerra, G. L. Malgioglio, A. Landi, A.; M. Renna, Models of Bifurcation and Gravity Induced Deflection in Wide Band Gap 4H-SiC Semiconductor Wafers. In Proceedings of the 2023 24th International Conference on Thermal, Mechanical and Multi-Physics Simulation and Experiments in Microelectronics and Microsystems (EuroSimE), Graz, Austria, 16–19 April 2023. http://dx.doi.org/10.1109/EuroSimE56861.2023.10100763

[16] V. Vinciguerra, M. Boutaleb, G. L. Malgioglio, A. Landi, F. Roqueta, M. Renna, Investigating the Occurrence of Bifurcation in Large Metalized Wafers using ANSYS Layered Shell Elements. In Proceedings of the 2023 24th International Conference on Thermal, Mechanical

and Multi-Physics Simulation and Experiments in Microelectronics and Microsystems (EuroSimE), Graz, Austria, 16–19 April 2023. http://dx.doi.org/10.1109/Euro-SimE56861.2023.10100793

[17] V. Vinciguerra, G. L. Malgioglio and M. Renna, "Finite Element Analysis of the Upsurge of Bifurcation during the Thinning Process of Large Semiconductor Wafers," 2024 25th International Conference on Thermal, Mechanical and Multi-Physics Simulation and Experiments in Microelectronics and Microsystems (EuroSimE), Catania, Italy, 2024, pp. 1-4, http://dx.doi.org/10.1109/Euro-SimE60745.2024.10491572

[18] V. Vinciguerra, G. L. Malgioglio and M. Renna, "Extension of the Equivalent Thickness Concept to the Bifurcation of Large Semiconductor Front Side Metal Taiko Wafer investigated by ANSYS Finite Element Analysis Methods," 2024 25th International Conference on Thermal, Mechanical and Multi-Physics Simulation and Experiments in Microelectronics and Microsystems (EuroSimE), Catania, Italy, 2024, pp. 1-5, http://dx.doi.org/10.1109/Euro-SimE60745.2024.10491503

[19] V. Vinciguerra, G. L. Malgioglio, A. Landi, S. Valastro, B. Cafra and M. Renna, "From Wafer Bifurcation to Warpage Die: a Correlation Method to determine the Warpage of a Metal-Coated Silicon Substrate," 2022 23rd International Conference on Thermal, Mechanical and Multi-Physics Simulation and Experiments in Microelectronics and Microsystems (EuroSimE), St Julian, Malta, 2022, pp. 1-6, http://dx.doi.org/10.1109/Euro-SimE54907.2022.9758875

[20] V. Vinciguerra, G. L. Malgioglio and A. Landi, "Models of Bifurcation in a Semiconductor Wafer: A Comparison of the Analytical Solution vs. the ANSYS Finite Element Analysis," 2022 23rd International Conference on Thermal, Mechanical and Multi-Physics Simulation and Experiments in Microelectronics and Microsystems (EuroSimE), St Julian, Malta, 2022, pp. 1-4, http://dx.doi.org/10.1109/Euro-SimE54907.2022.9758852

[21] V. Vinciguerra, A. Landi and G. L. Malgioglio, "Wafer Bifurcation as a Spontaneous Symmetry Breaking," 2022 23rd International Conference on Thermal, Mechanical and Multi-Physics Simulation and Experiments in Microelectronics and Microsystems (EuroSimE), St Julian, Malta, 2022, pp. 1-3, http://dx.doi.org/10.1109/Euro-SimE54907.2022.9758915

[22] M. Calabretta; A. Sitta; S. M. Oliveri; G. Sequenzia. Warpage Behavior on Silicon Semiconductor Device: The Impact of Thick Copper Metallization. Appl. Sci. 2021, 11, 5140. https://doi.org/10.3390/app11115140.

[23] G. Cheng, G. Xu, W. Gai and L. Luo, "Deep Understanding the role of Cu in RDL to Warpage by Exploring the Warpage Evolution with Microstructural Changes," 2018 IEEE 68th Electronic Components and Technology Conference (ECTC), San Diego, CA, USA, 2018, pp. 2416-2421, http://dx.doi.org/10.1109/ECTC.2018.00364

[24] G. G. Stoney, "The tension of metallic films deposited by electrolysis," Proceedings of the Royal Society of London. Series A, Containing Papers of a Mathematical and Physical Character, vol. 82, no. 553, 1909.

[25] G. C. Janssen, M. M. Abdalla, F. van Keulen, B. R. Pujada and B. van Venrooy, "Celebrating the 100th anniversary of the Stoney equation for film stress: Developments from polycrystalline steel strips to single crystal silicon wafers," Thin Solid Films,vol. 517, no. 6, pp. 1858-1867, 1 2009.

PCIM Europe 2024, 11– 13 June 2024, Nuremberg　　　DOI: 10.30420/566262132

# Maximum Junction Temperature Simulation and Validation for the Hot Spot in Multi-Chip SiC Power Module

Wonjin Cho[1], Byoungok Lee[1], Hansol Seo[1], Udaykumar Vangaveti[2]

[1] onsemi, Republic of Korea
[2] onsemi, the United States

Corresponding author:　Wonjin Cho, dylan.cho@onsemi.com
Speaker:　　　　　　　　Wonjin Cho, dylan.cho@onsemi.com

## Abstract

Thermal impedance ($Z_{th}$) of the power module derived by the relation of virtual junction temperature ($T_{vj}$) and corresponding power loss reflects the average junction temperature ($T_{j,avg}$) of the power semiconductor switch if multiple devices compose a single functional switch. However, the actual temperature across the individual devices can vary from the value represented in $T_{vj}$ of the switch due to uneven distribution of the temperature over the chips. This paper proposes localized $Z_{th}$ referred from the hot spot temperature of the switch and the use of proposed $Z_{th}$ for the inverter simulation to estimate the maximum allowable output power within the allowable range of the junction temperature ($T_j$). By applying localized $Z_{th}$, simulated $T_j$ shows ~10°C higher temperature compared with the result using the conventional $Z_{th}$ method in 250 kW automotive traction inverter simulation and empirical measurement, which can provide more practical feasibility analysis in the early stage of the system development.

## 1　Introduction

Because $T_j$ of the power semiconductor device is a dominant index that limits the output power of inverters, $T_j$ prediction has been used as a guidance of allowable operating conditions. As the relevant methods for power loss and corresponding $T_j$, various previous works have been made such as matching each element of power loss to the functional inverter model [1], processing $T_j$ calculation in decomposed power loss in the frequency domain [2], and numerical calculation of power loss based on reference measurement [3]. These predictive models provided calculated $T_j$ based on the transient power loss and thermal impedance in the Foster network model. Because thermal impedance is derived as a transfer function of the power loss toward transient $T_{vj}$, pre-measured data of power loss and $T_{vj}$ are required to characterize the Foster thermal model of the specific power module. According to AQG-324 guidelines for power modules [4], $T_{vj}$ of the switch from the case or coolant temperature is recommended to be characterized by an equivalent voltage drop associated with temperature. However, if multiple chips compose a single switch by connected in

parallel, $T_{vj}$ defined in AQG-324 only shows the average temperature across the chips. In the case that SiC device is adopted for high-power applications such as traction inverter modules for automotive, multiple SiC chips are connected in parallel to form a single switch. Many power module suppliers provide thermal impedance models for such high-power modules in the sense of the average junction temperature of the switch so that sufficient temperature margin should be considered for the safe operation of the system. This paper proposes thermal impedance based on the $T_{vj}$ at the hot spot across the chips and the corresponding result of inverter operation based on both simulation and experimental validation to enhance unknown temperature margins in system design.

## 2　Simulation models

### 2.1　Thermal Impedance Modelling

Thermal impedances are modeled using SiC power module, NVXR17S90M2SPC [5], designed for an automotive traction inverter powered by a 400 V battery with motor driving power of 250 kW and above. NVXR17S90M2SPC consists of 6

PCIM Europe 2024, 11– 13 June 2024, Nuremberg        DOI: 10.30420/566262132

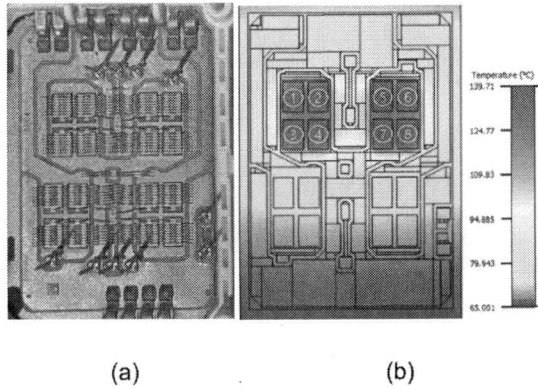

(a)                        (b)

**Fig. 1** Parallelly connected SiC module (a) configuration of a single phase of NVXR17S90M2SPC (b) FEM simulation model for a single switch.

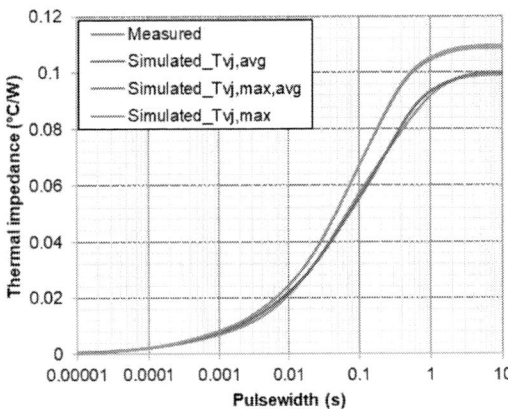

**Fig. 2** Modelled thermal impedances for different definitions of temperatures.

switches for a 3-phase full bridge, where each switch is composed of 8 SiC chips in parallel as shown in Fig. 1 (a). Thermal characteristics are modeled in the FEM simulation platform, where each chip numbered from 1 to 8 in Fig. 1 (b) is regarded as a thermal source of 85 W. Heat is transferred from each thermal source to a coolant that is the conventional ethylene-glycol mixture in 50% and 50% concentration and flows 10 l/m rate at 65°C. Because the massive amount of the heat is vertically dissipated to the coolant through the copper substrate and cooling fins, electrical connecting parts on the top surface such as bond wires, press-fit pins, and power terminal connections are not counted as meaningful thermal interfaces. From the temperature distribution illustrated in Fig. 1 (b), the average temperature of each chip distributes from 132 to 134°C across 8 chips, while the maximum temperature reaches about 139°C. To quantify thermal impedance associated with hot spot and average temperature, 3 kinds of temperatures are defined as follows:

1) $T_{vj,avg}$ is defined as the average temperature of a switch, which is the same as conventional $T_{vj}$ from the measurement of the voltage drop across the switch.
2) $T_{vj,max,avg}$ is defined as the average of the maximum temperature on each chip.
3) $T_{vj,max}$ is defined as the maximum temperature of all chips in the switch.

Simulated $Z_{th}$ using 3 kinds of temperature and associated power losses are plotted in Fig. 2. Measured $Z_{th}$ for product characterization for datasheet is also plotted on the same plane. The simulated $Z_{th}$ using $T_{vj,avg}$ well traces measured data from 10 μs to 10 s traces, which shows FEM simulations were done in appropriate conditions. Thermal impedances calculated using $T_{vj,max,avg}$ and $T_{vj,max}$

do not show meaningful differences in all pulse width ranges, which indicates that the temperature evenly distributes in the single chip but varies across the chips. The dominant difference of $Z_{th}$ characterized from $T_{vj,max,avg}$ and $T_{vj,max}$ is their levels at the steady state pulse width of 10 s, which shows ~10% higher value than the measured one. The other meaningful difference is that $Z_{th}$ value from $T_{vj,max}$ starts to diverge from the measured value at the pulse width of 2 ms, which indicates $Z_{th}$ difference happens at a close location from the junction. Because $Z_{th}$ gap is caused by the temperature difference on the surface of the chips, this simulation result coincides with physical insight, as well.

## 2.2 Inverter modeling

To investigate the feasibility of the power module with the effects of various definitions of thermal impedance, the inverter is modeled by the process described in Fig. 3. At the beginning stage of the modeling, the measured loss model of NVXR17S9 0M2SPC is imported in which each element of conduction and switching losses are characterized with respect to the current and $T_j$. Then, the inverter behavioral model is simulated using a modulation scheme, output current, power factor, bus voltage, modulation index, output frequency, and switching frequency. Each element of imported power loss is interpolated to be mapped at the corresponding inverter operation and results in the total power loss of the switch. The power loss in the switch increases $T_j$ by coupled with thermal impedance and increased $T_j$ induces more power loss. In this recursive feedback, averaged power loss and $T_{j,avg}$ are utilized to achieve the saturated value of $T_{j,avg}$ quickly, where the amount of

1047

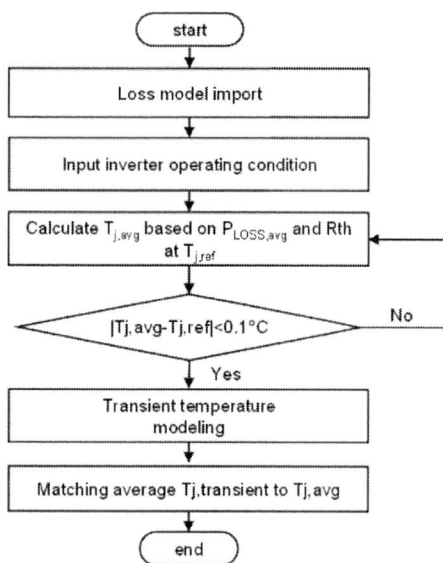

Fig. 3   Inverter modelling process.

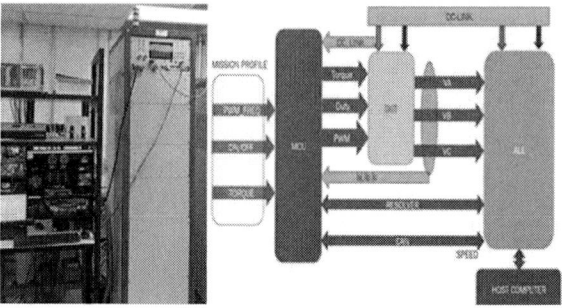

Fig. 4   Active load emulator and control flow process.

feedback for $T_{j,avg}$ is less than 0.1°C. Once $T_{j,avg}$ is saturated, transient $T_j$ is simulated on it by applying Foster thermal networks to the transient power loss. Because the transient power loss can be decomposed into many single rectangular pulses and each rectangular pulse can be represented as a linear combination of step functions with different delays and polarity of the step, transient $T_j$ can be represented as a linear combination of $\Delta T_{j,step}(t)$, which is step pulse response of Foster network expressed in (1).

$$\Delta T_{j,step}(t) = P \cdot \sum_n R_n \left(1 - e^{-\frac{t}{R_n C_n}}\right) \qquad (1)$$

, where P is the level of power loss in a step function, $R_n$ and $C_n$ are thermal resistance and capacitance at the n-th node of Foster network, respectively.

# 3   Verification and validation

## 3.1   Experimental configuration

The inverter system is configured employing the active load emulator (ALE) which provides equivalent load properties with an automotive traction motor and controllable software interface as shown in Fig. 4. The ALE equipment is capable of recirculating power up to 250 kW through the power module with a peak output current of 700 Arms. As the DUT part illustrated in the flow process in Fig. 4, the modified

NVXR17S90M2SPC module is used without gel filled on the chips and with black paint coated on the open surface after the gel is removed. Virtual junction temperature, $T_{vj}$, is measured by the IR camera with a resolution of 720 P at 60 Hz and tolerance of 5°C during inverter operation as demonstrated in Fig. 5. Output current from the inverter is set up to 450 Arms, where corresponding $T_{vj,max}$ is estimated as less than 120°C. Gate driving condition follows the recommendation described in the product datasheet, which was also referred to in characterizing the power loss model used in the inverter simulation.

## 3.2   Verification in the inverter operation

Simulated $Z_{th}$ from $T_{vj,avg}$ was verified by direct comparison with $Z_{th}$ curve from measured data for $T_{vj}$. However, $Z_{th}$ from $T_{vj,max}$ or $T_{vj,max,avg}$ cannot be derived from measured data because such localized temperatures do not appear in $T_{vj}$ measurement based on voltage drop depending on Tj. Even if such localized temperature is measured by the IR camera, temperature variation in a few ms range cannot be captured due to the limited frequency of the equipment. As an alternative method to verify $Z_{th}$ for localized temperature, the results from the inverter simulation and experimental operation are compared. Simulated $Z_{th}$ is applied to the inverter model described in section 2.2 and estimated $T_j$ is derived from the simulation. Applying the same operating conditions used in the inverter simulation listed in Table 1, a hardware experiment is performed using the ALE. Then, the associated temperature is measured on the chips of the DUT.

Obtained $T_{vj,avg}$ values from both simulation and experiment are plotted in Fig. 6, where simulated $T_{vj,avg}$ well follows the measured value within 3°C deviation. The maximum value of $T_{vj,avg}$ is 99°C at the output current of 450 Arms. Because measured $Z_{th}$ is applied to $T_{vj,avg}$ simulation, the result indicates the inverter model is simulated in a practical manner.

The simulation model for $T_{vj,max,avg}$ and $T_{vj,max}$ are

PCIM Europe 2024, 11– 13 June 2024, Nuremberg    DOI: 10.30420/566262132

**Fig. 5**    DUT configuration for thermal measurement.

| DC bus voltage (V) | 450 |
|---|---|
| Load current (Arms) | up to 450 |
| Power factor | 0.9 |
| Load frequency (Hz) | 100 |
| Switching frequency (kHz) | 10 |
| Coolant temperature (°C) | 65 |
| Coolant flow rate (*l*/m) | 10 |

**Table 1** Inverter operating condition for simulation and ALE experiment.

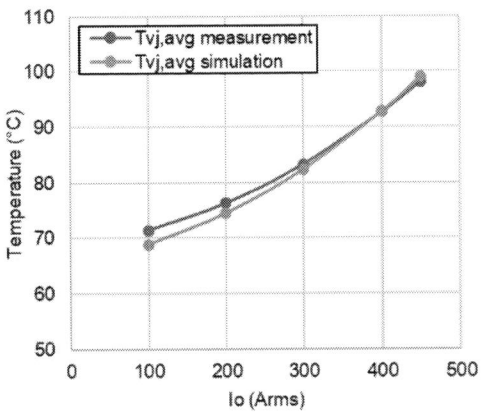

**Fig. 6**    $T_{vj,avg}$ comparison from simulation and experiment.

also verified by measuring the temperature on the hot spot, which is plotted in Fig. 7 with respect to the output current levels. Because the individual chip size of SiC device is not large enough to have meaningful temperature distribution on it, $T_{vj,max,avg}$ and $T_{vj,max}$ overlap in all operating current levels. The maximum temperature at the hot spot is 103°C and the gap of 4°C from $T_{vj,avg}$ should be considered in the system design.

### 3.3    Validation of system design

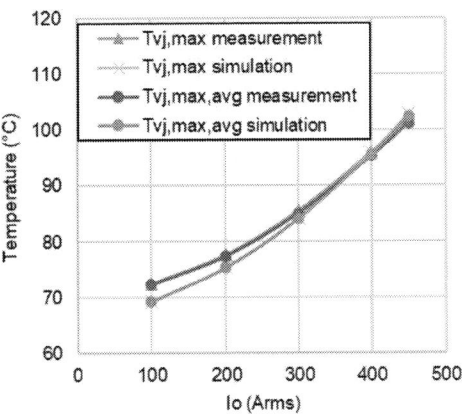

**Fig. 7**    $T_{vj,max,avg}$ and $T_{vj,max}$ comparison from simulation and experiment.

In addition to the geographical distribution of temperature across the chip, $T_j$ can vary in time as the power loss is attributed to the sinusoidal output current for AC motor operation. The fluctuation of $T_j$ depends on the output electrical frequency, where the amplitude of the temperature ripple increases with a lower output frequency. Thus, practical peak $T_j$ can be greater than $T_{vj,max}$ defined as the temperature at the hottest spot. As further investigation for the peak $T_j$, transient $T_{vj,max}$ is simulated using the same inverter operating condition described in Table 1, and plotted for the steady state condition in Fig. 8. Simulated $T_{vj}$ fluctuates at the frequency of 100 Hz and periodically hits 108°C, while time-averaged temperature corresponds to $T_{vj,max}$ of 103°C. This result indicates that even if the geographical distribution is counted on the maximum $T_j$ estimation using the more conservative definition of $Z_{th}$, additional $T_j$ increase must be considered due to the peak level in the thermal transient. The example shows 5°C increase due to temperature fluctuation but the gap can increase depending on the operating requirement of the inverter.

Since the maximally delivered output current of the inverter is limited by the allowable range of $T_j$, a more precise estimation for $T_j$ enhances the reliability of the inverter. To investigate the effect of thermal impedance proposed in this paper, the inverter operating condition in Table 1 can be recalled for NVXR17S90M2SPC where the maximum load current is determined by allowable $T_j$ limit of 175°C. The simulated $T_{vj,avg}$, the peak $T_{vj}$ ($T_{vj,peak}$) using $Z_{th1}$, and the maximum peak $T_{vj}$ ($T_{vj,max,peak}$) using $Z_{th2}$ are plotted in Fig. 9, where $Z_{th1}$ is the measured thermal impedance and $Z_{th2}$ is the modeled one from $T_{vj,max}$ at the hot spot. If $Z_{th1}$ is applied to the thermal calculation for the inverter

1049

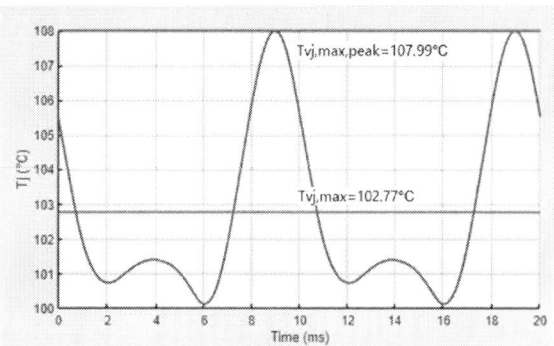

**Fig. 8** Transient junction temperature from the inverter simulation using geographical $T_{vj,max}$ definition.

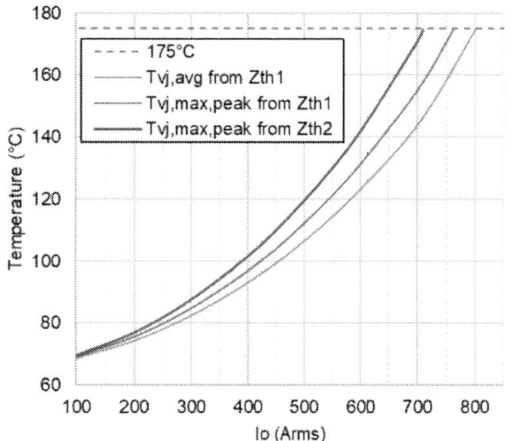

**Fig. 9** Maximum allowable current with respect to the thermal impedance models.

and only $T_{vj,avg}$ is considered as the performance liming factor in terms of temperature, the output current is allowed up to 800 Arms. Meanwhile, the output current is limited at 761 Arms if $T_{vj,peak}$ is counted with $Z_{th1}$, and 710 Arms if $T_{vj,max,peak}$ with $Z_{th2}$ are applied. The corresponding output powers of the inverter to 800, 761, and 710 Arms are 397, 377, and 352 kW, respectively. If the curve of $T_{vj,max,peak}$ from $Z_{th2}$ is extrapolated up to 800 Arms, temperature values become 197 and 217°C at 761 and 800 Arms, respectively. It indicates that $T_j$ far exceeds the maximum allowable limit of 175°C and may cause thermal damage to the device during inverter operation without considering the geographical distribution and transient fluctuation of the temperature.

# 4 Conclusions

As a more precise estimation method for junction temperature, a simulated thermal impedance for the hot spot was proposed in this paper. The proposed thermal impedance was applied to the inverter simulation to estimate junction temperature for the given operating condition. The estimated junction temperature was compared with the result from the inverter experiment at the same condition, which verified that the proposed method is practically effective. Considering thermal transient adding to the geographical distribution of temperature, the maximum junction temperature was estimated as a more reliable value for the system operation. Even the junction temperature can still have unknown distribution due to the dependency of thermal impedance on the power loss, and the specification tolerance of the power module, this proposed method aims to optimize system performance by mitigating uncertainty in junction temperature estimation. By doing so, it ensures that the system operates within desired specifications without falling below or exceeding them.

# 5 References

[1] U. Vangaveti, A. Sesha and D. Cho, "Measurement and Validation of Junction Temperatures of Chips in Automotive Traction Modules at Inverter Level," *PCIM Europe 2023; International Exhibition and Conference for Power Electronics, Intelligent Motion, Renewable Energy and Energy Management*, Nuremberg, Germany, 2023, pp. 1-7, doi: 10.30420/566091348.

[2] Z. Xu, Y. Zhang, H. Wang, X. Ge, Y. Liao and B. Yao, "A Novel Calculation Method for IGBT Junction Temperature Based on Fourier Transform," in *CPSS Transactions on Power Electronics and Applications*, vol. 8, no. 1, pp. 54-64, March 2023, doi: 10.24295/CPSSTPEA.2023.00006.

[3] M. Aydin and E. Beşer, "Power Loss and Thermal Temperature Calculation of IGBT Module by Mathematical Equations," *2023 International Conference on Power Energy Systems and Applications (ICoPESA)*, Nanjing, China, 2023, pp. 862-867, doi: 10.1109/ICoPESA56898.2023.10141374.

[4] ECPE Guideline AQG 324, Qualification of Power Modules for Use in Power Electronics Converter Units in Motor Vehicles. https://www.ecpe.org/research/working-groups/automotive-aqg-324/

[5] https://www.onsemi.com/products/discrete-power-modules/power-modules/silicon-carbide-sic-modules/NVXR17S90M2SPC.

# Integration of CFD Simulation Results in PLECS Using Lookup Tables

Simon Cepin[1], Holger Borcherding[1]

[1] University of Applied Sciences and Arts Lemgo, Germany

Corresponding author:    Simon Cepin, simon.cepin@th-owl.de
Speaker:                 Simon Cepin, simon.cepin@th-owl.de

## Abstract

In this paper an extended use of CFD simulation results in PLECS, an established simulation tool will be discussed. In the context of previous researches, a matrix to describe the characteristics of the developed thermal management was developed [1]. This description can be interpreted as a lookup table. This approach is used to extend the PLECS model by a dynamic heat sink. The thermal resistance is variable and depending on additional parameters, in this case the volume flow. In addition, control loops for the volume flow are considered. In further steps the consideration of additional losses that are required to cool the electrical parts can be added. Such as the power of the pump used for the coolant.

## 1 Introduction

The demand for WBG power semiconductors that embody efficiency and the ability to integrate into compact designs is relentlessly on the rise. Especially the heat sink is very important, as it is very complex due to the high integration level. Previous studies have shown that there are some promising approaches for the design of heat sinks. The basis for this publication is a liquid-cooled heat sink for double-sided cooled power semiconductors. In [1] a developed thermal management was simulated by the means of CFD. A Method to describe the thermal parameters based on fluid dynamic parameters has been defined. The resulting outcome was a matrix that describes thermal resistance and the occurring back pressure based on the volume flow of the coolant. The thermal management was verified at a later stage [2].

The next step is the utilization of the results. The origin of the investigations are the occurring losses in power electronics. To determine these losses, there are numerous tools available, which, among other things, provide the option to use heat sinks and calculate resulting temperatures of the semiconductors. In addition to tools like LTSpice and MATLAB Simulink, PLECS is often used for these investigations. In this tool, the losses are determined using lookup tables, as this leads to significantly reduced simulation time. The thermal parameters are represented by using thermal resistances and capacities. The approach of this paper is the use of lookup tables to describe the thermal parameters depending on the flow rate as there is no variable thermal resistor available in PLECS. This way, various arrangements can be compared with little time and effort. Moreover, this approach provides the opportunity to consider losses from an active cooling, such as a pump, or the limitations of an active cooling, like the maximum temperature and back pressure, in the simulation. This approach leads to a comprehensive view of the system and a more realistic assessment of the systems efficiency.

## 2 Implementation of the CFD results in PLECS

The following approach is considered on the basis of the software PLECS. The electrical and thermal libraries are used which leads to a model with an electrical and a thermal submodel. A controlled three-phased inverter consisting of three half bridge modules with an ohmic load represents the electrical submodel. The resulting losses represent the input variable for the thermal submodel. Furthermore, they are used for the calculation of the systems efficiency.

Likely there will be deviations compared to the CFD simulation. The reason is, that the thermal resistance was always related to the hottest chip. Therefore, an averaging of the CFD simulation results is necessary. The averaging must be reversed at the end, as this defines the temperature of the hottest chip and the survival of the module.

## 2.1 Modelling of a thermal resistance

This setup is the basis for all electrical-thermal simulations with PLECS and is also used in many current research projects [3]. However, there are other thermal components which are used for more complex thermal considerations. These include controlled thermal sources. The use of controlled sources allows very complex thermal behavior to be represented with them [4], [5]. However, components that are decisive for this case only exist as electrical components, these are variable thermal resistors and capacitors. At this point the chosen approach takes effect. By using controlled thermal sources and general control components, a variable resistance is simulated. Fig. 1 shows a variable thermal resistance consisting of a thermometer and a controlled heat flow source.

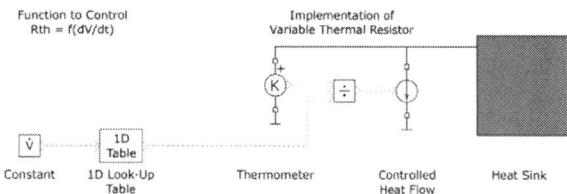

**Fig. 1** Implementation of variable thermal resistor and lookup table control

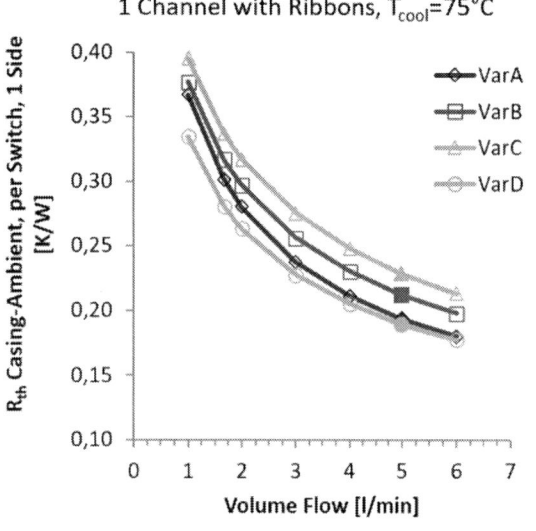

**Fig. 2** Conversion of CFD simulation results [1]: volume flow to thermal resistance

The model is extended by a 1D lookup table to calculate the thermal resistance according to the volume flow of the coolant. The use of a lookup table instead of a linear function provides two benefits. The simulation time is reduced and the results can be taken directly from the CFD simulation. The necessary matrices have been created within the framework of [1] which represent the thermal resistance and the resulting back pressure as a function of the coolants volume flow. These matrices fill the lookup table and thus are implemented in PLECS as you can see in Fig. 2 and Tab. 1.

**Tab. 1** 1D-lookup table matching Fig. 2

| Volume flow (in l/min) | Thermal resistance (in K/W) |
|---|---|
| 0 | Nan |
| 1 | 0,39 |
| 2 | 0,32 |
| 3 | 0,27 |
| 4 | 0,25 |
| 5 | 0,23 |
| 6 | 0,22 |
| 7 | Nan |

This simple examination shows some important limits. A necessary value for the numerical calculation is the thermal capacitance of the heat sink, alternatively of the electrical device on the heat sink. Otherwise, the model is containing an open loop and the resulting temperature would approach infinity immediately.

To test the concept, a simple thermal circuit consisting of a thermal capacitor and a thermal resistor was chosen. Then determining the controlled system, the thermal resistance is considered to be variable, therefore it is not part of the controlled system, but part of the manipulated variable. A controlled system was determined from the thermal capacity.

The controlled system for the cooling capacity and the resulting temperatures can be seen in Fig. 3 and Fig. 4.

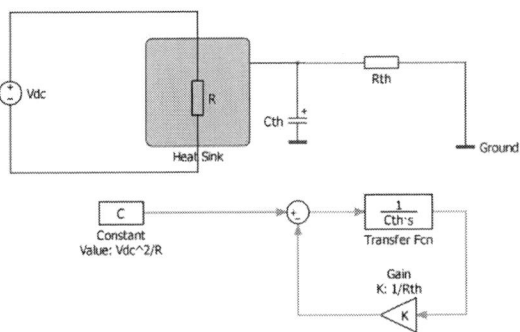

**Fig. 3** Controlled system for heat flow

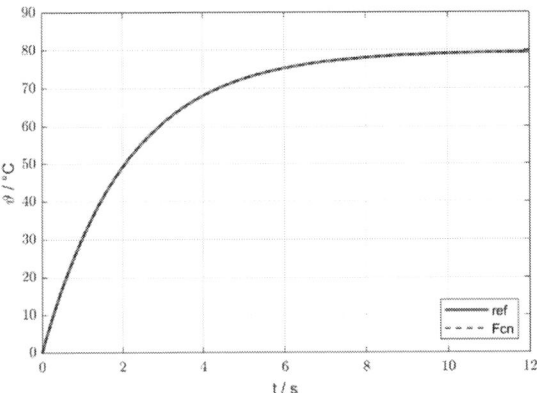

**Fig. 4** Step response of the controlled system

The variables derived from this for the controller design are shown in the Eq. 1 for the controlled system.

$$G_S(s) = \frac{1}{1/C_{th} \cdot s + 1} \quad (1)$$

## 2.2 Controller Design for dynamic cooling

Most cooling systems are switched on and work continuously during operation or have various discrete stages. Three operations are considered. The first operation is not controlled. The cooling system works continuously all the time with maximum power. The second and third operations contain a controller, whereas the second is a two-point controller which switches between zero and maximum power. The third is a continuous PI-controller.

The design of the PI controller is considered below and two approaches were discussed to set up the control loop. The first approach was to control the volume flow. Therefore, a transfer between heat flow and volume flow is necessary. The integration of a lookup table into the controlled system led to a significantly more complicated controller design.

For the second approach the control variable is set to be the required heat flow and not the volume flow. This approach needs other parts to implement the limits for the power, as these are nevertheless determined via the volume flow. This will be discussed later. Furthermore, the control loop is extended by a simplified behavior of the coolant pump. This is integrated as a dead time element for the needed heat flow, which is approximated as a PI element. The power dissipation of the DUT represents a disturbance variable. The reference variable is the temperature of the DUT.

$$G_{P1}(s) = \frac{1}{T_t \cdot s + 1} \quad (2)$$

$$G_R(s) = K_P \cdot \frac{T_n \cdot s + 1}{T_n \cdot s} \quad (3)$$

The controller is designed according to the optimal amount method. To apply the method, a guidance filter is added to the control loop, as it is shown in Fig. 5.

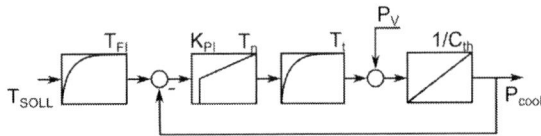

**Fig. 5** Control loop for the PI controlled system

An important part of the design are the limits of the controller with a corresponding anti-reset-windup method (short: ARW). The major problem for the ARW are the variable limits. Usually, the limits for are known and are specified directly. In this case the maximum and minimum heat flow depend on several parameters. The basis is described in Eq. 4.

$$Q = \Delta T / R_{th}(\dot{V}) \quad (4)$$

The temperature difference always has to be considered. The thermal resistance on the other hand would be constant for typical passive heat sinks. In this case it is variable and depends on the volume flow. The link between thermal resistance and volume flow is given in Tab. 1. This is sufficient, as the volume flow can assume minimum and maximum values so that the limits can be determined. The implementation in the model is shown in Fig. 6.

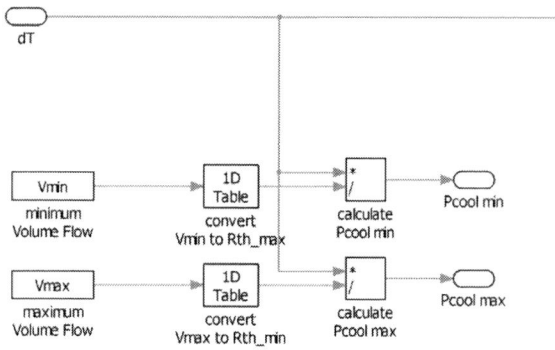

**Fig. 6** Calculation of the variable minimum and maximum heat flow

As a result, a PI controller with variable limits is implemented and the measured variable is compared to the mentioned continuous cooling the two-point controller which is shown in Fig. 7. The interesting part is the corresponding volume flow which is responsible for the slightly varying course of the temperature for the PI controller. The gradient changes as the controller reaches the maximum limit as shown in Fig. 8.

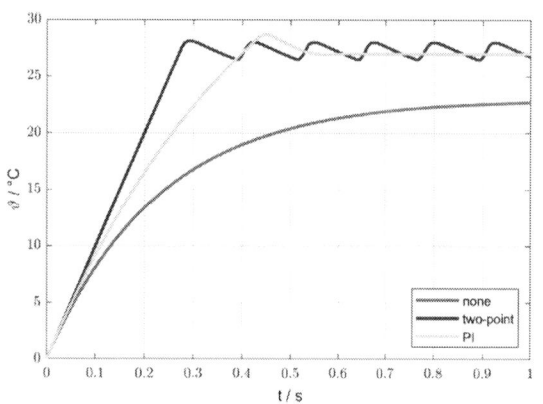

**Fig. 7** Temperature of none, two-point and PI controlled

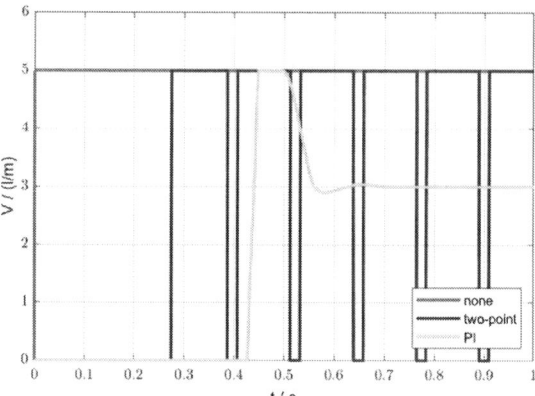

**Fig. 8** Volume flow of none, two-point and PI controlled (minimum = 0 l/min, maximum = 5 l/min)

## 3    Level of simulation

In addition to the thermal analysis, the use of lookup tables also offers the possibility of viewing the fluid dynamic results. The level of the simulation is variable. The first level is a simple distribution of the volume flow for different arrangements on the individual cooling surfaces [1]. In this case one of the control methods of chapter 2 are used. Consequently, it allows the determination of essential volume flow required to maintain the temperature below a maximum value.

Another result by using a lookup table is the calculation of the back pressure. This value is not essential for the first level. However, it provides very important information for the second level. The source and the lookup tables are shown in Fig.9 ad Tab. 2.

**Table 2**    2D-Lookup-Table matching Fig. 2 and 9

| Volume flow (in l/min) | Thermal resistance (in K/W) | Pressure drop (in mbar) |
|---|---|---|
| 0 | Nan | Nan |
| 1 | 0,39 | 2,7 |
| 2 | 0,32 | 5,7 |
| 3 | 0,27 | 11,3 |
| 4 | 0,25 | 18,7 |
| 5 | 0,23 | 27,9 |
| 6 | 0,22 | 39,0 |
| 7 | Nan | Nan |

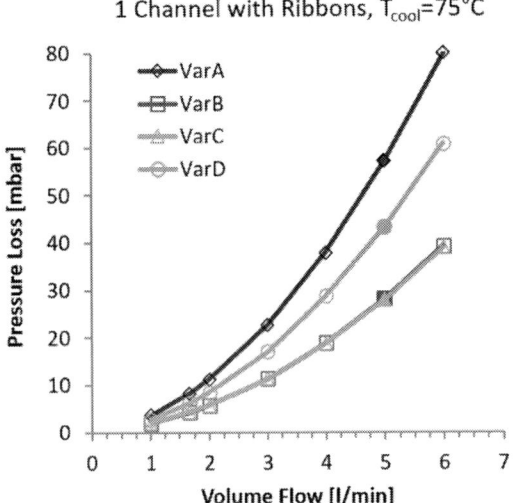

**Fig. 9** Conversion of CFD simulation results [1]; volume flow to back pressure

In this level, several cooling surfaces are interconnected, which leads to some interdependencies. For example, the temperature of the cooling fluent, the back pressure of the overall system and the volume flow influence each other. Two scenarios are used to explain this in detail, two cooling surfaces in series and two parallel. In series the volume flow has to be the same for all cooling surfaces. Therefore, the measurement value of the controller is not clear. The temperature of the cooling fluent for the second surface is increased due to the heat introduced by the first one. In parallel the temperature is equal for all surfaces. The volume flow across the two surfaces varies based on the characteristics of the associated cooling channels. For both scenarios the back pressure can be calculated to determine whether a permissible value is exceeded here. This value can be used similarly to the procedure in chapter 2 to extend the variable limits. A reference is given in [1] with two channels parallel and three surfaces in series in each channel.

The third level is the calculation of the power that is needed to provide the volume flow for a given back pressure. On the one hand, the result provides information as to whether controlled cooling makes sense for a given specific application. On the other hand, optimization of the cooling can be considered depending on the application.

The calculation and analyzation are discussed in the following chapter 4.

# 4 Implementation of cooling losses

PLECS is used due to fact that the occurring losses are calculated from lookup tables which leads to short simulation time. Although the focus is on the calculation of the temperature and back pressure, the losses in the semiconductors are also calculated, since these are responsible for the heating of the semiconductors. These losses are used to calculate the efficiency of the three-phased inverter.

To increase the degree of realism of the model, the power needed for cooling can be incorporated in a subsequent stage. The power for cooling depends on the pump used, the volume flow and the back pressure. Since the power has no benefit for the application, it is completely assigned to the losses in the efficiency analysis.

To calculate the power of the pump the volume flow and the back pressure are used. Therefore, the units are transferred to SI units. The efficiency of the pump is estimated with 85%.

$$[\dot{V}] = \frac{l}{min} = \frac{0,001}{60} \cdot \frac{m^3}{s} \tag{5}$$

$$[p] = bar = 100e3 \cdot \frac{kg \cdot m}{s^2} \tag{6}$$

$$P_{pump} = (\dot{V} \cdot p)/\eta \tag{7}$$

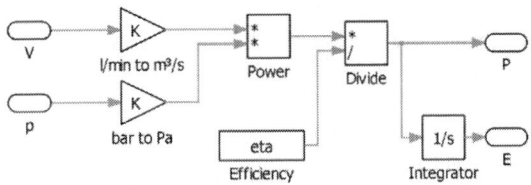

**Fig. 10** Implementation of the power calculation for the cooling pump

The corresponding power graphs are shown in Fig. 10. From this the mean values for each controller are calculated and listed in Tab. 3.

The power consumption of the PI controlled pump is three to four times lower compared to the uncontrolled pump. This is a significant improvement. Therefore, it can be said that the method offers potential for optimization of cooling systems. It should be noted that in some applications the lower output has a minor influence on the overall losses and efficiency.

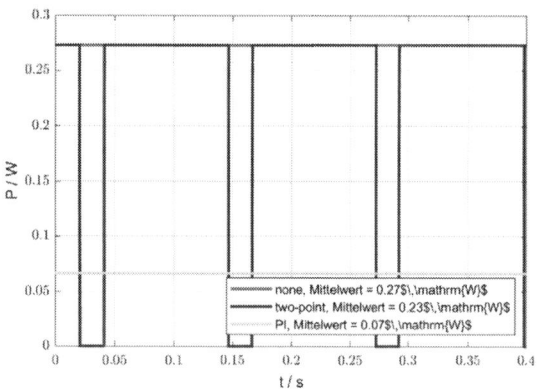

**Fig. 11** Power consumption of the cooling pump for different controller

**Tab. 3** Mean Values for Power consumption of cooling pump

| Controller | Power consumption |
|------------|-------------------|
| none       | 0,27 W            |
| Two-point  | 0,23 W            |
| PI         | 0,07 W            |

# 5 Conclusion

Within the scope of previous investigations, matrices were created by means of CFD simulation which represent the behavior of a cooling structure as a function of the volume flow [1]. These results serve as a basis for the design of a simulation model in PLECS, in which the static thermal components of the heat sink are replaced by variable components. For these components the thermal capacitance of the model is vital.

With the variable components a controller for the cooling system was designed. Two different approaches to implement the limits of the volume flow of the pump where mentioned, whereas the most promising one was designed and successfully modelled.

Furthermore, several levels of simulation were mentioned for further use of the implemented CFD results. One of these levels was elaborated to calculate the resulting power that is needed by the coolant pump to provide the needed volume flow.

The plan for the next steps is to integrate the back pressure into the variable limits. The controlled system will then be considered in more detail to model the internal structure of the module and thus the temperature of the SiC chip.

Following the finalization of the controller, it will be transferred to an overall model of a three-phase inverter and examined in terms of overall efficiency. This model can also provide information about the thermal stress on the module.

# 6 Acknowledgement

This paper is funded by the German Federal Ministry of Economic Affairs and Climate Action (BMWK) pursuant to a decision of the German Parliament in the project DCI4CHARGE (Extension of the DC-INDUSTRIE system concept for open low-voltage DC grids for bidi-charging applications), funding number: 01MV23005A and the German Federal Ministry of Education and Research (BMBF) pursuant to a decision of the German Parliament in the project UmSiChT (Inverter with silicon carbide based power electronics and double sided cooled power module), funding number: 16EMO0251

# References

[1] S. Cepin.: H. Borcherding.: Design of Thermal Management for Double-Sided Cooled SiC-Power Semiconductors, in PCIM 2022, 10-12 May 2022, Nuremberg, Germany.

[2] S. Cepin.: H. Borcherding.: Development of Heat Source for Performance Verification of Thermal Management for Double-Sided Cooled SiC-Power Semiconductors, in PCIM 2022, 9-11 May 2022, Nuremberg, Germany.

[3] Y. Yang.: M. Hefny.: K. Henneth.: A. Callegaro.: M. Goykhman.: A. Barronian.: A. Emadi.: A Fast and Accurate Thermal-Eletrical Coupled Model For SiC Traction Inverter, ITEC 2021, 21-25 June 2021, Chicago, IL, USA.

[4] A. Abubakar.: C. Klumpner.: P. Wheeler.: Analysis of Using a Thermoelectronic Module for Power Electronics Cooling, ONCON 2022, 9-11 December 2022.

[5] L. Graber.: J. Kim.: S. Pamidi.: Thermal Network Model for HTS Cable Systems and Components Cooled by Helium Gas, IEEE Transactions on Applied Superconductivity, 4 June 2016.

PCIM Europe 2024, 11– 13 June 2024, Nuremberg    DOI: 10.30420/566262134

# PCB Only Thermal Management Techniques for eGAN® FETs in a Half-Bridge Configuration.

Adolfo R. Herrera[1], Alejandro Pozo Arribas[1], Michael A. de Rooij[1]
[1] Efficient Power Conversion, U.S.A.

Corresponding author:    Michael de Rooij, michael.derooij@epc-co.com
Speaker:                         Adolfo R. Herrera, adolfo.herrera@epc-co.com

## Abstract

Cost constraints for low power GaN FET based converters limit the use of heatsink options for cooling, thus requiring the Printed Circuit Boards (PCB) to become the primary path for heat flux exchange. Various design options to minimize thermal resistance between the junction of a GaN FET and ambient are analyzed. Options include thicker copper, thermal vias to utilize the bottom-side copper for cooling, FET separation, and layout design to maximize heat-spreading are investigated. Proper implementation of these techniques can reduce thermal resistance from the junction to ambient ($R_{\theta JA}$) by up to 30% with little to no additional cost.

## 1    Introduction

Enhancement-mode gallium nitride eGaN FETs offer high power-density capabilities with ultra-fast switching and low on-resistance, all in a compact form factor [1]. However, the achievable power levels are limited by thermal limitations due to the extreme heat-flux densities. If not managed properly, the generated heat can result in excessive self-heating and elevated temperatures that compromise reliability and potential performance. In some applications, thermal management strategies are limited to just those that can be built using only the PCB. In these cases, proper thermal management is essential as devices such as eGaN FETs can easily rise above safe operating temperatures. Several effective design techniques can be implemented at the board-side for improved heat dissipation.

Power electronic devices in operation dissipate energy causing self-heating. The heat generated flows from the heat source, referred to as the device junction, towards colder bodies through two main parallel paths – to the printed circuit board (PCB) and to the case or backside of the device. The heat flux for each path is a result of the temperature difference and the thermal resistances encountered along the path, as shown in Eq. (1).

$$R_{\theta JX} = \frac{T_J - T_X}{P_V} \qquad (1)$$

Where $R_{\theta JX}$ is the thermal resistance from the junction to a reference location X; $T_J$ is the device junction temperature in steady state conditions; $T_X$ is the temperature at reference location X; and $P$ is the power dissipated in the device. The reference location commonly refers to the ambient environment (A), the board (B), or the backside of the device (C).

This paper details thermal management guidelines when designing lower power applications with GaN devices in which a heatsink cannot be used to cool the devices. Elements of the PCB such as the device size, copper layer thickness, PCB vias, and component layout on the board are studied to determine best design practices in a PCB-only cooling setting. Different PCB parameters for each of these elements are simulated in CFD to determine the effect on thermal resistance; the PCB parameters were then experimentally tested to validate the CFD simulation results.

## 2    Baseline and Methods

### 2.1    Thermal Resistance Network

Modeling the heat transfer characteristics of a PCB is complicated as conduction, convection, and radiation all combine to characterize the cooling on the board and any devices. However, because radiation can only be realistically affected by increasing the effective surface area of the PCB with a heatsink, only conduction and convection are modeled in the PCB thermal resistance network. The thermal resistance network will be modeled for two FETs in a half-bridge configuration. The half-bridge will be the focus for observing the thermal characteristics of a PCB-only cooling environment, but these heat-spreading techniques are broadly applicable to other topologies as well.

The thermal resistance circuit throughout an entire PCB contains too many inputs to be simply modeled, but by just examining the heat-generating FETs in a resistance model, a simplified network can be made as shown in Fig. 1. From Fig. 1, thermal resistances labeled with subscript C are resistances through the back of the die to ambient as there is no heatsink, they are connected to. Thermal resistances with subscript B are the resistance through the board all the way to ambient on either side. In a cooling application where there is no heatsink applied, the heat dissipated from the transistor to ambient is minimal due to the small area. As such, the main heat dissipation path from the transistor is through the PCB [2]. $R_{\theta JA}$ is the combined equivalent resistance of the network.

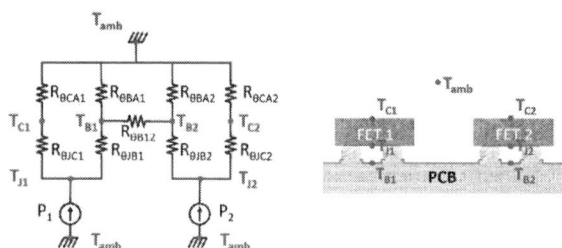

**Fig. 1** Simplified thermal resistance network for two transistors in a half-bridge.

## 2.2 Typical Board Studied

The size of the PCB is an important factor in its capacity for cooling. Smaller PCBs are much less effective at removing heat than larger ones, but there is a limit to the benefits of making a PCB larger relative to the device size. The relationship between board size and maximum temperature is shown in Fig. 2 for a fixed 3 W/cm$^2$ heat flux applied where the GaN devices would be placed. Maximum board temperature sharply rises as the PCB gets smaller, but there are diminishing returns in cooling as board size increases past a certain threshold for parts with the same power dissipation. Designing a large board for the sole purpose of increasing thermal performance will induce greater costs in the board while potentially penalizing electrical performance for very marginal thermal returns past a certain point.

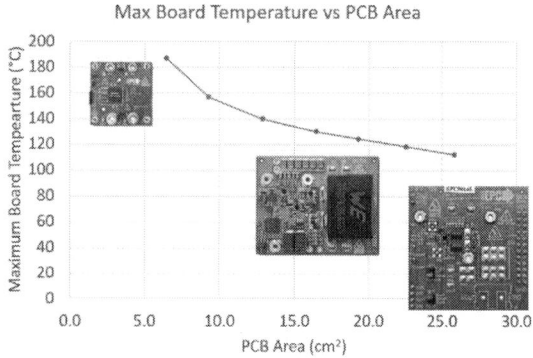

**Fig. 2** Effect of board size on temperature.

When discussing thermal management strategies for PCB-only cooling in the following sections, a standard 50 mm x 50 mm, 4-layer, board will be the baseline on which the strategies are examined. The typical board is shown in Fig. 3 bottom right. Copper thickness in the board will vary between 0.5 oz, 1 oz, 1.5 oz, 2 oz, and 3 oz copper. Note that 3 oz copper is only studied in simulation due to PCB manufacturing cost and feature size concerns. Some boards will contain vias near the FETs while others have a reduced amount or no vias. Half the boards will have decoupling capacitors between the FETs instead of being adjacent to them as is displayed in Fig. 3.

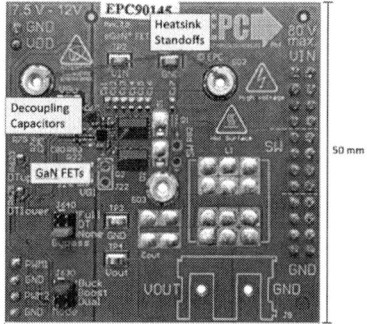

**Fig. 3** Image of a typical board used for this analysis.

## 2.3 FEA Approach

In this study, different PCB designs and their overall impact on thermal resistance from junction to ambient were simulated in 6SigmaET version 16.3, up to date versions of the software are now known as CelsiusEC [3]. 6SigmaET is a finite element analysis software specifically designed for simulations of thermal performance in electronic systems. Various elements of the standard PCB design being studied were modified to identify key variables that have the biggest impact on $R_{\theta JA}$. The simulations are set up so that the heat-generating

FETs are each given a power dissipation number in watts so that a steady state temperature can be reached. This temperature along with the given power dissipation numbers are plugged into Eq. (1) to obtain $R_{\theta JA}$. The individual FETs are modeled in CAD software then assigned material properties in the FEA software that match their physical properties in the real world. The PCBs are imported such that the model in the FEA software matches the amount of copper, the vias, and the thickness of the layers of the board in the real world. Simulation setups are later validated experimentally.

### 2.4 Experimental Validation

To validate the thermal resistance numbers obtained from the FEA simulations, the simulation setups were also run experimentally using the same board characteristics. The boards featured a half-bridge running the FETs activated in reverse-bias to allow for the current and power dissipation across the FETs to be accurately controlled. Three circuits were designed to allow for thermal imaging of configurations in which only the low-side is dissipating power, only the high-side is dissipating power, and when both FETs dissipate power. The power dissipation is calculated for each individual FET by multiplying the measured voltage across the part by the current running through the parts. The temperatures of the devices were measured both by a fiber optic thermal sensor beneath the die and with an IR camera. The dies were painted black to increase emissivity. The measured device temperature and calculated power dissipation are then input into Eq. (1) to calculate the thermal resistance of the part in the given board setup.

## 3 Device Size and Losses

### 3.1 Device Selection

For a given set of input/output voltages, load current, and max. operating temperature, the optimal converter design is the result of a complex balance between performance (efficiency/losses), cost, and size among other parameters. Understanding the relationships between them and implementing an optimal thermal management strategy are key to finding the best solution for each application.

Figure 4 shows an analysis of the power losses as a function of frequency for different eGaN FETs operating in a Buck converter configuration. The frequency can be considered as a good indicator of the physical size of the inductor and the overall converter [1]. In Fig. 4, the horizontal lines show

the max. power losses that would result in a maximum junction temperature of 100 °C, when the board described earlier is used. Figure 4(a) describes a low load scenario with only natural convection cooling; Fig. 4(b) shows a higher load scenario with 400 LFM airflow to cool the parts. For the low load scenario in Fig. 4(a), the minimum FET temperature rise is given by the lowest losses (point "1", EPC2204 at 250kHz), also resulting in the largest size. Increasing frequency allows a smaller inductor at the expense of a small increase in power losses. As an example, EPC2052 above > 625 kHz ("2"), shows lower losses than EPC2204, and lower cost given its smaller size. At higher load Fig. 4(b), the lowest losses are offered by EPC2218 switching at 250 kHz ("3"). Unfortunately, this solution would be very large and costly. Alternatively, EPC2204 operating at ~650 kHz ("4") allows a considerable reduction in size and cost. Moreover, with an optimal thermal management strategy, size and cost can be reduced even further by using EPC2044 above 1 MHz ("5"), while still meeting the max. junction temperature requirement. The next sections will discuss how to optimize thermal management.

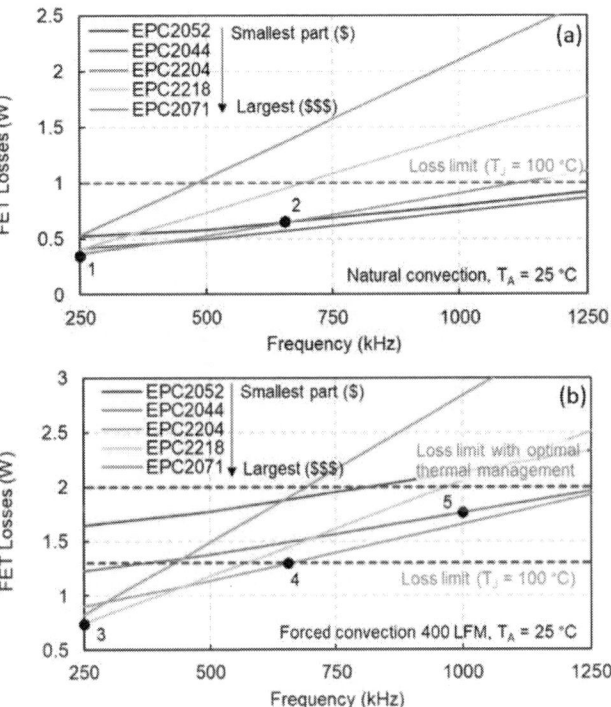

**Fig. 4** Power losses as a function of frequency for various eGaN FETs in a 48 V to 12 V Buck converter: (a) 2.2 µH inductance and 10 A load, (b) 2.2 µH inductance and 20 A load.

# 4 Copper Thickness

## 4.1 Advantages

The primary heat flux exchange mechanism on a PCB is to make use of the copper on the board itself for heat-spreading. Factors that affect heat-spreading are copper thickness, connection to the opposite side of the board and board area (in this narrative is fixed). Because there is no back-side cooling and the FET area is small compared to that of the PCB, almost all the heat from the FETs flows into the first copper layer of the PCB. Designing thicker copper traces for low electrical resistance also benefits thermal resistance and provides a high heat-conductance medium at each layer of the PCB. The effectiveness of increasing the thickness of copper layers is shown in Table 1 and Fig. 5.

| Thickness | ΔT Still Air [°C] | ΔT 400 LFM [°C] |
|---|---|---|
| 1oz Cu | 64 | 42 |
| 1.5oz Cu | 57 (-13%) | 39.5 (-6%) |
| 2oz Cu | 53 (-21%) | 37.5 (-12%) |
| 3oz Cu | 51 (-25%) | 35.5 (-18%) |

**Table 1** Temperature rise for different copper layer thicknesses under two cooling conditions using baseline PCB.

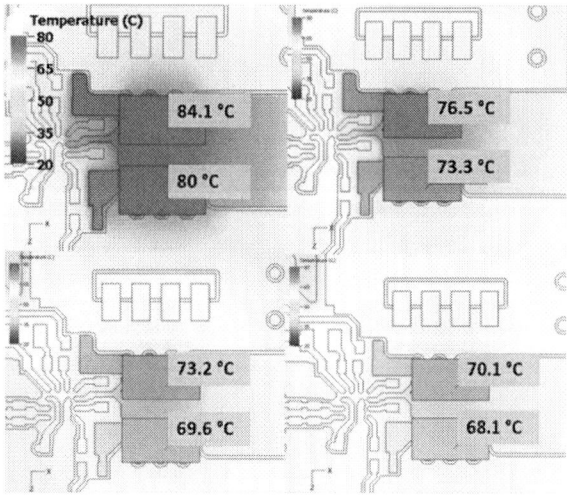

**Fig. 5** Simulation results for (a) 1 oz Cu, (b) 1.5 oz Cu, (c) 2 oz Cu, (d) 3 oz Cu.

Increasing the copper thickness from 1 oz to 2 oz results in a decrease in $R_{\theta JA}$ of 21%. Further increases in thickness beyond 2 oz copper result in less relative reduction in $R_{\theta JA}$. When airflow is considered as a cooling method, the decrease in $R_{\theta JMA}$ is lower relative to that of the natural convection condition – only decreasing about an additional 6% for each respective increase in copper thickness. This is because the forced convection is already effective at removing heat from the outer layers of the board, diminishing the maximum capability of their heatsinking and heat spreading.

## 4.2 Challenges

Copper thicknesses greater than 2 oz also create challenges in etching fine features into the PCB, overall cost, and PCB fabrication. When considering these challenges in combination with the lower relative level of $R_{\theta JA}$ reduction at greater thicknesses, designing boards with copper thickness greater than 2 oz should be limited to specific applications. Another drawback when designing thicker copper has to do with the buried layers as they have very little area of contact with the surrounding environment. The reduced convective area between the buried layers and the environment heavily limits their PCB cooling ability. This is especially the case in forced convection environments in which more heat transfer is occurring at the top and bottom layers. The buried layers in essence become additional small heatsinking components buried within the PCB, but because the buried layers are very thin compared to the overall thickness of the PCB that heatsinking impact is also limited. Nevertheless, the number of layers has a determinant influence on the electrical resistance of the board and thus should be designed based on electrical performance rather than on thermal considerations.

# 5 Using Vias to Reduce Thermal Resistance

## 5.1 Via Structure

The insulating dielectric layers of a PCB have low thermal conductivity, increasing the maximum temperature of the board and $R_{\theta JA}$ of the device. To solve this issue, vias can be implemented in the board to provide a heat path to the bottom layer and other buried layers – effectively increasing both the heatsinking capability and increasing the area for heat convection to both the top and bottom of the board. There are different configurations for vias that can be used in the pad to improve heat conduction, these are shown in Fig. 6. The most effective via configuration for reducing $R_{\theta JA}$ is to place them directly under the FET pads to maximize heat conduction away from the device. In applications where vias cannot be placed under the bumps for various reasons, placing them adjacent to the FET will also decrease $R_{\theta JA}$ but will not be as effective as under-bump vias. The vias should also be staggered when under the bump and not aligned in an equidistant rectangle to further improve heat spreading [4].

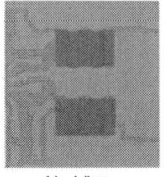 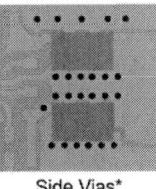 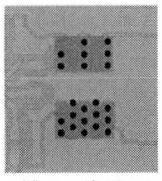

No Vias      Side Vias*      Vias under bump

**Fig. 6**    Different via configurations

The vias used in these boards are IPC4761 Type VII vias. These vias are tented over on both sides of the board and contain a non-conductive fill and then are plated over to seal them [5]. The minimum wall diameter of each via is 7.8 mils with a copper plating of 0.78 mil thickness and requiring an annular ring of at least 13.8 mils. The non-conductive fill serves two purposes: 1) allows the outer layers to be plated over effectively sealing the via closed, helping to prevent solder from draining down the via tub during reflow; and 2) it helps to reduce the effects of thermal cycling on the reliability of the board by having a better match between the coefficient of thermal expansion (CTE) of the via fill and the FR4 in the board [6]. These vias are usable in PCBs with copper thicknesses up to 2 oz, giving another reason for why the designed copper thickness of a PCB should not exceed that mark in these applications.

## 5.2  Via Effect on $R_{\theta JA}$

The cooling effect on the PCB of vias is heavily dependent on their proximity to the heat generating components. The most effective via configuration for reducing $R_{\theta JA}$ is to place them directly under the FET pads to maximize heat conduction away from the device. In applications where vias cannot be placed under the bumps for various reasons, placing them adjacent to the FET will also decrease $R_{\theta JA}$ making them not as effectively as under-bump vias. Figure 7 shows thermal simulations with various vias configurations implemented in the PCB. Including thermal vias in the board directly under the devices can reduce thermal resistance by up to 30% in both natural and forced convection conditions. Side vias also reduce thermal resistance but by about half the amount as under-bump vias. The images in Fig. 7 show the comparison of heat dispersion pattern around the FETs when vias are included or excluded. When no vias are designed into the board, the traces under the part become extreme hot spots relative to

the surrounding traces; the heat can only effectively escape the board from the top layer because there is relatively little heat flux to the other side. The side via and under-bump via images show the heat traveling through the "pipes" to the other side, resulting in a less extreme temperature difference around the part.

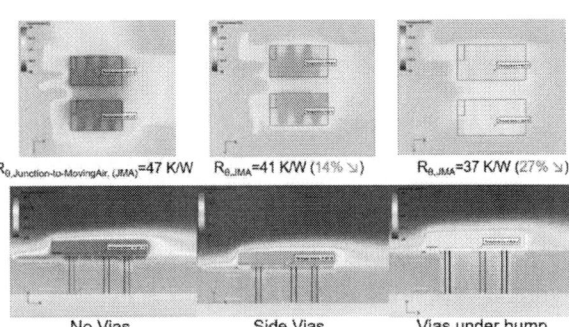

No Vias      Side Vias      Vias under bump

**Fig. 7**    PCB thermal performance comparison for different via configurations.

# 6  Layout of Components

## 6.1  Co-heating on a PCB

Typical GaN FET layouts [7, 8] require close proximity between the two devices in a half bridge configuration to ensure lowest loop inductance [9]. Heat-generating components on a PCB heat each other when placed in close proximity with each other. Due to this, two FETs in a half-bridge configuration will co-exchange heat flux which increases $R_{\theta JA}$ for each device. To approximate the effects of co-heating between two heat sources, a simple superposition can be used to analyze the combined effects of their temperature rise [10]. However, because there are nonlinear dependencies on temperature there will be a small error in the final temperature rise calculated. Figure 8 shows the superposition principle of two heat sources in simulation, note that the simulations are done in an environment with an ambient temperature of 20°C. With only the high-side FET active its temperature rise is 32.9°C; with the low-side FET active, co-heating raises the high-side FET by 22.4°C; adding the two temperature rises gives an estimated temperature rise of 55.3°C, which is close to the temperature rise of 52.3°C shown in Fig. 9(c). Similarly, for the low-side FET, the estimated combined temperature is 52.5°C which is slightly higher than the temperature rise shown of 49.5°C.

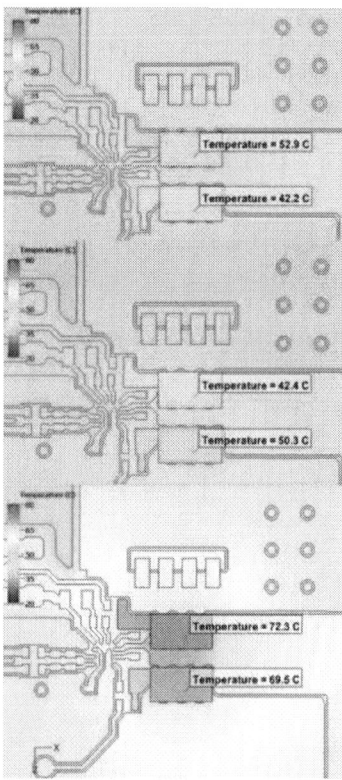

**Fig. 8** Principle of superposition with multiple heating elements.

Placing FETs close together will cause them to co-heat which increases thermal resistance and in turn temperature, reducing the ability to dissipate heat. Spacing out the FETs can reduce the effect of co-heating which in turn reduces $R_{\theta JA}$ without any penalty in loop inductance [11]. Figure 9 shows two different board layouts with different bus capacitor layout configurations intended for examining the effects of distance on co-heating. Keeping heat sources spread apart from each other as much as possible without sacrificing electrical performance is good practice for lowering the maximum temperature of the board. Figure 10 shows the differences in temperature rise for two FETs in a half-bridge configuration; changing to a center bus capacitor layout reduces $R_{\theta JA}$ by about 5% for both FETs.

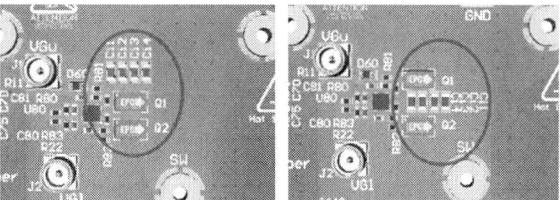

**Fig. 9** (a): Adjacent bus capacitor layout. (b): Center bus capacitor layout.

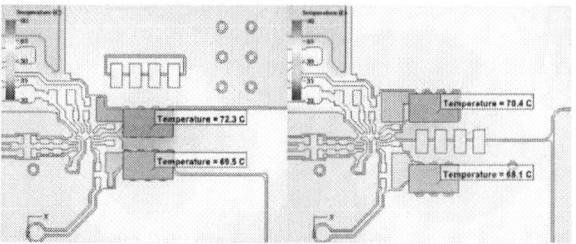

**Fig. 10** FEA simulation temperature results for (a) adjacent bus capacitor temperatures, (b) center bus capacitor temperatures.

# 7 Combining the Techniques

When combining the techniques outlined in this work, it is possible to reduce $R_{\theta JA}$ by close to 30%. Figure 11 shows a simulation of two boards and the temperature of each FET in a half bridge. The board in Fig. 11(a) has 0.5 oz copper, 4 layers, no vias near the FETs, and both FETs placed close to each other. The board in Fig. 11(b) has 2 oz copper, 4 layers, vias under the bump and close to the FETs, and has the FETs spread out with center configured bus capacitors to reduce co-heating. Making these changes to the board not only can reduce $R_{\theta JA}$ by over 35%, but also comes at little to no additional cost in manufacturing of the PCB.

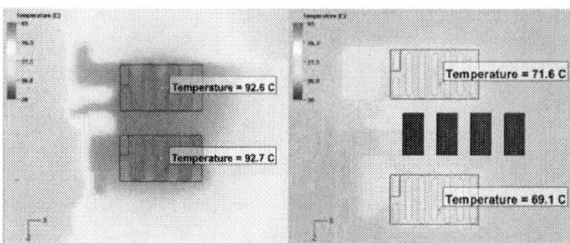

**Fig. 11** (a): Result with no thermal techniques used. (b): Result with thermal techniques used.

# 8 Experimental Results

The different PCB characteristics that maximize thermal performance without a heatsink were experimentally tested to verify the accuracy of the FEA simulations. A representative set of boards that isolate the different thermal techniques were tested to compare the results to those of the simulations. Each board was tested with a constant current of 0.75 A with temperature measurements being made five minutes after activating the FETs. Figure 12 shows the test setup. Results shown are for tests done in a natural convection setting with no airflow. Table 2 shows the temperature rise and thermal resistance results for the experimental tests by comparing the highest temperature and thermal resistance of each configuration. Note that thermal techniques were compared in isolation; for example, when comparing two copper thicknesses in PCBs then the vias and capacitor layouts are kept consistent between the different tested boards. The experimental results match closely with those done in the FEA simulations with a small error. Validation errors in the results could arise because of the difference in power dissipation between the simulations and the experiments. In the simulations each FET is dissipating exactly one watt while in the experiments the two FETs will each have a different power dissipation. This results in one FET contributing more to the heating of the pad and causing an imbalance between the two.

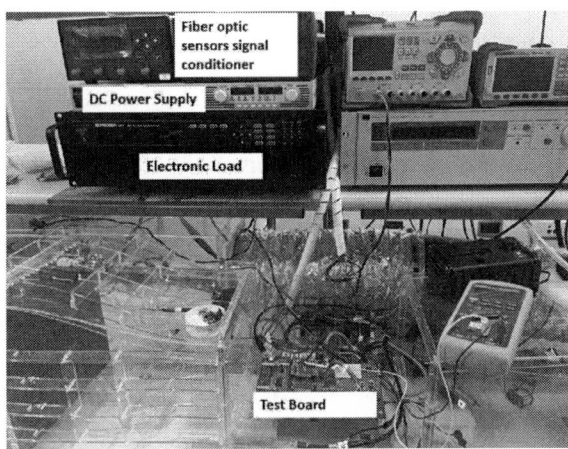

**Fig. 12**    Test setup for thermal measurements.

| Configuration | $\Delta T$ [°C] | $R_{\theta JA}$ [°C/W] |
|---|---|---|
| 0.5oz Cu | 53 | 46 |
| 2oz Cu | 44.6 (-19%) | 37.3 (-23%) |
| Min Vias | 60.4 | 48.9 |
| Side Vias | 49.6 (-18%) | 40.4 (-21%) |
| Vias in Pad | 43.8 (-27%) | 40.4 (-32%) |
| Adjacent Cap | 46.6 | 40.6 |
| Center Cap | 44.6 (-5%) | 37.3 (-9%) |

**Table 2**    Temperature rise and thermal resistance for the studied PCB configurations in isolation.

Figure 13(a) shows the results comparing a board with minimal thermal techniques used and Fig. 13(b) shows the results with all the thermal techniques being used. The experimental result shows high agreement with the simulation results shown in Fig. 11. The effectiveness of the heat spreading in Fig. 13(b) is evident compared to

that in Fig. 13(a), where the outer edges of the board are still relatively cool compared to the pad. The results show a satisfactory agreement with the FEA simulations and the efficacy of these PCB design techniques in reducing $R_{\theta JA}$.

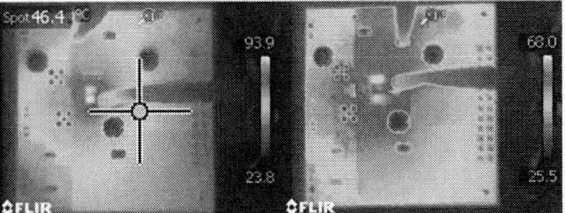

**Fig. 13**    (a) Experimental result with 0.5 oz Cu, minimal vias, and adjacent FETs; (b) Result with 2 oz Cu, vias under the FET, and spaced-out FETs.

# 9    Conclusion

## 9.1    Observations

The thermal techniques outlined allow for two FETs to each dissipate over 2 W of power with an ambient temperature of 25°C before the parts get too hot and there is a risk of failure. While this is not a very high-power result, it does show that with proper thermal design that GaN FETs can be cooled to a degree without necessitating a heatsink. Having an airflow of 400 LFM can increase the power dissipation from the FETs to over 3 W in each before reaching critical temperatures without the use of a heatsink.

## 9.2    Future Work

Other methods of reducing the maximum temperature of FETs on a PCB without a heatsink include thermal bridges and PCB substrates other than FR4. Thermal bridges are surface-mount chips made of highly thermally conductive materials that are also electrically isolating meant to give heat a pathway away from hot spots on a board [12]. There are applications with extreme hot spots on boards near relatively cool and isolated spots where thermal bridges could feasibly act as components that reduce the overall thermal resistance of the system. However, in applications where the hot spot is more disperse the thermal bridges may have minimal impact. Further work needs to be done to evaluate their effectiveness. AlN substrates for multilayer boards are another development designed to make cooling in a PCB much more effective, especially for wide bandgap semiconductors [13]. Their reliability, broad applicability, and cooling efficiency must be tested with GaN FETs, however.

## 9.3 Conclusions

An overview of PCB only thermal management strategies was presented to improve GaN FET cooling in those applications where heatsinking or external heat spreading cannot be used. These guidelines for maximizing board-side cooling recommend thicker copper layers, thermal vias under or near the devices, and spacing out the FETs to reduce co-heating. Combining these recommendations enable reductions of $R_{\theta JA}$ by over 35%, with little to no increase in overall system cost, also reducing peak operating temperatures. With reductions in thermal resistance using these techniques efficiency is improved, devices have longer lifespans, and higher power density can be achieved.

# 10 References

[1] A. Lidow, M. De Rooij, J. Strydom, D. Reusch, and J. Glaser, *GaN Transistors for Efficient Power Conversion*, 3rd ed. John Wiley & Sons, 2019. ISBN: 978-1119594147.

[2] Assaad Helou, "How2AppNote012 - How to Get More Power Out of an eGaN Converter.pdf," Efficient Power Conversion Corporation, Application note How2AppNote 012, 2021. [Online].

[3] Celsius EC Solver. Cadence, 2023. [Online]. Available: https://www.ca-dence.com/en_US/home/tools/system-analy-sis/thermal-solutions/celsius-ec-solver.html

[4] Y. Shen, H. Wang, F. Blaabjerg, H. Zhao and T. Long, "Thermal Modeling and Design Optimization of PCB Vias and Pads," in *IEEE Transactions on Power Electronics*, vol. 35, no. 1, pp. 882-900, Jan. 2020.

[5] IPC-4761: Design Guide for Protection of Printed Board Via Structures. IPC, July 2006.

[6] B. S. McCoy and M. A. Zimmermann, "Performance evaluation and reliability of thermal vias," *Nineteenth Annual IEEE Applied Power Electronics Conference and Exposition, 2004. APEC '04.*, Anaheim, CA, USA, 2004, pp. 1250-1256 vol.2.

[7] Efficient Power Conversion Corporation, "EPC90153: 80 V, 20 A Half-Bridge Development Board," [Online] Available: https://epc-co.com/epc/products/evaluation-boards/epc90153

[8] Efficient Power Conversion Corporation, "EPC90123: 100 V, 25 A Half-Bridge Development Board," [Online] Available: https://epc-co.com/epc/products/evaluation-boards/epc90123

[9] D. Reusch, J. Strydom, "Understanding the Effect of PCB Layout on Circuit Performance in a High Frequency Gallium Nitride Based Point of Load Converter," Applied Power Electronics Conference, APEC 2013, pp. 649–655, 16–21 March 2013.Observations and Conclusions

[10] "AN-2020 Thermal Design By Insight, Not Hindsight," Texas Instruments Application Report, Apr. 2013. [Online]. Available: http://www.ti.com/lit/an/snva419c/snva419c.pdf

[11] J. S. Glaser, A. Helou, "PCB Layout for Chip-Scale Package GaN FETs Optimizes Both Electrical and Thermal Performance," Applied Power Electronics Conference, APEC 2022, pp. 991–998, March 2022.

[12] Vishay Semiconductors, THJP datasheet, [Online]. Available: https://www.vishay.com/docs/60157/thjp.pdf. [Accessed March 2024].

[13] TDK Electronics, AIN Multilayer Substrate Design Rules, [Online]. Available: https://www.tdk-electronics.tdk.com/down-load/3094462/d40b91631de30e65bfe33a732352e4a4/design-rules-download.pdf. [Accessed March 2024]

PCIM Europe 2024, 11– 13 June 2024, Nuremberg    DOI: 10.30420/566262136

# From 4x to 3x STPAK® – What to be optimized for a more Compact EV Traction Inverter Solution

Vittorio Giuffrida[1], Simone Buonomo[1], Massimiliano Chiantello[1]

[1] STMicroelectronics, Italy

Corresponding author:     Vittorio Giuffrida, email: vittorio-mos.giuffrida@st.com
Speaker:                   Vittorio Giuffrida, email: vittorio-mos.giuffrida@st.com

## Abstract

The powertrain of full battery electric vehicle, mild-hybrid and plugin hybrid need the latest power technology and innovative package solution able to achieve the best-in-class performance. In this context the new generation of STMicroelectronics Silicon Carbide (SiC) MOSFETs properly meet this market requirements, leading to low energy consumption solutions for a sustainable future founded on decarbonization policies. The static features in combination with its temperature coefficient variation of this SiC MOSFETs generation can ensure high efficiency, reduced cooling effort and maximum power density as consequence.

Taking advantages of this new SiC generation plus those of STPAK®, that's the state-of-the-art discrete package with 2x dice in parallel, this paper illustrates how it is possible, from a technical point of view, moving from 4xSTPAK® to 3xSTPAK® solution in a traction inverter environment.

## 1  Introduction

Gradually SiC MOSFETs are replacing Si MOSFETs in a high demand application as the electrification of EV powertrain. To meet the high-power requirement for this application, the paralleling of SiC MOSFTEs has become an essential solution. However, it has some technical challenges because of current imbalance, different switching losses and so forth. In this paper the technical aspects behind moving from 4x STPAK® to 3x STPAK® solution are investigated deeply.

Devices with similar static characteristics such as on-state resistance (R) and threshold voltage ($V_{th}$) can help to avoid current unbalance in parallel solution. However, the selection of devices with very similar static characteristics is not enough to ensure balanced current levels in each die. More specifically, a proper design of the gate driving circuit and power loop is also required for current balancing [1]-[5].

Different power loop stray inductance ($L_{stray}$) leads to devices turning on at different rates and implies higher turn-off voltage overshot. The use of parallel discrete devices, paralleling Half-Bridge legs [1],

can lead to some benefits such as modularity and scalability of the inverter and power loop stray inductance optimization [1]. In other words, for power modules the stray inductance doesn't change moving from 8x to 4x dice paralleled solution, while for the STPAK® modular system, the power loop inductance can be optimized in relation to the number of legs in parallel. This paper illustrates why the only SiC MOSFET $R_{DS()}$ improvement is not enough to reduce SiC die area or number of STPAK® in parallel and why the system parasitic inductance optimization can help to maximize the performance when moving from 4x STPAK® (8x dice in parallel) to 3x STPAK® solution (6x dice in parallel).

The next ST SiC Generation introduces lower $R_{DS(on)}$ compared to the previous one, that means it can carry more current with the same die-size or theoretically the same current with smaller die-size for a given $R_{DS(on)}$. However, this last point is not always true. In fact, with lower die-size the thermal conductivity can change significantly, and this may need to be counterbalanced with further $R_{DS(on)}$ reduction.

1065

In this paper, three different solutions implemented with parallelled STPAK® are investigated, taking advantage of the latest SiC Generation with lower $R_{DS(on)}$. Moreover, the optimization of the system inductance is done to maximize the performance. Next, some technical concepts behind moving on reduced number of paralleled dice are done by maintaining the same thermal performance and the same current target. As anticipated, when moving from a 4x STPAK® to a 3x STPAK® solution, it is not enough having the same $R_{DS(on)}$ with 6x dice in parallel instead of 8x, because of two main factors:

- The equivalent $R_{th}$ is higher when moving from 4x to 3x devices.
- With no system and component layout modification, the switching speed needs to be reduced moving from 4x to 3x, leading to additional switching losses to be compensated.

Figure 1 tries to clarify the reasons behind these factors.

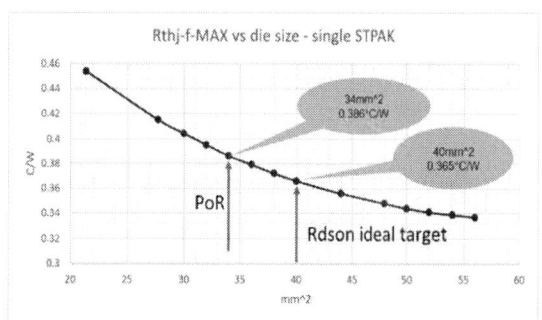

**Fig. 1**: $R_{thj\text{-}f\_MAX}$ profile vs die size of STPAK® - simulated by ST.

Assume to refer with "A" to the 4x STPAK® configuration and with "B" to the 3x STPAK® one. 40mm² is the SiC area of the next SiC Generation for single STPAK® of "B" configuration with equivalent $R_{DS(on)}$ of 34mm² current Generation in "A" one. But the equivalent $Rth_{j-f}(B)$ @ 3xSTPAK® is higher than $Rth_{j-f}(A)$ @ 4xSTPAK®, in fact we have:

$$Rth_{j-f}(B) = \frac{0.126^0 C}{W} > Rth_{j-f}(A) = \frac{0.096^0 C}{W} \quad (1)$$

In addition, with no system modification $Vds_{peak}(B) > Vds_{peak}(A)$ because of slightly higher total stray inductance of "B" configuration: connecting less STPAK® in parallel increases the effect of parasitic inductance, due to the higher number of paralleled stray inductive contributes. Furthermore, no gate driving resistance ($R_G$) modification (low equivalent $R_G$ for single STPAK® of 3x configuration) can lead to higher di/dt and consequently to $V_{ds}$ voltage exceeding the recommended operation conditon. This fact implies reduced speed and higher switching losses. That's $R_G$ gate driving of "B" configuration should be increased to counterbalance both the above-mentioned effects.

The paper is organized into three sections. In Section 1 the experimental measurements of 4x STPAK® solution have been performed. Section 2 shows the analysis and a proper gate driving setup for the 3xSTPAK® solution. In Section 3 the system inductance optimization of 3xSTPAK® has been proposed.

## 2   4xSTPAK® dynamic characterization

Half-bridge converter setup has been considered to perform dedicated Double Pulse Tests (DPTs). This converter is realized by putting in parallel four SiC devices, for both the low-side equivalent switch and the high-side one. The devices used as test vehicles are a 6.5 mΩ, 650 V SiC MOSFETs current Generation assembled in STPAK® package, with Kelvin Source (KS) terminal. The equivalent circuit is shown in Fig. 2.

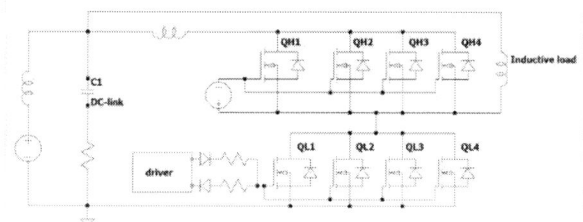

**Fig. 2**: Equivalent electric scheme for DPTs of four paralleled SiC MOSFETs.

The operative conditions are: $V_{bus}$ = 400 V, $I_{peak}$ = 1200 A, $V_{gs\_active}$ = [-5÷18] V, $V_{gs\_passive}$ = -5 V, $T_j$ = [25 − 175] °C.

First the $R_g$ setting has been done using the following criteria: $R_{goff}$ to achieve $V_{dspeak}$ on the active switch less than 650 V, while for $R_{gon}$ setting, $V_{peak\_diode}$ (on the complementary device) should be less than 600V. Decoupling resistor ($R_{dec}$) has been used to better balance gate voltage signal in each STPAK® device and to avoid potential very high frequency oscillations.

The devices under test (DUTs) have been properly selected with similar static parameters as reported. in Table I.

| STPAK® | $V_{th}[V]$ @ 1mA | $R_{dson}[m\Omega]$ @ 170A | $BV_{DS}[V]$ 1mA |
|---|---|---|---|
| QL1 | 2.82 | 7.2 | 893 |
| QL2 | 2.81 | 7.4 | 932 |
| QL3 | 2.82 | 7.0 | 910 |
| QL4 | 2.80 | 6.9 | 902 |

**Table I**: Static parameters of the DUTs.

A representative picture of the experimental setup is reported in Fig. 3.

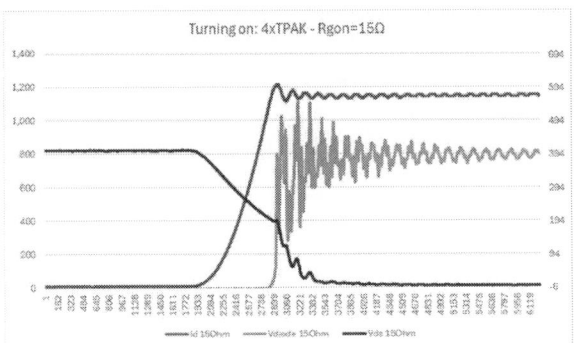

**Fig. 3** Test bench for DPT - experimental setup for STPAK® packaged devices.

The high side drain terminal is connected to DC+ bus bar, while the low side source is connected to DC- by a copper plate put very close to the STPAK® upper surface. This allows to minimize the overall inductive loop, as it results in $L_{stray}$ = 30 nH.

Fig. 4 and Fig.5 show respectively the turning on and turning off obtained with appropriate $R_g$ setting, done according to the above-mentioned operative conditions and criteria.

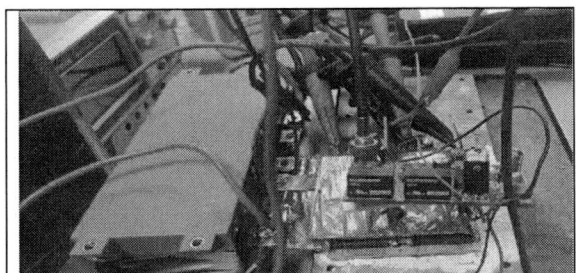

**Fig. 4:** 4xSTPAK® turning on with $R_{gon}$=15 Ω and $V_{peak\_diode}$ < 600 V.

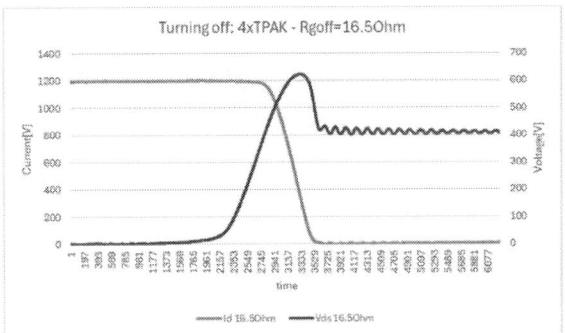

**Fig. 5:** 4xSTPAK® turning off with $R_{goff}$ =16.5 Ω and $V_{dspeak}$ < 650V.

A decoupling $R_{dec}$ = 1 Ω for each STPAK® has been used. This value is the minimum one to suppress the very high frequency oscillations phenomenon.

The experimental results are quantified in terms of switching losses at different load current levels, as reported in Fig. 6.

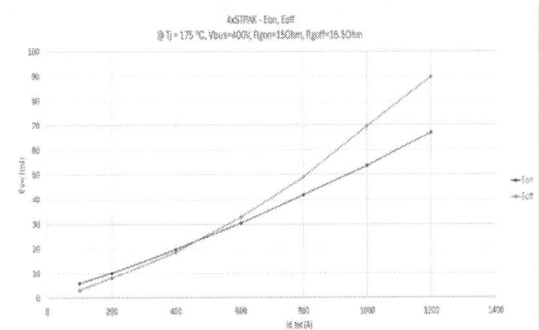

**Fig. 6**: 4xSTPAK® - Switching losses @ 175°C

From now on, we can estimate the traction inverter power losses of the 4xSTPAK® solution, given the operative conditions: $V_{bus}$ = 400V, Arms = 800 A, $f_{sw}$ = 8 kHz, $T_f$ = 60 °C and assuming $T_j$ ≤ 175 °C. Considering that, at $T_j$ = 175 °C the equivalent on-state resistance at switch level is $Rdson_{MAX}$ = $2.5m\Omega$, it possible to estimate the conduction losses:

$$P_{cond}(A) = Rdson * \frac{(RMS*1.414)^2}{4} = 798W \qquad (2)$$

While referring to Fig.6 the switching power losses are:

$$P_{sw}(A) = \frac{Eon+Eoff}{\pi} * fsw = 350W \qquad (3)$$

In this estimation the reverse recovery energy has been neglected.

$$P_{tot}(A) = P_{cond}(A) + P_{sw}(A) = 1148W \quad (4)$$

As reported in Eq. (1), for 4xSTPAK® we are considering $R_{thj\text{-}f} = 0.096$ ºC/W.

$$T_j(A) = 60^0C + 0.096 * P_{tot}(A) = 170^0C \quad (5)$$

The junction temperature esteem agrees with the assumption, i.e. $T_j \leq 175ºC$. Therefore, we can say that the 4xSTPAK® solution is able to meet the operative condition (800A, 8kHz, 400V).

## 3  3xSTPAK® dynamic characterization

In this case, the converter is realized by paralleling three SiC MOSFETs, for both the low-side and the high-side one. The devices used as test vehicles are a 4.88mΩ @ 25ºC for single STPAK®, 650V SiC MOSFETs next Generation.
The equivalent circuit is the same as shown in Fig. 2 but without "QH4" and "QL4" STPAK® devices. The experimental setup is the same as Fig 3. We start the analysis and the measurements with no $R_g$ modification, maintaining the same ones set in the 4xSTPAK® solution.

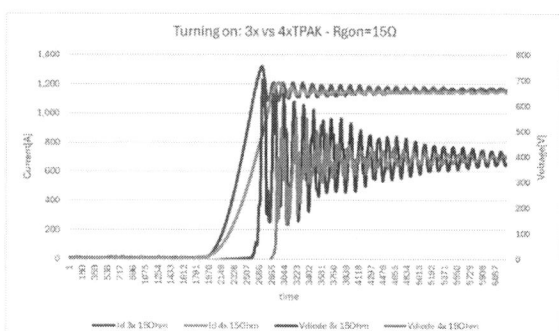

**Fig. 7:** Turning on: 4x vs 3xSTPAK® with $R_{gon}$=15 Ω.

As anticipated in the introduction, with no system modification and with the same $R_g$ the di/dt of 3xSTPAK® solution (dark blue waveform) is higher than the 4x one (light blue). This implicates a $V_{diode}$ exceeding the recommended voltage operative conditions, in fact $V_{diode} >> 600V$ (violet waveform). For this reason, the commutation speed must be reduced by setting Rg again according to the same above-mentioned criteria. In Fig. 8 the turning on waveform with Rgon = 27 Ω is shown.

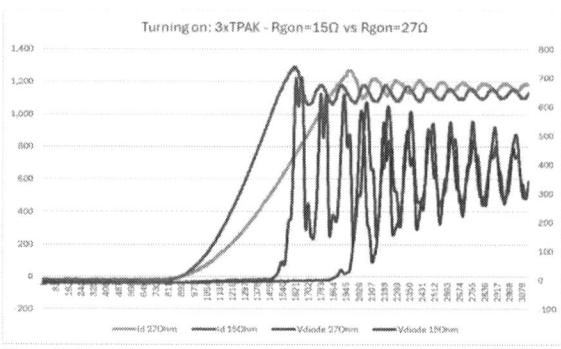

**Fig. 8:** Turning on: 3xSTPAK® with $R_{gon}$ = 15 Ω vs $R_{gon}$ = 27 Ω.

The increased $R_g$ of configuration "B" counterbalances the effects of no system modification reducing the $V_{peak\_diode}$ of the complementary switch as per recommendation. As a consequent drawback, higher switching losses have been measured, as shown in Fig. 9.

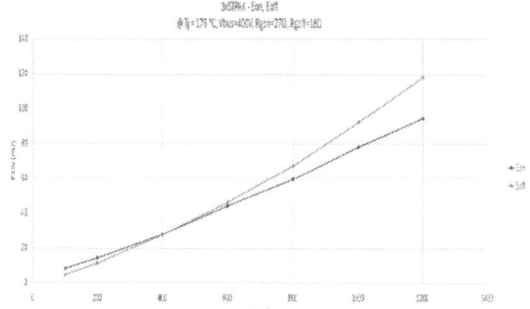

**Fig. 9**: 3xSTPAK® - Switching losses @ 175ºC

With this increased $R_g$, the 3xSTPAK® solution brings less current than the 4x one, in fact we have:

$$P_{sw}(B) > P_{sw}(A) \text{ up to } + 33\% \quad (6)$$

In addition, we have:

$$Rth_{j-f}(B) > Rth_{j-f}(A) \text{ up to } + 30\% \quad (7)$$

Basing on Eq. 6,7 results, we can estimate by how much current is reduced. Assuming the same thermal performance $T_j \leq 175ºC$ and the same equivalent $R_{ds(on)}$ at switch level between 3x and 4x STPAK® (2.5mΩ @ 175ºC) solutions, we have:

$$P_{tot}(A) = \frac{175^0C - 60^0C}{0.096} = 1197W \quad (8)$$

$$P_{tot}(B) = \frac{175^0C - 60^0C}{0.126} = 912W \qquad (9)$$

Therefore:

$$P_{tot}(B) = 0.76 * P_{tot}(A) \qquad (10)$$

For 4xSTPAK® solution the contribution of power losses is:

$$P_{cond}(A) = 0.7 * P_{tot}(A) \qquad (11)$$

$$P_{sw}(A) = 0.3 * P_{tot}(A) \qquad (12)$$

In the same way, given that 3xSTPAK® solution performs +33% of switching losses than the 4x one, we can assume the contribution will be:

$$P_{sw}(B) = 1.33 * 0.33 * P_{tot}(B) = 0.4 * P_{tot}(B) \qquad (13)$$

Consequently:

$$P_{cond}(B) = 0.6 * P_{tot}(B) = 547\ W \qquad (14)$$

At this point we can estimate the Arms current using the Eq. 2:

$$Arms(B) = \sqrt{4} * \frac{\sqrt{P_{cond}(B)}}{\sqrt{Rdson}} \frac{1}{1.414} = 660\ A \qquad (15)$$

In conclusion, due to higher $R_{thj-f}$ and higher $P_{sw}$, the Arms current is reduced of 17.5%.
As an alternative to this current estimation, we can also calculate the maximum achievable $R_{ds(on)}$ able to meet 800A which is the target condition.

Being

$$P_{sw}(B) = 1.33 * P_{sw}(A) = 465\ W \qquad (16)$$

Considering the Eq. 9 we have:

$$P_{cond}(B) = P_{tot}(B) - P_{sw}(B) = 446\ W \qquad (17)$$

Where:

$$Rdson = \frac{4 * P_{cond}(B)}{(800A * 1.414)^2} = 1.4\ m\Omega \qquad (18)$$

Even if we are saying to move on to the next SiC Generation, which guarantees 15% less $R_{ds(on)}$, the value coming out of Eq. 18 is not cost effective because it implicates more than 45% of sizing (SiC

MOSFET die area) than the current one used with 4xSTPAK®. This is the reason why a system inductance optimization is proposed to minimize the switching losses, trying to significantly reduce the die-size of 3xSTPAK® with the purpose of reaching the target of 800 A.

# 4    3xSTPAK® characterization with lower system inductance

In this case two separate DC+ and DC- bus bars have been used, as shown in Fig. 10. This connection simulates a modular approach in which two parallel branches, connected to the multiterminal DC-link, have been included. This strategy reduces the power loop stray inductance ($L_{stray}$). More specifically, it has been estimated a $L_{stray} \approx 15$ nH, that's half compared to the 4xSTPAK® solution.

**Fig. 10:** Experimental setup with modular connection approach.

Let's proceed again by selecting $R_g$ using the same criteria: $R_{goff}$ to have $V_{dspeak}$ on the active switch less than 650V, while for $R_{gon}$ setting, $V_{peak\_diode}$ has to be less than 600V.
For simplicity we only report $R_g$ setting during turning on (Fig. 11)

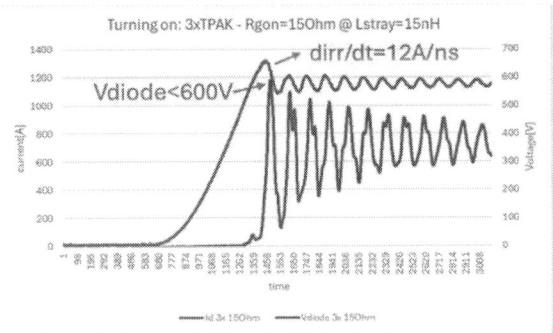

**Fig. 11:** Turning on: 3xSTPAK® with $R_{gon} = 15\ \Omega$ with $L_{stray} = 15$ nH

$$V_{peak\_diode} = 400\,V + 12\frac{A}{ns} * 15nH = 580\,V \quad (19)$$

While still respecting criteria recommendation, the reduced loop inductance allows to increase the di/dt and consequently to minimize the switching losses.

Fig. 12 shows the switching performance of the new 3xSTPAK® solution with $L_{stray}$ = 15 nH.

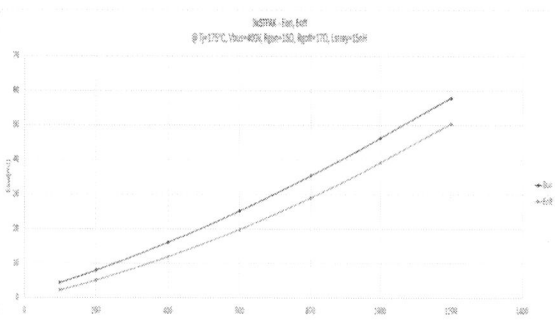

**Fig. 12**: 3xSTPAK® - Switching losses with $L_{stray}$=15nH @ 175°C.

As similarly done for the other cases, we can estimate again the maximum allowed phase current.

From Fig.12, in the worst case we have:

$$P_{sw}(B) = \frac{Eon + Eoff}{\pi} * fsw = 273W \quad (20)$$

$$P_{cond}(B) = P_{tot}(B) - P_{sw}(B) = 639W \quad (21)$$

Considering $R_{ds(on)}$ = 2.5mΩ, we obtain:

$$Arms(B) = \sqrt{4} * \frac{\sqrt{P_{cond}(B)}}{\sqrt{Rdson}}\frac{1}{1.414} = 715A \quad (22)$$

The switching improvement allows us to regain more than 8% of Arms current but, with the purpose of 800 A, this is not enough. At this point it is needed to recover a further 10% of Arms current with a further $R_{ds(on)}$ reduction. This is an iterative exercise where we increase the die-size, reducing gradually the equivalent $R_{thj-f}$.

The result with next SiC Generation leads to have $R_{ds(on)}$=2.08mΩ and $R_{thj-f}$=0.117°C/W. This combination allows us to meet the phase current target. In fact, we have:

$$P_{cond}(A) = 2.08m\Omega * \frac{(800A*1.414)^2}{4} = 665.4W \quad (23)$$

$$P_{sw}(B) = \frac{Eon + Eoff}{\pi} * fsw = 273W \quad (24)$$

Where:

$$\Delta T_{j-f} = \frac{Pcond + Psw}{0.117°C/W} = 109.8°C \quad (25)$$

Therefore:

$$T_j = 109.8°C + 60°C = 170°C \quad (26)$$

With the assumption that $T_j$ must be less than 175°C, we are able to meet 800 A.

# 5 Conclusions

This paper shows that using the next SiC Generation, even if it introduces a lower $R_{ds(on)}$ compared with the previous one, is not enough to reduce the number of devices in parallel because it is not a cost-effective solution. Therefore, despite having a SiC Generation with lower $R_{ds(on)}$, it is also necessary to have an optimization of the converter layout. Parasitic inductance minimization is useful to have benefits in terms of performance, cost, space, and weight. Table II summarizes the experimental results obtained comparing three different solutions based on paralleling STPAK® devices from STMicroelectronics. All these solutions allow to achieve the same target phase current (Arms = 800 A), with a maximum junction temperature $T_j$< 175°C.

| Configuration | Lstray | Arms | die-size | Number of paralleled dies at 3phase-Inverter level |
|---|---|---|---|---|
| 4xSTPAK® reference | 30nH | 800 A | B | 24 |
| 3xSTPAK® high L$_{stray}$ | 33nH | 800 A | B + 45% | 18 |
| 3xSTPAK® low L$_{stray}$ | 15nH | 800 A | B + 2% | 18 |

**Table II**: Comparison between three equivalent solutions.

# 6 References

[1] E. Alfonzetti, S. A. Rizzo, M. Pulvirenti and N. Salerno, "Power Loop Parasitics Impact on Paralleled Silicon CarbideMOSFETs," *PCIM Europe 2023; International Exhibition and*

*Conference for Power Electronics, Intelligent Motion, Renewable Energy and Energy Management*, Nuremberg, Germany, 2023, pp. 1-7.

[2] S. La Mantia, L. Abbatelli, C. Brusca, M. Melito and M. Nania, "Design Rules for Paralleling of Silicon Carbide Power MOSFETs," in *PCIM Europe 2017: International Exhibition and Conference for Power Electronics, Intelligent Motion, Renewable Energy and Energy Management*, edited by VDE (Nuremberg, Germany, 2017), pp. 1-6.

[3] A. Raciti, M. Melito, M. Nania and G. Montoro, "Effects of the Device Parameters and Circuit Mismatches on the Static and Dynamic Behavior of Parallel Connections of Silicon Carbide MOSFETs," *2018 IEEE Energy Conversion Congress and Exposition (ECCE)*, Portland, OR, USA, 2018, pp. 1846-1852.

[4] H. Li, S. Munk-Nielsen, X. Wang, R. Maheshwari, S. Bęczkowski, "Influences of Device and Circuit Mismatches on Paralleling Silicon Carbide MOSFETs," in *IEEE Transactions on Power Electronics*, vol. 31, no. 1, pp. 621-634, Jan. 2016

[5] J. -K. Lim, D. Peftitsis, J. Rabkowski, M. Bakowski and H. -P. Nee, "Analysis and Experimental Verification of the Influence of Fabrication Process Tolerances and Circuit Parasitics on Transient Current Sharing of Parallel-Connected SiC JFETs," in *IEEE Transactions on Power Electronics*, vol. 29, no. 5, pp. 2180-2191, May 2014.

PCIM Europe 2024, 11– 13 June 2024, Nuremberg    DOI: 10.30420/566262137

# A Multi-Objective Structural Optimization Method Based on Multi-Physics Simulations for Power Module

Baihan Liu[1] , Mengyao Du[1], Yiyang Yan[1], Jianwei Lv[1] , Cai Chen[1] , Yong Kang[1],
Yue Wu[2], Zhipeng He[2]

[1] Huazhong University of Science and Technology, China

[2] State Key Laboratory of HVDC, Electric Power Research Institute, CSG, China

Corresponding author:    Cai Chen, caichen@hust.edu.cn
Speaker:    Baihan Liu, loubeckham@hust.edu.cn

## Abstract

With the rise of wide-bandgap semiconductors SiC, power module packaging needs to be designed with all performance outstanding enough to give full play of advantages of SiC in harsh conditions. Due to the difficulty of taking into account all performance, packaging design of SiC power module faces great challenges. This study proposes a multi-objective structural optimization method based on Multiphysics simulation for power modules design, which combines Taguchi, TOPSIS, CRITIC, and entropy weight method to simplify the optimization process. Based on the proposed method, structural parameters of a SiC power module is optimized. According to the simulation results, a good effect of the compromise among the electrical, thermal, and mechanical performance is realized.

## 1 INTRODUCTION

Wide-bandgap semiconductors, such as SiC and GaN, have the advantages of high-temperature resistance, high breakdown field strength, fast switching speed, and high switching frequency, which allow more practical applications for high voltage, high temperature, and high frequency. The packaging design of SiC-based power modules, however, presents a challenge because it needs to ensure all performance indicators are excellent including electrical, thermal, mechanical, and other aspects of performance to adapt to harsh environments (high temperature, frequency, and switching speed) and operating loads (high voltage and current).

The package design of power modules is currently an iterative process conducted manually with the aid of little automation software, hard to consider every aspect of a matter. The reason is that packaging design requires weight and compromise between various performance indicators of power modules such as parasitic parameters, thermal resistance, and mechanical stress, which is very complex and time-consuming. First, many design criteria are contradictory, for example, thinning the ceramic layer of substrates is conducive to reducing thermal resistance, but it will increase the warping degree of substrates, making against the

reliability. Second, the mechanism of the influence of the structural parameters' change on all the performances is awfully complex, and if there are a large number of Parameter-Performance pairs, such research is extremely time-consuming. Third, the coupling between different physical fields increases the difficulty.

At present, research on the optimal design of power module structural parameters mainly focuses on reducing parasitic parameters [1], reducing thermal resistance [2], and improving reliability [3]. However, optimal design in these studies only aims at a single performance indicator and is mainly based on lumped parameter design or two-dimensional analysis, which has a limited effect on the optimization of structural parameters. The literature [4] proposed a multi-objective optimization design method based on the NSGA-II algorithm for two main failure mechanisms to improve the lifetime of IGBT power modules, but it requires the derivation of the analytical model, limiting the applied range. A multi-objective optimization method based on multi-physics coupled finite element simulation and MOGWO algorithm considering the electric field, thermal resistance, and mechanical stress is proposed in [5] to optimize the structural parameters of SiC power modules. However, the number of simulations depends on the number of algorithm iterations, increasing time cost.

In this study, a new structural optimization method is proposed, which reduces the number of simulations with orthogonal experimental design, simplifies the multi-index problem by the TOPSIS evaluation model based on the weighting method combining the CRITIC method and entropy weight method, and finds the structural parameter group with the best comprehensive/overall performance utilizing the Taguchi method.

# 2 MULTI-OBJECTIVE OPTIMIZATION DESIGN METHOD BASED ON THE TAGUCHI METHOD AND TOPSIS EVALUATION MODEL

## 2.1 Taguchi Method

Taguchi Method is an effective robust optimal design method applying the statistical methods and the design of experiment concepts, which is currently most widely used in construction design and the optimized analysis of the production process. Its core content is the Quality Loss Function $L(y)$ and the Signal-to-Noise Ratio $S/N$, and they are linked by Mean Square Deviation $MSD$, which is defined by the following formulas:

$$L(y) = (y - m)^2 \Rightarrow E[L(y)] = k(MSD) \quad (1)$$

$$S/N = -10\log(MSD) \quad (2)$$

$$MSD = \begin{cases} \dfrac{1}{n}\sum_{i}^{n}(y_i - m)^2 \\ \text{For one or more units;} \\ (\mu - m)^2 + \sigma^2 \\ \text{For a continuous function.} \end{cases} \quad (3)$$

where $k$ is a constant, $n$ is the number of units, $y_i$ is the quality level of products, $m$ is the target value, $\mu$ is the mean value of the quality level, and $\sigma^2$ is the variance of the quality level.

The degree of how close quality levels are to the target value is usually utilized to assess the quality of products, and any deviation means causing quality loss, which is what the quality loss function expresses. It can be inferred from formulas (1), (2), and (3) that in order to eliminate quality loss and achieve optimal design, $MSD$ needs to be equal to 0, that is, product quality is equal to the target

value. Therefore, according to the relationship between $L(y)$ and $S/N$, reducing quality loss is actually increasing the signal-to-noise ratio.

The basic process of the Taguchi Method is: first select factors and determine their levels; then establish an appropriate orthogonal array as the test table and conduct experiments according to it; finally, process the results to calculate their $S/N$ (signal-to-noise ratio) and select the combination of the factors' levels with the greatest $S/N$.

## 2.2 TOPSIS Evaluation Model

The conventional Taguchi method is generally applied to single-objective optimization problems. For multi-objective problems, such as the structural parameter optimization of power modules, introducing multi-objective optimization algorithm is usually necessary. In this study, TOPSIS evaluation method based on CRITIC-entropy weight method is used to simplify the multi-objective problem into a single-objective problem by scoring samples according to their performance indicators.

### 2.2.1 CRITIC-Entropy Weight Method

Both CRITIC method and entropy weight method are objective weighting methods, which determine weights of indicators only according to the intrinsic characteristics of data and avoid the interference of subjective factors.

The CRITIC method uses the independence coefficient and variation coefficient to express the conflict between indicators and information content respectively. The greater the independence coefficient, the smaller the correlation coefficient, the stronger the conflict between indicators, the less the repetition of the information reflected by indicators, and the greater the weight of the indicators; the greater the variation coefficient, the greater the standard deviation, the greater the difference between samples, the greater the information reflected by indicators, and the greater the weight.

In entropy weight method, entropy value is utilized to reflect the order degree of information. The smaller the entropy value, the greater the dispersion degree of indicators, the greater the usefulness of information, and the greater the weight. Fig. 1 shows the calculation steps and key formulas of the two methods.

CRITIC method has significant advantages due to taking into account both the correlation of indicators and the weight of information, but ignores the dispersion degree of indicators, which is exactly what entropy weight method is good at. The single evaluation criterion of entropy weight method, however, is easy to cause unrealistically high

PCIM Europe 2024, 11– 13 June 2024, Nuremberg        DOI: 10.30420/566262137

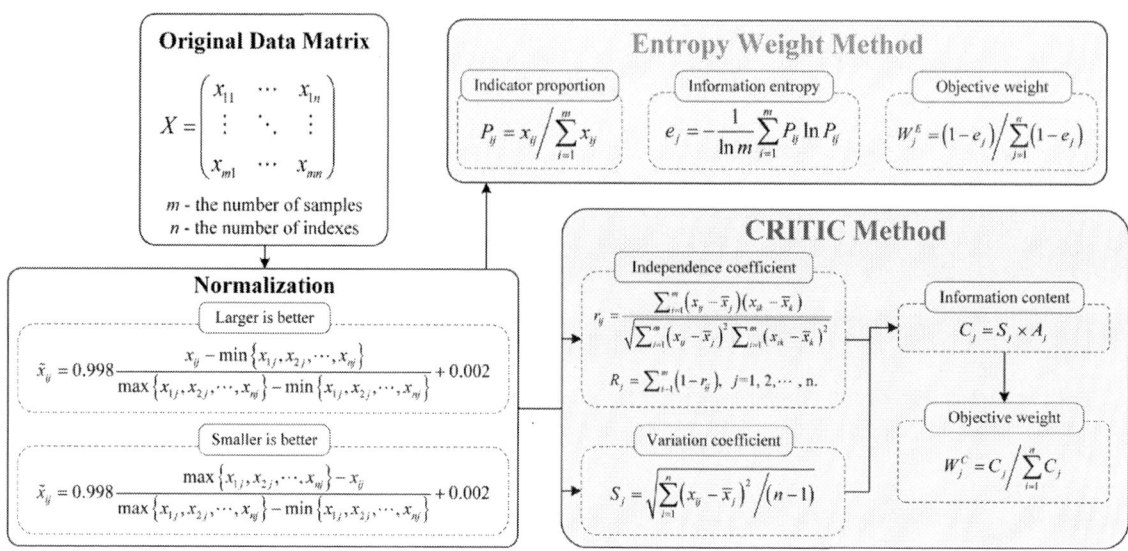

**Fig. 1**    Flow chart of CRITIC method and entropy weight method.

weight. In this study, the two weight methods are combined to complement each other to capture the diversity, dispersion, and correlation among different indicators. In CRITIC-entropy weight method, the final weights of indicators are determined according to the following formula:

$$W_j = \alpha \cdot W_j^C + \beta \cdot W_j^E \qquad (4)$$

where, $W_j^C$ and $W_j^E$ represent the weight of the $j$th indicator calculated by CRITIC method and entropy weight method respectively. $\alpha$ and $\beta$ are constant and their sum is 1.

### 2.2.2  TOPSIS method

TOPSIS method is a widely used multi-indicator comprehensive evaluation method, which scores the samples objectively based on data analysis. The core idea of TOPSIS method is to score and order the limited samples according to their closeness of the ideal target, so as to get the approximated ideal scheme. The flow chart of TOPSIS method based on CRITIC-entropy weight method is shown as Fig. 2.

### 2.3  Proposed Multi-Objective Optimization Design Method

Fig. 3 describes the implementation process of the proposed multi-objective optimization design method. The whole process contains three parts, Taguchi method, Multiphysics simulation, and evaluation model. Multiphysics simulation considers the coupling among electrical, thermal, and mechanical field. The circuit simulations give the power loss of chips to thermal simulations to get the temperature distribution, and thermal simulations prepare the thermal load for mechanical simulations to obtain the stress distribution.

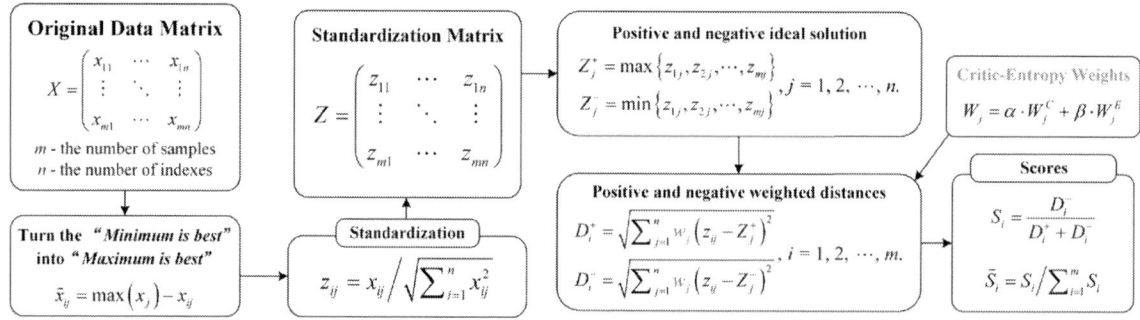

**Fig. 2**    Flow chart of TOPSIS method based on CRITIC-entropy weight method.

1074

PCIM Europe 2024, 11– 13 June 2024, Nuremberg          DOI: 10.30420/566262137

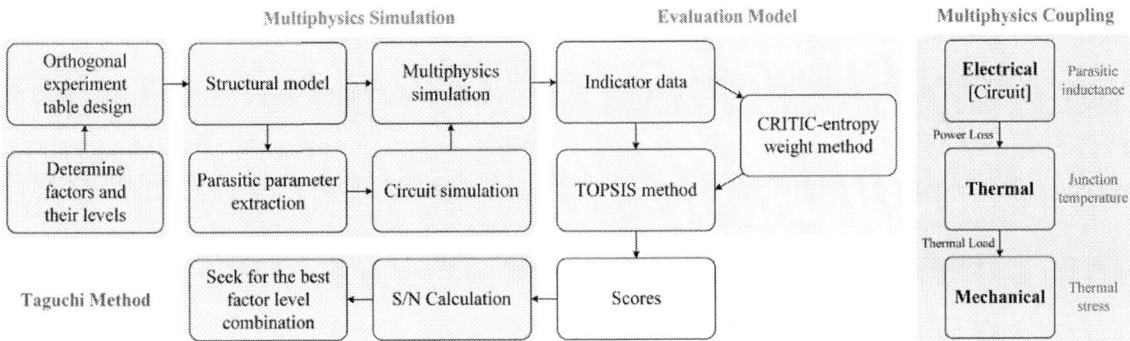

**Fig. 3**    Process of the proposed multi-objective structural optimization design method.

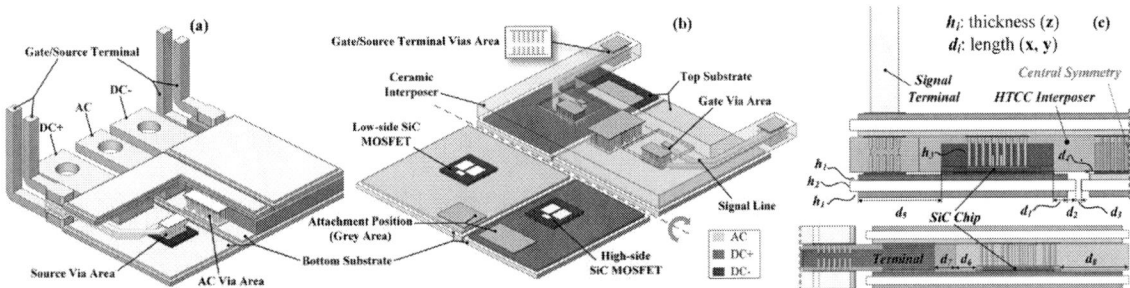

**Fig. 4**    Internal view of the power module based on island packaging structure. (a) Partial section view. (b) Internal view. (c) Meaning of factors.

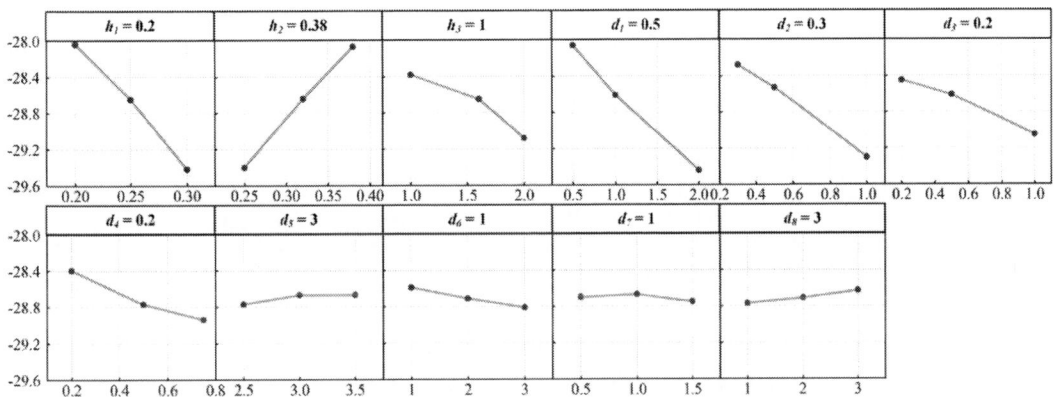

**Fig. 5**    Data processing result of the signal-to-noise ratios of all the factor levels.

## 3    Packaging Structural Optimal Design of a SiC Power Module

The proposed multi-objective optimization design method is employed to optimize the packaging structural parameters of a SiC-based power module as shown in Fig. 4. In the design, three thickness sizes and eight length sizes are chosen as the factors, and their levels are determined according to the size specifications and limitations provided by vendors. As to the experimental design, an $L^{27}(3^{13})$ orthogonal array is utilized to arrange factors and levels. Besides, power-loop parasitic inductance (PLIL), junction temperature (JT), maximum principal stress (MPS) of SiC-mosfet bare dies and AMB substrates, which are commonly emphasized in industry, are selected as the evaluation indicators.

1075

| Indicators | Best in samples | Worst in samples | Optimized values |
|---|---|---|---|
| MPS of Bare dies | 11.9942 | 48.40512 | 27.7601 |
| MPS of Substrate | 173.0723 | 272.2932 | 189.4401 |
| PLPI | 4.281 | 9.689 | 4.410 |
| JT | 248.381 | 249.384 | 248.722 |
| Scores | 0.046629 | 0.029674 | 0.046634 |

**Table 1** Indicators Data of the Optimized Power Module.

According to the experimental table, 27 sets of simulations are conducted, and samples are scored based on the indicator data from simulations by the proposed evaluation model. The S/N of factor levels is calculated utilizing (1) and the results are shown in Fig. 5. In accordance with the principle of maximizing S/N, the optimized structural parameter combination is easily determined. To verify the effect of optimal design, the power module with optimized structural parameters is modeled and simulated under the same conditions as other samples. As listed in Table I, each indicator of the power module after optimal design is very close to the best value of the corresponding indicator appearing in samples, and its score is the best. Hence, the proposed optimal design method plays a good compromise effect.

## 4    Conclusion and Future Work

In this study, a multi-objective structural optimization method based on Multiphysics simulation for power modules design is proposed, which combines Taguchi, TOPSIS, CRITIC, and entropy weight method to simplify the optimization process. The structural parameters of a SiC-based power module are optimized by the proposed optimal design method for better electrical, thermal, and mechanical performance. According to the results, the proposed optimal design method plays a good compromise effect among indicators.

## References

[1] Q. Le, T. Evans, S. Mukherjee, Y. Peng, T. Vrotsos, and H. A. Mantooth, "Response surface modeling for parasitic extraction for multi-objective optimization of multi-chip power modules (MCPMs)," in *2017 IEEE 5th Workshop on Wide Bandgap Power Devices and Applications (WiPDA)*, 2017, pp. 327–334. doi: 10.1109/WiPDA.2017.8170568.

[2] G. Tang, T. chong Chai, and X. Zhang, "Thermal Optimization and Characterization of SiC-Based High Power Electronics Packages With Advanced Thermal Design," *IEEE Transactions on Components, Packaging and Manufacturing Technology*, vol. 9, no. 5, pp. 854–863, 2019, doi: 10.1109/TCPMT.2018.2860998.

[3] H. Ren *et al.*, "High-Reliability Wireless Packaging for High-Temperature SiC Power Device Sintered by Novel Organic-Free Nanomaterial," *IEEE Transactions on Components, Packaging and Manufacturing Technology*, vol. 10, no. 12, pp. 1953–1959, 2020, doi: 10.1109/TCPMT.2020.3038430.

[4] B. Ji, X. Song, E. Sciberras, W. Cao, Y. Hu, and V. Pickert, "Multiobjective Design Optimization of IGBT Power Modules Considering Power Cycling and Thermal Cycling," *IEEE Transactions on Power Electronics*, vol. 30, no. 5, pp. 2493–2504, 2015, doi: 10.1109/TPEL.2014.2365531.

[5] W. Li, Y. Wang, Y. Ding, and Y. Yin, "Optimization Design of Packaging Insulation for Half-Bridge SiC MOSFET Power Module Based on Multi-Physics Simulation," *Energies*, vol. 15, no. 13, 2022, doi: 10.3390/en15134884.

PCIM Europe 2024, 11– 13 June 2024, Nuremberg   DOI: 10.30420/566262138

# Holistic Approach to Maximize Lifetime and Power Density in High Power Semiconductor Modules

Martin Schulz[1], Lukas Kreiner[1], Ralf Klemmer[1]

[1] Littelfuse, Germany

Corresponding author:   Martin Schulz, mschulz@littelfuse.com

## Abstract

The challenge in power electronics is, to achieve higher power throughput in smaller housings using less resources while increasing efficiency and reduce cost. As these targets have contradicting solutions, compromises need to be chosen. Higher currents increase the thermal stress in a given device, reducing its lifetime. To counteract, exchanging IGBTs using wide band gap SiC-MOSFETs is considered. However, the solution becomes more expensive in turn. Adapting an approach as done in press-pack-devices and allow electrically active heat sinks opens the path to massive improvement. The present work shows how this can be done in IGBT-based designs.

## 1 Comparing power modules and disc-style devices

Increasing power density always causes increased losses per area and thus typically higher temperatures. In turn, higher stress appears to interconnecting joints which recently lead to the replacement of soft-soldered connections by sintering [1]. Higher stability of the interconnecting technology can counteract the lifetime-reduction that results from higher temperature swings.

Another obvious solution countering stress due to temperature changes is improved cooling. Today, when building power semiconductors with insulating DCBs, the ceramic layers involved pose a physical limit to thermal transfer. A sketch of a power semiconductor module as it is built today is given in Fig. 1.

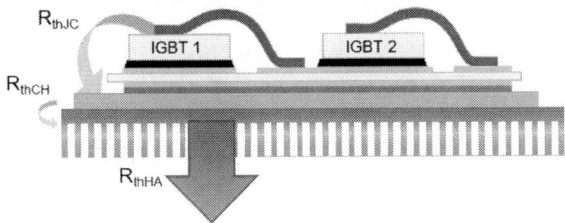

Fig. 1 Power semiconductor setup using DCBs for insulation

As a rule of thumb, any material that features electrical isolation also is a bad thermal conductor.
In contrast to power semiconductor modules, high-power disc-devices are often combined with electrically active cold-plates as higher performing cooling systems. A correlating stack-assembly along with a detailed view is given in Fig. 2:

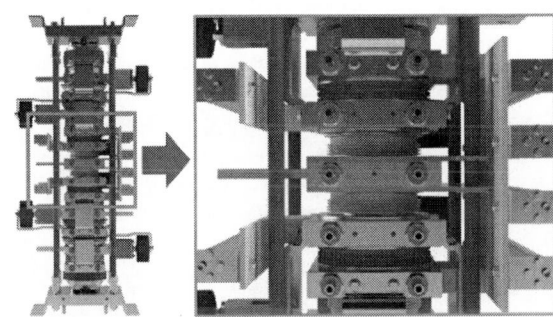

Fig. 2 3-Level IGBT-stack using electrically active cold-plates

The stack forms a 2400 A / 4500 V 3-level phase-leg [2]. The aluminum-made cold-plates are electrically active and additionally serve as terminals to the disc-devices and connecting point for snubber circuits, thus contributing to material savings and space reduction.

Consequently, a non-conducting cooling liquid must be used, typically a deionized water-glycol mixture.

1077

## 2 Feasibility study

To eliminate the dominating thermal resistance of the DCB's ceramic, a different solution was followed. In this, the IGBT-dies rated 1200 V/200 A are directly soldered to a suitable liquid cooled plate. This way, the cold-plate becomes the electric connection to the IGBT's collector and isolated islands in the design are solely required to mount the power-terminal at the emitter-side as well as the control-terminals. Using bond-wires to connect the individual parts in the setup was the obvious choice which led to a first device under test (DUT) as presented in Fig. 3.

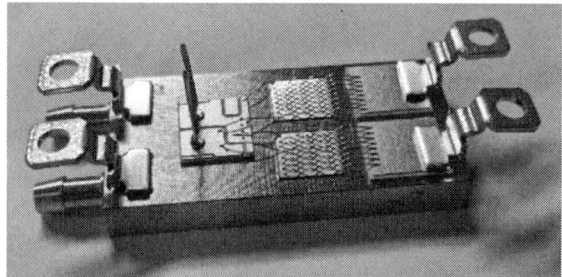

**Fig. 3** First device for feasibility study and thermal measurements

The die used in this first scheme is a 200 A IGBT with a size of approximately 200 mm². Eight bond-wires with 25 A current carrying capability were used per die to connect the emitters to the power terminals.

The device was blackened to conduct thermal measurements using infrared imaging. Current was applied from a laboratory supply, liquid cooling flowrate was set to about 6 L/min. Fig. 4 is a close-up view to the DUT within the setup.

**Fig. 4** DUT prepared for thermal imaging

The DUT is powered from a high-current, low-voltage source, capable to deliver 1 kA at 10 V. This way, safety precautions only are concerned with high temperatures but not with high voltages.
It was expected that the 200 A-chip remains well below it's thermal limits of 150°C even with much higher currents than 200 A being applied.

Conducting the experiment started with just 50 A and was repeated after increasing the current in 50 A-steps.

Reaching the rated chip current of 200 A, the die-temperature rose by about 70 K, leaving a margin for further 50 K before the maximum temperature of 150°C would be reached. However, with 25 A per bond wire another physical limit is reached as the bond wires' temperature already exceeded 400°C. The IR-image at 200 A per die can be seen in Fig. 5.

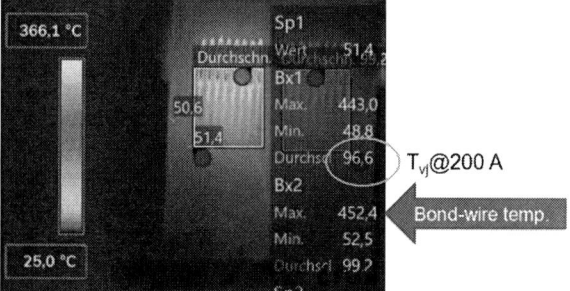

**Fig. 5** Thermal image taken at 200 A per die

Continuing the experiment by increasing the current, the bond wires fused at 250 A per die.

Very clearly, the assembly technology in such an approach becomes the weak link and a second approach was done using clip-soldered connections instead of bond wires.

This is only possible due to a new solderable chip-metallization applied to the emitter of the die.

Figure 6 is a photo of the second device, **Fig. 7** reveals the thermal development at 200 A per die, measured within the same setup as before.

**Fig. 6** DUT Version 2 with clip-soldered connections instead of bond wires

With this new construction, the chip temperature at 200 A reached 94°C while the connecting clip remained at a maximum temperature of 78°C and thus about 370°C below the bond wires' temperature in the previous design.

Operating temperature of $T_{vj}$ = 150°C was reached with a current exceeding 275 A.

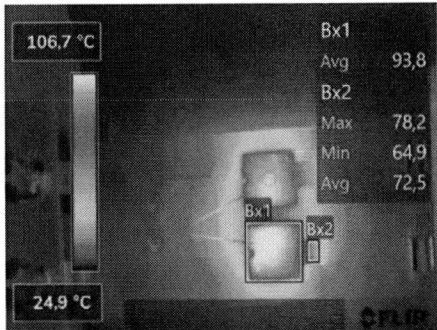

**Fig. 7** Thermal result at 200 A in the clip-soldered device

It is obvious, that the limitation imposed by using bond-wires has been removed. The clip's temperature remains well below 100°C which is about 350 K less than what was observed on the bond-wires.

As the IGBTs used in the two DUT were not identic, a direct comparison based on current only is not revealing the true picture. DUT1 was equipped with a chip designed for 200 A rated current and very low forward voltage of only 1.2 V at rated current. The transfer characteristic of the die used is displayed in Fig. 8.

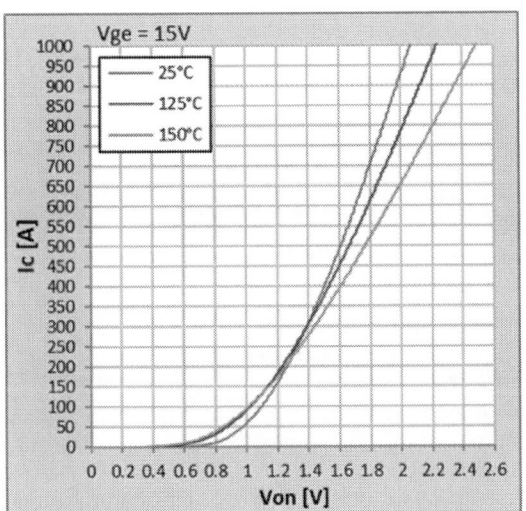

**Fig. 8** Transfer Characteristic of the newly developed 200 A-IGBT

A special feature of the chip is the high ratio between desaturating current and rated current. Typical IGBT-designs desaturate in a range of four to five times the rated current. This new chip won't desaturate below six to eight times it's nominal current. It becomes obvious, that exploiting this

feature to the full potential would lead to power loss densities easily exceeding 600 W/cm². This assumes operation at 650 A, resulting in 2 V forward voltage and 1300 W of losses on a 200 mm² die.

As this die does not feature a solderable front-side metallization, it was not an option to be used in the second DUT.

Instead, a standard die upgraded with a solderable metallization but a lower current rating of only 150 A rated current was used.

Besides the different chip area of only 183 mm², the forward voltage characteristic of the chip is different as is seen in Fig. 9.

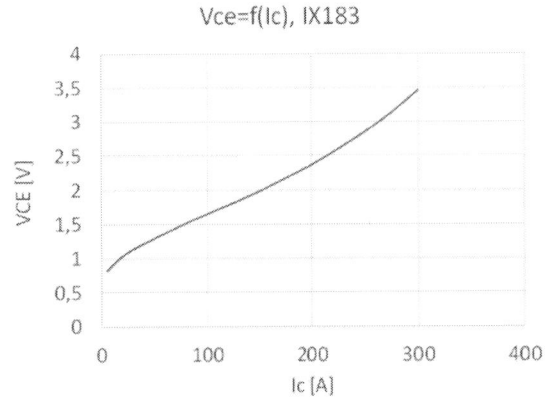

**Fig. 9** Forward characteristic of the IX183-IGBT

Therefore, a fair analysis can best be done considering power loss density [W/cm²] or be condensed in the thermal resistance calculated from both trials.

From the measurement done, a thermal transfer capability of the setup was calculated, allowing power-loss-densities of up to 380 W/cm² at a temperature swing of 100 K. This represents the difference between a maximum junction temperature of 175°C and an inlet temperature of the cooling liquid of 65°C.

For the newly designed chip with 2 cm² area, this allows to handle 760 W per die, leading to a maximum current of 450 A.

The 150 A /183 mm² standard die could be operated within it's thermal limits carrying up to 250 A.

## 3 Electric testing

Though the focus of the study was on the thermal performance, the basic electrical behaviour of the setup was also tested.

The result from a double-pulse-test is summarized in Fig. 10.

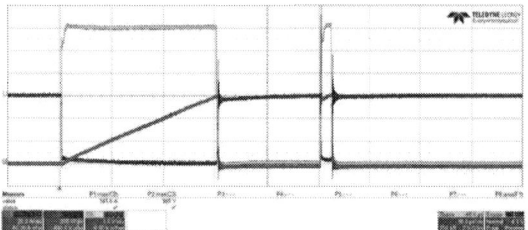

**Fig. 10** Results from the dynamic testing in double-pulse test equipment

The tests revealed a sufficiently clean switching but due to the nature of the test setup, no further details were investigated. Though the layout of the devices studied was not optimized for switching and a perfect switching behaviour was not expected, the results were still good enough to later be transferred into a potential series-development.

# 4    Cyclic load testing

The outstanding thermal performance is a key factor in increasing the lifetime of a power electronic component.

From the second design, several devices were subjected to a power cycling test ($PC_{sec}$) as defined in IEC 60749 [2].

The expectation was that the pad-and-clip assembly achieves a higher lifetime than a system using bond-wires as the failure mechanisms of bond-lift-off and bond-heel-crack are eliminated. However, as the pad is soldered to the chip's front-side, delamination of this interface is expected to happen eventually.

The test was conducted with an inlet temperature of 12°C. In a cycle of 4 seconds with 50% duty-cycle, the chip temperature swing observed was 90 K with a load current of 250 A.

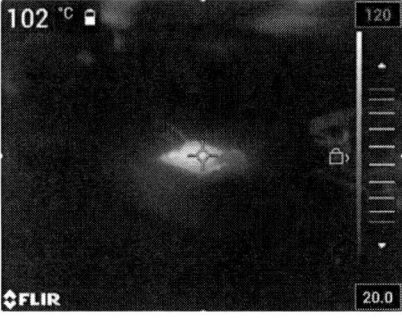

**Fig. 11** Thermal image of the chip during $PC_{sec}$-Test

As only very few DUTs were available, the chips were tested in a single arrangement.

Figure 11 reveals the outstanding thermal performance achieved and the low spreading of heat within the cold-plate.

The results from the long-term test conducted are summarized in the graph in Fig. 12.

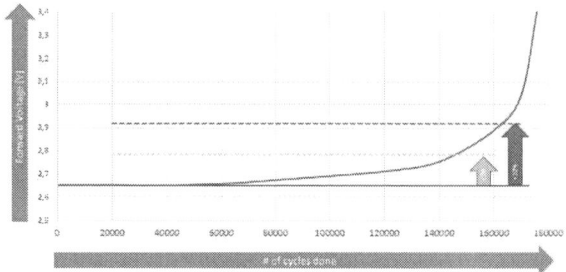

**Fig. 12** Results from $PC_{sec}$-Testing

It remains to be noted, that the IGBT in this test had a rated chip current of only 150 A, so despite staying within the given thermal limits for chip temperature, the chip was operated well beyond parameters that would be used in a real-world application.

The end-of-life criteria for this test is an increase of the forward voltage of 5%. This value was reached after roughly 145.000 cycles. Classical solder-bond-technology reaches about 80.000 cycles under these conditions. As further potential for improvement is seen in the chip metallization as well as in solder alloys and solder processes, achieving at least twice the power-cycling capability compared to solder-bond devices does seem reasonable.

# 5    Applications in focus and resulting benefits

From the structure chosen and power density achieved, it is obvious, that such a design is meant to operate in a high-power application.

Especially applications that already feature liquid cooling and demand high power throughput can benefit from a non-insulated power semiconductor arrangement.

Main targets for such an approach can be found in renewable energy generation in windmills or in the steel industry in metal melting or metal welding by induction heating.

With the heat sink forming the connection to the IGBT's collector, the scheme is a good choice for building single switches with high current carrying capabilities to replace existing designs based on current power modules with the same single-switch topology.

What could potentially be achieve can be estimated from the rendered image in Fig. 13.

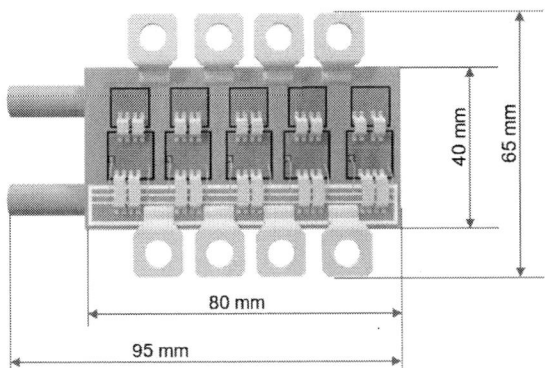

**Fig. 13** 1200 A-Device in non-isolated setup

Equipped with a 250 A-chipset, this version resembles a 1200 A single switch with an envelope of 123.5 cm³.

A half-bridge consisting of two such devices consumes about 250 cm³ of space. In comparison, common high-power modules used today demand up to 700 cm³ [3].

Besides single switch arrangements, the setup is also a good replacement when building bidirectional switches as used in DC-circuit breakers in battery-charging or UPS-systems as well as AC-circuit breakers.

A further benefit arising from the integrated liquid cooling is, that the necessary surrounding housing is no longer burdened with the high temperatures commonly seen in power semiconductors. This opens a path to use lower-grade plastics, potentially even materials that are easy to recycle, which will be a major topic in the years ahead.

In contrast to the classical approach, the terminals also benefit from the intense cooling, reducing the heat that is transferred to surrounding components like DC-link capacitors.

As for the use of resources, a half-bridge built from the device in Fig. 13 has a mass below 0.7 kg which is less than half the weight of a current design.

Sacrificing a fraction of the performance by replacing the copper heat sink using aluminum would allow for both, cost and weight reduction.

## Conclusions

The omnipresent trend of increasing power density in power semiconductors as they are built today starts reaching physical limits due to the isolation requirement. To push these limits further, new methods to extract heat from power semiconductors more efficiently need to be identified.

One way of doing so is direct liquid cooling combined with a suitable chip- and interconnection technology as presented.

The shift from classical isolated assemblies to non-isolated counterparts opens the door to increase power density by a factor of 10, compared to today's options.

## 6 Reference

[1] Karsten Guth et. al.
*New assembly and interconnects beyond sintering methods*
PCIM 2010, Nuremberg, Germany.

[2] Datasheet, XA2400GV45WT, Littelfuse

[2] International Standard IEC 60749-34

[3] Elaheh Arjmand
*Development of Cu-Cu Joining Technology by Laser Welding for Terminal Attach within Power Semiconductor Package*
PCIM 2023, Nuremberg, Germany

[4] Datasheet, FZ1200R12HE4, Infineon Technologies

PCIM Europe 2024, 11– 13 June 2024, Nuremberg          DOI: 10.30420/566262139

# Regulated High Density Switch Capacitor Topology

Pierrick Ausseresse, Josef Daimer[1] and Manfred Schlenk[2]

[1] Infineon Technologies AG, Germany

[2] Dr. Schlenk Consulting, Germany

Corresponding author:   Pierrick Ausseresse, Pierrick.ausseresse@infineon.com
Speaker:                        Pierrick Ausseresse, Pierrick.ausseresse@infineon.com

## 1    Abstract

Operating frequencies of around 100 to 200 kHz are commonly used for power of up to several kW.

As frequency increases towards MHz core losses increase greatly and force designers to limit flux density. This result in larger magnetic components with more turns per windings, limiting efficiency and density improvement that used to come with frequency increase.

To further increase power density in power supplies a shift from traditional inductor-based DC/DC converter to capacitor/inductor-based hybrids is happening. Capacitors are very effective at storing and transferring energy and their limitations occurs at much higher frequency than inductors. This means an increase of the upper frequency while offering higher efficiency and power density. Switched capacitor converter operate however at a constant step up/down ratio: regulation is barely possible.

To achieve regulation an inductor pre or post regulator is needed. The result is a dual stage DC/DC which increase costs and impact efficiency – even if power density goes up.

This paper will present a way to integrate a regulation block directly into the switch capacitor DC/DC, reducing cost and losses at the same time.

## 2    Introduction

Most DC/DC need to be able to output a regulated output. They operate either from a single-phase PFC pre-regulator with residual line ripple or from a non-constant source like a battery, both varying up to ±15%.

In the considered application a loosely regulated input voltage of about 160 V needs to be stepped down to a regulated voltage of 20 V at 3 A.

The first demonstrator was built using a state-of-the-art buck converter running at 100 kHz. Inductor size is 20*20*12=4.8 cm³ with a typical efficiency around 93% and a total power density of 10 W/cm³

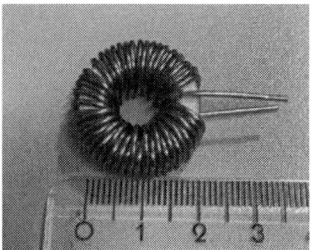

Fig. 1: 100 kHz, 160 V to 20 V buck inductor, around 5cm³

To increase power density an hybrid switch capacitor plus buck post regulator has already been used in the literature. One of the most succesfull is based on a bidirectionnal dickson charge pump.

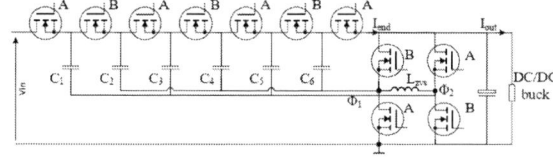

Fig. 2: non-regulated 180 V to 22 V Dickson switch capacitor. A low voltage buck regulates the Dickson output down to 20 V

Operation is very similar to the well-known diode-based Dickson charge pump, except that transistors are used to allow both step up and step down.

Such a system achieves an excellent power density of 50 W/cm³ due to the capacitor volume being much smaller than the single buck inductor. Due to the low remaining voltage difference, the regulating low voltage buck can be operated with higher frequency and smaller inductance, reducing its size manifold.

Still, this is a dual stage approach. Both stages must be able to handle the full output power of 60 W and losses multiplies. With density being the only advantage, the increase in cost is one of the main reasons such topology is not widely accepted.

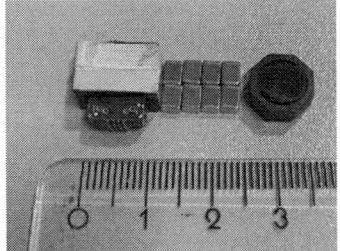

**Fig. 3:** power elements for 180 V to 22 V switch capacitor with 22 V to 20 V buck inductor. 1cm³

## 2 Switch capacitor regulation

Dickson charge pumps act like current multipliers. Each electrical charge going into one flying capacitor is also added to the output while being placed into the next flying capacitor. The number of charges flowing through the low voltage output is then directly proportional to the number of capacitors N:

$$I_{out} = (N + 1) * I_{in} \qquad (1)$$

During balancing, capacitor current is identical for both capacitors being balanced. At the same time, average current in each capacitor over a period must be zero. This allow to build following equation system, with Ick[A] being current into capacitor k during phase where all A transistors are active.

$$\begin{cases} I_{in} = I_{C1}[A] \\ I_{Ck}[A] = I_{Ck}[B] \\ I_{Ck}[B] = I_{C(k+1)}[B] \\ I_{C6}[A] = I_{end} \end{cases} \qquad (2)$$

The result being that end current of the Dickson charge pump is the same as input current and directly proportional to the output current.

$$I_{in} = I_{end} = \frac{I_{out}}{N + 1} \qquad (3)$$

Most power electronic engineers are used to build regulators that control the output current to regulate the output voltage.

Considering above equation, controlling the end current of the charge pump would have the same results - only with a larger DC gain.

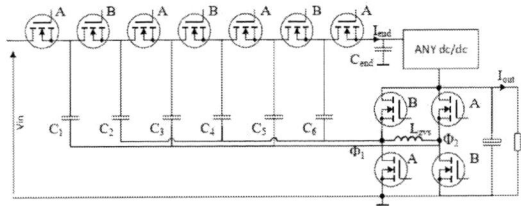

**Fig. 4:** Regulated Dickson charge pump using a low power dc/dc regulator.

Interestingly, it is possible to control the charge pump end current by placing a DC/DC converter.

In this case, the input current of the regulator becomes the control variable for the system output current. Note that any type of DC/DC works here but the type of DC/DC will impact the range where regulation is possible: using a buck controller will place a lower limit on the input regulation range, while a boost will enforce a higher limit. Even a linear regulator will work too.

The very positive effect is that the charge pump end current is equal to the input current, which is low due to the high voltage input, while the end voltage has been reduce trough the charge pump.

$$\begin{cases} I_{end=}{}^{I_{out}}/_{(N+1)} \\ V_{end} = V_{in} - N * V_{out} \end{cases} \qquad (4)$$

With both end current and end voltage being small, the power that need to be handled in the DC/DC is only a fraction of the total application power. This translates directly in costs reduction.

Minimizing the DC/DC power is also important for the total system efficiency. The charge pump and the DC/DC are operating not in series: this is not a dual stage solution, and this change the way to optimize.

To calculate losses each module efficiency must be pondered with the actual energy it is handling and the number of stages it goes through.

$$\begin{cases} P_{in} = {}^{P_{cp}}/_{\eta_{cp}} + {}^{P_{dc}}/_{\eta_{cp}} * \eta_{dc} \\ P_{cp} = P_{max} = 60W \\ P_{dc} = r * P_{cp} \text{ with } 0 < r < 1 \end{cases} \qquad (5)$$

Charge pumps being by nature very efficiency, to maximize efficiency is becomes important to minimize the losses in the DC/DC. As always, it is possible by mean of stage efficiency improvement. For this topology where DC/DC power is not constrained by output power, one more option is also valid: reducing the DC/DC power.

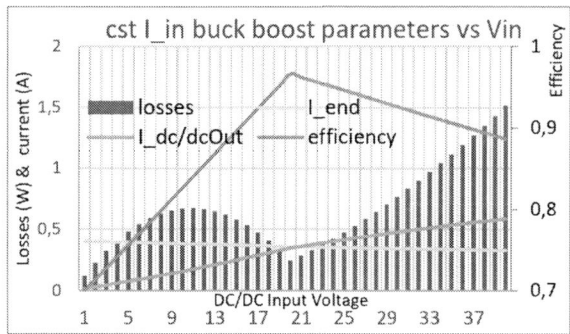

**Fig. 5:** Effect of input voltage on DC/DC losses for a regulated constant input current.

Using Fig.5, is becomes clear that there are two operating points where the losses can be considered negligible: with low input voltage where efficiency play little role on the losses, and with Vin=Vout where most converter achieve close to unity efficiency. The DC/DC that best match this range is the boost converter as it covers 0V input as well as 20V input. Over this range, losses go between 100 mW and 650 mW for an average of 500 mW. In total, regulation is brough to a 60 W fixed ratio charge pump with a low cost 5 W boost converter with only 500mW losses or 0.8% efficiency drop. This would have only been possible with a 99.2% or higher series 60W regulator in a two-stage system – with much higher price tag.

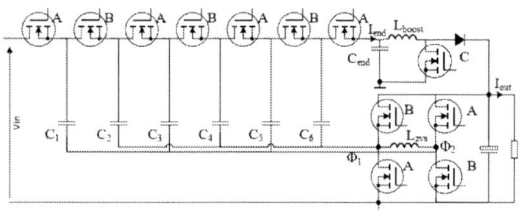

**Fig. 6:** Regulated Dickson charge pump using a low power boost regulator.

In a world where everything is a tradeoff, there are some limitations due to the hybrid construct – mainly range related. For proper operation, the range of the boost converter must still be respected.

$$0 < V_{end} < V_{out} = 20V \quad (6)$$

This in turn place a limit on the input voltage range.

$$N * V_{out} < V_{in} < (N + 1) * V_{out} \quad (7)$$

As a result soft start can be difficult, as is always the case with boost converter which suffer from

inrush currents. For input voltage higher than N*Vout (expected true for Vout=0 V) the system has low control on the output current. There are some mitigation strategies. For example, the charge pump could start together with the preregulator, reducing the amount by which the charge pump input voltage exceeds the desired range. Using a buck boost would also work but with extra costs.

For the application being tested with N=8 the output voltage can be regulated for any input between 160 V and 180 V, which is a relatively standard ripple after a PFC front end. Soft start wasn't an issue as by design the charge pump input voltage slowly rose from 0 V to its steady state.

**Fig. 7:** Hybrid regulated charge pump power passive components. Volume is around 0.7cm³

## 3 Practical results.

Three solutions were compared under similar conditions: a traditional buck converter, a fixed ratio charge pump (CP) plus low voltage buck and the presented hybrid solution.

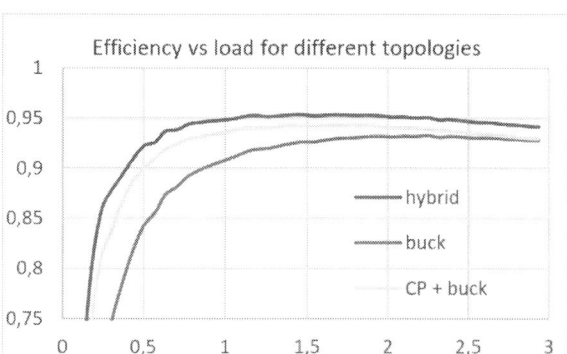

**Fig. 8:** Efficiency results for 3 different solutions.

Compared to a two-stage solution, the hybrid topology benefit from an almost constant efficiency improvement of 1% up to 1.2% over the whole load range. At the same time, it is about 20% smaller while using lower power components for lower cost, allowing improvement on size, cost, and efficiency at the same time – only trading off the input voltage regulation range. The single stage buck converter suffers from switching losses and performs poorly in the low load range (4% lower efficiency at 30% load) which is a known issue of high step-down ratio buck controller.

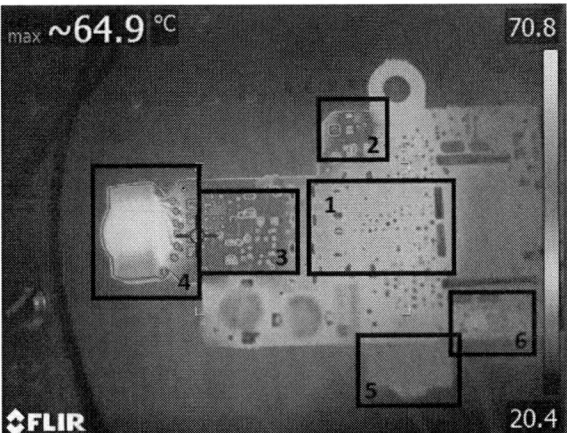

**Fig. 9:** Thermal results of the converter under full load.

Often overlooked, The ZVS inductor (4) is responsible of recycling the Coss energy. At 1MHz it circulates about 5W of power; an energy that would have been lost without it. As visible on the thermal picture it causes most of the system losses at 1.8 W, more than half of the total losses. In comparison, the Dickson chain of transistor (1) is virtually lossless (<100 mW) while the phase generation (3) and DC/DC (2) account for the remaining 1.5 W. Drivers (5+6) do not cause many losses due to the usage of GaN transistors.

Further design step will focus on optimizing the ZVS inductor. Measurement have shown high Rac – higher than what models predicted. Theoretical losses should have been below 1W.

## 4   Driving transistors.

Driving transistors is normally excluded from topologies presentation, but it wouldn't be fair when dealing with flying capacitor charge pumps.

Presented topology has 14 transistors, 11 of them not sharing a common source with any others. This is a challenge not only for providing isolated gates drives, but also for the supply of said gate drivers.

The solution is simpler than it seems and date far back in time: gate drive transformer – with a quirk. Instead of traditional high impedance wire wound

transformers, a low impedance embedded planar is used.

By using PCB planar magnetic it is possible to get rid of the limitation on the number of winding that traditional transformer have.

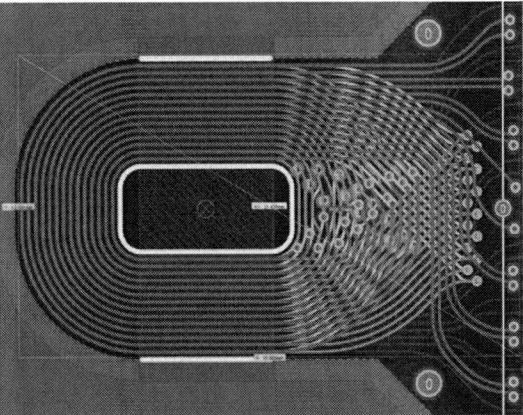

**Fig. 10:** 14 Windings planar gate drive transformer, with 4 turn per winding (R=3 Ω, L=12 µH)

On a regular transformer, those 28 contacts shared over two side of a coil former would need a contact area of 35 mm; larger than the whole set of power components that they are supposed to drive. Using planar, no dedicated contacts are needed and the whole gate drive transformer fits the footprint of an ELP14.5 ferrite. The latest generation present on the thermal picture is even smaller thanks to the use of 6 layers PCB and an ER9.5 core (5), the smallest available planar core on the market.

For Silicon FET a direct connection to the gate is possible, but not for GaN. GaN gate voltage is not rated symmetrically: transistors are fully ON at 5 V but the gate is damaged below -4 V. A transformer providing ±5V would exceed the gate capability, while driving ±4V would lead to high channel resistance.

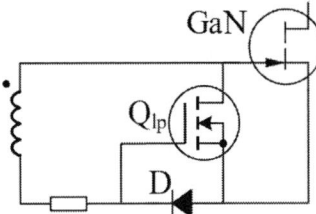

**Fig. 11:** Winding adapter circuit for GaN transistors

Using the circuit in figure 11 any negative voltage can be blocked without damaging the gate. For alternance with doted end positive, diode D conduct until the GaN transistor gate is fully charged and the transistor is ON. When voltage on the winding starts to drop it will first increase the low power transistor Qlp gate voltage, providing a short path to discharge the power transistor gate. Voltage on the doted end goes then negative, further charging Qlp gate and keeping the GaN transistor OFF with a low impedance path.

To drive pairs of transistors in half bridges, the doted end needs to be placed alternatively to the GaN gate or to the resistor – as usual with gate drive transformer half bridges. For a charge pump, that means grouping the "A" transistors against the "B" transistors.

There are a few reasons why such an old technology work so well with a groundbreaking topology. First, most flying capacitor work with 50% duty cycle. Varying duty cycle is a big problem when using gate drive transformer. Also, thanks to the high frequency a relatively low impedance per winding is acceptable. At 1 MHz, the previous transformer with 3 $\Omega$ of winding resistance only experience peak currents of 100mA – for losses below 30mW on the driving winding. Lastly, planar transformer with their extremely low leakage allows for excellent matching between all windings, enabling very tight timing control of many transistors at high frequency. For reference, the gate drive used has less than 4ns delay matching between all 12 "channels" – something that even high-end dedicated gate driver can find challenging to achieve.

## 5 Conclusion

An optimization way for high step-down ratio buck converter has been evaluated for a specific application. Excellent power density and efficiency improvements has been demonstrated with similar or slightly lower price point.

Even if the tests were conducted with relatively high voltage, it became clear that such topology is more likely to shine for high currents low voltage systems like point of load VRM. There, the current multiplying effect of the charge pump coupled with the regulation capability could provide very low regulated voltage at high current capability. Scaling for different input voltage is only a question of scaling the number of flying capacitors and could be the enabling factor for 48V DC links in tight clearance CPU power supply.

## References

[1] PhD student Andrew Stillwell with advisor R. Pilawa-Podgurski, "A Five-Level Flying Capacitor Multi-Level Converter with Integrated Auxiliary Supply and Start-Up" Grainger CEME

[2] Vincent Wai-Shan Ng Seth R. Sanders, "Switched Capacitor DC-DC Converter: Superior where the Buck Converter has Dominated"

[3] Owen Jong, Mohammed Ahmed, Yinsong Cai, Fred C. Lee and Qiang Li, "Multi Resonant Switched-Capacitor Converters for 48/1V Application"

[4] Gab-Su Seo, Ratul Das, and Hanh-Phuc Le, "A 95%-Efficient 48 V-to-1 V/10 A VRM Hybrid Converter"

[5] Michael Antivachis, Matthias Kasper, Dominik Bortis and Johann W. Kolar "Analysis of Capacitive Power Transfer GaN ISOP Multi-Cell DC/DC Converter Systems for Single-Phase Telecom Power Supply Modules"

PCIM Europe 2024, 11– 13 June 2024, Nuremberg          DOI: 10.30420/566262140

# Silicon Interposer as a Substrate for Power Modules with High Power Density and Superior Thermal Performance

Ömer Altan[1,2], Shuangyue Yang[1,2], Sebastian Tengvall[1], Junghyun Kang[1], Yasser Nour[1], Hoa Le Thanh[1], and Ahmed Ammar[1]

[1] Lotus Microsystems, Denmark
[2] Technical University of Denmark, Denmark

Corresponding author:  Ahmed Ammar, ammar@lotus-microsystems.com
Speaker:                Ahmed Ammar, ammar@lotus-microsystems.com

## Abstract

This paper presents a substrate technology for power modules based on a silicon interposer. The interposer is designed and processed with thick copper layers and high-aspect-ratio through-silicon vias (TSVs) [1], enabling its use in power applications. Components can be assembled on the substrate by means of conventional surface-mount technology (SMT) pick-and-place machines. Module encapsulation can then take place by molding of a desired molding compound. By applying the packaging technology to a power module design of a point-of-load buck converter for field programmable gate arrays (FPGAs) and server applications, and lead-frame packaging, the silicon substrate can achieve up to 25 % improvement in thermal performance compared to the laminate-substrate-based package [2]. That allows for higher power density and less derating for the same maximum allowable temperature on the application platform.

## 1   Introduction

In recent years, power modules have commonly been employed in the product development of different consumer and industrial applications. Thanks to their many advantages that include the reduced design time, space efficiency, and cost competitiveness. However, with the current technological advancements in feature-rich electronic devices that call for more power in less space, there is a persisting demand for increasing the power densities offered by power modules [3–5]. Assuming the same efficiency from the commonly employed topologies and components, increasing the power density results in additional power loss in the module package. The main challenge then becomes the thermal stress on the power module components, which results in higher parasitics as well as hot spots that exceed the maximum allowable temperature on the application platform.

## 2   Silicon Substrate

The typical substrate solutions for power modules are either a laminate substrate, or a lead-frame substrate. While the former has the advantages of low cost and possibility for embedding components within the substrate stack, it is limited with respect to the thermal performance it can provide [4]. On the other hand, the lead-frame substrate has better thermal capabilities, thanks to high copper density. Nevertheless, it has limited integration capabilities and routing densities.

Introducing the silicon substrate to the power module market can lead to differentiate from the state of the art. Regarding the level of integration, silicon has better integration capabilities with respect to both the active devices, e.g. power management integrated circuit (PMIC) [6], [7], as well as the passives by means of integrated passive devices (IPD) technologies [8]. With respect to thermal performance, the silicon material has up to 270 times the thermal conductivity of the different variants of the FR4 material in laminate substrates [9], [10]. Compared to the lead frame, which has superior capabilities with the vertical conduction of heat from the power devices to the application board, the silicon interposer allows for thermal distribution both vertically and laterally across the substrate. The results in a more even heat distribution between the power devices, and accordingly lower operating temperature.

1087

Silicon interposers are widely adopted as means for 2.5D and 3D packaging of dies processed in advanced technology nodes. Since they are used to carry and route high-bandwidth signals between different chiplets, they are typically processed on a thinned wafer with relatively thin copper layers, unfilled vias, and insulation materials relevant for logic-level voltages. In order to enable its use for power applications, the silicon interposer has to be reconfigured and processed. Lotus Microsystems has proprietary process technology for the development of silicon interposers for power applications. The technology features thicker copper layers, larger diameter TSVs fully filled with copper, as well as the ability to withstand higher voltages across the substrate.

## 3 Thermal Analysis

In this section, we demonstrate the performance of the power interposer in a switched inductor buck-type converter module that is shown in fig. 1. The module is employed for point-of-load conversion in several applications that include servers, storage, telecom and networking. The PMIC comes in a die-size-ball-grid-array (DSGBA) package and can support up to 6 A of output current. The PMIC is co-packaged with two surface-mount-device (SMD) inductors and two ceramic capacitors.

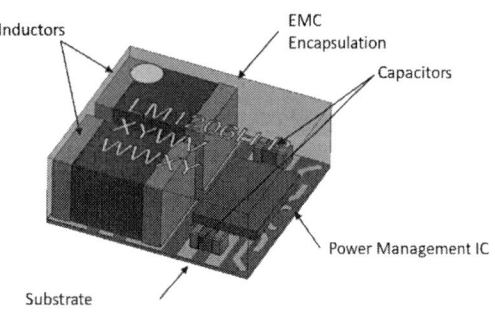

**Fig. 1** Buck-converter module for point-of-load conversion.

A thermal model is developed for the implementation of such module in the cases where the substrate is based on the conventional laminate substrate technology vs. the Lotus Power Interposer™. The thermal model assumes the assembly of the module on a 4-layer evaluative board PCB. Figure 2 shows a top and cross-section views of

the structure, with an indication of the thermal resistances between the power module components, the evaluation board, and ambient air.

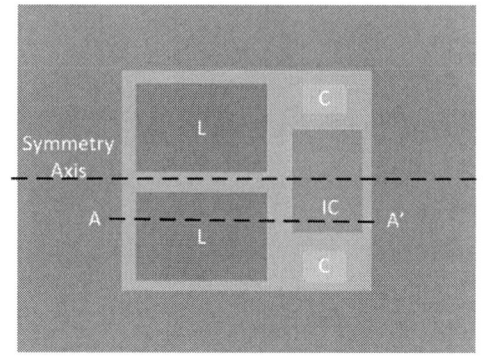

(a)

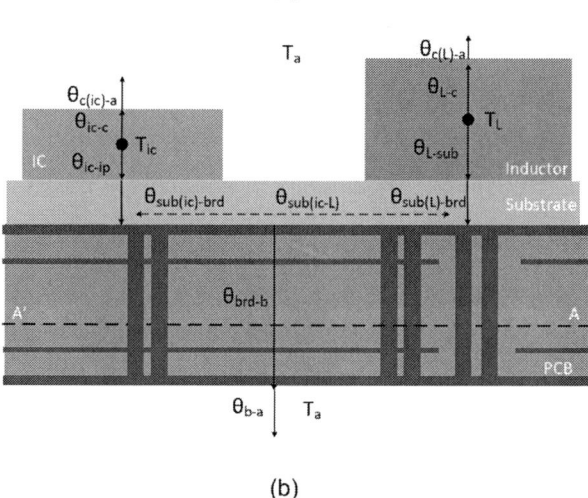

(b)

**Fig. 2** Power Module Structure (a) top view, (b) cross-section view.

Figure 3 shows a schematic of the thermal network corresponding to the structure illustrated in fig. 2 [11]. For simplicity, the model ignores a number of parameters including (1) the thermal coupling between the components on the evaluation board, (2) conduction from junction to case owing to the low thermal resistance value, and (3) power loss in the capacitors. The model is applied to one half of the module device from the symmetry axis shown in fig. 2 (a). Accordingly, the power loss is assumed only from one inductor and half the PMIC loss, whereas the thermal resistance of the PMIC is doubled.

PCIM Europe 2024, 11– 13 June 2024, Nuremberg    DOI: 10.30420/566262140

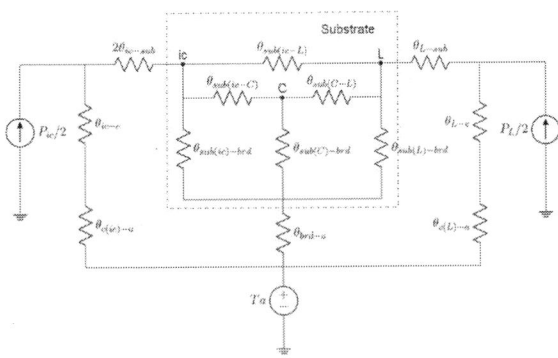

**Fig. 3** Schematic of the thermal network of the DUT evaluation board.

Figure 4 shows a schematic of the thermal network for the evaluation board, which is a 4-layer PCB with 2-oz copper thickness on the outer layers and 1-oz copper thickness on the inner layers. Convection is assumed on top and bottom of the PCB with an area of 37 cm². Four layers are connected to each other with 10 vias under the device area. Also, 6 extra vias under device area connects top and bottom copper layers. The thermal resistance of inductor ($\theta_{L-sub}$) is extracted from the datasheet by assuming the thermal resistance of the manufacturer 6-layer thermal test PCB as 25.4 ˚C/W.

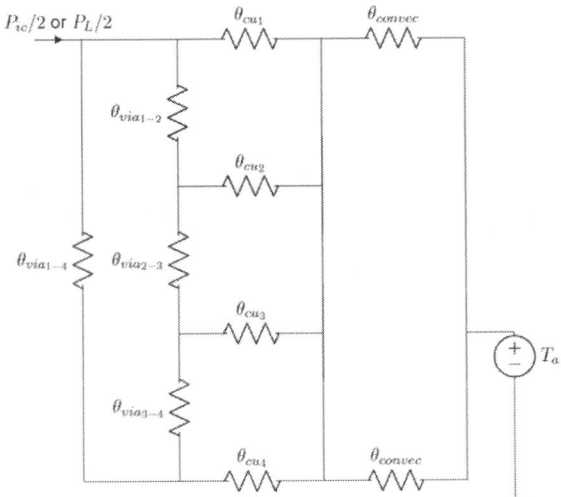

**Fig. 4** Schematic of thermal network of 4-layer PCB.

### 3.1 Substrate Thermal Model

Two substrates are designed and implemented for demonstrating the difference in thermal performance. The first is a silicon substrate processed using the Lotus Power Interposer™ technology ,

and the second is based on the conventional laminate substrate. The substrate thermal network is outlined in fig. 3.

For the laminate substrate case, a 200-μm-thick substrate is considered with 2 layers of 1-oz copper on each side. On the other hand, a silicon substrate with 100-μm thickness is considered with 2 metal layers of 1-oz copper each.

Table 1 shows a summary of the calculated values of the thermal resistances illustrated in fig. 3 and fig. 4.

| Parameter | Value [˚C/W] | |
|---|---|---|
| $\theta_{via1-4}$ | 30.9 | |
| $\theta_{via1-2}$, $\theta_{via3-4}$ | 4.6 | |
| $\theta_{via2-3}$ | 9.4 | |
| $\theta_{convec}$ | 25 | |
| $\theta_{ic-sub}$ | 10.1 | |
| $\theta_{L-sub}$ | 82.7 | |
| $\theta_{cu1}$, $\theta_{cu4}$ | 35.7 | |
| $\theta_{cu2}$, $\theta_{cu3}$ | 71.4 | |
| $\theta_{brd-a}$ | 27.1 | |
| $\theta_{c(ic)-a}$ | 31k | |
| $\theta_{c(sub)-a}$ | 13k | |
| | **FR4** | **Si** |
| $\theta_{sub(ic-C)}$ | 25.2 | 16.4 |
| $\theta_{sub(ic-L)}$ | 116.8 | 22.8 |
| $\theta_{sub(L-C)}$ | 733.7 | 25.6 |
| $\theta_{sub(ic)-brd}$ | 141.4 | 17 |
| $\theta_{sub(L)-brd}$ | 4.6 | 0.08 |
| $\theta_{sub(C)-brd}$ | 12 | 0.4 |

**Table 1** Summary of the values calculated for the thermal network resistances.

By applying the delta-wye conversion to the substrate thermal network and reducing the circuit in fig. 3, the operating temperatures for the PMIC and the inductors (the main sources of power loss) can be estimated using the following equations.

$$T_{ic\_FR4} = 35.4\,P_{ic} + 13.9\,P_L + T_a \qquad (1)$$

$$T_{L\_FR4} = 13.9\,P_{ic} + 56.6\,P_L + T_a \qquad (2)$$

$$T_{ic\_Si} = 26.7\,P_{ic} + 13.4\,P_L + T_a \qquad (3)$$

$$T_{L\_Si} = 13.4\,P_{ic} + 54.5\,P_L + T_a \qquad (4)$$

The analysis shows that the silicon interposer out-performs the laminate substrate and results in lower operating temperature for the power module components including the PMIC and the inductor. The thermal resistance between the components and the application board is reduced and a more even distribution of heat is achieved between the module components.

## 4 Experimental Results

The analytical model is verified by experimental results, where two devices under test (DUTs) employing the two versions of the substrate are implemented and assembled on similar evaluation boards. The two boards are fed into an efficiency characterization setup used for technology demonstration and competitive analysis of low-mid power dc-dc converters. A sense circuit measures the input and output voltage and current for each device, and the analog signals are fed into a high-resolution analog-to-digital converter (ADC). A microcontroller (MCU) processes each device's efficiency and writes the data on an on-board LCD display. The two DUTs are powered from a single source and loaded by similar loads. A thermal camera displays the difference in thermal performance. Figures 5 and 6 show a picture of the implemented setup and an illustration of the thermal performance difference between the two DUTs, respectively. Both figures demonstrate that for the same operational point and power loss for the two DUTs (~ 3.5 W), the device employing the Lotus Power Interposer™ has superior thermal performance over the conventional counterpart based on a laminate substrate, with more than 20 °C reduction in the max. operating temperature that results in an improvement of ~ 25% in thermal performance. That is thanks to the higher thermal conductivity of the silicon material over the FR4 employed in the laminate substrate, which results in lower thermal resistance from the power module components to the application board, as illustrated in section 3. That, in turn, allows for higher power

density and less thermal derating for the same maximum allowable temperature on the application platform.

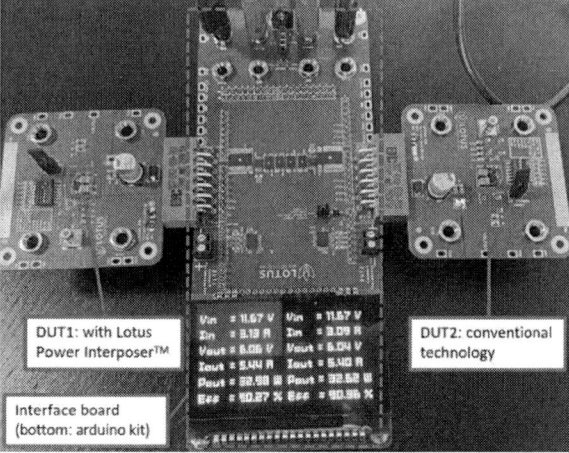

**Fig. 5** Picture of implemented characterization setup.

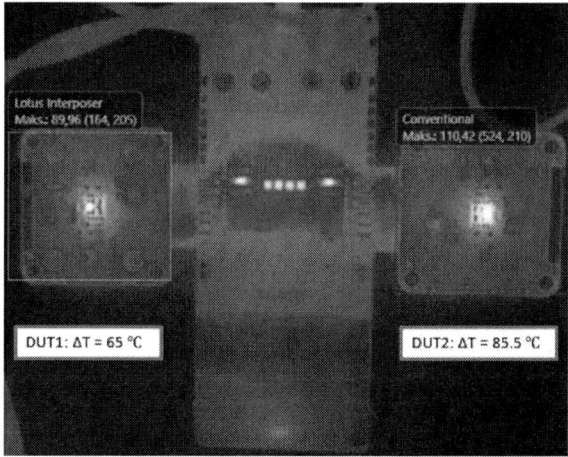

**Fig. 6** Illustration of the thermal performance of two DUTs employing the Lotus Power Interposer™ and the laminate substrate.

Table 2 shows a summary of the calculated and measured improvement in thermal performance obtained from the silicon substrate relative to the laminate counterpart for different operational points. The deviation between the calculated and measured results is a result of the simplicity of the thermal network model and abstraction of several parameters, as indicated in section 3. Nevertheless, both the analytical model and the measured results show substantial improvement in thermal performance using the Lotus Power Interposer™ across the power loss range.

| $P_{Loss}$ [W] | ΔT [%] | |
|---|---|---|
| | Calculated | Measured |
| 3.1 | 18.5 | 17.1 |
| 2.3 | 16.2 | 15.2 |
| 1.9 | 15.1 | 14.8 |
| 1.4 | 14.9 | 13.6 |

**Table 2** Summary of the improvement in the thermal performance for different operational points.

## 5 Conclusion

The silicon interposer as a substrate for power modules is presented. The technology has several advantages over the state-of-the-art substrates, which include the higher integration capability and better thermal performance. A buck-type converter is selected for a comparison between the silicon interposer package with the laminate substrate employed in land-grid-array (LGA) packages. By using the Lotus Power Interposer™, experimental results show an improvement of up to 25 % in thermal performance for the same power loss, with no reduction in efficiency. The results prove such a technology to be a primary candidate for system-in-package integration of power management components in feature-rich electronic devices that demand high-power-density solutions with superior thermal performance.

## References

[1] W. W. Shen and K. N. Chen, "Three-Dimensional Integrated Circuit (3D IC) Key Technology: Through-Silicon Via (TSV)," *Nanoscale Res Lett*, vol. 12, no. 1, pp. 1–9, Dec. 2017, doi: 10.1186/S11671-017-1831-4/FIGURES/16.

[2] F. Hou *et al.*, "Review of packaging schemes for power module," *IEEE J Emerg Sel Top Power Electron*, vol. 8, no. 1, pp. 223–238, 2019.

[3] Y. Yang, L. Dorn-Gomba, R. Rodriguez, C. Mak, and A. Emadi, "Automotive power module packaging: Current status and future trends," *IEEE Access*, vol. 8, pp. 160126–160144, 2020.

[4] J. Broughton, V. Smet, R. R. Tummala, and Y. K. Joshi, "Review of thermal packaging technologies for automotive power electronics for traction purposes," *J Electron Packag*, vol. 140, no. 4, p. 040801, 2018.

[5] M. Mahesh, K. Vinoth Kumar, M. Abebe, L. Udayakumar, and M. Mathankumar, "A review on enabling technologies for high power density power electronic applications," *Mater Today Proc*, vol. 46, pp. 3888–3892, Jan. 2021, doi: 10.1016/J.MATPR.2021.02.340.

[6] P. Vivet *et al.*, "3D advanced integration technology for heterogeneous systems," in *2015 International 3D Systems Integration Conference (3DIC)*, IEEE, 2015, pp. FS6-1.

[7] P. Coudrain *et al.*, "Active interposer technology for chiplet-based advanced 3D system architectures," in *2019 IEEE 69th Electronic Components and Technology Conference (ECTC)*, IEEE, 2019, pp. 569–578.

[8] H. T. Le *et al.*, "Fabrication of 3D air-core MEMS inductors for very-high-frequency power conversions," *Microsystems & Nanoengineering 2018 4:1*, vol. 4, no. 1, pp. 1–9, Jan. 2018, doi: 10.1038/micronano.2017.82.

[9] K. Azar and J. E. Graebner, "Experimental determination of thermal conductivity of printed wiring boards," Twelfth Annual IEEE Semiconductor Thermal Measurement and Management Symposium. Proceedings, Austin, TX, USA, 1996, pp. 169-182, doi: 10.1109/STHERM.1996.545107.

[10] Glassbrenner, C. J., & Slack, G. A. (1964). Thermal Conductivity of Silicon and Germanium from 3°K to the Melting Point. Physical Review, 134(4A), A1058. https://doi.org/10.1103/PhysRev.134.A1058

[11] Texas Instruments, 2013. An-2020 thermal design by insight, not hindsight. Tech. rep.

PCIM Europe 2024, 11– 13 June 2024, Nuremberg    DOI: 10.30420/566262141

# Analytical Modeling and Stability Characterization of a Damped Active EMI Filter for Single- and Three-Phase AC-DC Applications

Timothy Hegarty[1], Ashish Kumar[2], Yuetao Hou[1]

[1] Texas Instruments, Inc., USA
[2] Tau Motors, Inc., USA

Corresponding author and speaker: Timothy Hegarty, thegarty@ieee.org

## Abstract

A compact and efficient design of the electromagnetic interference (EMI) filter is critical to reach the full benefits of electrification in highly constrained operating environments, such as on-board chargers (OBCs) for battery electric vehicles (BEVs). Through miniaturization of toroidal-cored common-mode (CM) chokes, an active EMI filter (AEF) circuit for CM noise attenuation can substantially increase the gravimetric and volumetric power densities of the EMI filter implementation. This paper examines small-signal stability by deriving loop-gain expressions for a feedback-type, voltage-sense, current-inject (FB-VSCI) AEF circuit implemented using an integrated-circuit (IC) approach. The formulated expressions are subsequently confirmed by simulations and experimental validation of a two-stage EMI filter design suitable for a 22-kW universal-input OBC.

## 1 Introduction

Size reduction of EMI filter components is generally required to achieve the full benefits of electrification in automotive, enterprise, aerospace and other highly constrained system environments with strict power density requirements. Increased power levels and the number of fast-switching power devices in high-density AC-DC regulator systems make the electromagnetic environment (EME) more complex, with additional sources and victims of EMI present in applications such as OBCs. The OBC performs power factor correction (PFC), isolation and power regulation with power levels up to 19.2 kW or 22 kW for single-phase or three-phase inputs, respectively [1-3].

AEF circuits [4-8] now receive significant attention in these and other power-electronics constrained applications, owing to the attendant improvements in power density and cost of the EMI filter. An efficient, cost-effective and high-density filter is a key challenge in switching regulator design and is essential to effectually package the overall solution within demanding chassis-enclosed form factors.

A particular hallmark of CM EMI filters intended for grid-connected applications is their limited Y-capacitance due to touch-current safety requirements [8], leading to large-sized CM chokes to achieve a target corner frequency or a specific filter attenuation. This often results in unsatisfactory passive filter designs with bulky, heavy and expensive CM chokes that dominate the overall filter size.

However, space-limited power-electronics applications can now leverage commercially available active power-supply filter ICs [8] to reduce magnetic component and overall filter size by amplifying the effective value of Y-capacitance over a prescribed frequency range, thus helping to comply with applicable conducted emissions standards – for example, IEC 61851-21-1 is the relevant electromagnetic compatibility (EMC) product standard for OBCs [9]. As such, CM noise attenuation with AEF gives significant advancements in the volumetric (kW/l), gravimetric (kW/kg) and cost ($/kW) density metrics of the overall implementation.

However, loop stability is a key performance benchmark and must be fully understood and extensively detailed within the scope of a robust, reliable AEF design. The purpose and contribution of this paper lies in the derivation of expressions for the loop gain characteristic of a FB-VSCI AEF circuit. The expressions are essential to determine small-signal stability margins and extract component values for the requisite compensation and damping networks. Also included are simulation modeling and practical bench measurements of a two-stage, three-phase EMI filter to corroborate the theoretical findings and validate the proposed analytical approach.

# 2 Active EMI Filter for Grid-Tied Applications

## 2.1 FB-VSCI AEF Circuit

Shown in Fig. 1 are the schematics for a two-stage CLCL passive-only filter and an equivalent active filter design. The AEF IC for this three-phase application positions between the CM chokes, designated as $L_{CM1}$ and $L_{CM2}$ in Fig. 1, and provides a lower-impedance shunt path for CM noise currents to flow to chassis ground.

The approach aims for the reduction of the total filter volume, yet maintains low values of the low-frequency earth leakage current using an active circuit that shapes the frequency response of the inject capacitor – effectively multiplying its value at high frequencies. In turn, this amplified inject capacitance over the required frequency range is the key to lower CM choke inductances relative to the values of a passive filter with similar attenuation.

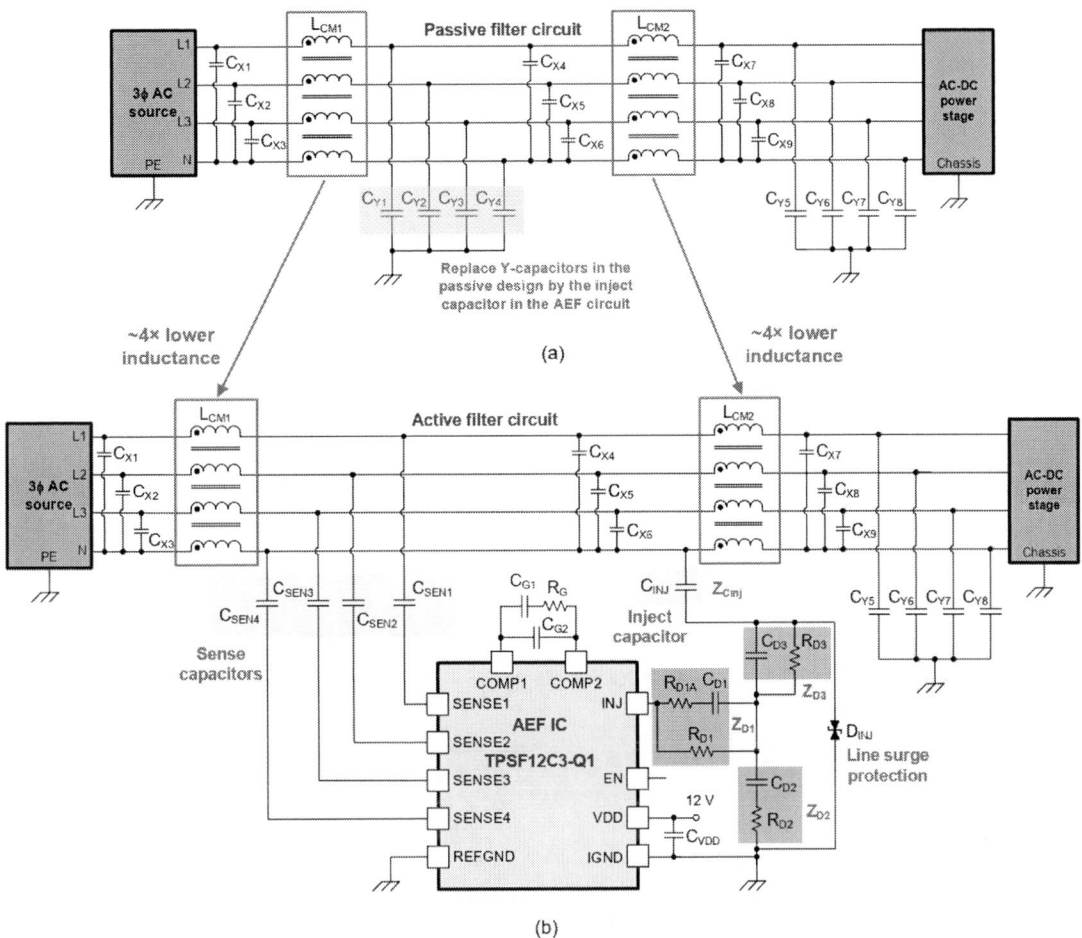

**Fig. 1** Schematic of a two-stage, three-phase passive filter (a) that is replaced by a corresponding VSCI AEF circuit (b) with reduced CM-choke inductance values.

The FB-VSCI AEF circuit in Fig. 1b leverages high-voltage Y-rated capacitors, highlighted in yellow, in combination with low-voltage active circuits for sensing and injection. This method provides a high level of integration and supports high density by eschewing magnetic components for noise sensing and injection. X-capacitors, designated as $C_{X4}$, $C_{X5}$ and $C_{X6}$ in Fig. 1 and positioned between the CM chokes, effectively provide a low-impedance path between the power lines from a CM noise standpoint, up to low-MHz frequencies (depending on the parasitic inductance of the X-capacitors). This allows current injection directly onto one power line using just one inject capacitor. Based on clearance spacing rules for the four power lines, neutral is often routed closest to the IC on the circuit board, and thus Fig. 1b shows it as the designated power line for injection.

Up to 25dB of CM noise attenuation is possible with this AEF circuit in the frequency range from 150 kHz to 3 MHz [5].

## 2.2 Functional Block Diagram

Depicted in the block diagram of Fig. 2, the AEF circuit rejects line-frequency AC voltage using a two-stage high-pass filter (HPF) sensing network, while amplifying the detected high-frequency CM noise and maintaining closed-loop stability using an external tunable damping circuit with impedance branches indicated as $Z_{D1}$, $Z_{D2}$ and $Z_{D3}$ (see also the components with subscript "D" reference designators in Fig. 2).

Components $R_G$, $C_{G1}$ and $C_{G2}$ connected between the COMP1 and COMP2 pins in Fig. 2 form an impedance $Z_G$, which behaves as a lead-lag network that sets the AEF amplification gain characteristic in association with the feedback impedance, denoted as $Z_F$. The output of the power amplifier (at the INJ pin) injects the required noise-cancelling signal back into the power lines through a damping network and a Y-rated inject capacitor $C_{INJ}$, typically set at 22 nF in three-phase applications with touch current specification of 3.5 mA$_{RMS}$ [8,10].

Inclusion of the damping network is to shape the amplifier-output-to-inject-capacitor transfer function and thus stabilize the LC resonant behavior that occurs between the CM choke inductances and the inject capacitance.

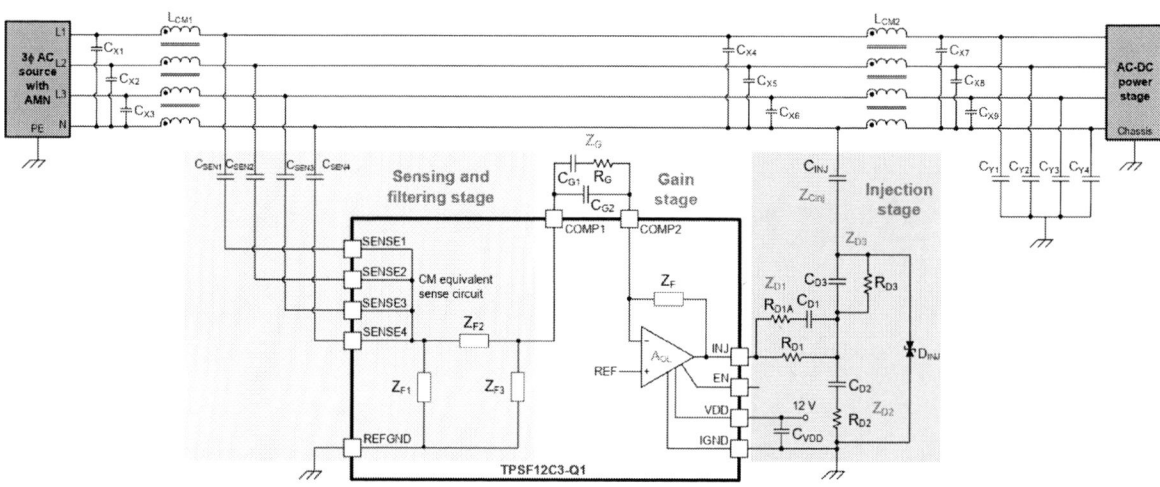

**Fig. 2**    CM-equivalent functional block diagram of the three-phase FB-VSCI AEF IC.

# 3    Loop Gain Derivation

## 3.1    CM Noise Model

Figure 3 gives a generalized CM circuit model with sensing, gain and inject stages and the equivalent circuits derived using an amplifier model as well as grid-side and regulator-side Thevenin-equivalent networks.

Impedances that appear symmetrically in each phase scale by a factor n for suitable representation in the CM model. More specifically, n = 2 applies for single-phase systems; n = 3 pertains to split-phase and three-phase three-wire (3p3w) systems; and n = 4 describes three-phase four-wire (3p4w) systems.

An equivalent voltage source and capacitive noise source impedance, denoted respectively as $V_{SW}$ and $C_{SW}$ in Fig. 3a, model the power-stage CM

noise source. Similarly, an equivalent CM voltage source and source impedance, designated as $V_{CM,grid}$ and $Z_{CM,grid}$ in Fig. 3a, represent the CM disturbance from the grid supply, which should ideally be zero in a balanced three-phase system.

Also included is a model for the line impedance stabilization network (LISN), often referred to as an artificial mains network (AMN) when connected at the AC input for EMI measurement. The LISN impedance elements in Fig. 3a also scale by the factor n.

The CM noise model includes the required regulator-side Y-capacitance $C_{Y2}$, as well as optional grid-side Y-capacitance $C_{Y1}$, which appears in parallel with the LISN for high-frequency filtering.

$Z_S$ in Fig. 3 is the effective CM impedance of the sense capacitor network, which is part of the HPF along with impedances $Z_{F1}$, $Z_{F2}$ and $Z_{F3}$ within the IC.

Furthermore, $Z_F$ and $Z_G$ denote the impedances of the amplifier gain stage, and $Z_{D1}$, $Z_{D2}$ and $Z_{iD3}$ interface the amplifier output to the filter injection node between the chokes. For simplicity, $Z_{iD3}$ combines damping impedance $Z_{D3}$ and inject capacitor impedance $Z_{CINJ}$, as they appear in series.

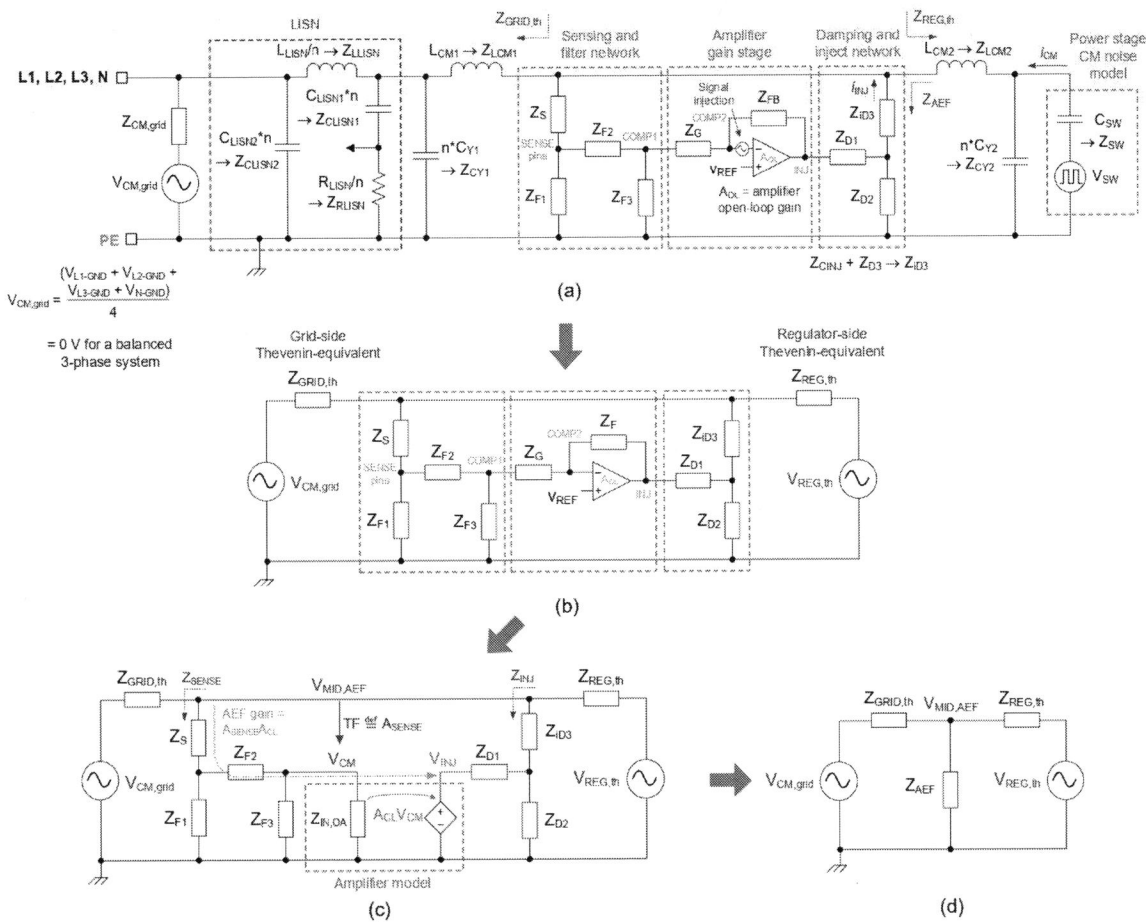

**Fig. 3**   Generalized CM noise model with LISN connected at the input (a); circuit reductions (b), (c) and (d).

## 3.2   Thevenin-Equivalent Impedances

The grid-side and regulator-side Thevenin-equivalent impedances, denoted respectively as $Z_{GRID,th}$ and $Z_{REG,th}$ in Fig. 3, incorporate the applicable passive filter components (CM chokes and Y-capacitors), source parasitic elements, and the measurement LISN if connected.

More specifically, Eq. (1) describes $Z_{GRID,th}$, which combines the choke inductance $L_{CM1}$, the grid-side Y-capacitors (if installed) and the components for the LISN as shown in Fig. 2, and includes the grid CM source impedance $Z_{CM,grid}$ if applicable. Meanwhile, Eq. (2) defines $Z_{REG,th}$, which includes the CM noise source impedance, the regulator-side Y-capacitors and the choke inductance $L_{CM2}$.

$$Z_{GRID,th}(s) = Z_{LCM1} +$$
$$Z_{CY1} \| \left( Z_{RLISN} + Z_{CLISN1} \right) \| \left( Z_{LLISN} + Z_{CLISN2} \| Z_{CM,grid} \right) \quad (1)$$

$$Z_{REG,th}(s) = Z_{LCM2}(s) + Z_{CY2}(s) \| Z_{SW}(s) \quad (2)$$

Equation (3) defines the Thevenin load impedance to the AEF circuit as the parallel combination of $Z_{GRID,th}$ and $Z_{REG,th}$ (see Fig. 3d).

$$Z_{LOAD,th}(s) = Z_{GRID,th}(s) \| Z_{REG,th}(s) \quad (3)$$

In addition, Eq. (4) gives the Thevenin regulator-side voltage source shown in Fig. 3b.

$$V_{REG,th}(s) = V_{SW}(s)\frac{Z_{CY2}(s)}{Z_{SW}(s) + Z_{CY2}(s)} \quad (4)$$

Figure 4 plots typical impedance magnitudes versus frequency. $Z_{GRID,th}$ is of lower value and thus determines the behavior of $Z_{LOAD,th}$ at low frequency, whereas $Z_{GRID,th}$ and $Z_{REG,th}$ are of similar value at higher frequency (above approximately 20kHz in this example), mainly because $L_{CM1}$ and $L_{CM2}$ are chosen to be of equal value.

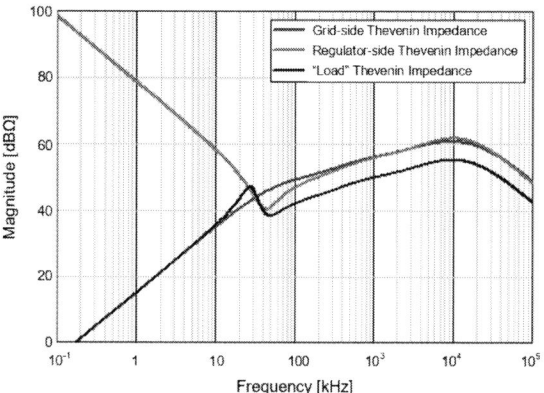

**Fig. 4** Typical plots of grid-side, regulator-side, and load Thevenin equivalent impedances.

### 3.3 AEF Amplifier Model

Using a conventional model for the amplifier, Eq. (5) expresses the open-loop gain as

$$A_{OL}(s) = -\frac{A_{DC}}{\left(1 + s/\omega_{p,dom}\right)\left(1 + s/\omega_{p,par}\right)} \quad (5)$$

where $A_{DC}$ is the DC gain and $\omega_{p,dom}$ and $\omega_{p,par}$ are dominant and parasitic poles of the amplifier response, respectively.

The amplifier model in Fig. 3c establishes the closed-loop (CL) gain, which is effectively the transfer function from the sensed and filtered CM disturbance at the COMP2 pin to the amplifier output at the INJ pin. This CL gain expresses as

$$A_{CL}(s) = -\frac{Z_F(s)}{Z_G(s)} \cdot \frac{1}{1 + \dfrac{\left(1 + Z_F(s)/Z_G(s)\right)}{A_{OL}(s)}} \quad (6)$$

The amplifier-stage input impedance shown in Fig 3c is given simply as

$$Z_{IN,OA}(s) = Z_G(s) + \frac{Z_F(s)}{1 + A_{OL}(s)} \quad (7)$$

### 3.4 Leveraging Design-Oriented Analysis

Shown in Fig. 5 is the circuit model for analysis and derivation of the loop gain, with the Thevenin-equivalent load impedance indicated explicitly.

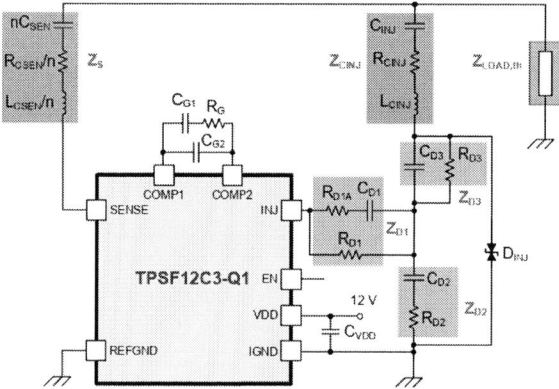

**Fig. 5** Circuit model with Thevenin-equivalent load impedance suitable for loop gain analysis.

The first step in the design-oriented analysis (DOA) is to derive an expression for the loop gain with *no load* and *no damping* (NLND), where impedances $Z_{LOAD,th}$ and $Z_{D2}$ in Fig. 5 are effectively open, and $Z_{D1}$ and $Z_{D3}$ are shorts.

As illustrated in Fig. 3a, we break the loop at the highest impedance point for signal injection, i.e. at the amplifier inverting input. Equation (8) conveniently expresses this NLND loop gain as

$$T_{NLND}(s) = A_{OL}(s) \cdot \frac{N_1(s) + N_2(s)}{D_1(s) + D_2(s)} \quad (8)$$

where $N_{1,2}(s)$ and $D_{1,2}(s)$ refer to numerator and denominator functions, respectively, as defined by Eq. (9) through (12) below.

Calculated using Mathcad [12] with values from a practical filter design [13], Fig. 6 plots the loop gain $T_{NLND}(s)$, which comprises a pair of complex-conjugate zeros and two real poles.

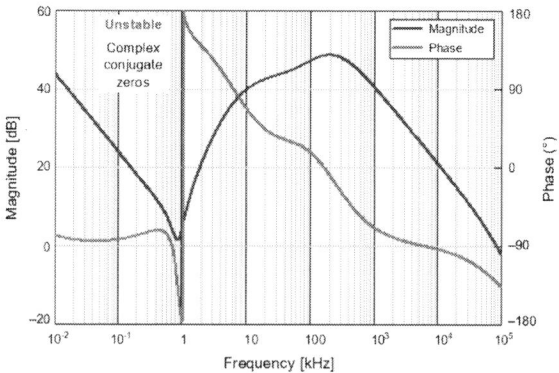

**Fig. 6** Loop gain $T_{NLND}(s)$ – no load, no damping.

$$N_1(s) = \left[ Z_{F3} \cdot Z_G + Z_{F2} \left( Z_{F3} + Z_G \right) \right] \cdot \left( Z_{CINJ} + Z_S \right) \tag{9}$$

$$N_2(s) = Z_{F1} \cdot \left[ Z_{F2} \cdot \left( Z_{F3} + Z_G \right) + Z_G \cdot \left( Z_{CINJ} + Z_S \right) + Z_{F3} \cdot \left( Z_G + Z_F + Z_{CINJ} + Z_S \right) \right] \tag{10}$$

$$D_1(s) = \left[ Z_{F3} \cdot \left( Z_G + Z_F \right) + Z_{F2} \cdot \left( Z_{F3} + Z_G + Z_F \right) \right] \cdot \left( Z_{CINJ} + Z_S \right) \tag{11}$$

$$D_2(s) = Z_{F1} \cdot \left[ Z_{F2} \cdot \left( Z_{F3} + Z_G + Z_F \right) + \left( Z_G + Z_F \right) \cdot \left( Z_{CINJ} + Z_S \right) + Z_{F3} \cdot \left( Z_G + Z_F + Z_{CINJ} + Z_S \right) \right] \tag{12}$$

As a convenient intermediate step, we avail of the EET [11] to obtain an expression for the loop gain *with load* and *no damping* (WLND). Availing of the open-circuit version of the EET, Eq. (13) defines the WLND loop gain as

$$T_{WLND}(s) = T_{NLND}(s) \cdot \left[ \frac{1 + \dfrac{Z_{N\_NLND\_to\_WLND}(s)}{Z_{LOAD,th}(s)}}{1 + \dfrac{Z_{D\_NLND\_to\_WLND}(s)}{Z_{LOAD,th}(s)}} \right] \tag{13}$$

where the quantity in brackets here represents the EET correction factor, and $Z_{LOAD,th}$ designates the open-circuit extra element. The nulling and driving-point impedances for the EET derive as follows

$$Z_{N\_NLND\_to\_WLND}(s) = Z_{N1}(s) \| Z_{N2}(s) \| Z_{N3}(s) \tag{14}$$

$$Z_{D\_NLND\_to\_WLND}(s) = Z_{CINJ}(s) \| \left( Z_S(s) + Z_{in\_FILT\_D}(s) \right) \tag{15}$$

where $Z_{N1}(s)$, $Z_{N2}(s)$ and $Z_{N3}(s)$ are given by Eq. (16) to (18), along with terms defined by Eq. (19) to (21) to simplify the expressions.

$$Z_{N1}(s) = Z_{CINJ}(s) \tag{16}$$

$$Z_{N2}(s) = Z_S(s) + Z_{in\_FILT\_N}(s) \tag{17}$$

$$Z_{N3}(s) = \text{Ratio\_}Z_1(s) \cdot Z_{N2}(s) \cdot \frac{Z_{CINJ}(s)}{Z_F(s)} \tag{18}$$

$$\text{Ratio\_}Z_1(s) = \frac{Z_G}{Z_{F1}} + \left( 1 + \frac{Z_{F2}}{Z_{F1}} \right) \cdot \left( 1 + \frac{Z_G}{Z_{F3}} \right) \tag{19}$$

$$Z_{in\_FILT\_N}(s) = Z_{F1} \| \left[ Z_{F2} + \left( Z_{F3} \| Z_G \right) \right] \tag{20}$$

$$Z_{in\_FILT\_D}(s) = Z_{F1} \| \left[ Z_{F2} + \left( Z_{F3} \| \left( Z_G + Z_F \right) \right) \right] \tag{21}$$

Figure 7 shows a typical plot of the magnitude and phase of $T_{WLND}(s)$, which includes a pair of right-half-plane (RHP) zeros, causing instability.

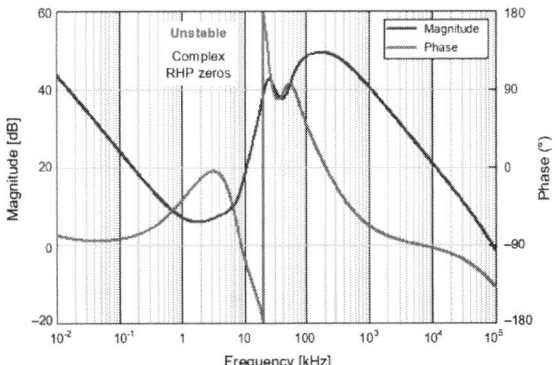

**Fig. 7**  Loop gain $T_{WLND}(s)$ – with load, no damping.

Next, we perform a star-delta transformation of the damping network elements from $Z_{D1}$, $Z_{D2}$ and $Z_{D3}$ to $Z_{dx}$, $Z_{dy}$ and $Z_{dz}$, as shown in Fig. 8.

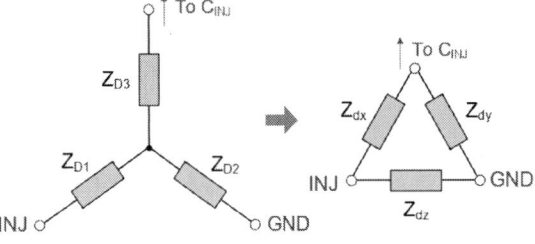

**Fig. 8**  Star-delta (Y-to-$\Delta$) transformation of the damping network impedances.

The delta-connected impedances in Fig. 8 derive as follows and facilitate the application of the EET to assimilate the effect of the damping network on the loop gain.

$$Z_{dx}(s) = \frac{Z_{D1} \cdot Z_{D2} + Z_{D2} \cdot Z_{D3} + Z_{D3} \cdot Z_{D1}}{Z_{D2}} \quad (22)$$

$$Z_{dy}(s) = \frac{Z_{D1} \cdot Z_{D2} + Z_{D2} \cdot Z_{D3} + Z_{D3} \cdot Z_{D1}}{Z_{D1}} \quad (23)$$

$$Z_{dz}(s) = \frac{Z_{D1} \cdot Z_{D2} + Z_{D2} \cdot Z_{D3} + Z_{D3} \cdot Z_{D1}}{Z_{D3}} \quad (24)$$

Applying the EET to the WLND gain expression from Eq. (13) establishes a final result *with load* and *with damping* (WLWD) that combines both the loading effect of $Z_{LOAD,th}$ and the damping effect of $Z_{D1}$, $Z_{D2}$ and $Z_{D3}$.

Using the short-circuit version of the EET, Eq. (25) defines the WLWD loop gain as

$$T_{WLWD}(s) = T_{WLND}(s) \cdot \left[ \frac{1 + \dfrac{Z_{dx}(s)}{Z_{N\_WLND\_to\_WLWD}(s)}}{1 + \dfrac{Z_{dx}(s)}{Z_{D\_WLND\_to\_WLWD}(s)}} \right] \quad (25)$$

where the quantity in brackets designates the EET correction factor with $Z_{dx}(s)$ classified as the short-circuit extra element.

The nulling and driving-point impedances derive respectively as follows.

$$Z_{N\_WLND\_to\_WLWD}(s) = \frac{Z_{N\_WLND\_to\_WLWD\_num}(s)}{Z_{N\_WLND\_to\_WLWD\_denom}(s)} \quad (26)$$

$$Z_{D\_WLND\_to\_WLWD}(s) = Z_{BC}(s) \| Z_{dy}(s) \quad (27)$$

The numerator and denominator terms in Eq (26) originate as follows.

$$Z_{N\_WLND\_to\_WLWD\_denom}(s) = \frac{Ratio\_Z_2(s)}{Z_{dy}(s)} + \frac{Ratio\_Z_1(s)}{Z_{dy}(s)} \cdot \frac{Z_S(s)}{Z_F(s)} + Y_N(s) \cdot \left(1 + \frac{Z_{CINJ}(s)}{Z_{dy}(s)}\right) \quad (28)$$

$$Z_{N\_WLND\_to\_WLWD\_num}(s) = 1 + Ratio\_Z_2(s) + Ratio\_Z_1(s) \cdot \frac{Z_S(s)}{Z_F(s)} + Y_N(s) \cdot Z_{CINJ}(s) \quad (29)$$

The following representations serve to streamline the previous expressions for loop gain.

$$Y_N(s) = \frac{Ratio\_Z_1(s)}{Z_F(s)} + \frac{Ratio\_Z_3(s)}{Z_{LOAD,th}(s)} \quad (30)$$

$$Ratio\_Z_2(s) = \frac{Z_G}{Z_F} + \frac{Z_{F2}}{Z_{F1}} \cdot \left(1 + \frac{Z_G}{Z_{F3}}\right) \quad (31)$$

$$Ratio\_Z_3(s) = Ratio\_Z_1(s) \cdot \frac{Z_S}{Z_F} + Ratio\_Z_2(s) \quad (32)$$

and, finally

$$Z_{BC}(s) = Z_{CINJ}(s) + Z_{LOAD,th}(s) \| \left(Z_S(s) + Z_{in\_FILT\_D}(s)\right) \quad (33)$$

## 3.5 Analytical Results

Based on the derived expressions above, Fig. 9 shows magnitude and phase plots for loop gain $T_{WLWD}(s)$.

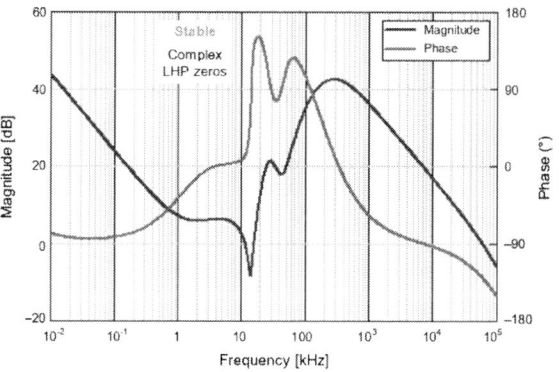

**Fig. 9** Loop gain $T_{WLWD}(s)$ – with load, with damping.

The damping network converts the complex RHP zeros to left-half-plane (LHP) zeros, stabilizing the system by preventing a negative gain margin from a phase drop to −180° or lower.

To avoid negative gain margin and thus instability, the phase characteristic should be well damped and not reach −180° at the resonant frequencies that typically occur within a range of approximately 2 kHz to 40 kHz.

# 4 Simulation and Experimental Results

## 4.1 Simulation Results

Shown in Fig. 10 is a SIMPLIS simulation result using the same three-phase filter design [12] as that used for the analytical results. The simulated plot demonstrates very close agreement with the calculated plot of Fig. 9, confirming the accuracy of the analytical approach presented in this paper.

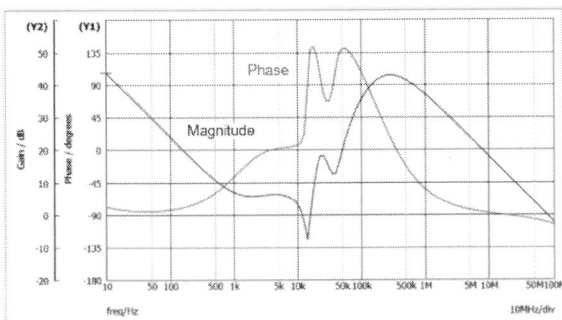

Fig. 10    Simulated loop gain (magnitude and phase).

## 4.2 Practical Results

Shown in Fig. 11 is a 3p4w AEF evaluation module (EVM) [13] rated at 32 $A_{RMS}$ per phase specifically for a 22-kW OBC application. The nanocrystalline CM chokes, manufactured by Würth Elektronik, each have a CM inductance of 0.75 mH at 10 kHz, and the Y-capacitors at the grid and regulator sides are 2.2 nF and 6.8 nF, respectively.

Fig. 11    EVM of a three-phase EMI filter with AEF suitable for a 22-kW OBC.

As shown in Fig. 12, we connect a Bode 100 vector network analyzer (VNA) from Omicron Labs, which also functions as a frequency response analyzer, to perform a practical measurement of the loop gain by injecting an isolated excitation signal at the inverting input of the AEF amplifier. See the schematic of Fig. 3a, which indicates the position of this signal source.

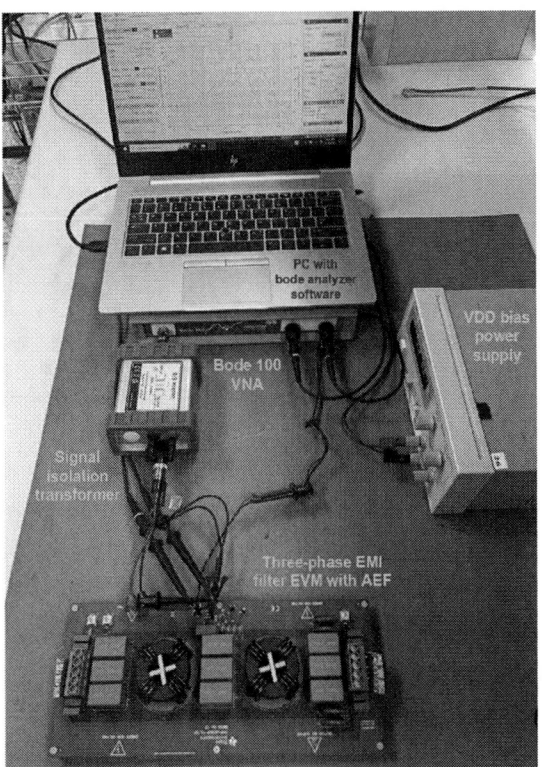

Fig. 12    Loop gain measurement setup with a Bode 100 VNA and signal injection transformer.

The TPSF12C3-Q1 AEF IC is specially modified for this test by using focused ion beam (also known as FIB) technology to remove the internally integrated feedback impedance $Z_F$ such that the excitation signal can be applied directly at the amplifier inverting input as needed. The components for impedance $Z_F$ are then connected externally on the board close to the IC (between the COMP2 and INJ pins, as indicated in Fig. 2).

The measured loop gain result in Fig. 13 largely aligns with the analytical and simulated results from Fig. 9 and Fig. 10, respectively.

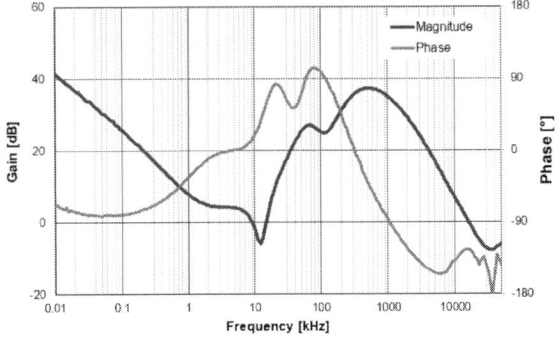

Fig. 13    Experimental loop gain result.

However, the loop crossover frequency in Fig. 13 occurs at 16 MHz, which is diminished relative to the predicted crossover of 60 MHz from Fig. 9. This relates to the capacitive parasitics inherent in the experimental setup. Essentially, breaking the loop and insertion of the excitation signal creates a capacitive loading effect at the high-impedance inverting input of the amplifier, thus causing the gain and phase to roll off earlier than expected.

In addition, the resonant behavior in the 10 kHz to 40 kHz range from Fig. 13 appears more damped than the predicted behavior in Fig. 9. This can be attributed to imprecisions in the resistive part of the modeled choke impedance [7], which relates to the frequency-dependent magnetic permeability of the nanocrystalline-core chokes.

# 5 Conclusions

With EMC being a key requirement in power-management designs, this paper details an analysis that centers on the stability of a CM AEF circuit with supplementary tunable damping network for applicability to single-phase, split-phase and three-phase applications. Careful manipulation of the basic loop-gain expression by applying the EET – first to include the effect of AEF circuit loading and second to incorporate a damping network – provides a rigorous derivation of the damped, loaded loop gain for stability analysis.

Close agreement of the formulated analytical results with simulation modeling and experimental loop measurements using a VNA substantiates the theoretical approach as described.

# 6 References

[1] H. Wouters and W. Martinez, "Bidirectional Onboard Chargers for Electric Vehicles: State-of-the-Art and Future Trends," *IEEE Transactions on Power Electronics*, vol. 39, no. 1, pp. 693-716, Jan. 2024, doi: 10.1109/TPEL.2023.3319996.

[2] R. Pradhan, N. Keshmiri and A. Emadi, "On-Board Chargers for High-Voltage Electric Vehicle Powertrains: Future Trends and Challenges," *IEEE Open Journal of Power Electronics*, vol. 4, pp. 189-207, Mar. 2023, doi: 10.1109/OJPEL.2023.3251992.

[3] I. Aghabali, J. Bauman, P. J. Kollmeyer, Y. Wang, B. Bilgin et al., "800-V Electric Vehicle Powertrains: Review and Analysis of Benefits, Challenges, and Future Trends," *IEEE Transactions on Transportation Electrification*, vol. 7, no. 3, pp. 927-948, Sept. 2021, doi: 10.1109/TTE.2020.3044938.

[4] P. Körner, P. Brockerhoff and F. Müller, "Analysis of Passive and Active EMI Filters for On-Board Chargers in Electric Vehicles," *2023 25th European Conference on Power Electronics and Applications* (EPE'23 ECCE Europe), Aalborg, Denmark, Sept. 2023, pp. 1-8, doi: 10.23919/EPE23ECCEEurope58414.2023.10264493.

[5] T. Hegarty, A. Kumar, R. Blattner and A. Obidat, "An Active EMI Filter for Common-Mode EMI Mitigation in High-Power AC Systems," *PCIM Europe 2023; International Exhibition and Conference for Power Electronics, Intelligent Motion, Renewable Energy and Energy Management*, Nuremberg, Germany, May 2023, pp. 1-6, doi: 10.30420/566091339.

[6] A. Kumar, Y. Hou, Y. Ramadass, T. Merkin, T. Hegarty et al., "An Active EMI Filter for High-Power Off-Line Applications," 2023 IEEE Applied Power Electronics Conference and Exposition (APEC), Orlando, FL, USA, 2023, pp. 2063-2067, doi: 10.1109/APEC43580.2023.10131427.

[7] T. Hegarty, "How Active EMI Filter ICs Reduce Common-Mode Emissions in Single- and Three-Phase Applications (Part 3): Modeling Nanocrystalline Chokes," *How2Power Today*, Mar. 2024.

[8] "TPSF12C1 and TPSF12C3 power-supply filter IC FAQs," Texas Instruments E2E™ design support forums, Aug. 2023.

[9] International Electrotechnical Commission, "Electric vehicle conductive charging system – Part 21-1: Electric vehicle on-board charger EMC requirements for conductive connection to an AC/DC supply," IEC 61851-21-1, first edition, June 2017.

[10] Underwriters. Laboratories, "UL 2202 – DC charging equipment for electric vehicles," third edition, Dec. 2022.

[11] "Techniques of Design-Oriented Analysis," ECEA 5706, University of Colorado Boulder.

[12] T. Hegarty, "TPSF12C1/3 Common-mode Active EMI Filter (AEF) Mathcad Design Tool," Texas Instruments, Feb. 2024.

[13] Texas Instruments EVM user's guide, "TPSF12C3EVM-FILTER active EMI filter evaluation module for three-phase systems," May 2024.

PCIM Europe 2024, 11– 13 June 2024, Nuremberg    DOI: 10.30420/566262143

# A Repetitive High Voltage Nanoseconds Pulse Generator: First Prototype Design and Tests Results

Serge Gavin[1], Simon Kissling[1], Bertrand Daout[2], Frédéric Castella[2], Mauro Carpita[1]

[1] HES-SO University of Applied Sciences of Western Switzerland, Switzerland
[2] Montena technology SA, Switzerland

Corresponding author:    Serge Gavin, serge.gavin@heig-vd.ch
Speaker:    Serge Gavin, serge.gavin@heig-vd.ch

## Abstract

In this paper a repetitive high voltage pulse generator is presented. The nominal output voltage of the high voltage pulses is 10kV. The main purpose is to create a pulsed electrical field to be used in scientific applications such as medicine, biology or food processing. In these cases, the pulses width should be as short as 100 ns with rising and falling times as shorter as possible. Moreover, for these applications, the pulsed electrical field is used in continuous processing, meaning that it needs to be repetitive and repeatable. In this case, the repetition rate is specified at 1 kHz.

## 1    Introduction

Pulsed electrical fields (PEF) are used in various applications such as electroporation in cellular biology, non-thermal plasma generation, and so forth [1]. These includes, for instance, air depollution, medical treatments of wounds or cancer, micro-organisms growth stimulation for food industry, etc.

Using very short repetitive pulses allow to achieve the required energy levels for these processes while using less average energy compared to conventional continuous processes. This assertion is valid as far as the pulse generation device efficiency is kept as high as possible.

In this paper a first prototype of a short, high voltage, repetitive pulse generator is presented. There are two aspects addressed in this paper; first of all, the pulse generation circuit is discussed, then the dedicated capacitor charger is presented.

## 2    Prototype specifications

The Table 1 presents the pulse generator specifications. The development project foreseen a twofold target: a first prototype, to validate the principle, and a second, higher voltage prototype which is under development and is foreseen in the future.

The first prototype, which is described in the present paper, has to produce a pulse of 10 kV with a duration of 100 ns. Rising and falling times have to be as short as possible, and in any case no longer than 10 ns. The main challenge is the repetition rate, which is specified at 1 kHz.

The second prototype that is foreseen will have an output pulse voltage of 30 kV while other parameters are the same.

| Parameter | | 1st prototype | Final prototype |
|---|---|---|---|
| Pulse voltage | $V_{pulse}$ | 10 kV | 30 kV |
| Pulse duration | $T_{pulse}$ | 100 ns | 100 ns |
| Rising/falling time | $T_r / T_f$ | < 10 ns | < 10 ns |
| Repetition rate | $f_{pulse}$ | 1 kHz | > 1 kHz |
| Load resistor | $R_{load}$ | 50 Ω | 50 Ω |
| Pulse power | $P_{pulse}$ | 2 MW | 18 MW |
| Average power | $P_{avg}$ | 200 W | > 1.8 kW |

**Table 1**    Pulse generator specifications

## 3    Prototype description

The pulse generator may be separated in two different parts; The pulse generation circuit [2] and the dedicated line charger [3].

### 3.1    Pulse generation circuit

To generate the pulse fulfilling the requirements of table 1 a method using a so-called pulse forming line (PFL) is used. The main principle is to charge a line which the characteristic impedance is equal to the load. Then a spark gap is used to discharge the stored energy into the load. If both impedances are matched, the resulting pulse is theoretically rectangular shaped with half the amplitude of the charging voltage at the input of the line [2]. The Fig. 1 shows a typical pulse generation circuit made of capacitors and inductors used as a pulse forming line. The L/C values are determined by the characteristic impedance needed, while the C value is directly chosen by the energy needed in the pulse, thus influencing the pulse length.

One of the drawbacks of this solution is the coupling resistance used between the high voltage power supply and the PFL. This is not suitable for a repetitive pulse circuit due to the dissipated power in this resistor. To overcome this, a specific line charger has been developed.

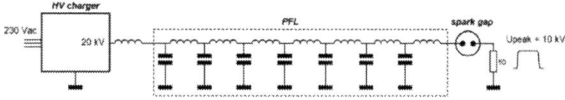

**Fig. 1** Pulse generation circuit for a pulse amplitude of 30 kV

## 3.2 Line charger circuit

As previously mentioned, the pulse is formed by using a PFL, requiring the input voltage of the line to be the double of the pulses amplitude. In this case, a charger with an output voltage of 20 kV is needed. To avoid a coupling resistor that causes a lower efficiency, the charger must be disconnected or switched off during the pulse.

Summarizing, some of the key points to address on the charger design are:

- Charger output has to be the double of pulses amplitude,

- to avoid any influence on the pulses shape, the charger as to be disconnected during the pulse,

- its charging time has to be lower than 1 ms to guarantee a repetition rate of 1 kHz,

- the output voltage has to be controlled and adjusted if needed.

Several solutions based on voltage multipliers [4] or flyback topologies [3] have already been investigated. In these cases, the charger has to be disconnected since its operation requires one or several output capacitors. These capacitors will influence the PFL and the pulse shape itself if the charger is not fully disconnected from the line.

The disconnection of the charger during the pulse requires a bulky element of multiple levels of semiconductors [5] that may be avoided by using a charger based on a converter topology that doesn't need any additional output capacitor.

## 4 Line charger prototype design

The charger has been designed according to the following requirements/characteristics:

- Input DC-link voltage of 400 V

- Full bridge inverter controlled by a microcontroller unit,

- High voltage, medium frequency transformer

- High voltage rectifier

A high-voltage measurement has been included, so the microcontroller is able to fully control the output voltage. The Fig. 2 shows a simplified schematic of the high voltage charger.

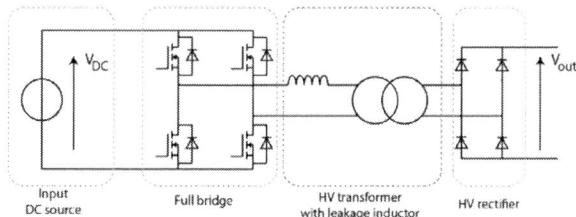

**Fig. 2** Line charger topology based on a full-bridge circuit

## 4.1 Line charger electronics

The full bridge inverter is based on a standard two legs phase shift control technique. That means the fundamental of the output voltage is controlled by the phase shift between the legs, each leg having a 50% duty cycle.

The Fig. 3 shows a picture of the charger electronics.

The chosen switching devices are GeneSiC 1200 V/71 A SiC MOSFET, mounted on a heatsink. Switching and sampling frequency have been chosen at 30kHz to stay below to the self-resonant frequency of the transformer.

**Fig. 3** Charger electronics PCB

The control is implemented in a TI microcontroller unit from the C2000 family (TMS320F28035). Thanks to its "Control Law Accelerator" (CLA) the controller and all the related time critical processes can be managed flawlessly.

On one side of the board one can find the power connections, DC-input and AC-output (not visible on Fig.3). On the other side of the board there are the coaxial connections used for:

- Line voltage measurement input,
- pulse triggering
- General purpose input

In addition, some optical fiber link and an isolated RS422 communication bus have also been implemented for supervising purposes.

## 4.2 Medium frequency high voltage transformer

A key element of the charger design is the high voltage transformer. It has been designed to withstand 20 kV at the secondary side but also to add some leakage inductor that allows to limit the current when the output voltage is low.

Due to the parasitic capacitor at the secondary, these medium frequency transformers have usually a self-resonance frequency that cannot be neglected. In this case, the self-resonance frequency has was measured at around 83 kHz.

The transformer has been built according to the following specifications:

- Number of turns: 14:540
- Nominal power: 400 VA
- Output voltage: 20 kV

The Fig. 4 shows a picture of the transformer

**Fig. 4**     Picture of the HV transformer

## 4.3 High voltage rectifier

A rectifier made of a series connected diodes is used at the secondary of the transformer. The diodes are rated for 5 kV. On the rectifier board, a voltage measurement has also been implemented.

The rectifier board is shown on Fig. 5

As the rectifier has to sustain the whole output voltage some millings have been made in the PCB design to increase the creepage distance and avoid any discharge.

The voltage measurement is made with a voltage resistive divider compensated with parallel capacitors to improve the dynamic response of the measurement.

**Fig. 4**     Picture of the rectifier board

## 5 Tests results

### 5.1 Charger test results

A prototype of the charger has been built and successfully tested.

The Fig. 5 shows some results of the charger tests. One can see on the oscilloscope screenshot that the charger output voltage reaches 20 kV in less than 1ms which is one of the mandatory conditions to achieve a repetition rate of 1 kHz. Moreover, the current in the transformer is also limited to less than 8 A even at the beginning of the charging process, thanks to the leakage inductance of the transformer.

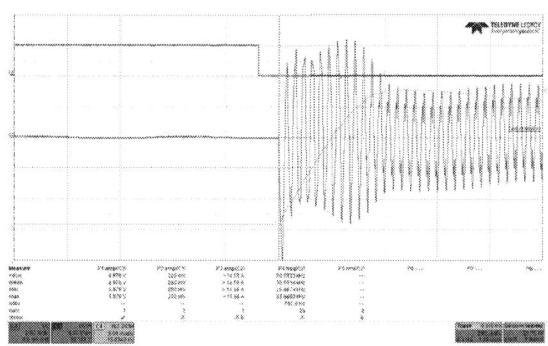

**Fig. 5** Charger tests results (200 us/div): Charger output voltage (green 5 kV/div), transformer primary current (purple 2 A/div) and charge starting signal (blue).

## 5.2 Complete pulse generator test results

In order to test the complete pulse generator, the charger, the pulse forming line, the spark gap unit and the triggering circuit have been assembled, as shown on Fig. 6 and Fig. 7.

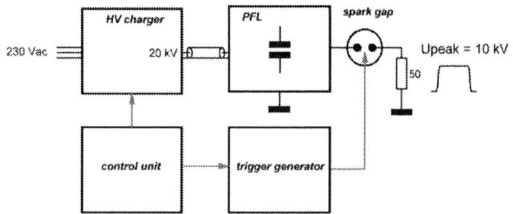

**Fig. 6** Whole pulse generator bloc schematic

**Fig. 7** Whole pulse generator assembly

As explained in point 3.2, due to the effect of the pulse forming line, the pulse amplitude will be half of the charger output voltage. The fig. 6 shows the result of a pulse made with a charging voltage of 24 kV.

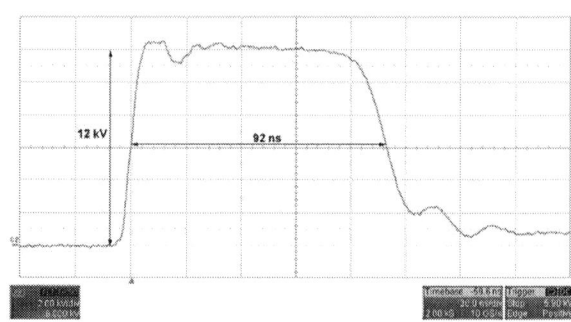

**Fig. 6** Pulse output voltage with an amplitude of 12 kV. The duration is 92 ns and rising time is <10 ns

## 6 Conclusion

In this paper a repetitive high voltage pulse generator is presented. The prototype that has been designed for pulse amplitude of 10 kV has been built and tested with pulses up to 12 kV, 100ns duration and >10 ns rising time. The instantaneous power of one pulse is then >2 MW as the load is 50 ohm.

Overall efficiency has been measured at 70% thanks to the charger technology avoiding any coupling resistance to charge de pulse forming line.

## References

[1] S. Kissling, S. Gavin, M. Carpita, "Power electronics for a pasteurization process working by electric resonance: First prototype experimental results", 17th European Conference on Power Electronics and Applications (EPE'15 ECCE-Europe), (2015)

[2] T. Huiskamp et al., "Design of a Subnanosecond Rise Time, Variable Pulse Duration, Variable Amplitude, Repetitive, High-Voltage Pulse Source", IEEE Transactions on Plasma Science, vol. 42, no. 1, 2014

[3] T. Huiskamp, et al., "A Solid-State 0–120 kV Microsecond Pulse Charger for a Nanosecond Pulse Source", IEEE Transactions on Plasma Science, vol. 41, no. 12, 2013

[4] M. Evans, B. Foy, D. Mager, R. Shapovalov and P.-A. Gourdain, "High voltage charging system for pulsed power generators", 2017

[5] S. Gavin, M. Carpita, "Asymmetric cascaded bridge converter for high voltage, high dynamics Power supply: small scale prototype test results", 5th International Conference on Power Engineering, Energy and Electrical Drives (POWERENG), 2015

PCIM Europe 2024, 11– 13 June 2024, Nuremberg     DOI: 10.30420/566262144

# Frequency Shift Keyed Dual Side Control of Inductive Power Transfer: An Application of Talkative Power Conversion

Julius Maximilian Placzek[1], Hamzeh Beiranvand[1,2], Marco Liserre[1,2,3]

[1] Kiel University, Germany
[2] Kiel Nano, Surface and Interface Science (KiNSIS), Germany
[3] Fraunhofer ISIT, Germany

Corresponding author:     Julius Maximilian Placzek, jmpl@tf.uni-kiel.de
Speaker:     Hamzeh Beiranvand, hab@tf.uni-kiel.de

## Abstract

Progress in control of Inductive Power Transfer (IPT) has resulted in Dual Side Controllers (DSC) which retain high efficiencies for a wide parameter range. These control schemes require coordination between the separate system sides. To avoid radio communication, hence extra system layers, and to enable underwater operation, communicationless and cooperative controllers have been proposed. They are significantly limited in which parameters can be tracked and generate additional design constraints. This paper instead applies the recent idea of Talkative Power Conversion (TPC) to IPT by proposing a low bandwidth Frequency Shift Keying (FSK) process. It is demodulated via the preexisting phase locked synchronization loop to implement communication of arbitrary information. The communication is demonstrated in experiment by utilising it for DSC of a Gallium Nitride (GaN) based Series-Series (SS) compensated 500 W 48 V IPT system with an Asymmetric Voltage Cancellation (AVC) waveform and a frequency shift of 8 Hz.

## 1 Introduction

For Series-Series (SS) compensated Inductive Power Transfer (IPT), increasingly sophisticated synchronous Dual Side Control (DSC) strategies have been proposed [1]–[6]. Usage of all degrees of freedom provided by the topology yields schemes such as Variable Frequency Triple Phase Shift Control (VFTPS) [4], provides control of losses and soft switching for a wide range of parameter variations, as well as bidirectional operation.

Multiparameter control of a plant via two entirely separate converters requires a method of coordination. Since DSC enables full local power control, this mainly concerns the outer loops which control efficiency and do not require low latency solutions [7]. A multitude of systems use commercial Radio Frequency (RF) communication systems [1]–[6]. For operation in RF dampening environments (water, living tissue, etc.) alternatives have been proposed, such as multiplexing based near field communication [8], [9] and communicationless DSC using estimator or parameter tracking methods [7], [10] such as the one shown in Fig. 1 a). These solutions typically either require significant additional hardware and offer fewer degrees of freedom than true communication.

For low power non-DSC systems such as medical implant telemetry and the "QI" standard for consumer IPT, direct modulation of the power flow has been widely adopted [11]–[14]. These low power solutions however are often focused on high data rates and operate at high khz or MHz frequencies [11], [14]. The Load Shift Keying (LSK) in QI typically uses additional switches on the die of the monolithic QI circuit and capacitors, which are small at this power level, to forcibly detune the power transfer process [13]. None of these schemes appear to translate well to medium and high power IPT, on the basis of added detuning, added components and other design assumptions.

In multireceiver IPT coordination, a frequency droop strategy was proposed in [15] as shown in Fig. 1 b), further improved to a frequency jitter strategy in [16] and the negligible impacts of sufficiently small frequency shifts upon power transfer shown in [17]. These strategies use smaller frequency variations

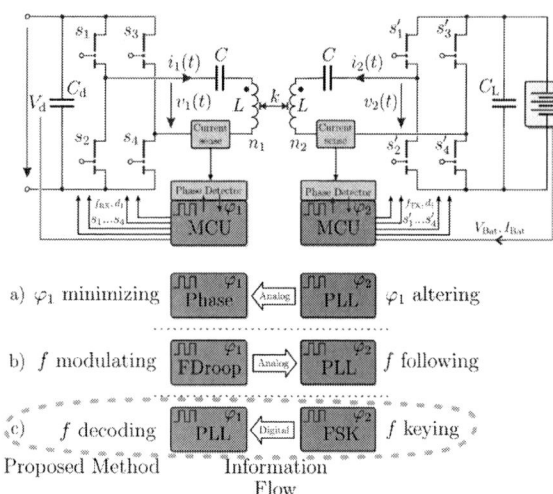

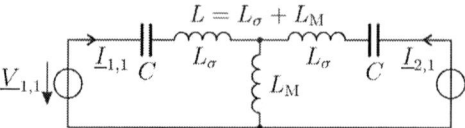

**Fig. 2:** Minimal representation of the SS compensated IPT topology.

**Fig. 1:** SS compensated system with controlled power conversion based on GaN High Electron Mobility Transistors (HEMT) with examples of preceding coordination schemes and the newly proposed TPC based digital arbitrary communication. a) The phase cooperative strategy, a simple but limited reverse channel [7]. b) An equivalent of the frequency droop forward channel first described in [15]. c) The method of implementing a reverse channel proposed in this paper.

of main power flow, i.e. 500 Hz of a 20 kHz switching frequency [16]. The signal is broadcast from a track system, directly feeds into the control loops of multiple receivers and is not digitally decoded.

This paper takes the idea of small frequency shifts in IPT system coordination and applies the principle of Talkative Power Conversion (TPC) [18], [19] - true digital communication via power flow without undue influence on it. To adress the challenge of DSC it demonstrates, that a reverse communication channel is as feasible as the forward one. For this, the synchronizing Phase Locked Loop (PLL) typical of advanced SS-IPT control is placed in the primary side instead of the secondary and acts as a receiver. Because the efficiency maximizing DSC loop as well as other potential data in the full application, such as battery State of Charge (SoC) and temperature are slowly varying quantities, a data rate on the order of tens of bit/s is sufficient. This is transmitted from the secondary side via Frequency Shift Keying (FSK) with a very low modulation depth. Binary words are encoded at a base rate of ca. 23 bit/s with a frequency shift of only 8 Hz out of a $f_1 = 52500$ Hz switching frequency. The rate of frequency change is limited to reduce PLL perturba-

tion and the resultant sequence is synthesized from discrete clock dividers. The Gallium Nitride (GaN) based experiment reaches DSC in 3.5 s with only 85 mV of additional ripple voltage, caused from residual system perturbation not rejected by the control loops.

## 2 Series-Series compensated IPT

The SS compensated topology has been widely considered in research on advanced IPT control schemes due to its well defined mathematical properties, low component count and compatibility with Voltage Source Converters (VSC) [1]–[6]. A minimal schematic of a fully controlled SS system with controlled voltage sources on either side is given in Fig. 2. Analysing the system under this equivalent, disregarding losses and winding ratio, leads to the following well known equations. A singular resonant frequency independent of the coupling factor $k$ can be found:

$$\omega_1 = 2\pi f_1 = \frac{1}{\sqrt{LC}}. \tag{1}$$

At this frequency $f_1$ lies the Constant Current (CC) mode of the SS topology. Considering possible implementations of the strategy with VSCs as in Fig. 1, the alternating voltages can be expressed as a product of the local DC voltages $V_1$, $V_2$ with the local complex modulation functions $\underline{M_1}(d_1)$ and $\underline{M_2}(d_2)$ consisting of magnitude and angular component, yielding the following perfectly symmetric current equations under the first harmonic approximation:

$$\underline{I}_{2,1} = j\frac{\underline{V}_{1,1}}{\omega_1 L_M} = j\frac{V_1 M_1(d_1)}{\omega_1 kL} \tag{2}$$

$$\underline{I}_{1,1} = -j\frac{\underline{V}_{2,1}}{\omega_1 L_M} = -j\frac{V_2 M_2(d_2)}{\omega_1 kL}. \tag{3}$$

The fundamental current of each side is coupled only to physical parameters of the resonant system and to the voltage of the respective other side. This decoupling enables the simple implementation of various control schemes based on PLLs, since it

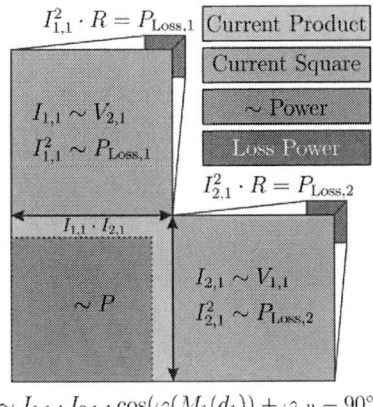

$$P \sim I_{1,1} \cdot I_{2,1} \cdot \cos(\varphi(\underline{M_1}(d_1)) + \varphi_{\text{pll}} - 90°)$$

**Fig. 3:** Transferred power and loss powers as a function of magnitudes and phase.

can synchronize on the local current without significant interference from the local power converter controlled by it.

## 2.1 Dual Side Control

DSC is the task of controlling power flow, conduction loss and soft switching from both system sides via multiple parameters distributed among the two power conversion systems. Fundamentally, four parameters can be used to achieve this. The equivalent AC output voltage of each converter, the phase angle between them and the switching frequency. Comprehensive in-depth optimizations of this process have been presented [4], [6], this paper limits itself to simple magnitude DSC of conductive losses as a test source of data for TPC.

Assuming that the system is in steady state, power will be $-P_1 = P_2 = P$ with the PLL angle $\varphi_{\text{pll}} \in [-90°, 90°]$ determining the power flow direction:

$$P_1 = -\frac{|\underline{V}_{1,1}||\underline{V}_{2,1}|}{\omega_1 kL}\cos(\varphi(\underline{M_1}(d_1)) + \varphi_{\text{pll}} - 90°). \quad (4)$$

The phase angle component $\varphi(\underline{M_1}(d_1))$ is an unwanted side effect of direct modulation with soft switching, further discussion can be found in [7].

The primary and secondary side converters can then use $|\underline{V}_{1,1}|$ and $|\underline{V}_{2,1}|$ respectively, for local control of the power flow. Assuming that losses are symmetric and too small to significantly change system transfer behavior, but large enough to degrade efficiency, the magnitude DSC condition is reached when:

$$|\underline{V}_{1,1}| = |\underline{V}_{2,1}| \rightarrow |\underline{I}_{1,1}| = |\underline{I}_{2,1}| \quad (5)$$

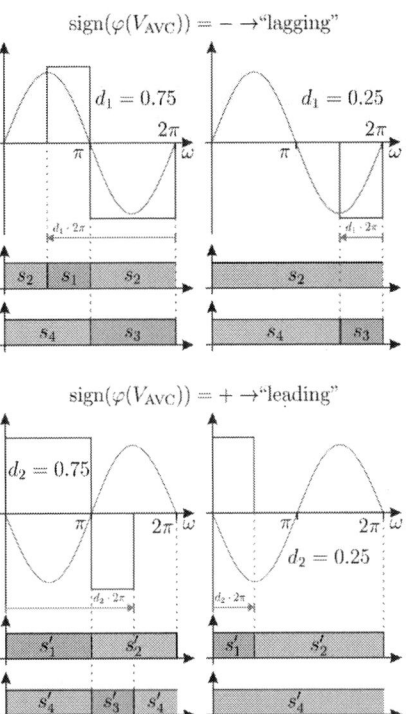

**Fig. 4:** Waveforms and switching pattern of the AVC process as aligned to the current under symmetric conditions.

because resistive $I^2R$ losses become minimal per unit of power, this is visualized in Fig. 3. For asymmetric coil systems additional considerations have to be made, a description is given in [6].

Here Asymmetric Voltage Cancellation (AVC) is considered, since [6], [7] have shown that it provides superior efficiency to phase shift modulation in a wide parameter range, despite increased harmonic losses. This is due to a reduction in switching events and improved power factor. AVC is shown in Fig. 4, the system sides use opposite symmetries to cancel out the undesired phase shift effects of directly modulated DSC [6], [7]. While AVC was shown to be incompatible with the PLL based phase cooperative control method in [7] (Fig. 1 a), the TPC based strategy demonstrated here, communicates arbitrary information and is not subject to the issues of the AVC phase function. The fundamental RMS magnitude behavior for a duty cycle of $d$ is independent of symmetry and can be derived via Fourier series for the half-bridge (HB) mode $0 < d \leq 0.5$ as:

$$\frac{|\underline{V}_{\text{HB},1}|}{V} = G_{\text{HB}} = \frac{\sqrt{2}}{\pi}\sin(\pi d) \quad (6)$$

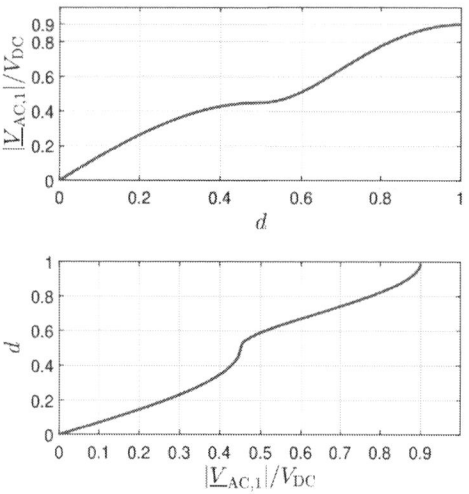

**Fig. 5:** Fundamental magnitude and reverse function for both symmetries of the AVC modulation process.

and the full-bridge (FB) mode $0.5 < d \leq 1$ as:

$$\frac{|\underline{V}_{\mathrm{FB},1}|}{V} = G_{\mathrm{FB}} = \frac{\sqrt{10 + 6\cos(2\pi d)}}{\pi\sqrt{2}}. \quad (7)$$

To close the magnitude DSC loop, the inverse function is also necessary, for the HB mode it can be given as:

$$d_{\mathrm{HB}} = \frac{1}{\pi}\arcsin\left(G_{\mathrm{HB}}\frac{\pi}{\sqrt{2}}\right). \quad (8)$$

For the FB mode the direct result of inverting the function proves nonsensical due to the ambiguities from periodicity in the cosine function, hence a sign inversion and an offset of $+\pi$ have to be introduced:

$$d_{\mathrm{FB}} = \frac{1}{2\pi}\left(\arccos\left(-\left(\frac{G_{\mathrm{FB}}^2\pi^2 - 5}{3}\right)\right) + \pi\right). \quad (9)$$

The $d$ to $G$ and reverse fundamental frequency functionalities are visualized in Fig. 5 and stored as Look-Up-Tables (LUT) in the control systems.

## 3 Proposed FSK TPC Scheme

An overview of the scheme is given in Fig. 6, the elements of which are explained in sequence in this section. The primary side will typically be connected to a fixed voltage source, such as an AC or DC grid, while the secondary operates on a globally unknown load or battery voltage. Therefore, the secondary side shall encode its local voltage $|V_2 \underline{M}_2(d_2)|$, as determined by a local power control loop, into a binary word. A synchronizing bit is added and the sequence is oversampled.

To avoid a large tracking error in the PLL, the rate of frequency change shall be limited. For this a discretized PT-1, implemented as an infinite impulse response filter, can be used.

The next challenge which has to be adressed, is the digital implementability of the strategy. Digital control systems such as the Microcontroller Units (MCU) used here, generate the switching signals by integer clock division. In this case the system clock is $f_{\mathrm{CLK}} = f_{\mathrm{CLK1}} = f_{\mathrm{CLK2}} = 120$ MHz and the target switching frequency $f_1 = 52.500$ Hz. This results in a non integer divider of $n^* = 2285.7$, which can only be approximated by $n_{\mathrm{a}} = 2285$ or $n_{\mathrm{b}} = 2286$. The minimum frequency step which can therefore be natively resolved by the control system becomes:

$$\frac{f_{\mathrm{CLK}}}{n_{\mathrm{a}}} - \frac{f_{\mathrm{CLK}}}{n_{\mathrm{b}}} = f_{\mathrm{a}} - f_{\mathrm{b}} \approx 23 \text{ Hz}. \quad (10)$$

Considering the nature of the receiving PLL, this issue however disappears. A PLL detects phase at the input and uses frequency for actuation. The PLL loop filter can have a bandwidth, so that any components significantly above the target bit frequency will be rejected. Therefore, at $f_1$ the PLL detects average phase and hence controls average frequency, being the average rate of phase change. This can be utilized to reconstruct the target average frequency $f^*$ from $f_{\mathrm{a}}$ and $f_{\mathrm{b}}$ by error summation and error minimizing decision, as is shown in Fig. 7. From a time domain perspective, the entire communication process is hidden in a small variation of an $1/f_{\mathrm{CLK}} = 8.33$ ns jitter of the switching frequency.

Since the topology is inherently bidirectional, the PLL can be placed in the primary side to oper-

**Tab. 1:** System parameters considered in this paper

| Parameter | Value |
|---|---|
| $V_1$ | 48 V |
| $V_2$ | 36 V |
| $k$ | 0.5 |
| $L$ | 18 µH |
| $C$ | 504 nF |
| $R_{\mathrm{L}}$ | 4.5 Ω |
| $C_{\mathrm{d}} = C_{\mathrm{L}}$ | 500 µF |
| $f_{\mathrm{I}}$ | 52 500 Hz |
| $f_{\mathrm{II}}$ | 52 508 Hz |
| $T_{\mathrm{bit}}$ | 43 ms |
| Power Device | GS 61008T |

PCIM Europe 2024, 11– 13 June 2024, Nuremberg · · · DOI: 10.30420/566262144

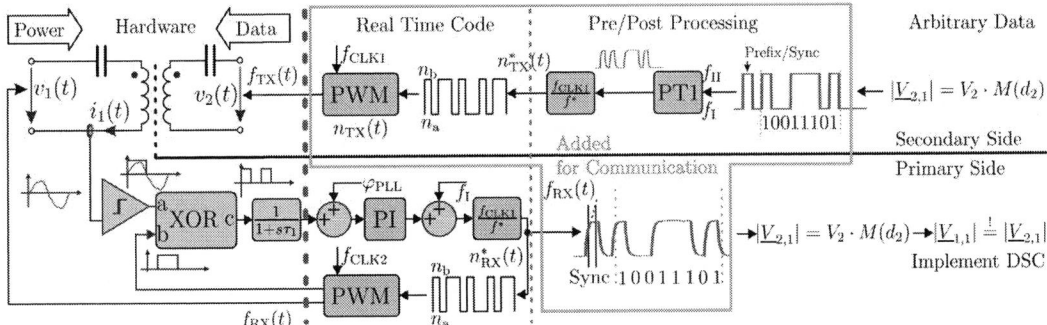

**Fig. 6:** An overview of the FSK process across all domains. All hardware and the majority of real time computations are necessary for power transfer. The main additions are non time critical code for communication encoding and decoding.

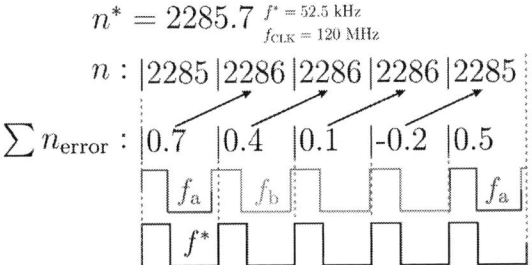

**Fig. 7:** The small $\Delta f$ shifts utilized for communication have to be translated into a sequence which can be generated by a typical digital PWM unit, which works by integer clock division. A signal, composed from the divider ratios $n_a$ and $n_b$ as selected by equivalent phase error accounting, has the correct average phase and frequency to implement $n^* \sim f^*$ when detected by a slow acting PLL.

ate as a "synchronized source" for forward power flow, instead of a "synchronous rectifier". The FSK sequence is tracked with a low error by the PLL, enabling a low interference on power flow. Remaining perturbations are further canceled by the local power control loop. The PLL also resolves the frequency from integer dividers, this is necessary to compensate for clock oscillator drift between the system sides, even in a fully communicationless case. Only the secondary side algorithm is added in terms of real time computation.

The receiver synchronizes a sampling pattern to $f_{RX}(t)$ and detects the sequence. The recovered voltage information is fed to the DSC control loop. As can be seen from Fig. 6, the signal processing hardware required for this scheme is minimal and identical to pure power transfer, a zero crossing detector, an XOR gate and a low pass filter. The PLL reference is directly sourced from the Pulse Width

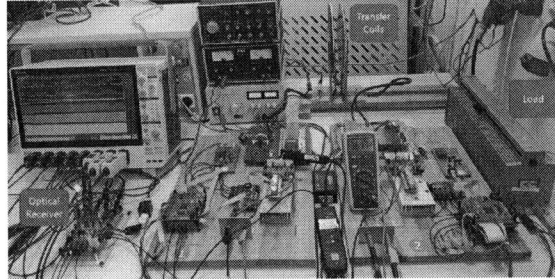

(a)

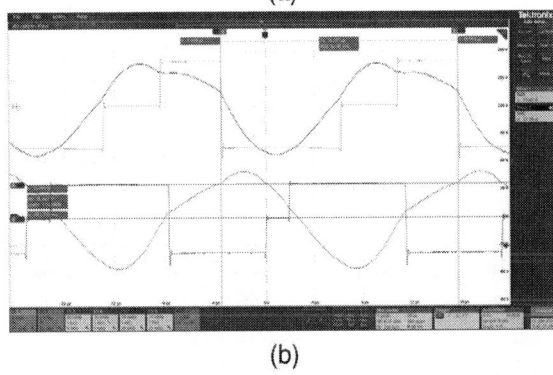

(b)

**Fig. 8:** Experimental setup: (a) Photo of the setup and (b) the AVC process on symmetric currents.

Modulation (PWM) unit. The main communication related processing happens either asynchronously, or at such a low frequency that it can be handled outside of interrupt service routines.

## 4 Experiment

The method is implemented on a 48 V 500 W GaN based testbed. The system parameters can be found in Tab. 1. The output voltage is reduced to $V_2 = 36$ V to simulate a low SoC battery charging from nominal DC link voltage $V_1 = 48$ V. This cre-

1109

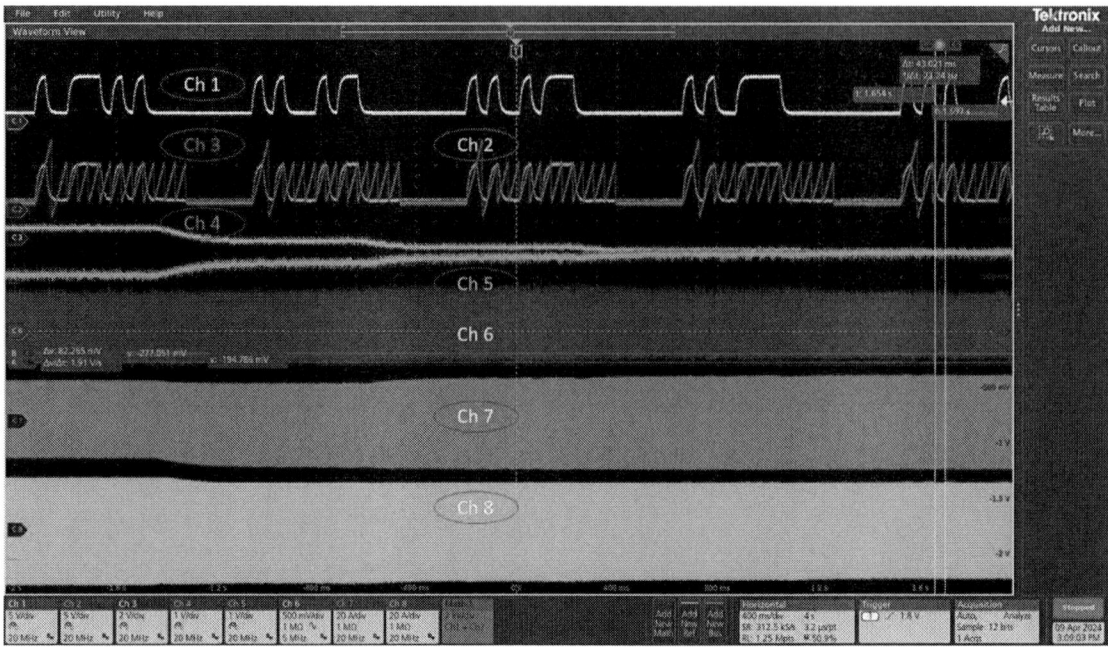

**Fig. 9:** Since this communication scheme is designed to generate small time domain impact, it is significantly difficult to visualize on an oscilloscope. Hence, signals Ch1-Ch5 from top to bottom are internal states of the MCUs, reproduced via PWM feeding low pass filters through optical fiber isolation. Ch1 shows the bit sequences computed by the secondary, with $\Delta f = 8$ Hz and ca. 43 ms bit duration. Ch2 reproduces the $f_{RX}$ of the PLL. Ch3 shows the synchronization scan and subsequent receiving process for the bit sequence. Ch4 and Ch5 show the $|V_d \underline{M}(d_1)|$ and $|V_L \underline{M}(d_2)|$ functions of primary and secondary respectively. The primary executes a filtered control step after every message is received, the secondary tracks to keep output power and reports the new state via the following message. Voltage DSC is achieved after 3.5 s. Ch6 shows the AC ripple of the output DC voltage as a direct measurement by differential probe. It can be seen that the communication process adds only ca. 85 mV of ripple. Ch7 and Ch8 show the envelope of $i_1(t)$ and $i_2(t)$ as a direct measurement via high speed current probe, again the action of TPC is barely visible and the DSC action correctly transitions the system currents to a balanced state.

ates a typical DSC challenge. The overall setup can be seen in Fig. 8 a) and the resonant waveforms of $v_1(t), i_1(t)$ and $v_2(t), i_2(t)$ in Fig. 8 b).

Figure 9 shows the low frequency waveforms during the DSC process. Ch1 to Ch5 are debugging signals taken by optical fiber from the MCU laboratory boards. They visualize quantities which are not properly detectable from the power flow by oscilloscope or broadband spectrum analyser, due to the very small variations. The frequency variation is reproduced by PLL in quasi steady state, without significant delay or distortion. The additional ripple voltage introduced by the process is ca. 85 mV and less than the counteracting ripple reduction achieved by DSC. Full DSC is achieved in 3.5 s after four messages have been received and processed. Impact on the system currents is marginal. It can be seen in the received frequency waveform on Ch2, that the DSC process in turn

equally generates little interference to FSK.

# 5 Conclusion

A low data rate FSK process implementing reverse arbitrary communication for SS compensated synchronous IPT was evolved. It results from the combination of previous cooperative and communicative approaches, which were either less capable, or generated significant other design constraints. It uses a very small modulation depth and data rate. According to the paradigm of TPC, it enables technically sufficient data transfer by power transfer, without undue influence on power transfer. The communication process is demonstrated in experiment by implementing magnitude DSC under an unsymmetrical DC voltage situation. DSC is reached in 3.5 s after four messages. It is thereby demonstrated, that such communication is technically practical in two ways: First, communication has marginal influence on power flow and power flow marginal influence on communication. Second, no hardware is added and the majority of computation is non time critical.

## Acknowledgment

Funded by the European Union - European Regional Development Fund (EFRE), the German Federal Government and the State of Schleswig-Holstein (LPW-E/1.1.2/1486).

## References

[1] T. Diekhans and R. W. De Doncker, "A Dual-Side Controlled Inductive Power Transfer System Optimized for Large Coupling Factor Variations and Partial Load," *IEEE Transactions on Power Electronics*, vol. 30, no. 11, pp. 6320–6328, Nov. 2015. DOI: 10.1109/TPEL.2015.2393912.

[2] Y. Li, J. Hu, F. Chen, Z. Li, Z. He, and R. Mai, "Dual-Phase-Shift Control Scheme With Current-Stress and Efficiency Optimization for Wireless Power Transfer Systems," *IEEE Transactions on Circuits and Systems I: Regular Papers*, vol. 65, no. 9, pp. 3110–3121, Sep. 2018. DOI: 10.1109/TCSI.2018.2817254.

[3] X. Zhang, T. Cai, S. Duan, H. Feng, H. Hu, *et al.*, "A Control Strategy for Efficiency Optimization and Wide ZVS Operation Range in Bidirectional Inductive Power Transfer System," *IEEE Transactions on Industrial Electronics*, vol. 66, no. 8, pp. 5958–5969, Aug. 2019. DOI: 10.1109/TIE.2018.2871794.

[4] Y. Liu, U. K. Madawala, R. Mai, and Z. He, "An Optimal Multivariable Control Strategy for Inductive Power Transfer Systems to Improve Efficiency," *IEEE Transactions on Power Electronics*, vol. 35, no. 9, pp. 8998–9010, Sep. 2020. DOI: 10.1109/TPEL.2020.2970780.

[5] M. Wu, X. Yang, W. Chen, L. Wang, Y. Jiang, *et al.*, "A Dual-Sided Control Strategy Based on Mode Switching for Efficiency Optimization in Wireless Power Transfer System," *IEEE Transactions on Power Electronics*, vol. 36, no. 8, pp. 8835–8848, Aug. 2021. DOI: 10.1109/TPEL.2021.3055963.

[6] S. Jia, C. Chen, S. Duan, and Z. Chao, "Dual-Side Asymmetrical Voltage-Cancelation Control for Bidirectional Inductive Power Transfer Systems," *IEEE Transactions on Industrial Electronics*, vol. 68, no. 9, pp. 8061–8071, Sep. 2021. DOI: 10.1109/TIE.2020.3016265.

[7] J. M. Placzek, H. Beiranvand, and M. Liserre, "Communicationless phase cooperative control of inductive power transfer using asymmetrical modulations," *IEEE Transactions on Power Electronics*, vol. 38, no. 6, pp. 7836–7847, 2023. DOI: 10.1109/TPEL.2023.3244673.

[8] M. Trautmann, B. Sanftl, R. Weigel, and A. Koelpin, "Simultaneous Inductive Power and Data Transmission System for Smart Applications," *IEEE Circuits and Systems Magazine*, vol. 19, no. 3, pp. 23–33, 2019, Conference Name: IEEE Circuits and Systems Magazine. DOI: 10.1109/MCAS.2019.2924508.

[9] J. Wu, C. Zhao, Z. Lin, J. Du, Y. Hu, and X. He, "Wireless power and data transfer via a common inductive link using frequency division multiplexing," *IEEE Transactions on Industrial Electronics*, vol. 62, no. 12, pp. 7810–7820, Dec. 2015. DOI: 10.1109/TIE.2015.2453934.

[10] R. Mai, Y. Liu, Y. Li, P. Yue, G. Cao, and Z. He, "An Active-Rectifier-Based Maximum Efficiency Tracking Method Using an Additional Measurement Coil for Wireless Power Transfer," *IEEE Transactions on Power Electronics*, vol. 33, no. 1, pp. 716–728, Jan. 2018. DOI: 10.1109/TPEL.2017.2665040.

[11] M. J. Karimi, A. Schmid, and C. Dehollain, "Wireless Power and Data Transmission for Implanted Devices via Inductive Links: A Systematic Review," *IEEE Sensors Journal*, vol. 21, no. 6, pp. 7145–7161, Mar. 2021. DOI: 10.1109/JSEN.2021.3049918.

[12] *Qi specification 1.3*, the Wireless Power Consortium, Inc., Jan. 2021.

[13] *Wlc1115, wireless charging ic (wlc) - transmitter 15w with integrated usb type-c pd controller*, WLC1115, Rev. C, Infineon, Datasheet: WLC1115, Wireless charging IC (WLC) - Transmitter 15W with integrated USB Type-C PD controller, Jul. 2023.

[14] P. A. Hoeher, "Fsk-based simultaneous wireless information and power transfer in inductively coupled resonant circuits exploiting frequency splitting," *IEEE Access*, vol. 7, pp. 40 183–40 194, 2019. DOI: 10.1109/ACCESS.2019.2907169.

[15] U. K. Madawala, M. Neath, and D. J. Thrimawithana, "A Power–Frequency Controller for Bidirectional Inductive Power Transfer Systems," *IEEE Transactions on Industrial Electronics*, vol. 60, no. 1, pp. 310–317, Jan. 2013, Conference Name: IEEE Transactions on Industrial Electronics. DOI: 10.1109/TIE.2011.2174537.

[16] M. Neath, U. Madawala, and D. Thrimawithana, "Frequency jitter control of a multiple pick-up Bidirectional Inductive Power Transfer system," in *2013 IEEE International Conference on Industrial Technology (ICIT)*, Feb. 2013, pp. 521–526. DOI: 10.1109/ICIT.2013.6505726.

[17] U. K. Madawala, M. J. Neath, and D. J. Thrimawithana, "The impact of variations in component values on power-frequency control of bidirectional Inductive Power Transfer systems," en, in *2012 IEEE International Symposium on Industrial Electronics*, Hangzhou, China: IEEE, May 2012, pp. 560–565. DOI: 10.1109/ISIE.2012.6237148.

[18] X. He, R. Wang, J. Wu, and W. Li, "Nature of power electronics and integration of power conversion with communication for talkative power," *Nature Communications*, vol. 11, no. 1, p. 2479, May 2020, Number: 1 Publisher: Nature Publishing Group. DOI: 10.1038/s41467-020-16262-0.

[19] M. Liserre, H. Beiranvand, Y. Leng, R. Zhu, and P. A. Hoeher, "Overview of talkative power conversion technologies," *IEEE Open Journal of Power Electronics*, vol. 4, pp. 67–80, 2023. DOI: 10.1109/OJPEL.2023.3237709.

PCIM Europe 2024, 11– 13 June 2024, Nuremberg    DOI: 10.30420/566262145

# Study of a Multi-Active Bridge Converter for a Domestic Electrical Grid

Abdennour Merrouche[1,2], Thierry Talbert[1], Daniel Matt[2], Thierry Martiré[2], Guillaume Pellecuer[2]

[1] PROMES-CNRS, UPR 8521, Perpignan University Via Domitia, France
[2] IES, UMR 5214, Montpellier University, France

Corresponding author:     Abdennour Merrouche, abdennour.merrouche@promes.cnrs.fr
Speaker:                  Abdennour Merrouche, abdennour.merrouche@promes.cnrs.fr

## Abstract

This article presents the design and development of a three-port, three-phase Multi-Active Bridge (MAB) converter adapted for domestic applications. The primary goal is to establish an autonomous electrical energy storage system utilizing photovoltaic (PV) sources and hydrogen as the storage medium. The converter is strategically designed to facilitate energy exchanges between different renewable energy sources and the electrical loads within the structure. Composed of three three-phase inverters and a transformer for magnetic coupling, each port enables bidirectional power flow and is galvanically isolated. Energy flow regulation is achieved through phase shifting of control commands among the three inverters. This type of converter enables enhanced flexibility and efficiency in integrating renewable energy sources with domestic energy systems, thereby advancing the transition towards sustainable energy applications.

## 1 Introduction

With the acceleration of climate change, the importance of prioritizing the use of renewable energy sources to reduce $CO_2$ emissions from fossil fuels has become imperative [1]. However, despite the considerable advantages of renewable energies, their intermittent nature, dependent on weather conditions, poses a major limitation. To overcome this intermittency, the use of energy storage means appears as a promising solution.

Currently, batteries are the most commonly used energy storage devices, particularly in the context of photovoltaic installations. However, it is essential to explore other energy storage alternatives in order to optimally and sustainably meet the increasing demand for renewable energy.

Hydrogen is emerging as a promising solution for energy storage, due to its remarkable physical properties as well as its non-toxic and non-polluting nature [2]. Furthermore, its abundant availability on Earth makes it an attractive candidate. The storage system would involve the use of an electrolyzer to produce hydrogen through electrolysis, as well as a fuel cell to regenerate electrical energy from the stored hydrogen.

In [3], a simulation of annual photovoltaic production and average annual consumption for four people shows that during the winter period, photovoltaic production is not sufficient to cover domestic consumption, while there is a surplus of production during the summer. The idea would therefore be to convert this surplus produced in summer into hydrogen, in order to generate electrical energy during the winter period. Hydrogen thus presents a viable solution for long-term storage.

Furthermore, the use of these different energy sources requires a power conversion system capable of interfacing these sources. To enable this interconnection, a multi-source converter is necessary. The two most commonly used structures are Fig. 1 : the first is the grid architecture using a DC voltage bus [4], where different conversion stages are applied for each source and the converters are connected to the common voltage bus. The second structure is the multiport structure. Among the multiport topologies, the Multi-Active Bridge (MAB) stands out as an extension of the dual-active bridge (DAB) [5]. The MAB consists of H-bridges

1113

magnetically coupled by a transformer. The advantages of this topology include high power density, bidirectional power transfer, and galvanic isolation between the sources.

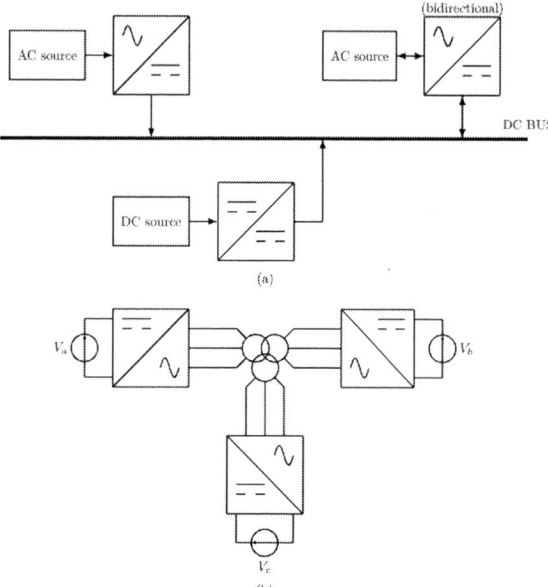

**Fig. 1:** (a) DC bus grid architecture, (b) Multiport structure.

In this study, the converter choice focused on a specific multiport converter: a three-port three-phase multi-active bridge.

# 2 Multi-Active Bridge : Theoretical Study

## 2.1 Structure of the converter

A demonstration of concept viability has been accomplished through the integration of three energy sources utilizing a specialized converter designed for this specific application. In Fig. 2, the schematic illustrates the interconnection of the energy sources via three-phase inverters (A, B and C) and a custom-designed transformer acting as a coupling element.

Each inverter generates a magnetic flux within the transformer. In a balanced state, the magnetic fluxes generated by the three inverters nullify each other, resulting in minimal power exchange. However, by fine-tuning the control parameters of the inverters to manipulate the phase difference between their control commands, we induce a magnetic imbalance, enabling the transfer of

energy from one source to another. This principle, known as "phase-shift" control, forms the basis of our approach.

The transformer plays a crucial role in the converter setup. Its successful implementation demands meticulous modeling and precise dimensioning.

## 2.2 Power Flow Expression

The multiport transformer consists of three legs, each containing three independent phases belonging to a different inverter. Thus, each inverter has one phase in one of the legs of the transformer (see Fig. 3). This configuration enables magnetic coupling between the inductances.

It is crucial to consider both the self-inductances and the mutual inductances to ensure an accurate representation of the system. In Fig. 4 the representation of the self and mutual inductance of phase 1 of inverter A can be observed.
The expression for the voltages and currents of the windings in Eq. (1) is:

$$
\begin{pmatrix} V_{1a} \\ V_{2a} \\ V_{3a} \\ \vdots \\ V_{3c} \end{pmatrix} = \begin{pmatrix} L_{1a} & M_{1a2a} & M_{1a3a} & \dots & M_{1a3c} \\ M_{2a1a} & L_{2a} & M_{2a3a} & \dots & M_{2a3c} \\ M_{3a1a} & M_{3a2a} & L_{3a} & \dots & M_{3a3c} \\ \vdots & \vdots & \vdots & \ddots & \vdots \\ M_{3c1a} & M_{3c2a} & M_{3c3a} & \dots & L_{3c} \end{pmatrix} \cdot \frac{d}{dt} \begin{pmatrix} I_{1a} \\ I_{2a} \\ I_{3a} \\ \vdots \\ I_{3c} \end{pmatrix}
$$

$$(1)$$

Thanks to the symmetrical properties of the self-inductance and the mutual-inductance matrix, we acquire a 9x9 inductance matrix, where each parameter is easily measurable.

The calculation of power flow between inverters, based on phase shift control, necessitates a comprehensive consideration of various factors. Notably, the inclusion of self-inductances and mutual inductances for each phase within the inverter. In the context of phase shift control, understanding and quantifying the self-inductances of each phase are essential as they directly impact the energy exchange dynamics. By integrating these considerations into the calculation framework, a more accurate depiction of power flow dynamics between inverters under phase shift control can be achieved, enabling effective power flow management strategies.

In the referenced study [6], it was demonstrated that the inductance matrix provides a

PCIM Europe 2024, 11– 13 June 2024, Nuremberg        DOI: 10.30420/566262145

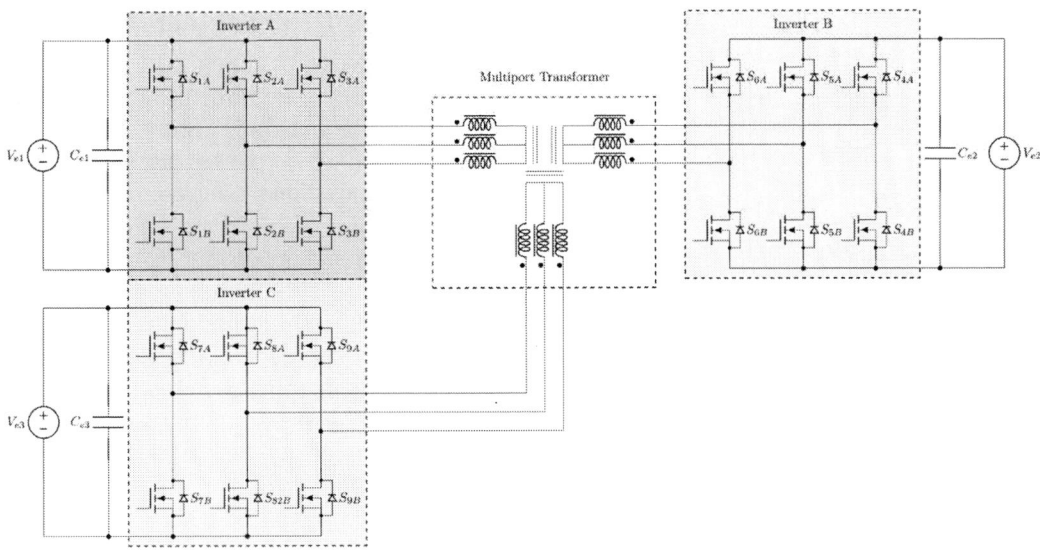

**Fig. 2:** Electrical circuit of the three-port three-phase multi-active bridge.

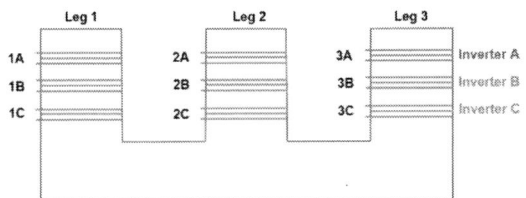

**Fig. 3:** Representation of the windings of the three-port transformer.

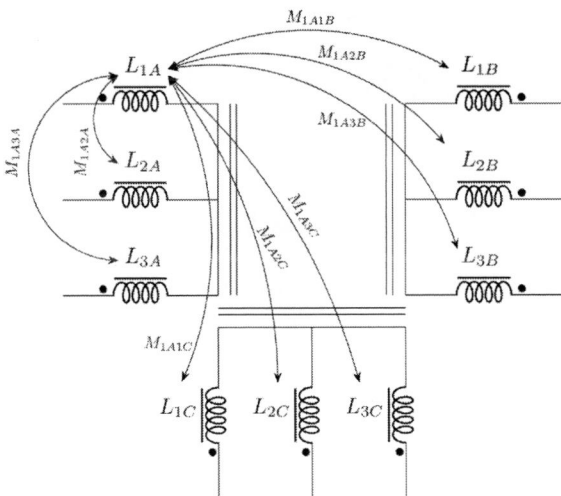

**Fig. 4:** Representation of the self and mutual inductance of phase 1 of inverter A.

method to establish a relationship between the magnetizing inductance and the leakage inductance inherent within the system. In references [7, 8], an N-branch transformer can be represented by an equivalent T-model, as illustrated in Fig. 5. The equivalent T-model [9,10] comprises a magnetic inductance, three leakage inductances, and an ideal transformer. The impedances and voltage sources of the inverters B and C are referred to the primary, which is in this case the inverter A. $L_f$ represents the Leakage inductance and $L_\mu$ the Magnetizing inductance. The equivalent source

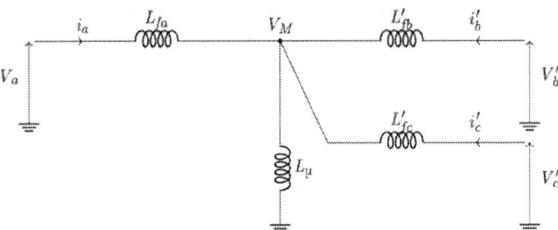

**Fig. 5:** T-model of the three-port transformer.

voltages at the B and C secondaries, as well as the related leakage inductances, are given by Eq. (2-7) [6].

$$K_{ab} = \frac{V_b}{V_a} = \frac{i_a}{i_b} = \frac{M_{C_{bc}}}{M_{C_{ac}}} \tag{2}$$

$$K_{ac} = \frac{V_c}{V_a} = \frac{i_a}{i_c} = \frac{M_{C_{bc}}}{M_{C_{ab}}} \tag{3}$$

1115

$$L_\mu = \frac{M_{C_{ac}}.M_{C_{ab}}}{M_{C_{bc}}} \quad (4)$$

$$L_{fa} = L_{C_a} - \frac{M_{C_{ac}}.M_{C_{ab}}}{M_{C_{bc}}} \quad (5)$$

$$L'_{fb} = \frac{L_{C_b} - \frac{M_{C_{ab}}.M_{C_{bc}}}{M_{C_{ac}}}}{K^2_{ab}} \quad (6)$$

$$L'_{fc} = \frac{L_{C_c} - \frac{M_{C_{ac}}.M_{C_{bc}}}{M_{C_{ab}}}}{K^2_{ac}} \quad (7)$$

$L_C$ "Cyclic self-inductance" represents the interaction between the self-inductance and the mutual inductances within the same inverter.

$M_C$ represents the cyclic mutual inductance, which considers the effect between the mutual inductances within the same leg and those of the same inverter.

If we consider the equivalent T-model, taking phase 1 of each inverter as a reference, it is possible to quickly calculate the power supplied by the three energy sources (see Eq. (8)(9)(10)). $V_M$ represents the midpoint potential, while $V_a$ denotes the input voltage of inverter A and serves as a reference. Additionally, $V_b$ and $V_c$ represent the input voltages of inverters B and C, respectively, phased by angles $\varphi_{ab}$ and $\varphi_{ac}$ relative to $V_a$. Specifically, $V_a = [V_M; 0°]$, $V_b = [V_M; \varphi_{ab}]$, and $V_c = [V_M; \varphi_{ac}]$.

$$P_a = - \frac{3V_m^2.(\frac{sin(\varphi_{ab})}{L'_{fb}} + \frac{sin(\varphi_{ac})}{L'_{fc}})}{2L_{fa}.w.\alpha} \quad (8)$$

$$P_b = - \frac{3V_m^2.(\frac{sin(\varphi_{ab})}{L'_{fb}} + \frac{sin(\varphi_{ac})}{L'_{fc}} - sin(\varphi_{ab}).\alpha)}{2L'_{fb}.w.\alpha} \quad (9)$$

$$P_c = - \frac{3V_m^2.(\frac{sin(\varphi_{ab})}{L'_{fb}} + \frac{sin(\varphi_{ac})}{L'_{fc}} - sin(\varphi_{ac}).\alpha)}{2L'_{fc}.w.\alpha} \quad (10)$$

With $\alpha = (\frac{1}{L_\mu} + \frac{1}{L_{fa}} + \frac{1}{L'_{fb}} + \frac{1}{L'_{fc}})$.

## 2.3  Multiport Transformer Modeling

Effective modeling of the magnetic coupler proves essential during the converter's design phase, as outlined in reference [11]. This modeling process serves two critical purposes: firstly, it validates the converter's operational principles, and secondly, it facilitates optimization of the coupler's geometry for the intended application. Employing an analytical modeling approach based on a geometry-dependent reluctance model is key to achieving accurate representation of the magnetic coupler's behavior. By leveraging such modeling techniques, insights into the dynamic interactions within the magnetic coupler can be gained. These insights enable fine-tuning of design parameters, ensuring optimal performance and efficiency. Furthermore, this modeling approach supports informed decision-making throughout the design process. As depicted in Fig. 6, the transformer comprises upper and lower plates and cylindrical-shaped legs. This symmetrical configuration ensures an even distribution of port powers [12], resulting in equally sized inductances.

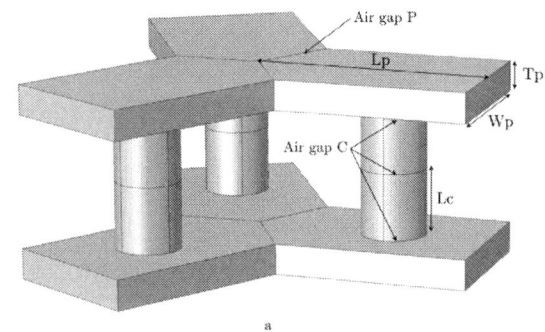

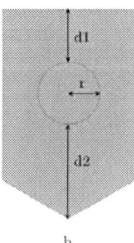

**Fig. 6:** 3D geometry of the three-port transformer.

The dimensions of the magnetic coupler are influenced by various parameters, including the length $L_p$ and width $W_p$ of both upper and lower plates, as well as the thickness $T_p$ and air gap $P$ between them. Additionally, factors such as the air gaps $C$ at the legs of the transformer, the length $L_C$ and radius $r$ of the transformer legs, and the distances $d1$ and $d2$ that define the position of the leg relative to the plate, play significant roles in determining its overall dimensions.

Taking into account a linear model for the magnetic behavior of the 3C90 material, characterized by a relative permeability (μr) of 2300, we can establish a straightforward reluctance model that relies on these parameters (see Fig. 7).

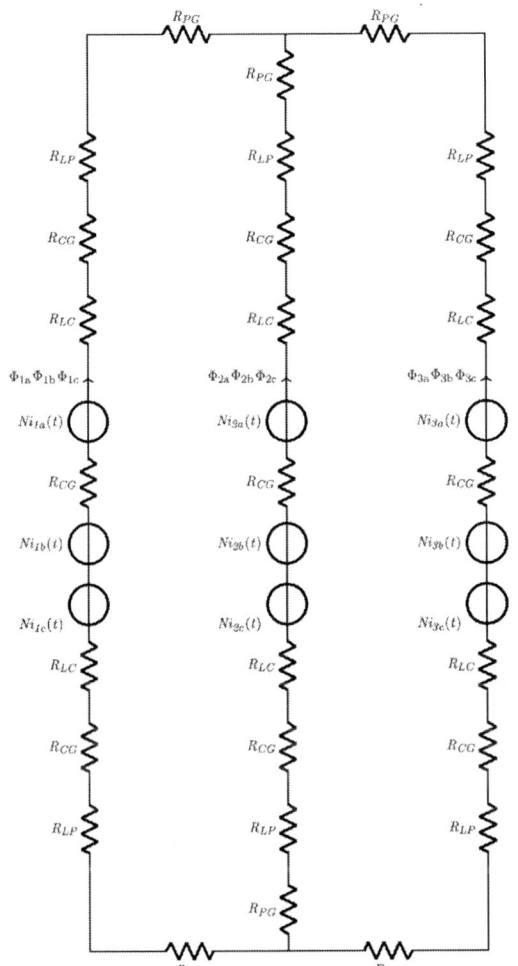

**Fig. 7:** Reluctance model circuit of the three-port transformer.

The reluctances $R_{PG}$ (plates air gap), $R_{LP}$ (plate), $R_{CG}$ (leg air gap), and $R_{LC}$ (leg) are defined by the following Eq. (11-14) :

$$R_{PG} = \frac{P}{\mu_0.S_{Pgap}} \quad (11)$$

$$R_{LP} = \frac{L_P - (d1 + r)}{\mu_0.\mu_r.S_P} \quad (12)$$

$$R_{CG} = \frac{C}{\mu_0.\pi r^2} \quad (13)$$

$$R_{LC} = \frac{L_C}{\mu_0.\mu_r.\pi r^2} \quad (14)$$

$S_{Pgap}$ and $S_P$ represent the cross-sectional area of the plate and the leg, respectively. The reluctance circuit, similar to an electrical circuit, enables the determination of equivalent reluctances seen from each source $N_i(t)$. This facilitates the determination of self-inductances $L_i$ and mutual inductances $M_{ij}$, and the establishment of the theoretical inductance matrix based on the geometric parameters of the transformer. We obtain the following 9x9 inductance matrix (in $\mu H$) :

$$\begin{pmatrix} 3.10 & -1.55 & -1.55 & 3.10 & -1.55 & -1.55 & 3.10 & -1.55 & -1.55 \\ -1.55 & 3.10 & -1.55 & -1.55 & 3.10 & -1.55 & -1.55 & 3.10 & -1.55 \\ -1.55 & -1.55 & 3.10 & -1.55 & -1.55 & 3.10 & -1.55 & -1.55 & 3.10 \\ 3.10 & -1.55 & -1.55 & 3.10 & -1.55 & -1.55 & 3.10 & -1.55 & -1.55 \\ -1.55 & 3.10 & -1.55 & -1.55 & 3.10 & -1.55 & -1.55 & 3.10 & -1.55 \\ -1.55 & -1.55 & 3.10 & -1.55 & -1.55 & 3.10 & -1.55 & -1.55 & 3.10 \\ 3.10 & -1.55 & -1.55 & 3.10 & -1.55 & -1.55 & 3.10 & -1.55 & -1.55 \\ -1.55 & 3.10 & -1.55 & -1.55 & 3.10 & -1.55 & -1.55 & 3.10 & -1.55 \\ -1.55 & -1.55 & 3.10 & -1.55 & -1.55 & 3.10 & -1.55 & -1.55 & 3.10 \end{pmatrix}$$

In the analysis of a magnetic circuit within the framework of the linear approximation of Hopkinson [13], it involves perfect magnetic circuits. These circuits are linear (with constant relative permeability μr) and free from magnetic leaks (where all the flux created by the windings appears in the magnetic circuit). Therefore, it is not possible to determine leakage inductances from this theoretical inductance matrix.

To complete the modeling of the transformer, finite element analysis allows for electromagnetic simulation of the transformer in 3D. Thanks to the software COMSOL Multiphysics we can obtain the magnetic induction in the transformer under nominal operating conditions (see Fig. 8).

The voltage applied across the terminals of the transformer windings is 33V at a frequency of 50kHz. The maximum inductance in the transformer is approximately 200mT, which falls within the linear zone of the $B(H)$ curve of the 3C90 material. This confirms that it is well below the saturation point, and that the transformer operates within the desired parameters and ensures its effective functionality.

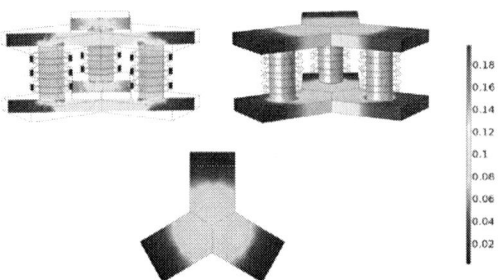

**Fig. 8:** 3D simulation of the magnetic induction (mT) of the three-port transformer.

This electromagnetic simulation also facilitates the extraction of the inductance matrix by employing the method of measuring the open-circuit voltage at each winding when a driving voltage is applied to the other winding. Equations (15) and (16) allow for the calculation of the self and mutual inductances.

$$L_i = \frac{1}{w} Im\left(\frac{V_{coil\,i}}{I_{coil\,i}}\right) \tag{15}$$

$$M_{ij} = \frac{1}{w} Im\left(\frac{V_{coil\,j}}{I_{coil\,i}}\right) \tag{16}$$

We obtain the following 9x9 inductance matrix (in $\mu H$) :

$$\begin{pmatrix} 3.2432 & -1.1384 & -1.1385 & 3.0203 & -1.1360 & -1.1360 & 2.9065 & -1.1361 & -1.1361 \\ -1.1384 & 3.2433 & -1.1387 & -1.1360 & 3.0206 & -1.1363 & -1.1361 & 2.9067 & -1.1364 \\ -1.1385 & -1.1387 & 3.2432 & -1.1360 & -1.1363 & 3.0204 & -1.1361 & -1.1364 & 2.9066 \\ 3.0203 & -1.1360 & -1.1360 & 3.1753 & -1.1348 & -1.1348 & 3.0177 & -1.1360 & -1.1360 \\ -1.1360 & 3.0206 & -1.1363 & -1.1348 & 3.1754 & -1.1351 & -1.1360 & 3.0180 & -1.1363 \\ -1.1360 & -1.1363 & 3.0204 & -1.1348 & -1.1351 & 3.1754 & -1.1360 & -1.1362 & 3.0178 \\ 2.9065 & -1.1361 & -1.1361 & 3.0177 & -1.1360 & -1.1360 & 3.2432 & -1.1384 & -1.1384 \\ -1.1361 & 2.9067 & -1.1364 & -1.1360 & 3.0180 & -1.1362 & -1.1384 & 3.2435 & -1.1387 \\ -1.1361 & -1.1364 & 2.9066 & -1.1360 & -1.1363 & 3.0178 & -1.1384 & -1.1387 & 3.2432 \end{pmatrix}$$

With this matrix, we can either determine the leakage inductances and calculate the power flow $P = f[\phi_{AB}, \phi_{Ac}]$ from Eq. (8-10) or simulate the converter on Ltspice by coupling the inductances using the coupling coefficient $k_{ij}$ Eq. (17).

$$k_{ij} = \frac{M_{ij}}{\sqrt{L_i L_j}} \tag{17}$$

## 2.4 Simulation of the Converter

With simulation we can accurately replicate and analyze the behavior of the converter. Through comprehensive simulation and modeling techniques, we can assess various performance metrics, including efficiency, voltage regulation, and power transfer characteristics.

We can compare the results obtained from different methods. Tables 1 and 2 present the power flow between the inverters for different phase-shifts, with Table 1 corresponding to a phase-shift of $\phi_{AB} = \phi_{AC} = -10°$, and Table 2 corresponding to a phase-shift of $\phi_{AB} = \phi_{AC} = -20°$. Comparing these results provides valuable insights on the power exchange dynamics between the inverters.

| | $P_A(W)$ | $P_B(W)$ | $P_C(W)$ |
|---|---|---|---|
| Theoretical | -1979 | 1320 | 495 |
| T-model | -1979 | 1320 | 495 |
| LTspice | -2048 | 1424 | 458 |
| COMSOL | -1976 | 1457 | 467 |

**Tab. 1:** Power flow for $\phi_{AB} = \phi_{AC} = -10°$

| | $P_A(W)$ | $P_B(W)$ | $P_C(W)$ |
|---|---|---|---|
| Theoretical | -4032 | 2395 | 1021 |
| T-model | -4032 | 2395 | 1021 |
| LTspice | -4165 | 2587 | 955 |
| COMSOL | -3907 | 2739 | 973 |

**Tab. 2:** Power flow for $\phi_{AB} = \phi_{AC} = -20°$

The results obtained from these different methods reveal that when negative phase-shifts are applied to inverters B and C, inverter A supplies power to inverters B and C. The negative sign of the power value indicates that it supplies power, whereas a positive value indicates that it receives power. This analysis sheds light on the directional flow of power within the system and underscores the significance of phase-shift control in regulating energy exchange between the inverters.

In the case of the Theoretical and the T-model we obtain the same results. Despite the similarity in magnitudes, differences emerge between the LTspice, COMSOL, Theoretical, and T-model results. This discrepancy stems from the theoretical and T-models not accounting for magnetic flux leakage, as they operate within the linear approximation of Hopkinson, assuming perfect magnetic circuits. Consequently, these models may not fully capture the complexities of real-world behavior, resulting in variations in the simulated results compared to those obtained from software simulations like LTspice and COMSOL.

Until this stage, all simulations have utilized a sinu-

PCIM Europe 2024, 11– 13 June 2024, Nuremberg          DOI: 10.30420/566262145

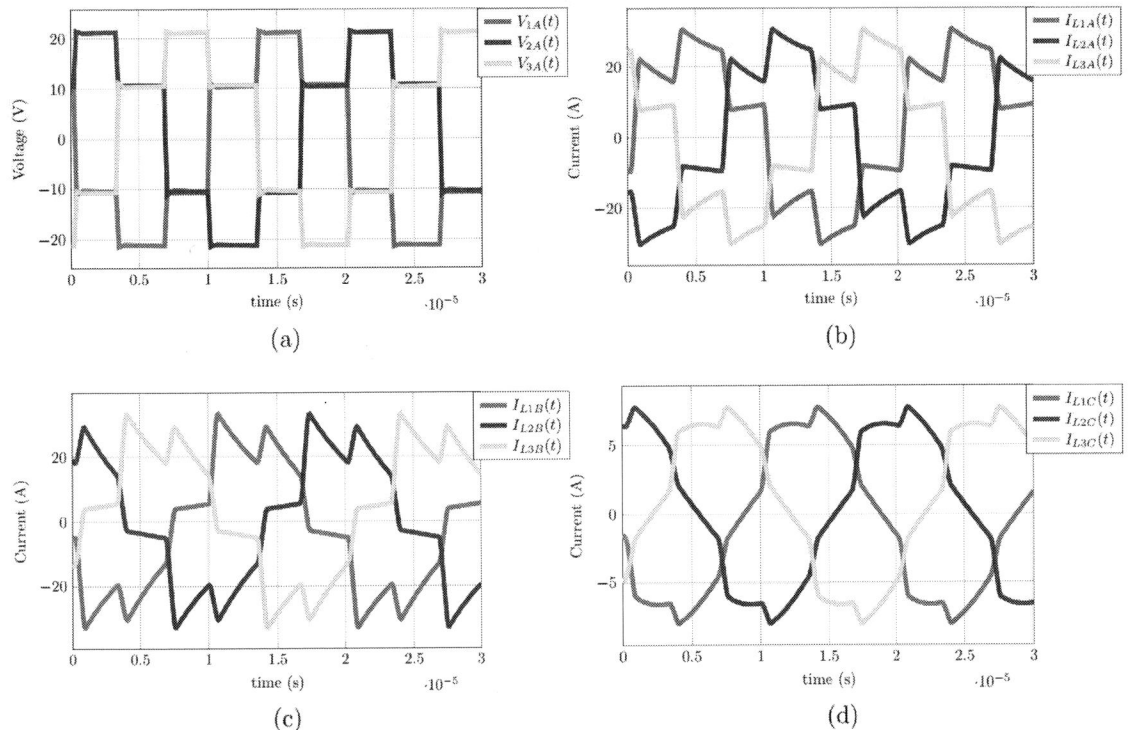

**Fig. 9:** Simulation results. (a) voltages of the three phases of inverter A. (b) currents in the windings of inverter A. (c) currents in the windings of inverter B. (d) currents in the windings of inverter C.

soidal waveform voltage. However, the converter will employ a six-step control. In Fig. 9, the waveforms and amplitudes of the voltage across the transformer windings, as well as the currents flowing through each transformer winding, are illustrated. The simulations conducted on LTspice using six-step control enable us to anticipate the behavior of the converter prior to experimental testing. This provides us with the opportunity to analyze the voltage and current waveforms for different phase-shifts.

## 3  Experimental Testing of the Converter

Figure 10 the transformer has been fabricated and assembled, featuring plates and legs with cylindrical shapes, constructed using 3C90 material. Within the transformer, the PCBs of the three inverters integrate the windings, each coil comprising $N = 2$ turns. These PCBs are arranged in a stacked configuration. The control board regulates the switches of the inverters, facilitated by rotary encoders, allowing for precise adjustment of the switching frequency and phase shifts $\phi_{AB}$ and $\phi_{AC}$. Experimental testing is conducted under

open-loop control conditions.

Following the completion of the converter, the actual inductance matrix was measured using an impedance analyzer. Subsequently, the measured matrix was compared with both the theoretical matrix extracted from the reluctance model and the finite element analysis matrix extracted from the electromagnetic simulation using COMSOL software. This comprehensive

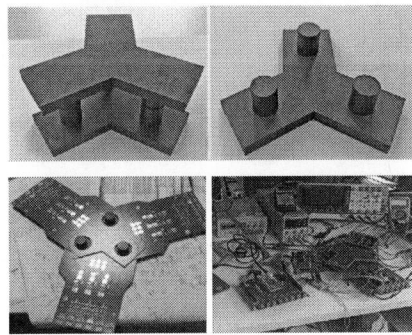

**Fig. 10:** Experimental setup of the three-port three-phase Multi-Active Bridge converter.

comparison enabled a thorough validation of

1119

the theoretical predictions and simulation results against empirical measurements. Such validation is essential for ensuring the accuracy and reliability of the theoretical and simulation-based approaches in predicting the transformer's behavior and performance under various operating conditions. We obtain the following 9x9 inductance matrix (in $\mu H$) :

$$
\begin{pmatrix}
3.32 & -0.91 & -0.96 & 3.00 & -1.08 & -1.10 & 2.86 & -1.10 & -1.12 \\
-0.93 & 3.55 & -1.22 & -1.05 & 3.24 & -1.35 & -1.08 & 3.08 & -1.37 \\
-0.99 & -1.23 & 3.76 & -1.09 & -1.36 & 3.41 & -1.10 & -1.39 & 3.28 \\
3.02 & -1.06 & -1.10 & 3.19 & -1.10 & -1.12 & 3.03 & -1.08 & -1.13 \\
-1.08 & 3.22 & -1.36 & -1.10 & 3.35 & -1.40 & -1.09 & 3.24 & -1.38 \\
-1.11 & -1.36 & 3.40 & -1.10 & -1.35 & 3.52 & -1.12 & -1.35 & 3.42 \\
2.91 & -1.08 & -1.13 & 3.00 & -1.12 & -1.15 & 3.22 & -1.11 & -1.14 \\
-1.10 & 3.12 & -1.39 & -1.12 & 3.20 & -1.39 & -1.07 & 3.40 & -1.38 \\
-1.13 & -1.39 & 3.30 & -1.15 & -1.41 & 3.39 & -1.10 & -1.35 & 3.68
\end{pmatrix}
$$

The initial tests of this converter were conducted with an input voltage of $V_{in} = 20V$ for each inverter, employing a switching frequency of $F_S = 50kHz$, and allowing for phase shifts $\phi_{AB}$ and $\phi_{AC}$ to vary within the range of $-10° \leq \phi \leq 13°$. These tests aimed to explore the converter's behavior under different phase-shift configurations and provide insights into its operational characteristics across a broad spectrum of operating conditions.

We can graphically represent the relationship between exchanged powers and phase shifts in Fig. 11. This graphical representation allows for a clear visualization of how changes in phase shift affect the power exchange dynamics between the inverters. By analyzing this evolution, we can gain valuable insights into the optimal phase-shift settings for maximizing power transfer efficiency and achieving desired operational performance.

The power mapping of the converter illustrates the validity of the Multi-Active Bridge's operating principle and showcases its capability to regulate energy exchange between different sources through the adjustment of phase shifts between the inverters. This validation underscores the effectiveness of the converter's design in facilitating dynamic control and efficient management of power flow, highlighting its potential for various applications requiring precise energy distribution and control.
In Fig. 12, the waveforms and amplitudes of the voltage across the transformer windings, as well as the currents, depict the maximum currents for each inverter, with phase shifts ranging from -10° to 13°. Each inverter is configured with a specific

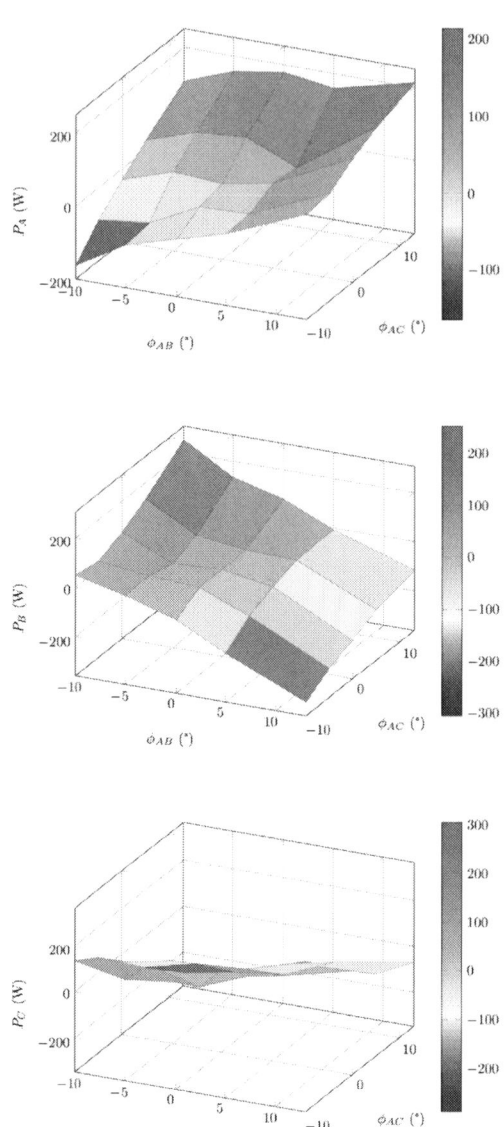

**Fig. 11:** Experimental power mapping of the three-port three-phase Multi-Active Bridge converter.

phase shift combination to achieve its peak current. For inverter A, the current reaches its maximum when both $\phi_{AB}$ and $\phi_{AC}$ are set to -10°. In the case of inverter B, the current peaks at $\phi_{AB} = 13$ and $\phi_{AC} = 10$. Similarly, for inverter C, the current reaches its peak at $\phi_{AB} = -10$ and $\phi_{AC} = 13$. This detailed analysis provides valuable insights into the optimal phase shift configurations for maximizing current flow in each inverter, facilitating informed decision-making during system optimization and

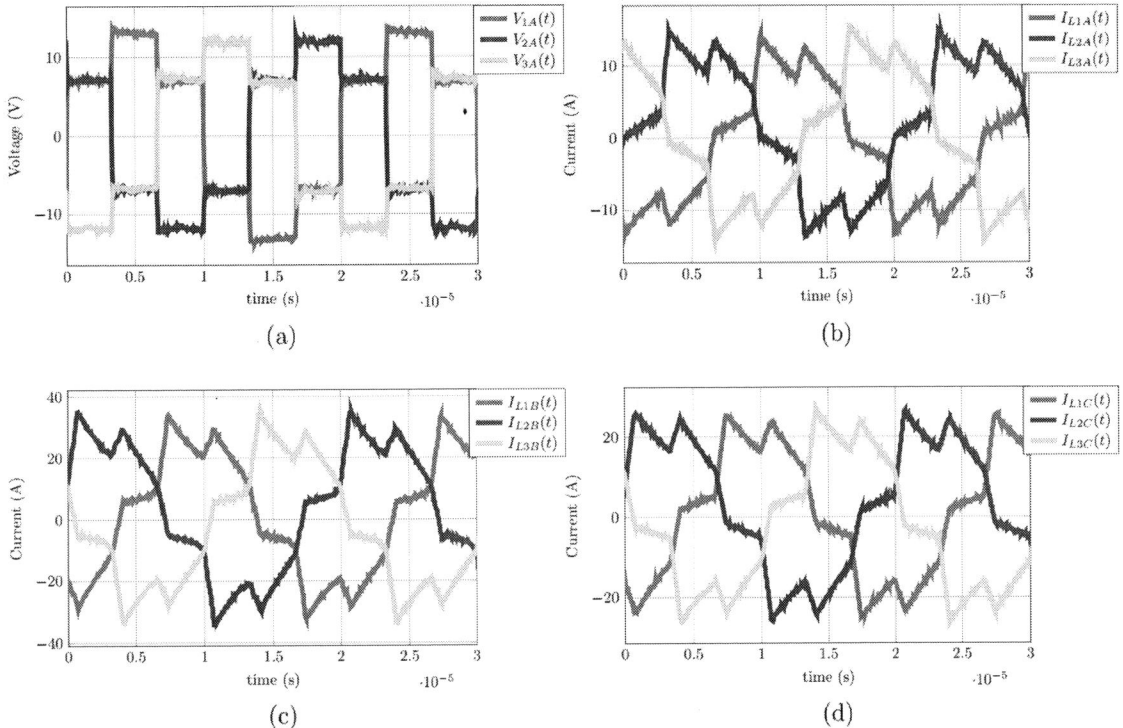

**Fig. 12:** Experimental results. (a) voltages of the three phases of inverter A. (b) currents in the windings of inverter A. (c) currents in the windings of inverter B. (d) currents in the windings of inverter C.

control design. The Multi-Active-Bridge relies heavily on the low values of leakage and magnetizing inductances of the transformer, typically in the microhenry range.

## 4   Conclusion

This study presents the design and implementation of a three-port, three-phase Multi-Active Bridge (MAB) converter, designed to enhance the integration of renewable energy sources into domestic electrical grids. Through theoretical insights, simulations, and practical experiments, we demonstrated the converter's efficient management of bidirectional energy flow, highlighting its potential to address renewable energy's intermittency. The converter's phase-shift control and magnetic coupling transformer prove effective in optimizing energy transfer, promising significant advancements in sustainable domestic energy management. Our work lays a foundation for future developments in smart grid technologies, emphasizing the importance of renewable energy integration for a sustainable energy future.

## 5   Acknowledgements

I would like to express my gratitude to Labex Solstice for providing the funding that supported the research presented in this article.

## References

[1] S. Adams and A. O. Acheampong, "Reducing carbon emissions: The role of renewable energy and democracy," Journal of Cleaner Production, vol. 240, p. 118245, Dec. 2019, doi: 10.1016/j.jclepro.2019.118245.

[2] O. Wildgruber, "Hydrogen as Energy Source: An Introduction," Energy Environment, vol. 17, no. 2, pp. 275–279, Mar. 2006, doi: 10.1260/095830506777070114.

[3] A. Merrouche et al., "Convertisseur multi-sources solaire pour pile à hydrogène," in Journées Nationales du Photovoltaïque 2022, Dourdan, France, Nov. 2022. Accessed: Apr. 08, 2024. [Online]. Available: https://hal.science/hal-03887957

[4] H. Tao, A. Kotsopoulos, J. L. Duarte, and M. a. M. Hendrix, "Family of multiport bidi-

rectional DC–DC converters," IEE Proceedings - Electric Power Applications, vol. 153, no. 3, pp. 451–458, May 2006, doi: 10.1049/ip-epa:20050362.

[5] H. Qin, H. Zhang, M. Liu, and C. Ma, "Comparison of Different Multi-winding Transformer Models in Multi-port AC-coupled Converter Application," in IECON 2021 – 47th Annual Conference of the IEEE Industrial Electronics Society, Toronto, ON, Canada: IEEE, Oct. 2021, pp. 1–6. doi: 10.1109/IECON48115.2021.9589283.

[6] G. Pellecuer, T. Martire, M. Petit and B. Loyer, "On-Board Power Management in a Marine Autonomous Surface Vehicle (ASV): Multi-Port Transformer Design," PCIM Europe 2022; International Exhibition and Conference for Power Electronics, Intelligent Motion, Renewable Energy and Energy Management, Nuremberg, Germany, 2022, pp. 1-9, doi: 10.30420/565822275.

[7] H. Tao, A. Kotsopoulos, J. L. Duarte and M. A. M. Hendrix, "A Soft-Switched Three-Port Bidirectional Converter for Fuel Cell and Supercapacitor Applications," 2005 IEEE 36th Power Electronics Specialists Conference, 2005, pp. 2487-2493,doi: 10.1109/PESC.2005.15819

[8] C. Sun, N. H. Kutkut, D. W. Novotny, et D. M. Divan, " General equivalent circuit of a multi-winding co-axial winding transformer ", in IAS

'95. Conference Record of the 1995 IEEE Industry Applications Conference Thirtieth IAS Annual Meeting, 1995, vol. 3, p. 2507-2514 vol.3. doi: 10.1109/IAS.1995.530622.

[9] J. G. Hayes, N. O'Donovan, and M. G. Egan, "The extended T model of the multi-winding transformer," in 2004 IEEE 35th Annual Power Electronics Specialists Conference (IEEE Cat. No.04CH37551), Aachen, Germany: IEEE, 2004, pp. 1812–1817. doi: 10.1109/PESC.2004.1355391.

[10] H. Shi-Ping, "Problems in Analysis and Design of Switching Regulators," California Institute of Technology, 1979.

[11] J. Yang, G. Buticchi, C. Gu, S. Gunter, H. Zhang, and P. Wheeler, "A Generalized Input Impedance Model of Multiple Active Bridge Converter," IEEE Trans. Transp. Electrific., vol. 6, no. 4, pp. 1695–1706, Dec. 2020, doi: 10.1109/TTE.2020.2986604.

[12] Neubert, M. (2020). Modeling, synthesis and operation of multiport-active bridge converters (Doctoral dissertation, Dissertation, Rheinisch-Westfälische Technische Hochschule Aachen, 2020).

[13] D. W. Jordan, "The Magnetic Circuit Model, 1850–1890: The Resisted Flow Image in Magnetostatics," The British Journal for the History of Science, vol. 23, no. 2, pp. 131–173, Jun. 1990, doi: 10.1017/S0007087400044733.

PCIM Europe 2024, 11– 13 June 2024, Nuremberg        DOI: 10.30420/566262146

# Fabrication Development for Gate Driver Embedded Double-Sided Cooling SiC Power Module for Electric Vehicle Application

Yuyang Wang[1], Riya Paul[1], Hao Chen[1], Cheng Tang[1], Anna Corbitt[1], Weiping Fu[1], H. Alan Mantooth[1]

[1] University of Arkansas, USA

Corresponding author: Yuyang Wang, yuyangw@uark.edu

Speaker: Anna Corbitt, amcorbit@uark.edu

## Abstract

High power-density packaging power modules are essential components within electric vehicle inverters. Heterogeneous packaging techniques significantly influence system performance, especially concerning the integration of gate drivers and decoupling capacitors. This paper presents a comprehensive approach to fabricating a double-sided cooling power integrated with gate drivers, decoupling capacitors, and a low temperature co-fired (LTCC) interposer. The proposed module's junction-to-case thermal resistance is 0.05 K/W and power commutation loop inductance is 12 nH. The fabrication process and electrical performance of the proposed module has been validated by Hi-Pot test and double pulse test.

## 1 Introduction

In the evolving field of power electronics, enhancing efficiency, reliability, and optimizing heat dissipation has become paramount. With the escalating demand for high-power electronic devices such as electric vehicles, the need for advanced cooling solutions has reached unprecedented urgency [1-6]. Addressing these challenges, double-sided cooling modules with outstanding heat dissipation capabilities, lower stray inductance in power circuitry, and higher power density design is required [7-11]. This paper delves into an in-depth exploration of the detailed fabrication process of a proposed double-sided cooling module. The first section introduces the fundamental structure of the proposed power module. The second section provides a comprehensive fabrication process for the proposed power module. The third section presents the electric performance characterization of the module.

## 2 Overview of Proposed 1.2 kV Heterogenous Packaging Power Module Fabrication

Fig. 1 shows a novel architecture for a half-bridge topology double-sided cooling SiC power module. This design integrates components such as gate drivers, decoupling capacitors, low-temperature co-fired ceramic (LTCC) interposer, and temperature sensors using a heterogeneous integration approach. In each switching position, two 1.2 kV SiC MOSFET (CPM312000013A) devices are connected in parallel. Integrating the gate driver board not only enhances the module's robustness but also shields the control circuitry from environmental interference, reducing gate loop inductance. Furthermore, integrating decoupling capacitors within the power module reduces loop inductance. The power loop inductance of the power

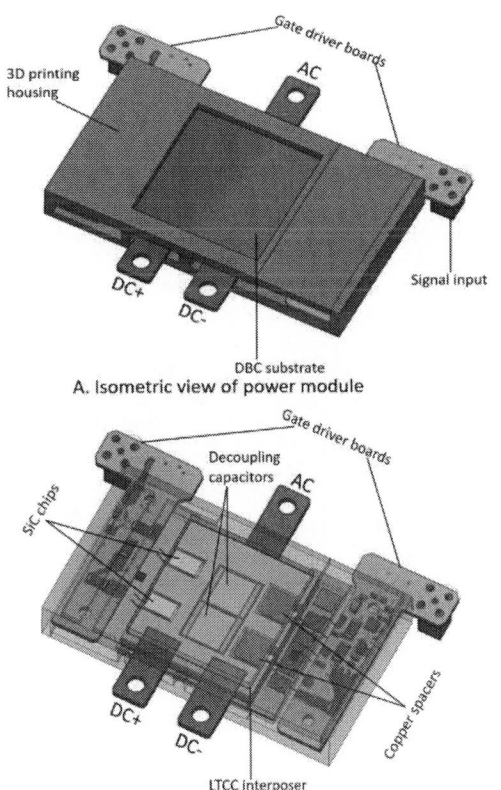

A. Isometric view of power module

B. Isometric view of power module with transparent housing

PCIM Europe 2024, 11– 13 June 2024, Nuremberg          DOI: 10.30420/566262146

module is only 1.76 nH when integrated with de-coupling capacitors which is validated in ANSYS Q3D.

**Fig. 1**    1.2 kV DSC power module integrated with decoupling capacitors, LTCC interposer, and gate driver boards.

## 3    Fabrication Flow of Double-sided Cooling Module

Fig. 2 illustrates the fabrication flow chart of the proposed power module.

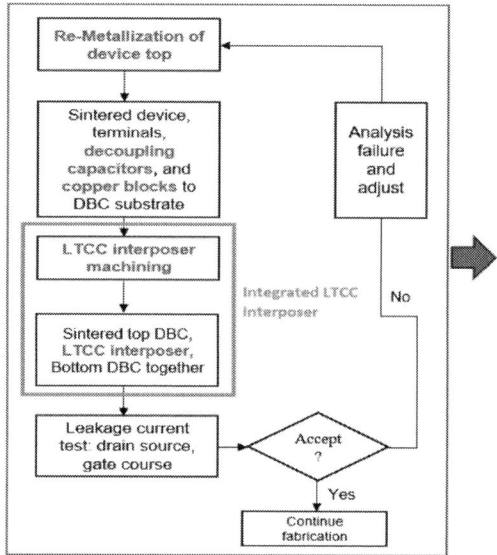

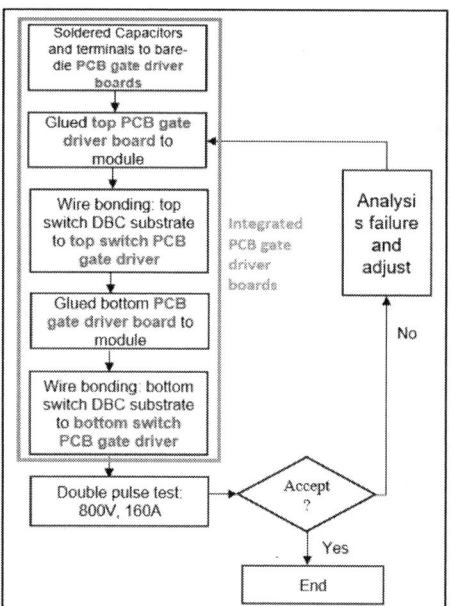

**Fig. 2**    Fabrication flow charts of the 1.2 kV DSC power module with integrated LTCC interposer, decoupling capacitors, and gate drivers

In Fig.2, the initial segment of the fabrication flow chart delineates the procedure for integrating decoupling capacitors and LTCC interposer into the module. The subsequent section of the flow chart outlines the process involving the preparation and integration of PCB gate driver boards.

Fig. 3 shows the specific fabrication process flow for the proposed power module, including steps such as re-metallization of the top source pads of devices, silver sintering process, wire bonding, LTCC interposer fabrication, encapsulation, and housing leads glue.

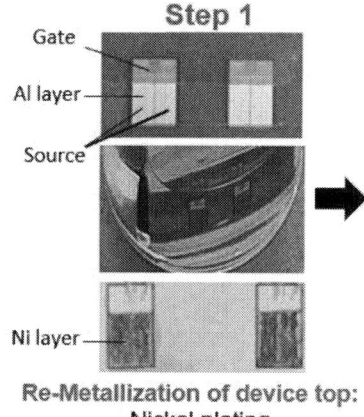

**Re-Metallization of device top:**
Nickel plating

**1st sintering:** Devices, decoupling capacitors, terminals, copper spacers, temp. sensor

1124

### Step 3

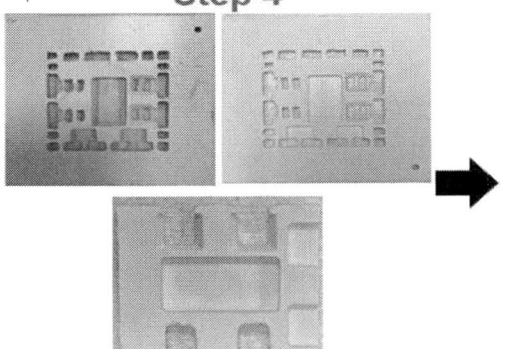

1st wire bonding:
Gate and Kelvin Source

### Step 4

LTCC interposer
machining:
9K7 LTCC interposer

### Step 5

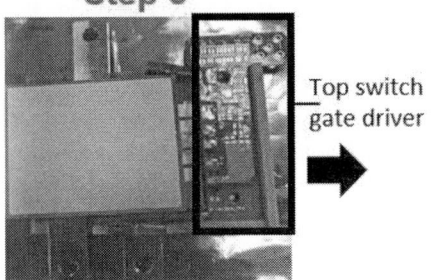

2nd Sintering: Bottom
DBC substrate,
LTCC interposer, top DBC
substrate

### Step 6

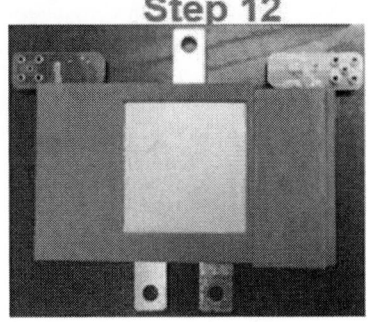

Top switch
gate driver

1st glue: top switch gate
driver, housing part A

### Step 7, 8

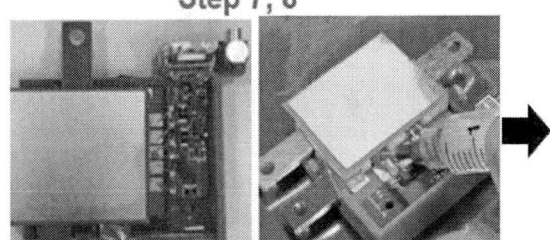

2nd wire-bonding: (top switch DBC
substrate to top switch gate driver),
1st encapsulation

### Step 9

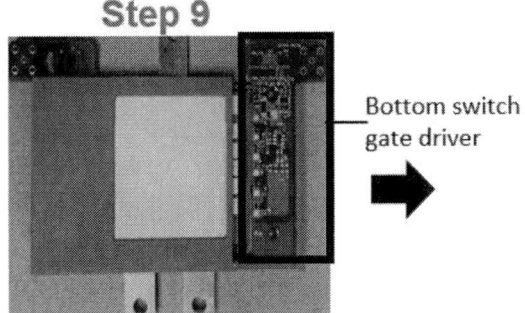

Bottom switch
gate driver

2nd glue: bottom switch gate
driver, housing part B

### Step 10,11

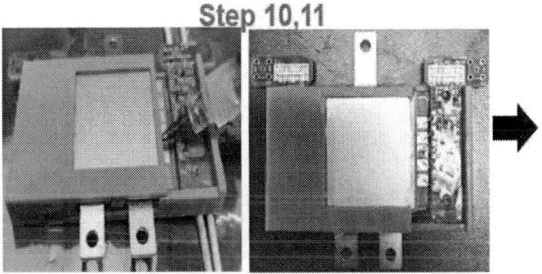

3rd wire-bonding: (bottom switch
DBC substrate to bottom switch gate driver),
2nd encapsulation

### Step 12

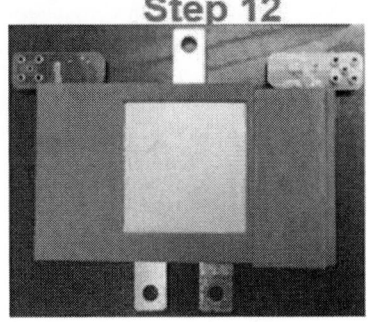

3rd glue: Housing part C

**Fig. 3** Specific fabrication process flow of the 1.2 kV DSC power module with integrated gate drivers and decoupling capacitors

Step 1: re-metallization of device top. The source surface of bare die is aluminum layer, which can't be sintered with silver. Electroless nickel plating was applied to re-metallization top metal pads. The thickness of the plated nickel layer is 2 μm. After wet plating, the chips were baked at 120°C for 2 hours to reduce the chip leaking current. Step 2: 1st silver sintering process. The DBC substrate was aligned by CNC machined graphite fixturing. Silver paste was utilized for chip attachment, copper terminals attachment, along with copper spacers and temperature sensor attachment. Two 2 kV, 0.056 μF high voltage capacitors were selected to reduce parasitic inductance in the power loop in the module. Subsequently, these parts were sintered at 210°C for 7 hours in N2 atmosphere. Step 3: 1st wire bonding. The 12-mil aluminum-wire Hesse wire bonder was applied to connect the gate and source signal loop. Step 4: LTCC interposer machining. Tormach CNC 1100 was utilized for machining the 14-layer laminated 9k7 low-temperature cofired at 950°C ceramic (LTCC) as insulation support. Step 5: 2nd sintering. The two DBC substrates were bonded together with LTCC interposer as medium layer. Step 6: 1st glue. The PCB top gate driver was integrated onto housing part A using instant adhesive Permabond 825. The integrated components require 24 hours room temperature to cure the bonding of glue. Step 7: 2nd wire bonding. The Hesse wire bonder was utilized to connect signal loop between the top switch DBC substrate and the PCB gate driver board with 12-mil aluminum wire. Step 8: 1st encapsulation. Kapton tape was used to seal the gaps in housing part A, preventing the encapsulant from leaking out. Soft silicone encapsulant R-2188 A and B were mixed in a 1:1 ratio and stirred for ten minutes. 50 mins vacuum degassing was required to debubble within module. Afterward, the assembly was baked at 60°C for 11 hours. Step 9: 2nd glue. The bottom switch DBC and PCB gate driver were integrated onto housing part B using instant adhesive Permabond 825. Step 10: 3rd wire bonding. The Bottom switch gate driver and DBC substrate were connected using the Hesse wire bonder. Step 11: 2nd encapsulation: similar to Step 8. Step 12 3rd glue: Integrate the final housing part C onto the module.

## 4 Characterization of Proposed Power Module

The insulation performance was validated by the Hi-Pot test with 81 nA Vds leakage current at 800 V and 1.2 nA Vgs leakage current at 15 V. The dynamic performance of the power module is verified by DC 800V and 164 A load current double pulse test, as shown in Fig. 4. These measurements are conducted using the Tektronix differential probes for voltage measurements, and the Rogowski coils for current measurement. In Channel 3, the drain-source voltage output displays voltage overshoot of 86V with Vgs 15V and gate resistor 0 Ω.

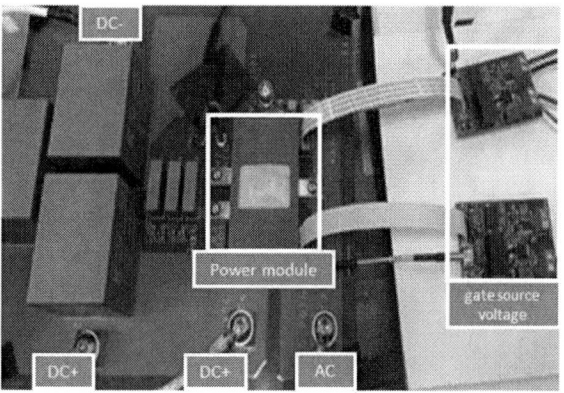

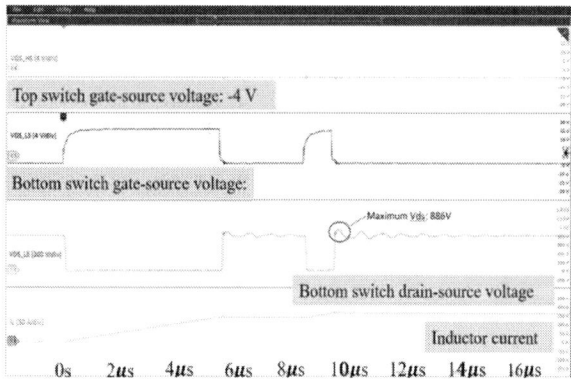

**Fig. 4** Double pulse test set up and the waveform at DC 800 V.

According to the experimental results, the overshoot voltage of the power module with integrated decoupling capacitors is 86 V, while the overshoot voltage of the power module without integrated decoupling capacitors is 310 V [12]. Compared with the power module without decoupling capacitors, the overshoot voltage of the power module with integrated decoupling capacitors is reduced by 72.3%. This shows that the power modules with integrated decoupling capacitors have better electrical performance.

# 5 Conclusion

This paper proposed a heterogeneous power module fabrication approach for integration gate drivers, decoupling capacitors, and low temperature co-fired (LTCC) interposer within module. The proposed fabrication process was validated successfully, and dynamic performance of module was characterized under 800 V DC voltage with 0 Ω gate resistance.

# References

[1] Paul, Riya, et al. "Fabrication of a double-sided cooled half-bridge silicon carbide power module for electric vehicles." PCIM Europe 2022; International Exhibition and Conference for Power Electronics, Intelligent Motion, Renewable Energy and Energy Management. VDE, 2022.

[2] She, Xu, et al. "Review of silicon carbide power devices and their applications." IEEE Transactions on Industrial Electronics 64.10 (2017): 8193-8205.

[3] Y. Xu, "Development of advanced SiC power modules", PhD dissertation, North Carolina State University, Ann Arbor, 2017.

[4] G. Bower, C. Rogan, J. Kozlowski, and M. Zugger, "SiC power elec- tronics packaging prognostics," in Proc. IEEE Aerosp. Conf., 2008, pp. 1–12, doi: 10.1109/AERO.2008.4526605.

[5] Chen, Hao, et al. "Design and optimization of SiC MOSFET wire bondless power modules." 2020 IEEE 9th international power electronics and motion control conference (IPEMC2020-ECCE Asia). IEEE, 2020.

[6] Yadlapalli, Ravindranath Tagore, et al. "A review on energy efficient technologies for electric vehicle applications." Journal of Energy Storage 50 (2022): 104212.

[7] C. Ding, H. Liu, K. D. T. Ngo, R. Burgos and G. -Q. Lu, "A Double-Side Cooled SiC MOSFET Power Module With Sintered-Silver Interposers: I-Design, Simulation, Fabrication, and Performance Characterization," in IEEE Transactions on Power Electronics, vol. 36, no. 10, pp. 11672-11680, Oct. 2021.

[8] F. Yang, Z. Wang, Z. Liang and F. Wang, "Electrical Performance Advancement in SiC Power Module Package Design With Kelvin Drain Connection and Low Parasitic Inductance," in IEEE Journal of Emerging and Selected Topics in Power Electronics, vol. 7, no. 1, pp. 84-98, March 2019.

[9] Yang, Yuhang, et al. "Automotive power module packaging: Current status and future trends." IEEE Access 8 (2020): 160126-160144.

[10] Chen, Zibo, and Alex Q. Huang. "High Performance SiC Power Module Based on Repackaging of Discrete SiC Devices." IEEE Transactions on Power Electronics (2023).

[11] J. Marcinkowski, "Dual-sided cooling of power semiconductor mod- ules," in Proc. PCIM Europe Conf., 2014, pp. 1179–1185.

[12] Paul, Riya, Ayesha Hassan, and H. Alan Mantooth. "A Double-sided Cooled Power Module with Embedded Decoupling Capacitors." IEEE Journal of Emerging and Selected Topics in Power Electronics (2024).

PCIM Europe 2024, 11– 13 June 2024, Nuremberg   DOI: 10.30420/566262147

# Printed Circuit Embedding of Prepackaged 150V Power MOSFETs in a Portable Welding Application

Thomas Gebhard [1], Franz Musil [2] , Manuel Schumann [3]

[1] Infineon Austria AG, Austria

[2] Fronius International GmbH, Austria

[3] Unimicron Germany GmbH, Germany

Corresponding author:   Thomas Gebhard, thomas.gebhard@infineon.com
Speaker:                Manuel Schumann, manuel.schumann@unimicron.de

## Abstract

This paper presents a full system implementation of 150V OptiMOS™ power MOSFETs embedded inside the printed circuit board (PCB) for a portable welding station "Fronius – Ignis Battery™ 150/750".

A significant reduction in electromagnetic conducted emission and voltage overshoot was observed for the synchronous buck stage. Different laminate materials were evaluated to optimize the thermal management. Several different qualification tests had been performed with pass, on the MOSFETs embedded inside the PCB, to demonstrate reliability readiness for volume production.

## 1   Outline

### 1.1   Introduction

The interface between semiconductor switches and the system becomes increasingly a bottleneck in electronic systems. Advanced semiconductor technologies hit the limit in terms of packaging within several different aspects. For low and ultra-low semiconductor voltages high current densities require shortest possible electrical interconnection between components on the PCB, to keep joule heating and stray inductances to a minimum.

**Fig. 1** Infineon's dual-phase power modules TDM22544D & TDM22545D based on a chip embedding package with a fully integrated power-stage.

Figure 1 shows a point of load system solution for computing recently launched by Infineon [1]. The semiconductors in this system are prepackaged inside laminate materials. There are several advantages with this approach. The thermal expansion coefficient between the printed circuit board (PCB) in the module and on the system-PCB are very similar, offering outstanding reliability characteristics. The vertical current flow through the semiconductor power switches minimizes joule heating in the substrate. The fully exposed pads on the top provide a direct thermal path to the heatsink for effective cooling.

Moving to systems with higher operation voltages, minimum creepage distance requirements can reduce the freedom for package fanout/ footprint and optimal arrangement on the PCB. The very harsh environment in an industrial environment, e.g. in a portable welding station "Fronius – Ignis Battery™ 150/750" (fig. 2) in terms of pollution and mechanical shock, requires special attention to reliability and rugged system design [2]. By embedding the semiconductor switches into the PCB those limitation can be overcome.

PCIM Europe 2024, 11– 13 June 2024, Nuremberg          DOI: 10.30420/566262147

**Fig. 2**    Fronius – Ignis Battery™ 150/75

In topologies with hard commutation, like in the synchronous buck converter, low parasitic inductances in the power loop are highly beneficial to enable fast switching to reduce switching losses on one hand and to keep voltage overshoot low on the other.

The portable welding station "Ignis Battery™ 150/750" from Fronius was selected as a perfect test vehicle to study PCB embedding of discrete power semiconductors. The following paragraphs are dedicated to describe the specifics of this application and the system implementation based on PCB-embedding of pre-packaged discrete power MOSFETs.

## 1.2   System description

Figure 3 shows a simplified diagram of the portable welding station shown in fig. 1. The system is powered by a Li-Ion battery with a nominal rating of 50.4V, with the usual variation depending on the charging condition of the battery.

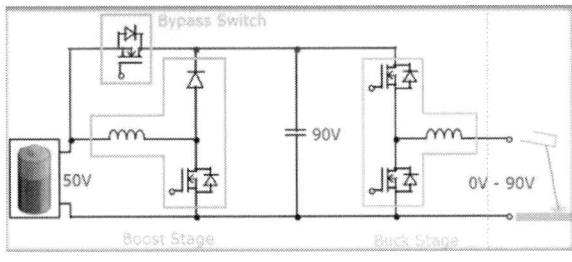

**Fig. 3**    Simplified electrical application diagram of the portable welding station shown in fig. 1

With a battery capacity of 750Wh up to thirty 2.5 mm electrodes can be welded with one battery charge. Alternatively, 33 minutes of continuous TIG welding can be performed at 100 A. A boost stage is active in the first phase of ignition which boosts the battery voltage up to 91V, for a maximum time of 2s, to establish a good initial ignition spark for welding (fig.4). After the ignition phase the current is regulated according to the target current set by the user. It should be noted that the maximum available steady state power is well above the standard characteristic curve (shown as grey line in fig. 4), which implies a corresponding performance of the converter. Chip embedding technology is advantageous here to enable compactness.

The different operation stages of the system are illustrated in fig. 4. As already mentioned in the introduction section, the synchronous buck converter is part of the investigation in this paper. This is particularly useful because, as can be seen from the area highlighted in blue in fig. 4, the buck stage is mainly operated during welding operation.

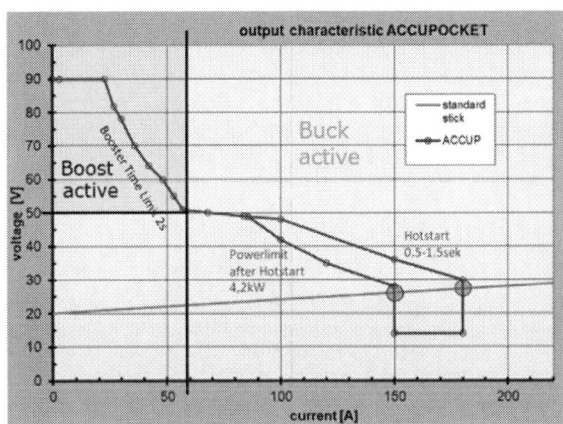

**Fig. 4**    Modes of operation: the boost stage comes active during the ignition of the spark for welding

## 1.3   Implementation

Figure 5 shows the arrangement of the power MOSFETs IPB072N15N3 in D²PAK on the standard PCB board. The blue box highlights the 6 parallel half bridges constituting the synchronous buck converter.

The PCB area of the synchronous buck MOSFETs alone occupies a relatively large area of 8x4.5cm². It is obvious that this PCB area can be effectively reduced by submersing the MOSFETs into the PCB as illustrated in fig 6.

PCIM Europe 2024, 11– 13 June 2024, Nuremberg    DOI: 10.30420/566262147

**Fig. 5** Standard system: top view on the power MOSFETs; the sync-buck part occupies a PCB area of 8cmx4.5cm

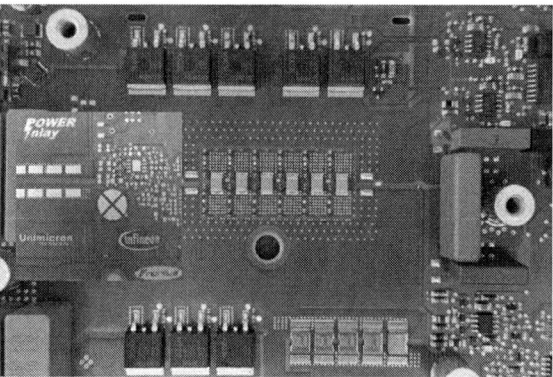

**Fig. 7** PCB-embedded system: top view on the power MOSFETs; the sync-buck part occupies a PCB area of 4cmx2cm

The embedding of traditional molded components is in principle possible, but this comes with the penalty that the overall thickness of the PCB would tremendously increase. On top of that the compatibility of materials in terms of thermal expansion coefficients and surface adhesion is not of the same quality compared to a setup where the MOSFETs are prepackaged within a laminate.

Another limitation that comes with molded packages is the solder die attach. Solder connections are not ideal in terms of thermal performance compared to a direct copper to copper plated interconnection.

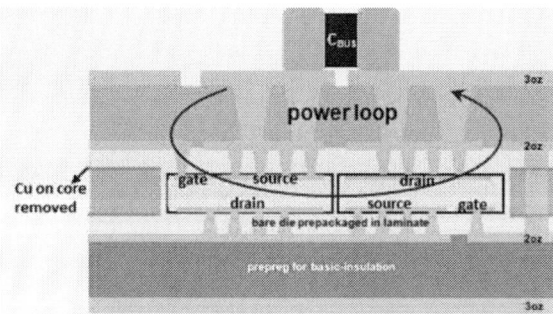

**Fig. 6** PCB embedding of discrete power MOSFETs: PCB layer stack, indicating the short power loop in the half-bridge closed by $C_{BUS}$

The prepackaged discrete power MOSFETs in this work were embedded inside the PCB with a symmetric layup as shown in fig. 6. The connections to

the MOSFETs are established by direct copper-to-copper connection on all interconnecting interfaces enabling high reliability, and thermal/ electrical conductivity.

As indicated in fig. 4, one layer of prepreg is designed as basic-insulation on the backside of the PCB where the heatsink is applied. In this way, the heat sink is not subject to potential, which has a favourable effect on EMI. It should be noted here that, as this is a battery-powered device, there is no earth potential.

Compared to a bare die a prepackaged MOSFET has the advantage that it can be handled in a standard PCB manufacturing line. Another advantage is that standard FR4 laminates can be used, allowing more flexibility for the PCB design. A main advantage for embedding is miniaturization. The PCB area occupied by the $D^2PAK$ is 8cm x 4.5cm versus 4cm x 2cm for the PCB-embedding version, i.e. an impressive reduction of the PCB real estate by a factor of 4.5.

Reducing the MOSFET area connecting to the thermal system might imply that heat spreading is less efficient, and that the hotspot temperature may rise. As it will be discussed in section 2.3 the opposite is the case, because thermal interconnections are significantly shortened and better in quality.

## 2 Results and Discussion

### 2.1 Reduction of voltage overshoot

When the buck-MOSFET in the circuit shown in Fig. 3 turns off, the di/dt of the drain current and the stray inductances of the drain leads cause a voltage spike (and therefore ringing) across the drain and the source potential. As the $D^2PAK$ have a stray inductance of approx. 8nH (including PCB tracks) this voltage spike can be seen clearly in the measurment in the first picture of fig. 8 in the orange line. The embedding build up has negligible stray inductance and so hardly no ringing occures.

For a fair comparison of course the same gate resistores have been used.

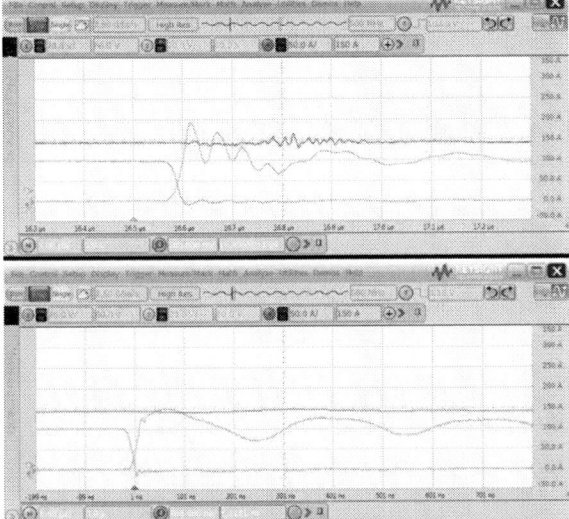

**Fig. 8** Voltage-overshoot: SMD (top) vs. "PCB-embedding" (bottom) -variant

## 2.2 Improvement of Electromagnetic Emission

As lower voltage overshoots and oscillations have already been demonstrated by measurement, the extent to which this has a positive effect on EMI compatibility was investigated.

Figure 9 shows the conducted emissions in constant current operation of 40A, as can be see there is a reduced emission up to 10dB, especially for frequencies higher than 3MHz. This is all the more astonishing as the battery charger must be connected to the grid for this test. In the lower frequency range of 150kHz to 1MHz both measurements appear quiet identical; this is understandable since the switching frequency of the power semiconductors was not changed.

The measurements were also repeated for other load points, up to the maximum operating point of 150A, where similar results were obtained.

A possible explanation for this improvement could be found in the very short interconnection to the bus-capacitor, suppressing high frequency components more effectively.

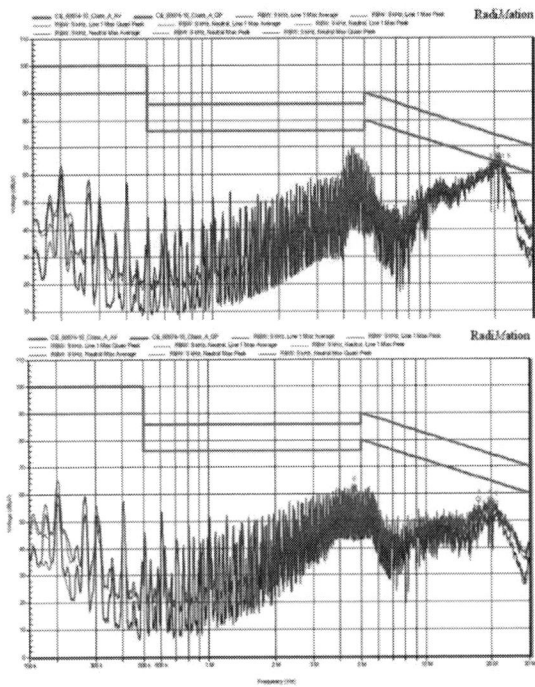

**Fig. 9** Up to 10dB conducted emission reduction: SMD-variant (top) vs. "PCB-embedding" variant (bottom)

Another factor might be that the overall volume of the power loop becomes small, leading to less coupling to the enviroment.

Last but not least, embedding the MOSFET inside the PCB could be used to contain radiative EMI.

## 2.3 Thermal Performance

The graph in fig. 11 compares the system temperature for one welding cycle for standard (0.44 W/K*m) versus thermally enhanced laminate (2.2 W/K*m) used on the PCB side where the heatsink is applied.

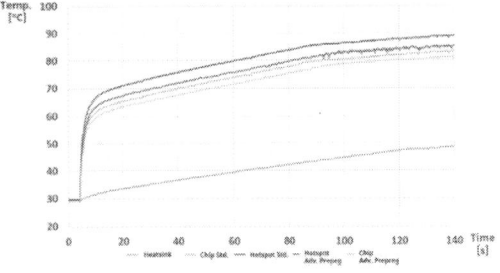

**Fig. 11** System temperature for one welding cycle comparing standard (Std) with thermally enhanced laminate (Adv. Prepreg)

PCIM Europe 2024, 11– 13 June 2024, Nuremberg    DOI: 10.30420/566262147

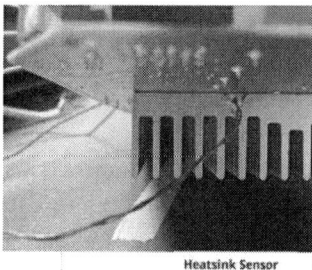

**Fig. 10**    Temperature measurement setup: thermocouple mounted on the heatsink on the backside of the PCB

The temperature of the hottest spot on the board is reduced by 5°C. Thermal images of the power MOSFETs are shown in fig. 12.

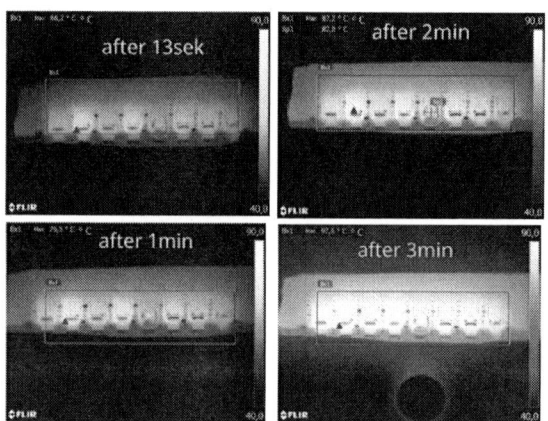

**Fig. 12**    Thermal images of the power MOSFETs after different times of operation

## 2.4   Reliability testing and qualification

A full product qualification was performed including 1000h H3TRB, 1000h HTGS, TC(-55/150°C) 3000x and IOL 15000x. A test card was used for this purpose as shown in fig. 13.

**Fig. 13**    Test card for reliability testing

A vibration test on the complete tool was performed, with a vibration frequency in the range of 5-35Hz with 5Hz steps (30 min per step). The system was fully functional afterwards.

## 3   Conclusion and outlook

This paper presents a full system implementation of 150V OptiMOS™ power MOSFETs embedded inside a PCB for a portable welding station "Fronius – Ignis Battery™ 150/750".

Different laminate materials were evaluated to optimize thermal management. Several different qualification tests had been performed with pass, on the MOSFETs embedded inside the PCB, to demonstrate reliability readiness for volume production.

A significant reduction in electromagnetic radiation emission and voltage overshoot was observed for the synchronous buck stage.

## References

Please follow international scientific citation rules.

[1]   https://www.infineon.com/cms/en/product/promopages/power-modules/

[2]   Fronius International GmbH „ Datasheet: Fronius – Ignis Battery 150/750". April 2024

PCIM Europe 2024, 11– 13 June 2024, Nuremberg    DOI: 10.30420/566262148

# Process Challenges and Progress Towards Direct Connection of Automotive Power Modules (TMM) to Heatsink

Indrajit Paul[1], Inpil Yoo[1], Sergio Savino[2], Gaetano Montalto[2], Ettore Chiacchio[2], Alessandro Tumminia[2], Francesco Salamone[2]

[1] STMicroelectronics, Munich, Germany

[2] STMicroelectronics, Catania, Italy

Corresponding author:    Indrajit Paul, indrajit.paul@st.com
Speaker:    Indrajit Paul, indrajit.paul@st.com

## Abstract

The direct connection of transfer molded modules (TMM) to a heatsink provides new opportunities for improving thermal performance and the supply chain at the system level. Two methods that could be used to perform direct connection are reflow soldering or sintering. This research focuses on the issues related to the mechanical robustness of TMM during the soldering process. The analysis of failed samples showed thermomechanical loading as the root cause of crack in EMC resin. To address this issue, a simulation-based sensitivity analysis was conducted to improve the design and material selection. The optimized assemblies were then subjected to thermal cycling tests, which confirmed their overall robustness. Additionally, the research considered other variables such as reflow equipment and soldering alloy selection.

## 1    Direct connection of moulded power modules

### 1.1    Introduction

There have been several recent developments in the power module for automotive traction inverter applications [1] including application of soldering/sintering process for direct connection of transfer molded power module to the heatsink [2]. Such an interconnection has been projected to have better thermal performance and lifetime stability as compared to thermal interface materials. Eventually manufacturing and supply chain simplification are considered as an added benefit of using such technology. A typical direct connected module system assembly is shown in Fig. 1. In this illustration, an off-the shelf molded power module has been soldered to a metal heatsink.

The key focus of this work was to understand the physical mechanisms leading to mechanical fracture of EMC during integration. We have investigated two different module designs and found high tensile stress in EMC as a key factor in limiting the application of direct connection of transfer molded power modules. To overcome these challenges

several design and material sensitivity analysis were performed.

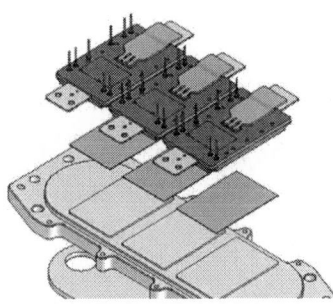

**Fig. 1** Direct connected transfer molded power module for automotive traction inverter

In the next section, we compare the different interconnection methods and compare the benefits and disadvantages. Also, we discuss the several materials and individual processes involved as well as the interactions during the assembly process.

1133

### 1.1.1 Existing methods (Indirect, direct connection)

The different methods of connecting module and heatsink are shown in Table 1, where we have highlighted the main feature. Thermal interface material (TIM) is commonly used indirect interconnection method where a layer of conductive material is placed between the module and the heatsink. To prevent mechanical squeezing (pumpout) during operation it is necessary to design a special clamping system. However, as the module is not mechanically connected there is no additional stress due to CTE mismatches with the heatsink.

Using direct connection method (solder or sinter) we can achieve an improved thermal behavior at the interconnection level. However, the attachement process could lead to severe mechanical stress in the power module. Using a direct connection method also limits the selection of solder component for other interconnections (e.g. die, clips, NTC etc). The biggest advantage of using direct interconnection is supply chain simplification and thermal performance.

| Method | Features |
|---|---|
| TIM | Clamping design, pumpout (-) |
| | No additional thermal stress (+) |
| | No constraint on solder (+) |
| | Reworkable solution (+) |
| Solder *or* Sinter** | No clamping design, compact (+) |
| | Improved thermal performance (+) |
| | Additional mechanical stress (-) |
| | Nonreworkable solution (-) |
| | Constrain solder selection (-) |

**Table 1** Comparison of existing module-heatsink connection methods (**: Not covered in this paper)

## 1.2 Material and process integration

Warpage and stress phenomenon during production and operation is well known for power module packaging. Several approaches are widely presented in the literature to optimize warpage/stress in baseplate based module [3]. In a TMM, presence of epoxy creates an additional complexity to the existing design and manufacturing approaches. This is caused due to the nonlinear material behaviour of EMC around glass transition temperature (Tg). Proper selection of epoxy resin is needed to avoid any long term issues related higher temperature operational capability of SiC technology [4]. The typical material behaviour of EMC with temperature is shown in Fig. 2.

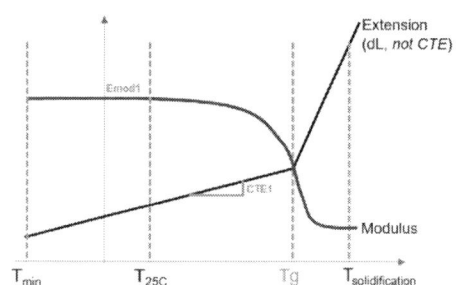

**Fig. 2** Temperature dependent material behaviour of EMC showing strong changes around Tg

Table 2 shows a list of materials used along with the typical coefficient of thermal expansion (CTE), glass transition temperature ($T_g$) and elastic modulus values.

| Material | Characteristic values | |
|---|---|---|
| AMB | Copper: 100 GPa, | CTE~17 ppm/K |
| | $Si_3N_4$: 300 GPa, | CTE~2.5 ppm/K |
| | AlN: 330 GPa, | CTE~4.6 ppm/K |
| Die | SiC: 501 GPa, | CTE~4.3 ppm/K |
| | Si: 162, GPa | CTE ~3 ppm/K |
| Epoxy | High Tg,CTE,** | CTE ~12 ppm/K |
| Heatsink | Aluminium: | CTE ~23 ppm/K |
| | Copper: | CTE ~17 ppm/K |
| | AlSiC: | CTE 7~ 9 ppm/K |
| Solder | Elastoviscoplastic | CTE~22ppm/K |

**Table 2** List of materials used in a typical power moulded module with direct connection to heatsink ($CTE_1$**: CTE below $T_g$)

Based on analytical models [6], we found some of the key factors that control the warpage (Fig. 3) and tensile stress generated during soldering process are as follows:

- AMB: Effective CTE, total thickness
- EMC: $CTE_1$ and $T_g$
- Heatsink: Thickness, CTE
- Soldering temperature (Solidus)

PCIM Europe 2024, 11– 13 June 2024, Nuremberg DOI: 10.30420/566262148

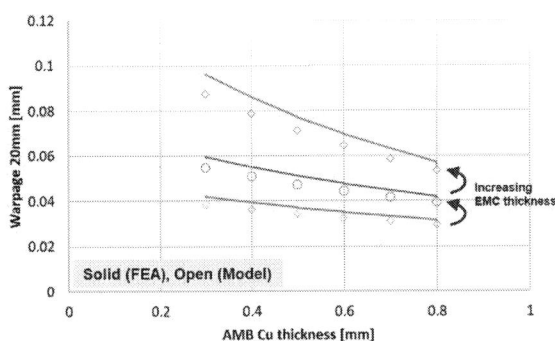

**Fig. 3** Influence of relative Cu to EMC thickness on the warpage (over 20mm length)

### 1.2.1 Warpage during soldering process

Moulded power module is constantly warping with changing temperature during the soldering process particularly beyond glass transition temperature ($T_g$). Fig. 4 shows the measured warpage of a power module.

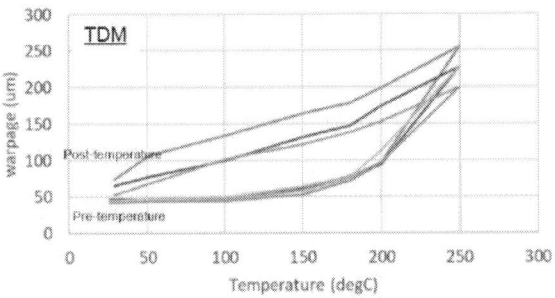

**Fig. 4** Warpage of the moulded power module as measured using TDM

The warpage behaviour of the module during soldering process could be explained using Fig. 5. Depending on the soldering temperature (for e.g. low temperature solder material) the power module can have a different stress condition at the point of joining. We also suggest that during cooling down the whole system (including module and heatsink) has a crying warpage resulting in a tensile stress on top side of EMC.

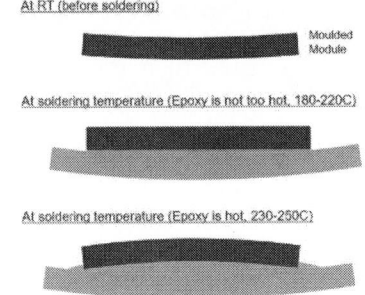

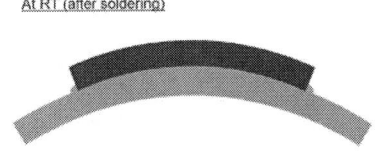

**Fig. 5** Warpage during soldering process (heating and cooling)

## 1.3 Integration challenges

### 1.3.1 Solder joint

During the initial phase of the study several soldering trials were carried out using AMB test vehicles to select the correct solder material and the nominal process conditions. A typical image showing different soldering joints is shown in Fig. 6. The top picture shows a good connection where a metal mesh-based solder material is used whereas we observed strong delamination without metal mesh.

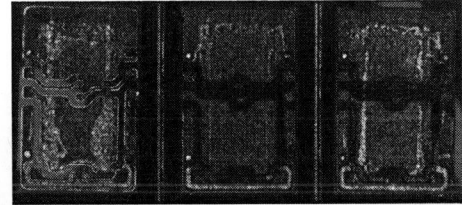

**Fig. 6** AMB samples with soldering on heatsink showing no delamination (top picture) and delamination (bottom picture)

The void observed in such an interconnection was very limited (<1%) as shown in Fig. 7. Based on similar studies we selected metal mesh-based solder material for connecting power modules to heatsink.

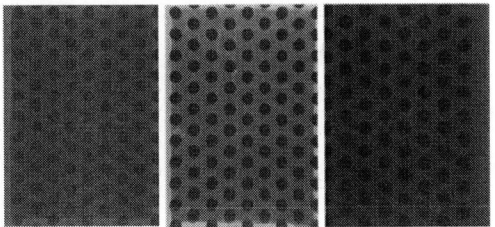

1135

**Fig. 7** X-ray images showing no voids in the AMB samples with soldering on heatsink

The initial results with attachment of power module using the metal mesh based solder showed good results however as mentioned during the previous section the dynamic warpage during soldering process could result in delamination (or unattached areas) as shown in Fig. 8. The delamination could also be due to improper surface treatment. Extensive study was not conducted as the delamination was solved using proper solder jig design.

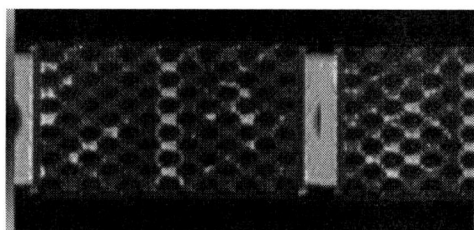

**Fig. 8** Possible delamination (or unattached areas) after soldering process

### 1.3.2 EMC crack

During the early phase of mechanical prototype development, we observed some of these issues as shown in Fig. 9. We observed cracks in EMC due to stress concentrations and solder delamination after reflow process. These observations were very systematic and the root cause for these cracks were CTE mismatch and design related aspects.

**Fig. 9** Failures observed during initial prototyping phase (cracks in EMC)

Further details on the mechanism leading to EMC cracks would be discussed in the next section.

## 2 Process modelling

### 2.1 Introduction

Direct connection approach necessitates additional thermal loading during the soldering/sintering process. Some of the additional parameters to be considered are [5]:

- Lower adhesion of EMC-Cu/Ag interfaces at high temperatures (e.g. reflow temperature)
- Crying warpage ➔ tensile stress in EMC at room temperature (long term creep)
- High stress concentration in EMC due to vertical connectors for signal pins

We have used finite element analysis (FEA) to simulate reflow process and estimate stresses in epoxy. Figure 10 shows the increased risk of delamination as we go higher with the soldering temperature (220°C➔260°C). It shows stress concentrations and high stress in several areas of EMC with low wall thickness. The stress calculation correlates well with the failure locations observed, as shown in Fig. 9.

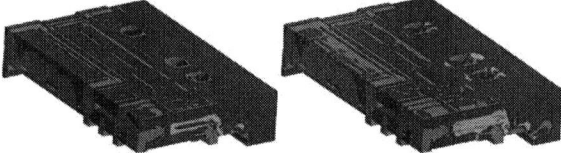

**Fig. 10** Tensile stress distribution in moulded power module at high temperatures (Left: 220°C, right: 260°C)

These high tensile stresses could also be interpreted as a possible delamination between the copper and EMC interface like the ones observed in SAM (Fig. 11).

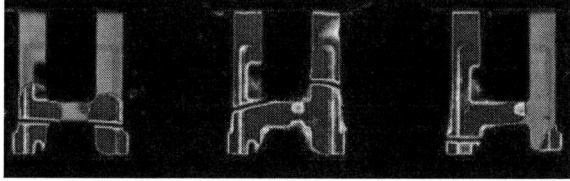

**Fig. 11** C-SAM images showing delamination at EMC interface with copper layer

In addition it was also observed that a large section of the EMC is in tensile stress on the top side as shown in Fig. 12 (dark yellow areas). This indicates that having design element with high stress sensitivity could result in EMC crack like Fig. 9 (right).

PCIM Europe 2024, 11– 13 June 2024, Nuremberg          DOI: 10.30420/566262148

**Fig. 12** Tensile stress in EMC during cooling down

## 2.2 Sensitivity analysis

### 2.2.1 Optimization

The optimization was done on a simplified element of the whole design as shown in Fig. 13 and used linear elastic material behaviour. The objective of this sensitivity was to identify the key parameters controlling the mechanical stresses in EMC during reflow process. One of our main observations was related to the prevalence of delamination at Cu-EMC vertical interfaces and stress due to bending (crying) during cooling down of the module in reflow process.

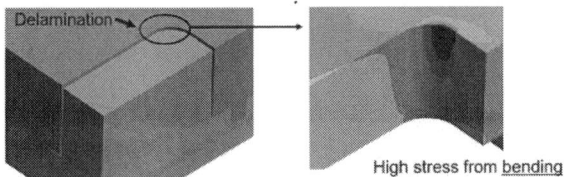

**Fig. 13** Model used for sensitivity analysis identifying delamination and CTE related warpage

### 2.2.2 Results

For the particular case as shown in Fig. 13, we identified the fillet radius and CTE1 of EMC as the key parameters after the sensitivity analysis (results shown in Fig. 14). In this particular study we used a specific EMC formulation with CTE1 lower than 10ppm/K. Such grades of EMC have a limited availablity in terms of commercial grades and eventually stress needs to be reduced through proper adjustment of the design parameters (for e.g. thickness). Other parameters such as the Tg needs to be properly matched as well. Further analysis was done to correlate the overall influence of material selection on warpage and stress.

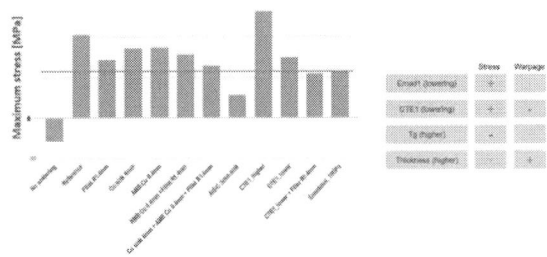

**Fig. 14** Sensitivity analysis showing the main influencing factors

# 3 Improvements and reliability testing

## 3.1 Cracks at room temperature

One of our initial observations was the presence of cracks after a short interval of time when the module in soldered conditions was kept at room temperature. We also observed early TC failure (<20cycles) which further indicated that high tensile stress in EMC is a major problem with direct connection of moulded power modules to heatsink. The failure pattern was similar to the one observed directly after soldering the module. So for the reliability testing we proposed a short cycle (10TC) and a longer cycle (>500TC) test to confirm the suitability of the whole design.

## 3.2 Improved design

Based on the simulation sensitivity analysis we increased the fillet radius and the EMC material. With this combination we were able to create functional prototypes without any crack on EMC or other internal components as shown in Fig. 15.

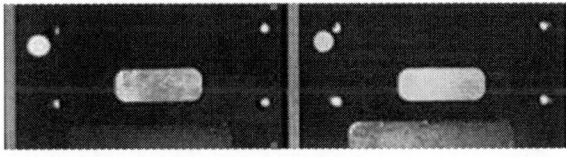

**Fig. 15** Mechanical prototypes with direct connection of moulded module with heatsink without any crack on EMC

## 3.3 TC testing

TC testing (both short and long) showed that the module once attached is well connected after the long cycle TC. There was no change observed in the solder coverage after long TC testing as shown in Fig. 16.

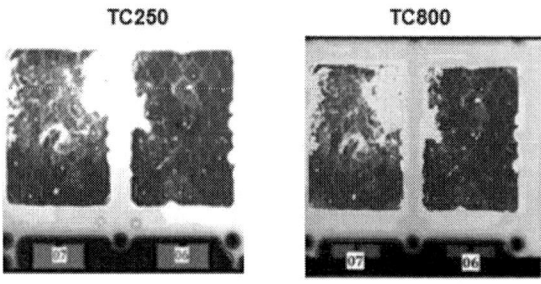

**Fig. 16** Solder joint connect at 250 and 800 TC cycles indicating limited delamination

## Conclusions

Direct connection of TMM for automotive inverter applications was presented in this paper. Mechanical crack in EMC material during solder reflow attachment of the module to heatsink was identified as one of the difficult challenges in the process integration. Various aspects of the mechanical crack were discussed and modelling techniques were used to identify an appropriate solution using a lot CTE mold compound and design change. TC performed on the improved design shows limited delamination during TC test. As a next step, process improvements are being made to increase the connection coverage of the module and eventually understand the impact on overall package performance.

## References

[1]. J. Reimers, L. Dorn-Gomba, C. Mak and A. Emadi, "Automotive Traction Inverters: Current Status and Future Trends," in IEEE Transactions on Vehicular Technology, vol. 68, no. 4, pp. 3337-3350, April 2019.

[2]. M. Fenech et al., "Power Package Attach by Silver Sintering – Process, Performance & Reliability," PCIM Asia 2020; International Exhibition and Conference for Power Electronics, Intelligent Motion, Renewable Energy and Energy Management, Shanghai, China, 2020, pp. 1-5.

[3]. J. Sommer, R. Bayerer, R. Tschirbs and B. Michel, "Base Plate Shape Optimisation for High-Power IGBT Modules," 5th International Conference on Integrated Power Electronics Systems, Nuremberg, Germany, 2008, pp. 1-4.

[4]. Q. Zou, G. Xie, G. Lei, Q. Ding and W. Zhan, "Development and Characterization of Epoxy Molding Compound with High Glass Transition Temperature," 2022 23rd International Conference on Electronic Packaging Technology (ICEPT), Dalian, China, 2022, pp. 1-6.

[5]. H. Pape, I. Maus, I. Paul, L. J. Ernst and B. Wunderle, "Fracture toughness characterization and modeling of interfaces in microelectronic packages - A status review," 2012 13th International Thermal, Mechanical and Multi-Physics Simulation and Experiments in Microelectronics and Microsystems, Cascais, Portugal, 2012, pp. 1/9-9/9.

[6] HSUEH, Chun-Hway. "Modeling of elastic deformation of multilayers due to residual stresses and external bending", Journal of Applied physics, 2002, 91. Jg., Nr. 12, S. 9652-9656.

PCIM Europe 2024, 11– 13 June 2024, Nuremberg    DOI: 10.30420/566262149

# Optimizing PCB Stackups for Enhanced GaN Transistor Performance in High-Power Applications

Philipp Czerwenka[1], Jan Frederik Wagenfeld[1], Gernot Schullerus[1]

[1] Electronics & Drives, Reutlingen University, Germany

Corresponding author:    Philipp Czerwenka, philipp.czerwenka@reutlingen-university.de
Speaker:    Philipp Czerwenka, philipp.czerwenka@reutlingen-university.de

## Abstract

This paper explores specialized PCB stackups to enhance GaN transistor performance in applications up to $10\,\mathrm{kW}$. Recognizing extensive prior research on GaN in high-power contexts, our study initially investigates layouts, which will be used for developing optimized stackups. Our objective is, to identify stackups that maximize thermal performance while minimizing parasitic effects. The analysis establishes insulated metal substrate and copper inlay PCB stackups with vertical layouts as promising options. These results enable a flexible integration of stackup designs in high-power GaN applications and their synergy with other design objectives.

## 1 Introduction

Gallium-Nitride (GaN) semiconductors offer reduced conduction and switching losses along with a smaller form factor compared to silicon semiconductors, owing to their material properties, like a significantly higher critical electrical field strength, and increased electron mobility when compared to silicon [1]. These properties enable the industry to build GaN power devices on a chip-scale level reducing device parasitics and therefore increasing switching speeds [2]. To harness these advantages in high-power applications, minimizing parasitics introduced by the printed circuit board (PCB) layout and maximizing thermal performance to handle the increased power density are crucial.

Numerous publications have addressed the challenge of designing PCB structures tailored towards the specific requirements of highly integrated GaN power electronics. The general consensus is, to use a vertical commutation loop for a half-bridge subcircuit with a local decoupling buffer capacitor [3–5]. Parasitic inductance values in the single digit nanohenry range are regularly reported, which are necessary for high signal quality switching operation. Additionally, novel layout concepts are also explored. In [6]

a vertical layout with transistors on opposite PCB sides is presented featuring a $28\,\%$ decrease in parasitic inductance value over comparable vertical designs. In [7] a vertical layout featuring interleaved structures for additional magnetic field cancellation effects is reported featuring an up to $35\,\%$ decrease in parasitic inductance. Furthermore, in [8] a GaN-based power stage build on a ceramic aluminium-nitride carrier is presented, yielding more dense integration.

In this publication we explore specialized PCB stackups to address requirements related to the parasitics as well as thermal requirements, using a half-bridge for voltages up to $600\,\mathrm{V}$ and currents up to $60\,\mathrm{A}$ as an example. Our goal is, to identify an advanced manufacturing process derived from traditional PCB technology to feature both the possibility to implement layout parasitic efficient designs and provide excellent thermal performance to aid with temperature management.

The paper is structured as follows: In Section 2 we provide a revision of the origins of layout parasitics as well as thermal analysis and the challenges associated with analytical methods. Section 3 presents the results of our simulation-based analysis of layout concepts for half-bridges given in the literature and stackup configurations for high power applications commonly used in the industry.

1139

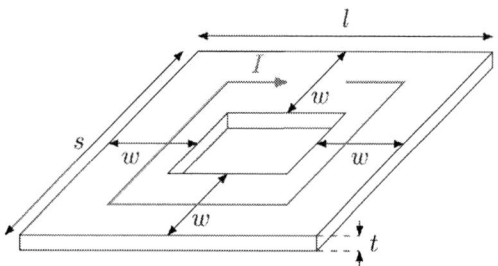

**Fig. 1:** Horizontal conductor loop, current path red

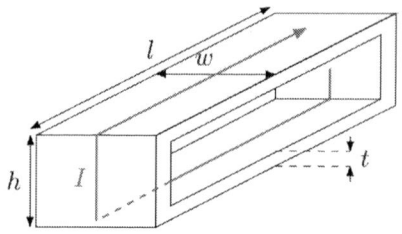

**Fig. 2:** Vertical conductor loop, current path red

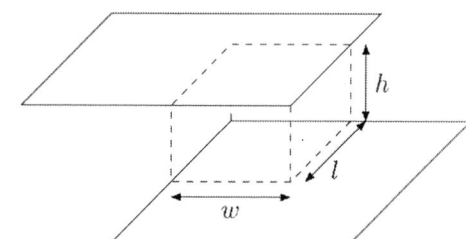

**Fig. 3:** Partially overlapping parallel layout geometries

In Section 4 the most promising combinations of layouts and stackups are further analyzed and manufactured as prototypes. Measurement results and methodologies are given. Section 5 concludes the paper.

## 2 Analytical Methods

### 2.1 Layout Parasitics in Half-Brigdes

In a PCB layout, every piece of conductor subjected to a current exhibits a magnetic field and therefore a self-inductance. For simple geometries, this inductance $L$ can be approximated analytically.

The inductance of a horizontal rectangular conductor loop as illustrated in Fig. 1 with dimensions $l$ and $s$, conductor width $w$, thickness $t$, $s \gg w$ and $w > t$ can be approximated by [9]

$$L = \frac{\mu_0 \mu_r}{\pi} \cosh^{-1}\left(\frac{s}{w}\right) l \ . \tag{1}$$

Similarly, the inductance of a vertical rectangular conductor loop (see Fig. 2) with length $l$, conductor width $w = 0.25\,\text{mm}$ to $50\,\text{mm}$, loop height $h = 0.1\,\text{mm}$ to $2\,\text{mm}$ and negligible conductor thickness $t$ can be approximated by [10]

$$L = \mu_0 \mu_r \frac{hl}{w}\left(\frac{1}{1 + \frac{h}{w}} + 0.024\right) \ . \tag{2}$$

For the analysis in the following sections, the paper uses a common strategy in parasitic analysis by combining the effects of all current path sections into a single inductance value for a given current path. In this analysis, we focus on the commutation loop inductance $L_{\text{PL}}$ of the half-bridge for high frequencies, which spans from the buffer decoupling capacitor across the high-side and low-side power transistor back to the buffer capacitor, as shown in Fig. 4.

As structures in the layout are differently charged due to the switching of voltage in the half-bridge circuit, electrical fields between the structures form and exhibit parasitic capacitances. Due to the nature of PCB manufacturing processes, lateral structures are preferred, resulting in large parallel planes in close proximity. There are capacitances between every permutation of circuit nets in the layout. An analytical estimate for such effects can be found when taking only the approximately homogeneous field between the parallel overlapping areas of two layout structures into account. For two rectangular geometries with side lengths $l$, $w$ and layer distance $h$ (Fig. 3) the parasitic capacitance $C$ can be approximated by

$$C = \varepsilon_0 \varepsilon_r \frac{wl}{h} \ . \tag{3}$$

In our analysis, we identified two capacitances as potentially disadvantageous to fast switching in a half-bridge circuit. $C_{\text{HSPCB}}$ and $C_{\text{LSPCB}}$ are parallel to the high-side and low-side transistors output capacitance respectively (see Fig. 4), causing an increase in output charge to be dissipated in the transistor's channel during each switching event. This negatively influences switching efficiency if the parasitic capacitance is too large, especially in soft-switching applications, where the majority of switching losses are caused by parasitic capacitance charging currents.

PCIM Europe 2024, 11– 13 June 2024, Nuremberg DOI: 10.30420/566262149

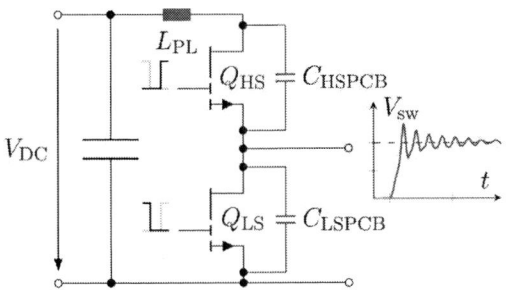

**Fig. 4:** Equivalent circuit of GaN-based half-bridge with decoupling capacitor showing how parasitic layout elements lead to ringing during switching

However, the accuracy of the presented approximations is severely limited by

- the increasing complexity of the commutation loop layouts and by

- the new layout structures introduced by advanced stackups.

A comparison given in Section 3 illustrates, that such analytical models are only accurate within the same order of magnitude of the simulation result and do not retain trends. This is due to the inherent simplifications of such models. Two particular simplifcations are pointed out below:

- Analytical models typically neglect vias, which on their own can influence the calculated inductance significantly.

- The interaction between the magnetic field of the currents, i.e. the mutual inductance between conductors also plays a major role in certain layouts [3] (see vertical layouts Section 3.1).

## 2.2 Thermal analysis

Thermal analysis is typically based on thermal equivalent diagrams, where the sources are given by the power loss $P_L$ and the temperature differences between different materials are modelled by thermal resistors $R_{th}$. Dynamic behavior can be included into such models by introducing thermal capacitances. In this work, only steady state considerations are given such that thermal capacitances are not considered.

The analytical thermal analysis is based on the thermal equivalent circuit given in Fig. 5 where a

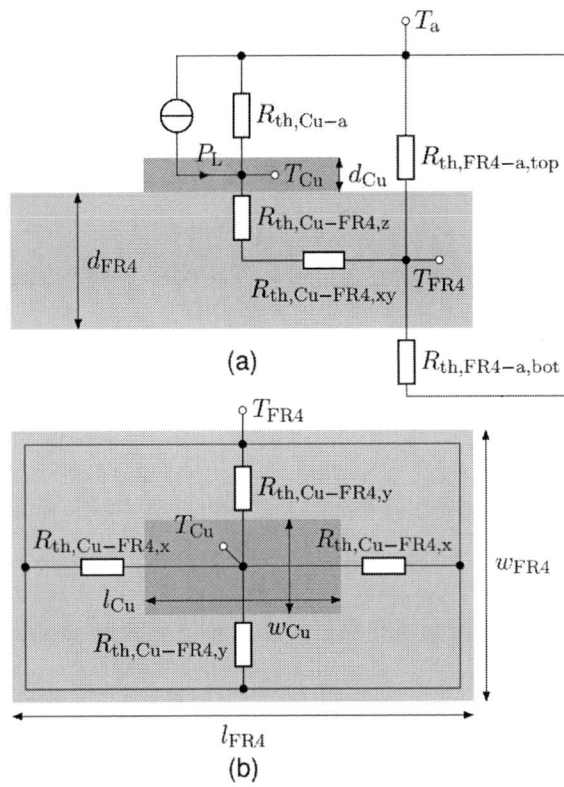

**Fig. 5:** Thermal equivalent diagram of a rectangular PCB structure for (a) vertical and (b) lateral considerations

copper foil with side lengths $l_{Cu}$, $w_{Cu}$, a thickness of $d_{Cu}$ and the temperature $T_{Cu}$ is placed on a FR4 substrate with side lengths $l_{FR4}$, $w_{FR4}$, a thickness $d_{FR4}$ and the temperature $T_{FR4}$. The thermal resistors $R_{th,Cu-FR4,(x,y,z)}$ model the thermal behavior inside the substrate whereas $R_{th,FR4-a,(top,bot)}$ describe the heat transport from to the top and bottom side of the substrate to ambient. Similarly, $R_{th,Cu-a}$ models the direct convection heat transport from copper to ambient with the temperature $T_a$.

The determination of the thermal resistances is done using the given geometries and considering conduction and convection. In the latter process, the Nusselt number $Nu$, the Prandtl number $Pr$ and the Grashof number $Gr$ [11] are used, where the latter depends on the temperature difference between the considered materials. Thus, the calculation of the thermal resistances requires an iterative procedure.

1141

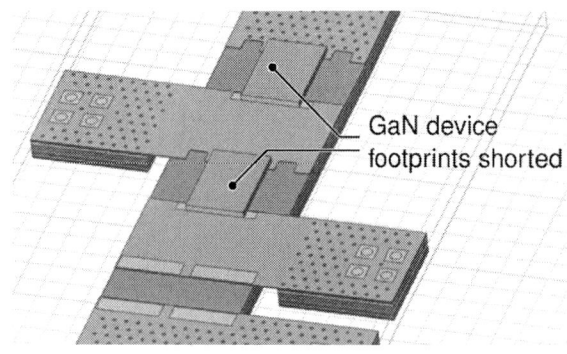

**Fig. 6:** Modified simulation model of the vertical 10-layer commutation loop in Ansys Q3D Extractor

Similar to the parasitics determination described in Section 2.1, vias are not considered. In addition, hot spots can not be easily modelled in such considerations, resulting in limited accuracy of such models. Nevertheless, these analytical approximations provide solutions which are valuable for simulation parameterizations and initial design considerations.

# 3 Layout and Stackup Analysis

The comparison of layouts and stackups in this paper is based on three criteria:

1. the parasitic element values,

2. the worst-case heating and

3. the expected ringing voltage.

The evaluation of these criteria is done on both, analytical solutions and simulations. The layout parasitics are modeled analytically as described in Section 2.1.

The worst-case heating is determined based on a scenario, where a current of $60\,\mathrm{A}$ is driven through the high-side and low-side transistor, respectively. The point of maximum temperature increase $\Delta T_{60\mathrm{A,max}}$ across the layout is used as criterion.

The ringing voltage which occurs during switching is an aggregate measurement of all parasitic element values. An analytical expression was developed for the maximum ringing voltage in [12]

$$V_{\mathrm{ring}} = V_{\mathrm{DC}} \left| \mathrm{sinc} \left( \frac{t_{\mathrm{r}}}{2\sqrt{L_{\mathrm{PL}}\left(C_{\mathrm{oss}} + C_{\mathrm{PCB}}\right)}} \right) \right| \quad (4)$$

where $V_{\mathrm{DC}}$ is the DC-supply voltage, $L_{\mathrm{PL}}$ represents the commutation loop inductance, the capacitance $C_{\mathrm{oss}}$ is the output capacitance of the considered GaN-device and $C_{\mathrm{PCB}}$ represents the capacitance $C_{\mathrm{HSPCB}}$ or $C_{\mathrm{LSPCB}}$ depending on whether the High-Side or the Low-Side switch is considered. The variable $t_{\mathrm{r}}$ is the rise time of the Drain-Source-Voltage. For the numerical evaluations, a GaN HEMT device GS-065-060 by GaNSystems [13] is assumed.

## 3.1 Analytical and Simulative Analysis of Layouts

We select four layout concepts from literature for our comparison [6, 7], illustrated in Fig. 7 as four layer designs for visual clarity. All layouts are analyzed using the same 10-layer stackup with $70\,\mu\mathrm{m}$ copper thickness (10L-Std) as shown in Fig. 8a, prioritizing comparability over realism. The 10L-Std stackup enables us to construct a layout with high copper cross section area, as well as small distances between layers. This will be the base case for a given layout structure and therefore emphasize the differences between the layout methods. All conductor layout widths where chosen to be the same width of $9\,\mathrm{mm}$.

The analytical and simulation results in Tab. 1 and Tab. 2, respectively (Designs A-D) confirm literature findings in that vertical layouts such as Design B and C exhibit smallest inductances, and thus low ringing voltage, due to magnetic field cancellation effects. We also confirm that increasing the copper content and copper surface area such as in Design A and C improves thermal performance. In all cases layout capacitances are negligible.

## 3.2 Analytical and Simulative Analysis of Stackups

We compare three specialized PCB stackups targeted at high power applications to the conventional multi-layer stackup mentioned above.

Insulated metal substrate stackups (IMS) are build on a metal core enabling high power density by high thermal conductivity. The analysis includes single layer (1L) and asymmetrical two layer (2L) IMS by Serigroup Srl. Single Layer IMS stackups are undesirable for practical applications as the load current passes through the IMS base, which is

PCIM Europe 2024, 11– 13 June 2024, Nuremberg    DOI: 10.30420/566262149

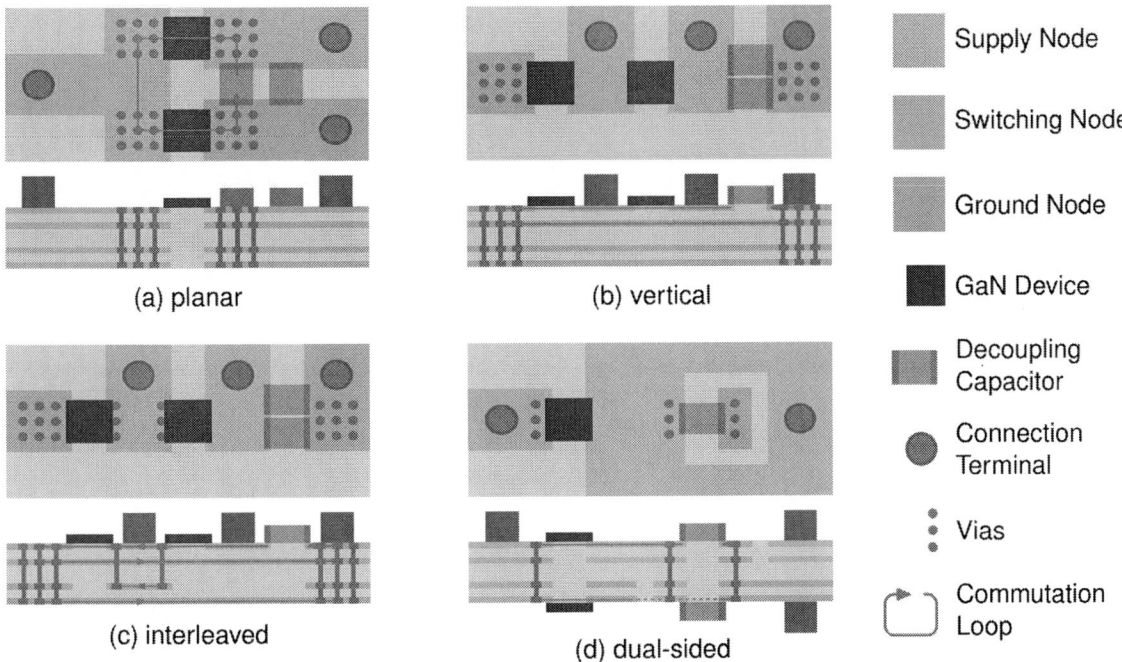

**Fig. 7:** Layout concepts included in comparison: (a) single-layer planar loop, (b) vertical loop with internal layer return current path, (c) vertical loop with additional interleaved structures from [7], (d) vertical loop with dual-sided switch placement from [6]

|  | Design A | Design B | Design C | Design D | Design E | Design F | Design G |
|---|---|---|---|---|---|---|---|
| Layout | planar | vertical | interleaved | dual-sided | vertical | vertical | vertical |
| Stackup | 10L-Std | 10L-Std | 10L-Std | 10L-Std | 1L-IMS | 2L-IMS | 4L-CuI |
| $L_{PL}$ | 10.87 nH | 1.33 nH | 0.81 nH | 5.65 nH | 3.57 nH | 1.81 nH | 0.91 nH |
| $C_{HS,PCB}$ | 41.6 fF | 5.4 fF | 53.9 fF | 17.6 fF | 8.1 fF | 8.1 fF | 53.8 fF |
| $C_{LS,PCB}$ | 41.6 fF | 48.93 pF | 565 pF | 17.6 fF | 9.89 pF | 18.86 pF | 41.74 pF |
| $\Delta T_{60A,max}$ | 22.6 K | 33.3 K | 62.8 K | 24.1 K | 25.1 K | 65.2 K | 43.6 K |
| $V_{ring}/V_{DC}$ | 4.76 % | 0.58 % | 0.25 % | 1.23 % | 1.97 % | 1.95 % | 0.96 % |

**Tab. 1:** Analytically determined parasitics and thermal results of high-power layout and stackup combinations

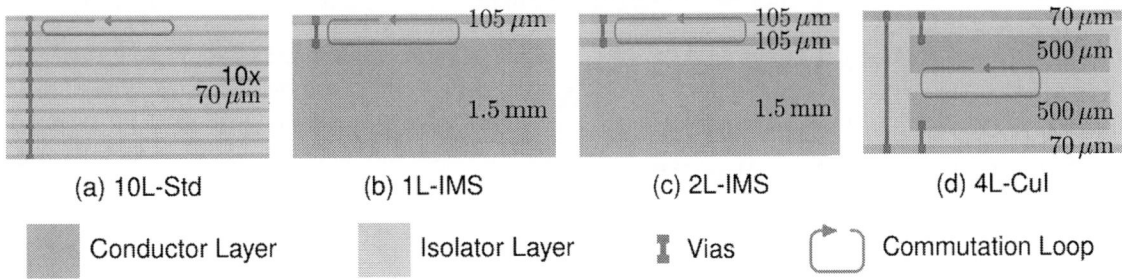

**Fig. 8:** Stackup concepts included in comparison, measurements denote conductor thicknesses: (a) 10-layer standard multi-layer PCB stackup, (b) single layer subtrate-contacted IMS stackup, (c) two-layer asymmetric IMS stackup, (d) four-layer stackup with internal copper inlays

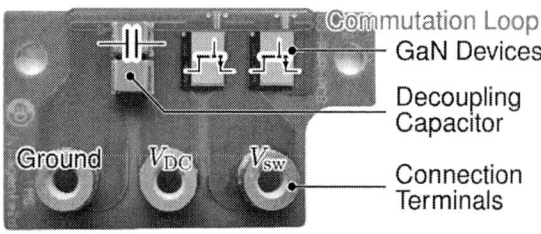

**Fig. 9:** Prototype of half-bridge with vertical layout and 2L-IMS stackup used for measurements

critical from an equipment safety standpoint where direct cooler attachments would be preferred. Symmetrical 2L-IMS stackups, where one layer of insulation and conductor are attached to both side of the metal substrate share the same drawbacks as single layer IMS and are therefore not included in this analysis.

Copper inlay stackups (CuI) are build by inserting high cross section area metal structures into a regular multi-layer PCB enabling high power density by high electrical conductivity. The analysis includes a four layer (4L) CuI configuration by Jumatech GmbH [14]. By having high cross section area conductors only on internal layers and thin copper on the outside, the outer layers can be structured by existing lithography processes which enables small fine-pitch components to be included in the design.

To ensure comparability, the vertical layout of Fig. 7b is chosen for this analysis. The simulation results in Tab. 2 (Designs E, F and G) show that the 2L-IMS exhibits superior parasitic characteristics over the CuI with a $28\%$ decrease in power loop inductance, owing to a reduced insulation thickness between its outer layers compared to the stackup core of the CuI. The 4L-CuI demonstrates superior thermal performance among all the compared stackups with a $48\%$ decrease in expected temperature rise over the 2L-IMS.

## 4 Prototype and Measurements

Based on the comparisons from Section 3.2, two prototype candidates, the 2L-IMS and 4L-CuI stackups, both with vertical Layout B, were chosen for a more detailed comparison in a realistic layout. Thus, for these two variants a real-world layout was designed and subsequently simulated for a physical analysis. Both layouts were designed

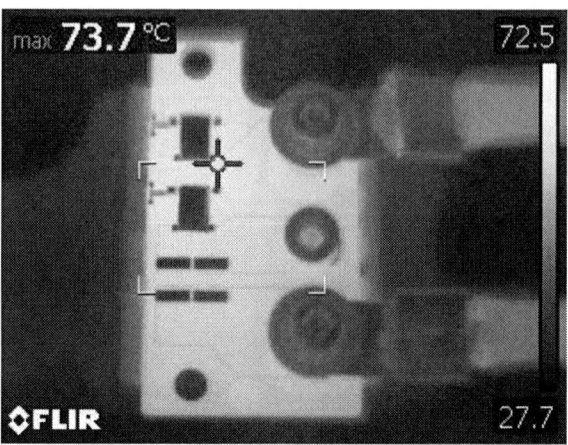

**Fig. 10:** Heat distribution of the prototype after $30\,\mathrm{min}$ at $60\,\mathrm{A}$ current.

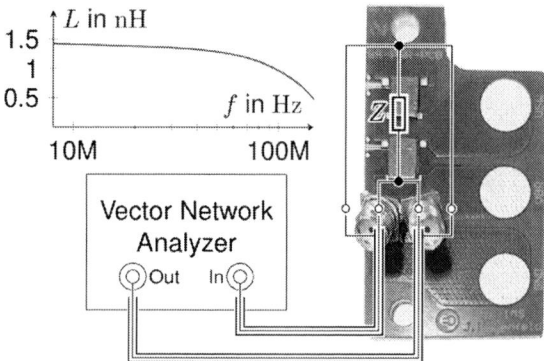

**Fig. 11:** Commutation loop impedance measurement setup using shunt-through vector network analysis, PCB without GaN devices mounted and footprints shorted

with the constraints of the two selected processes, keeping the layout as similar as possible. The simulation results in Tab. 3 confirm findings from Section 3.

Due to the lower parasitic inductance and an acceptable thermal behavior, the 2L-IMS stackup was subsequently chosen for prototype construction (see Fig. 9). Measurements of the constructed prototype are presented in Tab. 3. Parasitic measurements were conducted by impedance characterization and by transient measurement of the ringing frequency. Temperature distribution was analyzed at the defined worst cases using thermal imaging as shown in Fig. 10.

|  | Design A | Design B | Design C | Design D | Design E | Design F | Design G |
|---|---|---|---|---|---|---|---|
| Layout | planar | vertical | interleaved | dual-sided | vertical | vertical | vertical |
| Stackup | 10L-Std | 10L-Std | 10L-Std | 10L-Std | 1L-IMS | 2L-IMS | 4L-CuI |
| $L_{\mathrm{PL}}$ | 6.23 nH | 0.96 nH | 1.01 nH | 4.92 nH | 2.48 nH | 1.57 nH | 2.18 nH |
| $C_{\mathrm{HS,PCB}}$ | 418 fF | 65.9 fF | 11.1 fF | 177 fF | 28.8 fF | 19.4 fF | 142 fF |
| $C_{\mathrm{LS,PCB}}$ | 436 fF | 55.8 pF | 4.9 fF | 592 fF | 12.6 pF | 21.5 pF | 49.0 pF |
| $\Delta T_{\mathrm{60A,max}}$ | 45.1 K | 82.7 K | 42.5 K | 60.3 K | 56.4 K | 79.2 K | 32.2 K |
| $V_{\mathrm{ring}}/V_{\mathrm{DC}}$ | 3.72 % | 0.26 % | 0.11 % | 3.35 % | 1.50 % | 1.96 % | 1.14 % |

**Tab. 2:** Parasitic and thermal simulation results of high-power layout and stackup combinations

|  | Vertical 2L-IMS | | Vertical 4L-CuI | |
|---|---|---|---|---|
|  | Sim | Meas | Sim | Meas |
| $L_{\mathrm{PL}}$ | 1.56 nH | 1.60 nH | 1.94 nH | — |
| $\Delta T_{\mathrm{60A,max}}$ | 61.9 K | 48.7 K | 28.2 K | — |

**Tab. 3:** Parasitic and thermal analysis results of implemented prototypes, simulation (Sim), measurement (Meas), $T_{\mathrm{a}} = 25\,^{\circ}\mathrm{C}$

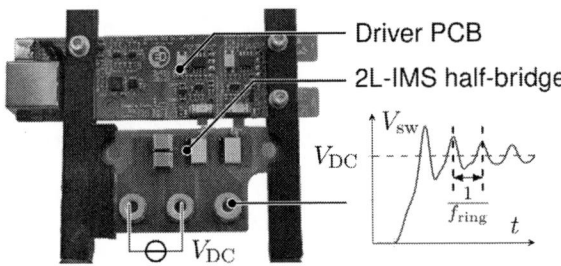

**Fig. 12:** Commutation loop impedance measurement setup using switching node ringing at $100\,\mathrm{kHz}$

For the measurement of loop impedance we use a vector network analyzer as shown in Fig. 11. All measurements were performed by shorting the power transistor footprints with copper shunts resulting in a loop similar to the modified 3D model for parasitic extraction as illustrated in Fig. 6. A two-port shunt type impedance measurement was performed to measure the impedance and as a result the inductance of the loop. Across a frequency range of $1\,\mathrm{MHz}$ to $100\,\mathrm{MHz}$ the parasitic inductance was found to be between $1\,\mathrm{nH}$ to $1.4\,\mathrm{nH}$.

The evaluation of the ringing was performed with the setup given in Fig. 12. Using a gate driver, the half-bridge was operated at a frequency of $100\,\mathrm{kHz}$ and the ringing overshoot of the switching node voltage was measured. The frequency of the ringing $f_{\mathrm{ring}}$ along with the output capacitance $C_{\mathrm{oss}}$ given in the transistors datasheet was used to determine the parasitic loop inductance value $L_{\mathrm{PL}}$ under the assumption of an undampened LC resonant tank circuit with

$$L_{\mathrm{PL}} \approx \frac{1}{4\pi^2 f_{\mathrm{ring}}^2 C_{\mathrm{oss}}}. \tag{5}$$

Across a switching voltage range of $5\,\mathrm{V}$ to $50\,\mathrm{V}$, where $C_{\mathrm{oss}}$ is largest, the parasitic inductance was found to be between $1.25\,\mathrm{nH}$ to $1.6\,\mathrm{nH}$, closely matching the simulation result.

## 5 Conclusion

In this paper we explored advanced PCB manufacturing stackups and layout techniques for power electronics applications using the example of a wide-bandgap semiconductor half-bridge commutation loop. Among the investigated layouts and stackups, a two-layer insulated metal substrate stackup as well as a four-layer copper inlay stackup in combination with vertical commutation loop layouts emerge as the most promising combinations. The IMS layouts achieves high performance by high thermal conductivity due to the heat spreading capabilities of the employed substrate, resulting in minimal layer distances and low parasitic values. The copper inlay layouts achieve high performance by high electrical conductivity featuring the least current-induced temperature increase. The prototype employing the 2L-IMS stackup demonstrates a minimal power loop inductance of approximately $1.6\,\mathrm{nH}$ with negligible layout capacitances, and it experiences a temperature increase of $48.7\,\mathrm{K}$ in the worst-case scenario at a current of $60\,\mathrm{A}$. The presented stackups are well suited for various high-power GaN applications, where more tightly integrated package solutions, such as ceramic carrier power modules, are not yet easily available.

## Acknowledgment

This work was supported by the German Federal Ministry of Education and Research, FH-Kooperativ Project 13FH063KX1 "Clean Motor Supply".

## References

[1] J. Millán, P. Godignon, X. Perpiñà, A. Pérez-Tomás and J. Rebollo, "A Survey of Wide Bandgap Power Semiconductor Devices," in *IEEE Transactions on Power Electronics*, vol. 29, no. 5, pp. 2155-2163, May 2014.

[2] A. Lidow, M. de Rooij, J. Strydom, D. Reusch and J. Glaser, *GaN Transistors for Efficient Power Conversion*, 3rd edition, Hoboken, USA, John Wiley & Sons, Inc., 2020.

[3] E. Persson, "Optimizing PCB Layout for HV GaN Power Transistors," in *IEEE Power Electronics Magazine*, vol. 10, no. 2, pp. 65-78, June 2023.

[4] D. Reusch and J. Strydom, "Understanding the Effect of PCB Layout on Circuit Performance in a High-Frequency Gallium-Nitride-Based Point of Load Converter," in *IEEE Transactions on Power Electronics*, vol. 29, no. 4, pp. 2008-2015, April 2014.

[5] K. Klein, E. Hoene and K. Lang, "Electromagnetic switching cell design and characterization for WBG power semiconductors," *CIPS 2020; 11th International Conference on Integrated Power Electronics Systems*, Berlin, Germany, 2020, pp. 1-7.

[6] B. Sun, K. L. Jørgensen, Z. Zhang and M. A. E. Andersen, "Research of Power Loop Layout and Parasitic Inductance in GaN Transistor Implementation," in *IEEE Transactions on Industry Applications*, vol. 57, no. 2, pp. 1677-1687, March-April 2021

[7] J. Hammer, I. G. Zurbriggen, M. Ali Saket and M. Ordonez, "Low Inductance PCB Layout for GaN Devices: Interleaving Scheme," *IEEE Applied Power Electronics Conference and Exposition*, Phoenix, AZ, USA, 2021, pp. 1537-1542.

[8] C. Kuring, M. Wolf, X. Geng, O. Hilt, J. Böcker et al., "GaN-Based Multichip Half-Bridge Power Module Integrated on High-Voltage AlN Ceramic Substrate," in *IEEE Transactions on Power Electronics*, vol. 37, no. 10, pp. 11896-11910, Oct. 2022.

[9] M. Di Paolo Emilio: *Edge Coupled Trace Inductance Calculator*, EEWeb, URL: https://www.eeweb.com/tools/edge-coupled-trace-inductance/, 22.09.2023.

[10] A. Letellier, M. R. Dubois, J. P. F. Trovão and H. Maher: "Calculation of Printed Circuit Board Power-Loop Stray Inductance in GaN or High di/dt Applications," *IEEE Transactions on Power Electronics*, vol. 34, no. 1, pp. 612-623, Jan. 2019.

[11] A. Griesinger: Wärmemanagement in der Elektronik - Theorie und Praxis, Springer Vieweg, Berlin, 2019.

[12] P. Czerwenka, J. Maier, T. Wolfer, G. Schullerus and E. Hennig, "Analytical Estimation of Parasitic Ringing Overvoltage in Fast Switching Half-Bridges," in *25th European Conference on Power Electronics and Applications*, Aalborg, Denmark, 2023, pp. 1-7.

[13] GaN Systems, *GS66516T Top-side cooled 650 V E-mode GaN transistor*, Datasheet, Toronto, 2021.

[14] M. Wölfel, "Drahtbeschriebene Leiterplatte oder Platine mit geätzten Leiterbahnen." European Patent EP 2 076 100 A1, July 1, 2009.

PCIM Europe 2024, 11– 13 June 2024, Nuremberg        DOI: 10.30420/566262150

# New Generation Ceramic Substrates – Key Components for Power Electronic Applications: Processing and Characterization

Dr. Stefanie Schindler[1], Dr. Tanja Einhellinger-Müller[1], Linda Schicker[1], Hans-Ulrich Völler[1]
[1] CeramTec GmbH, Germany

Corresponding author:    Dr. Stefanie Schindler, st.schindler@ceramtec.de
Speaker:                 Dr. Stefanie Schindler, st.schindler@ceramtec.de

## Abstract

Ceramic substrates are a key component for power electronic applications. Due to their thermal, mechanical, electrical, and chemical properties, the materials can be used for specific markets. The ceramic substrates are crucial materials for a wide range of applications. This presentation will compare different characterization methods for ceramic materials in terms of their suitability for different applications, as well as present challenges and requirements for the quality of ceramic substrates.

## 1    Introduction

### 1.1    Why advanced ceramics?

Advanced ceramics are used where other materials reach their limits: at extreme temperatures, under enormous mechanical stress, under highest current, with highest voltage applied – and even in the human body. They provide reliable solutions in all types of industrial production and high-tech applications. Technical ceramics keep us safely and comfortably where we need to go and provide us with clean energy using more efficient technologies. We are able communicate digitally on smaller devices and benefit from an improved quality of life.

The term "advanced ceramics" includes a variety of different ceramic materials – partly highly specialized – with unique mechanical, thermal, biologic-chemical and electric properties, as well as their combination. The ceramics can be classified into three large groups of materials: silicate ceramics, oxide ceramics and non-oxide ceramics. Silicate ceramics primarily consist of naturally occurring raw materials in combination with aluminum oxide. Oxide ceramics are based on materials including metal oxides. Non-oxide ceramics include materials based on carbon, nitrogen or silicon. Depending on the specific application and its requirements, the appropriate material is selected.

### 1.2    Ceramic substrates

A wide range of different key components for electronic applications, for example electronic circuit carriers, can be manufactured from ceramic substrates, which fulfil their function permanently and safely in various areas of applications. The major advantage of ceramic substrates lies in their excellent electrical, thermal, mechanical, isolating and chemical characteristics. Ceramic substrates are essential for the electronic and automotive industry, for renewable energy, and industrial applications. Their use depends on the reliability and thermal behavior of the several types, which renders them indispensable for e.g., power electronic applications in conjunction with Direct Copper Bonding (DCB) and Active Metal Brazing (AMB) (see Fig 1).

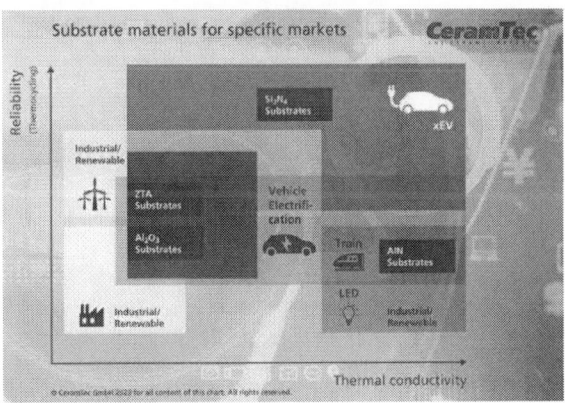

**Fig. 1**    Use of ceramic substrates subdivided by applications.

1147

Figure 2 shows an overview of the properties of the typical ceramics aluminum nitride, aluminum oxide, zirconium oxide, silicon carbide, and silicon nitride.

Aluminum oxide or alumina is the most commonly used substrate in high-performance electronics due to its excellent cost / performance ratio. It is characterized by its extremely high strength and offers consistently reliable and convincing performance even when subject to high thermal and electrical loads in terms of thermocycling capability, thermal shock resistance or flexural strength. This makes aluminum oxide ideal for power electronics in conjunction with direct copper bonding.

In contrast, aluminum nitride, for example, is a material that features a fascinating combination of very high thermal conductivity (170 W/mK) and excellent electrical insulation properties. This makes aluminum nitride predestinated for use in power and microelectronic applications, where high temperatures are generated such as in circuit carriers in semiconductors or in heat-sinks in LED lighting technology or high-power electronics. Achieving maximum high performance in minimum space technology inevitably means that power semiconductors (SiC and GaN) develop ever higher junction temperatures. To protect the components and connection technology, this heat must be dissipated quickly and reliably. This is where aluminum nitride material plays a significant role.

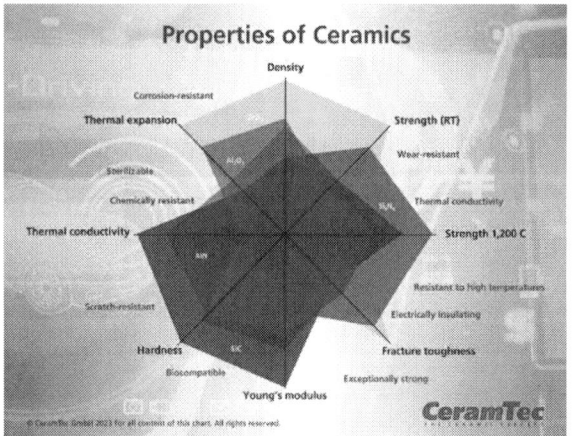

**Fig. 2**  Typical ceramic materials with their respective different excellent and unique properties.

# 2  Characterization methods

## 2.1  Bending strength

### 2.1.1  Theory

Strength is one important property of ceramic substrates. It describes the maximum stress that a material could withstand until it brakes under a specific load. Depending on the composition, the grain size of the raw materials and also the production process itself, the parameter varies statistically. The reliability of the material is determined by the distribution of defects and so-called critical defects according to the so-called weakest-link model (see Fig. 3).[1]

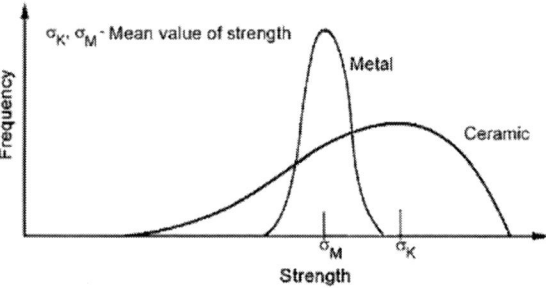

**Fig. 3**  Comparison of the strength distribution within batches of metal and ceramic material.[1]]

The parameter strength for ceramics can be divided into different types. The most crucial one is the bending strength $\sigma_M$. In addition, a ceramic material can also be characterized by the so-called compression strength. In comparison to the measurement of the bending strength, the influence of defects or inhomogeneities is small. The so-called tensile strength, however, describes the maximum mechanical tensile stress on the ceramic until the material breaks.

### 2.1.2  Different methods

The measurement results of the bending strength depend on a variety of parameters. This is the reason why it is indispensable to know the measurement conditions when different values are compared. The main influencing factors are sample geometry (*e.g.*, length, width, thickness), distance between sample and instrument, sample preparation, surface quality, velocity of the load, testing setup and testing method itself.[1]]

Commonly, there are three different test methods for the bending strength for ceramic substrates, namely the 3-point bending test, the 4-point bending test and the double ring method (ring-on-ring). The schematic setups for each method are shown in Fig. 4 to Fig. 6.

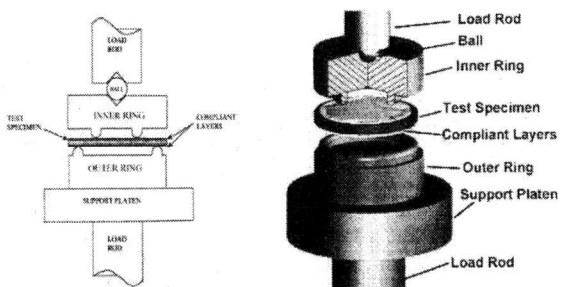

**Fig. 6** Schematic setup of equibiaxial testing in section and perspective view (double ring) according to ASTM 1499-09.[3]

For a 3-point bending test, the test specimen is placed onto two parallel supports and a third one in the middle, where the force is vertically introduced until the material fails. In a 4-point testing setup the test specimen is put under stress with two loading supports with a defined distance.

In contrast, the typical measurement setup for the double ring method is shown in Fig. 6. The test specimen is placed onto a support ring and a second load ring stresses the specimen until the material breaks. The diameter of both rings depends on the thickness of the material. The advantage of the double ring method is that the impact of the edge of the test specimen is negligible. Typically, the test specimen must be prepared by lasering in order to achieve the required dimensions.

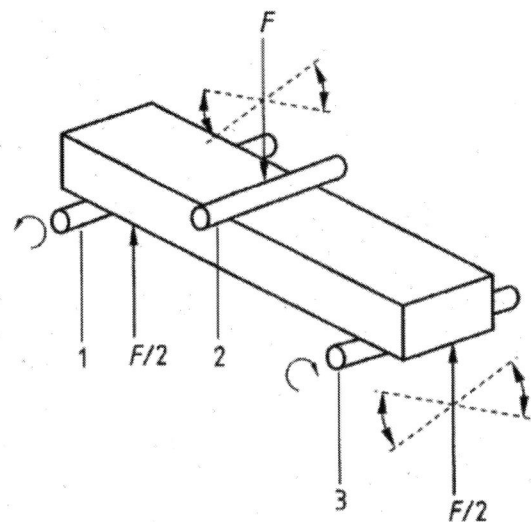

**Fig. 4** Schematic setup of 3-point testing according DIN-EN 843-1.[2]

### 2.1.3 Influence of testing methods

Figure 7 demonstrates, that the measurement method or measurement conditions influences the absolute values. The graph displays the bending strength results of 3-point bending, 4-point bending and double ring method, respectively. All test specimens of the as-fired ceramic substrate were prepared using fiber laser technology without any additional edge treatment.

The comparison is given as the difference of the individual bending strength results, whereas the 4-point testing is set as reference. All calculated values of bending strength are given as $\sigma_0$, which corresponds to a confidence level of 63.2 %. With respect to the reference (4-point testing), the 3-point method results are roughly 60 MPa higher, which corresponds to an increase of about 7.5 %. The double ring method, however, shows an increase of about 20 MPa compared to 4-point testing, which corresponds to an increase of 2.5 %.

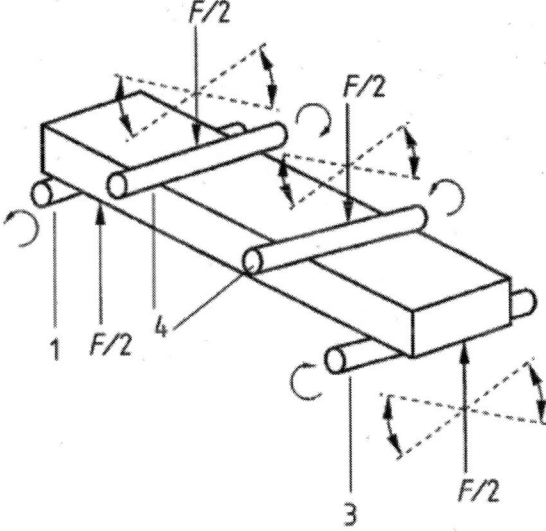

**Fig. 5** Schematic setup of 4-point testing according DIN-EN 843-1.[2]

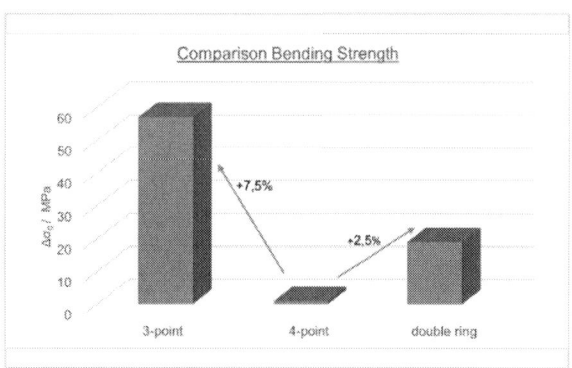

**Fig. 7** Bending strength measurement on a ceramic substrate using 3-point and 4-point testing according to DIN EN 843-1 and using double ring method according to ASTM1499-09, respectively.

### 2.1.4 Influence of testing speed

Not only the method itself has a huge impact on the test results, but also the individual measurement parameters, such as the testing speed.

Considering the 3-point measurement requirements according to DIN 843-1 and ISO 23242:2020, the testing speed must be selected so that the time to fracture is within 5 s to 15 s. The reason is that brittle ceramic material like ceramic substrates is sensible to slow crack growth in specific test environments which strongly influences the bending strength results. The use of low testing speeds leads to a bending of thin substrates which causes microcracks in the ceramic material. A lower force for the material breakdown is therefore required.

Figure 8 displays the bending strength test results performed with two different testing speeds with the same ceramic substrate. All test specimens were prepared identically without any edge treatment. When the testing speed is chosen significantly lower, which means that the material failure occurs later than within 15 s - for example 90 s - the bending strength $\sigma_0$ is about 70 MPa lower in comparison to a testing speed / time to failure of 10 s. This means that at lower testing speed, a lower force is needed until the fracture of the ceramic material occurs.

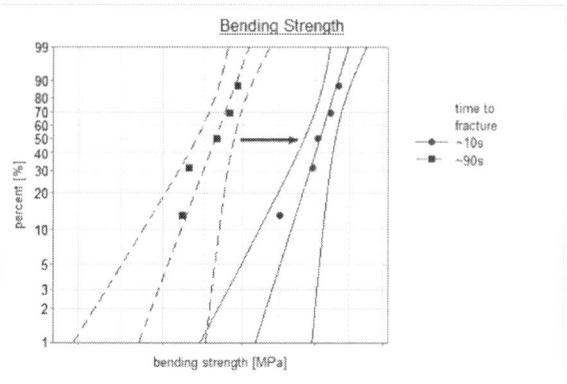

**Fig. 8** 3-point bending test on a ceramic substrate based on DIN EN 843-1. The data displayed as red points result from testing with testing speed / time to fracture of ≈90 s, the data in blue correspond to a test speed / time to fracture of ≈10 s.

## 2.2 Thermal conductivity

### 2.2.1 Theory

Thermal conductivity describes the fact how easily heat is transferred through a material. Heat is transported through a material as a result of a temperature gradient. Following the second law of thermodynamics, heat always flows in the direction of the lower temperature.[4]

The thermal conductivity is a material-specific property and can be calculated by the following equation:

$$\lambda(T) = a(T) * \rho(T) * c_p(T) \qquad (1)$$

$a(T)$ = thermal diffusivity [mm$^2$/s]

$\rho(T)$ = density [g/cm$^3$]

$c_p$ = specific heat capacity [J/gK)]

There are different possibilities how to measure the thermal conductivity on ceramics. One common method is the laser flash analysis based on the studies by Parker et al. in 1961.[5]

The material is heated from one side by an energy pulse or rather laser flash, which results in a time-dependent temperature rise, recorded by a detector on the opposite side. The higher the thermal diffusivity of the material, the faster the heat reaches the other side. Depending on the test requirements and the measurement setup, the test specimens must be prepared accordingly (see Fig. 9).[4]

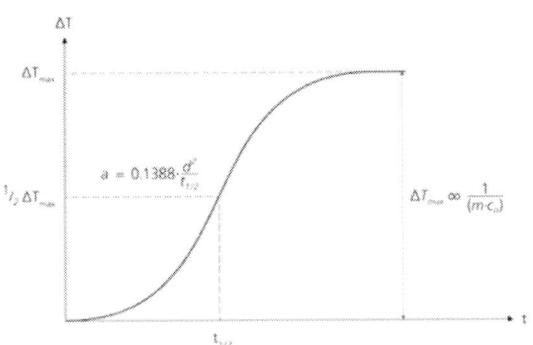

**Fig. 9** Typical schematic graph of the temperature increase on the sample backside.[4]

In general, ceramic materials have a lower heat conductivity in comparison to other materials like metals or metals compounds. Depending on the application, some ceramic substrates are used as thermal insulating material due to their relatively low thermal conductivity, whereas other ceramic materials with high thermal conductivities are used as heat conductors.

For example, aluminum nitride is a material that features an extremely interesting combination of very high thermal conductivity (170 W/mK) and excellent electrical insulation properties. This makes aluminum nitride ideal for use in power and microelectronic applications as the resulting heat must be dissipated quickly and reliably. Table 1 shows an overview of the thermal conductivity for different typical ceramic materials.

| Material | Thermal Conductivity [W/mK] |
|---|---|
|  |  |
| Aluminum Oxide | 22 |
| Aluminum Nitride | 170 |
| Zirconium Oxide | 1.5 |
| Zirconia Toughened Aluminum | 25 |
| Silicon Nitride | 80 |
|  |  |

**Table 1** Thermal conductivities of typical ceramic materials.

### 2.2.2 Influence of material thickness

Figure 10 underscores that the measurement conditions and preparation of the test specimen are important when specific values are compared. The graphs display the difference of the calculated thermal conductivity for three different ceramic substrate thicknesses, namely <0.5 mm, >0.5 mm and >1.0 mm. Each measurement was performed under the same conditions with the same ceramic material.

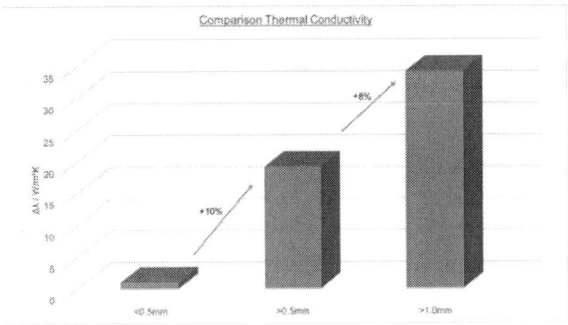

**Fig. 10** Measurement of thermal conductivity of a ceramic material of different thicknesses according to DIN EN 821-2.

# References

[1] Informationszentrum Technische Keramik (Hrsg): Brevier – Technische Keramik, Fahner Verlag, 2003.

[2] DIN EN 843-1:2008-08, Advanced technical ceramics – Mechanical properties of monolithic ceramics at room temperature – Part 1: Determination of flexural strength; German version EN 843-1:2006.

[3] ASTM C 1499-09 (2013), Standard Test Method for Monotonic Equibiaxial Flexural Strength of Advances Ceramics at Ambient Temperature.

[4] https://analyzing-testing.netzsch.com/de/produkte-und-loesungen/waermeleitfaehigkeitsbestimmung-und-temperaturleitfaehigkeitsbestimmung.

[5] W.J. Parker; R.J. Jenkins; C.P. Butler; G.L. Abbott (1961). "Method of Determining Thermal Diffusivity, Heat Capacity and Thermal Conductivity". Journal of Applied Physics.

PCIM Europe 2024, 11– 13 June 2024, Nuremberg    DOI: 10.30420/566262151

# AI-Enhanced Vacuum Reflow Oven: Precision Control for Reliable Large-Area Soldering

Chih Hui Lee [1*], Ta Chung Liu [2], Tsung I Lin [3], Shao-Yu Wang [4], Chung-Kun Yen [5,] Cheng-Tang Pan [6], Yow-Ling Shiue [7*]

[1] Institute of Biomedical Sciences, College of Medicine, National Sun Yat-sen University,Taiwan

[2] Mustec Corp., Taiwan

[3] Lighten Corp., Taiwan

[4] Department of Mechanical Engineering, National United University, Taiwan

[5] Department of Mechanical and Automation Engineering, I-Shou University, Taiwan

[6] Department of Mechanical and Electro-Mechanical Engineering, National Sun Yat-sen University, Taiwan

Department of Mechanical and Electro-Mechanical Engineering, National Sun Yat-Sen University, Taiwan
Institute of Advanced Semiconductor Packaging and Testing, College of Semiconductor and Advanced Technology Research, National Sun Yat-Sen University, Taiwan
Institute of Precision Medicine, National Sun Yat-Sen University, Taiwan

[7] Institute of Precision Medicine, College of Medicine, National Sun Yat-sen University, Taiwan

Corresponding author:    Yow-Ling Shiue, e-mail: juno2411@gmail.com
Speaker:                          Chih Hui Lee, e-mail: julia@mstc.com.tw

## Abstract

LED light source applications are becoming more and more widespread, and high-power components with large-area bonding (area > 20mm x 20mm) applications are growing rapidly in the automotive, aerospace, and other industries. In the past few years, silver sintering technology has become the subject of research for large-area bonding. However, this method is limited by the high cost of materials and the technical difficulty of the process environment. Tin-based solder paste bonding through proven reflow technology and lower material costs offers a promising approach for mass production, but the critical issue of void control should be addressed. In this study, high-quality bonding with void content of less than 5% was achieved in large bonding areas up to 40mm x 40mm. Key process parameters such as temperature and vacuum were optimized to provide guidance for actual production.

keywords：sing-chamber, large-area, void ration less, vacuum reflow oven.

## 1    Introduction

The replacement of traditional light sources by LEDs has emerged as prevailing trend. Driven by EU regulations and push for green energy, the lighting industry's development is predominantly centered around LED light sources [1].

With their high efficiency and environmental friendliness, LED have become the focal point for innovation in lighting technology. Apart from household lighting, the versatility of LEDs has led to the development of various product size and optical designs. Decentralized arrangements facilitate innovative optical designs, enhancing the application of high energy limps. High power chip integration technology offers notable advantages, including reduced time to market, enhanced performance, lower power consumption, improved signal integration, and cost effectiveness. Consequently, high power chips have garnered significant attention from the automotive industry, encroaching on some marker share [2].

The escalating development in this field has prompted component developers to continuously enhance the assembly and packaging technology of power electronic devices. One focal point is on improving soldering product quality by reducing solder void size relative to die size from 15% to 5% or less, aiming to achieve better reliability, heat dissipation, and conductivity. Additionally, there is a booming trend in the application of high-power chip packaging process is high stringent. According to the results discussed in related papers, solder joint failure is one of the significant factors affecting the reliability if electronic products [3].

The advent of large-coverage and high-performance power substrate/heatsink assemblies has brought significant attention to the forefront. Emphasizing bond strength, reliability, high electrical/thermal conductivity, and superior thermal reliability, the focus of research and development has intensified. However, this study is directed towards specialized lighting fixtures necessitating high power and intricate optical designs. The package dimensions notably surpass current standards, with the joint area of the power module reaching exceptionally large dimensions (40 mm x 40 mm). Solder joint characteristics prove exceedingly sensitive to the quality of electronic-grade packaging. Both the morphology of the intermetallic compound (IMC) layer and post-welding porosity play pivotal roles in determining the performance quality of forthcoming products. Meeting stringent void ratio requirements for large-area joints poses a recurring bottleneck in the packaging processes.

It is very difficult to implement traditional tunnel-type reflow ovens. This experiment will use the new single-chamber vacuum reflow soldering. Achieve the actual implementation effect of low void rate in a large area.

## 2 Research motivation

The challenge encountered in large area reflow soldering lies in the difficulty of eliminating voids caused by central bubbles. However, these voids often pose risks of product short circuits or burnouts [4]. Large voids can lead to deviations in thermal and electrical paths. There are various reasons for porosity in welded joints. In this experiment, we use a vacuum system into the process conditions to enhance it. The presence of defects can compromise mechanical robustness, thereby affecting the reliability and performance of electrically conductive solder joints. Nevertheless, the optimal void rate of products produced by traditional tunnel-type reflow ovens typically falls be-

tween 10% and 8%. This range remains substantial, especially considering the larger size of the experimental product, which measures 40mm x 40mm and is intended for larger, high-power special lamps. Under such circumstances, the utilization of products produced by traditional reflow furnaces may lead to decreased quality and potential reliability issues in product performance [5]. However, vacuum reflow furnace techniques have demonstrated significant improvements in reducing void size.

However, many numerous documents discuss the utilization of eutectic processes in the LED high-power packaging process, the direct Au/Sn eutectic bonding process is complex. Controlling the quality of the eutectic layer proves challenging and is easily influenced by the substrate's manufacturing state and bonding conditions. This complexity often results in the formation of voids within the eutectic bonding layer, thereby impacting thermal resistance and product performance [6].

## 3 Experiment

Although vacuum reflow soldering technology has been proven to significantly reduce void size [7], traditional reflow soldering ovens are commonly bulky, consume high energy, and have long product change times. They are typically used in large equipment. The performance of non-vacuum reflow soldering equipment often fails to meet the requirements for large area soldering and small cavity sizes.

In this experiment, a new vacuum reflow oven is utilized. Besides its vacuum function, it incorporates a multi-point AI temperature control thermal energy output model for product temperature feedback. Therefore, in a section where temperature control accuracy is high and vacuum can be controlled at will, this experiment can be flexibly established with a more diverse range of experimental contents.

### 3.1 Materials and product structures

The Sn-Ag-Cu (SAC) lead-free solder alloy has become a good joining material with low cost, good mechanical strength, and impact compared with Pb environ mental small. In the field of optoelectronics, no-clean flux has been developed in recent years. With the changing needs of products, after reflowing it to clean the flux maybe is not good.

This product uses PF602-P (SHENMAO) no-clean solder paste as the joint agent. The parameter recommendation curve is shown in Figure 1. The recommended reflow operating temperature for PF602-P solder paste is 165°C~185°C.

The LED light source is welded to the ceramic substrate. The overall heat dissipation and isolation design is combined with a copper plate under the large area ceramic substrate, with a joint area of 40mm x 40mm. The structure diagram is shown in Figure 2. The LED die bonding is based on SAC305 solder, which is not within the scope of this experiment and will not be described.

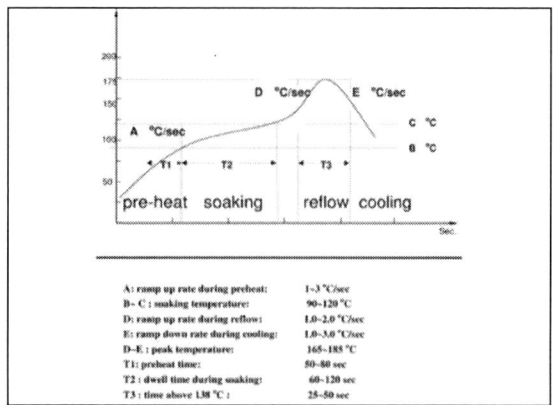

**Fig. 1**     Recommended temperature profile.

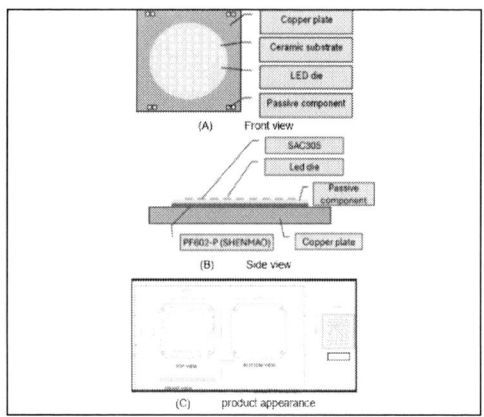

**Fig. 2**     Product structure diagram

## 3.2   Equipment introduction

The single-chamber vacuum reflow furnace is equipped with intelligent feedback for product temperature and fine tuning PID output. Despite have only one chamber, it can efficiently manage preheating/soaking temperature, reflow temperature, and vacuum pressure reduction rate. All processes are seamlessly executed within this single and cooling rates. Another notable feature is the ability to adjust vacuum settings, opening time, and duration based on the product's bubble condition, thereby achieving a relatively low void rate.

## 3.3   Experiment

### 3.3.1   Experiment process

The test sample to prepare, Sn-Ag-Cu (SAC305) lead-free solder paste (wet solder paste) is first printed on the LED die attach (DA) between the ceramic substrate and the die. The test samples then underwent a vacuum reflow process (using the SAC305 profile). The next main topic of discussion is to use low-temperature solder paste to join the ceramic substrate and the copper plate, using a steel plate with a thickness of 0.35mm and an area of 40mmx 40mm, and then performing a vacuum reflow process through the PF602-P profile. After passing through the vacuum reflow oven, use 2D X-RAY to measure the gap size. The experiment process as Fig.3.

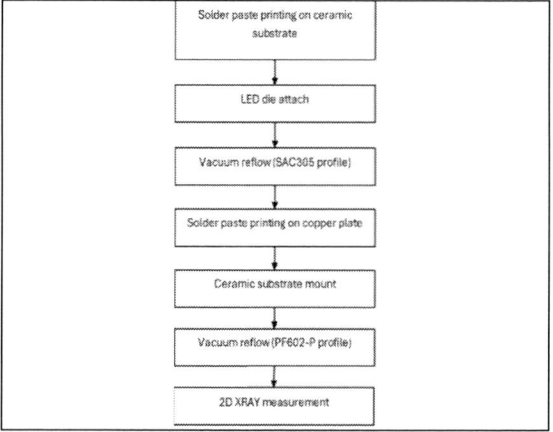

**Fig. 3**     Make a flowchart

### 3.3.2   Experimental parameters

The job of this experiment is to eliminate voids in large-area welding. The characteristic of the vacuum reflow oven we use is that it can feed back the temperature control system according to the actual temperature of the product. Finetune the temperature curve through intelligent temperature control and vacuum residence time adjustment. According to the reference, lower vacuum pressure is mentioned as the main reason for void reduction [8]. Therefore, the vacuum pressure of each set of experiments was controlled below 10torr. Higher reflow temperature and longer 10 Torr vacuum time have a greater impact on void reduction [9], so three main factors were: reflow peak temperature (170℃/180℃), residence time during peak temperature (10 sec/20 sec), and vacuum residence time (reflow time/soaking time + reflow time). Time, vacuum level, and solder paste thickness were set to constant values to plan the following experiments. There are 8 groups in total, and each group of experiments is performed 9 pcs.

### 3.3.3 Examine

The use of X-ray image analysis in the field of industrial production is affirmed, and it is crucial to evaluate the quality and reliability of many products. In this experiment, 2D X-ray can be used to detect voids and has reference value [10].

Image results of visually controlling void inspection die attachment using 2D x-ray imaging Fig.4.

The conditions for different groups are described below：

|   | Peak Temp. | Dewell Time during peak Temp. | Vacuum time |
|---|---|---|---|
| 1. | 170℃ | 10 sec | reflow time |
| 2. | 170℃ | 20 sec | reflow time |
| 3. | 180℃ | 10 sec | reflow time |
| 4 | 180℃ | 20 sec | reflow time |
| 5 | 170℃ | 10 sec | soaking time + reflow time |
| 6 | 170℃ | 20 sec | soaking time + reflow time |
| 7 | 180℃ | 10 sec | soaking time + reflow time |
| 8 | 180℃ | 20 sec | soaking time + reflow time |

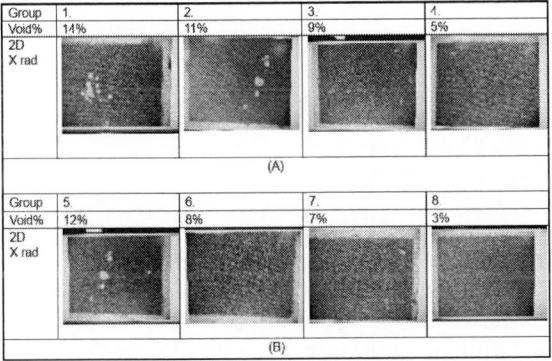

**Fig. 4** 8 sets of sampling 2D X-Ray photos; (A) is the void condition represented by the 1 to 4 groups only adding vacuum control in the reflow section. (B) is the void condition represented by the 5 to 8 groups that adding vacuum control in the reflow + soaking section.

### 3.3.4 Result

Fig5. (A) The group 1 to 4 on the vacuum during the reflow time. The performance of the voids rate of each group is as follows:

Group 1: the voids rate respectively 11.6%, 12.5%, 13.4%, 16%, 14.5%, 15%, 13.2%, 17.4% and 12.3%, the mean is 14%.

Group 2 : the voids rate respectively 10.7%, 11%, 9.8%,11.5%, 10.8%, 11.3%, 12%, 10% and 10% the mean is 11%.

Group 3 : the voids rate respectively 10%, 9%, 11%, 11%, 8% 9.6%, 8%,9% and 9% the mean is 9%.

Group 4 : the voids rate respectively 5%, 6%, 5%, 6%, 5%,6.8%, 5%, 4% and 3.4% the mean is 5%.

Fig 5 (B). The group 5 to 8 on the vacuum during the soaking time and reflow time. The performance of the voids rate of each group is as follows:

Group 5: the voids rate respectively 13%, 12%, 12.5%, 12%, 13%, 12.8%, 12%, 11% and 12%, the mean is 12%.

Group 6 : the voids rate respectively 8%, 6.6%, 8%, 8.4%, 9%, 6%, 8.8%, 8.8% and 8% the mean is 8%.

Group 7 : the voids rate respectively 6%, 6%, 6%, 7.6%, 7.8% 7%, 6.9%,6.8% and 6.7% the mean is 7%.

Group 8 : the voids rate respectively 3%, 3.1%, 2.8%, 2.9%, 2%, 2.8%, 2.6%, 2.4% and 2.4% the mean is 3%.

In this testing, we solely implement a vacuum environment during reflow soldering to facilitate the reflow of solder paste while expelling air bubbles. Unlike in a conventional reflow oven, where the product exits the preheat zone and the solder paste covers the gas, providing vacuum assistance as it enters the reflow section, the single chamber furnace used in the experiment streamlines this process. Here, the product doesn't need to flow out of the preheating zone; it simply waits for the solder paste to soften before receiving vacuum assistance. By adjusting the pressure difference to reduce trapped gas, we aim to minimize solder voids effectively.

Under the same temperature and dwell time conditions, the results obtained by turning on the vacuum solely during the reflow soldering time interval significantly differ from those achieved by applying vacuum during both the soaking time and reflow soldering Fig. 5 (C). The experiment revealed that the impact of maintaining a low-pressure environment and minimizing gaps during the soaking period is more pronounced.

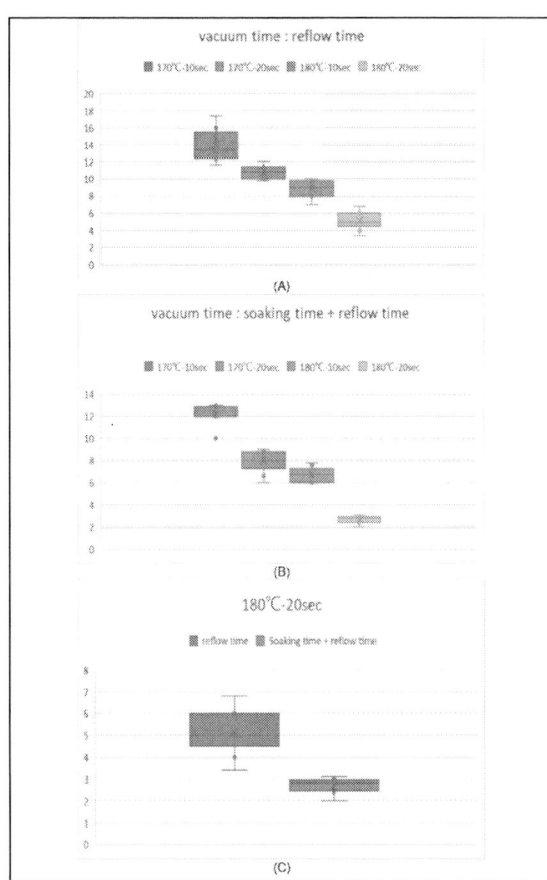

**Fig. 5** (A) In groups 1 to 4. First set of results showed that the average void ratio 14% (Blue box). When the vacuuming time dwell to 20 seconds, the void ratio 11% (orange box). And the temperature to 180°C and the residence time to 20 seconds can reduce the void ratio to 5% (yellow box).

(B) In groups 5 to 8, we increase the soaking time and provide a vacuum environment to allow the solder paste to reflow while squeezing out air bubbles. The fifth set of results shows that the average void ratio reaches 12% (blue box). When the vacuuming time was extended to 20 seconds, the results dropped significantly to 8% (orange box). Adjust the temper-nature to 180°C and the residence time to 20 seconds to reduce the void rate to 3% (yellow box).

(C) The experimental results show that the void rate can reach 5% or less at Group 4. When the vacuum

residence time is added at soaking time, the result is < 3%, which is the best effect.

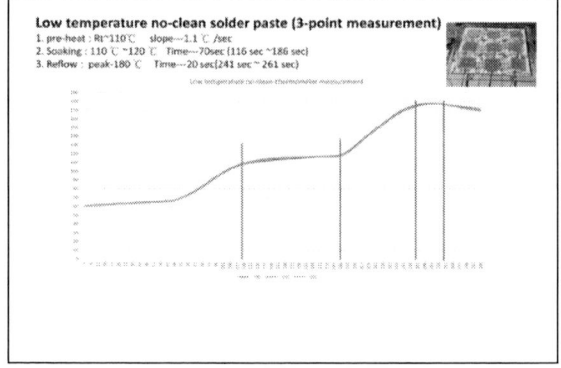

**Fig. 6** The temperature curve of the equipment used in this experiment for experimental testing.

The curve in Fig.6 clearly illustrates the performance of the AI single chamber vacuum reflow in terms of temperature control and accuracy. The temperature displayed in the figure represents the feedback of the actual temperature of the product within the chamber.

In this testing, a vacuum environment solder pastes in reflowing while expelling air bubbles. Unlike in a conventional reflow oven, where the product exits the preheat zone and the solder paste covers the gas, providing vacuum assistance as it enters the reflow section, the single-chamber furnace used in the experiment streamlines this process. Here, the product doesn't need to flow out of the preheating zone; it simply awaits the softening of the solder paste before receiving vacuum assistance. By adjusting the pressure difference to reduce trapped gas, our objective is to effectively minimize solder voids.

## 4 Conclusion

The difficulty in large-area bonding lies in effectively removing air bubbles from the middle. In this experiment, we boldly utilized a new vacuum reflow furnace to conduct a large-area bonding process and found that the conclusions obtained under this single-cavity temperature-controllable model were quite excellent.

In this study, the important factors that form reference to other literature DOE, the conclusions obtained by utilizing temperature, vacuum value, and vacuum time in this new vacuum reflow furnace are not only applicable but also indicate that maintaining the vacuum during the reflow time + soaking time can obtain the best result, with a voids rate of < 3%. It is a very positive discovery that such a result can be achieved in an area of 40mm x 40mm. In this experiment, it was observed that the new vacuum reflow oven exhibits precise temperature control characteristics and diversity of vacuum opening timing. As indicated by the AI-enhanced model discussed earlier, significant enhancements are evident in temperature control and feedback, resulting in improved solder paste melting and vacuum duration. Consequently, this amalgamation contributes to a reduction in void content to less than 3%. Moreover, it was discovered that AI can effectively regulate the PID value. Conducting experiments on products with more stringent electrical requirements is expected to yield favorable outcomes.

# References

[1] Implementing Directive 2005/32/EC of the European Parliament and of the Council With Regard to Ecodesign Requirements for NonDirectional Household Lamps, European Parliament, Brussel, Belgium, (EC) no. 244/2009, 2009.

[2] Maximilian Schmid, Andreas Zippelius, Alexander Hanß, Stephan Böckhorst, and Gordon Elger, "Investigations on High-Power LEDs and Solder Interconnects in Automotive Application: Part I—Initial Characterizatio" IEEE TRANSACTIONS ON DEVICE AND MATERIALS RELIABILITY, VOL. 22, NO. 2, JUNE 2022.

[3] Li Xunping, Zhou Bin, Yang Shaohua, En Yunfei," The Coupling Effect of Size Effect and Krikendall Voids on the Fracture Features of Ni/Sn3.0Ag0.5Cu/Cu Joint" 978-1-4799-4707-2/14/ ©2014 IEEE

[4] W. Gao, Q. Guo, Y. Peng, M. Ren, B. Zhang, and S. Cai, "Influence of solder layer void on thermal stability for power semiconductor device," in Proc. 20th Int. Conf. Electron. Packag. Technol. (ICEPT), Aug. 2019, pp. 1–4, doi: 10.1109/ICEPT47577.2019.245120.

[5] D. C. Katsis and J. D. van Wyk, "A thermal, mechanical, and electrical study of voiding in the solder die-attach of power MOSFETs," IEEE Trans. Compton., Package., Technol., vol. 29, no. 1, pp. 127–136, Feb. 2006, doi: 10.1109/TCAPT.2005.853301.

[6] Te-yuan Chung, Jian-Hong Jhang, Jing-Sian Chen, Yi-Chien Lo, Gwo-Herng Ho, Mount-Learn Wu, Ching-Cherng Sun," A study of large area die bonding materials and their cor-

responding mechanical and thermal properties", Microelectronics Reliability 52 (2012) 872-877

[7] Ningning Wang, Zongpeng He, Binbin Zhang, Rong An, "Research of Vacuum Soldering on 10# Steel and Thick Film Ceramic Substrates," 2016 17th International Conference on Electronic Packaging Technology 978-1-5090-1396-8/16©2016 IEEE

[8] Siang Miang Yeo, Ho-Kwang Yow, Senior Member, IEEE, and Keat Hoe Yeoh," Critical Threshold Limit for Effective Solder Void Size Reduction by Vacuum Reflow Process for Power Electronics Packaging," IEEE TRANSACTIONS ON COMPONENTS, PACKAGING AND MANUFACTURING TECHNOLOGY, VOL. 13, NO. 7, JULY 2023

[9] Siang Miang Yeo, Azman Mahmood, Shahrul Haizal Ishak," Vacuum Reflow Process Characterization for Void less Soldering Process in Semiconductor Pac," 38th International Electronics Manufacturing Technology Conference 2018

[10] Ondrej Kovac, Tomas Girasek, Alena Pietrikova," Image Processing of Die Attach's X-Ray Images for Automatic Voids Detection and Evaluation," 2016 39th International Spring Seminar on Electronics Technology (ISSE) 978-1-5090-1389-0/16 ©2016 IEEE

PCIM Europe 2024, 11– 13 June 2024, Nuremberg    DOI: 10.30420/566262153

# Corrosion-Compatible Drive Electronics for Electric Vehicles and Industrial Power Modules

Tom Petzold[1], Ronald Eisele[2], Markus Bast[1], Armin Hindel[1], Bennet Lorbeer[1], Knud Gripp[1], Sara Panahandeh[3], Corinna Grosse-Kockert[3], Daniel May[3], Mohamad Abo Ras[3], Anu Mathew[4], Susana Richter-Trummer[4], Rico Eichhorn[4], Sven Rzepka[4]

[1] Forschungs- und Entwicklungszentrum Fachhochschule Kiel GmbH, Germany
[2] Fachhochschule Kiel, Germany
[3] Berliner Nanotest und Design GmbH, Germany
[4] Fraunhofer-Institut für Elektronische Nanosysteme ENAS, Germany

Corresponding author: Tom Petzold, tom.petzold@fh-kiel.de[1]

## Abstract

One of the tasks of the research project is to develop large-area sintered joints on copper and aluminum water heat sinks in order to realize corrosion-compatible use in electric vehicles. To this end, strategies are being developed to equalize the strain differences of the different materials in the assembly and interconnection technology of the power modules, thus helping to reduce the Coefficients of Thermal Expansion (CTE) mismatch. This will help to extend the lifetime of power modules. Therefore, one goal is to conduct investigations with different materials and metallization between the power electronics and the heat sink, so that an adjustment of the CTE to the packaging and interconnection technology in the active (hot) area can be made. To achieve these goals, the development of suitable large-area sintering pastes and the implementation of necessary modifications to sintering tools and machines are required. Furthermore, the development of suitable failure analysis methods are helpful for the analysis of defects in the samples.

## 1   Introduction and motivation

With increasing power density in electronic drive modules in electric vehicles and industrial applications, good cooling is required, which can only be realized with liquid cooling. To withstand years of use, however, the liquid cooling circuit is not sufficiently protected against corrosion, which occurs due to the use of different materials. Typically, the cooled housing of the electric motor is made of aluminum and the cooler of the power module is made of copper. The aluminum of the e-machine would be the sacrificial anode in the water-driven bimetallic corrosion process and would rapidly corrode in a destructive manner. This will inevitably lead to the failure of the most cost-intensive assemblies (e-machine and motor-drive). To prevent corrosion in the cooling circuit, the materials must be harmonized. The aim of the project is to replace the classic soldered copper heat sink, often with nickel-plated surfaces,

by a sintered aluminum heat sink to achieve greater reliability. For this purpose, a suitable sintering tool and manufacturing processes must be adapted and verified. Large area sintering on liquid coolers and sintering on aluminum are not yet established solutions in today's manufacturing processes (see Fig. 1). But it has been shown in research projects that silver sintering on aluminum is generally possible. [1]

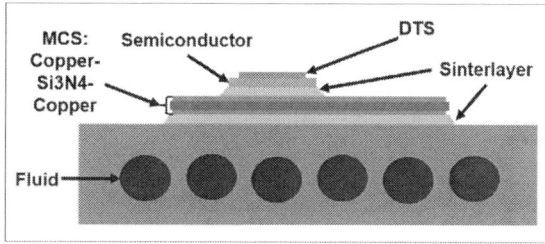

**Fig. 1** Schematic cross-section of an exemplary all-in-one-sintered module consisting of a semiconductor with die top system (DTS®) sintered to a direct bonded copper (MCS), which is sintered on a fluid cooler.

In contrast to soldering, sintering makes it possible to join the bond buffer, semiconductor, ceramic substrate and cooler in a single step. FEM (Finite Element Method) is used to simulate and validate different layer thicknesses of the required silver sintering paste in the module in order to find a functioning solution based on the CTE mismatch. The sintering layer thickness should be as low as possible for resource-saving production but must also ensure a durable connection due to the tension.

## 2 Large-Area sintering

### 2.1 Challenges

In addition to soldering, sintering is used to join modules on coolers. The advantage of sintering with silver pastes is the higher thermal conductivity of the joint, the lower layer thickness, higher mechanical strength and a high release temperature of the joint. The melting point of SAC (Sn, Ag, Cu) solder is approximately 221°C and that of silver 962°C. Sintering generally enables higher operating temperatures for the entire system. Due to the better thermal conductivity of the sintered compound, there is better heat transfer from power module to the cooler, which lowers the semiconductor temperature and increases the service life of the module. [2]
The joining of copper and Ag / AuNi coated materials with silver sintering is a well-known and proven process for joining semiconductors on active metal brazed substrates, for example. This requires an adjustment of the sintering parameters due to the larger thermal masses of the joining partners and a geometric adjustment of the sintering tool. In order to join power modules with a copper underside to aluminum coolers using proven processes, silver or copper plating is essential. A nickel coating is cheaper and serves as an oxide barrier but requires adjustments to the production process due to the material-specific properties.

### 2.2 Design of a power module with water cooler

The aim of the project is to build a functional power electronic demonstrator based on the DCM 1000 module by Semikron Danfoss with sintering technology (Fig. 2). So far, the original module is soldered by Danfoss. In this study, the solder layer is replaced by a sintered layer and a bottom-up all-in-one sintering process is developed.

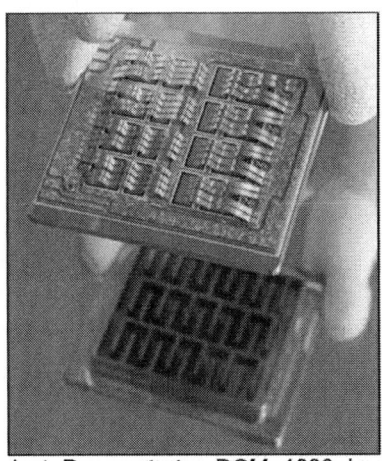

**Fig. 2** Project Demonstrator DCM 1000 large-area sintered on ShowerPower3D® from Semikron Danfoss

## 3 Large area, quasi hydrostatic sintering

### 3.1 Requirements of the sintering tool

The sintered connection between the cooler and the substrate requires a homogeneous sintering pressure over the surface of the substrate.
The use of a uniaxial sinter stamp in an all-in-one sintering process for a homogeneous pressure distribution proves to be difficult due to the height differences between the substrate, the semiconductors, gate resistors and sensors. The sintering tool must adapt flexibly to the different sizes of the components and compensate for this.
The quasi-hydrostatic sintering process, which has been established at Kiel University of Applied Sciences for many years, is suitable for all-in-on sintering on coolers. The soft stamp compensates for the height differences of the components and ensures uniform pressure on the substrate. Adjustments have to be made to the tool and the processes due to the geometries. The functional principle of the silicone pad as a sintering stamp is based on a quasi-hydrostatic pressure generation, which takes place through a pressure chamber with variable volume. The following figure shows a schematic representation of the sintering tool.

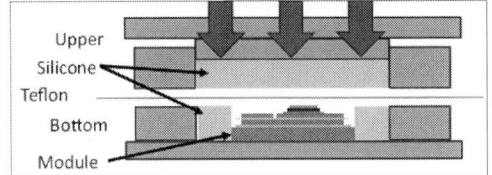

**Fig. 3** The sintering tool used in the project.

With quasi-hydrostatic sintering tools, the interior of the sintering tool is closed by the upper and bottom plungers so that the sintering pressure can be built up in the tool. By moving the tool together, the silicone of the sintering tool flows around the sintered material and transfers the force of the sintering press evenly to it.

In the project, the press of the project partner Budatec was used as the basic sintering press, as the SP300 press has all the necessary framework process parameters. Sintering parameters such as sintering under nitrogen, temperature 250°C, pressure 25 N/mm² with a holding time of 300 seconds can be generated. Thanks to the downstream cooling unit, the product can be cooled under process gas after the sintering process and removed without oxide.

In order to successfully sinter a substrate onto a cooler, the lower part of the sintering tool must be adapted to the geometric shape of the cooler. This is achieved by a negative mold of the lower shell. When designing the tool, care must be taken to ensure that distortion-free and damage-free results are achieved for the ceramic component. A real image of the lower shell can be seen in Fig. 4.

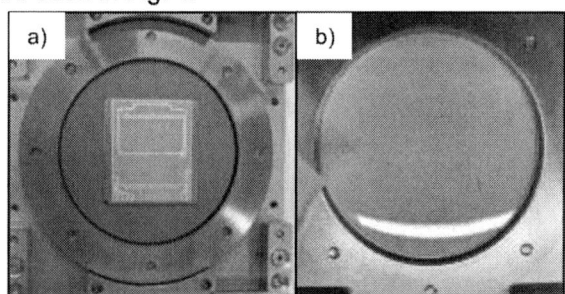

**Fig. 4 a)** Lower shell of the sintering tool **b)** lower shell with pressure-sensitive film on the right side

After loading and mounting the tool, it is placed on the conveyor belt of the SP300 sintering press. The sample is then exposed to the desired atmospheres in a cold state. Fig. 5 depicts this state. Once the process chamber is closed, nitrogen is introduced into the chamber to prevent oxidation on the sintered material. After the exclusion of oxygen, the sintering tool is transported to the back of the machine.

**Fig. 5** First state from the Budatec SP300 with sintering tool.

The sintering tool is then moved to the rear of the machine. The press then lifts the sintering tool with the preheated bottom hotplate out of the conveyor belt and presses it against the preheated top hotplate. As soon as the sintering tool comes into contact with the upper heating plate, the sintering tool is compressed, which closes the interior of the sintering tool. The heating of the sintering tool causes the silicone to expand and soften. Due to the thermal expansion of the silicone, the sintering force of the press must be continuously regulated. This effect is the desired quasi-hydrostatic behavior, which makes it possible to work with the same pressure ratios on all internal surfaces of the entire sintering tool. As soon as the interior of the sintering tool is completely filled with silicone, the pressure begins to rise due to the increasing pressing force until the sintering pressure in the interior (pressure chamber) of the sintering tool has built up. Fig. 4 b) shows the lower shell with an inserted substrate with cooler and a pressure-sensitive film (max. 50 N/mm²). The uniform red colouring shows the homogeneous pressure distribution when the tool is cold.

**Fig. 6** The sintering tool between the heating plates.

During the sintering process, the press permanently maintains the required contact pressure. Once the sintering process is complete, the pressing force is reduced and the silicone returns to its original shape. The hot sintering tool is then placed on the conveyor belt and transported to the coldplate.

## 3.2 Procedure for the sintering tests

### 3.2.1 Process determination

To enable large-area sintering of the metal ceramic substrate (MCS), substitute base plates made of OF-copper measuring 65 mm x 60 mm x 3 mm are used instead of coolers for process development. The size is similar to the dimensions and properties of the cooler of the power module. The substitute base plate, the sintered connection and the MCS are shown in Fig. 7.

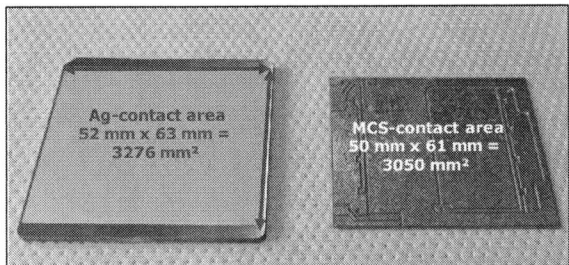

**Fig. 7** Size of the sinter pad and the MCS

Successful sintering consists of a drying process and a following sintering process. The sintering paste is first applied to the substitute base plate using a template and then dried. The drying process takes place in an oven and is determined by time, temperature and atmosphere. The drying of the sintering paste was controlled and optimized using thermal analytic methods. Pressure is added as a parameter in the sintering process.

To analyze the process, the samples are examined with a scanning acoustic microscope (SAM) for full-surface attachment. To verify the service life of the sintered joint, the samples are tested with a passive temperature cycle test of 1000 cycles (-40°C / 125°C - 30 min heating / 30 min cooling) and read-outs are carried out every 200 cycles using SAM analyses.

The sintering process is only considered good if the attachment area is > 90%. This sintering process now serves as a basis and the parameters are adapted to the individual changes in the samples.

### 3.2.2 Sinter layer thickness

The sinter layer thickness is crucial for the process and the power module. In conventional processes, this is ~100 µm before drying. Increasing the thickness of the sintered layer leads to better heat spreading and mechanical relief of the sintered layer in the finished structure. However, this also results in greater consumption of resources and makes the process less economical. The sinter paste is applied as thinly as possible and as thickly as necessary.

### 3.2.3 Sintering on cooler

In the previous investigations, optimum manufacturing processes were developed on the substitute base plate without a cooling structure. The developed sintering process and the adapted sintering tool are verified on the ShowerPower3D® heat sink from Semikron Danfoss, the quasi-hydrostatic sintering tool that was adapted and extensive shear tests were carried out. Shear bodies (Cu-Ag) with the dimensions 2.3 mm x 2.3 mm x 1 mm were used for the shear value test, evenly distributed and sintered over the surface of the cooler (Fig. 8).

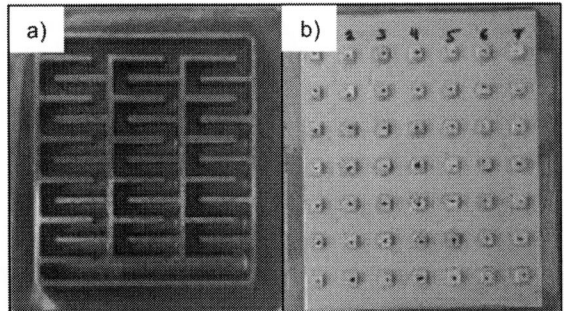

**Fig. 8 a)** Cooling structure of the ShowerPower3D® from Semikron Danfoss **b)** Shear bodies at the backside distributed over the surface of the cooler.

The shear values provide information about the homogeneous pressure and heat distribution. These are shown in the following diagrams. Each of the 7 rows results in a box plot (Fig. 9).

PCIM Europe 2024, 11– 13 June 2024, Nuremberg          DOI: 10.30420/566262153

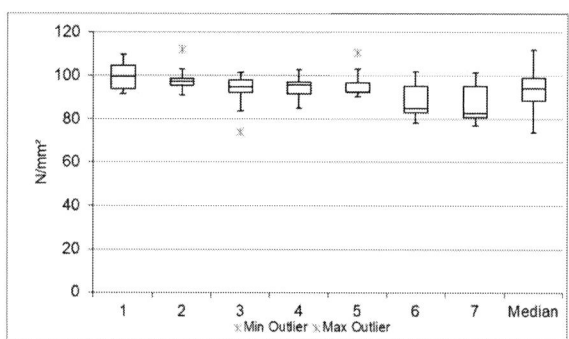

**Fig. 9** Box plot for the shear values from Fig. 8 b)

Sintering was carried out with a commercial reference paste at a pressure of 25 MPa and a temperature of 250°C for 5 minutes. The resulting median of the shear value is 94 N/mm² for all samples. The fracture code of all samples is an adhesion fracture. The individual shear values of the 49 positions are shown in Fig. 11 and show homogeneous sintering over the surface of the cooler.

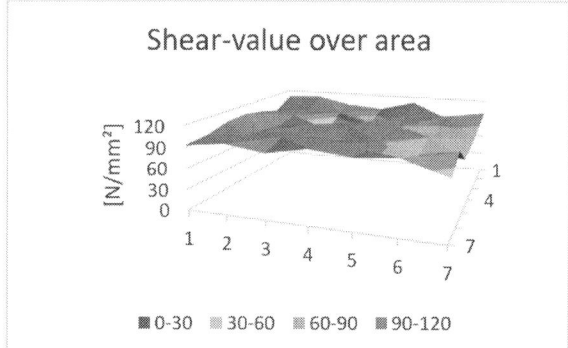

**Fig. 10** Presentation of the 49 shear values.

The MCS substrates were then sintered onto the heat sink and cycled. The results of passive temperature cycle tests will be shown in chapter 4.

### 3.3 Sintering on nickel

As described in chapter 1, the joining of not coated aluminum is not yet a conventional process. A cost-effective option is to coat the aluminum with nickel. The first examination with silver sintered joints on nickel coated aluminum substrates is a shear strength analysis.
Silver-coated aluminum shear bodies (2.3 mm x 2.3 mm x 1 mm) are used and sintered onto nickel-plated aluminum. The average of the shear strength is 48 N/mm² with a mixed fracture of the lower joining partner and the silver coating of the shear body. Furthermore, IGBTs were sintered onto the same samples and mandrel bending tests were carried out. The semiconductors

formed cracks and did not detach from the substrate, from which a successful connection can be qualitatively deduced. One of these samples can be seen in Fig. 11.

**Fig. 11** Mandrel bending test with nickel-plated aluminum substrate.

The joining process of the two previous tests was optimized and the shear strength was increased to ~ 66 N/mm². The fracture type changed to a cohesive fracture between the silver sinter paste and the shear body. In the combination of the same substrate with silver-plated copper shear bodies increases to ~ 85 N/mm². The fracture code changed to a cohesive fracture between the silver paste, the coating of the base plate and the base plate.

**Fig. 12** Shear craters, **a)** with AgCu shear body **b)** with AgAl shear body.

In the left picture from Fig. 12 a), cracks can be seen in the coating of the aluminum substrate, which highlights the greater shear values. The coating was torn by shearing off the shear bodies.

## 4 Simulation of reliability

Thermo-mechanical reliability analysis are investigated by finite element (FE) software ABAQUS release R2023. The thermo-mechanical simulation are performed on large area sintered samples to observe the mechanical behavior of the sample during the temperature cycle test. During this research, different designs of experiment has been conducted for various parameters such as aluminum and copper heatsink, thickness of sin-

1162

tered silver layer, heatsink thickness and sub-strates ($Al_2O_3$ and $Si_3N_4$). For the numerical analysis quarter symmetric FE model has been developed for computational efficiency and corresponding thermo-mechanical material property has been assigned to each layer shown in Fig. 13 respectively. A nickel layer is considered for aluminum heatsink.

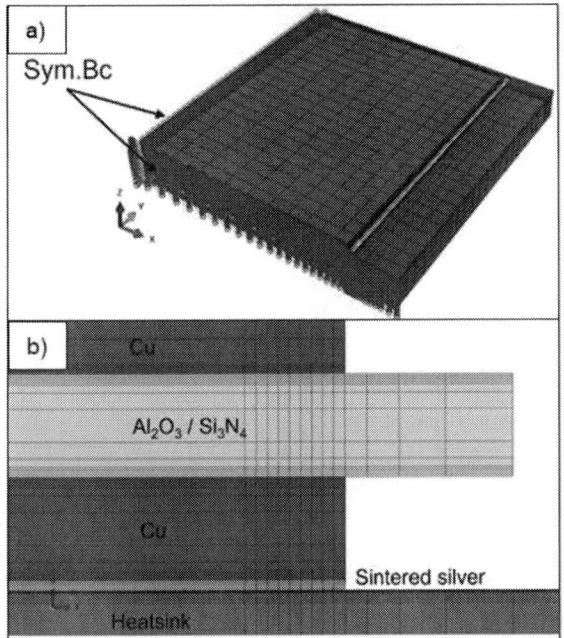

Fig. 13 FE model and material stack for finite element analysis. a) Quarter symmetry FE model and boundary condition, b) Material stack for numerical simulation

In passive temperature cycle test, uniform temperature is passively applied to whole module. The temperature profile used for thermo-mechanical simulation in given in Fig. 14. A stress-free temperature of 250°C considered before starting the thermal cycle which replicates the real condition of the samples for thermal shock test. The concept of result assessment is based on thermo-mechanical stresses and strains in the module. Initially four thermal cycles are performed for simulation and it has been observed that the cyclic equivalent plastic strain saturates after two cycles. Therefore, for further analysis only two thermal cycles are considered for mechanical analysis.

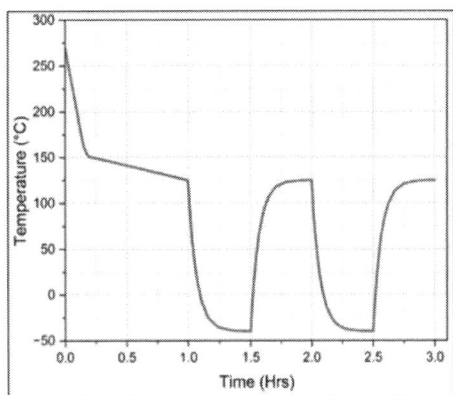

Fig. 14 Temperature profile for FE analysis

Higher cyclic equivalent plastic strain is expected to be at the outer periphery of the sintered silver layer and higher delamination of sintered silver layer is at the corner region where the cyclic equivalent plastic strain is higher. Lifetime of the module can be estimated by using Coffin-Manson equation, where $N_f$ is the number of mean cycles to failure, $\Delta\varepsilon_{pl}$ is the cyclic equivalent plastic strain, $C_1$ and $C_2$ are the Coffin- Manson coefficients. [3] [4]

$$N_f = C_1 \cdot (\Delta\varepsilon_{pl})^{-C_2} \quad (1)$$

Cyclic equivalent plastic strain values are weighted averaged with elemental volume to reduce the numerical singularity issues which will be used for estimating the lifetime of the module. The concept of result assessment is given in Fig. 15.

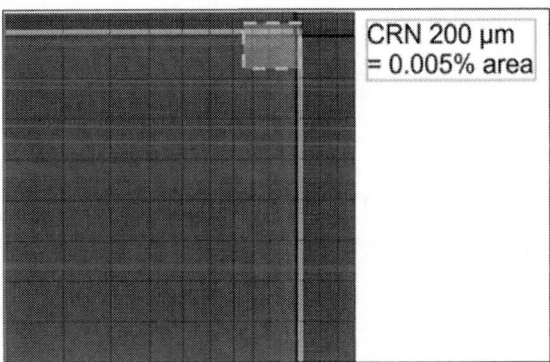

Fig. 15 Area considered (yellow box) for extracting discrete cyclic equivalent plastic strain value for lifetime estimation.

Warpage is the measure of out of plane displacement. Warpage has been observed on modules with $Si_3N_4$ on Al heatsink with different thickness at $\Delta T = 165$ K given in Fig. 16a. Concave warpage trend has been noticed in all modules. It has been also observed that the warpage decreases with increase in heatsink thickness because of bending stiffness. Bending stiffness is higher at higher thickness. Therefore, heatsink with higher thickness has lower warpage than other modules which is shown in Fig. 16b. The warpage along the diagonal of the modules is represented in Fig. 16b.

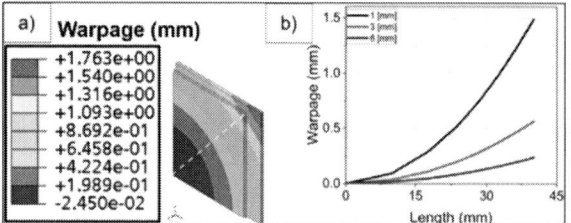

**Fig. 16** Warpage of $Si_3N_4$ substrate on Al heatsink with different thickness with sintered silver thickness of 110 µm at $\Delta T = 165$ K. **a)** Warpage of module with $Si_3N_4$ substrate on Al heatsink of thickness [1 mm]. **b)** Warpage along the diagonal (shown in yellow line) of modules with $Si_3N_4$ substrate on Al heatsink of different thickness.

Cyclic equivalent plastic strain (PEEQ) has been examined in the sintered silver layer and it has been noticed that the accumulation of strain increases with respect to increase in the thickness of heatsink. Fig. 17 gives the inference that the heatsink thickness has higher influence in the cyclic equivalent plastic strain accumulation which is at the interface between sintered silver layer and heatsink. Higher strain accumulation is observed at the corner region of the sintered silver where the delamination of sintered layer propagates. The propagation of delamination is denoted by red arrow in Fig. 17. The simulation results are validated with experimental results which is described in section 3.

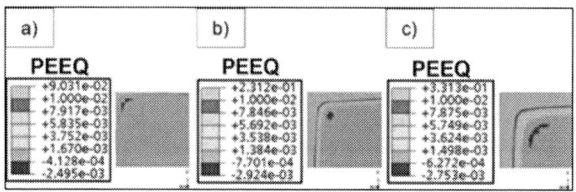

**Fig. 17** Cyclic equivalent plastic strain in sintered silver layer on Al heatsink with different thickness. The grey color on the image shows the region where PEEQ is greater than 1 %. **a)** 1 mm, **b)** 3 mm, **c)** 6 mm

The results from the DoE study showed that the module with Al heatsink has higher cyclic equivalent plastic strain than the module with Cu heatsink shown in Fig. 18. The results give an inference that the lifetime of modules with Cu heatsink is higher than that of Al heatsink. The study also shows that the increase in sintered silver thickness also enhances the lower strain accumulation in the sintered silver layer which leads to increase in the lifetime.

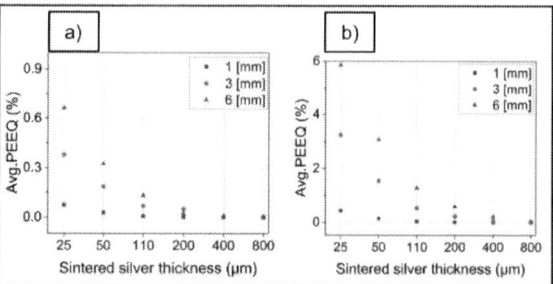

**Fig. 18** Cyclic equivalent plastic strain accumulation in sintered silver layer. Module with Cu heatsink is not scaled to 6 % to compare with Al heatsink as strain values are higher in these modules. **a)** Module with Cu heatsink. **b)** Module with Al heatsink

Warpage of the modules are also compared during this research. It has been observed that the modules with Al heatsink have higher warpage than copper heatsink which is due to the thermal mismatch. The warpage decreases with increase in heatsink thickness and sintered silver thickness from this investigation. The comparison of the warpage is shown in Fig. 19.

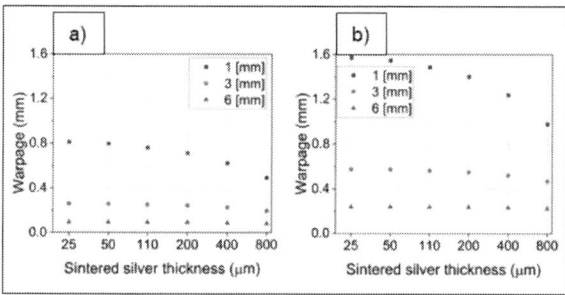

**Fig. 19** Comparison of warpage in modules with different heatsink and variation of thickness in heatsink and sintered silver layer. **a)** Module with copper heat sink. **b)** Module with Al heat sink.

# 5 Thermal, electrical and failure Analysis

## 5.1 Thermal and electrical conductivity of the sintered material

The pure sintered material was characterized thermally and electrically. The aim is to find sintering conditions that yield the best thermal and electrical, properties for the sintered material. For the measurements of thermal and electrical conductivity pure sintered stripes of width 4 mm, thickness 120 µm and length 40 mm were fabricated. They were sintered at 270°C for 300 s in the main phase. A thin teflon foil separated the sinter material stripes from the sinter soft tool. Thermal conductivity measurements were performed using Nanotests LaTIMA [4]. Electrical conductivity measurements were carried out using Keithley 6221/2182A current source and nanovoltmeter in pulse sweep mode. Fig. 21 shows the results of the in-plane thermal and electrical conductivity measurements of the sintered material as a function of sinter pressure. The results show that good thermal and electrical performance of the samples can be realized even with low pressure.

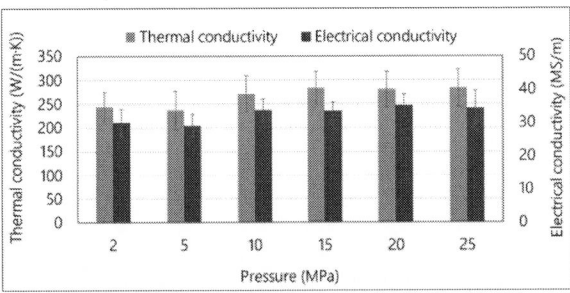

**Fig. 20** Thermal and electrical conductivity of sintered stripes, which were sintered using different pressures.

## 5.2 Failure analysis using C-SAM and pulsed infrared thermography

Passive temperature cycle tests (PCT) (-40°C / 125°C – 30 min heating / 30 min cooling, Fig. 21) were performed on MCS substrates ($Si_3N_4$) sintered on copper base plates.

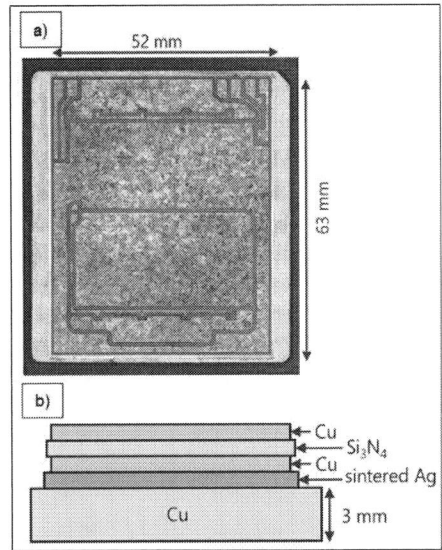

**Fig. 21 a)** top view from the setup, **b)** schematic of the samples (MCS on OF-copper plate) for the passive temperature cycle tests (not to scale).

The thickness of the applied sinter paste before drying was 100 µm. Failure analysis of the samples was performed before and after PCT by SAM [5] and Pulsed Infrared Thermography (PIRT) [6]. SAM and PIRT analysis have been performed at pristine state and at an interval of 200 cycles until 1000 cycles. Fig. 23 shows representative measurement results for samples sintered with a pressure of 12.5 MPa and 25 MPa and of a soldered sample.

Slight defects were observed by SAM and PIRT for all sintered and soldered samples at initial state. Samples sintered with a pressure of 25 MPa show less defects, than the other specimens. The two samples sintered with a pressure of 12.5 MPa displayed a high delamination of 13 % and 21% already after 200 cycles and of 26% an 38% after 100 cycles. After 1000 cycles, of the eight samples sintered with a pressure of 25 MPa, four samples showed a defect percentage of less than 6.1%, one sample showed a defect percentage of 13% and two showed a defect percentage of 70%. All of the five soldered samples also showed high defect densities of 18% - 21% after 1000 cycles which can be seen in Fig. 22.

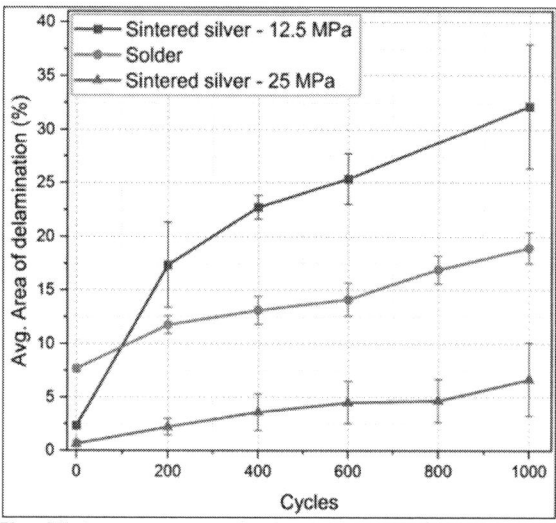

**Fig. 22** Average area of delamination of sintered silver samples with different sintering pressures and soldered samples. The error bars in the graph of sintered silver – 25 MPa and soldered samples are the standard deviation based on 5 samples each. The error bars in the graph of sintered silver – 12.5 MPa is the mean absolute deviation based on 2 samples.

As the number of cycles increases, delamination propagates from the corners to the center region, as shown in Fig. 23. The porosity increases with number of cycles, and the size of the pores also increases. Fig. 22 represents the consolidated results of delamination analysis on all samples with respect to thermal cycles. The graphs suggest that soldered samples have higher porosities than the samples sintered with a pressure of 25 MPa, which triggers the delamination in the solder layer.

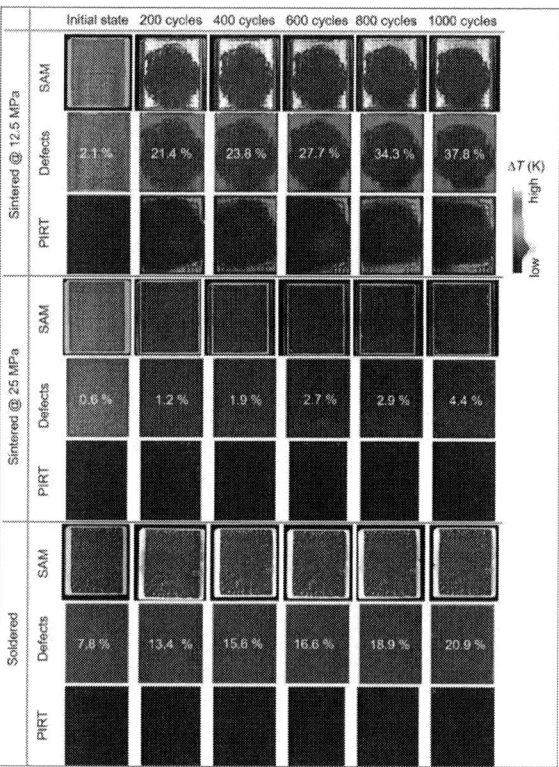

**Fig. 23** SAM and pulsed infrared thermography (PIRT) images of MCS substrates sintered (with 12.5 MPa and 25 MPa, with 100 μm initial thickness) or soldered to copper base plate after different numbers of thermal cycles. The defects were detected from the SAM images using ImageJ and confirmed by the PIRT images. PIRT images were flipped horizontally for a better comparison to SAM images.

## 6 Conclusion

Semiconductors and substrate were successfully sintered onto a cooler in an all-in-one sintering process. This was accomplished with the help of a quasi-hydrostatic sintering tool and adaptation of the process parameters. It was shown that large-area sintering of MCS's substrates with a size of 52 x 63 mm with silver sintering paste on a Shower-Power3D® by Semikron Danfoss copper heat sink is possible and represents a better connection than soldering. A successful connection after 1000 cycles was demonstrated using SAM and PIRT. The thermomechanical simulations have shown that the equivalent plastic strain increases with a reduction in the silver sintering layer thickness.

However, it was not possible to sinter on an aluminum heat sink. Further investigations are required for this.

## 7 Acknowledgement

The presented work has been performed within the project KoKo-Power funded under grant number 16ME0369 by the Federal Ministry of Education and Research (Germany). We acknowledge Marek Zajaczkowski (Nanotest) for TIFAS measurements.

## References

[1] C. Hennig et al., "Full-Aluminium-Powermodule," 2023 IMAPS Nordic Conference on Microelectronics Packaging (Nord-Pac), Oslo, Norway, 2023, doi: 10.23919

[2] Stefan Behrendt, "Analyse der thermischen Leistungsfähigkeit neuartiger Modulkonzepte unter Verwendung anorganischer Verkapse-lungsmaterialien", Martin-Luther-Universität Halle-Wittenberg, Germany 2022

[3] A. Mathew *et. al.*, "Investigation of reliability issues in sintered silver interconnected power devices and its lifetime prediction by FEM and experiment," in 12th International Conference on Integrated Power Electronics Systems, Berlin, Germany, 2022.

[4] M. Abo Ras, D. May, J. Heilmann, S. Rzepka, B. Michel and B. Wunderle, "Processing-structure-property correlations of sintered silver," in 15th IEEE Intersociety, Conference on Thermal and Thermomechanical Phenomena in Electronic Systems (ITherm), Las Vegas, USA, 2016.

[5] PVA TePla Scanning Acoustic Microscopy, "https://www.pvatepla-sam.com/produkte/sam-premium-line/," PVA TePla SAM, 11 10 2023. [Online]. Available: https://www.pvatepla-sam.com/produkte/sam-premium-line/. [Accessed 11 10 2023].

[6] S. Panahandeh, D. May, C. Grosse-Kockert, R. Schacht, M. Abo Ras and B. Wunderle, "Pulsed infrared thermal imaging as inline quality assessment tool," Microelectronics Reliability, vol. 142, p. 114910, 2023.

PCIM Europe 2024, 11– 13 June 2024, Nuremberg   DOI: 10.30420/566262154

# Evaluating the Safety Isolation of the Package in an Integrated Power Device

Thomas Anthony Capobianco[1]

[1] Power Integrations, United States of America

Corresponding author:   Thomas Anthony Capobianco, thomas.capobianco@power.com
Speaker:   Thomas Anthony Capobianco, thomas.capobianco@power.com

## Abstract

Power devices that integrate a high-voltage switch and provide internal safety-rated isolation enable compact power supply solutions for automotive applications. To meet safety requirements in an automotive application, manufacturers are required to demonstrate that in the event of a fault condition, the system will not cause hazard or injury to the end-user.

In this paper, a method for evaluating package safety isolation is described. An artificially induced catastrophic failure of the power switch is used to cause package damage in the vicinity of the high-voltage switch. Post-failure isolation tests demonstrate that the integrity of the isolation barrier within the semiconductor device is maintained.

## 1 Introduction

Performance, reliability, and safety are three factors that are considered in the use of any product. These are of great significance not only to the end-users, but more immediately to original equipment manufacturers (OEMs), semiconductor manufacturers, and automotive companies.

A conventional ICE-based automobile utilizes a 12 V battery to power its auxiliary electronics systems. With the emergence of hybrid and electric vehicles, the voltage range for batteries has increased to 400 V or even 800 V. Devices with an integrated high-voltage switch and internal isolation enable compact power supply design for these new vehicles and provide significant benefits. Isolation within the package eliminates the need for an optocoupler, an integrated solution that includes a high-voltage switch, a primary side controller, and a secondary side controller. The isolation greatly reduces component count and increases performance uniformity across production. These benefits simplify design and can increase reliability.

With the move to high-voltage systems, ensuring that an electrical fault will not cause shock, harm, or injury to the end-user is critical. In this paper, a method to evaluate package safety isolation in the event of catastrophic power-stage failure is proposed.

### 1.1 Industry standards

Safety standards apply to both system designs and integrated circuits. Specific standards are also used for qualifying automotive products. Tests associated with the isolation barrier are described below.

#### 1.1.1 UL 1577: UL standard for safety for optical isolators

UL 1577 [4] covers optical isolators, also known as optocouplers, and photocouplers. Aside from these, it also applies to non-optical devices that perform similar functions in terms of isolation and signal control. FluxLink™ technology employs magneto-inductive coupling (basically an isolation transformer forms the lead frame of the IC) used in the internal isolation of the InnoSwitch™ devices. Section 11 of the standard defines an important dielectric voltage-withstand test that is used to determine if a device is capable of withstanding — for 60 seconds without breakdown — a potential surge equal to the rated dielectric voltage. This is also commonly known as the High Potential (HiPot) test.

#### 1.1.2 IEC 60747-17: Magnetic and capacitive coupler for basic and reinforced insulation

In Section 6.4, isolation testing is described [3]. This test is used to verify the ability of a device to withstand the isolation test voltage. This test voltage can be a transient, repetitive, or continuous voltage under specified conditions. Passing criteria state that no external or internal flashover shall occur during testing. The unit must also pass post-testing isolation requirements. In this context, an appropriate definition of PASS / FAIL performance for post-stress testing is critical.

### 1.1.3 AEC-Q1000-007 Rev-B: Fault simulation and fault grading

Fault simulation is of great importance to automotive qualification. In Attachment 7 of AEC-Q1000 [1], the purpose of proper fault modeling is described. As stated, the fault simulation should expose any manufacturing defects. The procedure must describe the fault coverage provided to manage failure expectations. Test conditions also need to be appropriately set such that the relevant environmental activation conditions are present. These activation conditions may be temperature-, voltage-, current-, or frequency-related.

## 1.2 Integrated isolation within power devices

The use of integrated devices has numerous benefits in power supply design and manufacturing. These include less dependency on discrete components, simplicity and ease of design, scalability, and a reduction in design-to-production time.

InnoSwitch3-EP (Fig. 1) has the following functional blocks combined in a single IC: a 1700 V high voltage switch, driver, primary and secondary controllers, and protection circuitry. Secondary-side regulation is achieved using a proprietary feedback mechanism internal to the part — Flux-Link [2]. Having internal isolation eliminates the need for an optocoupler, which reduces component count and increases system reliability.

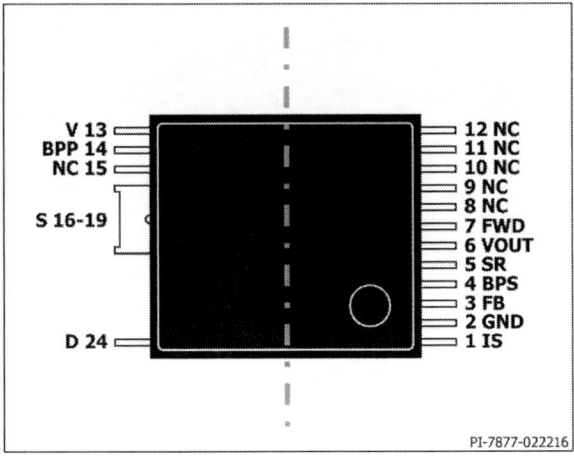

**Fig. 1** InnoSwitch3-EP pin configuration.

Because of the internal isolation feature, the InnoSwitch3-EP requires package safety isolation testing. The goal of this qualification was to demonstrate safety not only during normal operation but also in the event of a catastrophic failure.

## 2 Methodology

### 2.1 Pre-fault and post-fault HiPot test

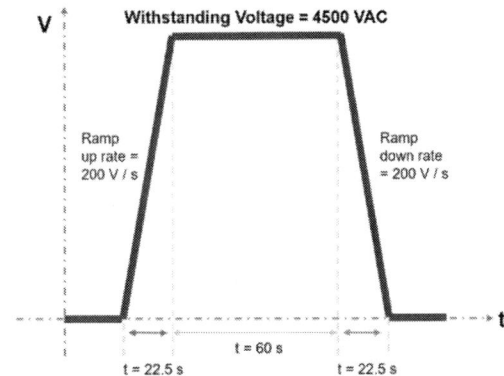

**Fig. 2** HiPot test (voltage vs. time).

In Fig. 2, a voltage versus time plot illustrates the pre-fault and post-fault HiPot voltage stress applied during testing.

For each test, all the pins on the primary side of the IC are shorted together. The pins on the secondary side of the part are also shorted together on the opposite side. This forms a two-terminal device that is subjected to the HiPot test. The test voltage is ramped up at a rate of 200 V / s until it reaches the desired withstand voltage of 4500 VAC. The test lasts for 60 seconds before the voltage is ramped down at a rate of 200 V / s.

A failure condition is defined as the part exceeding a primary-to-secondary current of 2.0 mA and/or arcing is observed.

According to the datasheet of the InnoSwitch3-EP part under test, the UL1577 isolation voltage is 4000 VAC (max) [2]. Tests were performed at 12.5% above the rated voltage.

## 2.2 Test set-up

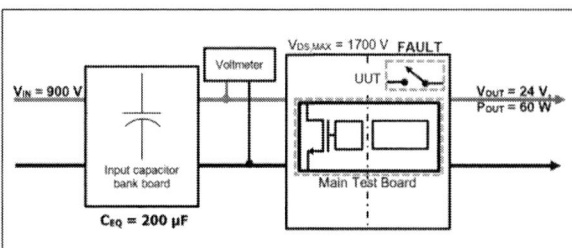

**Fig. 3** Test set-up block diagram.

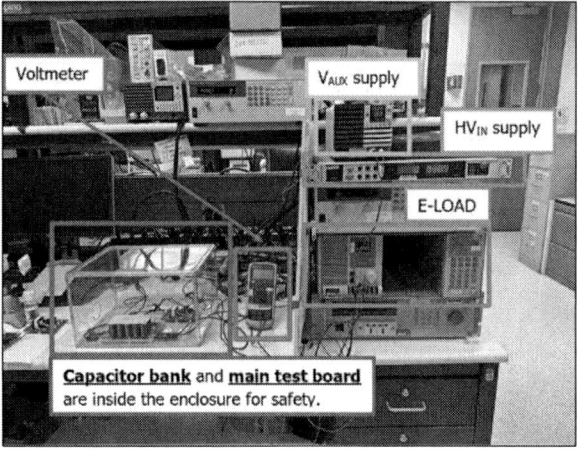

**Fig. 4** Test set-up.

The following equipment was used in the set-up as shown in Figs. 3 and 4.

1. High-voltage DC supply: Magna-Power SL1000-1.5 / UI
2. Low-voltage DC supply: Kikusui PWR801ML
3. DC Electronic Load: Chroma 63108 module in Chroma 6314 mainframe
4. Dielectric Withstand Tester: Associated Research, Inc. HYPOT® III
5. Voltmeter: Fluke 87 True RMS Multimeter
6. Soundproof box (optional)

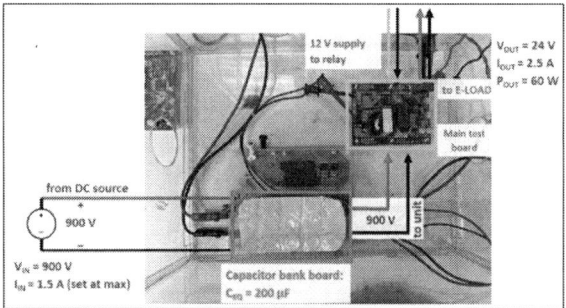

**Fig. 5** Test set-up shown inside the safety enclosure.

A capacitor bank with capacitance of 200 μF was used. This is 66% more than the 120 μF bulk capacitor for a typical 60 W design. This, along with the main test board, is placed inside a safety enclosure as shown in Fig. 5.

## 2.3 Evaluation board

A flyback converter that has an InnoSwitch3-EP part at its core was used as the evaluation platform. It is based on RDR-919Q [5], an automotive reference design from Power Integrations. This power supply has a rated output power of 60 W, and a nominal voltage of 24 V. External protection circuitry was removed, as well as the diode in the RCD primary clamp.

**Fig. 6** Main test board image.

As seen in Fig. 6, the InnoSwitch3-EP part is loaded on the main evaluation board. This part is PCB-mounted with male connector pins at the bottom-side. These mate with female socket pins on the evaluation board. The high-voltage DC supply is ramped up initially to 50 V.

**Fig. 7** Blue LED turns on at $V_{IN}$ = 50 V.

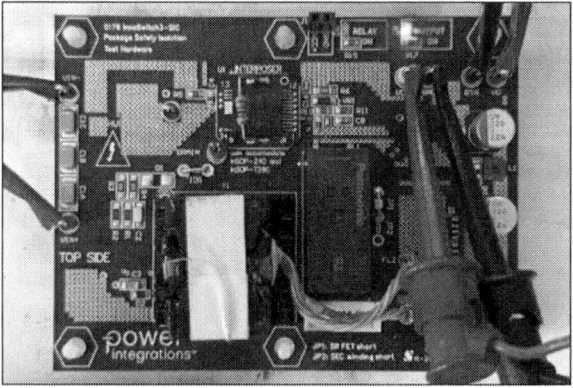

The blue LED indicates output voltage of 24 V as seen in Fig. 7. The input voltage is then ramped up further to 300 VDC.

PCIM Europe 2024, 11– 13 June 2024, Nuremberg      DOI: 10.30420/566262154

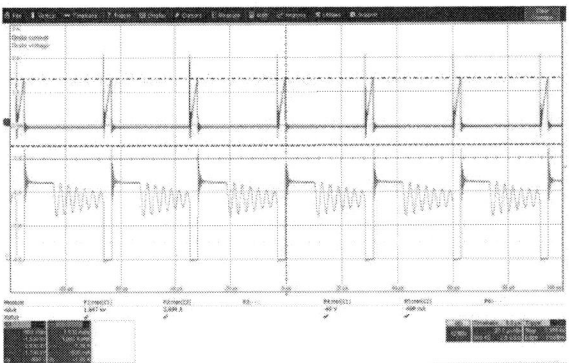

**Fig. 8** Drain voltage and current waveforms. Pink is the primary switch current (1 A per division); Brown is the $V_{DS}$ of the primary power switch (500 V per division); Time base is 20 us per division.

The electronic load is then turned on, and set to draw 2.5 A output current, allowing 60 W operation. The input voltage is again ramped up and set to 900 V.

Prior to part failure, the drain voltage is operating just below the stated 1700 V maximum device voltage specified for the device as shown in Fig. 8.

## 2.4 Inducing failure while part is in normal operation

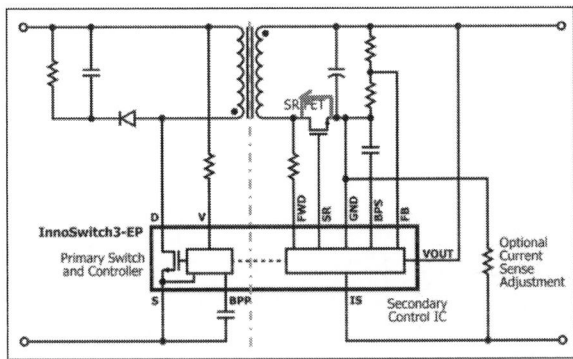

**Fig. 9** Simplified flyback schematic showing InnoSwitch3-EP.

A hard-short between the drain and source pins of the secondary side MOSFET using a relay was then introduced to induce part failure. At the onset of the fault, the electronic load will continue to deplete the output capacitor. The short circuit on the secondary rectifier appears as a short across the primary winding of the transformer which results in a rapid di/dt rise (limited only by the leakage inductance of the primary winding). The primary overcurrent limit then activates, but the very fast di/dt of the primary switch drain current will overshoot. The primary switch is turned off by the pro-

tection circuit resulting in a very high voltage appearing across the leakage inductance of the primary winding and causing catastrophic switch failure. This all happens quickly. Catastrophic failure and localized package damage in the vicinity of the high-voltage switch occur.

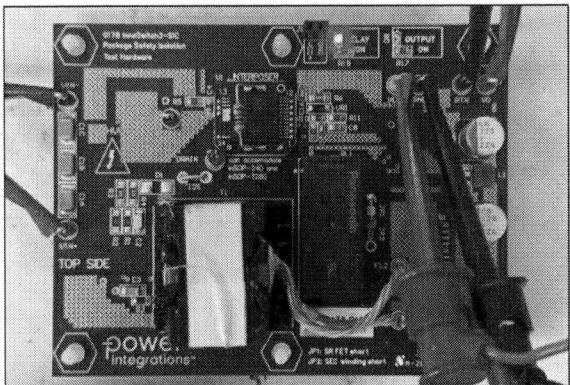

**Fig. 10** Red LED turns on when the short is performed.

When shorting the SR MOSFET, a red LED indicates that a secondary-side short has occurred. There is no new hazard associated with the short as the isolation is intact. This is shown in Fig. 10.

This test was repeated on three (3) different production lots, twenty (20) samples per lot, for a total of sixty (60) total samples tested.

## 3 Results

**Fig. 11** INN3949CQ, LOT: 03MCU738E, SN#4

1171

Figure 11 shows an InnoSwitch3-EP part after the catastrophic failure. A de-capped part is also shown: the damage on the package is near the high voltage DRAIN pin of the power switch.

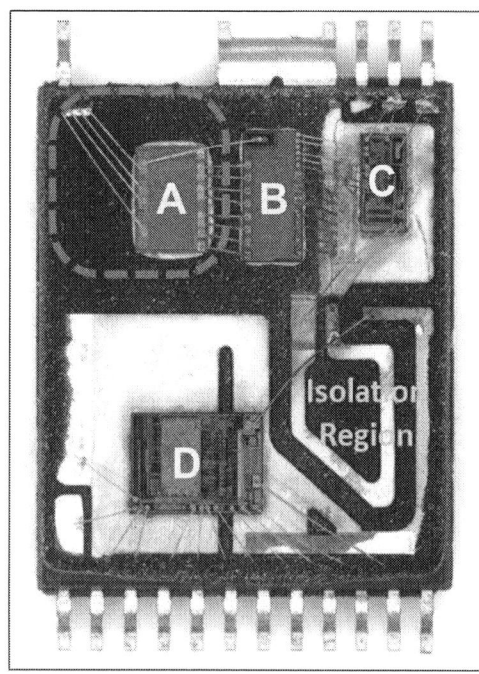

**Fig. 12** De-capped image of an InnoSwitch3

Figure 12 is a de-capped image of the InnoSwitch3-EP part. The different sections within the part are labelled from A to D. A is the high-voltage SiC device and B is the low-voltage MOSFET. Together, A and B form a cascode. C and D are the primary and secondary controllers, respectively, and FluxLink is realized using the lead frames in the area labelled as the Isolation Region.

All tested parts underwent post-fault HiPot tests. The setup and PASS/FAIL criteria were identical to those set during pre-fault HiPot tests. All parts passed post-fault HiPot tests. No arcing or flash-over was observed during the tests. Tables 1 to 3 show the results for each part from the three (3) different lots.

| INN3649C 03MCU738H 2306 H606 | Results |
|---|---|
| SN#1 | PASS |
| SN#2 | PASS |
| SN#3 | PASS |
| SN#4 | PASS |
| SN#5 | PASS |
| SN#6 | PASS |
| SN#7 | PASS |
| SN#8 | PASS |
| SN#9 | PASS |
| SN#10 | PASS |
| SN#11 | PASS |
| SN#12 | PASS |
| SN#13 | PASS |
| SN#14 | PASS |
| SN#15 | PASS |
| SN#16 | PASS |
| SN#17 | PASS |
| SN#18 | PASS |
| SN#19 | PASS |
| SN#20 | PASS |

**Table 1** Lot 1 results

| INN3949CQ 03MCU738E 2247 H907 | Results |
|---|---|
| SN#1 | PASS |
| SN#2 | PASS |
| SN#3 | PASS |
| SN#4 | PASS |
| SN#5 | PASS |
| SN#6 | PASS |
| SN#7 | PASS |
| SN#8 | PASS |
| SN#9 | PASS |
| SN#10 | PASS |
| SN#11 | PASS |

| SN#12 | PASS |
|---|---|
| SN#13 | PASS |
| SN#14 | PASS |
| SN#15 | PASS |
| SN#16 | PASS |
| SN#17 | PASS |
| SN#18 | PASS |
| SN#19 | PASS |
| SN#20 | PASS |

**Table 2** Lot 2 results

| INN3649C 03MCU738F 2247 H606 | Results |
|---|---|
| SN#1 | PASS |
| SN#2 | PASS |
| SN#3 | PASS |
| SN#4 | PASS |
| SN#5 | PASS |
| SN#6 | PASS |
| SN#7 | PASS |
| SN#8 | PASS |
| SN#9 | PASS |
| SN#10 | PASS |
| SN#11 | PASS |
| SN#12 | PASS |
| SN#13 | PASS |
| SN#14 | PASS |
| SN#15 | PASS |
| SN#16 | PASS |
| SN#17 | PASS |
| SN#18 | PASS |
| SN#19 | PASS |
| SN#20 | PASS |

**Table 3** Lot 3 results

# 4  Conclusion

Safety standards in electrical designs cover both system-level circuitry and component parts. The main intent of these standards is to act as safeguards in protecting end-users from harmful voltages and currents. The voltage-withstand test was used to evaluate the internal isolation of the integrated power device in a flyback converter. The evaluation consisted not only of this test but also included system fault modelling. The single fault was constructed to be not only one that was a critical worst-case fault, but also best represented what may be encountered in a real application. Results showed that the fault was "successful" in causing catastrophic damage to the package of the integrated device. Post-fault tests show that despite this, the strength of the isolation barrier of the integrated power device was preserved.

# References

[1] Fault Simulation and Fault Grading. AEC-Q100-007 Rev-B. Automotive Electronics Council. 2007.

[2] Power Integrations: InnoSwitch3-EP Family Off-Line CV/CC QR Flyback Switcher IC with Integrated Primary Switch, Synchronous Rectification and FluxLink Feedback. 2023. [accessed March 18, 2024]. https://www.power.com/sites/default/files/documents/innoswitch3-ep_family_datasheet.pdf

[3] International Standard Semiconductor devices – Part 17: Magnetic and capacitive coupler for basic and reinforced insulation. IEC 60747-17. International Electrotechnical Commission (IEC). 2020.

[4] Standard for Safety Optical Isolators, ANSI/UL 1577-2015. American National Standard. 2014.

[5] Power Integrations: Reference Design Report for a 60 W Isolated Flyback Power Supply Using InnoSwitch3-AQ INN3949CQ. [accessed March 18, 2024]. https://www.power.com/sites/default/files/documents/rdr-919q_60w_high_input_voltage_psu_automotive_innoswitch3-aq-1700v_sic.pdf

PCIM Europe 2024, 11– 13 June 2024, Nuremberg    DOI: 10.30420/566262155

# Flexible Control System for Modular One-Phase Interleaved GaN-based Totem Pole PFC using Real-Time Hardware

Oleksandr Solomakha[1], Siddhartha Menon[1], Manuel Rueß[1], Swapnil Sunil Roge[1], Dominik Koch[1], Ingmar Kallfass[1]

[1]Institute of Robust Power Semiconductor Systems, University of Stuttgart, Germany

Corresponding author: Oleksandr Solomakha, oleksandr.solomakha@ilh.uni-stuttgart.de

Speaker:        Oleksandr Solomakha, oleksandr.solomakha@ilh.uni-stuttgart.de

## Abstract

A flexible control system and modular GaN-based totem pole PFC setup is developed which is suitable to be applied in research and lab courses to attract students to power electronics. To implement the interleaved mode and the ability to quickly change the control strategy, the rapid control prototyping approach is used using dSPACE MicroLabBox. The developed modular 2.4 kW, 500 kHz GaN-based totem pole PFC setup is used as the power stage, which allows us to explore the characteristics of various GaN semiconductors during PFC system operation. The control system is implemented at CPU and FPGA levels using multirate sample time. The control system is also supplemented by a fault diagnostic system, a subsystem of time-triggered (background) tasks, and an efficiency calculation subsystem. A SCADA interface is developed for interactive management of the testing process and monitoring. The phase error between AC current and voltage in developed PFC system is less than 4.8%, power factor is 0.998, THD is less than 5.5%, which, taking into account the modularity of the developed system and the basic software, shows the possible hardware and software potential of the system to improve the PFC performance.

## 1 Introduction

The modular design of laboratory setups is a multifunctional solution for research and teaching tasks in the field of control systems and power stage of power electronics [1]. The choice of the totem pole PFC topology is motivated due to its widespread availability and inherent simplicity, which makes it much easier to achieve unity power factor. In order to minimize size and heat generation, modern developments are making use of GaN transistors [2-10]. To reduce THD, the interleaved mode is used [2-4, 11-14]. In order to achieve better PFC performance, different control strategies are applied. In most cases, the basis is the classical control structure [4-7], but predictive control [8, 9, 13] and multiphase control [14] are also used. Paper [2] provides the most extensive comparison of processors for PFC control and errors in phase correction – only processors of the latest series show acceptable results. The smallest phase error in closed-loop interleaved mode is 4.8°. It becomes obvious that for high-quality control and the opportunity to try different control strategies in the future,

it is necessary to use an FPGA in addition to a CPU. All of the above motivates to design a flexible control system with a CPU and FPGA with a modular power stage in order to be able to test totem pole PFC control strategies and characteristics of fast-switching GaN semiconductors.

## 2 Description of the PFC modular power stage

The test setup consists of a power stage and a control system. The power section is a modular totem pole PFC setup 2.4 kW @ 500 kHz, which consists of five types of plug-in circuit boards assembled on a motherboard (Fig. 1). The main PFC system parameters are shown in table 1. To meet these challenging specifications, GaN transistors were chosen due to their accelerated switching speeds and enhanced current-carrying capacity, while keeping switching losses to a minimum. The semiconductor board includes two GS66508T GaN devices with integrated drivers, configured as single half-bridges. To implement the topology under consideration, one half-bridge operates at a frequency of 50 Hz and Si-semiconductors are usually used. For the versatility of developing a modular system, the same GaN-based boards were used. Each driver is powered by a gate driver power board, which generates the alternating positive and negative voltage needed to switch the GaN devices. The two interleaved legs are connected in parallel via 68 μH inductors in each leg located on the PFC inductor board. On the same board there are sensors for measuring the input AC voltage, input AC current, and currents in each interleaved leg. The board with the output capacitor contains capacitors with a total capacity of 330 μF, as well as sensors for measuring the output dc voltage and dc current. For connection to the dSPACE MicroLabBox, all sensor signals are galvanically isolated and are brought to the range -10...+10 V. To suppress common-mode and differential noise, when PFC operates at high switching frequencies, an EMC filter board is installed at the input.

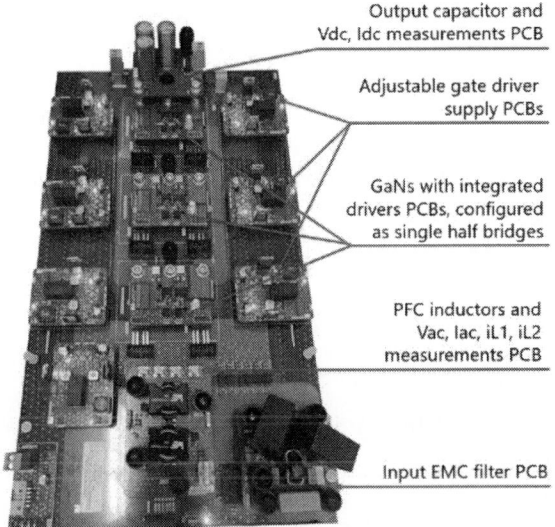

Output capacitor and Vdc, Idc measurements PCB

Adjustable gate driver supply PCBs

GaNs with integrated drivers PCBs, configured as single half bridges

PFC inductors and Vac, Iac, iL1, iL2 measurements PCB

Input EMC filter PCB

**Fig.1**— General view of the power stage

**Table 1:** Main PFC system parameters

| Parameter | Value |
|---|---|
| Output DC power | 2.4 kW @ 500 kHz |
| Output DC voltage | 400 Vdc |
| Input AC voltage | 230 Vac, one-phase |
| PWM frequency | 100...500 kHz |
| GaN semiconductor | GS66508T (650 V, 50 mΩ, 30 A, 6.1 nC) |
| Interleaved inductances | 68 μH (in each leg) |
| Output capacitor | 330 μF |

## 3 Description of the flexible PFC control system

The control system is built on a dSPACE MicroLabBox using the Rapid Control Prototyping principle. This made it possible to reduce development time and reuse the control system model from Matlab/Simulink. The software structure is shown in Fig. 2.

The dSPACE MicroLabBox hardware has CPU and FPGA levels. At the CPU level were placed: PFC control system, ADC conversion and signal normalizing system, fault diagnostic system, subsystem of time-triggered tasks, efficiency calculation subsystem. The structure of the PFC control system is shown in Fig. 2, and consists of a voltage controller, a current controller, and a PLL-based synchronization loop. A notch filter, tuned to 100 Hz, is implemented in the feedback loop of the voltage controller. It helps to stabilize the voltage controller removing certain harmonics from the Vdc feedback signal.

The voltage and current controllers use PI structure, with integrator reset by the start signal and integrator limitation using clamping method. Also, both controllers only work if the enable signal is present. For a smooth start of the system, an additional block with an increasing signal is used to generate a reference current. There are 5 variables that are transmitted from the CPU level to the FPGA using the built-in dSPACE interface— start signal, duty cycle, AC voltage, Interleaved mode activation, PWM frequency.

The ADC conversion and signal normalizing system includes several functions:

- reading the ADC channels of the dSPACE Microlabbox – input AC voltage, input AC current (common current), current through inductor L1 (1st interleaved leg), current through inductor L2 (2nd interleaved leg), DC voltage, DC current,

- signals normalizing by applying offsets and scales to each channel.

Fault Diagnostic System (FDS) consists of the following subsystems:

- fault detection systems in the electrical circuit: AC overvoltage, AC overcurrent (common input current), overcurrent in 1st interleaved leg, overcurrent in 2nd interleaved leg, DC overvoltage, DC overcurrent,

- ADC Fault Diagnostic, which can detect the following faults during ADC operation (for each channel): Burst Trigger overflow, No Burst Start, Data Lost, Store Error, Conversion Trigger Overflow,

- safety warning system.

PCIM Europe 2024, 11– 13 June 2024, Nuremberg            DOI: 10.30420/566262155

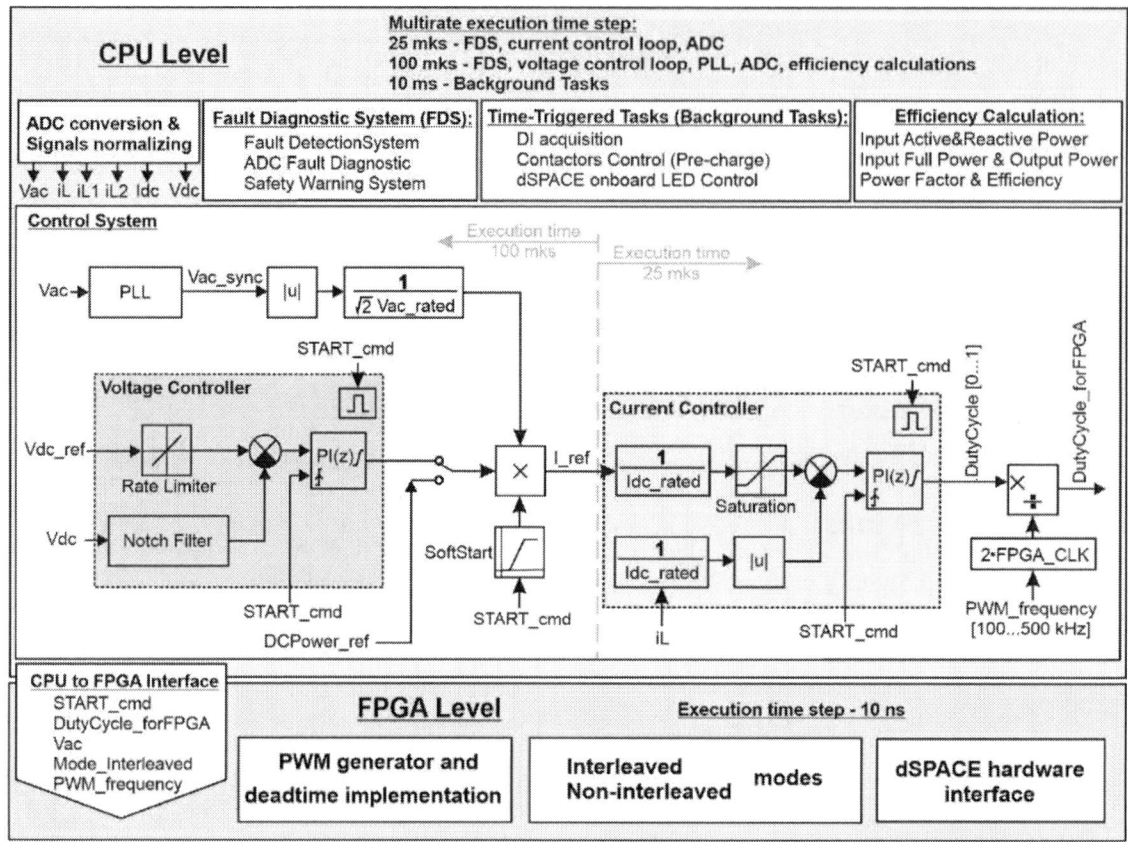

**Fig. 2:** Software structure

When a critical error is detected, the FDS generates a blocking signal, which removes the PWM pulses immediately and stops the control system.

In the subsystem of time-triggered tasks, tasks that were not time-critical were implemented— DI acquisition, contactors control and onboard LED control. The control system allows to send a start signal either by using a software button or by sending a DI signal.

In the efficiency calculation subsystem for online monitoring— input active, reactive and full power, output power, power factor and efficiency are calculated.

At the FPGA level, PWM control signals are generated in non-interleaved and interleaved modes, and dead-time is applied.

The developed structure allows testing PFC with a PWM frequency in the range of up to 500 kHz, testing various control strategies and parameters, optimizing PFC power circuit elements.

To optimize the use of real-time hardware resources, multi-rate was used.

On CPU level the following sample time values were used:

- 25 µs for ADC conversion of currents, current control loop, FDS,

- 100 µs for ADC conversion of voltages, voltage control loop, FDS,

- 10 ms for background tasks.

All software on the FPGA works with 10 ns sample time.

# 4    Description of the SCADA interface and test results

For interactive control of the test process and monitoring, a SCADA interface (Fig. 3) was developed using dSPACE ControlDesk. The SCADA interface consists of 4 tabs – Control, ADC, Controllers and Turnaround. The Control tab is the main one, it contains control and monitoring components. On Fig. 3 in the top graph we can see the synchronization voltage calculated by the PLL and the measured AC voltage at the PFC input. Measurements with an oscilloscope show a phase synchronization error of current and voltage less than 4.8%. In work [2], the same result was achieved on the C28346 300 MHz processor – the best result among 6 types of processors. Considering the modularity of the developed system and the basic software, this shows the possible hardware and software potential of the system for the improvement the performance of control strategies. In the

1176

PCIM Europe 2024, 11– 13 June 2024, Nuremberg          DOI: 10.30420/566262155

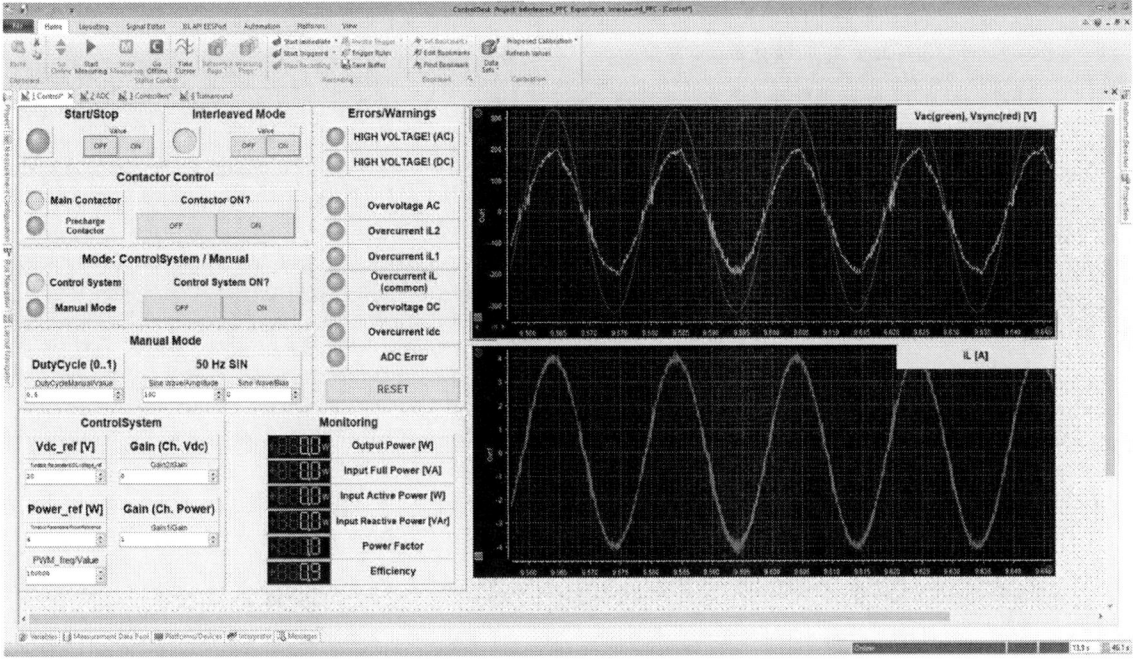

**Fig. 3:** SCADA interface

bottom graph on Fig. 3 the AC input current is shown. The upper and lower graphs are synchronized on the time scale, so we can see the synchronism of the input current and voltage, which shows the correct operation of the main PFC mode.

The ADC tab contains an interface for setting offset and gain for each ADC channel, as well as online graphs with all normalized ADC signals.

The Controllers tab contains an interface for setting the coefficients of current and voltage controllers, graphs of the reference values and feedback signals, and a graph of the output value of the controllers.

On Fig. 4 the graph of the voltage controller working out with different setpoints is presented. From the graph one can also evaluate the quality of the notch filter action, which significantly reduces the amplitude of ripples.

Fig. 5 shows a graph of how the current controller processes a reference signal.

In Fig. 6 a comparison of current THDi at different PWM frequencies is shown.

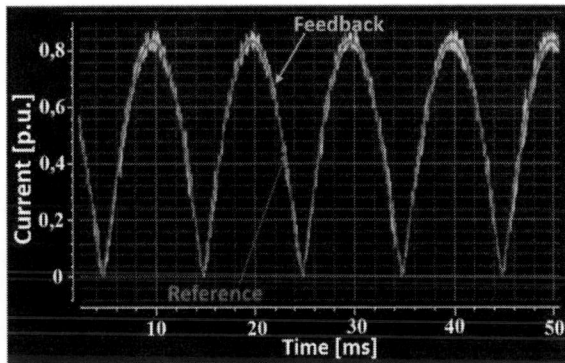

**Fig. 5:** Current controller performance

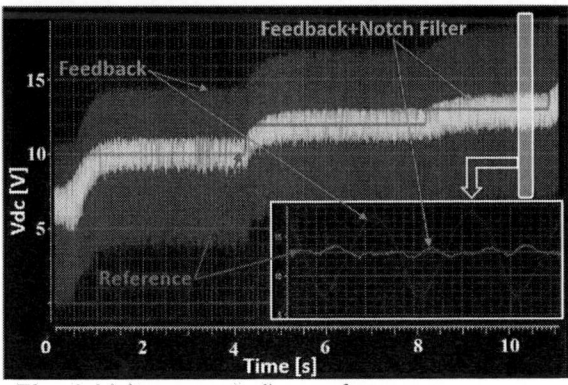

**Fig. 4:** Voltage controller performance

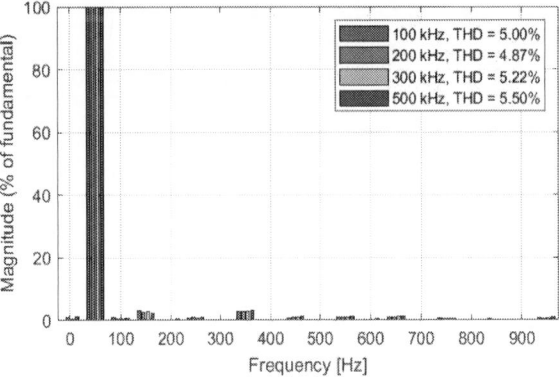

**Fig. 6:** THDi comparison

1177

Table 2 shows calculations of power factors calculated by THD and power ratio.

**Table 2:** Power factor calculations

| Test | PF (from THD) | PF (from [W/VA]) |
|------|---------------|------------------|
| 100 kHz | 0.998 | 0.996 |
| 200 kHz | 0.998 | 0.996 |
| 300 kHz | 0.998 | 0.995 |
| 500 kHz | 0.998 | 0.994 |

As can be seen from the testing results in Fig. 3-5, the described PFC control system is stable, has a flexible structure, and performs all the necessary functions.

Compared to [11], for the developed 2.4 kW PFC system, we obtained the same power factor PF = 0.998, as well as comparable THDi ≈ 5%, and in [11] THDi ≈ 3.5%.

# 5 Conclusions

A control system for modular 2.4 kW one-phase Interleaved GaN-based totem pole PFC using dSPACE MicroLabBox was developed. The control system is implemented at CPU and FPGA levels using multi-rate sample time. A rapid control prototyping approach was used for the development of a flexible control system. The phase error between current and voltage is less than 4.8%. Comparison with similar 2.4 kW PFC designs showed similar results in terms of power factor 0.998, and comparable THDi <5.5%. A custom PWM algorithm was developed on the FPGA because it was necessary to implement the interleaved mode and it gives opportunity to develop ultra-fast custom control design.

# 6 References

[1] Caldognetto T, Petucco A, Lauri A, Mattavelli P. A flexible power electronic converter system with rapid control prototyping for research and teaching. HardwareX. 2023 Mar 4;14:e00411. doi: 10.1016/j.ohx.2023.e00411. PMID: 36936810; PMCID: PMC10017425.

[2] Z. Liu, Z. Huang, F. C. Lee and Q. Li, "Digital-based interleaving control for GaN-based MHz CRM totem-pole PFC," 2016 IEEE Applied Power Electronics Conference and Exposition (APEC), Long Beach, CA, USA, 2016, pp. 1847-1852, doi: 10.1109/APEC.2016.7468119.

[3] Q. Ma, Q. Huang, R. Yu, T. Chen and A. Q. Huang, "Digital Interleaving Control for Two-Phase TCM GaN Totem-Pole PFC to Reduce Current Distortion," 2019 IEEE Energy Conversion Congress and Exposition (ECCE), Baltimore, MD, USA, 2019, pp. 3682-3688, doi: 10.1109/ECCE.2019.8912297.

[4] Y. Tang, W. Ding and A. Khaligh, "A bridgeless totem-pole interleaved PFC converter for plug-in electric vehicles," 2016 IEEE Applied Power Electronics Conference and Exposition (APEC), Long Beach, CA, USA, 2016, pp. 440-445, doi: 10.1109/APEC.2016.7467909.

[5] N. Korada and R. Ayyanar, "A 3kW, 500 kHz E-mode GaN HEMT based Soft-switching Totem-pole PFC," 2019 IEEE 7th Workshop on Wide Bandgap Power Devices and Applications (WiPDA), Raleigh, NC, USA, 2019, pp. 237-244, doi: 10.1109/WiPDA46397.2019.8998776.

[6] W. Zhao, B. Liu, S. Duan, W. Sheng, M. Wu and Z. Song, "A Digital Control Method with Feedforward Reconstruction for Improved Sampling Interference Suppression in GaN HEMTs based Totem-pole PFC Converters," 2018 1st Workshop on Wide Bandgap Power Devices and Applications in Asia (WiPDA Asia), Xi'an, China, 2018, pp. 1-6, doi: 10.1109/WiPDAAsia.2018.8734613.

[7] X. Chen, G. Son, F. Jin and Q. Li, "A Microcontroller-Based High Efficiency Critical Conduction Mode Control for GaN-Based Totem-Pole PFC," 2021 IEEE 22nd Workshop on Control and Modelling of Power Electronics (COMPEL), Cartagena, Colombia, 2021, pp. 1-7, doi: 10.1109/COMPEL52922.2021.9646009.

[8] Y. Liu, X. Huang, Y. Dou, M. Li, O. Ziwei and M. A. E. Andersen, "Adaptive dead time control for ZVS GaN based CRM Totem Pole PFC," 2020 IEEE 9th International Power Electronics and Motion Control Conference (IPEMC2020-ECCE Asia), Nanjing, China, 2020, pp. 438-442, doi: 10.1109/IPEMC-ECCEAsia48364.2020.9368082.

[9] H. Zhu, K. Wang, B. Li, X. Yang and Q. Chen, "Adaptive Tuning Method for ZVS Control in GaN-based MHz CRM Totem-Pole PFC Rectifier," 2021 IEEE Applied Power Electronics Conference and Exposition (APEC), Phoenix, AZ, USA, 2021, pp. 1837-1842, doi: 10.1109/APEC42165.2021.9487095.

[10] Z. Liu, F. C. Lee, Q. Li and Y. Yang, "Design of GaN-based MHz totem-pole PFC rectifier," 2015 IEEE Energy Conversion Congress

and Exposition (ECCE), Montreal, QC, Canada, 2015, pp. 682-688, doi: 10.1109/ECCE.2015.7309755.

[11] C. Zhang, K. Qu, B. Hu, J. Wang, X. Yin and Z. J. Shen, "A High-Frequency Dynamically Coordinated Hybrid Si/SiC Interleaved CCM Totem-Pole Bridgeless PFC Converter," in IEEE Journal of Emerging and Selected Topics in Power Electronics, vol. 10, no. 2, pp. 2088-2100, April 2022, doi: 10.1109/JESTPE.2021.3130083.

[12] M. -H. Park, J. Baek, Y. Jeong and G. -W. Moon, "An Interleaved Totem-Pole Bridgeless Boost PFC Converter with Soft-Switching Capability Adopting Phase-Shifting Control," in IEEE Transactions on Power Electronics, vol. 34, no. 11, pp. 10610-10618, Nov. 2019, doi: 10.1109/TPEL.2019.2900342.

[13] C. Zhang, X. Xie, K. Qu, B. Hu, Z. Li and J. Wang, "A Hybrid Si/SiC CCM Interleaved Totem-Pole Bridgeless PFC Converter with Coupled-Inductor and Hybrid-Frequency Interleaving Operation," 2021 IEEE 1st International Power Electronics and Application Symposium (PEAS), Shanghai, China, 2021, pp. 1-5, doi: 10.1109/PEAS53589.2021.9628585.

[14] G. Scarlatescu, T. K. Gachovska and T. Lipian, "A Four-Phase 5 kW Interleaved Totem-Pole PFC Platform Based on SiC FETs and Controlled by SA4041 Digital Power Processor," 2021 IEEE Energy Conversion Congress and Exposition (ECCE), Vancouver, BC, Canada, 2021, pp. 1840-1845, doi: 10.1109/ECCE47101.2021.9595232.

PCIM Europe 2024, 11– 13 June 2024, Nuremberg          DOI: 10.30420/566262156

# A Peak Current Mode Control Method for PFC

Bosheng Sun

Texas Instruments, USA

Corresponding author: Bosheng Sun, b-sun@ti.com

Speaker: Sheng-yang Yu, seanyu@ti.com

## Abstract

A peak current mode control method for a power factor correction (PFC) is introduced in this paper: the inductor peak current is modulated by a special sawtooth wave, the peak value of the sawtooth wave is a function of voltage loop output and PWM on time, and its magnitude linearly drops to 0 V at the end of the switching period. PWM turns on at the beginning of the switching cycle, and turns off when the inductor current value exceeds the sawtooth wave. This control method can achieve a unity power factor and low total harmonics distortion (THD) for both continuous conduction mode (CCM) and discontinuous conduction mode (DCM) operation, it also brings many advantages over the traditional average current mode control. The proposed peak current mode control method is proved through mathematical analysis, then verified on a PFC converter.

## 1 Introduction

Power factor correction (PFC) is used to force AC input current to follow AC input voltage to achieve high power factor. There are two traditional control methods for PFC: a constant ON time control for PFC operates at critical conduction mode (CRM), or average current mode control for PFC operates at continuous conduction mode (CCM) [1]. CRM PFC is good for low power application, while CCM is good for high power application. Although average current mode control has been proved very successful in CCM PFC, it has limitations.

The most common current sense method in average current mode control is to put a current shunt resistor at the ground return path to sense the inductor current. This current shunt resistor inevitably causes extra power loss, especially for high power application. Also, this current shunt sensing method becomes non-applicable in 2-phase interleaved PFC where each phase current needs to be sensed for current balancing purpose [2], and in dual boost semi-bridgeless PFC where not all the current go through the ground return path [3]. In these two cases, a current transformer (CT) instead of shunt resistor is used to sense current. The CT is put in series with the boost switch to sense the switching current, which is also the inductor charging current. Since the CT only sense the inductor charging current while average current mode control requires the average current information, a common practice is to sample at the middle of the CT output (the middle of the pulse-width modulation [PWM] on-time). This is because the middle point instantaneous current value equals the average inductor current value in CCM

mode (see Fig. 3). This method has fewer power losses than the current shunt, but it also has limitations: the duty cycle for PFC varies from 0% to 100%. When the duty cycle is small, the PWM on-time is small; therefore, it is difficult to sample exactly at the middle of the PWM on-time. Any sample position offset can cause feedback signal errors and deteriorate both total harmonic distortion (THD) and the power factor.

Peak current mode control [4], because it brings many advantages over average current mode control, is widely used in DC/DC converters. Employing peak current mode control in PFC also brings several practical advantages such as no need error-amplifier, multiplier and compensation network, input filter oscillation is virtually non-exist, etc. [5]. However, peak current mode control causes PFC input current distortion, especially at light load or high input voltage [5-7]. This is because PFC needs to control the average inductor current, the inductor peak current does not equal to average current, controlling the inductor peak current results in poor THD and a low power factor. To reduce THD, [5] proposed a few methods: (1) adding a variable DC offset to the sinusoidal reference, (2) predistorting the sinusoidal reference with a nonlinear network, and (3) converting the peak current control scheme into a simplified average current control scheme. [7] proposed another method of using a bi-edge modulation method with triangular wave as carrier. Although these methods can help reduce THD, they cannot totally remove the current distortion. In this paper, a new peak current mode control is proposed, the inductor current is modulated by a special sawtooth wave, the current distortion caused by peak current mode control can be completely eliminated, a unity power factor and low THD can be achieved.

## 2 Peak Current Mode Control for PFC in CCM

The proposed peak current mode control method is shown in Fig. 1. The inductor peak current is sensed through a CT which is put in series with the boost switch. PFC output voltage is sensed and compared to a reference. The output of the voltage loop is sent to a sawtooth wave generator.

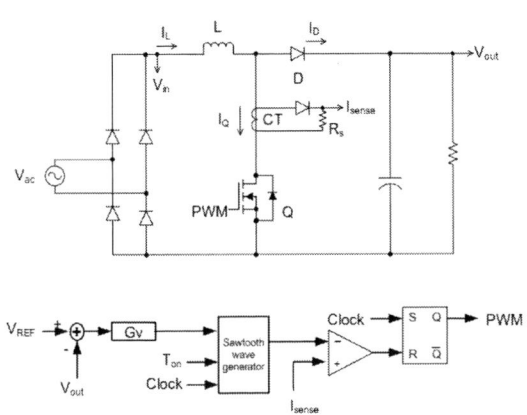

**Fig. 1** Proposed peak current mode control for PFC

The sawtooth wave generator generates a sawtooth wave. The sawtooth wave peak voltage ($V_{RAMP}$) starts at the beginning of each switching period, and its magnitude linearly drops to 0 V at the end of the switching period. The sensed current $I_{sense}$ is compared with this sawtooth wave. The boost switch (Q) turns on at the beginning of the switching period. Q turns off when $I_{sense}$ exceeds the sawtooth wave, as shown in Fig. 2. Fortunately, this kind of sawtooth wave and PWM generator already exists in almost all digital power controllers. These digital controllers have a peak current mode control module with programable slope compensation, it was original designed for DC/DC converter, now it can be used to implement this proposed control method for PFC. Programming the compensation with a slope of $V_{RAMP}/T$ (T is the switching period) generates the intended sawtooth wave.

To achieve a unity power factor, the peak value of the saw wave $V_{RAMP}$ is generated as:

$$V_{RAMP} = G_v * V_{out} + \frac{T_{on}*V_{out}*R_s}{2*L} \quad (1)$$

where $G_v$ is the voltage loop output, $V_{out}$ is the PFC output voltage, L is the inductance of the boost inductor, $R_s$ is the current-sense resistor at the current transformer output, and $T_{on}$ is the PFC PWM on-time. Since the PWM on-time is almost the same in two consecutive switching cycles, the $T_{on}$ information from the previous switching cycle is used to calculate the $V_{RAMP}$ value for this switching cycle.

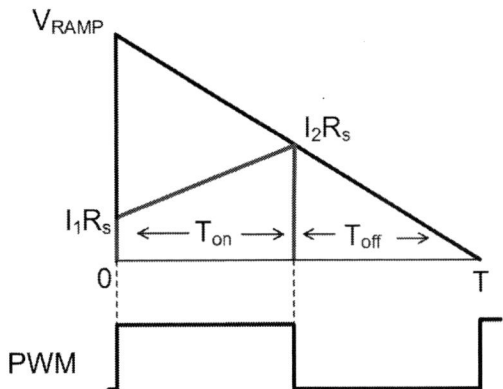

**Fig. 2** PWM waveform generation for the proposed method

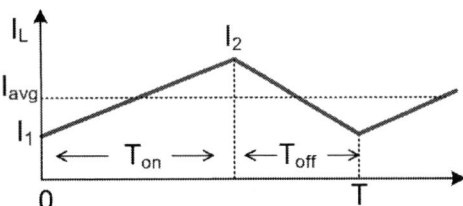

**Fig. 3** Inductor current waveform in CCM

From Fig. 2, during $T_{on}$ time, the input voltage applies to the inductor, causing the inductor current to rise from $I_1$ to $I_2$:

$$I_2 - I_1 = \frac{V_{in}*T_{on}}{L} \quad (2)$$

where $V_{in}$ is the PFC input voltage. From Fig. 3, the average inductor current in each switching cycle is:

$$I_{avg} = \frac{(I_1+I_2)}{2} \quad (3)$$

Substituting Eq. (2) into Eq. (3) results in Eq. (4):

$$I_{avg} = I_2 - \frac{V_{in}*T_{on}}{2*L} \quad (4)$$

From Fig. 2, Equation (5) is:

$$\frac{I_2 * R_s}{V_{RAMP}} = \frac{T_{off}}{T} \tag{5}$$

Eq. (6) applies to PFC operating at CCM in steady state:

$$\frac{T_{off}}{T} = \frac{V_{in}}{V_{out}} \tag{6}$$

Substituting Eq. (6) into Eq. (5) and solving for $I_2$ results in Eq. (7):

$$I_2 = V_{RAMP} * \frac{V_{in}}{R_s * V_{out}} \tag{7}$$

Substituting Eq. (1) and (7) into Eq. (4) results in Eq. (8):

$$I_{avg} = \frac{G_v}{R_s} * V_{in} + \frac{V_{in} * T_{on}}{2*L} - \frac{V_{in} * T_{on}}{2*L} = \frac{G_v}{R_s} * V_{in} \tag{8}$$

In Eq. (8), $G_v$ is the PFC voltage loop output, it is constant in steady state. $R_s$ is the current sense resistor at the current transformer output; therefore, $I_{avg}$ is proportional to $V_{in}$ and follows the shape of $V_{in}$. If $V_{in}$ is a sinusoidal wave, $I_{avg}$ will also be a sinusoidal wave, a unity power factor is achieved.

Compare to other peak current mode control methods, the current distortion is completely eliminated thanks to the special sawtooth wave. Compared to traditional average current-mode control, this method eliminates error-amplifier, multiplier, compensation network and the input voltage sensing circuit, resulting a lower system cost. The power losses caused by CT is also lower than the current shunt resistor. Since now the higher current ripple becomes acceptable, it is possible to reduce the boost inductance, thus a smaller size boost inductor can be used to increase power density and reduce cost.

To save system costs, some designers prefer combo control, where a single controller controls both PFC and the DC/DC controller. The combo controller can be put on either the primary or secondary side of the AC/DC power supply; each has its advantages and disadvantages. If the combo controller is chosen to be put on the primary side, the DC/DC output voltage and current information need to be sent to primary side across the isolation boundary, and the communication between the controller and host also needs to across the isolation boundary. If the combo controller is chosen to be put on the secondary side, because the conventional average current mode control method requires input AC voltage information, the input voltage must be sensed and used to modulate the current loop reference. Sensing the input voltage across the isolation boundary is a challenge.

In this peak current mode control method, only the switching current and $V_{out}$ information are needed. Since switching current is sensed by CT which provides isolation by itself, and a low-cost optocoupler can sense $V_{out}$ and send it to the secondary side, the PFC controller can be put on the secondary side of the AC/DC power supply and be combined with the DC/DC controller, which is also on the secondary side, to create a combo controller, which will significantly reduce system costs.

# 3 Peak Current Mode Control for PFC in CCM/DCM

The same algorithm can be extended to discontinuous conduction mode (DCM) operation. Fig. 4 shows the inductor current waveform in DCM. The inductor current drops to zero at the end of $T_{off}$ and stays at zero for the rest of period $T_{dcm}$, therefore:

$$T = T_{on} + T_{off} + T_{dcm} \tag{9}$$

The PWM waveform generator is the same as Fig. 2, but now the PWM off-time is $T_{off}$ + $T_{dcm}$, not $T_{off}$, as shown in Fig. 5.

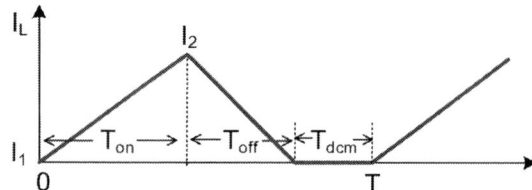

**Fig. 4**    Inductor current waveform in DCM

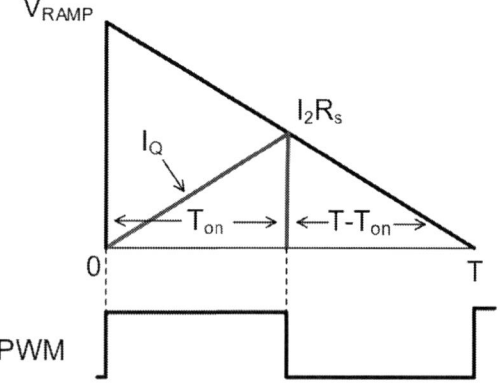

**Fig. 5**    PWM waveform generation for the proposed method in DCM

From Fig. 4, the average inductor current in DCM in one switching cycle is:

$$I_{avg} = \left(I_2 - \frac{V_{in} * T_{on}}{2 * L}\right) * \frac{T_{on} + T_{off}}{T} \qquad (10)$$

In steady state, inductor volt-second must be balanced in each switching cycle, resulting in Eq. (11):

$$V_{in} * T_{on} = (V_{out} - V_{in}) * T_{off} \qquad (11)$$

Solving for $T_{off}$ and substituting Eq. (10) results in Eq. (12):

$$I_{avg} = \left(I_2 - \frac{V_{in} * T_{on}}{2 * L}\right) * \frac{T_{on} * V_{out}}{T(V_{out} - V_{in})} \qquad (12)$$

From Fig. 5, Equation 13 is:

$$\frac{I_2 * R_s}{V_{RAMP}} = \frac{T - T_{on}}{T} \qquad (13)$$

Now the peak value of the sawtooth wave $V_{RAMP}$ is generated as:

$$V_{RAMP} = \left(\frac{G_v * V_{in} * T * (V_{out} - V_{in})}{T_{on} * V_{out}} + \frac{R_s * T_{on} * V_{in}}{2 * L}\right) * \frac{T}{T - T_{on}} \qquad (14)$$

Substituting Eq. (14) into Eq. (13) and solving for $I_2$ results in Eq. (15):

$$I_2 = \frac{G_v * V_{in} * T * (V_{out} - V_{in})}{R_s * T_{on} * V_{out}} + \frac{T_{on} * V_{in}}{2 * L} \qquad (15)$$

Substituting $I_2$ into Eq. (12) results in Eq. (16):

$$I_{avg} = \frac{G_v}{R_s} * V_{in} \qquad (16)$$

In Eq. (16), $G_v$ is constant in steady state, $R_s$ is the current-sense resistor at the current transformer output; therefore, $I_{avg}$ is proportional to $V_{in}$ and follows the shape of $V_{in}$. If $V_{in}$ is a sinusoidal wave, $I_{avg}$ will also be a sinusoidal wave, thus achieving a unity power factor.

Equations (10) through (16) are valid for both CCM and DCM, so if the sawtooth wave signal peak value is generated according to Eq. (14), then it is possible to achieve a unity power factor for both CCM and DCM operation.

Equation (1) is a special case of Eq. (14) where T = $T_{on}$ + $T_{off}$. For applications in which light loads (PFC will be in DCM mode at light load) THD and the power factor are not important, use Eq. (1) to simplify the implementation.

## 4 Test Results

The proposed control method is verified on a 360 W PFC controlled by Texas Instruments UCD3138 digital power controller. PFC output voltage is measured by analog-to-digital-converter (ADC), then it is compared to a digital reference,

the error goes to a proportional-integral (PI) voltage loop. A $V_{RAMP}$ value is calculated based on voltage loop output and $T_{on}$ time from previous switching cycle using either Eq. (1) or (14). This value is the starting point of the ramp, then its magnitude linearly drops to zero at the end of the switching cycle by programing the slope compensation rate of $V_{RAMP}/T$. The ramp value is converted to an analog signal through a digital-to-analog converter (DAC), then connects to the negative input of the internal peak current comparator. The sensed input current is compared to this ramp signal, PWM turns off when sensed input current exceeds the ramp signal. Fig. 6 shows the input current waveform with this peak current mode control, where a good sinusoidal current waveform is achieved.

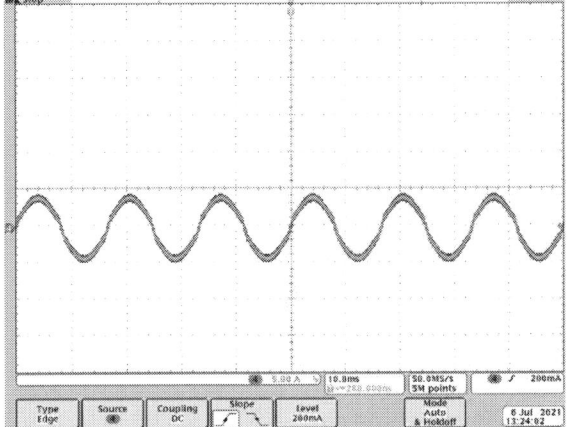

**Fig. 6** Input current waveform test results

## 5 Conclusion

By modulating the inductor peak current through a special sawtooth wave (the peak value of the sawtooth wave is a function of voltage loop output and PWM on-time, and its magnitude linearly drops to 0 V at the end of the switching period), a unity power factor and low THD can be achieved for PFC in both CCM and DCM operation. Compare to other peak current mode control methods, the current distortion is completely eliminated and a unity power factor can be achieved. Compare to traditional average current mode control, this method eliminates the current-error amplifier, multiplier and its compensation network, a smaller boost inductor with lower inductance can be used, these help to increase power density and reduces system cost. Also, since this method does not need input voltage sensing, it is possible to put the PFC controller on the secondary side of an AC/DC power supply to make a combo controller to further reduce system cost. Finally, it is easy to implement this control method with existing digital power controllers.

## References

[1] L. Dixon, "High Power Preregulator for Off-Line Power Supplies", Texas Instruments Power Supply Design Seminar SEM600, 1988.

[2] B. Sun, "Digital current balancing for an interleaved boost PFC", Texas Instruments Analog Design Journal, April 2013.

[3] L. Huber, Y. Jang and M. M. Jovanovic, "Performance Evaluation of Bridgeless PFC Boost Rectifiers," in IEEE Transactions on Power Electronics, vol. 23, no. 3, pp. 1381-1390, May 2008

[4] L. Dixon, "Current-Mode Control of Switching Power Supplies", Texas Instruments Power Supply Design Seminar SEM400, 1985.

[5] R. Redl and B. P. Erisman, "Reducing distortion in peak-current controlled boost power-factor correctors," in APEC'94, pp. 576–583.

[6] Z. Lai and K. M. Smedley, "A family of continuous-conduction-mode power-factor-correction controllers based on the general pulse-width modulator," in IEEE Transactions on Power Electronics, vol. 13, no. 3, pp. 501-510, May 1998.

[7] T. Jiang, P. Mao and S. Xie, "Analysis and improvement on input current of one-cycle controlled PFC converter," 2010 5th IEEE Conference on Industrial Electronics and Applications, Taichung, 2010, pp. 2094-2098

PCIM Europe 2024, 11– 13 June 2024, Nuremberg    DOI: 10.30420/566262157

# Adaptive Resonant Controller for a Three-Phase PFC Converter for an On-Board Charge Application

Rami Troudi[1], Kelly Ribeiro de Faria[1], Moctar Coulibaly[1]

[1] Valeo Powertrain Electrified Mobility, France

Corresponding author: Rami Troudi, rami.troudi@valeo.com
Speaker: Rami Troudi, rami.troudi@valeo.com

## Abstract

This paper presents the design methodology and the tuning of the current loop of a three-phase Power Factor Correction (PFC) converter used in the on-board charger application. The control loop is analyzed for different load conditions considering the main parameter variation, the boost inductance. The controller is proposed in the $\alpha\beta$ stationary frame with an adaptive resonant proportional compensator. Experimental results validate the controller performance for an 11 kW on-board charger.

## 1 Introduction

Three-phase PFC converters are used to increase the power and its control requires a voltage and current control loops. For on-board charge applications in Europe, single-phase mode operation is only possible for power up to 7.2 kW, therefore for higher power level; the converter needs to be connected in three-phase systems as shown in Fig. 1.

The need of increasing power density and efficiency leads to the selection of a three-phase PFC topology for power level application of 11 kW and 22 kW. In this context, there are many works talking about the control of single-phase PFC converters focusing on the bridgeless totem-pole converter [1]-[2], but only a few exploring the three-phase control for Electric Vehicle (EV) on-board charger application. The most common application references are usually from renewable energy and very high power domain [3], where the electrical parameters and thus the system dynamics differs a little bit from the medium power level application required for on-board chargers.

Therefore, this paper presents the entire design and tuning of the current loop for an 11 kW three-phase six-switch PFC boost converter, presented in Fig. 1.

The aim of the current loop is to regulate the current in the PFC choke. It should ensure a stable current waveform under different load conditions, PFC main parameter variation, grid impedance changes and possible resonances related to the AC filter connected to a very inductive grid. Furthermore, the current shall be smooth with low

THD (Total Harmonic Distortion) (standards IEC 61000-3-2 /3-12). Finally, the current control loop shall guarantee a very high power factor (PFC).

In this work, only the design and stability of the current loop system with the main parameter variable, $L_{pfc}$, inductance, is considered. In fact, the boost PFC inductance, $L_{pfc}$ as described in [4], changes as a function of the current among a wide range, and therefore the system stability needs to be evaluated in all operating conditions.

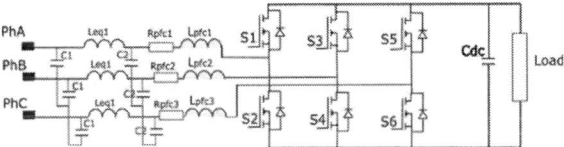

**Fig. 1** The three-phase six-switch PFC boost converter.

In terms of the control design, the rotating dq orthogonal frame is the most used. The stationary abc frame is also possible only requiring a small modification in the equivalent model to allow the decoupling of the phases; but the orthogonal $\alpha\beta$, with a proportional resonant compensator (PR) is the solution selected in this paper. More details about the frame selection can be seen in [5], but generally, any of the three frames can be used.

To present the control of the three-phase six-switch PFC converter, the first section of this paper will show the equivalent model of the system and its validation by using simulation and hardware in

1185

the loop comparisons. In section 3, we will treat the variation of the power parameters to perform the adaptive tuning of the PR controller. In section 4, we will present the experimental results to validate the controller proposed, and finally, we will discuss some improvements and further work.

## 2 Generic current loop design and model validation

Assuming the detailed architecture of a common PFC converter shown in Fig. 2 with six bidirectional switches, $S_i$ (i= 1,.., 6), three PFC chokes, $L_{pfci}$ (i= 1,.., 3), a three-phase AC filter, three Y-capacitors, $C_{fpfci}$ (i= 1,.., 3) and three equivalent inductive grid impedances, $R_{impGi}$- $L_{impGi}$ (i= 1,.., 3).

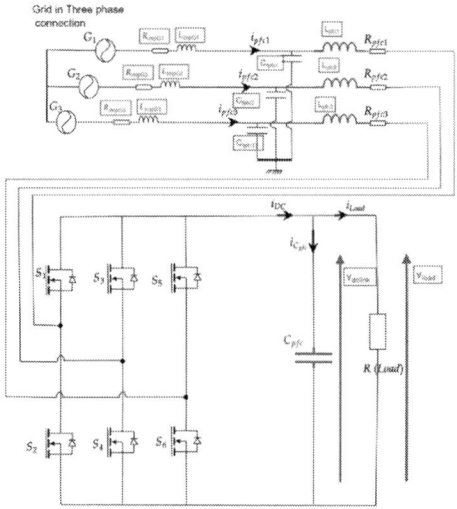

**Fig. 2:** Common bidirectional three-phase PFC converter.

The diagram of the current loop is presented in Fig. 3, with the PFC current plant, $H_{pfci}$, the AC voltage measurement filter, $H_{fpfcv}$, the AC current measurement, $H_{fpfci}$, the digital delay associated with the PWM modulation and the digital control computing time, $H_{pwmi}$, and the PR compensator, $H_{PRc}$.

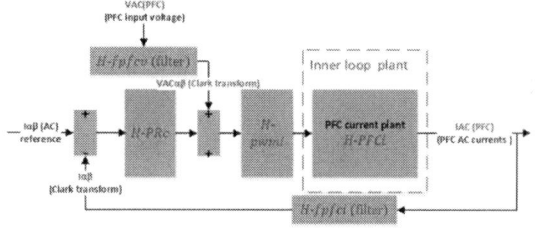

**Fig. 3:** Diagram of the current closed loop + feedforward function.

The AC voltage feed-forward is added to improve the system transient response time and tracking capabilities during all grid voltage conditions. In order to simplify this study, the Grid voltage is assumed clean and maintained to a constant value so the feed-forward function can be ignored. Thus, the current closed loop is given by the following expression (1).

$$H_{pfc_{iCL}}\ (s) = \frac{H_{PFCi}\ H_{pwmi}\ H_{PRc}}{1 + H_{PFCi}\ H_{pwmi}\ H_{PRc}\ H_{fpfci}} \qquad (1)$$

The voltage and the current measurement filters are considered as first order law pass filters. The current measurement is given by the following transfer function (2) with $f_{pfci}$, the filter cut-off frequency.

$$H_{fpfci}\ (s) = \frac{1}{\left(\dfrac{1}{2\pi f_{pfci}}\right)s + 1} \qquad (2)$$

The system plant $H_{pfci}$, duty cycle to current transfer function, can be computed using small signal approach.

The PR controller with damping factor is given in (3):

$$H_{PRc}\ (s) = K_p + \frac{2\ \omega_{cPR}\ K_R\ s}{s^2 + 2\xi\ \omega_{cPR}\ s + \omega_{0PR}^2} \qquad (3)$$

Note that $\omega_{cPR}$ designates the resonant cut-off frequency or the width of the resonant filter (rad/s) and $\omega_{0PR}$ designates the target reference frequency (rad/s). The PR controller is selected that the DC compensation/control network is moved to an AC control network. Thus, the control strategy is applied for both single-phase, three-phase systems and unbalanced grid system conditions. The PR compensator ensures very high gain and quality factor at the controlled frequency bandwidth ($\omega_{cPR}$). As a result, the steady state error is forced to a small value at the resonant frequency. Note, the easy tuning in the stationary reference (αβ) makes the PR controller more attractive in the control design of AC/DC converters in the electrical automotive industry. Furthermore, it is also possible to increase the bandwidth of the controller ($\omega_{cPR}$) and therefore improve the robustness in terms of variation in grid frequency.

Before proceeding with the design of an adaptive controller, time response analysis of the Hardware in the Loop (HIL) design and the theoretical model including the designed closed loop was done to carry out the model validation. The validation test conditions, considers that the voltage loop is open, the PR gains are maintained constant and the load is stepped from 11kW to 5kW. The comparison of the transient-response specifications of the equivalent model and the results from the HIL are given in Fig. 4 and Fig. 5, respectively.

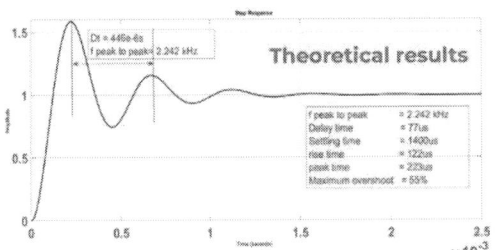

**Fig. 4:** Transient-response specifications of the designed closed loop Model

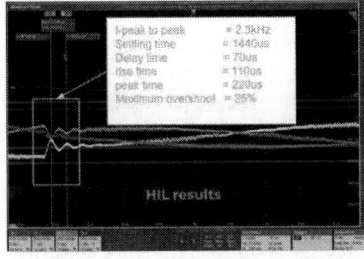

**Fig. 5:** Transient-response specifications of the HIL closed loop Model

The modelled PFC current loop is giving approximately the same dynamic results as in the HIL (Fig. 4 and Fig. 5). The changing of the PFC choke values changes the bandwidth/Gain of the closed-loop and can lead to instability, and that was verified theoretically and in the HIL.

The next study is using the Open Loop Gain (OLG) computed from (1) and is described by the equation in (4).

$$H_{pfc_{iOLG}}(s) = H_{PFCi} \ H_{pwmi} \ H_{PRc} \ H_{fpfci} \qquad (4)$$

Considering the PFC boost inductor variation as described in [4], the controller is initially tuned with fixed parameters: damping factor ξ=0.707, bandwidth of 5 rad/s and proportional ($K_p$) and resonant gains ($K_R$), of 1.82 and 165, respectively.

The Nichols chart of the OLG is presented in Fig. 6.

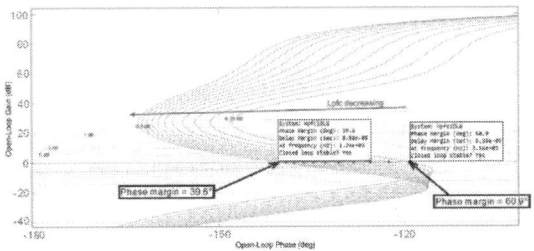

**Fig. 6:** Nichols chart: Minimum stability Margin variation of $H_{pfciOLG}$ assuming $L_{pfci}$ = [100μH - 360μH]

Note that while $L_{pfci}$ is changing from 100 μH to 360 μH, the phase margin is almost divided by 2 when the current reaches its maximum ($L_{pfc}$ = 100 μH). Another way of evaluating the impact of the inductance changing is to use the bode diagram analysis (Fig. 7).

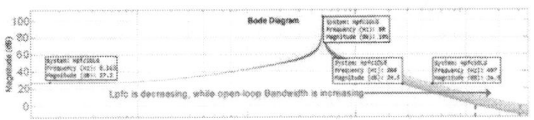

**Fig. 7:** Bandwidth variation of $H_{pfciOLG}$ assuming $L_{pfci}$ = [100μH - 360μH]

Therefore, the static gains implemented in the PR controller does not ensure good bandwidth and sufficient phase margin whatever the $L_{pfc}$ value.

The question arises as to how to maintain best phase margin and system bandwidth without being too much or too small during the constraint of PFC choke saturation.

## 3 Parameter selection criteria of the PR controller and the adaptive design

As mentioned in the previous section, the use of constant gains in the PR controller is not able to maintain the minimum phase margin and the desired bandwidth range for all the operating conditions of the converter. Therefore, in this section, we will perform the design and the tuning of the controller considering the variation of the main PFC inductor. The design process is taking into account the parameters in Table. 1.

| Stage | $V_{in}$ | Frequency | $P_{out}/\ I_{RMS}$ | L (µH) |
|---|---|---|---|---|
| AC/DC | 200-240 (AC) | 45-60 Hz | 7 kW/ 32 A | 100 - 360 |
| **Others:** | | | | |
| DC bus | 340-500Vdc | | Voltage ripple | 6 % |

**Table 1** Design parameters of the PFC converter

The inductance of the PFC chokes changes during all load conditions. In fact, the equivalent model of the inductor can be obtained as a function of the current and the coupling factor can be disregarded in order to treat the inductance variation per phase [4].

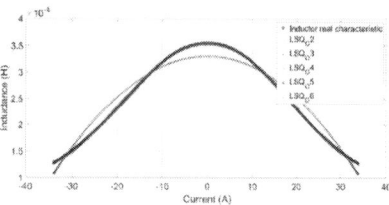

**Fig. 8:** Characteristic of the PFC choke during 1 cycle (50Hz)

This model can be simplified using Least Square approximation. In the application example presented in this paper, the inductance changes as illustrated in Fig. 8.

Using the PR controller in (2), we can define our controller specifications. It is desirable to obtain a high gain at the electrical grid frequency, $\omega_{OPR}$, to reduce the steady-state error and almost no gain outside the defined bandwidth $\omega_{cPR}$. This bandwidth is pre-defined and can be selected according to the frequency variations in the electrical grid.

As mentioned in [6], the proportional gain of the PR controller determines the dynamics of the system in a similar way while tuning a PI controller. Table 2 presents the specification for the current loop.

| Parameter | Value |
|---|---|
| Gain margin | 6 dB |
| Phase margin | > 45 ° |
| Bandwidth | [5-10] x λ $H_{PFCi}$ |
| Overshoot | < 10% |
| Peak gain at the grid frequency | 95 dB |

**Table. 2** Desired specification of the PFC current loop

The next step is to define the ranges of the proportional gains $K_p$ and $K_R$ in (2) to ensure a phase margin higher than 45° for all the load conditions and the gain margin above 6 dB for high stability. During this study, $\xi$ and ($\omega_{cPR}$) are assumed constant. We have considered the open-loop response of the system and the tuning of the proportional gains is based on the Ziegler-Nichols method.

Fig. 9 and 10 presents the impact of the gains $K_p$ and $K_R$, respectively, on the current loop while the PFC choke inductance changes between [360-100] µH. With the variation of the inductance, the gain $K_p$ can be increased to improve the bandwidth of the system, however the phase margin is decreased. Meanwhile, by decreasing the resonant gain $K_R$, the phase margin can be improved again and thus keep the stability of the current loop.

The tuning of the PR controller can then be resumed as the following: The bandwidth of the system can be improved by increasing the proportional gain $K_p$, but at the same time, the stability margin can be improved by decreasing $K_R$. Note that maximum bandwidth is essential to be able to follow all system dynamic transitions and maximum phase margin is mandatory for better current waveform. The low-steady error is ensured by higher $K_R$.

In order to obtain the system bandwidth at least 5 times higher than the constant time of the system plant, $H_{PFCi}$, depending on the minimum and maximum values of the PFC choke, we can determine the range of the proportional gains $K_p$ and $K_R$ to keep:

- A phase margin within [60° 90°] for better system stability;

- A bandwidth between [170 Hz (for better noise rejection) - 1273Hz (for better reference tracking)].

Note that the choice of a phase margin higher than 45°, i.e., 60° was to anticipate the impact of the delay existed in all PWM converters that can affect the desired phase margin.

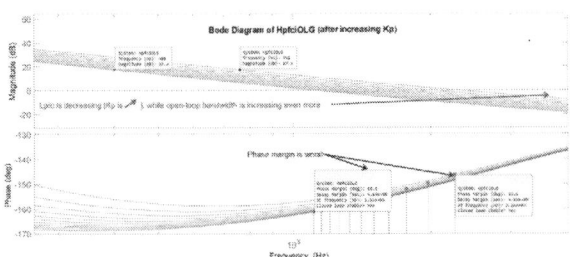

**Fig. 9:** Impact of the proportional gain $K_p$ on the current loop

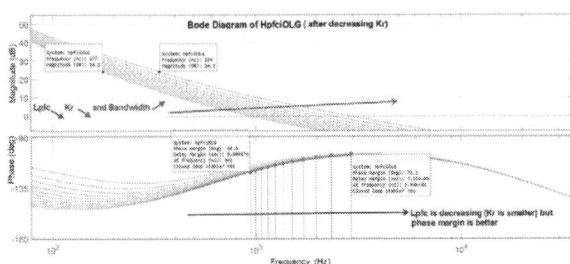

**Fig. 10:** Impact of the proportional gain $K_R$ on the current loop

As an example, Fig. 11 and Fig. 12 show the possible ranges for the proportional gains following the above design requirements.

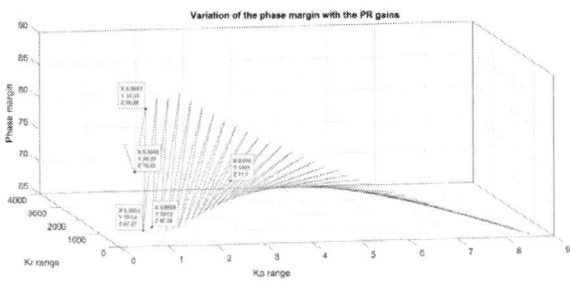

**Fig. 11:** Ranges of the proportional gains $K_R$ and $K_p$ for the minimum inductance (100 µH) to obtain phase margin > 60°.

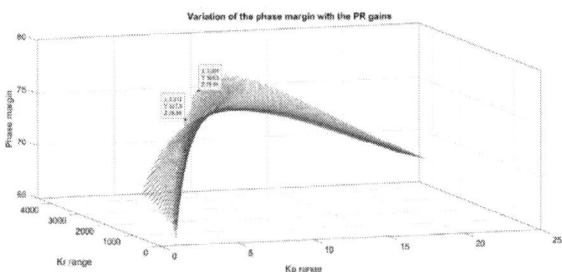

**Fig. 12:** Ranges of the proportional gains $K_R$ and $K_p$ for the minimum inductance (360 µH) to obtain phase margin > 60°.

Fig. 13 presents the evolution of the gains $K_p$ and $K_R$ implemented in the adaptive compensator to obtain the desired performance of the PFC converter for the entire range of the inductance values.

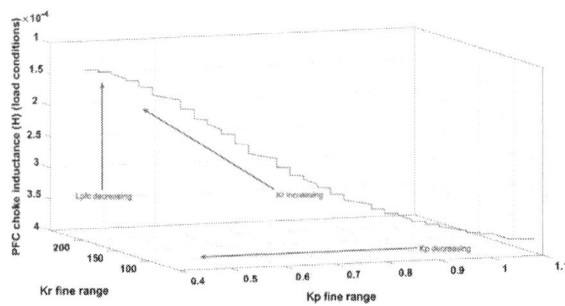

**Fig. 13:** The gains of the PR controller used to create the adaptive compensation based on the load condition variations.

Table 3 summarizes the effects of both proportional gains in the current loop.

| Actions | Phase margin | Bandwidth |
|---|---|---|
| Increase $K_{p\ (KR=0)}$ | Decreased | Increased |
| Increase $K_{R\ (Kp=0)}$ | Decreased | Increased |

**Table. 3:** Summary of the impact of the proportional gains $K_p$ and $K_R$ on the current loop.

# 4 Experimental validation of the controller

The implementation of the adaptive controller is based on the current measurement used to estimate the inductance value. The estimated inductance is then used in a lookup table to select the appropriate gains of the PR controller as defined in Fig. 13. But considering that the inductance behavior is known, the lookup table can be implemented based on the current measurement only.

Fig. 14 and Fig. 15 compare the static and adaptive gains for the three-phase PFC controller during a step load, while the power is increased from 2 kW to 11 kW. It is obvious that Fig. 14 shows distortion in the current waveforms at the maximum ranges. It can be explained by the fact that the bandwidth increases when the current is higher ($L_{PFCi}$ decreases). The waveform is not smooth due to the change of the phase margin during different load conditions. Meanwhile, the adaptive control in Fig. 15 and Fig. 16 present better performance during low and high load conditions, with better current waveform and consequently, lower THD.

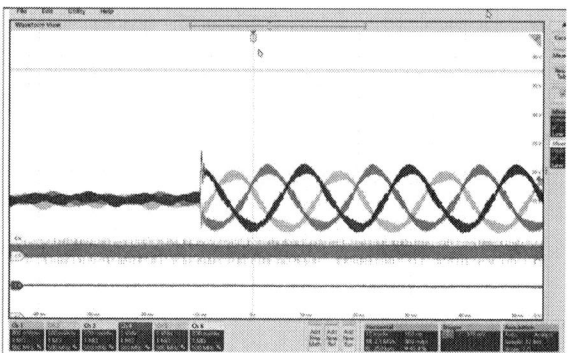

**Fig. 14**: AC current in the 3 phases during a transient response for a power step from 2 kW to 11 kW with the static/classic PR controller.

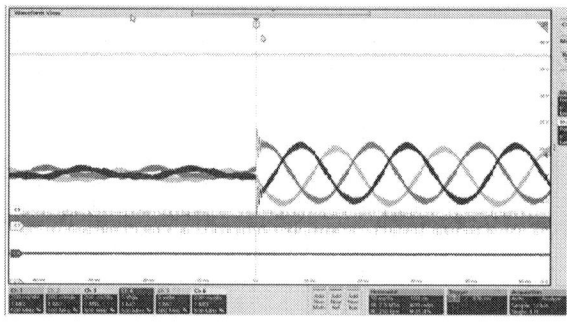

**Fig.15**: AC current in the 3 phases during a transient response for a power step from 2 kW to 11 kW with adaptive PR controller.

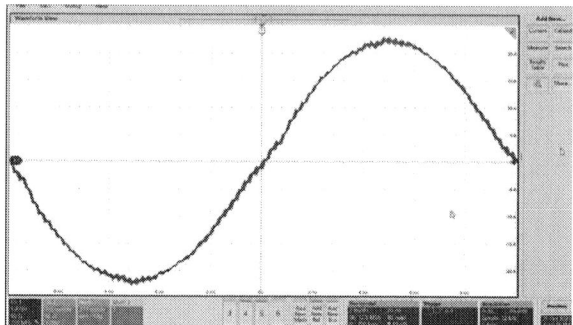

**Fig. 16**: Phase Current in the PFC chokes for maximum power (11 kW) with adaptive PR controller

# 5  Conclusion

This paper proposes an adaptive controller based on a resonant PR controller for a three-phase PFC converter. The results presented here, show the impact of the inductance variation in the margin of phase and the bandwidth response of the system that is hardly improved with a static tuning of the current controller. Therefore, the implementation of variable proportional gains in the PR controller allows improving both parameters. The effects of each one of the proportional gains are also highlighted concerning phase margin and bandwidth.

Although we have focused on the study of the controller tuning based on the variation of the boost inductor, this is not the only parameter affecting the system performance. As a matter of effect, the variation of the grid impedance and the main grid frequency should as well be considered during the tuning of the controller. However, even with these additional parameters, the methodology presented here can still be applied. Therefore, online grid impedance estimator can be used to update the compensator gains and notch filter should be deployed to mitigate the impact of the resonance related to the presence of X capacitors.

## Acknowledgment

The authors would like to thank the other members of system, control law and hardware teams from Valeo in Cergy-France for the help in providing the software and the support during the experimental tests.

## References

[1] J. W. -T. Fan, R. S. -C. Yeung and H. S. -H. Chung, "Optimized Hybrid PWM Scheme for

Mitigating Zero-Crossing Distortion in Totem-Pole Bridgeless PFC," in IEEE Transactions on Power Electronics, vol. 34, no. 1, pp. 928-942, Jan. 2019, doi: 10.1109/TPEL.2018.2819422.

[2] Q. Huang, Q. Ma, R. Yu, T. Chen, A. Q. Huang and Z. Liu, "Improved analysis, design and control for interleaved dual-phase ZVS GaN-based totem-pole PFC rectifier with coupled inductor," 2018 IEEE Applied Power Electronics Conference and Exposition (APEC), San Antonio, TX, USA, 2018, pp. 2077-2083,doi: 10.1109/APEC.2018.8341303.

[3] Mahajan A, Siedle C, Reichert S. Control of a Three Level Three Phase 20 kW Power Factor Correction (PFC) for Charging Applications. InPCIM Europe digital days 2021; International Exhibition and Conference for Power Electronics, Intelligent Motion, Renewable Energy and Energy Management 2021 May 3 (pp. 1-7). VDE.

[4] K. Ribeiro de Faria, L. Bendani and N. Bouzidi, "Design of a Four-Limb Coupled Inductor for a Three-phase Six-Switched Boost PFC Converter for EV Application," PCIM Europe 2023; International Exhibition and Conference for Power Electronics, Intelligent Motion, Renewable Energy and Energy Management, Nuremberg, Germany, 2023, pp. 1-6, doi: 10.30420/566091231.

[5] Divan, Israel & Brandao, Danilo & Matakas Junior, Lourenco & Simoes, Marcelo & Morais, L.M.F.. (2021). Analysis of Stationary- and Synchronous-Reference Frames for Three-Phase Three-Wire Grid-Connected Converter AC Current Regulators. Energies. 14. 8348. 10.3390/en14248348.

[6] Grid Converters for Photovoltaic and Wind Power Systems. Remos Teodorescu, Marco Liserre and Pedro Rodriguez. 2011, John Wiley & Sons, Ltd. ISBN: 978-0-470-05751-3.

PCIM Europe 2024, 11– 13 June 2024, Nuremberg    DOI: 10.30420/566262158

# Synthesis of a Field Oriented Control Algorithm by using two different Pole-Zero Compensation Approaches

Marco Denk[1], Felix Heigel[1], Johannes Schwarzkopf[2], Roman Filka[2]

[1] University of Applied Sciences Coburg, Germany
[2] Brose Fahrzeugteile SE & Co. KG, Würzburg, Germany

Corresponding author and Speaker:  Marco Denk, marco.denk@hs-coburg.de

## Abstract

This paper studies and compares two approaches to design and parametrize the q-current control loop of an electrical machine used in an electric vehicle application. Based on the identification of the parameters of the controlled system two control loop architectures were introduced and configured by pole-zero compensation. The stationary and dynamic behavior of both approaches were evaluated by simulations and verified on a motor test bench. Special focus is put on the drifting of motor parameters and their impact on the control stability and other control properties. It is found that both approaches differ from each other in their dynamic response behavior and their robustness against drifting motor parameters.

## 1    Introduction

### 1.1    Motor control system

Micromobility vehicles significantly contribute to an environmentally friendly, more flexible and socially inclusive way of traffic [1]. Besides low costs and high efficiency the control dynamic and the control robustness are very important requirements to the drivetrain. This paper investigates the control of a drive unit of micromobility vehicle that includes a 10 kW permanent magnet synchronous machine (PMSM) and a three-phase full bridge GaN voltage source inverter (VSI) as it can be seen in Fig. 1.

The PMSM has a nominal voltage of 48V and a nominal torque of 28 Nm. With field weakening the machine can be operated beyond their base speed of 2500 rpm and reach up to 4700 rpm. The stator winding is supplied by a three-phase GaN inverter [2]. Each topological switch consists of four single eGaN® FETs in parallel so that the inverter can deliver a peak output current of 200 $A_{RMS}$ and a continuous current of 150 $A_{RMS}$. To reach this, the GaN chips are attached to a heatsink with a forced air convection with 400 linear feet per minute. The inverter is controlled with a ST32K144 microcontroller, that is programmed in C++ and communicating with a motor control application tuning tool [3]. The microcontroller measures the three load currents with phase shunts and an ADC that is synchronized with the PWM signal. Moreover, the phase

voltages and the inverter temperatures are measured. In order to determine the rotor position of the electric machine a speed and position observer is used [4]. To control the speed and torque of electric machines, the field-oriented control (FOC) is the established state of art. It enables efficient motor drives with a fast dynamic response and high accuracy. The algorithms developed in this paper focus on controlling the torque of a PMSM. The desired torque $T_{req}$ can be specified by the user with a throttle. Based on this information the FOC adjusts the torque and keeps it constant as long as it is requested. One important use case is accelerating with a constant torque from a standstill position to a certain target speed with a good dynamic.

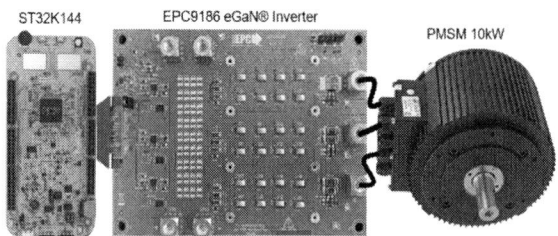

**Fig. 1**   Drivetrain of a 48V micromobility electric vehicle with GaN inverter and 10 kW electric machine.

To develop a torque control with good dynamic response behavior, this paper investigates the control circuit in detail and evaluates it on a motor test bench to pave the road for future measurements on a roller test bench. Special focus is put on the q-current control loop. The general structure of the control algorithm is given in Fig. 2. Based on the requested torque $T_{req}$ and the motor speed $\omega_{mech}$, the required d- and q-currents $I_{d,req}$ and $I_{q,req}$ are calculated with a maximum torque per ampere (MTPA) approach [5]. The requested currents are compared with the measured phase currents $I_{d,real}$ and $I_{q,real}$, which are transferred into the d/q- rotating frame with a α/β- and d/q-transformation [6].

The rotor position Θ is calculated with an observer, that is based on the measured and transformed phase currents $I_{d,real}$ and $I_{q,real}$ and the output voltages of the PI controllers [4]. The deviation between the required and measured d/q-currents are used as input for two PI controller. A feed-forward decoupling network [7] is used to ensure that the PI controller control the d- and the q-current independently from each other.

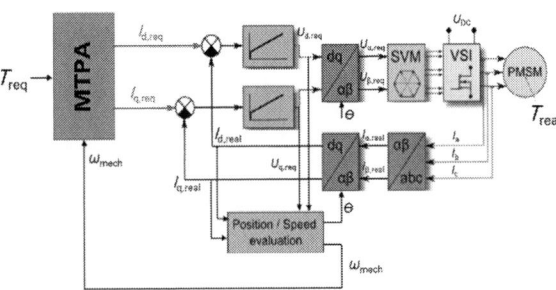

**Fig. 2** Field oriented control to operate a permanent magnet synchronous machine with desired torque $T_{req}$.

The control voltages $U_{d,req}$ and $U_{q,req}$ in the rotating frame can be derived from equation 1, where $L_d$ and $L_q$ are the stator inductances of the d- and q-axis and $R_S$ is the stator resistance. The quadrature and direct inductances are very similar, so the motor has a low saliency ($L_d \approx L_q$). Besides these motor parameters, equation 1 includes the electric frequency $\omega_{el} = 2\pi \cdot f_{el} = \omega_{mech}/p$ of the inverter output currents, where $p$ is the number of pole pairs, and the flux linkage of the rotor magnets $\Psi_f$.

$$
\begin{bmatrix} U_{d,req} \\ U_{q,req} \end{bmatrix} = \begin{bmatrix} R_S & -\omega_{el}L_q \\ \omega_{el}L_d & R_S \end{bmatrix} \cdot \begin{bmatrix} I_{d,req} \\ I_{q,req} \end{bmatrix}
$$

$$
+ \begin{bmatrix} L_d & 0 \\ 0 & L_q \end{bmatrix} \cdot \begin{bmatrix} \frac{d}{dt}I_{d,req} \\ \frac{d}{dt}I_{q,req} \end{bmatrix} + \begin{bmatrix} 0 \\ \omega_{el}\Psi_f \end{bmatrix} \tag{1}
$$

The output voltages of the PI controller are transformed back to the stator frame and used to calculate the PWM signals by space vector modulation. Since this paper focuses on the torque control of

the electric machine, the q-current control loop will be studied in detail. Equation 2 shows that the measured torque $T_{real}$ is a function of the measured q-current $I_{q,real}$, the flux linkage $\Psi_f$ of the PMSM and the number of pole pairs $p$. To control the torque, it is therefore sufficient to feedback the q-current as control parameter. A speed control operation would also be possible by adding an outer control loop [8]. In this application, driving with a constant speed is not relevant so far.

$$
T_{real} = \frac{3}{2} \cdot \Psi_f \cdot I_{q,real} \cdot p \tag{2}
$$

## 1.2 Research questions

Underneath the general FOC architecture different sub-variants appear when looking more precisely into the functional blocks. One essential difference can be found in the architecture and the parameterization of the PI controllers. Figures 3 and 4 show two commonly used architectures to control the d- or q-current in a FOC. The architecture in figure 3 consists of a PI controller R(s) with an integral component $T_N$ and a proportional gain $k_R$. These parameters need to be defined in a way to reach a desired behavior of the control loop. Especially an appropriate loop bandwidth and attenuation must be ensured. The PI controller is in series to the controlled system G(s) that is characterized by the ohmic-inductive behavior of the motor stator winding. In Laplace domain the controlled system for the q-current can be approximated by a first order element (PT1) that is parameterized with the time constant $T_S = L_q/R_S$ and the gain $k_S = 1/R_S$.

$$
G(s) = \frac{1}{L_q \cdot s + R_S} = \frac{k_S}{T_S \cdot s + 1} \tag{3}
$$

Applying a q-voltage $U_{q,req}$ to the stator winding of the motor G(s) leads to a q-current $I_{q,real}$, which is feed-back and subtracted from the requested current $I_{q,req}$. This closes the control loop. In chapter 2.1 this control circuit, called architecture 1, will be parameterized and investigated in detail.

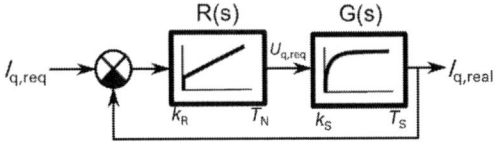

**Fig. 3** Control architecture 1 to control the q-current in a motor winding G(s) with a PI controller R(s).

The classical control architecture in figure 3 can be extended by an additional zero cancellation block $G_{ZC}(s)$ in the feed-forward path as it is shown in figure 4. Zero cancellation plays an important role in terms of compensation of a closed control loop system [8]. This kind of closed loop control circuit, which is called architecture 2, works very differently than architecture 1. The advantages and disadvantages of both architectures in the design and operating phase need to be clarified.

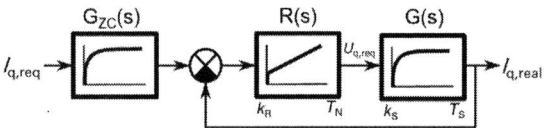

**Fig. 4** Control architecture 2 uses a zero-cancellation $R_{ZC}(s)$ in the feedforward path of the control loop.

This paper investigates both architectures in detail and compares their properties in terms of dynamic behavior and immunity against drifting parameter. The synthesis of the d-current control loop for field weakening would be very similar.

## 2 Synthesis of FOC Architectures

### 2.1 Architecture 1: Pole-zero-compensation in the open loop control circuit

The parameterization of the PI controller in architecture 1 is based on the transfer function of the open loop control system $F_{0,1}(s)$ as it is given in equation 4. It can be seen that the pole of the controlled system can be compensated with the zero of the PI controller by setting the integral component to $T_N = T_S$. Thanks to this compensation, neither the zero of the PI controller $R(s)$, nor the pole of the controlled system $G(s)$ have impact on the behavior of the closed control loop anymore.

$$F_{0,1}(s) = \underbrace{\frac{k_S}{T_S \cdot s + 1}}_{G(s)} \cdot \underbrace{k_R \cdot \frac{T_N \cdot s + 1}{T_N \cdot s}}_{R(s)} \quad (4)$$

After the compensation the transfer function of the closed loop circuit $F_{CL,1}(s)$ can be calculated with equation 5 where $T_N = T_S = L_q/R_S$ and $k_S = 1/R_S$. It can be seen that the closed control loop is a first order element. Laplace's final value theorem [9] shows that no steady-state deviation remains.

$$F_{CL,1}(s) = \frac{F_{0,1}(s)}{1 + F_{0,1}(s)} = \frac{1}{\frac{T_N}{k_R \cdot k_S} \cdot s + 1} \quad (5)$$

Figure 5 shows the p/z-diagram of architecture 1, where the pole of the controlled system $G(s)$ is compensated by the zero of the PI controller $R(s)$. All poles are in the left-half plane of the p/z-map, so the system is stable and the compensation is uncritical. After the compensation, the closed loop control circuit $F_{CL,1}(s)$ only has one pole on the left-half plane at $s_1 = - k_R/L_q$ (see equation 5).

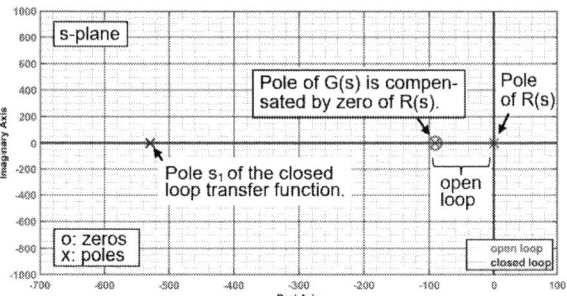

**Fig. 5** In architecture 1 the pole of $G(s)$ is compensated by the zero of $R(s)$ in the open loop circuit.

To identify the proportional gain factor $k_R$ of the PI controller, the frequency characteristic method is used [10]. Figure 6 shows the step response of the q-current control loop for the initial gain $k_{R1}$ and the optimized gain $k_{R2}$. The current is normalized to the nominal q-current $I_{q,nom}$. It can be seen that if the nominal q-current is requested $I_{q,req} = I_{q,nom}$ the step response shows a first order element characteristic (PT1) and reaches a stationary current value of $I_{q,req}/I_{q,nom} \approx 1$ (99%) after $\Delta t = 2.5$ ms.

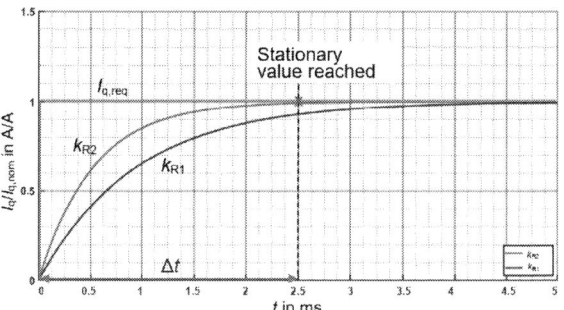

**Fig. 6** The step responses of architecture 1 with two different proportional gains $k_{Rx}$ shows a PT1 behavior.

### 2.2 Architecture 2: Zero-cancellation in the closed loop control circuit

In the second architecture the transfer function of the closed loop is calculated without any compensation measures in the open loop transfer function. To highlight the working principle of the compensation within the closed loop, the transfer function of the PI controller as given in equation 6 is used. It consists of the proportional component $k_P$ and an integration component $k_I$ as follows:

$$R(s) = k_R \cdot \frac{T_N \cdot s + 1}{T_N \cdot s} = \frac{k_P s + k_I}{s} \qquad (6)$$

Equation 7 shows the transfer function $F_{CL,2}(s)$ of architecture 2, that consists of the closed loop circuit transfer function $F_{CL,2}^*(s)$ and the feed-forward zero cancellation block $G_{ZC}(s)$. After some transformations it can be seen that the PI controller introduces a zero to the overall closed loop transfer function that is located at $s = -k_I/k_P$. This zero adds a derivative behavior to the control loop and creates an overshoot in the step response. To avoid this, the zero of the PI controller can be compensated. This compensation can be reached by introducing a zero cancellation block in the feed-forward path, as it can be seen in figure 4 and in equation 7. To compensate the zero of the PI controller, the zero-cancellation block has to be parameterized with the same parameters $k_P$ and $k_I$ as they are used in the controller.

$$F_{CL,2}(s) = G_{ZC}(s) \cdot \overbrace{\frac{R(s) \cdot G(s)}{1 + R(s) \cdot G(s)}}^{F_{CL,2}^*(s)} \qquad (7)$$

$$= \frac{1}{\frac{k_P}{k_I} s + 1} \cdot \frac{\frac{k_I}{L_q} \cdot \left(\frac{k_P}{k_I} \cdot s + 1\right)}{s^2 + \left(\frac{k_P + R_S}{L_q}\right)s + \frac{k_I}{L_q}}$$

$$= \frac{\frac{k_I}{L_q}}{s^2 + \left(\frac{k_P + R_S}{L_q}\right)s + \frac{k_I}{L_q}}$$

After the compensation the transfer function of the closed loop circuit has the form of a standard second order element $G_{PT2}(s)$ with a natural frequency of the closed loop system $\omega_0$ and the loop attenuation $\zeta$ as follows:

$$G_{PT2}(s) = \frac{\omega_0^2}{s^2 + (2 \cdot \zeta \cdot \omega_0)s + \omega_0^2} \qquad (8)$$

In figure 7 it can be seen that the compensation of the zero of the closed control loop with the pole of an additional zero-cancellation block happens in the left side of the p/z-diagram so it is also allowed and the system is stable. In contrast to the p/z-map

of architecture 1 (compare fig. 5) the closed loop transfer function of architecture 2 can have a double pole, two poles or a conjugated complex pole pair. The position of the poles defines the dynamic behavior of the control loop. Ziegler-Nichols optimum criteria specifies an attenuation of $\zeta = 0{,}707$ as good compromise between speed and stability [10]. Figure 7 shows that with a loop attenuation of $\zeta = 0{,}707$ and a frequency of $\omega_0 = 800~\text{s}^{-1}$ one complex conjugated pole pair at $s_{1/2}$ in the s-plane remains for the closed loop transfer function after the zero-cancellation.

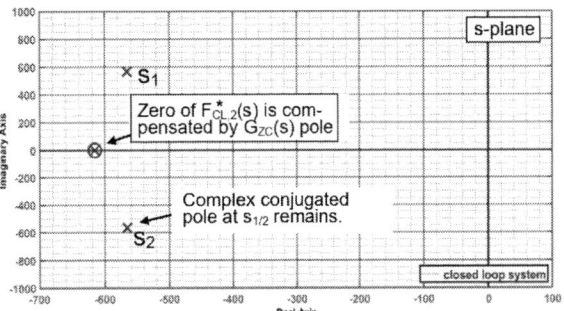

**Fig. 7** In architecture 2 the zero of $F_{CL,2}^*(s)$ is compensated by the pole of $G_{ZC}(s)$ in closed loop circuit.

The parameters of the PI controller can be tuned by matching the coefficients of the closed loop transfer function $F_{CL,2}(s)$ (see equ. 7) with those of a second order system as it is given in equation 8. This enables a simple tuning of the current control loop bandwidth $\omega_0$ and attenuation $\zeta$ to reach an anticipated control behavior. Figure 8 shows the step response of the parameterized closed control loop with and without zero-cancellation.

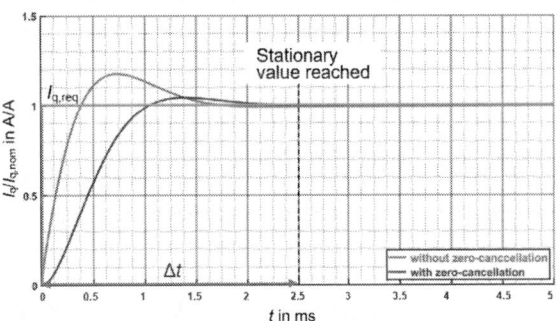

**Fig. 8** Step response of the closed loop control circuit with and without feed-forward zero-cancellation block.

It can be seen that with zero-cancellation block the control loop behaves like a PT2 element with the desired attenuation $\zeta = 0{,}707$ and loop bandwidth $\omega_0$. Without zero-cancellation block this simple adjustment of the control parameters is not possible.

## 2.3 Comparison of step responses

The pole-zero compensation in architecture 1 and the feed-forward zero-cancellation in architecture 2 work very differently and lead to different closed loop step response behaviors. Figure 9 shows the step responses of the finally parameterized control architectures 1 and 2 normalized to the nominal q-current $I_{q,nom}$. It can be seen that architecture 1 shows a PT1 and architecture 2 a PT2 behavior. In both architectures the q-current reaches a stationary value after $\Delta t = 2.5$ ms. The overshooting of architecture 2 can be used to improve the speed in the region close to the steady state current. At the beginning of the step response the PT2 behavior leads to a smaller gradient right after the input signal changes. This can offer advantages in subsequent signal processing blocks in the microcontroller and the dynamic operation of the traction inverter. The question is, how the dynamic behavior of both architectures is affected by drifting parameters in the controlled system and if both architectures are affected differently.

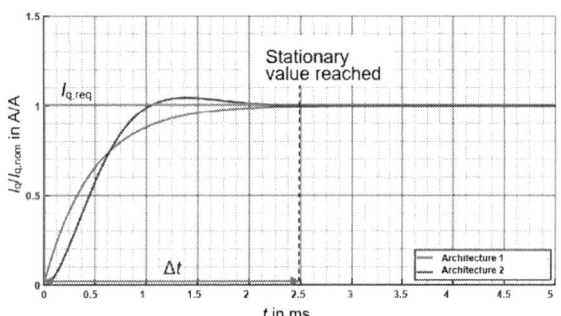

**Fig. 9** Comparison of the step responses of the finally configured and parameterized architectures 1 and 2.

# 3 Robustness against Parameter Drifting

## 3.1 Theoretical investigations

Besides the adjustment of good control properties in an initial design the robustness of a control loop against drifting parameters during the operating period is very important. This paper particularly investigates the impact of an increasing or decreasing stator resistance $R_S$ due to different motor temperatures. Moreover, in [11] it is found that the stator resistance can increase by 26% due to ageing. Adaptive methods to approximate the stator resistance with an observer and to use this value within the field oriented motor control exist and have the

potential to guarantee a similar control loop behavior independently from drifting parameters [12]. In addition to this it would be helpful to use a control architecture that is robust against drifting parameters on their own. To investigate this, both architectures are studied how they are affected by a drifting stator resistance. In architecture 1 the parameters of the PI controller $T_N$ and $k_R$ have been defined in direct relation to the parameters of the controlled system G(s) by choosing the integration time similar as the time constant of the stator winding $T_N = T_S$ and by identifying the proportional component $k_R$ with the Bode frequency characteristic method, where $k_S$ is design relevant. If $T_S$ and $k_S$ change due to drifting parameters and the parameters of the controller remain constant, the properties of the control circuit will be affected.

In figure 10 it can be seen that the pole of G(s) shifts to the left in the p/z-map with increasing stator resistance while the zero and the pole of R(s) remain unaffected. Consequently, the pole of G(s) moves away from the zero of the controller and is no longer fully compensated. Due to this not 100% working compensation the p/z-map of the closed loop consists of an additional zero and pole, that partially compensate each other. In addition, the pole on the left side shifts to the left and slows down the speed of the system (compare figure 5). To this end, in architecture 1 a drifting stator resistance changes the compensation and slows down the speed of the control loop.

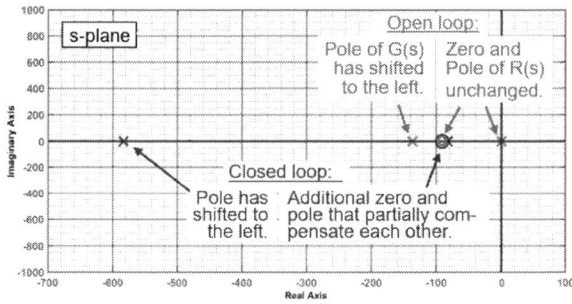

**Fig. 10** Poles and zeros of the open and closed loop of architecture 1 with 20% increased stator resistance.

In architecture 2 the cancellation of the zero of the closed loop transfer function $F^*_{CL,2}(s)$ by the pole of the zero-cancellation element $G_{ZC}(s)$ doesn't depend on the parameters $T_S$ and $k_S$ of the controlled system G(s), as it can be seen in equation 7. Instead, the cancellation only uses the constant parameters of the PI controller $k_P$ and $k_I$. The p/z-map in figure 11 shows that the zero of $F^*_{CL,2}(s)$ and the pole of $G_{ZC}(s)$ are not affected by drifting parameters and the compensation remains 100% working. Furthermore, it can be seen that the conjungated complex poles shift to the left with increasing stator resistance and slow down the system. The imaginary values of the poles are reduced, too, so overshooting will become less with increasing resistance. In summary, the compensation in architecture 2 is not affected by drifting parameters, but the speed of the control loop is slowed down, too.

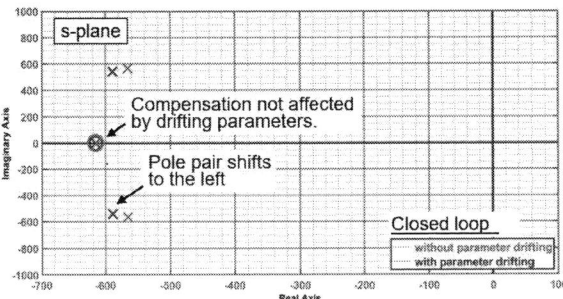

**Fig. 11** Impacts of an increasing stator resistance on the poles and zeros of the architecture 2 closed loop.

## 3.2 Simulation of drifting parameters

The impact of an 20% increased or decreased stator resistance on the step response of architecture 1 is shown in figure 12. It can be seen that an increase of $R_S$ slows down the speed of the closed control loop. A decrease of $R_S$ causes overshooting, which can be explained by the not 100% working compensation (compare figure 10).

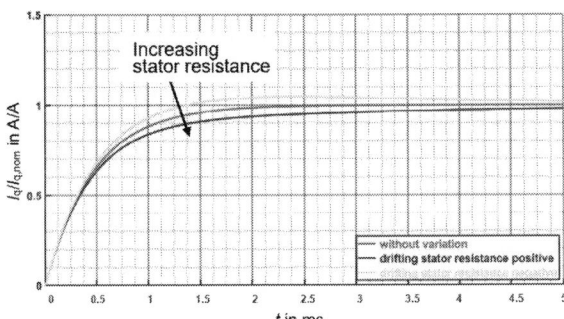

**Fig. 12** Impact of an increased or decreased stator resistance on the step response in architecture 1.

Figure 13 shows the step response of architecture 2 with a ±20% drifting stator resistance. Again, an increasing stator resistance slows down the velocity of the control circuit, but this impact is much smaller than in architecture 1 (compare figure 12). In addition, the impact of a decreasing stator resistance on the dynamic behavior is smaller and the overshooting remains virtually unchanged.

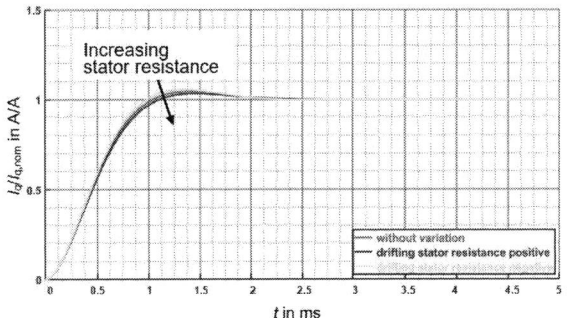

**Fig. 13** Impact of an increased or decreased stator resistance on the step response in architecture 2.

## 3.3 Evaluation by measurements

Due to the advantages of the zero-cancellation approach in terms of drifting parameters the control architecture 2 is programmed on the microcontroller of the drive unit and investigated on a motor test bench. In order to measure the q-current step response, the rotor was locked with a mechanical brake and the d-current was set to $I_{d,req} = 0$ A.

Figure 14 shows the measured step response of the q-current $I_{q,real}$ that is normalized to the continuous inverter output current $I_{q,nom}$. It can be seen that the measurement results are in good accordance with the simulation results (compare figure 8). The dynamic behavior of the step response can be described with a second order element and after a small overshooting ($\zeta = 0{,}707$) a stationary value is reached after about $\Delta t = 2{,}4$ ms. The delay time of about $t_{delay} = 100$ µs can be explained by the programmed slope in the requested q-current $I_{q,req}$ and processing times in the measuring circuit and the microcontroller. Figure 14 also shows the control voltage $U_{q,req}$, which is the output signal of the PI controller (compare figure 2).

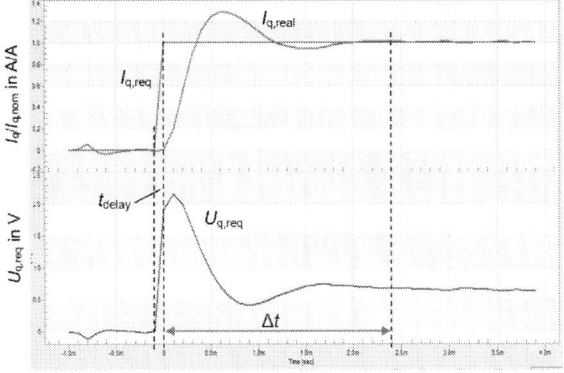

**Fig. 14** Measured step response of the q-current at room temperature after requesting $I_{q,req} = I_{q,nom}$.

To verify the effect of drifting parameters by measurements, the motor stator winding was heated-up to 85°C. Due to this the stator resistance increased by about 21%. The impact on the step response of the q-current is very small (compare figure 13) and the measuring result virtually looks like the measured system response at room temperature. This confirms the theoretical investigations and the robustness of architecture 2 against drifting parameters. To still prove the impact of drifting parameters on the control behavior of architecture 2, the stator resistance was increased by a factor of two by connecting three power resistors in series to the motor winding. The measuring result in figure 15 shows that this slows down the speed of the control circuit and reduces the overshooting.

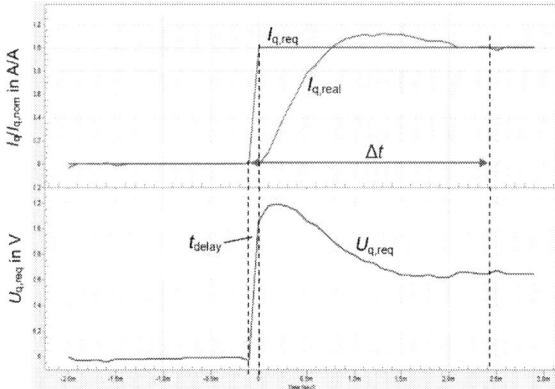

**Fig. 15** Measured step response of the q-current with an increased stator resistance by a factor of two.

## 4   Conclusions

This paper investigated two architectures to control the q-current in an electric machine. In architecture 1 the pole of the controlled system is compensated by the zero of the PI controller during the synthesis of the open loop control circuit. The proportional component of the PI controller is adjusted with the frequency characteristic method. Architecture 2 uses an additional zero-cancellation block in the feed-forward path to compensate the zero of the closed loop transfer function. The parameters of the PI controller are defined with the coefficient comparison method. The different control architectures and their different synthesis approaches lead to different dynamic behaviors. The step response of architecture 1 can be described with a first order element (PT1). Architecture 2 shows the behavior of a second order element (PT2), which can offer advantages in terms of speed and compatibility to

subsequent processing blocks. Moreover, the adjustment of a desired control loop behavior is easier so that architecture 2 shows advantages in the design phase. Special focus was put on the impact of drifting parameters during the operation phase, especially a changing stator resistance due to temperature and ageing. In theoretical investigations it was found that in architecture 1 the pole-zero compensation is worsened by drifting parameters, which has a significant impact on the control loop behavior. In contrast, the zero-cancellation in architecture 2 is not affected by an increasing or decreasing stator resistance. In both architectures the speed of the control circuit is slowed down by an increasing stator resistance, but in architecture 2 this effect is much smaller. To this end, architecture 2 offers advantages in the operating phase, too. Measurements on a motor test bench with a modified stator winding confirmed these findings. In future, the investigations should be extended to the impact of disturbances on the control loop and measurements on a roller test bench.

## References

[1] Badia, H.; Jenelius, E.: Shared e-scooter micromobility: review of use patterns, perceptions and environmental impacts, Transport reviews 2023, VOL. 43, NO. 5, DOI: 10.1080/01441647.2023.2171500

[2] Barba, V.; Musumeci, S.; Palma, M.: Enhanced Low-Voltage GaN FETs for e-Mobility Motor Control Improvements, AEIT International Conference on Electrical and Electronic Technologies for Automotive, Modena, Italy, 2023, pp. 1-6, DOI: 10.23919/AEITAUTOMOTIVE58986.2023.10217194.

[3] NXP Semiconductors User Guide, FreeMaster for Embedded Applications, Rev. 4.4, 03/2022, Document identifier: FMSTERUG

[4] Morimoto, S.; Kawamoto, K.; Sanada, M.; Takeda, Y.: Sensorless control strategy for salient-pole PMSM based on extended EMF in rotating reference frame, IEEE Transaction on Industrial Applications, vol. 38, no. 4, pp. 1054-1061, Jul./Aug. 2002

[5] Caruso, M.; Di Tommaso, A.O.; Miceli, R.; Nevoloso, C.; Spataro, C.; Trapanese, M.: Maximum Torque per Ampere Control Strategy for Low-Saliency Ratio IPMSMs, International Journal of renewable energy research, Vol.9, No.1, March, 2019

[6] Krause, P.; Wasynczuk, O.; Sudoff, S.: Analysis of Electric Machinery and Drive Systems, Engineering Physics, October 1995, DOI: 10.1109/9780470544167

[7] Milŏsević, M.: Decoupling Control of d and q Current Components in Three-Phase Voltage Source Inverter, Journal of Engineering, 2003, Article ID: 16197528

[8] Stulrajter, M.; Sustek, P.: Motor Control Application Tuning (MCAT) Tool for 3-Phase PMSM, NXP Application Note AN4642, Rev. 1, January 2013

[9] Nise, N.: Control Systems Engineering, John Wiley and Sons Ltd, December 2010, 6th edition, ISBN 9780470547564

[10] Ramdani, A.; Traiche, M.: PID controller compared with dynamic matrix control applied on disturbed complex system, International Journal of Digital Signals and Smart Systems, 2019, JDSSS.2019.103373.

[11] Ginzarly, R.; Hoblos, G.; Moubayed, N.: From Modeling to Failure Prognosis of Permanent Magnet Synchronous Machine, Journal of Applied Sciences, January 2020, DOI: 10.3390/app10020691

[12] Beguenane, R.; Ouhrouche, M. A.; Trzynadlowski, A. M.: Stator resistance tuning in an adaptive direct field-orientation induction motor drive at low speeds, 30th Annual Conference of IEEE Industrial Electronics Society *(IECON)*, Busan, Korea (South), 2004, DOI: 10.1109/IECON.2004.1433286

PCIM Europe 2024, 11– 13 June 2024, Nuremberg          DOI: 10.30420/566262160

# Average Current Mode Control and Its Loop Design

Feng Ji [1], Niklas Schwartz [2]
[1] Texas, Instruments, China
[2] Texas, Instruments, Germany

Corresponding author:    Feng Ji, feng-ji@ti.com
Speaker:                 Niklas Schwartz, n-schwarz@ti.com

## Abstract

Energy storage systems are becoming popular in these years. Multiphase interleaving and bi-directional operation are required in such applications. This paper will show the benefits of average current mode control in such applications. Models of the inner current loop and outer voltage loop are presented. Feedforward ramp is introduced and the benefits are analyzed. The compensation strategies for the current loop and voltage loop are suggested. A 48V to 12V/60A bi-directional DC/DC converter with the modified average current mode control is developed and introduced. Key waveforms are obtained to support the benefits. Simulation and bench test are performed to verify the model and compensation strategies.

## 1  Introduction

Energy storage systems are becoming popular in these years. To charge and discharge batteries, bi-directional operation is required. The energy storage system often goes up to hundreds of Amps. Multiphase interleaving is employed to share the power and reduce the current ripple to the output capacitors [1, 2].

Peak current mode control is widely adopted in switching mode power supplies. However, it also shows some problems, such as poor noise immunity and peak-to-average current error [3]. The current sharing accuracy will be limited by the inductor tolerance which is 20%~30% typically [4]. Current sharing is important for a balanced power dissipation in multiphase system. Also, it is quite noisy when it operates with high current and hard switching. Thus, peak current mode control is not the best choice in energy storage system.

Average current mode control solves the peak-to-average current error by regulating the average inductor current. Inductor tolerance will not affect the current regulation accuracy. So, accurate current regulation and sharing can be achieved with average current mode control. Also, it is less sensitive to noise.

These points make average current mode control ideal for energy storage system.

## 2  Current Loop Modelling

The control scheme of an average current mode controlled buck converter is shown in Fig. 1.

An outer voltage loop is utilized to regulate $V_{LV}$. And an inner current loop will regulate the inductor current.

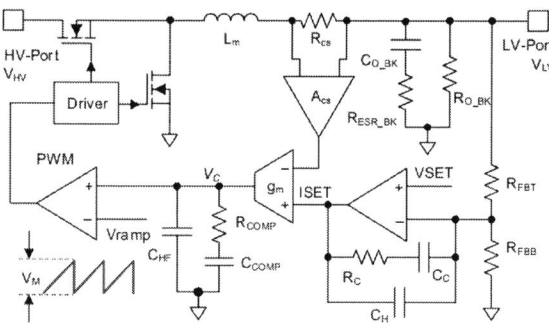

**Fig. 1**    Average current mode control scheme

The inner current loop control block diagram is shown in Fig. 2.

DIR controls the direction of the inductor current, that is in buck mode or boost mode.

1200

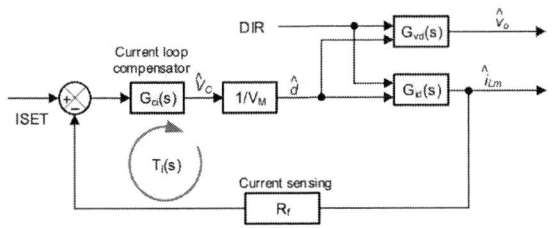

**Fig. 2** Control block diagram

From Fig. 2, the current loop open loop gain $T_i(s)$ can be found as,

$$T_i(s) = \frac{G_{ci}(s)G_{id}(s)R_f}{V_M} \quad (1)$$

Where

$$R_f = A_{CS}R_{CS} \quad (2)$$

The buck power plant current loop transfer function $G_{id\_BK}(s)$ can be written as,

$$G_{id\_BK}(s) = \frac{V_{HV}}{R_{O\_BK}} \frac{1 + \dfrac{s}{\omega_{Z\_il\_BK}}}{1 + \dfrac{s}{\omega_{0\_BK}Q_{BK}} + \dfrac{s^2}{\omega_{0\_BK}^2}} \quad (3)$$

Where

$$\omega_{Z\_il\_BK} = \frac{1}{R_{O\_BK}C_{O\_BK}} \quad (4)$$

$$\omega_{0\_BK} = \frac{1}{\sqrt{L_m C_{O\_BK}}} \quad (5)$$

$$Q_{BK} = \frac{\dfrac{1}{\omega_{0\_BK}}}{\dfrac{L_m}{R_{O\_BK}} + (R_{ESR\_BK} + R_S)C_{O\_BK}} \quad (6)$$

$C_{O\_BK}$ is total output capacitor in buck mode;

$R_{O\_BK}$ is total output load in buck mode;

$R_{ESR\_BK}$ is the equivalent series resistance of the output capacitor in buck mode;

$R_S$ is the total resistance along the current path.

The boost power plant current loop transfer function $G_{id\_BST}(s)$ can be written as,

$$G_{id\_BST}(s)$$

$$= \frac{2V_{LV}}{D'^3 R_{O\_BST}} \frac{1 + \dfrac{s}{\omega_{Z\_il\_BST}}}{1 + \dfrac{s}{\omega_{0\_BST}Q_{BST}} + \dfrac{s^2}{\omega_{0\_BST}^2}} \quad (7)$$

Where

$$\omega_{Z\_il\_BST} = \frac{2}{R_{O\_BST}C_{O\_BST}} \quad (8)$$

$$D' = \frac{V_{LV}}{V_{HV}} \quad (9)$$

$$\omega_{0\_BST} = \frac{D'}{\sqrt{L_m C_{OUT\_BST}}} \quad (10)$$

$$Q_{BST}$$

$$= \frac{\dfrac{D'}{\omega_{0\_BST}}}{\dfrac{L_m}{D'R_{O\_BST}} + \dfrac{R_S C_{O\_BST}}{D'} + R_{ESR\_BST}C_{O\_BST}} \quad (11)$$

$C_{O\_BST}$ is total output capacitor in boost mode;

$R_{O\_BST}$ is total output load in boost mode;

$R_{ESR\_BST}$ is the equivalent series resistance of the output capacitor in boost mode.

When we select the current loop cross over frequency at 1/6 of switching frequency, which is more than 10kHz typically, $G_{id\_BK}(s)$ can be simplified. For the numerator, $s/\omega_{Z\_il\_BK}$ dominates. And for the denominator, $s^2/\omega_{0\_BK}^2$ dominates. $G_{id\_BST}(s)$ can be simplified similarly.

$G_{id\_BK}(s)$ and $G_{id\_BST}(s)$ can be unified as,

$$G_{id}(s) \approx \frac{V_{HV}}{sL_m} \quad (12)$$

Substituting Eq. (12) into Eq. (1),

$$T_i(s) = \frac{V_{HV}}{sL_m} \frac{G_{ci}(s)R_f}{V_M} \quad (13)$$

It can be observed that the open loop gain $T_i(s)$ varies with $V_{HV}$.

By creating a feedforward ramp signal that is proportional to $V_{HV}$,

$$V_M = K_{FF}V_{HV} \quad (14)$$

We can find,

$$T_i(s) = \frac{1}{K_{FF}} \frac{G_{ci}(s)R_f}{sL_m} \quad (15)$$

Thus, we can get constant current loop gain regardless of voltage and load conditions. Also, single current loop compensation can be used for both buck and boost mode.

## 3 Current Loop Compensation

Equation (15) indicates that the current loop power plant is basically a first-order system. A Type-II compensator with gm amplifier as shown in Fig. 1 is adequate to stabilize the loop for both buck and boost mode operations.

The transfer function of the current loop compensator can be written as,

$$G_{ci}(s) = g_m Z_{COMP}(s) \qquad (16)$$

$g_m$ is the trans-conductance of the gm error amplifier;

$Z_{COMP}(s)$ is the impedance of the compensation network.

Considering $C_{HF} \ll C_{COMP}$, $Z_{COMP}(s)$ can be simplified as,

$$Z_{COMP}(s) = \frac{1 + \frac{s}{f_z}}{sC_{COMP}\left(1 + \frac{s}{f_p}\right)} \qquad (17)$$

Substituting Eq. (16), Eq. (17) into Eq. (15),

$$T_i(s) = \frac{R_f g_m}{sK_{FF}L_m} \frac{1 + \frac{s}{f_z}}{sC_{COMP}\left(1 + \frac{s}{f_p}\right)} \qquad (18)$$

The poles and zeros of the compensation stage are determined by,

$$f_{p0} = 0 \qquad (19)$$

$$f_p = \frac{1}{2\pi R_{COMP}C_{HF}} \qquad (20)$$

$$f_z = \frac{1}{2\pi R_{COMP}C_{COMP}} \qquad (21)$$

Considering the current loop for both buck mode and boost mode, the current loop crossover frequency $f_{CI}$ should be set to $f_{sw}/6$ [3].

Select the compensation network according to the following guidelines, then fine tune the network for optimal loop performance.

1. Set the current loop open loop gain to unity, that is $|T_i(2\pi i f_{CI})| = 1$,
2. The zero $f_z$ is placed at around 1/5 of target crossover frequency $f_{CI}$,
3. The pole $f_p$ is placed at approximately 1/2 of switching frequency $f_{sw}$,

Therefore, the compensation components can be derived as,

$$R_{COMP} = \frac{K_{FF}}{R_f g_m}|2\pi i f_{CI}L_m| \qquad (22)$$

$$C_{COMP} = \frac{1}{\left|2\pi i \frac{f_{CI}}{5}R_{COMP}\right|} \qquad (23)$$

$$C_{HF} = \frac{1}{\left|2\pi i \frac{f_{sw}}{2}R_{COMP}\right|} \qquad (24)$$

# 4 Voltage Loop Modelling

Figure 3 shows the outer voltage loop and the inner current loop.

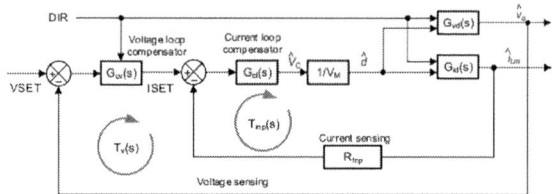

**Fig. 3** Outer loop and inner loop block diagram

As $n_p$ phases are employed for the inner loop. The equivalent circuit of a $n_p$ phase interleaving system can be obtained as Fig. 4.

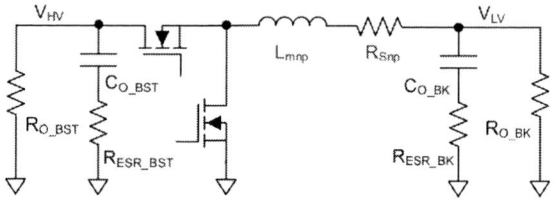

**Fig. 4** $n_p$ phase block diagram

The equivalent parameters of $n_p$ phase system can be summarized as,

$$L_{mnp} = \frac{L_m}{n_p} \qquad (25)$$

$$R_{Snp} = \frac{R_S}{n_p} \qquad (26)$$

$$R_{fnp} = \frac{R_f}{n_p} \qquad (27)$$

By using the equivalent parameters in Eq. (25), Eq. (26) and Eq. (27), $n_p$ phase transfer function of $G_{idnp\_BK}(s)$, $G_{idnp\_BST}(s)$, $T_{inp}(s)$ can be obtained from Eq. (3), Eq. (7) and Eq. (13).

$T_{inp}(s)$ can be found as,

$$T_{inp}(s) = \frac{V_{HV}}{sL_m}\frac{G_{ci}(s)R_{fnp}}{V_M} \qquad (28)$$

From Fig. 3, the voltage loop open-loop gain can be found as,

$$T_v(s) = G_{cv}(s)G_{vs}(s) \qquad (29)$$

ISET to output voltage ($v_{LV}$) transfer function can be found as,

$$G_{vs}(s) = \frac{1}{V_M}\frac{G_{ci}(s)G_{vd}(s)}{1 + T_{inp}(s)} \qquad (30)$$

As the outer loop crossover frequency is much lower than the inner loop, that is $1 \ll T_{inp}(s)$, Eq. (30) can be simplified as,

$$G_{vs}(s) = \frac{1}{V_M} \frac{G_{ci}(s)G_{vd}(s)}{T_{inp}(s)} \qquad (31)$$

Substituting Eq. (28) into Eq. (31),

$$G_{vs}(s) = \frac{G_{vd}(s)}{G_{idnp}(s)} \frac{1}{R_{fnp}} \qquad (32)$$

The buck converter duty cycle to output voltage transfer function $G_{vd\_BK}(s)$ is determined by,

$$G_{vd\_BK}(s) = V_{HV} \frac{1 + \dfrac{s}{\omega_{Z\_vl\_BK}}}{1 + \dfrac{s}{\omega_{0\_BK}Q_{BK}} + \dfrac{s^2}{\omega_{0\_BK}^2}} \qquad (33)$$

Where

$$\omega_{Z\_vl\_BK} = \frac{1}{R_{ESR\_BK}C_{O\_BK}} \qquad (34)$$

Substituting Eq. (33) into Eq. (32),

$$G_{vs\_BK}(s) = \frac{R_{O\_BK}}{R_{fnp}} \frac{1 + \dfrac{s}{\omega_{Z\_vl\_BK}}}{1 + \dfrac{s}{\omega_{Z\_il\_BK}}} \qquad (35)$$

Similarly, boost converter duty cycle to output voltage transfer function $G_{vd\_BST}(s)$ is determined by,

$$G_{vd\_BST}(s)$$
$$= \frac{V_{LV}}{D'^2} \frac{\left(1 + \dfrac{s}{\omega_{Z\_vl\_BST}}\right)\left(1 - \dfrac{s}{\omega_{RHPZ}}\right)}{1 + \dfrac{s}{\omega_{0\_BST}Q_{BST}} + \dfrac{s^2}{\omega_{0\_BST}^2}} \qquad (36)$$

Where

$$\omega_{Z\_vl\_BST} = \frac{1}{R_{ESR\_BST}C_{O\_BST}} \qquad (37)$$

$$\omega_{RHPZ} = \frac{R_{O\_BST}D'^2}{L_{mnp}} \qquad (38)$$

$C_{O\_BST}$ is output capacitor in boost mode;
$R_{O\_BST}$ is the output load in boost mode;
$R_{ESR\_BST}$ is the equivalent series resistance of the output capacitor in boost mode.
Substituting Eq. (36) into Eq. (32),

$$G_{vs\_BST}(s)$$
$$= \frac{R_{O\_BST}D'}{2R_{fnp}} \frac{\left(1 + \dfrac{s}{\omega_{Z\_vl\_BST}}\right)\left(1 - \dfrac{s}{\omega_{RHPZ}}\right)}{1 + \dfrac{s}{\omega_{Z\_il\_BST}}} \qquad (39)$$

Comparing Eq. (39) to Eq. (35), a right-half-plane zero can be found in the boost power stage.

# 5 Voltage loop Compensation

Similarly, A Type-II compensator is utilized to stabilize the voltage loop as shown in Fig. 1.

The transfer function of the voltage loop compensator can be written as,

$$G_{cv}(s) = \frac{Z_C(s)}{R_{FBT}} \qquad (37)$$

Where $Z_C(s)$ is the impedance of the outer loop compensation network. Considering $C_H \ll C_C$, $Z_C(s)$ can be simplified as,

$$Z_C(s) = \frac{1 + \dfrac{s}{f_{vz}}}{sC_C\left(1 + \dfrac{s}{f_{vp}}\right)} \qquad (38)$$

The poles and zeros of the compensation stage are determined by,

$$f_{vp0} = 0 \qquad (40)$$

$$f_{vp} = \frac{1}{2\pi R_C C_H} \qquad (41)$$

$$f_{vz} = \frac{1}{2\pi R_C C_C} \qquad (42)$$

The buck mode compensation is taken as an example. Substituting Eq. (37), Eq. (38) into Eq. (32),

$$T_v(s) = \frac{R_{O\_BK}}{R_{fnp}} \frac{1 + \dfrac{s}{\omega_{Z\_vl\_BK}}}{1 + \dfrac{s}{\omega_{Z\_il\_BK}}} \frac{1}{R_{FBT}} \frac{\left(1 + \dfrac{s}{f_{vz}}\right)}{sC_C\left(1 + \dfrac{s}{f_{vp}}\right)} \qquad (39)$$

It is advised that the outer voltage loop crossover frequency $f_{CV}$ should be one decade below that of the inner current loop crossover frequency $f_{CI}$.

And for the boost outer voltage loop crossover frequency should also be below 1/5 of the right-half-plane zero.

Select the compensation network according to the following guidelines, then fine tune the network for optimal loop performance.

1. Select $R_{FBT}$ based on the op amp bias current and power dissipation,
2. Set the voltage loop open loop gain to unity, that is $|T_v(2\pi i f_{CV})| = 1$,
3. The zero $f_{vz}$ is placed at around 1/5 of target crossover frequency $f_{CV}$,
4. The pole $f_{vp}$ is placed at approximately 10 times of $f_{CV}$,

Therefore, the buck mode compensation network can be derived as,

$$R_C = \frac{R_{FBT}}{\dfrac{R_{O\_BK}}{R_{fnp}} \left| \dfrac{1 + \dfrac{2\pi i f_{CV}}{\omega_{Z\_vl\_BK}}}{1 + \dfrac{2\pi i f_{CV}}{\omega_{Z\_il\_BK}}} \right|} \tag{43}$$

$$C_C = \frac{1}{\left| 2\pi i \dfrac{f_{CV}}{5} R_C \right|} \tag{44}$$

$$C_H = \frac{1}{|2\pi i \cdot 10 f_{CV} R_C|} \tag{45}$$

## 6 Experimental Validation

As shown in Fig. 5, a prototype is built based on LM5171. The prototype is designed for 48V to 12V/60A bidirectional power conversion. Dual phase interleaving is utilized for distributed power dissipation. The EVM also support 4-phase operation and constant current regulation. Feedforward ramp is also integrated in LM5171.

**Fig. 5** 48V to 12V/60A bi-directional converter

The design parameters are listed in Table 1.

| Parameter | Value |
|---|---|
| Typical HV port voltage | 48V |
| Typical LV port voltage | 14.5V |
| Max inductor average current | 30A |
| Total LV current | 60A |
| Switching frequency | 100kHz |

| | |
|---|---|
| Inductor $L_m$ | 4.7μH |
| Current sense resistor $R_{CS}$ | 1mΩ |

**Table 1** Design parameters

Other parameters are: $C_{O\_BK}$ = 800μF, $R_{O\_BK}$ = 0.24Ω, $R_{ESR\_BK}$ = 10mΩ, $C_{O\_BST}$ = 600μF, $R_{O\_BK}$ = 3.2Ω, $R_{ESR\_BST}$ = 10mΩ, $R_S$ = 10mΩ.

Accurate current sharing and 180° interleaving between 2 phases can be observed from Fig. 6.

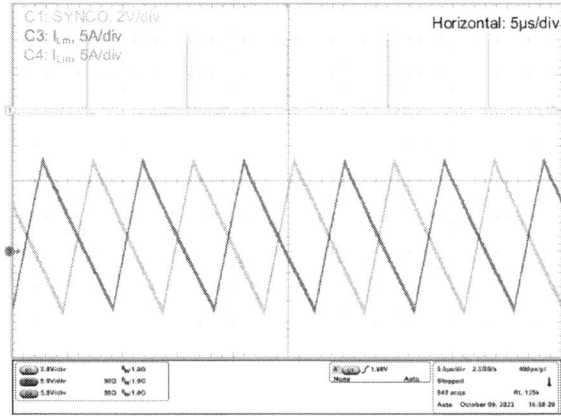

**Fig. 6** Accurate current sharing

Accurate and seamless bi-directional current regulation with ISET can be observed from Fig. 7.

There is a 1V offset voltage at ISET. The average inductor current can be calculated as $I_{Lm} = (V_{ISET} - 1V) \cdot 25A/V$.

When ISET is 1.8V, the converter operates in buck mode, and inductor current is regulated to 20A. When ISET is set to 0.2V, the current is regulated to -20A. The direction of the current flow changes within a few cycles.

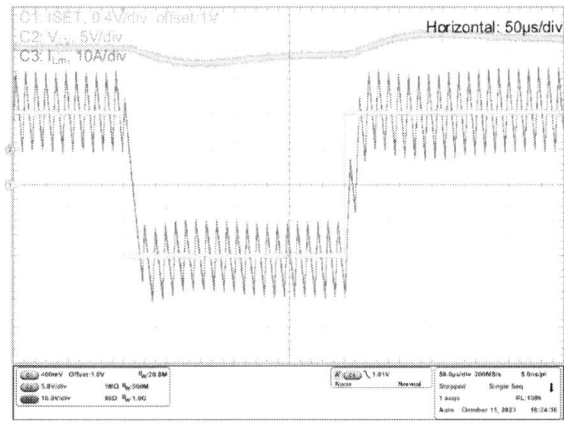

**Fig. 7** Bi-directional current regulation with ISET

The buck mode current loop gain $G_{id\_BK}(s)$, boost mode current loop gain $G_{id\_BST}(s)$ and unified current loop gain $G_{id}(s)$ are shown in Fig. 8. It can be seen that for 10kHz or above, $G_{id}(s)$ is a good approximation.

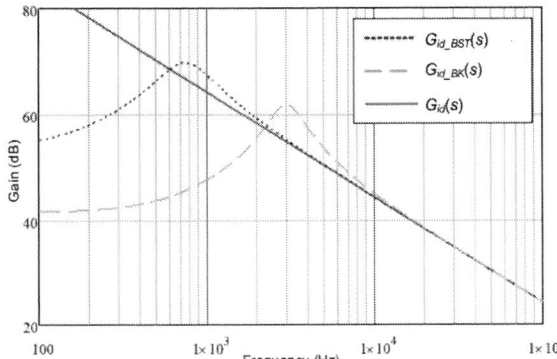

**Fig. 8**   Comparison of current loop gains

The buck mode efficiency over input voltage and load current is shown in Fig. 9. The buck mode output voltage is 14.5V. Both Diode Emulation Mode (DEM) and Forced PWM mode (FPWM) efficiency curve are shown.

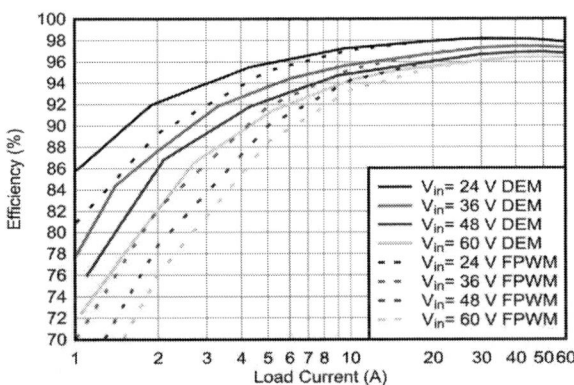

**Fig. 9**   Buck mode efficiency

Similarly, the boost mode efficiency is shown in Fig. 10. The boost mode output voltage is 50.5V.

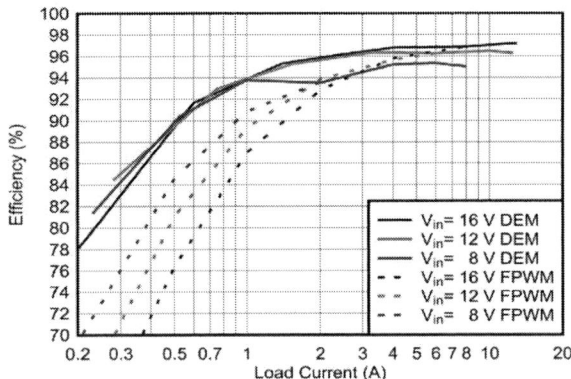

**Fig. 10**   Boost mode efficiency

The calculated current loop power plant $G_{id}(s)A_{CS}R_{CS}/V_M$, the current loop compensation stage $G_{ci}(s)$, and the resulting open-loop gain $T_i(s)$ are shown in Fig. 11.

The current loop crossover frequency is around 15kHz. $R_{COMP} = 3.65\text{k}\Omega$, $C_{COMP} = 15\text{nF}$, $C_{HF} = 1\text{nF}$ are selected for current loop compensation.

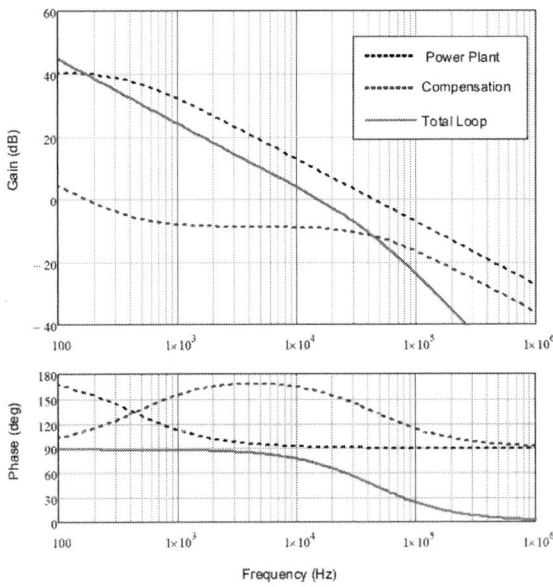

**Fig. 11**   Bode plots of the current loop

The calculated voltage loop power plant $G_{vs\_BK}(s)$, the voltage loop compensation stage $G_{cv}(s)$, and the resulting total open loop gain $T_{vs\_BK}(s)$ are shown in Fig. 12.

The voltage loop crossover frequency is around 1.5kHz. $R_C = 6.2\text{k}\Omega$, $C_C = 100\text{nF}$, $C_H = 1.5\text{nF}$ are selected for voltage loop compensation.

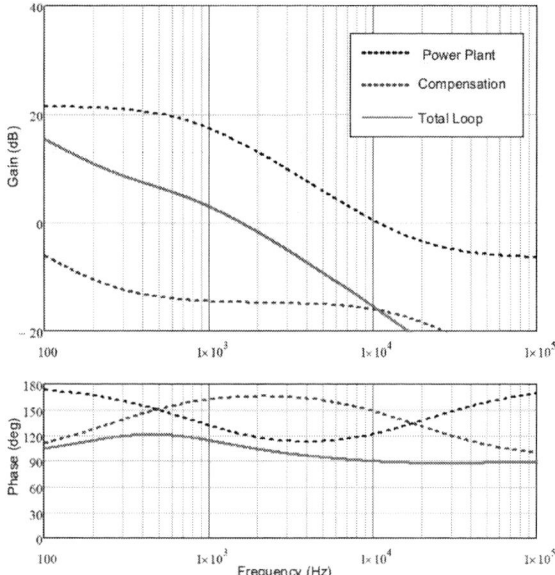

**Fig. 12**  Bode plots of the voltage loop

Figure 13 shows the simulated current loop open-loop gain $T_i(s)$. Figure 14 shows the measured voltage loop open-loop gain $T_v(s)$.

It can be found that the calculation, simulation and the bench test match well.

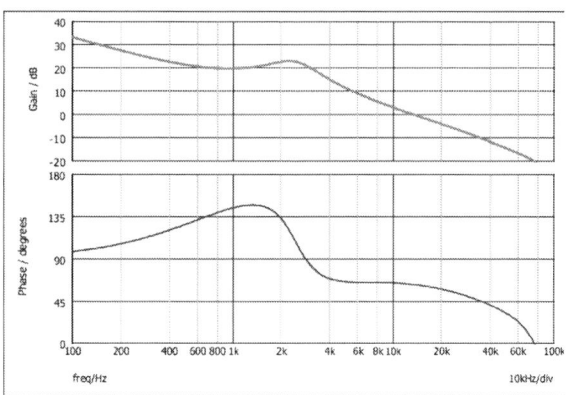

**Fig. 13**  Simulated current loop bode plots

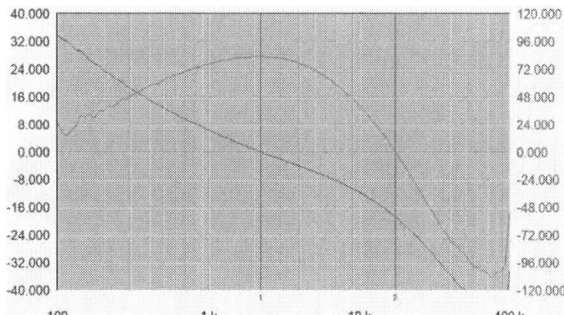

**Fig. 14**  Measured voltage loop bode plots

# 7  Conclusions

The system requirements of the energy storage system are analyzed, and benefits of average current mode control in such system are identified. The models of the inner current loop and outer voltage loop are presented. The benefit of feedforward ramp is analyzed. The compensation strategies for the current loop and voltage loop are suggested. A 48V to 12V/60A dual phase converter is developed and introduced. The compensation results are also shown to verify the model and the compensation strategies.

# References

[1] Xunwei Zhou, Pit-Leong Wong, Peng Xu, F. C. Lee and A. Q. Huang, "Investigation of candidate VRM topologies for future microprocessors," in IEEE Transactions on Power Electronics, vol. 15, no. 6, pp. 1172-1182.

[2] Garrett Roecker, "Average Current Mode Control of Bidirectional DCDC Systems," in Texas Instruments video library, 2017

[3] L. H. Dixon, "Average current-mode control of switching power supplies," in Unitrode Power Supply Design Seminar Handbook, 1990

[4] Youhao Xi, "Selecting a Bidirectional Converter Control Scheme," in Texas Instruments technical article, 2017

PCIM Europe 2024, 11– 13 June 2024, Nuremberg          DOI: 10.30420/566262161

# Novel Power Feed-Forward Regulation for Dual Stage PFC+DCDC Converters

Alfredo Medina-Garcia[1] , Martin Krueger[1], Cristina Martos-Contreras[1] , Pierrick Ausseresse[1], Josef Daimer[1] and Manfred Schlenk[2]

[1] Infineon Technologies AG, Germany

[2] Dr. Schlenk Consulting, Germany

Corresponding author: Alfredo Medina-Garcia, alfredo.medina-garcia@infineon.com

Speaker: Alfredo Medina-Garcia, alfredo.medina-garcia@infineon.com

## 1    Abstract

Current power adaptor and charger trends include supporting USB-PD [1], extended power range (5 V to 48 V), and at the same time overall size reduction. For output power greater than 75 W active power factor correction is required. In that case, a dual stage converter consisting of a boost PFC [2] followed by an isolated DC-DC converter is the most common used configuration. However, one of the biggest and most expensive components in the converter is the intermediate high voltage bulk capacitor between these two stages. This capacitor is used to hold the voltage ($V_{bus}$) within certain range for proper operation of the following DC-DC stage. While power feed-forward is a known method [3] to improve the Vbus control, the current methods come across with practical implementation issues. In this paper, an innovative and practical power feed-forward concept, from the DC-DC stage to the boost PFC stage to ensure the Vbus voltage stability during load transients, will be presented. It enables best-in-class regulated Vbus voltage even with very small bulk capacitor that ensures minimum adaptor size and lower cost. The concept has been implemented and tested in a 140 W prototype USB-PD EPR adaptor, the practical results will be presented and analyzed in this paper.

## 2    Introduction

Power supplies need to fulfill different standards and norms related to efficiency, stand-by power, power factor correction (PFC), safety, electromagnetic compatibility, etc. For usage with portable devices, having low weight and size is a very desired characteristic. On the other hand, the USB-PD standard [1] drives adaptors towards a universal power supply able to charge multiple devices. Adaptors with higher power level are able to cover a wider range of devices, for this reason USB-PD has been extended up to 240 W.

Power factor correction is required if the input power is over 75 W [4]. This ensures low reactive current with additional losses to be generated in the grid. For these cases a dual stage converter is

very common to address the wide AC input (90 V to 264 V) and output voltage range.
The first stage is typically a boost PFC and the

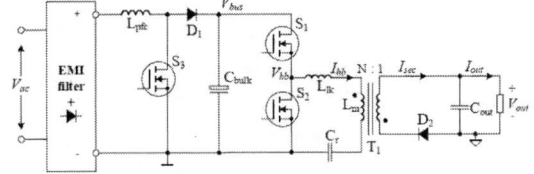

**Fig. 1**: Simplified schematic of a PFC boost converter followed by a hybrid-flyback stage [5-8]

second stage an isolated DC-DC converter.
Figure 1 shows a typical schematic of such application.

The first stage needs to regulate the bus voltage ($V_{bus}$) to ensure proper operation of the second stage. For example, during load transients at the output of the converter the Vbus voltage has to remain below the rating of the bulk capacitor ($C_{bulk}$) but above the minimum operating voltage of the DC-DC stage. At the same time, the boost PFC needs to ensure that the input current follows the AC voltage in shape and phase to fulfill the requirements of power factor and total harmonic distortion required by the regulation norms. [4].

Due to the sinusoidal input current there is a pulsating power from the first stage into Cbulk while the second stage will draw the equivalent average power. Under such conditions the Vbus voltage will experience a voltage ripple under static conditions given by equation (1):

$$V_{pp} = \frac{P_{in}}{2\pi f_{ac} C_{bulk} V_{bus}} \quad (1)$$

The frequency of the bus voltage ripple is twice the input line frequency ($f_{ac}$), i.e. 100 Hz for 50 Hz, this is shown in Figure 2.

Due to the nature of the boost PFC it is difficult to provide a control algorithm able to reject the natural ripple in Cbulk and, at the same time, react quickly to changes on Vbus due to the dynamic output load. Typical regulators for Vbus include a low pass filter or a notch filter with frequency rejection around 100 Hz and/or 120 Hz for 50 Hz and 60 Hz AC line frequency respectively.

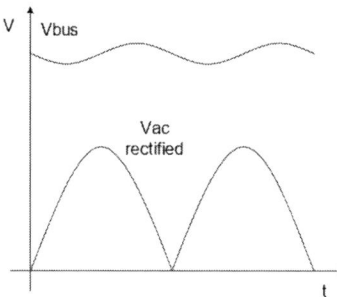

**Fig. 2**: Representative waveforms of the input and output voltage of the PFC boost stage

Most of the boost PFC controllers filter the measured bus voltage using a low pass filter with cut off frequency around 10 Hz, the result is compared with the reference voltage $V_{ref}$ and the resulting error signal is processed by the regulator.

The low pass filter, see Figure 3, helps to minimize the error signal ripple during steady state. However, it slows down the reaction to the dynamic load connected to Cbulk. For this reason, big and

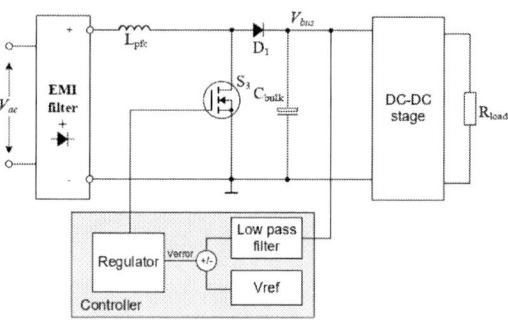

**Fig. 3**: Simplified schematic of a PFC boost converter and its control blocks

expensive bulk capacitors are typically used for PFC boost converters.

## 3 Applying known solutions

To improve the voltage regulation during dynamic load, the power measured at the load connected to the bulk capacitor, in this case the load of the DC-DC stage, is used as a feed-forward to the boost PFC controller. Considering such information, the PFC controller will react accordingly increasing or decreasing the power sent to the bulk capacitor even before the Vbus voltage drops or rise. This ensures the best reaction to load transients and can overcome the weakness of the slow loop required in PFC converters needed to provide good power factor.

This idea of feed-forward the power or current is not new [3], a diagram of such idea is shown in Figure 4.

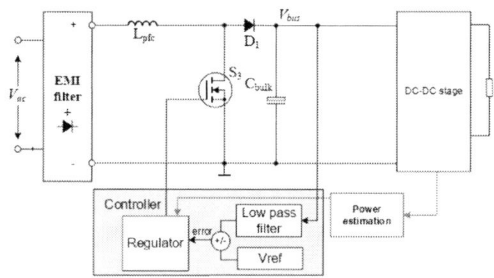

**Fig. 4**: Simplified schematic of a PFC boost converter including the control blocks required for power feed-forward in red

For a critical conduction mode (CrCM) PFC a control block diagram with power feed-forward is pre-

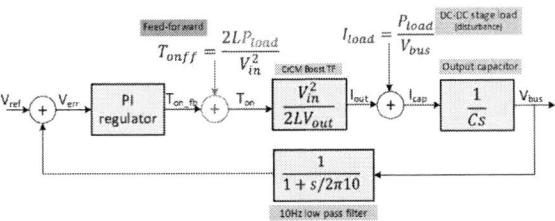

**Fig. 5**: PFC control block diagram including power feed-forward elements

sented in Figure 5. The feed-forward part is highlighted in red, the output current ($I_{out}$) is used to calculate the required $T_{on}$ of the PFC CrCM.

Nevertheless, in most of the methods proposed up to now, the absolute value of power is processed directly (instead of any relative power value). The main disadvantages of that method are: 1) Estimating the power taken by the DC-DC stage with relatively good accuracy is critical to avoid creating additional error on the power delivered by the first stage; 2) Real component values of the converter are required to calculate the feed-forward on-time ($T_{onff}$), e.g. the inductance value, which have tolerances and variations over temperature and life time; 3) The RMS value of the input voltage needs to be measured with good accuracy. Any error of the feed-forward $T_{onff}$ has to be corrected by the voltage feedback loop (i.e. the PI regulator). Due to the reasons listed previously, the concept is very limited and not very practical in most of the cases.

## 4 Proposed solution

To overcome the practical weakness of the previously described state-of-the-art concept, a new method, using the normalized change of power of the second stage ($\Delta P/P$) is proposed. For example, in case of a CrCM PFC using on-time control, this eliminates all three disadvantages of the previously described method. As shown in equation (2), using the normalized power change all tolerance contributors are cancelled.

$$\frac{\Delta P}{P} = \frac{2L\Delta P / V_{in}^2}{2LP / V_{in}^2} = \Delta T_{onff} / T_{onff} \qquad (2)$$

By feeding forward only the normalized power change we can simplify and avoid dependencies from system variables and components. Note, that this advantage also applies to several other converter topologies beyond CrCM PFC using on-time control. The actual output power of the PFC needs to change according to the normalized power

change, meaning it the current power will be multiplied by the power change given from the unify feedforward information. In the case of CrCM PFC, the $T_{on}$ can be used instead since it is proportional to the output power.

Nevertheless, this method requires changes on the PFC control. The PI compensator needs to be replaced by a linear filter and a different type of integrator given by Equation (2) and shown in Figure 6.

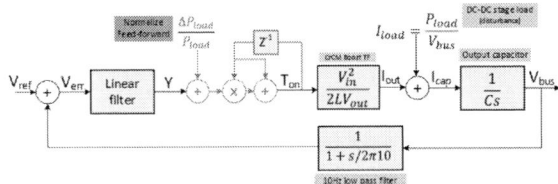

**Fig. 6**: PFC control block diagram with the improved power feed-forward elements

To better understand the working principle, we can take an example where the power taken by the second DC-DC stage experiences a dynamic load between 100 and 200 W. The resulting change of the internal variables over time is shown in Table 1. It can be seen how any change of power is di-

| Time interval | Y | $P_{load}$ (W) | $\Delta P/P_{load}$ | $T_{on}$ (µs) |
|---|---|---|---|---|
| 0 | 0 | 100 | 0 | 4 |
| 1 | 0 | 100 | 0 | 4+4*0 = 4 |
| 2 | 0 | 150 | 0.5 | 4+4*0.5 = 6 |
| 3 | 0 | 200 | 0.333... | 6+6*0.33...= 8 |
| 4 | 0 | 200 | 0 | 8+8*0 = 8 |
| 5 | 0 | 100 | -0.5 | 8-0.5*8= 4 |
| 6 | 0 | 100 | 0 | 4+4*0 = 4 |

**Table 1** Controller variables during dynamic load.

rectly tracked by the on-time which immediately provides the required output current $I_{out}$ considering a CrCM PFC control block shown in Figure 6.

In this simple example, the feed-forward tracks the dynamic load perfectly so that the liner filter output Y remains 0. The linear filter is still needed to compensate any source of error, drift and AC line changes.

## 5 Loop stability considerations

The improved power feed-forward method changes the open loop response analysis as indicated by the red marked elements in Figure 6. However, replacing the PI compensator by a different linear filter makes the open loop response itself and the related stability parameters remaining the same.

While using a feed-forward control, the frequency response and stability of the system does not change, the feed-forward offers various advantages. For example, the Vbus ripple can be filtered with lower band width which helps to improve the power factor or allows a lower bulk cap, or the combination of both.

## 6 Experimental results

To demonstrate the most important benefit i.e. a significant bulk capacitor size reduction, the novel controller concept has been implemented in a digital controller driving a 140 W power supply. The power supply system consists of two power stages: the first one is a critical conduction mode PFC and the second stage a hybrid-flyback converter [5-8] acting as an isolated DC-DC. The controller implements the control of both stages enabling this way easy information exchange within the controller.

To evaluate the effects of the new power feed-forward method a dynamic 10%/100% load test is performed. First, the system is tested without the power feed-forward features using an 82 µF bulk capacitor. Second, the system is tested with power feed-forward and same 82µF bulk capacitor. The result is shown in Figure 7. The upper part (a) of the figure shows the Vbus regulation without power feed-forward, it can be observed that, with the standard control, Vbus drop down to 346.6 V. When the power feed-forward mechanism is applied the Vbus voltage drop is reduced by 30V, this is shown in Figure 7.b where Vbus remains above 376.6 V.

Clearly, this should allow a reduction of the bulk capacitor. Therefore, the bulk capacitor of 82 µF is replaced by a 39 µF capacitor. The response of the PFC converter with power feed-forward is shown in Figure 8. On the top side (a) we can see the Vbus response to the dynamic load, it can be observed that even with such a small capacitor the Vbus voltage only drops down to 370 V which is even better than the original setup using 82 µF and having no feed-forward (346 V).

Figure 8.b shows how the PFC CrCM gate driver on-time perfectly tracks the dynamic load, increasing $T_{on}$ from 1.5µs to 5.5 µs within in 1ms.

The controller used is XDPS2221E from Infineon Technologies, which is under development and will be released by the end of 2024.

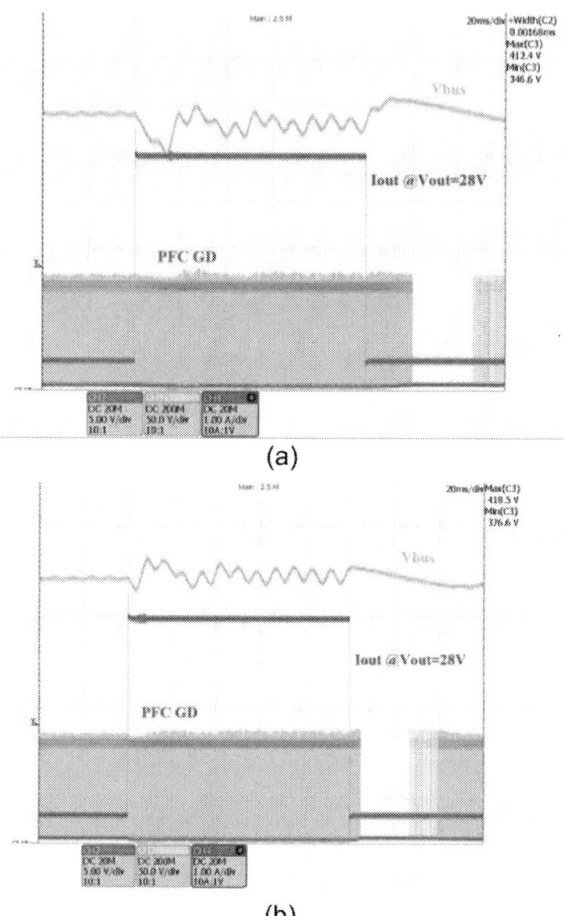

**Fig. 7**: Vbus regulation during dynamic load 14 W => 140 W => 14 W, Cbulk 82 µF. a) without power feed-forward, b) with power feed-forward.

## 7 Conclusions

For dual stage power supplies, i.e. PFC stage followed by an isolated DC-DC converter, with a bulk capacitor acting as buffer between both stages, the new proposed power feed-forward has been derived and demonstrated successfully. Unless the bulk capacitor value is limited by specific requirements like greater output voltage hold-up time, the new power feed-forward concept enables a significant reduction of the bulk capacitor, i.e. at least, by one half compared to state-of-the-art solutions. Being the bulk capacitor one of the bigger and most expensive components in such power supplies, the reduction greatly improves the cost and power density of the overall converter.

At the same time, the proposed concept solves known issues in current state-of-the-art solutions coming from tolerances on the power estimation and components values.

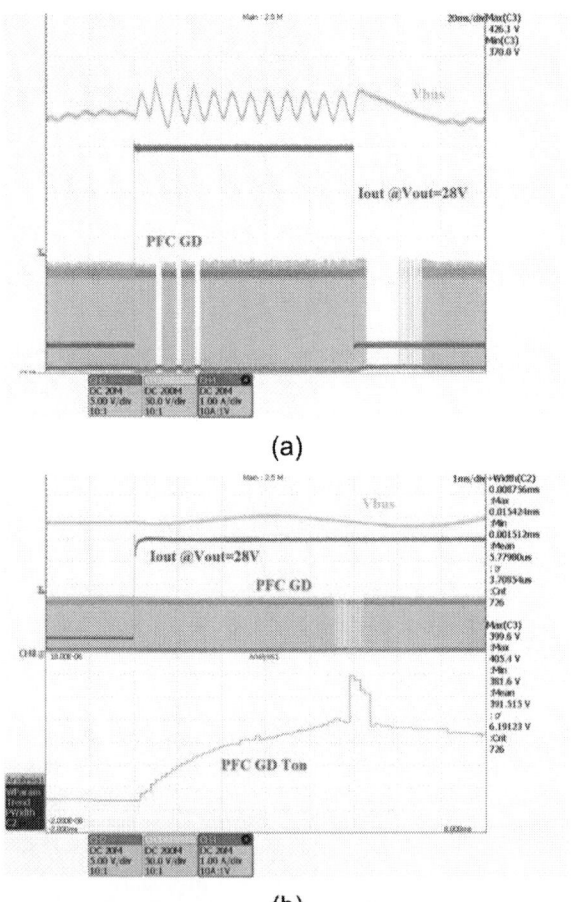

(a)

(b)

**Fig. 8**: a) Vbus regulation during dynamic load 14 W => 140 W => 14 W, Cbulk 39 µF with power feed-forward, b) On-time during load increase 14W => 140W.

# References

[1] www.usb.org. Universal Serial Bus Power Delivery Specification, revision 3.1, 26.May.2021.

[2] Nguyen Binh Nam *et al.*, "Design and implementation of digital controlled boost PFC converter under boundary conduction mode," *2017 IEEE 3rd International Future Energy Electronics Conference and ECCE Asia (IFEEC 2017 - ECCE Asia)*, Kaohsiung, 2017, pp. 1114-1119, doi: 10.1109/IFEEC.2017.7992197.

[3] R. Redl and N. O. Sokal, "Near-Optimum Dynamic Regulation of DC-DC Converters Using Feed-Forward of Output Current and Input Voltage with Current-Mode Control," in *IEEE Transactions on Power Electronics*, vol. PE-1, no. 3, pp. 181-192, July 1986, doi: 10.1109/TPEL.1986.4766303.

[4] IEC Electromagnetic compatibility (EMC) - Part 3-2: Limits - Limits for harmonic current emissions (equipment input current <= 16 A per phase), International Electrotechnical Commission, IEC 61000-3-2:2014.

[5] A. M. Garcia, M. J. Kasper, M. Schlenk and G. Deboy, "Asymmetrical Flyback Converter in High Density SMPS," PCIM Europe 2018; International Exhibition and Conference for Power Electronics, Intelligent Motion, Renewable Energy and Energy Management, 2018, pp. 1-5.

[6] A. Medina-Garcia, F. J. Romero, D. P. Morales and N. Rodriguez, "Advanced Control Methods for Asymmetrical Half-Bridge Flyback," in IEEE Transactions on Power Electronics, vol. 36, no. 11, pp. 13139-13148, Nov. 2021, doi: 10.1109/TPEL.2021.3077184.

[7] Medina-Garcia A., Schlenk M., Morales D.P., Rodriguez N., "Resonant Hybrid Flyback, a New Topology for High Density Power Adaptors," Electronics, vol. 7, 363,2018, doi: 10.3390/electronics7120363.

[8] B.Kohlhepp, M.Barwig, and T.Duerbaum,"Gan improves efficiency of an asymmetrical half-bridge pwm converter with synchronous rectifier," in PCIM Europe 2019; International Exhibition and Conference for Power Electronics, Intelligent Motion, Renewable Energy and Energy Management, 2019, pp. 1–8

PCIM Europe 2024, 11– 13 June 2024, Nuremberg    DOI: 10.30420/566262162

# 22 kW Bi-directional Wall-box Charger With 1200 V SiC MOSFET

Sanbao Shi[1] , Yi Zhang[1] ,Cheng Zhang[1] , Sidorov Vadim[2]

[1] Infineon Semiconductors (Shenzhen) Company Limited
[2] Infineon Technologies Austria AG

Corresponding author:   Sanbao Shi, simon.shi@infineon.com
Speaker:                Sanbao Shi, simon.shi@infineon.com

## Abstract

The demand for power converters with high power density, high efficiency, and more reliability is steadily rising to deal with issues such as environment pollution and sustainability. This demand is higher for automotive applications such as EV chargers, energy storage systems (ESS), onboard chargers (OBC), and vehicle-to-grid (V2G) systems.

This paper discusses a 22 kW wall-box charger system that can cater to this demand. The complete solution includes advanced 1200 V and 1700 V discrete SiC MOSFETs, a galvanically isolated gate driver, and AURIX™ microcontroller unit (MCU). Test results have shown that this innovative solution has several benefits including a wide range of output voltage (200 V – 900 V), compact size (650 x 290 × 200 mm), high power density, high efficiency (up to 96.5%), high level of waterproofing (grade IP65), and high reliability.

## 1  Introduction

The innovative 22 kW wall-box charger system includes an AC-DC part and a DC-DC part. In this system, a B6 full-bridge is applied to the AC-DC part thanks to a 1200 V discrete SiC MOSFET that enables the AC-DC part to be highly compact and efficient.

**Fig. 1**  Functional block diagram

Its phase-shift topology with electronic insulation, high efficiency, bi-directional power transmission, zero voltage switch (ZVS) soft-switching on, low electromagnetic interference (EMI), and wide range of output voltage is quite suitable for the DC-DC stage of automotive applications. In addition, the power conversion efficiency in forward direction is similar to that in the reverse direction. Intelligent, digitally controlled algorithms have been proposed to regulate the output voltage, control bi-directional power conversions, and achieve synchronous rectification. This innovative will have a very wide range of application prospects [1].

| Item | Parameter | Symbol | Min. | Typ. | Max. | Unit |
|---|---|---|---|---|---|---|
| AC side | Line voltage | $V_{line}$ | | 380 | | V |
| | Phase current | $I_{phase}$ | | 35 | | A |
| | Line frequency | $f_{line}$ | | 50 | | Hz |
| | Power factor | $\cos\theta$ | | ≥0.95 | | |
| | Total harmonic distortion | THD | | 3 | 5 | % |
| | Switching frequency | $f_{sw}$ | | 40 | | kHz |
| DC side | DC voltage (standard mode) | $V_{DC}$ | 200 | 530 | 750 | V |
| | DC voltage (high voltage mode) | $V_{DC}$ | 500 | 700 | 900 | V |
| | DC current | $I_{DC}$ | | | 50 | A |
| | Power | P | | | 22 | kW |
| | Switching frequency | $f_{sw}$ | | 60 | | kHz |

1212

| Other | Cooling | | Forced air | | |
|---|---|---|---|---|---|
| | Size | L x W x T | 650 x 290 × 140 | | mm |
| | Efficiency | η | 96 | | % |

**Table 1** Specifications of the 22 kW wall-box charger for forward mode

The key parameters and operating conditions for the system's forward mode (from AC to DC) are listed in Table 1. The parameters for the reverse working mode (from DC to AC) are similar.

The detailed functional block diagram is shown in Fig. 2. This power converter reference design consists of a 3-phase active front-end (AFE) converter and a 3-phase dual active bridge (DAB). It is capable of highly efficient bidirectional power transfer of up to 22 kW. The combination of 1200 V and 1700 V CoolSiC™ MOSFETs in a TO247-4 package [2] [3] with best-fitting EiceDRIVER™ 1ED compact gate driver ICs [4] unleashes the advantages of SiC technology, such as high-power conversion efficiency in both directions of energy flow, lower component part count, and enhanced system reliability.

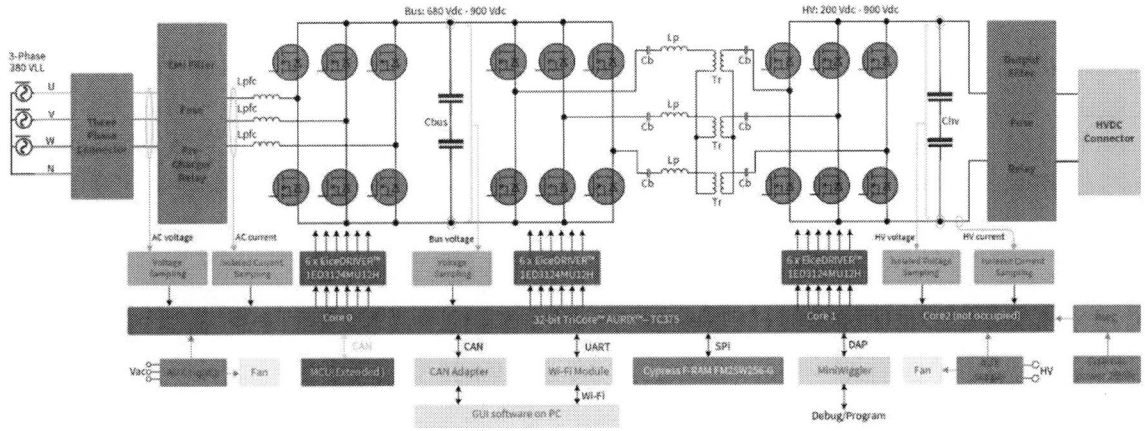

**Fig. 2** Detail functional block diagram of the 22 kW wall-box charger

## 2 System Design

### 2.1 AFE Control Theory

The AFE side has a full-bridge B6 topology. The PWM control algorithm uses a 3-phase boost converter with 120° phase shift, as shown in Fig. 3. There are two control loops in the boost topology, with the same parameters. The voltage loop is used to control the bus voltage and the current loop is used for power factor correction (PFC).

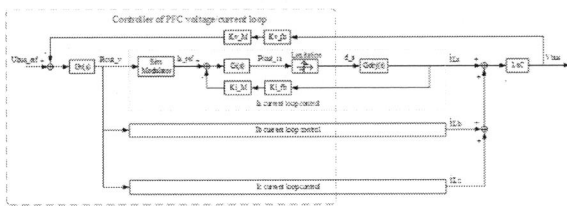

**Fig. 3** AFE control algorithm diagram

During the forward mode (from AC to DC), the voltage loop reference voltage is slightly higher than the bus voltage, to ensure that the voltage loop output, $P_{iout}$, is positive. The phase of the AC input voltage and AC input current is the same, therefore, the PF value is close to 1.

During reverse mode (from DC to AC), the voltage loop reference voltage is slightly lower than the bus voltage, to ensure that the voltage loop output, $P_{iout}$, is negative. The phase of the AC input voltage and AC input current is opposite, therefore, the PF value is close to -1.

### 2.2 DAB control method

The PWM control theory for the 3-phase DAB topology is single-phase shift (SPS), and the formula for output power is shown in Eq. (1).

$$P_O = \begin{cases} \dfrac{NV_{in}V_{out}}{2\pi f L}\dfrac{(4\pi-3\varphi)\varphi}{6\pi} & (0 \leq \varphi \leq \frac{\pi}{3}) \\ \dfrac{NV_{in}V_{out}}{2\pi f L}\dfrac{(18\pi\varphi-18\varphi^2-\pi^2)}{18\pi} & (\frac{\pi}{3} \leq \varphi \leq \frac{2\pi}{3}) \end{cases} \quad (1)$$

Here, $P_O$ is the output power, N is the transformer radio, $V_{in}$ is input voltage, $V_{out}$ is output voltage, $\varphi$ is the degree of phase shift, f is the PWM frequency, and L is the inductor value [5].

To enlarge the on/off scope of the MOSFET's soft switch, an innovative PWM control algorithm has been applied to the system, such as symmetrical and complementary PWM duty cycles for the bridge MOSFETs. This helps realize a wide output voltage range [6].

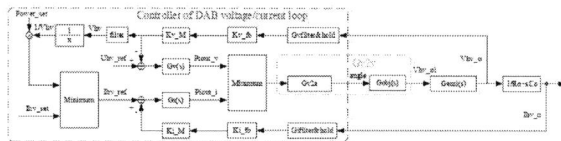

**Fig. 4** AFE control algorithm diagram

The AFE control algorithm diagram is shown in Fig. 4. As can be seen, there are two control loops, a voltage loop, and a current loop with double loop competition algorithm to realize the constant current mode and the constant power mode for charging the battery.

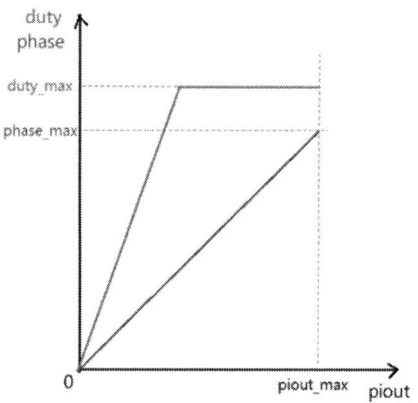

**Fig. 5** SPS and symmetric PWM algorithm

As shown in Fig. 5, the DAB loop control value, $P_{iout}$, will determine both the degree of the phase shift and the symmetric PWM duty cycle.

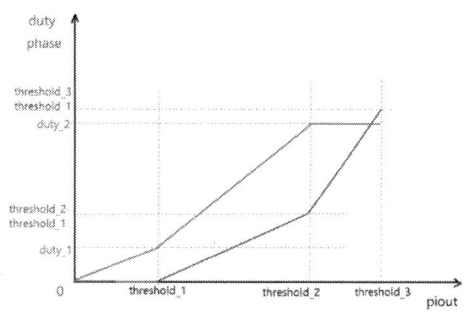

**Fig. 6** SPS and complementary PWM algorithm

As shown in Fig. 6, the DAB loop control value, $P_{iout}$, will determine both the degree of the phase shift and the complementary PWM duty cycle.

### 2.3 Mechanical and Thermal Design

Based on the analyses above, the 22 kW bi-directional wall-box charger has been designed, as shown in Fig. 7.

**Fig. 7** The 22 kW wall-box charger

To make the design compact and compatible with liquid and forced fan cooling, all the power components (MOSFET, transformer, inductor, electronic capacitors, etc.) were connected to a big heatsink. The thermal simulation in Fig. 8 shows that the temperature of the components was, basically, even.

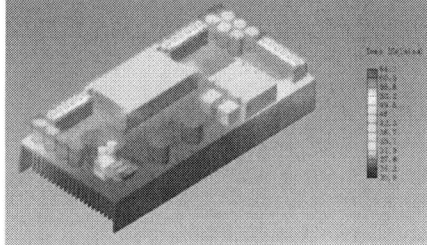

**Fig. 8** Result of a thermal simulation of the 22 kW system

## 3 Functional Test Result

### 3.1 Efficiency Test

The efficiency plots shown in Fig. 9 are based on measurements taken in the forward mode under different voltages at the HV DC side. The maximum current circulating through the DAB converter is limited due to the maximum current capability of the transformer. Therefore, the power of the converter is limited at low and high voltage. The maximum power of the converter can be achieved in the range of 380 to 550 V. A Yokogawa WT1800 high-precision power analyzer was used to capture the efficiency plots. The maximum efficiency of the reference design was 96.8% at $V_{DC}$ = 523 V, $P_{DC}$ = 14 kW.

PCIM Europe 2024, 11–13 June 2024, Nuremberg    DOI: 10.30420/566262162

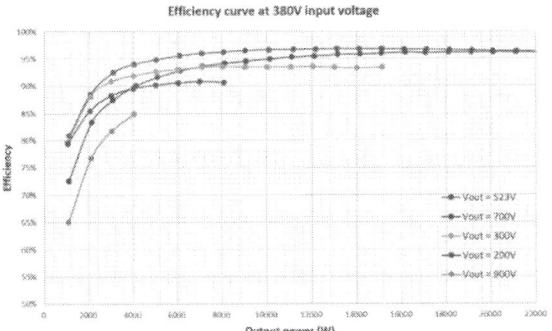

**Fig. 9** Efficiency results of the whole system

## 3.2 AFE Current and Voltage Test

Steady-state waveforms of the AC input current and voltage in forward mode at P = 22 kW, and inverse mode at P = 11 kW are shown in Fig. 10 and 11, respectively.

Channel C5 (grey) denotes the voltage in phase A (120 V/div). Channel C6 (violet) denotes the current in phase C (20 A/div). Channel C7 (red) denotes the current in phase A (20 A/div). Channel C8 (orange) denotes the current in phase B (20 A/div)

In both the tests, the HV DC voltage was equal to 530 V.

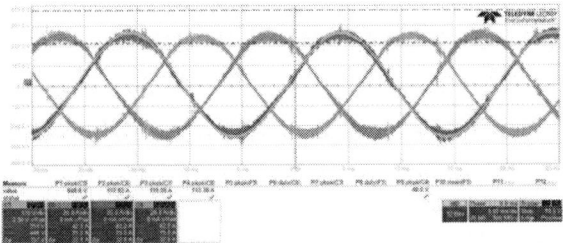

**Fig. 10** 3-phase AC input current, voltage, and total harmonic distortion (THD) in forward mode at 22 kW

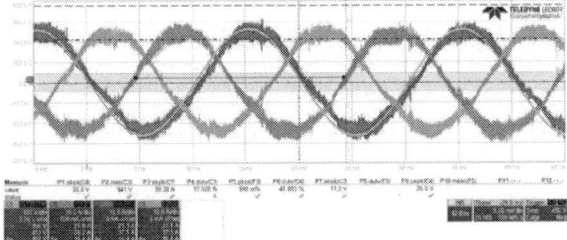

**Fig. 11** 3-phase AC input current and voltage in reverse mode at 11 kW

Figure 12 shows the steady-state waveforms of the AD-DC converter operating in forward mode as an active front-end converter at P = 22 kW, VDC = 530 V.

Channel C1 (yellow) denotes the gate-source voltage of the top switch in the leg (10 V/div). Channel C2 (red) denotes the drain-source voltage of the top switch in the leg (200 V/div). Channel C7 (red) denotes the gate-source voltage of the bottom switch in the leg (10 V/div).

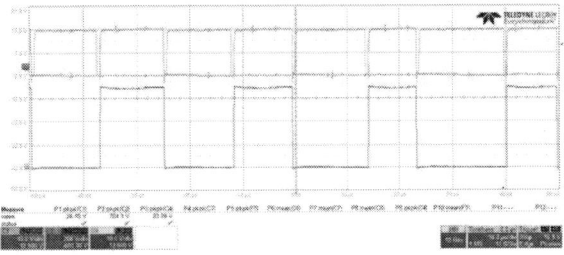

**Fig. 12** Steady-state waveforms of the AC-DC converter at P = 22 kW, VDC = 530 V

## 3.3 DAB current and voltage test

Figures 13 and 14 show the steady-state waveforms of transformer currents on the primary and secondary sides of the 3-phase DAB converter operating in the forward mode at P = 22 kW, VDC = 530 V.

Channel C6 (violet) denotes the current in phase C of the 3-phase DAB (20 A/div). Channel C7 (red) denotes the current in phase A of the 3-phase DAB (20 A/div). Channel C8 (orange) denotes the current in phase B of the 3-phase DAB (20 A/div)

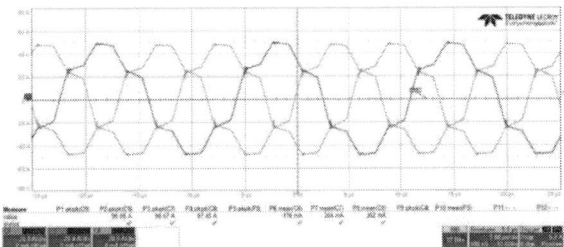

**Fig. 13** Steady-state waveforms of the currents on the primary side of the 3-phase DAB converter at P = 22 kW, VDC = 530 V

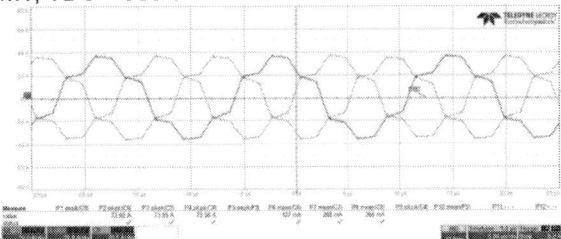

**Fig. 14** Steady-state waveforms of the currents on the secondary side of the 3-phase DAB converter at P = 22 kW, VDC = 530 V

Figures 15 and 16 show the steady-state waveforms of the switches' voltage on the primary and secondary sides of the 3-phase DAB converter operating in forward mode at P = 22 kW, VDC = 530 V.

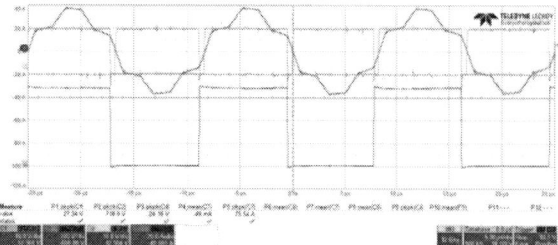

**Fig. 15** Steady-state waveforms of the current and voltage on the primary side of the 3-phase DAB converter at P = 22 kW, VDC = 530 V

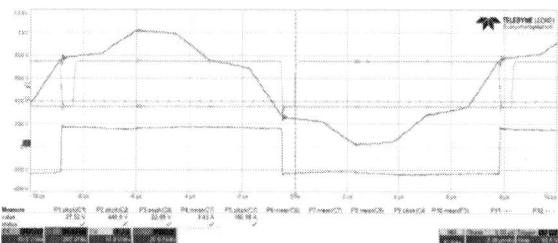

**Fig. 16** Steady-state waveforms of the current and voltage on the secondary side of the 3-phase DAB converter at P = 22 kW, VDC = 530 V

Channel C1 (yellow) denotes the gate-source voltage of the top switch (10 V/div). Channel C2 (pink) denotes the drain-source voltage of the top switch (10 V/div). Channel C4 (green) denotes the gate-source voltage of the bottom switch (10 V/div). Channel C7 (red) denotes the current of phase A (20 A/div).

## 4 Conclusion

A highly efficient, 22 kW bi-directional wall-box charger was presented in this paper. Its thermal design, functional design, and EMC design were discussed. The wide output voltage range implemented through a special software algorithm was also introduced. Thanks to the innovative 1200 V SiC MOSFET and isolated gate driver, the test results showed very good system performance including thermal efficiency, THD, and EMC. This can lead to a wide range of possible market opportunities.

## References

[1] R. Chen and S. Shi, "11 kW High-efficiency bidirectional CLLC converter with 1200 V SiC MOSFET," PCIM Asia 2021.

[2] Infineon Technologies datasheet: IMZ120R030M1H CoolSiC™ 1200 V SiC Trench MOSFET, 2020-12-11.

[3] B. Wu and S. Shi, "78 W auxiliary power supply for Infineon 1700 V silicon carbide MOSFET," PCIM Asia 2020.

[4] Infineon technology, datasheet: EiceDRIVER™ 1ED31xxMC12H Compact, , 2021-03-01.

[5] B. Zhao, Q. Song, W. Liu, and Y. Sun, "Overview of Dual-Active-Bridge Isolated Bidirectional DC–DC Converter for High-Frequency-Link Power-Conversion System, " IEEE Transactions on Power Electronics[J], vol. 29, no. 8, pp. 4091–4106, Aug. 2014.

[6] R. de Souza Coelho and A. L. Batschauer, "Analysis and Comparison of SPS, EPS and DPS Modulation Schemes for Dual Active Bridge Converter, " 2021 Brazilian Power Electronics Conference (COBEP), João Pessoa, Brazil, pp. 1-3, May. 2021.

PCIM Europe 2024, 11– 13 June 2024, Nuremberg          DOI: 10.30420/566262163

# Dynamic Switching Frequency Selection for Efficiency Optimization in On-Board Charger PFC Stage Based on Novel SiC MOSFET Power Module

Giuseppe Aiello [1], Dario Patti [1], Francesco Gennaro [1], Domenico Nardo [2],

[1] STMicroelectronics, Italy

[2] STMicroelectronics, Germany

Corresponding author:   Giuseppe Aiello, giuseppe.aiello01@st.com
Speaker:                 Giuseppe Aiello, giuseppe.aiello01@st.com

## Abstract

The paper deals with the control optimization of a PFC stage in bidirectional On-board-charger, **OBC**, application.   The implementation of a load-related switching frequency adaptation control is analyzed with efficiency estimation and experimental results.
The analysis consists of an evaluation of the optimization strategy proposed to improve the heavy load efficiency of a standard 2-level PFC.
Experimental validation has been carried out using a reference design based on 6-switch PFC stage with a novel highly integrated SiC MOSFET power module, The optimized control algorithm has been implemented on a 32-bit microcontroller with dedicated peripherals for digital power conversion.

## 1  Introduction

The increasing number of electric vehicles available and the reduction of the charging time make the design of On-Board Chargers (OBCs) increasingly challenging.    Size and weight constraint along with power conversion efficiency requirements are the design keys factors.   As the power level of the OBC increases while maintaining light and compact systems, the use of new topologies as well as new materials for power semiconductors, such as SiC and GaN, is mandatory.

Nowadays, 3-level SiC based power converter is one of the most popular architectures for high power application. Thanks to the best performance on high-power high-frequency operation, multilevel approach is typically preferred compared to the classic 2 level implementation.

On the other hand, in the case of cost effected design, the higher cost of multilevel solution is not suitable and specific countermeasures must be implemented on the control side to extend the efficiency of the 2-level topology.

## 2  PFC stage of OBC

The integration of a Power Factor Corrector (PFC) stage within an On-Board Charger (OBC) for electric vehicles (EVs), (Fig. 1), is a critical element in the pursuit of energy efficiency and compliance with regulatory standards for power quality. The primary function of the PFC stage is to enhance the power factor, which is a measure of how effectively incoming power is converted into useful work output. A high-power factor indicates efficient utilization of electrical power, where the power drawn from the grid is mostly converted into actual work with minimal reactive power, which does not perform any useful work but creates additional load on the power grid.

The PFC stage in an OBC ensures that the alternating current (AC) drawn from the power grid is as sinusoidal as possible and in phase with the voltage waveform. This is important because non-linear loads, such as the switching power supplies found in OBCs, can draw current in abrupt pulses rather than in a smooth sinusoidal manner. These pulses can introduce harmonics into the power system, leading to inefficiencies and potential interference with other equipment connected to the same power grid.

The PFC stage operates by using a boost converter topology, which consists of inductors, capacitors, and switching elements that adjust the input current to achieve the desired power factor. The boost converter increases the voltage to a level that is higher than the input AC voltage, while shaping the input current waveform to match the input voltage waveform.

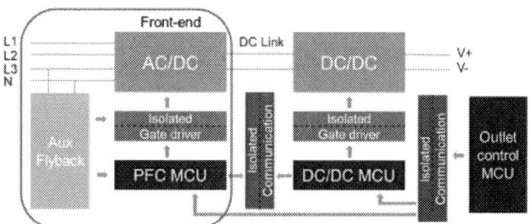

**Fig. 1 Typical architecture of OBC, focus on PFC stage.**

Moreover, the PFC stage contributes to the longevity and reliability of the OBC and the vehicle's battery system. By ensuring that the charger operates with minimal electrical noise and harmonics, the PFC stage helps to prevent undue stress on the electrical components of the OBC, thereby reducing the risk of premature failure. This reliability is crucial for electric vehicles, where the OBC is an integral part of the vehicle's daily operation and overall user experience.

Furthermore, the PFC stage plays a significant role in the bidirectional power flow capabilities of advanced OBCs, which not only charge the EV's battery but can also feed power back to the grid or power a home in a vehicle-to-grid (V2G) or vehicle-to-home (V2H) scenario.

In summary, the PFC stage of an OBC is a vital component that serves to optimize the power consumption from the grid, ensuring that the charging process is not only efficient but also harmonious with the power system infrastructure.

As the adoption of electric vehicles continues to grow, the role of the PFC stage in OBCs will become increasingly significant, underpinning the performance and sustainability of the EV ecosystem.

# 3 TOPOLOGY SELECTION FOR HIGH-POWER PFC STAGE

The selection of an appropriate topology for PFC stage of an OBC is a complex decision that hinges on multiple technical, economic, and regulatory considerations.

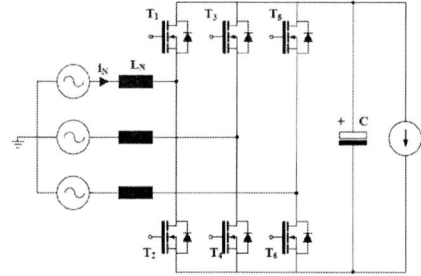

**Fig. 2 2-level three phase B6 PFC.**

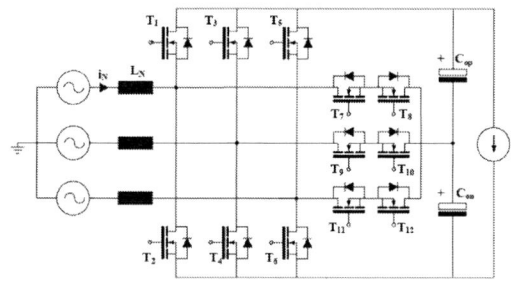

**Fig. 3 Multilevel T-type PFC.**

Choosing between two-level and three-level topologies, engineers must consider a several factors, including the power rating of the OBC, the desired power density, efficiency requirements, and cost constraints [1]. Each topology presents unique advantages and trade-offs that must be weighed against the specific demands of the electric vehicle (EV) charging application.

Two-level topologies, Fig. 2, are widely employed in OBC PFC stages due to their simplicity, proven track record, and lower component count, which can translate to cost and space savings.

The simplicity of the control scheme for two-level topologies is also a compelling advantage, making them a robust and straightforward solution for many OBC designs. However, two-level systems have limitations, particularly when dealing with higher power levels. The voltage stress on switches is equal to the DC link voltage, which can necessitate the use of components with higher voltage ratings, potentially increasing costs and reducing efficiency due to higher conduction losses.

1218

In contrast, three-level topologies, such as T-Type configuration shown in Fig. 3, offer several enhancements over their two-level counterparts, especially in applications requiring higher power and efficiency. By adding an extra level of voltage, these topologies effectively halve the voltage stress on each switch, allowing for the use of lower voltage-rated components that can switch faster and with less energy loss [2].

Nowadays, thanks to system improvements, OBC typically operate at high switching frequency, indeed, in that range, the best performance of multilevel converter is much visible, Fig. 4.

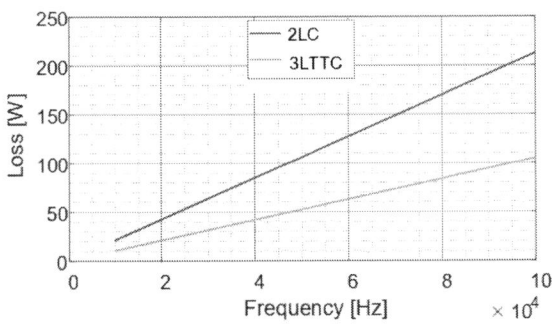

**Fig. 4 Switching losses comparison between 2L-B6 and 3L-TType.**

This reduction in voltage stress also leads to lower electromagnetic interference (EMI) and can improve the overall power quality. Furthermore, three-level topologies can achieve lower Total Harmonic Distortion (THD) in the input current, which is beneficial for meeting stringent grid compliance standards and for reducing the strain on the electrical infrastructure.

However, the benefits of three-level topologies come with increased complexity. The additional level of voltage requires more sophisticated control algorithms and a greater number of power electronic components, which can raise the cost and complexity of the OBC. The design of the gate drive circuitry also becomes more challenging, as it must manage the additional switches and ensure accurate timing to prevent shoot-through and other potentially damaging conditions. [3-4]

# 4 LOAD -DYNAMIC SWITCHING FREQUENCY ADAPTATION

In addition to the passive element size and weight, power quality parameters are strictly related to the switching harmonics content, thus high switching frequency is one of the main related designs parameters of a PFC application. Thanks to the Wide Band Gap, WBG, technologies, such as Silicon Carbide (SiC) power devices, the benefits of high frequency operation are much highlighted. In the case of high power OBCs, a crucial point is also the conduction losses [2]. Even in this case the multilevel topologies enable to extend the power range. Focusing on the topology efficiency behaviors, due to the very high RMS current, at high load operations the efficiency will dramatically decrease in 2-level B6 topology.

A load adaptive switching frequency control is beneficial to mitigate efficiency derating effect at high load operation by compensating the conduction losses increasing by reducing the switching losses contribution. The switching frequency range will be defined to prevent power quality issue and to guarantee the control bandwidth as well. As shown in Fig. 6, starting from the power where we have the nominal frequency (Fsw1) peak efficiency the switching frequency will be reduced (FswX) until heavy load configuration is reached. Thanks to this adaptation the slope of the efficiency curve compensated by the switching frequency reduction to guarantee the higher efficiency trend.

Focusing on OBC application this adaptation will be well fitted with the battery charging dynamic without big effort in terms of adaptive control execution routine that could be easily applied into the current control blocks of the OBC implementation in low cost 32bit microcontroller, as shown in Fig. 5 and Fig. 6.

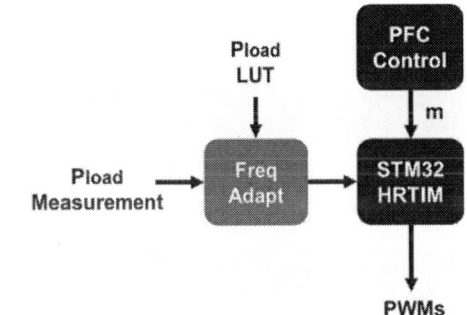

**Fig. 5 Frequency Adaptation control block based on STM32 High Resolution Timer**

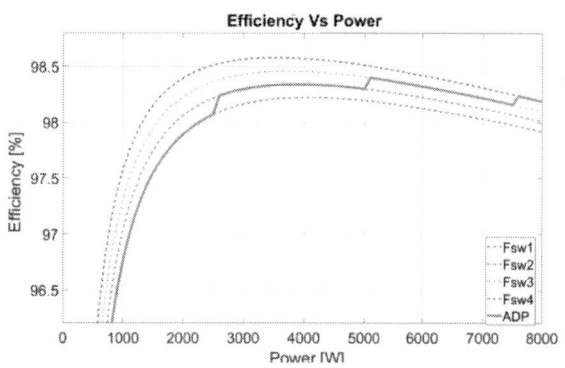

**Fig. 6 Efficiency estimation with Load-adaptive switching frequency control**

## 5 Experimental evaluation

An implementation and experimental evaluation of load adaptive switching frequency control is proposed. A real test bench is used to validate the impact on the performance of the PFC stage in a bidirectional OBC.

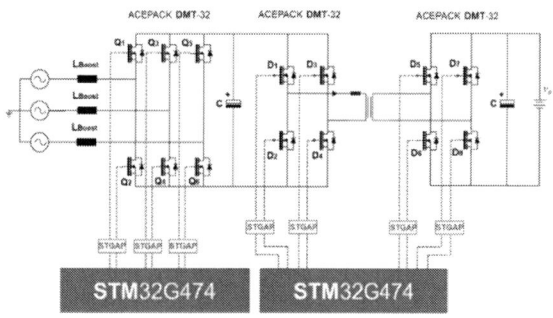

**Fig. 7 OBC test-bench reference design block diagram**

A dual stage B6 PFC and Dual Active Bridge DC-DC are used for this analysis.
Both stages are implemented in SiC power modules DMT32, that allow to optimize the system performance, reducing system size and power path as well, Table 1.

**Table 1 Power devices parameters**

| Parameter | Value | Unit |
|-----------|-------|------|
| VBR | 1200 | V |
| Ciss | 2086 | pF |
| Coss | 90 | pF |
| RDSon | 45 | Ω |

**Fig. 8 OBC test-bench STDES-BCBIDIR based on DMT32 SiC Power modules**

To implement the switching frequency adaptation algorithm, the High-Resolution Timer (HRTIM) embedded on STM32G4 MCU is used. The dynamic adaptation is implemented by using the internal synchronization stage of the HRTIM, any distortion on the modulation signal will be automatically guaranteed by the peripheral.

**Table 2 STDES-BCBIDIR specs**

| Parameter | Value | Description |
|-----------|-------|-------------|
| $V_{DC}$ | $720 - 880V$ | DC bus voltage range |
| $V_{BATT_{nom}}$ | $800V$ | Nominal battery voltage |
| $V_{AC}$ | $185 - 265\ V_{RMS}$ | AC main voltage range |
| $f_{SW}$ | $70kHz$ | PFC Nominal Sw. Freq. |
| $P_{max}$ | 13.2 kW | Max output power |
| $P_{nom}$ | 11kW | Nominal output power |

The experimental results of Fig. 9 shown the actual efficiency curve achieved by the adaptive control compared to three different fixed frequency operation.
As proposed, the benefits of low frequency operation are highlighted at mid/heavy load, instead to guarantee lower distortion at low load THDi the switching frequency will be increased (Fig. 10).

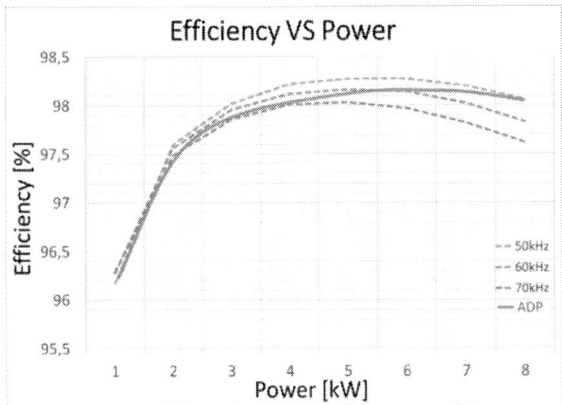

**Fig. 9 Efficiency measurement at 50kHz, 60kHz, 70kHz and Load-adaptive switching frequency control**

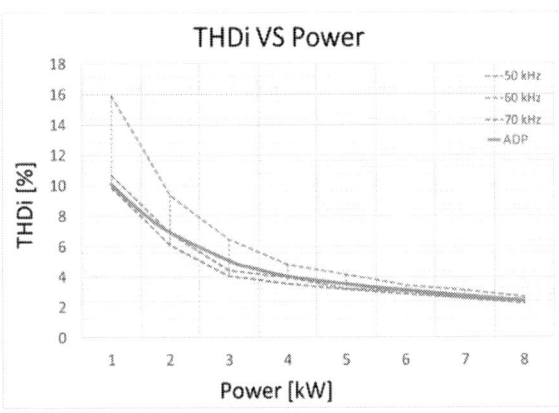

**Fig. 10 THD measurement at 50kHz, 60kHz, 70kHz and Load-adaptive switching frequency control**

## 6 Conclusion

An overview and a practical implementation of adaptive switching frequency control is introduced and evaluated. Several aspects are discussed with a focus on PFC stage for OBC. A description of the implementation validation on simulation was proposed to highlight the impact of the control on the actual efficiency results. A real platform was used to evaluate the effect of switching frequency adaptation on the overall efficiency performance of a DMT32 SiC power modules. The results confirm the theoretical aspects by considering the impact on a standard 2L configuration allows to obtain efficiency results typically achieved with multilevel configuration. The beneficial effect also enabling the possibility to implement cost effective 2-level

solution based on lower RDSon devices reducing the conduction losses contribution at heavy load with online frequency adaptation.

## References

[1] - Azurza, Jon & Zulauf, Grayson & Papamanolis, Panteleimon & Hobi, Simon & Miric, Spasoje & Kolar, Johann, "Three Levels Are Not Enough: Scaling Laws for Multilevel Converters in AC/DC Applications", IEEE Transactions on Power Electronics. PP. 1-1. 10.1109/TPEL.2020.3018857.

[2] - G. Aiello, F. Gennaro and M. Cacciato, "Advantages of SiC MOSFETs in High Frequency Bidirectional PFC Converters for Industrial Applications," PCIM Europe digital days 2021, Online, 2021, pp. 1-5.

[3] - Schweizer, Mario & Friedli, Thomas & Kolar, Johann., "Comparative Evaluation of Advanced Three-Phase Three-Level Inverter/Converter Topologies Against Two-Level Systems", IEEE Transactions on Industrial Electronics. 60. 5515-5527. 10.1109/TIE.2012.2233698.

[4] - Schweizer, Mario & Friedli, Thomas & Kolar, Johann. (2010). Comparison and Implementation of a 3-Level NPC Voltage Link Back-to-Back Converter with SiC and Si Diodes. 1527 - 1533. 10.1109/APEC.2010.5433434.

PCIM Europe 2024, 11– 13 June 2024, Nuremberg    DOI: 10.30420/566262164

# Design and Optimization of SiC-Based 11kW Motor Drive with High Efficiency

Ying Liu[1], Chen Wei[1], Jianlong Chen[1], Zongzeng Hu[1], Fulin Zhang[1], Haiming Zhan[1], Sidharth Gupta[2]

[1]Wolfspeed, China
[2]Wolfspeed, Germany

Corresponding author:    Ying Liu, Iris.Liu@Wolfspeed.com
Speaker:                           Ying Liu, Iris.Liu@Wolfspeed.com

## Abstract

This paper provides a 11kW motor drive with advanced Silicon Carbide (SiC) technology. A prototype of a three-phase, six-switch inverter is built and tested to compare the performance of the SiC MOSFET with the Si Insulated Gate Bipolar Transistor (IGBT) and illustrate the ability to drive an industrial motor at 3000 rpm and 11kW. >99% efficiency is demonstrated with the SiC based motor drive solution.

## 1    Introduction

IGBTs or Intelligent Power Module (IPMs) are widely utilized in industrial motor drives. However, to comply with the IE3 standard outlined in the global standard IEC 60034[1,2], the motor's efficiency needs to be enhanced from approximately 88% to 90% or higher. Achieving these new efficiency standards with IGBT-based inverters poses challenges [3]. Fig. 1 illustrates the block diagram of the motor drive system. Compared to silicon IGBTs, SiC MOSFETs offer superior performance in terms of coefficient of performance (COP), switching characteristics, conduction loss, and thermal properties. Additionally, SiC MOSFETs enable higher switching frequencies, which is proven to be effective in reducing the motor iron losses and contributing to reduced system noise. This paper presents and validates an advanced SiC-based 11kW three-phase 6-switch inverter with a sensorless field-oriented control (FOC) scheme.

To achieve a cost-effective, highly reliable, power-dense, and efficient design, several factors must be carefully considered. Ensuring compliance with

motor insulation requirements and EMI limits, the dv/dt of the switching nodes is a crucial consideration. Motor manufacturers typically advise against exceeding a dv/dt limit of approximately 5-10V/ns at the inverter terminals, particularly for 400V motors [5]. To leverage SiC technology, methods for adjusting the gate drive and achieving proper dv/dt are also explored. The paper will cover aspects such as hardware design, power device selection, fast overcurrent protection, and PCB layout considerations.

Fig. 2 shows the block diagram of the 11 kW motor drive converter using SiC MOSFETs, which will be presented in this paper.

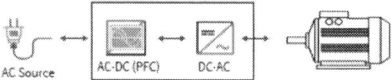

**Fig. 1** Block diagram of motor drive system

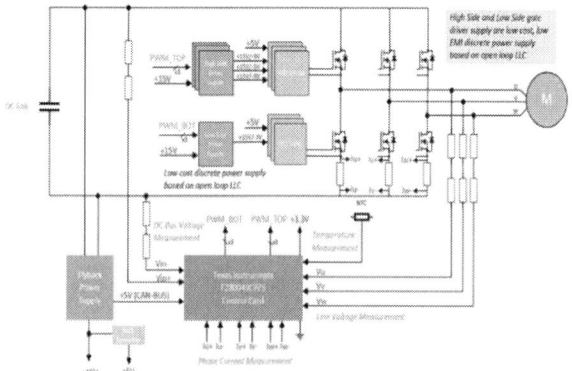

**Fig. 2** Block diagram of 11 kW motor drive Inverter

Table 1 shows its electrical and mechanical specification of the inverter, and table 2 shows the specification of the selected PMSM motor.

| DC Input Voltage Range | 550 - 850 Vdc |
|---|---|
| AC Output Voltage | 300 - 380 Vac |
| Rated Output Power | 11 kW |
| Switching Frequency | 16kHz /32kHz |
| Peak Efficiency | > 98.5% |
| Physical Dimensions (D x H) | Φ145 x 30 mm |

**Table 1** Design specifications of 11 kW Motor Drive

| Motor Type | Permanent Magnet Synchronous Motor |
|---|---|
| Rated Power | 20kW |
| Rated Voltage | 320V (L-L) |
| Rated Current | 38A (L-L) |
| Rated Speed | 3000r/min |

**Table 2** Specification of the selected PMSM motor

## 2 Key Components Selection and Hardware Design

In pursuit of high efficiency, high-frequency operation, and improved harmonic performance, this paper investigates SiC-based motor drives. Utilizing six SiC MOSFETs (Q1-Q6) with a 75mohm $R_{DS\_ON}$ and 1200V (C3M0075120K), the system operates at switching frequencies of 16kHz or 32kHz. Additionally, an auxiliary flyback power supply is designed to provide power for all ICs on the circuit boards. Detailed design and testing results will be presented herein.

The development and application of wide bandgap (WBG) power devices, such as SiC devices, have positioned themselves as promising alternatives to traditional Si devices in various applications. SiC MOSFETs offer numerous advantages in industrial motor drives compared to traditional IGBT solutions.

a) For SiC MOSFETs and IGBTs with the same current rating, SiC MOSFETs exhibit lower on-state voltage drop than IGBTs. Additionally, the on-state voltage drops of SiC MOSFETs demonstrates better thermal stability than IGBTs. In most cases, industrial drives operate below 80% of their rated power. This implies that a motor drive using SiC MOSFETs can achieve lower conduction losses and higher efficiency.

b) Without trailing current, turn-off switching loss of SiC MOSFETs is significantly less than that of IGBTs'. Apart from efficiency improvements, the smaller dead time in SiC solutions enables better Total Harmonic Distortion (THD) performance.

c) Motor drives using IGBTs typically operate at frequencies below 20kHz, where audible noise becomes a significant concern. SiC solutions operating at higher frequencies not only improves sampling accuracy and control bandwidth but also enables smaller and lighter motors with less windings.

d) Traditionally, industrial motor drives were designed with the motor and drive housed separately and connected by a long cable, leading to poor motor reliability or increased costs due to the requirements of output filters and additional power losses. However, SiC is stable at higher temperatures compared to IGBTs, enabling engineers to integrate drive electronics into the motor housing. This approach enables smaller input/output capacitors and results in a cost-effective solution.

## 3 Experimental results of 11kW motor drive

### 3.1 Prototype and dv/dt configuration

To validate the design, a digitally sensor-less controlled 11 kW motor drive inverter prototype was constructed and tested, as depicted in Fig. 3. With SiC, a high-power density of 22kW/l (D: 145mm,

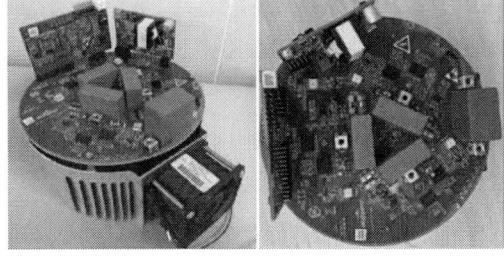

**Fig. 3** Photo of 11kW motor drive inverter

H: 30mm) is realized.

To comply with the motor insulation requirements and EMI limits of the converter, the switching speed of the MOSFET is restricted to 15V/ns. Fig. 4 displays the captured dv/dt test results under full load.

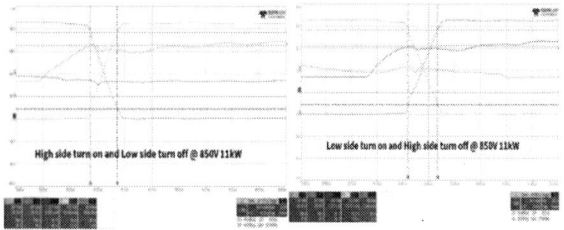

**Fig. 4** dv/dt test results under 850V and 11kW

## 3.2 Open loop test results

Fast and reliable overcurrent and short circuit protection are essential for safeguarding against catastrophic breakdown of the SiC MOSFET and enhancing system reliability. Therefore, the gate driver UCC21710 with a desaturation circuit is chosen. It incorporates the state-of-art protection features, including fast overcurrent and short circuit detection. Fig. 5 depicts the response time and captures the maximum current when a short circuit occurs.

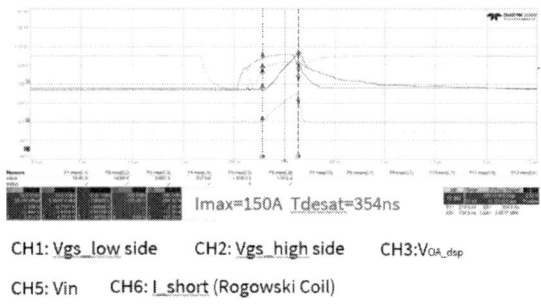

CH1: Vgs_low side    CH2: Vgs_high side    CH3: V$_{OA\_dsp}$

CH5: Vin    CH6: I_short (Rogowski Coil)

**Fig. 5** Desat performance and captured waveform

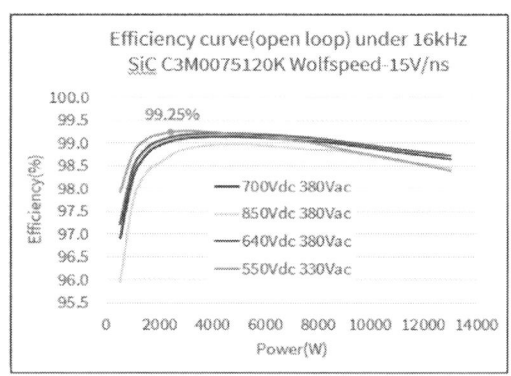

a)    Efficiency curve under 16kHz

Thanks to the superb performance of the SiC power devices (lower switching and conduction losses) and excellent thermal design, the prototype achieves over 99% peak efficiency with a switching frequency of 16 kHz under an input of 850 V$_{dc}$ and output of 380 V$_{ac}$, as illustrated in Fig. 6a). Specifically, the peak efficiency with the SiC device at 16kHz is 99.25%. Additionally, the peak efficiency at a switching frequency of 32kHz is approximately 98.8%, as depicted in Fig. 6b).

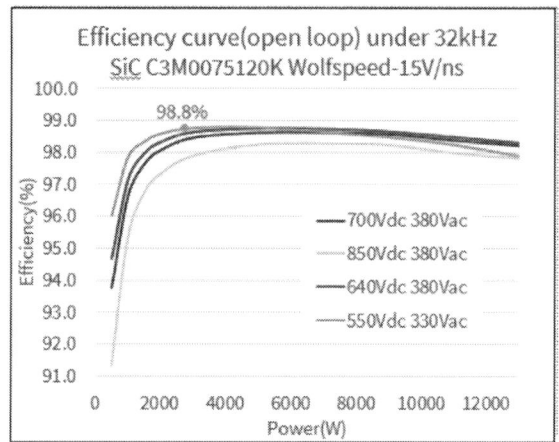

b)    Efficiency curve under 32kHz

**Fig. 6** Efficiency curve under different input voltage

Comparison testing with the popular IGBTs solution has been conducted using this prototype. A 25A 1200V IGBT was selected for the efficiency comparison testing. As illustrated in Fig. 7 a), the peak efficiency under 16kHz with the SiC solution is 1.5% higher than that of the IGBT solution. Furthermore, an efficiency improvement of up to 3% under 32kHz switching frequency is achieved compared to the IGBT solution, as shown in Fig. 7b).

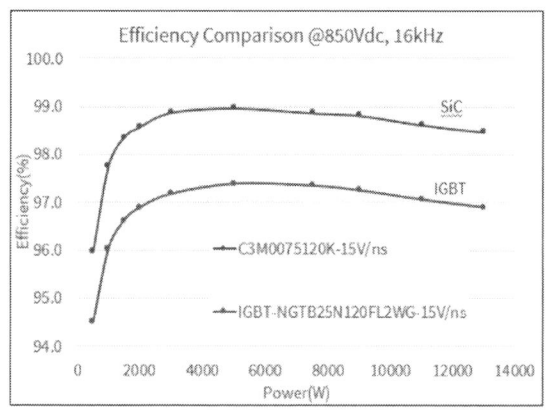

a)    Efficiency comparison curve under 16kHz

1224

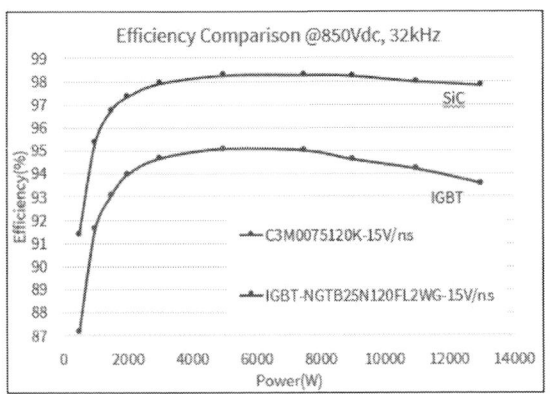

b) Efficiency comparison curve under 32kHz

**Fig. 7** Efficiency comparison with SiC and IGBT

## 3.3 Closed loop test results

Finally, the closed speed loop with sensor-less FOC control is tested based on the prototype. Fig. 8 and Fig. 9 show the captured key waveforms under full load of 11kW. From Fig. 9, it can be observed that the frequency of the phase current is approximately 200Hz. While the motor is running at a speed of 3000rpm, which aligns consistently with the set reference.

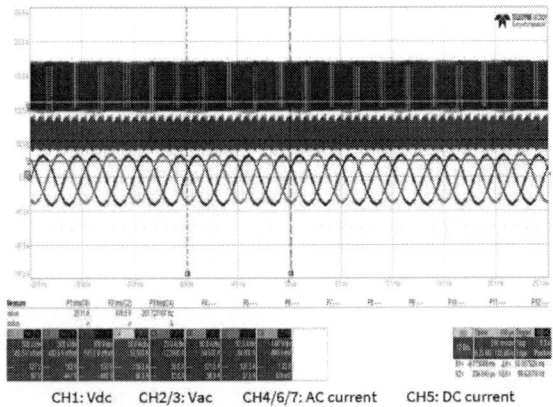

CH1: Vdc    CH2/3: Vac    CH4/6/7: AC current    CH5: DC current

**Fig. 8** Key waveform under closed loop

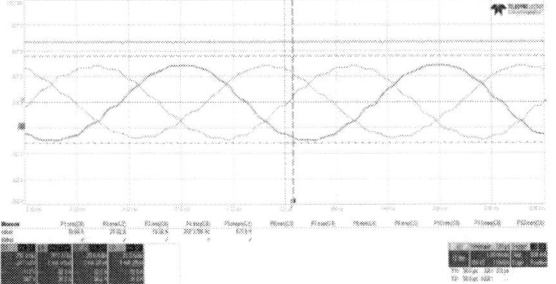

**Fig. 9** Phase current waveform under closed loop

Fig. 10 shows the efficiency curve and power loss of the inverter under 680Vdc and 3000rpm, the peak efficiency is above 99%.

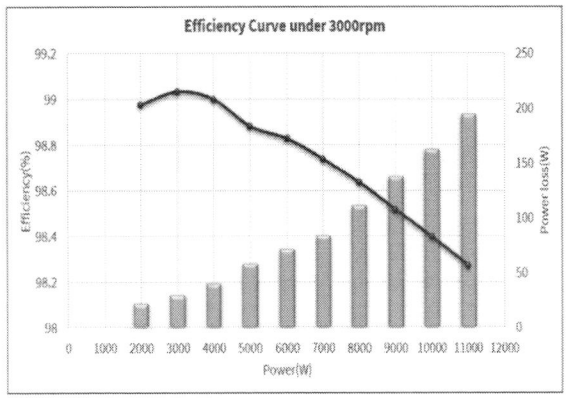

**Fig. 10** Efficiency and power loss test results

## 3.4 Thermal test results

Forced-air cooling was employed to cool down the power devices. Thermal testing was conducted at 850 Vdc input and 380Vac output under full load conditions (worst case scenario). K-type thermal couples and a data acquisition unit from Keysight Technologies Inc. (34972A) were utilized to measure the package temperature near the die of power switches.

| Devices | Heatsink Temperature (°C) | Calculated Power Loss(W) | Measured Case Temperature (°C) | Calculated Junction Temperature (°C) |
|---|---|---|---|---|
| | Input Voltage=850V, Output Voltage=380V (L-L), Po=11kW | | | |
| MOSFET Q1 | 64.72 | 21.7 | 69.67 | 93.54 |
| | Input Voltage=550V, Output Voltage=320V (L-L), Po=11kW | | | |
| MOSFET Q1 | 65.4 | 27.76 | 71.3 | 101.84 |

**Table 3** Thermal test results

The thermal test results are presented in Table 3. The junction temperature of the MOSFET is calculated based on the measured case temperature, thermal resistance, and estimated power loss of each MOSFET. It can be observed that the maximum junction temperature of the SiC MOSFET Q1 is 101.84°C at worse case, which is well below the rated maximum junction temperature of 175°C specified in the manufacturer's datasheet.

## 4 Summary

The paper investigates SiC-based motor drives to achieve high efficiency, high-frequency operations, and improved harmonic performance. The design approach and performance insights will prove beneficial to engineers interested in SiC and motor drive design.

# References

[1] G. Bucci, F. Ciancetta, E. Fiorucci and A. Ometto, "Uncertainty Issues in Direct and Indirect Efficiency Determination for Three-Phase Induction Motors: Remarks About the IEC 60034-2-1 Standard," in IEEE Transactions on Instrumentation and Measurement, vol. 65, no. 12, pp. 2701-2716, Dec. 2016, doi: 10.1109/TIM.2016.2599459.

[2] S. Deda and J. A. de Kock, "Induction motor efficiency test methods: A comparison of standards," 2017 International Conference on the Industrial and Commercial Use of Energy (ICUE), Cape Town, South Africa, 2017, pp. 1-6, doi: 10.23919/ICUE.2017.8067991.

[3] B. Blanque, J. I. Perat, P. Andrada and M. Torrent, "Improving efficiency in switched reluctance motor drives with online control of turn-on and turn-off angles," 2005 European Conference on Power Electronics and Applications, Dresden, Germany, 2005, pp. 9 pp.-P.9, doi: 10.1109/EPE.2005.219590.

[4] S. Yunus, W. Ming and C. E. Ugalde-Loo, "Efficiency Improvement Analysis of a SiC MOSFET-based PMSM Drive System with Variable Switching Frequency," 2021 23rd European Conference on Power Electronics and Applications (EPE'21 ECCE Europe), Ghent, Belgium, 2021, pp. 1-9, doi: 10.23919/EPE21ECCEEurope50061.2021.9570535.

[5] K. Vogel, et al. "Improve the efficiency in AC-Drives: New semiconductor solutions and their challenges", EEMODS 2016

PCIM Europe 2024, 11– 13 June 2024, Nuremberg    DOI: 10.30420/566262165

# Model Design Development for False Turn-on Characterization in SiC-Based Active T-Type Converter Considering All Parasitics

Amir Babaki[1], Sadegh Golsorkhi[1], Thomas Ebel[1], and Nicklas Christensen[2]

[1] Centre for Industrial Electronics (CIE), Institute of Electrical and Mechanical Engineering, University of Southern Denmark, Sønderborg, Denmark
[2] Danfoss Drives A/S, Gråsten, Denmark

Corresponding author:    Amir Babaki, amirbabaki@sdu.dk
Speaker:                         Amir Babaki, amirbabaki@sdu.dk

## Abstract

The dominant drawback of high-frequency high-voltage power converter is false turn-on of the OFF switches due to the existence of miller capacitor and high dv/dt. To ensure the secure converter operation, it is needed to investigate the comprehensive circuit model in case of false turn-on. This paper introduces a general circuitry method for investigating the gate-source induced voltage in SiC-MOSFET and Si-IGBT which are used in built T-type converter. This method is utilized to all possible scenarios to evaluate the effect of the parasitic parameters on the voltage amplitude across gate-source terminals of the switch during false turn-on. Simulation results and some standard experimental tests have been also carried out to verify the analytical methodology proposed in this paper.

## 1    Introduction

Nowadays, there is a promising attempt to enhance the performance of the power electronic switches, which are extensively used in high-voltage high frequency converters. Silicon Carbide (SiC) switches are being promising to fulfil the satisfactory performance in high frequency applications, where enables using smaller filter in converter [1]. So, this aims to potentially optimize the converter's power density. Nevertheless, lower turn-on threshold voltage ($V_{th}$) of SiC switches comparing Si ones makes the false turn-on occurrence more probable [2], which causes cross conduction of the converter leg. This issue can be trivialized somewhat by gate driver with active miller clamp feature, although it does not guarantee the reliable performance in some cases [3]. Additionally, this gate driver is specific to the switch characteristics which is more expensive than the common gate driver ICs. Using extra isolated voltage source which is connected to the gate-source capacitor during the switch turn-off helps to keep the actual voltage across the gate-source capacitor less than $V_{th}$. So, this may avoid turning-on of the switch which was supposed to be OFF during turning-on of the complimentary switch in the same leg [4]. Despite of the guaranteed safe performance,

the mentioned protection circuit is not cost wise. Before moving for other economic solution, some clarifications are required to evaluate the necessity of using any protection circuits. In fact, the issues associated with the SiC converters may be solved by utilizing optimum design of driving circuits [5]. Accordingly, an accurate comprehensive analytical model and its generalized one are needed to better characterization of the gate driver circuit as well as other related hardware design. The model presented in [6], which had been applied to half-bridge converter, is extended in this paper for active front-end converter. The hybrid T-type converter at the rectification side which contains of SiC MOSFETs and Si IGBTs is investigated. All possible parasitic component which can be modeled as stray inductances and switch intrinsic capacitors are considered in analytical model. The aim of this work is to optimize the gate driver parameters to have the safe switching without any practical issues which are imaginable in high-frequency, high-voltage applications. The simulation and experimental results have been also carried out to verify the investigated results of the parasitic component effect on system failure, which have been done the theory.

1227

## 2 Topology-based problem formulation

In order to figure out the false turn-on occurrence, an active T-type converter is investigated, which is shown with more details in Fig. 1(a). Three-level T-type converter has the best compromise between the high-power density and efficiency, as well as the possibility to reach the pure sinusoidal supply current with smaller input filter [7]. SiC MOSFETs are used as $Q_1$ and $Q_2$, while two back-to-back Si IGBT switches are employed as $Q_3$ and $Q_4$.

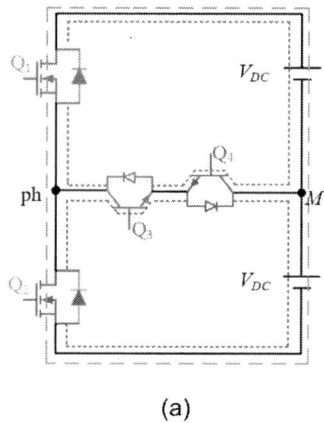

(a)

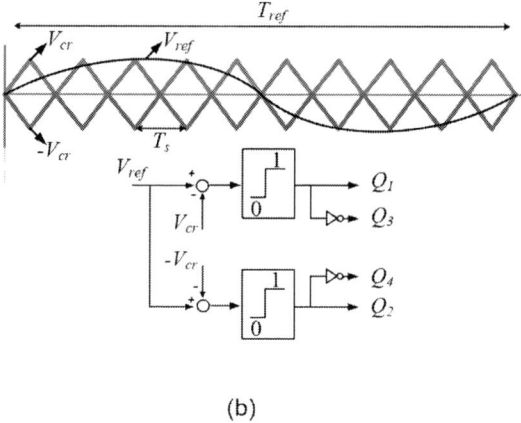

(b)

**Fig. 1** (a) Schematic of one phase in active T-type converter; (b) Switching modulation.

The zero-voltage level is provided by turning-on of the IGBTs. Generally, active T-type converter composes of two half bridges which are highlighted with red and blue color in Fig. 1(a). The upper side half bridge loop (red) includes complimentary switches $Q_1$ and $Q_3$, while the lower side half bridge (blue color) includes $Q_2$ and $Q_4$ as the complimentary set of switches. The modulation algorithm depicted in Fig. 1(b) shows that in positive half cycle of the reference waveform, the lower side switch of the leg ($Q_2$) is OFF, while switch $Q_4$ is ON. So, the current can be conducted in both direction through $Q_4$. The same situation is considered for lower half bridge. Accordingly, switch $Q_1$ is OFF during the negative half cycle of the reference waveform. Hence, the $Q_3$ is ON at this period and can conduct the bi-directional current. The deadtime is necessary to be consider for switching transient between $Q_1$ and $Q_3$ as well as $Q_2$ and $Q_4$ to protect both half bridges against shoot through current issue. Nevertheless, each switch inevitably experiences high Cdv/dt across its drain-source and the resulted induced gate-source voltage after turning on of the complimentary switch, which also cause unwanted shoot through current. This is because of Miller capacitor. In this converter, the Miller effect can be investigated in three cases: 1) upper side half bridge, where turning on of $Q_1$ affects false turn-on of $Q_3$; 2) lower side half bridge, where turning on of $Q_4$ affects false turn-on of $Q_2$; 3) converter leg, where turning on of $Q_1$ affects false turn-on of $Q_2$. Among them, case 3 is the most extreme one, which means it has the highest possibility of false turning-on. As a result, it is just needed to monitor the gate-source voltage oscillations of switch $Q_2$ during the transient turning on of switch $Q_1$ which are related to case 3.

## 3 Model development

$Q_2$ is supposed to be OFF in theory during turning on of $Q_1$. However, findings in this paper show that the induced voltage across the gate-source capacitor of $Q_2$ may even exceed the turn-on threshold voltage in some cases, which results in current cross conduction issue. For demonstrative purposes, the parasitic capacitances of switches $Q_2$ is displayed in Fig. 2(a) together with the simplified circuit considering all parasitic inductances as Fig. 2(b). The DC link in case of $Q_1$ turning on can be modeled by a sequential pulsed voltage in series with switch's on resistance ($R_{on,Q1}$).

In Fig. 2, when $Q_2$ is in its off state ($V_{drv,2}$ is grounded), turning on of the switch $Q_1$ results in a sudden voltage level change at the drain node of $Q_2$. Source of $Q_2$ is connected to ground through the leakage inductance, $L_l$. High voltage change rate(dv/dt), sensed at $Q_2$ drain node, leads to a current flow through the gate-drain capacitor, $C_{gd2}$. This is because before turning on of the $Q_1$, $Q_3$ and $Q_4$ conduct the current. So, the drain voltage ($V_{D2}$) equals $V_{dc}$. Once after turning on of $Q_1$, drain voltage is jumping to $2V_{dc}$. So, this results in a current flow in the loop containing the gate resistance of $Q_2$, which is depicted as $R_{G2}$ in Fig. 2(b). The turn-on time of $Q_1$, which is related to dv/dt, changes by turn-on gate resistor in practice. However, in the proposed model, the value of $R_{total}$ (see Fig. 2(b)) is varied to adjust dv/dt rate. In the case that the resulted voltage across resistor $R_{G2}$ exceed threshold, then the switch will be pushed to conduct the current which is leading significant switching loss.

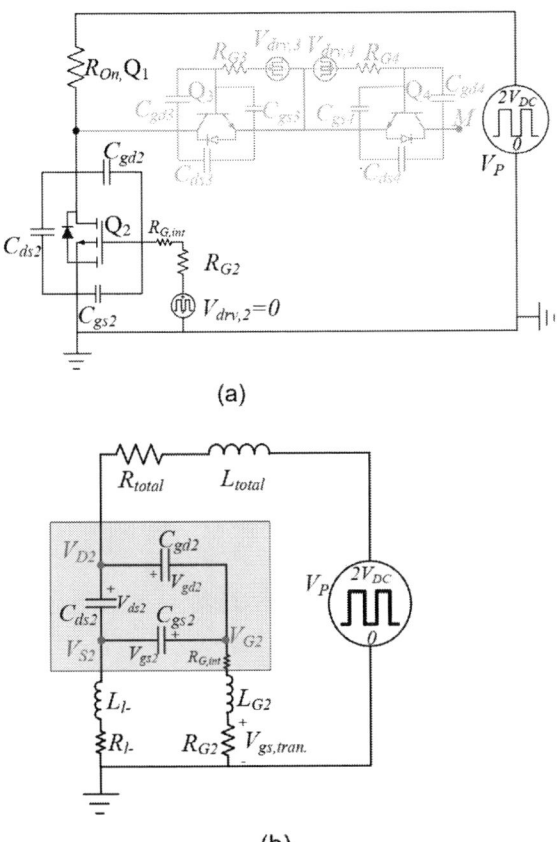

(a)

(b)

**Fig. 2** (a)Equivalent circuit for $Q_2$ false turn on demonstration by turning on of $Q_1$,(b) simplified circuit considering all parasitic capacitances and stray inductances.

This paper aims to analytically find which specific parameter has the most sensible effect on false turn-on. This can be found by assessing the relation of the natural frequency($\omega_n$) and damping ratio($\zeta$) in term of the switch's intrinsic capacitors. From the first order model which is the simplest one, the transfer function for the $V_{gs2}$ versus $V_{ds2}$ is calculated as:

$$\frac{V_{gs2}(s)}{V_{sd2}(s)} = \frac{sR_{G2}C_{gd2}}{1+sR_{G2}(C_{gd2}+C_{gs2})} \quad (1)$$

According to (1), it is found that lower $C_{gd2}$ as well as higher $C_{gs2}$ decreases the possibility to false turning-on of $Q_2$. Of course, this model neglects the circuit's stray inductance as well as the drain-source capacitor, $C_{ds2}$, which can play a significant role in this regard. From analyzing the converter leg loop (brown color in Fig. 1(a)) and using KCL for the nodes in Fig. 2(b), the function of $V_{ds2}$ and $V_{gs2}$ are obtained in term of the intrinsic capacitors. Moreover, these functions' responses with respect to the input step voltage are demonstrated in Figs. 3(a) and (b), respectively. The values of the system parameters which are used in analysis, simulation and experiment are specified in Table 1 with their definition.

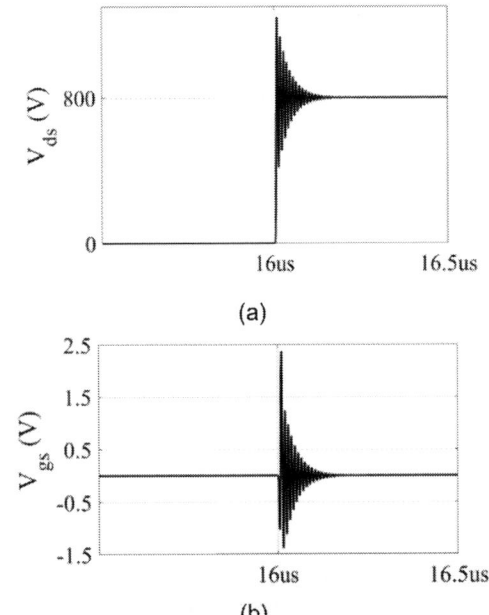

(a)

(b)

**Fig. 3** Step response of (a) $Q_2$ drain-source voltage, and (b) $Q_2$ gate-source voltage during $Q_1$ turning on.

As shown in Fig. 3(b), gate-source voltage ($V_{gs}$) jumps to around 2.4 V during the transient time of $Q_1$ turn-on. The turn-on threshold voltage ($V_{th}$) of the actual used switch is 2.5V, which guarantees the safe performance of the converter. The values for the parasitic capacitances are extracted from the switch datasheet that has been used in practice. The leakage inductance values are also determined by simulating the actual power module with the same size and material in Ansys Q3D.

**Table 1** Parameters of the circuit model in analysis, simulation, and experiment.

| Parameter | Definition | Value | Parameter | Definition | Value |
|---|---|---|---|---|---|

| $L_{total}$ | total loop stray inductance | 15nH | $R_{total}$ | total loop resistance | 12nH |
|---|---|---|---|---|---|
| $L_{l-}$ | Leg inductance under switch $Q_2$ | 2nH | $R_{l-}$ | Leg resistance under switch $Q_2$ | 1.1m$\Omega$ |
| $L_{G2}$ | Gate on/off stray inductance | 0.5nH | $R_{G2}$ | Gate turn on resistance | 15.6$\Omega$ |
| $C_{gd2}$ | Gate-Drain capacitor for SiC | 10pF | $C_{gs2}$ | Gate-Source capacitor for SiC | 1300pF |
| $C_{ds2}$ | Drain-Source capacitor for SiC | 50pF | $R_{on,Q1}$ | SiC MOSFET on resistance | 75m$\Omega$ |

## 4 Analytical/Simulation result

The general form of $V_{ds}$ in Laplace area can be mentioned as (2) regarding to the introduced equivalent circuit model:

$$V_{ds}(s) = \frac{(s+z_1)(s+z_2)(s^2+2\alpha s+\beta)}{s(s+p_1)(s+p_2)(s^2+2\zeta\omega_n s+\omega_n^2)}. \quad (2)$$

Two factors, $\zeta$ and $\omega_n$, represent the convergency speed of voltage to steady state and the oscillation situation. These parameters are mainly affected by parasitic components (intrinsic capacitances $C_{gs2}, C_{gd2}, C_{ds2}$) of the switches as well as stray inductances existed in the circuit ($L_{l-}, L_{G2}, L_{total}$). Each parameter's impact is then compared with others. In total, the effect of parasitic capacitances on the natural frequency and damping ratio is more considerable comparing to the impact of the parasitic inductances. Fig. 4 illustrates both the simulation results and the ones in theory, which have a good agreement with each other. Referring to Fig. 4(a), $C_{ds2}$ is inversely proportional to the induced gate-source voltage, which is not considered in the frequency-response formulation about the first order model equivalent. $C_{gd}$ significantly influences the induced gate-source voltage and the resulted false turn-on possibility. The induced voltage is linearly proportional to the gate-drain capacitor. This is in a good agreement with the finding in the simple first order model (see Eq.(1)). It is noteworthy that a wide range of $C_{gd}$ has been investigated in theory and simulation to ensure lack of bifurcation, which is not happened in practice. Moreover Fig. 4(c) reveals that an increase in the value of $C_{gs}$ results in induced gate-source voltage reduction and false turn-on percentage. This is also understand-

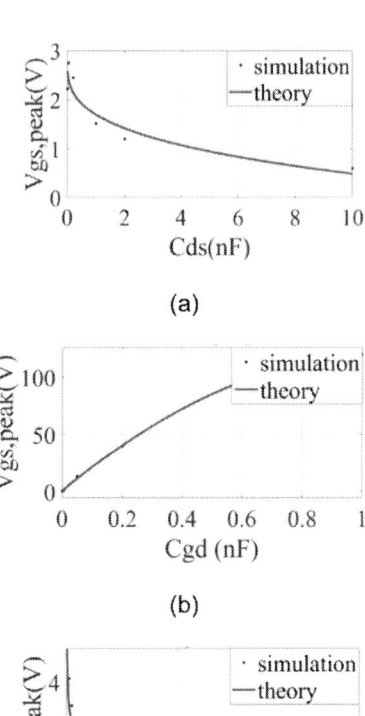

(a)

(b)

(c)

**Fig. 4** Effect of each parasitic capacitance of the switch on the induced gate-source voltage; (a) $C_{ds}$, (b) $C_{gd}$, and (c) $C_{gs}$.

able with respect to Eq. 1 relating to the first order model. Generally, increasing capacitance will increase switching loss. So, to get rid of false turn-on occurrence and reaching to the minimum possible switching loss in a high dv/dt and high frequency power converter, $C_{gs}$ and $C_{ds}$ need to be selected, compromisingly.

## 5 Experimental Verification

Hybrid three-phase active T-type converter has been built up, which employs SiC MOSFETs for the legs and back-to-back Si IGBTs for zero voltage level, which the general view is shown in Fig. 5. Standard Double Pulse Test (DPT) is carried out in experiment to validate the abovementioned results and also avoid any switches damage in case of false turn on occurrence during the test. The Tektronix AFG 31000 Double Pulse generator is used for making only two pulses with adjustable

PCIM Europe 2024, 11– 13 June 2024, Nuremberg          DOI: 10.30420/566262165

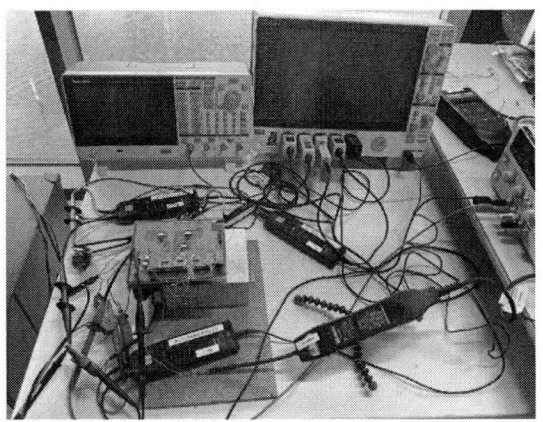

**Fig. 5** The experimental setup for result verification by DPT test.

duty cycle depending on the switch current rate. Fig. 6(b) depicts the inductor (*L*) current during the double pulse test for case 3(Fig. 6(a)), which was the only case simulated and analyzed as the critical case.

Fig. 7 illustrates the observed gate-source voltage of $Q_2$ with the possibility of false turn-on as well as the gate-source of the switch tested by double pulses in case 3. The enlarged view of the voltage in two turning on transients are also shown for better clarification. Reasonable agreement is obtained by comparing the analytical/simulation model with the experimental results. The peak value of $V_{gs}$, observed in both experimental and analytical/simulation, is around 2.4V. Analytically, $V_{gs}$ reaches to the minimum value for the first time after 100 ns, slightly preceding the experimental timing, suggesting that the experimental outcome exhibits a slightly slower response compared to the analytical one. This discrepancy can be attributed to increased circuit parasitic resulting from the adoption of new fabrication technology and layout design. Nonetheless, despite this difference, the experimental waveforms exhibit the same trends that reasonably align with the analytical/simulation results. As it can be seen in Fig. 7(a), voltage is induced in both transient from ON to OFF and vice versa. Nevertheless, false turning on during falling edge does not make cross conduction problem for the circuit due to the deadtime between switching state changes. So, rising edge

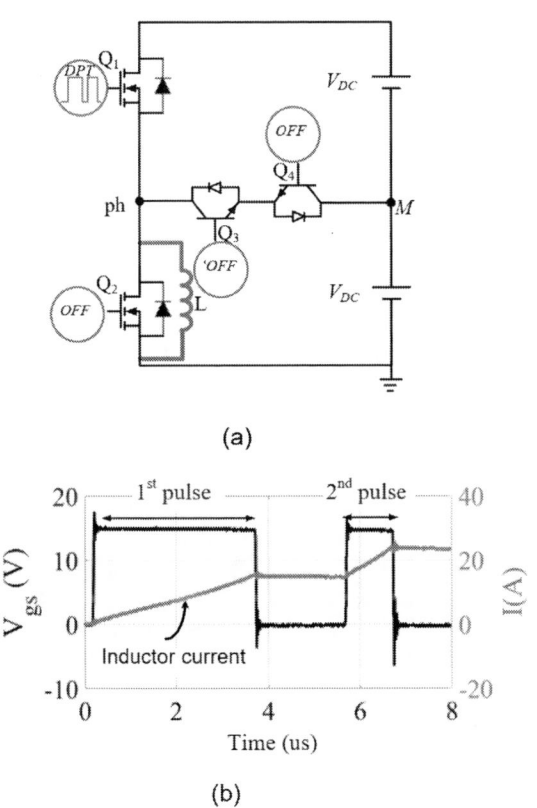

(a)

(b)

**Fig. 6** (a) circuit model of case 3 under DPT test; (b) Inductor current and $Q_1$ gate-source voltage in DPT.

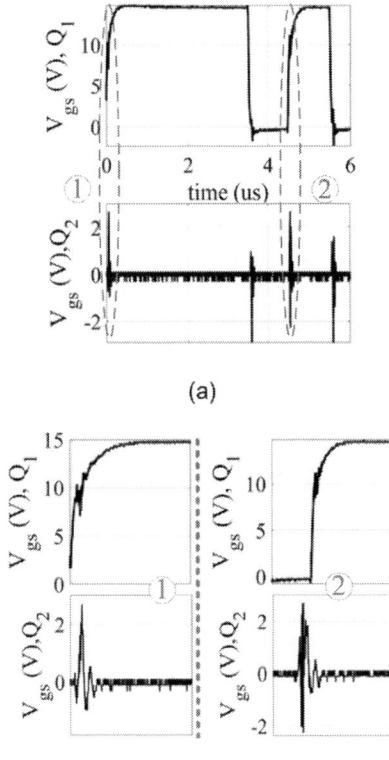

(a)

(b)

1231

**Fig. 7** (a) Induced gate-source voltage of $Q_2$ in DPT test of case 3, (b) enlarged view at first and second $Q_1$ turn-on transient.

of the pulses (pointed by red color dash loop) are only necessary to be monitored. Of course, in real practice, the amplitude of induced gate-source voltage and also voltage oscillation across the switches will be reduced in half. This is because unlike the DPT test here, no inductance is placed in parallel to the switch. So, the corresponding freewheeling diode does not conduct the current during turning off of complimentary switch. Accordingly, the step voltage that is sensed across the switch in case 3 is $V_{DC}$ (400V) instead of $2V_{DC}$ (800V), which was shown in Fig. 3(a).

## 5.1 dv/dt investigation

The effect of dv/dt can be investigated by changing the turn on time of the switch. This is done by changing gate turn on resistance. dv/dt increases with lower values of $R_{G,on}$. Fig. 8 shows the enlarged view of the gate-source voltage in experiment like the one in Fig. 7, while $R_{G,on}$ is reduced to 11.5Ω from nominal value (15.6Ω). Comparing the results in Fig. 8, it is obvious that the induced gate-source voltage across $C_{gs}$ of the OFF switch($Q_2$) has higher overshoot for $R_{G,on}=11.5\Omega$ during turning on of the complimentary switch($Q_1$) which is approximately 3.2V. This means the possibility for false turning on increases with higher voltage change rate resulted by lower gate resistor. The voltage overshoot is lower than threshold voltage for $R_{G,on}=15.6\Omega$ and higher than it for $R_{G,on}=11.5\Omega$ which is out of safety region. This experiment can be carried out to recognize the upper bound of the eligible dv/dt regarding the resulted induced voltage that may cause false turn on.

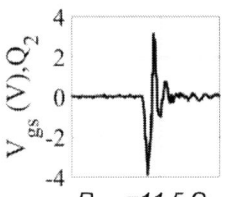 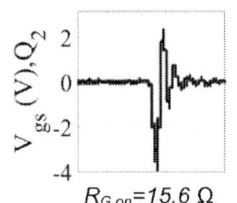

$R_{G,on}=11.5 \Omega$      $R_{G,on}=15.6 \Omega$

**Fig. 8** Comparison result of the induced gate-source voltage of switch $Q_1$ in different $R_{G,on}$.

## 6  CONCLUSION

False turn-on possibility for SiC MOSFETs in T-type converter has been analytically evaluated in this paper. The parasitic components including intrinsic capacitor and layout leakage inductances have been also considered in the proposed model. The significant impact of the parasitic capacitors on the induced gate-source voltage, its damping ratio and the voltage oscillation frequency has been claimed in both theory and simulation. The results obtained beforehand in theory are then verified by the experimental test using DPT approach. The obtained experimental results have a good qualitatively alignment with the findings in both theory and simulation. The method reported here can be applied to model false turn-on all types of power converters which are imposed of multiple Wide Band Gap (WBG) switches with high voltage changes rate (high dv/dt).

## Acknowledgment

This work was supported by The Energy Technology Development and Demonstration Program (EUDP) funds via the project with ref.: 64020-2075.

## References

[1] Yaakub, Muhamad Faizal, et al. "Silicon carbide power device characteristics, applications and challenges: an overview." International Journal of Power Electronics and Drive Systems 11.4 (2020): 2194.

[2] Ahmed, Md Rishad, Rebecca Todd, and Andrew J. Forsyth. "Predicting SiC MOSFET behavior under hard-switching, soft-switching, and false turn-on conditions." IEEE Transactions on Industrial Electronics 64.11 (2017): 9001-9011.

[3] Haihong, Qin, et al. "An overview of SiC MOSFET gate drivers." 2017 12th IEEE Conference on Industrial Electronics and Applications (ICIEA). IEEE, 2017.

[4] Gao, Feng, et al. "A gate driver of SiC MOSFET for suppressing the negative voltage spikes in a bridge circuit." *IEEE Transactions on Power Electronics* 33.3, pp. 2339-2353, 2017.

[5] Mocevic, Slavko, et al. "Design of a 10 kV SiC MOSFET-based high-density, high-efficiency, modular medium-voltage power converter." IEnergy 1.1 (2022): 100-113.

[6] Khanna, Raghav, et al. "An analytical model for evaluating the influence of device parasitics on Cdv/dt induced false turn-on in SiC MOSFETs" *28th Annual IEEE Applied Power Electronics Conference and Exposition (APEC)*, 2013.

[7] Christensen, Nicklas. "Demonstration of High-Power Density kW Converters utilizing Wide-Band Gap Devices", 2019.

PCIM Europe 2024, 11– 13 June 2024, Nuremberg  DOI: 10.30420/566262166

# Efficiency Investigations of an Auxiliary Resonant Commutated Pole (ARCP) Inverter

Markus Zocher[1], Norbert Grass[1], Ralph Kennel[2]

[1] Institute ELSYS, Nuremberg Institute of Technology, Germany
[2] HLU, Technical University of Munich, Germany

Corresponding author:   Markus Zocher, markus.zocher@th-nuernberg.de
Speaker:   Markus Zocher, markus.zocher@th-nuernberg.de

## Abstract

This work presents a method to measure the losses of an Auxiliary Resonant Commutated Pole (ARCP) inverter in periodic operation. Conduction losses and the switching loss contributions are identified for different semiconductor technologies. The results show that SiC-MOSFETs are especially suited for the ARCP topology, since the majority of switching losses are eliminated. When using IGBTs, the turn-on losses are reduced significantly, but the turn-off losses only decrease little. Using this method, different ARCP control algorithms can be verified and compared.

## 1   Introduction

Power electronic inverters are widely used in motor and grid applications to transform direct current (DC) into alternating current (AC) or vice versa. The majority of these are built based on the hard-switching two-level topology which requires the least number of active components and is therefore often seen as beneficial from cost-perspective. However, such a hard-switching topology does not only limit the switching frequency due to the switching losses but also requires a large filtering effort to fulfil electromagnetic compliance (EMC) requirements.

The Auxiliary Resonant Commutated Pole (ARCP) topology shown in Fig. 1 is an extension to the two-level inverter which allows soft-switching. With the help of the resonant inductance $L_r$ an oscillation circuit with the output capacitances of the main transistors $S_{1,2}$ is created. The resonance circuit is activated by the auxiliary IGBTs $S_{A1,2}$. The topology has been invented independently by De Doncker [1] and McMurray [2]. Traditionally, the ARCP has been mainly researched with large resonant circuits including an additional external resonance capacitance $C_{r,ext}$ and a resonance period of many microseconds. This relaxes control requirements but also limits the modulation range [3]. In recent years the ARCP is also considered for small resonant circuits which keep the additional cost, weight, and volume for auxiliary transistors and the resonant components small and enables higher switching frequencies.

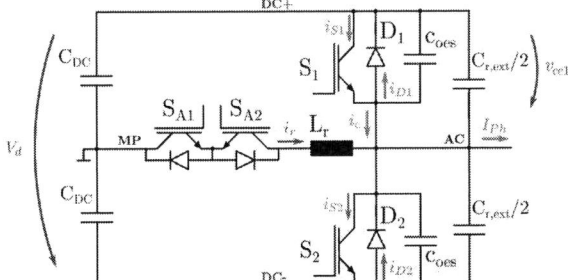

**Fig. 1.**   Schematic of half-bridge with ARCP extension and connection to DC link

From an EMC perspective the ARCP topology can be seen as an active electromagnetic interference (EMI) filter which shapes the inverter output voltage directly at the source of disturbance and also happens to decrease the switching losses [4].

Small resonant circuits require a more advanced control to achieve soft-switching - in particular when nonlinearities of the resonant components cannot be neglected and the power semiconductor switching behavior cannot be assumed as ideal [5]. The effectiveness of the adaptive control algorithms [6] is tested practically in this paper regarding the achievable savings in switching losses.

The paper is structured as follows. Chapter 2 provides an overview of the principle of operation assuming ideal switching behavior. Afterwards an overview of the tested semiconductor devices and typical aspects of their switching behavior is provided in chapter 3. The proposed measurement setup is presented in chapter 4. The following chapters provide the results for two different methods to determine the average switching losses. In chapter 5 a DC current measurement is used to determine the average conduction and switching losses. In chapter 6 oscilloscope measurements help to evaluate the loss distribution among the main and auxiliary semiconductors and the resonant inductor. The measurements are conducted for one SiC-MOSFET and different IGBT generations used as main transistors $S_{1,2}$.

## 2 ARCP principle

Two typical commutations are shown in Fig. 2. Between $t_1$ and $t_5$ the positive output current $I_{Ph}$ is commutated from $D_2$ to $S_1$, and between $t_6$ and $t_8$ back from $S_1$ to $D_2$. The load at the inverter output is seen as inductive causing the output current $I_{Ph}$ to remain constant during a switching action.

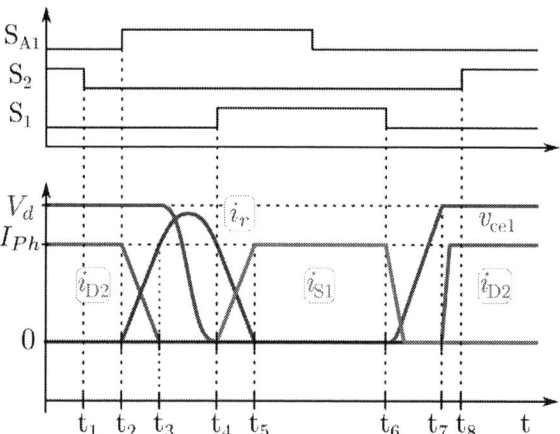

**Fig. 2.** Waveforms of ARCP commutations

The first commutation from diode to transistor is supported by the auxiliary leg. After turning off the main transistor $S_2$, the turn-on of the auxiliary transistor $S_{A1}$ at $t = t_2$ leads to an increase of the resonant current $i_r$ which takes over the current from diode $D_2$. When this current has reached the level of the output current $I_{Ph}$ at $t = t_3$, the resonant circuit between $L_r$ and $C_r = 2 \cdot c_{oes} + C_{r,ext}$ is able to oscillate. When the transistor $S_1$ is turned on after a half oscillation period at $t = t_4$ where the voltage $v_{ce1}$ has a minimum close to zero the turn-on losses are practically zero and soft-switching is achieved.

The complementary commutation does not always require the auxiliary leg. After $S_1$ is turned off at $t = t_6$, the output current automatically charges the resonant capacitance. After the diode $D_2$ is conducting again, the transistor $S_2$ can be turned on softly at $t = t_8$. The turn-off from $S_1$ is softened by the resonant capacitance which limits the slope of $v_{ce1}$. At smaller output current levels both commutations are supported by the auxiliary leg.

## 3 Devices under test

The semiconductor devices used as main transistors are summarized in Table 1. In this work they will be referenced by their generation.

| Generation | Device | $v_{max}(V)$ | $I_{nom}(A)$ |
|---|---|---|---|
| IGBT2 (NPT) | SKM200GB 125D | 1200 | 150 |
| IGBT3 (Trench) | SKM100GB 176D | 1700 | 75 |
| IGBT4-Fast (Trench) | SKM100GB 12T4 | 1200 | 100 |
| SiC-MOS-FET (Rohm 2nd Gen.) | SKM350MB 120SCH15 | 1200 | 350 |

**Table 1** Overview of the used main semiconductors

To calculate the energies that are dissipated during turn-on and turn-off it is useful to correct the electrically measured turn-on and turn-off energies by the energy stored in the output capacitance in the off-state according to (1) and (2).

$$E_{off} = \int v_{ds}(t) \cdot i_d(t)\, dt - E_{Coss} \quad (1)$$

$$E_{on} = \int v_{ds}(t) \cdot i_d(t)\, dt + E_{Coss} \quad (2)$$

$$E_{Coss} = \int_{v_{ds}=0}^{v_{ds}=V_d} v_{ds} \cdot c_{oss}(v_{ds})\, dv_{ds} \quad (3)$$

The uncorrected turn-off losses measured with the IGBT4-Fast and SiC-MOSFET modules are depicted in Fig. 3 with and without an external capacitance. At the IGBT (a) an external capacitance will only slightly decrease the turn-off losses and not offer real ZVS at turn-off. Also the energy stored in $c_{oes}$ is low compared to the measured turn-off losses. With the SiC-MOSFET (b) the main part of the measured turn-off energy at small currents is not dissipated during turn-off but stored in $c_{oss}$.

PCIM Europe 2024, 11– 13 June 2024, Nuremberg          DOI: 10.30420/566262166

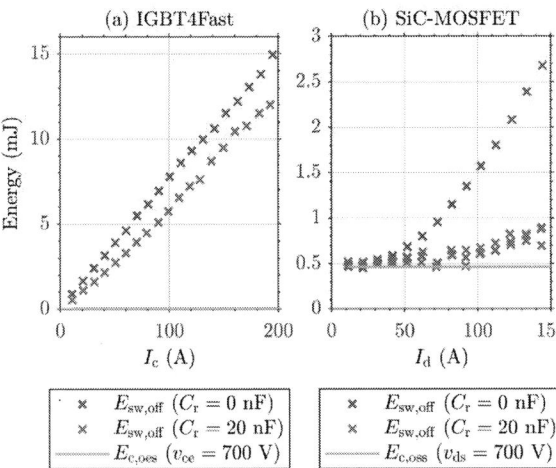

**Fig. 3.** Measured turn-off energies with external capacitance $C_r = 20\,nF$ and without, stored energy in chip internal output capacitance, comparison of (a) IGBT4-Fast, (b) SiC-MOSFET module

## 4 Proposed measurement setup

A direct efficiency measurement of power converters by measuring input and output power often leads to a large error, since the transferred power is magnitudes above the power losses, especially when high-efficient resonant topologies are investigated. To measure losses more accurately without using calorimetric techniques, a full-bridge (FB) setup with two ARCP legs and a mainly inductive load is proposed as a power in the loop (PIL) configuration and displayed in Fig. 4.

The converter drives a sinusoidal current $I_{Ph}$ into the inductive load, consisting of the ideal inductor $L_L$ and the ohmic resistor $R_L$ representing the losses of the load inductor. The FB setup offers the advantage, that the high frequency current ripple in the load is minimal, when a suitable modulation is used like shown in Fig. 5. In FB mode the dutycycle of the two ARCP half-bridge legs is alternated sinusoidally with the fundamental frequency and a phase shift of 180°. The PWM of both legs is symmetric to the center of the pulses. The inverter output voltages $v_{Ph,n}$ are referenced to the DC link midpoint.

In FB mode the load inductor is placed like shown in Fig. 4, in half-bridge (HB) mode the load inductor is assumed to be connected between one HB output and the DC link midpoint. The voltage across the inductive load is calculated in FB and HB mode in (4) and (5). The load current $I_{Ph}$ is approximately found by neglecting $R_L$ and integration of $v_L$ (6). As a reference the setup is also used in hard-switching mode without external resonant capacitances and auxiliary transistors.

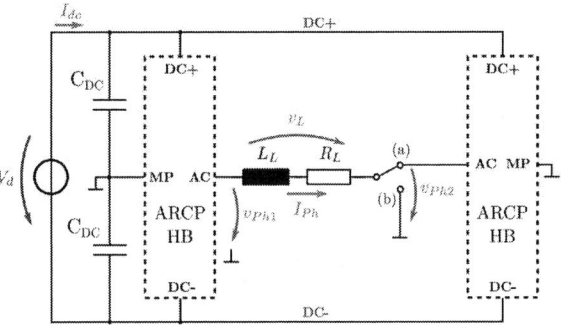

**Fig. 4.** ARCP full-bridge setup with inductive load, switch position (a) full-bridge mode, (b) half-bridge mode

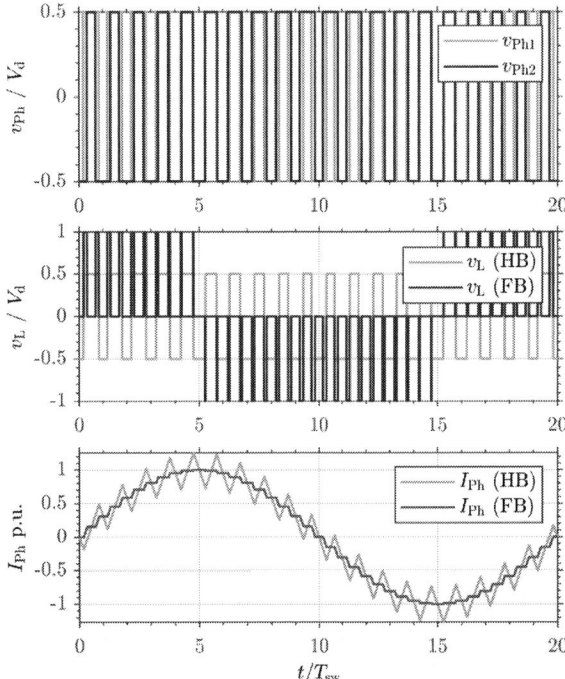

**Fig. 5.** Ideal Waveform comparison of periodic operation with half-bridge and full-bridge connection of the load inductor

$$v_{L,FB}(t) = v_{Ph1}(t) - v_{Ph2}(t) \qquad (4)$$

$$v_{L,HB}(t) = v_{Ph1}(t) \qquad (5)$$

$$I_{Ph}(t) = \frac{1}{L_L} \int_0^t v_L(\tau)\, d\tau \qquad (6)$$

As Fig. 5 depicts, the current ripple in FB mode is low compared to the HB mode. This simplifies the loss estimation inside the inductive load since losses due to the high frequent ripple which can be difficult to estimate are significantly reduced. As

1235

the ratio between fundamental period and switching period will be at least a magnitude beyond the ideal example of Fig. 5, the current ripple becomes even more neglectable.

Fig. 6 shows a practical measurement in periodic operation in FB soft-switching mode with IGBT4-Fast modules as main transistors. To ensure the correct timing of the auxiliary transistors for soft-switching the adaptive lookup table control method from [6] is used. The resonant current $i_r$ exceeds the main transistor current $i_c$ during the commutations. The auxiliary leg is usually active in one direction. Near the zero-crossings of $I_{Ph}$ both commutations are supported with a resonant current. At $t = 5\,\mathrm{ms}$, two switching actions are displayed in detail. The commutation from diode to transistor, shown in the middle diagram of Fig. 6 is supported by the auxiliary leg leading to almost zero turn-on energy. The complementary commutation from transistor to diode, shown in the bottom diagram of Fig. 6 leads to turn-off losses that also could be seen in Fig. 3 (a).

Fig. 7 shows a periodic operation at the same operational point and control setup with SiC-MOSFET modules as main transistors.

In comparison to the IGBT4-Fast the SiC-MOSFET has a significant larger chip internal output capacitance with an energy-based equivalent value of $\overline{c_{oss}} \approx 2.91\,\mathrm{nF}$. This allows to operate the ARCP without an external resonant capacitor.

Since the resonant capacitance is small compared to the resulting resonant capacitance of the IGBT4-Fast measurement setup, the peak resonant current $i_r$ only slightly exceeds the main transistor drain current $i_d$. When looking at switching actions in detail, it can be seen, that the measured turn-on energy would be negative since the energy stored in the internal output capacitance is released leading to a negative transistor current in the middle diagram of Fig. 7. The measured turn-off energy on the other hand includes the charging of this capacitance leading to a larger measured turn-off energy than the really dissipated turn-off energy. This again shows the importance to correct the measured switching energies according to (1) and (2).

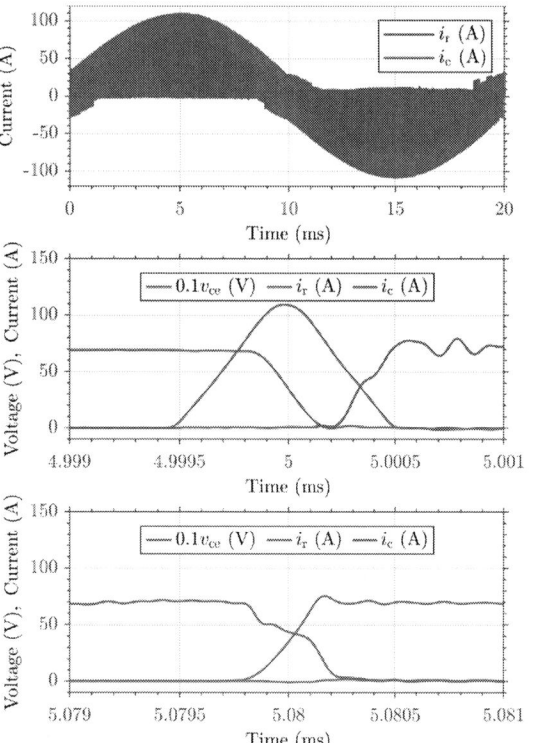

**Fig. 6.** Measurement of periodic operation at $V_d = 700\,\mathrm{V}$, $\hat{I}_{Ph} = 70\,\mathrm{A}$, $L_r = 1.51\,\mathrm{\mu H}$, $C_{r,ext} = 10\,\mathrm{nF}$ with IGBT4-Fast module and adaptive current-based control algorithm, $f_1 = 50\,\mathrm{Hz}$, $f_{sw} = 10\,\mathrm{kHz}$, zoom on two switching actions in lower diagrams

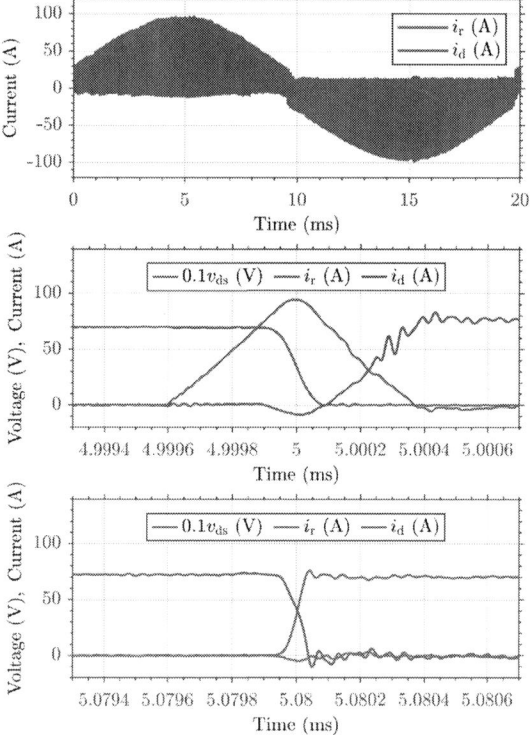

**Fig. 7.** Measurement of periodic operation at $V_d = 700\,\mathrm{V}$, $\hat{I}_{Ph} = 70\,\mathrm{A}$, $L_r = 1.51\,\mathrm{\mu H}$, $C_{r,ext} = 0\,\mathrm{nF}$ with SiC-MOSFET module and adaptive periodic Lookup table, $f_1 = 50\,\mathrm{Hz}$, $f_{sw} = 10\,\mathrm{kHz}$, zoom on two switching actions in lower diagrams

# 5 Losses and efficiency from DC measurement

According to the setup from Fig. 4 the total losses are obtained with relatively high accuracy from DC measurements and represented by

$$P_{V,\text{total}} = V_d \cdot I_{dc} \qquad (7)$$

Without saturation effects the losses in the inductive load can be generally approximated by (8) where $I_{\text{Ph},\nu}$ is the RMS-value of the load current at the harmonic $\nu$ and $R_L(\nu f_1)$ represents the frequency dependent losses in the inductive load. In general, the frequency dependent ohmic resistance reflects the increased winding losses of the load inductor due to skin and proximity effects. Also the core losses of the load inductor will increase with frequency at the same current amplitude due to higher hysteresis and eddy current losses at higher harmonics.

$$P_{V,\text{load}} = \sum I_{\text{Ph},\nu}^2 \cdot R_L(\nu f_1) \qquad (8)$$

As shown in Fig. 5, in FB mode higher order current harmonics can be neglected in good approximation. Therefore, only the fundamental current harmonic at $\nu = 1$ has to be considered which simplifies the calculation of (8).

After subtracting (8) from (7), the losses of the two half-bridges are retained. By varying the amplitude of the output current $\hat{I}_{\text{Ph}}$ and the switching frequency $f_{\text{sw}}$, the conduction losses and average switching energies can be obtained according to (9).

$$P_{V,\text{HB}}(V_d, \hat{I}_{\text{Ph}}, f_{\text{sw}}) = \frac{1}{2}\left(P_{V,\text{total}} - P_{V,\text{load}}\right)$$

$$\overset{!}{=} P_{V,\text{cond}}(\hat{I}_{\text{Ph}}) + \overline{E_{\text{sw,HB}}(V_d, \hat{I}_{\text{Ph}})} \cdot f_{\text{sw}} \qquad (9)$$

At the hard-switching inverter, the average switching energy $\overline{E_{\text{sw,HB}}}$ is the sum of the turn-on and turn-off energy from the transistor and the reverse recovery energy of the diode. In the ARCP mode the average switching energy also includes the switching and conduction losses in the auxiliary leg. All presented losses are also dependent on the ambient and chip-temperature and should be evaluated at the expected temperatures in the later application.

To separate the conduction losses and switching energy properly, all measurements should be carried out at the same semiconductor junction temperature. This can be challenging since the losses themselves are highly dependent on the output current.

Therefore, the loss evaluation and comparison in this chapter is based on a junction temperature close to 25 °C or only slightly above. This temperature is ensured by recording the losses directly after turning on the inverter at the desired operational point. This method has the additional advantage that the semiconductors' datasheet values for conduction losses and switching energies usually given for 25 °C can be used for further verification.

Fig. 8 and Fig. 9 display the influence of the output current amplitude and switching frequency to the inverter losses using IGBT4-Fast and SiC-MOSFET modules for the main transistors $S_{1..4}$. The offset of a linear fit curve at $f_{\text{sw}} = 0$ kHz represents the conduction losses in the main transistors and diodes $P_{V,cond}$. The slope is representing the average switching energy $\overline{E_{\text{sw,HB}}}$.

In ARCP mode and at high output current amplitudes the switching losses decrease by approximately 30 % when used with a Trench IGBT module (Fig. 8) and reduced by more than 80 % when used with a SiC-MOSFET module (Fig. 9), compared to the hard-switched configuration.

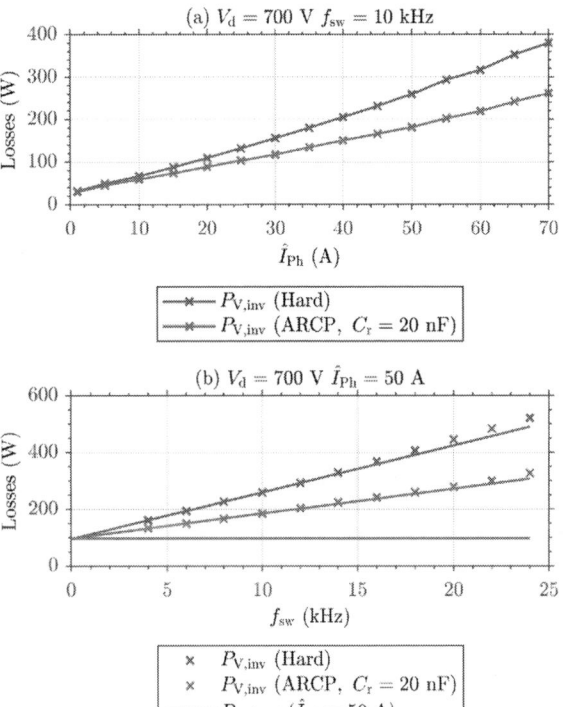

**Fig. 8.** Inverter Losses in hard-switching and ARCP mode with IGBT4-Fast module depending on (a) output current amplitude, (b) switching frequency

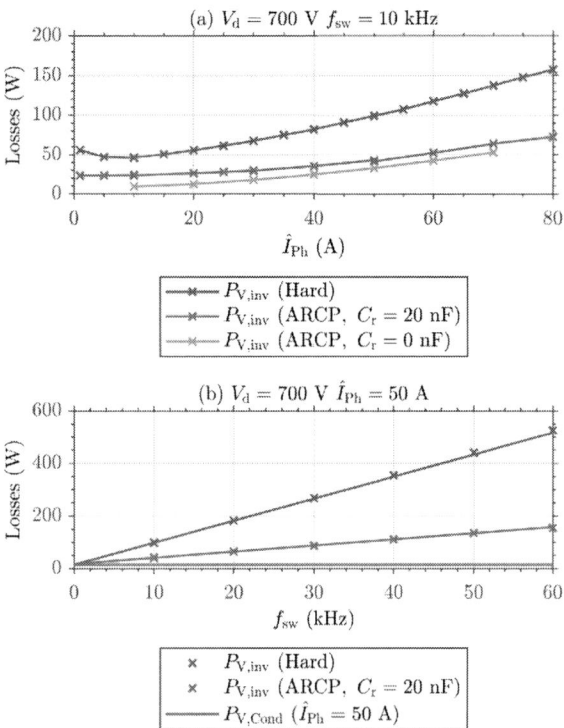

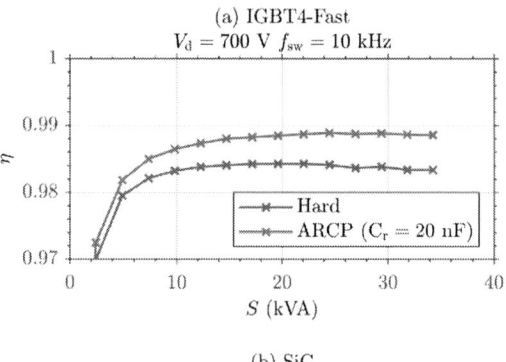

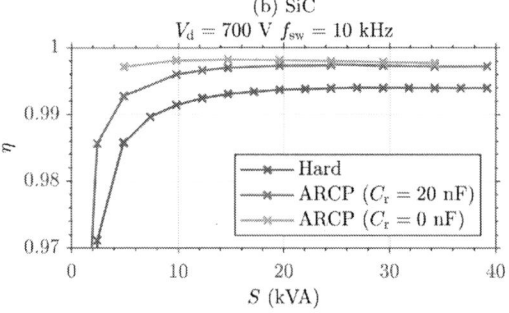

**Fig. 9.** Inverter Losses in hard-switching and ARCP-mode with SiC-MOSFET module depending on (a) output current amplitude, (b) switching frequency

This results in an increased efficiency of the ARCP mode as Fig. 10 shows. Although a switching frequency of $f_{sw} = 10$ kHz is low for a SiC-MOSFET inverter and only the semiconductor and auxiliary leg losses are included, it can be seen, that especially the combination of the SiC-module in ARCP-mode with a maximum efficiency of 99.8 % is very close to an ideal converter.

The higher efficiency of the operation with the SiC-MOSFET module results from lower conduction and switching losses as Fig. 11 depicts in an exemplary operational point. Since a MOSFET does not have a forward saturation voltage the conduction losses are low compared to the IGBT-module. This difference is enhanced by the fact that the investigated operational point is below the SiC-MOSFETs nominal current.

The switching losses of the hard-switched SiC-module are also low compared to the hard-switched IGBT-module. When the ARCP mode is used the conduction losses of IGBT and SiC variants remain equal. When the IGBT-module is operated in ARCP mode the switching losses are decreased by 47 % compared to the hard-switching mode. The switching losses of the hard-switching SiC variant are even cut by 72 % when operated in ARCP mode.

This reduction in conduction and switching losses of SiC-MOSFETs in combination with the ARCP topology is responsible for the high efficiency shown in Fig. 10 (b).

**Fig. 10.** Comparison of efficiency in hard-switching and ARCP mode at $V_d = 700$ V, $f_{sw} = 10$ kHz with usage of (a) IGBT4-Fast and (b) SiC-module

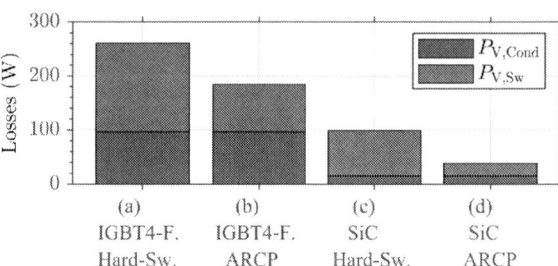

**Fig. 11.** Comparison of conduction ($P_{V,Cond}$) and switching losses ($P_{V,sw}$) per half-bridge of IGBT4-Fast and SiC-MOSFET in hard-switching and ARCP mode, $V_d = 700$ V, $f_{sw} = 10$ kHz, $I_{Ph} = 50$ A ($L_r = 1.51$ µH, $C_r = 20$ nF)

# 6 Oscilloscope measurements for loss distribution

To investigate the distribution of the switching energies, oscilloscope measurements of voltages and currents are carried out. Modern digital oscilloscopes offering a "Segmented trigger" function make it simple to record all switching actions of a single fundamental period $T_1 = 1/f_1$ in a sufficient sample rate. In the hard-switching mode currents and voltages of the main semiconductors are recorded. In the ARCP mode additionally the voltages and currents of the auxiliary IGBTs and diodes and the resonant inductor are captured. All measurements are executed at a junction temperature of 25 °C. After processing the measured voltages and currents, the distribution of the average switching energy is found. The measurements are conducted with all IGBT and MOSFET modules summarized in Table 1.

For the different main transistors, switching modes, and combinations the following diagrams show the average values of the switching loss contribution. The sum of the IGBT turn-off ($E_{\text{off}}$) and main diode reverse recovery losses ($E_{\text{RR}}$) are given in blue color. The IGBT turn-on losses ($E_{\text{on}}$) are drawn in red. The ARCP modes additionally have losses in the resonant inductor ($E_{\text{Lr}}$), turn-on losses of the auxiliary IGBT ($E_{\text{AuxIgbt}}$), and reverse recovery losses of the auxiliary diode ($E_{\text{AuxD}}$). The deviation between the average switching energy determined from the DC measurement according to (9) and the oscilloscope measurement are displayed as "Add. Losses".

The results for the IGBT4-Fast module in Fig. 12 show, that the main IGBT turn-on losses ($E_{\text{on}}$) are reduced significantly when the hard-switching and ARCP modes are compared. The turn-off and reverse recovery losses ($E_{\text{off}}$ + $E_{\text{RR}}$) are only slightly reduced and contribute the dominant part of the average switching energy in the ARCP modes. This is caused by the slow turn-off from IGBTs under ZVS condition with a parallel capacitance as can be seen from Fig. 6 and Fig. 3. This behavior has also been reported in [7].

By increasing the resonant capacitance, the turn-off energies will decrease, but the losses in the resonant inductor $L_{\text{r}}$ will increase through the higher resonant current. The losses in the auxiliary semiconductors are small but will also increase with the amplitude of the resonant current. This leads to an almost constant total average switching energy and implies that a further increase of $C_{\text{r}}$ is not favorable for switching loss reduction.

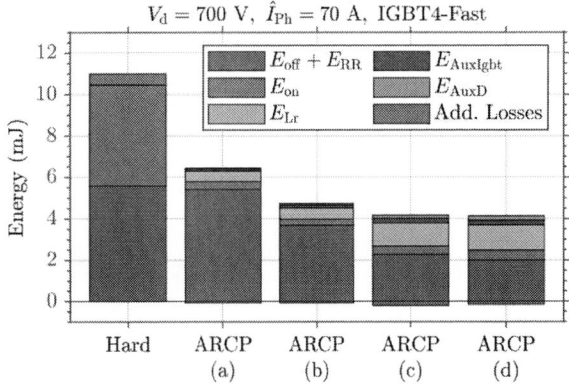

**Fig. 12.** Loss distribution with IGBT4-Fast module, hard-switching and ARCP mode, $L_{\text{r}} = 1.51\ \mu\text{H}$, variation of $C_{\text{r}} = [0\ \text{nF}, 20\ \text{nF}, 66\ \text{nF}, 99\ \text{nF}]$ and dead-time $t_{\text{dead}} = [0.5\ \mu s,\ 1\ \mu s,\ 2\ \mu s, 2\ \mu s]$

The additional losses show that DC and the oscilloscope measurements match very well with a deviation of less than 5 %.

In Fig. 13 the loss distribution results for an IGBT3 module is shown. Since this module is from a higher voltage class (1700 V), the switching energies are generally higher and especially the turn-off energy is not changed significantly by a resonant capacitance. The higher voltage blocking capability comes at the expense of higher charge carrier lifetimes which are required to keep the forward saturation voltage and thus the conduction losses in an acceptable region. The results from Fig. 13 further show that the investigated small resonant circuits are more interesting for IGBTs of the 1200 V class, at least from an efficiency point of view.

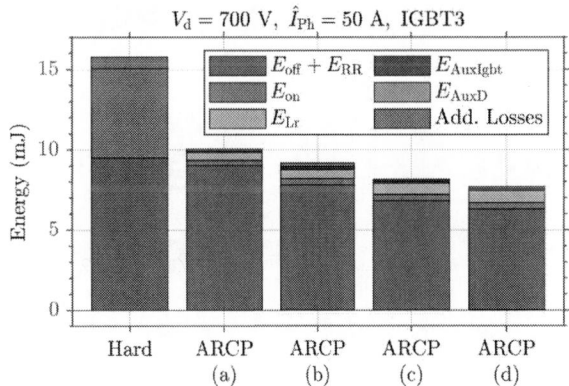

**Fig. 13.** Loss distribution with IGBT3 module (1700 V class), hard-switching and ARCP mode, variation of $C_r = [0\ \text{nF}, 10\ \text{nF}, 20\ \text{nF}, 30\ \text{nF}]$, dead-time $t_{\text{dead}} = 1\ \mu s$, $L_{\text{r}} = 1.51\ \mu\text{H}$

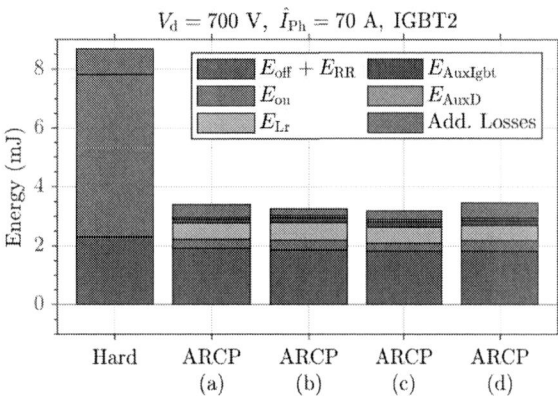

**Fig. 14.** Loss distribution with IGBT2 module, hard-switching and ARCP mode, $L_r = 1.51\,\mu H$, variation of $C_r = [10\,nF, 20\,nF, 20\,nF, 20\,nF]$ and dead-time $t_{dead} = [0.5\,\mu s, \; 0.5\,\mu s, 1\,\mu s, 2\,\mu s]$

In Fig. 14 the results for an IGBT2-module can be found. In contrast to the other IGBT modules processed in trench technology, this IGBT generation is based on the Non-Punch-Through (NPT) technology. This technology has generally smaller turn-off losses making the relative savings of the average switching energy in the ARCP mode higher and leading to a reduction of approximately 60 % in comparison to the hard-switching mode. These lower turn-off losses come at the expense of a higher forward voltage of this IGBT generation of $v_{ce,sat} = 3.3\,V$ at nominal current [8] in comparison to $v_{ce,sat} = 2.3\,V$ of the IGBT4-Fast module [9]. The dead-time has a minor influence on the average switching energies. The deviation of the DC and oscilloscope measurements represented as "Add. Losses" in Fig. 14 is slightly higher compared to the other IGBT generations but is still smaller than 15 %.

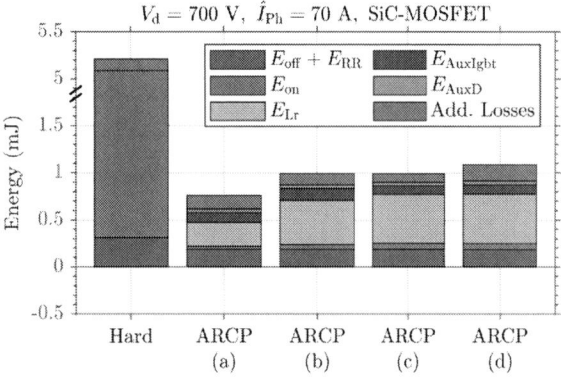

**Fig. 15.** Loss distribution with SiC module, comparison of hard-switching and ARCP mode without additional capacitance $C_{r,ext} = 0\,nF$, $t_{dead} = 0.5\,\mu s$, variation of $L_r = [0.85\,\mu H, 1.03\,\mu H, 1.51\,\mu H, 2.67\,\mu H]$

Fig. 15 shows the loss distribution of a SiC-MOSFET module in hard-switched and ARCP mode. The relative saving in switching energy is above 80 % and very large compared to the investigated IGBT modules. This results from the fact, that the turn-off from the unipolar MOSFET semiconductor is generally much faster compared to a bipolar IGBT since there is no charge carrier recombination and hence, no tail current. The chip internal output capacitance which is large compared to an IGBT is sufficient to achieve soft-switching. The main losses now come from the resonant inductor which is also varied in Fig. 15. The inductor with $L_r = 0.85\,\mu H$ is realized as toroidal air coil, avoiding core losses, and resulting in a lower effective switching energy. The higher inductances are realized with toroidal cores of a high frequency, low permeability ($\mu_r = 10$) iron powder material resulting in slightly higher inductor losses.

# 7 Conclusion

The proposed methods allow the loss measurement of an ARCP inverter without the need for a calorimetric measurement setup. The DC loss measurement is suited very well to obtain the total losses of different operational points with little effort and covers all inverter losses. To gain more information on the loss distribution this measurement can be accompanied by an oscilloscope measurement containing all switching actions in one fundamental period. Both measurements lead to similar results regarding the average switching energy, which helps to built confidence regarding the measurement accuracy.

The results further show that the ARCP is especially beneficial when used with SiC-MOSFETs where the switching losses are reduced remarkably, also when combined with a small resonant circuit. When using IGBTs, the turn-on losses are also greatly reduced, whereas the turn-off losses are reduced more or less - depending on the IGBT technology.

Therefore, the ARCP extension enhanced the Figure of Merit from SiC-MOSFET based hard-switching inverters by utilizing the output capacitance to eliminate the switching losses and shape the output voltage.

# 8 References

[1] R. W. De Doncker and J. P. Lyons, "The auxiliary resonant commutated pole converter," in *Conference Record of the 1990 IEEE Industry Applications Society Annual Meeting*, Seattle, WA, USA: IEEE, 1990, pp. 1228–1235. doi: 10.1109/IAS.1990.152341.

[2] W. McMurray, "Resonant snubbers with auxiliary switches," in *Conference Record of the IEEE Industry Applications Society Annual Meeting*, San Diego, CA, USA: IEEE, 1989, pp. 289–834. doi: 10.1109/IAS.1989.96608.

[3] R. Teichmann and H. Gueldner, "Analysis of transfer ratio limitations in auxiliary resonant commutated pole converters," in *7th IEEE International Power Electronics Congress. Technical Proceedings. CIEP 2000 (Cat. No.00TH8529)*, Acapulco, Mexico: IEEE, 2000, pp. 15–20. doi: 10.1109/CIEP.2000.891385.

[4] A. Charalambous, X. Yuan, and N. McNeill, "High-Frequency EMI Attenuation at Source With the Auxiliary Commutated Pole Inverter," *IEEE Trans. Power Electron.*, vol. 33, no. 7, pp. 5660–5676, Jul. 2018, doi: 10.1109/TPEL.2017.2743041.

[5] M. Zocher, N. Grass, and R. Kennel, "Auxiliary Resonant Commutated Pole Inverter (ARCPI) Operation Using online voltage measurements," in *2022 IEEE Applied Power Electronics Conference and Exposition (APEC)*, Houston, TX, USA: IEEE, Mar. 2022, pp. 1592–1597. doi: 10.1109/APEC43599.2022.9773560.

[6] M. Zocher, N. Grass, and R. Kennel, "Adaptive Control Methods for an Auxiliary Resonant Commutated Pole Inverter," presented at the PCIM Europe 2023; International Exhibition and Conference for Power Electronics, Intelligent Motion, Renewable Energy and Energy Management, Nuremberg, Germany: VDE VERLAG GMBH, May 2023. Accessed: Dec. 13, 2023. [Online]. Available: https://ieeexplore.ieee.org/document/10173339

[7] M. Helsper, F. W. Fuchs, and M. Munzer, "Analysis and comparison of planar- and Trench-IGBT-Modules under ZVS and ZCS switching conditions," in *2002 IEEE 33rd Annual IEEE Power Electronics Specialists Conference. Proceedings (Cat. No.02CH37289)*, Cairns, Qld., Australia: IEEE, 2002, pp. 614–619. doi: 10.1109/PSEC.2002.1022521.

[8] "Semikron SKM200GB125D." [Online]. Available: https://www.semikron-danfoss.com/dl/service-support/downloads/download/semikron-datasheet-skm200gb125d-22890620.pdf

[9] "Semikron SKM100GB12T4." [Online]. Available: https://www.semikron-danfoss.com/dl/service-support/downloads/download/semikron-datasheet-skm100gb12t4-22892020.pdf

PCIM Europe 2024, 11– 13 June 2024, Nuremberg    DOI: 10.30420/566262167

# A Novel Hybrid Two-Stage AC-DC Converter with Soft-Switched CCM PFC Stage for EVs Charging Applications

Lei Wang[1], Sinan Li [1]

[1] The University of Sydney, Australia

Corresponding author:    Sinan Li, sinan.li@sydney.edu.au
Speaker:                        Lei Wang, lwan6253@uni.sydney.edu.au

## Abstract

Totem-pole rectifiers are widely adopted in electric vehicles (EVs) on-board chargers. Conventional totem-pole rectifiers suffer from either high conduction losses in triangular current mode (TCM), or high switching losses in continuous conduction mode (CCM). To tackle this challenge, this paper presents a novel hybrid two-stage solution where the totem-pole rectifier can achieve CCM and soft switching simultaneously. Our new solution is achieved by smartly introducing current from the back-end LLC stage into the front-end totem-pole PFC stage, creating a Zero Voltage Switching (ZVS) condition for the PFC stage despite operating in the CCM. A scaled-down 100-W EVs charging prototype is developed to validate the proposed concept.

## 1    Introduction

A high efficiency and high power-density on-board charger (OBC) plays a critical role in the performance of electric vehicles (EVs) [1], [2]. The conventional OBC has a two-stage architecture, comprising a Power Factor Correction (PFC) rectifier followed by a DC/DC isolated converter [3]. With the advancement of wide band gap power devices, the OBC's efficiency and power density have seen notable improvements, leading to the adoption of the bridgeless totem-pole PFC combined with an LLC DC/DC converter as the mainstream OBC solution [4]. While the switching frequency of an LLC converter is typically in the range of several kilo- or mega-Hertz, that of the PFC stage typically is less than 100 kHz. The reason for PFC stage's low switching frequency is that it can't operate efficiently at a very high operating frequency: when operating in continuous conduction mode (CCM), the PFC stage has lower conduction losses due to lower inductor current ripple but higher switching losses due to hard switching; whereas in critical conduction mode (CRM) and triangular current mode (TCM), the PFC stage has lower switching losses due to the zero-voltage switching capability but higher conduction losses and magnetic losses due to higher inductor current ripple [5].

Recently, a semi-single-stage bridgeless PFC architecture has been proposed [6]. It has almost the same architecture as the conventional two-stage AC/DC converters, but it can simultaneously

achieve full ZVS and CCM for the front-end PFC stage which the two-stage counterpart cannot. The new architecture is based on the idea of "borrowing" current from the back-end DC/DC stage to establish the ZVS condition for the front-end PFC stage. Fig. 1 shows the semi-single-stage bridgeless Star PFC architecture from [6], which enables "current borrowing" from the back-end stage to the front-end stage when $Q_{aux}$ is turned off. This concept can also be extended to Fig. 2, named as semi-single-stage bridgeless Delta PFC architecture here, which similarly allows for "current borrowing" when $Q_{aux}$ is turned off. However, all the validation of the semi-single-stage bridgeless PFC architecture relies on an Asymmetrical Half Bridge (AHB) Flyback converter as the back-end DC/DC stage [6], [7], which is unsuitable for high-power applications, such as EVs OBC.

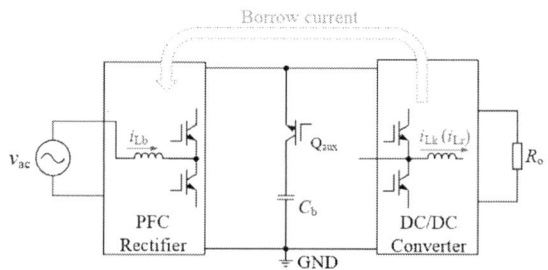

**Fig. 1**    Semi-single-stage bridgeless Star PFC architecture.

1242

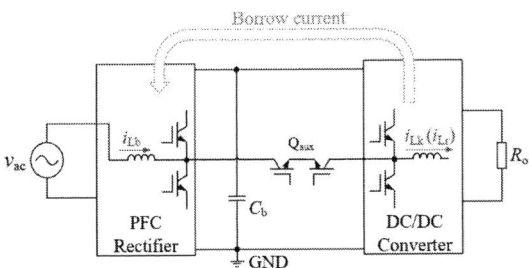

**Fig. 2** Semi-single-stage bridgeless Delta PFC architecture.

Motivated by the application of the semi-single-stage bridgeless PFC architecture in high-power scenarios, this paper presents an AC/DC converter based on the configuration of a totem-pole PFC rectifier and an LLC converter, alongside the concept of the semi-single-stage bridgeless Star PFC architecture, which uses one fewer switch than the Delta PFC architecture. In this paper, the proposed converter is named hybrid AC/DC converter. This proposed solution aims to minimize conduction, magnetic, and switching losses simultaneously, effectively addressing the challenges posed by traditional approaches in EVs OBC.

# 2 Characteristics of Proposed Hybrid AC/DC Converter

## 2.1 Challenge in the Conventional Two-stage AC/DC Converter

Fig. 3 illustrates the conventional two-stage AC/DC converter, consisting of a bridgeless totem-pole PFC rectifier, an LLC DC/DC converter, and a DC-bus capacitor $C_b$. $C_b$ is typically large enough to decouple the high-frequency dynamics of both stages, enabling nearly independent operation. In this setup, $Q_1$ and $Q_2$ constitute the high-frequency bridge leg, while $Q_3$ and $Q_4$ form the low-frequency bridge leg. Notably, $Q_5$ and $Q_6$ are turned on alternately, each with a 50% duty cycle.

Exploring the design challenges of the conventional two-stage AC/DC converter, Fig. 4(a) and Fig. 4(b) display the typical operational waveforms of the totem-pole PFC rectifier and the LLC converter, respectively. We assume a CCM of operation for the totem-pole PFC rectifier. This paper focuses on the totem-pole PFC in CCM, as it is widely adopted for medium to high power applications. Observations from Fig. 4(a) and Fig. 4(b) reveal that while the LLC converter can achieve full ZVS, the totem-pole PFC fails to realize full ZVS in either the positive or negative cycle of $v_{ac}$. Thus,

for the conventional two-stage AC/DC converter, despite the totem-pole PFC's low conduction and magnetic losses in CCM, its switching losses are significantly high.

## 2.2 Implementation and Advantages of the Proposed Hybrid AC/DC Converter

The proposed two-stage AC/DC converter topology is depicted in Fig. 5. This design integrates a bridgeless totem-pole PFC rectifier, an LLC DC/DC converter, and an auxiliary switch ($Q_{aux}$) positioned between the intermediate DC Bus and the DC-bus capacitor ($C_b$). Compared with the traditional two-stage AC/DC converter design, our approach only adds an auxiliary switch. This facilitates a partial hybridization of both the PFC and LLC stages by cutting the connection between DC Bus and $C_b$ though turning off $Q_{aux}$, i.e., "current borrowing" process, thereby enabling complete ZVS turn-on of the PFC stage, even in CCM.

The operation of the proposed converter is almost identical to that of the conventional two-stage converter, except the PFC and LLC operating at the same switching frequency, synchronized at $T_{sw\_0}$, as illustrated in Fig. 6, along with the integration of the auxiliary switch $Q_{aux}$'s functionality. When $Q_{aux}$ is turned off at $T_{sw\_0}$, it introduces a new operating state in which the LLC inductor current $i_{Lr}$ contributes to the commutation process of the PFC stage. This process, as "current borrowing", involves the front-end side borrowing current from the back-end side.

Fig. 7 illustrates this new state (at $T_{sw\_0}$ in Fig. 6) and states around it through the converter's four key operating stages:

(i) Stage I: PFC rectifier operates in the $(1-d_{PFC})T_{sw}$ phase, with $Q_1$ and $Q_4$ turned on, causing $i_{Lb}$ to decrease linearly; LLC operates in the phase with $Q_5$ turned on, leading to $i_{Lm}$ being increased linearly; $Q_{aux}$ is constantly turned on, and $V_{Bus}$ is equal to $V_{Cb}$.

(ii) Stage II: Due to the turn-off of $Q_{aux}$ at $T_{sw\_0}$, the differential current ($i_{Lr} - |i_{Lb}|$) of $i_{Lr}$ and $|i_{Lb}|$ at $T_{sw\_0}$ discharges/charges the parasitic capacitor $C_{oss}$ of $Q_2$, $Q_3$, $Q_6$, and $Q_{aux}$ (provided that $i_{Lr} - |i_{Lb}| > 0$). In addition, $i_{Lr} - |i_{Lb}| = i_1+i_2+i_3+i_4$, where $i_1$, $i_2$, and $i_4$ represent the discharging current of $C_{oss}$ of $Q_2$, $Q_3$, and $Q_6$, while $i_3$ represents the charging current of $C_{oss}$ of $Q_{aux}$.

(iii) Stage III: Upon completion of the charging and discharging process of

$C_{oss}$, the parallel diodes of $Q_2$, $Q_3$, and $Q_6$ become forward biased, resulting in the clamping of $V_{Bus}$ to ground, thereby creating ZVS conditions for all switches.

(iv)   Stage IV: $Q_1$ is turned off, and $Q_2$ and $Q_6$ are turned on under ZVS. Consequently, PFC rectifier shifts its phase from $(1-d_{PFC})T_{sw}$ to $d_{PFC}T_{sw}$, causing $i_{Lb}$ to increase linearly. Simultaneously, the LLC converter transitions to a phase characterized by a linear decrease in $i_{Lm}$.

Due to the partial hybridization of the PFC and LLC stages shown in Fig. 7, where the PFC and LLC function almost as a single stage without $C_b$, the proposed converter is named the hybrid AC/DC converter (an application of semi-single-stage bridgeless Star PFC architecture [6]). Furthermore, the proposed hybrid converter exhibits the same current waveforms as the conventional two-stage converter, evident from Fig. 4 and Fig. 6. This resemblance facilitates the simplification of the control scheme for the proposed converter, allowing for rapid implementation based on the conventional two-stage converter's framework.

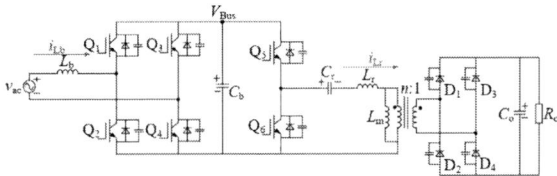

**Fig. 3**   The conventional two-stage AC/DC converter topology.

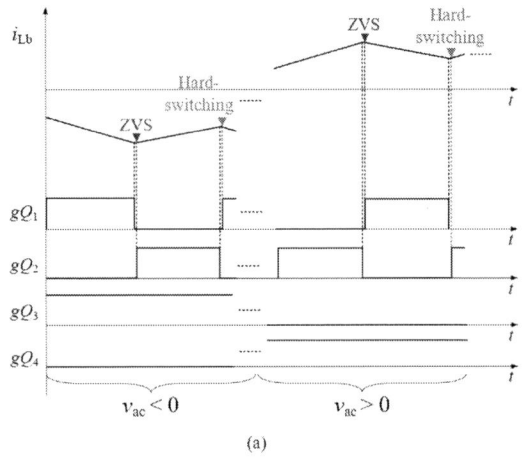

(a)

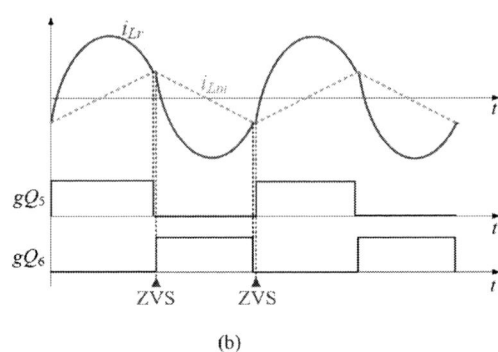

(b)

**Fig. 4**   Operating current waveforms of conventional two-stage AC/DC converter: (a) Totem-pole PFC rectifier (b) LLC DC/DC converter.

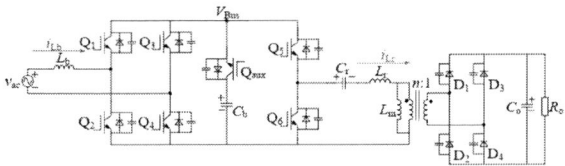

**Fig. 5**   The proposed hybrid two-stage AC/DC converter.

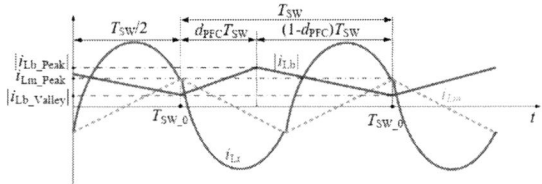

**Fig. 6**   Operating current waveforms of the proposed converter.

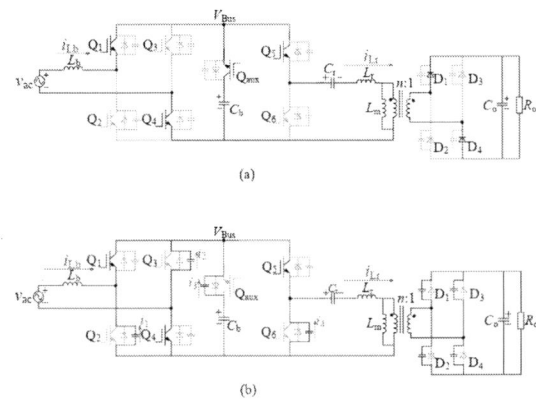

(a)

(b)

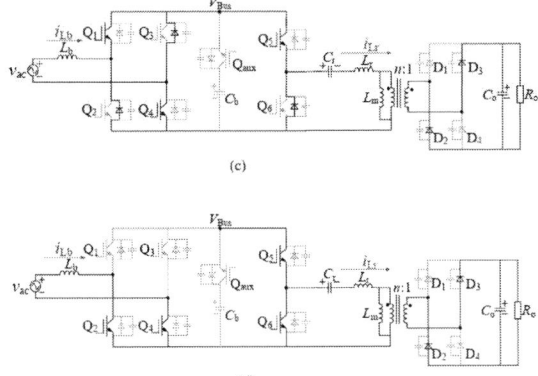

**Fig. 7** Key Operating stages at the positive line cycle: (a) Stage I (b) Stage II (c) Stage III (d) Stage IV.

## 3 Theoretical Analysis of ZVS Conditions

Considering that (i) LLC inherently achieves ZVS, and (ii) PFC rectifier's $Q_2$ ($Q_1$) cannot achieve ZVS turn-on at positive (negative) line cycle in the conventional AC/DC converter and (iii) the proposed converter introduces an additional switch $Q_{aux}$, only the ZVS conditions of $Q_1$(at negative line cycle), $Q_2$ (at positive line cycle), and $Q_{aux}$ need to be discussed.

### 3.1 ZVS Conditions for PFC Rectifier

Achieving ZVS for $Q_2$ during the positive line cycle and $Q_1$ during the negative line cycle is critical to the efficiency of the PFC rectifier, as it enables full ZVS turn-on. Corresponding ZVS conditions are determined at the switching instant $T_{sw\_0}$, as illustrated in Fig. 6 and Fig. 7, where the difference between $i_{Lr}$ (or $i_{Lm}$) of the LLC converter and $i_{Lb}$ of the PFC rectifier determines the ZVS capability. Apparently, a greater difference allows for a larger current to be "borrowed" from the back-end side to the front-end side to achieve ZVS.

#### 3.1.1 ZVS Conditions in the View of Energy

As for the LLC converter, it's observed that $i_{Lr}$ matches $i_{Lm}$ at $T_{sw\_0}$ if the switching frequency aligns with or falls below the resonant frequency of $L_r$ and $C_r$, as illustrated in Fig. 6 and Fig. 8(a) [8]. The energy stored in the inductor and transformer of LLC can be expressed as:

$$E_{LLC}\big|_{t=T_{sw\_0}} = \frac{1}{2}(L_m + L_r)i_{Lr}^2\big|_{t=T_{sw\_0}} = \frac{1}{2}(L_m + L_r)i_{Lm}^2\big|_{t=T_{sw\_0}}. \quad (1)$$

At switching frequency above resonant frequency (see Fig. 8(b)) [8], despite a divergence between $i_{Lr}$ and $i_{Lm}$, the inductor and transformer energy storage can be approximately formulated by (2), which is identical to (1). This approximation in (2) is justified by the significant magnitude of $L_m$ to $L_r$, and the comparable magnitude of $i_{Lr}$ to $i_{Lm}$.

$$E_{LLC}\big|_{t=T_{sw\_0}} = \left(\frac{1}{2}L_m i_{Lm}^2 + \frac{1}{2}L_r i_{Lr}^2\right)\bigg|_{t=T_{sw\_0}} = \frac{1}{2}(L_m + L_r)i_{Lm}^2 + \frac{1}{2}L_r(i_{Lr}^2 - i_{Lm}^2)$$

$$\approx \left\{\frac{1}{2}(L_m + L_r)i_{Lm}^2\right\}\bigg|_{t=T_{sw\_0}} \quad (2)$$

The stored energy can be utilized to discharge/charge the parasitic capacitors parallel to the switches (e.g., as shown in Fig. 7(b), the parasitic capacitors parallel to $Q_2$, $Q_3$, $Q_6$ and $Q_{aux}$), thus establishing ZVS conditions, only when the current $i_{Lm}$ exceeds $i_{Lb}$. Consequently, the energy used to accomplish ZVS can be described in (3). On the other hand, the energy requisite for achieving ZVS conditions is contingent upon the energy needed to fully charge and discharge the parasitic capacitors, as calculated in (4), derived from the energy stored in four parasitic capacitors illustrated in Fig. 7(b). To achieve ZVS, the magnitude of energy represented by (3) must exceed that of (4).

$$E_{ZVS}\big|_{t=T_{sw\_0}} = \frac{1}{2}(L_m + L_r)(i_{Lm}^2 - i_{Lb}^2)\big|_{t=T_{sw\_0}}. \quad (3)$$

$$E_{req} = 4 \times \frac{1}{2}C_{oss}(V_{Cb})^2. \quad (4)$$

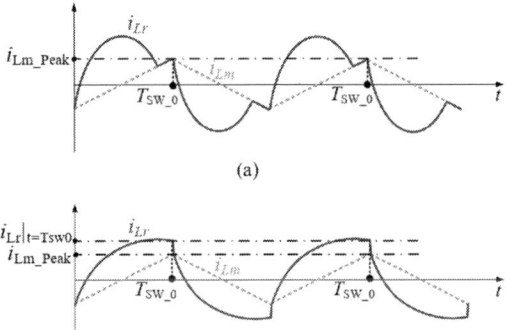

**Fig. 8** Current waveforms of the LLC converter: (a) switching frequency below resonant frequency (b) switching frequency above resonant frequency.

#### 3.1.2 Simplified ZVS Conditions in the View of Current

To further simplify the ZVS analysis, the ZVS criterion can be simplified that the value of (3) is greater than zero, under the assumption that $C_{oss}$ is negligible, leading to:

$$i_d = (i_{Lm} - |i_{Lb}|)|_{t=T_{sw\_0}} = i_{Lm\_Peak} - |i_{Lb\_Valley}| > 0, \quad (5)$$

where $i_d$ represents the difference between $i_{Lm}$ and $|i_{Lb}|$ at $T_{sw0}$, and $i_{Lm\_Peak}$ and $\max(|i_{Lb\_Valley}|)$ can be further derived as shown in (6) and (7). Here, $n$, $V_o$, $\Delta i_{ac}$, $i_{ac(Peak)}$, $v_{ac(Peak)}$ represent the turns ratio of the transformer, the output voltage, the ripple factor (in percentage) of $i_{ac}$, the instantaneous peak value of $i_{ac}$, the instantaneous peak value of $v_{ac}$.

$$i_{Lm\_Peak} = \frac{n V_o T_{sw}}{8 L_m}. \quad (6)$$

$$\max(|i_{Lb\_Valley}|) = (1 - \frac{1}{2}\Delta i_{ac})i_{ac(Peak)} = \frac{(1 - \frac{1}{2}\Delta i_{ac})V_o I_o}{\frac{1}{2}v_{ac(Peak)}}. \quad (7)$$

In (7), the absolute value of $i_{Lb\_Valley}$ reaches its maximum at the peak of $v_{ac}$, indicating the most difficult point in a line cycle to achieve ZVS. By combining (5), (6) and (7), the minimum value of $i_d$ is obtained:

$$i_{d\_min} = \min\left\{(i_{Lm} - |i_{Lb}|)|_{t=T_{sw\_0}}\right\} = \frac{n V_o T_{sw}}{8 L_m} - \frac{(1 - \frac{1}{2}\Delta i_{ac})V_o I_o}{\frac{1}{2}v_{ac(Peak)}}. \quad (8)$$

As long as $i_{d\_min}$ exceeds 0 A, the PFC stage in the proposed converter is approximately considered capable of achieving full ZVS turn-on. The magnitude of $\Delta i_{ac}$ directly influences $i_{d\_min}$: a larger $\Delta i_{ac}$ results in a smaller $i_{d\_min}$. Given that the value of $\Delta i_{ac}$ is contingent upon the inductance of $L_b$, selecting an appropriate $L_b$ value is crucial to ensure the ZVS operation of the PFC rectifier.

### 3.2 ZVS Conditions for $Q_{aux}$

$Q_{aux}$ can achieve ZVS turn-on through one of two mechanisms: (i) utilizing $|i_{Lb\_Peak}|$, if $Q_2$ (at positive line cycle) or $Q_1$ (at negative line cycle) in PFC stage turns off before $Q_6$ in LLC stage, or (ii) utilizing $-i_{Lm\_Peak}$, if $Q_6$ in LLC stage turns off before $Q_2$ (at positive line cycle) or $Q_1$ (at negative line cycle) in PFC stage.

Fig. 9 demonstrates examples for both scenarios, i.e., the utilization of $|i_{Lb\_Peak}|$ or $-i_{Lm\_Peak}$, during a positive line cycle: (i) In Fig. 9(a), when $Q_2$ is turned off, positive $i_{Lb}$ is directed to the diode of $Q_1$ and $Q_{aux}$, establishing ZVS conditions for both $Q_1$ and $Q_{aux}$; (ii) In Fig. 9(b), when $Q_6$ is turned off, the negative $i_{Lr}$ ($i_{Lm}$) is directed to the diode of $Q_{aux}$, facilitating ZVS conditions for $Q_{aux}$. In both scenarios, the absolute values of $i_{Lb}$ and $i_{Lm}$ reach their peak when the corresponding switch is turned off.

Given that the magnitudes of the absolute value of $i_{Lb\_Peak}$ and $i_{Lm\_Peak}$ are sufficiently large, the ZVS turn-on for $Q_{aux}$ is inherently guaranteed.

In summary, all switches in the proposed converter can realize fully ZVS turn-on.

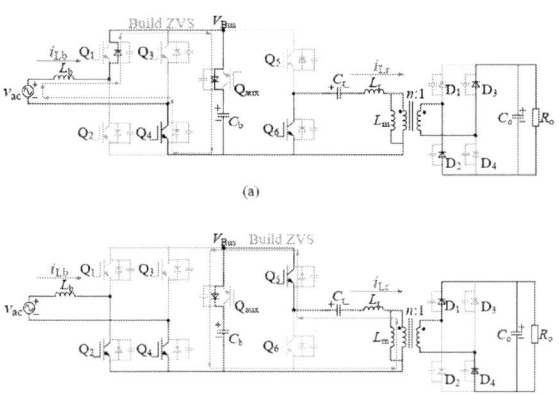

**Fig. 9** ZVS process for $Q_{aux}$ at a positive line cycle: (a) Utilization of $|i_{Lb\_Peak}|$ ($Q_2$ turns off before $Q_6$) (b) Utilization of $-i_{Lm\_Peak}$ ($Q_6$ turns off before $Q_2$).

## 4 Experiment Verification

To evaluate the performance of the proposed hybrid converter, experimental tests were conducted on a scaled-down prototype with key parameters: (i) Input voltage $v_{ac(RMS)}$ = 110 V, $f_{ac}$ = 60 Hz (ii) output voltage $V_o$ = 100 V, (iii) rated output power $P_o$ = 100 W.

Fig. 10 and Fig. 11 display the current and voltage waveforms at the input and output under full and half load conditions, respectively. In each scenario, the current remains sinusoidal while the output voltage maintains stability, demonstrating the essential functions of AC input current shaping and output voltage regulation.

Fig. 12 presents the key switching logic, maintaining consistency with logic shown in Fig. 6 and Fig. 7. Furthermore, Fig. 13 and Fig. 14 show critical switching waveforms near the peak of $v_{ac}$, demonstrating the successful achievement of ZVS for both PFC stage and $Q_{aux}$, in agreement with the analysis conducted in Section 3. Specifically, (i) Fig. 13 shows that at the falling edge moment (i.e., $T_{sw0}$) of $v_{ds(Q2)}$, the difference current $i_{Lr} - i_{Lb} > 0$ and inductor current $i_{Lb} > 0$, verifying that $Q_2$'s ZVS condition can be fulfilled even when PFC stage is operating in CCM. (ii) Fig. 14 shows that $Q_{aux}$ achieves ZVS turn-on utilizing $i_{Lb\_Peak}$. Therefore, the full ZVS capability of the proposed converter is confirmed experimentally.

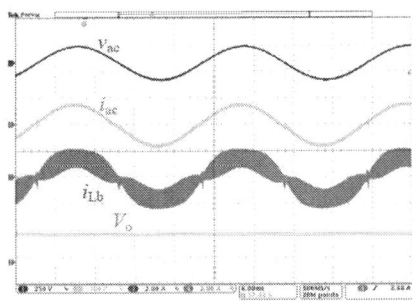

**Fig. 10** Operating waveforms at 110-V input and 100-W output.

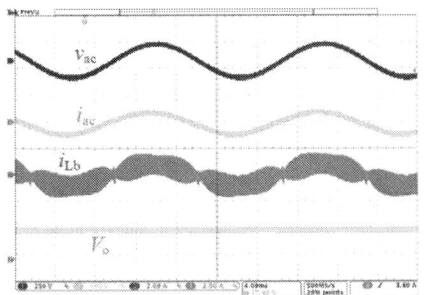

**Fig. 11** Operating waveforms at 110-V input and 50-W output.

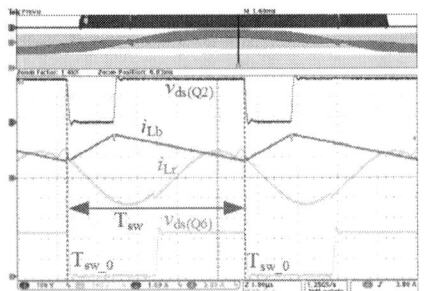

**Fig. 12** Switching waveforms at 110-V input and 100-W output.

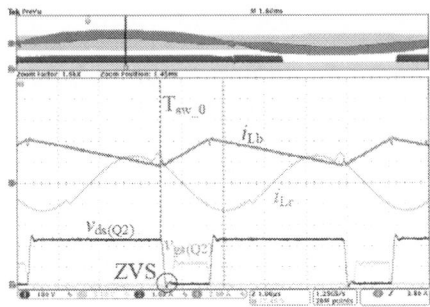

**Fig. 13** Switching waveforms of PFC stage at 110-V input and 100-W output.

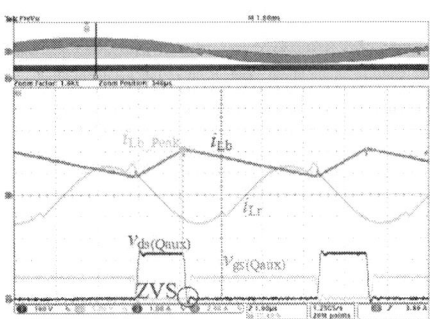

**Fig. 14** Switching waveforms of $Q_{aux}$ at 110-V input and 100-W output.

# 5 Conclusion

The proposed hybrid two-stage AC/DC converter for the on-board chargers (OBC) presents a significant advancement over conventional converters. By enabling simultaneous achievement of CCM and ZVS in the PFC stage, the converter effectively minimizes conduction loss, magnetic loss, and switching loss. The control strategy employed is also notably simple, further emphasizing its potential for practical applications. The successful demonstration of the proposed converter in a scaled-down 100-W prototype further validates its viability.

# References

[1] Khaligh and M. D'Antonio, "Global Trends in High-Power On- Board Chargers for Electric Vehicles," in IEEE Transactions on Vehicular Technology, vol. 68, no. 4, pp. 3306-3324, April 2019.

[2] M. R. Khalid, I. A. Khan, S. Hameed, M. S. J. Asghar and J. -S. Ro, "A Comprehensive Review on Structural Topologies, Power Levels, Energy Storage Systems, and Standards for Electric Vehicle Charging Stations and Their Impacts on Grid," in IEEE Access, vol. 9, pp. 128069-128094, 2021.

[3] R. Hou and A. Emadi, "A Primary Full-Integrated Active Filter Auxiliary Power Module in Electrified Vehicles with Single-Phase Onboard Chargers," IEEE Transactions on Power Electronics, vol. 32, no. 11, pp. 8393-8405, 2017.

[4] W. Liu, A. Yurek, B. Sheng, Y. Chen, Y. -F. Liu and P. C. Sen, "A Single Stage 1.65kW AC-DC LLC Converter with Power Factor Correction (PFC) for On-Board Charger (OBC) Application," 2020 IEEE Energy Conversion

Congress and Exposition (ECCE), Detroit, MI, USA, 2020, pp. 4594-4601.

[5] Marxgut, F. Krismer, D. Bortis, and J. W. Kolar, "Ultraflat Interleaved Triangular Current Mode (TCM) Single-Phase PFC Rectifier," IEEE Trans. Power Electron., vol. 29, no. 2, pp. 873–882, 2014.

[6] L. Wang and S. Li, "Bridgeless Star Power Factor Correction Architecture," in *IEEE Transactions on Power Electronics*, doi: 10.1109/TPEL.2024.3362409.

[7] L. Wang, H. Li and S. Li, "A Soft-Switched CCM Bridgeless PFC Rectifier for 240-W USB-PD EPR Wall Adapter," 2023 IEEE Energy Conversion Congress and Exposition (ECCE), Nashville, TN, USA, 2023, pp. 2263-2268.

[8] Texas Instruments, " Designing an LLC Resonant Half-Bridge Power Converter," in Texas Instruments Literature Number SLUP263, 2010.

PCIM Europe 2024, 11– 13 June 2024, Nuremberg    DOI: 10.30420/566262168

# A Method for Tuning Leakage Inductance in Transformers

Rosemary O'Keeffe, Michael Dunleavy, Cathal Sheehan

Bourns Inc., Ireland

Corresponding author: Rosemary O'Keeffe, rosemaryokeeffe@bourns.com

Speaker: Rosemary O'Keeffe, rosemaryokeeffe@bourns.com

## Abstract

Being able to control the value of the Leakage Inductance in a transformer in production is becoming critical in automotive and industrial applications. This is especially true in designs for a dual active bridge topology. The coupling factor K can be calculated using FEA which takes into account the dimensions in three axes of the winding distribution and the distance between the coupled coils. However being able to guarantee the values of these variables in a manufacturing environment is not at all practical. In this paper, a novel method for tuning the leakage inductance to the target value with a tolerance of less than 10% during assembly is presented. Ansys Maxwell simulations, as well as sample design and testing are discussed here and the results show how it is possible to manually tune leakage using a two bobbin configuration.

## 1 Introduction

For many applications the leakage inductance is a critical factor in the operation of the transformer. This is particularly true in dual active bridge circuits where the leakage inductance is the main source responsible for energy transfer. This is due to the power transfer inductance of the transformer being a sum of the leakage inductance and additional primary and secondary inductances [1]. It is becoming a critical factor for many applications in power and it is often the case that the tolerances for leakage can be less than 10%. For manufacturers, this poses a real challenge since so many factors can affect the leakage. Leakage in a transformer is due to the imperfect coupling between the primary and secondary coils. The coupling factor k is given by [2]:

$$k = \frac{M}{\sqrt{L_p L_s}} \qquad (1)$$

where M is the mutual inductance and $L_p$ and Ls are the self-inductances in the primary and the secondary respectively.

Differences in each of these factors from one sample to another would only cause minor differences in the leakage and when the range is not small then these differences can be ignored. However, it has been seen that this range is becoming more and more critical in the design. A method for tuning the leakage in the device after winding is discussed. This method allows the leakage to be tuned in each device so that it falls within a narrow range and allows for each transformer to operate correctly. The design, implementation, and test of a transformer with tuned leakage inductance is presented here. This solution includes a winding configuration to initial set up the leakage within the tuneable range. In this case the primary and secondary windings are wound side by side and wound from the inner bobbin onto the outer bobbin. Then cable ties are used to move the coils in relation to each other and tune the leakage. The transformer is intended to be used in a DC-to-DC dual active bridge circuit and requires high power, 13kW, and current, 250A, operation. For this design, the leakage current needed to be designed around 1.35µH with only a +10% and -5% tolerance.

This paper is divided as follows, Initial design, Ansys Maxwell Simulations, Comparison of Modelling and Test Results and Conclusions. In the first section various schemes and configurations are considered and modelled using Ansys. The results are then considered and the solution for tuning is discussed. A transformer sample using the novel method is built and the results of tuning are provided. The differences between the model and test results are then discussed. Finally possible solutions for scaling up to a large volume manufacturing environment are provided. Figure 2 shows the design using Ansys, as well as the sample built to confirm the results. A blow up of the bobbin design is presented in Fig. 1. Figure 3 shows how the movement of the coils was simulated in Ansys by moving the interior coils by 1mm and testing the leakage, inductance and coupling coefficient.

## 2 Initial Design

The transformer discussed here was designed for use in a 13kW DCDC dual active bridge (DAB) fig. 1. A DAB provides galvanic isolation between the high and low sides of the converter as well as bidirectional power flow [3]. This type of circuit is particularly sensitive to leakage. The high power and current requirements meant that a large PQ101 core and multistranded litz were required. For this design, the leakage was required to be 1.35 with tolerances of +10 and -5%. Typically, for a transformer tolerances of up to 20% could be expected in the leakage inductance. For this reason, a design where leakage could be tuned during manufacture had to be considered.

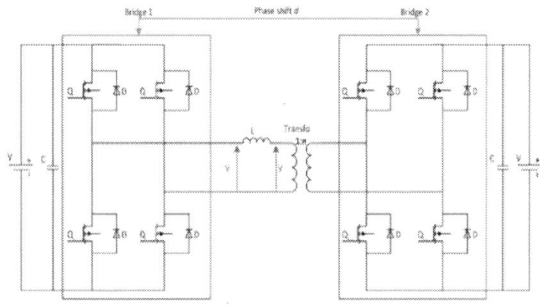

**Fig. 1:** DCDC Dual Active Bridge

For the initial design, a number of winding configurations were considered to reach the leakage requirements. This was difficult since the leakage was required to be high, but the winding window size restricted the distance which could be achieved from typical winding. To get within the initial value a side-by-side winding was chosen. This type of winding increases the distance between the winding and decreases the coupling. With this type of winding on a single bobbin with no overlap of the windings a leakage of approx. 1.7uH was achieved. The initial results were from Ansys simulations which will be discussed in the next section. The first design was a single bobbin with side-by-side windings and was simulated as in fig. 2.

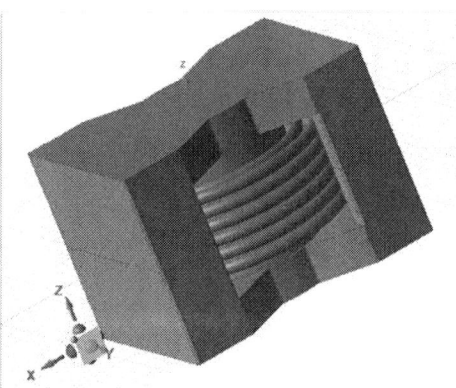

**Fig. 2:** Initial Design from Ansys

## 3 Ansys Maxwell Simulations

Ansys Maxwell was used to simulate the initial transformer design. Both 2D and 3D models were created, and Eddy Current and Transient Solvers were used to analyze the design. In this paper the Eddy current analysis will be focused on. The transient analysis was used to determine the losses of the core and the windings and are outside the scope of this paper. For the 2D analysis cylindrical about z was used to reduce the geometry to 2D since this provides a good approximation when using a PQ core. The model was set up first with a typical winding of primary on the inside (closest to the centre leg of the core) and the secondary wrapped on top of the primary. The leakage was calculated in Ansys using the formula in [2]

$$L_l{}^*(1-K^2) \qquad (2)$$

where $L_l$ is the magnetizing inductance and K is the coupling coefficient. This formula is recommended by Ansys for calculating the leakage inductance of a transformer.

The secondary was then moved so that the coils of the secondary and primary did not overlap completely, and this was moved in increments of 1 coil width until the overlap was minimised and the winding window was completely full. From this, it was discovered that the leakage could not be achieved with such a configuration. Next a side-by-side configuration was considered. In this case, the primary windings were situated as shown in fig. 3. Here the primary windings are identified in purple. This winding significantly reduced the leakage and was the initial design which was created in the laboratory. From the physical device it was discovered that the leakage was still not within the limits and a design which allowed for some overlap between primary and secondary would be necessary.

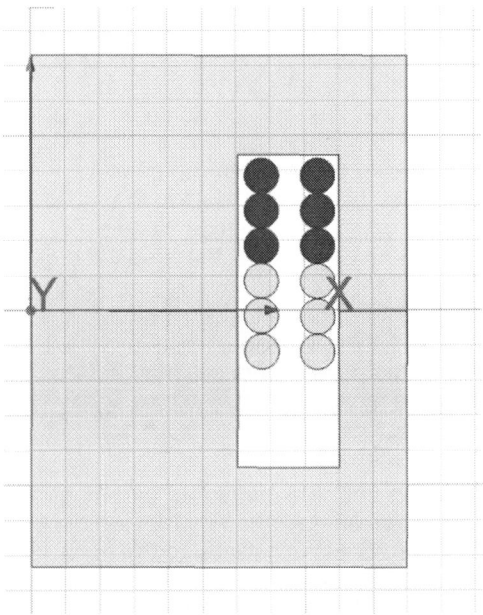

**Fig. 3:** Initial Design with primary indicated by purple

The side-by-side winding with the coils moved so that some overlap was achieved was simulated in Ansys and it was found that this could created the leakage inductance required. However, it was found that in a real device the leakage was subject to big changes due to the slight differences in winding tightness, spacing

of windings and other factors that cannot be fully controlled when winding. To account for this, a number of designs were considered. Eventually a design was modelled in Ansys where the windings were moved in relation to each other and the effect on the leakage was observed.

In a model this type of design is very simple but to create a working device which could be tuned in such a way in a factory setting was a more difficult prospect. A number of designs were considered. Taking into account the small window size and the large size of the winding space became an issue. Initially a design where the only moving part was the coil was considered. However, this was an issue since the winding was not only moved but also deformed by the mechanism. Creating a consistent design in the factory was of a major

**Fig. 4:** Design with coils moving

concern so this type of design was non-ideal. This design with moving coils is shown in fig. 4.

The next solution was looking at how to move all the coils as a single unit without compressing the coils. To do this, a two-bobbin design (fig. 5). This was a unique solution but required a carefully designed bobbin configuration which would hold the coils exactly in place. The winding was also more complicated due to the side-by-side winding design. The primary winding is wound onto half the inner bobbin and then the first few windings on the secondary is wound on the rest of the bobbin. The outer bobbin is then placed on top, and the windings are continued onto the outer bobbin. The cores are then placed on the bobbin and the transformer is ready to be tested.

The initial leakage is measured using Wayne Kerr 6500B and then the outer bobbin is moved to increase or decrease the leakage as needed. In this way the leakage can be tuned to within very tight limits for use in a DAB transformer and stiff tolerances can be respected. This system could also be used to tune the capacitance which would decrease as the coils are moved further from each other. This is a typical trade off between capacitance and leakage. For this design, the major concern was the leakage but a system like this could also be used to get a compromise between leakage and capacitance.

Currently the design is fully manually controlled and so the movement of the coils are not as precise and the tuning requires human intervention which increases the cost and the time of the production of the transformer. However, since the coil only needs to be moved in one direction, it is possible that the process could be automated and the movement halted when the desired leakage is achieved. This would require a machine to press or pull the outer coil with the leakage being measured. Some code could be written to stop the movement of the coil when the desired leakage is achieved and the bobbins could then be held in place.

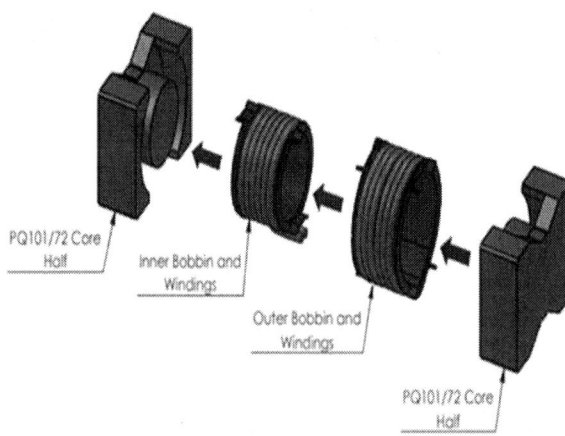

**Fig. 3:** Expanded 2 bobbin design

# 4 Comparison of Modelling and Test Results

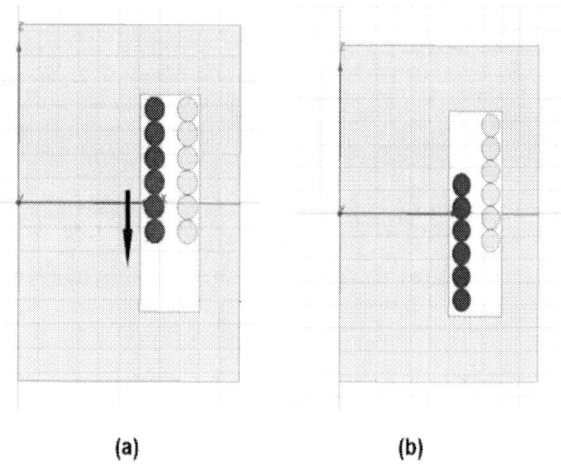

(a)                              (b)

**Fig. 4:** Initial (a) and Final (b) Coil Positions in Ansys Simulations. Arrow Indicates Direction from Position 1 to Position 15

From the initial modelling results it was determined that the leakage could be tuned by moving the coils in the manner described in this paper. However, to consider the effect of differences in the physical design the transformer was tested with as close a comparison as possible. In the model the coils are moved from position 1 to 15 in precise increments of 1mm to show how minor changes affect the leakage. In the device under test (DUT) the movements are matched to this, however the movements are not as tightly controlled as was possible in the model. The results are shown in table 1.

As is seen in both the measured and modelled results the leakage is tuned by moving the coils on the outer bobbin in relation to the inner bobbin. The model shows a difference of 0.16μH between the starting and finishing positions. In the DUT the drop is more significant 0.28μH. This is useful for tuning the device. The larger difference could be due to slight differences in the initial and finishing positions due to spaces in the model between the windings. In the DUT the windings are in direct contact with each other and the distance from starting the finishing position would be increased due to this.

| Position | Coupling Coefficient | Leakage Inductance (µH) | |
|---|---|---|---|
| | Simulated | Simulated | Measured |
| 1 | 0.9494 | 1.77 | 1.6 |
| 2 | 0.9498 | 1.76 | 1.58 |
| 3 | 0.9528 | 1.75 | 1.56 |
| 4 | 0.953 | 1.74 | 1.54 |
| 5 | 0.9532 | 1.73 | 1.53 |
| 6 | 0.954 | 1.71 | 1.52 |
| 7 | 0.9542 | 1.7 | 1.51 |
| 8 | 0.9551 | 1.69 | 1.48 |
| 9 | 0.9608 | 1.68 | 1.46 |
| 10 | 0.9628 | 1.66 | 1.43 |
| 11 | 0.9649 | 1.65 | 1.41 |
| 12 | 0.9669 | 1.64 | 1.39 |
| 13 | 0.9688 | 1.63 | 1.36 |
| 14 | 0.97 | 1.62 | 1.35 |
| 15 | 0.9715 | 1.61 | 1.32 |

**Table 1:** Measured and Simulated Leakage Inductance as the Coils are Moved

The initial condition simulated in the model has a higher leakage inductance than that of the DUT. The magnetizing inductance of the model is 170µH and the DUT magnetizing inductance, as measured, is 225µH. This accounts for some of the differences in the leakage between the model and the DUT. The leakage is a function of the magnetizing inductance as shown in formula (2).

The results show how the coupling coefficient is effected by the movement of the coils and this also affects the capacitance of the devices. As the leakage is decreased, the coupling coefficient is increased, and the capacitance is increased. Essentially what is happening is the two coils can be considered as two plates of a capacitor so the more they overlap the better the coupling and the higher the capacitance. Capacitance is often also considered a crucial factor in the design of a transformer, however, for this paper the focus was on the leakage inductance and so the capacitance was not discussed in detail here.

In this paper a manual method for tuning the leakage of a transformer was discussed. However, it has been considered that his design could be adapted to an automated process in a factory with some minor changes. As a manual process this could only be considered for small volumes. For larger volumes a device to move the coils while the leakage is tested is envisioned and this could be added to an industrial setting. In many cases, larger leakage tolerances can be accepted, and this system would not be required.

However, it has been observed that as transformer and circuit design continue to advance reducing tolerance limits is being more important. For example, if the leakage inductance is not within acceptable limits circuits can see ringing and capacitors need to be used to reduce this. This can cause the circuit to become large and complex and costly. In the future, it is possible that methods such as that discussed in this paper will become more usual to allow for high efficiency circuits.

# 5 References

[1] S. D. R. W. P. M. M. S. S. Yin, "Impact of the Transformer Magnetizing Inductance on the Performance of the Dual-Active Bridge Coverter," in *2021 IEEE 22nd Workshop on Contol and Modelling of Power Electronics (COMPEL)*, Cartagena, Columbia, 2021.

[2] R. R. X. Y. N. Y. Y. Y. Fengjuan Wang, "A transformer with high coupling coefficient and small area based on TSV," *Integration,* vol. 81, pp. 211-220, 2021.

[3] Imperix, "Dual Active Bridge Control," 2024. [Online]. Available: https://imperix.com/doc/implementation/dual-active-bridge-control.

PCIM Europe 2024, 11– 13 June 2024, Nuremberg        DOI: 10.30420/566262169

# Low Cost High Density 300W/20V AC/DC Converter Enabled by GaN Power ICs

Tom Ribarich[1], Xiucheng Huang[2]

[1]Navitas Semiconductor, USA
[2]Navitas Semiconductor, China

Corresponding Author:  Tom Ribarich, tom.ribarich@navitassemi.com

## Abstract

A low cost 300W high-density AC-to-DC converter has been designed and demonstrated to achieve >96% peak efficiency and 270cc cased size. The circuit topologies include a 2-ph interleaved PFC input stage, an LLC dc-dc stage, and a synchronous rectification output stage. The design includes GaN Power ICs and off-the-shelf controllers running at 250kHz. This new design has resulted in a cased power density of 1.1 W/cc. The waveforms demonstrate zero-voltage switching in all stages and the performance results show very high efficiency and acceptable component temperatures.

## 1   Introduction

Designers are continuously looking for converter topologies that have higher efficiency, smaller size, and lower cost. PFC front-end solutions today include 1-ph boost, 2-ph interleaved boost, and bridgeless totem-pole. 1-ph CrCM boost has high peak currents, large conduction losses, high differential EMI noise, and a large dc bus output ripple. Totem-pole has emerged as a possible solution for higher efficiency, but is complex, has high EMI noise, and is expensive. 2-ph interleaved achieves lower peak currents, less EMI filtering, high efficiency, smaller DC bus capacitance, and low cost. The circuit topology selected for this 300W GaN-based design includes (Figure 1) a 2-ph interleaved PFC front end, a half-bridge LLC dc-dc stage, and an SR output stage. A CrCM 2-ph interleaved PFC front end was selected due to low cost and high efficiency. The LLC step down converter was selected due to high-efficiency, zero-voltage switching, and very suitable for operating at high frequencies. The SR output stage is necessary to rectify the high output currents (15A) and to minimize conduction losses and thermals.

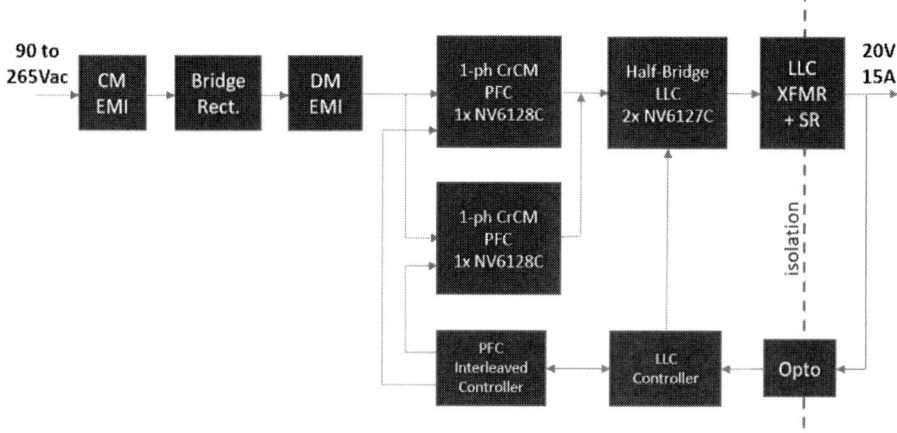

**Fig. 1**, GaN-based 300W AC-DC power supply block diagram.

1254

## 2   2-ph Interleaved PFC Stage

Several solutions can be used for the input PFC stage including single phase and dual phase interleaved boost topologies. The dual phase interleaved topology consists of two separate boost converters that are switched 180deg out of phase with each other (Figure 2). Interleaved control generates lower input and output ripple currents versus single phase solutions. This results in less EMI filtering, lower EMI filter losses, and lower output capacitance. Additional benefits include high efficiency, better heat spreading, low cost, and can be realized using off-the-shelf components.

The 2-ph interleaved PFC stage for this design consists of (Figure 3) two boost inductors, two GaN Power ICs, two boost diodes, and a 2-ph interleaved PFC controller. The NV6128C GaN Power IC integrates a low capacitance and low RDSon GaN power FET together with the gate drive circuit into a 6x8 QFN package. Additional functions such as turn-on dV/dt control and a linear regulator are also included in the GaN IC. The key advantages of the 2-ph interleaved PFC topology driven by GaN Power ICs are high frequency operation enabling small boost inductor size, low EMI noise, small DC bus capacitance, and high system efficiency.

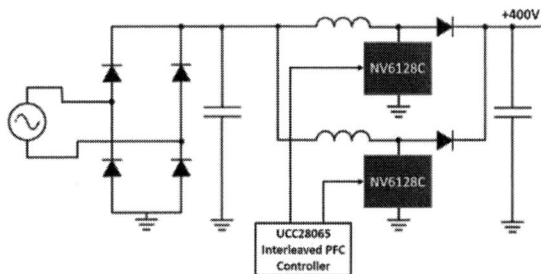

**Fig. 3**, 2-ph interleaved PFC stage using 2x NV6128C GaN Power ICs.

## 3   LLC DC-DC Stage

The LLC topology was selected for the dc-dc stage due to high frequency, high efficiency, and ZVS operating mode. The NCP13992AB controller was selected due to off-the-shelf availability and 500 kHz maximum frequency capability (Figure 4). The NV6127C GaNFast Power ICs have been selected for the LLC half-bridge due to low capacitance and low Qg for high-frequency operation, fast start-up during soft-start and burst modes, high dV/dt immunity, and integrated gate driver and regulator for robust and accurate gate voltage control.

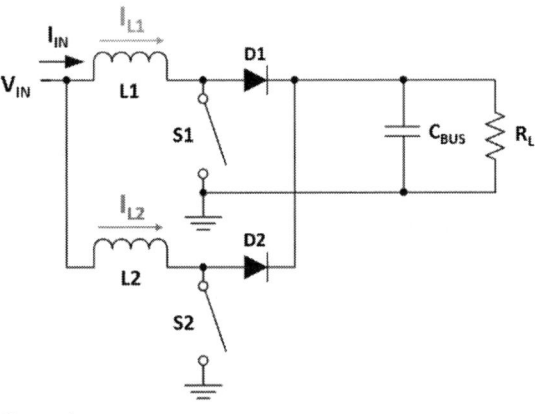

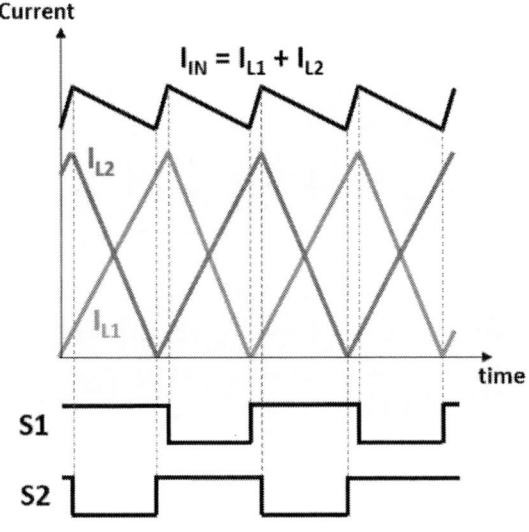

**Fig. 2**, 2-ph interleaved PFC stage simplified schematic (upper) and switching waveforms (lower).

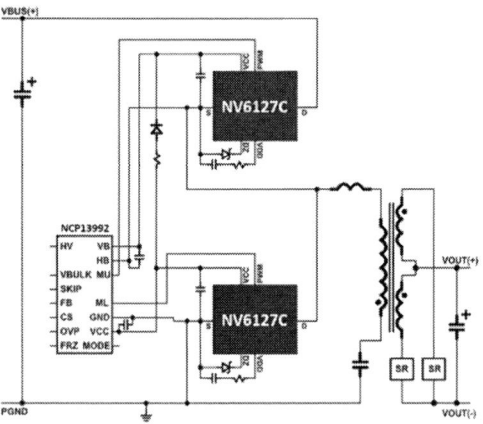

**Fig. 4**, LLC dc-dc stage basic circuit schematic using NV6127C GaNFast Power ICs.

## 4  AC-DC Converter Prototype

The design approach implemented for the AC-DC converter prototype (Figure 5) includes a single mother board, bobbin-based PFC inductors, and an LLC planar transformer. The LLC planar transformer includes the SR controller and silicon FETs mounted together on the planar PCB to reduce insertion losses and to maintain a low board profile (23.5mm). The GaN Power ICs are mounted on the bottom side together with the controllers for easy PCB layout and excellent bottom side conductive cooling to the housing.

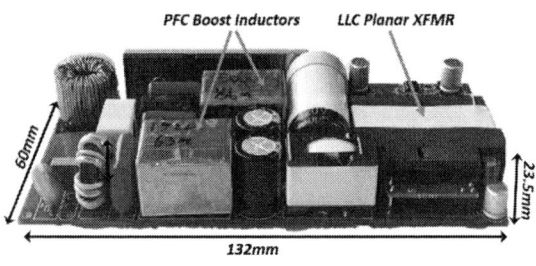

**Fig. 5**, 300W AC-DC converter PCBA (186cc).

## 5  Circuit Waveforms

The complete power supply was designed, built and tested for performance. The 2-ph interleaved PFC and LLC waveforms are shown in Figure 6. The PFC waveforms are at 115Vac and 230Vac input, at 100% load, and operating at a frequency from 125 to 150 kHz at the peak of the AC line. The LLC half-bridge voltage and current waveforms show ZVS operation at an operating frequency of 250 kHz. The overall efficiency curves show a peak efficiency > 96% at 230Vac input.

### Full Load@115Vac

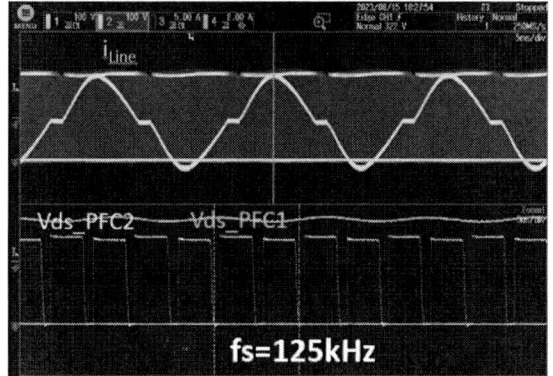

### Full Load@230Vac

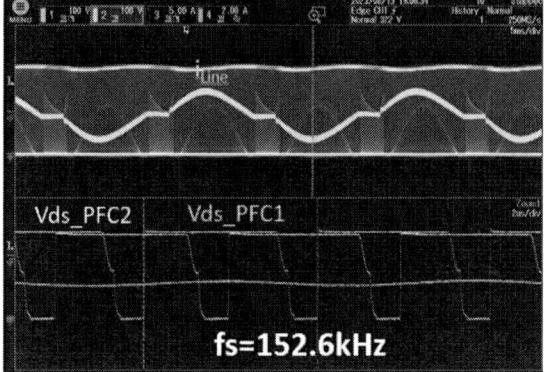

PCIM Europe 2024, 11– 13 June 2024, Nuremberg          DOI: 10.30420/566262169

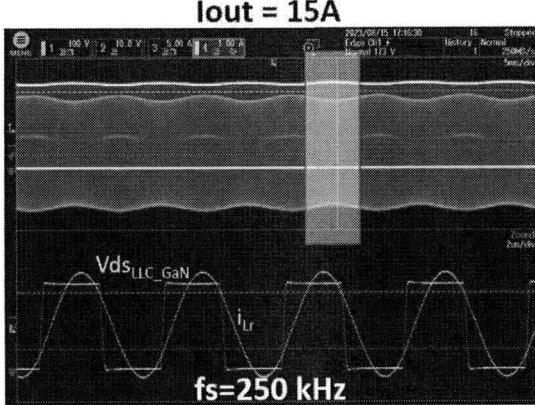

**Fig. 6**, PFC waveforms (upper & middle), and LLC waveforms (lower) during full load conditions.

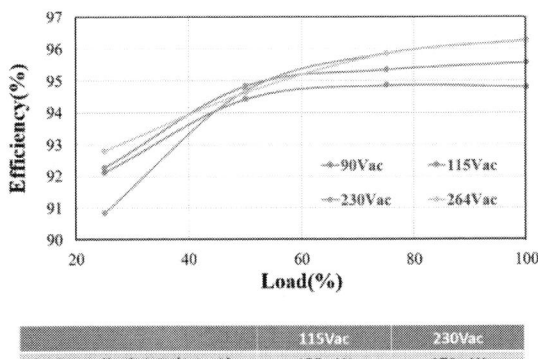

| | 115Vac | 230Vac |
|---|---|---|
| Pin @Standby (19.5V/1.6mA) | 155mW | 179mW |

**Fig. 7**, Max load efficiency (upper), efficiency vs. line/load (middle), and standby power (lower).

## 6 Efficiency & Thermals

The power supply was tested for overall performance including efficiency and thermals. The efficiency curves show peak efficiencies of approximately 94.8% at 90 Vac, 95.6% at 115 Vac and 96.3% at 230 Vac (Figure 7). Low standby power measurements are also shown (179mW at 230Vac/19.5V/1.6mA). Thermal measurements (Figure 8) were taken during 90 Vac and full load conditions, and show acceptable component temperatures necessary for manageable thermals during cased conditions. The hot spots on the top side (103.4C) and bottom side (106C) are due to heat flow from the input bridge rectifier and can be reduced with proper thermal management (TIM + copper shielding).

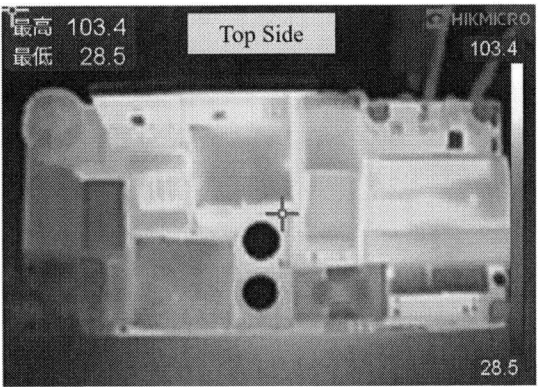

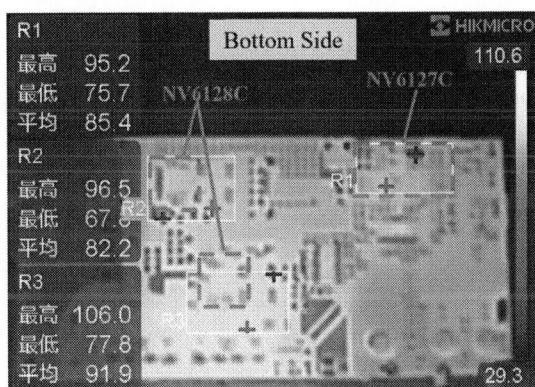

**Fig. 8**, Top side and bottom side thermal images during 90 Vac and full load conditions.

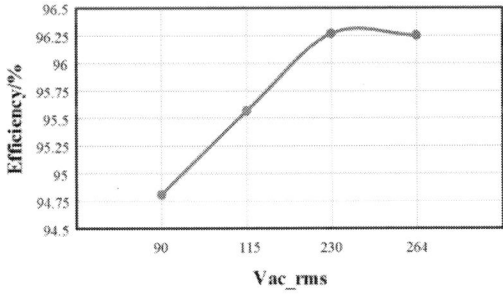

1257

## 7 EMI Measurements

Conducted EMI measurements were taken (Figure 9) for both 115Vac and 230Vac full-load conditions. AV and PK results are well below their limits with a minimum of 10dB of margin to accommodate for design tolerances.

### 115Vac

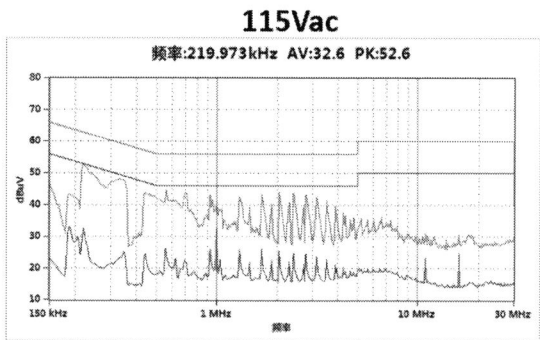

### 230Vac

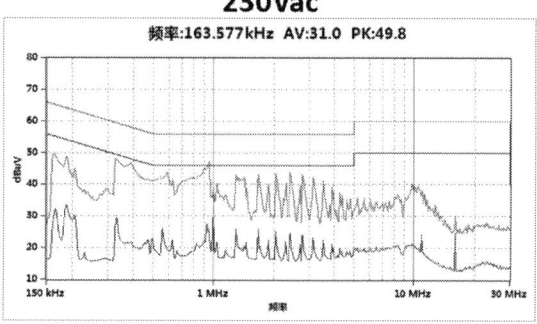

**Fig. 9**, Conducted EMI measurements at 115Vac (upper) and 230Vac (lower) at full load conditions.

## 8 Conclusions and Future Work

A 300W AC-DC operating at 250 kHz has been designed and demonstrated to show that a high density is achievable with readily available components and controllers. GaN Power ICs have enabled higher switching frequencies resulting in a dramatic reduction of the size of the magnetic components. The performance data shows high efficiency (>96% peak efficiency at 230Vac input) which is necessary to achieve a high power density with acceptable thermals. Future work includes thermal management inside cased housing, radiated EMI compliance, and voltage transient/surge testing.

## References

[1]. B. Yang, F. C. Lee, A. J. Zhang, and G. Huang. "LLC resonant converter for front end DC/DC conversion." in Proc. IEEE APEC, 2002, pp. 1108-1112.

[2]. B. Lu, W. Liu, Y. Liang, F. C. Lee, and J. D. Van Wyk. "Optimal design methodology for LLC resonant converter." In Proc. IEEE APEC, 2006, pp. 533-538.

[3]. D. Fu, B. Lu, and F. C. Lee. "1MHz high efficiency LLC resonant converters with synchronous rectifier." In Proc. IEEE PESC, 2007, pp. 2404-2410.

[4]. Y. C. Li, F. C. Lee, Q. Li, X. Huang and Z. Liu, "A novel AC-to-DC adaptor with ultra-high power density and efficiency," in IEEE APEC, 2016, pp. 1853-1860.

PCIM Europe 2024, 11– 13 June 2024, Nuremberg          DOI: 10.30420/566262170

**pcim**
EUROPE

# 25kW Grid-Tied Bi-directional T-Type Inverter with High-Efficiency and High-Power Density Using SiC MOSFETs

Tamanna Bhatia[1], Jun Zhang[1], Sumana Ghosh[1], Frank Wei[2]

[1] Wolfspeed, USA
[2] Wolfspeed, China

Corresponding author:     Tamanna Bhatia, Tamanna.Bhatia@wolfspeed.com
Speaker:                          Tamanna Bhatia, Tamanna.Bhatia@wolfspeed.com

## Abstract

In the past decade, solar installations have experienced substantial expansion, primarily driven by their myriad benefits, such as economical operation, scalability, flexible installation and more. At this growth rate, there is a demand for these systems to handle higher power, be more energy efficient, smaller in size and deliver better power quality. In this paper, we will discuss how using SiC MOSFETs help to reduce the size of the bi-directional DC-AC inverter stage by increasing the switching frequency while still maintaining peak efficiency of 99.2% and a power density of greater than 6kW/L.

## 1   Introduction

The solar energy system architecture can be divided into three main types of systems namely the central-type, string-type and micro systems. Central-type systems risk single point failure and have higher DC transmission losses, whereas the string-type system's modular structure provides redundancy to the system and they're easy to service. The string type system finds many applications in commercial builds and residential homes because of the wider power range, 10kW – 200kW. The string inverter is much easier to scale to meet wilder power requirements desirable for this segment [1]. Fig. 1 shows the energy storage architecture around a solar panel.

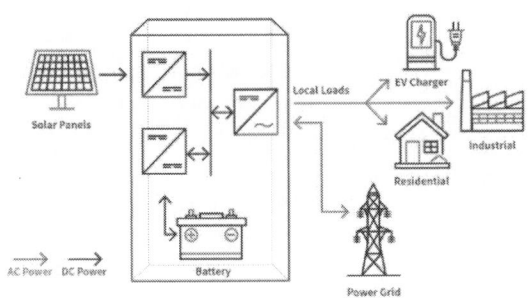

**Fig 1:** Solar Energy Storage System Architecture

The output power fluctuation of the renewable energy sources and increasing power demand on the electric grid bought by the EV transformation puts increased stress onto the existing grid. Improving the current grid stability is very much needed. The bi-directional converter is very crucial to help manage the grid since it is able transfer the energy between the renewable energy source, storage battery (which includes EV battery) and electric grid depending on the demand. It can reduce the grid stress by releasing the energy from storage to the grid or store the energy when grid supply is plenty.

For the inverter stage in a bi-directional system, the two-level three-phase inverter is very popular due to its simplicity and ease of control. Since there are only two voltage levels, 2-level inverters relatively have higher output THD. Multilevel converters help to solve this problem of high THD by employing multiple input voltage levels and can operate at higher switching frequencies, high voltage and higher power without compromising on efficiency [2]. This also helps to reduce the size of the system by having to use a smaller EMI filter. T-type converter has a simple structure and lower cost compared to other multi-level topologies due to lesser number of power devices than other multi-level topologies like the I-type and active-NPC. SiC MOSFETs in T-type inverters can improve the efficiency, power density and provide power flow in both directions [3]. In the following section a comparison of SiC

1259

and Si IGBT based T-type inverter has been presented, followed by the 25kW SiC based prototype. In the final paper, detailed design and hardware results along with efficiency comparison will be presented.

## 2 Key Components and Hardware Design

### 2.1 Design Specifications

The design specifications are like the range needed for solar string inverters with 800-1000V DC bus and 20-30kVA power.

| Parameter | Target Value |
|---|---|
| Input Voltage | 800Vdc nominal (600-1kV) |
| Output Voltage | 380-480Vac |
| Rated Power | 25kVA @ 400V AC L-L |
| Peak Efficiency | 99.2% @800V Vin |
| Output Current | 36Arms |
| Switching Frequency | 50kHz |
| Power Density | >6kW/L |
| Modulation technique | Space Vector PWM |

**Table 1**: Design Specifications

### 2.2 Power Devices

The two modes for this inverter design are the inverter mode and the PFC mode as seen in fig. 2.

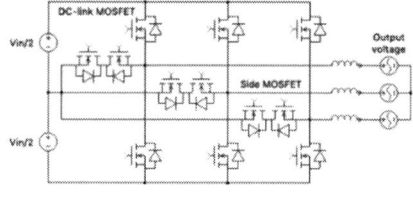

(a)

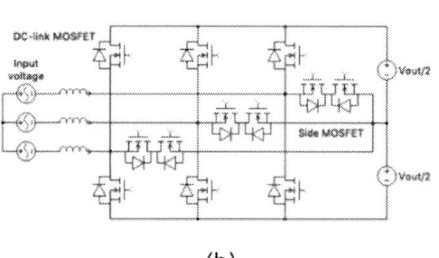

(b)

**Fig. 2:** (a) Inverter Mode (b) PFC Mode

In this section two device technology implementations of the same design will be discussed. IGBT device technology has low "on-state" static losses but have slow switching performance and therefore much higher dynamic losses. This limits their application to low switching frequencies which require bulky and large magnetics.

| Power Device | | DC-link MOSFET: 32mΩ 1200V SiC<br><br>Side MOSFET: 25mΩ 650V SiC | DC-link IGBT: 50A 1200V Si<br><br>Side IGBT: 50A 650V Si |
|---|---|---|---|
| Switching Frequency | | 64kHz | 24kHz |
| Heatsink Temperature | | 85°C | 85°C |
| $R_{th,ch}$ | | 0.5 °C/W | 0.5 °C/W |
| Efficiency* | | **98.9%** | **98.8%** |
| DC-Link MOSFET | Switching Losses | 75.79 W | 127.14 W |
| | Conduction Losses | 129.17 W | 104.16 W |
| | Total Losses | 204.96 W | 231.3 W |
| Side MOSFET | Switching Losses | 0 W | 0 W |
| | Conduction Losses | 70.1 W | 55.5 W |
| | Total Losses | 70.1 W | 55.5 W |
| Total Losses | | 275.06 W | 286.8 |
| Tj of DC-Link Device | | 134.9 °C | 134.7 °C |
| Tj of Side Device | | 97.7 °C | 91.8 °C |

**Table 2**: Comparing power loss for SiC MOSFET and Si IGBT at 25kW: Inverter Mode

| Power Device | | DC-link MOSFET: 32mΩ 1200V SiC<br><br>Side MOSFET: 25mΩ 650V SiC | DC-link IGBT: 50A 1200V Si<br><br>Side IGBT: 50A 650V Si |
|---|---|---|---|
| Switching Frequency | | 64kHz | 24kHz |
| Heatsink Temperature | | 85°C | 85°C |
| $R_{th,ch}$ | | 0.5 °C/W | 0.5 °C/W |
| Efficiency* | | **99.06%** | **98.7%** |
| DC-Link MOSFET | Switching Losses | 0 W | 0 W |
| | Conduction Losses | 115.86 W | 138.55 W |

| | | | |
|---|---|---|---|
| | Total Losses | 115.86 W | 138.55 W |
| Side MOSFET | Switching Losses | 50.16 W | 97.44 W |
| | Conduction Losses | 67.34 W | 99.13 W |
| | Total Losses | 117.5 W | 196.6 W |
| Total Losses | | 233.36 W | 335.12W |
| Tj of DC-Link Device | | 114.3 °C | 115.1 °C |
| Tj of Side Device | | 109.1 °C | 125.4 °C |

**Table 3**: Comparing power loss for SiC MOSFET and Si IGBT at 25kW: PFC Mode

From Table 2 and 3, it is noted that even though the switching frequency for Si IGBTs is 24kHz which is much lower than that of SiC MOSFETs the power loss for both inverter and PFC mode are higher than that of SiC MOSFETs. This clearly shows that the inverter will have much higher efficiency with SiC MOSFETs even after including other losses in the system. Fig. 3 shows a weight and size comparison between the 2 chokes that would be needed for switching frequency of 64kHz for SiC and 24kHz for IGBTs. SiC solution provides higher efficiency, lower magnetics cost, size and weight and therefore higher power density solution for bi-directional grid-tied inverters.

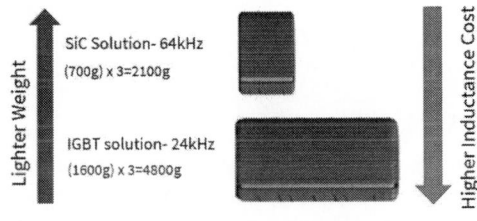

**Fig. 3:** Weight of PFC choke comparison for SiC and IGBT switching frequency

A switching frequency of 50kHz is selected to optimize between the switching losses and magnetics components selection.

## 2.3 Gate Drive

The gate drive design is crucial for working with SiC MOSFETs to fully utilize the benefit of fast switching speeds. The T-type Inverter topology has 12 switch positions and each switch requires a gate bias power supply. The gate-source voltage ($V_{GS}$) rating for Wolfspeed SiC is +15V/-3V. The gate-bias power supply is a discrete implementation of a push-pull converter in this design to reduce the cost than selection of the shelves DC-DC modules.

To further reduce the cost of these power supplies and space on the board a "common-drain" topology of the middle switches is used in this T-type Inverter. This helps to use same bias power supply generator for the high-side switch Q1 and one of the middle switches Q3 for a single phase as shown in fig 4. All the other middle switches Q2_A, Q2_B and Q2_C share the same gate power supply and all the low-side switches Q4_A, Q4_B and Q4_C share the gate power supply. This reduces the total number of gate bias power supplies to 5 for the entire board instead of 12 separate gate power supplies for each switch.

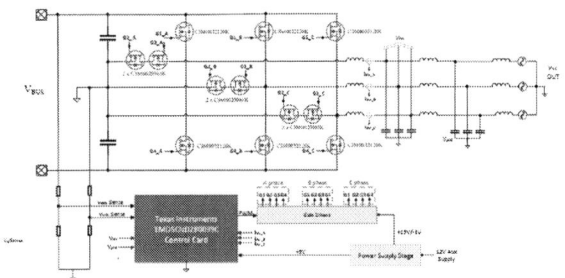

**Fig. 4:** (a) Inverter Mode (b) PFC Mode

For hard-switched topologies like the inverter and PFC, it is also very important to limit the cross-talk between the half-bridge switches. Gate-drive IC UCC5350MC from Texas Instruments is used in this application for all switches for Miller-Clamp to prevent parasitic turn-on of switches.

## 2.4 Hardware Implementation

The 25kW T-type inverter's hardware implementation is shown in Fig. 5. The DC bulk capacitors are split across the two sides of the board for DC+ and DC- voltage rails. For each phase there are PFC chokes, LC EMI filters, phase relay and inrush current protection mechanism also marked in Fig. 5. The 4 switches per phase are placed in a way to minimize commutation loop, optimize thermal balancing between switches and minimize the length of high noise switching node. The control implementation is using TI's TMDSCND2800039C control card.

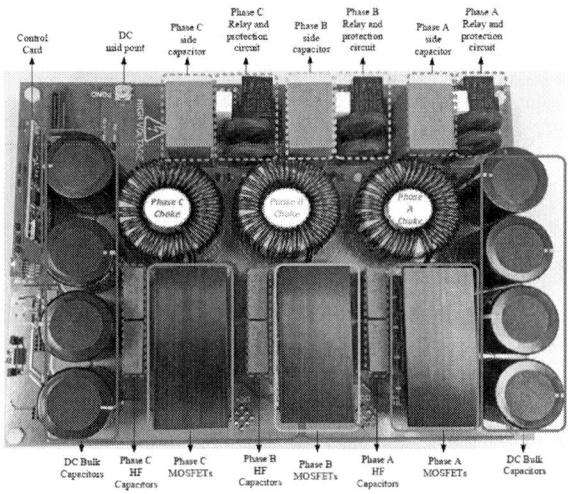

**Fig. 5:** Top view of the 25kW T-type Inverter board

The dimensions of the board are 320.5mm x 235.5mm x 50mm.

# 3 Test Results

## 3.1 Inverter Mode

An all SiC design for the T-type Inverter helps us to achieve 99.2% peak efficiency. Using the Space-Vector Modulation technique PWM technique for Inverter control the output Voltage and currents across the 3 phases are very balanced and symmetrical as seen in fig. 6. The waveform shown is at $V_{in}$= 800V and $V_{out}$ = 400Vl-l at 25kVA.

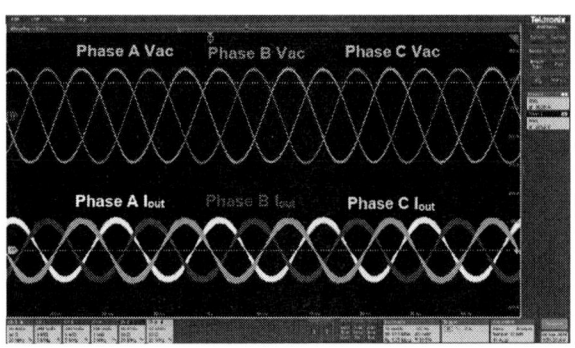

**Fig. 6:** Inverter output waveforms 25kVA

The phase A MOSFET gate-source voltage Vgs and drain-source voltage Vds are shown in fig. 7. The maximum negative Vgs spikes seen in the image is about -7.8V which is lower than the

maximum rating of dynamic Vgs -8V for C3M00320120K and C3M0025065K.

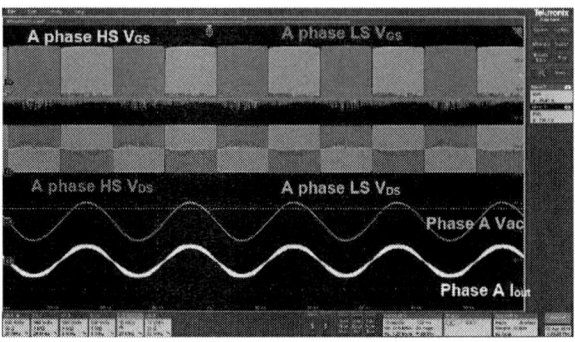

**Fig. 6:** Phase A High-Side and Low-Side MOSFET waveforms

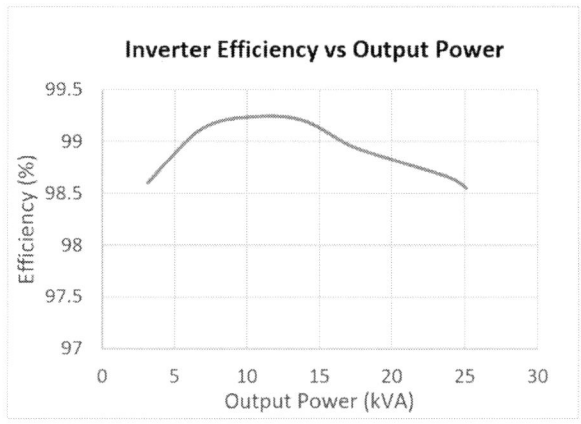

**Fig. 7:** Inverter Mode Efficiency at $V_{in}$= 800V and $V_{out}$ = 400Vl-l and $F_{sw}$= 50kHz

## 3.2 PFC Mode

The efficiency in PFC mode is slightly lower than the inverter more as work is still ongoing to optimize the current control loop in PFC mode. As can be seen in fig. 8, the output voltage ripple is <10% but there still some harmonic distortion in the input current.

The peak efficiency in PFC mode 99% with a full load efficiency of above 98.5% as seen in fig.9.

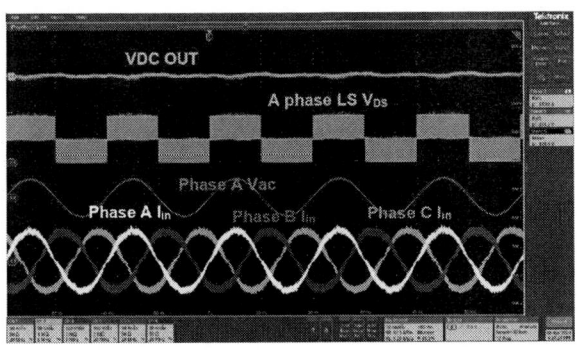

**Fig. 8:** PFC waveforms 25kVA

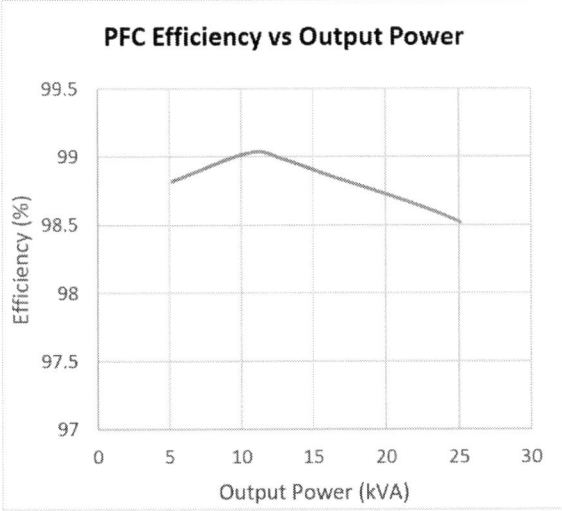

**Fig. 9:** PFC Mode Efficiency at Vin= 800V and Vout = 400VI-I and $F_{sw}$= 50kHz

Please note, all efficiency numbers do not include the auxiliary and cooling power. The auxiliary and fan power is about 30W in this design.

# 4 Summary

An all SiC T-type bi-directional Inverter is studied in this paper which achieves very high efficiency, high-power density, uses higher switching frequency than traditional IGBT approach and therefore is less bulky due to reduced magnetics size. This approach also talks about how to save some space and cost on the board by "common-drain" side MOSFETs approach. This design should be helpful for engineers who want to design for Solar inverters, uninterruptible power supplies and other inverter topologies with SiC.

# 5 References:

[1] Shehadeh, S.H., Aly, H.H., El-Hawary, M.E.: 'Photovoltaic inverters technology'. IEEE 28th Canadian Conf. on Electrical and Computer Engineering (CCECE), Halifax, NS, 2015, pp. 436–443

[2] M. Schweizer, I. Lizama, T. Friedli, and J. W. Kolar, "Comparison of the chip area usage of 2-level and 3-level voltage source converter topologies," in IECON 2010 - 36th Annual Conference on IEEE Industrial Electronics Society, 2010, pp. 391–396

[3] M. Schweizer, J. W. Kolar, "Design and Implementation of a Highly Efficient Three-Level T-Type Converter for Low-Voltage Applications", IEEE Transactions on Power Electronics, Vol. 28, No. s2, pp. 899-907, February 2013

[4] M. Gu, P. Xu, L. Zhang and K. Sun, "A SiC-based T-type three-phase three-level gridtied inverter," 2015 IEEE 10th Conference on Industrial Electronics and Applications (ICIEA), Auckland, New Zealand, 2015, pp. 1116-1121, doi: 10.1109/ICIEA.2015.7334274

[5] S. Wei, F. He, L. Yuan, Z. Zhao, T. Lu and J. Ma, "Design and implementation of high efficient two-stage three-phase/level isolated PV converter," 2015 18th International Conference on Electrical Machines and Systems (ICEMS), Pattaya, Thailand, 2015, pp. 1649-1654, doi: 10.1109/ICEMS.2015.7385305.

PCIM Europe 2024, 11– 13 June 2024, Nuremberg    DOI: 10.30420/566262171

# Cost-Effective Efficiency Enhancement in AC-DC Converters: A Study Across the Full Load Cycle

Sebastian Gick [1], Markus Pfeifer[2], Sebastian Nielebock[2], Mark-M. Bakran [1]

[1] University of Bayreuth, Centre for Energy Technology, Department of Mechatronics, Germany
[2] Siemens AG, Germany

Corresponding author:    Sebastian Gick, sebastian.gick@uni-bayreuth.de
Speaker:                 Sebastian Gick, sebastian.gick@uni-bayreuth.de

## Abstract

This paper addresses the need for cost-effective efficiency enhancement in AC to DC conversion, with a focus on industrial variable frequency drive systems. An examination of diverse AC-DC converter topologies and power semiconductor options is conducted. Switching losses of the various device combinations are measured in a double pulse test. With the use of simulation models semiconductor losses are analyzed. Thereby the die area of the converters is optimized and the converter efficiency is determined. The study encompasses the B6 converter, T-Type converter, and a Vienna rectifier. The investigation reveals that efficiency over the full load cycle can be drastically improved compared to the reference converter without increased semiconductor cost.

## 1    Introduction

Variable frequency drives (VFDs) are the state of the art in modern industry, enabling precise motor speed control and consequent energy savings, especially during partial load operation. In contrast, modern insulated-gate bipolar transistor (IGBT)-based AC-DC converters used for these industrial drive systems typically exhibit their highest efficiency in the medium load range [1]. This results in a low efficiency at partial load where a significant portion of the operating life of the converter takes place. This results in a low overall efficiency even though the nominal efficiency of the AC-DC converter is $> 98\%$.

Previous research has examined this low efficiency at light load. [1] challenged this problem by applying silicon carbide (SiC)/ IGBT4 hybrid switches. This adds complexity to the system, resulting in higher cost and higher susceptibility to failures. Other contributions to the field often neglect cost and solve the problem by employing wide bandgap semiconductors for the entire converter, massively increasing the semiconductor cost as shown in this paper. The focus of this paper is on cost-effective measures to increase partial load efficiency of AC-DC converters.

The outcomes of this study promise to enhance the selection and deployment of AC-DC converters in industrial VFDs, thereby augmenting energy efficiency whilst staying cost competitive with today´s converter solutions.

Chapter 2 defines which converter topologies are suited for the study. Chapter 3 describes the double pulse test setup. Chapter 4 presents the simulation model that is used for calculation loss data. Chapter 5 introduces the cost model used to evaluate semiconductor costs for the converter. The results of the study are given in chapter 6.

## 2    Investigated converters

The state-of-the-art solution for AC-DC rectification of the European 400 V grid voltage is the B6 converter (Fig. 1). Its widespread use can be explained primarily by its simplicity requiring only six switches with corresponding drivers and a simple 2-level modulation scheme. In industry IGBTs are commonly used as switches for the B6 converter, resulting in a converter with high output power at moderate cost. However, this comes at the expense of partial load efficiency. The forward voltages of the IGBTs and diodes lead to high conduction losses that result in only moderate partial load efficiency.

Furthermore, the utilization of a B6 converter leads to significant switching losses due to the high commutation voltage for each device and the inherent high switching losses of IGBTs caused by the tail

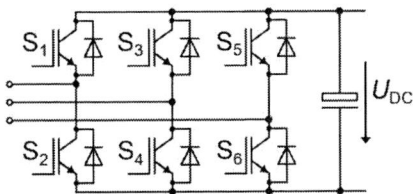

**Fig. 1:** B6 converter

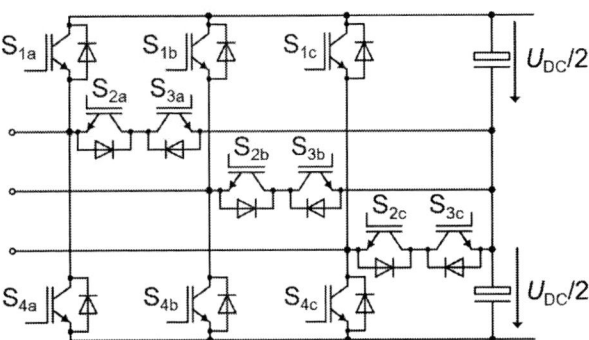

**Fig. 2:** TNPC converter

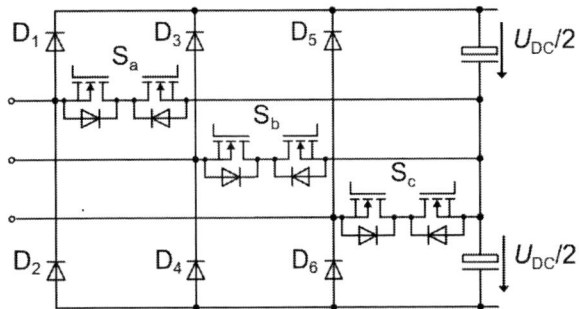

**Fig. 3:** Vienna rectifier

current and reverse recovery of the antiparallel silicon diodes. These high switching losses also lead to poor efficiency at partial load. Hence other topolgies that come with lower switching losses are being considered.

Multilevel converters can be used to minimize switching losses due to the lower harmonic content in the output waveform and due to the reduced commutation voltage for each switch. However, they have the disadvantage of high conduction losses because there is at least one additional switch/diode in the conduction path compared to the B6 converter [2].

These additional conduction losses are outweighed by the lower switching losses when higher switching frequencies or higher DC voltages are used. When the converter is used in the 400 V low voltage AC grid these higher conduction losses have to be taken into account when selecting the converter topology. Therefore, a 3-level converter with a small number of switches in the conduction path is preferred. In addition, for industrial use, high part count topologies such as the (active) neutral point clamped converter are not favorable due to higher cost and higher susceptibility to failures.

In [2], a full IGBT T-Type converter (TNPC) is compared with an IGBT B6 converter and an IGBT neutral point clamped (NPC) converter. The comparison is done for full load and various switching frequencies. It is shown that the IGBT T-Type converter exhibits superior efficiency over the B6 converter and NPC converter even down to 5 kHz switching frequency at rectifier operation.

Therefore, the T-Type topology is selected as the most promising 3-level converter (Fig. 2). The B6 and T-Type converters can be used as bi-directional rectifiers, which is critical in certain industrial applications because it allows the recuperated power from a connected drive to be fed back into the grid. Other applications like pump drives do not require bidirectional power flow and therefore unidirectional rectifiers are possible. The simplest unidirectional

rectifier is the diode full bridge. It is not examined because the line current is highly distorted.

When the application allows an unidirectional rectifier the Vienna rectifier is also a promising converter topology (see Fig. 3). It has the same advantages as the T-Type converter, which means high efficiency, sinusoidal input current and low input current distortion but it comes with even higher power density and therefore lower semiconductor cost [3], [4]. In addition, the Vienna rectifier is reliable, and the drive circuit is even simpler than a B6 converter, because only three gate drivers are required.

For these reasons, the three topologies B6 converter, T-Type converter and Vienna rectifier are investigated. The choice of the semiconductor type remains free.

Today the B6 IGBT converter is commonly chosen because of its simplicity and reliability. Therefore, the reference for comparing different converter variations is a B6 converter based on Infineon's IGBT4 technology. Although the IGBT4 is low in semiconductor cost, there are other devices that offer superior efficiency but at a higher cost that need to be considered. On the one hand, a newer IGBT generation can be selected that offers lower switching losses and lower conduction losses. However, the problem of the diode-like behavior with high forward voltage even at low current remains which has a strong impact on the partial load efficiency. Therefore, devices with no forward voltage drop are also considered.

Common DC-Link voltages for active rectification from the 400 V AC Grid are in the range of 650 V-800 V, consequently 1.2 kV switching devices are required when using a B6 converter. Only SiC metal oxide semiconductor field-effect transistors (MOSFETs) are capable to provide the high blocking voltage and an ohmic output characteristic that is needed for an efficient B6 converter. However, this comes with a prohibitively high semiconductor cost. The B6 SiC MOSFET converter is nevertheless used as an efficiency and cost benchmark in this study because it is expected to have the highest efficiency and also the highest cost of the converters studied.

The semiconductors in the T-branch of the T-Type converter or Vienna rectifier only perceive half the voltage stress compared to that of B6 switches or the switches in the main branch. Therefore, only 650 V devices are necessary which allows the application of silicon superjunction (SJ) MOSFETs, gallium nitride (GaN) high-electron-mobility-transistors (HEMTs) and lower threshold voltage 650 V IGBTs which all come with lower cost than SiC MOSFETs. Therefore, mixed topologies of 1.2 kV IGBTs and these 650 V devices in the T-branch are very attractive because they are offering a compromise between high efficiency and low cost.

The implementation of wide bandgap devices in the T-branch comes with the advantage of lower current stress of the T-branch compared to the main switches at unity power factor AC-DC rectification. As a result, less die area is required for the T-branch than for the main branch, making the use of SiC in the T-branch attractive for applications where conduction losses outweigh switching losses.

Focusing on mixed topologies has the disadvantage that the switching losses of the semiconductors can no longer be estimated from data sheet values. The turn-on losses of one device are dependent on the reverse recovery behavior of the corresponding switch in the switching cell. Turn-off losses are higher when the corresponding diode shows significant forward recovery [5]. Therefore, extensive double pulse measurements are necessary. The switches evaluated in the double pulse test are listed in Tab. 1.

For the B6 topology all 1.2 kV switches in Tab. 1

| Device | Type | $U_{BD}$ |
|---|---|---|
| IKW40N120T2 | IGBT4 | 1.2 kV |
| IKY50N120CH7 | IGBT7 | 1.2 kV |
| IMZA120R014M1H | SiC FET | 1.2 kV |
| IKZ75N65EL5 | L5 IGBT | 650 V |
| IMZA65R107M1H | SiC FET | 650 V |
| IPZA65R018CFD7 | SJ FET | 650 V |

**Tab. 1:** Investigated switches for the use in the various converter topologies

are evaluated. For the 3-level converters the IGBT4 is always kept as the main switch to allow a comparison. As T-switches the 650 V IGBTs and SiC MOSFET are evaluated. The SJ MOSFET can only be used for the Vienna rectifier. Because of its poor body diode, it cannot be used in any switching cell with two hard switching active switches.

## 3 Double pulse test setup

The reference for the comparison is an IGBT4 B6 converter. The IGBTs are packaged in an EconoPACK™ 3 module from Infineon. The converter is assumed to deliver a maximum peak AC input current of $I_p = 180$ A and comes with a stray inductance of $L_\sigma = 44$ nH in the commutation cell. For the double pulse test, transistors with smaller die area in TO packages are utilized. This scaled double pulse test allows for a simple setup and measurement. Consequently, the stray inductance of the switching cell has to be scaled as well. Scaling down of the maximum double pulse current $I_{p,DP}$ is done by the die area ratio of the IGBT4 to the die area in the application. Scaling of the double pulse stray inductance $L_{\sigma,DP}$ is than done according to (1).

$$L_{\sigma,DP} = L_\sigma \cdot \frac{I_p}{I_{p,DP}} \qquad (1)$$

The double pulse tests are therefore done with a scaled down maximum current of $I_{p,DP} = 39$ A and a scaled stray inductance of $L_\sigma = 200$ nH (refer to Fig. 4).

Gate resistor optimization for the B6 switches is done at a DC-Link voltage of 800 V until a peak voltage of 1.1 kV over the 1.2 kV devices is reached.

For the 3-level topologies 400 V DC-Link voltage is used because the T-Type converter operates with only half the voltage in the two DC-Link capacitors. For the 650 V devices a maximum voltage of 600 V is permitted. For the 1.2 kV devices in the T-Type

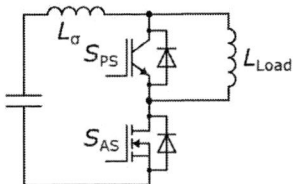

**Fig. 4:** Example schematic of the double pulse test setup. For characterizing switches for the 3-level topologies only half the DC-Link voltage is used.

converter a maximum voltage of 1.1 kV is also permitted.

To allow for comparability, the main switch of the T-Type converter is always the IGBT4 with its co-packaged diode. For the Vienna rectifier switching is also done against the co-packaged diode of the IGBT4. The investigated combinations of switches are listed in Tab. 3.

For 3-level topologies a 650 V device is combined with a 1.2 kV device in the same commutation cell. This leads to an effect that should be highlighted. In Fig. 5 the turn on of the 650 V SiC MOSFET against the 1.2 kV IGBTs is depicted.

The anti parallel diode of the IGBTs is a 1.2 kV diode commutating against only 325 V DC-Link voltage, which allows the complementary switch to be turned on quickly. In fact, the SiC MOSFET can be turned on so quickly that the diode does not pick up any voltage when the SiC MOSFET is turned on. All the voltage is picked up by the stray inductance $L_\sigma$ of the commutation cell (refer to Fig. 4). This leads to a turn on of the SiC MOSFET without carrying any significant current, resulting in a virtually lossless turn-on of the MOSFET. The current then commutates from the diode into the MOSFET. The

diode picks up the the voltage during the reverse recovery phase. In combination with the low switching and reverse revovery losses of SiC devices, the SiC MOSFET only percieves a small portion of the total switching losses of the T-Type converter. The majority of the switching losses are dissipated in the silicon of the main switch.

A similar effect is also observed in the SJ FET and the L5 IGBT double pulse tests (see Fig 6). Therefore, their turn-on energy is also effectively zero.

The total high temperature switching losses of all investigated device combinations are shown in Fig. 7 for a junction temperature of $T_J = 150\,^\circ$C. The left side of the legend lists the switches for the B6 configurations. These double pulse measurements are taken at 650 V DC-Link voltage, the nominal DC-Link voltage of the B6 converter. It shows that by using the IGBT7 instead of the IGBT4 the hot switching loss in the B6 converter can already be reduced by 41%. By using a SiC MOSFET instead of an IGBT4 the switching loss can be reduced by 73%.

The switching losses of the switch combinations for the 3-level configurations are listed on the right. They are measured at 325 V DC-Link voltage as this is the nominal DC-Link voltage of the two DC-Links of the three level converters. For these combinations due to using two different switches, two switching events are possible depending on the current direction of the phase current. Either the 650 V switch is the active switch and the 1.2 kV switch is the passive switch or the 650 V switch is the passive switch and the 1.2 kV switch is the active switch. The loss curve shown is the average of the curves for both possible switching events.

The hot switching loss of the 3-level switching cells is between the loss of the 1.2 kV SiC MOSFET and the IGBT7. Therefore, the SiC MOSFET B6 converter seems to come with the lowest losses for

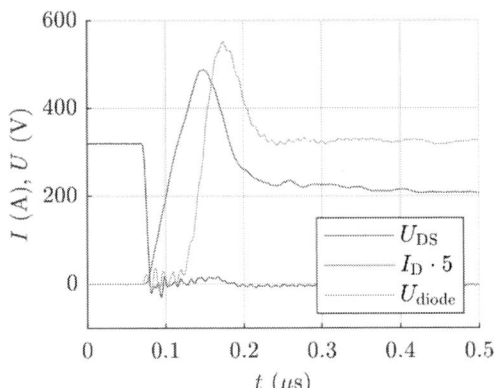

**Fig. 5:** Turn-on event of the 650 V SiC MOSFET against the 1.2 kV co-packaged diode of the IGBT4.

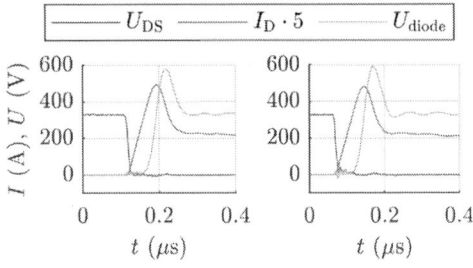

**Fig. 6:** Turn-on event of the SJ MOSFET (left) and the L5 IGBT (right) against the diode of the IGBT4.

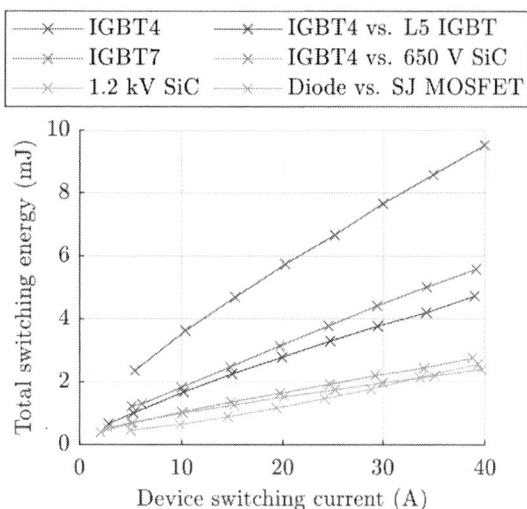

**Fig. 7:** Measured total switching losses at $T_J = 150\,°C$. Left legend entries measured at 650 V DC-Link voltage, right entries measured at 325 V.

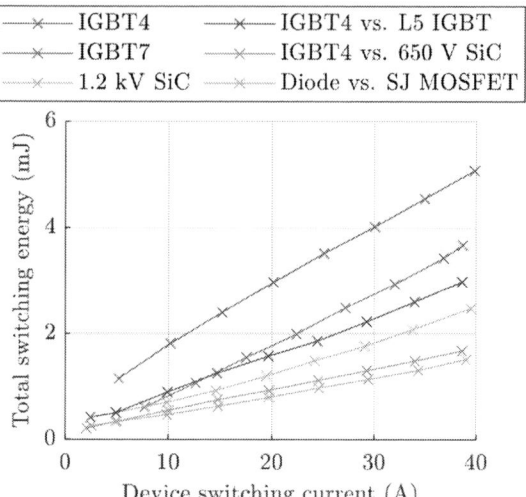

**Fig. 8:** Measured total switching losses at $T_J = 25\,°C$. Left legend entries measured at 650 V DC-Link voltage, right entries measured at 325 V.

each switching event.

More relevant for the partial load operation are the cold switching losses, as the semiconductor temperature decreases at reduced load. At lower temperatures IGBTs have an advantage because their switching energy has a higher temperature dependency than the switching energy of SiC MOSFETs. Fig. 8 shows the switching losses for a junction temperature of $T_J = 25\,°C$. Because of the lower temperature the IGBT4 switched against the 650 V SiC or SJ MOSFET offers the lowest switching energy. The SJ MOSFET switching against the copackaged diode of the IGBT4 experiences even lower losses than the SiC MOSFET in this combination. This effect is already described in [6].

## 4 Simulation model

An electrothermal PLECS simulation model is created to calculate the efficiency of the different converter variants. This requires the modeling of semiconductor losses. Conduction losses and switching losses are implemented as lookup table in a PLECS thermal model for each switch. The switching losses are taken from the double pulse measurements. Conduction losses taken from the output characteristics given in the data sheets.

For a fair comparison of different semiconductors and different topologies in a simulation model the same thermal boundary conditions must be used for all converter variants. In the simulation, the junction to case thermal resistance ($R_{th,JC}$) and the

heatsink to ambient thermal resistance ($R_{th,CA}$) are modeled as they are the most dominant thermal resistors in the application. A rule has to be defined for the value of these resistances for the different converters.

It is assumed, that in the application the power semiconductors are placed in an appropriately sized module, which in turn is placed on an appropriately sized air heat sink. Therefore, the the thermal resistance is dependent on the area of the module and therefore dependent on the installed die area. Therefore, the thermal resistances are all modeled as area specific thermal resistances. As data base for the specific thermal resistances the die area of a EconoPACK™ 3 module and the installed heat sink is utilized. Following that, the area specific thermal resistances for the simulation model are calculated by (2) and (3) using the full die area of the simulated converter $A_{die,converter}$. The die area of the transistors from Tab. 1 is determined by breaking the TO package and measuring the die area with a caliper.

$$R_{th,JC} = \frac{27\,\mathrm{K\,mm^2/W}}{A_{die,converter}} \quad (2)$$

$$R_{th,CA} = \frac{94.3\,\mathrm{K\,mm^2/W}}{A_{die,converter}} \quad (3)$$

The next step is to find the maximum output power of each converter so that the efficiency of all converters can be compared at the same load point. For the B6 converter the maximum output power can be found simply by iterating through a PLECS simulation model increasing the output power till maximum junction temperature of the switches is reached. At this maximum output power, the semiconductors are thermally fully utilized.

For the 3-level topology, an additional step must be taken because the die area of the T-branch and the main branch must be sized independently, as they experience different current stress. This would lead to either the T-switches or the main switches not reaching the maximum junction temperature and therefore these semiconductors would not be thermally fully utilized. By sizing the die areas independently the ideal die area ratio for the three level converters can be found.

The PLECS switch thermal model does not inherently allow for simple modification of the die area of a given switch. Rather, a combination of changing the thermal model of the switch and the $R_{\mathrm{th,JC}}$ has to be performed, which is time-consuming and error-prone. Therefore, a workaround is found to size the switches of the T-branch independently from the main branch switches.

As shown in Fig. 9, a parallel path to each T-switch is added which allows a specific ratio of the T-current $I_{\mathrm{T}}$ to bypass the switches of the T-branch. The bypassed current $I_{\mathrm{BP}}$ is described by (4). $k_{\mathrm{par}}$ is defied as the ratio of the changed die area of the analyzed switch in the T-branch $A_{\mathrm{T}}$ and the area of the physical die $A_{\mathrm{die}}$ (5). This way the current stress of the modeled T-switches can be changed to any arbitrary value without changing the current stress of the main switches. Thus, the die area can be effectively changed.

$$I_{\mathrm{BP}} = 1 - \frac{1}{k_{\mathrm{par}}} \cdot I_{\mathrm{T}} \qquad (4)$$

$$k_{\mathrm{par}} = \frac{A_{\mathrm{T}}}{A_{\mathrm{die}}} \qquad (5)$$

For example, if half the current of the T-switches is bypassed, the T-branch acts as a parallel connection of two dies. Half of the current is passed through the modeled die and half through the parallel path, a fictitious die. This is the same behavior as it is for a parallel connection of 2 dies.

The die than experiences half the losses of 2 parallel die. Therefore, the losses of the fictitious parallel die $P_{\mathrm{loss,add}}$ are added afterwards via a thermal heat source. The added losses are described by (6).

$$P_{\mathrm{loss,add}} = (k_{\mathrm{par}} - 1) \cdot P_{\mathrm{loss}} \qquad (6)$$

This method can be used for any parallel connection of $k_{\mathrm{par}}$ dies and therefore allows precise setting of the die area of the T-branch.

An electrothermal simulation model is set up for each topology. A simplified PLECS model for the T-Type converter is depicted in Fig. 9. The 3 phase input current of the converter is set by AC current sources. This models a connection to the grid via a 3 phase choke, neglecting the small additional losses due to ripple current on the AC lines. Symmetrical space vector modulation is used to modulate the B6 converter output voltage with unity power factor. For the T-Type converter 3-level space vector modulation is used. For the Vienna rectifier a digital space vector modulation based on [7] is implemented.

The nominal grid voltage of 400 V is used in the simulation. For dimensioning of the die area of the T-branch of the 3-level converters there is also accounted for an undervoltage operation at 80% grid voltage. A nominal DC-Link voltage of 650 V is chosen. Switching frequency for the B6 converter is set to 8 kHz. Because the 3-level converters achieve the same output current ripple for half the switching frequency, their switching frequency is set to 4 kHz. To reduce complexity, the simulation does not model any transient thermal behavior.

The thermal resistors are placed in the simulation model as depicted in Fig. 9. Based on this model an iteration is performed to find the maximum

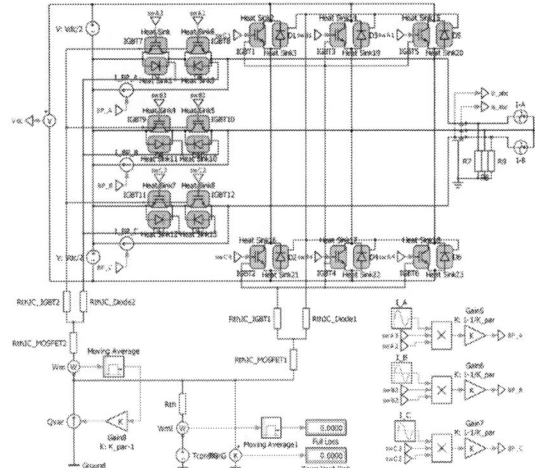

**Fig. 9:** Simplified PLECS simulation model for the T-Type converter.

output power of the investigated converter. The maximum output power is reached when the switches reach a junction temperature of 150 °C, regardless whether the actual devices allow a higher junction temperature. This eliminates the differences caused from different maximum junction temperatures permitted by the manufacturer of different devices. When the maximum output power is found, all load points of the different converters can be described as a portion of the maximum output power. In this way, different converters with different total die areas can be compared for maximum full load efficiency.

## 5 Cost model

A very simple cost model is used to compare the semiconductor cost of different topologies applying different semiconductors. The semiconductor price is expressed as an arbitrary monetary unit (MU). For each semiconductor material, an area-specific cost is assumed that is independent of the voltage class or switch type. The cost rates for Si and SiC are defined in Tab. 2.

| Material | Cost factor |
|---|---|
| Silicon | 1 MU/mm$^2$ |
| Silicon carbide | 6 MU/mm$^2$ |

**Tab. 2:** Area specific cost for each semiconductor material

## 6 Results

The maximum output power and die area for the different topologies are determined by the PLECS simulation. The die are and output power are used to calculate the specific semiconductor cost of the converters. In Tab. 3 the converter topologies and the results are listed.

The two reference converters are the B6 IGBT4 converter and the B6 SiC MOSFET converter. The IGBT4 B6 converter comes with low semiconductor cost, although not the lowest, and the SiC converter comes with the highest cost. In this application, the semiconductor cost for the SiC converter is 230% of the IGBT4 converters cost, which states the point that a full SiC solution is still cost prohibitive today. The IGBT7, as a newer generation of IGBT, allows for lower cost than the IGBT4 B6 converter. In terms of semiconductor cost, the 3-level converters outperform their 2-level counterparts.

The Vienna rectifier topology is particularly notable because of its drastically reduced semiconductor

| Devices | Topology | Specific cost |
|---|---|---|
| IGBT4 | B6 | 17.9 MU/kW |
| IGBT7 | B6 | 16.1 MU/kW |
| 1.2 kV SiC MOSFET | B6 | 41.4 MU/kW |
| IGBT4 + L5 IGBT | T-Type | 15.2 MU/kW |
| IGBT4 + SiC MOSFET | T-Type | 30.4 MU/kW |
| 1.2 kV Si diode + SJ MOSFET | Vienna | 13.8 MU/kW |
| 1.2 kV Si diode + L5 IGBT | Vienna | 11.3 MU/kW |

**Tab. 3:** Investigated converters, their maximum output power and specific semiconductor cost

cost compared to the IGBT4 B6 converter. The Vienna rectifier with L5 IGBTs only comes at 63% of the semiconductor cost of the B6 IGBT4 converter. This cost savings come from the elimination of the IGBTs as the main switches and the decreased conduction and switching loss due to the Vienna rectifier topology.

This advantage comes with the major disadvantage that the Vienna rectifier does not allow power to be fed back into the grid. In applications where this is not required, the Vienna rectifier clearly outperforms all other converters in terms of cost.

The focus of this paper is also to compare the efficiency of the different converters. For this purpose a simulation is made for ten steps of input power ($P_{in}$). The calculated losses of the converters ($P_{loss}$) are then used to calculate the efficiency of the converters $\eta$ with the help of (7).

$$\eta = \frac{P_{in} - P_{loss}}{P_{in}} \tag{7}$$

In Fig. 10 the efficiency curves of the converters are depicted. Again the SiC B6 converter and the IGBT4 B6 are used as the reference. They have the highest/ lowest efficiency of all investigated converters.

Switching out the IGBT4 for a newer generation IGBT, the IGBT7, already improves the efficiency of the B6 converter drastically. Especially at light load, high loss savings of up to 45% are achieved. The next higher efficiency class are the TNPC consisting of the IGBT4 and the L5 IGBT and the Vienna rectifier consisting of the 1.2 kV diode and the L5 IGBT. These converters benefit from the

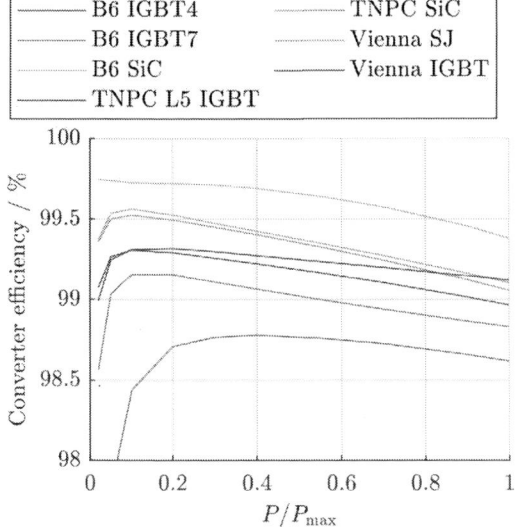

**Fig. 10:** Efficiency curves of the investigated converters. The B6 IGBT4 converter and B6 SiC MOSFET converter are the reference converters.

lower switching loss of the 3-level topology and do not have a significant increase in conduction loss due to the low forward voltage of the L5 IGBT. The L5 IGBT is an IGBT which is optimized for low forward voltage drop gives therefore an advantage in conduction losses.

The highest efficiency using silicon devices is achieved by the T-Type converter consisting of the IGBT4 and 650 V SiC MOSFETs and the Vienna rectifier consisting of the 1.2 kV diode and the silicon SJ MOSFET. These converters benefit from the lack of a forward voltage drop of the MOSFETs, which then provides a major efficiency advantage at light load due to very low conduction losses. At full load, the MOSFET has no conduction loss advantage over the IGBTs which results in a similar efficiency at full load for converters with MOSFET and IGBT T-branches.

## 7 Conclusion and outlook

This paper compares several converter topologies with different power semiconductors for use in industrial drive systems. The focus is on cost effective measures to improve the full load efficiency of today's converters. Therefore, the switching losses for the used semiconductor combinations are measured in a double pulse test setup. A simulation model is presented that is used to calculate converter die area, maximum output power, and converter efficiency.

For bidirectional operation maximum efficiency is achieved by the B6 SiC MOSFET converter. The T-Type converter with SiC MOSFET T-branch offers the next best efficiency but still comes with 172% the cost of the reference converter. The best cost efficiency tradeoff is offered by the IGBT T-Type converter. For unidirectional converters, the Vienna rectifier with silicon SJ MOSFETs as the T-switch combines the highest efficiency over the full load cycle of all silicon converters with one of the lowest semiconductor cost of all the converters studied.

## Acknowledgments

This work was supported by the Technology Alliance Oberfranken (TAO).

## References

[1] M. Makoschitz and S. Biswas, "Light load efficient silicon power converters based on wide bandgap circuit extensions," *Applied Sciences*, vol. 10, no. 14, p. 4730, 2020. DOI: 10.3390/app10144730.

[2] M. Schweizer, I. Lizama, T. Friedli, and J. W. Kolar, "Comparison of the chip area usage of 2-level and 3-level voltage source converter topologies," *IECON 2010 - 36th Annual Conference on IEEE Industrial Electronics Society*, pp. 391–396, 2010. DOI: 10.1109/IECON.2010.5674994.

[3] J. W. Kolar and F. C. Zach, "A novel three-phase utility interface minimizing line current harmonics of high-power telecommunications rectifier modules," *IEEE Transactions on Industrial Electronics*, 1997.

[4] S. D. Round, P. Karutz, M. L. Heldwein, and J. W. Kolar, "Towards a 30 kw/liter, three-phase unity power factor rectifier," *2007 Power Conversion Conference - Nagoya*, 2007.

[5] J. Lutz, H. Schlangenotto, U. Scheuermann, and R. de Doncker, *Semiconductor Power Devices*. Cham: Springer International Publishing, 2018. DOI: 10.1007/978-3-319-70917-8.

[6] H. Gui, Z. Zhang, R. Ren, R. Chen, J. Niu, *et al.*, "Sic mosfet versus si super junction mosfet - switching loss comparison in different switching cell configurations," *2018 IEEE Energy Conversion Congress and Exposition (ECCE)*, pp. 6146–6151, 2018.

[7] R. Burgos, R. Lai, and Y. Pei, "Space vector modulation for vienna-type rectifiers based on the equivalence between two- and three-level converters: A carrier-based implementation," *2007 IEEE Power Electronics Specialists Conference*, 2007.

PCIM Europe 2024, 11– 13 June 2024, Nuremberg    DOI: 10.30420/566262172

# Next Generation Power Module with Parallel Connected SiC MOSFETs for BEV Traction Inverters

Kohei Tanikawa[1], Oji Sato[1], Kotaro Shibata[1], Masashi Hayashiguchi[1], Tomohiro Yasunishi[1], Daiki Ikeda[1], Takara Kosaka[1], Tomoki Fujimura[1], Hiroto Sakai[1], Yuta Okawauchi[1] and Kenji Hayashi[1]

[1] ROHM CO., Ltd., Japan

Corresponding author: Kohei Tanikawa, Kohei.tanikawa@mnf.rohm.co.jp
Speaker: Kohei Tanikawa, Kohei.tanikawa@mnf.rohm.co.jp

## Abstract

We show a simple technique for suppression of self-excited oscillation in parallel connected multiple Silicon Carbide (SiC) MOSFETs. Here, we focus on the source-to-source inductance in adjacent chips ($L_{ss}$) which is considered one of the most critical parameters stabilizing the power devices connected in parallel. By introducing copper clips and optimizing layout, $L_{ss}$ is completely well controlled. As a result, a strong dependency of oscillation behavior to $L_{ss}$ value is observed. In addition, a highly efficient switching operation and marked power cycling durability is demonstrated by our newly developed SiC power module, mainly target to traction inverter, "TRCDRIVE pack™". Note that "TRCDRIVE pack™" is a trademark or a registered trademark of ROHM Co., Ltd.. This study gives an important knowledge to draw SiC potential at the maximum and to design next generation SiC-based battery electric vehicle traction inverters.

## 1    Introduction

Replacing petrol vehicles with battery electric vehicles (BEV) is accelerating to realize a carbon neutral society around the world. In this market, there is a huge SiC device demand for next-generation power saved traction inverters because of its high breakdown voltage, low resistance, and low switching loss. Furthermore, downsizing of power module with power devices such as SiC MOSFETs is strongly required, leading to space, weight and power saving for BEV applications. On the other hand, SiC crystal consists of more lattice defects than those of Silicon (Si), giving rise to the limitation of the size for SiC device. Therefore, handling multiple SiC devices in parallel is crucial key technology to meet high power demands. However, it is said that this configuration has a potential for inducing unintended oscillation, namely, parallel oscillation. Very recently, a guideline is illustrated for avoiding parallel oscillation in power module [1]. This study implies that the oscillation is suppressed as smaller $L_{ss}$ or larger gate-to-gate inductance in adjacent chips ($L_{gg}$). It should be noted that an increase in $L_{gg}$ negatively effects the

switching speed, while a reduction in $L_{ss}$ has little effect on the switching characteristic.

In this article, the newly developed next generation SiC power module is proposed. We demonstrate reducing $L_{ss}$ value and suppressing parallel oscillation by tuning the shape of the copper clip in this SiC-based power module. Moreover, this module shows very low inductance, generating highly efficient switching performance. An outstanding power cycling durability compared to the conventional case type power module is also verified.

## 2    TRCDRIVE pack™

We have so far developed SiC-based power modules for BEV traction inverters. Figure 1 exhibits our newly developed next generation SiC-based power module, "TRCDRIVE pack™" which consists of half-bridge circuit. SiC MOSFET dies are bonded to the substrate with silver sintered layer, which results in low thermal resistance [2]. To achieve lower total module resistance, inductance, higher power density and reliability, copper clips are introduced instead of multiple wires or ribbons [3]. In addition, press-fit terminals are applied to gate, drain sense, source sense and thermistor terminals. These technologies contribute to balancing an electric current that flows into each SiC MOSFET die and to downsizing power module.

Here, the stability of the switching operation, switching performance and reliability are discussed for this power module with SiC MOSFET dies in parallel.

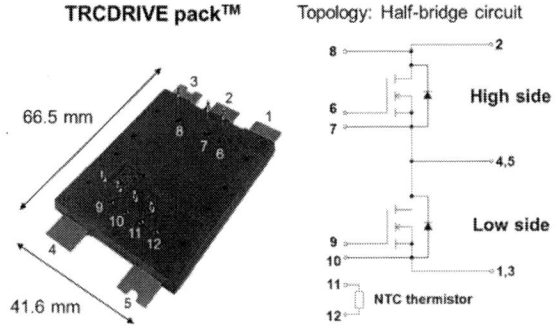

**Fig. 1** TRCDRIVE pack™

# 3 Suppression of parallel oscillation

To observe switching waveforms, the double pulse tests (DPTs) were performed under the conditions listed in Table 1. The gate source voltage ($V_{GS}$) switching waveforms at the turn-on for both high and low side SiC MOSFETs are displayed in Fig. 2. We can see an ideal switching waveform for the low side. In contrast, a clear oscillation was observed for the high side.

| $V_{DS}$ (V) | $I_{DS}$ (A) | $T_j$ | $V_{GS}$ (V) |
|---|---|---|---|
| 600 | 400 | RT | 18/0 |

**Table 1** The test conditions of the DPTs

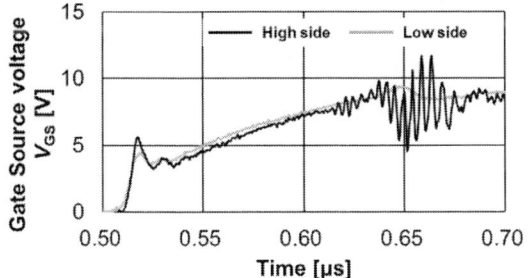

**Fig. 2** Switching waveforms for both high and low side

We infer that this difference in phenomenon derives from the design of copper clip. For high side, copper clips are arranged for each SiC MOSFET die, respectively (Type A), whereas one large copper clip connect SiC MOSFET dies directly for low side. Consequently, $L_{ss}$ for low side is about 65% smaller than that for high side. To overcome this issue, the isolated copper clips in the high side are connected in the same way as the low side, leading to the dramatic reduction in $L_{ss}$ (see Type B in Fig. 3). As a result, a smooth $V_{GS}$ switching waveform was confirmed as shown in Fig.4. Namely, the suppression of parallel oscillation has been achieved for power module with parallel connected SiC MOSFETs. These results are consistent with the previous study [1].

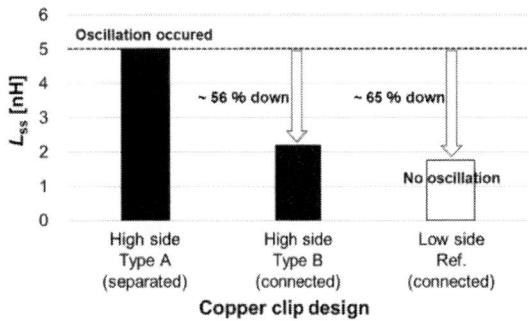

**Fig. 3** $L_{ss}$ versus copper clip structure for both high and low side

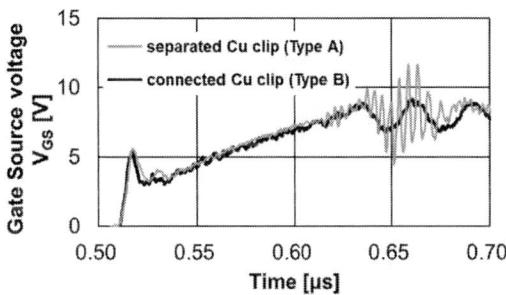

**Fig. 4** Switching waveforms versus copper clip structure for high side

# 4 Switching Characteristics

It is widely acknowledged that the magnitude of the surge voltage $V_{surge}$ at turn-off switching is proportional to the stray inductance $L_{stray}$ and switching frequency as given by Eq. (1).

$$V_{surge} = -L\frac{di}{dt} \qquad (1)$$

Therefore, reducing the $L_{stray}$ is essential for preventing the MOSFET from exceeding the rated voltage and enabling high-speed switching, one of the main characteristics of SiC MOSFETs [4,5].

By optimizing copper clip shape and inner layout, the very low stray inductance of 5.7 nH is obtained for TRCDRIVE pack™. This value is approximately 40% lower than that of our conventional case type module called "Gtype".

To evaluate the switching losses, DPTs were applied again for both modules. The test conditions are listed in Table 2. The switching speed and losses were adjusted by changing the external gate resistor $R_g$. Figure 5 gives the relationship between the $V_{surge}$ and the total switching losses.

| $V_{DS}$ (V) | $I_{DS}$ (A) | $T_j$ | $V_{GS}$ (V) |
|---|---|---|---|
| 600 | 400 | RT | 18/0 |

**Table 2** The DPTs conditions

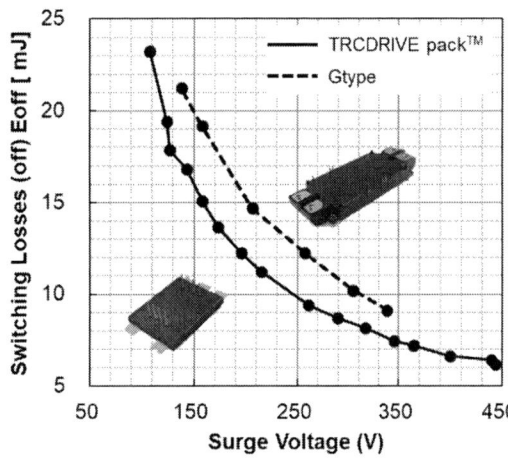

**Fig. 5** Switching Losses versus Surge Voltage

Generally, module users need to ensure that the surge voltage peak never exceeds the absolute maximum voltage of the module under worst case conditions. If, for example, a surge voltage up to 300 V can be allowed in a certain application, the switching loss values for TRCDRIVE pack™ were found to be about 10% ~ 20% smaller compared to that for Gtype. This result clearly demonstrates that low $L_{stray}$ of the module is a key factor not only to gain safety margin with respect to the drain-to-source peak voltage but also to enhance the efficiency by reducing switching losses.

In the next step, the energy consumption rate (ECR) was evaluated for TRCDRIVE pack™ with 4th generation SiC MOSFETs and Gtype with IG-BTs by means of an electric motor test bench under a worldwide harmonized light vehicles test cycle (WLTC). The representative test conditions are listed in Table 3. The other detail conditions including methods are described in the previous study [6,7].

| Power supply voltage $V_{DC \text{ or } DS}$ (V) | 800 | |
|---|---|---|
| Switching frequency $f_{sw}$ (kHz) | 10 | |
| Dead time $T_d$ (μs) | 2 | |
| Base-emitter or Gate-source voltage $V_{BE \text{ or } GS}$ (V) | + 18 / - 4 | + 18 / 0 |
| Gate resistance $R_{gon}/R_{goff}$ (Ω) | 3.3 / 3.3 | 6.8 / 6.8 |
| DUT | Gtype | TRCDRIVE pack™ |
| | IGBT | 4th Gen SiC |
| | | |

**Table 3** Test conditions

The results of ECR are depicted in Fig. 6. The ECR of the TRCDRIVE pack™ with 4G-SiC is improved by 9.8% compared to that of Gtype with IGBT. This suggests that TRCDRIVE pack™ with 4G-SiC brings benefits for BEV application as it extends the driving distance for a given battery capacity or allows battery size reduction without sacrificing driving distance.

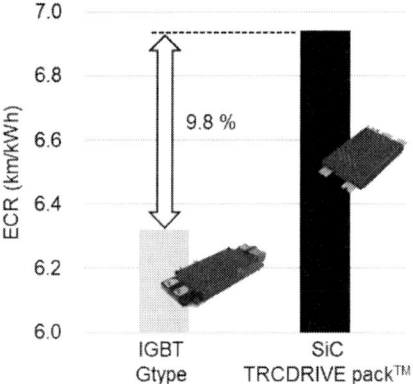

**Fig. 6** The measured ECR for both modules

Thus, the superiority of TRCDRIVE pack™ with SiC MOSFETs to Gtype with IGBT in switching performance is definitely verified.

## 5 Power Cycling Capability

In this section, we examine power cycling durability for TRCDRIVE pack™. Table 4 summarizes the power cycling test conditions. The failure criteria of power cycling test are defined as a 5% rise in $V_{DS}$ based on the initial value ($V_{DS}$ Ratio) according to the AQG-324.

| $\Delta T_j$ (K) | $T_{jmin}$ (°C) | $T_{jmax}$ (°C) | $T_{on}$ (s) | n (pcs) |
|---|---|---|---|---|
| 100 | 75 | 175 | 2 | 6 |

**Table 4** Power cycling test conditions

The power cycling diagrams are indicated in Fig. 7. The average end-of-life is calculated at 776 kcycs. These experimental results are also fitted into a Weibull distribution by plotting ln(-ln(1-F)) as a function of number of cycles (see Fig. 8). The estimated lifetime at failure rate of 1% is found to be over 250 kcycs which is around 17 times higher than that for the existing our case type module. It can be speculated that this long operation life of TRCDRIVE pack™ derives from the copper clip, silver sintering and transfer molding technology.

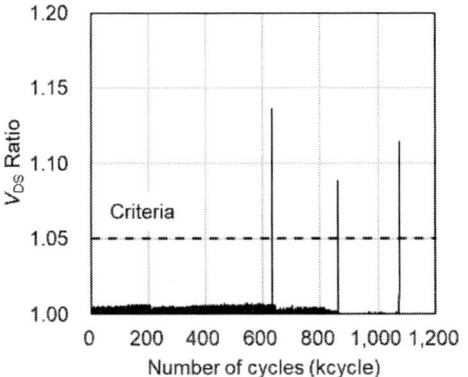

**Fig. 7** Power cycling diagram for TRCDRIVE pack™

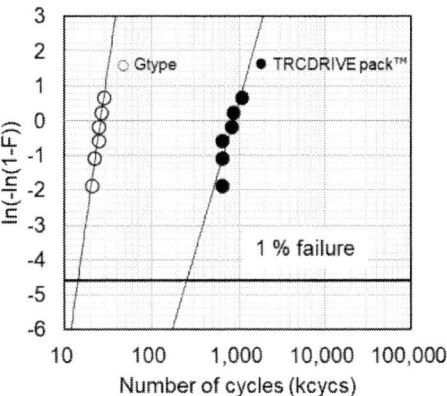

**Fig. 8** Weibull distribution for power cycling test

## 6 Conclusion

We developed a new state-of-the-art SiC based power module, a good benchmark for BEV traction inverter, bring the best out of SiC die performance, that it should be. Firstly, a simple method to design a power module inner structure with parallel connected SiC MOSFETs shrinking the risk of oscillations was proposed. By connecting the SiC MOSFET dies via copper clip directly, great reduction in $L_{ss}$ was confirmed. Note that, $L_{ss}$ is one of the most important parameters involved in the circuit stability. Consequently, an ideal switching waveform was observed, meaning that suppression of the self-excited oscillations was achieved. Furthermore, a high switching performance and power cycling durability compared to the conventional case type module was clearly demonstrated by our newly developed power module with copper clip technology, TRCDRIVE pack™.

These results will accelerate the adoption of SiC MOSFETs in the field of BEV traction inverters.

## References

[1] H. Sakai *et al.*, "Simplified Open-Loop Transfer Functions to Analyze Influential Parasitic Parameters for Oscillation Caused by Parallel Connected Transistors" Proceedings of the 35th International Symposium on Power Semiconductor Devices and ICs.

[2] K. Wakamoto *et al.*, "Degradation Mechanism of Silver Sintering Die Attach Based on Thermal and Mechanical Reliability Testing" IEEE Transactions on Components, Packaging and Manufacturing Technology, vol. 13, no. 2, pp. 197-210, 2023.

[3] Kyaw Ko Lwin *et al.*, "Copper Clip Package for high performance MOSFETs and its optimization" 2016 IEEE 18th Electronics Packaging Technology Conference, pp.123-128, 2016.

[4] P. Beckedahl *et al.*, "400A, 1200V SiC power module with 1nH commutation inductance", CIPS 2016.

[5] S. Hain *et al.*, "The Effect of Different Stray Inductances on the Performance of Various Types of IGBTs – Is Less Always Better?", 2015 17th European Conference on Power Electronics and Applications (EPPE`15 ECCE-Europe), PCIM Europe 2016, Nuremberg, Germany.

[6] H. Umegami *et al.*, "Performance Comparison of Si IGBT and SiC MOSFET Power Module Driving IPMSM or IM under WLTC" World Electr. Veh. J. 2023, 14, 112.

[7] K. Shibata *et al.*, "Automotive Traction Inverter using the 4th Generation SiC MOSFET Power Module", PCIM Europe 2021, Nuremberg, Germany.

PCIM Europe 2024, 11– 13 June 2024, Nuremberg    DOI: 10.30420/566262174

# Investigation of Common Source Feedback in SiC Power Modules Regarding Performance and Short Circuit Robustness

Dominik Alexander Ruoff[1], Zong Xern Sim[1], Burkhard Ulrich[2]

[1] Robert Bosch GmbH, Germany
[2] Reutlingen University, Germany

Corresponding author:    Dominik Alexander Ruoff, dominikalexander.ruoff@de.bosch.com
Speaker:    Dominik Alexander Ruoff, dominikalexander.ruoff@de.bosch.com

## Abstract

SiC power modules are crucial in the automotive industry due to their high efficiency, but the change from Si to SiC brings new challenges regarding the short-circuit withstand time (SCWT). This paper investigates the influence of a common source feedback gate topology on short-circuit behavior. Implementing source feedback enhances the short-circuit withstand time but comes at the cost of increased switching losses. A more balanced trade-off between robustness and performance can be achieved by combining a well-defined common source feedback with an increasing gate-source voltage. This article investigates the concept using simulations, followed by characterization tests on a prototype commutation cell.

## 1 Introduction

The global automotive industry has experienced a significant shift towards electromobility in recent years, leading to a surge in demand for power electronics with higher efficiency. Wide band gap (WBG) devices such as SiC MOSFETs have lower specific on-resistance and perform better under partial load conditions than Si IGBTs [1]. As a result, SiC MOSFETs have emerged as a promising alternative to conventional Si transistors.

However, SiC MOSFETs exhibit reduced short-circuit robustness when compared to IGBTs. Specifically, the short-circuit withstand time of SiC MOSFETs is only one-fourth that of conventional IGBTs under similar short-circuit conditions [2]. SCWT is the most extended period for which the MOSFET is under a short-circuit event before experiencing degradation or physical failure [2]. This phenomenon is caused by higher saturation current and temperature, and it is crucial to turn off the MOSFET before damage occurs.

Multiple approaches are known to improve the short-circuit robustness of SiC MOSFETs. Thicker gate oxide will increase the maximum short-circuit energy threshold and lengthen the short-circuit transient at the cost of a slower device switching speed due to higher threshold voltage and gate

capacitance [3]. An alternative approach involves lowering the source doping concentration, which can effectively double the SCWT while causing a marginal increase in the on-resistance [4]. These methods, however, require structural redesigning of the MOSFET, which is costly and time-consuming. Paper [5] investigates the influence of various circuit characteristics on short-circuit behavior. It was found that source inductance reduces the device's short-circuit current and temperature but increases switching losses.

This paper examines the influence of source inductance and source resistance on the saturation current and short-circuit energy and presents a novel approach to improve the SCWT, but also addressing the increased switching losses.

## 2 Test Case

### 2.1 Test Configurations

A half-bridge configuration, as shown in Figure 1, is used for the investigation, as such a topology is commonly used in electric vehicles to control three-phase motors. The low-side (LS) switch acts as the device under test (DUT), while the high-side (HS) switch serves either as a passive switch during double pulse test or is deliberately shorted during short-circuit test, which is shown in Figure 2(a). In this case, the pulse duration of DUT corresponds to a short-circuit period.

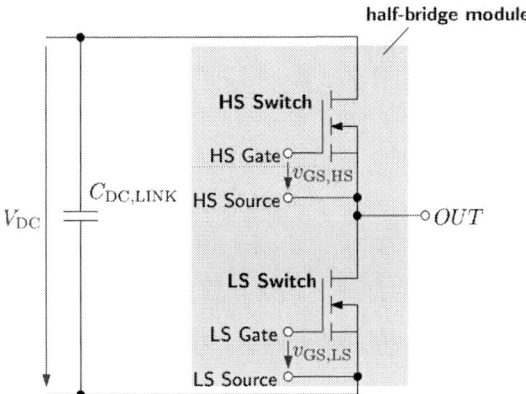

**Fig. 1:** Half-bridge Test Configuration

The double pulse test setup is illustrated in Figure 2(b). The HS switch is permanently turned off by applying a gate-source voltage below its turn-on threshold, thus acting as a freewheeling diode, while a load inductance is connected in parallel to the HS switch. The LS switch control signal will then be used to adjust the test current value for measuring the turn-on and turn-off switching energy. In both test configurations the DUT's drain current and drain-source voltage are evaluated. In addition, the gate voltage can be either connected to the DUT's Kelvin Source terminal or a source terminal, as depicted in Figure 2.

### 2.2 Source Feedback and Kelvin Source Configuration

To turn on a MOSFET, a certain charge must be supplied to the gate to raise the gate-source voltage $V_{GS}$ above its threshold value $V_{GS,TH}$. Typically, a voltage source based gate driver is used for this purpose. A gate resistor $R_G$ is added in series in the gate loop to control overvoltages and limit the current of gate drivers. In this way, the speed of the turn-on and turn-off processes can be adjusted (i.e., slowed down) to reduce the voltage overshoot between the drain and source during switching events. The reference potential for gate control voltage can be either connected to a Kelvin Source terminal or at the external source terminal. The difference in connectivity is illustrated in Figure 3.

The Kelvin Source (SK) connection isolates the gate current from the power current path, increasing switching speed and reducing switching losses. The effective gate source voltage is expressed as

$$V_{GS} = V_G - I_G \cdot R_G \tag{1}$$

where $V_G$ is the gate voltage and $I_G$ the gate cur-

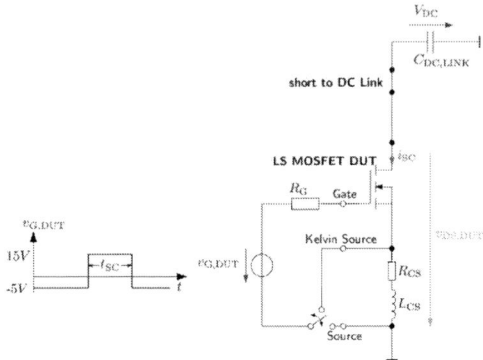

(a) Short-circuit test Setup

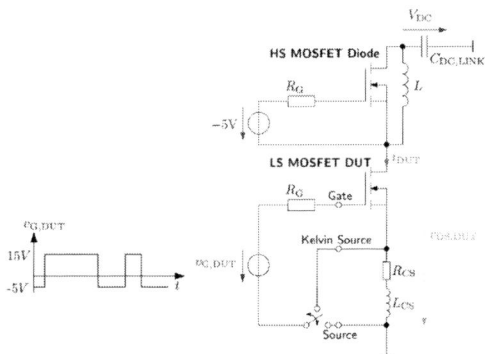

(b) Double-pulse test setup

**Fig. 2:** Comparison of switching events between SK and SFB configuration

rent. In the SFB circuit, the gate loop includes the source inductance $L_{CS}$ and resistance $R_{CS}$. These parasitic components arise from the inductance and resistance of the connection between a chip and the copper traces on the substrate, typically wire bonds. During a current transient, a voltage is induced through $L_{CS}$ and counteracts the applied gate driver voltage $V_G$, reducing the effective voltage observed at the gate-source terminal. This relationship can be expressed as

$$V_{GS} = V_G - I_G \cdot R_G - V_{CS} \tag{2}$$

where $V_{CS}$ is the voltage induced across common source impedance and is a function of drain current:

$$V_{CS} = R_{CS} \cdot I_D + L_{CS} \cdot \frac{di_D}{dt} \tag{3}$$

According to Eq. (2) and Eq. (3), the introduction of a SFB connection affects the device's switching transient by counteracting the gate driver imposed $v_{GS}$ by the drain current and its rate of change dur-

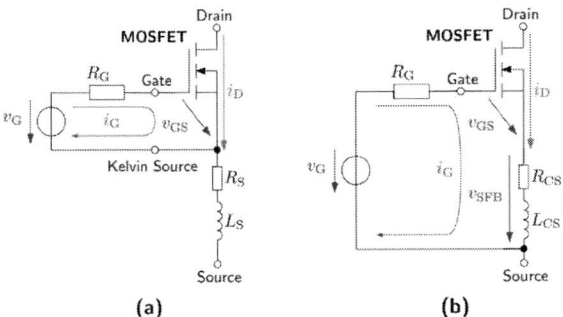

**Fig. 3:** (a) Kelvin Source and (b) Source Feedback connection

ing the switching transients. Therefore, the switching process of the device can be influenced by appropriate adjustment of the additional parameters $L_{CS}$ and $R_{CS}$. The influence of these two parameters will be investigated in the following sections.

## 3 Simulation Analysis

In this investigation, the source inductance $L_{CS}$ and resistance $R_{CS}$ are simulated using a simplified half-bridge circuit in a SPICE tool. The influence of the source inductance on the short-circuit behavior is first examined by varying $L_{CS}$ within the range of 0 to 2 nH. The corresponding simulation outcomes are presented in Fig. 4. The increase in $L_{CS}$ reduces the slew rate of short-circuit current $I_{SC}$, yielding a delayed current peak during turn-on and a reduced voltage overshoot during turn-off event. The slowdown of the short-circuit current also reduces the short-circuit energy loss, allowing the MOSFET to be turned on for a more extended period before reaching the critical short-circuit energy threshold.

To evaluate the effect of $R_{CS}$ on the short-circuit performance, in a second simulation series, the value of $R_{CS}$ is varied within the range of 0 to 2 mΩ. The simulation outcomes are presented in Fig. 5. At the beginning of a short-circuit event, the voltage drop across $R_{CS}$ is low, causing a minimal feedback impact. The rate of change of short-circuit current remains relatively consistent across different $R_{CS}$ values. As the short-circuit current increases, the voltage drop across $R_{CS}$ also increases, intensifying the feedback effect, while the influence of $L_{CS}$ decreases. Larger $R_{CS}$ values result in smaller drain currents, improving short-circuit robustness while reducing operational efficiency.

## 4 Measurement Results

### 4.1 Measurement Setup

A practical test setup for evaluation has been built around a test commutation cell (TCC), which is designed to emulate the behavior of actual power modules. The TCC implements a half-bridge module using two paralleled SiC semiconductor chips, directly bonded on a PCB, for each topological switch of the half-bridge. The TCC is designed to allow either a SK or a SFB connection for the DUT. The ratings of the used devices as DUT are summarized in Table 1. For the short-circuit test, the HS switch on the TCC is replaced by a direct short. Figure 6 illustrates the short-circuit setup using the TCC and a DC-link capacitor. For a double-pulse test, the HS switch is turned off and acts as a diode. A load inductor is connected across the HS switch to emulate the behavior of an inductive load in practical applications.

**Tab. 1:** Rating of 1200 V Bosch SiC MOSFET

| $V_{DS}$ | $I_{DS}$ | $R_{DS(on)}$ | $V_{GS}$ |
|---|---|---|---|
| 1200 V | 200 A | 10.6 mΩ | −5 V...18 V |

### 4.2 Short-Circuit Measurement Results

Short-circuit tests are performed on the TCC with both SK and SFB connections. Figure 7 shows the measured waveforms of the short-circuit current, drain-source voltage, and gate-source voltage. Due to source feedback, the voltage overshoot (orange curve) and peak current (red curve) during the short-circuit are reduced. This results in reduced energy dissipation, thereby enabling an extended duration for the short-circuit $t_{SC}$.

In the experiment, the short-circuit time $t_{SC}$ for both configurations is gradually increased until a short-circuit energy $E_{SC}$ of 1.1 J is reached. The results are plotted in Fig. 8. Assuming the maximum short-circuit energy is 1 J, the SCWT is increased from 1.1 µs to 1.6 µs with SFB. Table 2 shows the summary of the short-circuit measurement comparison.

| | $I_{sat}$ | $t_{SC}$ | $E_{SC}$ |
|---|---|---|---|
| SK | 3227 A | 1.1 µs | 1.058 J |
| SFB | 2207 A | 1.6 µs | 1.095 J |

**Tab. 2:** Short-circuit measurement results

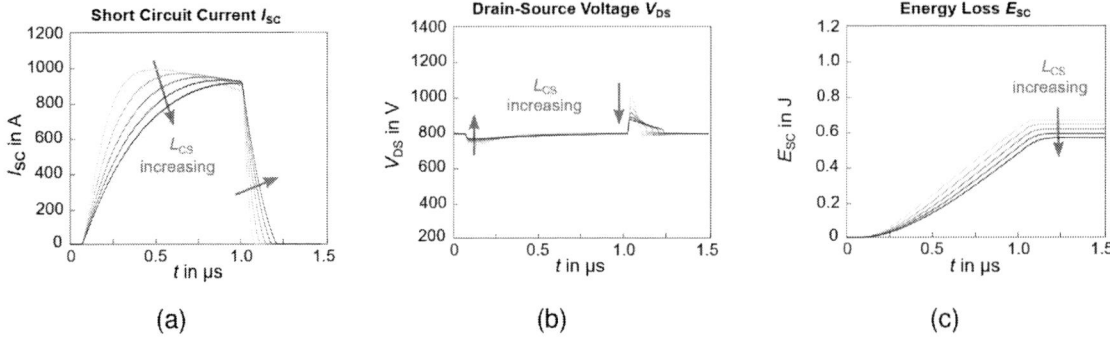

**Fig. 4:** Effect of $L_{CS}$ on short-circuit (a) current $I_{SC}$; (b) voltage $V_{DS}$; (c) energy loss $E_{SC}$.

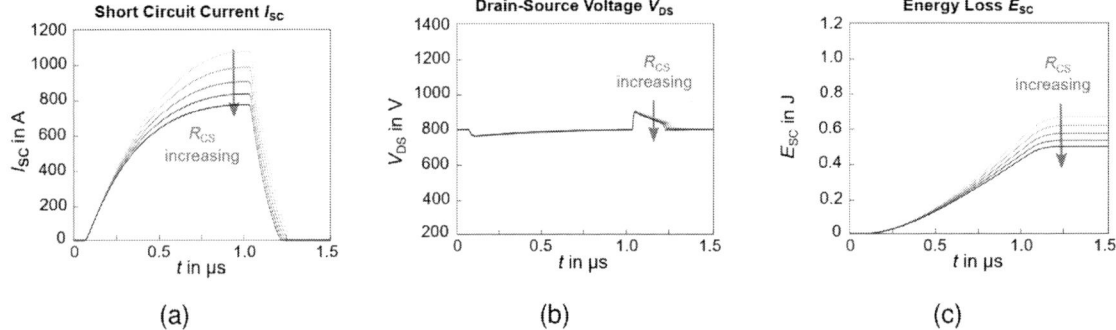

**Fig. 5:** Effect of $R_{CS}$ on short-circuit (a) current $I_{SC}$; (b) voltage $V_{DS}$; (c) energy loss $E_{SC}$.

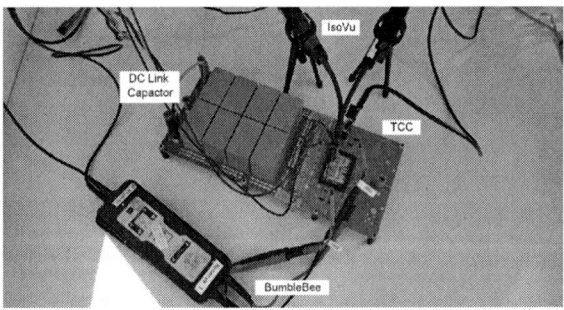

**Fig. 6:** TCC setup for dynamic measurements

### 4.3 Double-Pulse Measurement Results

The measurement results for turn-on and turn-off switching events are shown in Fig. 9. Due to source feedback, the current commutation speed decreases, and a delay in the drop of $V_{DS}$ is introduced, leading to higher losses than in a SK connection. In Table 3, the results of the measured switching losses are summarized.

An increase of $22.6\,\%$ in switching losses is observed when changing from a SK to a SFB connection. At first glance, this appears to be a major drawback of a SFB connection. Therefore, careful consideration must be given when incorporating an SFB connection into an application.

|  | $E_{on}$ | $E_{off}$ | $E_{tot}$ |
|---|---|---|---|
| SK | $12.8\,mJ$ | $8.9\,mJ$ | $21.7\,mJ$ |
| SFB | $17.1\,mJ$ | $9.5\,mJ$ | $26.6\,mJ$ |

**Tab. 3:** Double-pulse measurement results per die

## 5 Performance Investigation

In a traction inverter application, the power module is designed for a specific performance using exactly the required SiC semiconductor area to have a reliable and cost-efficient system design when considering conduction losses, switching losses, and thermomechanical boundary conditions. Besides the performance, it is crucial to achieve the required $t_{scwt}$ to allow the system to safely turn off half-bridge short-circuits and avoid safety hazards like thermal incidents and unwanted motor torque. Between these requirements, a trade-off must be found. This relationship is illustrated in Fig. 10. The green (solid) line shows the value of $R_{on}$ against the applied gate-voltage $V_{GS}$. The yellow (dashed)

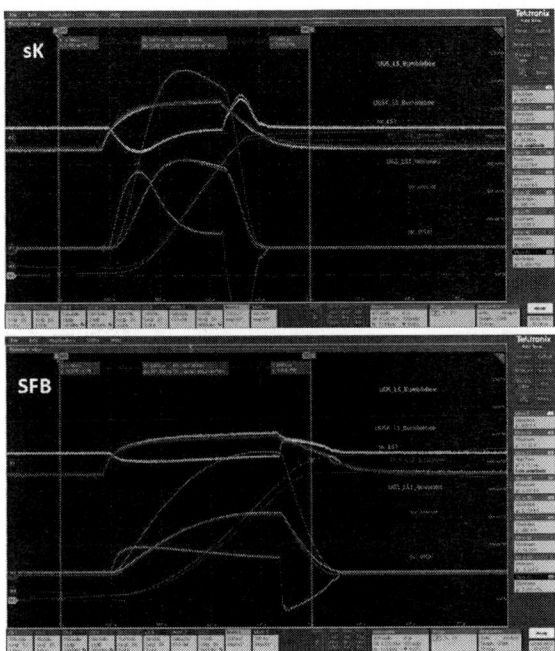

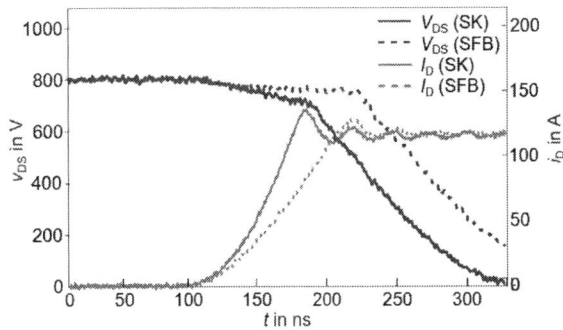

(a) Turn-on event per die.

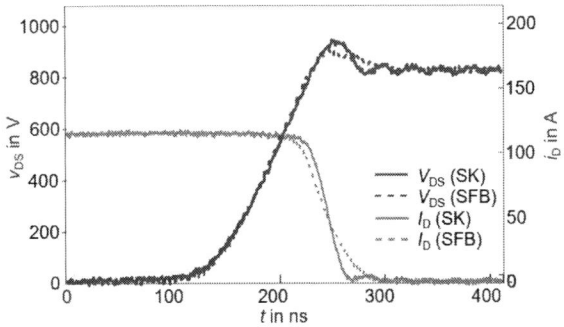

(b) Turn-off event per die.

**Fig. 9:** Comparison of switching events between SK and SFB configuration.

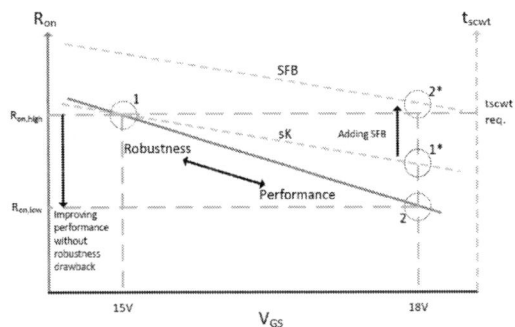

**Fig. 7:** Short-circuit measurement with SK and SFB configuration. $V_{GS} = 15\,\text{V}, V_{DS} = 800\,\text{V}$

lines show the achievable $t_{scwt}$ against $V_{GS}$ for two different source connections (upper trace with SFB and lower trace without SFB). In a typical application, the driver $V_{GS}$ is the primary parameter, allowing a shift between performance and robustness. This voltage is set within the allowed limits to the value achieving the required $t_{scwt}$, resulting in a particular $R_{on}$. For example, in operating point 1 (cf. Fig. 10), $V_{GS}$ is set to $15\,\text{V}$ to achieve the required $t_{scwt}$. This operating point may not be the point where the actual SiC technology provides the best $R_{on}$ and therefore the lowest conduction losses.

If $V_{GS}$ is increased from $15\,\text{V}$ to $18\,\text{V}$ without a SFB,

**Fig. 10:** Trade off between performance and short-circuit robustness

the device operates at point 2 with a decreased $R_{on}$ value, indicating lower conduction losses. However, the allowable $t_{scwt}$, which is indicated by point 1*, violates the required short-circuit withstand time, thus operating the device at this point with $V_{GS} = 18\,\text{V}$ is not permissible. If a SFB is added, the $t_{scwt}$ line is shifted up and the device would operate in point 2*, achieving an improved $t_{scwt}$ again. The aforementioned explanation indicates that a performance

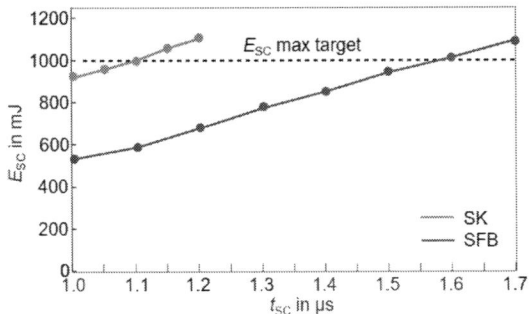

**Fig. 8:** Comparison of short-circuit energy between SK and SFB configuration

improvement using a SFB is possible, as will be derived using a numerical example in the following. If a MOSFET is used in a inverter application, the switching losses $P_{SW}$ and conduction losses $P_{cond}$ yield a total losses of:

$$P_{tot} = P_{cond} + P_{sw} \qquad (4)$$

Considering the thermal resistance $R_{th}$ of the power module, the resulting temperature swing $dT$ is given by:

$$dT = P_{tot} \cdot R_{th} \qquad (5)$$

Introducing a common source feedback increases $P_{sw}$, but $P_{cond}$ can be reduced by using an increased $V_{GS}$. A better performance is achieved if the overall losses $P_{tot}$ decrease. Table 4 shows example values from a typical inverter application used for the calculation.

| $P_{sw}$ | $P_{cond}$ | $R_{th}$ | $dT$ |
|---|---|---|---|
| 33 W | 77 W | 0.86 K/W | 95 K |

**Tab. 4:** Performance parameter of an example application per die

From the measurements with SFB, a rise of $P_{sw}$ by $22.6\%$ is observed while at the same time the short-circuit withstand time increases from $1.1\,\mu s$ to $1.6\,\mu s$. For the MOSFET in use, it is established that the $R_{on}$ value decreases by $6.2\%$ while the $t_{scwt}$ decreases by $3.5\%$ per volt increase in $V_{GS}$. The static voltage drop across the common source resistance of $R_{CS} = 2\,m\Omega$ at the rated current of $200\,A$ is $V_{Rcs} = 0.4\,V$. This results in an effective increase of the gate-source voltage by $dV_{GS} = 2.6\,V$, when compared to SK connection. Considering these results, a numerical summary is given in Table 5 for a particular application comparing a Kelvin Source connection operating at $V_{GS} = 15\,V$ to a common source feedback connection operating at $V_{GS} = 18\,V$.

| | $P_{cond}$ | $P_{sw}$ | $dT$ | $t_{scwt}$ |
|---|---|---|---|---|
| $SK_{15V}$ | 77 W | 33 W | 95 K | $1.1\,\mu s$ |
| $SFB_{18V}$ | 64.6 W | 40.5 W | 90.4 K | $1.4\,\mu s$ |

**Tab. 5:** Parameter Comparison SK and SFB configuration

It is shown that the total losses decrease by $4.5\%$, and therefore, the SFB connection overcompensates the $t_{scwt}$ drawback of increased $V_{GS}$. This

provides the freedom to decrease the chip size, draw more current out of the system, or achieve the required short-circuit withstand time.

# 6 Conclusion

In this work, the influence of the source inductance $L_{CS}$ and the source resistance $R_{CS}$ on the short-circuit robustness and losses of SiC MOSFETs in a half-bridge configuration for traction inverters is investigated. $L_{CS}$ primarily affects the dynamic current rise, whereas $R_{CS}$ affects mainly the static current. Incorporating a well-defined source feedback can improve the short-circuit withstand time while allowing the application to use a higher gate-source voltage of the inverter switches and can, therefore, reduce the overall losses of the system.

# References

[1] A. Nisch, M. Heller, W. Wondrak, A. Bucher, C. Hasenohr, *et al.*, "Simulation and measurement-based analysis of efficiency improvement of sic mosfets in a series-production ready 300 kw / 400 v automotive traction inverter," in *2020 22nd European Conference on Power Electronics and Applications (EPE'20 ECCE Europe)*, 2020, P.1–P.10. DOI: 10.23919/EPE20ECCEEurope43536.2020.9215765.

[2] J. Sun, H. Xu, X. Wu, and K. Sheng, "Comparison and analysis of short-circuit capability of 1200v single-chip sic mosfet and si igbt," in *2016 13th China International Forum on Solid State Lighting: International Forum on Wide Bandgap Semiconductors China (SSLChina: IFWS)*, IEEE, 2016, pp. 42–45.

[3] J. An, M. Namai, H. Yano, N. Iwamuro, Y. Kobayashi, and S. Harada, "Methodology for enhanced short-circuit capability of sic mosfets," in *2018 IEEE 30th International Symposium on Power Semiconductor Devices and ICs (ISPSD)*, IEEE, 2018, pp. 391–394.

[4] A. Bolotnikov, P. A. Losee, R. Ghandi, A. Halverson, and L. Stevanovic, "Optimization of 1700v sic mosfet for short circuit ruggedness," in *Materials Science Forum*, Trans Tech Publ, vol. 963, 2019, pp. 801–804.

[5] L. Xu, D. Xiaochuan, Z. Hao, W. Yi, L. Xuan, *et al.*, "Analysis of short-circuit behavior for sic mosfets with various circuit characteristics," in *2020 17th China International Forum on Solid State Lighting and 2020 International Forum on Wide Bandgap Semiconductors China*, IEEE, 2020, pp. 54–57.

PCIM Europe 2024, 11– 13 June 2024, Nuremberg      DOI: 10.30420/566262175

# HybridPACK™ Drive Power Module with SiC- MOSFET's and Monolithic RC- Snubber Chips for Optimized Power Density

Andre Uhlemann, Nikolaj Gorte, Andreas Groove, Thomas Hunger
Infineon Technologies AG, Germany

Corresponding author and poster Speaker: Andre Uhlemann, andre.uhlemann@infineon.com

## Abstract

The snappiness of the parasitic body diodes of SiC- MOSFETs lead to high voltage overshoots, often exceeding the physical limits of the chip technology. The impact of monolithic RC-snubber chips paralleled to MOSFET half bridges were investigated in HybridPACK™ Drive power modules. Due to the filter-effect no oscillation effects occur and the EMI-amplitude is reduced by 30 dB. For diode recovery operation a significant overvoltage reduction is found. Hence, lower $R_{G,on}$-resistors can be used which results in lower total dynamic losses (26%).This saving-effect is significant and manifests itself in 8% higher RMS-currents or in 12% lower junction temperatures.

## 1   Introduction

The constant demand for increasing the efficiency of the power electronic components utilized for traction in electrical vehicles (EV), especially for battery EV, lead to a strong push of SiC-MOSFETs. SiC offers faster switching and lower static and dynamic losses compared with the well-established Si-IGBT solutions. However, fast SiC switching can suffer on ringing and pronounced voltage overshoot phenomena mainly caused by the high current transients together with the circuit parasitics [1]. In our paper we focus on the influence of integrated **RC-S**nubber (RCS) chips which are a monolithic integration as a series connection of a resistor and a capacitor in one Silicon die in terms of switching performance of the body diode. The advantage of different snubber concepts is demonstrated in [2]. The paper is organized as follows: section 2 introduces the concept of the RC snubber chips and the assembly in power module demonstrators. In section 3 the dynamic behavior is studied followed by simulating parameter variation of RCS in part 4. Section 5 discusses the implications on inverter operation in terms of total power output and EMC.

## 2   RC-Snubber single chip

### 2.1   Basic Principle

In our investigation RCS-chips manufactured by Fraunhofer IISB, Erlangen, Germany, were utilized. The capacitor in series with a resistor is realized on a processed silicon substrate here the bulk silicon of thickness t and chip area A provides the resistance. The capacitor is formed utilizing an array of deep trenches etched into the surface of the silicon wafer and subsequently filled with dielectrics and a highly doped polysilicon finally forming a MOS capacitance [3]. **Fig. 1** shows a schematic cross section of an RC snubber equipped with a bondable front side aluminum metallization and a solderable rear side allowing for an ease integration into a power module.

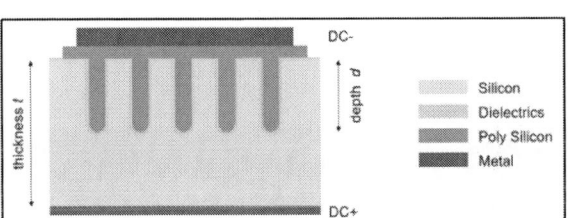

**Fig. 1**   Schematic cross section of a RC-snubber chip.

The diagrams in **Fig. 2** show typical design values for the resistance and capacitance value for a chip of 0.25 cm² deduced from simple geometrical considerations. For the resistance the silicon without trenches is considered. The capacitance is simply calculated for a trench having a diameter of 4 μm and an isolating dielectric of thickness 1 μm and $\varepsilon_r = 4$, neglecting the contribution of the planar parts, space charge region effects and more sophisticated dielectric stacks. Resistance and capacitance values in the range of several $\Omega$ and nF are achievable.

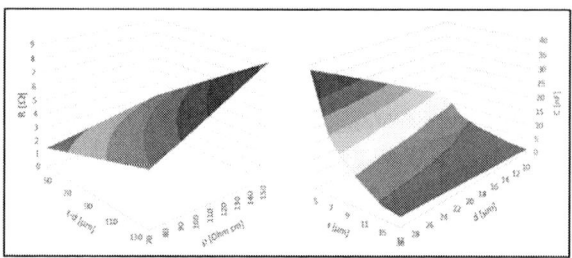

**Fig. 2** Design values of a hypothetical RC snubber chip having an area of A = 0.25 cm². Left: Resistance as a function of the trenchless part of silicon thickness t-d and the resistivity of the base material, right: capacitance in dependence of trench separation and trench depth.

## 2.2 Power Module Demonstrators and Double Pulse Setup

**Fig. 3** shows the overview on a half bridge in a HybridPACK™ Drive power module equipped with the SiC MOSFET devices and the RCS-chips placed between both DC terminals. The RCS parameters are: ESR = 3,8 Ω and ESC = 4,6 nF.
(**E**quivalent **S**eries **R**esistance: ESR)
(**E**quivalent **S**eries **C**apacitance: ESC)

**Fig. 4** shows schematically the measurement setup used for dynamic characterization. Rogowski coils were used to individual determine the current flowing both through the MOSFET in diode mode and the RCS- chips. It is necessary to measure the snubber current separately by a Rogowski-coil placed under the bond wires. With these measurements the MOSFET current and the relevant energy losses can be calculated.

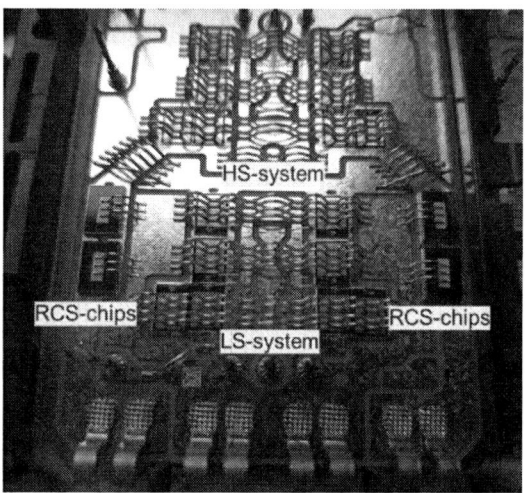

**Fig. 3** Photograph of a half bridge power module with 750V SiC- MOSFET Gen2 devices as High Side (HS) and Low Side (LS) switches and RCS- chips placed between the DC-terminals

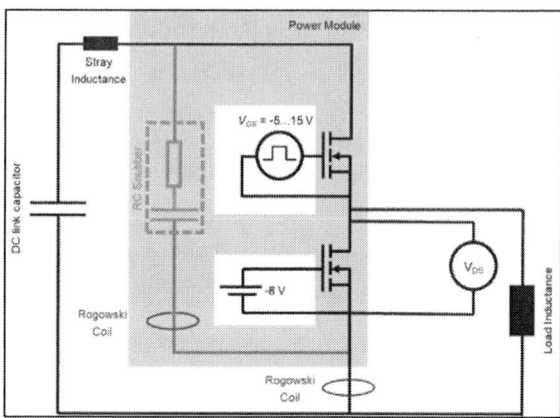

**Fig. 4** Circuit diagram for SiC- MOSFET body diode characterization. The placement of the RCS as well as the Rogowski coils for current measurements are shown. The system stray inductance is 14,7nH.

# 3 Investigation of SiC- MOSFETS with RC- Snubber Chips

## 3.1 Dynamic Measurements & working principle

The power modules are characterized by using a double pulse procedure. In **Fig. 5** typical diode recovery waveforms are shown. The drain current ($I_D$) and the source-drain voltage ($V_{DS}$) are plotted versus time. This is done for a power module with activated (red curves) and deactivated RCS (black curves).

For measurements with activated RCS, no significant oscillation occurs and the overvoltage is kept below 450 V. The peak current, however, is higher than the measurement without RCS. This surplus current comes from the discharging snubber capacitor.

The measurements with deactivated snubber were performed at the same power module but with detached bond wires. The measurement results show strong oscillations both in $V_{DS}$ and $I_D$ as well as a significant higher overvoltage reaching 700 V. This effect is also visible in **Fig. 6** for different switching speeds and load currents. It can be clearly seen that the power module with RCS-chips can be switched significantly faster than the same module without RCS- chips (comparison at overvoltage $V_{max}$ = 700 V). Concerning the voltage overshoot during turn-off operation we found a different picture. Due to the small ESC value of the RCS, the overvoltage reduction is relatively small (≤ 4%).

For a higher overvoltage reduction, capacitors with higher capacitance values are needed.

At the passive switch -side the snubber chips impact the oscillation effect and the high di/dt- vales. It leads to a significant damping effect analog to the critical damping case. Due to this damping effect, the signal heights, the slopes (di/dt) and the over voltages ($L_s$*di/dt) are reduced.

In **Fig. 7** the dynamic losses are compared in dependence of the presence of the RCS. The overvoltage limit is set to 700 V for both configurations, see **Fig. 6**. The gate turn-on resistors $R_{G,on}$ are chosen accordingly.

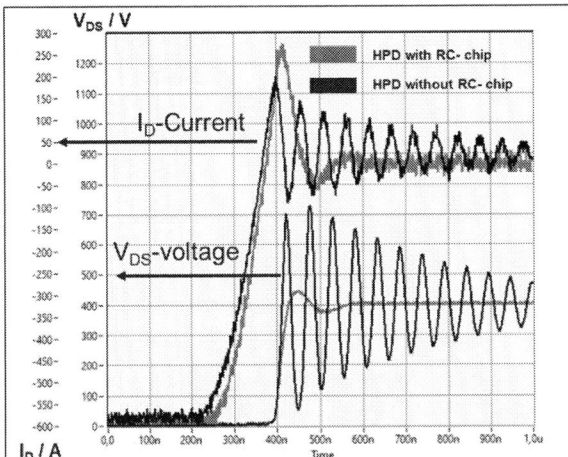

**Fig. 5** Transients of $I_D$ and $V_{DS}$ during diode recovery operation with (red) and without (black) RCS-chips at $V_{DC}$ = 400 V.

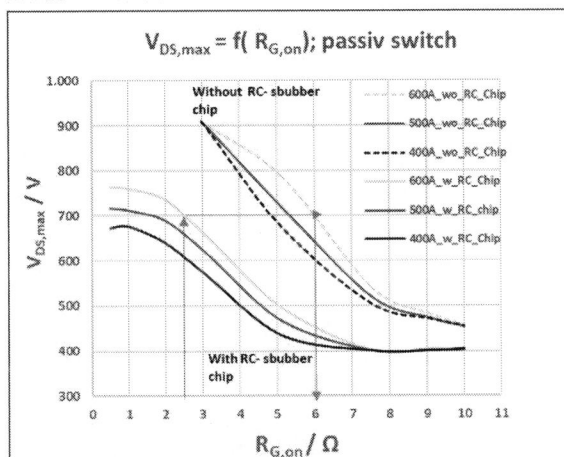

**Fig. 6** Maximum overvoltage $V_{DS,max}$ of passive switch during diode recovery with (w) and without (wo) RCS- chips as function of $R_{G,on}$.

The power module with activated RCS can be turned-on with $R_{G,on}$ = 2,4 Ω. With deactivated RCS the turn-on must be slowed down with a $R_{G,on}$ = 6,2 Ω. This leads to a five times lower turn-on energy. But higher switching speed also implies higher dynamic losses in the parasitic body diode.

The chosen turn-off resistor is almost the same for both configurations. Regarding the accumulated total switching losses, the RCS leads to an energy reduction of 26%.

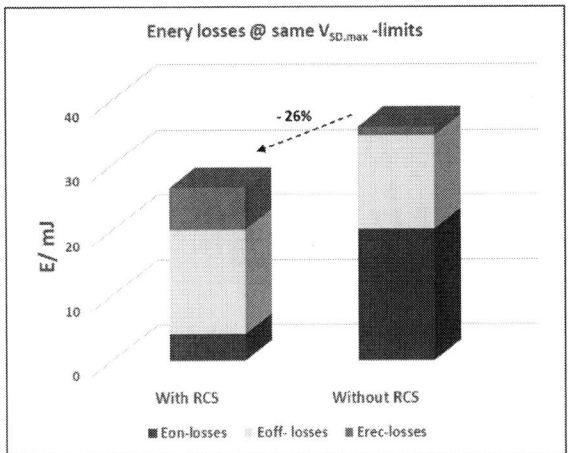

**Fig. 7** Dynamic losses with and without RCS-chips targeting the same $V_{DS}$-limits. Tj=175°C; $V_{DC}$ = 400 V, $I_{load}$ = 400 A; $V_{GS}$= -5/+15V

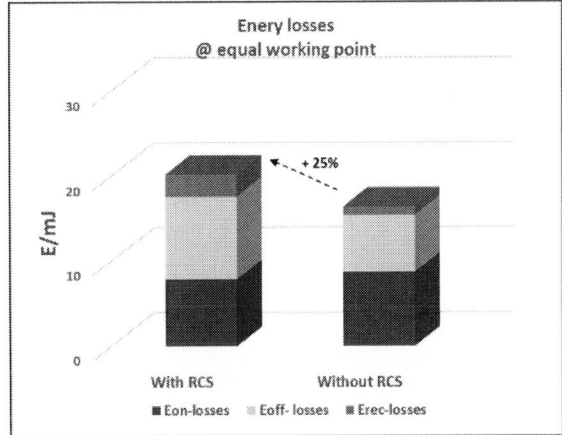

**Fig. 8** Dynamic losses for equal driving conditions: Tj=175°C; $V_{DC}$ = 400 V, $I_{load}$ = 400 A; $R_{G,on}$ =5,1 Ω and $R_{G,off}$ = 1 Ω; $V_{GS}$= -5/+15V

**Fig. 8**. The presence of the RCS leads to higher $E_{off}$ and $E_{rec}$. This is caused by higher current under- and overshoots leading to higher energy losses, see **Fig. 5** again. Both effects are hence caused by discharging and charging snubber capacitors.

Thus, for improving the snubber efficiency it is mandatory to reduce the snubber peak current. This will be investigated in the next chapter via simulation.

# 4 Electrical Simulation of Snubber behaviour

## 4.1 Setting up the model

To set up a proper simulation model, a full package model of the electrical parasitic elements is generated first. The model is analyzed in Ansys Q3D Extractor. All electrical parasitic elements are hence considered. The snubber capacitors are modelled as blocks of non-conducting material equipped with metal plates at the front and rear side forming the terminals which are then available in compact model for the SPICE simulator (**Fig. 9**).

The parasitic compact model is then imported into a SPICE-environment and connected to inner parts like chip models and snubber capacitors, and to the outer periphery like gate-driver, DC-link, probes etc. A double pulse setup similar to **Fig. 4** is created in the simulator (**Fig. 10**).

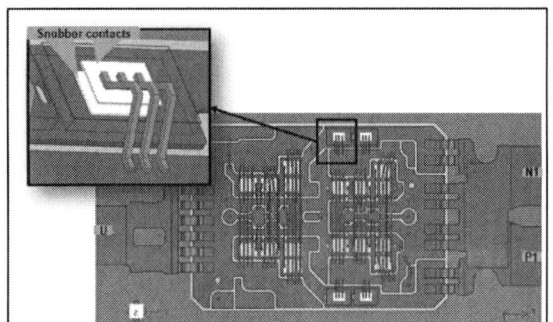

**Fig. 9** 3D-model to calculate the compact model of electrical relevant elements.

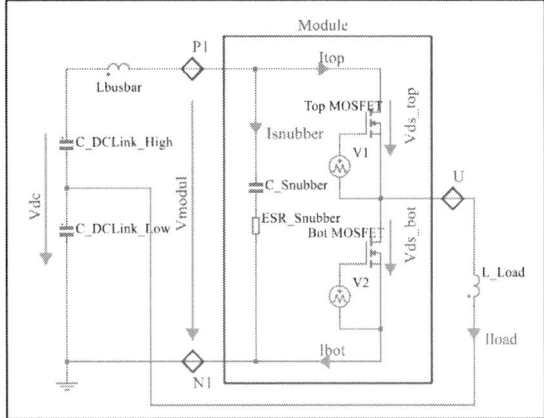

**Fig. 10** Circuit diagram to define the measurements

## 4.2 Circuit simulation

An example showing the turning on the bottom switch is shown in **Fig 11**.

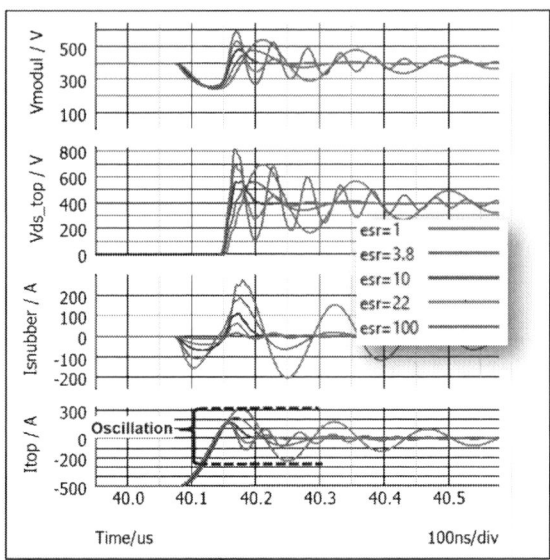

**Fig. 11** Waveforms at bottom switch turn-on with influence of the snubber-ESR in $\Omega$.

By varying the electrical series resistance of the snubber capacitor it can be seen, that there is an optimum ESR-value to achieve a minimum of oscillations amplitude. Let's define the oscillation as the difference between the maximum and the minimum of the first oscillation period of the recovery current $I_{top}$ in **Fig. 12**. For ESR = 1 $\Omega$ and ESR = 100 $\Omega$ the oscillation value is very high. For ESR = 10 $\Omega$ there seems to be an optimum, because the oscillation shows only a slight overshoot.

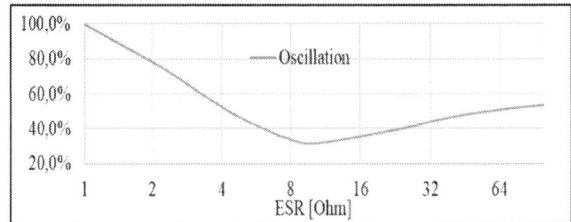

**Fig. 12** Oscillations amplitude as function of ESR normalized to value at ESR = 1 $\Omega$.

The losses in dependence of ESR are displayed in **Fig. 13**. $E_{on}$ and $E_{off}$ relates to the switching MOSFET (here the bottom device) whereas $E_{rec}$ is related to the top diode (in freewheeling mode). The total value is given as $E_{tot}$. Interestingly, the minimum of $E_{tot}$ seems to occur at larger values than the ESR determined for minimizing the oscillation amplitude.

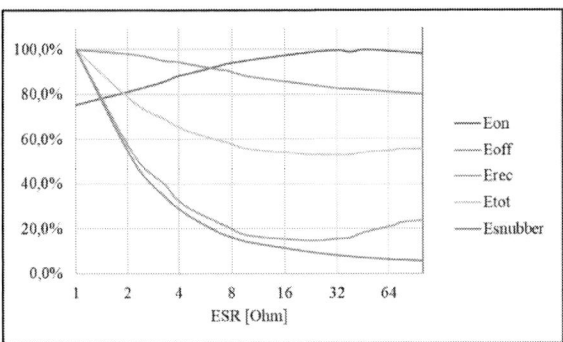

**Fig. 13** Normalized energies as function of ESR during bottom switch turn-on.

# 5 Inverter operation

## 5.1 Description of the test setup

The following measurement results are based on an inverter measurement close to the application. The measurement was carried out using the following equipment: DC source with a power range up to 30 kW, an Infineon HybridPACK™ Drive SiC-G2 Evaluation Kit, an IR (InfraRed)-camera and a specially designed PCB in order to obtain IR images from the chip surface. An oscilloscope was used to record the voltages and currents of the module during operation. The SiC-G2 Evaluation Kit typically includes a standard cooler, DC-Link capacitor, logic board and a driver board for controlling the module. Two modules were used for the measurement, one with additional RCS-chips ($R_{G,on}$: 1,65 Ω / $R_{G,off}$: 1,65 Ω) and one without RCS- chips ($R_{G,on}$: 5,95 Ω / $R_{G,off}$: 2,35 Ω). The setup as well as a photograph of the substrate specially prepared for IR- thermography are shown in **Fig. 14**.

The thermal behavior of the module was investigated using an IR- camera and an IR- PCB. This way, the thermal behavior of the module was observed during operation. A passive inductive load was used instead of a motor.

## 5.2 Electrical measurement results

The measurement results show the electrical behavior of the module with and without RC-snubber chips. In order to be able to compare the results, the modules (with and without RCS-chips) were operated under the same conditions (Tj, DC- link voltage, $I_D$- current and the same $V_{DS, max}$- level). The switching curves of the module without RC-snubber exhibit the typical oscillating behavior of the SiC- module. The switching behavior with the RC- snubber is smoother but also faster, which results in fewer switching losses and more power can be obtained from the module, see **Fig. 15**.

**Fig. 14** Experimental setup (left) and detail of substrate prepared for IR thermography (right).

In this measurement, the total switching energy from $E_{on}$, $E_{off}$ and $E_{rec}$ was reduced by up to 32 %; the total current (Ic) of the module/system was used to calculate the energies.

For modules with RCS, the usage of total current instead of the chip related current leads to lower $E_{off}$ and $E_{rec}$- losses.

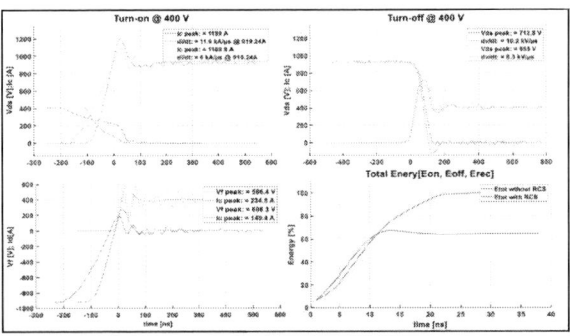

**Fig. 15** Turn-on behavior and diode recovery@ passive switch (Left hand side). Turn- off behaviour and total energy plotted vs. time (right hand side). With RCS = blue line, Without RCS= yellow line.

The smooth switching behavior of the module has a positive effect on the electromagnetic compatibility (EMC) of the module and thus on the whole system. The results of the conducted EMC measurement show that the module with RC- snubber chips has a reduction in conducted interference of approx. 30 dB in the 22 MHz range (setup according to the CISPR 25 [4]). This has a positive effect on the entire system, see **Fig. 16**.

The red curve shows the frequency behavior of the module without RC- snubber, the blue curve shows the frequency behavior of the module with the RCS- chips. The module's smooth switching behavior significantly enhances the electromagnetic compatibility (EMC) performance, thereby benefiting the overall system.

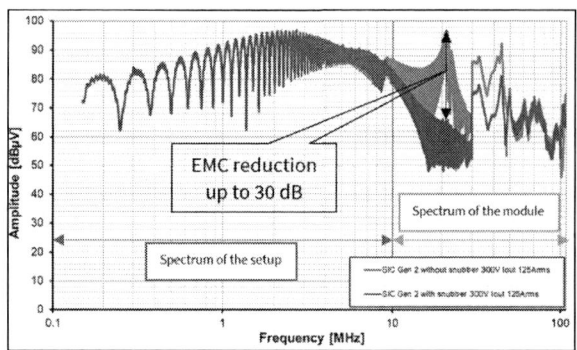

**Fig. 16** EMC spectrum of a module with (blue) and without (red) RC-snubber chips

The results obtained from the conducted EMC measurement demonstrate that the module, integrated with RC-snubber chips operating within the 22 MHz frequency range, experiences a substantial decrease in conducted interference, estimated at approximately 30 dB. This reduction brings about advantageous consequences throughout the entire system.

### 5.3 IR Measurement Results

The IR images (**Fig. 17**) show the thermal behavior of the module (MOSFET chip) as well as the RCS- chips. The images illustrate the heat distribution in the module as well as the heat distribution on the RC-snubber chips. Even though the SiC chips have a temperature above 100°C, the RC-snubber chips only get slightly warmer than the DBC of the module.

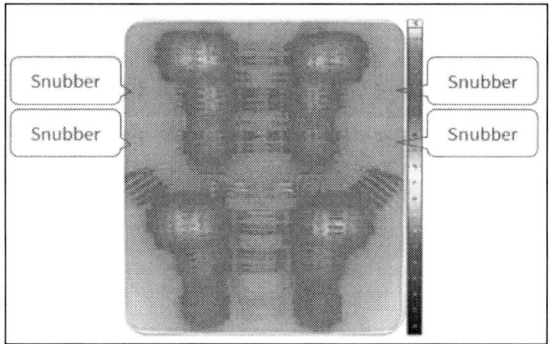

**Fig. 17** Thermal behaviour of the module/system

In **Fig. 18**, the Y-axis ordinate shows the temperature of the MOSFET and the RCS-chips. The abscissa represents the normalized current. Comparing the performance of the modules under the same conditions ($T_{vj}$ = 200 °C), it can be seen that the module with RCS can drive around 8% more power than the module without RCS. This results in a temperature difference of up to 12%.

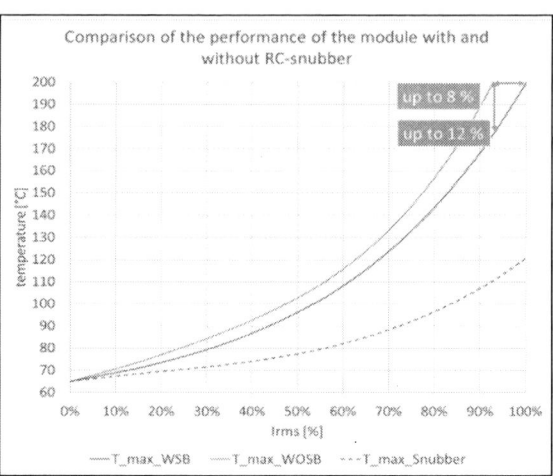

**Fig. 18** Comparison of the performance of the modules with and without RCS in inverter operation.

## 6 Summary

Monolithic RC- snubber chips were tested together with SiC- MOSFETs. Due to the RC- snubber chips the EMI- amplitude was reduced by 30dB and a significant $V_{max}$-reduction could be observed for diode recovery. Hence, lower $R_{G,on}$- resistors could be used which results in lower total dynamic losses (26%). This saving- effect is significant and manifests itself in 8% higher RMS- currents or in 12% lower junction temperatures. These findings can help to increase the RMS- output current or to reduce the active chip area for cost saving. For snubber -chip optimization, higher ESR values up to 10Ω were recommended and motivated by simulation results.

## 7 References

[1] R. Bayerer, D. Domes: *"Power circuit design for clean switching"* in 2010, 6th International Conference on Integrated Power Electronic Systems, 2010 pp. 1- 6

[2] M. Schlüter, M. Pfost "A comprehensive analytical description of asymmetric active snubber" IET Power Electronic 2022,

[3] T. Becker et al., Proc. CIPS 2022, VDE Verlag Offenbach Berlin, 61

[4] CISPR 25: https://www.vde-verlag.de/iec-normen/250584/cispr-25-2021.html [19.03.2024]

## Acknowledgement:

This Topic is supported by the national funded- project trustAE

PCIM Europe 2024, 11– 13 June 2024, Nuremberg    DOI: 10.30420/566262176

# Robust Auxiliary Power Supply for EVs Based on Innovative STi²GaN 650V IC

Federica Cammarata[1], Filippo Scrimizzi[1], Andrea Russo[1], K. Kamran[1], Claudia Malannino[1]

[1] STMicroelectronics, Italy

Corresponding author: Federica Cammarata, federica.cammarata@st.com

Speaker: Federica Cammarata, federica.cammarata@st.com

## Abstract

Among the main building blocks of EV architecture, a relevant position is covered by the emergency DCDC converter, crucial for provisioning power to a wide variety of low voltage subsystems. For this purpose, STMicroelectronics proposes its innovative STi²GaN 650V flyback IC in a closed control loop for HV-LV conversion. Working at 300kHz switching frequency, by autonomously tuning the duty cycle, the system can deliver up to 68W output power and guarantee 87.2% efficiency in a wide loading range. The strength of the system is its outstanding robustness in sustaining very high RMS input voltage stress (battery overvoltage) for long time window.

## 1    Introduction

The fast developing of PHEV and EV markets is moving the design and development of power devices towards high density of integration for electronic solutions that can satisfy stringent automotive requirements in terms of efficiency, compactness and reliability. Beyond the well-known Si-based technology, today the research is focusing on alternative WBG technologies able to withstand higher power delivering (Silicon Carbide), or higher switching frequency (Gallium Nitride) [1]. Thanks to the strong features characterizing GaN, this technology results highly suitable for high frequency and high efficiency power conversion domain [2], [3]. In this direction, STMicroelectronics is releasing a new family of Intelligent & Integrated products based on monolithic GaN solutions tailored for both HV and LV systems in EVs market with HV insulated DC-DC converters, on-board chargers and low voltage DC-DC converters applications' target.

STi²GaN IC solution allows to integrate in a monolithic chip the driver circuitry, with some protections embedded, and the GaN power stage thus minimizing the stray inductances deriving from a

discrete architecture and allowing a safest high frequency working mode.

Among the main building blocks of an EV system, there is the Auxiliary DC-DC converter whose fundamental role is to guarantee, starting from the high voltage battery, the desired DC supply voltage, required by the main safety critical bocks of the car, when the main low voltage battery is down.

For this purpose, a 650V e-GaN IC properly designed for automotive applications is proposed to realize a closed control loop isolated HV to LV DC-DC flyback converter which ensures 15-18V output voltage starting from the main 400V HV battery. The main advantage introduced by the proposed IC, compared to the already available solution, is the extremely high level of integration that the monolithic nature of the device allows thus resulting in heavy reduction of the parasitic inductances characterizing a discrete solution, with not negligible benefits such as:

• Reduced gate-source loop ringing which allows higher switching frequencies and improved reliability;

- Reduced oscillation peaks, meaning lower failure probability and possibility to go to higher supply voltages;
- No need of gate resistances and thus higher switching speed with lower switching losses.

The power HEMT is a 650V e-Mode Gallium Nitride HemT device with 190mΩ or 410mΩ typ. RDSon. The IC has an internal clock generation circuit embedded which guarantees adjustable PWM frequency up to 500kHz by connecting a dedicated RC network to a specific pin. It is also based on a current control loop to properly tune the duty cycle according to the target output voltage. Moreover, logic level inputs compatibility, adjustable soft start, turn on slew rate tuning and maximum on time protections integrated with overtemperature, overcurrent and overload protections are the main features of such smart GaN solution device.

Being its target application an emergency power supply system, it is expected to be in OFF state most of the time. This is the reason why the IC implements a standby mode operation to reduce power consumption, autonomously turning off some not needed internal blocks when working in open load condition.

To obtain a high-power density of this solution, the passive parts have also been optimized with special focus on the transformer.

Able to deliver up to 68W output power, but at the same time properly optimized for open load operation, the converter, with its compact design and simplicity of implementation, is characterized by 87.2% peak efficiency. Moreover, the strong point of the system is its outstanding robustness and reliability in terms of high input voltage withstanding capability.

## 2 Monolithic GaN-based IC overview

Based on the concept of heterojunction [2], interface between two semiconductors (GaN and AlGaN) with dissimilar bandgaps but the same crystal structure, by default, Gallium Nitride is a normally-ON technology, but several methods to make it normally off already exist. Among the e-mode GaN, intelligent and integrated GaN devices are based on p-GaN gate normally off solution [4]. It is realized by inserting an (Al)GaN layer with p-type doping between the AlGaN barrier and the

Schottky gate contact Fig. 1. The effect of the introduction of the p-GaN layer is an increase in the band diagram with the resulting depletion of the 2DEG even at zero gate bias.

Based on Gallium Nitride technology, the integrated power device features a 650V maximum rated voltage.

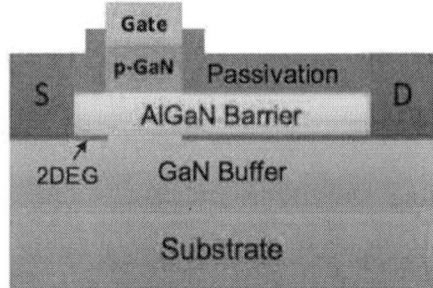

**Fig. 1**   p-GaN HEMT structure.

Both 12V Enhancement HEMT and 12V Depletion HEMT plus Resistor and P GaN capacitors and 650V Enhancement HEMT basic technology block are used in the monolithic integrated circuit (IC). The power switch is a lateral enhancement-mode GaN FET with a current sensing cell used for overcurrent detection. The block diagram of the integrated device is reported in Fig. 2.

Thanks to the monolithic solution, the stray inductances of the power can be reduced thus avoiding the dangerous voltage gate and drain ringing during the switching transients [5]. The power SoC arrangement combines features such as variable current mode PWM controller, soft start, programmable turn-on dv/dt, and thermal diagnostic. Moreover, being housed in a QFN package with double side cooling, the IC ensures a good thermal dissipation.

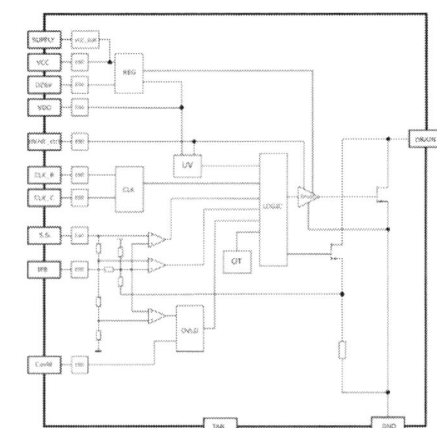

**Fig. 2**   Block diagram of the IC.

# 3 Closed control loop flyback converter topology and design

As already known, a flyback converter can be simply realized by adopting a low side switching device, a transformer with properly selected turns ratio and a diode on the secondary side [4]. To obtain a compact-size high-voltage Flyback auxiliary converter directly connected up to 400V voltage battery bus, it is designed based on the introduced integrated and intelligent GaN IC, which, by embedding the control and driving circuitry, is useful to create an extremely miniaturized and compact closed loop solution based on current feedback.

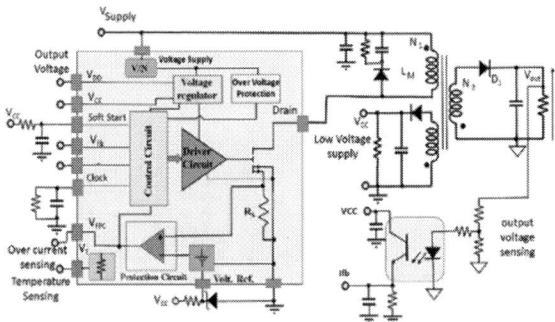

**Fig. 3** Block diagram of flyback converter based on GaN monolithic IC.

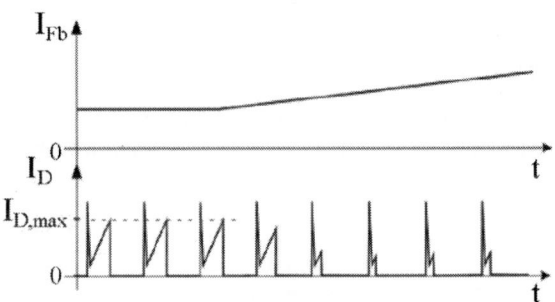

**Fig. 4** Feedback current diagram.

In particular, the developed architecture exploits a TS431 voltage reference on the secondary side and an optocoupler to realize the isolated feedback network starting from a voltage divider on the converter output, see Fig. 3. The alternative solution to optocoupler for real automotive environment is the use of isolator for the car platform.

The working principle of the IC is a current regulation feedback realized by using a current comparator. Without feedback current the circuit operates

up to the ID,max value while, by increasing the feedback current, the drain current starts to decrease, see Fig. 4.

A third winding of the transformer is used to generate the aux voltage to supply the low voltage VCC pin of the part (VCC,max=30V).

The presence of an internal shunt resistance allows to implement a lossless current sensing cell. Moreover, an external RC network makes the application frequency tuning possible, a dedicated resistor is connected to dV/dt pin in order to regulate the turn off drain-source slope and an external RC network is also used to set the soft start transient.

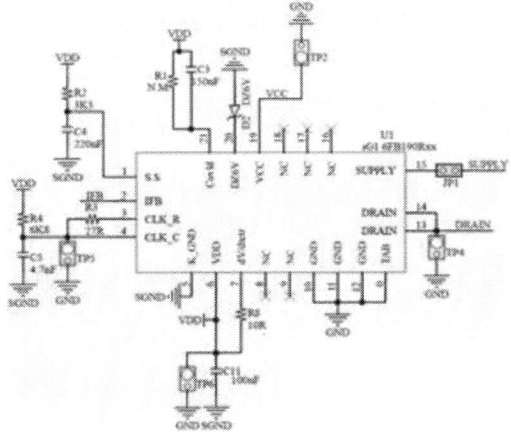

**Fig. 5** Device supplying network.

## 3.1 Emergency power supply specifics

Considering the target application an emergency power supply for EVs, the DCDC converter must be compliant with some specifics especially related to the capability in withstanding a desired output power profile. In particular, Fig. 6 shows an example of the time related power profile of a typical application.

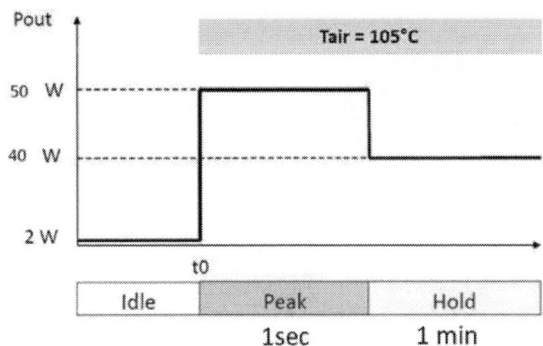

**Fig. 6** Typical AUX converter Power specific.

The converter remains in an idle state for 99% of its lifetime providing 2W output power; then it must be able to provide a peak of about 50W for a limited timeframe (1 second) while it has to be able to manage 40W for a longer period (about 1 minute). Moreover, the complete solution will deal with an ambient temperature equal to 105°C.

## 4 Flyback Transformer Design

With integration of GaN HEMT into the STi²GaN650V controller IC, it can achieve switching frequency up to 500kHz. The possibility of high switching frequency opens the door to reducing the size of magnetics, further enhancing the power density of the solution. To achieve this, planar magnetic technology is employed, leveraging its low profile, reliable reproducibility, excellent control over parasitics, and cost-effectiveness. Additionally, it offers the opportunity to integrate the transformer into the main converter PCB, leading to a more integrated and compact design.

The planar transformer comprises flat copper sheets wound around a rigid or flex PCB but can also be a hybrid. They are inserted in a low-profile magnetic core made from a "soft" ferrite. Their copper tracks have rectangular cross-sections, different from conventional wire-wound components that have circular cross-sections [6].

In this design, a custom core shape is utilized to tailor the design for maximum optimization. The rectangular-shaped core is adopted with rounded corners on the center leg, ensuring a more even current density in the rounded corner of the windings. The core fully encloses the windings, maximizing inductance with a limited number of turns while also making the transformer less susceptible to radiated emissions. With a noticeably low height (8.3mm) as shown in Fig. 7, the overall volume of the transformer is almost 50% less than the conventional wire-wound counterpart.

An interleaved layer arrangement has been implemented, resulting in significantly lower leakage inductance (0.6µH) compared to the traditional wire-wound counterpart. This arrangement is crucial for reducing leakage inductance in high input voltage designs, which in turn helps to minimize voltage spikes during the turn-off event of MOSFETs and reduces snubber loss. This approach is particularly important for managing high input voltage designs. The transformer specifications are shown in Table 1 below:

**Table 1**: Specifications of Planar Transformer

| Power | 65W |
|---|---|
| Turn ratio (Prim/Sec/Aux) | 5.3:1:3.6 (16T/3T/5T) |
| No. of Layers/Copper thickness | 8 Layers/ 2Oz |
| Magnetizing Inductance (Lm) | 300µH |
| Total Leakage Inductance | 0.6µH (sec & aux shorted) |
| Primary DC Resistance | 0.390Ω |
| Secondary DC Resistance | 0.016Ω |
| Primary AC/DC Res. ratio | <6 @ 300kHz |
| Secondary AC/DC Res. ratio | <4 @ 300kHz |

Through iterative FEM analysis, the core and winding geometry is further optimized, specifically focusing on the placement of the airgap and windings to achieve lower AC resistance across the 500kHz range. The ferrite material used is Ferroxcube 3C97 (MnZn) offering very low losses and high saturation in desired frequency range. Also, it provides flat power loss density over wide temperature range (from 60 to 150 °C). The flux distribution, as depicted in Fig. 8, is consistently even, ensuring controlled core loss and maintaining a sufficient margin from saturation.

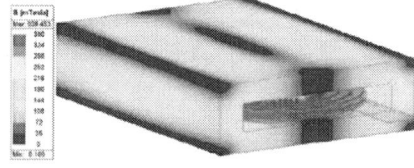

**Fig. 8**   Magnetic Field Distribution in Core

When striving for design compactness, the thermal integrity becomes even more crucial at the system level. Therefore, it is important to accurately estimate the losses stemming from the transformer, particularly the copper and core

**Fig. 7**   Prototype of Planar Transformer.

losses, so that suitable measures may be taken to limit temperature rise. Due to the improved design of the transformer, it minimally contributes to core and copper losses. The worst-case total losses estimated by FEM Analysis amount to 1.5W, comprising 0.9W of copper losses and 0.6W of core losses. As shown below core loss has been estimated also mathematically evaluating the peak flux density and extracting the correspondent power over volume value from 3C97 datasheet.

$$I_{PRI,PEAK} = \left(\frac{I_{OUT} \times K}{1 - D_{MAX}}\right) + \left(\frac{V_{IN,MIN} \times D_{MAX}}{2 \times L_{PRI} \times F_{sw}}\right) = 2.12A$$

$$\Delta I_{PRI} = \left(\frac{V_{IN,MIN} \times D_{MAX}}{L_{PRI} \times F_{sw}}\right) = 0.433A$$

$$B_{MAX} = \frac{n \times I_{MAX} \times \mu_0}{lg} = 237mT$$

$$\hat{B} = B_{MAX} \times \frac{\Delta I}{I_{PRI,PEAK}} = 56mT$$

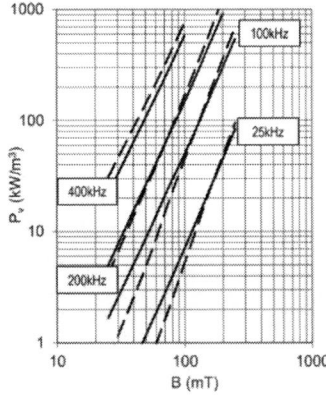

Fig. 9 Power loss as a function of peak flux density

Core Loss=Pv (kW/m$^3$) x Ve (m$^3$)

Core Loss≈170 (kW/m$^3$) x 3504.9(m$^3$) x 10$^{-9}$

Core Loss≈ 600 mW

Copper losses are established considering both AC and DC contributions. DC losses can be evaluated taking into account the values of $I_{DC}$ and $R_{DC}$ of primary, secondary and auxiliary windings while the AC ones can be properly estimated through Dowell's equations.

Thanks to the large surface area of the planar transformer, it provides an efficient path for heat dissipation caused by transformer losses, therefore, improving the overall thermal integrity of the system.

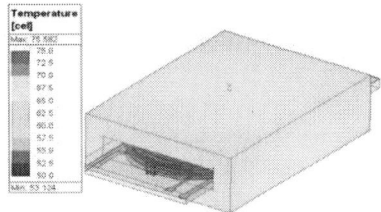

Fig. 10 Temperature Distribution of Maximum Load

The temperature profile of the transformer shown in Figure 9 reveals a smooth distribution over the core surface. The average core temperature is approximately 65 °C which occurs when mapping the maximum core and copper losses at ambient temperature of 25 °C.

## 5 Testing results

The converter, whose demo-board is shown in Fig.11, was evaluated at 300kHz switching frequency with different power delivery requests fixed by a constant current load on the output. The desired switching frequency was set by using R= 6.8k and C= 4.7nF network on oscillator pin.

The converter was tested both in DCM and CCM operation thus increasing the output power.

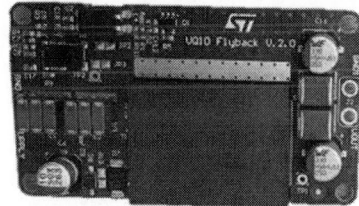

Fig. 11 Picture of the flyback converter demo-board

Moreover, at the startup phase the converter works in burst mode operation, meaning that when the load is low the switch doesn't turn ON at every switching cycle, see Fig. 12. During the light/open load condition, to reduce power consumption, the device turns off some internal blocks, keeping ON only the necessary ones. In such conditions the overall measured power consumption was 600mW.

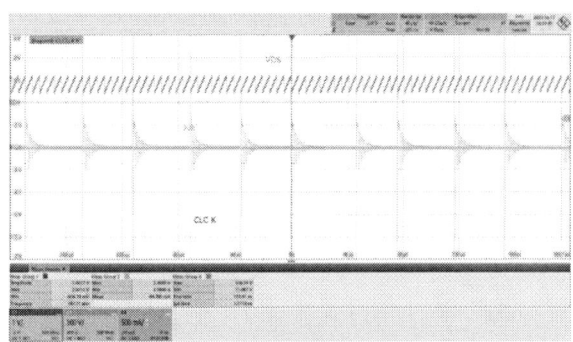

**Fig. 12** Burst mode operation.

Fig. 13 shows the capability of the system to manage hard switching conditions at 400V input voltage and 65W output power. In particular, the blue curve represents the internal clock signal, while the light blue one is the Vds of the device and the yellow one is the drain current with the typical flyback current shape.

Fig. 14 shows the dynamic load variation from 1A to 3A when the DC/DC converter realizes the 400V/15V conversion resulting in a very low (~450mV) output voltage drop.

Finally, Fig.15 shows the switching node zoom details at 4A load when the bridge connection among the transformer and the drain pin of the device, used for the current measurement, is removed. By this connection improvement the final Vds voltage spike is reduced up to 508V.

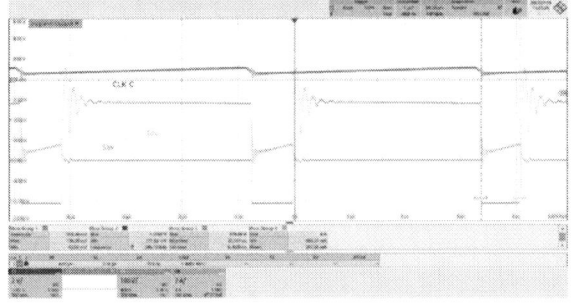

**Fig. 13** Switching waveforms with Vin=400V, Vout=15V, Fsw= 300kHz, Iout= 5A

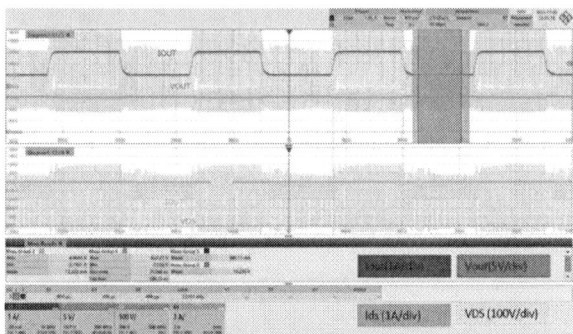

**Fig. 14** Dynamic load regulation from 1A to 3A.

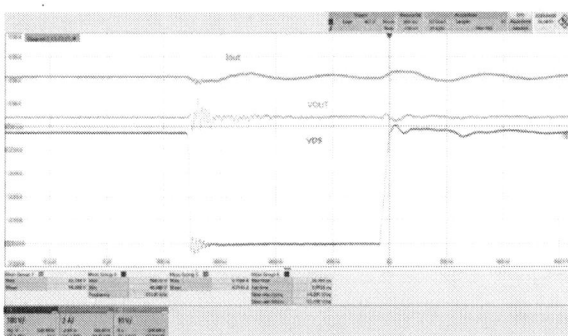

**Fig. 15** Switching node zoom at Iout= 4A with improved connection.

# 6 Voltage stress test

Based on the real use of the device which is directly connected to the high voltage battery of the EV (400V) and considering the battery level fluctuations which could induce battery voltage enhancement up to 500V RMS value, the device has been heavily stressed in such real operating conditions by testing it with 500V supply voltage both at room temperature and 105°C for more than two hours testing time.

Fig. 16 shows the switching waveforms during the test. In particular, the blue line is the drain-source voltage of the power stage, the purple curve is the internal clock of the device and the yellow one represents the current flowing through the primary winding of the transformer. By analyzing the VDS curve it's possible to recognize the hard switching condition and to note the absence of heavy peaks and oscillations thanks to the integrated nature of the device. For this purpose, a 91V Zener diode was selected as clamping network in order to not exceed the absolute maximum rating of the component.

Those tests emphasize the absolute robustness and reliability of the device when used in hard conditions like the real AUX DC-DC converter ones. Moreover, considering that the realized flyback topology is characterized by RCD clamping network and diode-based rectification on the secondary side, Fig.17 highlights the excellent performance of the device in terms of efficiency in a wide output power range. In particular, the efficiency is equal to 87.2% when the power delivery is 68W and the application works in CCM and fixed frequency mode (300kHz). Some additional design changes could be added in order to further improve the efficiency of the system by paying an extra cost in terms of complexity and higher number of components of the application. Different topologies such as quasi resonant or active clamp flyback can help in such direction.

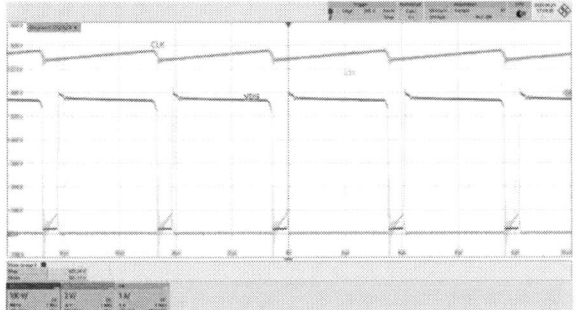

**Fig. 16** Switching waveforms in 500V voltage stress conditions.

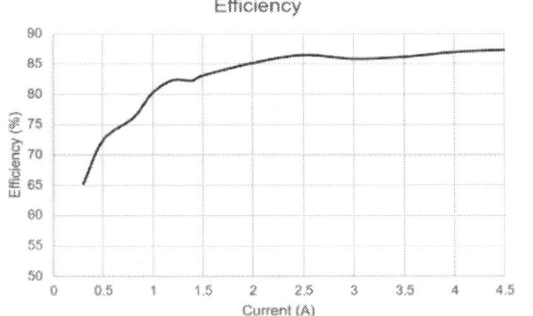

**Fig. 17** Efficiency at 300kHz, Vin=400V.

## 7 Conclusion

The aim of the paper is to provide an innovative solution for high frequency power converter applications tailored for the Automotive field based on HV monolithic GaN power stage with integrated driver/controller and protections. The system is a closed loop flyback converter based on voltage divider plus reference circuit feedback to provide

400V/15V typical voltage conversion. The proposed STi²GaN 650V IC for automotive applications allows to achieve higher power density and very good efficiency results at high switching frequency compared to the conventional available Si based converters. The monolithic system on chip solution minimizes the impact of the stray inductances in the switching behavior and improves the efficiency performance. One of the main strengths of the device is its capability in withstanding very high supply voltage at hot temperatures for long testing time guaranteeing a very high level of reliability for the final application.

## References

[1]. N. Kaminski, O. Hilt, "SiC and GaN devices – wide bandgap is not all the same," Volume8, Issue3, Special Issue: Power Semiconductor Devices and Integrated Circuit, May 2014.

[2]. A. Lidow, M. De Rooij, J. Strydom, D. Reusch, J. Glaser. GaN Transistors for Efficient Power Conversion, 3rd ed.; John Wiley & Sons: Hoboken, NJ, USA, 2019.

[3]. J. Xu, L. Gu, Z. Ye, S. Kargarrazi and J. M. Rivas-Davila, "Cascode GaN/SiC: A Wide-Bandgap Heterogenous Power Device for High-Frequency Applications," in IEEE Transactions on Power Electronics, vol. 35, no. 6, pp. 6340-6349, June 2020, doi: 10.1109/TPEL.2019.2954322.

[4]. S. Musumeci, E. Armando, F. Mandrile, F. Scrimizzi, G. Longo and C. Mistretta, "Experimental Evaluation of an Enhanced GaN-Based Non-Symmetric Switching Leg Integrated Module for Synchronous Buck Converter Applications," 2021 23rd European Conference on Power Electronics and Applications (EPE'21 ECCE Europe), Ghent, Belgium, 2021, pp. 1-10, doi:

10.23919/EPE21ECCEEurope50061.2021.95 70541.

[5]. Scrimizzi F. et al.: The GaN Breakthrough for Sustainable and Cost-Effective Mobility Electrification and Digitalization. Electronics 2023, 12, 1436.

[6]. Z. Ouyang, O. C. Thomsen and M. A. E. Andersen, "Optimal Design and Tradeoff Analysis of Planar Transformer in High-Power DC–DC Converters," in IEEE Transactions on Industrial Electronics, vol. 59, no. 7, pp. 2800-2810, July 2012

[7]. A. A. Mohammed and S. M. Nafie, "Flyback converter design for low power application," 2015 International Conference on Computing, Control, Networking, Electronics and Embedded Systems Engineering (ICCNEEE), Khartoum, Sudan, 2015, pp. 447-450, doi: 10.1109/ICCNEEE.2015.7381410.

# Impact of Various Silicon Diodes on the Hybrid Switch Inverter

Michael Walter ⊚ , Mark-M. Bakran ⊚

University of Bayreuth, Centre for Energy Technology - ZET, Germany

Corresponding author:     Michael Walter, michael.walter@uni-bayreuth.de
Speaker:                  Michael Walter, michael.walter@uni-bayreuth.de

## Abstract

The primary objective of this study is to improve the efficiency of a traction hybrid switch inverter designed for operation at a DC-link voltage of 400V. This improvement is achieved by integrating discrete 650V Si-IGBTs and freewheeling diodes alongside 650V SiC-MOSFETs in a parallel configuration. Due to interlock time the MOSFET switches against the silicon diode which increases the turn-on losses of the MOSFET that lowers the efficiency. To address this issue, the study delves into an analysis of the switching losses when various silicon diodes are employed. Following this analysis, the research evaluates the efficiency of the system during the Worldwide Harmonized Light Vehicles Test Cycle (WLTP) driving cycle

## 1  Introduction

In the pursuit of optimising the cost-effectiveness of electric vehicle (EV) traction systems, a number of strategies are being investigated. One prominent strategy involves the utilisation of low-cost IGBTs to fabricate a full-Si inverter. This solution is economically appealing; however, it introduces a trade-off in the form of heightened inverter losses. In order to achieve a desired vehicle range, there is an inherent requirement for more battery capacity. Conversely, the adoption of silicon carbide (SiC) metal-oxide-semiconductor field-effect transistors (MOSFET) semiconductors, though they carry higher initial costs, offers notable advantages. Inverters utilising SiC MOSFETs have consistently demonstrated superior efficiency in contrast to silicon Insulated-Gate Bipolar Transistor (IGBT), as evidenced by recent studies. Consequently, they require less battery capacity. [1]–[3].

In the evolving field of power electronics, achieving an optimal balance between cost and efficiency remains a pivotal challenge, especially in the context of electric vehicle traction systems. In response to this challenge, the hybrid-switch topology inverter has emerged as an innovative solution. This design strategy combines the strengths of Si-IGBTs and SiC-MOSFETs. Si-IGBTs are particularly favoured in numerous applications due to their cost-effectiveness. The economic appeal of these

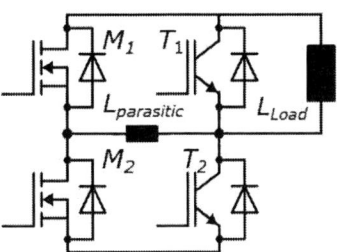

**Fig. 1:** Electrical schematic of the double-pulse measurement setup.

devices does not come at the expense of performance, as they exhibit good conduction characteristics at high current. This makes them a suitable choice for operations under high-load scenarios. Conversely, SiC-MOSFETs, despite their higher cost, offer a number of advantages. They are characterised by reduced switching losses, which becomes increasingly important for efficiency in traction applications. Additionally, their conduction characteristics becomes advantageous under lower load currents, providing a complement to the characteristics of Si-IGBTs. By merging these two distinct sets of advantages, the hybrid-switch topology offers a pathway to exploit the best of both worlds. An electrical schematic of this topology is depicted in Fig. 1. The parasitic inductance amounts $L_{\text{parasitic}} \approx 20\,\text{nH}$. Previous research has not only identified the potential of this topology but has also delved deeper into its intricacies. In particular, numerous studies have focused on the parallel operation of IGBTs and SiC-MOSFETs, with a particular emphasis on delayed switching techniques, in order to ensure performance across a broad operational spectrum. [4], [5].

## 2 Control Strategy

A number of control strategies have been described in the literature, with the majority focusing on delaying the switching process between MOSFETs and IGBTs, and vice versa, across the entire load current range. This paper presents an alternative approach, which separates the control scheme into an MOSFET operation mode at low current and an IGBT operation mode at high current. In the MOSFET operation mode, only the MOSFET is the active switch, while the IGBT remains passive. In IGBT operation mode, the IGBT switches, while the MOSFET supports the IGBT in conduction mode. Therefore, the MOSFET turns on after the hard switching process of the IGBT is completed, as illustrated in Fig. 2. The hybrid switch operating in this manner has been demonstrated to be the most cost-effective solution in comparison to other control strategies presented in [3].

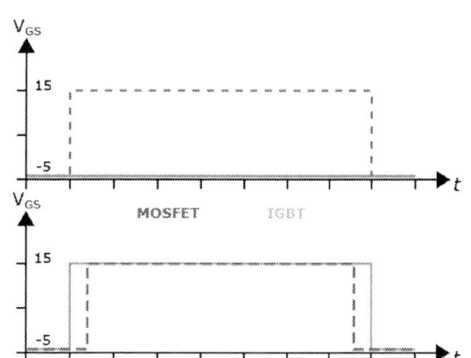

**Fig. 2:** Top: Control scheme in MOSFET operation mode. Bottom: Control scheme in IGBT operation mode.

SiC-MOSFETs and IGBTs differ in their output characteristics. SiC-MOSFETs exhibit an ohmic output characteristic, which means that, relative to their surface area, they experience lower conduction losses at lower current levels compared to IGBTs. However, as the current magnitude increases, the bipolar output characteristic of IGBTs, along with the Si-diode, provides an advantage in terms of conduction losses.

Figure 3 illustrates the probability of occurrence of the load current at a typical driving cycle. It can

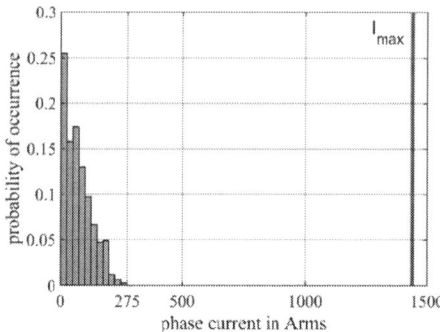

**Fig. 3:** Probability of occurrence of the phase current in a typical driving cycle.

be observed that the driving cycle is significantly influenced by partial load in comparison to the peak load current of a $600\,\mathrm{kW}$ inverter. By using the best cost chip size ratio $A_{\mathrm{SiC}}/(A_{\mathrm{SiC}} + A_{\mathrm{Si}})$ [3], only MOSFET operation is active during driving cycle.

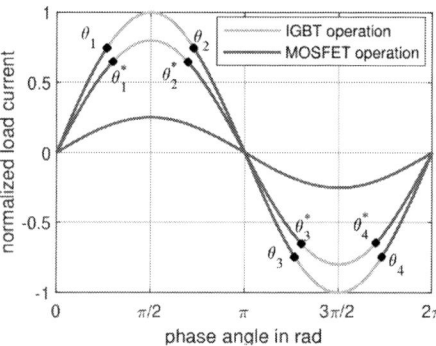

**Fig. 4:** Fundamental control scheme for the transition from MOSFET operation to IGBT operation. The transition occurs within the sine wave.

A strategic approach is employed in the hybrid inverter to optimize the benefits of the distinct output characteristics. In particular, the MOSFET is responsible for driving the inverter when the load current is at a low level within the overall load current waveform. This decision is based on the MOSFET's more favourable output characteristics under these conditions. In contrast, the IGBT is engaged when the load current reaches higher levels, allowing the system to exploit the advantages associated with its bipolar output characteristic. The operational concept underlying this strategy, depicted in detail in Fig. 4, involves a transition from MOSFET operation mode to IGBT operation mode within the sinusoidal load current. The MOSFET operates at a constant RMS current within the load current. The calculation of the phase angle for the transition to IGBT operation is based on Eq. (1).

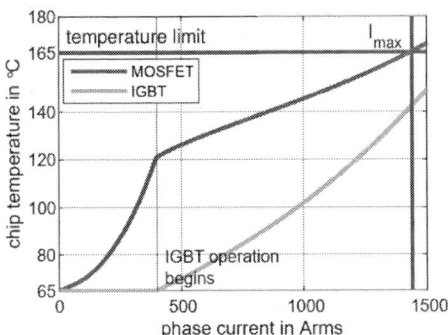

**Fig. 5:** Steady state temperature of MOSFET and IGBT in continuous operation mode.

$$\sqrt{\frac{4}{2\pi} \cdot \int_{\varphi}^{\varphi+\theta} \left[ \hat{I} \cdot \sin(x - \varphi) \right]^2 \mathrm{d}x} = I_{\mathrm{RMS}}; \quad (1)$$

$$\theta - \frac{1}{2} \cdot \sin(2\theta) = \frac{I_{\mathrm{RMS}}^2 \cdot \pi}{\hat{I}^2}; \quad \theta \in [0, \frac{\pi}{2}] \quad (2)$$

The resolution of Eq. (2) results in Eq. (3). These calculated phase angles play a pivotal role in determining the precise point of transition from MOSFET operation to IGBT operation.

$$\theta_1 = \theta; \; \theta_2 = \pi - \theta; \; \theta_3 = \pi + \theta; \; \theta_4 = 2\pi - \theta \quad (3)$$

The RMS current limit of the MOSFET is derived from its steady-state temperature. Initially, the steady-state temperature limit of the MOSFET is set at $120\,°C$ [3] to simplify parameters. The calculated RMS current is also used for the driving cycle. Figure 5 depicts the steady-state temperature of the semiconductors during continuous operation. Until the MOSFET reaches the temperature limit of $120\,°C$, only MOSFET operation is active. As the phase load current increases further, both the MOSFET and the IGBT become active.

## 3 Measurements

The primary objective of this paper is to optimize the silicon diode within the hybrid switch inverter. This is achieved by cutting the bond wires of the silicon diode and attaching a new diode to the IGBT. Three different silicon diodes are used: the standard diode from a 200A IGBT, together with two diodes of different current ratings - one with a nominal current of 150A (referred to as "diode A") and the other with a current rating of 80A (referred to as

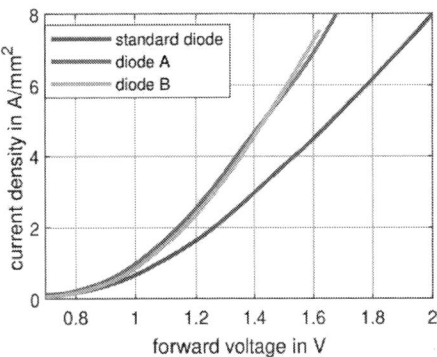

**Fig. 6:** Specific conduction characteristics of the diodes used in the experiment.

"diode B"). Figure 6 shows the normalised conduction characteristics of the diodes used. It can be seen that the standard diode differs in technology from diode A and diode B. Diode A and B have the same technology, but differ in current rating and therefore in chip area, as shown in table 1. The strategy is to minimise the presence of bipolar carriers in the silicon diode, thereby reducing both the turn-on losses of the MOSFET and the turn-off losses of the silicon diode.

**Tab. 1:** Area and nominal current of the used silicon diodes

|  | stand. diode | device A | device B |
|---|---|---|---|
| Area in $\mathrm{mm}^2$ | 35 | 19 | 11 |
| $I_{\mathrm{nominal}}$ in A | 200 | 150 | 80 |

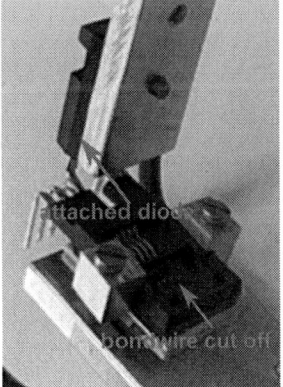

**Fig. 7:** IGBT with removed mould, cut off bond wires to silicon diode and new attached diode.

The electrical schematic of the hybrid switch half bridge is shown in Fig. 1, while Fig. 7 shows the IGBT with the bond wires removed and the newly attached silicon diode. This novel configuration was subjected to double pulse test measurements to extract the switching losses. In particular the turn-on losses of the active MOSFET and the turn-off losses of the passive silicon diode. The gate resistance of the semiconductor devices is chosen to ensure optimum switching speed without exceeding the voltage rating of $650\,\mathrm{V}$.

The turn-on losses of the active MOSFET are shown in Fig. 8. The losses shown are scaled to the required chip area to achieve an output power of $600\,\mathrm{kW}$. It can be seen that the turn-on losses with standard diode, the diode B and with body diode have almost similar values. However, the turn-on losses of the MOSFET can be reduced by using different silicon diodes or just the SiC body diode. On the other hand, the turn-on losses with diode A are significantly higher. This discrepancy can be attributed to the reduced switching speed achievable in this configuration, necessitated by the need to stay within the voltage limit of $650\,\mathrm{V}$. Consequently, the turn-on losses for diode A are increased by about 50%. Figure 9 shows the measured transients of the turn-on process of the active MOSFET with different diodes. The parasitic inductance between SiC and Si half bridge leading to a slight inductive decoupling effect, refer to [6]. The scaling of the chip size relative to the application differs between the referenced paper and this paper. Therefore the small parasitic inductance $L_{\mathrm{parasitic}} \approx 20\,\mathrm{nH}$ shows slight decoupling effects.

In contrast, Fig. 10 displays the turn-off losses of the different diodes. The reverse recovery losses are also scaled to the required chip area to provide

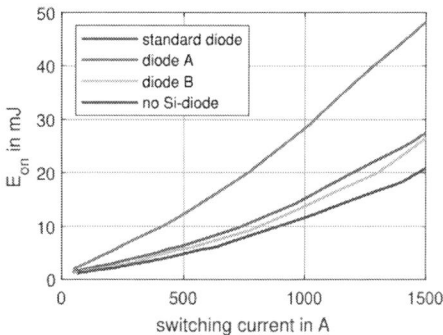

**Fig. 8:** scaled turn on losses of the active MOSFET with different silicon diodes.

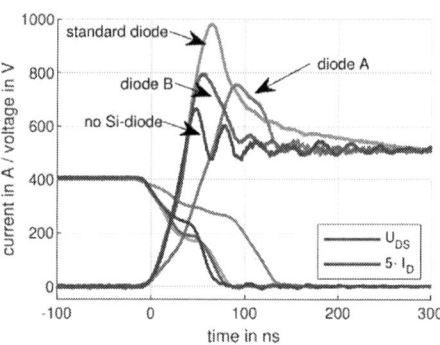

**Fig. 9:** turn-on transients of MOSFET with different diodes at $25\,^{\circ}\mathrm{C}$.

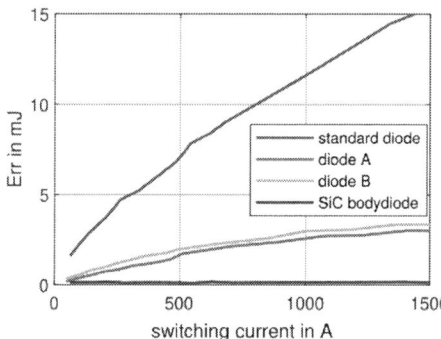

**Fig. 10:** scaled reverse recovery losses of different Si-diodes with active MOSFET.

an output power of $600\,\mathrm{kW}$. It is evident that the standard silicon diode incurs the highest turn-off losses, followed by diode B and diode A. Interestingly, the turn-off losses of the silicon diodes A and B are almost identical. This observation is surprising, as it might be expected for the device A diode to have higher turn-off losses compared to the device B, due to the larger chip size area. However, this can be attributed to the influence of switching speed and the lower current slope during the turn-off of the diode A. This results in almost equivalent turn-off losses for both diodes. The turn-off losses of the SiC body diode are very low compared to the silicon diode due to the lack of bipolar charge carriers. The passive turn-off losses mainly consist of capacitive losses of the MOSFET. The turn-off transients of the diodes are shown in Fig. 11

## 4   Simulation Results

After measuring the switching losses of the different hybrid switch variants, the steady-state temperature and therefore the maximum output power of the inverter is investigated. The modulation index is set to 1 in each case, and the power factor is varied

PCIM Europe 2024, 11– 13 June 2024, Nuremberg        DOI: 10.30420/566262177

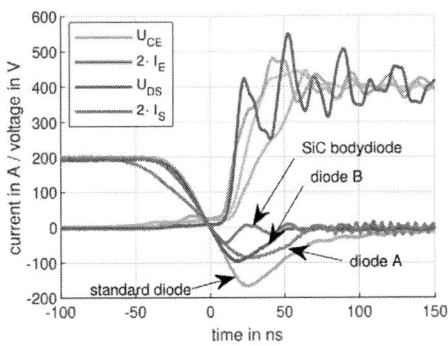

**Fig. 11:** turn-off transients of diodes at $25\,°C$.

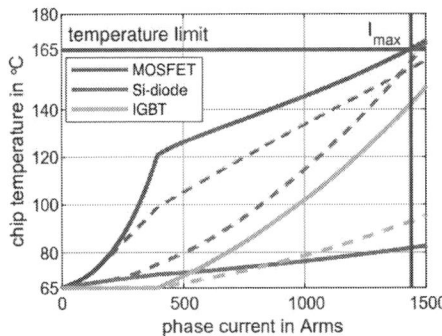

**Fig. 12:** Steady state temperature of MOSFET, Si-diode (standard), and IGBT in continuous operation mode. (Solid line) Power factor of 1; (dashed line) Power factor of -1.

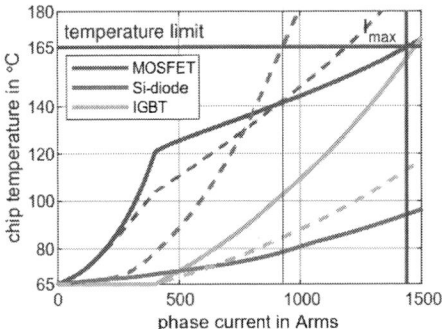

**Fig. 13:** Steady state temperature of MOSFET, Si-diode (diode A), and IGBT in continuous operation mode. (Solid line) Power factor of 1; (dashed line) Power factor of -1.

to $\cos\varphi = 1$ and $\cos\varphi = -1$ to cover the cases of full motor power and full regenerative power. The thermal parameters include a coolant temperature of $T_{\text{Coolant}} = 65\,°C$ and specific thermal resistances of $r_{\text{th-SiC}} = 20\,\frac{\text{K·mm}^2}{\text{W}}$ for SiC and $r_{\text{th-Si}} = 30\,\frac{\text{K·mm}^2}{\text{W}}$ for silicon. The difference in specific thermal resistance can be attributed to heat spread effects in discrete modules for SiC and silicon. A discrete SiC module has a smaller semiconductor chip size area compared to an equivalent module with silicon semiconductors, leading to lower thermal resistance for SiC. For example, a discrete SiC module contains $50\,\text{mm}^2$ of SiC area, while a silicon module contains $155\,\text{mm}^2$ of silicon area.

Another point to note is that the area ratio between SiC MOSFET and IGBT is chosen depending on the output power (600 kW) of the inverter and the used control scheme to achieve cost optimization, as shown in [3]. However, the drivetrain cost can be further optimized by adjusting the temperature limit for the transition from MOSFET operation to IGBT operation. This optimization is not the focus of this paper; the temperature limit is primarily kept constant at $120\,°C$.

Figure 12 depicts the steady-state temperature in continuous operation mode of the MOSFET, the standard Si-diode, and the IGBT. The solid line represents the temperature over the phase load current at full motor power with a power factor of 1, while the dashed lines show the temperature at full regenerative power.

At full motor power, the MOSFET reaches the temperature limit of $165\,°C$ at the maximum load current of a $600\,\text{kW}$ inverter. Similarly, in the case of full regenerative power, the silicon diode reaches the temperature limit at the maximum load current of a $600\,\text{kW}$ inverter. This configuration enables symmetrical motor and regenerative output power.

The subsequent configuration depicts the case with diode A, as seen in Fig. 13. During full motor operation, the MOSFET reaches the temperature limit at the maximum load current. Comparatively, the IGBT exhibits a higher temperature compared to the configuration with the standard diode. This increase is due to the necessity of reducing the switching speed with the device A diode, similar to the MOSFET. This reduction in switching speed leads to higher switching losses of the IGBT, similar to Fig. 8. During regenerative operation, the silicon diode reaches its temperature limit before reaching the maximum phase load current due to its smaller chip size area. This limitation restricts the regenerative power to $380\,\text{kW}$.

The configuration with the device B diode exhibits similar behavior, as shown in Fig. 14. In full motor operation, the MOSFET reaches the temperature limit at the maximum phase load current. During regenerative operation, the device B diode, with its

1301

smaller chip size area, further limits the regenerative power to $300\,\text{kW}$.

The final configuration, depicted in Fig. 15, features the hybrid switch without a Si diode. In order to achieve an output power of $600\,\text{kW}$ while maintaining an unmodified SiC-MOSFET to IGBT area ratio, the MOSFET area must be relatively large compared to the other configurations. It can also be seen that the IGBT is not well utilized. The relatively large SiC-MOSFET area provides a regenerative output power of $300\,\text{kW}$.

Table 2 presents a summary of the previously discussed results. The first column displays the SiC area ratio of a specific hybrid switch configuration in comparison to the full SiC inverter. At the case with silicon diode, approximately one-third of the SiC area is required in comparison to a full-SiC inverter. The silicon area ratio increases slightly from the standard diode to diode A and diode B, because the MOSFET is subjected to greater stress in backward conduction mode and requires a larger chip area to provide the output power. In the absence of a silicon diode, the SiC area must be increased, because only the SiC-MOSFET conducts in reverse. The second column presents a description of the silicon diode ratio relative to the hybrid switch configuration with the standard diode. It can be observed that with the diode B, the diode area is significantly reduced, although this is accompanied by a reduction in the regenerative output power.

Following the simulation of the steady-state continuous operation mode of the inverter, the drive cycle of the different configurations must be evaluated. Figure 16 presents the overall losses, normalized to a full-Si inverter, of the hybrid switch inverter when

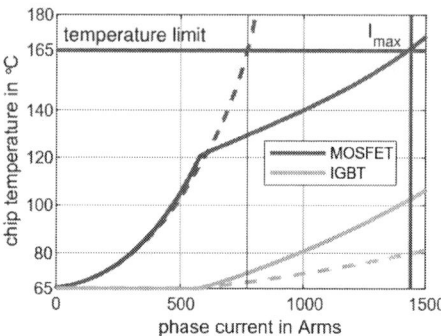

**Fig. 15:** Steady state temperature of MOSFET and IGBT without a Si-diode in continuous operation mode. (Solid line) Power factor of 1; (dashed line) Power factor of -1.

**Tab. 2:** Important parameters of the different inverter configurations

|  | $\dfrac{A_{\text{SiC}}}{A_{\text{full-SiC}}}$ | $\dfrac{A_{\text{Si-diode}}}{A_{\text{standard}}}$ | $P_{\text{reg.}}$ in kW |
|---|---|---|---|
| stand. diode | 0.33 | 1 | 600 |
| diode A | 0.35 | 0.4 | 380 |
| diode B | 0.36 | 0.24 | 300 |
| no Si-diode | 0.49 | 0 | 320 |

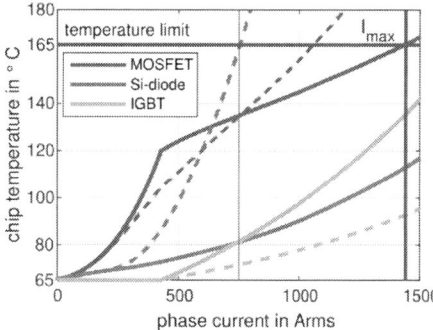

**Fig. 14:** Steady state temperature of MOSFET, Si-diode (diode B), and IGBT in continuous operation mode. (Solid line) Power factor of 1; (dashed line) Power factor of -1.

utilizing different silicon diodes during the Worldwide Harmonized Light Vehicles Test Procedure (WLTP). Additionally, the performance of the full-Si and the full-SiC inverter is shown.

The full-Si inverter exhibits the highest losses during the driving cycle, primarily due to high switching losses and bipolar conduction characteristics. The losses of the hybrid switch inverter with the standard diode are considerably lower than those of the full-Si inverter, but they can be further reduced by employing alternative diodes. With diode A, the switching losses of the active MOSFET are increased, while the turn-off losses of the silicon diode are decreased, resulting in a net reduction of losses. With the use of diode B, it is possible to reduce the losses even further.

The most intriguing comparison is between the configuration with no silicon diode and the full-SiC inverter. It can be observed that the hybrid switch inverter exhibits lower overall losses during the driving cycle in comparison to the full-SiC inverter. Despite the hybrid inverter exhibiting doubled conduction losses due to the halved SiC-MOSFET area, the switching losses are lower. This phenomenon can be attributed to the high part-load operation of the inverter, which results in the dominance of capacitive switching losses. These losses are lower in this hybrid switch configuration compared to the full-SiC inverter, leading to lower overall losses. The enlarged capacitive losses during MOSFET switching of the hybrid switch inverter due to the passive silicon area are considered in the simulation. It is important to note that the turn-off losses of the SiC body diode are included in the "sw MOSFET" field of Fig. 16.

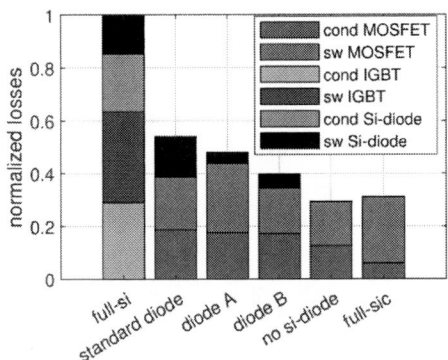

**Fig. 16:** Normalized losses at the driving cycle of different inverter configurations relative to the full-Si inverter.

$$c_{\text{drivetrain}} = 6 \cdot (A_{\text{SiC}} \cdot c_{\text{SiC}} + A_{\text{Si}} \cdot c_{\text{Si}})$$
$$+ 6 \cdot c_{\text{pack}} \cdot \left(\frac{A_{\text{SiC}}}{50\,\text{mm}^2} + \frac{A_{\text{Si}}}{155\,\text{mm}^2}\right) \quad (4)$$
$$+ \frac{1}{\eta} \cdot (C_{\text{batt}} \cdot b_{\text{k}})$$

In addition of evaluating the losses, another factor that must be considered is the cost of the drivetrain. The drivetrain consists of the battery, the inverter, and the electric motor. The cost of the electric motor is neglected because this term would vanish in the cost comparison. The cost of the battery is a function of its size, $C_{\text{batt}}$, and the cost per kWh of battery capacity, represented by the parameter $b_{\text{k}}$. The cost of the inverter is dependent on the required semiconductor area and the packaging costs of the system. The required battery size is influenced by the efficiency of the inverter needed to achieve a given vehicle range. Equation (4) illustrates the comprehensive cost equation utilized to determine the cost of the drivetrain. The battery size and cost per kWh will be interpreted as one

parameter $(C_{\text{batt}} \cdot b_{\text{k}})$ with the unit relative currency, as shown in Eq. (4).

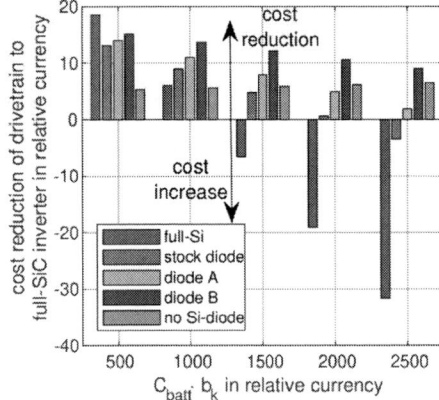

**Fig. 17:** Cost reduction of the drivetrain with different inverter configurations relative to a drivetrain with a full-SiC inverter. The cost ratio per area chip size amounts to $c_{\text{SiC}}/c_{\text{Si}} = 10$.

Figure 17 illustrates the cost reduction in relative currency for the different inverter configurations compared to a full SiC inverter. The cost ratio per unit chip area is $c_{\text{SiC}}/c_{\text{Si}} = 10$. At $(C_{\text{batt}} \cdot b_{\text{k}}) = 500$, all configurations are more cost-effective than the full SiC inverter. It can be observed that the most cost-effective inverter is the full-Si inverter. This can be attributed to the relatively small costs of the battery package, which has a significant impact on the overall drivetrain costs. As the battery cost increases further, this factor becomes more relevant in the cost analysis, making a high degree of inverter efficiency more important. Consequently, the full-Si inverter becomes disadvantageous with a higher battery cost.

Nevertheless, hybrid configuration A, B, and without a silicon diode are advantageous compared to the full-SiC inverter across the full range of the parameter $(C_{\text{batt}} \cdot b_{\text{k}})$ as shown in this figure. While the inverter without a silicon diode has better efficiency than the configuration with diode B, the hybrid B with a silicon diode is more cost-effective due to the smaller SiC area chip size. It is important to note that the configuration without a silicon diode is not the most cost-effective option, due to the significant SiC area that is required.

Another adjustable parameter is the current limit for the transition from MOSFET operation to IGBT operation. Figure 18 depicts the temperature of the MOSFET and IGBT with varying temperature limits for the transition to IGBT operation for the

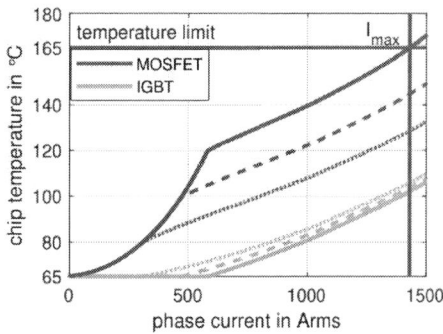

**Fig. 18:** Steady state temperature curves of MOSFET and IGBT in continuous operation mode without a Si-Diode and power factor of 1. The current limit for transition from MOSFET-only operation has changed. The chip size area remains the same.

configuration without a silicon diode with a power factor of 1. The temperature limit, and therefore the current for the transition, is set to $80\,^\circ\mathrm{C}$, $100\,^\circ\mathrm{C}$, and $120\,^\circ\mathrm{C}$. The chip size of the configuration remains constant in all cases.

It can be observed that modifying the transition current has a minor impact on the IGBT temperature, whereas it has a pronounced effect on the MOSFET temperature curves. By reducing the transition current, neither the MOSFET nor the IGBT reaches the maximum temperature limit of $165\,^\circ\mathrm{C}$ at maximum phase current. This indicates that the inverter could potentially provide more output power than required, which could result in a reduction in the size of the semiconductors.

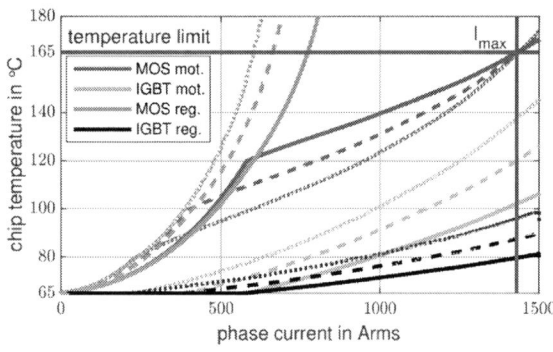

**Fig. 19:** Steady state temperature curves of MOSFET and IGBT in continuous operation mode without a Si-Diode. The current limit for transition from MOSFET-only operation and the chip area have changed.

The results of reducing the chip size area can be observed in Fig. 19. The SiC area to IGBT area

ratio is held constant for cost optimization as published in [3]. It is evident that in motoric operation, all MOSFET temperature curves intersect at the maximum output current. This indicates that all configurations provide the proposed motoric output power precisely.

Furthermore, by reducing the transition current and consequently the chip size area, the IGBT temperature increases, allowing for more utilization of the IGBT. Further reduction of the transition to IGBT operation could lead to better exploitation of the IGBT, aligning its temperature curve with that of the MOSFET at the maximum temperature limit. Nevertheless, such an adjustment results in IGBT operation during the WLTP driving cycle, see Fig. 3. This would reduce the efficiency during the driving cycle, which would impact the cost analysis. This consideration is not within the scope of this paper, as finding the global optimum with all parameters affecting the hybrid inverter will be the subject of a future study.

Nevertheless, reducing the chip size to enhance the inverter's utilization in the motoric state has the consequence of limiting the maximum regenerative output power. It is evident that the temperature limit of the MOSFET in the regenerative state with a power factor of -1 reduces the maximum phase current of the inverter and consequently the regenerative power.

**Tab. 3:** Summarized parameters for the configuration shown in Fig. 19

| temp. limit IGBT operation begins | | |
|---|---|---|
| $80\,^\circ\mathrm{C}$ | $100\,^\circ\mathrm{C}$ | $120\,^\circ\mathrm{C}$ |
| $\dfrac{A_{\mathrm{SiC}}}{A_{\mathrm{full-SiC}}}$ | | |
| 0.36 | 0.41 | 0.49 |
| $P_{\mathrm{reg.}}$ in kW  250 | 275 | 320 |
| efficiency $\eta$ in %  98.88 | 98.93 | 98.97 |

Table 3 presents the results of the configuration without a silicon diode. It can be observed that the SiC area can be reduced by 25% by lowering the transition temperature, or equivalently, the transition current. Additionally, the regenerative power decreases from $320\,\mathrm{kW}$ to $250\,\mathrm{kW}$. Changing the chip size area also alters the efficiency of the inverter, which consequently impacts the cost analysis.

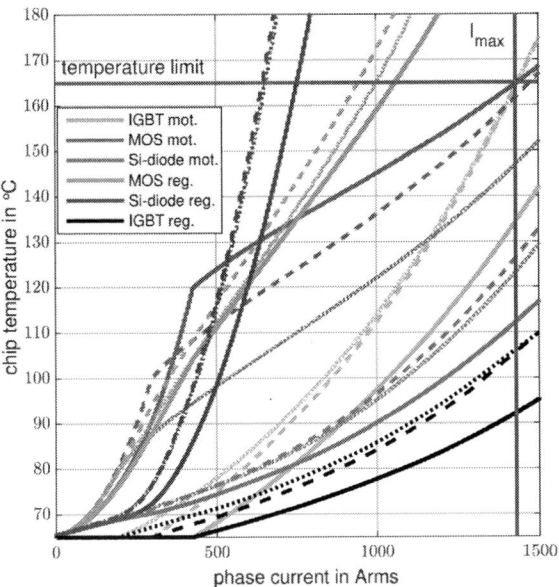

**Fig. 20:** Steady state temperature curves of MOSFET, diode B, and IGBT in continuous operation mode. Current limit for transition to IGBT operation has changed

Figure 20 presents the temperature curves of the MOSFET, IGBT, and Si-diode with diode B. As previously mentioned, the temperature limit for transitioning to IGBT operation and the current limit have been reduced. During the regenerative state, the regenerative power is now constrained by the Si-diode.

**Tab. 4:** Summarized parameters for the configuration shown in Fig. 20

| | temp. limit IGBT operation begins | | |
| --- | --- | --- | --- |
| | 80 °C | 100 °C | 120 °C |
| $\dfrac{A_{\text{SiC}}}{A_{\text{full-SiC}}}$ | 0.31 | 0.31 | 0.36 |
| $P_{\text{reg.}}$ in kW | 275 | 275 | 300 |
| efficiency $\eta$ in % | 98.56 | 98.56 | 98.62 |

Table 4 presents the results of the configuration with diode B. It can be observed that reducing the transition temperature, or equivalently, the transition current, can result in a 14% reduction in the SiC area. Additionally, the regenerative power decreases from 300 kW to 275 kW. Furthermore, the efficiency is negatively impacted by using a smaller

chip size area.

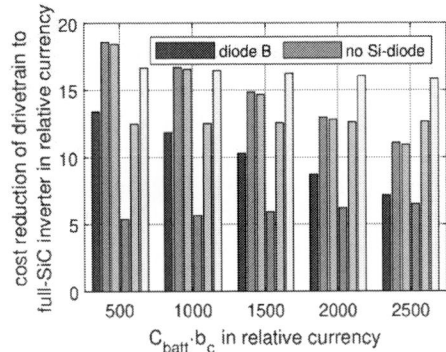

**Fig. 21:** Cost comparison to a full-SiC inverter with different current limits and chip areas. Lighter facecolor shading indicates a lower transition current.

In Fig. 21, another cost comparison between the diode B and the no Si-diode inverter is presented based on the cases depicted in Figs. 19 and 20. The cost ratio of SiC and Si per unit area amounts 10. At low battery costs, the diode B inverter exhibits advantages compared to the no Si-diode inverter. This can be attributed to the fact that in this scenario, chip size is more crucial for the cost analysis. However, as battery costs increase, the no-diode configuration becomes more cost advantageous.

## 5 Outlook

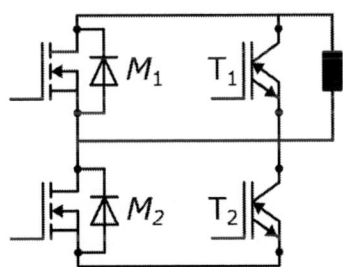

**Fig. 22:** Electrical schematic of a hybrid half bridge consisting of MOSFET and RCDC-IGBT with highly modulating diode.

As previously stated, the regenerative power of this shown approach is constrained. This limitation could potentially lead to complications in the event of a failure of the drivetrain, where the inverter transitions to a secure state designated as active short circuit (ASC), as referenced in [7]. In this state, either the high-side switches or the low-side switches are activated, and the semiconductor components must conduct excess current. ASC may result in damage or destruction of the inverter due to the limited reverse conduction capability.

One potential solution to this issue is the use of a diode with adjustable parameters, such as the charge carriers, as described in [8]. This device can be employed to reduce the bipolar charge carriers of the diode, thereby improving the switching performance of the MOSFET due to the absence of bipolar charge carriers. Additionally, in the event of a failure state in the drivetrain, the diode can be utilized to provide more regenerative power.

A novel approach to combine the advantages of both scenarios is the hybrid switch inverter, which employs a combination of MOSFET and RCDC-IGBT with highly modulating diode behavior. An electrical schematic of this approach can be seen in Fig. 22.

# 6 Conclusion

This paper examines the hybrid switch inverter with different silicon diodes to lowering losses at a typical driving cycle and reduce the cost of the drivetrain. It is observed that due to the fact that the MOSFET switches against the silicon diode during turn-on, the MOSFET losses are increased in comparison to a pure SiC MOSFET half bridge. The objective of this paper is to optimize the silicon diode in order to reduce the turn-on losses of the active MOSFET and the turn-off losses of the Si-diode by decreasing the presence of bipolar charge carriers. Four distinct hybrid configurations are contrasted with a full-Si and full SiC inverter. These hybrid configurations utilize the standard diode of the IGBT, diode A and diode B, which share the same diode technology but differ in current rating, in conjunction with a hybrid inverter without silicon diode. The results indicate that by optimizing the hybrid configuration, the losses can be reduced for a typical driving cycle. Furthermore, the absence of a silicon diode permits the reduction of the hybrid inverter's losses to a level below that of a full silicon carbide inverter. Additionally, the cost of the drivetrain can be reduced to below that of either a full silicon or full silicon carbide inverter.

# References

[1] M. Nitzsche, C. Cheshire, M. Fischer, J. Ruthardt, and J. Roth-Stielow, "Comprehensive comparison of a sic mosfet and si igbt based inverter," in *PCIM Europe 2019*, pp. 1–7.

[2] F. Kayser, F. Pfirsch, F. -J. Niedernostheide, R. Baburske, and H. -G. Eckel, "Novel si-sic hybrid switch and its design optimization path," in *2022 IEEE 34th International Symposium on Power Semiconductor Devices and ICs (ISPSD)*, 2022, pp. 225–228.

[3] M.Walter and Mark-M. Bakran, "Hybrid-switch-inverter - a new approach reducing the system cost of the electric powertrain," in *PCIM Europe 2023*, pp. 2340–2349. DOI: 10.30420/566091324.

[4] M. Rahimo, F. Canales, R. A. Minamisawa, C. Papadopoulos, U. Vemulapati, *et al.*, "Characterization of a silicon igbt and silicon carbide mosfet cross-switch hybrid," *IEEE Transactions on Power Electronics*, vol. 30, no. 9, pp. 4638–4642, 2015.

[5] R. E. Mathieson, P. D. Judge, and S. Finney, "Si/sic hybrid switch for improved switching and part-load performance," in *2020 IEEE 21st Workshop (COMPEL)*, 2020, pp. 1–7.

[6] M. Walter and M.-M. Bakran, "Optimization of the hybrid-switch inverter by decoupling sic and si," in *2023 11th International Conference on Power Electronics and ECCE Asia (ICPE 2023 - ECCE Asia)*, IEEE, 22.05.2023 - 25.05.2023, pp. 1835–1842. DOI: 10.23919/ICPE2023-ECCEAsia54778. 2023.10213959.

[7] A. Lanzafame, L. D. Tornello, G. Scelba, E. Venuti, A. Raffa, *et al.*, "Experimental test setup for thermal stress analysis of sic devices under active short circuits," in *2022 IEEE Energy Conversion Congress and Exposition (ECCE)*, 2022, pp. 1–6. DOI: 10.1109/ECCE50734.2022.9947533.

[8] D. Werber, F. Pfirsch, T. Gutt, V. Komarnitskyy, C. Schaeffer, *et al.*, "6.5kv rcdc: For increased power density in igbt-modules," in *2014 IEEE 26th International Symposium on Power Semiconductor Devices & IC's (ISPSD)*, 2014, pp. 35–38. DOI: 10.1109/ISPSD.2014.6855969.

PCIM Europe 2024, 11– 13 June 2024, Nuremberg    DOI: 10.30420/566262178

# Advanced Pulse Sequence for Saliency-based High-Accurate Rotor Position Estimation of Railway Traction Locomotive Motors

Eduardo Rodriguez Montero[2], Markus Vogelsberger[1], Wolfram Teppan[3], Hans Ertl[2], Thomas Wolbank[2]

[1] ALSTOM Transport Austria GmbH, Rolling Stock Components-DRIVES/Platform, Austria

[2] TU Wien, Institute of Energy Systems and Electrical Drives, Austria

[3] LEM Advisory Services SA, Switzerland

Corresponding author:    Markus Vogelsberger, markus.vogelsberger@alstomgroup.com
Spealer:                 Markus Vogelsberger, markus.vogelsberger@alstomgroup.com

## Abstract

In recent years, railway industry has evinced interest in encoderless traction drive systems, stimulated by the multiple advantages that encoder removal involves. Yet, to achieve sustained zero-speed encoderless control, motor spatial saliencies are required. Particularly, standard railway traction induction motors used in locomotives often exhibit a saliency usually named intermodulation in literature, which allows to track the rotor position with remarkably high accuracy at both low and high loads. This paper presents a method to extract saliencies using advanced pulsing and compares it with the classical INFORM pulse sequence. Experimental validation using a standard 1.6 MW locomotive traction motor is provided.

## 1    Introduction

Large induction motor drives used in railway traction, such as those in charge of locomotive propulsion, are typically equipped with current transducers and a shaft encoder. In recent years, railway businesses (industry and operators) have evinced interest in encoderless traction drive systems, stimulated by the reliability increase that encoder removal involves, in addition to the cost volume, and weight savings. Thus far, the only choice to control an AC motor at near-zero electrical frequency is by extraction of motor spatial saliencies [1]. This can be done using voltage pulse methods (e.g. [3]-[9]), which rely on the application of transient test voltage vectors during which the resulting transient stator currents are evaluated. In these transient currents (current derivatives), motor spatial saliencies are reflected.

The standard design of induction motors used for propulsion of locomotives has a positive by-product effect in the context of spatial saliencies. These locomotive motors sometimes develop unintentionally an intermodulation saliency, whose intensity increases as loading increases. As published in [3], the intermodulation angle can be translated into a rotor angle with remarkably high accuracy. The accuracy of the rotor angle estimation does not degrade as the torque increases, in contrast to the well-known decrease of flux angle estimation

quality that occurs when employing the saturation saliency, since a phase shift arises due to a large share of saturation being caused by stator leakage flux [2], making flux vector and saturation vector being spatially misaligned.

In order to extract multiple motor spatial saliencies, an advanced voltage step excitation sequence is used in this paper. This sequence has already been published in [4] where current derivative sensors were employed. However, this paper does not use current derivative sensors. Instead, standard locomotive current transducers are used to measure the stator currents of the high-power railway motor. Therefore, to acquire a multi-saliency-vector (saliency vector), phase current derivatives are computed using linear regression during the available voltage steps. These phase current derivatives are processed to form an offsetless saliency vector (average transient inductance effect is cancelled using special current slope vector combinations). After saliency decoupling, the intermodulation saliency angle can be tracked, thus enabling to estimate the rotor position with very high accuracy at low and high loads [3].

This paper is arranged as follows: Section 2 will explain the proposed saliency extraction and pulse application methods. Section 3 will shortly describe the test stand and test motor. Section 4 will provide experimental validation measured on a 1.6

1307

MW railway locomotive induction motor. Section 5 will provide the conclusions.

## 2 Proposed Spatial Saliency Extraction and Voltage-Pulsing Methods

The meticulous design of traction motors used in locomotives sometimes ends up in the creation of a tiny by-product: a spatial saliency named intermodulation in literature (e.g. [3]). This saliency in practice has an unnoticeable impact in the fundamental-wave behavior of the motor. However, it is possible to acquire the intermodulation saliency vector using short voltage pulses and analyzing the resulting transient currents. As explained in [3], the intermodulation saliency angle allows to track the rotor position with remarkably high accuracy at any load, especially at high loads.

### 2.1 Proposed Pulse Application Method

Traditionally, test voltage pulses used to follow the well-known triple-pulse-sequence 1-2-3 used in the classical INFORM [5]. Each sequence consisted of four active inverter states and, during the two middle active states ($t_V$ pulse duration) of each of the pulse sequences, the current derivative (di/dt) was computed. In [5], the classical INFORM and an enhanced pulsing sequence were investigated. Minimum current distortion was achieved by the application of a fifth active state during each sequence designed to compensate induced voltage and stator-resistance-based current distortion. In this paper, current distortion reduction is achieved by a decrease in the voltage pulse amount without adversely affecting saliency extraction quality.

A generic representation of Space-Vector Pulse Width Modulation (SVPWM) including the triple-pulse-sequence is shown in Fig. 1. Note that 1-2-3 denote sequence subindex and A-B-C three-phase sub-index in the following. The length of SVPWM period and pulse period is equal and defined as $t_{PWM}$.

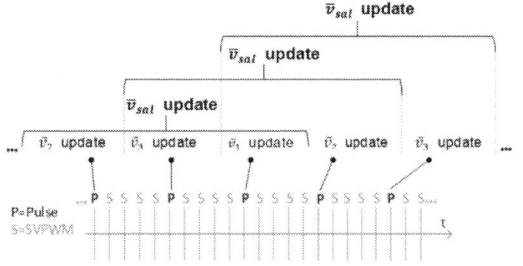

**Fig. 1**   Generic schematic representation of triple-pulse-sequence and SVPWM.

Fig. 2 shows an example of the three-phase currents during classical INFORM sequence-1. The period length is $t_{PWM}$, $t_V$ is the pulse duration of the two middle actives states and $t_V/2$ is the pulse duration of the first and last actives state. $t_{0,INF}$ is the inactive state duration.

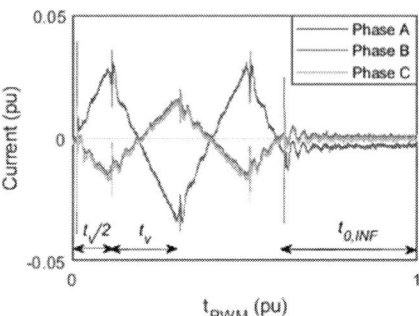

**Fig. 2**   Example of transient currents during classical pulsing U-sequence.

In this work, the voltage pulses used to excite the locomotive motor follow the sequence of [4], named Advanced-Pulsing in this work. In contrast to [4], this work presents a new saliency acquisition method using only standard railway current sensors and no current derivative sensors. In this paper, di/dt are estimated via least-square linear regression.

An example of the resulting three-phase currents during Advanced-Pulsing U-sequence is shown in Fig. 3. The period length is $t_{PWM}$ and $t_V$ is the pulse duration of the middle actives state. $t_V/2$ is the pulse duration of the first and last actives state and $t_{0,INF}$ is the inactive state duration.

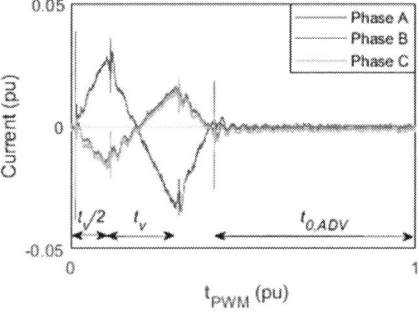

**Fig. 3**   Example of transient currents during Advanced-Pulsing U-sequence.

Table 1 shows the pulsing sequences for classical (INFORM) and Advanced-Pulsing schemes. The bold pulses given in Table 1 highlight when a di/dt is computed.

| Pulsing | Sequence | Test voltage vectors |
|---|---|---|
| Classical INFORM [5] | 1 | $011 \to \mathbf{100} \to 011 \to \mathbf{100} \to 000$ |
| | 2 | $101 \to \mathbf{010} \to 101 \to 010 \to 000$ |
| | 3 | $110 \to \mathbf{001} \to 110 \to 001 \to 000$ |
| Advanced-Pulsing [4] | 1 | $011 \to \mathbf{100} \to 011 \to \mathbf{000}$ |
| | 2 | $101 \to \mathbf{010} \to 101 \to \mathbf{000}$ |
| | 3 | $110 \to \mathbf{001} \to 110 \to \mathbf{000}$ |

**Table 1** Pulse sequence and test voltage vectors for classical pulsing and Advanced-Pulsing methods.

Table 2 shows the voltage vectors and phase-phase voltages of a 2-level 3-phase inverter.

| Voltage vector | $U_{AB}$ | $U_{BC}$ | $U_{CA}$ |
|---|---|---|---|
| 100 | $U_{DC}$ | 0 | $-U_{DC}$ |
| 110 | 0 | $U_{DC}$ | $-U_{DC}$ |
| 010 | $-U_{DC}$ | $U_{DC}$ | 0 |
| 011 | $-U_{DC}$ | 0 | $U_{DC}$ |
| 001 | 0 | $-U_{DC}$ | $U_{DC}$ |
| 101 | $U_{DC}$ | $-U_{DC}$ | 0 |
| 000 | 0 | 0 | 0 |
| 111 | 0 | 0 | 0 |

**Table 2** Voltage vectors and phase-phase voltages of a 2-level 3-phase inverter.

The duration of each test voltage vector is given in Table 3. $t_V$ is the duration of the active voltage vector, regardless of the sequence number, and $t_0$ is the voltage vector duration of the inactive voltage vectors. In line with Table 1, a bold duration denotes when a di/dt is computed.

| Pulsing | Test voltage vectors duration |
|---|---|
| Classical INFORM [5] | $t_V/2 \to \mathbf{t_V} \to \mathbf{t_V} \to t_V/2 \to t_{0,INF}$ |
| Advanced-Pulsing [4] | $t_V/2 \to \mathbf{t_V} \to t_V/2 \to \mathbf{t_{0,ADV}}$ |

**Table 3** Test voltage vector duration for classical (INFORM) and Advanced-Pulsing methods.

Therefore, under classical-INFORM pulsing, it follows that $t_{PWM} = 3 \cdot t_V + t_{0,INF}$ and under the proposed Advanced-Pulsing it follows that $t_{PWM} = 2 \cdot t_V + t_{0,ADV}$.
Thanks to the reduction in active switching duration during Advanced-Pulsing compared to classical pulsing, current distortion will also be reduced. Besides, the consequent reduction in the amount of switching events per pulse period will lead to

lower switching losses in Advanced-Pulsing than in INFORM-pulsing.

## 2.2 Proposed Saliency Extraction

The transient leakage inductance of N-salient induction motors can be modelled as in (1)-(3).

$$L_A = L_0 + \sum_{i=1}^N L_{m,i} \cdot sin\left(\gamma_i\right) \quad (1)$$

$$L_B = L_0 + \sum_{i=1}^N L_{m,i} \cdot sin\left(\gamma_i + 2 \cdot \pi/3\right) \quad (2)$$

$$L_C = L_0 + \sum_{i=1}^N L_{m,i} \cdot sin\left(\gamma_i + 4 \cdot \pi/3\right) \quad (3)$$

Being $L_0$ the average HF inductance, $L_{m,i}$ the saliency amplitude of the $i_{th}$ saliency and $\gamma_i$ the angle of the $i_{th}$ saliency.

As explained in Fig. 1 and Table 1-3, the motor is excited with three 1-2-3 sequences for both the proposed and classical pulsing methods.

### 2.2.1 Classical Pulsing

Using the phase current derivatives caused by each classical pulse sequence, $\bar{v}_1$, $\bar{v}_2$ and $\bar{v}_3$ vectors are calculated according to (4)-(6). $U_{DC}$ is the DC-link voltage and $a = \exp\left(j \cdot 2 \cdot \pi/3\right)$.

$$
\begin{aligned}
\bar{v}_1 = \frac{2}{3} \cdot (&\left(\frac{di_{A|100}}{dt} - \frac{di_{A|011}}{dt}\right) \\
+ &\left(\frac{di_{B|100}}{dt} - \frac{di_{B|011}}{dt}\right) \cdot a \\
+ &\left(\frac{di_{C|100}}{dt} - \frac{di_{C|011}}{dt}\right) \cdot a^2)
\end{aligned} \quad (4)
$$

$$
\begin{aligned}
\bar{v}_2 = \frac{2}{3} \cdot (&\left(\frac{di_{A|010}}{dt} - \frac{di_{A|101}}{dt}\right) \\
+ &\left(\frac{di_{B|010}}{dt} - \frac{di_{B|101}}{dt}\right) \cdot a \\
+ &\left(\frac{di_{C|101}}{dt} - \frac{di_{C|101}}{dt}\right) \cdot a^2)
\end{aligned} \quad (5)
$$

$$
\begin{aligned}
\bar{v}_3 = \frac{2}{3} \cdot (&\left(\frac{di_{A|001}}{dt} - \frac{di_{A|110}}{dt}\right) \\
+ &\left(\frac{di_{B|001}}{dt} - \frac{di_{B|110}}{dt}\right) \cdot a \\
+ &\left(\frac{di_{C|001}}{dt} - \frac{di_{C|110}}{dt}\right) \cdot a^2)
\end{aligned} \quad (6)
$$

Solving (4)-(6) with the help of (1)-(3) and assuming $L_0 \gg L_{m,i}$ results in (7)-(9).

$$\bar{v}_1 \approx 2 \cdot U_{DC} \cdot \left(\frac{4}{3 \cdot L_0} + \sum_{i=1}^N \frac{2 \cdot L_{m,i}}{3 \cdot L_0^2} \cdot e^{j \cdot \left(\gamma_i - \frac{\pi}{2}\right)}\right) \quad (7)$$

$$\bar{v}_2 \approx 2 \cdot U_{DC} \cdot \left( \frac{4}{3 \cdot L_0} + \sum_{i=1}^{N} \frac{2 \cdot L_{m,i}}{3 \cdot L_0^2} \cdot e^{j \cdot \left( \gamma_i + 4 \frac{\pi}{3} - \frac{\pi}{2} \right)} \right) \quad (8)$$

$$\bar{v}_3 \approx 2 \cdot U_{DC} \cdot \left( \frac{4}{3 \cdot L_0} + \sum_{i=1}^{N} \frac{2 \cdot L_{m,i}}{3 \cdot L_0^2} \cdot e^{j \cdot \left( \gamma_i + 2 \frac{\pi}{3} - \frac{\pi}{2} \right)} \right) \quad (9)$$

As (7)-(9) show, the offset part inversely proportional to $L_0$ forms a zero sequence. On the other side, the $i_{th}$ saliency part, proportional to $L_{m,i}$, forms a positive sequence. Thus, a multiple-saliency vector can be obtained when applying Clarke transformation. The multiple-saliency vector resulting in classical pulsing ($\bar{v}_{sal|INF}$) is shown in (10). In (10), the term $\sum \bar{e}_{HF}$ is added, accounting for the random/HF(High Frequency) noise arising in $\bar{v}_{sal|INF}$.

$$\bar{v}_{sal|INF} \approx \sum_{i=1}^{N} A_{i|INF} \cdot e^{j \cdot \left( \gamma_i - \frac{\pi}{2} \right)} + \sum \bar{e}_{HF} \quad (10)$$

The intermodulation saliency part is decoupled using feedforward compensation from $\bar{v}_{sal|INF}$ [6]. The result (assuming perfect decoupling) is $\bar{v}_{1|INF}$.

$$\bar{v}_{1|INF} \approx A_{1|INF} \cdot e^{j \cdot \left( \gamma_1 - \frac{\pi}{2} \right)} + \sum \bar{e}_{HF} \quad (11)$$

Being

$$A_{i|INF} = \frac{2 \cdot L_{m,i}}{3 \cdot L_0^2} \quad (12)$$

### 2.2.2 Advanced-Pulsing

The calculation of $\bar{v}_1$, $\bar{v}_2$ and $\bar{v}_3$ when using the described Advanced-Pulsing scheme is done according to (13)-(15).

$$\bar{v}_1 = \frac{2}{3} \cdot \left( \left( \frac{di_{A|100}}{dt} - \frac{di_{A|000}}{dt} \right) \right.$$
$$+ \left( \frac{di_{B|100}}{dt} - \frac{di_{B|000}}{dt} \right) \cdot a \quad (13)$$
$$\left. + \left( \frac{di_{C|100}}{dt} - \frac{di_{C|000}}{dt} \right) \cdot a^2 \right)$$

$$\bar{v}_2 = \frac{2}{3} \cdot \left( \left( \frac{di_{A|010}}{dt} - \frac{di_{A|000}}{dt} \right) \right.$$
$$+ \left( \frac{di_{B|010}}{dt} - \frac{di_{B|000}}{dt} \right) \cdot a \quad (14)$$
$$\left. + \left( \frac{di_{C|101}}{dt} - \frac{di_{C|000}}{dt} \right) \cdot a^2 \right)$$

$$\bar{v}_3 = \frac{2}{3} \cdot \left( \left( \frac{di_{A|001}}{dt} - \frac{di_{A|000}}{dt} \right) \right. \quad (15)$$

$$+ \left( \frac{di_{B|001}}{dt} - \frac{di_{B|000}}{dt} \right) \cdot a$$
$$+ \left( \frac{di_{C|001}}{dt} - \frac{di_{C|000}}{dt} \right) \cdot a^2 \right)$$

Solving (13)-(15) with the help of (1)-(3) and assuming $L_0 \gg L_{m,i}$ results in (16)-(18).

$$\bar{v}_1 \approx U_{DC} \cdot \left( \frac{4}{3 \cdot L_0} + \sum_{i=1}^{N} \frac{2 \cdot L_{m,i}}{3 \cdot L_0^2} \cdot e^{j \cdot \left( \gamma_i - \frac{\pi}{2} \right)} \right) \quad (16)$$

$$\bar{v}_2 \approx U_{DC} \cdot \left( \frac{4}{3 \cdot L_0} + \sum_{i=1}^{N} \frac{2 \cdot L_{m,i}}{3 \cdot L_0^2} \cdot e^{j \cdot \left( \gamma_i + 4 \frac{\pi}{3} - \frac{\pi}{2} \right)} \right) \quad (17)$$

$$\bar{v}_3 \approx U_{DC} \cdot \left( \frac{4}{3 \cdot L_0} + \sum_{i=1}^{N} \frac{2 \cdot L_{m,i}}{3 \cdot L_0^2} \cdot e^{j \cdot \left( \gamma_i + 2 \frac{\pi}{3} - \frac{\pi}{2} \right)} \right) \quad (18)$$

As in (7)-(9), (16)-(18) contain a zero-sequence component (offset) and positive sequence saliency vectors. Thus, a multiple-saliency vector can be obtained when applying Clarke transformation to (16)-(18). The multiple-saliency vector resulting from Advanced-Pulsing $\bar{v}_{sal|ADV}$ is shown in (19). In (19), the term $\sum \bar{e}_{HF}$ is added, accounting for the resulting random/HF noise arising in $\bar{v}_{sal|ADV}$.

$$\bar{v}_{sal|ADV} \approx \sum_{i=1}^{N} A_{i|ADV} \cdot e^{j \cdot \left( \gamma_i - \frac{\pi}{2} \right)} + \sum \bar{e}_{HF} \quad (19)$$

The intermodulation saliency harmonic is decoupled using feedforward compensation from $\bar{v}_{sal|ADV}$ [6]. The result (assuming perfect decoupling) is $\bar{v}_{1|ADV}$.

$$\bar{v}_{1|ADV} \approx A_{1|ADV} \cdot e^{j \cdot \left( \gamma_1 - \frac{\pi}{2} \right)} + \sum \bar{e}_{HF} \quad (20)$$

Being

$$A_{i|ADV} = A_{i|INF}/2 \quad (21)$$

As mathematically derived in (11) and (20), the decoupled intermodulation vectors $\bar{v}_{1|ADV}$ and $\bar{v}_{1|INF}$ show equal noise content. However, the magnitude of the saliency-containing vector in Advanced-Pulsing results half of the classical-INFORM according to (21). This is due to the non-deterministic addition of random/HF noise in the phase current measurement and di/dt calculation. However, given the noise suppression characteristic of linear regression, the random/HF noise coming along with di/dt is very low. Thus, saliency extraction using Advanced-Pulsing can be guaranteed.

# 3 Test Bench

A standard 1.6 MW locomotive traction motor (8000 Nm rated torque) is investigated regarding the applicability of classical INFORM pulsing (INF) and Advanced-Pulsing (ADV) methods described above. The motor exhibits a significant intermodulation saliency used to compute a rotor position as explained in [3]. Fig. 4 shows a typical electric locomotive driven by this type of motor. For experimental validation, a traction motor is loaded using an additional loading machine mechanically coupled to the test motor in the lab.

**Fig. 4**  Example of locomotive driven by four traction motor units.

# 4 Experimental Results

## 4.1 Multiple-Saliency Vector

The experimental multiple-saliency vectors ($\bar{v}_{sal|INF}$ and $\bar{v}_{sal|ADV}$) are shown in Fig. 5 at 100% rated torque (8000 Nm) and very low speed.

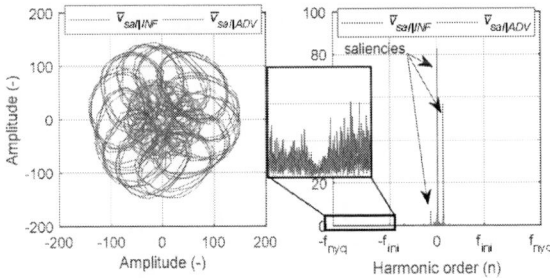

**Fig. 5**  Experimental $\bar{v}_{sal|INF}$ and $\bar{v}_{sal|ADV}$ in time domain (left) and FFT in frequency domain (right).

Fig. 5 reveals that the saliency amplitudes in $\bar{v}_{sal|INF}$ are twice as big as in $\bar{v}_{sal|ADV}$ as predicted by (21). Besides, a zoom-in view of the negative-sequence HF content of $\bar{v}_{sal|ADV}$ and $\bar{v}_{sal|INF}$ is also provided. The frequency window of the zoom-in view spans from $-f_{nyq}$ (negative

Nyquist frequency of pulsing frequency) to $f_{ini}$, where $f_{ini}$ illustrates the defined begin of the considered HF noise.
In this HF negative frequency interval, it can be observed that $\bar{v}_{sal|INF}$ contains a slightly higher average noise floor and the highest HF amplitude compared to $\bar{v}_{sal|ADV}$. This is discussed following subsection.

## 4.2 Noise and SNR

To illustrate the noise characteristic of $\bar{v}_{sal|ADV}$ and $\bar{v}_{sal|INF}$, Fig. 6 shows the average absolute noise in the interval $-f_{nyq}$ to $f_{ini}$ and $f_{ini}$ to $f_{nyq}$ of $\bar{v}_{sal|ADV}$ and $\bar{v}_{sal|INF}$ for several torque levels and very low shaft speeds, calculated according to (22). $n$ denotes the number of discrete harmonics sampled from $-f_{nyq}$ to $f_{ini}$. The mean value of 20 FFTs has been done to generate Fig. 6.

$$e_n = \frac{1}{2n} \cdot \left( \sum_{-f_{nyq}}^{-f_{ini}} FFT(\bar{v}_{sal}) + \sum_{f_{ini}}^{f_{nyq}} FFT(\bar{v}_{sal}) \right) \quad (22)$$

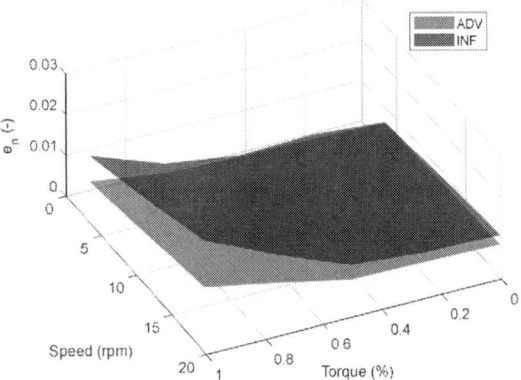

**Fig. 6**  $e_n$ for classical-INFORM and Advanced-Pulsing methods.

Besides, $e_n$ can be related to the intermodulation saliency amplitude in a quasi-Signal-to-Noise-Ratio (quasi-SNR) that only considers noise in frequency interval $-f_{nyq}$ to $f_{ini}$ and $f_{ini}$ to $f_{nyq}$. The quasi-SNR helps to assess whether the Advanced-Pulsing or INFORM pulsing present better saliency signal extractability.

$$quasi - SNR = A_1/e_n \quad (23)$$

The measured experimental quasi-SNR of the classical-INFORM and Advanced-Pulsing methods is shown in Fig. 7.

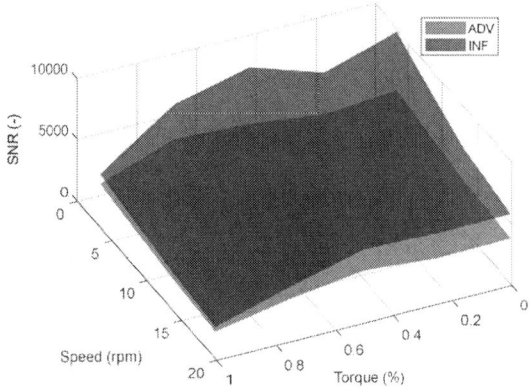

**Fig. 7**   Experimental quasi-$SNR$ for classical-IN-FORM and Advanced-Pulsing methods.

As observed in Fig. 7, the calculated quasi-$SNR$ is larger when using INFORM pulsing at low torque and near-zero speeds than when using Advanced-Pulsing. However, the calculated quasi-$SNR$ using Advanced-Pulsing becomes similar to the quasi-$SNR$ using INFORM pulsing towards very high loads and higher speeds. This might be explained by the lower current ripple and the lower current deviation during excitation periods generated by Advanced-Pulsing. Therefore, is spite of reduced active states, the quasi-$SNR$ when using Advanced-Pulsing results nearly as high as the quasi-$SNR$ when using INFORM-Pulsing if operating the motor at high torques and increasing shaft speeds. This implies a high applicability of the proposed Advanced-Pulsing regarding saliency extraction and ultimately rotor position estimation.

## 4.3  THD

The Total Harmonic Distortion (THD) is calculated for classical-INFORM and Advanced-Pulsing methods according to (24). Phase-A current is measured at 500 kHz sampling frequency. Thus, the Nyquist frequency (nyq) is 250 kHz. Pulse sequences are injected every 5 SVPWM periods (see Fig. 1).

$$THD = \frac{\sqrt{\sum_{n=2}^{nyq} I_n^2}}{|I_1|} \cdot 100 \tag{24}$$

Where $I$ is the current amplitude and $I_1$ the fundamental-wave current amplitude.

The ratio of THD between ADV and INF ($THD_{ADV/INF}$) is calculated according to (25):

$$THD_{ADV/INF} = (1 - \frac{THD_{ADV}}{THD_{INF}}) \cdot 100 \tag{25}$$

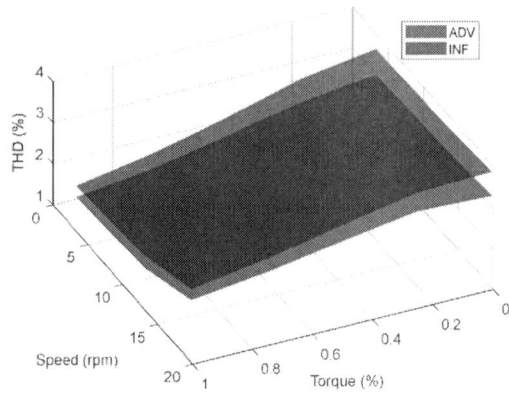

**Fig. 8**   Experimental $THD(\%)$ for classical-IN-FORM and Advanced-Pulsing methods.

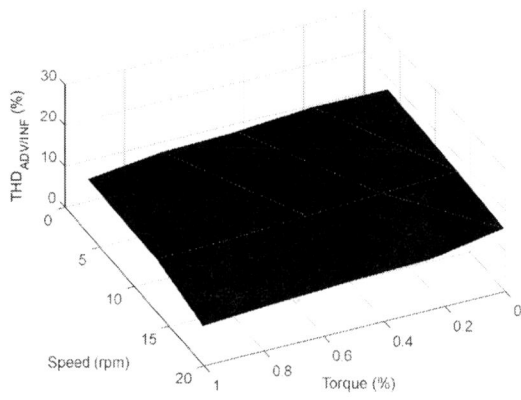

**Fig. 9**   Experimental $THD_{ADV/INF}(\%)$.

As expected by the improved pulsing sequences, the proposed Advanced-Pulsing method shows between 10% and 20% lower current harmonic distortion than classical INFORM pulsing. This implies a slight improvement in acoustic noise and drive losses.

## 4.4  Rotor Position Performance

The angle of the intermodulation saliency vector is transformed into a rotor position after saliency decoupling as explained in [3]. The rotor position deviation, utilizing a shaft encoder as reference, is shown for an 8000 Nm (1 pu) step using the Advanced-Pulsing in Fig. 9 and classical INFORM pulsing in Fig. 10.

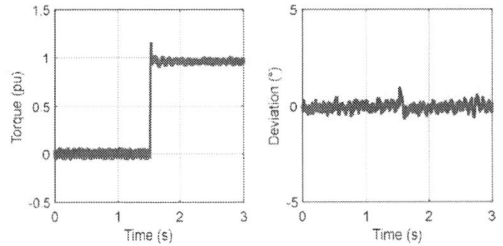

**Fig. 9** Experimental torque (left) and rotor position deviation using Advanced-Pulsing.

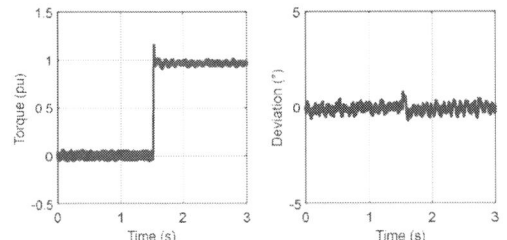

**Fig. 10** Experimental torque (left) and rotor position deviation using INFORM-Pulsing.

As observed in Fig. 9 and Fig. 10, the rotor position using both pulsing methods presents very high precision, showing maximum deviation of ±1° (mechanical degrees). No significant difference in the deviation angle between both methods is observed (specially at no-load). This is due to the use of least-square linear regression to calculate current derivatives in both pulsing methods. As a result, HF noise is strongly filtered and, although the quasi-SNR in Advanced-Pulsing and classical-pulsing differs at no-load, this difference does not reach a sufficient level to be noticed in the accuracy of the position estimates. The position deviation seen in in Fig. 9 and Fig. 10 is rather originated by the discrete nature of the saliency vector, the saliency harmonic products [6], the non-exact saliency decoupling and the spatial harmonic order of the control saliency (in this paper, intermodulation). Such characteristics are present in the same way for both investigated pulsing methods, since the saliency vector was acquired under each method using the vector sum of the three pulsing sequences.

Thus, it can be stated that the proposed Advanced-Pulsing involves an improvement in current distortion without losing rotor position estimation accuracy compared to the classical pulsing method.

## 5 Conclusions

An advanced pulsing method for saliency-based rotor position estimation in sensorless motor drives using standard current sensors has been investigated. The method was mathematically and experimentally compared with the classical pulsing method. The proposed advanced pulsing method was found to show lower current harmonic distortion and similar rotor position estimation precision compared to the classical pulsing method. Experimental measurements collected at a test multi-salient railway traction motor demonstrated the improvement in current distortion using the proposed method without compromising rotor position estimation accuracy. Experimental rotor position results as well as THD and SNR results were provided up to the 8000 Nm motor rated torque.

## 6 Acknowledgement

The authors want to thank ALSTOM, especially Mr. H. Mannsbarth (head of Components-Bogies/Drives department in global Rolling Stock Platform/Components division) and Mr. Cedric Zanutti (head of R&D/Technology in RSC-Bogies/Drives) for their generous support, research/development funding and project supervision. Furthermore, the authors want to thank colleagues from ALSTOM group (Mr. Cepak, Mr. Ganster and Mr. E. Moser) for their feedback and great support.

The authors further are very indebted to LEM Company (especially Mr. A. Hürlimann/Chairman of Board of Directors, Dr. W. Teppan & Mr. J. Burk) and OMICRON Lab Company (especially Mr. F. Hämmerle, Mr. M. Pfitscher and Mr. T. Schuster) for the cooperation and generous support.

## References

[1] J. Holtz and H. Pan, "Acquisition of rotor anisotropy signals in sensorless position control systems," IEEE Trans. Ind. Appl. , Sept./Oct. 2004, Vol. 40, No 5, pp. 1379–1387.

[2] F. Briz, M. W. Degner, A. Diez and R. D. Lorenz, "Measuring, modeling, and decoupling of saturation-induced saliencies in carrier-signal injection-based sensorless AC drives," in IEEE Transactions on Industry Applications, vol. 37, no. 5, pp. 1356-1364, Sept.-Oct. 2001.

[3] E. Rodriguez Montero, M. Vogelsberger and T. Wolbank, "Robust Saliency-Based Speed Sensorless Control of Induction Machines under Overload Operation," 2022 International Conference on Electrical Machines (ICEM), Valencia, Spain, 2022, pp. 586-591.

[4] T. M. Wolbank and J. L. Machl, "Influence of inverter-nonlinearity and measurement setup on zero speed sensorless control of AC machines based on voltage pulse injection," 31st Annual Conference of IEEE Industrial Electronics Society, 2005. IECON 2005., Raleigh, NC, USA, 2005.

[5] E. Robeischl and M. Schroedl, "Optimized IN-FORM measurement sequence for sensorless PM synchronous motor drives with respect to minimum current distortion," in IEEE Transactions on Industry Applications, vol. 40, no. 2, pp. 591-598, March-April 2004.

[6] E. Rodriguez Montero, M. Vogelsberger and T. Wolbank, "Analysis and Compensation of Saliency Harmonic Components in Multi-Salient Induction Motors for Encoderless Speed Control," 2023 IEEE International Electric Machines & Drives Conference (IEMDC), San Francisco, CA, USA, 2023, pp. 1-7.

[7] D. Hind, C. Li, M. Sumner and C. Gerada, "Realising robust low speed sensorless PMSM control using current derivatives obtained from standard current sensors," 2017 IEEE International Electric Machines and Drives Conference (IEMDC), Miami, FL, USA, 2017, pp. 1-6.

[8] M. X. Bui, "Sensorless Position Estimation, Parameter Identification and Control Integration for Permanent Magnet Synchronous Machines using Current Derivative Measurements," 2018 International Power Electronics Conference (IPEC-Niigata 2018 -ECCE Asia), Niigata, Japan, 2018, pp. 4174-4180.

[9] J. -Y. Chen, K. -Y. Hung, G. -R. Chen and S. -C. Yang, "Integration of Measurement Vector Insertion With Discontinuous PWM to Improve Saliency-Based Sensorless Drive Position Estimation," in IEEE Transactions on Power Electronics, vol. 37, no. 12, pp. 15283-15296, Dec. 2022.

[10] J. P. Degel, S. Haehnlein, C. Kloeffer and M. Doppelbauer, "A moving least-square approach for current slope estimation in an inverter fed IPMSM using field programmable gateway arrays," 2020 International Conference on Electrical Machines (ICEM), Gothenburg, Sweden, 2020, pp. 1033-1039.

PCIM Europe 2024, 11– 13 June 2024, Nuremberg    DOI: 10.30420/566262179

# Optimized Half-Bridge Gate-Drive with low Time-Skew for RC-IGBTs and SiC-MOSFET Dead-Time Control

Jan Fuhrmann[1], Faiq Siddiqui[1], Till-Mathis Ploetz[2], Hans-Guenter Eckel[1]

[1] University of Rostock, Germany
[2] Rostocker Kompetenzzentrum für Leistungselektronik GmbH, Germany

Corresponding author:    Jan Fuhrmann, jan.fuhrmann@uni-rostock.de
Speaker:    Jan Fuhrmann, jan.fuhrmann@uni-rostock.de

## Abstract

To enhance the efficiency of reverse-conducting IGBTs and silicon carbide MOSFETs, advanced gate drive units can be employed to minimize operational losses. This entails the crucial task of current direction detection in the gate drive system, ensuring optimal semiconductor operation. If necessary, these units insert desaturation pulses just before switching, reducing switching losses. However, the introduction of desaturation pulses also raises the risk of overlapping switching, as the most effective desaturation pulse must be timed very precisely relative to the opposite switch's turn-on. To address this challenge, a sophisticated gate drive unit is proposed in this paper, emphasizing minimal time-skew and precise timing.

## 1   Introduction

Advanced gate drivers offer a significant opportunity to reduce losses in semiconductor devices by enabling sophisticated switching techniques such as desaturation pulses for reverse-conducting IGBTs and extremely brief lock times during freewheeling in SiC MOSFETs. In both scenarios, semiconductors engage in rapid switching with minimal lock times within a half-bridge configuration, demanding precise timing from the half-bridge drivers. Additionally, current direction detection within the gate drive is imperative.

This paper introduces a gate drive unit that excels in timing precision and incorporates a straightforward current detection feature on each gate drive unit.

The first section elucidates a switching pattern designed for reverse-conducting IGBTs, integrating a desaturation pulse via a compact FPGA. The second section outlines the development of the gate drive unit, offering insights into jitter and switching measurements for reverse-conducting IGBTs. The final section provides a glimpse of how this gate drive unit can be adapted with minor modifications to achieve extremely short lock times in SiC-MOSFET modules, thereby mitigating reverse-recovery losses.

A reverse-conducting IGBT integrates both a diode and an IGBT into a single semiconductor chip, offering distinct advantages, particularly in diode behavior. In this semiconductor design, the diode component occupies a larger area compared to conventional modules that feature a ratio of two IGBTs to one diode. This design characteristic contributes to the reduction of temperature swings in both semiconductors, especially when utilized in rectifier applications with very low frequencies, such as those encountered in double-fed induction generators.

The biggest disadvantage is the mixture of IGBT and diode with reduced diode performance when the gate of the IGBT is opened. This leads to a lowered plasma profile and higher conduction losses. Morover the switching losses of the diode increases with closed IGBT gate due to the big diode area. This requires an advanced gate drive which turns off the IGBT gate in diode mode and inserts a desaturation pulse, by opening the IGBT gate shortly before commutation. This concepts are described in [1], [2], [3]. Also concepts with optimized reverse-conducting IGBT, which are a comprimise especially for the diode behavior, are described in literature [4], [5]

PCIM Europe 2024, 11– 13 June 2024, Nuremberg    DOI: 10.30420/566262179

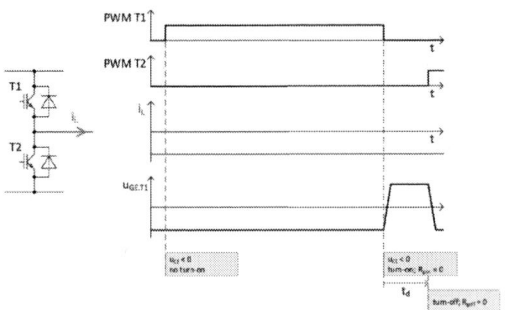

(a) Diode mode with turned off gate and desaturation pulse.

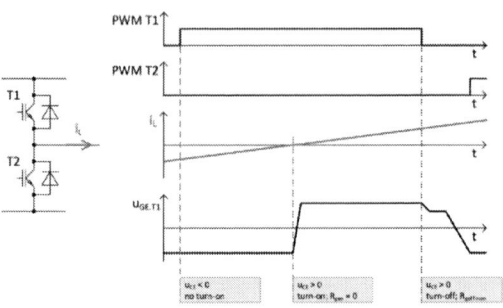

(b) Commutation from diode to IGBT with immediate turn on of the gate to avoid IGBT blocking.

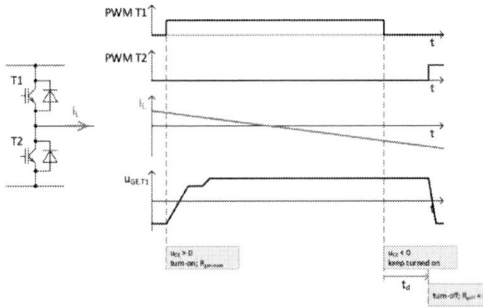

(c) Current commutation from IGBT to diode without turning off the gate to avoid too many switching events.

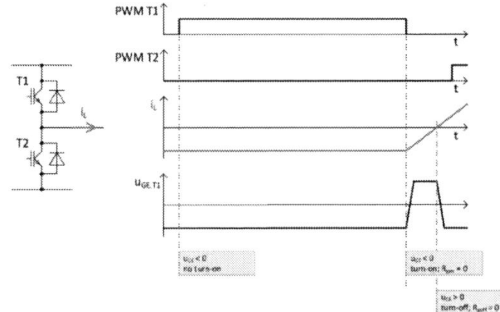

(d) Current commutation from diode to IGBT during the desaturation pulse, here an immediate turn off is required to avoid a shoot through.

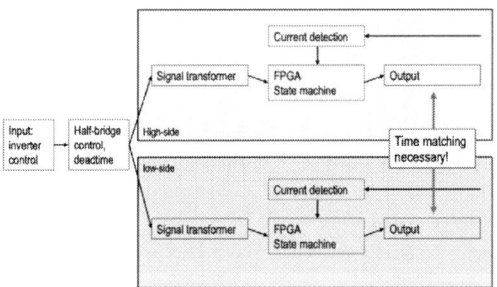

(e) Signal path and the required timing of the half-bridge gate drive.

(f) 3D model of the realized gate driver.

**Fig. 1:** Switching pattern for an reverse conduction IGBT with a current detection, critical timing path within the half-bridge gate drive unit and a picture showing the proposed gate drive.

## 2 Simplified Pulse Pattern for RC-IGBT with Desaturation Pulse

Our gate drive implementation utilizes a straightforward current detection method, incorporating a basic clamping diode chain to measure the collector-emitter voltage. This approach leverages the properties of pn-junctions, resulting in detection thresholds at both $0.7\,\text{V}$ and $-0.7\,\text{V}$. The diode chain contributes to this stability by introducing a current, which maintains a consistent collector-emitter volt-

age and facilitates reliable current detection. Furthermore, this voltage measurement serves a secondary purpose in short-circuit detection.

The current detection mechanism enables the development of a simple control pattern for the IGBT gate. In Figure 1a, we illustrate the standard diode mode operation where the gate is turned off. This results in a good diode characteristic with a high plasma concentration and low on-state losses. During turn-on, the collector-emitter voltage is inher-

1316

ently less than −0.7 V, due to the conducting free-wheeling diode. This condition is detected, preventing gate turn-on. On the flip side, during turn-off, a desaturation pulse, timed to match the lock time, is introduced. This gate switching occurs with a minimal gate resistor since there's no need for a semiconductor switching process. By opening the gate channel, the plasma concentration in the semiconductor decreases, leading to higher on-state losses, albeit negligible for brief durations, and reduced switching losses due to the removal of less plasma. It is imperative to close the gate before the opposing semiconductor activates and commutates the diode. Upon gate closure, the plasma begins to rebuild, underscoring the necessity for precise timing between the desaturation pulse and the onset of the opposite turn-on state.

In cases where a change in current direction is detected within the diode mode, as depicted in Figure 1b, the gate is promptly activated to prevent IGBT blocking. Without opening the IGBT gate, the IGBT would commence blocking the DC link voltage, leading to an erroneous output voltage vector that is unacceptable for inverter control. In such instances, the conventional IGBT turn-on process is inadequate, necessitating a minimal gate resistor to facilitate the IGBT gate charging process.

Conversely, a reverse current transition, as shown in Figure 1c, is disregarded to avoid unnecessary gate switching events, which could increase the switching frequency, potentially straining the gate drive and elevating losses. Fortunately, the additional losses in the diode during this short period are negligible.

The most critical situation arises when a current direction change occurs from the diode to the IGBT during the desaturation/lock time at the end of the switching cycle. As presented in Figure 1d, a transition happens during the desaturation pulse. In such cases, immediate action is required to prevent a bridge short-circuit, as the IGBT must be turned off earlier. If this transition is detected, the desaturation pulse is promptly terminated by deactivating the gate.

To achieve precise timing in this application, a half-bridge gate-drive unit with excellent synchronization between both channels is required. This synchronization is essential to enable desaturation pulses to occur in close proximity to the turn-on of the opposite IGBT, responsible for commutating the freewheeling diode. This level of timing precision

must remain stable across a broad range of operating temperatures and be particularly accurate at the semiconductor gate.

Ensuring this precision encompasses various aspects, including data transmission from the half-bridge input, the generation of lock times, the operation of the state machine within the FPGA, and the gate drivers. All these components contribute to the entire signal path, as illustrated in Figure 1e. Therefore, nearly every element within the half-bridge gate drive system plays a role in maintaining timing consistency and minimizing jitter, necessitating careful consideration in component selection.

# 3 Realized Half-Bridge Gate Drive Unit for RC IGBTs

In this section, we delve into the details of our implemented half-bridge gate drive unit. We elucidate the state machine embedded within the FPGA, delineate the selected hardware components starting from the input stage and culminating in the two bridge outputs of the gate driver. Additionally, we provide comprehensive measurements pertaining to this gate driver, with particular emphasis on timing and the histogram depicting the delays of each component.

## 3.1 State Machine for Controlling the RC IGBT

The state machine depicted in Figure 2 delineates the switching behavior of an RC IGBT, encompassing two distinct on-state modes alongside a desaturation pulse for the diode. Additionally, it incorporates a desaturation monitor essential for short-circuit detection. The state machine comprises three steady states highlighted in orange and four transition states denoted in blue.

The steady states are as follows:

– Off-state: In this state, the gate-source voltage ($V_{GE}$) is set to −15 V, coupled with a minimal $R_{G\text{-}OFF}$ to mitigate the risk of parasitic turn-on.

– On-state (with $V_{GE}$=−15 V): This state is characterized by a continuation of the −15 V $V_{GE}$ voltage, with a minimal $R_{G\text{-}OFF}$ ensuring optimal conduction characteristics for the diode.

– On-state (with $V_{GE}$=+15 V): Here, the $V_{GE}$ voltage is set to +15 V, accompanied by an IGBT $R_{G\text{-}ON}$, facilitating the conventional operation of the IGBT.

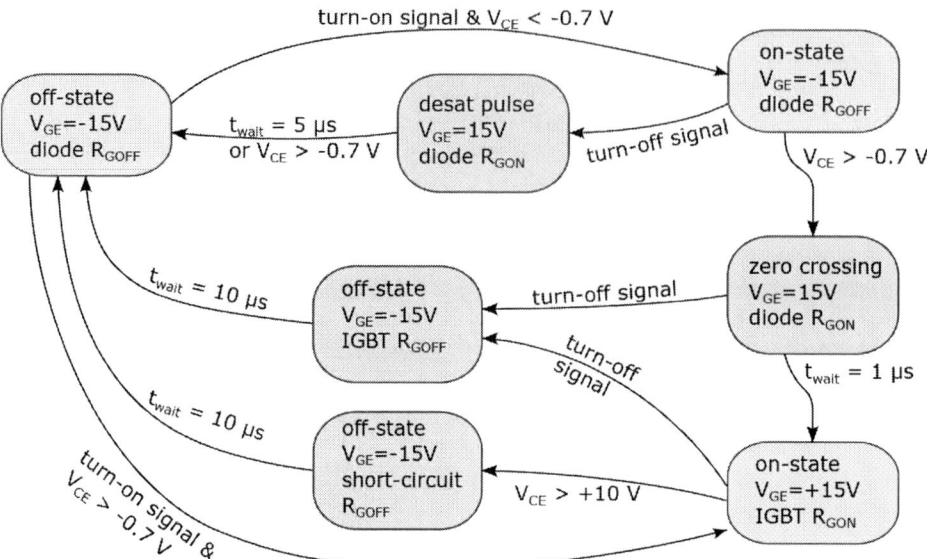

**Fig. 2:** State machine of the realized gate driver with three steady states (orange) and four transition states (blue). The conditions for switching the states are given at the lines between the states.

These states collectively govern the operational behavior of the RC IGBT, ensuring precise control over its switching dynamics and safeguarding against potential failure modes.

The four transition states play a crucial role in seamlessly connecting all steady states in a coherent manner.

The gate drive initiates in the off-state and transitions to the on-state upon receipt of the turn-on signal transmitted via the coreless transformer. At this juncture, the comparator within the diode chain measures the collector-emitter voltage, discerning between positive and negative collector current. Given the bipolar nature of the device, the detection threshold is set at –0.7 V. If the voltage falls below this threshold, indicating negative current flow, the diode region is activated, necessitating the gate channel to remain closed. The on-state of the diode is achieved directly.

Conversely, if the current is positive, signifying the need to turn on the IGBT, this action is executed utilizing $R_{G\text{-}ON}$ and a driving voltage of 15 V. The turn-off process is initiated either through a command relayed via the coreless transformer or prompted by a desaturation event. In the event of a short-circuit, the gate driver promptly detects the desaturation and proceeds to swiftly turn off the fault current, employing a large gate resistor to preempt overvoltage. Notably, this necessitates the inclusion of an additional turn-off MOSFET, a requirement obviated

with the implementation of active clamping. Under normal operational conditions, turn-off is facilitated by a suitable gate resistor, followed by a transition to the diode off-state after a predetermined delay, orchestrated by the state machine.

The diode turn-off operation is executed by the opposing IGBT, and to mitigate switching losses, a desaturation pulse is strategically incorporated. During this process, the gate is opened, causing a reduction in plasma concentration. Shortly before commutation commences, the gate is promptly closed. This action, typically performed after a predefined interval, involves the utilization of a small gate resistor to expedite the charging and discharging of the gate. Alternatively, an alternative approach involves triggering gate closure upon zero crossing of the current during the desaturation pulse, facilitating immediate gate discharge.

Another critical transition occurs from diode mode to IGBT mode in the on-state. Upon the collector-emitter current surpassing –0.7 V, the state machine swiftly switches the diode $R_{G\text{-}ON}$ to an open gate configuration to avert IGBT blocking. After a brief duration, the gate drive seamlessly transitions to disengage the IGBT $R_{G\text{-}ON}$.

## 3.2 Realized Half-Bridge Gate Drive Unit

As demonstrated earlier, the gate-drive unit necessitates highly precise timing, starting from the input signal and extending to the gate switching, with a compact FPGA responsible for generating the

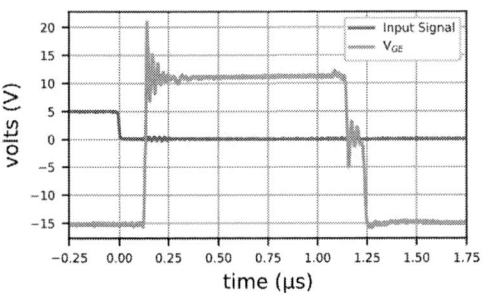

**Fig. 3:** Desaturation pulse which was insert by the gate driver after the turn-off signal was received. The gate is opened for circa 1 $\mu$s to lower the plasma concentration of the diode to reduce switching losses.

| Resource | Detail |
|---|---|
| Total LUT4s | 130 out of 5280 (2%) Logic LUT4s: 91 Ripple Logic: 36 LUT4s Feed thru LUT4s: 3 |
| Slice Registers | 73 out of 5280 (1%) |
| Phase-Locked Loop (PLL) | Input clock: 12MHz Output clock: 200MHz |
| Input/Output sites | 16 out of 39 (41%) |
| Number of I/O registers | 11 out of 117 (9%) |

**Tab. 1:** High-side and low-side FPGA resource utilization for one state machine; with the low usage of the components one FPGA could be suitable to drive the whole half-bridge driver.

switching pattern for driving the reverse-conducting IGBT. To minimize jitter, component selection is crucial, especially concerning delay and time skew. This aspect is well-handled when employing standard high-speed devices.

However, the primary challenge lies in transmitting the input signal to the isolated side of the gate drives. Traditional optical methods, such as optocouplers and optical fibers, introduce delays and jitter on the order of several hundreds of nanoseconds, rendering them impractical for this gate drive application. A more suitable solution involves using a coreless transformer with high blocking capabilities. In this case, we utilized an ISO7762FDW coreless transformer. For the power supply, a half-bridge supply module from Murata (MGJ6D05H24MC) is employed, and the FPGA chosen is a Lattice iCE device.

The main components of this gate drive are:

- Power supply:
  Murata MGJ6D05H24MC

- Data transmission:
  Texas Instruments ISO7762FDW

- FPGA: Lattice Semiconductors ICE40UP5K-SG48I

In the initial concept, a single FPGA at the input stage was employed to receive input signals for the half-bridge and generate two switching signals for the high- and low-side components. Additionally, it integrated a dead-time mechanism to prevent shoot-through. At the heart of the driver's operation, another FPGA orchestrated the state machine

functionality. However, there is potential to streamline this setup further to optimize space and reduce costs, a topic explored in the subsequent discussion.

Figure 3 illustrates a desaturation pulse within the gate drive. Following the reception of the turn-off signal, the FPGA initiates a desaturation pulse to mitigate switching losses in the diode by decreasing plasma concentration.

### 3.3 Gate Driver Costs

Currently, our setup utilizes three FPGAs to accommodate two state machines and generate the necessary input signals. Each FPGA requires additional supply voltages of 1.8 V and 3.3 V. This incurs additional costs estimated as follows:

- 30 € for 3x FPGA

- 10 € for 2x transmitter

- 10 € for 2x additional output bridges

Consequently, our prototype costs approximately 50 € more than a conventional half-bridge gate drive setup. However, by employing a smaller and single FPGA, costs could be reduced to around 30 €. This optimization holds the potential for significant cost savings without compromising functionality.

### 3.4 FPGA Resource Utilization

For the IGBT gate driver, the primary state machine responsible for half-bridge control was realized using Lattice's ICE40UP5K FPGA, harnessing its capabilities for efficient control and signal processing.

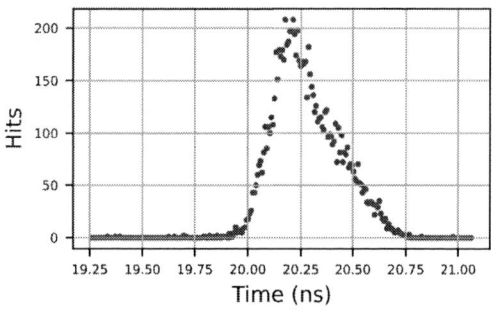

(a) Transmission time of the input signal to the FPGA via the coreless transformer

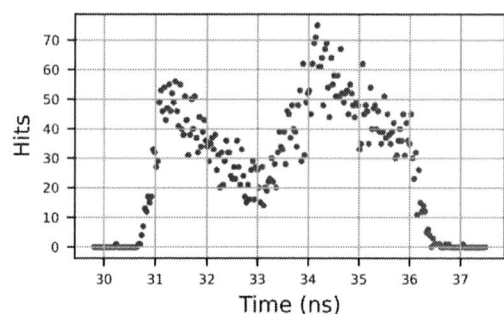

(b) Transmission time from the input signal to the FPGA and trough the FPGA

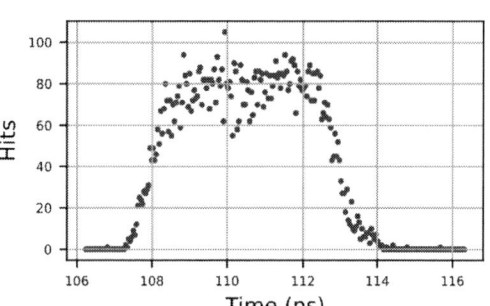

(c) Transmission time to the driving MOSFET of the output half bridge

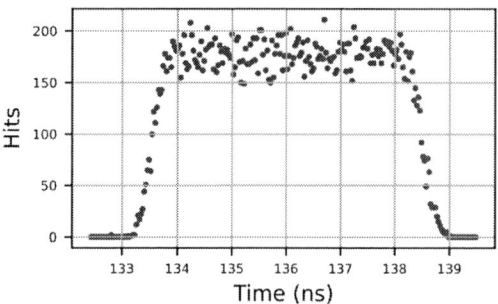

(d) Scatter plot of the delay between turn-off signal and beginning of the desaturation pulse, see Figure 3

**Fig. 4:** Measurement of time delays for different signal paths from the input to the driving MOSFET at the output of the gate driver. For the whole signal path from the input to the output a delay of 135.5 ns ± 3.5 ns was reached.

Monitoring resource utilization within the FPGA platform is paramount, as it significantly impacts the system's performance and scalability. While the Lattice Radiant software generates a comprehensive report on resource utilization, here we provide a concise breakdown of the FPGA's resource consumption in Table 1.

The complexity of the state machine required to govern the switching behavior of the RC gate driver is relatively low, rendering the FPGA utilized in the current setup over-designed. Only a few percent of the look-up-tables are needed to describe the state machine and also a lot of free pins of the package are available. While this level of FPGA capacity is acceptable for an initial prototype, there exists an opportunity to optimize and downscale in subsequent iterations to achieve cost savings.

### 3.5 Timing of the gate driver components

In addition to the FPGA state machine, the selection of components for precise timing is central to the functionality of this gate driver. Consequently, extensive measurements pertaining to the timing precision of our proposed gate driver have been conducted. These measurements serve to validate the accuracy and reliability of our design, ensuring its suitability for demanding applications.

In Figure 4, four histograms depicting timing are presented, illustrating the time delay between the input and various points within the signal chain.

The first measurement, depicted in Figure 4a, showcases the transmission time of the coreless transformer, which transmits the half-bridge input signal after the FPGA adds dead-time to the high- and low-sides. An average delay of 20.1 ns is observed, with a deviation of ±1.1 ns.

Figure 4b illustrates the timing from the input through the FPGA. The input signal, post-coreless transformer, is sampled and transmitted by the FPGA, resulting in an average delay of 33.5 ns with a wider spread. This spread is attributed to the discretization of the FPGA.

The FPGA's state machine requires a few clock cycles for processing, leading to an overall delay of 111 ns. Additionally, the gate driver for the output half-bridge introduces a small delay, but the overall jitter remains relatively low at $\pm 5$ ns.

The final sub-figure, see Figure 4d, presents the complete timing from input to gate signal, including the timing of the desaturation pulse (cf. Figure 3). After the turn-off signal, the state machine adds a desaturation pulse with an overall delay of 135.5 ns and a variation of $\pm 3.5$ ns. These results underscore the rapid data transmission and exceptional timing precision achieved by the gate driver.

## 4 Summary and Outlook

This paper introduced a half-bridge gate driver characterized by precise timing and minimal delay. A remarkable overall delay of 135.5 ns with a variation of $\pm 3.5$ ns has been achieved.

Furthermore, the half-bridge gate driver leverages an FPGA to govern a state machine allowing the operation of an RC IGBT in an advanced mode aimed at reducing diode on-state and reverse-recovery losses. This is achieved through the implementation of a straightforward state machine that employs the collector-emitter voltage as the primary means of current direction detection.

The gate drive strategy ensures that the IGBT channel remains closed when current is being conducted by the diode. During turn-off, it introduces a desaturation pulse to lower plasma concentration just before current commutation occurs. The precision timing required for these operations is effectively attained with the proposed gate driver, thus offering a comprehensive solution for optimizing IGBT performance and minimizing losses.

The current iteration of the gate driver utilizes three FPGAs: one for managing the input stage and two for controlling the high and low side IGBTs of the half-bridge configuration. However, there's room for improvement by consolidating all functions onto a single FPGA. This consolidation promises a more streamlined design and a reduction in component count. Moreover, it would significantly shrink the board footprint and decrease interconnect lengths, thereby enhancing signal integrity and achieving a more compact overall design.

Transitioning to a single FPGA also underscores the importance of resource management. Essential resources such as Look-Up Tables (LUTs), logic elements, and memory blocks must be carefully managed to ensure optimal performance and functionality. By efficiently allocating these resources, we can maximize the capabilities of the FPGA while maintaining the desired level of functionality and performance.

## References

[1] X. Huang, C. Ling, X. You, and T. Q. Zheng, "Research of the loss and gate desaturation control for RC-IGBT used in vehicle power converters," in *IEEE International Conference on Industrial Technology (ICIT)*, 2017, pp. 201–206. DOI: 10.1109/ICIT.2017.7913083.

[2] D. Lexow, Q. T. Tran, and H.-G. Eckel, "Performance modulation of reverse conducting IGBT through gate-voltage adjustment in diode conduction mode," in *21st European Conference on Power Electronics and Applications,*, 2019. DOI: 10.23919/EPE.2019.8914784.

[3] D. Lexow and H.-G. Eckel, "Boosting Pilot-Diode Reverse-Conducting IGBTs Turn- ON and Reverse-Recovery Losses with a Simple Gate-Control Technique," in *24th European Conference on Power Electronics and Applications*, 2022, pp. 1–10.

[4] M. Rahimo, C. Papadopoulos, C. Corvasce, and A. Kopta, "An Optimized Plug-In BIGT with No Requirements for Gate Control Adaptations," in *PCIM Europe 2017. International Exhibition and Conference for Power Electronics, Intelligent Motion, Power Quality*, 2017, pp. 16–18.

[5] Q. T. Tran, F. J. Niedernostheide, F. Pfirsch, A. Mauder, R. Baburske, and H.-G. Eckel, "RC-GID IGBT-A novel reverse-conducting IGBT with a gate voltage independent diode characteristic and low power losses," in *Proceedings of the International Symposium on Power Semiconductor Devices and ICs*, vol. 2021-May, 2021, pp. 347–350. DOI: 10.23919/ISPSD50666.2021.9452199.

PCIM Europe 2024, 11– 13 June 2024, Nuremberg          DOI: 10.30420/566262180

# Design of a Traction Inverter Based on PCB-Embedded GaN Devices

Luca Bongiovanni[1], Maurizio Tranchero[1], Claudio Romano[1], Paolo Santero[1]

[1] Ideas & Motion s.r.l., Italy

Corresponding author:     Maurizio Tranchero, maurizio.tranchero@ideasandmotion.com
Speaker:                  Maurizio Tranchero, maurizio.tranchero@ideasandmotion.com

## Abstract

This paper summarizes the design of a highly integrated power stage for a traction inverter based on PCB-embedded GaN transistors, compatible with Hybrid-PACK Drive power modules from Infineon. Starting from the modeling of the transistors to estimate the power losses, the requirements to properly manage heat dissipation are derived and are translated in specifications to guide the design, the PCB stack-up, and the layout. Special care has been devoted to the gate driving circuitry, integrating several solutions to evaluate GaN transistors and at the same time comparing the behavior of commercial core-less hall-effect current sensors with more standard hall-effect devices.

## 1  Introduction

On the road for the electrification of electric vehicles, *Wide Band-Gap* (WBG) devices are key to achieve efficient and compact devices. SiC transistor have been used extensively in traction applications, whilst the adoption of GaN-based solution is still limited. In the realm of the HiEFFI-CIENT funded project [1], one of the objective was to evaluate the use of GaN transistors in traction applications. To achieve this, PCB embedding [2] technology has been used to reduce the thermal impedance between the junction and the liquid cooler.

This paper describes the criteria and the solution implemented to design a traction inverter to be used in the automotive field, based on GaN transistors. More specifically, the targets are

1. to present a novel inverter family taking advantage of GaN technology and PCB embedding methodology optimized for HV traction motor, and

2. to decrease time-to-market through efficient design space exploration with the usage of high-level internal simulation environment, enabling evaluate power losses of an inverter for a full *Worldwide Harmonized Light Vehicles* (WLTP) cycle (30 minutes) within $< 5\,\text{s}$ computing time in a standard laptop.

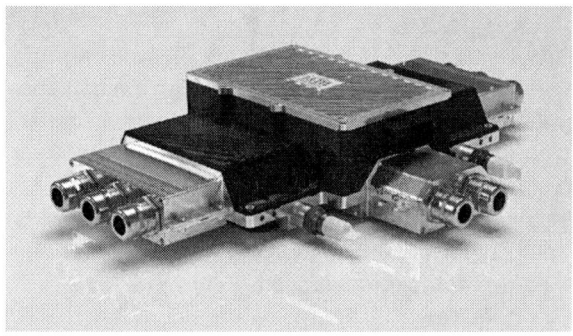

**Fig. 1:** DIDIMO_HV400 a dual inverter developed at Ideas & Motion.

The paper is organized as follows: Sec. 2 describes the high-level considerations made to shape the initial design; Sec. 3 describes the inverter architecture and each board composing the unit; Sec. 4 provides some initial achievement from the very first prototype; and Sec. 5 draws final conclusion.

## 2  High-Level Design

### 2.1  Mechanical Constraints

Mechanical constraints for this design derives from the DIDIMO_HV inverter family developed at I&M in the last years [3] (Fig. 1). This inverter family is based on a highly modular idea: keeping all the functionalities separated and well-defined in order to make them easy to substitute and update depending on the target application.

All the units developed so far have been based

PCIM Europe 2024, 11– 13 June 2024, Nuremberg    DOI: 10.30420/566262180

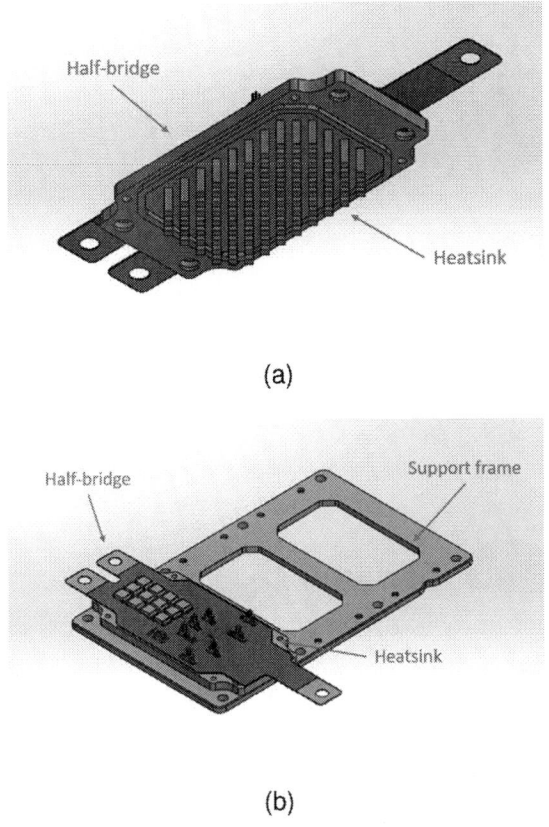

(a)

(b)

**Fig. 2:** The half-bridge stage, the heat sink, and the support frame.

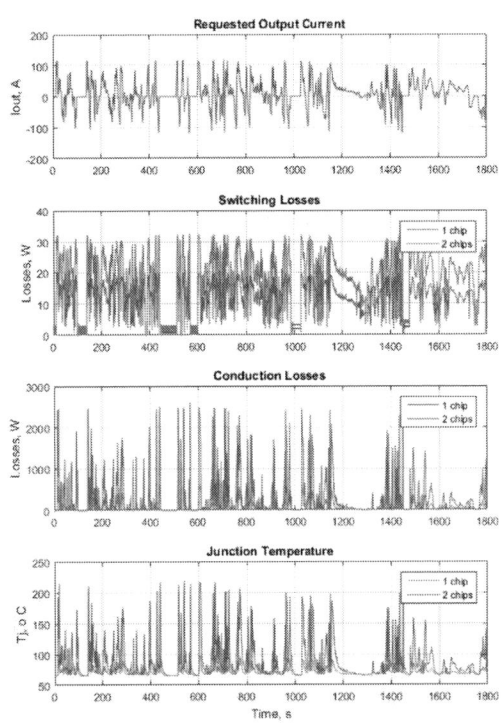

**Fig. 3:** WLTP simulation for our GaN power stage.

on *Hybrid-Pack Drive* (HPD) power modules [4]. This because it represents a well-established power module architecture for automotive applications used by several manufacturers and became a *de facto* standard for high-power inverters. To reuse all the mechanical design done with the previous devices, it has been decided to design something compatible with HPD modules. GaN devices available on HPD power modules are few. GaN Systems few years ago did some investigation on this field [5], but they are no longer available for purchasing. Building a similar device would have high costs and hence was not a viable solution for this project.

To achieve similar thermal performances and reduce manufacturing costs, in the HiEFFICIENT project [1], thanks to one partner [6], it was possible to access to the chip-embedding technology [2]. This technology enables the inclusion of bare dice into the fabrication process of a *Printed Circuit Board* (PCB) and helps reducing the dimension

of a given solution (package is not present) and the thermal impedance between the junction and the coolant. The final shape of the PCB depends on the designer's need, hence the idea was to make a HPD-compatible module to have a one-to-one replacement in the inverter architecture and hence easing the performance comparison.

To further reduce the thermal resistance between the junction and the heat-sink, the PCB has been decided to use sintering on the base-plate in direct contact with the coolant (Fig. 2).

## 2.2 System-Level Simulations

The target of the demonstrator was to drive 100 A continuously, after discussion with the manufacturer, the choice fell on some bare dice from GaN Systems, each of them has $25\,m\Omega$ RDSon. To decide how many devices were needed to meet the requirements, the evaluation of these devices in the DIDIMO_HV [3] cooling system was necessary. To do this it has been developed a surrogate model to simulate the system behavior under realistic conditions.

1323

For this reason, the device and its substrate have been described into the SimPLE modeling environment [7], an internal high-level simulator.

The simulator requires to define the components in an object-oriented fashion, i.e., defining a class for each of them, i.e.,

- the power switches, including all the functional parameters of the devices;

- the package where the switches are installed, describing the stray elements given by coupling and inductive paths;

- the topology implemented;

- the gate driver characteristics.

The parameters for the transistors have been extracted from the datasheet to correctly populate the model, whilst those of the substrate have been estimated using manufacturer information and geometrical considerations. Thanks to this information, SimPLE is able to extract the key parameters of the system and run the electro-thermal model of the system under test to extract power losses and estimate the junction temperature.

Since look-up tables and simple approximated formulae are used for calculation, the execution of the model is fast (less than 5 s to simulate an inverter executing the WLTP [8] cycle), this makes the tool well-suited to design-space exploration and hence obtain an optimized design.

An example of the output a simulation performed with this tool provides is reported in Fig. 3. It shows the results of a WLTP simulation: the top pane shows the current profile requested by the application; power losses are reported in the two central panes, separated as switching and conduction losses; and the estimated junction temperature of the devices in the lower pane.

The simulations performed with this method enabled to decide the number of dice in parallel to be used. Indeed, looking at Fig. 3 it is possible to see how much conduction and power losses are changing depending on the number of devices in parallel. This has a direct impact on the junction temperature and it is evident that a single chip is not enough to guarantee to operate in the allowed temperature range for these devices, for this reason it was decided to go for the integration of two devices in parallel.

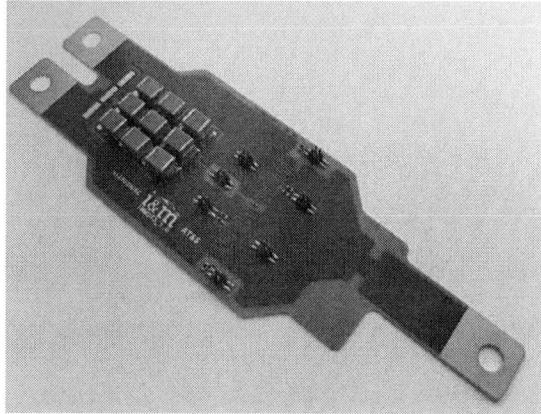

**Fig. 4:** DC-link directly mounted on the half-bridge stage.

# 3   Architecture

To improve modularity and scalability of the application, the inverter has been split into three different boards implementing the following functionalities

- the power stage

- the gate driver

- the control board

Following sections will introduce each of these subsystems and provide details about how they have been designed and which functionalities have been included.

## 3.1   Power Stage

The Power Stage is designed to be fully compatible, in terms of physical dimensions, with the HPD power modules from Infineon. It is composed of three main parts: the half-bridge stage, the support frame, and the heat sink (Fig. 2b).

The half-bridge stage is a single phase power stage developed using the PCB embedding technology in which the GaN die is embedded in the PCB inner layer. The heat sink is an aluminum frame, with a fin structure to increase the dissipation area towards the liquid coolant, directly sintered with the PCB. The support frame, made of aluminum, is used to fix the three modules in a way that matches the form factor of the HPD.

Embedding power semiconductor devices into printed circuit boards (PCBs) offers numerous advantages over traditional packaging methods. By integrating the semiconductor dies directly into the

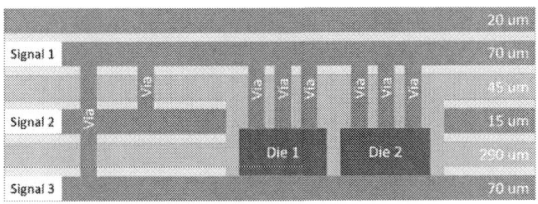

**Fig. 5:** The PCB stack-up.

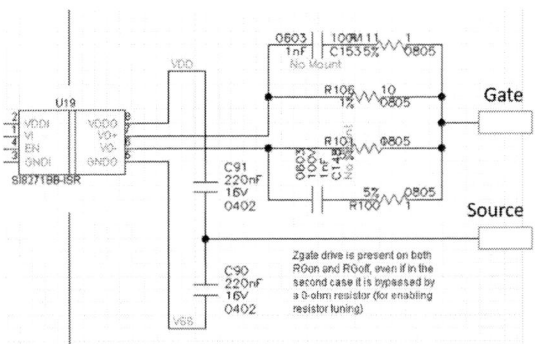

**Fig. 6:** Detail of the gate-driving circuit.

circuit board, the size of converters can be significantly reduced. This reduction in size leads to shorter current loops, which in turn decreases interconnection resistances and parasitic inductances. Consequently, system-level efficiency is improved as both conduction and switching losses are minimized. Additionally, utilizing thick copper substrates facilitates efficient heat dissipation, thanks to the low thermal resistance they provide.

Three of these modules are needed for one complete three-phase power stage. The single module has four GS-065-060-2-D2 GaN with a Vds of 650 V and Imax of 60 A: two in parallel for the low-side and two for the high-side. The embedding technology offers the possibility of providing a distributed DC-link capacitor directly on the module, reducing significantly the inductance with respect the traditional technology.

The chosen embedding technology consists of a three-layer PCB: the top layer is used to carry the current and to contact, with apposite vias, the embedded GaN; the inner layer can be used for signals while the bottom layer is used as a support layer, directly in contact with the GaN die that it is placed between the bottom and the inner layers.

### 3.2 Gate-Driver Board

The *Gate-Driver Board* (GDB) aims at interfacing correctly the high-voltage power stage with the low-voltage controller. This is implemented in two ways:

- Adapting control signals (both driving current and voltage levels) to the power transistors and high-voltage feedbacks to the voltage range used in the controller domain.

- Providing the proper electrical insulation required to accomplish safety rules to be granted in a high voltage application [9].

#### 3.2.1 Control Signals

The power stage is based on GaN-System's bare dice [10], embedded on a PCB. These devices require specific signal levels to correctly operate (i.e., $V_{GS,max}=[-10\,V:7\,V]$), unfortunately standard microcontrollers cannot operate to these voltage. Moreover, the port of a microcontroller cannot provide the current to drive these devices quick enough and this will automatically result in increasing the switching losses reducing the efficiency of the energy conversion. For these reasons, a gate driver is inserted to adapt the low-voltage signals coming from the logic domain into those compatible with the power transistors. Starting from the reference design GaN-System provides [11], it has been decided to use the same gate driver from Skyworks [12]. Two separated gate resistors have been used to switch on and off the power device with different speeds. Moreover, an RC network has been added in parallel to both gate resistances to bypass them during the initial phase of the switching and hence reducing the switching time, keeping limited the resulting overvoltage [13]. Figure 6 shows the resulting schematic.

#### 3.2.2 Analog Measurements

From the HV domain several signals have to be brought back to the microcontroller, crossing the insulation barrier, namely

1. The AC currents, i.e., the measurement of the output currents sunk by the load. Two of them are required for our control scheme, but for safety reason all the three phases are monitored either with a hall-effect sensor equipped with a concentrator [14] or using a core-less alternative [15]. Galvanic insulation between the sensing element and the reading circuitry is ensured by sensor technology itself that does not require to be in contact with the HV domain.

2. The DC voltage (VDC), a high-voltage divider

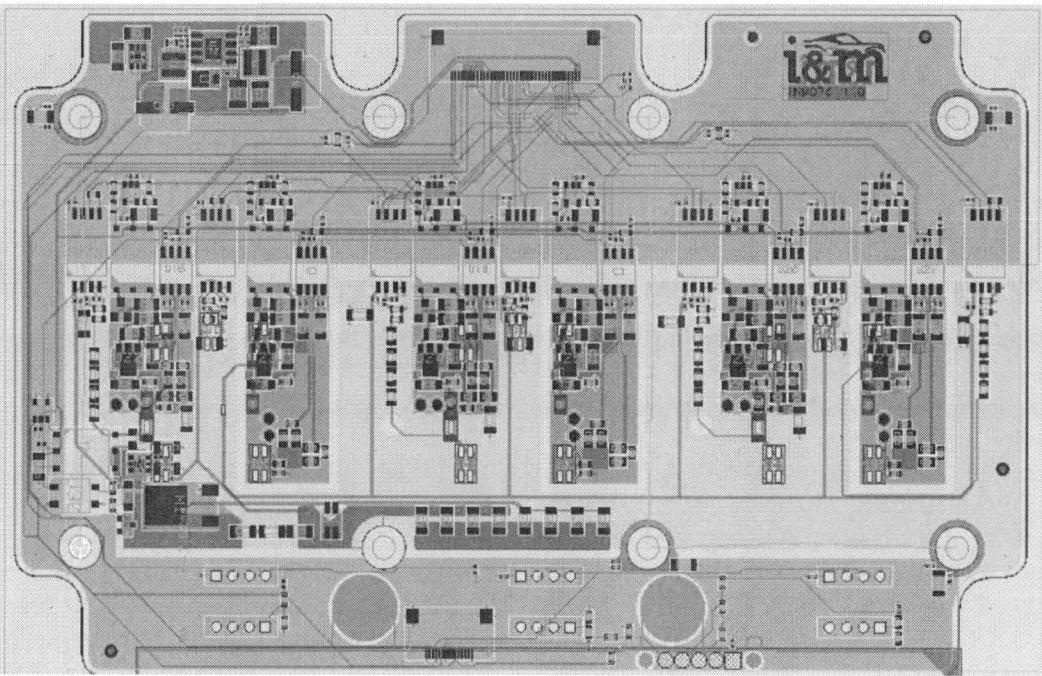

**Fig. 7:** Layout of the gate-driver board.

is used to adapt the DC bus voltage to a level compatible with the reading circuitry and then insulated by means of an isolated amplifier.

3. The AC voltages, implemented as the DC voltage monitor and used to improve the control in case of sensor-less activation [16] and measure the rotor temperature using the back EMF [17].

Since coming from the HV domain, these signals have been adapted in amplitude to be suitable to the dynamic input range of the controller and then insulated to keep the required safety level.

### 3.2.3 Electrical Insulation and Safety

This inverter will operate at 400 V (nominal) and hence electrical insulation between the HV domain and the microcontroller has to be granted for safety reason [9]. Both clearance and creepage distances have to be calculated to ensure, not only the functional insulation, but also the basic insulation between live parts. (Double insulation is not required since all metallic parts are connected to chassis and electric vehicles are equipped with an insulation monitor [9], [18].)

### 3.2.4 PCB Layout

Special care has been devoted to the layout of the GDB (Fig. 7). The approach followed is the one proposed in [19]:

1. To keep commutation loops as small as possible, the top was used as routing layer and the first inner layer devoted to the source connection on the switching nodes.

2. Planes have been kept as small as possible to avoid coupling to the other boards of the inverter and then generate lower EM noise.

3. Each switching node has been isolated from the others and special care has been devoted to avoid any overlapping between areas at different potential.

4. The low-voltage domain uses the first inner layer as ground plane.

### 3.3 Control Board

The control board is the brain of the traction inverter, it is designed to be used in various applications, thanks to the high performance and configurability of its microcontroller (Aurix TC38x [20]). The applications for which the board is designed are the single, double and three-level voltage inverters, and single and double current inverters. More in details, the *Generic Timer Module* (GTM) of the TC38x can be configured to synchronously drive

PCIM Europe 2024, 11– 13 June 2024, Nuremberg    DOI: 10.30420/566262180

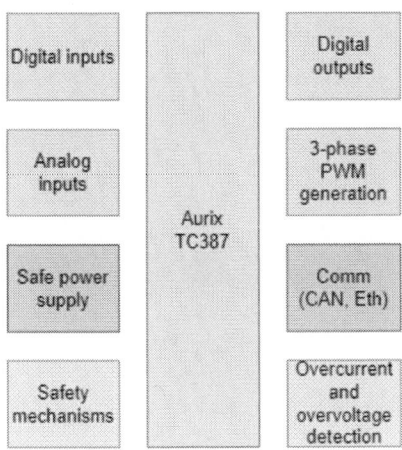

**Fig. 8:** Block diagram of the control board.

the three-phase bridge driver inputs (up to 18 in case of three-level inverter) and acquire the phases current.

The control board interfaces the gate driver board (two gate drivers for the double inverter) and the outside of the inverter. The gate driver interface includes all the IO necessary to command the phase drivers and to acquire the phase current sensors. The interface towards the outside consist of the outputs of the motor (position sensor and temperature) and generic inputs and outputs that allow the board to acquire external signals and drive external loads. The control board is designed to address safety applications: the TC38x microcontroller coupled with its power supply is able to reach the ASIL-D level [21]. This capability is beneficial in EV applications, in case of malfunctions the control board is able to trigger different countermeasures like fast discharge, three-phase bridge in freewheeling or in active short circuit.

## 4   The Prototype

The first samples received of the power stage received were damaged. Indeed, during the sintering process, due to a wrong setting, the PCB were exposed to a pressure higher than needed, resulting in an almost systematic damage on the connection between one of the two gates and trace connecting to the external world. The gate connection, in a large part of the samples, is shorted to the source. This result probably is due to the high technological ambition of the project: the combination of the sintering process and the PCB-embedding of dice required much more effort to be put in place and

fine tuned.

The project is still ongoing, hence another run would be necessary to have some specific data and hence perform a characterization of the system.

Despite the power stage, the other boards are fully operating, hence once the new batch of power stages will be available, it will be possible to complete the testing (Fig. 9).

## 5   Conclusion

This paper presented the design of a traction inverter based on PCB-embedded GaN devices. An intial phase devoted to design exploration has been done into an internal high-level simulator, based on analytical and empirical data. The results of these simulations helped defining the power stage structure and organization.

The actual design has then been separated into multiple boards to keep the main inverter functionalities separated and being able to target different applications maintaining a high level of modularity and scalability. The power stage has been designed to keep mechanical interface compatible with HybridPACK Drive power modules to reuse existing cooling circuits and rails.

The initial testing of the power boards is affected by issues on process reliability for the sintering step in the assembly. For this reason the power stage cannot be tested at the moment of writing this paper. The remaining boards (i.e., control and gate-driver board) are fulfilling the expectation and operate correctly. We are looking forward to the next batch of power modules to complete the testing of the whole inverter.

## Acknowledgment

This work was supported by HiEFFICIENT project. This project has received funding from the ECSEL Joint Undertaking (JU) under grant agreement no. 101007281. The JU receives support from the European Union's Horizon 2020 research and innovation programme and Austria, Germany, Slovenia, Netherlands, Belgium, Slovakia, France, Italy, and Turkey.

## References

[1]   "Highly Efficient and Reliable Electric Drivetrains Based on Modular, Intelligent and Highly Integrated Wide Bandgap Power Electronics Modules." (2024), [Online]. Available: https://www.hiefficient.eu/.

**Fig. 9:** Prototype assembly of the power stage and the gate-driver board.

[2] C. Buttay, C. Martin, F. Morel, R. Caillaud, J. Le Leslé, *et al.*, "Application of the PCB-Embedding Technology in Power Electronics - State of the Art and Proposed Developments," in *2018 Second International Symposium on 3D Power Electronics Integration and Manufacturing*, 2018, pp. 1–10.

[3] Ideas & Motion. "DIDIMO_HV Product Page: A Dual Inverter for High Voltage Applications." (2017), [Online]. Available: https : / / www . ideasandmotion . com / products / e - mobility - controllers/dual-400v/.

[4] Infineon. "Assembly instructions for the Hybrid-PACK Drive." (2024), [Online]. Available: https : //www.infineon.com/.

[5] GaN Systems. "650V 300A 3-Phase GaN Power Module with External Gate Driver Board." (2021), [Online]. Available: https : // gansystems . com / wp - content / uploads / 2021 / 10 / GS - EVx - 3P - 650V300A - SP1x - Technical - Manual - Rev - 211013.pdf.

[6] *Advanced Technologies & Solutions*, 2024.

[7] L. Giraudi, M. Tranchero, C. Romano, and P. Santero, "A Fast and Reliable Simulator for the Evaluation of Losses in Power Devices based on a Mixed Analytical and Empirical Model," in *Thermal Investigation of ICs and Systems, 2023. THERMINIC 2023*, 2023.

[8] UNECE. "Proposal for a new global technical regulation on the Worldwide harmonized Light vehicles Test Procedure (WLTP)." (2014), [Online]. Available: https://unece.org/DAM/trans/doc/2014/wp29/ECE-TRANS-WP29-2014-027e.pdf.

[9] International Electrotechnical Commission, "Insulation coordination for equipment within low-voltage supply systems," International Electrotechnical Commission, Standard, 2020.

[10] GaN Systems. "GS-065-060-5-T-A Automotive 650 V GaN E-mode transistor Datasheet." (2022), [Online]. Available: https://gansystems.com/wp-content/uploads/2022/01/GS-065-060-5-T-A-DS-Rev-220127.pdf.

[11] GaN Systems. "Design with GaN Enhancement mode HEMT." (2018), [Online]. Available: https://gansystems.com/wp-content/uploads/2018/02/GN001_Design_with_GaN_EHEMT_180228.pdf.

[12] Skyworks. "Si827x Data Sheet." (2022), [Online]. Available: https://www.skyworksinc.com/-/media/Skyworks/SL/documents/public/data-sheets/Si827x.pdf.

[13] F. Stormer, H.-G. Eckel, F. Pfirsch, and F.-J. Niedernostheide, "Switching Behavior of SiC-MOSFETs in High Power Modules," in *PCIM Europe 2018; International Exhibition and Conference for Power Electronics, Intelligent Motion, Renewable Energy and Energy Management*, 2018, pp. 969–974.

[14] LEM. "HAH3DR 1200-S07/SP3, Open Loop Hall Effect Technology." (2024), [Online]. Available: https://www.lem.com/en/product-list/hah3dr-1200s07sp3.

[15] Allegro Microsystems. "Coreless, High Precision, Hall-Effect Current Sensor IC with Common-Mode Field Rejection, Overcurrent and Overtemperature Detection." (2024), [Online]. Available: https://www.allegromicro.com/en/products/sense/current-sensor-ics/sip-package-zero-to-thousand-amp-sensor-ics/acs37610.

[16] G. Liu, C. Cui, K. Wang, B. Han, and S. Zheng, "Sensorless control for high-speed brushless dc motor based on the line-to-line back emf," *IEEE Transactions on Power Electronics*, vol. 31, no. 7, pp. 4669–4683, 2016. DOI: 10.1109/TPEL.2014.2328655.

[17] D. Fernandez, D. Reigosa, T. Tanimoto, T. Kato, and F. Briz, "Wireless permanent magnet temperature & field distribution measurement sys-tem for ipmsms," in *2015 IEEE Energy Conversion Congress and Exposition (ECCE)*, 2015, pp. 3996–4003. DOI: 10.1109/ECCE.2015.7310224.

[18] R. Foley, R. Nagappala, G. Ressler, P. Andres, and B. Martel, "Application of Insulation Standards to High Voltage Automotive Applications," vol. 2, Apr. 2013. DOI: 10.4271/2013-01-1528.

[19] E. Persson, "Optimizing PCB Layout for HV GaN Power Transistors," *IEEE Power Electronics Magazine*, vol. 10, no. 2, pp. 65–78, 2023. DOI: 10.1109/MPEL.2023.3275311.

[20] Infineon. "AURIX™ Family – TC38xQP." (2024), [Online]. Available: https://www.infineon.com/cms/en/product/microcontroller/32-bit-tricore-microcontroller/32-bit-tricore-aurix-tc3xx/aurix-family-tc38xqp/.

[21] International Standard Organization, "Road Vehicles — Functional Safety," International Standard Organization, Standard, 2011.

PCIM Europe 2024, 11– 13 June 2024, Nuremberg    DOI: 10.30420/566262181

# Optimizing Electric Vehicle Performance with GaN Design

**Authors** : Andrew Patterson, David Green, Ahmed Nejim

**Silvaco UK Ltd.**

**PCIM Speakers**:    Andrew Patterson,  _andrew.patterson@silvaco.com_
                     Dr. David Green,  _david.green@silvaco.com_

## Abstract

A rapidly growing number of consumers are preferring electric vehicles as their primary mode of transport, in order to support the global drive to reduce fossil-fuel burning. These consumers are regularly finding that the vehicle manufacturer claims on driving-range from a full battery, performance, re-sell value, re-cycling, repair, insurance and other everyday criteria are unfortunately not being met. Many global companies and industries are part of the complex Electric Vehicle (EV) supply chain,   including power electronics semiconductor designers. This paper focuses on the improvements that can be applied to GaN semiconductor design specifically,   as a future automotive technology to improve the efficiency and performance of the power MOSFETs used in electric vehicle traction drives, on-board chargers, and DC-DC Inverters.   Semiconductor device simulations are used to analyze the tradeoffs that can be made at the device and atomistic level, to improve conductivity, breakdown voltage performance, temperature effects and switching efficiency.

## Problem Statement

Today's EV charging infrastructure allows for an average vehicle travel of 160 km between charges, according to Deloitte (DTTL). Consumer expectation, after years of using internal combustion engines, averages at 320 km between refueling – two times more. Power inverters are at the heart of the electronic subsystems of Electric Vehicles, used to transform DC voltage levels from the battery in order to drive a wide range of vehicle accessories,   as well of course to provide AC power for the vehicle electric motor / powertrain.   Anything that can be done to reduce inverter and semiconductor device power-loss, will help to preserve battery energy, and also reduce internal self-heating and wasted energy through heat.

Manufacturers generally test their vehicles under ideal conditions, but in real-life use consumers will make use of devices which consume a lot of battery energy,   such as cabin heaters, air-conditioning, heated-seats, de-misters and much more. Consumers expect to be able to operate these devices, as well as achieve good driving range, when using the vehicle in different weather conditions. Manufacturers will be very interested in any design approaches allowing optimization of the available battery energy.

Vehicle and environmental temperature changes significantly affect semiconductor performance, but are regularly not reported in EV test figures. Performance at extreme temperature points needs to be modelled to give a complete picture. Among the many challenges such as the battery technology and traction improvement, the underlying power semiconductors remain a fundamental area of EV R&D for the foreseeable future (1).

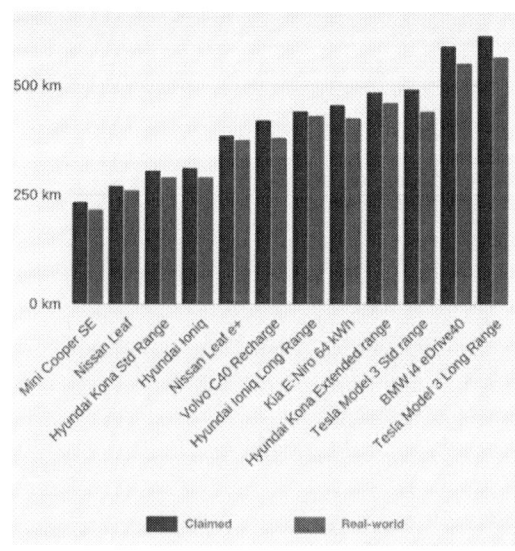

**Fig. 1**    Electric Vehicle Range : Actual v Claimed

# 1    Solution Objectives

Gallium Nitride (GaN) semiconductor switches are emerging as a superior technology to the more widely used Silicon Carbide (SiC) devices in EV Inverters. GaN device parameters of interest, which directly affect performance include:

- Switching Speed
- Channel Conductivity (Gds)
- Channel Capacitance
- Device Weight
- GaN material characteristics
- HEMT Channel Geometry

The analysis described here set out to investigate how much improvement could be achieved by adjusting the typical design parameters. This performance analysis made use of both 2D and 3D simulations, and made use of a GaN High Electron Mobility Transistor (HEMT) model. The overall performance enveloped was explored using "Design of Experiments" (DoE) simulation techniques.

## 1.1    GaN Building Block

GaN devices have the capability to be used for very high efficiency inverter design (in the range 98-99%) and handling up to tens of kilowatts of power (2). Through CMOS integration, operating frequencies of 40–75 MHz can be reached. This can substantially reduce the size and weight of associated inverter components, such as inductors and capacitors, thus resulting in a more compact design. Technology Computer Aided Design (TCAD) simulation of Wide-Bandgap power semiconductor including GaN HEMTS allows designers to explore the critical performance dependencies identified here, including device geometry, blocking voltage and breakdown, Ron, Transconductance, Transient Switching , Parasitics, as well as self-heating and proximity effects.

The preferred solution for power applications is the "normally-off" GaN design, and the pGaN HEMT is shown in Figure 2. The top panel shows the interruption of the 2DEG charge channel below the gate contact at zero or low gate bias. This explains the normally off condition.The bottom panel shows how the design of the field shield metal plates protects the gate stack from the high drain potential under blocking conditions.

This geometry was set up in the simulator, in order to then further explore parameter changes in the available design space.

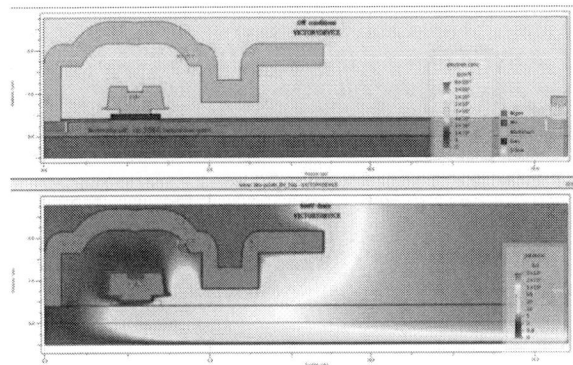

**Fig 2** : Normally Off GaN Design

All the volume data obtained in TCAD simulations are difficult or impossible to measure directly from the real device. The simulator-provided information provides an essential insight into the device performance, and allows an investigation into the device performance limits and technology limitations. The alternative would be real-device tests at very high currents and voltages – both expensive and potentially hazardous.

## 1.2    Design Space

Product design is a tough task faced by all OEMs and their supply chain. Sometimes, design criteria are conflicting, and navigating the design space to discover the best compromise is a multi-disciplinary activity. For semiconductor device and circuit design, the effort starts by parameterizing all the process steps and the subsequent device response. Numerical design tools are well-suited to this task, allowing a full flow from concept through to manufacture/fabrication.

Naturally, the device manufacturing process (Fab) will have rules that cannot be violated either due to the available capabilities or due to significant loss of yield. These must be appreciated by the technology designers and respected in the Design of Experiment (DoE) configuration.

Figure 3 shows the scenario for the simulation flow process. The effects of GaN device variations are collected and presented, and the effect on overall design performance judged. Among the many parameters to consider one could list the configuration of the gate contact, the field plates, the length of the channel and the design of the stack. Add to that the intentional doping and the diffusion to create the pGaN region under the gate.

The impact of these technological variations on the device performance as well as the circuit performance can be obtained within a Design Technology Co-Optimization flow (DTCO). A seamless approach driven by a global script serves this purpose well.

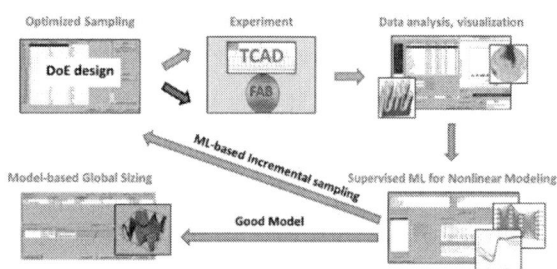

**Fig 3.** GaN Simulation Flow Process

When the DoE is coupled with machine learning algorithms, then a frugal design of experiments (DoE) data set can be generated that effectively covers all the available design space. Predictions on product yield can also be made.

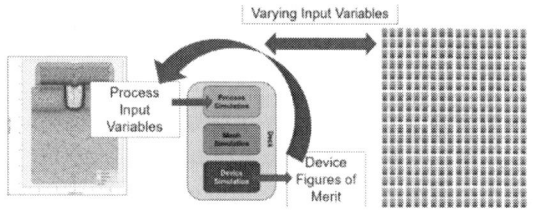

**Fig 4.** Generating a DOE matrix

A typical design cycle would involve identifying a starting flow based on the engineer's best guess. This is then expanded into a large matrix of simulations as shown in figure 4. The large number of data are needed to explore all possible combinations of technology parameters in search of the most optimal parameters. Finally, generating a

parametric relational model that can be understood by a wide number of technology engineers and stakeholders beyond the specialist design community represents a significant value added.

Silvaco has developed a tool flow that allows TCAD simulation data to be made available for the engineers involved in device fabrication. The tool chain is illustrated in Figure 5 below, and is currently in use at large semiconductor facilities. The combination of data from DoE and the AI analytics is expected to improve the manufacturing yield, and improve understanding of yield variation based design variants.

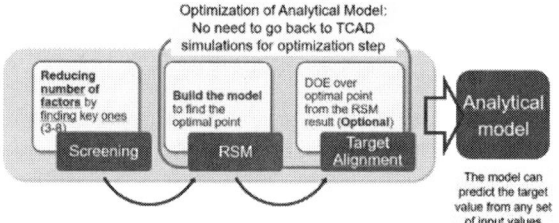

**Fig 5.** Analytical Model Creation from DoE

## 1.3 Innovations

By parameterizing device design and the underlying physics of the device response, complex novel power electronic GaN based devices can emerge, such as the tri-gate GaN HEMT device that was published in 2021 by Yunwei Ma et. al. (4)

Figure 6 shows a TCAD simulation of a 3D tri-gate JHEMT. Two cutplanes showing the barrier height and the potential profile under a bias of 1000 V on the drain contact. This design produces a good blocking voltage without compromising the device on-resistance. Such devices contribute to faster switching-times and improved charging for electric vehicles.

It remains to be seen if more complex designs can be produced with adequate yield to match the price point needed for EV manufacturers.

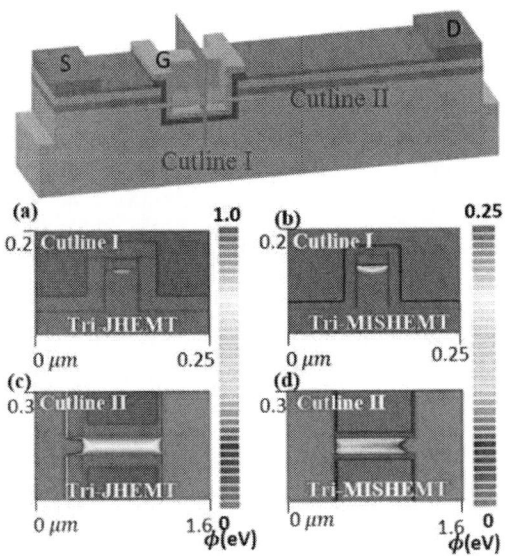

**Fig 6** : TCAD Simulation of JHEMT

## 1.4 Breakdown Voltage and Temperature

Reliability is a key requirement in automotive electronics. Semiconductors often have to operate in a harsh environment, with wide ambient temperature swings, and immunity to high voltages / electric field breakdown effects. When it comes to breakdown voltage, improvements can be made by using a GaN substrate, instead of Silicon (Si). Blocking voltages for GaN-on-Si are of the order of 1 kV, whereas GaN-on-GaN can achieve 94 kV (6).

The higher cost of manufacture of a GaN-on-GaN solution means that SiC technology remains dominant today in automotive EV components, but it is expected there will be a steady transition to GaN based technologies as energy conservation becomes even more important.

GaN does not perform as well as Silicon Carbide (SiC) for Thermal Conductivity – but if less heat is being generated through device power loss, the cooling requirement is reduced.

TCAD simulation allows the material and geometric properties of the GaN HEMT to be modelled and voltage breakdown and temperature performance analyzed on a design workstation. Breakdown voltage performance, which will depend on trench dimensions, insulator thickness, and differ-

ent material doping strategies is vital to understand, and adjustments can yield significant performance and reliability increases.

## 1.5 Higher Frequency Operation

GaN devices with integrated gate drivers can switch at up to 105 V/ns, resulting in an 82% reduction in switching losses when compared to Silicon Carbide technology (5). Energy is lost in inverters during half-bridge switching, when both semiconductors are in an off-state – the so-called dead time. Making this as short as possible, reduces device energy loss. In a GaN HEMT device, there is no P-channel / N-channel junction within the lateral structure of a GaN device, meaning there is no body-diode and associated reverse-recovery losses. Silicon Carbide devices include a body diode in their structure, and associated reverse-recovery charge during switching.

GaN devices have extended the operating frequency of EV inverters far beyond what is possible with SiC technology based designs. Figure 7 below shows a comparison of the operating frequencies of different technologies and materials.

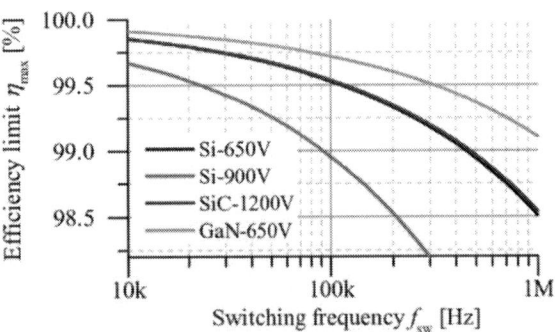

**Fig 7** : Efficiency Comparison based on switching frequency (7)

## 2   Impact on EV Design

For the last century, motor vehicles powered by internal combustion engines made use of low-voltage systems to drive all the many vehicle functions. A single voltage rail of 12v from an alternator/generator has served the industry well for decades. The higher voltages needed in EVs need to be handled differently – energy starts from the DC battery - typically 350-450 volt DC multi-cell units, and transformed for other consumers in the vehicle. These will include a 3-phase AC supply for the traction-motor, as well as conventional 12 or 24 volt DC rails used for multiple vehicle accessories. Fine-tuning this energy conversion can greatly improve and preserve the available battery energy.

## 2.1 Traction Motor Power Supply

The AC Motor, built into the powertrain of the EV, is typically an induction motor, as the magnetic field control circuits are simpler to control than the alternative Synchronous AC motors. Today's EV traction motors will have a typical power requirement of 10-15kW, energy which has to be converted from the DC battery source. Many different inverter topologies have been well proven over decades, and fine-tuning the switch performance and GaN semiconductors offers additional savings that are now worth harvesting. Figure 8 below shows a standard three-phase switch arrangement for controlling the traction motor.

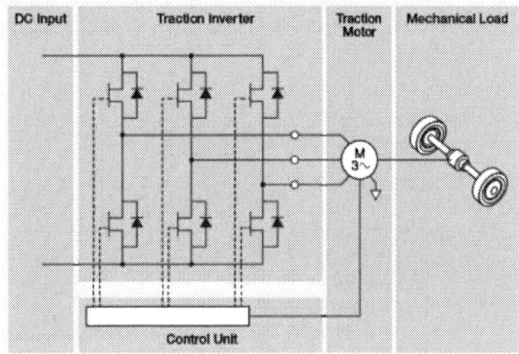

**Fig 8**. Traction Motor DC to AC Inverter Topology(5)

Operating the inverter switches at high frequencies has both advantages but also new challenges. High-speed switching kilowatts of power means significant electromagnetic radiation (EM), so the EV inverters and their control units must be adequately shielded. As the frequency goes up, so does emitted power through radiation. However an advantage of these high frequencies, is that the magnetic cores and components that form part of the inverter, can be made smaller, and not necessarily be iron-cored. This can save both weight and cost in a design.

## 2.2 EV Temperature Performance

The automotive environment for its electronic devices is typically -40 C to +105 C, and component performance will vary significantly over this range. The battery energy source in particular is significantly dependent on temperature. The Lithium-Ion batteries used in today's electric vehicle suffer performance degradation at both low and high temperatures. At low temperatures, the chemical reactions inside the battery slow down. It is estimated that there is a 10-15 mile range reduction in EVs when operating in extreme low temperatures.

High temperatures are also hazardous for EV batteries, where thermal run-away can occur. Battery Management ECU / control units (BMS) ensure that each cell is managed, and cell charge or discharge currents controlled, according to working temperature.

Figure 9 below shows an optimum temperature operating point between 25-40 Celsius, with degradation outside this range. (8)

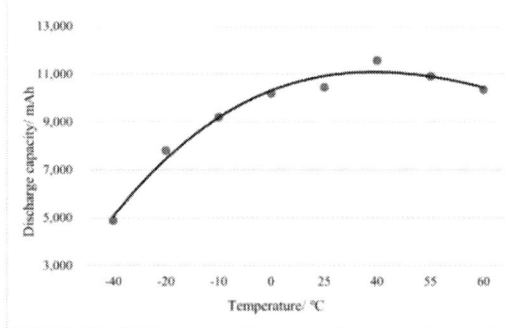

**Fig 9** : Battery variation with Temperature

TCAD simulations allow operating temperatures to be specified, so that the changes in overall circuit performance dependent on GaN devices can be predicted. Properties such as thermal and electrical conductivity of GaN vary significantly with ambient temperature. Figure 10 shows the variation in thermal conductivity, important when considering heat-extraction from the device under power load. This is one area where Silicon Carbide performs better.

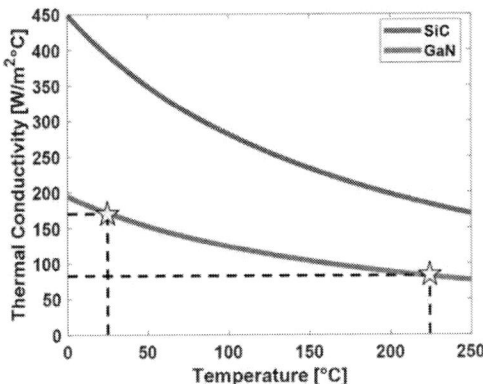

**Fig 10.** Variation of Thermal Conductivity of SiC/GaN with Temperature

# 3 Conclusions

Stored energy in an Electric Vehicle battery is a very precious commodity. This energy has to provide maximum driving range, as well as provide power for vehicle occupant comfort and driver-information systems. Designers are looking to optimize every aspect of the available battery energy, and the use of numerical TCAD simulations for power semiconductor design has shown that further improvements for the electronic sub-systems performance are possible, and are being achieved. Models based on carrier transport physics can improve GaN HEMT device performance, and improve the efficiency of the power modules used in Electric Vehicles. These simulations are predictive, and provide valuable insight early in the design-cycle, saving expensive re-designs or iterations later in the life-cycle. Furthermore, when parameterized, the design space can be investigated to obtain the optimal design compromise for the target application. Advanced simulation tools incorporating machine learning algorithms can also take manufacturability issues into the optimization flow linking the device designer directly to the fab and product criteria.

# References

1. Samuel, Edward & Osaka, Motohisa. (2024). Power Electronics in Electric Vehicles: A Comprehensive Overview. 10.13140/RG.2.2.15320.67840.)
2. Matteo Meneghini et. al., "GaN-based power devices: Physics, reliability, and perspectives" J. Appl. Phys. 130, 181101 (2021)
3. Gurpinar et al., "600 V normally-off p-gate GaN HEMT based 3-level inverter," in 2017 IEEE 3rd International Future Energy Electronics Conference and ECCE Asia (IFEEC 2017—ECCE Asia) (IEEE, 2017), pp. 621–626.
4. Ma, Yunwei, et. al. 2021. "Kilovolt Tri-Gate GaN Junction HEMTs with High Thermal Stability". United States. https://www.osti.gov/servlets/purl/1817457
5. Tektronix Report on Measurement and Testing of Traction Motors
6. Potential for GaN in Electric Vehicle Power Electronics, ID Tech Research March 2023.
7. Efficiency Comparison of semiconductor materials, ResearchGate, 2015.
8. The Influence of Temperature on the Capacity of Lithium Ion Batteries with Different Anodes, School of Mechanical Engineering, Nantong University, China. 2021
9. Modeling GaN on SiC: Temperature Dependent Properties, JetCool 2021

PCIM Europe 2024, 11– 13 June 2024, Nuremberg          DOI: 10.30420/566262182

# Fast Analytical Calculation of the Magnetic Field in Permanent Magnet Synchronous Machines with Flux Barriers Including Saturation

Martin Ackermann[1], José-Luis Marqués[2], Claus Hillermeier[1]

[1] University of the Bundeswehr Munich, Chair for Automation and Control, Munich, Germany
[2] University of the Bundeswehr Munich, Chair for Mathematics, Munich, Germany

Corresponding author:     Martin Ackermann, Martin.Ackermann@unibw.de
Speaker:                  Martin Ackermann, Martin.Ackermann@unibw.de

## Abstract

For a Permanent Magnet Synchronous Machine with concentrated windings and stator flux barriers including magnetic saturation, a fast but accurate analytical calculation of the magnetic field distribution and of the resulting torque is presented. The main idea is to use the highly accurate Subdomain Model to model the linear field behavior and to incorporate magnetic saturation as effective virtual currents obtained from a magnetic Equivalent Circuit Model. The comparison of the obtained results with Finite Element Analysis shows a good precision of the presented method, suitable for control applications.

## 1   Introduction

In Permanent Magnet Synchronous Machines (PMSM), the working harmonic of the coils interacts with the magnetic field of the magnets producing an effective torque, whereas the other harmonics cause torque ripple and losses. One approach for improving the spectrum of the magnetic field is to insert magnetic flux barriers (FBs) in the stator as shown in Fig. 1. FBs have been investigated in several publications where some advantages have been found, including increased electromagnetic torque density [1], increased efficiency by reducing losses like eddy currents (especially within the magnets) [2], weight advantage [3] and the possibility to use the space of the FBs for other purposes like cooling [4].

Up to now, there exists no accurate analytical model of the magnetic field and the produced torque in the presence of stator FBs. However, fast and accurate analytical models are mandatory for high performance control of such PMSMs. That is why in this paper the well investigated Subdomain Model (SDM) for common PMSMs [5]–[8] is adapted to PMSMs with stator FBs as shown in Fig. 1. Since in the SDM infinitely high permeable ferromagnetic material is assumed, it is enhanced by an equivalent circuit model (ECM) to incorporate the effect of magnetic saturation by virtual currents.

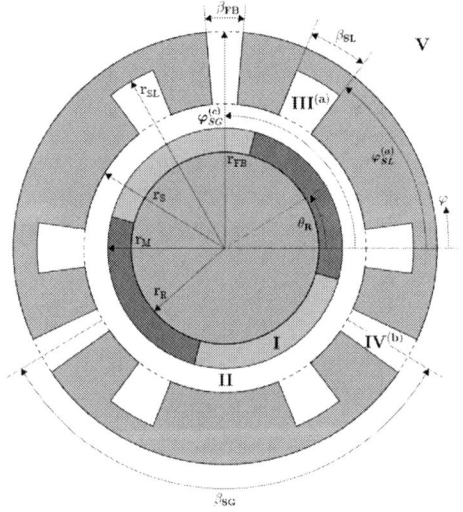

**Fig. 1:** PMSM containing stator FBs (subdomain $IV^{(b)}$). The angular position of each FB $\varphi_{FB}^{(b)}$ is defined analogously to the slot position $\varphi_{SL}^{(a)}$.

In the literature some approaches combining SDM and ECM with virtual currents already exist. For instance, Wu *et al.* modeled the saturation effect for open circuit [9] as well as on-load case [10] for a PMSM using virtual current sheets on the stator surface. Dianati *et al.* [11] used a similar method to calculate the saturation effect in a magnetic geared surface permanent magnet machine. Yin *et al.* [12] calculated virtual current sheets for modeling parallel slots under consideration of saturation.

As described, in this work the magnetic saturation is incorporated by virtual currents to the SDM as well. The original contribution is the introduction of a methodology to calculate these virtual currents by an ECM in advance and storing the results in a lookup-table (LUT) for subsequent application. Hence this whole procedure is well suited for real time computation of the torque during motor operation and can be applied for high performance control.

## 2 Subdomain Model with FBs

As already described in the introduction, there exists no version of the SDM for a PMSM with stator FBs. However, there is a similar machine type: flux switching permanent magnet machines. Using [13] as basis, combining it with the rotor geometry of PMSMs [5] and replacing the stator magnets by air, the SDM can be efficiently adapted to a PMSM with FBs in the stator, as shown in Fig. 1. In this section the SDM is derived in a compact formulation to present in integrated notation all successive steps until incorporating the magnetic saturation effect in the analytical modeling.

For the PMSM it is assumed that the periodically arranged magnets display a magnetization with only a radial component $M_r$. This gives the following Fourier expansion

$$\mu_0 M_r(\varphi) = \sum_{k \in \mathbb{N}} (\mu_0 M_k) \cos\left(k(\varphi - \theta_R)\right) \quad (1)$$

$$= \sum_{k \in \mathbb{N}} \frac{1}{r_R} \left[ -a_k^{(M)} \sin(k\varphi) + c_k^{(M)} \cos(k\varphi) \right]$$

where $M_k \propto B_{rem}$, $\theta_R$ is the rotor angular position,

$$a_k^{(M)} = -r_R \mu_0 M_k \sin(k\theta_R),$$
$$c_k^{(M)} = +r_R \mu_0 M_k \cos(k\theta_R). \quad (2)$$

Due to the choice of the cosine function for the magnetization Fourier expansion, only Fourier coefficients are nonzero with index $k = (1, 3, 5, \ldots)\tilde{p}$, where $\tilde{p}$ is the number of pole pairs.

For a PMSM of long axial extension, such that the contributions at both ends of the motor can be neglected, the magnetic field $\vec{B}$ displays no dependence on the axial coordinate $z$. Due to Maxwell equation $\mathrm{div}\, \vec{B} = 0$, the magnetic field can be described as the curl of a vector potential $\vec{A}$, $\vec{B} = \mathrm{curl}\, \vec{A}$, where $\vec{A}$ is aligned along the axial direction. Its single component $A_z$ is only a function

of the radial coordinate $r$ and the tangential coordinate $\varphi$: $\vec{A} = A_z(r, \varphi)\, \vec{e}_z$. $\vec{B} = \mathrm{curl}\, \vec{A}$ implies that the magnetic field in radial ($B_r$) and tangential ($B_\varphi$) direction is given by derivatives of $A_z$ according to

$$B_r = \frac{1}{r} \frac{\partial A_z}{\partial \varphi},$$
$$B_\varphi = -\frac{\partial A_z}{\partial r}. \quad (3)$$

The solution for $A_z$ results from the Maxwell equation for magnetostatics $\mathrm{curl}\, \vec{B} = \mu_0(\vec{J} + \mathrm{curl}\, \vec{M})$, where the sources for the magnetic field $\vec{B}$ are either some magnetization $\vec{M}$ or some external current density $\vec{J}$. Based on these assumptions, the Maxwell equation to be solved is given by

$$\left(-\mathrm{curl}\, \vec{B}\right)_z = +\Delta A_z = \frac{\mu_0}{r} \frac{\partial M_r}{\partial \varphi} - \mu_0 J_z. \quad (4)$$

The solution of this equation is achieved by using different Fourier expansions for each of the five different subdomains shown in Fig. 1: subdomain I for the rotor magnets with a given nonzero magnetization, subdomain II for the airgap, subdomain III[a] for slot $a$ with slot index $a$ running from 1 to the total slot number $n_{sl}$, where each slot contains a constant current density $J_z^{(a)}$, IV[b] for the FBs with index $b$ running from 1 to $n_{FB}$ and subdomain V for the external surrounding air.

The solutions for the different subdomains have to fulfill boundary conditions at each interface, either to some high permeable material or between two of the subdomains. At each interface constraints are given by the continuity of the normal component of field $\vec{B}$ and of the tangential component of field $\vec{H}$ at both sides of that interface.

In all subdomains, the general solution of the homogeneous Laplacian equation, (4) without magnetization or current density, is given by

$$\sum_{k \in \mathbb{N}} \left[ a_k^{(SD)} \left(\frac{r}{r_{ref}}\right)^k \cos(k\varphi) + b_k^{(SD)} \left(\frac{r}{r_{ref}}\right)^{-k} \cos(k\varphi) \right.$$
$$\left. + c_k^{(SD)} \left(\frac{r}{r_{ref}}\right)^k \sin(k\varphi) + d_k^{(SD)} \left(\frac{r}{r_{ref}}\right)^{-k} \sin(k\varphi) \right].$$

$r_{ref}$ denotes some appropriate constant reference radial coordinate. Hence that general solution is the sum of four different contribution series, each one weighted by different Fourier coefficients $\left\{ a_k^{(SD)}, b_k^{(SD)}, c_k^{(SD)}, d_k^{(SD)} \right\}$, where SD is the subdomain index I-V.

Using this general solution in combination with a particular solution for the magnets [8] and slots [14]

| Vector potential | General solution of homogeneous equation | Particular solution of inhomogenous equation |
|---|---|---|
| $A_z^{(I)}$ (Magnets) | $\displaystyle\sum_{k\in\mathbb{N}}\left(\left(\frac{r}{r_R}\right)^k + \left(\frac{r}{r_R}\right)^{-k}\right)\left[a_k^{(I)}\cos(k\varphi) + c_k^{(I)}\sin(k\varphi)\right]$ | $\displaystyle\sum_{k\in\mathbb{N}}\frac{r}{r_R}\frac{k+\left(\frac{r}{r_R}\right)^{-k-1}}{k^2-1}\left[a_k^{(M)}\cos(k\varphi) + c_k^{(M)}\sin(k\varphi)\right]$ |
| $A_z^{(II)}$ (Airgap) | $\displaystyle\sum_{k\in\mathbb{N}}\left[a_k^{(II)}\left(\frac{r}{r_S}\right)^k\cos(k\varphi) + b_k^{(II)}\left(\frac{r}{r_S}\right)^{-k}\cos(k\varphi)\right.$ $\left. + c_k^{(II)}\left(\frac{r}{r_S}\right)^k\sin(k\varphi) + d_k^{(II)}\left(\frac{r}{r_S}\right)^{-k}\sin(k\varphi)\right]$ | – |
| $A_z^{(III,a)}$ (Slots) | $\displaystyle\sum_{n\in\mathbb{N}}a_n^{(III,a)}\left(\left(\frac{r}{r_{SL}}\right)^{\frac{n\pi}{\beta_{SL}}} + \left(\frac{r}{r_{SL}}\right)^{-\frac{n\pi}{\beta_{SL}}}\right)\times$ $\times\cos\left(\frac{n\pi}{\beta_{SL}}(\varphi - \varphi_{SL}^{(a)})\right) + a_0^{(III,a)}$ | $\dfrac{\mu_0 J_z^{(a)}}{2}\left(r_{SL}^2\ln(r) - \dfrac{r^2}{2}\right)$ |
| $A_z^{(IV,b)}$ (Flux Barriers) | $\displaystyle\sum_{m\in\mathbb{N}}\left[a_m^{(IV,b)}\left(\frac{r}{r_{FB}}\right)^{\frac{m\pi}{\beta_{FB}}} + b_m^{(IV,b)}\left(\frac{r}{r_S}\right)^{\frac{m\pi}{\beta_{FB}}}\right]\times$ $\times\cos\left(\frac{m\pi}{\beta_{FB}}(\varphi - \varphi_{FB}^{(b)})\right) + a_0^{(IV,b)} + b_0^{(IV,b)}\ln(r)$ | – |
| $A_z^{(V)}$ (External Air) | $\displaystyle\sum_{t\in\mathbb{N}}\left[a_t^{(V)}\left(\frac{r}{r_{FB}}\right)^{-t}\cos(t\varphi) + c_t^{(V)}\left(\frac{r}{r_{FB}}\right)^{-t}\sin(t\varphi)\right]$ | – |

**Tab. 1:** General and particular solutions for the vector potential of each subdomain under consideration of the boundary conditions to the high permeable material. The magnetic field can be determined according to (3).

the vector potential for each subdomain can be described. Assuming an infinite permeability of the iron, fields enter or leave orthogonally such high permeable materials. In a first step the boundary conditions based on this assumption are incorporated into each subdomain, whereby some of the four unknown Fourier coefficients can be eliminated. The resulting equations are given in Tab. 1 and are explained in the following: for the magnets (subdomain I) the boundary condition at the rotor interface $r = r_R$ to the high permeable iron

$$H_\varphi^{(I)}\Big|_{r=r_R} = 0 \quad \forall \quad 0 \leq \varphi < 2\pi$$

is applied, reducing the number of independent Fourier coefficients in the resulting equation to two. The boundary condition of the air gap (subdomain II) to the stator interface at $r = r_S$ is not considered yet, since it will be applied as integral expression later. Incorporating the boundary conditions at each slot's (subdomain III$^{(a)}$) surface to the high permeable material

$$H_\varphi^{(III,a)}\Big|_{r=r_{SL}} = 0 \quad \forall \quad \varphi_{SL}^{(a)} \leq \varphi \leq \varphi_{SL}^{(a)} + \beta_{SL}$$

$$H_r^{(III,a)}\Big|_{\varphi=\varphi_{SL}^{(a)}} = 0 \quad \forall \quad r_S \leq r \leq r_{SL}$$

$$H_r^{(III,a)}\Big|_{\varphi=\varphi_{SL}^{(a)}+\beta_{SL}} = 0 \quad \forall \quad r_S \leq r \leq r_{SL}$$

only one Fourier coefficient remains unknown. As it can be seen in the resulting equation, the constant of the slot's vector potential $a_0^{(III,a)}$ is considered,

although it does not contribute to the magnetic field due to (3). However it is useful for determining the inductances and back-EMFs of the coils [6].

By considering the boundary conditions for the FBs IV$^{(b)}$ to the high permeable stator material

$$H_r^{(IV,b)}\Big|_{\varphi=\varphi_{FB}^{(b)}} = 0 \quad \forall \quad r_S \leq r \leq r_{FB}$$

$$H_r^{(IV,b)}\Big|_{\varphi=\varphi_{FB}^{(b)}+\beta_{FB}} = 0 \quad \forall \quad r_S \leq r \leq r_{FB}$$

only two of the four Fourier coefficients remain unknown (Tab. 1). Besides constant vector potential $a_0^{(IV,b)}$ in the resulting equation, a second constant with radial dependency, $b_0^{(IV,b)}\ln(r)$, based on a more general solution of the Laplacian equation needs to be considered. This is required to incorporate correctly the effect of the FBs. For the external air, subdomain V, only the two Fourier coefficients with $r^{-t}$ dependency are nonzero, since the magnetic field for $r \to \infty$ is vanishing.

In the next step the aim is to impose continuity at the interfaces between two subdomains. The resulting equations can be put into the form of a linear equation system for determining all unknown Fourier coefficients as function of the rotor angle and the current densities within each slot or coil. By that, the magnetic field within each subdomain can be calculated.

The continuity between the magnets and the airgap

| Equation | Resulting boundary condition |
|---|---|
| $BC_{1,1}$ | $\dfrac{1}{\mu_0}\displaystyle\int_0^{2\pi} B_\varphi^{(II)}\Big|_{r=r_S}\cos(k\varphi)\,d\varphi = \dfrac{1}{\mu_0}\sum_{a=1}^{n_{sl}}\int_{\varphi_{SL}^{(a)}}^{\varphi_{SL}^{(a)}+\beta_{SL}} B_\varphi^{(I,a)}\Big|_{r=r_S}\cos(k\varphi)\,d\varphi + \dfrac{1}{\mu_0}\sum_{b=1}^{n_{FB}}\int_{\varphi_{FB}^{(b)}}^{\varphi_{FB}^{(b)}+\beta_{FB}} B_\varphi^{(IV,b)}\Big|_{r=r_S}\cos(k\varphi)\,d\varphi$ |
| $BC_{1,2}$ | $\dfrac{1}{\mu_0}\displaystyle\int_0^{2\pi} B_\varphi^{(II)}\Big|_{r=r_S}\sin(k\varphi)\,d\varphi = \dfrac{1}{\mu_0}\sum_{a=1}^{n_{sl}}\int_{\varphi_{SL}^{(a)}}^{\varphi_{SL}^{(a)}+\beta_{SL}} B_\varphi^{(I,a)}\Big|_{r=r_S}\sin(k\varphi)\,d\varphi + \dfrac{1}{\mu_0}\sum_{b=1}^{n_{FB}}\int_{\varphi_{FB}^{(b)}}^{\varphi_{FB}^{(b)}+\beta_{FB}} B_\varphi^{(IV,b)}\Big|_{r=r_S}\sin(k\varphi)\,d\varphi$ |
| $BC_{2,1}$ | $\displaystyle\int_{\varphi_{SL}^{(a)}}^{\varphi_{SL}^{(a)}+\beta_{SL}} A_z^{(II)}\Big|_{r=r_S}\cos\left(\dfrac{n\pi}{\beta_{SL}}(\varphi-\varphi_{SL}^{(a)})\right)d\varphi = \int_{\varphi_{SL}^{(a)}}^{\varphi_{SL}^{(a)}+\beta_{SL}} A_z^{(III,a)}\Big|_{r=r_S}\cos\left(\dfrac{n\pi}{\beta_{SL}}(\varphi-\varphi_{SL}^{(a)})\right)d\varphi$ |
| $BC_{2,2}$ | $\displaystyle\int_{\varphi_{SL}^{(a)}}^{\varphi_{SL}^{(a)}+\beta_{SL}} A_z^{(II)}\Big|_{r=r_S}d\varphi = \int_{\varphi_{SL}^{(a)}}^{\varphi_{SL}^{(a)}+\beta_{SL}} A_z^{(III,a)}\Big|_{r=r_S}d\varphi$ |
| $BC_{3,1}$ | $\displaystyle\int_{\varphi_{FB}^{(a)}}^{\varphi_{FB}^{(b)}+\beta_{FB}} A_z^{(II)}\Big|_{r=r_S}\cos\left(\dfrac{m\pi}{\beta_{FB}}(\varphi-\varphi_{FB}^{(b)})\right)d\varphi = \int_{\varphi_{FB}^{(b)}}^{\varphi_{SL}^{(a)}+\beta_{SL}} A_z^{(IV,b)}\Big|_{r=r_S}\cos\left(\dfrac{m\pi}{\beta_{FB}}(\varphi-\varphi_{FB}^{(b)})\right)d\varphi$ |
| $BC_{3,2}$ | $\displaystyle\int_{\varphi_{FB}^{(b)}}^{\varphi_{FB}^{(b)}+\beta_{FB}} A_z^{(II)}\Big|_{r=r_S}d\varphi = \int_{\varphi_{FB}^{(b)}}^{\varphi_{FB}^{(b)}+\beta_{FB}} A_z^{(IV,b)}\Big|_{r=r_S}d\varphi$ |

**Tab. 2:** Resulting boundary conditions at the inner stator interface ($r = r_S$) between the airgap, slots and FBs.

| Equation | Resulting boundary condition |
|---|---|
| $BC_{4,1}$ | $\dfrac{1}{\mu_0}\displaystyle\int_0^{2\pi} B_\varphi^{(V)}\Big|_{r=r_{FB}}\cos(t\varphi)\,d\varphi = \dfrac{1}{\mu_0}\sum_{b=1}^{n_{FB}}\int_{\varphi_{FB}^{(b)}}^{\varphi_{FB}^{(b)}+\beta_{FB}} B_\varphi^{(IV,b)}\Big|_{r=r_{FB}}\cos(t\varphi)\,d\varphi$ |
| $BC_{4,2}$ | $\dfrac{1}{\mu_0}\displaystyle\int_0^{2\pi} B_\varphi^{(V)}\Big|_{r=r_{FB}}\sin(t\varphi)\,d\varphi = \dfrac{1}{\mu_0}\sum_{b=1}^{n_{FB}}\int_{\varphi_{FB}^{(b)}}^{\varphi_{FB}^{(b)}+\beta_{FB}} B_\varphi^{(IV,b)}\Big|_{r=r_{FB}}\sin(t\varphi)\,d\varphi$ |
| $BC_{5,1}$ | $\displaystyle\int_{\varphi_{FB}^{(b)}}^{\varphi_{FB}^{(b)}+\beta_{FB}} A_z^{(V)}\Big|_{r=r_{FB}}\cos\left(\dfrac{m\pi}{\beta_{FB}}(\varphi-\varphi_{FB}^{(b)})\right)d\varphi = \int_{\varphi_{FB}^{(b)}}^{\varphi_{FB}^{(b)}+\beta_{FB}} A_z^{(IV,b)}\Big|_{r=r_{FB}}\cos\left(\dfrac{m\pi}{\beta_{FB}}(\varphi-\varphi_{FB}^{(b)})\right)d\varphi$ |
| $BC_{5,2}$ | $\displaystyle\int_{\varphi_{FB}^{(b)}}^{\varphi_{FB}^{(b)}+\beta_{FB}} A_z^{(V)}\Big|_{r=r_{FB}}d\varphi = \int_{\varphi_{FB}^{(b)}}^{\varphi_{FB}^{(b)}+\beta_{FB}} A_z^{(IV,b)}\Big|_{r=r_{FB}}d\varphi$ |

**Tab. 3:** Resulting boundary conditions at the outer stator interface ($r = r_{FB}$) between the FBs and surrounding air.

at $r = r_M$ can be simply imposed by

$$BC_{0,1}: \quad B_r^{(I)}\Big|_{r=r_M} = B_r^{(II)}\Big|_{r=r_M}$$

$$BC_{0,2}: \quad H_\varphi^{(I)}\Big|_{r=r_M} = H_\varphi^{(II)}\Big|_{r=r_M}.$$

From these two continuity conditions, four separate equations are obtained: two for the coefficients $\{a_k^{(I)}, a_k^{(II)}, b_k^{(II)}\}$ of the cosine functions and two for the coefficients $\{c_k^{(I)}, c_k^{(II)}, d_k^{(II)}\}$ of the sine functions (see also Tab. 4). The consideration of the continuity between the airgap, slots and FBs at $r = r_S$ is more complex, since multiple boundary conditions have to be incorporated. The continuity of the tangential magnetic field

$$H_\varphi^{(II)}\Big|_{r=r_S} = H_\varphi^{(III,a)}\Big|_{r=r_S} \quad \forall \quad \varphi_{SL}^{(a)} \leq \varphi \leq \varphi_{SL}^{(a)} + \beta_{SL}$$

$$H_\varphi^{(II)}\Big|_{r=r_S} = H_\varphi^{(IV,b)}\Big|_{r=r_S} \quad \forall \quad \varphi_{FB}^{(b)} \leq \varphi \leq \varphi_{FB}^{(b)} + \beta_{FB}$$

$$H_\varphi^{(II)}\Big|_{r=r_S} = 0 \quad \text{elsewhere}$$

can be applied by integrating $H_\varphi$ multiplied either by cosine or by sine functions along $\varphi$, as shown in Tab. 2 ($BC_{1,1}$ and $BC_{1,2}$).

The second set of equations is based on the continuity of the vector potential

$$A_z^{(II)}\Big|_{r=r_S} = A_z^{(III,a)}\Big|_{r=r_S} \quad \forall \quad \varphi_{SL}^{(a)} \leq \varphi \leq \varphi_{SL}^{(a)} + \beta_{SL}$$

for each slot and

$$A_z^{(II)}\Big|_{r=r_S} = A_z^{(IV,b)}\Big|_{r=r_S} \quad \forall \quad \varphi_{FB}^{(b)} \leq \varphi \leq \varphi_{FB}^{(b)} + \beta_{FB}$$

for each FB. These boundary conditions can be formulated as an integral over the interface between slot and airgap or FB and airgap. While the Fourier coefficients of the harmonics are extracted by multiplying a cosine function with periodicity half of a slot ($BC_{2,1}$) or FB ($BC_{3,1}$), the vector potential's constants of the slots ($BC_{2,2}$) and FBs ($BC_{3,2}$) are extracted without a trigonometric function.

In an analogous way the boundary conditions at the interface of the outer stator $r = r_{FB}$ can be applied, the resulting equations are shown in Tab. 3 ($BC_4$-$BC_5$). Solving these integrals, as well as the ones shown in Tab. 2 for the continuity at the stator interface ($r = r_S$), and rearranging the resulting equations in combination with the ones based on the continuity between subdomain I and II ($BC_{0,1}$,

| Unknown Fourier coefficients of $\vec{C}_F$ | Number of unknown Fourier coefficients | Boundary conditions | Resulting number of equations |
|---|---|---|---|
| $\{a_k^{(I)}, c_k^{(I)}\}$ | $2n_h$ | $\{BC_{0,1}, BC_{0,2}\}$ | $4n_h$ |
| $\{a_k^{(II)}, b_k^{(II)}, c_k^{(II)}, d_k^{(II)}\}$ | $4n_h$ | $\{BC_{1,1}, BC_{1,2}\}$ | $2n_h$ |
| $\{a_0^{(III,a)}\}$ | $n_{sl}$ | $\{BC_{2,2}\}$ | $n_{sl}$ |
| $\{a_n^{(III,a)}\}$ | $n_{sl}n_h$ | $\{BC_{2,1}\}$ | $n_{sl}n_h$ |
| $\{a_0^{(IV,b)}, b_0^{(IV,b)}\}$ | $2n_{fb}$ | $\{BC_{3,2}, BC_{5,2}\}$ | $2n_{fb}$ |
| $\{a_m^{(IV,b)}, b_m^{(IV,b)}\}$ | $2n_{fb}n_h$ | $\{BC_{3,1}, BC_{5,1}\}$ | $2n_{fb}n_h$ |
| $\{a_t^{(V)}, b_t^{(V)}\}$ | $2n_h$ | $\{BC_{4,1}, BC_{4,2}\}$ | $2n_h$ |

**Tab. 4:** Unknown Fourier coefficients $\vec{C}_F$ for each subdomain of the resulting equation system (5), corresponding number of unknown Fourier coefficients, the boundary conditions and the thereby extracted number of equations. $n_h$ is the number of Fourier coefficients which are taken into account for each subdomain.

$BC_{0,2}$), a linear equation system can be created

$$\mathbf{M}\vec{C}_F = \vec{f}(\theta_R, \vec{J}_z). \tag{5}$$

The matrix $\mathbf{M}$ contains the geometry depending parts of the equations, $\vec{C}_F$ all unknown Fourier coefficients of each subdomain and $\vec{f}(\theta_R, \vec{J}_z)$ effectively the particular solutions as the sources of the magnetic field. For a better overview, all required Fourier coefficients in each subdomain (collected together in vector $\vec{C}_F$) are listed in Tab. 4, totaling $8n_h + n_{sl}(1 + n_h) + 2n_{fb}(1 + n_h)$, with $n_h$ representing the maximal number of harmonics being considered. The number of applied boundary conditions (also listed in Tab. 4) matches this number of unknown Fourier coefficients thus allowing an unequivocal solution of vector $\vec{C}_F$ for a given rotor's magnet arrangement and stator's current distribution. By solving (5) the magnetic field within each subdomain can be determined by inserting $\vec{C}_F$ in the equations shown in Tab. 1 and derivating the vector potential according to (3).

## 3 Modeling magnetic saturation

A second option to determine the magnetic flux distribution in an electrical machine is the ECM. The idea for modeling the magnetic saturation is to develop both a linear and a nonlinear ECM, whereby virtual currents can be determined and stored in a LUT. These virtual currents are then incorporated into the SDM.

### 3.1 Linear ECM

Since a reluctance can be seen as a magnetic flux tube, an electrical machine as shown in Fig. 1 can be separated into these tubes. With the Magneto Motive Force (MMF) of the coils and magnets an ECM can be developed for one segment (see Fig.

2 and 3), for simplification in Cartesian coordinates. The complete ECM is built by connecting all segments.

The MMF for each coil is defined by

$$\mathcal{F}_C^{(c)} = n_{turn}i^{(c)}, \tag{6}$$

with $n_{turn}$ as the number of winding turns. The MMF for the magnets is equal to

$$\mathcal{F}_G^{(c)} = \frac{B_{rem}h_{mag}}{\mu_0\beta_G} \int_{\varphi_{SG}^{(c)}+\frac{\beta_{FB}}{2}}^{\varphi_{SL}^{(cr)}+\beta_{SL}} \text{sign}(\vec{M}(\varphi, \theta_R))d\varphi,$$

$$\mathcal{F}_M^{(c)} = \frac{B_{rem}h_{mag}}{\mu_0\beta_M} \int_{\varphi_{SL}^{(cr)}+\beta_{SL}}^{\varphi_{SL}^{(cl)}} \text{sign}(\vec{M}(\varphi, \theta_R))d\varphi,$$

$$\mathcal{F}_G'^{(c)} = \frac{B_{rem}h_{mag}}{\mu_0\beta'_G} \int_{\varphi_{SL}^{(cl)}}^{\varphi_{SG}^{(c+1)}-\frac{\beta_{FB}}{2}} \text{sign}(\vec{M}(\varphi, \theta_R))d\varphi, \tag{7}$$

with $B_{rem}$ and $h_{mag}$ as remanence and thickness of the magnets. $\beta_G$, $\beta_M$ and $\beta'_G$ are the angle lengths of the integration paths. $c_r$ and $c_l$ denote the right and left slot of segment $c$. As it can be seen in (7), the equivalent MMF sources of the magnets as sketched in Fig. 2 are calculated by integrating over the magnetization distribution (1). The reason for that is to handle in a simple way the continuous angle relation of stator and rotor by a static network. The reluctances are depending on the relative permeability, length and width of its material. For example, reluctance $\mathcal{R}_{S,M}^{(c)}$, the stator tooth of segment $c$, can be calculated by

$$\mathcal{R}_{S,M}^{(c)} = \frac{h_S}{\mu_0\mu_{r,S,M}^{(c)}\ell w_M} \tag{8}$$

with $\mu_{r,S,M}^{(c)} = $ const. and having high values in the linear case. The other reluctances sketched in

PCIM Europe 2024, 11– 13 June 2024, Nuremberg — DOI: 10.30420/566262182

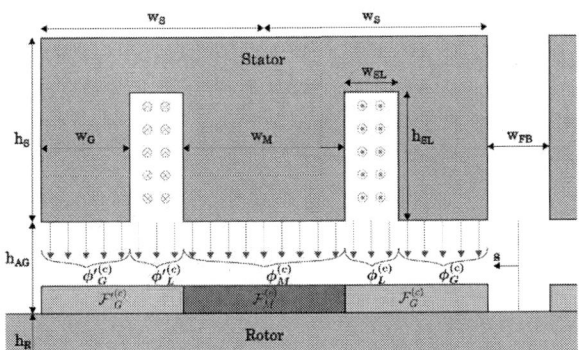

**Fig. 2:** Segment of an electrical machine with FBs in Cartesian coordinates.

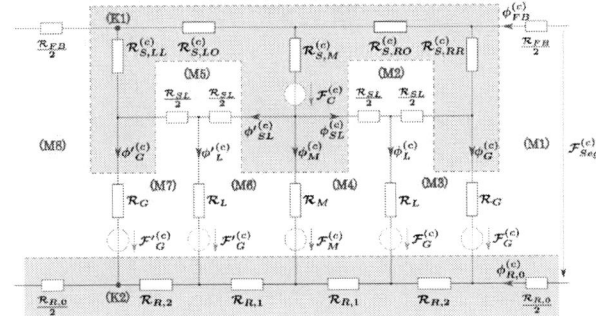

**Fig. 3:** ECM of one segment $c$ in Cartesian coordinates.

Fig. 3 are analogously defined, based on the geometric relations shown in Fig. 2. By defining the linearly independent fluxes in combination with Kirchhoff's current and voltage law (all relevant fluxes $\phi$, nodes $K$ and meshes $M$ are exemplary marked in Fig. 3), a linear equation system can be set up and efficiently written in Matrix notation

$$\mathbf{R}\vec{\Phi} = \vec{F}, \qquad (9)$$

where matrix $\mathbf{R}$ contains the reluctances based on Kirchhoff's equations, $\vec{\Phi}$ the unknown magnetic flux flowing through each reluctance and $\vec{F}$ the MMF source of the magnets and the coil within each segment (see Fig. 3). By solving (9), the flux distribution within the magnetic network is known.

### 3.2 Nonlinear ECM

To account for the effect of magnetic saturation in the above presented ECM, some modifications are required. Firstly, the reluctances of stator iron are now dependent on the magnetic field. For example, the permeability of reluctance $\mathcal{R}_{S,M}^{(c)}$ in (8) is not constant anymore, but a function of the magnetic field flowing through it: $\mu_{r,S,M}^{(c)}\left(B_{r,S,M}^{(c)}\right)$. The redefined reluctances of the stator are substituted into the equation system (9). Since $\mu_r(B)$ is a nonlinear relation, the equation system to calculate the flux distribution becomes nonlinear as well:

$$\vec{\Phi} = \mathbf{R_{Sat}}^{-1}(\vec{\Phi})\,\vec{F} \qquad (10)$$

Matrix $\mathbf{R_{Sat}}$ now depends on the flux flowing through stator iron and therefore on the solution vector $\vec{\Phi}$. Having solved that nonlinear equation system, the magnetic field in the air gap of a segment can be calculated on the basis of the corresponding fluxes:

$$\overline{B}_{SG}^{(c)}(s) = \begin{cases} 0 & \text{for} \quad s_{SG}^{(c)} \le s < s_{SG}^{(c)} + \frac{w_{FB}}{2} \\ \frac{\phi_G^{(c)}}{w_G\ell} & \text{for} \quad s \in Z_G \\ \frac{\phi_L^{(c)}}{w_{SL}\ell} & \text{for} \quad s \in Z_L \\ \frac{\phi_M^{(c)}}{w_M\ell} & \text{for} \quad s \in Z_M \\ \frac{\phi_L'^{(c)}}{w_{SL}\ell} & \text{for} \quad s \in Z_L' \\ \frac{\phi_G'^{(c)}}{w_G\ell} & \text{for} \quad s \in Z_G' \\ 0 & \text{for} \quad s_{SG}^{(c+1)} - \frac{w_{FB}}{2} \le s < s_{SG}^{(c+1)} \end{cases} \qquad (11)$$

where $s = \varphi\, r_S$ and $Z_G,\ Z_L,\ Z_M,\ Z_L',\ Z_G'$ denote the flux zones according to Fig. 2. For a better understanding, all magnetic fields calculated by the ECM are marked with an overline. The magnetic field within the whole machine in the air gap is then determined by

$$\overline{B}(s) = \begin{cases} \overline{B}_{SG}^{(1)}(s) & \text{for} \quad s_{SG}^{(1)} \le s < s_{SG}^{(2)} \\ \overline{B}_{SG}^{(2)}(s) & \text{for} \quad s_{SG}^{(2)} \le s < s_{SG}^{(3)} \\ \vdots & \vdots \\ \overline{B}_{SG}^{(c)}(s) & \text{for} \quad s_{SG}^{(c)} \le s < s_{SG}^{(c+1)} \\ \vdots & \vdots \\ \overline{B}_{SG}^{(n_{sg})}(s) & \text{for} \quad s_{SG}^{(n_{sg})} \le s < s_{SG}^{(1)} \end{cases} \qquad (12)$$

where $s_{SG}^{(c)}$ is the begin of segment $c$ in Cartesian coordinates.

### 3.3 Virtual saturation currents

This subsection describes the method for determining the virtual saturation currents. These virtual currents are calculated by the introduced ECM, the results are stored in a LUT and can then be incorporated into the SDM. The assumption is made that the saturation effects are rotating synchronously with the rotor and the corresponding stator currents.

1341

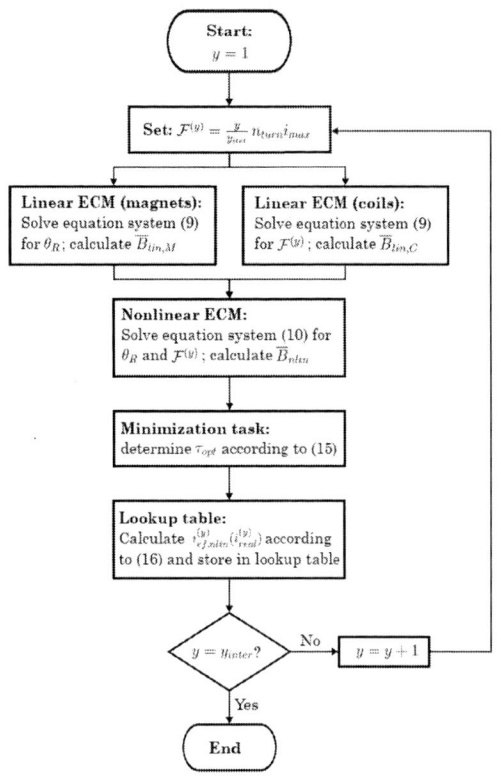

**Fig. 4:** Flowchart for determining the LUT $i_{eff,nlin}(i_{real})$ with $y_{inter}$ different intervals.

Therefore, these effects can be calculated with a fixed, arbitrarily chosen rotor angle $\theta_R$. The LUT has entries for different amplitudes of currents (or MMF's). For its computation $n_{turn}i_{max}$ needs to be defined, the MMF which corresponds to the machine specific maximum current. As shown by the flowchart in Fig. 4, this maximal coil MMF is divided into $y_{inter}$ intervals. For each MMF amplitude $\mathcal{F}^{(y)}$, the magnetic voltage for all coils can be defined according to their phases:

$$\mathcal{F}_C^{(c)} = \mathcal{F}^{(y)} \cos\left( \tilde{p}\theta_R + \frac{2(P_{coil}^{(c)} - 1)}{n_{phase}}\pi + \frac{\pi}{2} \right) \quad (13)$$

where $P_{coil}^{(c)} \in \{1, 2, \ldots, n_{phase}\}$ assigns a phase to coil $c$. As it can be seen in (13), the electrical phase shift is set to $+\frac{\pi}{2}$, which corresponds to an operating point with no d-current in d/q coordinates. To analyze other operating points with $|i_d| > 0$, the phase shift can be set accordingly. With the rotor angle $\theta_R$, the MMF of the magnets $\mathcal{F}_G^{(c)}$, $\mathcal{F}_M^{(c)}$ and $\mathcal{F}_G'^{(c)}$ for each segment $c$ can be calculated by (7). In the next step, the linear ECM (9) with the influence of the coils and magnets separately, as well as the nonlinear ECM (10) for coils and mag-

nets simultaneously, is solved on the basis of the defined MMF sources. Thus, the magnetic field $\overline{B}_{lin,C}$, $\overline{B}_{lin,M}$ and $\overline{B}_{nlin}$ can be determined by (11) and (12) for each case.

Now, the field calculated by linear ECM can be adjusted to the nonlinear ECM by using a correction factor $\tau$, which is determined by minimizing the following cost function

$$K(\tau) = \int_0^{w_{EM}} \left( \overline{B}_{nlin} - \left( \tau \overline{B}_{lin,C} + \overline{B}_{lin,M} \right) \right)^2 ds, \quad (14)$$

where $w_{EM}$ is the air gap's circumference length, $\overline{B}_{lin,C}$ and $\overline{B}_{lin,M}$ are the fields produced by the coils and magnets, calculated by linear ECM, and $\overline{B}_{nlin}$ is the overall field (by coils *and* magnets) calculated by nonlinear ECM. Assuming the same currents flowing through the coils for linear as well as nonlinear case the solution of optimization problem (14) is given by

$$\frac{\partial K}{\partial \tau} = \int_0^{w_{EM}} \tau \overline{B}_{lin,C}^2 - \overline{B}_{lin,C}\overline{B}_{nlin} - \overline{B}_{lin,C}\overline{B}_{lin,M} \, ds = 0$$

$$\tau_{opt} = \frac{\int_0^{w_{EM}} (\overline{B}_{nlin} - \overline{B}_{lin,M})\overline{B}_{lin,C} \, ds}{\int_0^{w_{EM}} \overline{B}_{lin,C}^2 \, ds}. \quad (15)$$

$\tau_{opt}$ can be interpreted as the necessary reduction of the coil's field for a given external current amplitude to model the effect of saturation as close as possible, nevertheless independent of the actual rotor angular position $\theta_R$. Since, in the linear case, the field produced by the coils is proportional to the coil currents, an effective saturation current can be determined, describing in the linear ECM framework the reduction of the magnetic field due to nonlinear saturation effects:

$$i_{eff,nlin}(i_{real}) = \tau_{opt}i_{real}. \quad (16)$$

As final step, the input of the SDM is no longer the real current $i_{real}$ flowing physically within a coil, but the current $i_{eff,nlin}$ which adjusts the linear model to the nonlinear behaviour of the saturation effect. As described in the beginning of this section, it can be calculated offline and in advance by gradually increasing $i_{real}$ in small steps (see Fig. 4), starting from zero currents, and storing $i_{eff,nlin}(i_{real})$ from equation (16) in a LUT.

Summarizing, this approach can be used to calculate efficiently the field distribution and thereby the produced torque taking magnetic saturation into account. Once $i_{eff,nlin}(i_{real})$ is calculated and stored during initialization, the implementation of the saturation effect causes nearly no extra computation

**Tab. 5:** Important parameters of the studied machine.

| Parameter | Value | Unit |
|---|---|---|
| Number of Phases ($m$) | 5 | 1 |
| Number of winding turns ($n_{turn}$) | 100 | 1 |
| Number of pole pairs ($\tilde{p}$) | 6 | 1 |
| Number of flux barriers ($n_{FB}$) | 5 | 1 |
| Flux barrier width ($\beta_{FB}$) | 7.5 | ° |
| Slot width ($\beta_{SL}$) | 11.15 | ° |
| Stator outer radius ($r_{FB}$) | 95 | mm |
| Stator inner radius ($r_S$) | 61.75 | mm |
| Active rotor length ($\ell$) | 90 | mm |
| Air gap width ($r_S - r_M$) | 0.75 | mm |
| Magnet thickness ($r_M - r_R$) | 7.85 | mm |
| Coil inductance ($L^{(c)}$) | 4.5 | mH |
| Amplitude PM flux linkage ($\hat{\Phi}_M^{(c)}$) | 213 | mWb |

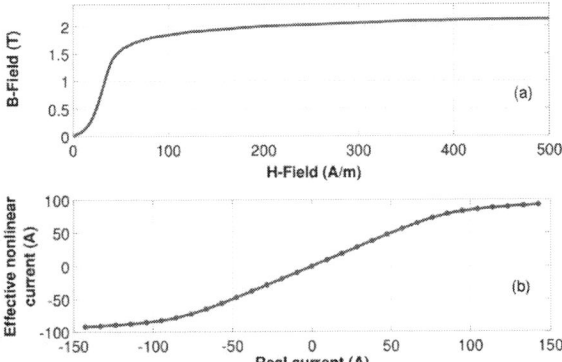

**Fig. 5:** (a) B-H curve of the stator ferromagnetic material in the analyzed machine. (b) Effective nonlinear current against real (physical) current.

effort during online simulation e.g. for control applications.

## 4 Modeling results

For validation purposes the developed nonlinear SDM is used to calculate the air gap flux and torque of a representative 5-phase PMSM with FBs and compared to corresponding FEA's. The qualitative structure of the discussed machine is the one shown in Fig. 1, its most relevant parameters are listed in Tab. 5. Ferromagnetic (nonlinear) material of the iron is assumed, whose B-H curve is shown in Fig. 5a). In Fig. 5b) the effective nonlinear current $i_{eff,nlin}$ against physical current $i_{real}$ can be seen, calculated using the method presented in the section above.
Fig. 6b) and Fig. 6c) show the radial and tangential magnetic field distribution in on-load case at a fixed time step, calculated by the developed nonlinear

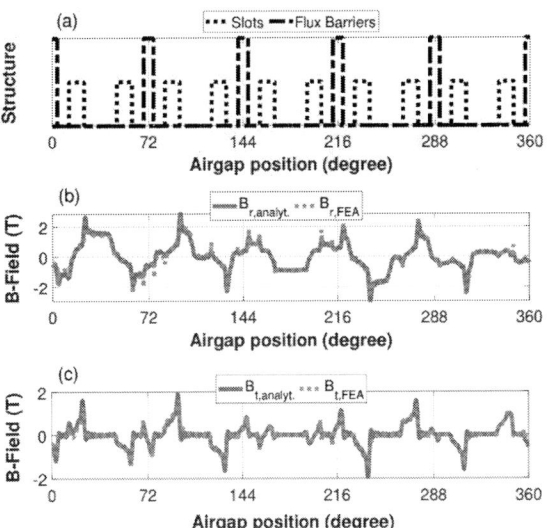

**Fig. 6:** Magnetic field of magnets and coils ($\hat{i} = 120A$) within the airgap. (a) Slot and FB structure. (b) Radial and (c) tangential component of the magnetic field calculated by the nonlinear SDM (red) and by nonlinear FEA (green).

SDM. The amplitude of the currents $\hat{i}$ is set to 120A. In both cases the magnetic field components are compared to nonlinear FEAs. They show a high accordance and can therefore be used for an accurate determination of the torque. This can be efficiently done by the Maxwell stress tensor [5]

$$ T = \frac{\ell r_S^2}{\mu_0} \int_0^{2\pi} B_r^{(II)} B_\varphi^{(II)} \, d\varphi, \qquad (17) $$

integrating over the interface between air gap and stator, where $\ell$ denotes the active rotor length.
In Fig. 7-9 the calculated torque for increasing stator currents is shown. Again, the developed nonlinear SDM is compared to corresponding nonlinear FEA's. Furthermore the torque determined by the common linear SDM, derived in section 2, is also drawn for comparison. At first it can be seen that the difference between the linear SDM to the nonlinear SDM and FEA significantly increases with higher current amplitudes, due to the growing influence of the saturation effect. While the nonlinear SDM and FEA show a high accordance at a current amplitude of 100A (Fig. 8), a slightly increased difference can be noticed at the current amplitudes of 80A (Fig. 7) and 120A (Fig. 9). Also it can be seen that the nonlinear SDM overestimates the torque ripple at a current amplitude of 120A. These slight

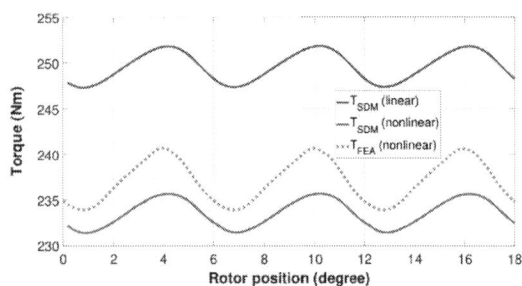

**Fig. 7:** Torque determined by linear SDM, nonlinear SDM and nonlinear FEA ($\hat{i} = 80\text{A}$).

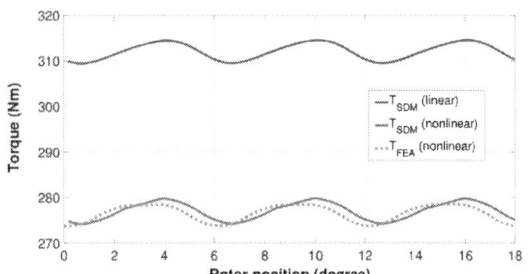

**Fig. 8:** Torque determined by linear SDM, nonlinear SDM and nonlinear FEA ($\hat{i} = 100\text{A}$).

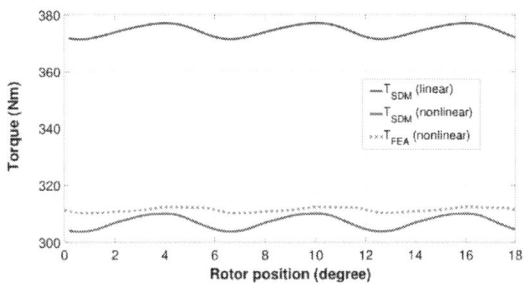

**Fig. 9:** Torque determined by linear SDM, nonlinear SDM and nonlinear FEA ($\hat{i} = 120\text{A}$).

deviations are due to the approximative character of the presented method for incorporating the magnetic saturation. Since the developed model takes magnetic saturation into account with minimal computational effort, it is – based on the representative accuracy shown in Fig. 7-9 – most suitable for control applications, where a fast and efficient prediction of the torque with sufficient accuracy is desirable.

## 5 Conclusion

In this paper a self contained derivation of a complete analytical modeling of a PMSM with FBs has been developed combining both SDM and ECM. The model is able to calculate the magnetic field

in the air gap and the resulting torque, taking into account the nonlinear magnetic saturation of stator material. The main idea of incorporating the saturation effect, which is hard to directly include in SDM, is to use virtual currents determined in a simplified ECM. The analytical calculations reproduce in all considered cases the results of time consuming FEAs with sufficient accuracy, however running several orders of magnitude faster. Thus, the model is suitable for a real time computation of the torque during operation and can be applied for high performance motor control.

## Acknowledgment

This research is funded by dtec.bw – Digitalization and Technology Research Center of the Bundeswehr which we gratefully acknowledge. dtec.bw is funded by the European Union – NextGenerationEU.

## References

[1] J. W. Gerold and D. Gerling, "An equivalent winding factor larger than 1 by using flux barriers in the stator," in *2019 IEEE International Electric Machines and Drives Conference (IEMDC)*, IEEE, 2019. DOI: 10.1109/iemdc.2019.8785148.

[2] G. Dajaku and D. Gerling, "Low costs and high-efficiency electric machines," in *2012 2nd International Electric Drives Production Conference (EDPC)*, IEEE, 2012. DOI: 10.1109/edpc.2012.6425093.

[3] S. Spas, G. Dajaku, and D. Gerling, "Comparison of PM machines with concentrated windings for automotive application," in *2014 International Conference on Electrical Machines (ICEM)*, IEEE, 2014. DOI: 10.1109/icelmach.2014.6960458.

[4] A. Nollau and D. Gerling, "Novel cooling methods using flux-barriers," in *2014 International Conference on Electrical Machines (ICEM)*, IEEE, 2014. DOI: 10.1109/icelmach.2014.6960354.

[5] B. Ackermann and R. Sottek, "Analytical modeling of the cogging torque in permanent magnet motors," *Electrical Engineering*, vol. 78, no. 2, pp. 117–125, 1995. DOI: 10.1007/bf01245643.

[6] T. Lubin, S. Mezani, and A. Rezzoug, "IMPROVED ANALYTICAL MODEL FOR SURFACE-MOUNTED PM MOTORS CONSIDERING SLOTTING EFFECTS AND ARMATURE REACTION," *Progress In Electromagnetics Research B*, vol. 25, pp. 293–314, 2010. DOI: 10.2528/pierb10081209.

[7] L. J. Wu, Z. Q. Zhu, D. Staton, M. Popescu, and D. Hawkins, "An improved subdomain model for predicting magnetic field of surface-mounted permanent magnet machines accounting for tooth-tips," *IEEE Transactions on Magnetics*, vol. 47, no. 6, pp. 1693–1704, 2011. DOI: 10.1109/tmag.2011.2116031.

[8] Z. Zhu, D. Howe, E. Bolte, and B. Ackermann, "Instantaneous magnetic field distribution in brushless permanent magnet DC motors. i. open-circuit field," *IEEE Transactions on Magnetics*, vol. 29, no. 1, pp. 124–135, 1993. DOI: 10.1109/20.195557.

[9] L. J. Wu, Z. Li, X. Huang, Y. Zhong, Y. Fang, and Z. Q. Zhu, "A hybrid field model for open-circuit field prediction in surface-mounted PM machines considering saturation," *IEEE Transactions on Magnetics*, vol. 54, no. 6, pp. 1–12, 2018. DOI: 10.1109/tmag.2018.2817178.

[10] L. J. Wu, Z. Li, D. Wang, H. Yin, X. Huang, and Z. Q. Zhu, "On-load field prediction of surface-mounted PM machines considering nonlinearity based on hybrid field model," *IEEE Transactions on Magnetics*, vol. 55, no. 3, pp. 1–11, 2019. DOI: 10.1109/tmag.2018.2890244.

[11] B. Dianati and I. Hahn, "Nonlinear modeling of MGSPMs based on hybrid subdomain and magnetic equivalent circuitry," *IEEE Transactions on Magnetics*, vol. 58, no. 1, pp. 1–10, 2022. DOI: 10.1109/tmag.2021.3128739.

[12] H. Yin, L. Wu, Y. Zheng, and Y. Fang, "Magnetic field prediction in surface-mounted PM machines with parallel slot based on a nonlinear subdomain and magnetic circuit hybrid model," in *2019 IEEE International Electric Machines and Drives Conference (IEMDC)*, IEEE, 2019. DOI: 10.1109/iemdc.2019.8785318.

[13] Y. Fang, J. Ji, and W. Zhao, "Modeling of fault-tolerant flux-switching permanent-magnet machines for predicting magnetic and armature reaction fields," *CES Transactions on Electrical Machines and Systems*, vol. 6, no. 4, pp. 413–421, 2022. DOI: 10.30941/CESTEMS.2022.00053.

[14] A. Bellara, Y. Amara, G. Barakat, and B. Dakyo, "Two-dimensional exact analytical solution of armature reaction field in slotted surface mounted PM radial flux synchronous machines," *IEEE Transactions on Magnetics*, vol. 45, no. 10, pp. 4534–4538, 2009. DOI: 10.1109/tmag.2009.2021527.

PCIM Europe 2024, 11– 13 June 2024, Nuremberg　　DOI: 10.30420/566262183

# Modeling and Stability Analysis of LCL filtered VSCs in Interleaved Topology

Adeel Jamal [1], Gerd Griepentrog [2]

[1,2] Technical University of Darmstadt, Germany

Corresponding author:　　Adeel Jamal, adeel.jamal@tu-darmstadt.de
Speaker:　　　　　　　　　Adeel Jamal, adeel.jamal@tu-darmstadt.de

## Abstract

As the integration of renewable energy resources into the European power grid accelerates, high-power grid-connected Voltage Source Converters (VSCs) are becoming increasingly pivotal. This paper delves into the modeling of interleaved VSCs and analyzes two different control schemes, utilizing converter-current and grid-current feedback methodology. Through the application of the Nyquist Stability Criterion (NSC), the stability of the two control schemes is analyzed, identifying the critical frequency essential for system stability. Performance validation of the two control schemes is presented using an experimental testrig.

## 1 Introduction

The European power grid, particularly in Germany, is witnessing an increasing integration of renewable energy resources. This transformation is significant, as demonstrated by the fact that renewable net energy generation contributed to 46.7% of the total energy mix in the European Union (EU) and 60.3% in Germany in 2022 [1]. Such a shift necessitates the operation of high-power VSCs. However, the design and availability of high-power VSCs are hampered by challenges related to the blocking voltages and conduction losses of power semiconductor switches.

Moreover, limited switching frequency in high-power VSCs requires large filters to adhere to stringent grid regulations [2]. This increases the system space [3], reduces power density, and escalates the overall system and platform costs [4]. One potential solution to increase the power is to interleave two VSCs, which can effectively double the output current and hence increase the power transfer capacity of the system at higher efficiency [5]. Interleaving brings several benefits, including an increased power capacity [6], improved reliability at the power system network node where the Interleaved VSC system (IVS) is connected [5], and reduced size of the filters due to the lower Total Harmonic Distortion (THD) of the output current [6],[7],[5]. It is

important to note that cost increases very steeply for high-power filters as the manufacturing process gets complex with the comparatively high price for handling, shipping, and installation. Power quality is also improved if the right choice of modulation technique is made. Contrasting space vectors for each of the VSC's Space Vector Modulators (SVM) can bring about the cancellation of certain harmonics from the output voltage and improve the line current quality [7],[8]. It can ensure better power quality and reduce harmonic distortion [5]. The LCL filter is a prevalent choice for grid-connected converter designs, primarily due to its compact size and strong attenuation of high-frequency switching ripples, which provides an advantage over the L-filter [[9]-[10]]. However, controlling an LCL-filtered converter poses a challenge due to the availability of grid- and converter-side currents for feedback, a complexity absent in L-filtered converters where both currents are always equal. Despite these challenges, it remains advantageous to regulate the output current of an LCL-filtered grid converter using a single current feedback loop for its simplicity, while still retaining effectiveness [[11]-[12]]. Previous studies have proposed control methodologies for LCL-filtered VSCs, such as the state-space control [13] and an observer-based current controller. Despite the rising importance of IVS, comprehensive literature on potential control configurations remains scant. This paper aims to bridge this gap by introducing and evaluating two distinct control schemes. The focus of this paper centers first on the modeling of LCL filters discussed in section 2 and then control of two interleaved LCL-filtered VSCs is presented in detail in section 3.1 and 3.2. The stability analysis of the two control configurations assessed using NSC is illustrated for three different LCL filter variations using Nyquist plots in section 4.

## 2 System Description and Modeling

Figure 1 shows two three-level T-type inverter (3LT2I) interleaved via their respective LCL filters. Each filter consists of a converter side inductor $L_i$ & a grid side inductor $L_g$. They are connected to the grid via grid impedance denoted by $R_{gd}$ & $L_{gd}$. The design can incorporate either Grid-Current feedback (GCF) or Converter-Current feedback (CCF) to regulate the two types of output current as illustrated in Fig. 1. The respective transfer function (TF) of converter side current ($i_i$) to the converter's output voltage ($v$) and converter's grid-side current ($i_g$) to converter's output voltage ($v$) can be derived to obtain eq. (1) and eq. (2).

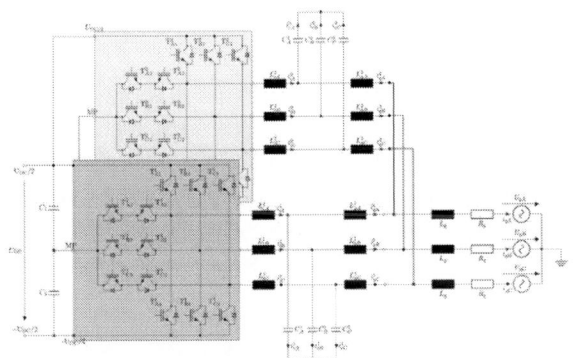

**Fig. 1:** $3LT^2I$ inverter interleaved at the output of LCL filter

$$\frac{i_i(s)}{v(s)} = \frac{1}{sL_i} \frac{s^2 + \frac{1}{L_gC}}{s^2 + \frac{L_i+L_g}{L_iL_gC}} \tag{1}$$

$$\frac{i_i(s)}{v(s)} = \frac{1}{sL_i} \frac{\frac{1}{L_gC}}{s^2 + \frac{L_i+L_g}{L_iL_gC}} \tag{2}$$

The two equations can be simplified by defining the resonance frequency ($\omega_r$) and anti-resonance frequency ($\omega_o$) of the LCL filter as follows:

$$\omega_r = \sqrt{\frac{L_i + L_g + L_{gd}}{L_i(L_g + L_{gd})C}}, f_r = \frac{\omega_r}{2\pi} \tag{3}$$

$$\omega_r = \sqrt{\frac{1}{(L_g + L_{gd})C}}, f_o = \frac{\omega_o}{2\pi} \tag{4}$$

Other TFs such as $\frac{i_c(s)}{i_i(s)}$ and $\frac{i_g(s)}{i_c(s)}$ can be obtained from eq.(1) and eq.(2) as:

$$\frac{i_c(s)}{i_i(s)} = \frac{s^2}{s^2 + \omega_o^2} \tag{5}$$

$$\frac{i_g(s)}{i_c(s)} = \frac{\omega_o^2}{s^2} \tag{6}$$

Each feedback option can be fed back to a proportional-resonant (PR) controller for eliminating steady-state current tracking error at fundamental frequency or alternatively PI controller can be utilized in the rotational reference frame (dq-coordinates). The TF of PR controller is given in equation (7) where $K_p$ is the proportional gain, $K_i$ is the resonant gain and $\omega_b$ is the fundamental frequency. VSC can be modelled as a delay function since sampling, computation of the controller equations for output reference variable and modulator introduces a delay of $1.5T_s$, whereas $T_s$ corresponds to one sampling/switching period.

$$G_c(s) = K_p + \frac{K_i s}{s^2 + \omega_b^2} \tag{7}$$

$$G_d(s) = e^{-1.5T_s} \tag{8}$$

## 3 Control Design Techniques and Stability Analysis

The VSI output power regulation can be achieved by controlling either grid-side or converter-side currents. Therefore current sensing for converter-current or grid-current can be done to enable feedback control. GCF control directly regulates grid current, enabling control of low-order current harmonics with Harmonic Controllers in parallel to the main control loop, albeit at the expense of additional grid-current measurement costs. CCF control, utilizing existing current sensors for over-current protection, avoids extra sensor costs. However, it indirectly regulates injected grid current, the harmonic information at the point of common coupling (PCC) is lost and therefore the harmonic controllability is limited. Controller designs using Grid-current feedback and Converter-current feedback will be described in the following sections.

### 3.1 Controller design using Grid-current feedback

Figure 2 shows the GCF control scheme (GCF-CS) where inner control loop regulates $i_g^1$ and $i_g^2$ of the LCL filters. Total current injected into the grid is fed back to an outer control loop which ensures $i_{gd}$ follows $i_{gd}^*$. $k^1$ and $k^2$ are the load sharing factors that can be chosen between 0 and 1 to non-uniformly share the load between the VSCs. One of the advantages of the design is $i_{gd}$ can be regulated even when one of the VSCs is not in operation. Efficiency of the whole system in case of low power demand can be increased by running only one VSC to ensure that the VSC in operation is running up to its rated power. One of the downsides of the design is the need for additional current sensors to measure the grid-side currents of each of the converter. The converter-side currents need to be measured anyways to provide over-current protection to individual half-bridges of which the converter is composed of. The outer loop control can be based on either P or PI controller. The open loop transfer function (OLTF) of the GCF-CS is

$$T_{i_g} = F_{gd}(s)[k^1 \cdot \frac{F_g^1(s)G_d^1(s)G_{i_g}^1(s)}{1 + F_g^1(s)G_d^1(s)G_{i_g}^1(s)} + k^2 \cdot \frac{F_g^2(s)G_d^2(s)G_{i_g}^2(s)}{1 + F_g^2(s)G_d^2(s)G_{i_g}^2(s)}] \quad (9)$$

### 3.2 Controller design using Converter-current feedback

Figure 3 shows the CCF-CS for the feedback of converter current $i_i$. This scheme is more practical as mostly VSC manufacturers provide the current sensing of the converter output current built-in to the hardware so extra current sensors can be avoided in this control scheme design[14], [15]. Only one additional sensor needs to be added for measuring the total current injected into the grid ($i_{gd}$). Load sharing is possible by selecting the respective constants $k_1$ and $k_2$. Since total injected grid current is regulated in the slower outer control loop, high-order harmonics cannot be influenced. The indirect method of regulating grid current (i.e. controlling $i_i^1$ and $i_i^2$) may result in severe distortion

when exposed to grid-voltage harmonics. This issue arises from the unrestricted flow of harmonic currents through the capacitor [16], [17].

$$T_{i_i} = F_{gd}(s)[k^1 \cdot \frac{F_i^1(s)G_d^1(s)G_{i_i}^1(s)}{1 + F_i^1(s)G_d^1(s)G_{i_i}^1(s)} \cdot \frac{i_c^1(s)}{i_i^1(s)} \cdot \frac{i_g^1(s)}{i_c^1(s)} + k^2 \cdot \frac{F_g^2(s)G_d^2(s)G_{i_g}^2(s)}{1 + F_g^2(s)G_d^2(s)G_{i_g}^2(s)} \cdot \frac{i_c^1(s)}{i_i^1(s)} \cdot \frac{i_g^1(s)}{i_c^1(s)}] \quad (10)$$

## 4 Stability Analysis Using Nyquist Stability Criterion

Three variations depending on the parameters of the inductance and capacitance of the LCL filter as detailed in Tab. 1 for GCF-CS and CCF-CS has been studied.

| Variations of LCL filter | Anti-resonance frequency $\omega_0$ | Resonance frequency $\omega_r$ |
|---|---|---|
| (a) C=30$\mu F$ | 1.62 kHz | 1.92 kHz |
| (b) C=5$\mu F$ | 3.98 kHz | 4.71 kHz |
| (c) C=10$\mu F$ | 2.81 kHz | 3.33 kHz |

**Tab. 1:** LCL filter variations with three different capacitance; $L_i = 800\mu F$, $L_g = 220\mu F$ and $L_{gd} = 100\mu F$

It is observed that CCF interleaved topology has inherent damping characteristics as also confirmed in single converter current-feedback based converters [18] but for ensuring the stability of the system against weak grid conditions, additional damping function has to integrated for GCF interleaved topology.

| System | Gain Margin (dB) | Phase Margin | Stability |
|---|---|---|---|
| GCF-CS (a) | 45.16 | 122.4 | Unstable |
| GCF-CS (b) | 3.8 | -102.24 | Stable |
| GCF-CS (c) | -9.25 | 51.48 | Unstable |
| CCF-CS (a) | -8 | -179 | Unstable |
| CCF-CS(b) | 6.01 | 180 | Stable |
| CCF-CS(c) | 1.89 | -8.4 | Weakly stable |

**Tab. 2:** Comparison of GM and PM for different filter variations of GCF-CS and CCF-CS

Using the open-loop transfer functions eq. (9),(10), Nyquist plots are obtained as shown in Fig. 4. Sys-

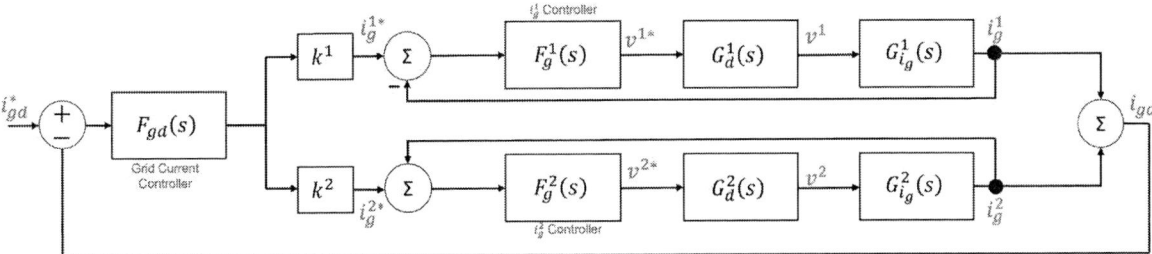

**Fig. 2:** Controller Design using grid-current feedback

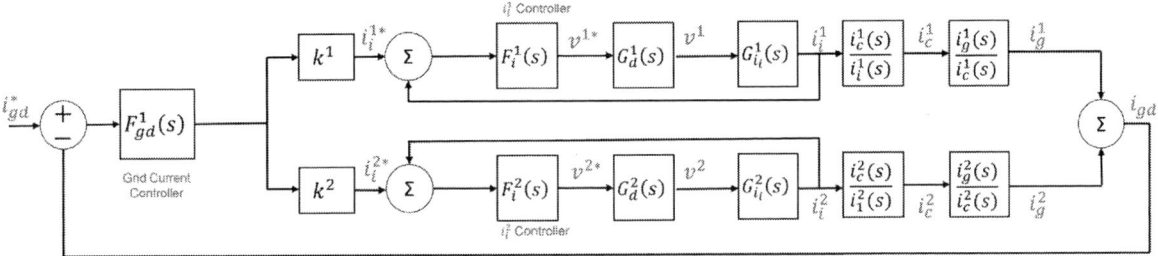

**Fig. 3:** Controller design using converter-current feedback

tems with greater gain and phase margins can withstand greater changes in system parameters before becoming unstable [19]. After analyzing the Nyquist plots and deriving phase margins (PM) and gain margins (GM) from them, stability is assessed. All of the PM and GM for each of the control scheme for three different filter variations are given in Tab. 2. It can be concluded that the closed-loop system for GCF-CS is stable if $\omega_r > \omega_{cr}$, whereas $\omega_{cr}$ is the critical frequency. Critical frequency [20] seperates the stable and unstable regions of the GCF-CS and CCF-CS systems. The value of critical frequency is only affected by the phase lag of the time delay in the digital control system and it is $f_s/6$. Similarly, closed-loop system for CCF-CS is stable if $\omega_r < \omega_{cr}$. It is also noted that the system becomes highly prone to instability close to the critical frequency if the controller gains are not chosen carefully as in the case of GCF-CS variation(c). The stable region of GCF-CS for interleaved converters is $(f_s/6, f_s/2)$ and the unstable region is $(0, f_s/6)$ while for CCF-CS, the stable region for the interleaved converters is $(0, f_s/6)$ and unstable region is $(f_s/6, f_s/2)$.

## 5 Results and Discussion

The test rig shown in Fig. 6 has been used for the implementation and performance validation of the proposed control schemes. The main inverter KEBA SOTL.610 houses two thee-phase t-type

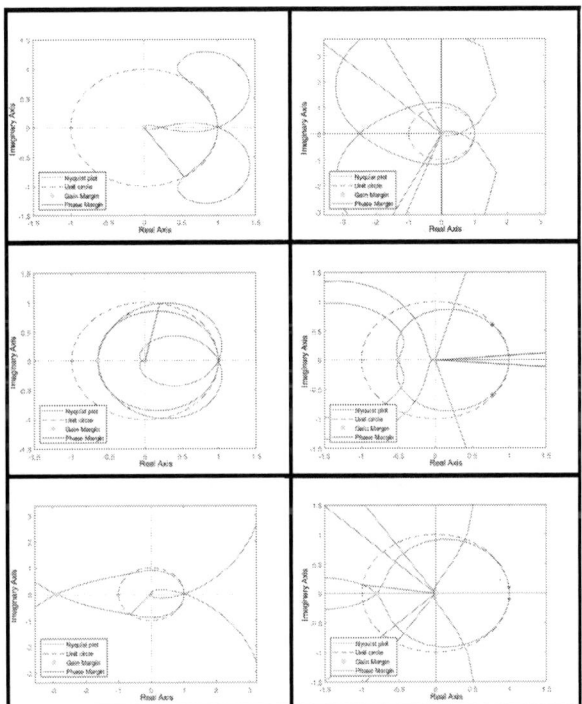

**Fig. 4:** Nyquist plots of GCF-CS(a)[top-left], GCF-CS(b)[center-left], GCF-CS(c)[bottom-left], CCF-CS(a)[top-right], CCF-CS(b)[center-right], GCF-CS(c)[bottom-right]

power stages and each power stage operates independently. Each power stage is connected to three phase LCL filter with $L_i = 800\mu F$, $L_g = 220\mu F$, and $C_f = 5\mu F$. Delta point of LCL filter is not grounded. EMI filter is installed after the LCL filter to filter out noise frequencies in the higher frequency bandwidth range. Xilinx's Ultrascale+ SoC-FPGA platform is deployed for implementing controller, generating PWMs, and for voltage and current sensors' signals post-processing [21]. PI controller is implemented for the control of grid current with $k_p = 0.2$ and $k_i = 15$. Fig. 5 shows the experimental results of implementing the grid-current feedback control according to Fig. 2. Interleaved VSCs in the experimental testrig are programmed as grid-following converter therefore, after the Phase locked loop (PLL) is done with synchronizing with the grid voltage waveform, at $t = 40ms$, contactor is turned on thereby connecting the grid to the VSCs. The reference currents are zero, therefore controller regulates the current at 0A. At $t = 90ms$, $i_{gd,d}$ is set to 30A and at $t = 0.13s$, $i_{gd,q}$ is set to 30A. It can be easily observed that the current injected into the grid i.e. $i_{gd}$ follows $i_{gd}^*$ with a very slight overshoot. Voltage at the point of common coupling is displayed on the top of Fig.5. Similar results are obtained for converter-current feedback control scheme in case of variation (b). Both of the control schemes are stable for $C_f = 5\mu F$ as it was obtained theoretically in Sec. 3.1 and Sec. 3.2.

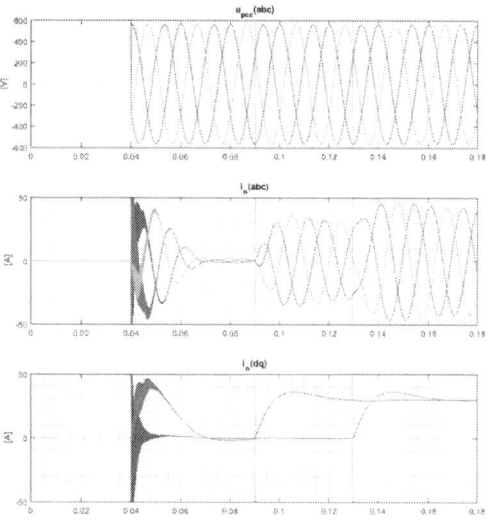

**Fig. 5:** Experimental results for Grid-current feedback control

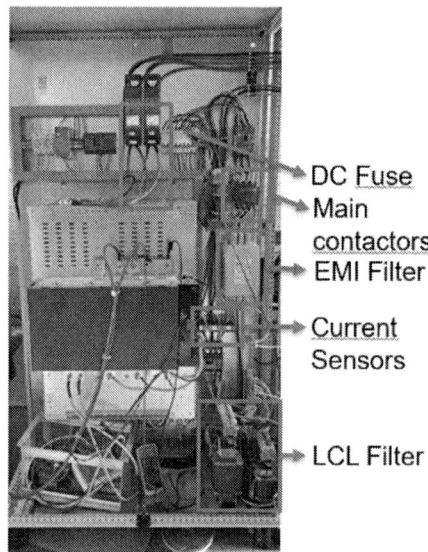

**Fig. 6:** Experimental test rig hosting interleaved converter with LCL filter

# 6 Conclusion

This paper analyzes two potential control schemes for interleaved VSCs consisting of LCL filters. The stability of control schemes is analyzed by applying NSC on the OLTF of three different variations of GCF-CS and CCF-CS. It is found for CCF-CS, if the resonant frequency $\omega_r$ of the interfacing LCL filter is below a certain critical frequency $\omega_{cr}$, then system is stable and outer loop control can be designed for stationary accurate tracking of reference injected grid current. While for GCF-CS, the system is only stable if $\omega_r > \omega_{cr}$. Any time delay compensation technique can be further employed to improve the control performance and widen the stable region for CCF-CS.

# References

[1] B. ENTSO-E Brussels, "Statistical factsheet 2022 [provisional values as of june 2023], technical report, www.entsoe.eu, 2022."

[2] "Technical guidline: Generating plants connected to the medium-voltage network, bdew bundesverband der energie- und wasserwirtschaft e.v," 2008.

[3] J. Muhlethaler, M. Schweizer, R. Blattmann, J. W. Kolar, and A. Ecklebe, "Optimal design of lcl harmonic filters for three-phase pfc rectifiers," *IEEE Transactions on Power Electronics*, vol. 28, no. 7, pp. 3114–3125, 2013. DOI: 10.1109/TPEL.2012.2225641.

[4] M. Liserre, R. Cárdenas, M. Molinas, and J. Rodriguez, "Overview of multi-mw wind turbines and wind parks," *IEEE Transactions on Industrial Electronics*, vol. 58, no. 4, pp. 1081–1095, 2011. DOI: 10.1109/TIE.2010.2103910.

[5] J. Birk and B. Andresen, "Parallel-connected converters for optimizing efficiency, reliability and grid harmonics in a wind turbine," in *2007 European Conference on Power Electronics and Applications*, IEEE, 2007. DOI: 10.1109/epe.2007.4417318.

[6] G. Gohil, L. Bede, R. Teodorescu, T. Kerekes, and F. Blaabjerg, "Design of the trap filter for the high power converters with parallel interleaved vscs," 2014, pp. 2030–2036. DOI: 10.1109/IECON.2014.7048781.

[7] G. Gohil, L. Bede, R. Maheshwari, R. Teodorescu, T. Kerekes, and F. Blaabjerg, "Parallel interleaved vscs: Influence of the pwm scheme on the design of the coupled inductor," in *IECON 2014 - 40th Annual Conference of the IEEE Industrial Electronics Society*, 2014, pp. 1693–1699. DOI: 10.1109/IECON.2014.7048730.

[8] M. Abbes, I. Mehouachi, and S. Chebbi, "Circulating current reduction of a grid-connected parallel interleaved converter using energy shaping control," *Electric Power Systems Research*, vol. 170, pp. 184–193, 2019. DOI: https://doi.org/10.1016/j.epsr.2019.01.020.

[9] M. Liserre, F. Blaabjerg, and S. Hansen, "Design and control of an lcl-filter-based three-phase active rectifier," *IEEE Transactions on Industry Applications*, vol. 41, no. 5, pp. 1281–1291, Sep. 2005. DOI: 10.1109/tia.2005.853373.

[10] Z. Xin, P. C. Loh, X. Wang, F. Blaabjerg, and Y. Tang, "Highly accurate derivatives for lcl-filtered grid converter with capacitor voltage active damping," *IEEE Transactions on Power Electronics*, vol. 31, no. 5, pp. 3612–3625, 2015.

[11] C. Zou, B. Liu, S. Duan, and R. Li, "Influence of delay on system stability and delay optimization of grid-connected inverters with lcl filter," *IEEE Transactions on Industrial Informatics*, vol. 10, no. 3, pp. 1775–1784, 2014.

[12] J. Yin, S. Duan, and B. Liu, "Stability analysis of grid-connected inverter with lcl filter adopting a digital single-loop controller with inherent damping characteristic," *IEEE Transactions on Industrial Informatics*, vol. 9, no. 2, pp. 1104–1112, 2012.

[13] F. M. Mahafugur Rahman, J. Kukkola, V. Pirsto, M. Routimo, and M. Hinkkanen, "State-space control for lcl filters: Comparison between the converter and grid current measurements," in *2019 IEEE Energy Conversion Congress and Exposition (ECCE)*, IEEE, Sep. 2019. DOI: 10.1109/ecce.2019.8913204.

[14] T. Abeyasekera, C. M. Johnson, D. J. Atkinson, and M. Armstrong, "Suppression of line voltage related distortion in current controlled grid connected inverters," *IEEE Transactions on Power Electronics*, vol. 20, no. 6, pp. 1393–1401, 2005.

[15] Q. Yan, X. Wu, X. Yuan, and Y. Geng, "An improved grid-voltage feedforward strategy for high-power three-phase grid-connected inverters based on the simplified repetitive predictor," *IEEE Transactions on Power Electronics*, vol. 31, no. 5, pp. 3880–3897, 2015.

[16] W. Yao, X. Wang, P. C. Loh, X. Zhang, and F. Blaabjerg, "Improved power decoupling scheme for a single-phase grid-connected differential inverter with realistic mismatch in storage capacitances," *IEEE Transactions on Power Electronics*, vol. 32, no. 1, pp. 186–199, 2016.

[17] G. Escobar, M. J. Lopez-Sanchez, D. F. Balam-Tamayo, J. A. Alonzo-Chavarria, and J. M. Sosa, "Inverter-side current control of a single-phase inverter grid connected trough an lcl filter," in *IECON 2014-40th Annual Conference of the IEEE Industrial Electronics Society*, IEEE, 2014, pp. 5552–5558.

[18] Y. Tang, P. C. Loh, P. Wang, F. H. Choo, and F. Gao, "Exploring inherent damping characteristic of lcl-filters for three-phase grid-connected voltage source inverters," *IEEE Transactions on Power Electronics*, vol. 27, no. 3, pp. 1433–1443, 2011.

[19] N. S. Nise, *Control systems engineering*. John Wiley & Sons, 2020.

[20] S. G. Parker, B. P. McGrath, and D. G. Holmes, "Regions of active damping control for lcl filters," *IEEE Transactions on Industry Applications*, vol. 50, no. 1, pp. 424–432, 2013.

[21] A. Jamal and G. Griepentrog, "Using system-on-chip boards for the deployment of controller for verification and prototyping," in *2022 24th European Conference on Power Electronics and Applications (EPE'22 ECCE Europe)*, IEEE, 2022, pp. 1–8.

PCIM Europe 2024, 11– 13 June 2024, Nuremberg          DOI: 10.30420/566262185

# Enhancing Safety and Efficiency for Isolated PLC I/O Designs with SPI Daisy Chain

Travis Lenz[1]

[1] Skyworks Solutions, Inc., United States

Corresponding author:    Travis Lenz, travis.lenz@skyworskinc.com
Speaker:                 Travis Lenz, travis.lenz@skyworskinc.com

## Abstract

Programmable Logic Controllers have the difficult task of interfacing with dozens of input and output devices spread out across an industrial or manufacturing facility. The modularity of PLCs and the hazardous nature of the factory environment impose challenging requirements on the efficiency and safety capabilities of these input and output channels beyond the IEC 61131-2 specification for PLC design. This paper will introduce an architecture that utilizes SPI daisy chaining to combine up to 128 input channels or 128 output channels along with integral safety, robustness, and diagnostic features to enhance PLC I/O design.

## 1    Introduction

Programmable Logic Controllers (PLCs) are a specialized form of microprocessor-based controller that differ from general-purpose computers in their expandable input and output interfaces, user-friendly programming languages designed for logic and switching, high-reliability to minimize downtime, and ruggedized design to withstand harsh environments. PLCs have become the de facto replacement to older, electromechanical designs based on relays and feature an architecture consisting of a core processing unit (CPU), power supply unit (PSU), input/output (I/O) interface, communications interface, and a programming interface [1].

These PLC platforms must be able to withstand a wide range of harsh operating conditions such as high humidity and temperature, operate 24/7 for years on end, tolerate a wide range of power sources and unknown I/O connections, and do so in a confined physical space all while conforming to the governing standard of IEC 61131. Yet with the advancement of sensor and actuator technology that interface with PLCs, the demand for larger and faster control systems, and the proliferation of low-cost and powerful microprocessing components, these requirements and the complexity of PLC designs grow even steeper.

PLC designs are improving to accommodate the demand of an evolving industrial ecosystem with the expansion of programming capabilities that can incorporate general purpose function blocks such as those written in an embedded C language [2]. This has in part been enabled by the increased processing power of low-cost microprocessors often used in PLC designs. To take advantage of this increased resource, the PLC I/O channel hardware must also evolve to avoid bottlenecks and deliver enhanced performance to the sensors and actuators on the other end of the PLC program.

These I/O channel improvements which are beginning to appear across the industry include: higher density to interface with more devices while consuming the same or less board space and other resources, intelligent feedback that provides actionable insights for complex control schemas, and greater physical and electrical reliability to further improve uptime and reduce the cost of lifetime maintenance. This paper will survey these advancements in PLC I/O design and present a solution that meets the requirements of high I/O channel density, robust internal protection, and actionable diagnostics feedback from I/O to controller.

## 2    I/O Device Communication Interfaces

The primary mode of achieving higher density in the I/O channel design for PLCs is via integration of discrete components. The traditional approach to handling 24V digital PLC inputs is to use one discrete optocoupler per channel to provide channel-to-channel isolation and signal conditioning, as introduced in [3] and [4]. Digital PLC outputs have been designed in a similar manner, typically pairing an optocoupler with a discrete MOSFET to provide 0.5 A to 2 A of current to the load. In the last two decades, semiconductor manufacturers have begun integrating multiple isolated input or output channels into a single package to achieve higher density. This integration reduces bill of material cost and board space, enhances the isolation barrier with CMOS technology, and adds safety and protection features not found in optocoupler-based designs.

The innovation of these integrated input and output devices has birthed a wide variety of device types with different features and CPU interfaces. There are multiple factors that limit the number of channels integrated into a single I/O semiconductor device, including package heat dissipation, modularity of I/O design, and the burden of communication interface to the CPU. The following subsections will focus on different types of CPU communication interfaces with I/O devices and the advantages and disadvantages of each.

### 2.1 Parallel I/O Architecture

The incumbent approach of using one optocoupler per I/O channel requires a dedicated GPIO pin for each input or output. This approach, called parallel I/O, can be seen in Fig. 1. This architecture is parallel in nature because data from each channel can be processed by the CPU in parallel, or simultaneously. This also allows devices connected to different ground potentials to interface with the same controller, as the inputs and outputs of each optocoupler are isolated from one another. As a result, many of the modern integrated I/O devices utilize a similar approach where each input has its own dedicated isolation channel to replace the optocoupler.

The parallel interface approach requires a unique processor GPIO pin to be dedicated to each input channel, resulting in the following relationship between input or output channels and GPIO pins:

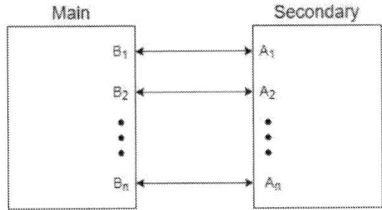

**Fig. 1**     Parallel communication interface

$$\frac{channels_{tot}}{pins_{par}} = 1 \qquad (1)$$

or

$$pins_{par} = channels_{tot} \qquad (2)$$

As the number of required channels increases to 32, 64, 128 or more, pin consumption quickly becomes a bottleneck in the design as few practical CPUs can support this many dedicated GPIO connections.

### 2.2 SPI Architecture

As a result of this design constraint, many I/O devices have evolved to utilize isolated 4-wire serial peripheral interfaces (SPI) such as the one introduced in [4]. The high-level interface is demonstrated in Figure 2 below.

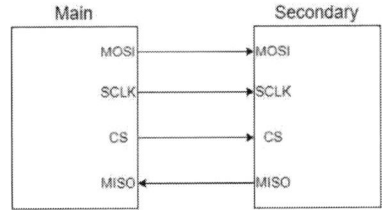

**Fig. 2**     4-wire SPI interface

While there are many advantages to an SPI interface, the main advantage over a parallel interface is the reduction of CPU resources required to process individual I/O channel statuses. In an SPI network, the CPU can issue a read or write command to a target secondary device by asserting that device's chip select (CS) pin. If the CS pin is unasserted, the secondary device will ignore the data transmitted on the main data out line (MOSI). As a result, adding devices to an existing SPI network requires only one additional pin to function as a unique CS for that device. Equation (3) shows the relationship between number of devices on a single SPI bus ($n$) and required CPU GPIO pins ($p$).

$$pins_{SPI} = 3 + n \qquad (3)$$

If we factor in the number of I/O channels per device, we get a general relationship between pins and channels.

$$n = \frac{channels_{tot}}{channels_{n}} \qquad (4)$$

yielding

$$pins_{SPI} = 3 + \frac{channels_{tot}}{channels_n} \quad (5)$$

Although the number of channels per device can vary, the most common devices in the industry offer 4 to 8 channels per SPI device. Choose a 4-channel SPI device used in a system with 32 I/O channels to compute the ratio of pins required to implement a parallel interface compared with the SPI interface:

$$\frac{pins_{SPI}}{pins_{par}} = \frac{3 + \frac{32}{4}}{32} = \frac{11}{32} = .344 \quad (6)$$

In the example above, the SPI interface reduces the GPIO pin cost by 65%, with the savings increasing exponentially with total channel count and channels per device. This reduction in CPU pin usage is a major factor in driving integrated I/O devices to use SPI interfaces. However, as industrial control systems continue to expand in device count, even adding a single additional GPIO pin per device can limit the number of channels in a system.

## 2.3 Daisy Chain SPI Architecture

In order to eliminate this extra GPIO pin per device, newer SPI-based device interfaces have adopted an architecture known as daisy chaining, shown in Fig. 3.

This architecture has been demonstrated to address complex cabling and scalability challenges in PLC I/O design with only a modest reduction in data throughput [5]. In [6], an SPI daisy chain system was implemented with a clever technique for isolating the CPU from the high-potential devices in the daisy-chain, thus enabling devices at different potentials to utilize the same bus and preserving the advantage of the optocoupler approach above.

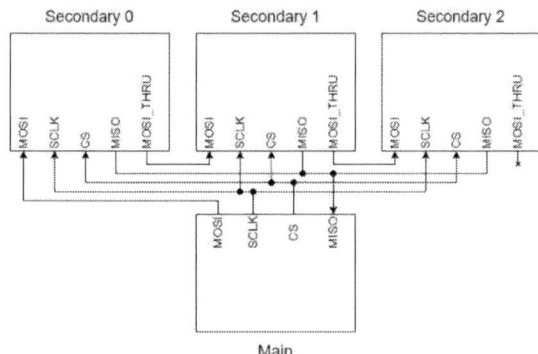

**Fig. 3**     4-wire SPI daisy chain interface

The SPI daisy chain architecture can have different signal connection schemes; some utilize a direct connection from the secondary data out pin (MISO) to the next device's secondary data in (MOSI), while others use a special data out pin (MISO_THRU or similar) that connects to the next device's MOSI pin. In either case, the CPU only needs 4 pins to interface with all devices in the daisy chain:

$$pins_{dc} = 4 \quad (7)$$

Using the same example as in Eq. (6), the ratios between parallel interface, SPI interface, and SPI daisy chain interface are recalculated below:

$$\frac{pins_{dc}}{pins_{par}} = \frac{4}{channels_{tot.}} = \frac{4}{32} = .125 \quad (8)$$

$$\frac{pins_{dc}}{pins_{SPI}} = \frac{4}{3 + \frac{channels_{tot}}{channels_n}} = \frac{4}{8} = .500 \quad (9)$$

The SPI daisy chain architecture provides 50% pin savings over the traditional SPI architecture, and an astounding 87.5% pin savings over the parallel I/O architecture. These savings increase with the total system channel count.

As shown in the calculations above, the advantage of the daisy-chain architecture provides exponential CPU resource savings as the number of I/O channels increases. In review, the utilization of a daisy-chained SPI architecture for PLC I/O design provides a major advantage towards the goal of supporting larger and more complex I/O networks while maintaining or reducing the GPIO resource burden on the CPU.

# 3 Integrated Protection and Diagnostic Mechanisms

The capacity to interface with many I/O devices is only beneficial if the channels can tolerate the electrical stress induced by the load connection. Digital output devices may be designed to drive up to 2 A of current per channel and must tolerate a temporary surge current of twice the rated current per IEC 61131-2 [7]. These digital outputs are often connected to inductive loads such as relays or coils and therefore must be able to demagnetize the inductive load and handle overcurrent situations safely to avoid catastrophic damage to the output switch, as shown in Fig. 4 [8].

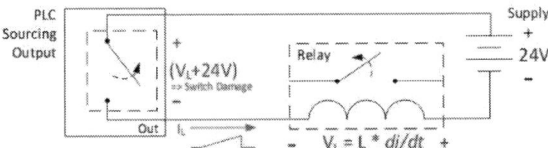

**Fig. 4**     Inductive kickback in PLC output channel

This section will examine various methods of kickback protection, overcurrent protection, and diagnostic reporting that enable PLC output devices to tolerate potentially dangerous load conditions in the industrial environment.

## 3.1 Traditional Approach to Kickback Protection

There are several commonly used approaches to clamp inductive kickback and discharge inductor current. For discrete solutions such as the optocoupler design discussed in [3], kickback protection must also be implemented with discrete components and is typically implemented with back-to-back diodes in the form of a passive clamp as shown in Fig. 5 below.

This approach is simple and relatively inexpensive, but it comes at the cost of larger BOM, more board space, and a fixed clamp voltage that may not be suitable for all inductive loads. Integrated output devices have evolved to either utilize the internal switching device itself to function as a demagnetization clamp or to utilize an integrated demagnetization clamp in parallel. Figures 6 and 7 demonstrate these approaches, respectively.

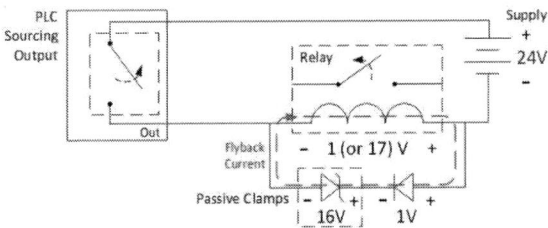

**Fig. 5**     Kickback protection by passive clamp

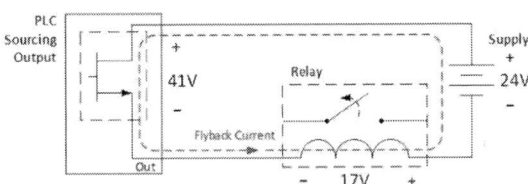

**Fig. 6**     Kickback protection by avalanche breakdown

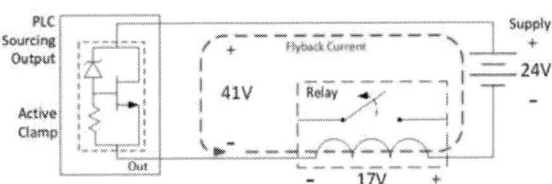

**Fig. 7**     Kickback protection by integrated clamp

Both approaches provide design flexibility due to the removal of external BOM components and reduced board space. However, there are drawbacks to each.

Clamping by avalanche breakdown requires the switching device to be able to tolerate a large Vds voltage equal to the supply voltage plus the inductor kickback voltage. These devices must therefore be fabricated using a high-voltage process and can drive up the cost of production. During demagnetization, the switch will dissipate power equal to the breakdown voltage times the inductor current, resulting in very high-power dissipation that induces significant heating. This can result in thermal degradation and reduce the lifetime of the device.

Demagnetizing the inductive load in a separate, integrated clamp alleviates this issue by providing a separate demagnetization current path. This is an effective approach that is used in many integrated PLC output devices in the industry today. The primary drawback of this approach is applying a fixed demagnetization voltage to all inductive loads, resulting in high power dissipation and heat generation in the clamping device.

## 3.2 Kickback Protection with Integrated Multi-Voltage Smart Clamp

In [9], a clever solution has been presented to detect inductive kickback and enable a freewheeling current path using integrated logic. Such a solution is only useful in an H-bridge configuration and still presents the challenge of discharging the inductor rapidly, since the inductor coils will discharge at a slow rate in the proposed solution. In [8], an integrated multi-voltage clamp is introduced that addresses the previous shortcomings by utilizing a passive clamp to limit the initial kickback voltage, a smart detection scheme to trigger activation of the active clamp, and multiple clamp voltages to optimize power dissipation versus demagnetization time. The high-level architecture of this approach is shown in Fig. 8 below.

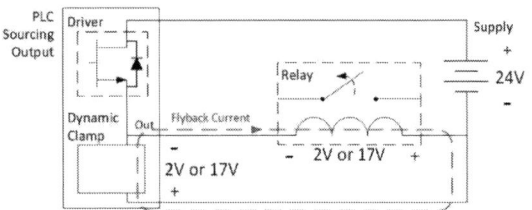

**Fig. 8**    Multi-voltage clamp for kickback protection

This approach is an evolution of the aforementioned solutions and enables the PLC output device to drive loads up to the rated load current in IEC 61131-2 without risk of damage to the device during demagnetization. The smart detection scheme senses the output current and determines if the clamp should operate in low-voltage mode to limit thermal dissipation or high-voltage mode to minimize discharge time.

## 3.3 Overcurrent Protection Through Power Estimation

With any high-current output, there is a possibility of overcurrent (or overload) conditions such as short-circuit events. Most PLC output devices utilize thermal sensing to determine when an overcurrent condition has occurred, as shown in Fig. 9.

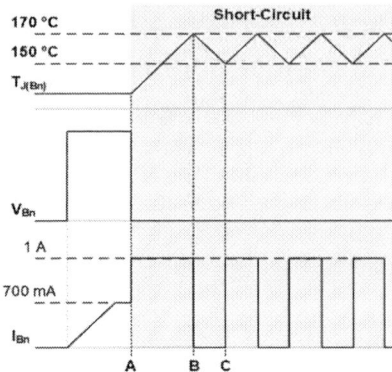

**Fig. 9**    Thermal sensing for overcurrent detection

In the figure, $T_J$ represents the junction temperature of the output device, $V_{Bn}$ represents the output channel voltage, and $I_{Bn}$ represents the output channel current. As a result of a short-circuit event at time A, the junction temperature quickly rises until the rising overtemperature threshold is crossed, leading to an output shutdown at time B. After sufficient cooling time, the falling overtemperature threshold is crossed, and the channel enables once again. This cycle induces a high average temperature in the switching device and can lead to thermal degradation and reduced device lifetime.

An alternative approach is introduced in [8] that uses a power estimation technique to determine when an overcurrent condition is present. Upon detection, the output current is limited via impedance control for a duration of 150 µs, after which if the overcurrent condition remains, the output channel shuts down for 500 ms. After the shutdown period expires, the device sends short pulses to the output channel to determine if the overcurrent condition is still present. If it is, the channel is shut down for 500 ms once again. This behavior is demonstrated in Fig. 10.

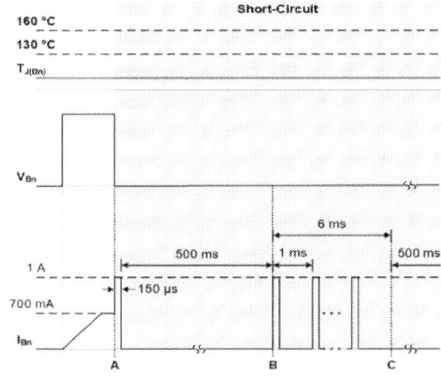

**Fig. 10**    Power estimation for short-circuit protection

The advantage of this approach is greatly reduced thermal dissipation in the output device. As a result, the device can withstand an overcurrent condition for an indefinite amount of time without overheating and causing thermal degradation of the output.

## 3.4 Diagnostic Reporting

In an environment where uptime is critical and system complexity is on the rise, detecting and resolving PLC fault conditions such as the overcurrent event discussed in the previous section becomes paramount [2]. In [10], a multistep process is introduced for debugging a PLC fault with the goal of diagnosing the problem as fast and accurately as possible. In the same paper, however, a lack of professional skills of technical personnel in the industrial environment is also identified as a major challenge. As such, utilizing binary LED indicators to deduce the origin of an I/O device fault can be a difficult and unreliable way to debug a PLC system.

With the advent of integrated I/O devices, intelligent sensors and diagnostic reporting systems can be integrated into the devices to provide actionable feedback to the PLC's human-machine interface (HMI). According to [11], a safety PLC (s-PLC) which lacks sufficient application diagnostic information (such as fault signals reported back from the I/O devices) to achieve the desired safety

integrity level (SIL) must use external sensors and actuators to test the I/O circuitry, or equipment operators must conduct manual tests to correct operation. In addition to reducing costs by eliminating redundant test circuits or operator maintenance, actionable diagnostics significantly reduce debug time by relaying fault states such as overcurrent fault, overtemperature fault, and open-circuit detection directly to the operator.

## 4 System Design Example: Useless Machine Demonstration

The following design demonstrates an effective implementation of integrated PLC I/O devices to address the challenges of I/O channel density, enhanced physical robustness and electrical reliability, and providing actionable diagnostic feedback for improved system fault detection and resolution. Figure 11 shows a high-level system block diagram for the Useless Machine demonstration, which emulates the famous "useless machine" electronics project but utilizes a set of PLC I/O devices and a Python-based GUI for real-time control.

This emulated PLC system design includes five components: the power supply which powers all field-side components, the MCU which functions as the PLC's CPU, the I/O devices which provide the interface to the input and output loads, the input and output loads themselves, and a PC which

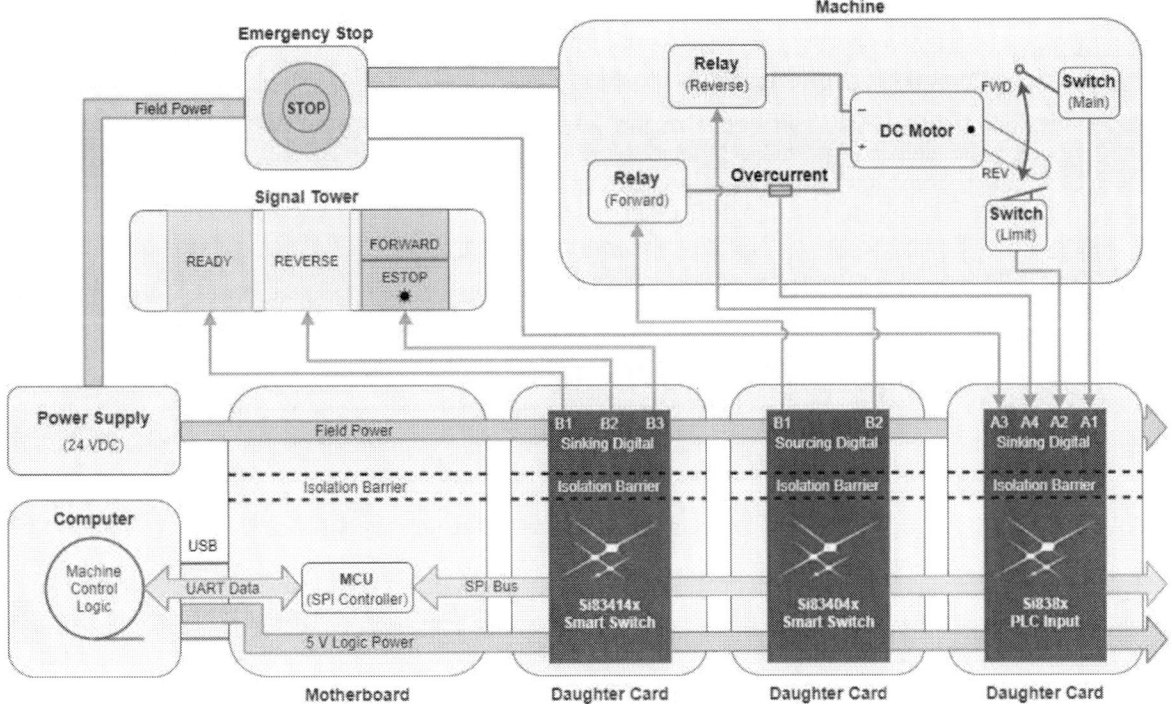

Fig. 11    Useless Machine system block diagram

functions as the HMI or programming interface for the PLC. The Skyworks Si834x Isolated Smart Switch is used for the PLC output channels, while the Skyworks Si838x PLC Input device is used for the PLC input channels.

## 4.1 State Machine Operation

The Useless Machine demonstration implements a basic truth table that translates the four input states into a set of output states that dictate the operation of the motor and the color and duty cycle of the LED signal tower. The simplified state machine is shown in Fig. 12 below.

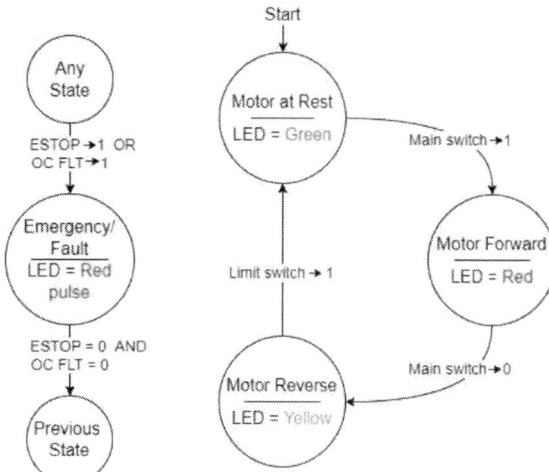

**Fig. 12** Useless Machine simplified state machine

## 4.2 Implementation of SPI Daisy Chain Interface for I/O Communication

In order to minimize the GPIO pin usage of the MCU for communication to the PLC I/O devices, an SPI daisy chain architecture was used as discussed in Section 2.3. With a total of 16 I/O channels supported by 3 devices, this architecture yields the following theoretical ratios:

$$\frac{pins_{dc}}{pins_{par}} = \frac{4}{channels_{tot.}} = \frac{4}{16} = .25 \qquad (10)$$

$$\frac{pins_{dc}}{pins_{SPI}} = \frac{4}{3 + \frac{channels_{tot}}{channels_n}} = \frac{4}{6} = .667 \qquad (11)$$

Although only 9 channels are utilized in the design, the ability to expand to up to 128 input or output channels using just four GPIO pins enables expansion of the existing design to accommodate a large-scale I/O control schema.

## 4.3 Real-time Loop Control and Actionable Diagnostic Feedback

The Useless Machine demonstration is controlled in real time by the GUI running on the PC to provide visual feedback of the input switch states and output device statuses. This adds significant loop time, as all SPI data must be transacted to the GUI via USB. Even so, the average loop time measured from a rising or falling edge on any input channel to the rising or falling edge of a corresponding output channel (e.g. main switch toggle to FWD relay asserted) is just 0.765 ms, fast enough to handle a low-speed actuator such as the DC motor in this design. Pushing the loop control algorithm to the MCU and providing asynchronous diagnostic data back to the PC – comparable to real PLC design – could reduce the loop time by two orders of magnitude or more while maintaining the flow of actionable insight to the operator.

In addition to a tight control loop that monitors the input states and services the outputs, the PLC system also monitors diagnostic data from the Si834x output devices to determine if a fault condition has occurred on any of the output channels. These diagnostics include a system-wide active-low fault signal, open-wire detection on each output, overcurrent detection and shutdown on each channel, and overtemperature warning and shutdown. This information is displayed in the GUI in a manner comparable to a PLC HMI.

In contrast to the PLC fault debug approach described in [10], debug of such reported fault conditions enables the user or operator to quickly identify not only the I/O device where the fault originated, but also the nature of the fault and the specific device that may need service or replacement. This cost-saving information can only be gathered by the PLC system if an SPI daisy chain is utilized for device communication, freeing up valuable GPIO ports to make use of the additional diagnostic feedback signals. Additionally, the integration of output channel sensors and comparators enables the detection of such fault conditions that can be stored and communicated via an advanced digital control block within the devices, an advantage this type of design has over traditional discrete optocoupler design.

## 4.4 Safe and Reliable Relay Actuation with Multi-Voltage Clamp

The Si834x output device utilizes the multi-voltage smart clamp introduced in Section 3.2. This demagnetization clamp design optimizes inductor discharge time versus power dissipation in the clamping device. In Fig. 13 below, a scope capture of the multi-voltage clamp discharging a 200 mH inductive relay coil demonstrates the operation during demagnetization.

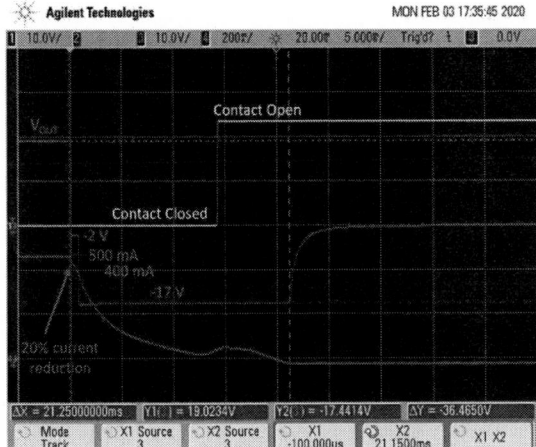

**Fig. 13** Demagnetization of 200 mH inductor using Si834x integrated multi-voltage smart clamp

When the output channel is turned off at the first cursor, the channel voltage (in blue) drops from 24 V to -2 V to begin discharging the inductor slowly while the clamp current level (in pink) is still high. Once the current drops below a 400 mA threshold, the clamp voltage increases to a magnitude of -17 V to rapidly discharge the remaining current and reliably actuate the relay contactor. This method of operation enables the inductor current to drop by 20% – from 500 mA to 400 mA – while in low-voltage clamping mode, reducing the total energy stored in the inductor by 36% while dissipating 900 mW in the clamp for 3 ms, resulting in just 3 mJ of energy consumed in the clamp. The result of this low-voltage clamp state is nearly 35% reduction in average total power dissipation in the multi-voltage clamp when compared to a static -21 V clamp.

## 5 Conclusion

This paper surveyed various approaches to modern PLC I/O design in order to tackle the challenges of higher I/O channel count, robust and reliable electrical operation of channels for long-term durability, and reporting of actionable diagnostics to improve ease of debug and reduce the cost of maintenance. A system design called the Useless Machine was introduced which utilizes an SPI daisy-chained architecture to interface with sixteen input and output channels. The design features device self-protection mechanisms such as an integrated multi-voltage clamp and power estimator for overcurrent detection and reports a multitude of fault conditions for fault detection and origination. This design demonstrates a real-time control system (RTCS) that services four input switch states and actuates two motor relays and a signal tower to indicate device state. In conclusion, the advantages of isolated PLC I/O device features as identified and discussed in Sections 1, 2, and 3 of this paper were proven in concept by this Useless Machine system design.

## References

[1] Bolton W. Programmable Logic Controllers. Sixth Edition. Oxford, UK: Newnes; 2015.

[2] Sehr M, Lohstroh M, Weber M, Ugalde I, Witte M, et al. Programmable Logic Controllers in the Context of Industry 4.0. IEEE Trans. Ind. Inform. 2020;17(5):3523-3533

[3] Yang V, Mu S, Hartmann D. PLC DCS Analog Input Module Design Breaks Barriers in Channel-to-Channel Isolation and High Density. 2016;50(12):1-5.

[4] Gao M. Product How-to: Isolated PLC digital inputs for industrial control. EDN. 2013 [accessed 28 March 2024].

[5] Coskun O, Egeli E, Tarcan E, Kurtulus I, Tilmaz G. Analysis and Implementation of a Daisy-Chain Serial Peripheral Interface Bus for a Communication Network with Multiple PLC Modules. Paper presented at: ELECO 2023. 14th Int. Conf. Elect. Electron. Eng. 30 November – 2 December 2023; Bursa, Turkey.

[6] Wang X, Zhang H, Zhang L, Zhang J, Hao Y. A daisy-chain SPI interface in a battery voltage monitoring IC for electric vehicles. Paper presented at: ICSICT 2014. 12th IEEE Int. Conf. Solid-State Integr. Circuit Technol. 28-31 October 2014; Guilin, China.

[7] International Standard IEC 61131: Programmable Controllers. 3rd edition. Geneva, Switzerland: IEC; 2017.

[8] Skyworks Solutions, Appl. Note 1212. Advanced Load Driving Considerations for Skyworks Smart Switches. 2023 [accessed 28 March 2024].

[9] Sun J, Zhang B, Wang H, Ming X, Xiao K, et al. An Inductive Kickback Absorption Scheme Without Power Zener and Large Capacitor. IEEE Trans. Ind. Electron. 2010;58(2):709-716.

[10] Song Q. Research on Engineering Problems in PLC Control System. Paper presented at: ICICIP 2018. 9th Int. Conf. on Intell. Control and Inf. Process. 09 – 11 November 2018; Wanzhou, China.

[11] Rastocny K, Zdansky J, Balak J, Holecko P. Effects of diagnostic on the safety of a control system realised by safety PLC. Paper presented at: 2016 Elektro. 14 July 2016; Strbske Pleso, Slovakia.

# AUTHOR INDEX

Abbas, Khizra ............... 764
Ackermann, Martin............... 1336
Aiello, Giuseppe ............... 1217
Akbari, Saeed............... 2094
Akturk, Akin ............... 739
Alauzet, Louis............... 2811
Albert, Tianlong............... 1759
Alfonso, Irene Maria Torres............... 2503
Alfonzetti, Emanuela ............... 1844
Allioua, Abdelmoumin............... 2128
Ammar, Ahmed ............... 1087
Appleby, Matthew............... 3276
Arai, Nobuhide ............... 298
Araujo, Lucas............... 1673
Arnaudov, Dimitar............... 2268
Askan, Kenan............... 1545
Aspalter, Paul............... 2258
Augustin, Tim ............... 3086
Aunon, Fernando ............... 1467
Ausseresse, Pierrick............... 1082
Austrup, Isabel............... 2956
Babaki, Amir............... 1227
Bagheribavaryani, Mohammadreza ............... 1418
Baharizadeh, Mehdi............... 378
Bai, Yeriel............... 1804
Baker, Nick............... 1923
Bándy, Kristóf............... 403, 2566
Barcelos, Renan Pillon............... 264
Barón, Kevin Muñoz ............... 1978
Barth, Henry............... 2838
Basso, Christophe............... 3096
Bastawros, Adel............... 440, 1951
Batista, Emmanuel............... 2394
Baudais, Briac............... 3187
Behrendt, Stefan ............... 361
Beiranvand, Hamzeh............... 1105
Beyerle, Raphael............... 958
Bhatia, Tamanna ............... 1259
Bicer, Ekin Alp ............... 40
Bimmel, Luc............... 3206
Blechinger, Christoph ............... 1717
Block, Marius ............... 2217
Bockholt, Yannick............... 3334
Böhning, Lukas............... 2208
Boldyrjew-Mast, Roman............... 723
Bosnjic, Zlatko............... 1788
Boutry, Arthur............... 1878
Bouzerd, Souhila............... 581

Branas, Christian............... 2286
Brandl, Anja Katerina............... 1613
Breidenstein, Daniel............... 1634
Bürger, Matthias............... 863
Cairnie, Mark............... 599
Calmels, Alain............... 3305
Cammarata, Federica............... 1289
Campos, Adriana............... 2663
Cannone, Marco............... 502
Capobianco, Thomas Anthony ............... 1168
Çay, Yunus............... 3247
Cepin, Simon............... 1051
Chaisakdanugull, Chanuch............... 3067
Chatroux, Daniel............... 2278
Chatterjee, Bhaskar............... 774
Chen, Mengxing............... 424
Cherief, Wahid............... 1910
Cho, Wonjin Dylan............... 1046
Choo, Vin Loong............... 1775
Chorfi, Ilias............... 2175
Cinik, Sadik............... 2453
Colak, Baris............... 490
Colomer, Pau............... 456
Conilh, Christophe ............... 2227
Corbitt, Anna............... 135, 1123, 1821
Croston, Jose Andres Aguilar ............... 3150
Curbow, Austin............... 1475
Cusumano, Andrea............... 1627
Czerwenka, Philipp............... 1139, 3034
Daire, Baptiste............... 3110
Dasch, Michael............... 1907
Davoodi, Hossein............... 1013
Debbadi, Karthik............... 2963
Deboy, Gerald............... 15
Dedew, Mohamed Lemine ............... 34
Delaforge, Timothé............... 1797
Denk, Marco............... 1192
Despesse, Ghislain............... 797
Diz, Sergio De Lopez............... 411
Do, Nguyen Nghia............... 1428
Dresel, Lars............... 2737
Du, Xinyuan............... 1987
Duijsen, Peter Van............... 1658, 2248, 2657, 3213
Dumollard, Yannick............... 1751
Dupont, Max............... 93
Dusmez, Serkan............... 383, 2334, 3060
Eichler, Felix............... 3020
Eyama, Takaaki............... 56

Fabian, Benjamin ............................................. 190
Fenske, Florian .............................................. 3390
Fey, Justin ................................................... 1902
Fleck, Soenke ................................................. 338
Förster, Nikolas ............................................. 3237
Fotteler, Oleg ............................................... 3328
Fräger, Lukas ................................................. 926
Frank, Michael ................................................ 754
Frank, Wolfgang .............................................. 1770
Frei, Steffen ......................................... 2478, 3007
Fuchs-Gade, Jannik ........................................... 2632
Fuhrmann, Jan ................................................ 1315
Gackowski, Bartosz ........................................... 1504
Gandluru, Veera Bharath Chandra Reddy ........................ 2167
Gavin, Serge ................................................. 1101
Gebhard, Thomas .............................................. 1128
Gebhardt, Mathias ............................................ 2769
Gellman, Ziv .................................................. 608
Gendrin, Martin ............................................... 909
Ghanbari, Alireza Ramezan .................................... 3175
Ghosh, Priyanka .............................................. 1523
Gick, Sebastian .............................................. 1264
Gioda, Alexis ................................................ 3400
Girgin, Mehmet Oguz .......................................... 3353
Giuffrida, Simone ............................................. 248
Giuffrida, Vittorio .......................................... 1065
Gleissner, Michael ........................................... 2803
Goff, Gregoire Le ............................................ 3160
Gomez, Antonio Miguel Munoz ................................... 625
Gottardo, Davide ............................................. 2461
Gragger, Johannes ............................................ 2104
Graham, Robert ............................................... 1410
Groon, Fabian ................................................ 3380
Groos, Gerhard ................................................ 986
Guan, Jiajia ......................................... 2591, 3395
Gudala, Bhavana .............................................. 2524
Guiot, Eric .................................................. 1604
Gunes, Ekrem R. .............................................. 3221
Gupta, Gaurav ................................................. 534
Gürlek, Yavuz ................................................. 745
Haake, Daniel ................................................ 2538
Haas, Tobias ......................................... 2326, 3017
Haehre, Karsten ............................................... 214
Haensel, Stefan ............................................... 230
Hanf, Michael ......................................... 351, 571
Harmand, Thomas .............................................. 2138
Hasegawa, Kazunori ........................................... 3002
Hauenschild, Philipp ......................................... 1969
Hegarty, Timothy ............................................. 1092
Hegde, Niranjan .............................................. 1374
Heimler, Patrick ............................................. 1955
Hellinger, Rolf ................................................. 1

Hepp, Maximilian ............................................. 3045
Herrera, Adolfo .............................................. 1057
Herrmann, Clemens ............................................. 731
Hertline, Joseph ............................................. 1886
Herzog, Fabian ............................................... 3136
Hirao, Takashi ............................................... 1007
Hironaka, Yoichi .............................................. 699
Hoffmann, Lennart ............................................ 3264
Horat, Andreas ................................................ 480
Hornbuckle, Malachi .......................................... 2724
Hosseinzadehlish, Mana ............................... 1402, 1610
Hu, Jhih-Cheng ................................................ 791
Huber, Jonas .................................................. 254
Huerner, Andreas .............................................. 681
Huselstein, Jean-Jacques ..................................... 2547
Husev, Oleksandr .............................................. 893
Igartuburu, Daniel San Laureano .............................. 2303
Imai, Ayano ................................................... 180
Ippisch, Matthias ............................................ 2638
Irifune, Hiroyuki ............................................ 2028
Jahn, Simon ................................................... 883
Jamal, Adeel ................................................. 1346
Jappe, Tiago ................................................. 2843
Jegal, Junhyeok .............................................. 1590
Jha, Kunal ................................................... 2930
Jia, Minli ................................................... 2730
Jo, David .................................................... 1732
Jones, Jeremy ................................................ 1031
Kaiser, Jeremias ............................................. 1538
Kampert, Erik ................................................ 2342
Kanatzar, Paul ............................................... 1361
Kangjia, He ................................................... 62
Karout, Mohammed Amer ........................................ 1835
Kasko, Igor .................................................. 1991
Kato, Koji ................................................... 1368
Kaufmann-Bühler, Marius ...................................... 2400
Kawabata, Junya .............................................. 2049
Keilmann, Robert ............................................. 2972
Kempitiya, Asantha ............................................ 497
Klever, Severin .............................................. 1561
Knappstein, Lukas ............................................ 1745
Knecht, Martin ............................................... 3142
Koch, Jan-Niklas ............................................. 2240
Koczy, Dawid ................................................. 1651
Kohlhepp, Benedikt ........................................... 2316
Koi, Kenichi ................................................... 67
Kono, Hiroshi ................................................ 2022
Kopischke, Ruben ............................................. 2796
Körner, Patrick ............................................... 615
Kragl, Robert ................................................ 1385
Kreppel, Thomas .............................................. 2416
Krigar, Tim ................................................... 174

Kroics, Kaspars.....................................................510
Kugener, Jeff.............................................3315, 3318
Kurukuru, Varaha Satya Bharath .........................875
Kuzmanoska, Sara............................................2745
Ladentin, Kevin................................................1964
Lambert, Adrien................................................1574
Langfermann, Sascha.........................................1516
Lavery, Melanie................................................1485
Lee, Chih Hui...................................................1152
Lee, Jongmu.....................................................1712
Lee, Kihyun ...........................................1724, 1737
Lemaitre, Damien..............................................2596
Lenz, Travis.....................................................1352
Lenzen, Patrick...................................................903
Leung, Wing Tai...................................................74
Liao, Xinyuan.....................................................322
Lim, Alex.........................................................2937
Lindner, Lars....................................................2370
Lippold, Florian................................................2981
Liu, Baihan......................................................1072
Liu, Iris...........................................................1222
Liu, Yusi..........................................................3181
Lottis, Christian.........................................3347, 3358
Lotz, Marc René................................................1457
Lu, Juncheng..............................................19, 837
Lucia, Oscar.............................................2448, 2513
Lutzen, Hauke....................................................976
Lv, Jianwei......................................................1872
Ma, Kwokwai....................................................2778
Machtinger, Katharina........................................2119
Madloch, Sonja...................................................369
Maheshwari, Ramkrishan.....................................3118
Mai, Annette......................................................284
Maier, Jannik...................................................1642
Mandrioli, Riccardo...........................................2576
Mannen, Tomoyuki..............................................831
Mari, Jorge.......................................................843
Marie, Alexandre...............................................2819
Martano, Emanuele............................................2874
Martínez, Alfonso..............................................2359
Masuda, Akiyoshi..............................................1018
Mauromicale, Giuseppe.......................................2751
Mazzer, Simone.........................................2162, 2532
McRae, Tim......................................2190, 2364, 3042
Medina-Garcia, Alfredo.......................................1207
Meligy, Ahmed.................................................1495
Menzel, Steffen...................................................933
Merrouche, Abdennour........................................1113
Minamisawa, Renato Amaral.................................2036
Mirkovic, Nikola...............................................2488
Mo, Xianghao...................................................2386
Mochizuki, Yo....................................................870

Mönch, Stefan....................................................167
Mueller, Lukas..................................................2425
Mühlfeld, Christian............................................3340
Muralikrishna, Ajay Krishna Voppu........................2886
Nachete, Idriss.................................................2408
Nakako, Hideo.....................................................49
Nawaz, Muhammad............................................2013
Nehmer, Dominik.................................................24
Neira, Sebastian................................................2700
Neuner, Matthias...............................................3296
Nikiforidis, Ioannis......................................916, 2718
Nkembi, Armel Asongu........................................2469
Oberdieck, Karl.................................................1828
O'Keeffe, Rosemary............................................1249
Olalla, David....................................................1814
Ong, Shu Ee.....................................................2942
Orlando, Stefano........................................1434, 2673
Otori, Daichi.....................................................518
Otte, Raphael...................................................2234
Ouhab, Merouane.........................................589, 2948
Owzareck, Michael.............................................3371
Palma, Marco....................................................1568
Panchal, Pranav..................................................315
Paradkar, Sachin Shridhar...................................2627
Patterson, Andrew.............................................1330
Paul, Indrajit....................................................1133
Peng, Hujun......................................................665
Petzold, Tom....................................................1158
Pham, Thanh-Toan.............................................2786
Philippe, Antoine........................................1441, 1449
Phung, Thanh Hai.......................................1555, 2612
Piccioni, Andrea...............................................2680
Piepenbrock, Till................................................463
Poller, Tilo......................................................2914
Porpora, Francesco..............................................222
Pouresmaeil, Mobina...........................................419
Prince, Aswathy M..............................................2850
Rabay, Battist..................................................2082
Radix, Bryan....................................................2112
Radomsky, Lukas...............................................3286
Randerath, Joschka............................................3256
Raßmann, Rando...............................................2350
Rauh, Michael....................................................690
Rebenklau, Lars................................................2088
Reddy, Niranjan Suravarapu..................................2273
Rehlaender, Philipp............................................2686
Reimann, René..................................................2377
Reiner, Richard..................................................557
Reißenweber, Lukas.......................................447, 525
Reitz, Niclas....................................................1393
Ren, Linhao.....................................................1917
Ren, Xufu........................................................803

| | |
|---|---|
| Rendek, Karol | 2831 |
| Reymond-Laruina, Frédéric | 103 |
| Rezaeizadeh, Amin | 1686 |
| Ribarich, Tom | 1254 |
| Ribeiro, Kelly | 2294 |
| Rillo, Oriol Subirats | 290 |
| Ringelmann, Tim | 2708 |
| Rodrigues, Luis Alves | 635 |
| Rodriguez, Manuel Escudero | 812 |
| Rodruigez, Manuel Escudero | 3077 |
| Rosensaft, Boris | 2620 |
| Rudzki, Jacek | 1942 |
| Ruoff, Dominik | 1277 |
| Ruppert, Lukas | 1703 |
| Sakai, Junya | 2006 |
| Salomez, Florentin | 2431 |
| Samura, Koki | 2764 |
| Sankari, Rasched | 197 |
| Sawada, Takashi | 161 |
| Schindler, Stefanie | 1147 |
| Schindler, Tobias | 140 |
| Schmidhuber, Michael | 2438 |
| Schmidt, Matthias | 397 |
| Schmidt, Paul | 1999 |
| Schmitz, Laurids | 2995 |
| Schnell, Raffael | 855 |
| Schnitzler, Ruben | 1851 |
| Schulte, Felix | 3364 |
| Schulz, Martin | 1077 |
| Schwab, Stefan | 343 |
| Schwarz, Niklas | 1200 |
| Scuto, Alfio | 2921 |
| Seber, Elizabeth | 888 |
| Sekar, Ajith Kumar | 2041 |
| Sen, Gokhan | 784 |
| Seo, Hansol | 1896 |
| Sheikhan, Alireza | 549, 2606 |
| Shi, Sanbao | 1212, 3029 |
| Sifoune, Sarah | 390 |
| Singer, Mehyeddine | 2309 |
| Solomakha, Oleksandr | 1174 |
| Somarin, Hasan Mousavi | 2494 |
| Sos, Carlos Costas | 205 |
| Sousa, Gean | 2557 |
| Srikrishna, N. H | 3269 |
| Steenbock, Liska | 3169 |
| Steiner, Felix | 1891 |
| Stone, David A. | 949 |
| Subotic, Stefan | 274 |
| Sugie, Hisashi | 1765 |
| Sun, Qing | 2791 |
| Suzuki, Keita | 2053 |

| | |
|---|---|
| Syed, Hadiuzzaman | 2758 |
| Talits, Kevin | 997 |
| Tan, John Emmanuel | 150 |
| Tanikawa, Kohei | 1272 |
| Tarmoom, Ehab | 942 |
| Tekir, Bünyamin | 672 |
| Tengvall, Sebastian | 1380 |
| Thamm, Merlin | 1532 |
| Thekemuriyil, Tanya | 645, 2986 |
| Thirukoluri, Rajani Kumar | 1865 |
| Thomas, Mark | 564 |
| Thönnessen, André | 1584 |
| Tigira, Sandu | 3130 |
| To, Pham Ha Trieu | 707 |
| Tobler, Stefan | 472 |
| Tokorozuki, Takeshi | 849 |
| Torrisi, Marco | 822 |
| Tranchero, Maurizio | 1322 |
| Troudi, Rami | 1185 |
| Tuncay, Sebnem | 1858 |
| Uemura, Hirofumi | 433 |
| Ueno, Masaki | 2909 |
| Ugur, Abdulkerim | 3229 |
| Uhlemann, Andre | 1283 |
| Urbaneck, Daniel | 2152 |
| Varadarajan, Kamal | 1598 |
| Vemulapati, Umamaheswara Reddy | 1025 |
| Vinciguerra, Vincenzo | 1039 |
| Vobecky, Jan | 1002 |
| Vogelsberger, Markus | 1307 |
| Vogt, Michael | 2866 |
| Vuletic, Radovan | 305 |
| Walter, Michael | 1297 |
| Wang, Hamlin | 2185 |
| Wang, Hao | 2693 |
| Wang, Lei | 1242 |
| Wang, Lisheng | 2060 |
| Wang, Qilei | 113 |
| Wang, Rui | 84 |
| Wang, Yushi | 966 |
| Watanabe, Hiroki | 3125 |
| Weckbrodt, Julien | 121 |
| Wei, Frank | 2146 |
| Wei, Suhang | 2074 |
| Weihe, Sven | 330 |
| Wen, Jin | 2182, 3103 |
| Wessel, Wilfried | 655 |
| Wietschel, Martin | 7 |
| Wille, Christopher | 128 |
| Winkler, Paul | 1511 |
| Xie, Dong | 2880 |
| Xie, Luhong | 1930 |

Yadav, Sachin...................................................2583
Yan, Xingda ......................................................1680
Yan, Yiyang.......................................................1809
Ye, Yijun ...........................................................2826
Ye, Zhong ...............................................1621, 2646
Yoshida, Satoshi...................................................543
Yoshioka, Kentaro.............................................2067
Yu, Renze ............................................................717
Yu, Sean ....................................................1180, 2518
Yu, Sheng-Yang .................................................2200
Zeng, Chenhang.................................................1693
Zhang, Chi .........................................................1781
Zhang, Hongpeng ................................................238
Zhang, Huaiyuan................................................2901
Zhang, Yi ...........................................................1936
Zhao, Yue ...........................................................3197
Zheng, Zexiang ..................................................2860
Zhu, Shiwu.........................................................2893
Zipperstein, David .............................................2651
Zipprich, Robert.................................................1664
Zocher, Markus...................................................1233

**VDE VERLAG GMBH**
Bismarckstr. 33
P.O.B. 12 01 43
10625 Berlin, Germany

ISBN 978-1-7138-9966-2